Encyclopedia of VIROLOGY
THIRD EDITION

EDITORS-IN-CHIEF

Dr BRIAN W J MAHY
and
Dr MARC H V VAN REGENMORTEL

AMSTERDAM • BOSTON • HEIDELBERG • LONDON • NEW YORK • OXFORD
PARIS • SAN DIEGO • SAN FRANCISCO • SINGAPORE • SYDNEY • TOKYO
Academic Press is an imprint of Elsevier

ELSEVIER

ACADEMIC PRESS

Academic Press is an imprint of Elsevier
Linacre House, Jordan Hill, Oxford, OX2 8DP, UK
525 B Street, Suite 1900, San Diego, CA 92101-4495, USA

Copyright © 2008 Elsevier Inc. All rights reserved

The following articles are US government works in the public domain and are not subject to copyright:
Bovine Viral Diarrhea Virus, Coxsackieviruses, Prions of Yeast and Fungi, Human Respiratory Syncytial Virus, Fish Rhabdoviruses, Varicella-Zoster Virus: General Features, Viruses and Bioterrorism, Bean Common Mosaic Virus and Bean Common Mosaic Necrosis Virus, Metaviruses, Crimean-Congo Hemorrhagic Fever Virus and Other Nairoviruses, AIDS: Global Epidemiology, Papaya Ringspot Virus, Transcriptional Regulation in Bacteriophage.

Nepovirus, Canadian Crown Copyright 2008

No part of this publication may be reproduced, stored in a retrieval system or transmitted in any form or by any means electronic, mechanical, photocopying, recording or otherwise without the prior written permission of the publisher

Permissions may be sought directly from Elsevier's Science & Technology Rights Department in Oxford, UK: phone (+44) (0) 1865 843830; fax (+44) (0) 1865 853333; email: permissions@elsevier.com. Alternatively you can submit your request online by visiting the Elsevier web site at (http://elsevier.com/locate/permission), and selecting *Obtaining permission to use Elsevier material*

Notice
No responsibility is assumed by the publisher for any injury and/or damage to persons or property as a matter of products liability, negligence or otherwise, or from any use or operation of any methods, products, instructions or ideas contained in the material herein. Because of rapid advances in the medical sciences, in particular, independent verification of diagnoses and drug dosages should be made

British Library Cataloguing in Publication Data
A catalogue record for this book is available from the British Library

Library of Congress Catalog Number: 200892260

ISBN: 978-0-12-373935-3

For information on all Elsevier publications
visit our website at books.elsevier.com

PRINTED AND BOUND IN SLOVENIA
08 09 10 11   10 9 8 7 6 5 4 3 2 1

Working together to grow
libraries in developing countries

www.elsevier.com | www.bookaid.org | www.sabre.org

ELSEVIER   BOOK AID International   Sabre Foundation

Encyclopedia of Virology

Third Edition

EDITORS-IN-CHIEF

Brian W J Mahy MA PhD ScD DSc
Senior Scientific Advisor,
Division of Emerging Infections and Surveillance Services,
Centers for Disease Control and Prevention,
Atlanta GA, USA

Marc H V Van Regenmortel PhD
Emeritus Director at the CNRS,
French National Center for Scientific Research,
Biotechnology School of the University of Strasbourg,
Illkirch, France

ASSOCIATE EDITORS

Dennis H Bamford, Ph.D.
Department of Biological and Environmental Sciences
and Institute of Biotechnology, Biocenter 2,
P.O. Box 56 (Viikinkaari 5),
00014 University of Helsinki,
Finland

Charles Calisher, B.S., M.S., Ph.D.
Arthropod-borne and Infectious Diseases Laboratory
Department of Microbiology, Immunology and Pathology
College of Veterinary Medicine and Biomedical Sciences
Colorado State University
Fort Collins
CO 80523
USA

Andrew J Davison, M.A., Ph.D.
MRC Virology Unit
Institute of Virology
University of Glasgow
Church Street
Glasgow G11 5JR
UK

Claude Fauquet
ILTAB/Donald Danforth Plant Science Center
975 North Warson Road
St. Louis, MO 63132

Said Ghabrial, B.S., M.S., Ph.D.
Plant Pathology Department
University of Kentucky
201F Plant Science Building
1405 Veterans Drive
Lexington
KY 4050546-0312
USA

Eric Hunter, B.Sc., Ph.D.
Department of Pathology and Laboratory Medicine, and
Emory Vaccine Center
Emory University
954 Gatewood Road NE
Atlanta Georgia 30329
USA

Robert A. Lamb, Ph.D., Sc.D.
Department of Biochemistry,
Molecular Biology and Cell Biology
Howard Hughes Medical Institute
Northwestern University
2205 Tech Dr.
Evanston
IL 60208-3500
USA

Olivier Le Gall
IPV, UMR GDPP, IBVM,
INRA Bordeaux-Aquitaine, BP 81,
F-33883 Villenave d'Ornon Cedex
FRANCE

Vincent Racaniello, Ph.D.
Department of Microbiology
Columbia University
New York, NY 10032
USA

David A. Theilmann, Ph.D., B.Sc., M.Sc
Pacific Agri-Food Research Centre
Agriculture and Agri-Food Canada
Box 5000, 4200 Highway 97
Summerland
BC V0H 1Z0
Canada

H. Josef Vetten, Ph.D.
Julius Kuehn Institute, Federal Research Centre for
Cultivated Plants (JKI)
Messeweg 11-12
38104 Braunschweig
Germany

Peter J Walker, B.Sc., Ph.D.
CSIRO Livestock Industries
Australian Animal Health Laboratory (AAHL)
Private Bag 24
Geelong
VIC 3220
Australia

PREFACE

This third edition of the *Encyclopedia of Virology* is being published nine years after the second edition, a period which has seen enormous growth both in our understanding of virology and in our recognition of the viruses themselves, many of which were unknown when the second edition was prepared. Considering viruses affecting human hosts alone, the worldwide epidemic of severe acute respiratory syndrome (SARS), caused by a previously unknown coronavirus, led to the discovery of other human coronaviruses such as HKU1 and NL63. As many as seven chapters are devoted to the AIDS epidemic and to human immunodeficiency viruses. In addition, the development of new molecular technologies led to the discovery of viruses with no obvious disease associations, such as torque-teno virus (one of the most ubiquitous viruses in the human population), human bocavirus, human metapneumovirus, and three new human polyomaviruses.

Other new developments of importance to human virology have included the introduction of a virulent strain of West Nile virus from Israel to North America in 1999. Since that time the virus has become established in mosquito, bird and horse populations throughout the USA, the Caribbean and Mexico as well as the southern regions of Canada.

As in the two previous editions, we have tried to include information about all known species of virus infecting bacteria, fungi, invertebrates, plants and vertebrates, as well as descriptions of related topics in virology such as antiviral drug development, cell- and antibody-mediated immunity, vaccine development, electron microscopy and molecular methods for virus characterization and identification. Many chapters are devoted to the considerable economic importance of virus diseases of cereals, legumes, vegetable crops, fruit trees and ornamentals, and new approaches to control these diseases are reviewed.

General issues such as the origin, evolution and phylogeny of viruses are also discussed as well as the history of the different groups of viruses.

To cover all these subjects and new developments, we have had to increase the size of the Encyclopedia from three to five volumes.

Throughout this work we have relied upon the 8th Report of the International Committee on Taxonomy of Viruses published in 2005, which lists more than 6000 viruses classified into some 2000 virus species distributed among more than 390 different genera and families. In recent years the criteria for placing viruses in different taxa have shifted away from traditional serological methods and increasingly rely upon molecular techniques, particularly the nucleotide sequence of the virus genome. This has changed many of the previous groupings of viruses, and is reflected in this third edition.

Needless to say, a work of this magnitude has involved many expert scientists, who have given generously of their time to bring it to fruition. We extend our grateful thanks to all contributors and associate editors for their excellent and timely contributions.

Brian W J Mahy
Marc H V van Regenmortel

HOW TO USE THE ENCYCLOPEDIA

Structure of the Encyclopedia

The major topics discussed in detail in the text are presented in alphabetical order (see the Alphabetical Contents list which appears in all five volumes).

Finding Specific Information

Information on specific viruses, virus diseases and other matters can be located by consulting the General Index at the end of Volume 5.

Taxonomic Groups of Viruses

For locating detailed information on the major taxonomic groups of viruses, namely virus genera, families and orders, the Taxonomic Index in Volume 5 (page...) should be consulted.

Further Reading sections

The articles do not feature bibliographic citations within the body of the article text itself. The articles are intended to be a first introduction to the topic, or a 'refresher', readable from beginning to end without referring the reader outside of the encyclopedia itself. Bibliographic references to external literature are grouped at the end of each article in a Further Reading section, containing review articles, 'seminal' primary articles and book chapters. These point users to the next level of information for any given topic.

Cross referencing between articles

The "See also" section at the end of each article directs the reader to other entries on related topics. For example. The entry *Lassa, Junin, Machupo and Guanarito Viruses* includes the following cross-references:

See also: Lymphocytic Choriomeningitis Virus: General Features.

CONTRIBUTORS

S T Abedon
The Ohio State University, Mansfield, OH, USA

G P Accotto
Istituto di Virologia Vegetale CNR, Torino, Italy

H-W Ackermann
Laval University, Quebec, QC, Canada

G Adam
Universität Hamburg, Hamburg, Germany

M J Adams
Rothamsted Research, Harpenden, UK

C Adams
University of Duisburg–Essen, Essen, Germany

E Adderson
St. Jude Children's Research Hospital, Memphis, TN, USA

S Adhya
National Institutes of Health, Bethesda, MD, USA

C L Afonso
Southeast Poultry Research Laboratory, Athens, GA, USA

P Ahlquist
University of Wisconsin – Madison, Madison, WI, USA

G M Air
University of Oklahoma Health Sciences Center, Oklahoma City, OK, USA

D J Alcendor
Johns Hopkins School of Medicine, Baltimore, MD, USA

J W Almond
sanofi pasteur, Lyon, France

I Amin
National Institute for Biotechnology and Genetic Engineering, Faisalabad, Pakistan

J Angel
Pontificia Universidad Javeriana, Bogota, Republic of Colombia

C Apetrei
Tulane National Primate Research Center, Covington, LA, USA

B M Arif
Great Lakes Forestry Centre, Sault Ste. Marie, ON, Canada

H Attoui
Faculté de Médecine de Marseilles, Etablissement Français Du Sang, Marseilles, France

H Attoui
Université de la Méditerranée, Marseille, France

H Attoui
Institute for Animal Health, Pirbright, UK

L Aurelian
University of Maryland School of Medicine, Baltimore, MD, USA

L A Babiuk
University of Alberta, Edmonton, AB, Canada

S Babiuk
National Centre for Foreign Animal Disease, Winnipeg, MB, Canada

A G Bader
The Scripps Research Institute, La Jolla, CA, USA

S C Baker
Loyola University of Chicago, Maywood, IL, USA

T S Baker
University of California, San Diego, La Jolla, CA, USA

J K H Bamford
University of Jyväskylä, Jyväskylä, Finland

Y Bao
National Institutes of Health, Bethesda, MD, USA

M Bar-Joseph
The Volcani Center, Bet Dagan, Israel

H Barker
Scottish Crop Research Institute, Dundee, UK

A D T Barrett
University of Texas Medical Branch, Galveston, TX, USA

J W Barrett
The University of Western Ontario, London, ON, Canada

T Barrett
Institute for Animal Health, Pirbright, UK

R Bartenschlager
University of Heidelberg, Heidelberg, Germany

N W Bartlett
Imperial College London, London, UK

S Basak
University of California, San Diego, CA, USA

C F Basler
Mount Sinai School of Medicine, New York, NY, USA

T Basta
Institut Pasteur, Paris, France

D Baxby
University of Liverpool, Liverpool, UK

P Beard
Imperial College London, London, UK

M N Becker
University of Florida, Gainesville, FL, USA

J J Becnel
Agriculture Research Service, Gainesville, FL, USA

K L Beemon
Johns Hopkins University, Baltimore, MD, USA

E D Belay
Centers for Disease Control and Prevention, Atlanta, GA, USA

M Benkő
Veterinary Medical Research Institute, Hungarian Academy of Sciences, Budapest, Hungary

M Bennett
University of Liverpool, Liverpool, UK

M Bergoin
Université Montpellier II, Montpellier, France

H U Bernard
University of California, Irvine, Irvine, CA, USA

K I Berns
University of Florida College of Medicine, Gainesville, FL, USA

P Biagini
Etablissement Français du Sang Alpes-Méditerranée, Marseilles, France

P D Bieniasz
Aaron Diamond AIDS Research Center, The Rockefeller University, New York, NY, USA

Y Bigot
University of Tours, Tours, France

C Billinis
University of Thessaly, Karditsa, Greece

R F Bishop
Murdoch Childrens Research Institute Royal Children's Hospital, Melbourne, VIC, Australia

B A Blacklaws
University of Cambridge, Cambridge, UK

C D Blair
Colorado State University, Fort Collins, CO, USA

S Blanc
INRA–CIRAD–AgroM, Montpellier, France

R Blawid
Institute of Plant Diseases and Plant Protection, Hannover, Germany

G W Blissard
Boyce Thompson Institute at Cornell University, Ithaca, NY, USA

S Blomqvist
National Public Health Institute (KTL), Helsinki, Finland

J F Bol
Leiden University, Leiden, The Netherlands

J-R Bonami
CNRS, Montpellier, France

L Bos
Wageningen University and Research Centre (WUR), Wageningen, The Netherlands

H R Bose Jr.
University of Texas at Austin, Austin, TX, USA

H Bourhy
Institut Pasteur, Paris, France

P R Bowser
Cornell University, Ithaca, NY, USA

D B Boyle
CSIRO Livestock Industries, Geelong, VIC, Australia

C Bragard
Université Catholique de Louvain, Leuven, Belgium

J N Bragg
University of California, Berkeley, Berkeley, CA, USA

R W Briddon
National Institute for Biotechnology and Genetic Engineering, Faisalabad, Pakistan

M A Brinton
Georgia State University, Atlanta, GA, USA

P Britton
Institute for Animal Health, Compton, UK

J K Brown
The University of Arizona, Tucson, AZ, USA

K S Brown
University of Manitoba, Winnipeg, MB, Canada

J Bruenn
State University of New York, Buffalo, NY, USA

C P D Brussaard
Royal Netherlands Institute for Sea Research, Texel, The Netherlands

J J Bugert
Wales College of Medicine, Heath Park, Cardiff, UK

J J Bujarski
Northern Illinois University, DeKalb, IL, USA and Polish Academy of Sciences, Poznan, Poland

R M Buller
Saint Louis University School of Medicine, St. Louis, MO, USA

J P Burand
University of Massachusetts at Amherst, Amherst, MA, USA

J Burgyan
Agricultural Biotechnology Center, Godollo, Hungary

F J Burt
University of the Free State, Bloemfontein, South Africa

S J Butcher
University of Helsinki, Helsinki, Finland

J S Butel
Baylor College of Medicine, Houston, TX, USA

M I Butler
University of Otago, Dunedin, New Zealand

S Bühler
University of Heidelberg, Heidelberg, Germany

P Caciagli
Istituto di Virologia Vegetale – CNR, Turin, Italy

C H Calisher
Colorado State University, Fort Collins, CO, USA

T Candresse
UMR GDPP, Centre INRA de Bordeaux, Villenave d'Ornon, France

A J Cann
University of Leicester, Leicester, UK

C Caranta
INRA, Montfavet, France

G Carlile
CSIRO Livestock Industries, Geelong, VIC, Australia

J P Carr
University of Cambridge, Cambridge, UK

R Carrion, Jr.
Southwest Foundation for Biomedical Research, San Antonio, TX, USA

J W Casey
Cornell University, Ithaca, NY, USA

R N Casey
Cornell University, Ithaca, NY, USA

S Casjens
University of Utah School of Medicine, Salt Lake City, UT, USA

R Cattaneo
Mayo Clinic College of Medicine, Rochester, MN, USA

D Cavanagh
Institute for Animal Health, Compton, UK

A Chahroudi
University of Pennsylvania School of Medicine, Philadelphia, PA, USA

S Chakraborty
Jawaharlal Nehru University, New Delhi, India

T J Chambers
Saint Louis University School of Medicine, St. Louis, MO, USA

Y Chang
University of Pittsburgh Cancer Institute, Pittsburgh, PA, USA

J T Chang
Baylor College of Medicine, Houston, TX, USA

D Chapman
Institute for Animal Health, Pirbright, UK

D Chattopadhyay
University of Calcutta, Kolkata, India

M Chen
University of Arizona, Tucson, AZ, USA

J E Cherwa
University of Arizona, Tucson, AZ, USA

V G Chinchar
University of Mississippi Medical Center, Jackson, MS, USA

A V Chintakuntlawar
University of Oklahoma Health Sciences Center, Oklahoma City, OK, USA

W Chiu
Baylor College of Medicine, Houston, TX, USA

J Chodosh
University of Oklahoma Health Sciences Center, Oklahoma City, OK, USA

I-R Choi
International Rice Research Institute, Los Baños, The Philippines

P D Christian
National Institute of Biological Standards and Control, South Mimms, UK

M G Ciufolini
Istituto Superiore di Sanità, Rome, Italy

P Clarke
University of Colorado Health Sciences, Denver, CO, USA

J-M Claverie
Université de la Méditerranée, Marseille, France

J R Clayton
Johns Hopkins University Schools of Public Health and Medicine, Baltimore, MD, USA

R J Clem
Kansas State University, Manhattan, KS, USA

C J Clements
The Macfarlane Burnet Institute for Medical Research and Public Health Ltd., Melbourne, VIC, Australia

L L Coffey,
Institut Pasteur, Paris, France

J I Cohen
National Institutes of Health, Bethesda, MD, USA

J Collinge
University College London, London, UK

P L Collins
National Institute of Allergy and Infectious Diseases, Bethesda, MD, USA

A Collins
University of Wisconsin School of Medicine and Public Health, Madison, WI, USA

D Contamine
Université Versailles St-Quentin, CNRS, Versailles, France

K M Coombs
University of Manitoba, Winnipeg, MB, Canada

J A Cowley
CSIRO Livestock Industries, Brisbane, QLD, Australia

J K Craigo
University of Pittsburgh School of Medicine, Pittsburgh, PA, USA

M St. J Crane
CSIRO Livestock Industries, Geelong, VIC, Australia

J E Crowe, Jr.
Vanderbilt University Medical Center, Nashville, TN, USA

H Czosnek
The Hebrew University of Jerusalem, Rehovot, Israel

T Dalmay
University of East Anglia, Norwich, UK

B H Dannevig
National Veterinary Institute, Oslo, Norway

C J D'Arcy
University of Illinois at Urbana-Champaign, Urbana, IL, USA

A J Davison
MRC Virology Unit, Glasgow, UK

W O Dawson
University of Florida, Lake Alfred, FL, USA

L A Day
The Public Health Research Institute, Newark, NJ, USA

J C de la Torre
The Scripps Research Institute, La Jolla, CA, USA

X de Lamballerie
Faculté de Médecine de Marseille, Marseilles, France

M de Vega
Universidad Autónoma, Madrid, Spain

P Delfosse
Centre de Recherche Public-Gabriel Lippmann, Belvaux, Luxembourg

B Delmas
INRA, Jouy-en-Josas, France

M Deng
University of California, Berkeley, CA, USA

J DeRisi
University of California, San Francisco, San Francisco, CA, USA

C Desbiez
Institut National de la Recherche Agronomique (INRA), Station de Pathologie Végétale, Montfavet, France

R C Desrosiers
New England Primate Research Center, Southborough, MA, USA

A K Dhar
Advanced BioNutrition Corp, Columbia, MD, USA

R G Dietzgen
The University of Queensland, St. Lucia, QLD, Australia

S P Dinesh-Kumar
Yale University, New Haven, CT, USA

L K Dixon
Institute for Animal Health, Pirbright, UK

C Dogimont
INRA, Montfavet, France

A Domanska
University of Helsinki, Helsinki, Finland

L L Domier
USDA–ARS, Urbana, IL, USA

L L Domier
USDA-ARS, Urbana-Champaign, IL, USA

A Dotzauer
University of Bremen, Bremen, Germany

T W Dreher
Oregon State University, Corvallis, OR, USA

S Dreschers
University of Duisburg–Essen, Essen, Germany

R L Duda
University of Pittsburgh, Pittsburgh, PA, USA

J P Dudley
The University of Texas at Austin, Austin, TX, USA

W P Duprex
The Queen's University of Belfast, Belfast, UK

R E Dutch
University of Kentucky, Lexington, KY, USA

B M Dutia
University of Edinburgh, Edinburgh, UK

M L Dyall-Smith
The University of Melbourne, Parkville, VIC, Australia

J East
University of Texas Medical Branch – Galveston, Galveston, TX, USA

A J Easton
University of Warwick, Coventry, UK

K C Eastwell
Washington State University – IAREC, Prosser, WA, USA

B T Eaton
Australian Animal Health Laboratory, Geelong, VIC, Australia

H Edskes
National Institutes of Health, Bethesda, MD, USA

B Ehlers
Robert Koch-Institut, Berlin, Germany

R M Elliott
University of St. Andrews, St. Andrews, UK

A Engel
National Institutes of Health, Bethesda, MD, USA and
D Kryndushkin
National Institutes of Health, Bethesda, MD, USA

J Engelmann
INRES, University of Bonn, Bonn, Germany

L Enjuanes
CNB, CSIC, Madrid, Spain

A Ensser
Virologisches Institut, Universitätsklinikum, Erlangen, Germany

M Erlandson
Agriculture & Agri-Food Canada, Saskatoon, SK, Canada

K J Ertel
University of California, Irvine, CA, USA

R Esteban
Instituto de Microbiología Bioquímica CSIC/University de Salamanca, Salamanca, Spain

R Esteban
Instituto de Microbiología Bioquímica CSIC/University of Salamanca, Salamanca, Spain

J L Van Etten
University of Nebraska–Lincoln, Lincoln, NE, USA

D J Evans
University of Warwick, Coventry, UK

Ø Evensen
Norwegian School of Veterinary Science, Oslo, Norway

D Falzarano
University of Manitoba, Winnipeg, MB, Canada

B A Fane
University of Arizona, Tucson, AZ, USA

R-X. Fang
Chinese Academy of Sciences, Beijing, People's Republic of China

D Fargette
IRD, Montpellier, France

A Fath-Goodin
University of Kentucky, Lexington, KY, USA

C M Fauquet
Danforth Plant Science Center, St. Louis, MO, USA

B A Federici
University of California, Riverside, CA, USA

H Feldmann
National Microbiology Laboratory, Public Health Agency of Canada, Winnipeg, MB, Canada

H Feldmann
Public Health Agency of Canada, Winnipeg, MB, Canada

F Fenner
Australian National University, Canberra, ACT, Australia

S A Ferreira
University of Hawaii at Manoa, Honolulu, HI, USA

H J Field
University of Cambridge, Cambridge, UK

K Fischer
University of California, San Francisco, San Francisco, CA, USA

J A Fishman
Massachusetts General Hospital, Boston, MA, USA

B Fleckenstein
University of Erlangen – Nürnberg, Erlangen, Germany

R Flores
Instituto de Biología Molecular y Celular de Plantas (UPV-CSIC), Valencia, Spain

T R Flotte
University of Florida College of Medicine, Gainesville, FL, USA

P Forterre
Institut Pasteur, Paris, France

M A Franco
Pontificia Universidad Javeriana, Bogota, Republic of Colombia

T K Frey
Georgia State University, Atlanta, GA, USA

M Fuchs
Cornell University, Geneva, NY, USA

S Fuentes
International Potato Center (CIP), Lima, Peru

T Fujimura
Instituto de Microbiología Bioquímica CSIC/University of Salamanca, Salamanca, Spain

R S Fujinami
University of Utah School of Medicine, Salt Lake City, UT, USA

T Fukuhara
Tokyo University of Agriculture and Technology, Fuchu, Japan

D Gallitelli
Università degli Studi and Istituto di Virologia Vegetale del CNR, Bari, Italy

F García-Arenal
Universidad Politécnica de Madrid, Madrid, Spain

J A García
Centro Nacional de Biotecnología (CNB), CSIC, Madrid, Spain

R A Garrett
Copenhagen University, Copenhagen, Denmark

S Gaumer
Université Versailles St-Quentin, CNRS, Versailles, France

R J Geijskes
Queensland University of Technology, Brisbane, QLD, Australia

T W Geisbert
National Emerging Infectious Diseases Laboratories, Boston, MA, USA

E Gellermann
Hannover Medical School, Hannover, Germany

A Gessain
Pasteur Institute, CNRS URA 3015, Paris, France

S A Ghabrial
University of Kentucky, Lexington, KY, USA

W Gibson
Johns Hopkins University School of Medicine, Baltimore, MD, USA

M Glasa
Slovak Academy of Sciences, Bratislava, Slovakia

Y Gleba
Icon Genetics GmbH, Weinbergweg, Germany

U A Gompels
University of London, London, UK

D Gonsalves
USDA, Pacific Basin Agricultural Research Center, Hilo, HI, USA

M M Goodin
University of Kentucky, Lexington, KY, USA

T J D Goodwin
University of Otago, Dunedin, New Zealand

A E Gorbalenya
Leiden University Medical Center, Leiden, The Netherlands

E A Gould
University of Reading, Reading, UK

A Grakoui
Emory University School of Medicine, Atlanta, GA, USA

M-A Grandbastien
INRA, Versailles, France

R Grassmann
University of Erlangen – Nürnberg, Erlangen, Germany

M Gravell
National Institutes of Health, Bethesda, MD, USA

M V Graves
University of Massachusetts–Lowell, Lowell, MA, USA

K Y Green
National Institutes of Health, Bethesda, MD, USA

H B Greenberg
Stanford University School of Medicine and Veterans Affairs Palo Alto Health Care System, Palo Alto, CA, USA

B M Greenberg
Johns Hopkins School of Medicine, Baltimore, MD, USA

I Greiser-Wilke
School of Veterinary Medicine, Hanover, Germany

D E Griffin
Johns Hopkins Bloomberg School of Public Health, Baltimore, MD, USA

T S Gritsun
University of Reading, Reading, UK

R J de Groot
Utrecht University, Utrecht, The Netherlands

A J Gubala
CSIRO Livestock Industries, Geelong, VIC, Australia

D J Gubler
John A. Burns School of Medicine, Honolulu, HI, USA

A-L Haenni
Institut Jacques Monod, Paris, France

D Haig
Nottingham University, Nottingham, UK

F J Haines
Oxford Brookes University, Oxford, UK

J Hamacher
INRES, University of Bonn, Bonn, Germany

J Hammond
USDA-ARS, Beltsville, MD, USA

R M Harding
Queensland University of Technology, Brisbane, QLD, Australia

J M Hardwick
Johns Hopkins University Schools of Public Health and Medicine, Baltimore, MD, USA

D Hariri
INRA – Département Santé des Plantes et Environnement, Versailles, France

B Harrach
Veterinary Medical Research Institute, Budapest, Hungary

P A Harries
Samuel Roberts Noble Foundation, Inc., Ardmore, OK, USA

L E Harrington
University of Alabama at Birmingham, Birmingham, AL, USA

T J Harrison
University College London, London, UK

T Hatziioannou
Aaron Diamond AIDS Research Center, The Rockefeller University, New York, NY, USA

J Hay
The State University of New York, Buffalo, NY, USA

G S Hayward
Johns Hopkins School of Medicine, Baltimore, MD, USA

E Hébrard
IRD, Montpellier, France

R W Hendrix
University of Pittsburgh, Pittsburgh, PA, USA

L E Hensley
USAMRIID, Fort Detrick, MD, USA

M de las Heras
University of Glasgow Veterinary School, Glasgow, UK

S Hertzler
University of Illinois at Chicago, Chicago, IL, USA

F van Heuverswyn
University of Montpellier 1, Montpellier, France

J Hilliard
Georgia State University, Atlanta, GA, USA

B I Hillman
Rutgers University, New Brunswick, NJ, USA

S Hilton
University of Warwick, Warwick, UK

D M Hinton
National Institutes of Health, Bethesda, MD, USA

A Hinz
UMR 5233 UJF-EMBL-CNRS, Grenoble, France

A E Hoet
The Ohio State University, Columbus, OH, USA

S A Hogenhout
The John Innes Centre, Norwich, UK

T Hohn
Basel university, Institute of Botany, Basel, Switzerland

J S Hong
Seoul Women's University, Seoul, South Korea

M C Horzinek
Utrecht University, Utrecht, The Netherlands

T Hovi
National Public Health Institute (KTL), Helsinki, Finland

A M Huger
Institute for Biological Control, Darmstadt, Germany

L E Hughes
University of St. Andrews, St. Andrews, UK

R Hull
John Innes Centre, Colney, UK

E Hunter
Emory University Vaccine Center, Atlanta, GA, USA

A D Hyatt
Australian Animal Health Laboratory, Geelong, VIC, Australia

T Hyypiä
University of Turku, Turku, Finland

T Iwanami
National Institute of Fruit Tree Science, Tsukuba, Japan

A O Jackson
University of California, Berkeley, CA, USA

P Jardine
University of Minnesota, Minneapolis, MN, USA

J A Jehle
DLR Rheinpfalz, Neustadt, Germany

A R Jilbert
Institute of Medical and Veterinary Science, Adelaide, SA, Australia

P John
Indian Agricultural Research Institute, New Delhi, India

J E Johnson
The Scripps Research Institute, La Jolla, CA, USA

R T Johnson
Johns Hopkins School of Medicine, Baltimore, MD, USA

W E Johnson
New England Primate Research Center, Southborough, MA, USA

S L Johnston
Imperial College London, London, UK

A T Jones
Scottish Crop Research Institute, Dundee, UK

R Jordan
USDA-ARS, Beltsville, MD, USA

Y Kapustin
National Institutes of Health, Bethesda, MD, USA

P Karayiannis
Imperial College London, London, UK

P Kazmierczak
University of California, Davis, CA, USA

K M Keene
Colorado State University, Fort Collins, CO, USA

C Kerlan
Institut National de la Recherche Agronomique (INRA), Le Rheu, France

K Khalili
Temple University School of Medicine, Philadelphia, PA, USA

P H Kilmarx
Centers for Disease Control and Prevention, Atlanta, GA, USA

L A King
Oxford Brookes University, Oxford, UK

P D Kirkland
Elizabeth Macarthur Agricultural Institute, Menangle, NSW, Australia

C D Kirkwood
Murdoch Childrens Research Institute Royal Children's Hospital, Melbourne, VIC, Australia

R P Kitching
Canadian Food Inspection Agency, Winnipeg, MB, Canada

P J Klasse
Cornell University, New York, NY, USA

N R Klatt
University of Pennsylvania School of Medicine, Philadelphia, PA, USA

R G Kleespies
Institute for Biological Control, Darmstadt, Germany

D F Klessig
Boyce Thompson Institute for Plant Research, Ithaca, NY, USA

W B Klimstra
Louisiana State University Health Sciences Center at Shreveport, Shreveport, LA, USA

V Klimyuk
Icon Genetics GmbH, Weinbergweg, Germany

N Knowles
Institute for Animal Health, Pirbright, UK

R Koenig
Biologische Bundesanstalt für Land- und Forstwirtschaft, Brunswick, Germany

R Koenig
Institut für Pflanzenvirologie, Mikrobiologie und biologische Sicherheit, Brunswick, Germany

G Konaté
INERA, Ouagadougou, Burkina Faso

C N Kotton
Massachusetts General Hospital, Boston, MA, USA

L D Kramer
Wadsworth Center, New York State Department of Health, Albany, NY, USA

P J Krell
University of Guelph, Guelph, ON, Canada

J Kreuze
International Potato Center (CIP), Lima, Peru

M J Kuehnert
Centers for Disease Control and Prevention, Atlanta, GA, USA

R J Kuhn
Purdue University, West Lafayette, IN, USA

G Kurath
Western Fisheries Research Center, Seattle, WA, USA

I Kusters
sanofi pasteur, Lyon, France

I V Kuzmin
Centers for Disease Control and Prevention, Atlanta, GA, USA

M E Laird
New England Primate Research Center, Southborough, MA, USA

R A Lamb
Howard Hughes Medical Institute at Northwestern University, Evanston, IL, USA

P F Lambert
University of Wisconsin School of Medicine and Public Health, Madison, WI, USA

A S Lang
Memorial University of Newfoundland, St. John's, NL, Canada

H D Lapierre
INRA – Département Santé des Plantes et Environnement, Versailles, France

G Lawrence
The Children's Hospital at Westmead, Westmead, NSW, Australia and
University of Sydney, Westmead, NSW, Australia

H Lecoq
Institut National de la Recherche Agronomique (INRA), Station de Pathologie Végétale, Montfavet, France

B Y Lee
Seoul Women's University, Seoul, South Korea

E J Lefkowitz
University of Alabama at Birmingham, Birmingham, AL, USA

J P Legg
International Institute of Tropical Agriculture, Dar es Salaam, Tanzania,
UK and
Natural Resources Institute, Chatham Maritime, UK

P Leinikki
National Public Health Institute, Helsinki, Finland

J Lenard
University of Medicine and Dentistry of New Jersey (UMDNJ), Piscataway, NJ, USA

J C Leong
University of Hawaii at Manoa, Honolulu, HI, USA

K N Leppard
University of Warwick, Coventry, UK

A Lescoute
Université Louis Pasteur, Strasbourg, France

D-E Lesemann
Biologische Bundesanstalt für Land- und Forstwirtschaft, Brunswick, Germany

J-H Leu
National Taiwan University, Taipei, Republic of China

H L Levin
National Institutes of Health, Bethesda, MD, USA

D J Lewandowski
The Ohio State University, Columbus, OH, USA

H-S Lim
University of California, Berkeley, Berkeley, CA, USA

M D A Lindsay
Western Australian Department of Health, Mount Claremont, WA, Australia

R Ling
University of Warwick, Coventry, UK

M L Linial
Fred Hutchinson Cancer Research Center, Seattle, WA, USA

D C Liotta
Emory University, Atlanta, GA, USA

W Ian Lipkin
Columbia University, New York, NY, USA

H L Lipton
University of Illinois at Chicago, Chicago, IL, USA

A S Liss
University of Texas at Austin, Austin, TX, USA

J J López-Moya
Instituto de Biología Molecular de Barcelona (IBMB), CSIC, Barcelona, Spain

G Loebenstein
Agricultural Research Organization, Bet Dagan, Israel

C-F Lo
National Taiwan University, Taipei, Republic of China

S A Lommel
North Carolina State University, Raleigh, NC, USA

G P Lomonossoff
John Innes Centre, Norwich, UK

M Luo
University of Alabama at Birmingham, Birmingham, AL, USA

S A MacFarlane
Scottish Crop Research Institute, Dundee, UK

J S Mackenzie
Curtin University of Technology, Shenton Park, WA, Australia

R Mahieux
Pasteur Institute, CNRS URA 3015, Paris, France

B W J Mahy
Centers for Disease Control and Prevention, Atlanta, GA, USA

E Maiss
Institute of Plant Diseases and Plant Protection, Hannover, Germany

E O Major
National Institutes of Health, Bethesda, MD, USA

V G Malathi
Indian Agricultural Research Institute, New Delhi, India

A Mankertz
Robert Koch-Institut, Berlin, Germany

S Mansoor
National Institute for Biotechnology and Genetic Engineering, Faisalabad, Pakistan

A A Marfin
Centers for Disease Control and Prevention, Atlanta, GA, USA

S Marillonnet
Icon Genetics GmbH, Weinbergweg, Germany

G P Martelli
Università degli Studi and Istituto di Virologia vegetale CNR, Bari, Italy

M Marthas
University of California, Davis, Davis, CA, USA

D P Martin
University of Cape Town, Cape Town, South Africa

P A Marx
Tulane University, Covington, LA, USA

W S Mason
Fox Chase Cancer Center, Philadelphia, PA, USA

T D Mastro
Centers for Disease Control and Prevention, Atlanta, GA, USA

A A McBride
National Institutes of Health, Bethesda, MD, USA

L McCann
National Institutes of Health, Bethesda, MD, USA

M McChesney
University of California, Davis, Davis, CA, USA

J B McCormick
University of Texas, School of Public Health, Brownsville, TX, USA

G McFadden
University of Florida, Gainesville, FL, USA

G McFadden
The University of Western Ontario, London, ON, Canada

D B McGavern
The Scripps Research Institute, La Jolla, CA, USA

A L McNees
Baylor College of Medicine, Houston, TX, USA

M Meier
Tallinn University of Technology, Tallinn, Estonia

P S Mellor
Institute for Animal Health, Woking, UK

X J Meng
Virginia Polytechnic Institute and State University, Blacksburg, VA, USA

A A Mercer
University of Otago, Dunedin, New Zealand

P P C Mertens
Institute for Animal Health, Woking, UK

T C Mettenleiter
Friedrich-Loeffler-Institut, Greifswald-Insel Riems, Germany

H Meyer
Bundeswehr Institute of Microbiology, Munich, Germany

R F Meyer
Centers for Disease Control and Prevention, Atlanta, GA, USA

P de Micco
Etablissement Français du Sang Alpes-Méditerranée, Marseilles, France

B R Miller
Centers for Disease Control and Prevention (CDC), Fort Collins, CO, USA

C J Miller
University of California, Davis, Davis, CA, USA

R G Milne
Istituto di Virologia Vegetale CNR, Torino, Italy

P D Minor
NIBSC, Potters Bar, UK

S Mjaaland
Norwegian School of Veterinary Science, Oslo, Norway

E S Mocarski
Emory University School of Medicine, Atlanta, GA, USA

E S Mocarski, Jr.
Emory University School of Medicine, Emory, GA, USA

V Moennig
School of Veterinary Medicine, Hanover, Germany

P Moffett
Boyce Thompson Institute for Plant Research, Ithaca, NY, USA

T P Monath
Kleiner Perkins Caufield and Byers, Menlo Park, CA, USA

R C Montelaro
University of Pittsburgh School of Medicine, Pittsburgh, PA, USA

P S Moore
University of Pittsburgh Cancer Institute, Pittsburgh, PA, USA

F J Morales
International Center for Tropical Agriculture, Cali, Colombia

H Moriyama
Tokyo University of Agriculture and Technology, Fuchu, Japan

T J Morris
University of Nebraska, Lincoln, NE, USA

S A Morse
Centers for Disease Control and Prevention, Atlanta, GA, USA

L Moser
University of Wisconsin – Madison, Madison, WI, USA

B Moury
INRA – Station de Pathologie Végétale, Montfavet, France

J W Moyer
North Carolina State University, Raleigh, NC, USA

R W Moyer
University of Florida, Gainesville, FL, USA

E Muller
CIRAD/UMR BGPI, Montpellier, France

F A Murphy
University of Texas Medical Branch, Galveston, TX, USA

A Müllbacher
Australian National University, Canberra, ACT, Australia

K Nagasaki
Fisheries Research Agency, Hiroshima, Japan

T Nakayashiki
National Institutes of Health, Bethesda, MD, USA

A A Nash
University of Edinburgh, Edinburgh, UK

N Nathanson
University of Pennsylvania, Philadelphia, PA, USA

C K Navaratnarajah
Purdue University, West Lafayette, IN, USA

M S Nawaz-ul-Rehman
Danforth Plant Science Center, St. Louis, MO, USA

J C Neil
University of Glasgow, Glasgow, UK

R S Nelson
Samuel Roberts Noble Foundation, Inc., Ardmore, OK, USA

P Nettleton
Moredun Research Institute, Edinburgh, UK

A W Neuman
Emory University, Atlanta, GA, USA

A R Neurath
Virotech, New York, NY, USA

M L Nibert
Harvard Medical School, Boston, MA, USA

L Nicoletti
Istituto Superiore di Sanità, Rome, Italy

N Noah
London School of Hygiene and Tropical Medicine, London, UK

D L Nuss
University of Maryland Biotechnology Institute, Rockville, MD, USA

M S Oberste
Centers for Disease Control and Prevention, Atlanta, GA, USA

W A O'Brien
University of Texas Medical Branch – Galveston, Galveston, TX, USA

D J O'Callaghan
Louisiana State University Health Sciences Center, Shreveport, LA, USA

W F Ochoa
University of California, San Diego, La Jolla, CA, USA

M R Odom
University of Alabama at Birmingham, Birmingham, AL, USA

M M van Oers
Wageningen University, Wageningen, The Netherlands

M B A Oldstone
The Scripps Research Institute, La Jolla, CA, USA

G Olinger
USAMRIID, Fort Detrick, MD, USA

K E Olson
Colorado State University, Fort Collins, CO, USA

A Olspert
Tallinn University of Technology, Tallinn, Estonia

G Orth
Institut Pasteur, Paris, France

J E Osorio
University of Wisconsin, Madison, WI, USA

N Osterrieder
Cornell University, Ithaca, NY, USA

S A Overman
University of Missouri – Kansas City, Kansas City, MO, USA

R A Owens
Beltsville Agricultural Research Center, Beltsville, MD, USA

M S Padmanabhan
Yale University, New Haven, CT, USA

S Paessler
University of Texas Medical Branch, Galveston, TX, USA

P Palese
Mount Sinai School of Medicine, New York, NY, USA

M A Pallansch
Centers for Disease Control and Prevention, Atlanta, GA, USA

M Palmarini
University of Glasgow Veterinary School, Glasgow, UK

P Palukaitis
Scottish Crop Research Institute, Invergowrie, Dundee, UK

I Pandrea
Tulane National Primate Research Center, Covington, LA, USA

O Papadopoulos
Aristotle University, Thessaloniki, Greece

H R Pappu
Washington State University, Pullman, WA, USA

S Parker
Saint Louis University School of Medicine, St. Louis, MO, USA

C R Parrish
Cornell University, Ithaca, NY, USA

R F Pass
University of Alabama School of Medicine, Birmingham, AL, USA

J L Patterson
Southwest Foundation for Biomedical Research, San Antonio, TX, USA

T A Paul
Cornell University, Ithaca, NY, USA

A E Peaston
The Jackson Laboratory, Bar Harbor, ME, USA

M Peeters
University of Montpellier 1, Montpellier, France

J S M Peiris
The University of Hong Kong, Hong Kong, People's Republic of China

P J Peters
Centers for Disease Control and Prevention, Atlanta, GA, USA

M Pfeffer
Bundeswehr Institute of Microbiology, Munich, Germany

H Pfister
University of Köln, Cologne, Germany

O Planz
Federal Research Institute for Animal Health, Tuebingen, Gemany

L L M Poon
The University of Hong Kong, Hong Kong, People's Republic of China

M M Poranen
University of Helsinki, Helsinki, Finland

K Porter
The University of Melbourne, Parkville, VIC, Australia

A Portner
St. Jude Children's Research Hospital, Memphis, TN, USA

R D Possee
NERC Institute of Virology and Environmental Microbiology, Oxford, UK

R T M Poulter
University of Otago, Dunedin, New Zealand

A M Powers
Centers for Disease Control and Prevention, Fort Collins, CO, USA

D Prangishvili
Institut Pasteur, Paris, France

C M Preston
Medical Research Council Virology Unit, Glasgow, UK

S L Quackenbush
Colorado State University, Fort Collins, CO, USA

F Qu
University of Nebraska, Lincoln, NE, USA

B C Ramirez
CNRS, Paris, France

A Rapose
University of Texas Medical Branch – Galveston, Galveston, TX, USA

D V R Reddy
Hyderabad, India

A J Redwood
The University of Western Australia, Crawley, WA, Australia

M Regner
Australian National University, Canberra, ACT, Australia

W K Reisen
University of California, Davis, CA, USA

T Renault
IFREMER, La Tremblade, France

P A Revill
Victorian Infectious Diseases Reference Laboratory, Melbourne, VIC, Australia

A Rezaian
University of Adelaide, Adelaide, SA, Australia

J F Ridpath
USDA, Ames, IA, USA

B K Rima
The Queen's University of Belfast, Belfast, UK

E Rimstad
Norwegian School of Veterinary Science, Oslo, Norway

F J Rixon
MRC Virology Unit, Glasgow, UK

Y-T Ro
Konkuk University, Seoul, South Korea

C M Robinson
University of Oklahoma Health Sciences Center, Oklahoma City, OK, USA

G F Rohrmann
Oregon State University, Corvallis, OR, USA

M Roivainen
National Public Health Institute (KTL), Helsinki, Finland

L Roux
University of Geneva Medical School, Geneva, Switzerland

J Rovnak
Colorado State University, Fort Collins, CO, USA

D J Rowlands
University of Leeds, Leeds, UK

P Roy
London School of Hygiene and Tropical Medicine, London, UK

L Rubino
Istituto di Virologia Vegetale del CNR, Bari, Italy

R W H Ruigrok
CNRS, Grenoble, France

C E Rupprecht
Centers for Disease Control and Prevention, Atlanta, GA, USA

R J Russell
University of St. Andrews, St. Andrews, UK

B E Russ
The University of Melbourne, Parkville, VIC, Australia

W T Ruyechan
The State University of New York, Buffalo, NY, USA

E Ryabov
University of Warwick, Warwick, UK

M D Ryan
University of St. Andrews, St. Andrews, UK

E P Rybicki
University of Cape Town, Cape Town, South Africa

K D Ryman
Louisiana State University Health Sciences Center at Shreveport, Shreveport, LA, USA

K D Ryman
Louisiana State University Health Sciences Center, Shreveport, LA, USA

K H Ryu
Seoul Women's University, Seoul, South Korea

M Safak
Temple University School of Medicine, Philadelphia, PA, USA

M Salas
Universidad Autónoma, Madrid, Spain

S K Samal
University of Maryland, College Park, MD, USA

J T Sample
The Pennsylvania State University College of Medicine, Hershey, PA, USA

C E Sample
The Pennsylvania State University College of Medicine, Hershey, PA, USA

R M Sandri-Goldin
University of California, Irvine, Irvine, CA, USA

H Sanfaçon
Pacific Agri-Food Research Centre, Summerland, BC, Canada

R Sanjuán
Instituto de Biología Molecular y Cellular de Plantas, CSIC-UPV, Valencia, Spain

N Santi
Norwegian School of Veterinary Science, Oslo, Norway

C Sarmiento
Tallinn University of Technology, Tallinn, Estonia

T Sasaya
National Agricultural Research Center, Ibaraki, Japan

Q J Sattentau
University of Oxford, Oxford, UK

C Savolainen-Kopra
National Public Health Institute (KTL), Helsinki, Finland

B Schaffhausen
Tufts University School of Medicine, Boston, MA, USA

K Scheets
Oklahoma State University, Stillwater, OK, USA

M J Schmitt
University of the Saarland, Saarbrücken, Germany

A Schneemann
The Scripps Research Institute, La Jolla, CA, USA

G Schoehn
CNRS, Grenoble, France

J E Schoelz
University of Missouri, Columbia, MO, USA

L B Schonberger
Centers for Disease Control and Prevention, Atlanta, GA, USA

U Schubert
Klinikum der Universität Erlangen-Nürnberg, Erlangen, Germany

D A Schultz
Johns Hopkins University School of Medicine, Baltimore, MD, USA

S Schultz-Cherry
University of Wisconsin – Madison, Madison, WI, USA

T F Schulz
Hannover Medical School, Hannover, Germany

P D Scotti
Waiatarua, New Zealand

B L Semler
University of California, Irvine, CA, USA

J M Sharp
Veterinary Laboratories Agency, Penicuik, UK

M L Shaw
Mount Sinai School of Medicine, New York, NY, USA

G R Shellam
The University of Western Australia,
Crawley, WA, Australia

D N Shepherd
University of Cape Town, Cape Town, South Africa

N C Sheppard
University of Oxford, Oxford, UK

F Shewmaker
National Institutes of Health, Bethesda, MD, USA

P A Signoret
Montpellier SupAgro, Montpellier, France

A Silaghi
University of Manitoba, Winnipeg, MB, Canada

G Silvestri
University of Pennsylvania, Philadelphia, PA, USA

T L Sit
North Carolina State University, Raleigh, NC, USA

N Sittidilokratna
Centex Shrimp and Center for Genetic Engineering and Biotechnology, Bangkok, Thailand

M A Skinner
Imperial College London, London, UK

D W Smith
PathWest Laboratory Medicine WA, Nedlands, WA, Australia

G L Smith
Imperial College London, London, UK

L M Smith
The University of Western Australia,
Crawley, WA, Australia

E J Snijder
Leiden University Medical Center, Leiden, The Netherlands

M Sova
University of Texas Medical Branch – Galveston, Galveston, TX, USA

J A Speir
The Scripps Research Institute, La Jolla, CA, USA

T E Spencer
Texas A&M University, College Station, TX, USA

P Sreenivasulu
Sri Venkateswara University, Tirupati, India

J Stanley
John Innes Centre, Colney, UK

K M Stedman
Portland State University, Portland, OR, USA

D Stephan
Institute of Plant Diseases and Plant Protection, Hannover, Germany

C C M M Stijger
Wageningen University and Research Centre, Naaldwijk, The Netherlands

L Stitz
Federal Research Institute for Animal Health, Tuebingen, Gemany

P G Stockley
University of Leeds, Leeds, UK

M R Strand
University of Georgia, Athens, GA, USA

M J Studdert
The University of Melbourne, Parkville, VIC, Australia

C A Suttle
University of British Columbia, Vancouver, BC, Canada

N Suzuki
Okayama University, Okayama, Japan

J Y Suzuki
USDA, Pacific Basin Agricultural Research Center, Hilo, HI, USA

R Swanepoel
National Institute for Communicable Diseases, Sandringham, South Africa

S J Symes
The University of Melbourne, Parkville, VIC, Australia

G Szittya
Agricultural Biotechnology Center, Godollo, Hungary

M Taliansky
Scottish Crop Research Institute, Dundee, UK

P Tattersall
Yale University Medical School, New Haven, CT, USA

T Tatusova
National Institutes of Health, Bethesda, MD, USA

S Tavantzis
University of Maine, Orono, ME, USA

J M Taylor
Fox Chase Cancer Center, Philadelphia, PA, USA

D A Theilmann
Agriculture and Agri-Food Canada, Summerland, BC, Canada

F C Thomas Allnutt
National Science Foundation, Arlington, VA, USA

G J Thomas Jr.
University of Missouri – Kansas City, Kansas City, MO, USA

J E Thomas
Department of Primary Industries and Fisheries, Indooroopilly, QLD, Australia

H C Thomas
Imperial College London, London, UK

A N Thorburn
The University of Melbourne, Parkville, VIC, Australia

P Tijssen
Université du Québec, Laval, QC, Canada

S A Tolin
Virginia Polytechnic Institute and State University, Blacksburg, VA, USA

L Torrance
Scottish Crop Research Institute, Invergowrie, UK

S Trapp
Cornell University, Ithaca, NY, USA

S Tripathi
USDA, Pacific Basin Agricultural Research Center, Hilo, HI, USA

E Truve
Tallinn University of Technology, Tallinn, Estonia

J-M Tsai
National Taiwan University, Taipei, Republic of China

M Tsompana
North Carolina State University, Raleigh, NC, USA

R Tuma
University of Helsinki, Helsinki, Finland

A S Turnell
The University of Birmingham, Birmingham, UK

K L Tyler
University of Colorado Health Sciences, Denver, CO, USA

A Uchiyama
Cornell University, Ithaca, NY, USA

C Upton
University of Victoria, Victoria, BC, Canada

A Urisman
University of California, San Francisco, San Francisco, CA, USA

J K Uyemoto
University of California, Davis, CA, USA

A Vaheri
University of Helsinki, Helsinki, Finland

R Vainionpää
University of Turku, Turku, Finland

A M Vaira
Istituto di Virologia Vegetale, CNR, Turin, Italy

N K Van Alfen
University of California, Davis, CA, USA

R A A Van der Vlugt
Wageningen University and Research Centre, Wageningen, The Netherlands

M H V Van Regenmortel
CNRS, Illkirch, France

P A Venter
The Scripps Research Institute, La Jolla, CA, USA

J Verchot-Lubicz
Oklahoma State University, Stillwater, OK, USA

R A Vere Hodge
Vere Hodge Antivirals Ltd., Reigate, UK

H J Vetten
Federal Research Centre for Agriculture and Forestry (BBA), Brunswick, Germany

L P Villarreal
University of California, Irvine, Irvine, CA, USA

J M Vlak
Wageningen University, Wageningen, The Netherlands

P K Vogt
The Scripps Research Institute, La Jolla, CA, USA

L E Volkman
University of California, Berkeley, Berkeley, CA, USA

J Votteler
Klinikum der Universität Erlangen-Nürnberg, Erlangen, Germany

D F Voytas
Iowa State University, Ames, IA, USA

J D F Wadsworth
University College London, London, UK

E K Wagner
University of California, Irvine, Irvine, CA, USA

P J Walker
CSIRO Australian Animal Health Laboratory, Geelong, VIC, Australia

A L Wang
University of California, San Francisco, CA, USA

X Wang
University of Wisconsin – Madison, Madison, WI, USA

C C Wang
University of California, San Francisco, CA, USA

L-F Wang
Australian Animal Health Laboratory, Geelong, VIC, Australia

R Warrier
Purdue University, West Lafayette, IN, USA

S C Weaver
University of Texas Medical Branch, Galveston, TX, USA

B A Webb
University of Kentucky, Lexington, KY, USA

F Weber
University of Freiburg, Freiburg, Germany

R P Weir
Berrimah Research Farm, Darwin, NT, Australia

R A Weisberg
National Institutes of Health, Bethesda, MD, USA

W Weissenhorn
UMR 5233 UJF-EMBL-CNRS, Grenoble, France

R M Welsh
University of Massachusetts Medical School, Worcester, MA, USA

J T West
University of Oklahoma Health Sciences Center, Oklahoma City, OK, USA

E Westhof
Université Louis Pasteur, Strasbourg, France

S P J Whelan
Harvard Medical School, Boston, MA, USA

R L White
Texas A&M University, College Station, TX, USA

C A Whitehouse
United States Army Medical Research Institute of Infectious Diseases, Frederick, MD, USA

R B Wickner
National Institutes of Health, Bethesda, MD, USA

R G Will
Western General Hospital, Edinburgh, UK

T Williams
Instituto de Ecología A.C., Xalapa, Mexico

K Willoughby
Moredun Research Institute, Edinburgh, UK

S Winter
Deutsche Sammlung für Mikroorganismen und Zellkulturen, Brunswick, Germany

J Winton
Western Fisheries Research Center, Seattle, WA, USA

J K Yamamoto
University of Florida, Gainesville, FL, USA

M Yoshida
University of Tokyo, Chiba, Japan

N Yoshikawa
Iwate University, Ueda, Japan

L S Young
University of Birmingham, Birmingham, UK

R F Young, III
Texas A&M University, College Station, TX, USA

T M Yuill
University of Wisconsin, Madison, WI, USA

A J Zajac
University of Alabama at Birmingham, Birmingham, AL, USA

S K Zavriev
Shemyakin and Ovchinnikov Institute of Bioorganic Chemistry, Russian Academy of Sciences, Moscow, Russia

J Ziebuhr
The Queen's University of Belfast, Belfast, UK

E I Zuniga
The Scripps Research Institute, La Jolla, CA, USA

CONTENTS

Editors-in-Chief	v
Associate Editors	vii
Preface	ix
How to Use the Encyclopedia	xi
Contributors	xiii

VOLUME 1

A

Adenoviruses: General Features	B Harrach	1
Adenoviruses: Malignant Transformation and Oncology	A S Turnell	9
Adenoviruses: Molecular Biology	K N Leppard	17
Adenoviruses: Pathogenesis	M Benkő	24
African Cassava Mosaic Disease	J P Legg	30
African Horse Sickness Viruses	P S Mellor and P P C Mertens	37
African Swine Fever Virus	L K Dixon and D Chapman	43
AIDS: Disease Manifestation	A Rapose, J East, M Sova and W A O'Brien	51
AIDS: Global Epidemiology	P J Peters, P H Kilmarx and T D Mastro	58
AIDS: Vaccine Development	N C Sheppard and Q J Sattentau	69
Akabane Virus	P S Mellor and P D Kirkland	76
Alfalfa Mosaic Virus	J F Bol	81
Algal Viruses	K Nagasaki and C P D Brussaard	87
Allexivirus	S K Zavriev	96
Alphacryptovirus and *Betacryptovirus*	R Blawid, D Stephan and E Maiss	98
Anellovirus	P Biagini and P de Micco	104
Animal Rhabdoviruses	H Bourhy, A J Gubala, R P Weir and D B Boyle	111
Antigen Presentation	E I Zuniga, D B McGavern and M B A Oldstone	121
Antigenic Variation	G M Air and J T West	127
Antigenicity and Immunogenicity of Viral Proteins	M H V Van Regenmortel	137

Antiviral Agents H J Field and R A Vere Hodge	142
Apoptosis and Virus Infection J R Clayton and J M Hardwick	154
Aquareoviruses M St J Crane and G Carlile	163
Arboviruses B R Miller	170
Arteriviruses M A Brinton and E J Snijder	176
Ascoviruses B A Federici and Y Bigot	186
Assembly of Viruses: Enveloped Particles C K Navaratnarajah, R Warrier and R J Kuhn	193
Assembly of Viruses: Nonenveloped Particles M Luo	200
Astroviruses L Moser and S Schultz-Cherry	204

B

Baculoviruses: Molecular Biology of Granuloviruses S Hilton	211
Baculoviruses: Molecular Biology of Mosquito Baculoviruses J J Becnel and C L Afonso	219
Baculoviruses: Molecular Biology of Sawfly Baculoviruses B M Arif	225
Baculoviruses: Apoptosis Inhibitors R J Clem	231
Baculoviruses: Expression Vector F J Haines, R D Possee and L A King	237
Baculoviruses: General Features P J Krell	247
Baculoviruses: Molecular Biology of Nucleopolyhedroviruses D A Theilmann and G W Blissard	254
Baculoviruses: Pathogenesis L E Volkman	265
Banana Bunchy Top Virus J E Thomas	272
Barley Yellow Dwarf Viruses L L Domier	279
Barnaviruses P A Revill	286
Bean Common Mosaic Virus and Bean Common Mosaic Necrosis Virus R Jordan and J Hammond	288
Bean Golden Mosaic Virus F J Morales	295
Beet Curly Top Virus J Stanley	301
Benyvirus R Koenig	308
Beta ssDNA Satellites R W Briddon and S Mansoor	314
Birnaviruses B Delmas	321
Bluetongue Viruses P Roy	328
Border Disease Virus P Nettleton and K Willoughby	335
Bornaviruses L Stitz, O Planz and W Ian Lipkin	341
Bovine and Feline Immunodeficiency Viruses J K Yamamoto	347
Bovine Ephemeral Fever Virus P J Walker	354
Bovine Herpesviruses M J Studdert	362
Bovine Spongiform Encephalopathy R G Will	368
Bovine Viral Diarrhea Virus J F Ridpath	374
Brome Mosaic Virus X Wang and P Ahlquist	381
Bromoviruses J J Bujarski	386

Bunyaviruses: General Features *R M Elliott*	390
Bunyaviruses: Unassigned *C H Calisher*	399

C

Cacao Swollen Shoot Virus *E Muller*	403
Caliciviruses *M J Studdert and S J Symes*	410
Capillovirus, Foveavirus, Trichovirus, Vitivirus *N Yoshikawa*	419
Capripoxviruses *R P Kitching*	427
Capsid Assembly: Bacterial Virus Structure and Assembly *S Casjens*	432
Cardioviruses *C Billinis and O Papadopoulos*	440
Carlavirus *K H Ryu and B Y Lee*	448
Carmovirus *F Qu and T J Morris*	453
Caulimoviruses: General Features *J E Schoelz*	457
Caulimoviruses: Molecular Biology *T Hohn*	464
Central Nervous System Viral Diseases *R T Johnson and B M Greenberg*	469
Cereal Viruses: Maize/Corn *P A Signoret*	475
Cereal Viruses: Rice *F Morales*	482
Cereal Viruses: Wheat and Barley *H D Lapierre and D Hariri*	490
Chandipura Virus *S Basak and D Chattopadhyay*	497
Chrysoviruses *S A Ghabrial*	503
Circoviruses *A Mankertz*	513
Citrus Tristeza Virus *M Bar-Joseph and W O Dawson*	520
Classical Swine Fever Virus *V Moennig and I Greiser-Wilke*	525
Coltiviruses *H Attoui and X de Lamballerie*	533
Common Cold Viruses *S Dreschers and C Adams*	541
Coronaviruses: General Features *D Cavanagh and P Britton*	549
Coronaviruses: Molecular Biology *S C Baker*	554
Cotton Leaf Curl Disease *S Mansoor, I Amin and R W Briddon*	563
Cowpea Mosaic Virus *G P Lomonossoff*	569
Cowpox Virus *M Bennett, G L Smith and D Baxby*	574
Coxsackieviruses *M S Oberste and M A Pallansch*	580
Crenarchaeal Viruses: Morphotypes and Genomes *D Prangishvili, T Basta and R A Garrett*	587
Crimean–Congo Hemorrhagic Fever Virus and Other Nairoviruses *C A Whitehouse*	596
Cryo-Electron Microscopy *W Chiu, J T Chang and F J Rixon*	603
Cucumber Mosaic Virus *F García-Arenal and P Palukaitis*	614
Cytokines and Chemokines *D E Griffin*	620
Cytomegaloviruses: Murine and Other Nonprimate Cytomegaloviruses *A J Redwood, L M Smith and G R Shellam*	624
Cytomegaloviruses: Simian Cytomegaloviruses *D J Alcendor and G S Hayward*	634

VOLUME 2

D

Defective-Interfering Viruses *L Roux*	1
Dengue Viruses *D J Gubler*	5
Diagnostic Techniques: Microarrays *K Fischer, A Urisman and J DeRisi*	14
Diagnostic Techniques: Plant Viruses *R Koenig, D-E Lesemann, G Adam and S Winter*	18
Diagnostic Techniques: Serological and Molecular Approaches *R Vainionpää and P Leinikki*	29
Dicistroviruses *P D Christian and P D Scotti*	37
Disease Surveillance *N Noah*	44
DNA Vaccines *S Babiuk and L A Babiuk*	51

E

Ebolavirus *K S Brown, A Silaghi and H Feldmann*	57
Echoviruses *T Hyypiä*	65
Ecology of Viruses Infecting Bacteria *S T Abedon*	71
Electron Microscopy of Viruses *G Schoehn and R W H Ruigrok*	78
Emerging and Reemerging Virus Diseases of Plants *G P Martelli and D Gallitelli*	86
Emerging and Reemerging Virus Diseases of Vertebrates *B W J Mahy*	93
Emerging Geminiviruses *C M Fauquet and M S Nawaz-ul-Rehman*	97
Endogenous Retroviruses *W E Johnson*	105
Endornavirus *T Fukuhara and H Moriyama*	109
Enteric Viruses *R F Bishop and C D Kirkwood*	116
Enteroviruses of Animals *L E Hughes and M D Ryan*	123
Enteroviruses: Human Enteroviruses Numbered 68 and Beyond *T Hovi, S Blomqvist, C Savolainen-Kopra and M Roivainen*	130
Entomopoxviruses *M N Becker and R W Moyer*	136
Epidemiology of Human and Animal Viral Diseases *F A Murphy*	140
Epstein–Barr Virus: General Features *L S Young*	148
Epstein–Barr Virus: Molecular Biology *J T Sample and C E Sample*	157
Equine Infectious Anemia Virus *J K Craigo and R C Montelaro*	167
Evolution of Viruses *L P Villarreal*	174

F

Feline Leukemia and Sarcoma Viruses *J C Neil*	185
Filamentous ssDNA Bacterial Viruses *S A Overman and G J Thomas Jr.*	190
Filoviruses *G Olinger, T W Geisbert and L E Hensley*	198
Fish and Amphibian Herpesviruses *A J Davison*	205
Fish Retroviruses *T A Paul, R N Casey, P R Bowser, J W Casey, J Rovnak and S L Quackenbush*	212

Fish Rhabdoviruses *G Kurath and J Winton*	221
Fish Viruses *J C Leong*	227
Flaviviruses of Veterinary Importance *R Swanepoel and F J Burt*	234
Flaviviruses: General Features *T J Chambers*	241
Flexiviruses *M J Adams*	253
Foamy Viruses *M L Linial*	259
Foot and Mouth Disease Viruses *D J Rowlands*	265
Fowlpox Virus and Other Avipoxviruses *M A Skinner*	274
Fungal Viruses *S A Ghabrial and N Suzuki*	284
Furovirus *R Koenig*	291
Fuselloviruses of Archaea *K M Stedman*	296

G

Gene Therapy: Use of Viruses as Vectors *K I Berns and T R Flotte*	301
Genome Packaging in Bacterial Viruses *P Jardine*	306
Giardiaviruses *A L Wang and C C Wang*	312

H

Hantaviruses *A Vaheri*	317
Henipaviruses *B T Eaton and L-F Wang*	321
Hepadnaviruses of Birds *A R Jilbert and W S Mason*	327
Hepadnaviruses: General Features *T J Harrison*	335
Hepatitis A Virus *A Dotzauer*	343
Hepatitis B Virus: General Features *P Karayiannis and H C Thomas*	350
Hepatitis B Virus: Molecular Biology *T J Harrison*	360
Hepatitis C Virus *R Bartenschlager and S Bühler*	367
Hepatitis Delta Virus *J M Taylor*	375
Hepatitis E Virus *X J Meng*	377
Herpes Simplex Viruses: General Features *L Aurelian*	383
Herpes Simplex Viruses: Molecular Biology *E K Wagner and R M Sandri-Goldin*	397
Herpesviruses of Birds *S Trapp and N Osterrieder*	405
Herpesviruses of Horses *D J O'Callaghan and N Osterrieder*	411
Herpesviruses: Discovery *B Ehlers*	420
Herpesviruses: General Features *A J Davison*	430
Herpesviruses: Latency *C M Preston*	436
History of Virology: Bacteriophages *H-W Ackermann*	442
History of Virology: Plant Viruses *R Hull*	450
History of Virology: Vertebrate Viruses *F J Fenner*	455

Hordeivirus J N Bragg, H-S Lim and A O Jackson	459
Host Resistance to Retroviruses T Hatziioannou and P D Bieniasz	467
Human Cytomegalovirus: General Features E S Mocarski Jr. and R F Pass	474
Human Cytomegalovirus: Molecular Biology W Gibson	485
Human Eye Infections J Chodosh, A V Chintakuntlawar and C M Robinson	491
Human Herpesviruses 6 and 7 U A Gompels	498
Human Immunodeficiency Viruses: Antiretroviral Agents A W Neuman and D C Liotta	505
Human Immunodeficiency Viruses: Molecular Biology J Votteler and U Schubert	517
Human Immunodeficiency Viruses: Origin F van Heuverswyn and M Peeters	525
Human Immunodeficiency Viruses: Pathogenesis N R Klatt, A Chahroudi and G Silvestri	534
Human Respiratory Syncytial Virus P L Collins	542
Human Respiratory Viruses J E Crowe Jr.	551
Human T-Cell Leukemia Viruses: General Features M Yoshida	558
Human T-Cell Leukemia Viruses: Human Disease R Mahieux and A Gessain	564
Hypovirulence N K Van Alfen and P Kazmierczak	574
Hypoviruses D L Nuss	580

VOLUME 3

I

Icosahedral dsDNA Bacterial Viruses with an Internal Membrane J K H Bamford and S J Butcher	1
Icosahedral Enveloped dsRNA Bacterial Viruses R Tuma	6
Icosahedral ssDNA Bacterial Viruses B A Fane, M Chen, J E Cherwa and A Uchiyama	13
Icosahedral ssRNA Bacterial Viruses P G Stockley	21
Icosahedral Tailed dsDNA Bacterial Viruses R L Duda	30
Idaeovirus A T Jones and H Barker	37
Iflavirus M M van Oers	42
Ilarvirus K C Eastwell	46
Immune Response to Viruses: Antibody-Mediated Immunity A R Neurath	56
Immune Response to Viruses: Cell-Mediated Immunity A J Zajac and L E Harrington	70
Immunopathology M B A Oldstone and R S Fujinami	78
Infectious Pancreatic Necrosis Virus Ø Evensen and N Santi	83
Infectious Salmon Anemia Virus B H Dannevig, S Mjaaland and E Rimstad	89
Influenza R A Lamb	95
Innate Immunity: Defeating C F Basler	104
Innate Immunity: Introduction F Weber	111
Inoviruses L A Day	117
Insect Pest Control by Viruses M Erlandson	125
Insect Reoviruses P P C Mertens and H Attoui	133
Insect Viruses: Nonoccluded J P Burand	144

Interfering RNAs	K E Olson, K M Keene and C D Blair	148
Iridoviruses of Vertebrates	A D Hyatt and V G Chinchar	155
Iridoviruses of Invertebrates	T Williams and A D Hyatt	161
Iridoviruses: General Features	V G Chinchar and A D Hyatt	167

J

Jaagsiekte Sheep Retrovirus	J M Sharp, M de las Heras, T E Spencer and M Palmarini	175
Japanese Encephalitis Virus	A D T Barrett	182

K

Kaposi's Sarcoma-Associated Herpesvirus: General Features	Y Chang and P S Moore	189
Kaposi's Sarcoma-Associated Herpesvirus: Molecular Biology	E Gellermann and T F Schulz	195

L

Lassa, Junin, Machupo and Guanarito Viruses	J B McCormick	203
Legume Viruses	L Bos	212
Leishmaniaviruses	R Carrion Jr, Y-T Ro and J L Patterson	220
Leporipoviruses and Suipoxviruses	G McFadden	225
Luteoviruses	L L Domier and C J D'Arcy	231
Lymphocytic Choriomeningitis Virus: General Features	R M Welsh	238
Lymphocytic Choriomeningitis Virus: Molecular Biology	J C de la Torre	243
Lysis of the Host by Bacteriophage	R F Young III and R L White	248

M

Machlomovirus	K Scheets	259
Maize Streak Virus	D P Martin, D N Shepherd and E P Rybicki	263
Marburg Virus	D Falzarano and H Feldmann	272
Marnaviruses	A S Lang and C A Suttle	280
Measles Virus	R Cattaneo and M McChesney	285
Membrane Fusion	A Hinz and W Weissenhorn	292
Metaviruses	H L Levin	301
Mimivirus	J-M Claverie	311
Molluscum Contagiosum Virus	J J Bugert	319
Mononegavirales	A J Easton and R Ling	324
Mouse Mammary Tumor Virus	J P Dudley	334
Mousepox and Rabbitpox Viruses	M Regner, F Fenner and A Müllbacher	342
Movement of Viruses in Plants	P A Harries and R S Nelson	348

Mumps Virus *B K Rima and W P Duprex*	356
Mungbean Yellow Mosaic Viruses *V G Malathi and P John*	364
Murine Gammaherpesvirus 68 *A A Nash and B M Dutia*	372
Mycoreoviruses *B I Hillman*	378

N

Nanoviruses *H J Vetten*	385
Narnaviruses *R Esteban and T Fujimura*	392
Nature of Viruses *M H V Van Regenmortel*	398
Necrovirus *L Rubino and G P Martelli*	403
Nepovirus *H Sanfaçon*	405
Neutralization of Infectivity *P J Klasse*	413
Nidovirales *L Enjuanes, A E Gorbalenya, R J de Groot, J A Cowley, J Ziebuhr and E J Snijder*	419
Nodaviruses *P A Venter and A Schneemann*	430
Noroviruses and Sapoviruses *K Y Green*	438

O

Ophiovirus *A M Vaira and R G Milne*	447
Orbiviruses *P P C Mertens, H Attoui and P S Mellor*	454
Organ Transplantation, Risks *C N Kotton, M J Kuehnert and J A Fishman*	466
Origin of Viruses *P Forterre*	472
Orthobunyaviruses *C H Calisher*	479
Orthomyxoviruses: Molecular Biology *M L Shaw and P Palese*	483
Orthomyxoviruses: Structure of Antigens *R J Russell*	489
Oryctes Rhinoceros Virus *J M Vlak, A M Huger, J A Jehle and R G Kleespies*	495
Ourmiavirus *G P Accotto and R G Milne*	500

VOLUME 4

P

Papaya Ringspot Virus *D Gonsalves, J Y Suzuki, S Tripathi and S A Ferreira*	1
Papillomaviruses: General Features of Human Viruses *G Orth*	8
Papillomaviruses: Molecular Biology of Human Viruses *P F Lambert and A Collins*	18
Papillomaviruses of Animals *A A McBride*	26
Papillomaviruses: General Features *H U Bernard*	34
Paramyxoviruses of Animals *S K Samal*	40
Parainfluenza Viruses of Humans *E Adderson and A Portner*	47
Paramyxoviruses *R E Dutch*	52
Parapoxviruses *D Haig and A A Mercer*	57

Partitiviruses of Fungi *S Tavantzis*	63
Partitiviruses: General Features *S A Ghabrial, W F Ochoa, T S Baker and M L Nibert*	68
Parvoviruses of Arthropods *M Bergoin and P Tijssen*	76
Parvoviruses of Vertebrates *C R Parrish*	85
Parvoviruses: General Features *P Tattersall*	90
Pecluvirus *D V R Reddy, C Bragard, P Sreenivasulu and P Delfosse*	97
Pepino Mosaic Virus *R A A Van der Vlugt and C C M M Stijger*	103
Persistent and Latent Viral Infection *E S Mocarski and A Grakoui*	108
Phycodnaviruses *J L Van Etten and M V Graves*	116
Phylogeny of Viruses *A E Gorbalenya*	125
Picornaviruses: Molecular Biology *B L Semler and K J Ertel*	129
Plant Antiviral Defense: Gene Silencing Pathway *G Szittya, T Dalmay and J Burgyan*	141
Plant Reoviruses *R J Geijskes and R M Harding*	149
Plant Resistance to Viruses: Engineered Resistance *M Fuchs*	156
Plant Resistance to Viruses: Geminiviruses *J K Brown*	164
Plant Resistance to Viruses: Natural Resistance Associated with Dominant Genes *P Moffett and D F Klessig*	170
Plant Resistance to Viruses: Natural Resistance Associated with Recessive Genes *C Caranta and C Dogimont*	177
Plant Rhabdoviruses *A O Jackson, R G Dietzgen, R-X Fang, M M Goodin, S A Hogenhout, M Deng and J N Bragg*	187
Plant Virus Diseases: Economic Aspects *G Loebenstein*	197
Plant Virus Diseases: Fruit Trees and Grapevine *G P Martelli and J K Uyemoto*	201
Plant Virus Diseases: Ornamental Plants *J Engelmann and J Hamacher*	207
Plant Virus Vectors (Gene Expression Systems) *Y Gleba, S Marillonnet and V Klimyuk*	229
Plum Pox Virus *M Glasa and T Candresse*	238
Poliomyelitis *P D Minor*	243
Polydnaviruses: Abrogation of Invertebrate Immune Systems *M R Strand*	250
Polydnaviruses: General Features *A Fath-Goodin and B A Webb*	257
Polyomaviruses of Humans *M Safak and K Khalili*	262
Polyomaviruses of Mice *B Schaffhausen*	271
Polyomaviruses *M Gravell and E O Major*	277
Pomovirus *L Torrance*	283
Potato Virus Y *C Kerlan and B Moury*	288
Potato Viruses *C Kerlan*	302
Potexvirus *K H Ryu and J S Hong*	310
Potyviruses *J J López-Moya and J A García*	314
Poxviruses *G L Smith, P Beard and M A Skinner*	325
Prions of Vertebrates *J D F Wadsworth and J Collinge*	331
Prions of Yeast and Fungi *R B Wickner, H Edskes, T Nakayashiki, F Shewmaker, L McCann, A Engel and D Kryndushkin*	338

Pseudorabies Virus *T C Mettenleiter*	342
Pseudoviruses *D F Voytas*	352

Q

Quasispecies *R Sanjuán*	359

R

Rabies Virus *I V Kuzmin and C E Rupprecht*	367
Recombination *J J Bujarski*	374
Reoviruses: General Features *P Clarke and K L Tyler*	382
Reoviruses: Molecular Biology *K M Coombs*	390
Replication of Bacterial Viruses *M Salas and M de Vega*	399
Replication of Viruses *A J Cann*	406
Reticuloendotheliosis Viruses *A S Liss and H R Bose Jr.*	412
Retrotransposons of Fungi *T J D Goodwin, M I Butler and R T M Poulter*	419
Retrotransposons of Plants *M-A Grandbastien*	428
Retrotransposons of Vertebrates *A E Peaston*	436
Retroviral Oncogenes *P K Vogt and A G Bader*	445
Retroviruses of Insects *G F Rohrmann*	451
Retroviruses of Birds *K L Beemon*	455
Retroviruses: General Features *E Hunter*	459
Rhinoviruses *N W Bartlett and S L Johnston*	467
Ribozymes *E Westhof and A Lescoute*	475
Rice Tungro Disease *R Hull*	481
Rice Yellow Mottle Virus *E Hébrard and D Fargette*	485
Rift Valley Fever and Other Phleboviruses *L Nicoletti and M G Ciufolini*	490
Rinderpest and Distemper Viruses *T Barrett*	497
Rotaviruses *J Angel, M A Franco and H B Greenberg*	507
Rubella Virus *T K Frey*	514

S

Sadwavirus *T Iwanami*	523
Satellite Nucleic Acids and Viruses *P Palukaitis, A Rezaian and F García-Arenal*	526
Seadornaviruses *H Attoui and P P C Mertens*	535
Sequiviruses *I-R Choi*	546
Severe Acute Respiratory Syndrome (SARS) *J S M Peiris and L L M Poon*	552
Shellfish Viruses *T Renault*	560
Shrimp Viruses *J-R Bonami*	567

Sigma Rhabdoviruses	*D Contamine and S Gaumer*	576
Simian Alphaherpesviruses	*J Hilliard*	581
Simian Gammaherpesviruses	*A Ensser*	585
Simian Immunodeficiency Virus: Animal Models of Disease	*C J Miller and M Marthas*	594
Simian Immunodeficiency Virus: General Features	*M E Laird and R C Desrosiers*	603
Simian Immunodeficiency Virus: Natural Infection	*I Pandrea, G Silvestri and C Apetrei*	611
Simian Retrovirus D	*P A Marx*	623
Simian Virus 40	*A L McNees and J S Butel*	630
Smallpox and Monkeypox Viruses	*S Parker, D A Schultz, H Meyer and R M Buller*	639
Sobemovirus	*M Meier, A Olspert, C Sarmiento and E Truve*	644
St. Louis Encephalitis	*W K Reisen*	652
Sweetpotato Viruses	*J Kreuze and S Fuentes*	659

VOLUME 5

T

Taura Syndrome Virus	*A K Dhar and F C T Allnutt*	1
Taxonomy, Classification and Nomenclature of Viruses	*C M Fauquet*	9
Tenuivirus	*B C Ramirez*	24
Tetraviruses	*J A Speir and J E Johnson*	27
Theiler's Virus	*H L Lipton, S Hertzler and N Knowles*	37
Tick-Borne Encephalitis Viruses	*T S Gritsun and E A Gould*	45
Tobacco Mosaic Virus	*M H V Van Regenmortel*	54
Tobacco Viruses	*S A Tolin*	60
Tobamovirus	*D J Lewandowski*	68
Tobravirus	*S A MacFarlane*	72
Togaviruses Causing Encephalitis	*S Paessler and M Pfeffer*	76
Togaviruses Causing Rash and Fever	*D W Smith, J S Mackenzie and M D A Lindsay*	83
Togaviruses Not Associated with Human Disease	*L L Coffey,*	91
Togaviruses: Alphaviruses	*A M Powers*	96
Togaviruses: Equine Encephalitic Viruses	*D E Griffin*	101
Togaviruses: General Features	*S C Weaver, W B Klimstra and K D Ryman*	107
Togaviruses: Molecular Biology	*K D Ryman, W B Klimstra and S C Weaver*	116
Tomato Leaf Curl Viruses from India	*S Chakraborty*	124
Tomato Spotted Wilt Virus	*H R Pappu*	133
Tomato Yellow Leaf Curl Virus	*H Czosnek*	138
Tombusviruses	*S A Lommel and T L Sit*	145
Torovirus	*A E Hoet and M C Horzinek*	151
Tospovirus	*M Tsompana and J W Moyer*	157
Totiviruses	*S A Ghabrial*	163

Transcriptional Regulation in Bacteriophage R A Weisberg, D M Hinton and S Adhya 174
Transmissible Spongiform Encephalopathies E D Belay and L B Schonberger 186
Tumor Viruses: Human R Grassmann, B Fleckenstein and H Pfister 193
Tymoviruses A-L Haenni and T W Dreher 199

U

Umbravirus M Taliansky and E Ryabov 209
Ustilago Maydis Viruses J Bruenn 214

V

Vaccine Production in Plants E P Rybicki 221
Vaccine Safety C J Clements and G Lawrence 226
Vaccine Strategies I Kusters and J W Almond 235
Vaccinia Virus G L Smith 243
Varicella-Zoster Virus: General Features J I Cohen 250
Varicella-Zoster Virus: Molecular Biology W T Ruyechan and J Hay 256
Varicosavirus T Sasaya 263
Vector Transmission of Animal Viruses W K Reisen 268
Vector Transmission of Plant Viruses S Blanc 274
Vegetable Viruses P Caciagli 282
Vesicular Stomatitis Virus S P J Whelan 291
Viral Killer Toxins M J Schmitt 299
Viral Membranes J Lenard 308
Viral Pathogenesis N Nathanson 314
Viral Receptors D J Evans 319
Viral Suppressors of Gene Silencing J Verchot-Lubicz and J P Carr 325
Viroids R Flores and R A Owens 332
Virus Classification by Pairwise Sequence Comparison (PASC) Y Bao, Y Kapustin and T Tatusova 342
Virus Databases E J Lefkowitz, M R Odom and C Upton 348
Virus Entry to Bacterial Cells M M Poranen and A Domanska 365
Virus Evolution: Bacterial Viruses R W Hendrix 370
Virus-Induced Gene Silencing (VIGS) M S Padmanabhan and S P Dinesh-Kumar 375
Virus Particle Structure: Nonenveloped Viruses J A Speir and J E Johnson 380
Virus Particle Structure: Principles J E Johnson and J A Speir 393
Virus Species M H V van Regenmortel 401
Viruses and Bioterrorism R F Meyer and S A Morse 406
Viruses Infecting Euryarchaea K Porter, B E Russ, A N Thorburn and M L Dyall-Smith 411
Visna-Maedi Viruses B A Blacklaws 423

W

Watermelon Mosaic Virus and Zucchini Yellow Mosaic Virus	H Lecoq and C Desbiez	433
West Nile Virus	L D Kramer	440
White Spot Syndrome Virus	J-H Leu, J-M Tsai and C-F Lo	450

Y

Yatapoxviruses	J W Barrett and G McFadden	461
Yeast L-A Virus	R B Wickner, T Fujimura and R Esteban	465
Yellow Fever Virus	A A Marfin and T P Monath	469
Yellow Head Virus	P J Walker and N Sittidilokratna	476

Z

Zoonoses	J E Osorio and T M Yuill	485

Taxonomic Index   497

Subject Index   499

T

Taura Syndrome Virus

A K Dhar, Advanced BioNutrition Corp, Columbia, MD, USA
F C Thomas Allnutt, National Science Foundation, Arlington, VA, USA

© 2008 Elsevier Ltd. All rights reserved.

Glossary

Internal ribosomal entry site (IRES) The sequence in a RNA transcript that allows internal entry of the ribosomal assembly for translation in a cap-independent manner.
Quasispecies Due to the error-prone replication of RNA, a specific RNA virus exists as a steady state or equilibrium mixture of mutant sequences clustered around a master sequence or sequences.
Specific pathogen free (SPF) Shrimp lines that have been bred in captivity and certified to be free of a specific disease or diseases.

Introduction

Taura syndrome disease (TSD) was first described as a shrimp disease in 1992 from Pacific white shrimp (*Penaeus vannamei*) farms located along the mouth of the Taura River basin in Guayas Province, Ecuador. Initially the disease was incorrectly attributed to the toxic effects of fungicides used in a near by banana plantation. The etiologic agent was later proven to be of viral origin. The virus was named Taura syndrome virus (TSV) based on the location where the disease was first recorded. Movement of TSV-contaminated shrimp stocks led to the rapid spread of the disease throughout the Americas between 1992 and 1996. Since the initial discovery of TSD, the economic losses to the shrimp farming industry in the Western Hemisphere have been enormous. From the Americas, the disease spread to Taiwan in 1999, and subsequently TSV epizootics have been recorded in China, Thailand, Korea, and Indonesia.

TSV particles are nonenveloped and icosahedral, measuring 31–32 nm in diameter (**Figure 1**), and have a buoyant density of $1.338 \, \text{g ml}^{-1}$. The viral capsid contains three major (VP1, VP2, and VP3) and one minor (VP0) polypeptide. TSV genome is a linear, positive-sense, single-stranded RNA of approximately 10.2 kb. Nascent viral RNA synthesis occurs in intracellular vesicle-like membranes within the cytosol of the host cells. In infected cells, mature virions are also seen in close association with proliferating membranes. Currently, TSV has been classified as an unassigned member of the family *Dicistroviridae*.

History and Classification

TSD, caused by the TSV, is listed as a reportable disease by the Office Internationale des Epizooties (OIE), the international organization responsible for ensuring worldwide information transparency by dissemination of the current status of important animal diseases. TSD etiology was initially very controversial causing uncertainty that contributed to TSV spread through the Western Hemisphere between 1992 and 1996. The resulting epizootic cost the shrimp farming industry in the Americas an estimated US$1.2–2 billion. In 1999, TSD was reported from Taiwan and since then, the disease has spread to China, Thailand, Korea, and Indonesia. The disease has already caused considerable economic losses in Asia.

Based on clinical and histological data, researchers from Ecuador recognized a new syndrome in *P. vannamei* shrimp farms located along the mouth of the Taura River basin (Guayas Province, Ecuador) during the summer of 1992. The cause of TSD was initially attributed to toxicity by two systemic fungicides, Tilt (Propiconazole, Ciba-Geigy) and Calixin (Tridemorph, BASF) that were being sprayed on local banana plantations to control black leaf wilt disease or Singatoka negra. It was hypothesized that fungicide-contaminated runoff resulted in mortalities in shrimp farms. During the 2 years following its discovery, TSD spread west through Ecuador's most concentrated shrimp-farming region, southward into Northern Peru, and northward to the Pacific Coast of Colombia. These latter regions had neither banana plantations nor application of the suspected fungicides. Meanwhile, attempts to induce TSD experimentally by waterborne, oral, or injection delivery of Tilt, Calixin, or Benlate O.D. (Benomyl, DuPont, a widely used banana fungicide in Ecuador) failed. During May 1994, two separate outbreaks of TSD occurred among cultured *P. vannamei* in Oahu (Hawaii, USA), representing the first TSD incursion into the United States. Subsequently, the induction of TSD clinical signs and histological lesions was demonstrated by experimental challenge of healthy shrimp *per os*, and the viral etiology was established. The virus was named Taura syndrome virus.

Based on morphology and biophysical properties, TSV was then presumed to be a member of the family *Picornaviridae*. Availability of TSV genome sequence data and phylogenetic analysis using the capsid gene sequence of TSV and picornavirus superfamily viruses, revealed two major clusters (**Figure 2**). One cluster contains mammalian picornaviruses and picornavirus superfamily viruses infecting insects while the other cluster contains the plant picornaviruses. Within the first cluster, there were two subclusters. First subcluster contains insect picornavirus superfamily viruses with a dicistronic genome. The second subcluster contains two groups: one group includes sacbrood virus (SBV) of honeybee and infectious flacherie virus (IFV) of silkworm that share more similarities with mammalian than insect picornaviruses and the

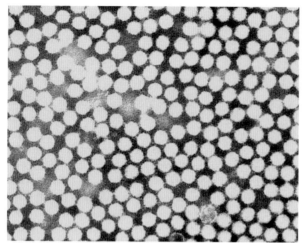

Figure 1 Transmission electron micrograph of CsCl gradient-purified and negatively stained (with 2% PTA) TSV particles. The virus particles are nonenveloped and icosahedral, measuring 31–32 nm in diameter.

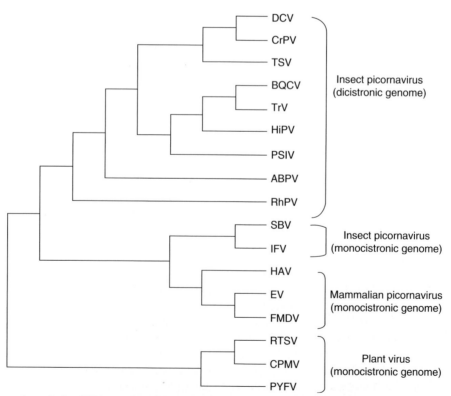

Figure 2 Phylogenetic analysis of TSV and other RNA viruses infecting insects, plants, and mammalian hosts. Reproduced from Dhar AK, Cowley JA, Hasson KW, and Walker PJ (2004) Genome organization, biology and diagnosis of Taura syndrome virus and yellowhead virus of penaeid shrimp. *Advances in Virus Research* 63: 353–419, with permission from Elsevier.

other group is the mammalian picornaviruses (**Figure 2**). The picornavirus superfamily viruses with dicistronic genomes were subsequently classified under a new family, *Dicistroviridae*, of which TSV is a new but unassigned member.

To date, the genomes of RNA viruses infecting a wide range of insect species have been sequenced. These include cricket paralysis virus (CrPV, AF218039), drosophila C virus (DCV, AF014388), acute bee paralysis virus (ABPV, NC002548), black queen cell virus (BQCV, AF183905), sacbrood virus (SBV, AF092924), rhopalosiphum padi virus (RhPV, AF022937), plautia stali intestine virus (PSIV, AB006531), triatoma virus (TrV, AF178440), himetobi P virus, (HiPV, AB017037), infectious falcherie virus (IFV, AB000906), homalodisca coagulata virus-1 (HoCV-1, DQ288865), aphid lethal paralysis virus (ALPV, AF536531), Kashmir bee virus (KBV, AY275710), solenopsis invicta virus-1 (SINV-1, AY634314), and TSV of shrimp (AF277675). The genome organization of IFV and SBV is similar to mammalian picornaviruses. They contain a single open reading frame (ORF) with the capsid genes located at the 5′ end and nonstructural genes at the 3′ end. In contrast, the genomes of CrPV, DCV, ABPV, BQCV, RhPV, PSIV, TrV, HiPV, HoCv-1, ALPV, KBV, SINV-1, and TSV contain two ORFs (ORF1 and ORF2) separated by an intergenic region (IGR). At the 5′ end, ORF1 encodes the nonstructural genes, and at the 3′ end ORF2 encodes the capsid proteins (as seen for TSV in **Figure 5**). All show greater sequence similarity to each other than to mammalian picornaviruses. To date, eight of these viruses (ALPV, BQCV, CrPV, DCV, HiPV, PSIV, RhPV, and TrV) have been assigned into the newly created family *Dicistroviridae*, genus *Cripavirus*, with CrPV as the type species. SINV-1 is tentatively assigned to this genus, while APBL and TSV are net assigned to any genus but placed in the *Dicistroviridae* family.

Host Range and Clinical Signs

Species of penaeid shrimp differ in their susceptibility to TSV. *Penaeus vannamei*, the principal penaeid species cultivated in the Americas, and *P. schmitti* are highly susceptible to TSV. TSV causes serious disease in postlarval (PL), juvenile, and adult *P. vannamei*, but has not been reported in PL smaller than the PL11 stage (11 day old). Other American penaeid species, such as *P. stylirostris*, *P. setiferus*, *P. duorarum*, and *P. aztecus*, are generally less susceptible to TSV. Among the Eastern Hemisphere penaeid species, TSV infection has been reported in *P. monodon*, *P. japonicus*, and *P. chinensis*.

To combat catastrophic losses caused by TSV in *P. vannamei*, TSV-resistant lines of specific pathogen-free (SPF) *P. vannamei* have been developed by selective breeding. However, due to the ability of the virus to mutate rapidly, less susceptible or resistant species and selected breeding lines may succumb to TSD as more virulent strains emerge. This was observed during TSV epizootics in Mexico, in 1999, when TSV-tolerant *P. stylirostris* showed high mortality due to emergence of a virulent strain of TSV.

The clinical signs of acute-phase TSV infection in *P. vannamei* include lethargy, anorexia, opaque musculature, ataxic swimming behavior, flaccid bodies, soft cuticles, and chromatophore expansion resulting in reddening of the uropods, appendages, and body. These shrimp are typically in late pre-molt or early post-molt stage. Cumulative mortalities during acute-phase infection may reach 95%. In a pond where a TSV outbreak occurs, dead or dying shrimp float on the watersurface, attracting predatory birds (seagulls, cormorants, terns, etc.). Individuals that survive acute-phase infection develop grossly visible, multifocal, melanized lesions on the cephalothorax, tail, and appendages (**Figure 3**). These lesions are characteristic of the transition phase of infection. Chronically

Figure 3 *Penaeus vannamei* shrimp displaying typical clinical signs of Taura syndrome disease at the end of the acute phase. Multifocal, melanized lesions on the thorax and tail are visible (indicated by arrows) in TSV-infected shrimp.

infected shrimp shed the cuticle with melanized foci and display normal behavior but can remain asymptomatic carriers of TSV.

Pathogenesis

TSD pathogenesis consists of three overlapping yet clinically and histologically distinct phases: a peracute to acute phase, a short transition phase, and a persistent chronic phase.

Acute Phase

The acute phase begins approximately 24 h post exposure and persists for approximately 7 days. The acute phase is characterized by the rapid development of severe, multifocal cuticular epithelial necrosis and high mortalities (**Figure 4(b)**). Target tissues of TSV infection include the cuticular epithelium of the foregut, gills, appendages, hindgut, and general body cuticle. Lesions sometimes extend into underlying subcuticular connective tissue and striated muscle. In severe cases, the antennal gland, hematopoietic tissue, testes, and ovaries may also be infected. Acutely infected epithelial cells detach from the underlying stroma, assume a spherical shape, and this is followed by cell lysis and release of virions into the circulatory system. Infected cells display highly basophilic pyknotic and karyorrhectic nuclei and marked cytoplasmic eosinophilia, with variably staining and sized cytoplasmic inclusion bodies. These histological lesions are collectively called 'peppered' or 'buckshot-laden' lesions, and are considered pathodiagnostic for acute phase of TSD (**Figure 4(b)**).

Transition Phase

Shrimp surviving acute-phase infection enter a brief transitional phase for approximately 5 days before entering into chronic phase. Appearance of multifocal melanized lesions on the cephalothorax and tail, with declining mortalities, are characteristic of the transition phase (**Figure 3**). Histologically, transition-phase animals present infrequent, scattered, epithelial lesions, normally appearing lymphoid organ (LO) arterioles (tubules) that display a TSV-positive signal by *in situ* hybridization (ISH), and the onset of spheroid development within the LO. Transition-phase animals are lethargic, anorexic, start molting, and shed melanized exoskeleton.

Chronic Phase

During the chronic phase, mortality ceases and animals resume normal behavior. The chronic phase may last for 8–12 months in experimentally infected *P. vannamei*. The histological hallmark of chronic phase infection is the appearance of LO spheroids (**Figure 4(d)**). Occasionally, ectopic spheroids are also found associated with tegmental glands located within connective tissues of the cephalothorax and appendages. Spheroids consist of phagocytic semigranular and granular hemocytes undergoing apoptosis.

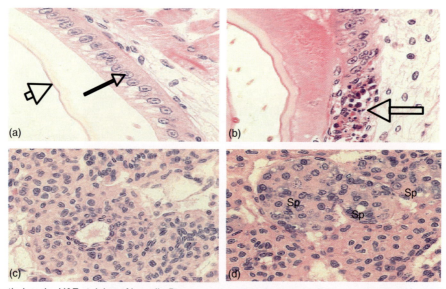

Figure 4 Histopathology by H&E staining of juvenile *Penaeus vannamei* infected with Taura syndrome virus. (a) Normal stomach cuticular epithelium (thin arrow) and cuticle (thick arrow); (b) acute-phase TSV infection evident within the stomach cuticular epithelium (arrow) with nuclear pyknosis, karyorrhexis, and cytoplasmic eosinophilia clearly visible; (c) normal lymphoid organ (LO) showing a prominent arteriole with centralized lumen; and (d) chronic TSV infection visible inside the LO, displays prominent spheroids (Sp). All panels are at ×600 magnification.

Genetics

Genome Organization and Gene Expression

The TSV genome consists of a linear, positive-sense, single-stranded, polyadenylated RNA of 10 205 nucleotides (nt). It has a 417 nt 5′ untranslated region and a 224 nt 3′ untranslated region. There are two ORFs. ORF1 encodes nonstructural proteins, while ORF2 encodes structural proteins (**Figure 5**). ORF1 is 6324 nt long, encoding a 2107 amino acid (aa) polyprotein with a molecular mass of 234 kDa. ORF2 is 3036 nt long, encoding a 1011 aa long polypeptide with a molecular mass of 112 kDa. Nonstructural ORF1 proteins contain conserved helicase (NTP-binding protein), protease, and RNA-dependent RNA polymerase (RdRp) motifs. The RNA helicase consensus sequence (Gx_4GK) is at aa 752–758 and has similarity with the cognate domains of other *Dicistroviridae* and insect picornaviruses. The protease domain in the TSV ORF1-encoded polypeptide, located between aa 1380 and 1570, shows similarity with the 3C proteases of the *Dicistroviridae*, *Picornaviridae*, *Sequiviridae*, and *Comoviridae*. The TSV protease motif is partially conserved, with glycine replaced by cysteine. Similar to picornaviruses, the His-Asp-Cys catalytic triad in the protease domain is conserved in TSV. The RdRp domain, located at the 3′ end of ORF1, contains eight conserved motifs analogous to those in picornaviruses of insects, mammals, and plants.

The N-terminal region of the ORF1-encoded polypeptide contains a stretch of 65 aa (166–230) that shows similarity to the 'inhibition of apoptosis' proteins (IAP) found in mammals, yeast, insects, and some DNA viruses. No other RNA viruses are known to encode polypeptides containing such a motif. It is unknown if this IAP motif of TSV plays a role during the long-term chronic-phase infection in evading the host immune system or facilitates virus replication.

There is an IGR of 207 nt between the two ORFs in the TSV genome. The IGR region contains an internal ribosomal entry site (IRES) that allows cap-independent translation of the ORF2 capsid proteins. Functional IRES elements have also been identified in the IGRs of CrPV and PSIV. In mammalian cells infected with picornaviruses, structural and nonstructural proteins are produced in equimolar amounts. However, in insect viruses, like DCV, structural proteins are produced in excess. This is due to IRES-mediated translation of DCV coat proteins. Translation of the two distinct polyproteins in DCV (ORF1 and ORF2) appears to be independently controlled. This is in contrast to picornaviruses encoding a single ORF in which a single polyprotein is processed post-translationally to generate both the structural and nonstructural proteins. The cap-independent translation in PSIV ORF2 has been demonstrated *in vitro* using a rabbit reticulocyte lysate system. In CrPV, the initiation codon for IRES-mediated translation has been identified as CCU, whereas, the initiation codon in PSIV and RhPV is CUU. The CCU/CUU triplets have been identified as part of the inverted repeat sequence of the IRES elements that form RNA pseudoknot structures essential for IRES activity. In TSV, there is an in-frame methionine upstream of ORF2 although N-terminal sequencing of the VP2 capsid protein has revealed alanine at the terminal position (ANPVEIDNFDTT). The alanine codon is preceded by proline (CCU) and methionine (AUG) codons (MPANPVE). Site-specific mutation in TSV has demonstrated that conversion of the AUG to UUG has no effect on capsid protein expression. Additionally, cell-free expression in the absence of initiator methionine tRNA or in the presence of ribosome inhibitor is possible with TSV. These results indicate that the IRES element in TSV IGR is functional and that the TSV intergenic AUG codon does not act as a translation initiation site. TSV has an unusual side stem–loop structure in its IGR-IRES region that is involved in this cap-independent translation. The presence of a methionine codon in the IGR-IRES has been shown to be unimportant to translation; that is, TSV translation of ORF2 does not require initiator tRNA and relies instead on the IRES element for translation initiation.

The TSV genome is transcribed as a single ∼10 kb transcript and does not produce subgenomic RNA for translation of the capsid genes. This distinguishes TSV from many positive-strand RNA viruses (e.g., *Caliciviridae* and *Togaviridae*) in which capsid proteins encoded in the 3′ end of the genome are generally translated from subgenomic RNA.

Genetic Diversity

The isolate that caused the first TSV epizootics in the USA (Hawaii) was subsequently characterized using biological, serological, and molecular data as Serotype A. In 1999, significant mortalities in a TSV-tolerant strain of the Western blue shrimp (*P. stylirostris*) occurred at shrimp farms in México and there were concerns that a new TSV

Figure 5 Genome organization of Taura syndrome virus (TSV). The conserved helicase, protease, and the RNA-dependent RNA polymerase (RdRp) domains are indicated in open reading frame 1 (ORF1). IGR, intergenic region; UTR, untranslated region.

strain might have emerged. These shrimp showed acute- or chronic-phase TSV infections by staining with H&E, and were virus-positive by reverse transcriptase polymerase chain reaction analysis (RT-PCR) and by ISH. However, these animals were TSV-negative by immunohistochemical (IHC) analysis using a TSV-specific monoclonal antibody (mAb) specific to the VP1 capsid gene.

Nucleotide sequences of capsid genes (VP1 and VP2) of the Mexican isolates revealed nonconservative amino acid replacements (such as S→A, polar uncharged to nonpolar hydrophobic in VP1, and Q→K, polar uncharged to positively charged in VP2) compared to the Hawaiian isolate. Phylogenetic analysis showed that the Mexican isolate falls into a different clade. The Mexican isolate was classified as TSV serotype B.

In 2001, significant mortalities of *P. vannamei* occurred at shrimp farms in Belize, resulting from TSV epizootics. Using OIE diagnostic methods and sequence analysis, the Belize isolate was shown to represent a new serotype, TSV serotype C (**Table 1**).

In 2004, a TSV epizootic occurred in some farms in Texas, USA, near shrimp-processing plants that process frozen shrimp from Southeast (SE) Asia. Phylogenetic analysis using the VP1 capsid gene sequence revealed that TSV Texas isolate groups with the isolates from SE Asia (Thailand, Taiwan, and China, **Figure 6**). The spread of TSV to Asia most likely resulted from the introduction of TSV-infected *P. vannamei* from South America. The first TSV epizootic in Asia occurred in Taiwan in 1998, and subsequently spread to other Asian countries. It is possible that the Texas isolate was introduced through imported shrimp from SE Asia. Based on the phylogenetic analysis, the Asian isolates have been grouped tentatively as serotype D (**Figure 6**). Further characterization (biological and serological) might reveal that this group represents more than one serotype.

Overall, the TSV isolates show very high similarity (∼95%) in nucleotide sequence of the VP1 gene among the isolates and most of the variations are located in the C-terminal end of the VP1 capsid gene. In CrPV, the type species of the genus *Cripavirus*, family *Dicistroviridae*, the crystal structure of the capsid gene has revealed that the C-terminal end of the VP3 gene (homologous to TSV VP1) is a helical extension on the virion surface. Therefore, this region is less constrained by structural requirements allowing mutations to accumulate. These mutations have serious implications for antibody-driven diagnosis and potentially modulate the virulence of the newly evolved strain(s).

Evolution and Spread

RNA viruses exist as 'quasispecies' in nature due to the high mutation frequency caused by the lack of effective proofreading activity of RdRp. The presence of a large array of related genotypes in the population allows an RNA virus to evolve and adapt to changes in the host and the environment. The spread and evolution of TSV appears to provide an example of this process. In México, until about the mid-1990s, *P. vannamei* was the principal species for shrimp farming. When TSV epizootics in México reached a peak in 1996, farmers started switching from TSV-susceptible *P. vannamei* to TSV-tolerant *P. stylirostris*. This resulted in the decline of TSV epizootics and, by 1998, shrimp production in México stabilized. In 1999, a new strain of TSV evolved in México that caused significant mortalities in previously tolerant *P. stylirostris*. The replacement of *P. vannamei* with *P. stylirostris* in shrimp farms in México might have contributed to the development of a virulent strain of TSV as the virus adapted to a new host species.

The transportation of live postlarval and adult shrimp (*P. vannamei*) from South America to Asia facilitated TSV dissemination in Asia and in 1998, the first TSV outbreak was reported in Taiwan. From Taiwan, TSV spread to other Asian countries. In Thailand, *P. vannamei* cultivation commenced in late 1990. Due to the ban on cultivation of native *P. monodon* species in freshwater areas in Thailand in 2000 and widespread slow growth in cultivated *P. monodon* in 2001 and 2002, farmers adopted *P. vannamei* as an alternative species. During this time, almost 70% of the broodstock in Thailand was imported from China. In 2003, the first TSV outbreak was confirmed in Thailand. In 2004, TSV epizootics occurred in farms in Texas, USA, located proximal to shrimp-processing plants that import frozen and bait shrimp from SE Asia. Phylogenetic analysis, using nucleotide sequence data from the TSV capsid gene VP1, showed that the isolates from Thailand, China, Taiwan, and Texas formed a single clade (**Figure 6**). As naïve shrimp populations are exposed to TSV, virus/host selection results in the emergence of more virulent strain(s) with devastating consequences. The

Table 1 List of TSV serotypes identified based on biological, genomic, and serological properties

Serotype	Place of origin	Year	Species of origin
TSV serotype A	Hawaii, USA	1994	Penaeus vannamei
TSV serotype B	Sinaloa, México	1998	Penaeus stylirostris
TSV serotype C	Belize	2001	Penaeus vannamei
TSV serotype D	Asia	1999–2005	Penaeus vannamei, Penaeus monodon, Metapenaeus ensis

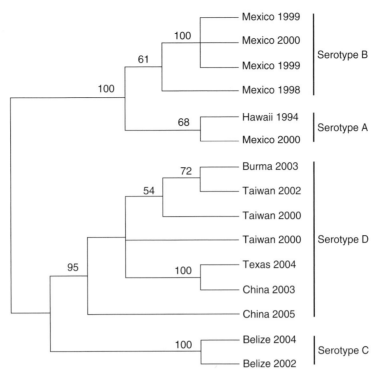

Figure 6 Maximum likelihood phylogenetic tree of TSV isolates collected from different geographical areas. The phylogenetic tree was constructed using the phylogenetic analysis using parsimony (PAUP) method, and numbers on the branches represent bootstrap values. The origin and the year of collection of these isolates are indicated.

availability of TSV genomic information has provided valuable insight into the origin, evolution, and spread of TSV worldwide.

Dissemination

TSV is transmitted by cannibalism of infected moribund or dead shrimp by healthy shrimp in a pond or in an experimental tank. The virus can remain infectious in decaying shrimp carcasses for up to 3 weeks and these may serve as a source of infection. In addition, chronically infected shrimp harbor infectious TSV within the LO and hemolymph for 8–12 months post infection, representing a source for renewed outbreaks in ponds, canals, or adjacent estuaries. In experimentally induced TSV epizootics, waterborne transmission has been reported to occur up to 48 h following peak mortality.

Mechanical transmission of TSV can occur via an aquatic insect (water boatmen, *Trichocorixa reticulate*) living in ponds and via seagulls (*Larus atricilla*). Infective TSV has been demonstrated in the seagull feces collected from the banks of ponds where TSV outbreaks have occurred.

Since the initial outbreak of Taura syndrome in Ecuador, the geographic range of the disease has spread and TSV is now considered endemic along the mid-American Pacific Coast, ranging from northern Peru through to México. Wild postlarvae and broodstock showing acute and/or chronic disease have been collected in Ecuador, Honduras, El Salvador, and Southern México. Therefore, wild postlarvae and broodstock are potential sources of dissemination for TSV. The movement of TSV-infected live postlarvae and adult shrimp has been the principal means by which TSV spread into shrimp farming nations within the Western Hemisphere and Asia. Export of frozen commodity shrimp and bait shrimp from infected areas also may have contributed to the spread of TSV across the continents.

Diagnosis

Taura syndrome diagnosis and the detection of TSV can be conducted by observation of clinical signs, bioassay, histopathology, enzyme-linked immunosorbent assay (ELISA), ISH using nonradiolabeled probes, and reverse transcription-polymerase chain reaction (RT-PCR).

As no immortalized shrimp cell lines are currently available, detection of infectious virus relies on *in vivo* bioassays. The bioassay protocol involves exposing SPF juvenile *P. vannamei* to TSV-suspect shrimp either by *per os* challenge or by injection of a tissue extract.

TSV-positive shrimp are identified by gross signs and histopathology.

A standard diagnostic tool for identifying TSV-induced pathology involves H&E histology of Davidson's alcohol formalin acetic acid fixative (AFA)-preserved shrimp. The presence of acute-phase lesions in cuticular epithelia is considered to be a definitive diagnosis of TSV infection (**Figure 4**).

Digoxigenin-labeled cDNA probes have been developed for TSV detection by ISH. Positive ISH reactions are identified by the presence of a blue-black precipitate within the cytoplasm of TSV-infected cells. ISH has greater sensitivity than H&E staining, as TSV can be detected in shrimp with mild acute infections and in asymptomatic carriers.

ELISA-based dot blot detection of TSV capsid protein using a TSV-specific monoclonal antibody is available and has been modified for application to IHC detection of TSV in histological sections. IHC has advantages over ISH in that it is a more economical and rapid assay with similar sensitivity. However, IHC using monoclonal antibodies can fail to detect some TSV strains due to epitope drift, as discussed above.

An RT-PCR method has been described for TSV detection in hemolymph, providing a nonlethal diagnostic method that is rapid and has a greater sensitivity than other methods. Recently, real-time RT-PCR methods using SYBR Green dye and TaqMan probes have been developed for rapid detection and quantification of TSV. The real-time RT-PCR is highly sensitive, specific, detects subclinical infections, and has a high throughput potential. The drawback with the real-time RT-PCR is its high cost and the need for significant technical expertise. However, combining real-time PCR with conventional diagnosis might be acceptable for large-scale shrimp farm operations to screen broodstock and field samples for TSV.

Control/Prevention

The current TSV control methods emphasize biosecurity at the farm level and the utilization of SPF, TSV-resistant shrimp for stocking. This approach is preventative and has considerable success in the control of TSD on carefully managed farms. Since the immune system of shrimp is relatively primitive, the development of a traditional vaccination procedure for control of TSV is not likely. Experimentation with RNA interference and the expression of antisense RNA against the TSV capsid gene in transgenic shrimp show promise in the control of the disease but currently have only been tested at the laboratory level. Also, disease protection by nonspecific dsRNA indicates that shrimp have a fairly strong innate immune response that could be exploited.

Acknowledgment

The authors thank Dr. Kenneth W. Hasson, Texas Veterinary Medical Diagnostic Laboratory (TVMDL), College Station, Texas for providing the histopathology and electron microscopy photographs.

See also: Picornaviruses: Molecular Biology; White Spot Syndrome Virus; Yellow Head Virus.

Further Reading

Bonami JR, Hasson KW, Mari J, Poulos BT, and Lightner DV (1997) Taura syndrome of marine penaeid shrimp: Characterization of the viral agent. *Journal of General Virology* 78: 313–319.

Cevallos RC and Sarnow P (2005) Factor-independent assembly of elongation-competent ribosomes by an internal ribosome entry site located in an RNA virus that infect penaeid shrimp. *Journal of Virology* 79: 677–683.

Dhar AK, Cowley JA, Hasson KW, and Walker PJ (2004) Genome organization, biology and diagnosis of Taura syndrome virus and yellowhead virus of penaeid shrimp. *Advances in Virus Research* 63: 353–419.

Dhar AK, Roux MM, and Klimpel KR (2002) Quantitative assay for measuring the Taura syndrome virus (TSV) and yellow head virus (YHV) load in shrimp by real-time RT-PCR using SYBR Green chemistry. *Journal of Virological Methods* 104: 69–82.

Hasson KW, Lightner DV, Mohney LL, Redman RM, Poulos BT, and White BL (1999) Taura syndrome virus (TSV) lesion development and the disease cycle in the Pacific white shrimp *Penaeus vannamei*. *Diseases of Aquatic Organisms* 36: 81–93.

Hasson KW, Lightner DV, Poulos BT, *et al.* (1995) Taura syndrome in *Penaeus vannamei*: Demonstration of a viral etiology. *Diseases of Aquatic Organisms* 23: 115–126.

Hatakeyama Y, Shibuya N, Nishiyama T, and Nakashima N (2004) Structural variant of the intergenic internal ribosomal entry site elements in dicistroviruses and computational search for their counterparts. *RNA* 10: 779–786.

Mari J, Poulos BT, Lightner DV, and Bonami JR (2002) Shrimp Taura syndrome virus: Genomic characterization and similarity with members of the genus *Cricket paralysis-like viruses*. *Journal of General Virology* 83: 915–926.

Nielsen L, Sang-oum W, Cheevadhanarak S, and Flegel TW (2005) Taura syndrome virus (TSV) in Thailand and its relationship to TSV in China and the Americas. *Diseases of Aquatic Organisms* 63: 101–106.

Office International des Epizooties (OIE) Fish Disease Commission (2003) Manual of diagnostic manual for aquatic animal diseases. ch 4. 1. 1, Taura Syndrome. http://www.oie.int/eng/normes/fmanual/A_summry.htm (accessed August. 2007).

Robalino J, Browdy CL, Prior S, *et al.* (2004) Induction of antiviral immunity by double-stranded RNA in a marine invertebrate. *Journal of Virology* 78: 10442–10448.

Robles-Sikisaka R, Hasson KW, Garcia DK, *et al.* (2002) Genetic variation and immunohistochemical differences among geographic isolates of Taura syndrome virus of penaeid shrimp. *Journal of General Virology* 83: 3123–3130.

Srisuvan T, Pantoja CR, Redman CR, and Lightner DV (2006) Ultrastructure of the replication site in Taura syndrome virus (TSV)-infected cells. *Diseases of Aquatic Organisms* 73: 89–101.

Tang KFJ and Lightner DV (2005) Phylogenetic analysis of Taura syndrome virus isolates collected between 1993 and 2004 and virulence comparison between two isolates representing different genetic variants. *Virus Research* 11: 69–76.

Yu CI and Somg YL (2002) Outbreak of Taura syndrome in Pacific white shrimp *Penaeus vannamei* cultured in Taiwan. *Fish Pathology* 35: 21–24.

Taxonomy, Classification and Nomenclature of Viruses

C M Fauquet, Danforth Plant Science Center, St. Louis, MO, USA

© 2008 Elsevier Ltd. All rights reserved.

Glossary

Classification The categorization of organisms into defined groups on the basis of identified characteristics.
Family A category in the virus taxonomic classification of related organisms, comprising one or more genera.
Genus A category in the virus taxonomic classification of related organisms, comprising one or more species.
Nomenclature The assigning of names to organisms in a scientific classification system.
Order A category in the virus taxonomic classification of related organisms, comprising one or more families.
Species A virus species is a polythetic class of viruses that constitutes a replicating lineage and occupies a particular ecological niche.
Taxonomy The science of classifying plants, animals, and microorganisms into increasingly broader categories based on shared features. The practice or principles of classification.

Introduction

Virus taxonomy is a very important but controversial field of science. It was ranked as the first constraint for the modern development of virus databases, and the exponential increase in virus sequencing is worsening the situation. However, substantial progress has been made particularly in the last 20 years, both on the conceptual framework and practical implication of virus taxonomy. The International Committee on Taxonomy of Viruses (ICTV) is the only committee of the Virology Division, of the International Union of Microbiological Societies (IUMS), in charge of that task since 1966, for the international virology community. Virus Taxonomy Reports have been published regularly by ICTV and they became the reference in virus taxonomy and nomenclature. This article aims at providing some historical information about the establishment and changes in virus taxonomy and describes the current status of virus classification, nomenclature, and orthography.

There is no such thing as a 'natural' or a 'biological' classification; by essence any classification is an arbitrary human invention and viruses are no exception. The question really is: "How can we classify viruses in such way that it makes sense and is useful to many scientists?" The need for virus classification is not only supported by the common human need for organization, but also as a scientific tool to compare viruses and extrapolate useful information from one virus to another and from one family to another. Crucial biological information can be extrapolated directly from human viruses like picornaviruses to plant comoviruses and vice versa, when the classification indicates many structural and genomic common characteristics. When a newly discovered virus is assigned to a particular taxon, this virus can immediately be granted a number of *a priori* properties that only need confirmation and that have an immediate impact on specific virological studies. Furthermore, although it is not clearly stated by ICTV that the current classification is thought to reflect virus evolution, it is accepted that virus taxonomy is aiming at this objective and could become a tool by itself to study and evaluate virus evolution.

Virus nomenclature cannot be dissociated from classification. There must be a coherent system for naming viruses accompanied with a system for classifying viruses. Furthermore, using correct orthography and typography of virus taxa is not simply an excercise meant to complicate the task of virologists but is based on rules that help scientists to extract useful information from what is written down. It is therefore important to establish and follow guidelines for orthography, nomenclature, and classification of viruses. For all these reasons, virus classification and nomenclature have always been very controversial and have led to passionate discussions over the past four decades.

Historical Background of Virus Taxonomy

The first evidence for the existence of viruses was shown by Beijerinck in 1898, but it was not until the 1920s that virologists began to classify viruses. The first system referred to pathogenic properties of animal and human viruses and to symptoms for plant viruses. For example, viruses sharing the pathogenic properties causing hepatitis (e.g., hepatitis A virus, hepatitis B virus, yellow fever virus, Rift Valley fever virus) were grouped together as 'the hepatitis viruses'. In 1939, Holmes published a classification of plant viruses dependent on host reactions and differential host species using a binomial–trinomial nomenclature based on the name of the infected plant. It was only in the 1950s, with the utilization of the electron microscope, that the first real virus classification was established. Naturally the shape and size of virus particles became major criteria for virus classification. Because of that powerful and rapid technology the number of newly discovered viruses increased rapidly and several hundreds of new viruses were listed in a short period of time.

In 1966 in Moscow, at the International Congress for Microbiology, 43 virologists created the International Committee on Nomenclature of Viruses (ICNV) with the aim of developing a worldwide recognized taxonomy and nomenclature system for all viruses. The name of the ICNV was changed in 1974 to the more appropriate ICTV. The ICTV, which is the only committee of the Virology Division of the International Union of Microbiological Societies, is now recognized as the official international body that decides on all matters related to taxonomy and nomenclature of viruses.

Since 1966, virologists have agreed that all viruses isolated from different organisms should be classified together in a unique system, separate from that of microorganisms such as fungi, bacteria, and mycoplasma. However, there has been much controversy on how to achieve this aim. Lwoff, Horne, and Tournier in 1962 proposed the adoption of a system classifying viruses into subphyla, classes, orders, suborders, and families. It was also proposed that the hierarchical classification would be based on the type of nucleic acid (DNA or RNA), the strandedness (single = ss or double = ds), the presence or absence of an envelope, the capsid symmetry, the type of replication cycle (with or without an RNA intermediate for DNA viruses), and the number of genome segments. This hierarchical classification system has never been recognized by the ICTV, but most of the criteria used to demarcate the major classes of viruses formed the basis of the universal taxonomy system now in place, and all published ICTV reports have used this scheme with only minor changes.

It is only in the last 15 years that a hierarchical classification level higher than the family was proposed and accepted. A first order, *Mononegavirales*, was accepted in 1990, and the orders *Caudovirales* and *Nidovirales* were adopted in 1996. In 2005, ICTV considered introducing four new orders namely *Picornavirales*, *Herpesvirales*, *Reovirales*, and *Retrovirales*, and these may become accepted in the near future.

It is important to note that the species category for viruses was only adopted in 1991. From then onwards viruses were assigned to species or tentative species. In addition, a list of species demarcation criteria has been established for each family. It is anticipated that by 2010 the ICTV will have introduced species criteria for all viruses and that some level of homogeneity will have been reached, although it is perfectly acceptable to have different sets of criteria for different families of viruses.

Since the establishment of the ICTV, a total of eight virus taxonomic reports have been published. At the first meeting of the Committee in Mexico City in 1970, two families with corresponding two genera and 24 floating genera were adopted to begin grouping the vertebrate, invertebrate, and bacterial viruses together, and in addition, 16 plant virus 'groups' were introduced. Although virologists working with vertebrate viruses had assigned viruses to genera and families for many years, plant virologists until 1993 used the term 'group' to designate viruses with similar properties. It was only in 1995 that the ICTV adopted a uniform system for all viruses, encompassing 2644 assigned viruses. The *Eighth ICTV Report on Virus Taxonomy*, published in 2005 describes a Universal Virus Classification that comprises 3 orders, 73 families, 9 subfamilies, 287 genera, and 5450 viruses belonging to 1950 species (**Table 1**).

Over the past four decades the number of classified viruses, as well as the number of each type of taxa, has increased exponentially and continues to grow (**Figure 1**). Because DNA sequencing has become a routine technique, it seems likely that the number of recognizes viruses and viral taxa will continue to grow exponentially. Furthermore, virus genome sequences provide qualitative and quantitative criteria for defining the molecular variability of viruses that are useful for classification purposes. Sequencing will also permit identification and classification of many viruses that are difficult to isolate and characterize by other methods.

Organization and Structure of ICTV

The ICTV is the only Committee of the Virology Division of the International Union of Microbiological Societies. It is a non-profit organization composed of volunteered virologists from many countries who make decisions on virus names and taxa through a democratic process. The ICTV operates through subcommittees and study groups consisting of more than 500 virologists with expertise in human, animal, insect, protozoa, archaea, bacteria, mycoplasma, fungi, algae, and plant viruses.

Taxonomic proposals are initiated and formulated by study groups or by single individuals. The proposals are examined, offered to public scrutiny, accepted by the relevant subcommittee and presented for approval by the Executive Committee of the ICTV. All decisions are ratified by postal vote, where all members of the ICTV and more than 50 national microbiological societies are represented. Presently, there are 75 study groups working in concert with six subcommittees: one each for the vertebrate, invertebrate, plant, bacterial, and fungal viruses and one for the virus ICTV DataBase (ICTVdB). The ICTV does not impose taxa but ensures that all propositions are compatible with the International Code for Virus Classification and Nomenclature for accuracy, homogeneity, and consistency. The ICTV regularly

Table 1 The order of presentation of viruses in the eighth ICTV report

Order	Family	Subfamily	Genus	Type species	Host
The DNA viruses					
The dsDNA viruses					
Caudovirales	*Myoviridae*		"T4-like viruses"	*Enterobacteria* phage T4	B
			"P-like viruses"	*Enterobacteria* phage P	B
			"P-like viruses"	*Enterobacteria* phage P	B
			"Mu-like viruses"	*Enterobacteria* phage Mu	B
			"SP1-like viruses"	*Bacillus* phage SP1	B
			"ϕH-like viruses"	*Halobacterium* phage ϕH	Ar
	Siphoviridae		"λ-like viruses"	*Enterobacteria* phage λ	B
			"T1-like viruses"	*Enterobacteria* phage T1	B
			"T5-like viruses"	*Enterobacteria* phage T5	B
			"L5-like viruses"	*Mycobacterium* phage L5	B
			"c2-like viruses"	*Lactococcus* phage c2	B
			"ψM1-like viruses"	*Methanobacterium* phage ψM1	Ar
			"ϕC31-like viruses"	*Streptomyces* phage ϕC31	B
			"N15-like viruses"	*Enterobacteria* phage N15	B
	Podoviridae		"T7-like viruses"	*Enterobacteria* phage T7	B
			"P2-like viruses"	*Enterobacteria* phage P2	B
			"ϕ29-like viruses"	*Bacillus* phage ϕ29	B
			"N4-like viruses"	*Enterobacteria* phage N4	B
	Tectiviridae		*Tectivirus*	*Enterobacteria* phage PRD1	B
	Corticoviridae		*Corticovirus*	*Pseudoalteromonas* phage PM2	B
	Plasmaviridae		*Plasmavirus*	*Acholeplasma* phage L2	B
	Lipothrixviridae		*Alphalipothrixvirus*	*Thermoproteus tenax* virus 1	Ar
			Betalipothrixvirus	*Sulfolobus islandicus* filamentous virus	Ar
			Gammalipothrixvirus	*Acidianus* filamentous virus 1	Ar
	Rudiviridae		*Rudivirus*	*Sulfolobus islandicus* rod-shaped virus 2	Ar
	Fuselloviridae		*Fusellovirus*	*Sulfolobus* spindle-shaped virus 1	Ar
			Salterprovirus	His1 virus	Ar
	Guttaviridae		*Guttavirus*	*Sulfolobus newzealandicus* droplet-shaped virus	Ar
	Poxviridae	*Chordopoxvirinae*	*Orthopoxvirus*	Vaccinia virus	V
			Parapoxvirus	Orf virus	V
			Avipoxvirus	Fowlpox virus	V
			Capripoxvirus	Sheeppox virus	V
			Leporipoxvirus	Myxoma virus	V
			Suipoxvirus	Swinepox virus	V
			Molluscipoxvirus	Molluscum contagiosum virus	V
			Yatapoxvirus	Yaba monkey tumor virus	V
		Entomopoxvirinae			

Continued

Table 1 Continued

Order	Family	Subfamily	Genus	Type species	Host
	Asfarviridae		Asfivirus	African swine fever virus	V, I
	Iridoviridae		Iridovirus	Invertebrate iridescent virus 6	I
			Chloriridovirus	Invertebrate iridescent virus 3	I
			Ranavirus	Frog virus 3	V
			Lymphocystivirus	Lymphocystis disease virus 1	V
			Megalocytivirus	Infectious spleen and kidney necrosis virus	V
	Phycodnaviridae		Chlorovirus	Paramecium bursaria Chlorella virus 1	Al
			Coccolithovirus	Emiliania huxleyi virus 86	Al
			Prasinovirus	Micromonas pusilla virus SP	Al
			Prymnesiovirus	Chrysochromulina brevifilum virus PW1	Al
			Phaeovirus	Ectocarpus siliculosus virus 1	Al
			Raphidovirus	Heterosigma akashiwo virus 01	Al
	Baculoviridae		Nucleopolyhedrovirus	Autographa californica multiple nucleopolyhedrovirus	I
			Granulovirus	Cydia pomonella granulovirus	I
	Nimaviridae		Whispovirus	White spot syndrome virus 1	I
	Herpesviridae	Alphaherpesvirinae	Simplexvirus	Human herpesvirus 1	V
			Varicellovirus	Human herpesvirus 3	V
			Mardivirus	Gallid herpesvirus 2	V
			Iltovirus	Gallid herpesvirus 1	V
		Betaherpesvirinae	Cytomegalovirus	Human herpesvirus 5	V
			Muromegalovirus	Murid herpesvirus 1	V
			Roseolovirus	Human herpesvirus 6	V
		Gammaherpesvirinae	Lymphocryptovirus	Human herpesvirus 4	V
			Rhadinovirus	Saimiriine herpesvirus 2	V
			Ictalurivirus	Ictalurid herpesvirus 1	V
	Adenoviridae		Mastadenovirus	Human adenovirus C	V
			Aviadenovirus	Fowl adenovirus A	V
			Atadenovirus	Ovine adenovirus D	V
			Siadenovirus	Frog adenovirus	V
			Rhizidiovirus	Rhizidiomyces virus	F
	Polyomaviridae		Polyomavirus	Simian virus 40	V
	Papillomaviridae		Alphapapillomavirus	Human papillomavirus 32	V
			Betapapillomavirus	Human papillomavirus 5	V
			Gammapapillomavirus	Human papillomavirus 4	V
			Deltapapillomavirus	European elk papillomavirus	V

		Epsilonpapillomavirus	Bovine papillomavirus 5	V
		Zetapapillomavirus	Equine papillomavirus 1	V
		Etapapillomavirus	Fringilla coelebs papillomavirus	V
		Thetapapillomavirus	Psittacus erithacus timneh papillomavirus	V
		Iotapapillomavirus	Mastomys natalensis papillomavirus	V
		Kappapapillomavirus	Cottontail rabbit papillomavirus	V
		Lambdapapillomavirus	Canine oral papillomavirus	V
		Mupapillomavirus	Human papillomavirus 1	V
		Nupapillomavirus	Human papillomavirus 41	V
		Xipapillomavirus	Bovine papillomavirus 3	V
		Omikronpapillomavirus	Phocoena spinipinnis papillomavirus	V
		Pipapillomavirus	Hamster oral papillomavirus	V
Polydnaviridae		Bracovirus	Cotesia melanoscela bracovirus	I
		Ichnovirus	Campoletis sonorensis ichnovirus	I
Ascoviridae		Ascovirus	Spodoptera frugiperda ascovirus 1a	I
Unassigned		Mimivirus	Acanthamoeba polyphaga mimivirus	Pr, V
The ssDNA viruses				
Inoviridae		Inovirus	Enterobacteria phage M13	B
		Plectrovirus	Acholeplasma phage L51	B
Microviridae		Microvirus	Enterobacteria phage φX174	B
		Chlamydiamicrovirus	Chlamydia phage 1	B
		Bdellomicrovirus	Bdellovibrio phage MAC1	B
		Spiromicrovirus	Spiroplasma phage 4	B
Geminiviridae		Mastrevirus	Maize streak virus	P
		Curtovirus	Beet curly top virus	P
		Topocuvirus	Tomato pseudo-curly top virus	P
		Begomovirus	Bean golden yellow mosaic virus	P
Circoviridae		Circovirus	Porcine circovirus-1	V
		Gyrovirus	Chicken anemia virus	V
Unassigned		Anellovirus	Torque teno virus	V
Nanoviridae		Nanovirus	Subterranean clover stunt virus	P
		Babuvirus	Banana bunchy top virus	P
Parvoviridae	Parvovirinae	Parvovirus	Minute virus of mice	V
		Erythrovirus	Human parvovirus B9	V
		Dependovirus	Adeno-associated virus 2	V
		Amdovirus	Aleutian mink disease virus	V
		Bocavirus	Bovine parvovirus	V
	Densovirinae	Densovirus	Junonia coenia densovirus	I
		Iteravirus	Bombyx mori densovirus	I
		Brevidensovirus	Aedes aegypti densovirus	I
		Pefudensovirus	Periplaneta fuliginosa densovirus	I

Continued

Table 1 Continued

Order	Family	Subfamily	Genus	Type species	Host
The DNA and RNA reverse transcribing viruses					
	Hepadnaviridae		Orthohepadnavirus	Hepatitis B virus	V
			Avihepadnavirus	Duck hepatitis B virus	V
	Caulimoviridae		Caulimovirus	Cauliflower mosaic virus	P
			Petuvirus	Petunia vein clearing virus	P
			Soymovirus	Soybean chlorotic mottle virus	P
			Cavemovirus	Cassava vein mosaic virus	P
			Badnavirus	Commelina yellow mottle virus	P
			Tungrovirus	Rice tungro bacilliform virus	P
	Pseudoviridae		Pseudovirus	Saccharomyces cerevisiae Ty1 virus	F, P
			Hemivirus	Drosophila melanogaster copia virus	F, I
			Sirevirus	Glycine max SIRE1 virus	P
	Metaviridae		Metavirus	Saccharomyces cerevisiae Ty3 virus	F, P, I
			Errantivirus	Drosophila melanogaster Gypsy virus	I
			Semotivirus	Ascaris lumbricoides Tas virus	I
	Retroviridae	Orthoretrovirinae	Alpharetrovirus	Avian leukosis virus	V
			Betaretrovirus	Mouse mammary tumor virus	V
			Gammaretrovirus	Murine leukemia virus	V
			Deltaretrovirus	Bovine leukemia virus	V
			Epsilonretrovirus	Walleye dermal sarcoma virus	V
			Lentivirus	Human immunodeficiency virus 1	V
		Spumaretrovirinae	Spumavirus	Simian foamy virus	V
The RNA viruses					
The dsRNA viruses					
	Cystoviridae		Cystovirus	Pseudomonas phage φ6	B
	Reoviridae		Orthoreovirus	Mammalian orthoreovirus	V
			Orbivirus	Bluetongue virus	V, I
			Rotavirus	Rotavirus A	V, I
			Coltivirus	Colorado tick fever virus	V, I
			Seadornavirus	Banna virus	V, I
			Aquareovirus	Aquareovirus A	V
			Idnoreovirus	Idnoreovirus 1	I
			Cypovirus	Cypovirus 1	I
			Fijivirus	Fiji disease virus	P, I
			Phytoreovirus	Wound tumor virus	P, I
			Oryzavirus	Rice ragged stunt virus	P, I
			Mycoreovirus	Mycoreovirus 1	F
	Birnaviridae		Aquabirnavirus	Infectious pancreatic necrosis virus	V
			Avibirnavirus	Infectious bursal disease virus	V

	Totiviridae	Entomobirnavirus	Drosophila X virus	I
		Totivirus	Saccharomyces cerevisiae virus L-A	F
		Giardiavirus	Giardia lamblia virus	Pr
		Leishmaniavirus	Leishmania RNA virus 1–1	Pr
	Partitiviridae	Partitivirus	Atkinsonella hypoxylon virus	F
		Alphacryptovirus	White clover cryptic virus 1	P
		Betacryptovirus	White clover cryptic virus 2	P
	Chrysoviridae	Chrysovirus	Penicillium chrysogenum virus	F
	Hypoviridae	Hypovirus	Cryphonectria hypovirus 1	F
	Unassigned	Endornavirus	Vicia faba endornavirus	P

The negative-stranded ssRNA viruses

Mononegavirales

	Bornaviridae	Bornavirus	Borna disease virus	V
	Rhabdoviridae	Vesiculovirus	Vesicular stomatitis Indiana virus	V, I
		Lyssavirus	Rabies virus	V
		Ephemerovirus	Bovine ephemeral fever virus	V, I
		Novirhabdovirus	Infectious hematopoietic necrosis virus	V
		Cytorhabdovirus	Lettuce necrotic yellows virus	P, I
		Nucleorhabdovirus	Potato yellow dwarf virus	P, I
	Filoviridae	Marburgvirus	Lake Victoria marburgvirus	V
		Ebolavirus	Zaire ebolavirus	V
	Paramyxoviridae			
Paramyxovirinae		Rubulavirus	Mumps virus	V
		Avulavirus	Newcastle disease virus	V
		Respirovirus	Sendai virus	V
		Henipavirus	Hendra virus	V
		Morbillivirus	Measles virus	V
Pneumovirinae		Pneumovirus	Human respiratory syncytial virus	V
		Metapneumovirus	Avian metapneumovirus	V
	Unassigned	Varicosavirus	Lettuce big-vein associated virus	P
		Ophiovirus	Citrus psorosis virus	P
	Orthomyxoviridae	Influenzavirus A	Influenza A virus	V
		Influenzavirus B	Influenza B virus	V
		Influenzavirus C	Influenza C virus	V
		Thogotovirus	Thogoto virus	V, I
		Isavirus	Infectious salmon anemia virus	V, I
	Bunyaviridae	Orthobunyavirus	Bunyamwera virus	V, I
		Hantavirus	Hantaan virus	V
		Nairovirus	Dugbe virus	V, I
		Phlebovirus	Rift Valley fever virus	V, I
		Tospovirus	Tomato spotted wilt virus	P, I
	Unassigned	Tenuivirus	Rice stripe virus	P, I
	Arenaviridae	Arenavirus	Lymphocytic choriomeningitis virus	V

Continued

Table 1 Continued

Order	Family	Subfamily	Genus	Type species	Host
			Deltavirus	Hepatitis delta virus	V
The positive-stranded ssRNA viruses					
	Leviviridae		Levivirus	Enterobacteria phage MS2	B
			Allolevivirus	Enterobacteria phage Qβ	B
	Narnaviridae		Narnavirus	Saccharomyces 20S narnavirus	F
			Mitovirus	Cryphonectria mitovirus 1	F
	Picornaviridae		Enterovirus	Poliovirus	V
			Rhinovirus	Human rhinovirus A	V
			Cardiovirus	Encephalomyocarditis virus	V
			Aphthovirus	Foot-and-mouth disease virus	V
			Hepatovirus	Hepatitis A virus	V
			Parechovirus	Human parechovirus	V
			Erbovirus	Equine rhinitis B virus	V
			Kobuvirus	Aichi virus	V
			Teschovirus	Porcine teschovirus	V
	Unassigned		Iflavirus	Infectious flacherie virus	I
	Dicistroviridae		Cripavirus	Cricket paralysis virus	I
	Marnaviridae		Marnavirus	Heterosigma akashiwo RNA virus	F
	Sequiviridae		Sequivirus	Parsnip yellow fleck virus	P
			Waikavirus	Rice tungro spherical virus	P
	Unassigned		Sadwavirus	Satsuma dwarf virus	P
			Cheravirus	Cherry rasp leaf virus	P
	Comoviridae		Comovirus	Cowpea mosaic virus	P
			Fabavirus	Broad bean wilt virus 1	P
			Nepovirus	Tobacco ringspot virus	P
	Potyviridae		Potyvirus	Potato virus Y	P
			Ipomovirus	Sweet potato mild mottle virus	P
			Macluravirus	Maclura mosaic virus	P
			Rymovirus	Ryegrass mosaic virus	P
			Tritimovirus	Wheat streak mosaic virus	P
			Bymovirus	Barley yellow mosaic virus	P
	Caliciviridae		Lagovirus	Rabbit hemorrhagic disease virus	V
			Norovirus	Norwalk virus	V
			Sapovirus	Sapporo virus	V
			Vesivirus	Vesicular exanthema of swine virus	V
	Unassigned		Hepevirus	Hepatitis E virus	V
	Astroviridae		Avastrovirus	Turkey astrovirus	V
			Mamastrovirus	Human astrovirus	V
	Nodaviridae		Alphanodavirus	Nodamura virus	I
			Betanodavirus	Striped jack nervous necrosis virus	V
	Tetraviridae		Betatetravirus	Nudaurelia capensis β virus	I

Unassigned	Omegatetravirus	Nudaurelia capensis ω virus	I
Luteoviridae	Sobemovirus	Southern bean mosaic virus	P
	Luteovirus	Barley yellow dwarf virus-PAV	P
	Polerovirus	Potato leafroll virus	P
	Enamovirus	Pea enation mosaic virus-1	P
Unassigned	Umbravirus	Carrot mottle virus	P
Tombusviridae	Dianthovirus	Carnation ringspot virus	P
	Tombusvirus	Tomato bushy stunt virus	P
	Aureusvirus	Pothos latent virus	P
	Avenavirus	Oat chlorotic stunt virus	P
	Carmovirus	Carnation mottle virus	P
	Necrovirus	Tobacco necrosis virus A	P
	Panicovirus	Panicum mosaic virus	P
	Machlomovirus	Maize chlorotic mottle virus	P
Nidovirales			
Coronaviridae	Coronavirus	Infectious bronchitis virus	V
	Torovirus	Equine torovirus	V
Arteriviridae	Arterivirus	Equine arteritis virus	V
Roniviridae	Okavirus	Gill-associated virus	I
Flaviviridae	Flavivirus	Yellow fever virus	V, I
	Pestivirus	Bovine viral diarrhea virus 1	V
	Hepacivirus	Hepatitis C virus	V
Togaviridae	Alphavirus	Sindbis virus	V, I
	Rubivirus	Rubella virus	V
Unassigned	Tobamovirus	Tobacco mosaic virus	P
	Tobravirus	Tobacco rattle virus	P
	Hordeivirus	Barley stripe mosaic virus	P
	Furovirus	Soil-borne wheat mosaic virus	P
	Pomovirus	Potato mop-top virus	P
	Pecluvirus	Peanut clump virus	P
	Benyvirus	Beet necrotic yellow vein virus	P
Bromoviridae	Alfamovirus	Alfalfa mosaic virus	P
	Bromovirus	Brome mosaic virus	P
	Cucumovirus	Cucumber mosaic virus	P
	Ilarvirus	Tobacco streak virus	P
	Oleavirus	Olive latent virus 2	P
Unassigned	Ourmiavirus	Ourmia melon virus	P
	Idaeovirus	Rasberry bushy dwarf virus	P
Tymoviridae	Tymovirus	Turnip yellow mosaic virus	P
	Marafivirus	Maize rayado fino virus	P, I
	Maculavirus	Grapevine fleck virus	P
Closteroviridae	Closterovirus	Beet yellows virus	P
	Ampelovirus	Grapevine leafroll-associated virus 3	P
	Crinivirus	Lettuce infectious yellows virus	P

Continued

Table 1 Continued

Order	Family	Subfamily	Genus	Type species	Host
	Flexiviridae		Potexvirus	Potato virus X	P
			Mandarivirus	Indian citrus ringspot virus	P
			Allexivirus	Shallot virus X	P
			Carlavirus	Carnation latent virus	P
			Foveavirus	Apple stem pitting virus	P
			Capillovirus	Apple stem grooving virus	P
			Vitivirus	Grapevine virus A	P
			Trichovirus	Apple chlorotic leaf spot virus	P
	Barnaviridae		Barnavirus	Mushroom bacilliform virus	F

Unassigned viruses

Unassigned Vertebrate Viruses					V
Unassigned Invertebrate Viruses					I
Unassigned Prokaryote Viruses					B
Unassigned Fungus Viruses					F
Unassigned Plant Viruses					P

The subviral agents: Viroids, satellites and agents of spongiform encephalopathies (prions)

Order	Family	Subfamily	Genus	Type species	Host
Viroids	Pospiviroidae		Pospiviroid	Potato spindle tuber viroid	P
			Hostuviroid	Hop stunt viroid	P
			Cocadviroid	Coconut cadang-cadang viroid	P
			Apscaviroid	Apple scar skin viroid	P
			Coleviroid	Coleus blumei viroid 1	S
	Avsunviroidae		Avsunviroid	Avocado sunblotch viroid	P
			Pelamoviroid	Peach latent mosaic virus	P
Satellites					
Vertebrate Prions					V
Fungi prions					F

Virus hosts: Al, Algae; Ar, Archaea; B, Bacteria; F, Fungi; I, Invertebrates; P, Plants; Pr, Protozoa; V, Vertebrates.

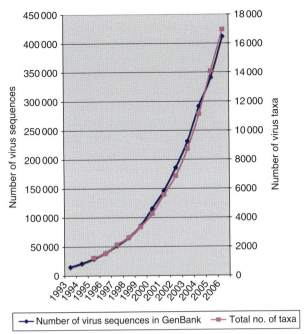

Figure 1 Number of virus taxa (including isolates) and virus sequences stored at GenBank since 1993.

publishes reports describing all existing virus taxa and containing a complete list of classified viruses with their abbreviations. The ICTV published its Eighth Report in 2005. An internet website is also maintained where all new taxonomic proposals are loaded and where the most important information relative to virus taxonomy is made available and updated regularly. The increasing number of virus species and virus strains being identified, along with the explosion of data on many descriptive aspects of viruses and viral diseases, particularly sequence data, has led the ICTV to launch an international virus database project (ICTVdB) and a Taxonomic Proposal Management System specifically to handle taxonomic proposals.

Polythetic Classification and Demarcation Criteria

There are currently two systems in use for classifying organisms: the Linnean and the Adansonian systems. The Linnean system is the monothetic hierarchical classification applied by Linnaeus to plants and animals, while the Adansonian is a polythetic hierarchical system. Although convenient to use, the Linnean system has shortcomings when applied to the classification of viruses because there is no obvious reason to privilege one criterion over another. The Adansonian system considers all available criteria at once and makes several classifications, taking the criteria successively into consideration. Criteria leading to the same classifications are considered correlated and are therefore not discriminatory. Subsequently, a subset of criteria is considered, and the process is repeated until all criteria can be ranked to provide the best discrimination of the species. Furthermore, qualitative and quantitative data can be simultaneously considered when building such a classification. In the case of viruses, the method is not used on a systematic basis, although it has been shown that at least 60 characters are needed for a complete virus description (**Table 2**).

The increasing number of reported viral nucleic acid sequences allows the construction of phylogenetic trees based on a single gene or a group of genes. Sequence comparisons by themselves have not satisfactorily provided a clear classification of all viruses together but are widely used at the order, family, and genus levels. Recently the National Center for Biotechnology Information (NCBI) in Washington developed a system of pairwise sequence comparisons (the so-called PASC system) between viral sequences which allows a new virus to be assigned to known taxa. It seems probable that, in future, virus classification will make increasing use of sequence data.

For more than 40 years, the ICTV has been classifying viruses essentially at the family and genus levels using a nonsystematic polythetic approach. Viruses are first clustered in genera and then in families. A subset of characters including physicochemical, structural, genomic, and biological criteria is then used to compare and group viruses. This subset of characters may change from one family to another according to the availability of the data and depending on the importance of a particular character for a particular family. Obviously, there is no homogeneity in this respect in the current virus classification system, and virologists weigh the criteria in a subjective process. Nevertheless, over time, there has been a great stability of the current classification at the genus and family levels. It is also clear that hierarchical classification above the family level will encounter conflicts between phenotypic and genotypic criteria and that virologists may have to reconsider the entire classification process in order to progress at this level.

Virus Taxa Descriptions

Virus classification continues to evolve with the technologies available for describing viruses. The first wave of descriptions, those before 1940, took into account mostly the visual symptoms of viral diseases along with modes of viral transmission. A second wave, between 1940 and

Table 2 Virus family descriptors used in virus taxonomy

I. Virion properties
  A. Morphology properties of virions
    1. Virion size
    2. Virion shape
    3. Presence or absence of an envelope and peplomers
    4. Capsomeric symmetry and structure
  B. Physical properties of virons
    1. Molecular mass of virions
    2. Buoyant density of virions
    3. Sedimentation coefficient
    4. pH stability
    5. Thermal stability
    6. Cation (Mg^{++}, Mn^{++}) stability
    7. Solvent stability
    8. Detergent stability
    9. Radiation stability
  C. Properties of genome
    1. Type of nucleic acid – DNA or RNA
    2. Strandedness – single stranded or double stranded
    3. Linear or circular
    4. Sense – positive, negative, or ambisense
    5. Number of segments
    6. Size of genome or genome segments
    7. Presence or absence and type of 5′-terminal cap
    8. Presence or absence of 5′-terminal covalently linked polypeptide
    9. Presence or absence of 3′-terminal poly(A) tract (or other specific tract)
    10. Nucleotide sequence comparisons
  D. Properties of proteins
    1. Number of proteins
    2. Size of proteins
    3. Functional activities of proteins (especially virion transcriptase, virion reverse transcriptase, virion hemagglutinin, virion neuraminidase, virion fusion protein)
    4. Amino-acid-sequence comparisons
  E. Lipids
    1. Presence or absence of lipids
    2. Nature of lipids
  F. Carbohydrates
    1. Presence or absence of carbohydrates
    2. Nature of carbohydrates
II. Genome organization and replication
    1. Genome organization
    2. Strategy of replication of nucleic acid
    3. Characteristics of transcription
    4. Characteristics of translation and post-translational processing
    5. Site of accumulation of virion proteins, site of assembly, site of maturation and release
    6. Cytopathology, inclusion body formation
III. Antigenic properties
    1. Serological relationships
    2. Mapping epitopes
IV. Biological properties
    1. Host range, natural and experimental
    2. Pathogenicity, association with disease
    3. Tissue tropisms, pathology, histopathology
    4. Mode of transmission in nature
    5. Vector relationships
    6. Geographic distribution

1970, brought together an enormous amount of information from studies of virion morphology (electron microscopy, structural data), biology (serology and virus properties), and physicochemical properties of viruses (nature and size of the genome, number and size of viral proteins). The impact of descriptions on virus classification has been particularly influenced by electron microscopy and the negative-staining technique for virions in the 1960s and 1970s. With this technique, viruses could be identified from poorly purified preparations of all tissue types and information about size, shape, structure, and symmetry could be quickly provided. As a result, virology progressed simultaneously for all viruses infecting animals, insects, plants, and bacteria. Since 1970, the virus descriptors list has included genome and replication information (sequence of genes, sequence of proteins), as well as molecular relationships with virus hosts.

The most recent wave of information used to classify viruses is virus genome sequences. Genome sequence comparisons are becoming more and more prevalent in virus taxonomy as exemplified by the presence of a significant number of phylogenetic trees in the *Eighth ICTV Report*. Some scientists promote the concept of quantitative taxonomy, aimed at demonstrating that virus genome sequences contain all the coding information required for all the biological properties of the viruses. This is in complete agreement with the polythetic concept of virus species definition if one considers that the unique sequence of a genome contains in fact all the information of the virus to perform all the steps of its replication cycle with structural and nonstructural genes and all of its biological functions. A good example of quantitative taxonomy is the re-classification of flaviviruses from the genus *Flavivirus* in the family *Togaviridae* into the new family *Flaviviridae* based upon sequencing of the yellow fever virus genome and comparisons with the gene sequence arrangement of members of the genus *Alphavirus* in the family *Togaviridae*. Another recent example is the merging of the genera *Rhinovirus* and *Enterovirus* in the family *Picornaviridae*, based on the fact that phylogenetic trees and pairwise comparisons did not support the continued distinction between the two genera.

There is a correlative modification of the list of virus descriptors, and **Table 2** lists the family and genus descriptors which are used in the current ICTV report. **Table 2** lists 45 different types of properties where each property (e.g., morphology) can take on different individual states (e.g., filamentous, icosahedral, etc.). A universal lists of virus descriptors has been established which is used by the ICTVdB. It contains a common set of descriptors for all viruses and subsets for specific viruses in relation to their specific hosts (human, animal, insect, plant, and bacterial).

The Order of Presentation of the Virus Classification

Currently, and for practical reasons only, virus classification is structured according to the 'Order of Presentation of Viruses' indicated in **Table 1**. The presentation of virus orders, families, and genera in this particular order reflects convenience rather than any hierarchical or phylogenetic consideration. The Order of Presentation of Viruses follows four criteria: (1) the nature of the viral nucleic acid, (2) the strandedness of the nucleic acid (single stranded (ss) or double stranded (ds)), (3) the use of a reverse transcription process (DNA or RNA), and (4) the sense of gene coding on the encapsidated genome (positive, negative, or ambisense). These four criteria give rise to six clusters comprising the 86 families and unassigned genera (genera without a designated family). Within each cluster, families and unassigned genera have been listed according to their possible affinities. For example, the families *Picornaviridae, Dicistroviridae, Sequiviridae, Comoviridae,* and *Potyviridae* are listed one after another because they share a number of similarities in their genome organization and sequence relatedness and they may form the basis for a proposed order in the future.

Table 3 List of criteria demarcating different virus taxa

I. Order
  Common properties between several families including:
    Biochemical composition
    Virus replication strategy
    Particle structure (to some extent)
    General genome organization
II. Family
  Common properties between several genera including:
    Biochemical composition
    Virus replication strategy
    Nature of the particle structure
    Genome organization
III. Genus
  Common properties within a genus including:
    Virus replication strategy
    Genome size, organization, and/or number of segments
    Sequence homologies (hybridization properties)
    Vector transmission
IV. Species
  Common properties within a species including:
    Genome arrangement
    Sequence homologies (hybridization properties)
    Serological relationships
    Vector transmission
    Host range
    Pathogenicity
    Tissue tropism
    Geographical distribution

A New Virus Taxon: The Virus Species

For many years, virologists debated the existence of virus species which was a very controversial issue and a series of definitions surfaced at regular intervals but none was adopted. However, in 1991, the ICTV Executive Committee accepted the species concept and the adopted definition is "A virus species is a polythetic class of viruses that constitutes a replicating lineage and occupies a particular ecological niche." This simple definition has already and will continue to have a profound effect on virus classification. In the *Eighth Report of the ICTV*, the 'List of Species' and the 'List of Tentative Species' are accompanied by a 'List of Species Demarcating Criteria' provided for each genus. Naturally, this list of criteria should follow the polythetic nature of the species definition, and more than one criterion should be used to determine a new species. It is obvious that most criteria are shared among the different genera, within and across families. These shared criteria include host range, serological relationships, vector transmission type, tissue tropism, genome rearrangement, and sequence homology (**Table 3**). However, while the nature of the criteria is similar, the levels of demarcation clearly differ from one family to another. This may reflect differences in appreciation from one family to another, but most likely reflects the differential ranking of a particular criterion in different families. The huge differences in sequence homologies (up to 30%) among lentivirus nucleoprotein sequences may not have the same biological significance as small differences for potyvirus capsid protein sequences (0–10%), and therefore universal levels of sequence identity for similar genes may not exist for viruses. However, it is important to note that the nature of the demarcating criteria at the genus level will probably not change since they have passed the test of years. Despite the fact that they were mostly established using biochemical and structural criteria, most of them have remained valid when correlated with genome organization and sequence data.

A Uniform Nomenclature for All Virus Taxa

Nomenclature is tightly associated with classification, in the sense that the taxonomic names indicate, to some extent, the nature of the taxa. Similarly for viruses, the ICTV has set rules for virus nomenclature and the orthography of taxonomic names that are regularly revised and improved. The international virus species names end in 'virus', international genus names in

'...virus', international subfamily names in '...virinae', international family names in '...viridae', and international order names in '...virales'. In formal taxonomic usage, the virus order, family, subfamily, genus, and species names are printed in italics (or underlined) and the first letter is capitalized. For all taxa except species, new names are created following ICTV guidelines. Because of the difficulty in creating new official international names for virus species, it has been decided in 1998 by the ICTV to use the existing English vernacular virus names. However, to differentiate virus species names from virus names it has also been decided that their typography would be different, that is, the species names would be italicized, and the first letter of the name capitalized while the virus names would not. In addition ICTV had created an additional category called 'Tentative Species Names' to accommodate viruses that seemed to belong to a new species, but did not have enough data to support this decision; it was also a way to 'reserve' a name already used in literature. In 2005, ICTV decided to replace this category by 'Unassigned Viruses' in the genus.

Latinized binomials for virus names have been supported by animal and human virologists of the ICTV for many years but have never been implemented. Their recommendation was in fact withdrawn from ICTV nomenclature rules in 1990, and consequently, such names as *Herpesvirus varicella* or *Polyomavirus hominis* should not be used. For several years, plant virologists adopted a different nomenclature, using the vernacular name of a virus but replacing the word 'virus' by the genus name: for example, *Cucumber mosaic cucumovirus* and *Tobacco mosaic tobamovirus*. This system is called 'the non-latinized binomial system', although the binomial order is the opposite of the typical latinized binomial system where the genus name ends with the virus name. Though this usage is favored by many scientists, and examples of such a practice can be found for human, animal, and insect viruses (e.g., *Human rhinovirus*, *Canine calicivirus*, and *Acheta densovirus*), it has not yet been adopted as a universal system by the ICTV; however, it has been decided that each study group would decide what is best for the viruses they deal with and the new names would have to be ratified through a formal taxonomic proposal by the ICTV.

In formal usage, the name of the taxon precedes the name of the taxonomic unit: for example, "the family *Picornaviridae*" or "the genus *Rhinovirus*". In informal vernacular usage, virus order, family, subfamily, genus, and species names are written in lower case roman script; they are not capitalized or italicized (or underlined) – for example 'animal reoviruses'. To avoid ambiguous identifications, it has been recommended to journal editors that published virological papers follow ICTV guidelines for proper virus identification and nomenclature and that viruses should be cited with their full taxonomic terminology when they are first mentioned in an article, for example, order *Mononegavirales*, family *Paramyxoviridae*, subfamily *Pneumovirinae*, genus *Pneumovirus*, species *Human respiratory syncytial virus*.

A Universal Classification System

The present universal system of virus taxonomy is set arbitrarily at the hierarchical levels of order, family, subfamily, genus, and species. Lower hierarchical levels, such as suborder, subgenera, and subspecies, may be considered in the future if need arises. Hierarchical levels under the species level such as strains, serotypes, variants, and pathotypes are established by international specialty groups and/or by culture collections, but not by the ICTV.

Species

The species taxon is always regarded as the most important taxonomic level in classification but has proved difficult to apply to viruses. In 1991, the ICTV accepted the definition of species, stated above, proposed by Marc van Regenmortel. The major advantage of this definition is that it can accommodate the inherent variability of viruses and is not dependent on the existence of a unique set of characteristics. Members of a polythetic class are defined by more than one property and no single property is absolutely indispensable. Thus, in each family, it might be possible to determine the set of properties of the class 'species' and thus to verify if the family members are representatives of the class 'species' or if they belong to a different taxonomic level.

Many practical matters are related to the definition of a virus species. These include (1) homogeneity of the different taxa across the classification, (2) diagnostic-related matters, (3) virus collections, (4) evolution studies, (5) biotechnology, (6) sequence database projects, (7) virus database projects, (8) publication matters, and also (9) intellectual property rights.

Genera

There is no formal ICTV definition for a genus, but it is commonly considered as "a population of virus species that share common characteristics and are different from other populations of species." Although this definition is somewhat elusive, this level of classification seems stable and useful. Some genera have been moved from one family to another (or from one family to an unassigned genus status such as the genus *Hepevirus*) over the years, but the composition and description of these genera has remained very stable. The characters defining a genus are different from one family to another. The use of subgenera has been abandoned in current virus classification.

Families

Genera are usually clustered in families, and most of the time, when a new genus, obviously not belonging to any existing family, is created, virologists also create a new family. Even after the creation of the ICTV, plant virologists have continued to classify plant viruses in 'groups', refusing to place them in genera and families. This position was mostly caused by a refusal to accept a binomial nomenclature. However, because of obvious similarities, plant reoviruses and rhabdoviruses had been integrated into the families *Reoviridae* and *Rhabdoviridae* (**Table 1**). Plant virologists subsequently accepted in 1995 the placing of plant viruses into species, genera, and families. The number of unassigned genera is regularly decreasing with time; the most recent clustering is the creation of the family *Flexiviridae* with the genera *Potexvirus*, *Carlavirus*, *Mandarivirus*, *Foveavirus*, *Capillovirus*, *Allexivirus*, *Vitivirus*, and *Trichovirus*. However, there are still 22 unassigned genera that do not belong to any family. Their presence originates mostly from the preference of plant virologists for accumulating data on virus species and genera before clustering genera in families. The unassigned genus status is now being used by animal virologists as a convenient temporary classification status. Examples are the unassigned genera *Anellovirus*, that is close to the family *Circoviridae* but different enough to be separated, the previously unassigned genus *Cripavirus* that has been upgraded to full family status (*Dicistrovirida*), and another unassigned genus *Iflavirus* that has been created to accommodate new picorna-like viruses that are not typical picornaviruses.

Orders

As mentioned above, the higher hierarchical levels for virus classification are extremely difficult to establish. To date only three orders have been accepted: *Caudovirales*, *Mononegavirales*, and *Nidovirales*. The first order, *Mononegavirales*, was established in 1990 and comprises the nonsegmented ssRNA negative-sense viruses, namely, the families *Bornaviridae*, *Filoviridae*, *Paramyxoviridae*, and *Rhabdoviridae*. This order was formed because of the great similarity between these families over many criteria, including their replication strategies. A second order, *Caudovirales*, contains all families of dsDNA phages possessing a tail, including the *Myoviridae*, *Podoviridae*, and *Siphoviridae*. The order *Nidovirales* comprises the families *Coronaviridae*, *Arteriviridae*, and *Roniviridae* and was created because it was clear that the viruses belonging to these families share many properties and yet are so different that they cannot be placed together in the same family. Many members of the ICTV advocate the creation of many more orders, and as a matter of fact four new orders encompassing the families *Herpesviridae*, *Picornaviridae*, *Reoviridae*, and *Retroviridae* have been proposed and provisionally named but it has been decided to proceed cautiously in this area so as to avoid creation of short-lived orders. The creation of formal taxa higher than the orders – for example kingdoms, classes, and subclasses – has not been considered by the ICTV.

See also: Nature of Viruses; Phylogeny of Viruses; Virus Classification by Pairwise Sequence Comparison (PASC); Virus Databases; Virus Species.

Further Reading

Adams MJ, Antoniw JF, Bar-Joseph M, et al. (2004) The new plant virus family *Flexiviridae* and assessment of molecular criteria for species demarcation. *Archives of Virology* 149: 1045–1060.

Dolja VV, Boyko VP, Agranovsky AA, and Koonin EV (1991) Phylogeny of capsid proteins of rod-shaped and filamentous RNA plant viruses: Two families with distinct patterns of sequence and probably structure conservation. *Virology* 184: 79–86.

Dolja VV and Koonin EV (1991) Phylogeny of capsid proteins of small icosahedral RNA plant viruses. *Journal of General Virology* 72: 1481–1486.

Fauquet CM, Mayo MA, Maniloff J, Desselberger U, and Ball LA (eds.) (2004) *Virus Taxonomy: Eigth Report of the International Committee on Taxonomy of Viruses*, p.1258. San Diego, CA: Elsevier Academic Press.

Francki RIB, Milne RG, and Hatta T (1985) *Atlas of Plant Viruses*. Boca Raton, FL: CRC Press.

Harrison BD, Finch JT, Gibbs AJ, et al. (1971) Sixteen groups of plant viruses. *Virology* 45: 356–363.

Koonin EV (1991) The phylogeny of RNA-dependent RNA polymerases of positive-strand RNA viruses. *Journal of General Virology* 72: 2197–2206.

Mayo MA and Horzinek MC (1998) A revised version of the International Code of Virus Classification and Nomenclature. *Archives in Virology* 143: 1645–1654.

Van Regenmortel MHV (1990) Virus species, a much overlooked but essential concept in virus classification. *Intervirology* 31: 241–254.

Van Regenmortel MHV, Bishop DHL, Fauquet CM, Mayo MA, Maniloff J, and Calisher CH (1997) Guidelines to the demarcation of virus species. *Archives of Virology* 142: 1505–1518.

Relevant Website

http://www.ictv.ird.fr – ICTV; Taxonomic Proposal Management System.

Tenuivirus

B C Ramirez, CNRS, Paris, France

© 2008 Elsevier Ltd. All rights reserved.

Glossary

RNA silencing Evolutionarily conserved mechanism in many eukaryotes to target and degrade aberrant RNA molecules. It constitutes an antiviral defense in plants and insects.

Introduction

The tenuiviruses were first described in the Fifth Report of the International Committee on Taxonomy of Viruses in 1982 as nonenveloped plant viruses possibly possessing a negative single-stranded (ss) RNA genome. Their name comes from the Latin 'tenuis', (thin, fine, weak) which refers to the structure of the viral particle as seen by electron microscopy (**Figure 1**). Epidemics of rice stripe virus (RSV) and rice hoja blanca virus (RHBV) cause important yield losses in rice-growing areas of Asia and the former USSR and of tropical America, respectively. Tenuiviruses exhibit unique properties that make them different from other plant viruses. Some properties of tenuiviruses are the following:

1. The peculiar flexuous viral particles have a thread-like morphology and can adopt circular forms (**Figure 1(a)**).
2. The viruses are persistently transmitted by a particular species of planthopper in a circulative, propagative manner. For some of the members of the genus, it has been demonstrated that the virus multiplies both in the host plant and in the insect vector. Multiplication of the virus in the vector may have deleterious effects on the insect. The viruses can be transmitted transovarially by viruliferous female planthoppers to their offspring, and through sperm from viruliferous males.
3. The genome of tenuiviruses is multisegmented and composed of ssRNAs that have either negative or ambisense polarity (**Figure 2**).
4. An RNA-dependent RNA polymerase (RdRp) is associated with the viral particle.
5. A nonstructural protein of 16–22 kDa accumulates in large amounts in infected plants, forming large inclusions (**Figure 1(b)**).
6. It has been observed for some tenuiviruses that the mRNAs are synthesized via cap-snatching (**Figure 3**).
7. Tenuiviruses infect plants of the family Poaceae.

Type Species and Other Species in the Genus

Rice stripe virus (RSV) is the type species of the genus *Tenuivirus*. Other species in the genus are *Echinochloa hoja blanca virus* (EHBV), *Maize stripe virus* (MSpV), *Rice grassy stunt virus* (RGSV), *Rice hoja blanca virus* (RHBV), and *Urochloa hoja blanca virus* (UHBV). Tentative species are Brazilian wheat spike virus (BWSpV), European wheat striate mosaic virus (EWSMV), Iranian wheat stripe virus (IWSV), rice wilted stunt virus, and winter wheat mosaic virus.

Virion Properties

Morphology

The particles also referred as ribonucleoproteins (RNPs) are thin filaments that may appear circular or spiral shaped (**Figure 1(a)**). The RNPs are 3–10 nm in diameter, with lengths proportional to the sizes of the RNAs they contain. No envelope has been observed.

Physical and Physicochemical Properties

RNP preparations can be separated into four or five components by sucrose density gradient centrifugation. The buoyant density of the RNP in CsCl when centrifuged to equilibrium is 1.282–1.288 g cm^{-3}. RNA constitutes 5–12% of the particle weight.

Components

The tenuivirus genome is composed of 4–6 noncapped ssRNA segments; the approximate size of the genome of the type member RSV is 17 kbp (**Figure 2**). The largest segment (RNA1, ~9 kbp) of RSV, MSpV, and RHBV is of negative polarity and encodes the RdRp. Segments 2 (RNA2, 3.3–3.6 kbp), 3 (RNA3, 2.2–2.5 kbp), and 4 (RNA4, 1.9–2.2 kb) of RSV, MSpV, and RHBV are ambisense. Segment 5 (1.3 kbp) detected in virions of MSpV and of EHBV is of negative polarity. A fifth RNA segment has also been reported for some isolates of RSV. RGSV RNA1, 2, 5, and 6 are homologous to RNA1, 2, 3, and 4, respectively, of other tenuiviruses, whereas RNA3 (3.1 kbp) and 4 (2.9 kbp) are ambisense and unique to RGSV. Subgenomic RNAs (sgRNAs) of different sizes and of either polarity are detected; they serve as mRNA for the synthesis of the viral proteins (**Figure 3**).

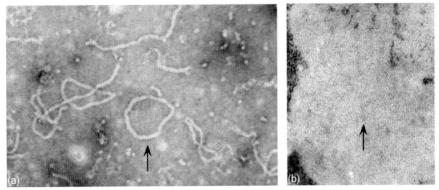

Figure 1 (a) Micrograph of purified ribonucleoproteins (RNPs) of rice hoja blanca virus (RHBV): An RNP is indicated by the arrow. (b) Ultrathin section of a rice leaf cell infected with RHBV. The arrow indicates viral inclusion bodies inside the nucleus with cytopathic effects.

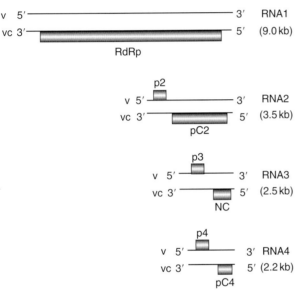

Figure 2 Schematic representation of the genome organization of RSV. RdRp, RNA-dependent RNA polymerase; NC, nucleocapsid protein; p4, nonstructural protein, which accumulates in viral inclusion bodies; v, viral sense RNA; vc, viral complementary RNA.

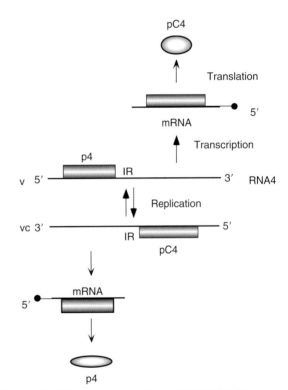

Figure 3 Schematic representation of the replication, transcription, and translation strategies of the ambisense RNA segments of tenuiviruses. Cap and nonviral nucleotides of host origin used as primers by the RdRp as a result of cap-snatching are represented by '——•'. IR, intergenic region.

The RNPs contain a nucleocapsid (NC) protein of 34–35 kDa, and small amounts of a 230 kDa protein which is co-purified with RNPs of RSV, RHBV and RGSV. This protein is the RdRp, associated with filamentous RNPs. The RNA polymerase activity of RHBV is capable of replicating and transcribing the RNA segments *in vitro*.

Genome Organization, Encoded Proteins, Replication and Transcription

The 5' and 3' terminal sequences (for about 20 nt) are complementary to each other: they can base-pair and give rise to circular RNPs. The terminal 8 nt (5' ACACAAAG) and their complement are conserved between tenuiviruses and viruses of the genus *Phlebovirus* of the family *Bunyaviridae*. Several RNA segments encode two proteins in an ambisense arrangement (**Figure 2**). The RdRp and NC protein (32 Kda) are encoded by the viral complementary (vc)RNA1 and vcRNA3, respectively. The p3 (24 Kda) encoded by the viral (v)RNA3 is a suppressor of RNA silencing as demonstrated for RHBV in both plant and insect hosts. A major nonstructural protein (p4; 22 Kda) that accumulates in infected plants is encoded by vRNA4. The sizes of other RSV proteins shown in **Figure 2** are 23 kDa (p2), 94 kDa (pC2), 24 kDa (p3), and 32 kDa (pC4).

For MSpV, RHBV, and RSV, the mRNAs are synthesized via cap-snatching (**Figure 3**). The 5′ end of the mRNAs contains 10–17 nonviral nucleotides and are capped, these extra sequences are derived from host cell mRNAs that are taken or snatched by the RdRp and used as primers to initiate mRNA synthesis. Intergenic noncoding regions located between the open reading frames (ORFs) can in certain cases adopt hairpin structures. The cap-snatching mechanism has been observed for mRNA synthesis of influenza viruses and viruses of the families *Bunyaviridae* and *Arenaviridae*.

Antigenic Properties

The NC proteins of RSV and MSpV are serologically related, as are the NC and p4 proteins of RSV and RGSV. Likewise, the NC proteins of RHBV, EHBV, and UHBV are serologically related. The NC protein of RSV reacts weakly with antibodies made to virion preparations of RGSV or RHBV.

Biological Properties

Host Range

Plant hosts of tenuiviruses all belong to the family Poaceae.

Cytopathology

In infected plants large inclusion bodies (**Figure 1(b)**) are observed which contain the major nonstructural protein p4.

Symptoms

Infected plants exhibit chlorotic stripes and yellow stippling on the leaf blade. Rice plants infected with RHBV at an early stage of development are stunted and may develop necrosis.

Insect Vector and Kind of Transmission

Each virus species is transmitted by a particular species of planthopper in a circulative, propagative manner and can be transmitted transovarially by viruliferous female planthoppers to their offspring, and through sperm from viruliferous males. The principal vectors of the species are *Laodelphax striatellus* for RSV, *Tagosodes cubanus* for EHBV, *Peregrinus maidis* for MSpV, *Nilaparvata lugens* for RGSV, *Tagosodes orizicolus* for RHBV, and *Caenodelphax teapae* for UHBV. Known vectors of the tentative species are *Sogatella kolophon* for BWSpV, *Javesella pellucida* for EWSMV, and *Ukanoes tanasijevici* for IWSV.

Criteria for Species Demarcation and Phylogenetic Relationships between Species in the Genus

The criteria for species demarcation are given as follows:

1. Vector specificity: transmission by different species of vector.
2. Different sizes and/or numbers of RNA segments.
3. Host range: abilities to infect different key plant species.
4. Amino acid (aa) sequence identity of less than 85% between corresponding gene products.
5. Nucleotide sequence identity of less than 60% between corresponding noncoding intergenic regions.

An example of species discrimination is that between RSV and MSpV. RSV is transmitted by *L. striatellus* and infects 37 graminaceous species including rice and wheat. MSpV is transmitted by *P. maidis* and infects maize, occasionally sorghum, and a few other graminaceous plants but not rice or wheat. RSV isolates have genomes of four RNA segments and the MSpV genome consists of five segments.

An example of difficult species demarcation is the group of the hoja blanca viruses (RHBV, EHBV, and UHBV). They have different vectors, different sizes, and numbers of RNA segments, different hosts, and the nucleotide sequence identity of their intergenic regions is less than 60%. However, the aa sequences of the four proteins on RNAs 3 and 4 are about 90% identical between RHBV, EHBV, and UHBV. Since four of five criteria are met, these viruses could be considered as distinct species that possibly separated recently and are now diverging with little field contact between them.

Phylogenetic analysis of the sequence data from RNA3 and RNA4 of the tenuiviruses shows that RHBV, EHBV, and UHBV are related and form a group distinct from RSV and MSpV (**Figure 4**).

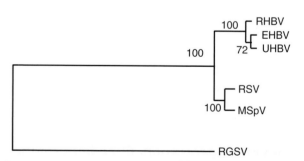

Figure 4 Phylogenetic tree showing the relationships between tenuiviruses on the basis of the nucleotide sequences of the ORFs on RNAs 3 and 4 (RNAs 5 and 6 for RGSV). Shown is the percentage bootstrap support at each node. RHBV, *Rice hoja blanca virus;* EHBV, *Echinochloa hoja blanca virus;* UHBV, *Urochloa hoja blanca virus;* RSV, *Rice stripe virus;* MSpV, *Maize stripe virus;* RGSV, *Rice grassy stunt virus.*

Relation to Other Taxa

Viruses of the genus *Tenuivirus* share several properties with those of the genus *Phlebovirus* of the family *Bunyaviridae*. The multisegmented genomes of tenui- and phleboviruses contain negative-sense and ambisense components. The 5′ and 3′ terminal complementary sequences of viruses of either genus can base-pair and could give rise to circular RNPs. The terminal 8 nt (5′ ACACAAAG) and their complement are conserved between tenui- and phleboviruses. Synthesis of mRNA by the viral RdRp follows a cap-snatching mechanism for viruses of the two genera. Tenuiviruses and most genera in the family *Bunyaviridae* infect their insect vectors as well as their plant hosts. The different number of genome components and the apparent lack of an enveloped viral particle distinguish tenuiviruses from viruses in the family *Bunyaviridae*.

Acknowledgments

The author is grateful to F. Morales for the electron microscope pictures and to Anne-Lise Haenni for useful discussions and critical reading of the manuscript.

See also: Bunyaviruses: General Features.

Further Reading

de Miranda JR, Muñoz M, Wu R, and Espinoza AM (2001) Phylogenetic placement of a novel tenuivirus from the grass *Urochloa plantaginea*. *Virus Genes* 22: 329–333.

de Miranda JR, Muñoz M, Wu R, Hull R, and Espinoza AM (1996) Sequence of rice hoja blanca tenuivirus RNA-2. *Virus Genes* 12: 231–237.

de Miranda JR, Ramirez BC, Muñoz M, *et al.* (1997) Comparison of Colombian and Costa Rican strains of rice hoja blanca tenuivirus. *Virus Genes* 15: 191–193.

Estabrook EM, Suyenaga K, Tsai JH, and Falk BW (1996) Maize stripe tenuivirus RNA 2 transcripts in plant and insect hosts and analysis of pvc2, a protein similar to the Phlebovirus virion membrane glycoproteins. *Virus Genes* 12: 239–247.

Falk BW and Tsai JH (1998) Biology and molecular biology of viruses in the genus Tenuivirus. *Annual Review of Phytopathology* 36: 139–163.

Haenni A-L, de Miranda JR, Falk BW, *et al.* (2005) *Tenuivirus*. In: Fauquet CM, Mayo MA, Maniloff J, Desselberg U, and Ball LA (eds.) *Virus Taxonomy: Eighth Report of the International Committee on Taxonomy of Virus*, pp. 717–723. San Diego, CA: Elsevier Academic Press.

Huiet L, Feldstein PA, Tsai JH, and Falk BW (1993) The maize stripe virus major non-capsid protein messenger RNA transcripts contain heterogeneous leader sequences at their 5′ termini. *Virology* 197: 808–812.

Nguyen M, Ramirez BC, Golbach R, and Haenni A-L (1997) Characterization of the *in vitro* activity of the RNA-dependent RNA polymerase associated with the ribonucleoproteins of rice hoja blanca tenuivirus. *Journal of Virology* 71: 2621–2627.

Ramirez BC, Garcin D, Calvert LA, Kolakofsky D, and Haenni A-L (1995) Capped non-viral sequences at the 5′ end of the mRNA of rice hoja blanca virus RNA4. *Journal of Virology* 69: 1951–1954.

Ramirez BC and Haenni A-L (1994) Molecular biology of tenuiviruses, a remarkable group of plant viruses. *Journal of General Virology* 75: 467–475.

Ramirez BC, Lozano I, Constantino L-M, Haenni A-L, and Calvert LA (1993) Complete nucleotide sequence and coding strategy of rice hoja blanca virus RNA4. *Journal of General Virology* 74: 2463–2468.

Ramirez BC, Macaya G, Calvert LA, and Haenni A-L (1992) Rice hoja blanca virus genome characterization and expression *in vitro*. *Journal of General Virology* 73: 1457–1464.

Shimizu T, Toriyama S, Takahashi M, Akutsu K, and Yoneyama K (1996) Non-viral sequences at the 5′-termini of mRNAs derived from virus-sense and virus-complementary sequences of the ambisense RNA segments of rice stripe tenuivirus. *Journal of General Virology* 77: 541–546.

Toriyama S, Kimishima T, and Takahashi M (1997) The proteins encoded by rice grassy stunt virus RNA5 and RNA6 are only distantly related to the corresponding proteins of other members of the genus Tenuivirus. *Journal of General Virology* 78: 2355–2363.

Toriyama S, Kimishima T, Takahashi M, Shimizu T, Minaka N, and Akutsu K (1998) The complete nucleotide sequence of the rice grassy stunt virus genome and genomic comparisons of the genus Tenuivirus. *Journal of General Virology* 79: 2051–2058.

Tetraviruses

J A Speir and J E Johnson, The Scripps Research Institute, La Jolla, CA, USA

© 2008 Elsevier Ltd. All rights reserved.

Introduction

The *Tetraviridae* are a family of viruses with nonenveloped $T=4$ particles that package single-stranded, positive-sense RNA genomes and infect only a single order of insects, the Lepidoptera (moths and butterflies). This family stands out from the rest of the known viruses based on these properties alone. It is the only RNA virus family with a host range restricted to insects and the only family with nonenveloped icosahedral $T=4$ particles. The unusual symmetry of their capsids has become the basis for their family name (from the Greek *tettares*, four).

The first report of a larvae disease later to be attributed to a tetravirus was from South Africa in 1941. Researchers Tooke and Hubbard described a devastating annual mortality of the emperor pine moth, *Nudaurelia cytherea capensis*, such that the dead larvae formed a carpet beneath the pine trees (*Pinus radiata*). The outbreaks have now reduced the larvae population on the pine plantations to a great extent. Five different viruses were found in the infected larvae that were named nudaurelia α, β, γ, δ, and ε, with β being the most abundant. The nudaurelia β virus (NβV) was extensively studied and became the type member of a new virus group called the nudaurelia β-like viruses (betatetraviruses) that was officially recognized in 1982. These studies showed NβV had a monopartite genome and the first known $T=4$ icosahedral capsid. The most recently characterized betatetravirus is from the USA, named Providence virus, which was discovered by Pringle, Ball, and colleagues in 2003. Importantly, this is the first tetravirus that replicates in cell culture, greatly aiding the study of tetravirus molecular biology (see below).

While the betatetraviruses were the first tetraviruses discovered and form the majority of the family, Hendry and colleagues isolated another virus from *Nudaurelia* larvae in 1985, the nudaurelia ω virus (NωV), which was physically similar to NβV but encapsidated a second genomic RNA. NωV became the type member for a second genus within the *Tetraviridae*, the ω-like genus (omegatetraviruses). A second omegatetravirus was isolated by Hanzlik and colleagues in 1993 from Australia, the helicoverpa armigera stunt virus (HaSV), and a third was discovered in 2005 by Yi and colleagues from the Yunnan Province of China, the dendrolimus punctatus tetravirus (DpTV). Both HaSV and DpTV are closely similar to NωV (see below). Currently, the family consists of three omegatetraviruses, and nine betatetraviruses, as well as eight other possible members (**Table 1**). The tetravirus family has grown little in recent years, with only a few new members identified in the last decade. This is due to their limited host range, the difficult field collections of the insect hosts due to the locations where they thrive, and the difficulty or impossibility of rearing the host insects in the laboratory. Also, several aspects of their life cycle remain poorly characterized due to their inability to grow in cell culture despite extensive efforts covering a wide variety of cell types. In spite of these drawbacks, studies of the tetraviruses have expanded to a great extent. It is possible to raise HaSV from laboratory-grown insects, Providence virus is the first tetravirus found to replicate in cell culture, and virus-like particles (VLPs) can be produced in a baculovirus expression system for NωV, HaSV, and TaV. These systems have allowed for advances in detailed biophysical descriptions of these capsids, including a crystal structure of NωV, and new data on replication, pathology, and the biotechnological potential of this unique virus family.

Table 1 Members of the *Tetraviridae*

Virus	Acronym	Family of Lepidoptera host	Geographic location
Betatetraviruses			
Nudaurelia capensis β virus[a]	NβV	Saturniidae	South Africa
Antheraea eucalypti virus[b]	AeV	Saturniidae	Australia
Dama trima virus	DtV	Limacodidae	Malaysia
Dasychira pudibunda virus[c]	DpV	Lymantriidae	UK
Philosamia cynthia X ricini virus	PxV	Saturniidae	UK
Providence virus	PrV	Noctuidae	USA
Pseudoplusia includens virus	PiV	Noctuidae	USA
Thosea asigna virus[d]	TaV	Limacodidae	Malaysia
Trichoplusia ni virus	TnV	Noctuidae	USA
Omegatetraviruses			
Nudaurelia capensis ω virus[a]	NwV	Saturniidae	South Africa
Dendrolimus punctatus tetravirus	DpTV	Lasiocampidae	China
Helicoverpa armigera stunt virus	HaSV	Noctuidae	Australia
Unassigned possible members[e]			
Acherontia atropas virus	AaV	Sphingidae	Canary Islands
Agraulis vanillae virus	AvV	Nymphalidae	Argentina
Callimorpha quadripuntata virus	CqV	Arctiidae	UK
Eucocytis meeki virus	EmV	Cocytiidae	Papua New Ginea
Euploea corea virus	EcV	Danadidae	Australia/Germany
Hypocritae jacobeae virus	HjV	Arctiidae	UK
Lymantria ninayi virus	LnV	Lymantriidae	Papua New Ginea
Nudaurelia ε virus[f]	NεV	Saturniidae	South Africa

[a]Type virus for genus.
[b]Serological evidence shows identity to NβV.
[c]Host renamed *Calliteara pudibunda*.
[d]Host renamed *Setothosea asigna*.
[e]Viruses showing serological relationship to a known betatetravirus (excluding NεV), but otherwise uncharacterized.
[f]NεV resembles the tetraviruses but is serologically unrelated to any known member.
Reproduced by permission of Elsevier.

Genome Organization, Replication, and Capsid Assembly

General Characteristics of Tetravirus Genomic RNAs

The main characteristic separating the betatetraviruses and omegatetraviruses is the RNA encapsidated by the particles in each genus. All tetraviruses analyzed for RNA content share the presence of a large genomic RNA strand of about 5.3–6.5 kbp in length and a smaller genomic or subgenomic RNA strand of about 2.5 kbp. In the monopartite β-like viruses, both the replicase and capsid protein genes are on the large genomic RNA strand (~6.5 kbp). In the bipartite ω-like viruses, the replicase and capsid protein genes are split between the large (~5.3 kbp) and small (~2.5 kbp) genomic RNA strands, respectively.

Although the betatetraviruses are monopartite, the capsids often package two RNA species. A subgenomic mRNA of 2.5 kbp coding for the capsid protein is also encapsidated with the genomic RNA strand in NβV, PrV, TaV, and TnV. Thus, virus particles from both genera package two RNA molecules totaling ~7.8–9.0 kbp. Experimental evidence also suggests that the ω-like viruses package both genomic RNAs in a single particle, as is the case for the nodaviruses, the only other small spherical RNA animal viruses with a bipartite genome. This contrasts with plant viruses having multipartite genomes, which package single genomic RNA strands in different particles.

Analysis of the HaSV genomic RNAs revealed 5' cap structures and distinctive tRNA-like structures at the 3' termini, which are not polyadenylated. A tRNA-like structure also exists at the 3' terminal of the NβV RNA, and no β-like virus genome has been found to be polyadenylated. However, both the HaSV and NβV RNAs could be polyadenylated, indicating no terminal blockage. Therefore, all tetravirus genomic RNAs are likely to have unblocked, tRNA-like 3' termini and 5' caps, although the existence of the cap has not been experimentally tested with any β-like virus.

β-Like Genome Organization and Replication Model

The complete nucleotide sequence of the NβV genome (6625 nt) was elucidated in 1998, and partial cDNA clones of the TaV and PrV genomes have been analyzed more recently (2001 and 2003, respectively). All three show nearly identical organization. The single genomic RNAs are dicistronic and have overlapping open reading frames (ORFs) (**Figure 1**). The 5' ends encode the replicase and the 3' ends encode the capsid protein genes, which have extensive overlap with the replicase. The NβV replicase ORF contains 5778 nt encoding 1925 amino acids with a calculated molecular weight of 215 kDa. It contains the three domains characteristic of viruses in the α-like RNA virus superfamily (the N-terminal methyltransferase domain, the nucleotide-binding or helicase domain, and the RNA-dependent RNA polymerase (RdRp) domain), but lacks the motifs for the papainlike protease and proline-rich hinge domains that are implicated in autoprocessing of the replicase proteins of some +RNA viruses of vertebrates. Finally, the genomic RNA appears to be the mRNA for the replicase, since its initiation codon is the first located at the 5' end of the genomic RNA. Translation studies of other β-like virus genomic RNAs have also produced large polypeptides, supporting this scenario.

The sequence of the putative TaV replicase shows a novel permutation among RNA viruses, and together with birnaviruses and an unassigned possible tetravirus (Euprosterna elaeasa virus) defines a unique, ancient lineage of RdRp's. None of the other sequenced tetravirus replicases have a permutation. The TaV replicase lacks both the methyltransferase and helicase domains and the RdRp has little homology with other RNA virus replicases, which as a group have a universal sequence conservation. In 2002, Gorbalenya and colleagues discovered that the catalytic palm subdomain of the TaV RdRp had permuted the conserved 'A–B–C' motif to 'C–A–B' by moving the C motif, which has two Asp residues critical for activity, 110 residues upstream in the sequence. Strikingly, structural modeling demonstrated that this permutation can be accommodated in the conserved RdRp structures by simply rearranging the connectivity of three loops, all of which are positioned away from the active site. Thus, the TaV RdRp is likely to have a conserved fold but with unique connectivity, which might be a major event in the evolution of polymerases.

The 3' proximal ORF of NβV has 1836 nt and encodes the 66 kDa capsid protein precursor (named α). The capsid protein ORF overlaps the replicase ORF by more than 99% and is in the +1 reading frame relative to the replicase. The capsid protein gene is apparently translated through the production of the previously mentioned 2.5 kbp subgenomic RNA (**Figure 1**), yet replication experiments designed to detect double-stranded RNA (dsRNA) intermediates do not show evidence for a separate complex for the β-like subgenomic RNA indicating that both genomic and subgenomic RNAs are produced from a single full length (–)RNA template. One copy of the genomic (+)RNA, one copy of the subgenomic (+)RNA, and 240 copies of the expressed capsid protein precursor assemble to form an ~400 Å diameter icosahedral $T = 4$ virion (**Figure 2**). The capsid protein is labeled a precursor at this stage because each copy executes an autocatalytic cleavage to produce a large protein (β, ~60 kDa) and a small protein (γ, ~6 kDa), but only after the capsid is assembled (**Table 2**). Both the β and γ polypeptides remain part of the virion after cleavage.

The capsid protein precursors in TaV and PrV may be significantly larger than that of NβV. An additional codon

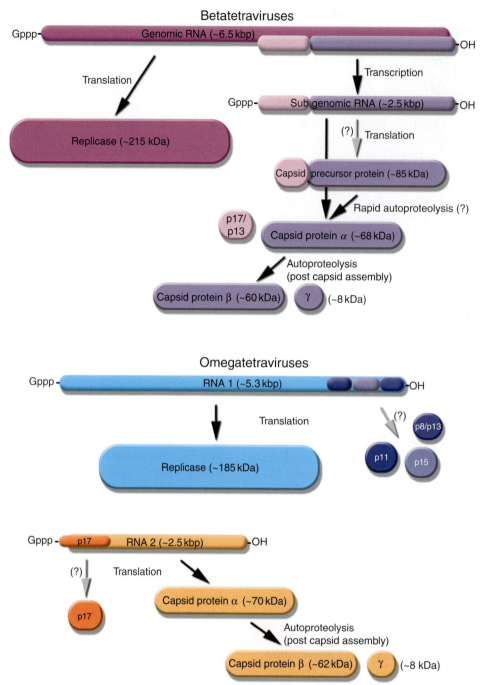

Figure 1 Genome organization and replication strategy of the tetraviruses. Each diagram represents the consensus of available data, but some group members may have slightly different organizations and steps from those shown. The individual events and gene products are discussed in the text. The gray arrows with question marks represent little, if any, translation of those gene products or an unverified event. While autocatalytic processing of the 68–70 kDa capsid protein is nearly identical, significant differences exist between the two groups in their pathways and their potential products. The β-like group utilizes a subgenomic RNA for translation of the capsid precursor protein, which may have two autoproteolytic processing events. The ω-like group has additional ORFs that overlap the end of the replicase gene with unknown functions. Both groups encode 13–17 kDa proteins on the small RNAs in addition to the capsid protein, but it is an additional ORF that would be translated as part of a polyprotein in the β-like group and an overlapping ORF that would be translated separately in the ω-like group.

exists upstream from the known N-terminus of the capsid proteins found in the TaV and PrV virions, yet the extra codon is transcribed to their subgenomic RNAs (**Figure 1**). The additional codons would add approximately 17 and 13 kDa to the TaV and PrV capsid protein precursors, respectively. Since the assembled capsids do not have a protein of this size, these precursors possibly undergo two post-assembly, autocatalytic cleavages, with the

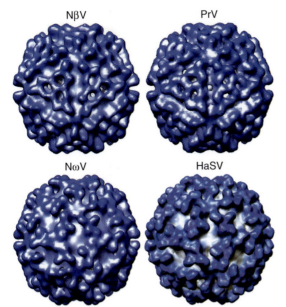

Figure 2 Cryoelectron microscopy reconstructions of the icosahedral tetravirus virions. All four structures are at ~25–30 Å resolution and are viewed down their icosahedral twofold axes. The capsids are ~400 Å in diameter and have $T = 4$ quasi-icosahedral symmetry. The capsid morphologies within each group are closely similar (β-like at top, ω-like at bottom), but differ between the groups. Most notably, within the triangular facets, the β-like viruses have a pitted surface where the ω-like surface is filled in. TaV also displays the same pitted surface in negative-stain EM photographs. The recent crystal structure determination of PrV showed this is mainly due to different rotations of the Ig-like domains relative to the β-barrels when compared to the subunits in the NωV crystal structure (see **Figure 3**). This supports the suggested role of the Ig-like domain in determining the restrictive host cell specificity of these viruses by displaying distinctly different surfaces to cellular receptors.

additional N-terminal cleavage producing polypeptides corresponding to the extra codons (p17, p13). Evidence for this came from expressing the entire capsid protein precursor of TaV in a recombinant baculovirus system as a single polypeptide (p17 + α). TaV VLPs were produced and contained only polypeptides of 58 and 7 kDa, suggesting that the additional 17 kDa had been previously removed by auto-proteolysis. However, it has not yet been shown that the extra codons are translated *in vivo* as part of the capsid protein precursor, and neither p17 nor p13 has been found in assembled particles (authentic or VLPs) of either virus.

ω-Like Genome Organization and Replication Model

The complete sequences of both RNA1 and RNA2 from HaSV and DpTV were determined in 1995 and 2005, respectively, and the complete RNA2 sequence from NωV was initially determined in 1992 and amended in 2005. Partial data for the RNA1 sequence of NωV are also available (**Table 2**). Like the β-like group, the RNAs of the ω-like group have near-identical organization. The length of RNA1 is 5312 nt in HaSV and 5492 in DpTV. Both RNAs contain a single large ORF covering over 90% of their total length, and three smaller ORFs located within the 3′ portion of the large ORF (**Figure 1**). While the three smaller ORFs are in the same reading frame, they are out of frame with the large ORF.

The large ORFs encode the viral replicases with calculated molecular weights of 187 and 180 kDa for HaSV and DpTV, respectively. Like the replicases of the β-like viruses, they also contain the N-terminal methyltransferase domain, the nucleotide-binding or helicase domain, and the RdRp domain, but lack the motifs for the papain-like protease and proline-rich hinge domains. The domains share from 63% to 76% identity between the HaSV and DpTV replicases, and about 36% identity with the same domains from the β-like replicases. The HaSV and NωV replicases are more closely related, sharing close to 90% identity. RNA1 self-replication has been detected before RNA2 replication has been observed, and protein expressed from HaSV RNA1 in a baculovirus system is able to replicate RNA2 transcripts at very low levels.

The three small ORFs encode proteins of 11, 15, and 8 kDa in HaSV and 11, 15, and 13 kDa in DpTV. In both genomes, the stop codon for p11 is located immediately to the 5′ side of the p15 ORF, allowing these two ORFs to form a single 26 kDa protein if the stop codon was suppressed. Expression of these proteins would almost certainly require a subgenomic RNA due to the long 5′ leading sequence; however, the existence of this RNA has not been confirmed. The amino acid sequences of all three ORFs show no discernible relationship to other proteins; therefore, no putative functions can be assigned.

The RNA2 lengths in NωV, HaSV, and DpTV are 2445, 2478, and 2490 nt, respectively. Each has a long (~280 nt) 5′ lead sequence followed by two overlapping genes starting at the first two AUG codons. The first ORF encodes a 17 kDa protein (p17) and overlaps the larger second ORF that encodes the 70–71 kDa capsid protein precursor (**Figure 1**). The initiation codon for p17 is in a poor context, suggesting that it will be translated inefficiently by scanning ribosomes. However, *in vitro* translation studies of both NωV and HaSV RNA2's revealed both gene products were produced. The function of p17, if any, remains unclear, but preliminary experimental observations have supported a regulatory role in genomic RNA replication, and/or a function analogous to that of the movement proteins of plant viruses.

The close relationship between the ω-like viruses is seen in their capsid protein precursors (α). The amino acid lengths of the precursors are 644, 647, and 643 for NωV, HaSV, and DpTV, respectively, and they share

Table 2 Properties of tetravirus capsid proteins, genomic (g), and subgenomic (sg) RNAs[a]

	Beta (Da)	Gamma (Da)			GenBank accession number(s)
β-Like virus			*gRNA (bases)*	*sgRNA (bases)*	
NβV	58 448[b]	7975[b]	6625[b]	2656[b]	AF102884
PrV	60 649[b]	7343[b]	6400	2500	AF548354
TaV	58 327[b]	6781[b]	6500	2500	AF282930, AF062037
AeV			Present		
TnV	67 000–68 000		5865[c]		
DtV	62 000–66 000		Present		
PxV	62 400		Present		
DpV	66 000		5555[c]		
PiV	55 000		5865[c]		
ω-Like virus			*gRNA1 (bases)*	*gRNA2 (bases)*	
NωV	62 019[b]	7817[b]	5300	2445[b]	S43937[d]
HaSV	63 378[b]	7309[b]	5312[b]	2478[b]	U18246, L37299
DpTV	62 107[b]	7636[b]	5492[b]	2490[b]	AY594352, AY594353
Unassigned possible members			*gRNA or RNA1*	*sgRNA or RNA2*	
NεV	61 000				

[a]Blank fields represent undetermined or unavailable data.
[b]Data derived from sequence analysis.
[c]Converted from RNA molecular weight using average of 324 Da/nt.
[d]Sequence was amended in 2005 by GenBank accession numbers DQ054382 and DQ054403 to determine the full 5′ terminal sequence of RNA2.
Reproduced by permission of Elsevier.

Table 3 Percentage of sequence identity and similarity between the capsid proteins of six tetraviruses (three ω-like: HaSV, NωV, DpTV; three β-like: PrV, NβV, TaV). The similarity score is in parentheses

	NωV	*HaSV*	*DpTV*	*PrV*	*NβV*	*TaV*
NωV		66 (72)	87 (93)	37 (46)	29 (35)	27 (29)
HaSV	66 (72)		66 (78)	38 (47)	29 (36)	27 (33)
DpTV	87 (93)	66 (78)		37 (55)	25 (39)	26 (41)
PrV	37 (46)	38 (47)	37 (55)		27 (35)	28 (35)
NβV	29 (35)	29 (36)	25 (39)	27 (35)		35 (43)
TaV	27 (29)	27 (33)	26 (41)	28 (35)	35 (43)	

between 66% and 86% sequence identity (**Table 3**). One copy of the genomic (+)RNA1, one copy of the genomic (+)RNA2, and 240 copies of the expressed capsid protein precursor assemble to form an ∼400 Å diameter icosahedral $T = 4$ virion (**Figure 2**). As in the β-like viruses, each copy of α undergoes an autocatalytic cleavage to produce a large protein (β, ∼62–63 kDa) and a small protein (γ, ∼7–8 kDa) after the capsid is assembled (**Table 2**), and both the β and γ polypeptides remain part of the virion after cleavage. Both genomic RNAs are likely to be replicated from separate (−)RNA templates based on larvae infected with NωV having separate dsRNAs corresponding to each of the genomic RNAs. RNAs 1 and 2 represent the complete genomic information required for virus growth based on studies of HaSV assembly and replication in plant and insect cells.

Providence Virus Replication in a Fat Body Cell Line

PrV is currently the only tetravirus shown to replicate in cell culture. The virus was discovered in 2003 as a persistent infection of a midgut cell line (MG8) derived from the corn earworm (*Helicoverpa zea*). In addition to the *H. zea* midgut, PrV was also shown to infect the *H. zea* fat body cell line (FB33) at low doses. A time course study of the PrV infection cycle was conducted using the FB33 cells. Input virus genomic and subgenomic (+)RNA is detected inside the cells at 6 h post infection (h.p.i.). Viral RNA synthesis is first detected at 18 h.p.i. and significantly increases between 18 and 24 h.p.i. At 36 h.p.i., total viral RNA accumulation reaches its maximum level, and (−)RNA also reaches its peak level, which is about 10-fold lower than that of corresponding (+)RNA.

The PrV β capsid protein (~60 kDa) appears at 6 h.p.i. along with the genomic and subgenomic RNAs. The β protein begins increasing in copy number after about 24 h.p.i., and after 36 h the uncleaved capsid protein precursor (α) also accumulates, slightly lagging behind the appearance of the subgenomic RNA that is thought to drive its translation. Although the α protein accumulated in infected cells, it is not detected in purified virions. In contrast, p13 translation was not detected during the time course experiment and it is not found in purified virions. However, proteolytic processing and degradation of this precursor could be extremely rapid as observed with the baculovirus-expressed TaV precursor.

To determine if assembled virions could be seen in infected cells, both MG8 and FB33 cells infected with PrV for 3 days were sectioned and examined by TEM. Both cell lines contained virus particles that were ~400 Å in diameter. The particles were seen only in the cytoplasm and often associated with one or more unidentified cytoplasmic vesicles that appear to be induced by the infection. In a few percent of the FB33 cells with very high virus accumulation, large crystalline particle arrays appeared within the cytoplamic structures.

Capsid Assembly

PrV particles are seen only in the cytoplasm of sectioned fat body cells at all time points measured; thus, it is highly probable they assemble in the cytoplasm and the cytoplasmic structures produced by the infection may aid in the process. In lieu of cell culture systems, assembly of VLPs has been achieved using the capsid proteins from NωV, HaSV, and TaV (the only β-like virus for which successful VLP production has been reported). The VLPs will assemble upon expression of only the capsid protein precursor in heterologous expression systems, namely yeast, plant protoplasts, and baculoviruses. Considerable yields of VLPs can be purified from the baculovirus expression systems and this has enabled a number of biophysical and biochemical studies of tetravirus capsids, particularly those of NωV and HaSV (see below). Like native virions, the VLP capsid protein subunits undergo autocatalytic cleavage from α to β and γ peptides, can specifically package viral RNA, and can bind specifically to a larval midgut cell receptor.

Capsid Structure and Dynamics

High-Resolution Capsid Structure

Both X-ray crystallography and cryoelectron microscopy image reconstruction (cryoEM) have been used to examine the different structural states of tetravirus capsids. The crystal structure of authentic NωV was determined in 1996 to 2.8 Å resolution, and is currently the only tetravirus crystal structure that has been published. Recently, the authors' laboratory has also determined the crystal structure of purified HaSV VLPs, and found it closely similar to that of NωV. The NωV crystal structure showed that the capsid is icosahedral, 420 Å in diameter, and composed of 240 capsid proteins arranged with $T=4$ quasi-symmetry. Thus, the 60 icosahedral asymmetric units each have four subunits, named A–D, in unique chemical environments (**Figure 3(a)**). The 644 amino acid capsid subunit has three domains: an exterior immunoglobulin (Ig)-like fold, a central β-barrel in tangential orientation, and an interior helix bundle (**Figure 3(b)**). The helix bundle is created by the N- and C-termini of the polypeptide before and after the β-barrel fold, and the Ig-like domain is an insert between the E and F β-strands of the barrel. The exterior Ig domain is unique in nonenveloped viruses and is expected to be involved in cell binding and virus tropism, and has the least conserved sequence among tetravirus subunits. The interior helix domain is anticipated to interact with the RNA genome and may have a role in capsid polymorphism (see below). All 240 of the NωV capsid subunits have undergone autoproteolysis between residues N570 and F571 after assembly, leaving the cleaved portion of the C-terminus, called the γ peptide, associated with the capsid. The quasi-symmetry is controlled by a molecular switch (**Figures 3(a) and 3(b)**). In NωV, a segment of the γ peptide (residues 608–641) is only ordered in the C and D subunits, and functions as a wedge between the ABC and DDD morphological units to prevent curvature and create a flat contact. The interface between two ABC units is bent partly due to the lack of ordered γ peptide. Importantly, the subunit structure and auto-proteolysis revealed a strong relationship between tetraviruses and the $T=3$ insect nodaviruses. Indeed, the β-barrel folds and autocatalytic sites superimpose with little variation, and the γ peptide in the nodaviruses is also involved in quasi-equivalent switches.

The crystal structure of PrV, the first high-resolution structure of a β-like virus, was recently determined in the authors' laboratory (data not shown). The PrV subunit retains the three domains and a superimposable autoproteolysis site (all four subunits show a clear break at the cleavage site), but the Ig domain and the first and last ordered residues of the termini in the helical domain have different positions and structure compared to NωV. The different Ig domain position was first observed in the cryoelectron microscopy reconstruction of NβV as a distinctly different surface structure compared to the ω-like particle surfaces (**Figure 2**). The use of subunit termini in PrV is more like that seen in the nodaviruses. The quasi-symmetry switch in PrV has swapped elements compared to the omegatetraviruses, and represents a new and unique mixture of structural features among the insect viruses.

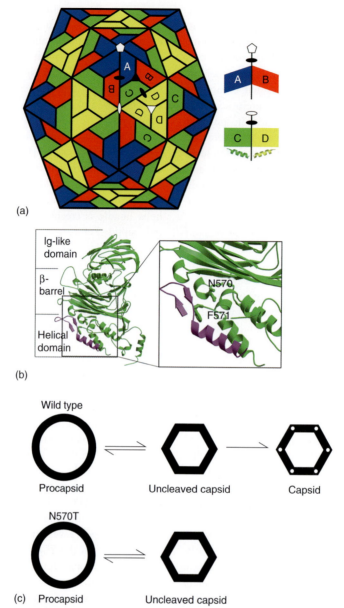

Figure 3 Structure and dynamics of the NωV capsid. (a) The subunit organization in the NωV $T = 4$ capsid crystal structure. There are four subunits in the asymmetric unit (A in blue, B in red, C in green, and D in yellow). Icosahedral symmetry axes are marked with white symbols, quasi-symmetry axes are marked with black symbols. Quasi-twofold dimer interfaces occur with the bent A–B or flat C–D conformations (right). The well-ordered C-terminal helices in the C–D subunits comprise a molecular switch that stabilizes the flat interface. (b) Ribbon diagram of the C subunit illustrates the three domain tertiary structure present in all 240 subunits (exterior of capsid on top, interior on bottom). Maturation cleavage occurs within the helical domain between residues Asn570 and Phe571. The molecular switch, present only in the C or D subunits, is highlighted in magenta. (c) In the wild-type virus, the procapsid converts to uncleaved capsid with the drop of pH, followed by a maturation cleavage, which locks the particle in capsid conformation after ∼10% of the subunits are cleaved. In the cleavage-defective mutant (N570T), particle conformational change becomes reversible.

Capsid Maturation, Auto-Proteolysis, and Large-Scale Transitions

NωV and HaSV coat proteins expressed in the baculovirus system assemble into round and porous intermediate VLPs that are 480–490 Å in diameter, called procapsids, when purified at neutral pH (pH 7.6). The procapsid coat proteins remain in the uncleaved α precursor form until a reduction in pH. When exposed to acidic conditions (pH 5.0), procapsids underwent a large-scale structural rearrangement (maturation) to the

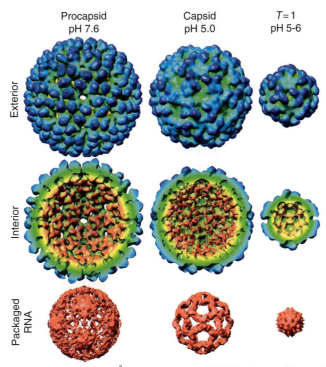

Figure 4 CryoEM reconstructions between 23 and 27 Å resolution of three HaSV polymorphic capsid forms. The polymorphic form and its associated pH are given at the top. Three views of each form are shown with radial coloring from red on the interior to blue on the exterior. The procapsid and capsid diameters are similar to those of NωV (490 and 430 Å, respectively). Note the well-ordered dodecahedral RNA structure in the mature capsid, which has been observed in several insect viruses. The $T = 1$ form has so far been found only in HaSV purifications as an extra peak in the sucrose gradients, and only appears after the acid maturation step (i.e., not found at pH 7.6). The diameter of the $T = 1$ particles is 260 Å and the protein shell is composed only of a truncated version of the β subunits (55 kDa).

smaller, angular mature capsid, where α begins to cleave into the β and γ polypeptides (**Figures 3(c)** and **4**). Small-angle X-ray scattering (SAXS) demonstrated that the pH-induced conformational change occurs in less than 100 ms in NωV, while the autocatalytic cleavage event had a half-life of hours. The relationship of the conformational change and cleavage was also examined using SAXS, which suggested that the conformational change was initially reversible until approximately 10% of the subunits had cleaved, and then the particles were 'locked' in the capsid conformation (**Figure 3(c)**). Finer pH titrations of NωV revealed intermediate capsid states at pH 5.8 and 5.5 (data not shown). Presumably similar intermediates also exist for HaSV.

Both ANS (1-anilino-8-naphthalene sulfonate) and the thiol-reactive fluorophore, maleimide-ANS (MIANS), have been bound to procapsids to further characterize the conformational transition to capsid. Absorption of the covalently attached MIANS (forms covalent thiol-ether linkage with cysteines) situated at subunit interfaces can report on environment changes as it is buried during maturation. Unexpectedly, acidification of the labeled particles trapped them at an intermediate stage of maturation and the autocatalytic cleavage was inhibited. The labeled VLPs showed no cleavage after 4 h of incubation at pH 5.0 buffer while wild-type particles were fully cleaved. A cryoEM image reconstruction of the MIANS-labeled VLPs at pH 5.0 revealed the particles are 440 Å in diameter, having only progressed about halfway along the trajectory between procapsid and capsid. This demonstrates that cleavage depends on the conformational change and is not triggered by just lowering the pH, and provides support for the strategy of targeting subunit interfaces with antiviral compounds as a means of disrupting the viral life cycle.

Cleave Site Mutation Allows Reversible Transition and Procapsid Structure Analysis

Mutagenesis studies have shown the asparagine at the scissile bond (N570) in NωV is required for cleavage. Substituting threonine for asparagine (N570T) results in assembly of less stable particles, but they do not cleave and continue to undergo the conformational change reversibly as a function of pH (**Figure 3(c)**). CryoEM structures at 8 Å resolution of cleaved and N570T VLPs showed the role of cleavage in stabilizing the capsid form. There are more extensive inter-subunit contacts between the C-termini and the neighboring N-termini in cleaved particles that results in more ordered C-terminal helices, which accounts for the added stability.

The N570T mutant was also used to study the structure of the uncleaved procapsid state. An 11 Å resolution cryoEM structure of N570T procapsids displays a morphology indistinguishable from procapsids formed by wild-type subunits. The higher resolution showed, at pH 7.6, that portions of the C-terminal regions of the γ polypeptides of the A and B subunits, which are not ordered in the capsid form, are ordered and contribute to a second molecular 'wedge' that prevents hinging between the A and B subunits. It makes the quasi-twofold interactions at the A–B and C–D interfaces closely similar in the procapsid, producing its round morphology and consistent with its role as an assembly intermediate during NωV particle maturation.

Pathobiology

Host Range

Lepidopteran insects are the only known hosts for tetraviruses. No replication in other animals has been detected. The hosts are confined largely to the Noctuidae, Saturniidae, and Limacodidae families within the Lepidoptera order. The range of species that they are able to infect differs between the viruses. HaSV appears unable to infect species outside the Heliothinae subfamily, while TnV and DpV are able to infect insects outside the family of their nominal host.

Transmission and Symptoms

Horizontal transmission via ingestion by larvae has been demonstrated for several tetraviruses with a widely varying range of symptoms. Only slight growth retardation is observed using high doses of TnV, but NβV infection of larvae is lethal, causing cessation of feeding, and moribund, discolored, and flaccid appearance at 7–9 days post infection. A dependence upon the larval life stage has also been seen. HaSV is highly active against neonate larvae in the first three instars of development, needing as little as 5000 particles to cause them to stunt and cease feeding within 24 h and to die within 4 days. In contrast, no detectable symptoms occur in later larval development, signifying a high degree of developmental resistance. It has not been demonstrated whether adults or pupae are capable of being infected by tetraviruses. Vertical transmission of tetraviruses is also believed to occur, but the evidence remains undefinitive.

Histopathology and Tropism

Natural infection by tetraviruses appears to be exclusively in the larval midgut, although PrV has been cultured in fat body cells. In a definitive experiment, Northern blots of RNA extracted from HaSV-infected larvae showed viral RNA only in the midgut tissue, even after virus was injected into the hemocoel.

In a quantitative study of HaSV infection of its host, the African or cotton bollworm, supplying the virus to larvae *per os* during the first three instars of development resulted in an infection restricted to three of the four midgut cell types. The virus initiated infection in closely situated foci that expanded and converged with others over time. In response, the midgut cells increased their rate of sloughing and apoptosis to an extent that incapacitated the midgut due to loss of all but a few cells. Infection of older larvae results in only sparsely situated foci, which fail to expand and eventually disappear presumably due to cell sloughing. These data indicate that cell sloughing is an immune response existing throughout development, but older larvae have an additional mechanism for increased resistance to infection.

Biotechnology

Application as Biopesticides

Small RNA viruses offer attractive properties as biological control agents for pests: high specificity, high efficacy, and potentially self-replicating. A notable drawback is the difficulty in producing large quantities of infectious virions, but the simplicity of the tetraviral genomes makes production of virions in nonhost cells feasible. Agrobacterium-mediated transgenesis has been used to place the genes required for HaSV assembly into tobacco plants, which resulted in stunted, HaSV-infected larvae when they fed on the modified plants. Expanding on this result, Hanzlik and colleagues conducted the first field trial of HaSV as a control agent in 2005 on a sorghum crop in Queensland, Australia. The semipurified HaSV preparation reduced the larval population by 50% at both 3 and 6 days post application (dpa), which was equivalent to a commercial baculovirus preparation. Notably, the amount of HaSV on the sorghum heads increased at 5.5 dpa, probably reflecting dispersal of newly produced virus from cadavers and frass. These results indicate that HaSV could be used effectively as a biopesticide.

Application as Chemical Platforms

A study in 2006 demonstrated that the Ig domains of NωV particles could be modified to possibly alter its tropism and/or to carry specific chemical tags. A $(His)_6$ tag was inserted into the GH loop (between A378 and G379) of a capsid precursor construct that was then expressed in a baculovirus system. The His-tagged p70 self-assembled into VLPs that had similar morphological and RNA encapsidation properties as wild-type VLPs. Two assays using paramagnetic nickel beads confirmed that multiple affinity tags were present on the surface of the modified VLPs and could bind the beads. Thus, the GH loop of NωV appears to be a suitable modification site for further biotechnology applications.

See also: Cryo-Electron Microscopy; Nodaviruses; Virus Particle Structure: Nonenveloped Viruses; Virus Particle Structure: Principles.

Further Reading

Canady MA, Tsuruta H, and Johnson JE (2001) Analysis of rapid, large-scale protein quaternary structural changes: Time resolved X-ray solution scattering of nudaurelia capensis ω virus. *Journal of Molecular Biology* 311: 803–814.

Christian PD, Murray D, Powell R, Hopkinson J, Gibb NN, and Hanzlik TN (2005) Effective control of a field population of *Helicoverpa armigera* by using a small RNA virus Helicoverpa armigera stunt virus (*Tetraviridae, Omegatetravirus*). *Journal of Economic Entomology* 98: 1839–1847.

Gorbalenya AE, Pringle FM, Zeddam J-L, et al. (2002) The palm subdomain-based active site is internally permuted in viral RNA-dependent RNA polymerases of an ancient lineage. *Journal of Molecular Biology* 324: 47–62.

Hanzlik TN and Gordon KHJ (1997) The *Tetraviridae*. *Advances in Virus Research* 48: 101–168.

Helgstrand C, Munshi S, Johnson JE, and Liljas L (2004) The refined structure of nudaurelia capensis ω virus reveals control elements for a T = 4 capsid maturation. *Virology* 318: 192–203.

Johnson JE and Reddy V (1998) Structural studies of nodaviruses and tetraviruses. In: Miller LK and Ball LA (eds.) *The Insect Viruses*, pp. 171–223. New York: Plenum.

Lee KL, Tang J, Taylor D, Bothner B, and Johnson JE (2004) Small compounds targeted to subunit interfaces arrest maturation in a nonenveloped, icosahedral animal virus. *Journal of Virology* 78: 7208–7216.

Maree HJ, Van der Walt E, Tiedt FA, Hanzlik TN, and Appel M (2006) Surface display of an internal His-tag on virus-like particles of nudaurelia capensis omega virus (NωV) produced in a baculovirus expression system. *Journal of Virological Methods* 136: 283–288.

Pringle FM, Johnson KN, Goodmann CL, McIntosh AH, and Ball LA (2003) Providence virus: A new member of the *Tetraviridae* that infects cultured insect cells. *Virology* 306: 359–370.

Taylor DJ, Krishna NK, Canady MA, Schneemann A, and Johnson JE (2002) Large-scale, pH-dependent, quaternary structure changes in an RNA virus capsid are reversible in the absence of subunit autoproteolysis. *Journal of Virology* 76: 9972–9980.

Theiler's Virus

H L Lipton and S Hertzler, University of Illinois at Chicago, Chicago, IL, USA
N Knowles, Institute for Animal Health, Pirbright, UK

© 2008 Elsevier Ltd. All rights reserved.

Glossary

Cardiovirus Genus in the family *Picornaviridae*.
Demyelination Pathological process of damage to the proteolipid coat around nerve fibers (axons).
L* Virus-encoded protein that is translated in alternative reading frame to that of the polyprotein.
Theilovirus Species within the genus *Cardiovirus*.
Theiler's murine encephalomyelitis viruses (TMEVs) Abbreviation for mouse Theiler's viruses.

History, Geographic Distribution, and Host Range

The mouse encephalomyelitis viruses are enteric pathogens of mice. Discovered by Max Theiler in the early 1930s and originally called murine polioviruses, these agents are commonly referred to as Theiler's murine encephalomyelitis viruses (TMEVs). Theiler initially recovered isolates from mice with spontaneous paralysis housed at the time in the animal colony at the Rockefeller Institute; subsequently, TMEVs were found in virtually all nonbarrier mouse colonies, where they caused asymptomatic intestinal infections. While TMEVs are widely distributed in the world, their host range is narrow, including mice and rats. Serological evidence indicates that *Mus musculus*, the feral house mouse, is the natural host, but several other species of voles and possibly rats may also serve as hosts. As is the case for other picornaviruses, following peripheral routes of infection, TMEV spreads to the central nervous system (CNS) producing encephalitis or poliomyelitis (spontaneous paralysis). In the older literature, the incidence of spontaneous paralysis was reported to be approximately one paralyzed animal per 1000–5000 colony-bred mice. Since TMEVs may go undetected unless appropriate serological tests are performed, these agents are a potential hazard for investigators using mice in biomedical research.

In recent years, this group of viruses has assumed additional importance, because TMEV infection in mice provides one of the few available experimental animal models for multiple sclerosis. TMEV-induced demyelinating disease in mice is a relevant model for multiple sclerosis because: (1) chronic pathological involvement is virtually limited to the CNS white matter; (2) myelin breakdown is accompanied by mononuclear cell inflammation; (3) demyelination results in clinical disease, for example, spasticity, from involvement of upper motor neuron pathways; (4) myelin breakdown is in part immune-mediated; and (5) the disease is under multigenic control with a strong linkage to the major histocompatibility complex (MHC) gene H2D.

Classification and Serologic Relationships

Based on the complete nucleotide sequence and genome organization, TMEVs have been classified in the genus *Cardiovirus* of the family *Picornaviridae* along with encephalomyocarditis virus (EMCV). EMCV and TMEV (or Theilovirus) constitute separate species within the genus *Cardiovirus*. Polyclonal antisera show no cross-neutralization between TMEV and EMCV; however, since the coat proteins of the two *Cardiovirus* species share ~50% identity of their amino acid sequences, cross-reactions are observed with antibody binding assays such as enzyme-linked immunosorbent assay (ELISA) and complement fixation. Phylogenetic analysis of the complete P1 capsid-coding region has revealed three major clades: TMEV, Vilyuisk human encephalomyelitis virus (VHEV), and the Theiler's-like virus (TLV) of rats (**Figure 1**). These three clades presumably represent three distinct serotypes of the species *Theilovirus*; however, comprehensive serological studies remain to be performed.

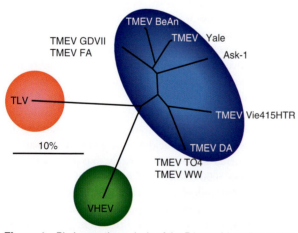

Figure 1 Phylogenetic analysis of the P1 capsid-coding region of the Theiloviruses showing three clades, presumably representing three distinct serogroups: TMEV, Vilyuisk human encephalomyelitis virus (VHEV), and Theiler's-like virus of rats (TLV).

Physical Properties

Because the TMEVs do not have an envelope they are insensitive to chloroform, ether, nonionic detergents, such as deoxycholate, NP40 and Tween-80, and the ionic detergent sodium dodecyl sulfate, but are inactivated by 0.3% formaldehyde and HCl 0.1 mol l^{-1}. TMEVs are rapidly destroyed at temperatures over 50 °C and lose some infectivity upon lyophilization. Purified virions can be stored for long periods of time at −70 °C without loss of infectivity, but slowly lose infectivity on storage at −20 °C. Enteroviruses require stability at low pH to pass through the acidic conditions of the stomach. TMEVs are stable over the entire pH range from 3 to 9.5, but in contrast to the other cardioviruses, such as EMCV, they are not highly thermolabile in the presence of 0.1 mol l^{-1} chloride or bromide in the pH range 5–7.

Virion Structure

Picornavirions have a relative molecular mass of ~8.5 × 10^6 Da, of which 30% is RNA and 70% protein. Theiler's virions are 30 nm in diameter and their structure reveals only a spherical virion by negative staining (**Figure 2(a)**). Virions have a sedimentation coefficient of 150 S by velocity centrifugation in sucrose and a buoyant density of 1.34 g ml^{-1} by isopycnic centrifugation in cesium salts.

The three-dimensional structures of the GDVII, BeAn, and DA strains have been determined at ~3 Å resolution by X-ray crystallography (**Figures 2(b)** and **2(c)**). The overall architecture is quite similar to that of other picornaviruses and closely resembles that of Mengo virus, another member of the EMCV species. Each virion is composed of 12 pentamers, with each pentamer made up of five protomers, each containing a single copy of the four capsid proteins. Each of the three major capsid proteins VP1, VP2, and VP3 consist of a wedge-shaped eight-stranded antiparallel β-barrel. The N-termini of the capsid proteins form an extensive, intertwined network on the inner surface of the protein shell. The loops connecting the β-strands form the outer surface features of the protein shell and provide surface differences with Mengo virus. A 25 Å depression on the virion surface at the junction between VP1 and VP2 along the twofold axis, termed the pit, is believed to be the viral receptor and is a broad depression. The TMEV surface structures are differentiated from that of Mengo virus in having: (1) a larger VP1 CD double loop, with loop I containing an extra five residues, that is shifted more toward the VP2 EF puff at the twofold axis, while loop II is directed more toward the fivefold axis; (2) an 11-residue insertion in the VP2 EF puff forming a double loop in which the inserted loop interacts with

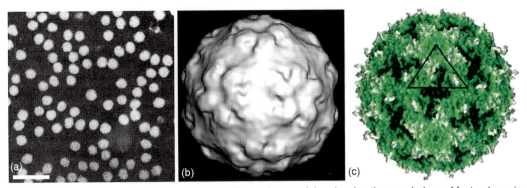

Figure 2 Theiler's virions. (a) Negative staining of purified BeAn virus particles showing the morphology of featureless virus profiles due to the impermeability of the capsid to the electron dense stain. Scale = 100 μm. (b) Cyroelectron micrograph of a GDVII virion at 20 Å resolution showing the star-shaped profile of the plateau around the fivefold axes. (c) GRASP image of a BeAn virion showing a topographical relief map of the VP1 and VP2 surface loops in white. The superimposed triangle represents a protomer with its apex pointing at the fivefold axis. The darkened area within the outlined iscosahedral unit is a surface depression of the pit or receptor binding site. This depression extends across the twofold axis.

the VP1 FMDV GH loop; and (3) the tip of the VP3 knob (a loop inserted in βE) points directly outward on the rim of the pit.

Properties of the Genome

The genetic component of the TMEV virion is a single-stranded, positive-polarity RNA molecule 8100 nt in size that has a sedimentation coefficient of 35 S. The virion RNA has a virally encoded 20 amino acid protein, VPg, covalently linked to the 5′ uridine, and a poly(A) tract on the 3′ end. The complete genomes of the GDVII, DA, BeAn strains, as well as TLV have been sequenced. With the notable absence of a poly(C) tract in the 5′ untranslated region, the organization and sequence of the TMEV genome is remarkably similar to that of EMCV. The polyprotein of the prototypic BeAn strain initiates at the AUG codon at nucleotide 1065 and extends for 6909 nt (or 2303 codons), ending at the single UGA termination triplet at base 7972 (**Figure 3**). The polyprotein-coding region is flanked by 5′ and 3′ untranslated sequences of 1064 and 125 nt, respectively. In BeAn the 5′ untranslated region contains a stretch of 11 pyrimidines interrupted by a single purine before the AUG at nucleotide 1065. In picornaviruses, the 5′ untranslated region mediates cap-independent translation and serves as an internal ribosome entry site (IRES). The TMEV 5′ untranslated sequences have been predicted to form stable secondary structures, which in the 500 nt upstream of the authentic AUG (at 1065) are nearly identical to those predicted for EMCV and foot-and-mouth disease virus. In BeAn, eight AUGs precede the initiator AUG, but none of them has an optimum Kozak context sequence. Hence, it could be argued that selection of the authentic initiator AUG after binding of ribosomes to TMEV RNA does not involve internal ribosome binding. However, BeAn nucleotides ∼500–1065 determine a structure that serves as an IRES in bicistronic mRNAs both *in vitro* (BHK-21 cells). A poly(A) tail of indeterminate length is present on the 3′ end of the viral genome.

Translation: Polyprotein Processing and Final Gene Products

The 12 TMEV gene products are the result of post-translational processing of a 2303 amino acid polyprotein (mol. weight of 256 kDa) (**Figure 3**). The polyprotein of GDVII virus contains no insertions or deletions; however, two VP1 amino acids are deleted in DA, TO4, and WW viruses. The processing scheme follows a standard L-4-3-4 picornavirus polypeptide arrangement, that is, the leader protein (L), four capsid proteins in part one (P1) of the genome, three proteins in P2, and four proteins in P3 (**Figure 3**). The coding limits of individual viral proteins have been predicted by analogy with those of EMCV since the only confirmation to date of a deduced sequence is that of the N terminus of 1D. The eight amino acids flanking the putative cleavage sites are highly conserved for the two viruses. All of the cleavage sites in the polyprotein except for two, 1A/1B and 2A/2B, are processed by the viral protease 3C. The TMEV 3C protease therefore processes Q-C, as well as Q-S and Q-A, dipeptides and, in addition, the E-N dipeptide at the 1D/2A cleavage. However, only 6 of 8 Q-G, 2 of 13 Q-S, and 1 of 7 Q-A dipeptides in the polyprotein are cleaved by 3C, indicating that involvement of secondary, tertiary, or both types of structure is also important for recognition of these particular dipeptides. The tetrapeptide NP/GP at 2A/2B which is the primary cleavage site in the cardiovirus polyprotein is cleaved autocatalytically, as it is in EMCV.

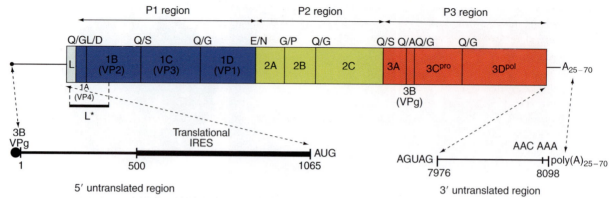

Figure 3 TMEV 8 kbp genome showing 5′ and 3′ untranslated regions (enlarged) and the polyprotein which is superimposed over the 7 kbp nucleotide open reading frame (ORF) and contains the final 12 viral gene products. The 5′ end of the viral RNA is not capped but has protein 3B (VPg) covalently attached to the first nucleotide. A ~500 base internal ribosome entry site (IRES) is present in the 5′ untranslated region immediately upstream of the authentic AUG initiation codon for the polyprotein. The 3′ untranslated region ends in a poly (A) tail. The ORF is divided into L, P1 (blue) encoding the four capsid proteins, P2 (yellow) three nonstructural proteins, and P3 (red) four nonstructural proteins, including the 3C cysteine protease which makes all but two of the cleavages in the polyprotein, and the 3D, the RNA-dependent RNA polymerase. Cleavage dipeptides are shown above viral protein boundaries. Finally, the location of the L* protein that is translated out-of-frame is shown near the N-terminus of the polyprotein.

Table 1 Functional activities of the TMEV nonstructural proteins

Nonstructural protein	Activities
Leader (L)	Binds to Ran, inhibiting import and export of proteins to and from the nucleus, including those involved in the innate interferon response
L*	Out-of-frame 19-kDa protein initiated by AUG or ACG codons immediately downstream of the authentic AUG for polyprotein translation; has a putative role in TMEV persistence
2A	A C-terminal tetrapeptide NP/GP responsible for autocatalytic cleavage of 2A/2B, the primary cleavage in the cardiovirus polyprotein
2B	Possible role in cell membrane permeability
2C	A multifunctional protein involved in viral RNA replication, including ATPase, GTPase, membrane binding, and RNA binding activities
3A	Inhibition of ER-to-Golgi transport that results in reduction of host cell cytokine secretion
3B	Covalently linked to the 5′-terminal uridine of the single-stranded RNA and primer for the viral RNA-dependent RNA polymerase
3C	Cysteine-reactive protease responsible for all of the monomolecular and bimolecular cleavages in the polyprotein except for the assembly-dependent maturation cleavage of VP0 (1A/1B) and autocatalytic cleavage at 2A/2B
3D	RNA-dependent RNA polymerase responsible for viral RNA replication

Cleavage of the polyprotein gives rise to three primary products, the first of which (116 kDa) contains the leader protein (8 kDa), the P1 capsid proteins, and the first P2 protein 2A (15 kDa). Thus, the initial precursor released from the polyprotein is like that of the other cardioviruses but differs from that of other groups of picornaviruses. The capsid proteins are arranged in the following order: 1A (VP4; 7 kDa), 1B (VP2; 29 kDa), 1C (VP3; 25 kDa), and 1D (VP1; 30 kDa). The second processing precursor (2BC) is 51 kDa and gives rise to 2B (14 kDa) and 2C (37 kDa). The third or C-terminal precursor protein is 88 kDa and is processed into the four mature proteins 3A (10 kDa), 3B (2 kDa), 3C (24 kDa), and 3D (52 kDa). The functions of the nonstructural proteins are shown in **Table 1**. Another protein, termed L*, translated out-of-frame, is initiated 22 nt downstream of the authentic AUG in the polyprotein, giving rise to an 18 kDa protein that was originally shown to be involved in the TMEV persistent infection.

Viral RNA Replication

Cardiovirus replication is similar to that of other picornaviruses. Following entry, VPg is removed from the 5′ end of the RNA, and the RNA is directly translated using cellular factors. The viral genomic RNA (plus-sense), is then transcribed into negative-strand copies, and each one is used as template for reiterative synthesis of ~25–50 plus-sense RNAs identical to the genome of the virus. An RNA structure located within the VP2 sequence of TMEV is necessary for RNA replication, and is referred to as the *cis*-acting replicative element (CRE). The CRE functions as a template for uridylylation of viral protein 3B (VPg) forming VPgpUpU$_{OH}$ which primes positive-strand RNA synthesis. Viral protein 3D is the RNA-dependent RNA polymerase which catalyzes both of these processes.

Viral Genomic Determinants in Pathogenesis

The existence of two naturally occurring TMEV neurovirulence groups provides a useful system for investigating molecular pathogenesis. Moreover, the difference in neurovirulence in terms of mean 50% lethal dose between high and low neurovirulence strains is of the order of 10^5 plaque forming units. With the advent of reverse genetics the generation of full-length cDNA clones of high neurovirulence GDVII and low virulence DA and BeAn viruses was feasible. Viral RNAs transcribed from cDNAs are infectious upon transfection of mammalian cells. To identify pathogenetic determinants, for example, those of neurovirulence and persistence, recombinant viruses between parental cDNAs have been assembled and point mutations introduced into wild-type strains for analysis in cell culture and mice.

Neurovirulence and persistence have been mapped to the L-P1 regions encoding the L and L* proteins and the capsid proteins. The results suggest that the mechanism for the pathogenetic properties of neurovirulence and persistence involve several determinants, including an exclusive role for the capsid in neurovirulence and the out-of-frame L* protein and capsid (exterior surface of the virion) in persistence. The latter observation suggested that receptor-mediated events may be involved in persistence; it was demonstrated that persistence required the ability of low neurovirulence strains to bind to its sialic acid co-receptor. The L* protein was initially recognized because only the low neurovirulence strains have an out-of-frame AUG start codon. However, the ACG codon in high neurovirulence strains was then shown to initiate L* translation, and chimeric TMEVs that contain a high neurovirulence L protein in a low neurovirulence background were shown to persist in the mouse CNS as well as the low neurovirulence parental virus. The role of the L* protein in persistence will require further study.

Transmission and Tissue Tropism

TMEVs are transmitted by the fecal–oral route and are separable into two biological groups based on neurovirulence (**Table 2**). The first group, consisting of three isolates, GDVII, FA, and ASK-I, possesses high neurovirulence, causing a rapidly fatal encephalitis in mice. Some 10–15 isolates, including viruses recovered from the CNS of spontaneously paralyzed mice and feces of asymptomatic mice, form a second, low neurovirulence group. Experimentally, the low neurovirulence strains produce poliomyelitis (early disease) followed by demyelinating disease (late disease). When mice are inoculated intracerebrally with cell culture-adapted low neurovirulence strains, the poliomyelitis phase is attenuated (subclinical); whereas, brain-derived stocks of these viruses produce both disease phases.

Pathogenesis

Very little information is available about the pathogenesis of TMEV infection following peripheral routes of infection, including the oral route. In general, isolates from either of the neurovirulence groups do not readily produce CNS disease following peripheral routes of inoculation, with the exception of TO(B15), a mutant selected for its invasiveness from the intestinal tract. When mice are inoculated intracerebrally with the high neurovirulence strains, virus replicates widely in the brain and spinal cord, causing encephalitis or encephalomyelitis. Thus,

Table 2 Two TMEV neurovirulence groups

Characteristic	High neurovirulence	Low neurovirulence
Isolates (strains)	GDVII, FA, ASK-1	DA, BeAn 8386, Yale, WW, TO4, TO(B15), Vie415HTR
Disease	Encephalomyelitis	Poliomyelitis/demyelination
Incubation period	1–10 days	7–20 days/>30 days
CNS target cell	Cortical and motor neurons	Motor neurons/macrophages and oligodendrocytes
LD$_{50}$	1–10 PFU	>10^5 PFU
Persistent infection	No	Yes
Temperature sensitive	No	Yes
Carbohydrate co-receptor	Heparan sulfate	α2,3-linked sialic acid
Differences in virion structure	Only in the Cα chain of the VP2 puff B	

neurons as well as glial cells (astrocytes and oligodendrocytes) become infected in the cerebral cortex, hippocampus, basal ganglia, thalamus, brainstem, and spinal cord. Affected mice develop hunched posture and hind-limb paralysis. A rapid demise is the result of widespread cytolytic infection. The following sections focus on the pathogenesis of the biphasic disease produced by the low neurovirulence strains, which provides an experimental model system for human demyelinating disease, multiple sclerosis.

Clinical Features of Infection

The intracerebral route of inoculation maximizes the incidence of neurological disease. Following intracerebral inoculation, the low neurovirulence strains produce a distinct biphasic CNS process in susceptible strains of mice, which is characterized by poliomyelitis during the first 2 weeks post infection, followed by chronic, inflammatory demyelination that begins during the second or third week post infection and becomes clinically manifest between 1 and 3 months post infection. Some investigators who inoculated lower amounts of virus observed an even later onset of demyelinating disease. Mice with poliomyelitis develop flaccid paralysis, usually of the hind limbs; only one limb may be affected or paralysis may involve both hind limbs and spread to involve the forelimbs, occasionally leading to death. In contrast to the fatal outcome of paralysis produced by the Lansing strain of human poliovirus type 2, complete recovery from TMEV-induced poliomyelitis is usually seen. Residual limb deformities may be seen as the result of extensive anterior horn cell infection and severe paralysis (early disease).

Gait spasticity is the clinical hallmark of the demyelinating (late) disease. Late disease is first manifest by slightly unkempt fur and decreased activity, followed by an unstable, waddling gait. Subsequently, generalized tremulousness and ataxia develop, and the waddling gait evolves into overt paralysis. Incontinence of urine and priapism are seen. As the disease advances, prolonged extensor spasms of the limbs occur spontaneously followed by difficulty in righting. Extensor spasms can be induced by turning an animal over on its back by quickly rotating the tail. Weight loss occurs during late disease. The clinical manifestations of late disease are progressive and lead to an animal's demise in several to many months post infection. A functional motor assay using a rotorod apparatus has been used to quantitate motor function in this disease.

Pathogenesis and Histopathology

Motor neurons in the brainstem and spinal cord are the main targets of infection during poliomyelitis, but sensory neurons and astrocytes are also infected. TMEVs do not replicate in endothelial and ependymal cells. A brisk microglial reaction is elicited, with the appearance of numerous microglial nodules, particularly in the anterior gray matter of the spinal cord. Examples of neuronophagia are quite frequent at this time, but little lymphocytic response is seen. The poliomyelitis phase lasts 1–4 weeks, after which time little residual gray matter involvement is inapparent other than for astrocytosis.

As early as 2 weeks post infection, inflammation of the spinal leptomeninges begins to appear, followed by involvement of the white matter. Initially, the inflammatory infiltrates are almost exclusively composed of lymphocytes, but at later times plasma cells and macrophages become numerous (**Figure 4(a)**). The influx of macrophages is in close temporal and anatomic relationship with myelin breakdown. Both light microscopic and ultrastructural studies show that myelin breakdown is related to the presence of macrophages (**Figure 4**), which either actively strip myelin lamellae from otherwise normal-appearing axons or are found in contact with myelin sheaths undergoing vesicular disruption. Foci of inflammation and myelin destruction extend from the perivascular spaces into the surrounding white matter, leading to sharply demarcated plaques of demyelination (**Figure 4(b)**). The ultrastructure studies during the initial phase of myelin breakdown have not shown alterations in oligodendroglia, which are in close apposition with naked but otherwise normal axons; however, oligodendrocytopathology has been observed later although it appears to be a minor contributor to demyelination.

Sites of TMEV Persistence

TMEV persistence clearly involves active virus replication, since infectious virus can be readily isolated from the CNS of infected mice and high viral genome copy numbers are present. *In situ* hybridization has revealed two populations of CNS cells differing in the number of viral genomes. Virus replication in the majority of these cells (>90%) appears to be restricted, as they contain <500 viral genomes. A small percentage of CNS cells contain >1500 genomes, possibly as many as 10^4–10^5, and are probably productively infected. The absolute number of viral genomes as determined by *in situ* hybridization is probably only an approximation. Restricted virus production has been demonstrated in macrophages isolated from the CNS of diseased mice; therefore, macrophages appear to be the primary target for persisting virus. It is also possible that some of the cells with restricted infection are astrocytes. The kinetics of virus replication in the CNS cells with restricted infection remains to be elucidated – the length of the replicative cycle and whether the cells are lysed or continue to produce infectious virus for longer times is not known. TMEV infection *in vitro*

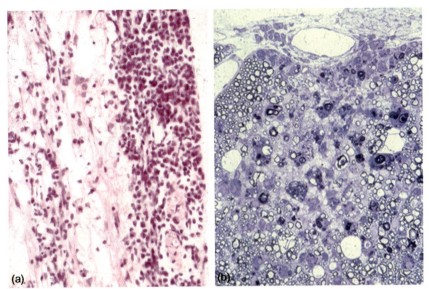

Figure 4 Histopathology of TMEV-induced demyelinating disease (late disease) in an SJL mouse. (a) H&E-stained longitudinal section (6 μm) of the spinal cord showing intense mononuclear inflammatory cell infiltrates in the leptomeninges (right) and parenchyma where many vacuolated macrophages are observed. (b) Toluidine blue-stained, Epon-embedded coronal section (1 μm) of a spinal cord showing a discrete plaque-like area of demyelination (many naked axons present) and lipid-laden macrophages at the cord margin and surrounded by normal myelinated axons. Magnification ×400 (a); ×300 (b).

only occurs in monocytes once they have differentiated into macrophages. Virus production in infected macrophage cell lines is restricted but viral translation, polyprotein processing, and assembly of virion intermediates (protomers and pentamers) as well as virions appear to be normal. Ultimately infected macrophages undergo apoptosis, which is caspase-9 and -3 dependent, consistent with activation of the mitochondrial pathway. In contrast, oligodendrocytes appear to be productively infected, since ultrastructural analysis has shown crystalline arrays of virions in oligodendrocytes in demyelinating lesions in mice. Oligodendrocytes may correspond to the CNS cells containing large numbers of viral genomes by *in situ* hybridization. These data suggest that a cytolytic infection of oligodendrocytes contributes to demyelination along with immune-mediated mechanisms of myelin damage.

Immune Response

During the first week, TMEV-infected mice mount a virus-specific humoral immune response that reaches a peak by 1–2 months post infection and is sustained for the life of the host. Neutralizing and other virus-specific antibodies have been measured. The majority of the antiviral IgG response in persistently infected, susceptible mice is IgG2a subclass, with little antiviral IgM detected by day 14 post infection, whereas IgG1 antiviral antibodies appear to predominate in resistant as well as in immunized mice. Murine CD4+ T cells of the Th1 subset mediate delayed-type hypersensitivity (DTH) and regulate IgG2a production via interferon-γ production, whereas CD4+ Th2 cells regulate IgG1 and IgE production via interleukin 4. Thus, the predominant IgG2a antiviral response in susceptible mice may be an *in vivo* measure of preferential stimulation of a Th1-like pattern of cytokine synthesis. Virus-specific CD8+ cytolytic T-cell responses have also been shown to help in virus clearance during the acute phase of the infection, and may be responsible for the resistance of C57BL/6 mice to the demyelinating disease.

Infected, susceptible strains of mice also produce substantial levels of virus-specific CD4+ T-cell responses. T-cell proliferation and DTH appear by 2 weeks post infection and remain elevated for at least 6 months. Both DTH and T-cell proliferation have been shown to be specific for TMEV and mediated by CD4+ class II restricted T cells. A temporal correlation has also been found between the onset of demyelination and the appearance of virus-specific CD4+ T-cell responses, as well as for high levels of virus-specific DTH. High levels of the Th1 cytokines interferon-γ and TNF-α also correlate temporally with the evolution of demeylination. DTH and CD4+ T-cell proliferative responses in infected and immunized susceptible SJL mice are directed toward immunodominant regions (peptides) in each of the three major coat proteins. T-cell responses to these epitopes in VP1 and VP2 are believed to participate in the immunopathology.

Although mice mount virus-specific humoral and cellular immune responses early in the infection and peak CNS virus titers fall by 100 000-fold, TMEVs somehow evade immune clearance to persist at low infectious levels indefinitely. Extraneural persistence has not been observed. Current dogma holds that humoral immunity is more important than cellular immunity in clearing infections by nonenveloped viruses, such as picornaviruses. Evidence has been presented for a role for both neutralizing antibodies and cytolytic CD8+ T cells in TMEV clearance. The precise mechanism by which TMEVs evade immune surveillance is not known but does not appear to involve antigenic variation. Although complement and virus-antibody deposition in the CNS parenchyma has detected extracellular transport of virus as infectious virus antibody complexes, viral aggregates, or as virus contained within cellular membranes, are a possible means whereby virus could be protected from TMEV-specific immune responses and continue to infect other cells. This is an area for further study to enable a better understanding of how TMEV evade immune surveillance.

Immune-Mediated Mechanism of Demyelination

Appropriately timed immunosuppression can prevent the clinical signs and pathological changes of TMEV-induced demyelinating disease, indicating that the immune response participates in myelin breakdown. A number of different immunosuppressive modalities have proved to be effective, including cyclophosphamide, antilymphocyte serum, antitumor necrosis factor antibodies, and monoclonal anti-IA, CD4+, and CD8+ antibodies. If given too early in the course of early disease, the infection in neurons is potentiated and results in encephalitis and a high mortality rate. Thus, immunosuppression may be most effective when administered after the first week of infection. The incidence of demyelinating disease is increased in SJL mice infected with a dose of virus that normally produces a low incidence of disease and adoptively immunized with TMEV VP2-specific T-cell line. This observation supports a role for CD4+ T cells in mediating TMEV-induced demyelinating disease.

The effector mechanism by which a nonbudding virus, such as TMEV, might lead to immune-mediated tissue injury is unknown. Because TMEV antigens have been primarily found in macrophages, it has been proposed that myelin breakdown results from an interaction between virus-specific T cells trafficking into infected areas of the CNS and the virus. Thus, myelinated axons may be nonspecifically damaged as a consequence of a virus-specific immune response, that is, an 'innocent bystander' response. In this circumstance, cytokines produced by MHC class II-restricted, TMEV-specific T_{DTH} cells primed by interaction with infected macrophages lead to the recruitment and activation of additional macrophages in the CNS, resulting in nonspecific macrophage-mediated demyelination. This hypothesis is consistent with the CNS pathology observed in mice exhibiting TMEV-induced demyelinating disease and the fact that antigen-specific T cells and T-cell lines have been shown to cause bystander CNS damage via macrophage activation in other model systems. Alternatively, in the case of extensive infection of oligodendrocytes, demyelination might result from immune injury to these myelin-maintaining cells expressing TMEV antigens in conjunction with H-2 class I determinants. CD8+ T cells would then be the likely T cells to kill infected oligodendrocytes; however, widespread degeneration of oligodendrocytes has not been observed.

Acknowledgments

The authors thank Jan-Yve Sgro for providing the GRASP image of a BeAn virion and Paul Chipman for the cryoelectron micrograph of a GDVII virion.

See also: Cardioviruses.

Further Reading

Adami C, Pritchard AE, Knauf T, Luo M, and Lipton HL (1998) Mapping a determinant for central nervous system persistence in the capsid of Theiler's murine encephalomyelitis virus (TMEV) with recombinant viruses. *Journal of Virology* 71: 1662–1665.

Borson NO, Paul C, Lin X, et al. (1997) Brain-infiltrating cytolytic T lymphocytes specific for Theiler's virus recognize H2D^b molecules complexed with a viral VP2 peptide lacking a consensus anchor residue. *Journal of Virology* 71: 5244–5250.

Dethlefs S, Brahic M, and Larsson-Sciard EL (1997) An early, abundant cytotoxic T-lymphocyte response against Theiler's virus is critical for preventing viral persistence. *Journal of Virology* 71: 8875–8878.

Grant RA, Filman OJ, Fujinami RS, Icenogle JP, and Hogle JM (1992) Three-dimensional structure of Theiler virus. *Proceedings of the National Academy of Sciences, USA* 89: 2061–2065.

Jelachich ML, Bramlage C, and Lipton HL (1999) Differentiation of M1 myeloid precursor cells into macrophages results in binding and infection by Theiler's murine encephalomyelitis virus (TMEV) and apoptosis. *Journal of Virology* 73: 3227–3235.

Kumar ASM, Reddi HV, Kung A, and Lipton HL (2004) Virus persistence and disease in a mouse model of multiple sclerosis requires binding to the sialic acid co-receptor. *Journal of Virology* 78: 8860–8867.

Lobert PE, Escriou N, Ruelle J, and Michiels T (1999) A coding RNA sequence acts as a replication signal in cardioviruses. *Proceedings of the National Academy of Sciences, USA* 96: 11560–11565.

Luo M, He C, Toth KS, Zhang CX, and Lipton HL (1992) Three-dimensional structure of Theiler's murine encephalomyelitis virus (BeAn strain). *Proceedings of the National Academy of Sciences, USA* 89: 2409–2413.

Paul S and Michiels T (2006) Cardiovirus leader proteins are functionally interchangeable and have evolved to adapt to virus replication fitness. *Journal of General Virology* 87: 1237–1246.

Penna-Rossi C, Delcroix M, Huitinga I, et al. (1997) Role of macrophages during Theiler's virus infection. *Journal of Virology* 71: 3336–3340.

Peterson JO, Waltenbaugh C, and Miller SD (1992) IgG subclass responses to Theiler's murine encephalomyelitis virus infection and immunization suggest a dominant role for Th1 cells in susceptible mouse strains. *Immunology* 75: 652–658.

Roussarie J-P, Ruffie C, and Brahic M (2007) The role of myelin in Theiler's virus persistence in the central nervous system. *PLoS Pathogens* 3(2): e23(doi:10.1371/journal.ppat.0030023).

Simas JP and Fazakerley JK (1996) The course of disease and persistence of virus in the central nervous system varies between individual CBA mice infected with the BeAn strain of Theiler's murine encephalomyelitis virus. *Journal of General Virology* 77: 2701–2711.

Trottier M, Wang W, and Lipton HL (2001) High numbers of viral RNA copies in the central nervous system during persistent infection with Theiler's virus. *Journal of Virology* 75: 7420–7428.

Yauch RL, Palma JP, Yahikozawa H, Chang-Sung K, and Kim BS (1998) Role of individual T-cell epitopes of Theiler's virus in the pathogenesis of demyelination correlates with the ability to induce a Th1 response. *Journal of Virology* 72: 6169–6174.

Tick-Borne Encephalitis Viruses

T S Gritsun and E A Gould, University of Reading, Reading, UK

© 2008 Elsevier Ltd. All rights reserved.

Taxonomy, Nomenclature, and Phylogenetic Relationships

'Tick-borne encephalitis antigenic complex' was the original term for viruses now classified as the mammalian tick-borne flaviviruses. Together with the seabird-associated tick-borne flaviviruses they comprise one ecological group in the genus *Flavivirus*, family *Flaviviridae*. The genus contains two other groups, namely, the mosquito-borne flaviviruses and the flaviviruses with no known arthropod vectors. Other flaviviruses are referred to as nonclassified flaviviruses. The prototype TBEV is a human and animal pathogen. The earliest TBEV isolate, known as Russian Spring and Summer Encephalitis virus, was isolated in 1937 in far-East Asia. Currently, the TBEV (currently classified as virus species in the group of mammalian tick-borne flaviviruses) is subdivided into three subtypes, far-Eastern, Siberian, and West European reflecting their antigenic, phylogenetic, and geographic relationships. Other antigenically related but distinct tick-borne flaviviruses include Omsk hemorrhagic fever virus, Powassan virus (POWV), Langat virus, Kyasanur Forest disease virus, Alkhurma virus, Louping ill virus, Spanish sheep encephalomyelitis virus, Turkish sheep encephalomyelitis virus, and Greek goat encephalomyelitis virus. Each of them may cause encephalitis and/or hemorrhagic disease in humans, farmed, domestic, and wild animals. Other related mammalian tick-borne flaviviruses, not known to cause human or animal disease, include Royal Farm virus, Karshi virus, Gadgets Gully virus, and Kadam virus. The viruses associated primarily with seabirds and their ticks are Saumarez Reef virus, Meaban virus, and Tyuleniy (Three Arch) virus. They are not recognized pathogens but they show interesting geographic dispersion that reflects the flight patterns of the birds with which they are associated. The phylogenetic relationships between the TBEVs are illustrated in **Figure 1**.

Virions

Infectious (mature) virions are spherical particles (50 nm) with a relatively smooth surface and no distinct projections. They have an electron-dense core (30 nm) surrounded by a lipid membrane. The core consists of positive-polarity genomic RNA (~11 kbp) and capsid protein C (12K). The lipid membrane incorporates an envelope glycoprotein (E, 53K) and a membrane glycoprotein (M, 8K). The immature (intracellular) virions contain a precursor membrane protein (prM, 18K), the cleavage of which occurs in the secretory pathway during egress of virions from infected cells.

Virions sediment at about 200S and their buoyant density is 1.19 g ml^{-1}, although there is significant heterogeneity among them. Electron micrographs frequently revealed virions in association with cellular membranes, probably explaining this heterogeneity. They are most stable at pH 8.0, although TBEV virions remain infectious in normal human gastric juice at pH 1.49–1.80. As with all enveloped viruses, infectivity of TBEVs decays rapidly at temperatures above 40 °C. Most flaviviruses agglutinate erythrocytes but a few nonagglutinating strains of TBEVs have been described. Most of the physical characteristics of the TBEVs were established using virus isolated from mammalian cells. However, after adaptation to ticks, they become less cytopathic for cultured cells and laboratory animals. These virions do not move toward the cathode in rocket immunoelectrophoresis and they have reduced hemagglutinating activity. These phenotypic characteristics may be reversed following re-adaptation to mammalian cells and host-selection of virions with a shifted net surface charge due to a single amino acid substitution in the E glycoprotein.

The E glycoprotein mediates virus binding to cellular receptors and thereby directly affects virus host range, virulence, and immunological properties by inducing

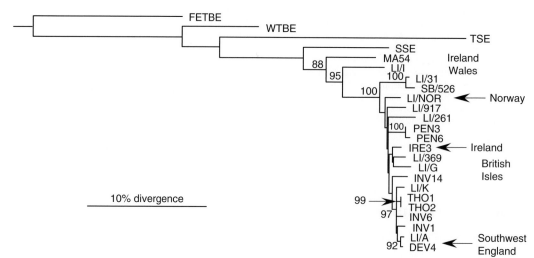

Figure 1 Maximum likelihood phylogenetic tree of the E gene from 24 tick-borne flaviviruses. Branch lengths are drawn to scale and all nodes supported by more than 75% bootstrap support are indicated. The tree is rooted with the sequence from FETBE virus, Sofjin (RSSE) strain. The three main populations of virus in the British Isles are indicated, along with those viruses secondarily introduced into Ireland and Norway, and the viruses found in the south-west of England. Reproduced from McGuire K, Holmes EC, Gao GF, Reid HW, and Gould EA (1998) Tracing the origins of Louping ill virus by molecular phylogenetic analysis. *Journal of General Virology* 79: 981–988.

protective antibodies. X-ray crystallography has revealed the E-protein ectodomain (N-terminal 395 amino acids) as homodimers folded in a 'head-to-tail' manner (**Figure 2(a)**) and orientated parallel to the membrane surface. It contains three structural domains, each based on β-sheets: the central domain I, the dimerization (fusion) domain II, and receptor domain III (dI, dII, and dIII in **Figure 2**). The C-terminal 101 residues of the E protein form a stem-anchor region consisting of two stem α-helices and two transmembrane α-helices that anchor the E protein into the lipid bilayer (**Figure 3**). Domain II contains a hydrophobic fusion peptide consisting of 13 residues that is highly conserved between all flaviviruses. It is located on the tip of domain II and plays a central role in fusion of the virion membrane to cellular endosomal membranes resulting in release of virion RNA into the cytoplasm (**Figure 2**).

A three-dimensional image reconstruction of the virion surface reveals a protein shell composed of 90 E-dimers organized into a 'herringbone' configuration; three quasi-parallel E-dimer molecules make up the main structural asymmetric unit of the shell (**Figures 4(b)** and **4(c)**). The fivefold symmetry axes that appear as holes are generated by appropriate positioning of five domain IIIs and their lateral surface is accessible to cellular receptors and neutralizing antibodies. The M protein protrudes through holes formed between dimerization domains of E molecules. The transmembrane regions of the M and E proteins traverse the external but not the internal layer of the lipid membrane and there is no direct contact with the nucleocapsid (**Figure 3**).

In contrast with those of many other enveloped viruses, the flavivirus capsid is poorly organized and appears to be positioned randomly in the core. In solution, the capsid protein appears as dimer molecules with four helices (**Figure 5**). Two C-terminal helices fold into the positively charged interface interacting with viral RNA; the residues on the opposite side of these helices are hydrophobic and support dimerization. Two of the internal α-helices form capsid-dimers, are hydrophobic, and are required for contact with the virion membrane.

TBEV Life Cycle

Virus Entry into the Cells

Cellular receptors for tick-borne flaviviruses

The virus life cycle commences with the attachment of virions to specific receptors on clathrin-coated pits on the cell surface. Flaviviruses infect a wide variety of primary and continuous cells, derived from mammalian, avian, and arthropod tissues. It is not yet clear if the different flaviviruses use similar or different cell surface receptors. Cellular heparan sulfate molecules are involved in TBEV attachment to cells. Nevertheless, cells that lack heparan sulfate are still sensitive to virus infection. For TBEV, a number of different proteins have been associated with cell receptor activity, including the high-affinity laminin receptor and α1β3-integrin that also recognizes laminin as a natural ligand.

The E protein mediates attachment and entry of virions into cells

The immunoglobulin-like folding of the E-protein domain III implies that it is involved in cell attachment,

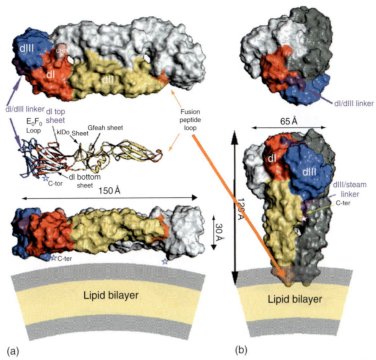

Figure 2 Three-dimensional analysis of TBEV envelope (E) glycoprotein (a) at neutral pH (in mature virion) and (b) acidic pH (postfusion conformation). The view from above and lateral view correspond to the top and bottom rows. Protein domains dI, dII, and dIII are colored in red, yellow, and blue, respectively. Ribbon diagram corresponds to the lateral view of the E protein. Reproduced from Bressanelli S, Stiasny K, Allison SL, et al. (2004) Structure of a flavivirus envelope glycoprotein in its low-pH-induced membrane fusion conformation. *EMBO Journal* 23: 728–738.

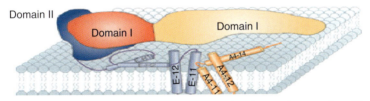

Figure 3 Schematic presentation of flavivirus structural glycoproteins E and M on the membrane surface. The coloring of the domains I, II, and III is as on **Figure 2**. Two α-helices of the stem-anchor region and two α-helices of the transmembrane domain of the E protein are depicted as blue- and colored cylinders. One α-helix of the stem region and two α-helices of the transmembrane domains of the M protein are depicted as orange-colored cylinders. Reproduced from Mukhopadhyay S, Kuhn RJ, and Rossmann MG (2005) A structural perspective of the flavivirus life cycle. *Nature Reviews Microbiology* 3: 13–22, with permission from Macmillan Magazines Limited.

mediated by high-affinity cellular receptors. The receptors on the surface of vertebrate and invertebrate cells are different and probably recognized by different residues on domain III. However, domains I and II on the surface of the E protein also appear to play a role in virus–cell interactions. Domain I carries N-linked carbohydrates, which are recognized by cell surface lectins. Domain II is enriched with patches of basic positively charged surface residues that have been shown to mediate virion attachment to the cellular, negatively charged, heparan sulfate; domains I and III might also contribute to this type of interaction.

Virus attachment to the cell receptors initiates receptor-mediated endocytosis, followed by fusion of virion and endosomal membranes (**Figures 4(c)–4(e)**). Exposure of the attached virus to the endosomal acid pH converts the herringbone configuration into a fusogenic structure resulting in particle expansion and eventually in the re-assembly of the E dimers into vertical trimers (**Figures 2** and **4**). It is believed that during this process domains I and II flex relative to each other, allowing domain II and the stem region to move toward the endosomal membrane facilitating trimerization of domain II; the fusion tripeptide is embedded into the target membrane followed by formation of a pre-fusion intermediate (**Figure 6**). Trimerization spreads from fusion peptides to domain I; meanwhile domain III rotates relative to domain II pushing the stem region back toward the fusion peptide. This refolding

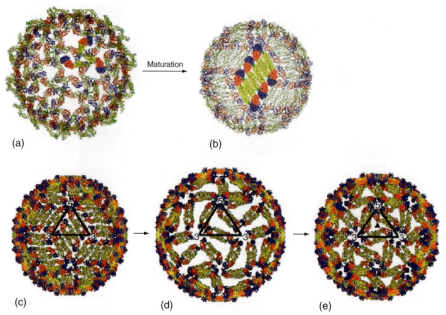

Figure 4 Different conformations of E protein during the transition from immature (a) into the mature (b) particles and during pH-dependent fusion of mature particles (c), through the putative intermediates T = 3 (d and e). Domains I, II, and III are depicted with the same colors as those on the **Figure 2**. Reproduced from Ma L, Jones CT, Groesch TD, Kuhn RJ, and Post CB (2004) Solution structure of Dengue virus capsid protein reveals another fold. *Proceedings of the National Academy of Sciences, USA* 101: 3414–3419, with permission from National Academy of Sciences USA.

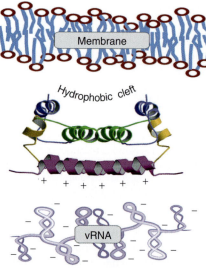

Figure 5 Model for the molecular interactions of flavivirus capsid dimers with genome RNA and virion membrane. Reproduced from Ma L, Jones CT, Groesch TD, Kuhn RJ, and Post CB (2004) Solution structure of Dengue virus capsid protein reveals another fold. *Proceedings of the National Academy of Sciences, USA* 101: 3414–3419, with permission from National Academy of Sciences USA.

brings two membranes together, resulting in the formation of the 'hemifusion stalk' with only the proximal lipid leaflets fused. The 'zipping' up of stems followed by the migration of transmembrane domains eventually leads to fusion of the distal lipid outlets and formation of a fusion pore. In the final low-pH conformation, domains I and III appear to move to the tip of the vertical trimer whereas the fusion peptide becomes embedded in the membrane eventually being juxtaposed to the transmembrane domains (**Figures 2**(b) and **6**). Five trimers form the fusion pore; the stem region appears to play an essential role in promoting and stabilizing trimer assembly. Although the E-protein domains shift and rotate during exposure to acidic pH, each retains the neutral-pH conformation, the essential feature of class II fusion molecules that do not require major molecular rearrangements at acidic pH.

Strategy of the TBEV Genome

Translation and processing of virus proteins

After uncoating, the RNA is translated into a polyprotein of ~3400 residues from one open reading frame (ORF) that is co-translationally translocated and anchored in the endoplasmic reticulum (**Figure 7**). It is then processed by cellular signalases and viral serine protease producing three structural (capsid, prM, and E) and seven nonstructural (NS1 through NS5) proteins. The gene order, protein molecular masses, and their membrane localization are shown in **Figure 7**. Viral protease activity is provided by the N-terminal domain (~180 residues) of the NS3 protein in association with cofactor activity of the NS2B protein that probably anchors the NS3 into

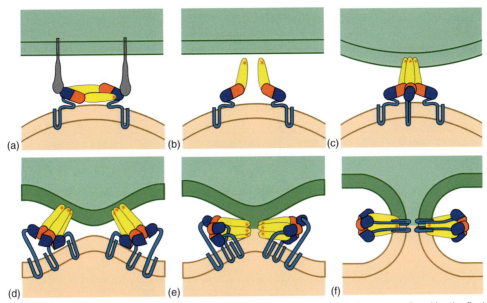

Figure 6 Tentative mechanism for the fusion of viral (brown) and endosomal (green) membranes mediated by the flavivirus E protein following interaction with cellular receptor (gray) (a). Acidic pH triggers the movement of domain II (with fusion peptide on the top) toward the endosomal membrane (b) with its subsequent trimerization and insertion into the endosomal membrane (c). Trimerization spread toward domains I and III (d) causing the C-terminal part of E protein to fold backwards toward the fusion peptide. The trimerization between stem-anchor and domain II (e) results in partial fusion (e) and eventually in the formation of a fusion pore (f). Reproduced with permission from Modis Y, Ogata S, Clements D, and Harrison SC (2005) Variable surface epitopes in the crystal structure of Dengue virus type 3 envelope glycoprotein. *Journal of Virology* 79: 1223–1231.

the membrane. The ORF is flanked by 5′- and 3′-untranslated regions (UTRs; ~130 and 700 bp, respectively) that contain signals essential to initiate translation and replication. The 5′-UTR preceding the ORF is capped by $m^7GpppAmpN_2$ and initiates virus translation *in vitro* although an interactive effect from the 3′-UTR has also been demonstrated. Computer-assisted prediction has demonstrated folding of the 5′-UTR into a Y-shaped structure, conserved between all flaviviruses; mutations in this structure appear to impact on virus translation and replication. The translation of flavivirus proteins is probably carried out in specialized smooth membrane structures, called convoluted membranes, derived from proliferation of the rough endoplasmic reticulum in response to flavivirus infections.

Replication of viral RNA

The initiation of viral replication (synthesis of negative strand on positive strand from the 3′-UTR) requires direct interaction between the 5′- and 3′-UTRs resulting in the formation of a double-stranded RNA (dsRNA) stem and circularization of the virus genome. A terminal 3′ long stable hairpin (3′ LSH) structure has been revealed in the 3′-UTR that, together with the 5′-Y-shaped-structure and dsRNA circularization stem, forms a complex promoter required to initiate virus RNA replication. Additional predicted conserved secondary RNA structures in the 3′-UTR might function as replication enhancers, essential for efficient RNA synthesis and virus transmission. The synthesis of genome-size dsRNA (replicative form, RF) accomplishes the first stage of viral replication.

Flavivirus RNA replication is semiconservative and asymmetric, with an average of one nascent positive-sense single-stranded RNA (ssRNA) molecule (genomic) per negative-sense template, with no free minus-strands and with 10–100 times excess of positive-strand relative to negative-strand RNA synthesis. The replicative intermediate presents a partially double-stranded RNA with nascent, and displaced, plus-sense ssRNA molecules undergoing elongation. Viral RNA capping of the 5′-UTR occurs on the displaced plus-strand RNA; the trifunctional NS3 and NS5 proteins provide nucleoside triphosphatase and guanylyl/methyltransferase activities, respectively.

Formation of the replication complex involves nonstructural proteins (**Figure 8**) most of which are bi- and even trifunctional. The C-terminal domain of NS5 protein acts as a viral RNA polymerase and the C-terminal domain of NS3 as a helicase; both NS5 and NS3 proteins interact with 3′-LSH to initiate virus replication. The functions of other nonstructural proteins are less precisely identified. The NS1 glycoprotein forms membrane-associated hexamers that dissociate into dimers. This protein is translocated in the endoplasmic reticulum and secreted together with virions in mammalian but not in mosquito cells; its anchoring into the membranes is mediated by

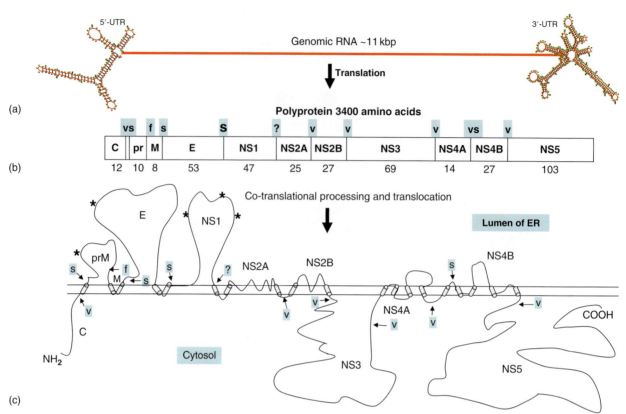

Figure 7 Genome strategy of TBEV. (a) Genomic RNA is presented as a solid line on the top; the 5′- and 3′-UTR are depicted in the predicted conformations according to Gritsun TS, Venugopal K, Zanotto PM, et al. (1997) Complete sequence of two tick-borne flaviviruses isolated from Siberia and the UK: Analysis and significance of the 5′- and 3′-UTRs. *Virus Research* 49: 27–39. (b) Translation and co-translational processing of flavivirus polyprotein. The flavivirus polyprotein is depicted as a bar, with specified individual proteins and their molecular masses (numbers below the bar). (c) Membrane topology of viral proteins in relation to the lumen of the ER and cytoplasm after the completion of co-translational processing and translocation. Adapted from Westaway EG, Mackenzie JM, and Khromykh AA (2003) Kunjin RNA replication and applications of Kunjin replicons. *Advances in Virus Research* 59: 99–140. Transmembrane domains are shown as cylinders. Glycosylation is indicated as (*). The polyprotein is processed by ER signalases (s) and viral-specific protease NS2B-NS3 (V). The cleavage of M from prM is carried out by the Golgi protease, furin (f) and cleavage between NS1 and NS2A is by an unknown (?) ER protease.

glycosyl-phosphatidylinositol. The NS1 protein induces the production of protective antibodies that also participate in complement-mediated lysis of infected cells, implying a role in immunopathology. The NS2A protein is found in association with the NS3 helicase domain, the NS5 protein, and the 3′-LSH. It may be involved in trafficking of viral RNA between the translation, replication, and virion assembly sites. The NS2B protein is a membrane-anchored cofactor of the serine proteinase, NS3. It also has membrane-permeability modulating activity possibly related to replication. The hydrophobic NS4A protein in conjunction with the NS1 protein probably anchors the polymerase complex to cell membranes and is involved in virus-induced membrane rearrangements that compartmentalize virus translation, replication, and assembly processes (**Figure 8**).

Replication of flavivirus RNA is associated with host membrane 75–100 nm vesicular packets enclosed in the second outer membrane and connected to convoluted membranes, the sites of virus translation. The vesicular packets proliferate in the perinuclear region in response to virus infection and are probably derived from the *trans*-Golgi network. The architecture of the replication complex in relation to vesicular packets and convoluted membranes is illustrated in **Figure 9**. The replicative forms and replication complexes are enclosed within vesicular packets, whereas nascent ssRNA is externally located. In addition, a variety of cellular proteins associates with the 5′-UTR, 3′-UTR, and the RNA polymerase; they probably facilitate assembly of the replication complex and trafficking of the viral RNA/polymerase into the appropriate cellular compartment for replication.

Assembly, Maturation, and Release of Virions

The sequence of molecular events during virion morphogenesis is not completely understood. On completion of post-translational processing, the prM and E proteins rapidly form heterodimeric complexes; the

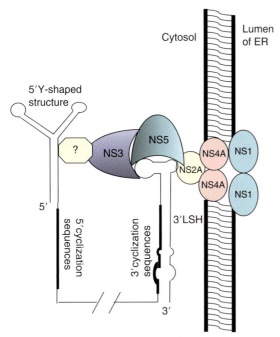

Figure 8 Representation of the assembly of the membrane-anchored polymerase complex. The conformations of 3′LSH and 5′ Y-shaped structure are shown. This stage probably precedes the cyclization of the virus genome mediated by direct interaction between the inverted complementary 5′- and 3′-cyclization sequences (thick lines); this is followed by the initiation of minus-strand RNA synthesis. Adapted from Westaway EG, Mackenzie JM, and Khromykh AA (2002) Replication and gene function in Kunjin virus. *Current Topics in Microbiology and Immunology* 267: 323–351.

chaperone-like function of the prM protein is essential for correct folding of E protein. Heterodimers are rapidly assembled into immature particles or, in the absence of capsid, into virus-like particles that also accumulate during infection. Virion packaging is coupled with RNA replication; only actively replicating RNA is encapsidated. The orientation of structural proteins in the lipid membrane suggests that assembly is mediated by capsid budding through the endoplasmic reticulum. However, assembled virus capsids have never been visualized, neither have budding particles nor specific release mechanisms been identified.

Assembly of structural proteins in the viral lipid membrane initially results in the formation of immature (intracellular) virions that are noninfectious, resistant to acidic pH, and unable to be structurally rearranged prior to fusion. Image reconstruction has defined the different organization of E/prM proteins compared with E/M proteins on the surface of immature and mature virions, respectively (**Figure 4**). Sixty projections were identified on the surface of immature virions making them look larger (60 nm) in comparison with smooth 50 nm mature virus particles. Each spike is formed by three heterodimeric E/M molecules. In immature particles the E protein points away from the membrane with the fusion peptide at its extremity. Virion maturation is accompanied by cleavage of prM to M, mediated by the cellular protease furin in the *trans*-Golgi network at acidic pH, during the transportation of virions to the cell surface. PrM-to-M cleavage results in the dissociation of E/M trimeric spikes with the formation of E/E homodimers, probably following shift of the stem-transmembrane domain relative to the ectodomain. In mature particles, the E ectodomain lies on the virus membrane, with the fusion peptide embedded in the cavity between domains I and III of the neighboring dimer. Formation of the herringbone building unit and release of the glycosylated pre-peptide then follows. Eventually, *trans*-Golgi-derived vesicles packed with mature virions follow the host secretory pathway, fuse with the plasma membrane, and release mature virions.

Pathogenesis of TBEV

Incidence of Tick-Borne Encephalitis

Classical tick-borne encephalitis (TBE) is contracted by humans when they are fed upon by infected *Ixodes* spp. ticks. Forested areas across the Northern Hemisphere, with thick moist undergrowth and abundance of small wild animals, provide ideal habitats for ticks and for TBEV. Long-term survival of the virus occurs mainly in ticks, which remain infected throughout their 2–5-year life cycle. Efficient virus transmission between infected nymphs and noninfected larvae occurs on the forest animals when these ticks co-feed. There is no requirement for viremia, since the virus can be transmitted to the larvae in animals that are not susceptible to the virus. Moreover, nonviremic transmission between the ticks can occur even on immune animals. Cells that migrate between the skin surface and local lymph nodes (e.g., dendritic cells) become infected and are then imbibed by the feeding noninfected larvae. Nonviremic transmission provides an efficient mechanism for long-term virus survival in the absence of overt disease. At the same time, susceptible animals such as sheep, goats, horses, pigs, dogs, grouse, and even humans may inadvertently become infected by ticks either in the forests or by ticks carried away from the forests. Under these circumstances the vertebrate host may become viremic and the virus may then be transmitted to a noninfected tick that subsequently transmits the virus to another vertebrate host when the infected tick feeds for a second time. Such situations arise on the sheep-rearing moorlands in the British Isles, Norway, Spain, Turkey, Greece, and other parts of Europe. In some situations sheep/goat and grouse populations may be severely affected.

The incidence of human TBE varies from year to year in different areas. The Urals and Siberia annually

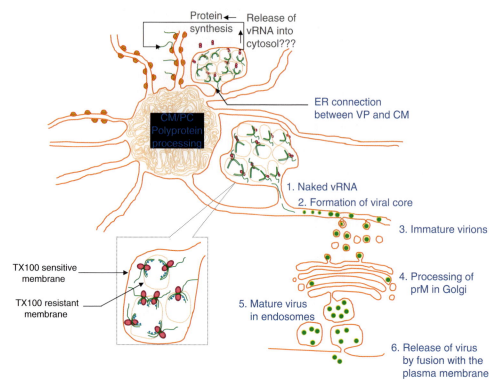

Figure 9 Proposed model of flavivirus replication in connection with cellular membranous structures. Translation of viral proteins is associated with CMs whereas replication is carried out within VPs. The dsRNA of RF in association with proteins of the polymerase complex (RC) is anchored to the internal leaflet of VP membrane whereas newly synthesized plus ssRNA is released from the RC. Reproduced with permission from Uchil PD and Satchidanandam V (2003) Architecture of the flaviviral replication complex. Protease, nuclease, and detergents reveal encasement within double-layered membrane compartments. *Journal of Biological Chemistry* 278: 24388–24398.

record the highest number of hospitalized cases. These numbers have risen from 700 to 1200 cases per year in the 1950s and 1960s to greater than 11 000 per year in the 1990s, following *Perestroika* and initial breakdown of the infrastructure. The incidence in Europe is lower but nonetheless significant, with about 3000 cases per year. Although it can affect people of all ages, the highest incidence occurs among the most active groups, that is, 17–40 year olds.

The incidence of clinical disease in endemic regions is dependent on the frequency of forest visits, the density of ticks in different years, the concentration of virus in ticks, and the virulence of circulating strains. Taking all these factors into account, it was estimated that one clinical case must occur for every 100 people bitten by ticks, and this correlates with the observation that in regions where more than 60% of ticks carry virus, about 1.4% of people develop TBE after being bitten.

Variety of Clinical Manifestations

TBEV is recognized as a dangerous human pathogen, causing a variety of clinical manifestations, although asymptomatic infections constitute about 70–95% of all TBEV infections. Symptoms include (1) mild or moderate fever with complete recovery of patients (about 90% of all clinical manifestations), (2) subacute encephalitis with nearly complete recovery or residual symptoms that may or may not disappear with time, (3) severe encephalitis associated with irreversible damage to the central nervous system (CNS), resulting in disability or death (about 4–8%), and (4) slow (months to decades) progressive or chronic encephalitis (1–2%).

Initially in the 1940–50s it was believed that the same virus produced encephalitis over the whole of Europe and Asia. However, with the improvement of serological diagnosis and the advent of phylogenetic analysis, three subtypes of TBEV were identified.

Human infections with far-Eastern TBEV subtype viruses produce the most severe form of CNS disorder, with a tendency for the patient to develop focal meningoencephalitis or polyencephalitis, and case–fatality rates between 20% and 40%. Siberian strains are often associated with high prevalence of the nonparalytic febrile form, with case–fatality rates rarely exceeding 6–8%. There is a tendency for some patients to develop chronic TBE, predominantly in association with Siberian TBEV strains. The disease produced by West European TBEV strains is biphasic, with fever during the first phase and neurological disorders of differing severity during the second phase, which occurs in 20–30% of patients, with a case–fatality rate of 1–2%. While Louping ill virus and the related Spanish, and Turkish/Greek viruses produce fatal encephalitis in animals, they are only rarely associated with disease in humans, probably primarily because of the lower (in comparison with TBEV) rates of human exposure to infected ticks.

Biphasic milk fever is an unusual form of TBE originally observed in western Russia mainly associated with the consumption of goat's milk. West European strains of TBEV were isolated from unpasteurized goat's milk and from patients. Later, similar outbreaks of biphasic milk fever were reported in central Europe and Siberia. The apparent difference in clinical manifestations of TBE contracted by tick-bite or by the alimentary route may be explained by differences in the type of immunological response.

Human Disease Produced by Other Tick-Borne Flaviviruses

POWV also produces encephalitic infections in humans although not on an epidemic scale. POWV circulates in Russia, the USA, and Canada, where it may cause human encephalitis with a high incidence of neurological sequelae and up to a 60% case–fatality rate. In North America POWV has diverged to produce a closely related deer tick virus (DTV) that has a predilection for different rodent species. Langat virus has been isolated in Malaysia and Thailand; there are no registered cases of human disease associated with this virus. However, when Langat virus was used as a live, attenuated vaccine in human trials in Russia, one in 10 000 patients developed encephalitis.

Three tick-borne flaviviruses cause hemorrhagic disease rather than encephalitis in humans, namely Omsk hemorrhagic fever virus, Kyasanur Forest disease virus, and Alkhurma virus. As yet there is no explanation for these differences in virulence characteristics. Epidemic foci of Omsk hemorrhagic fever virus in the Omsk and Novosibirsk regions of western Siberia usually follow epizootics in muskrats (*Ondatra zibethica*) from which local hunters become infected by the virus when they handle infected animals. The most marked pathological signs of the disease are focal visceral hemorrhages in the mucus of the nose, gums, uterus, and lungs. Convalescence is usually uneventful without residual effects; fatal cases have been recorded, but rarely (0.5–3%).

Kyasanur Forest disease virus was first recognized in 1957 in India where it caused hemorrhagic disease among monkeys and humans, frequently with fatal outcome. It is believed that perturbation of regions of the forest for land development led to increased exposure of monkeys and humans to ticks carrying the virus. Subsequently, it has become evident that this virus circulates throughout western India and this may explain the close antigenic and genetic link with Alkhurma virus, which was isolated from fatal human cases of hemorrhagic fever in Saudi Arabia. Recent evidence has demonstrated the presence of Alkhurma virus in ticks associated with camels. Whether or not this or a closely related virus circulates in Africa remains to be determined.

Prevention and Control of TBE

Vaccination is the most efficient method available for preventing TBE in enzootic regions. Currently two inactivated vaccines are commercially available. The European vaccine has significantly reduced the annual incidence of TBE in Austria and Germany. A similar vaccine, based on a far-Eastern strain of TBEV, has been used successfully in Russia to immunize at-risk laboratory personnel. Other preventive measures routinely employed in TBE endemic areas are (1) education of residents in methods of avoiding tick-bites (when visiting tick-infested areas, wearing appropriate clothing, regularly inspect for feeding ticks, report tick-bite to medical authorities); (2) treat cats and dogs with acaricides; (3) clear thick, moist vegetation areas from around houses; and (4) spray acaricides in forested areas, close to habitation.

Human trials of a live, attenuated vaccine for Langat virus produced an unacceptably high incidence of TBE (1/10 000). However, these trials demonstrated the higher protection efficiency of live, attenuated vaccines in comparison with inactivated vaccines. Currently safer strategies to produce live, attenuated vaccine are being developed such as (1) RNA- or DNA-based vaccines; (2) engineering TBEV mutants with multiple attenuating mutations or large deletions within their genomes, resulting in the loss of neuroinvasiveness; and (3) engineering chimeric yellow fever virus vaccine containing substituted TBEV E and M proteins.

Currently there are no safe, effective antivirals for TBEV infections, but there is promising progress in tests with small-interfering RNAs (siRNAs). These short molecules, ~21 nt long, bind to homologous regions of viral mRNA, thus interfering with the replication cycle. Another antiviral strategy with which rapid progress is being made is based on the design of molecules that can target specific regions within viral replicative enzymes. If the twentieth century was significant for its development of effective vaccines, the twenty-first century might be recognized for its development of effective antivirals.

See also: Japanese Encephalitis Virus; West Nile Virus.

Further Reading

Aberle JH, Aberle SW, Kofler RM, and Mandl CW (2005) Humoral and cellular immune response to RNA immunization with flavivirus replicons derived from tick-borne encephalitis virus. *Journal of Virology* 79: 15107–15113.

Anderson R (2003) Manipulation of cell surface macromolecules by flaviviruses. *Advances in Virus Research* 59: 229–274.
Bressanelli S, Stiasny K, Allison SL, et al. (2004) Structure of a flavivirus envelope glycoprotein in its low-pH-induced membrane fusion conformation. *EMBO Journal* 23: 728–738.
Gelpi E, Preusser M, Garzuly F, Holzmann H, Heinz FX, and Budka H (2005) Visualization of Central European tick-borne encephalitis infection in fatal human cases. *Journal of Neuropathology and Experimental Neurology* 64: 506–512.
Gritsun TS, Lashkevich VA, and Gould EA (2003) Tick-borne encephalitis. *Antiviral Research* 57: 129–146.
Gritsun TS, Nuttall PA, and Gould EA (2003) Tick-borne flaviviruses. *Advances in Virus Research* 61: 317–371.
Gritsun TS, Tuplin AK, and Gould EA (2006) Origin, evolution and function of flavivirus RNA in untranslated and coding regions: Implications for virus transmission. In: Kalitzky M and Borowski P (eds.) *Flaviviridae: Pathogenesis, Molecular Biology and Genetics*, pp. 47–99. Norwich, UK: Horizon Scientific Press.
Heinz FX and Allison SL (2003) Flavivirus structure and membrane fusion. *Advances in Virus Research* 59: 63–97.
Heinz FX, Stiasny K, and Allison SL (2004) The entry machinery of flaviviruses. *Archives of Virology Supplement* 18: 133–137.
Lindenbach BD and Rice CM (2001) *Flaviviridae*: The viruses and their replication. In: Knipe DM and Howley PM (eds.) *Fields Virology*, 4th edn., vol. 1, pp. 991–1042. London: Lippincott Williams and Wilkins.
Markoff L (2003) 5′- and 3′-noncoding regions in flavivirus RNA. *Advances in Virus Research* 59: 177–228.
Mukhopadhyay S, Kuhn RJ, and Rossmann MG (2005) A structural perspective of the flavivirus life cycle. *Nature Reviews Microbiology* 3: 13–22.
Rey FA, Heinz FX, Mandl C, Kunz C, and Harrison SC (1995) The envelope glycoprotein from tick-borne encephalitis virus at 2 Å resolution. *Nature* 375: 291–398.
Uchil PD and Sachidanandam V (2003) Architecture of the flaviviral replication complex. Protease, nuclease, and detergents reveal encasement within double-layered membrane compartments. *Journal of Biological Chemistry* 278: 24388–24398.
Westaway EG, Mackenzie JM, and Khromykh AA (2003) Kunjin RNA replication and applications of Kunjin replicons. *Advances in Virus Research* 59: 99–140.

Tobacco Mosaic Virus

M H V Van Regenmortel, CNRS, Illkirch, France

© 2008 Elsevier Ltd. All rights reserved.

The Beginnings of Virology

Tobacco mosaic virus occupies a unique place in the history of virology and was in the forefront of virus research since the end of the nineteenth century. It was the German Adolf Mayer, working in the Netherlands, who in 1882 first described an important disease of tobacco which he called tobacco mosaic disease. He showed that the disease was infectious and could be transmitted to healthy tobacco plants by inoculation with capillary glass tubes containing sap from diseased plants. Although Mayer could not isolate a germ as the cause of the disease, he did not question the then prevailing view that all infectious diseases were caused by microbes and he remained convinced that he was dealing with a bacterial disease.

About the same time, in St. Petersburg, Dmitri Ivanovsky was studying the same disease and he reported in 1892 that when sap from a diseased tobacco plant was passed through a bacteria-retaining Chamberland filter, the filtrate remained infectious and could be used to infect healthy tobacco plants. Ivanovsky was the first person to show that the agent causing the tobacco mosaic disease passed through a sterilizing filter and this gave rise to the subsequent characterization of viruses as filterable agents. A virology conference was held in 1992 in St. Petersburg to celebrate the centenary of this discovery. Although Ivanovsky is often considered one of the fathers of virology, the significance of his work for the development of virology remains somewhat controversial because all his publications show that he did not really grasp the significance of his filtration experiments. He remained convinced that he was dealing with either a small bacterium or with bacterial spores and never appreciated that he had discovered a new type of infectious agent.

Following in the footsteps of Mayer, Beijerinck in Delft, Holland, again showed in 1898 that sap from tobacco plants infected with the mosaic disease was still infectious after filtration through porcelain filters. He also demonstrated that the causative agent was able to diffuse through several millimeters of an agar gel and he concluded that the infection was not caused by a microbe.

Beijerinck called the agent causing the tobacco mosaic disease a *contagium vivum fluidum* (a contagious living liquid), in opposition to a *contagium vivum fixum*. In those days the term *contagium* was used to refer to any contagious, disease-causing agent, while the term *fixum* meant that the agent was a solid particle or a cellular microbe. On the basis of his filtration and agar diffusion experiments, Beyerinck was convinced that the agent causing tobacco mosaic was neither a microbe nor a small particle or corpuscle (meaning a small body or particle from the Latin *corpus* for body). He proposed instead that the disease-causing agent, which he called a virus, was a living liquid containing a dissolved, nonparticular and noncorpuscular entity.

Lute Bos has claimed that Beijerinck's introduction in 1898 of the unorthodox and rather odd concept of a *contagium vivum fluidum* marked the historic moment when virology was conceived conceptually. However, it is clear that his definition of a virus as a soluble, living agent not consisting of particles certainly does not correspond to our modern view of what a virus is. It is Beijerinck's insistence that a virus is not a microbe and his willingness to challenge the then widely held view that all infectious diseases are caused by germs that make many people regard his contribution to the development of virology as more important than that of Ivanovsky. In 1998, a meeting held in the Netherlands commemorated the centenary of the work of Beijerinck on TMV. About the same time a meeting was held in Germany to honor the contribution of Loeffler and Frosch who, also in 1898, had shown that the agent causing the foot-and-mouth disease of cattle was able to pass through a Chamberland-type filter, in the same way as TMV. In addition the German workers also established that their disease-causing agent was not able to go through a finer grain Kitasato filter, from which they concluded that the agent, which was multiplying within the host, was corpuscular and not soluble as claimed by Beijerinck. Loeffler, however, continued to believe that his pathogen was a very small germ or spore, invisible in the light microscope, that was unable to cross the small pores of a Kitasato filter.

There is no doubt that none of these 'fathers' of virology realized the nature of the filterable pathogenic agents they were investigating and that the term virus which they used did not have the meaning it has today. It took another 30 years before chemical analysis eventually revealed what viruses are actually made of.

Physical and Chemical Properties of TMV

A major change in our perception of what viruses are occurred in 1935 when Stanley, working at the Rockefeller Institute in Princeton, obtained needle-like crystals of TMV that were infectious and consisted of protein. He used methods that were being developed at the time for purifying enzymes and was greatly helped by a bioassay developed by Holmes in 1929 that made it possible to quantify the amount of TMV present in plant sap. In this assay, extracts containing the virus are rubbed on the leaves of *Nicotiana glutinosa* tobacco plants which leads to a number of necrotic lesions proportional to the amount of virus present.

Stanley's demonstration that TMV was a crystallizable chemical substance rather than a microorganism, a discovery that earned him the Nobel Prize in 1946, had a profound impact on the thinking of biologists because it suggested that viruses were actually living molecules able to reproduce themselves. For a while, it seemed that viruses closed the gap between chemistry and biology and might even hold the key to the origin of life.

Stanley had originally described TMV as a pure protein, but in 1936 Bawden and Pirie showed that the virus contained phosphorus and carbohydrate and actually consisted of 95% protein and 5% RNA. TMV thus became the first virus to be purified and shown to be a complex of protein and nucleic acid. Following the development of ultracentrifuges in the 1940s, ultracentrifugation became the standard method for purifying TMV and many other viruses.

In 1939, TMV became the first virus to be visualized in the electron microscope and in 1941, TMV particles were shown to be rods about 280 nm long and 15 nm wide. Subsequently, X-ray analysis established that the rods were hollow tubes consisting of a helical array of 2130 identical protein subunits with 16(1/3) subunits per turn and containing one molecule of RNA deeply embedded in the protein subunits at a radius of 4 nm (**Figure 1**). The length of the TMV particle is controlled by the length of the RNA molecule which becomes fully coated with protein and is thereby protected from nuclease attack.

The viral protein subunits can also aggregate on their own, without incorporating RNA, to form rods very similar to TMV particles but in this case the rods are of variable length (**Figure 2**). It is possible to degrade the virus particles with acetic acid or weak alkali and to obtain in this way dissociated protein subunits and viral RNA, the latter being rapidly degraded by nucleases. In 1956 Gierer and Schramm in Tübingen, Germany, and Fraenkel-Conrat in Berkeley, California, showed that if the viral RNA was obtained by degrading the particles with phenol or detergent, intact RNA molecules were obtained that were infectious and could produce the same disease as intact virus. This was the first demonstration that the nucleic acid component of a virus was the carrier of viral infectivity and that it possessed the genetic information that coded for the viral coat protein.

In 1960 the TMV coat protein was the first viral protein to have its primary structure elucidated when its sequence was determined simultaneously in Tübingen and in Berkeley and found to consist of 158 amino acids. Subsequently, the coat protein sequence of numerous TMV mutants obtained by treating the virus with the mutagenic agent nitrous acid was also determined and these sequence data helped to establish the validity of the genetic code that was being elucidated in the early 1960s in Nirenberg's and Ochoa's laboratories. One of the changes induced by the action of nitrous acid on viral RNA is the nucleotide conversion of cytosine (C) to uracil (U). This changes the codon CCC for the amino acid proline to the codons UCC or CUC for serine and leucine, respectively. In addition, these two codons can be further converted to UUU which codes for phenylalanine. When the coat protein sequences of various TMV mutants were determined, it was found that

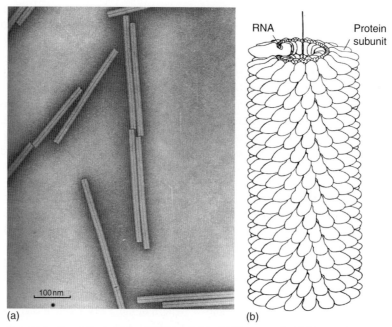

Figure 1 (a) Electron micrograph of negatively stained TMV particles. (b) Diagram of a TMV particle showing about one-sixth of the length of a complete particle. The protein subunits form a helical array with 16(1/3) units per turn and the RNA is packed at a radius of about 4 nm from the helix axis. From Klug A (1999) The tobacco mosaic virus particle: Structure and assembly. *Philosophical Transactions of the Royal Society*: *Biological Sciences* 354: 531–535.

most of the exchanged amino acids could be attributed to C→U nucleotide conversions, which confirmed the validity and universality of the proposed genetic code.

A particularly interesting mutation present in mutant Ni 1927 was the exchange proline to leucine at position 156 in the sequence which was found to greatly decrease the chemical stability of the virus. It had been known for many years that the virus was resistant to degradation by the enzyme carboxypeptidase and that this enzyme released only about 2000 threonine residues from each virus particle. The C-terminal residue in the coat protein is in fact threonine and the enzyme degradation data were interpreted at the time to mean that the virus particle contained 2000 subunits. However, when the proline at position 156 was replaced by leucine in the mutated protein, the enzyme was able to degrade the protein far beyond the C-terminal residue. The location of a proline near the C-terminus together with the presence of an acetyl group on the N-terminus are actually responsible for the remarkable resistance of the virus to exopeptidase degradation. The same exchange at position 156 in mutant Ni 1927 also led to an interesting discovery in immunology (see below).

Self-Assembly of TMV Particles

Dissociated coat protein molecules of TMV are able to assemble into different types of aggregates depending on the pH and ionic strength (**Figure 2**). The disk aggregate is a two-layer cylindrical structure with a sedimentation coefficient of 20 S, where each layer consists of a ring of 17 subunits compared with the 16(1/3) molecules present in each turn of the assembled helix. The disks are able to form stacks which can be either polar or nonpolar, depending on the relative orientation of adjacent disks (**Figure 2**). Short stacks of disks were shown to be nonpolar by immunoelectron microscopy, using monoclonal antibodies that reacted with only one end of the viral helix but were able to bind to both ends of the stacked disks. Whereas short polar stacks of disks can be transformed via lock-washer intermediates into helices (**Figure 2**), this is not possible for nonpolar stacks which cannot be incorporated as such during the elongation process that leads to the formation of RNA-containing virus particles.

It is generally accepted that initiation of TMV assembly from its RNA and protein components requires a 20 S protein aggregate, which could be either a disk or a short, lock-washer type proto-helix. The surface of this aggregate constitutes a template which is recognized by a specific viral RNA sequence. The nucleation of the assembly reaction was found to occur by a rather complex process involving a hairpin structure on the viral RNA, known as the origin of assembly and located about 1000 nucleotides from the 3′ end, which is inserted through the central hole of the 20 S aggregate. The nucleotide sequence of the stem–loop hairpin is responsible for the preferential incorporation of viral RNA rather than foreign RNA. Elongation occurs by addition of more protein subunits which pull up more RNA through the central hole. This

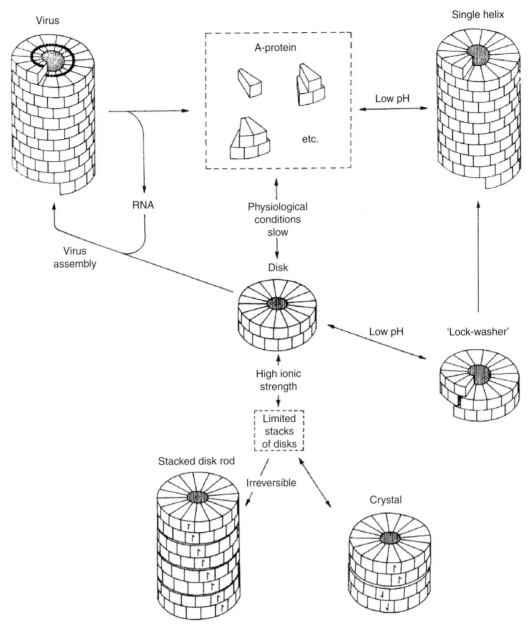

Figure 2 Diagram showing the different polymorphic aggregates of TMV coat protein obtained at different pH values and ionic strengths. Note that both polar and nonpolar stacked disks can be obtained. The 'lock-washer' is not well defined and represents a metastable transitory state. Adapted from Durham ACH and Klug A (1972) Structures and roles of the polymorphic forms of tobacco mosaic virus protein. *Journal of Molecular Biology* 67: 315–332, with permission from Elsevier.

process requires that the 5' tail of the RNA loops back down the central hole and is slowly 'swallowed up' while the 3' tail continues to protrude freely from the other end. This unexpected mechanism was visualized by electron micrographs taken in Hirth's laboratory in Strasbourg which revealed how particles were growing with both 3' and 5' RNA tails protruding initially from the same end of the particle. The elongation is thus bidirectional and occurs faster along the longer 5' tail by incorporation of 20 S aggregates and more slowly toward the 3' end through incorporation of 4 S protein. The elucidation of the assembly process was greatly facilitated by the sequencing of the viral RNA which was completed in 1982. TMV RNA consists of 6395 nucleotides and was the first genome of a plant virus to be sequenced completely.

Virus Disassembly

Since TMV particles are extremely stable *in vitro*, it was not at all obvious how the RNA managed to be released from the particles in order to start the virus replication cycle. Using a cell-free translation system, Wilson discovered that the disassembly of TMV particles is initiated when the end of the particles containing the 5′ terminus of the RNA becomes associated with ribosomes. This leads to viral subunits becoming dislodged from the particles while the 5′ terminal open reading frame in the RNA is being translated by the ribosomes, a process known as co-translational disassembly. This mechanism allows the coat protein subunits to protect the RNA from enzymatic degradation until the particle has reached a site in the infected cell where translation can be initiated.

Antigenicity of TMV

The antigenic properties of TMV have been studied extensively for more than 60 years and these studies have given us much information on how antibodies recognize proteins and viruses. TMV is an excellent immunogen and antibodies to the virus are readily obtained by immunization of experimental animals. When the sequence of TMV coat protein became available, it was possible for the first time to locate the antigenic sites or epitopes of a viral protein at the molecular level. Initial studies focused on two antigenic regions of the coat protein, the C-terminal region (residues 153–158) situated at the surface of virus particles and the disordered loop region (residues 103–112) located in the central hole of the particles and accessible to antibodies only in dissociated protein subunits (**Figure 3**). The C-terminal hexapeptide coupled to bovine serum albumin was used by Anderer in Tübingen to raise antibodies and the resulting antiserum was found to precipitate the virus and neutralize its infectivity. Since both natural peptide fragments and synthetic peptides were used in this work, Anderer and his colleagues should be credited with the discovery that synthetic peptides can elicit antibodies that neutralize the infectivity of a virus. Only when similar results were obtained with animal viruses 15 years later, did the potential of peptides for developing synthetic vaccines become clear.

It has been known since the 1950s that intact TMV particles and dissociated coat protein subunits harbor different types of epitopes recognized by specific antibodies. Certain epitopes present only on virions are constituted by residues from neighboring subunits that are recognized as a single entity by certain antibodies; other epitopes arise from conformational changes in the protein that result from intersubunit bonds. Both these types of epitopes which depend on the protein quaternary structure and are absent in dissociated protein subunits have been called neotopes. Another type of epitope known as cryptotope occurs on the portion of the protein surface that is buried in the polymerized rod and becomes accessible to antibodies only in the dissociated subunits.

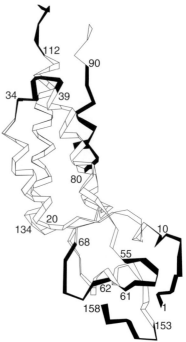

Figure 3 Backbone of the TMV coat protein subunit based on crystallographic data. Residues 94–106 have been omitted because this region, located on the inside of the particle, is disordered. The N and C termini of the protein are located on the outer surface of TMV particles. The position of seven continuous epitopes (residues 1–10, 34–39, 55–61, 62–68, 80–90, 108–112, and 153–158) is indicated by solid lines. Reproduced from Al Moudallal Z, Briand JP, and Van Regenmortel MHV (1985) A major part of the polypeptide chain of tobacco mosaic virus is antigenic. *EMBO Journal* 4: 1231–1235, with permission from Nature Publishing Group.

The mapping of neotopes and cryptotopes on the surface of TMV coat protein was greatly simplified once monoclonal antibodies (Mabs) became available. Many continuous epitopes were identified at the surface of dissociated coat protein subunits by measuring the ability of peptides to react with Mabs or with antisera (**Figure 3**). These epitopes were found to correspond to regions of the protein shown by X-ray crystallography to possess a high segmental mobility. This correlation between antigenicity and mobility along the peptide chain was also found to exist in other proteins and was used to develop algorithms for predicting the location of epitopes in proteins from their primary structure.

The surface of the protein subunits accessible at either end of the virus particle is different and the one located near the 5′ terminus end of the RNA harbors the two helices corresponding to residues 73–89 and 115–135 (**Figure 3**). Many Mabs specific for this surface have been obtained and in addition to reacting with both ends of nonpolar stacked disks, they were found to block TMV disassembly by sterically preventing the interaction between RNA and ribosomes. It has been suggested that if such antibodies could be expressed in plants, they might be able to control viral infection.

Another interesting immunological phenomenon was discovered when TMV antibodies were analyzed for their ability to react with certain TMV mutants. It was found that all rabbits immunized with TMV induced the formation of heterospecific antibodies, that is, antibodies that were unable to react with the TMV immunogen but recognized the mutant Ni 1927 quite well. When all the antibodies in a TMV antiserum capable of reacting with TMV were removed by cross-absorption with the virus, it was found that the depleted antiserum still reacted strongly with this mutant which had a single proline→leucine exchange. Apparently the removal of the proline at position 156 exposes binding sites for both peptidases and antibodies that are normally out of reach in the wild-type protein structure. The induction of heterospecific antibodies by immunization with TMV is another illustration of the difference between the antigenicity and immunogenicity of proteins. TMV particles possess the immunogenic capacity of eliciting heterospecific antibodies that react with the Ni 1927 mutant but do not have the antigenic capacity of reacting with these antibodies. The reverse situation where an antigenic peptide or protein is able to react with a particular antibody but is unable to induce the same type of antibody when used as immunogen is a more commonly observed phenomenon which greatly hampers the development of synthetic peptide vaccines.

Many of the viruses that are currently classified in the genus *Tobamovirus* were initially considered to be strains of TMV on the basis of similar particle morphology and ability to cross-react with TMV antibodies. Antigenic relationships between different tobamoviruses were quantified using a parameter known as the serological differentiation index (SDI) which is the average number of twofold dilution steps separating homologous from heterologous antiserum titers. A close correlation exists between the antigenic distance of two viruses expressed as SDI values and the degree of sequence difference in their coat proteins. In general, when two viruses differ antigenically by an SDI value larger than 4, they are considered to belong to separate species. This is valid only if relatedness is measured with polyclonal antisera containing antibodies to a range of different epitopes since comparisons made with Mabs specific for a single epitope will emphasize antigenic similarities or differences depending on which particular Mab is used.

Relationships between individual tobamoviruses are nowadays usually assessed by comparisons between viral genomes. Phylogenetic studies have shown that tobamoviruses are very ancient and co-evolved with their angiosperm hosts, which means that they are at least 120 million years old.

Replication and Cell-to-Cell Movement of TMV

TMV-infected tissues contain four viral proteins: the 126 kDa and 183 kDa proteins of the replicase complex, the 30 kDa movement protein, and the 17.6 kDa coat protein. TMV replication is initiated by the translation of the viral RNA to produce the replicase proteins and this leads to the synthesis of minus-sense and plus-sense copies of the RNA. Translation of the viral coat protein gene then occurs leading to the assembly of progeny virus particles from full-length genomic RNA and coat protein. The translation of the coat protein and movement protein is controlled by the production of two separate subgenomic mRNAs, a mechanism later found to occur in many other viruses but first demonstrated with TMV.

The mechanism that allows plant viruses to move from cell to cell in their hosts has remained a mystery for many years. The rigid cellulose-rich walls of plant cells impede intercellular communication which occurs only through tubular connections known as plasmodesmata. However, plasmodesmata are too narrow to allow the passage of virus particles. Studies with the 30 kDa TMV movement protein showed that this protein changed the size exclusion limit of the plasmodesmata, allowing the virus to move through them in the form of a thin, less than 2 nm wide, ribonucleoprotein complex composed of the genomic RNA and the movement protein. Movement proteins have subsequently been found in many other plant viruses and shown to possess RNA-binding properties.

Studies with TMV also revealed how plant viruses encode a suppressor to combat the post-transcriptional gene silencing reaction that plants use to fight virus infection.

Biotechnology Applications of TMV

TMV-resistant plants have been obtained by transforming them with a DNA copy of the TMV coat protein gene. This coat protein-mediated resistance is due to the ability of transgenically expressed coat protein to interfere in transgenic cells with the disassembly of TMV particles. Coat protein-mediated resistance has subsequently been obtained with many positive-sense RNA plant viruses.

A portion of the TMV RNA leader sequence, called omega, was shown to enhance the translation of foreign genes introduced into transgenic plants. This translational enhancer has been incorporated successfully in many gene vectors for a variety of applications.

In another type of application, TMV particles have been used as surface carriers of foreign peptide epitopes for constructing recombinant vaccines and producing them in plants. It was found that the N- and C-termini of the TMV coat protein as well as the surface loop area corresponding to residues 59–65 were able to accept foreign peptide fusions without impairing the ability of the resulting chimeric virus to infect plants systemically. Several experimental vaccines against viral and parasitic infections that are based on genetically engineered TMV particles produced in tobacco are currently under development.

In conclusion, it is clear that a series of historical accidents together with the fact that large amounts of remarkably stable TMV particles could be obtained from infected plants, allowed this virus to play a major role in the development of virology. In addition, studies of TMV also contributed significantly to the advancement of molecular biology and to our understanding of the physicochemical and antigenic properties of macromolecules.

See also: Antigenicity and Immunogenicity of Viral Proteins; Nature of Viruses; Plant Virus Vectors (Gene Expression Systems); *Tobamovirus*; Vaccine Production in Plants; Viral Suppressors of Gene Silencing.

Further Reading

Al Moudallal Z, Briand JP, and Van Regenmortel MHV (1985) A major part of the polypeptide chain of tobacco mosaic virus is antigenic. *EMBO Journal* 4: 1231–1235.

Bos L (1999) Beijerinck's work on tobacco mosaic virus: Historical context and legacy. *Philosophical Transactions of the Royal Society, London, Series B* 534: 675–685.

Calisher CH and Horzinek MC (eds.) *100 Years of Virology. The Birth and Growth of a Discipline*, pp. 1–220. Vienna: Springer.

Creager ANH (ed.) (2002) *The Life of a Virus. Tobacco Mosaic Virus as an Experimental Model*, pp. 1–398. Chicago, IL: University of Chicago Press.

Durham ACH and Klug A (1972) Structures and roles of the polymorphic forms of tobacco mosaic virus protein. *Journal of Molecular Biology* 67: 315–332.

Harrison BD and Wilson TMA (1999) *Tobacco Mosaic Virus: Pioneering Research for a Century. Philosophical Transactions of the Royal Society, London, Series B* 354: 517–685.

Hirth L and Richards KE (1981) Tobacco mosaic virus: Model for structure and function of a simple virus. *Advances in Virus Research* 26: 145–199.

Mahy BWJ and Lvov DK (eds.) (1993) *Concepts in Virology. From Ivanovsky to the Present*, pp. 1–438. Langhorne, PA: Harwood Academic Publishers.

Scholthof K-BG (2004) Tobacco mosaic virus: A model system for plant biology. *Annual Review of Phytopathology* 42: 13–34.

Scholthof K-BG, Shaw JG, and Zaitlin M (1999) *Tobacco Mosaic Virus: One Hundred Years of Contributions to Virology*, pp. 1–256. St. Paul, MN: American Phytopathologial Society Press.

Van Helvoort T (1991) What is a virus? The case of tobacco mosaic disease. *Studies in History and Philosophy of Science* 22: 557–588.

Van Regenmortel MHV (1999) The antigenicity of TMV. *Philosophical Transactions of the Royal Society, London, Series B* 354: 559–568.

Van Regenmortel MHV and Fraenkel-Conrat H (eds.) (1986) *The Plant Viruses. Vol. 2: The Rod-Shaped Plant Viruses*, pp. 1–180. New York, NY: Plenum.

Tobacco Viruses

S A Tolin, Virginia Polytechnic Institute and State University, Blacksburg, VA, USA

© 2008 Elsevier Ltd. All rights reserved.

Glossary

Transient virus vector The use of a virus, such as tobacco mosaic virus, for the expression of a foreign gene following insertion of the sequence at a site in the viral genome.

Virus vector A biological organism, such as an aphid, leafhopper, thrips, whitefly, or fungus, that is capable of specific horizontal transmission of a virus from one plant to another.

Introduction

Tobacco has played a role in agricultural history and in the history of virology. Twenty plant virus names begin with tobacco, indicating that tobacco was a host of great agricultural and economic importance, was susceptible to a large number of viruses, and was the object of early virus research. Early in the last century, when viruses were first discovered and named, research programs emphasized controlling diseases of tobacco, many of which were caused by viruses. This article discusses the role of the tobacco in virology, viruses as causes of tobacco diseases, and some of the major milestones in virology achieved with tobacco viruses.

Tobacco is in the genus *Nicotiana* (family Solanaceae), created in 1565 and named after a French promoter of tobacco, Jean Nicot von Villemain. Although the genus is quite diverse and contains about 100 species, only two species have been extensively cultivated as commercial crops. *Nicotiana rustica* was originally grown by Native Americans in the eastern United States, and was the first tobacco species introduced to England and Portugal. *Nicotiana tabacum* was grown in Mexico and South America, and was introduced to the early Spanish explorers. With the cultivation of the species by the Virginia colony in Jamestown, *N. tabacum* became the preferred tobacco type, and is the predominant species today.

Tobacco is the most widely grown nonfood crop in the world, and is thought to have been used by people in the Americas for smoking and chewing since 1000 BC. It played a major role in the colonization of the Americas by Europeans, and was an economic driver for nearly four centuries. Although always a controversial and exotic commodity, tobacco became established in Europe before coffee, tea, sugar, and chocolate. In spite of the human health risks of tobacco products, tobacco continues to be grown throughout the world.

Three major types of the *N. tabacum* predominate, each with different characteristics depending on the final use of the tobacco product. The most common type is Virginia or flue-cured tobacco, which is used for cigarette, pipe, and chewing tobacco. It is grown worldwide, with the USA and China as major producers. Burley tobacco, the second-most popular type, is air-cured and used in chewing tobacco, and is flavored and blended for American-type cigarettes and pipe tobacco. Burley tobacco is also widely grown, but in not as many countries as is Virginia tobacco. A third type, Oriental or Turkish tobacco, is a small-leafed tobacco and is sun-cured and used in English blends. The name is derived from the eastern Mediterranean area where it is grown. Oriental cultivars Samson from Turkey and Xanthium from Greece have been widely used in virus research.

Viruses as Causal Agents of Tobacco Diseases

There are 20 viruses that were either first isolated from tobacco or contain tobacco as the host in their name (TEV, TWV) (**Table 1**). Chronologically these reports span over a century. Seven viruses naturally infect tobacco, but were first isolated from other crops and named from them (**Table 2**). The viruses are discussed individually, but some are synergistic in mixed infections and cause symptoms more severe than either virus alone. Satellites viruses and RNAs associated with tobacco viruses were the first to be recognized as subviral entities (**Table 3**). The greatest recent new activity with tobacco viral diseases has been in understanding the complexity of the viruses with similarities to tobacco leaf curl virus (family *Geminiviridae*) and their worldwide distribution.

Tobacco in Virology History

Tobacco was used as a host plant for virology research leading to many fundamental discoveries, only a few of which are highlighted here. *Nicotiana* species widely used for virus propagation and host range tests are *N. clevelandii*, *N. tabacum*, *N. glutinosa*, *N. megalosiphon*, *N. sylvestris*, and, more recently, *N. benthamiana*. These species are reported as being experimentally susceptible to 267, 231, 201, 93, 57, and 141 viruses, and insusceptible to 80, >200, 175, 35, 35, and 40 viruses, respectively. Responses of tobacco species and cultivars to mechanical inoculation are used for virus and strain identification. Cytopathic effects and inclusion bodies induced by plant viruses were first observed by light microscopy with X-bodies of TMV and nuclear and cytoplasmic inclusions of TEV and PVY. Details of plant virus replication processes were advanced by studies using tobacco leaf protoplasts. Tobacco was also the first plant to be genetically transformed by recombinant DNA techniques, and the first in which pathogen-derived resistance was demonstrated by the insertion of the coat protein gene from TMV. Tobacco transformed with noncoding viral sequences of TEV led to the discovery of RNAi and viral-induced gene silencing (VIGS).

Viruses Causing Economically Important Diseases of Tobacco

Tobamoviruses

The causal agent of a mosaic disease of tobacco, tobacco mosaic virus (TMV, genus *Tobamovirus*) became the first virus to be purified, crystallized, and characterized. Milestones in virology history that have been accomplished with TMV have been well documented, and will not be repeated here. The virus has a rigid, rod-shaped particle constructed of a single capsid protein arranged helically around a single strand of RNA encoding four proteins, functions of which are well characterized in viral replication and movement. Transient vectors can be constructed from TMV, permitting insertion and expression of foreign genes for new products, or for labeling virus for pathogenicity and cellular biology experiments.

The mosaic disease occurs worldwide wherever tobacco is grown. The stability of the virus in tobacco products, its highly infectious nature, and efficient inoculation by mechanical abrasion contributes to the difficulty tobacco farmers have had in controlling this virus. In certain cultivars of burley and ornamental tobaccos, the *N* gene for resistance has provided durable resistance for nearly 75 years, considered remarkable for an RNA virus, as resistance-breaking strains have not developed. Use of the *N* gene in Virginia or flue-cured tobacco, however, has not been successful because of poor-quality traits associated with linked genes. Efforts to resolve this using hybrids have only been moderately successful because systemic necrosis can develop in field-infected plants and cause more serious loss than that caused by mosaic.

Other tobamoviruses infecting tobacco are tomato mosaic virus (ToMV) and tobacco mild green mosaic virus (TMGMV). ToMV and TMV are seed-borne on tomato seed, but reportedly not on tobacco seed. However, production of tobacco seedlings in float trays in

Table 1 Chronology of discovery and naming of tobacco viruses

Virus name	Virus acronym	Genus/family	Year described	Crop of origin	Country	Investigator	Vector taxon
Tobacco mosaic virus	TMV	*Tobamovirus*	1892, 1896 1914	Tobacco Tobacco	USSR USA	Iwanowski Allard	None
Tobacco etch virus	TEV	*Potyvirus/Potyviridae*	1921	*Datura* spp.	USA	Blakeslee	Aphid
Tobacco ringspot virus	TRSV	*Nepovirus / Comoviridae*	1927	Tobacco	USA (VA)	Fromme	Nematode
Tobacco mild green mosaic virus	TMGMV	*Tobamovirus*	1929	*N. glauca*	Canary Islands	McKinney	None
Tobacco rattle virus	TRV	*Tobravirus*	1931	Tobacco	Germany	Boning	Nematode
Tobacco leaf curl virus	TLCV	*Begomovirus / Geminiviridae*	1931	Tobacco	Tanzania	Storey	Whitefly
Tobacco necrosis virus	TNV	*Necrovirus Tombusviridae*	1935	Tobacco	UK	Smith & Bald	Fungus
Tobacco streak virus	TSV	*Ilarvirus / Bromoviridae*	1936	Tobacco	USA (WI)	Johnson	Thrips
Tobacco yellow dwarf virus	TYDV	*Mastrevirus/ Geminiviridae*	1937	Tobacco	Australia	Hill	Leafhopper
Tobacco mottle virus + Tobacco vein distorting virus	TMoV + TVDV	*Umbravirus + Luteovirus*	1946	Tobacco	Zimbabwe	Smith	Aphid
Tobacco stunt virus	TStV	*Varicosavirus*	1950	Tobacco	Japan	Hidaka	*Olpidium brassicae*
Tobacco wilt virus	TWV	*Potyvirus/Potyviridae*	1959	*Solanum jasminoides*	India		Aphid
Tobacco bushy top virus	TBTV	*Umbravirus*	1962	Tobacco	Rhodesia		Aphid
Tobacco yellow vein virus + Tobacco yellow vein assistor virus	TYV + TYVAV	*Umbravirus + Luteoviridae*	1972	Tobacco	Malawi	Adams & Hull	Aphid
Tobacco vein mottling virus	TVMV	*Potyvirus/Potyviridae*	1972	Tobacco	USA (NC)	Gooding	Aphid
Tobacco necrotic dwarf virus	TNDV	*Luteoviridae*	1977	Tobacco	Japan	Kubo	Aphid
Tobacco apical stunt virus	TASV	*Begomovirus/ Geminiviridae*	1999	Tobacco	Mexico	Brown	Whitefly
Tobacco curly shoot virus	TCSV	*Begomovirus/ Geminiviridae*	2002	Tobacco	China (Yunnan)	Zhou	Whitefly

Table 2 Additional viruses reported to naturally occur in tobacco

Virus name	Virus acronym	Genus/family	Year described	Crop of origin	Country	Investigator	Vector taxon
Cucumber mosaic virus	CMV	Cucumovirus/ Bromoviridae	1916	Cucumber	USA	Doolittle Jagger	Aphid
Tomato mosaic virus	ToMV	Tobamovirus	1919	Tomato	USA	Clinton	None
Tomato spotted wilt virus	TSWV	Tospovirus/ Bunyaviridae	1930	Tomato	India	Samuel	Thrips
Alfalfa mosaic virus	AMV	Alfamovirus/ Bromoviridae	1931	Alfalfa	USA	Weime	Aphid
Potato virus Y	PVY	Potyvirus/Potyviridae	1931	Potato	USA	Smith	Aphid
Eggplant mottled dwarf virus	EMDV	Nucleorhabdovirus/ Rhabdoviridae	1969	Eggplant	Italy	Martelli	Cicadellid leafhopper
Eggplant mosaic virus – tobacco strain	EMV-T	Tymovirus/ Tymoviridae	1996	Tobacco	Brazil	Ribiero	Chrosomelid beetle

Table 3 Satellite viruses and nucleic acids associated with tobacco viruses

Satellite name	Helper virus	Year described	Crop of origin	Investigator – country	Characteristics
Tobacco necrosis satellite virus	TNV	1960	Tobacco	Kassanis – UK	18 nm, $T=1$
Tobacco mosaic satellite virus	TMV	1986	N. glauca	Dodds – US	17 nm, $T=1$; 1059 nt
Tobacco bushy top virus satellite RNA	TBTV	2001	Tobacco	China	Linear, 650–800 nt
Tobacco ringspot virus satellite RNA	TRSV	1972	Bean	Schneider	Circular, 359 nt, viroid-like
Cucumber mosaic virus satellite RNA	CMV	1976	Tobacco	Kaper – Italy	Linear, 330–390 nt
Tobacco leaf curl virus DNA β	TLCV	2003	Tobacco	Zhou – China	Circular, 1300–1350 nt
Tobacco curly shoot virus DNA β	TCSV	2005	Tobacco	Li – China	Circular

greenhouses suggests anecdotally that TMV-infested seed may provide inoculum source for seedling infections. Instances of spread of ToMV from tomato seedlings to tobacco seedlings have been observed. ToMV causes more severe symptoms than does TMV. TMGMV naturally infects *N. glauca*, and is often associated with an icosahedral satellite virus, satTMV (**Table 3**).

Bromoviruses

Certain viruses of tobacco are the type members of three genera of the family *Bromoviridae*, namely cucumber mosaic virus (CMV, genus *Cucumovirus*), alfalfa mosaic virus (AMV, genus *Alfamovirus*), and tobacco streak virus (TSV, genus *Ilarvirus*). These viruses continue to occur in tobacco worldwide. Disease outbreaks are sporadic, but are of economic significance in fields infected at a high incidence. All three viruses have very wide host ranges, are known to infect a number of weeds and native plants naturally, and are seed transmitted in many of their host plants. CMV and AMV are transmitted in a stylet-borne, nonpersistent manner by many species of aphids. TSV, however, is spread by contact and through pollen carried by thrips.

Common properties include a tripartite, single-stranded RNA (ssRNA) genome, and encapsidation of a fourth subgenomic RNA encoding the capsid protein. Particles have icosahedral symmetry, but AMV forms multiple bacilliform particles and TSV forms quasi-icosahedrons. CMV capsids are of $T=3$ symmetry.

Cucumber mosaic and alfalfa mosaic viruses

Symptoms caused by CMV are similar to those of TMV. CMV is an important tobacco pathogen in Asia and in certain European countries. In recent surveys in Greece, burley, Virginia, and oriental types all had a high incidence of CMV, even up to 100% infected plants in some fields. Symptoms of AMV on tobacco are a bright yellowing of leaf areas, broad rings, and mosaic on young leaves. AMV incidence was lower, but was highest in tobacco growing near alfalfa fields, a source of the virus.

Many fundamental discoveries of genome structure, organization, and encapsidation occurred with AMV and CMV in the 1960s and 1970s using tobacco as the main test species. Sucrose density-gradient centrifugation enabled separation of the multiple particles of AMV, leading to experiments to demonstrate the requirement of RNA1, RNA2, and RNA3 plus either RNA4 or free coat protein for infectivity. Genetic experiments were conducted by mixed inoculations with RNA species from biologically distinct strains. Similar experiments were conducted with CMV, which required separation on cesium chloride gradients based on density to demonstrate separate encapsidation of RNA1, RNA2, and RNA3 + RNA4 in a third capsid. Infectivity required RNAs 1, 2, and 3, but not RNA4 or coat protein. A fifth, small RNA species, initially termed RNA5 or CARNA5 (CMV-associated RNA), was later shown to be the first satellite RNA species. These satellite RNAs can either ameliorate or exacerbate symptoms in crops. Severe lethal necrosis symptoms are associated with CMV satellite RNA in tomato in Italy and tobacco in India.

Tobacco streak virus

TSV is rarely reported on tobacco, but is an important pathogen of many annual and vegetatively propagated crops and wild species worldwide. In tobacco, TSV causes necrotic lines along veins of leaves as virus moves systemically from the point of inoculation. Symptoms develop on a few leaves, and then plants recover. Incidence of infected tobacco plants is usually low. Thrips are recognized as a vector of TSV. TSV belongs to subgroup 1 of the genus *Ilarvirus*, which is divided into six subgroups based on serological relatedness.

Potyviruses

Three viruses of the genus *Potyvirus* (family *Potyviridae*), tobacco etch virus (TEV), potato virus Y (PVY), and tobacco vein mottling virus (TVMV), are among the most important viral pathogens of tobacco. All are aphid transmitted, with flexuous filamentous particles $c.\ 13 \times 750$ nm containing an ssRNA genome of about 10 kb. Successful aphid transmission depends upon a DAG sequence in the N-terminus of the coat protein and a viral sequence encoding a helper component (HC). The genome is expressed as a polyprotein, which is cleaved by viral-encoded proteases (Pro) into a single structural and several nonstructural proteins. Both TEV and PVY occur nearly worldwide and have wide host ranges. Diseases caused by potyviruses are managed in part through resistant cultivars such as TN86, a burley tobacco resistant to TVMV, TEV, and PVY. Several burley varieties have partial or high resistance to one or more viruses, as well as resistance to fungi and nematodes.

Tobacco etch virus

TEV has a rather narrow host range, but is found in solanaceous weeds including *Datura* sp., from which it was first isolated in the 1920s. The virus is common only in North and South America. Isolates vary in symptom severity and ability to cause necrosis or severe versus mild etch in tobacco. TEV is also known to cause severe symptoms in pepper and tomato crops in the USA, Central America, and the Caribbean. The few isolates that have been sequenced show very little diversity, and are indistinguishable serologically.

TEV has also been a model virus in the elucidation of many host–virus interactions and viral genome expression. The 5′ leader of TEV genomic RNA is used in many constructs for genetic engineering, as it contains an internal ribosome entry site and directs efficient translation of uncapped mRNA. TEV was the first virus demonstrated to encode proteases, including the autocatalytic small nuclear inclusion proteinase and a protease linked to helper component (HC-Pro).

Transgenic resistance to TEV was the first to be due to RNA interference and gene silencing, as nontranslatable sequences of the capsid protein-encoding region of the viral genome were more effective than the capsid protein translatable sequences. This observation led to the discovery RNA interference as well as the silencing suppression function by the HC-Pro region of the genome.

Potato virus Y

PVY, first isolated from potato in the 1930s as one of a complex of viruses associated with the century-old potato degeneration, is one of the most studied plant viruses and is the type member of the genus *Potyvirus*. PVY remains an important pathogen of potato and also of tobacco. Severe epidemics of PVY in tobacco have been reported recently in China. In Greece, PVY is one of the most prevalent viruses isolated from seedbeds and fields. Symptoms associated with PVY are vein clearing and/or top necrosis in oriental tobacco, and yellowing with veinal necrosis in Virginia and burley tobacco. PVY has a wide host range among the Solanaceae, including 70 species of *Nicotiana* and is also found in annual and perennial weeds in 15 or more plant families. A number of strains have been described both on potato and on tobacco, distinguishable by inoculation to differential hosts and by sequence analysis.

Other potyviruses

TVMV was not described until the 1970s, when it emerged as a new virus in the burley tobacco-growing regions of the USA, namely Kentucky, North Carolina, Tennessee, and Virginia. Severe losses were recorded until resistance to this virus, and partial resistance to PVY and TEV, was incorporated into burley tobacco cultivars. A virus isolated from tobacco in Nigeria was identified as being closely related to the potyvirus pepper veinal mottle virus, and both reacted with TVMV antibody in electron microscope serology, but not enzyme-linked immunosorbent assay (ELISA). However, this reaction alone is not sufficient to prove virus identity. Only one additional potyvirus, tobacco wilt virus (TWV), has been reported from tobacco, and it is only known in India.

Tospoviruses

Tomato spotted wilt virus (TSWV), a member of the genus *Tospovirus* (family *Bunyaviridae*), is a serious pathogen on tobacco in many regions of the world. This thrips-transmitted virus has a very wide host range infecting over 800 species in 90 or more plant families, both monocots and dicots, and causes significant yield losses in a number of economically important crops. The virus has spherical, membrane-bound particles, 80–120 nm, with a three-segment, ambisense genome encapsided in the same particle. Symptoms in tobacco infected early are severe leaf necrosis and stunting. Several management approaches have involved weed reservoir and vector management. In Georgia in the southeastern US, successful management has been accomplished with compounds to induce resistance combined with imidochloprid insecticide applications in seedling trays and at transplanting. Yield losses were reduced significantly.

Geminiviruses (Family *Geminiviridae*)

Mastreviruses

Tobacco yellow dwarf virus (TYDV) was described in Australia in 1937, and later determined to be one of only three dicot-infecting members of the genus *Mastrevirus* (family *Geminiviridae*). It is yet to be recorded outside of Australia where it causes severe dwarfing and downward leaf curl in tobacco, and a lethal, necrotic disease in bean (syn. bean summer death virus). The virus is transmitted persistently by the brown leafhopper *Orosius argentatus*, but not by other leafhoppers nor mechanically. TYDV also infects several summer annual and autumn–spring growing plants, providing a continuous succession of host plants for the virus, and a source of virus for tobacco. Like other mastreviruses, TYDV has a single-stranded DNA genome, single component, and has coding regions in both virion sense and complementary sense sequences.

Begomoviruses

The tobacco leaf curl disease was first described in Tanzania in the early 1930s, in which infected tobacco plants were severely crinkled, curled, and dwarfed. The disease was successfully transmitted by the whitefly *Bemisia tabaci*, but not mechanically, suggesting that a virus was the causal agent. Similar disease symptoms in tobacco were noted as early as 1902 in South Africa, in 1912 in the Netherlands East Indies, in the 1930s in India, and in Brazil and Venezuela prior to 1950. Positive association of the tobacco leaf curl disease with a begomovirus was not demonstrated until DNA analyses and polymerase chain reaction (PCR) became available. The name tobacco leaf curl virus (TLCV, genus *Begomovirus*, family *Geminiviridae*) is now accepted for this virus.

However, new viruses are being described worldwide in association with severely diseased tobacco. There are numerous recent reports of TLCV-like viruses from Asia, including Japan, China, and India, as well as from Dominican Republic and Cuba. Those viruses are now accepted as members of the genus *Begomovirus*: tobacco leaf curl Japan virus (TbLCJV), tobacco leaf curl Kochi virus (TbLCKoV), tobacco leaf curl Yunnan virus (TbLCYNV), tobacco leaf curl Zimbabwe virus (TbLCZV), and tobacco curly shoot virus (=tobacco leaf curl – China, TbCSV). Sequences have also been reported for tobacco leaf curl virus – Karnataka 1 and 2 from India. Tentative begomovirus members include tobacco apical stunt virus, tobacco leaf curl India virus, and tobacco leaf rugose virus – Cuba.

All TLCV-like viruses have a single-stranded, monopartite DNA genome, and have only the DNA-A component. TbLCYNV was the first virus to be shown to have a DNAβ component, a type of satellite DNA widely associated with the virus in the Yunnan Province of China and with TbCSV (**Table 3**). Like other satellites, DNA β molecules have no sequences in common with their helper virus, DNA-A, which replicates independently of DNA β. With TbCSV, DNA β affects pathogenicity, with increasing symptom severity in the presence of specific molecules.

Viruses Causing Diseases of Tobacco of Lesser or Minor Importance

Nepoviruses

Tobacco ringspot virus (TRSV) caused a severe disease of tobacco in Virginia in the 1920s. Symptoms initially were severe necrotic rings with green centers, and a pattern of necrotic lines described as an 'oak-leaf' pattern. Tobacco plants expressed symptoms on three to four leaves developing following inoculation, but not on later-developing leaves. This apparent recovery was the first phenomenon to be described as acquired immunity. In recent research, recovery from symptoms induced by TRSV was the first to be attributed to plant defenses and to involve RNAi. The ssRNA genome is bipartite and encapsidated in separate icosahedral particles, typical of members of the family *Comoviridae*. Additional small RNAs found in the 1970s to also be encapsidated in TRSV particles were the first satellite RNAs to be discovered (**Table 3**) and later shown to have hammerhead, ribozyme-like structures capable of self-cleavage. These structures are also found in certain viroids.

TRSV is the type species of the genus *Nepovirus*. It is classified in subgroup A of this genus based on RNA2 size and sequence, and on serological relationships. The genus is named from the property of being nematode transmitted (by *Xiphenema* spp.) and having polyhedral particles. The virus is spread naturally in North America by these nematodes as well as by possible arthropod vectors. It has a very wide host range and is found naturally in many weeds and woody species. Additionally, TRSV has been disseminated worldwide in vegetative propagules of perennial crops and ornamentals, but does not appear to spread in the absence of the nematode vector.

Artichoke yellow ringspot virus (AYRSV) is known to infect tobacco naturally, causing chlorotic blotches, rings, and lines, and to spread naturally in Greece, Italy, and Eurasia. It is reported to spread via pollen to seeds or seed parent plants of *N. clevelandii*, *N. glutinosa*, and *N. tabacum*. Serological evidence indicates that AYRSV is a distinct species, and RNA2 properties place it in the subgroup C of nepoviruses, which includes tomato ringspot and cherry leaf roll viruses.

Tobraviruses

Tobacco rattle virus (TRV, genus *Tobravirus*), first described on tobacco in the 1930s, is now seldom reported on tobacco crops. Symptoms of rattle on tobacco are systemic necrotic flecks and line patterns, and death or stunting of shoots. It was known as *Mauche, ratel, Streifen und Kräuselkrankheit* in Germany, its geographic origin. Distribution today is limited to Europe, Japan, New Zealand, and North America. The host range of TRV is very wide, as it infects over 400 species of both dicots and monocots, many of which are weeds and other wild plants. TRV causes important diseases of potato, pepper, and various ornamentals.

Biological vectors of TRV are nematodes in the genera *Paratrichodorus* and *Trichodorus* (Trichodoridae). *Nicotiana clevelandii* is recommended as a propagative host for TRV, and as a trap plant for testing nematode transmission characteristics. Transmissibility is mediated by read-through proteins encoded on RNA-2 of the bipartite, ssRNA genome which extend from the surface of rigid rod-shaped particles of various sizes. This region of RNA-2 has been exploited for expression of introduced genes, and the virus is now widely used as a virus-induced gene silencing (VIGS) vector for genetic experiments.

Necroviruses

Tobacco necrosis virus (TNV) is no longer associated with diseases of tobacco, but rather with bean stipple streak and tulip necrosis diseases. TNV is classified in the genus *Necrovirus* (family *Tombusviridae*). It is one of the few viruses transmitted by the fungus *Olpidium brassicae* and often occurs in irrigated soils and in greenhouses. Virus particles adsorb to the surface of resting spores, on which they do not survive for long periods. Symptoms on tobacco are localized necrosis, with very little systemic movement within infected plants. Electron microscopy of purified TNV revealed 26–28 nm isometric particles of $T=3$ symmetry. Some cultures also had 17 nm particles, which were a satellite virus (satTNV) encoding a capsid protein that assembled into a $T=1$ particle. satTNV is found in Europe and North America in association with bean, but not with tobacco.

Varicosaviruses

Tobacco stunt virus (TStV) is reported only from Japan, where it was found in the 1950s to cause severe stunting and necrosis in tobacco. This virus was among the first to be shown to be transmitted by a fungus, *O. brassicae* (order Chytridiales), in which it is borne internally in resting spores and persists for 20 years or more. Once soil is infested with spores containing virus, tobacco planted into the soil will become infected and show symptoms within 2 weeks. The RNA genome of TStV is encapsided as two segments in straight, rod-shaped, rather labile, particles. The genome is now classified as being negative sense and single-stranded, with genome organization like that of rhabdoviruses. Taxonomically, TStV is in the genus *Varicosavirus*, whose name comes from the enlarged (varicose) veins induced in lettuce by the virus now known as lettuce big-vein associated virus. Recent molecular analyses suggest that TStV is a strain of this virus, rather than a distinct species.

Luteoviruses and Umbraviruses

Diseases caused by luteoviruses alone

Tobacco necrotic dwarf virus (TNDV) of the genus *Enamovirus* (family *Luteoviridae*) causes severe disease symptoms in some tobacco fields in Japan. It is transmitted by the aphid *Myzus persicae* in a persistent manner, but not mechanically. Early infection causes stunting and premature yellowing or death of young leaves of tobacco, but distribution is apparently limited.

Tobacco rosette and other diseases with dependent transmission

An aphid-transmitted disease known as tobacco rosette was widespread in Zimbabwe in sub-Saharan Africa in the 1930s. Smith in 1946 discovered that the sap-transmissible tobacco mottle virus (TMoV) could only be transmitted by aphids if plants were co-infected with tobacco vein distorting virus (TVDV), making tobacco rosette the first disease complex associated with dependent transmission. TVDV, a member of the genus *Enamovirus*, is the assistor virus for TMoV, a presumed member of the genus

Umbravirus based on sap transmissibility and other biological properties. Sequence data of one of the two recognized strains of TMoV confirmed this assignment in 2001.

Two other diseases of tobacco have complex etiology presumed to involve an umbravirus–luteovirus complex. Tobacco yellow vein virus (TYVV), a tentative species of the genus *Umbravirus*, and tobacco yellow vein assistor virus, a tentative member of the family *Luteoviridae*, were described as causing a tobacco disease with chlorotic vein-banding and leaf malformation symptoms in Malawi. Tobacco bushy top disease occurring in Rhodesia since the early 1960s is known to be caused by TVDV and a different umbravirus, tobacco bushy top virus (TBTV), and is characterized by pale green, mottling, and extreme shoot proliferation. TBTV and TVDV have recently been detected in China in tobacco showing severe witches' broom symptoms, the etiology of which had been presumed to be phytoplasma. Sequence data from these viruses confirm their classification. The significance of this observation is that this is the first incidence of an umbravirus–luteovirus complex reported in China, and first outside of sub-Saharan Africa.

Tymoviruses

A virus was isolated from naturally infected tobacco in Brazil in 1996, and described as eggplant mosaic virus–tobacco strain (EMV-T), the first report of a virus of the genus *Tymovirus* (family *Tymoviridae*) from tobacco. The virus caused mosaic symptoms, but did not reduce plant productivity. It was experimentally transmitted by sap and by the chrysomelid beetle, *Diabrotica speciosa*. As this is the only report of its incidence in tobacco, EMV-T is of only minor importance in tobacco in a limited geographic region. EMV is a pathogen of eggplant and other solanaceous vegetables, and was limited in distribution to Trinidad and Tobago. It is closely related to Andean potato latent virus, but is considered a distinct species. Genomic relationships between EMV and EMV-T have not been reported.

Nucleorhabdoviruses

Eggplant mottled dwarf virus (EMDV, genus *Nucleorhabdovirus*, family *Rhabdoviridae*) was found in tobacco plants in Greece. Plants showed severe stunting, leaf deformation, and vein clearing, and were not infected by other common tobacco viruses. Incidence was at less than 0.01%, mainly in plants near the edges of the field suggesting a weed reservoir for the virus, which has a cicadellid leafhopper vector, *Agallia vorobjevi*. EMDV has been known in the Mediterranean area since 1969 and was reported from tobacco in Italy in 1996 as tobacco vein yellowing virus, now a synonym of EMDV. EMDV infects other solanaceous crops, and appears to be spreading to a wider geographical range from Spain to the Middle East.

See also: Alfalfa Mosaic Virus; Cucumber Mosaic Virus; Emerging and Reemerging Virus Diseases of Plants; History of Virology: Plant Viruses; *Ilarvirus*; Plant Resistance to Viruses: Engineered Resistance; Plant Virus Diseases: Economic Aspects; Potato Virus Y; Ribozymes; Satellite Nucleic Acids and Viruses; Tobacco Mosaic Virus; Tomato Spotted Wilt Virus; Vector Transmission of Plant Viruses; Viral Suppressors of Gene Silencing; Virus Induced Gene Silencing (VIGS).

Further Reading

Adams MJ and Antoniw JF (2005) DPVweb: An open access Internet resource on plant viruses and virus diseases. *Outlooks on Pest Management* 16: 268–270.

Adams MJ, Antoniw JF, and Fauquet CM (2005) Molecular criteria for genus and species discrimination within the family. *Potyviridae Archives for Virology* 150: 459–479.

Brunt AA, Crabtree K, Dallwitz MJ, Gibbs AJ, Watson L, and Zurcher EJ (eds.) (1996 onward) Plant viruses online: Descriptions and lists from the VIDE Database. Version: 20 Aug. 1996. http://image.fs.uidaho.edu/vide/ (accessed May 2007).

Chatzivassiliou IK, Efthimiou K, Drossos E, Papadopoulou A, Poimenidis G, and Katis NI (2004) A survey of tobacco viruses in tobacco crops and native flora in Greece. *European Journal of Plant Pathology* 110: 1011–1023.

Cui XF, Xie Y, and Zhou XP (2004) Molecular characterization of DNAβ molecules associated with tobacco leaf curl Yunnan virus. *Journal of Phytopathology* 152: 647–650.

Harrison BD and Wilson TMA (1999) Milestones in the research on tobacco mosaic virus. *Philosophical Transactions of the Royal Society of London, Section B* 345: 521–529.

Lindbo JA and Dougherty WG (2005) Plant pathology and RNAi: A brief history. *Annual Review of Phytopathology* 43: 191–204.

Mansoor S, Briddon RW, Zafir Y, and Stanley J (2003) Geminivirus disease complexes: An emerging threat. *Trends in Plant Science* 8: 128–134.

Paximadis M, Idris AM, Torres-Jerez I, Villareal A, Rey MEC, and Brown JK (1999) Characterization of tobacco geminiviruses in the Old and New World. *Archives of Virology* 144: 703–717.

Ratcliff F, Martin-Hernandez AM, and Baulcombe DC (2001) Tobacco rattle virus as a vector for analysis of gene function by silencing. *Plant Journal* 25: 237–245.

Sasaya T, Ishikawa K, Kuwata S, and Koganezawa H (2005) Molecular analysis of coat protein coding region of tobacco stunt virus shows that it is a strain of *Lettuce big-vein virus* in the genus Varicosavirus. *Archives of Virolology* 150: 1013–1021.

Shew HD and Lucas GB (eds.) (1991) *Compendium of Tobacco Diseases*, 96pp. St. Paul, MN: APS Press.

Simon AE, Roossinck MJ, and Havelda Z (2004) Plant virus satellite and defective interfering RNAs: New paradigms for a new century. *Annual Review of Phytopathology* 42: 415–437.

Syller J (2003) Molecular and biological features of umbravirus, the unusual plant viruses lacking genetic information for a capsid protein. *Physiological and Molecular Plant Pathology* 63: 35–46.

Xie Y, Jiang T, and Zhou X (2006) Agroinoculation shows tobacco leaf curl Yunnan virus is a monopartitie begomovirus. *European Journal of Plant Pathology* 115: 369–375.

Relevant Website

http://www.dpvweb.net – DPVWeb Home Page.

Tobamovirus

D J Lewandowski, The Ohio State University, Columbus, OH, USA

© 2008 Elsevier Ltd. All rights reserved.

Glossary

Origin of assembly Stem–loop structure that is the site of initiation of virion assembly.

Pseudoknot An RNA structure with base pairing between a loop and other regions of the RNA.

Introduction

Early research in the late 1800s on the causal agent of the mosaic disease of tobacco led to the discovery of viruses as new infectious agents. Thus tobacco mosaic virus (TMV), the type species of the genus *Tobamovirus*, became the first virus to be discovered, and subsequently has had a significant role in many fundamental discoveries in virology. The first quantitative biological assay for plant viruses was the use of *Nicotiana glutinosa* plants, which produce necrotic local lesions when inoculated with TMV and many tobamoviruses. The resistance gene *N* that confers this hypersensitive response-type resistance to TMV was the first resistance gene against a plant virus to be cloned and characterized. TMV was the first virus to be purified and crystallized, which led to the discovery of the nucleoprotein nature of viruses and determination of the atomic structure of the coat protein and the virion. TMV was the first virus to be visualized in the electron microscope, confirming the predicted rigid rod-shaped virions. The genetic material of TMV was shown to be RNA, a property previously thought to be restricted to DNA. The first viral protein for which an amino acid sequence was determined was the coat protein of TMV. TMV was the first virus to be mutagenized and the subsequent determination of coat protein sequences from a number of strains and mutants helped to establish the universality of the genetic code. Methods of infecting plant protoplasts with viruses were developed with the tobacco–TMV system, creating a synchronous system to study events in the infection cycle. The TMV 30 kDa protein was the first viral protein shown to be required for virus movement.

Taxonomy and Classification

The genus *Tobamovirus* has not been assigned to a family. Currently, there are 23 recognized species within the genus *Tobamovirus* (**Table 1**). Several recently sequenced viruses are tentative members of new species (**Table 1**). Although tobamoviruses comprise one of the more intensively studied plant virus genera, taxonomy has often been confusing. Historically, plant viruses with rigid virions of approximately 18×300 nm^2 and causing various diseases were all designated strains of TMV. Thus, many viruses originally referred to as TMV strains are now recognized as belonging to separate species. For example, the tobamovirus that was referred to as the tomato strain of TMV, and is approximately 80% identical to TMV at the nucleotide sequence level, is actually tomato mosaic virus. One criterion for distinguishing the members of separate tobamovirus species is a nucleotide sequence difference of at least 10%.

Virus Structure and Composition

Tobamovirus virions are straight tubes of approximately 18×300 nm^2 with a central hollow core 4 nm in diameter. Virion composition is approximately 95% protein and 5% RNA. For TMV, approximately 2100 subunits of a single coat protein are arranged in a right-handed helix around a single genomic RNA molecule, with each subunit associated with three adjacent nucleotides. Protein–protein associations are the essential first event of virion assembly. Coat protein subunits assemble into several types of aggregates. Coat protein monomers and small heterogeneous aggregates of a few subunits are collectively referred to as 'A-protein'. The equilibrium between A-protein and larger aggregates is primarily dependent upon pH and ionic strength. Purified coat protein and viral RNA can assemble into infectious particles *in vitro*. Larger aggregates are disks composed of two individual stacked rings of coat protein subunits, and protohelices. Protohelices contain approximately 40 coat protein subunits arranged in a spiral around a central hollow core, similar to the arrangement within the virion. A sequence-specific stem–loop structure in the RNA, the origin of assembly (OAS), initiates encapsidation and prevents defective packaging that could result from multiple independent initiation events on a single RNA molecule. Virion assembly initiates as the primary loop of the OAS is threaded through a coat protein disk or protohelix with both ends of the RNA trailing from one side. The conformation of the coat protein protohelix changes as the RNA becomes embedded within the groove between the two layers of subunits. Elongation is

Table 1 Definitive and tentative members of the genus *Tobamovirus*

Cucumber fruit mottle mosaic virus	CFMMV
Cucumber green mottle mosaic virus	CGMMV
Frangipani mosaic virus	FrMV
Hibiscus latent Fort Pierce virus	HLFPV
Hibiscus latent Singapore virus	HLSV
Kyuri green mottle mosaic virus	KGMMV
Maracuja mosaic virus	MaMV
Obuda pepper virus	ObPV
Odontoglossum ringspot virus	ORSV
Paprika mild mottle virus	PaMMV
Pepper mild mottle virus	PMMoV
Ribgrass mosaic virus	RMV
Sunn-hemp mosaic virus	SHMV
Sammons' Opuntia virus	SOV
Tobacco latent virus	TLV
Tobacco mild green mosaic virus	TMGMV
Tobacco mosaic virus	TMV
Tomato mosaic virus	ToMV
Turnip vein-clearing virus	TVCV
Ullucus mild mottle virus	UMMV
Wasabi mottle virus	WMoV
Youcai mosaic virus	YoMV
Zucchini green mottle virus	ZGMMV
Tentative members	
Brugmansia mild mottle virus	
Cucumber mottle virus	
Cactus mild mottle virus	
Streptocarpus flower break virus	
Tropical soda apple mosaic virus	

bidirectional, proceeding rapidly toward the 5′ end of the RNA as the RNA loop is extruded through the elongating virion and additional coat protein disks are added. There is disagreement about the mechanism of elongation toward the 3′ terminus of the RNA, but it appears that this slower process involves the addition of smaller protein aggregates.

Subgenomic mRNAs containing the OAS are encapsidated into shorter virions that are not required for infectivity. The OAS is located within the open reading frame (ORF) for the movement protein of most tobamoviruses. The level of accumulation of a particular subgenomic mRNA containing the OAS determines the relative proportion of that particular virion species. Thus, all tobamovirus virion populations contain a small percentage of movement protein subgenomic mRNAs. In some tobamoviruses, including cucumber green mottle mosaic virus, hibiscus latent Singapore virus (HLSV), kyuri green mottle mosaic virus, maracuja mosaic virus, sunn-hemp mosaic virus, zucchini green mottle mosaic virus, and cactus mild mottle virus, the OAS is located within the coat protein ORF. Thus, these so-called subgroup 2 tobamoviruses produce a significant proportion of small virions that contain coat protein subgenomic mRNA. Hybrid nonviral RNAs containing an OAS will also assemble with coat protein into virus-like particles of length proportional to that of the RNA.

Genome Organization

The genome of tobamoviruses consists of one single-stranded positive-sense RNA of approximately 6300–6800 nt (**Figure 1(a)**). There is a methylguanosine cap at the 5′ terminus, followed by an AU-rich leader 55–75 nt in length. The 3′ nontranslated end of the RNA consists of sequences that can be folded into a series of pseudoknot structures, followed by a tRNA-like terminus. The hibiscus-infecting tobamoviruses HLSV and hibiscus latent Fort Pierce virus contain a polyA stretch between the 3′ end of the coat protein ORF and the tRNA-like structure. The tRNA-like terminus can be aminoacylated *in vitro*, and in most cases specifically accepts histidine. The exception is the 3′ terminus of SHMV, which accepts valine and appears to have arisen by a recombination event between a tobamovirus and a tymovirus.

Four ORFs that are contained within all tobamovirus genomes (**Figure 1(a)**) correspond to the proteins found in infected tissue. Two overlapping ORFs begin at the 5′ proximal start codon. Termination at the first in-frame stop codon produces a 125–130 kDa protein. A 180–190 kDa protein is produced by readthrough of this leaky termination codon approximately 5–10% of the time. The remaining proteins are expressed from individual 3′ co-terminal subgenomic mRNAs, from which only the 5′ proximal ORF is expressed (**Figures 1(c) and 2(a)**). The next ORF encodes the 28–34 kDa movement protein, which has RNA-binding activity and is required for cell-to-cell movement. The 3′ proximal ORF encodes a 17–18 kDa coat protein. A subgenomic mRNA containing an ORF for a 54 kDa protein that encompasses the readthrough domain of the 180–190 kDa ORF has been isolated from infected tissue, although no protein has been detected.

Within the protein-coding regions of the genome, there are nucleotide sequences that also function as *cis*-acting elements for subgenomic mRNA synthesis, virion assembly, and replication. Gene expression from subgenomic mRNAs is regulated both temporally and quantitatively. The movement protein is produced early and accumulates to low levels, whereas the coat protein is produced late and accumulates to high levels. The regulatory elements for subgenomic mRNA synthesis are located on the genome-length complementary RNA overlapping the upstream ORF (**Figure 1(a)**). There is limited (40%) sequence identity between the TMV movement protein and coat protein subgenomic promoters. The TMV movement protein subgenomic promoter is located upstream of the movement protein ORF, flanking the transcription initiation site. Unlike the movement protein subgenomic promoter, full activity of the coat protein subgenomic promoter requires sequences within the coat protein ORF.

Viral Proteins

The tobamovirus 125–130/180–190 kDa proteins are involved in viral replication, gene expression, and movement. Both are contained in crude replicase preparations, and temperature-sensitive replication-deficient mutants map to these ORFs. The 125–130/180–190 kDa proteins contain two functional domains common to replicase proteins of many positive-stranded RNA plant and animal viruses (**Figure 1(b)**). The N-terminal domain has methyltransferase and guanylyltransferase activities associated with capping of viral RNA. The second common domain is a proposed helicase, based upon conserved sequence motifs. The readthrough domain of the 180–190 kDa protein has sequence motifs characteristic of RNA-dependent RNA polymerases. Both proteins are necessary for efficient replication, although the TMV 126 kDa protein is dispensable for replication and gene expression in protoplasts. The 125–130 kDa protein (or sequences within this region) of the 180–190 kDa protein are required for cell-to-cell movement. Additionally, these multifunctional proteins are symptom determinants, as mutations in mild strains map to these ORFs.

The 28–34 kDa movement protein has a plasmodesmatal binding function associated with its C-terminus and a single-stranded nucleic acid-binding domain associated with the N-terminus. The movement protein–host interaction determines whether the virus can systemically infect some plant species. Although principally a structural protein, the coat protein is also involved in other host interactions. Coat protein is required for efficient long-distance movement of the virus. Coat protein is also a symptom determinant in some susceptible plant species and an elicitor of plant defense mechanisms in other plant species.

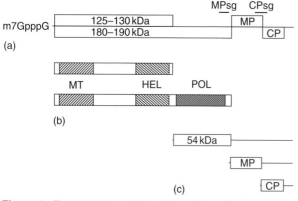

Figure 1 Tobamovirus genome organization and gene expression strategy. (a) Tobamovirus genome organization. ORFs designated as open boxes. Nontranslated sequences designated as lines; positions of subgenomic promoters are marked. (b) Nonstructural proteins involved in tobamovirus replication. Functional domains shared with other viruses within the 'alphavirus supergroup' are designated as hatched boxes. (c) Subgenomic mRNAs with 5′ proximal ORF labeled. MP, movement protein; CP, coat protein; MT, methyltransferase; HEL, helicase; POL, polymerase; MPsg, MP subgenomic mRNA; CPsg, CP subgenomic mRNA.

Interactions between Viral and Host Proteins

Available evidence suggests that the interactions of viral proteins with host factors are important determinants of viral movement and host ranges. Amino acid substitutions in the movement protein and 125–130/180–190 kDa

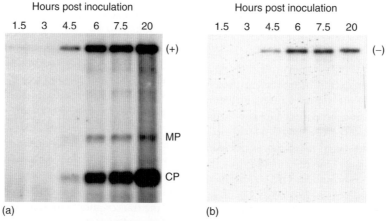

Figure 2 Time course of accumulation of TMV positive-stranded (a) and negative-stranded (b) RNAs in tobacco protoplasts. Total RNA was extracted from tobacco suspension cell protoplasts transfected with TMV *in vitro* transcripts at the time points indicated and analyzed by Northern blot hybridization. (+), TMV genomic RNA; MP, movement protein subgenomic RNA; CP, coat protein subgenomic RNA; (−), negative-stranded complement to the genomic RNA.

proteins can alter the movement function in different hosts. Some viruses, including tobamoviruses, can assist movement of other viruses that are incapable of movement in a particular plant species. These interactions suggest that there are more precise associations of viral proteins with host factor(s) than with viral RNA. Additionally, precise coat protein–plant interactions are required for movement to distal positions within the plant. The helicase domain of the 130–190 kDa proteins elicits the *N* gene-mediated resistance in *N. glutinosa*.

Virus Replication

Virions or free viral RNA will infect plants or protoplasts. Because tobamoviruses have a genome consisting of messenger-sense RNA that is infectious, one of the first events is translation of the 5′-proximal ORFs to produce the proteins required for replication of the genomic RNA and transcription of subgenomic mRNAs. When virions are the infecting agent, the first event is thought to be co-translational disassembly, in which the coat protein subunits at the end of the virion surrounding the 5′ end of the RNA loosen, making the RNA available for translation. Ribosomes then associate with the RNA, and translation of the 126/183 kDa ORFs is thought to displace coat protein subunits from the viral RNA. After the formation of an active replicase complex, a complementary negative-strand RNA is synthesized from the genomic positive-strand RNA template. Negative-strand RNA serves as template for both genomic and subgenomic mRNAs. Negative-strand RNA synthesis ceases early in infection, while positive-strand RNA synthesis continues. This results in an asymmetric positive- to negative-strand RNA ratio. Early in infection, genomic RNA functions as template for negative-strand RNA synthesis and as mRNA for production of the 126/183 kDa proteins. Later in the infection cycle, most of the newly synthesized genomic RNA is encapsidated into virions. Subgenomic mRNAs transcribed during infection function as mRNA for the 3′ ORFs. Within cells of an infected leaf, replication proceeds rapidly between approximately 16 and 96 h post infection within a cell, then ceases. Even though the infected cells become packed with virions, these cells remain metabolically active for long periods. During the early stages of infection of an individual cell, the infection spreads through plasmodesmatal connections to adjacent cells. This event requires the viral movement protein that modifies plasmodesmata to accommodate larger molecules and the 126/183 kDa proteins. Movement through plasmodesmata does not require the coat protein. A second function of the movement protein appears to be binding to the viral RNA to assist its movement through the small plasmodesmatal openings. The movement protein also appears to associate with the cytoskeleton. As the virus spreads from cell to cell throughout a leaf, it enters the phloem for rapid long-distance movement to other leaves and organs of the plant, a complex process that requires the coat protein.

cis-Acting Sequences

The 5′ nontranslated region contains sequences that are required for replication. This region is an efficient translational leader. The 3′ nontranslated region contains *cis*-acting sequences that are involved in replication. Certain deletions within the pseudoknots are not lethal, but result in reduced levels of replication. Exchange of 3′ nontranslated elements between cloned tobamovirus species has resulted in some lethal and nonlethal hybrids, suggesting a requirement for sequence specificity and/or secondary structure. The 3′ nontranslated region appears to be a translational enhancer, both in the viral genome and when fused to heterologous reporter mRNAs. Sequences encoding the internal ORFs for the movement and coat proteins are dispensable for replication. Duplication of the subgenomic promoters results in transcription of an additional new subgenomic mRNA. Heterologous tobamovirus subgenomic promoters inserted into the viral genome are recognized by the replicase complex and transcribed. Foreign sequences inserted behind tobamovirus subgenomic mRNA promoters have been expressed to high levels in plants and protoplasts.

Satellite Tobacco Mosaic Virus

Satellite tobacco mosaic virus (STMV), a tobamovirus-dependent satellite virus, was isolated from *Nicotiana glauca* plants infected with tobacco mild green mosaic virus (TMGMV). The STMV genome consists of one single-stranded positive-sense RNA of 1059 nt. The 240 3′ nucleotides share approximately 65% sequence identity with TMGMV and TMV, contain pseudoknot structures, and have a tRNA-like terminus. No sequence similarity with any tobamovirus exists over the remainder of the genome. Two overlapping ORFs that are expressed in *in vitro* translation reactions are present in the genomic RNA of most STMV isolates. The 5′-proximal ORF encodes a 6.8 kDa protein that has not been detected *in vivo* and is not present in all STMV isolates. The second ORF encodes a 17.5 kDa coat protein that is not serologically related to any tobamovirus coat protein. The 17 nm icosahedral virions are composed of a single STMV genomic RNA encapsidated within 60 STMV coat protein subunits. Replication of natural populations of STMV is supported by other tobamoviruses, but at lower levels than with the natural helper virus, TMGMV. The host range of STMV parallels that of the helper virus.

See also: Plant Virus Diseases: Economic Aspects; Satellite Nucleic Acids and Viruses; Tobacco Mosaic Virus.

Further Reading

Gibbs AJ (1977) Tobamovirus group. CMI/AAB Descriptions of Plant Viruses, No. 184. http://www.dpvweb.net/dpv/showadpv.php?dpvno=184 (accessed July 2007).

Hull R (2002) *Matthews' Plant Virology*, 4th edn., New York: Academic Press.

Lewandowski DJ (2005) *Tobamovirus* genus. In: Fauquet CM, Mayo MA, Maniloff J, Desselburger U, and Ball LA (eds.) *Virus Taxonomy, Classification and Nomenclature of Viruses: Eighth Report of the International Committee on Taxonomy of Viruses*, pp. 1009–1014. San Diego, CA: Elsevier Academic Press.

Pogue GP, Lindbo JA, Garger SJ, and Fitzmaurice WP (2002) Making an ally from an enemy: Plant virology and the new agriculture. *Annual Review of Phytopathology* 40: 45–74.

Scholthof KB (2005) Tobacco mosaic virus, a model system for plant virology. *Annual Review of Phytopathology* 42: 13–34.

Scholthof KBG, Shaw JG, and Zaitlin M (eds.) (1999) *Tobacco Mosaic Virus: One Hundred Years of Contributions to Virology*. St. Paul, MN: APS Press.

Van Regenmortel MHV and Fraenkel-Conrat H (eds.) (1986) *The Plant Viruses, Vol. 2: The Rod-Shaped Plant Viruses*. New York: Plenum.

Tobravirus

S A MacFarlane, Scottish Crop Research Institute, Dundee, UK

© 2008 Elsevier Ltd. All rights reserved.

Taxonomy and Characteristics

The genus *Tobravirus* is comprised of three species, the type species *Tobacco rattle virus* (TRV) together with *Pea early-browning virus* (PEBV) and *Pepper ringspot virus* (PepRSV), which was previously referred to as the CAM strain of TRV. The genus has not been assigned to a virus family. Tobraviruses have a genome of two, positive-sense, single-strand RNAs that are packaged separately into rod-shaped particles. In some situations, the larger genomic RNA (RNA1) can cause a systemic infection in the absence of the second, smaller RNA (RNA2) and without the formation of virus particles. Tobraviruses are transmitted between plants by root-feeding nematodes of the genera *Trichodorus* and *Paratrichodorus*, and in some plant species are also seed transmitted.

Virus Particle Production and Structure

Tobravirus RNA1 is encapsidated into the L (long) particle with a length of 180–215 nm, depending on virus species, and RNA2 is encapsidated into S (short) particles which range in length from 46 to 115 nm, depending on virus isolate (**Figure 1**). Both L and S particles have an apparent diameter of 20–23 nm, depending on the technique used to examine them. *In vitro* translation experiments using RNA extracted from purified virus preparations, as well as studies with the TRV SYM isolate, which has an unusual genome structure, showed that some, if not all, tobravirus subgenomic (sg)RNAs are also encapsidated, in particles of various lengths. The tobraviruses encode a single coat protein (CP), molecules of which assemble in a helical arrangement around a central cavity with a diameter of 4–5 nm, and with a distance of 2.5 nm between successive turns of the helix.

In vitro reconstitution experiments suggested that virus particle formation initiates at the 5′ end of the viral RNA, although the encapsidated sgRNAs do not carry the 5′ terminal part of the virus genomic RNAs. Peptide mapping showed that the major antigenic regions of the CP are, in descending order of strength, the C-terminal 20 amino acids (aa), 5 aa in the central region of the protein and 5 aa at the N-terminus. This and other spectroscopic analyses suggest that the N- and C-termini are exposed on the outer surface of the particle, while the central region is exposed in the central canal (where interactions of the CP with the viral RNA take place). The C-terminal domain appears to be unstructured and

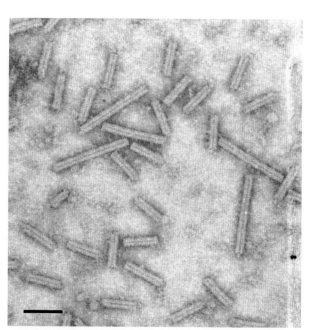

Figure 1 Electron micrograph of long and short particles of TRV isolate RH. Scale = 100 nm.

plays a role in interactions with other virus proteins that are involved in nematode transmission of the virus. Particularly with TRV, there is significant amino acid sequence difference between the CP of different virus isolates, resulting in many different serotypes of this virus.

M-Type and NM-Type Infections

In early studies with tobraviruses (usually with TRV), the infectivity of fractionated and purified L and S particles was examined, showing that L particles were infectious (producing local and systemic symptoms on particular hosts) but that S particles were not. However, plants infected with L particles did not contain virus particles, whereas plants infected with both L and S particles did contain (both) virus particles. This was explained by sequencing, which revealed that RNA1 encodes proteins for RNA replication and movement, whereas RNA2 encodes the CP. Because infections with RNA1 do not produce virus particles, they were very difficult to maintain by repeat inoculation of sap extracts, and were referred to as nonmultiplying (NM-type) infections. Infections derived from both RNA1 and RNA2 and producing virus particles were easily passaged and were referred to as multiplying (M-type) infections. Subsequently, it was found that extraction with phenol of RNA from NM-infected plants in fact produced highly infectious preparations.

Early observations also suggested that NM-type infections caused more severe symptoms than M-type infections and moved only slowly up the plant, giving rise to the theory that unencapsidated RNA1 could spread systemically only from cell to cell via plasmodesmata. However, experiments with TRV and PEBV carrying defined mutations in the CP gene demonstrated unequivocally that unencapsidated virus RNAs moved very rapidly via the vascular system in both *Nicotiana benthamiana* and *Nicotiana clevelandii* plants. In addition, with wild-type (encapsidated) virus, systemic infection always included both RNA1 and RNA2 but with the (unencapsidated) CP mutants RNA2 occasionally became separated from RNA1 and was not detected in some systemic leaves. One PEBV mutant carried only a small (28 aa) deletion at the C-terminus of the CP but did not produce virus particles or indeed any detectable CP in infected plants. Nevertheless, this mutant moved systemically as quickly as did wild-type virus and in this case RNA2 did not become separate from RNA1. This suggests, perhaps, that some form of CP but not incorporated into particles, which can be present at very low levels, ensures the coordinated transport of RNA1 and RNA2 through the phloem.

In another study, plants were infected with TRV RNA1 and two types of RNA2, one wild type and the second encoding a CP in which the 15 aa at the C-terminus were replaced with three nonviral residues. Both RNA2 species were encapsidated by the CP that they encoded but neither appeared to be encapsidated by the CP from the other RNA2. Apparently, although RNA1 is encapsidated *in trans*, RNA2 is only encapsidated *in cis*. The mechanism for this is not known but may be linked to the role of CP in the coordinated systemic movement of RNA1 and RNA2.

Genome Structure and Expression

The viral RNAs have a 5' methylated guanosine cap. The 3' end of the RNAs is not polyadenylated but folds into a tRNA-like pseudoknot structure that, in contrast to some other virus RNAs, cannot be aminoacylated. RNA1 is from 6.8 to 7 kb in size, and RNA2 varies between isolates from 1.8 to ∼4 kb (**Figure 2**). Complete sequences have been obtained for RNA1 from four isolates of TRV, two isolates of PEBV, and one isolate of PepRSV. In addition, complete sequences have been obtained for RNA2 from 15 isolates of TRV, three isolates of PEBV, and one isolate of PepRSV. (Information on RNA sequences can be obtained from the eighth ICTV report.) Other than the sequencing, almost no further molecular studies have been carried out on PepRSV.

RNA Sequences

The larger RNA (RNA1) is highly conserved in nucleotide sequence between different isolates of the same virus (e.g., 99% identity between RNA1 of TRV isolates SYM and ORY), but there is much less identity between isolates of the different tobravirus species (e.g., 62% identity between RNA1 of TRV isolate SYM and PEBV isolate SP5). The smaller RNA (RNA2) varies considerably both in terms of overall nucleotide sequence identity as well as protein-coding capacity between isolates of the same

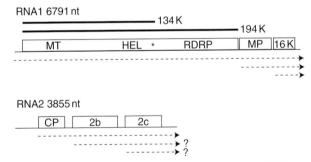

Figure 2 Genome diagram and expression strategy of TRV isolate PpK20. Open boxes denote virus genes. The solid lines above the RNA1 genes shows the location of the two, overlapping replicase genes. The 134K protein contains methyltransferase (MT) and helicase (HEL) motifs. The C-terminal part of the 194K protein contains an RNA-dependent RNA polymerase (RDRP) motif. The asterisk denotes the 134K gene termination codon where readthrough translation to produce the 194K protein occurs. The movement protein (MP) gene is also known as the 29K or 1a gene. The cysteine-rich 16K gene is also known as the 1b gene. For RNA2, CP denotes the coat protein gene. The dashed lines beneath each RNA denote the 3' co-terminal subgenomic RNAs that are known or suspected to exist.

virus (TRV or PEBV), although phylogenetic analysis shows that the CPs of the different tobraviruses, which are encoded by RNA2, are more closely related to each other than to CPs from viruses in other genera. The tobraviruses are most closely related to tobamoviruses, in terms of particle structure, overall gene organization, and viral protein sequence homologies.

RNA1 has a 5′ noncoding region (NCR) of between 126 and 202 nt, and a 3′ NCR of 459–255 nt. RNA2 has a much larger 5′ NCR of 470–710 nt, and a 3′ NCR of 780–392 nt. There is almost no sequence homology between the 5′ NCRs of the tobraviruses; however, the 25 nt at the 3′ terminus of the TRV and PEBV RNAs are identical and there is 70% identity over the next 140 nt. Consequently, when recombinants were made between the RNA2 of TRV PpK20 and PEBV SP5, the replicase complex of both viruses could replicate RNA2 molecules carrying the 3′ NCR from either virus. However, RNA2 carrying the TRV 5′ NCR could only be replicated by a TRV replicase complex, and RNA2 carrying the PEBV 5′ NCR could only be replicated by a PEBV replicase complex.

Expression of Virus Genes

The 5′ proximal gene of RNA1, encoding viral replicase proteins, is expressed by direct translation of the genomic RNA. The two other RNA1-encoded genes (1a and 1b) are expressed from sgRNAs that start with the sequence AUA (within a conserved motif GCAUA) and that are co-terminal with the 3′ end of the genomic RNA.

The 5′ proximal gene of RNA2 of almost all tobravirus isolates encodes the CP. However, the CP gene is located at least 470 nt downstream of the 5′ terminus of RNA2, and this region contains numerous (at least six) AUG codons upstream of the translation initiation codon of the CP gene, that are inhibitory to CP expression. Experiments to delete parts of the 5′ NCR of PEBV RNA2 showed that only a very small amount of CP is translated *in vitro* from full-length (genomic) RNA2 but that translation is increased 25-fold when the 5′ NCR is removed. Thus, even though the CP gene is the 5′ proximal gene on RNA2, it (and the other genes on RNA2) is expressed from an sgRNA. Uniquely, the CP gene of TRV SYM is located near to the 3′ end of RNA2, downstream of another gene. The sgRNA for this CP is much shorter than genomic RNA2 and is encapsidated into a clearly identifiable VS (very short) virus particle.

A stem–loop sequence is found upstream of tobravirus sgRNA start sites, and is particularly conserved for the CP sgRNA. The introduction of mutations into this structure (8 bp stem, 4 nt loop) upstream of the PEBV CP gene showed that the stem–loop forms an essential part of the sgRNA promoter, and that the structure of the stem rather than its actual sequence is important for sgRNA synthesis *in planta*. The promoter regions for the TRV and PEBV CP genes could be used as a cassette and moved to other locations in RNA2 to express nonviral genes.

Recombination in Tobravirus RNAs

The large variation in the overall length of RNA2 from different tobravirus isolates reflects the fact that most are recombinant molecules, where the 3′ part of RNA2 has been replaced by sequences from the 3′ part of RNA1. The recombination junction can occur at any position in RNA2 that is downstream of the CP gene, and the recombinant may retain none, some, or even both of the other genes (2b and 2c) that might be considered RNA2 specific. The region of RNA1 that is transferred to RNA2 always includes the 3′ NCR, often includes the 1b gene immediately upstream of the 3′ NCR, and may occasionally include part of the 1a gene that is located upstream of the 1b gene. This means that many of these recombinant isolates carry two, probably both functional, copies of the 1b gene, one on RNA1 and the second on RNA2. The mechanism for recombination is not known but is speculated to be caused by template switching by viral replicase from RNA1 to RNA2 during minus-strand RNA synthesis. Sequences that resemble the 5′ terminus of tobravirus genomic RNAs (rich in A and U residues) are often found at or near the recombination junction, and conceivably could facilitate the recombination event. It is also not completely clear when tobravirus recombination occurs, though the feeling is that repeated passage of virus by mechanical inoculation in glasshouse-grown plants, bypassing the nematode transmission process, either encourages recombination or reduces selection against the survival of recombinants. Nevertheless, at least one known recombinant isolate, TRV PaY4, which was cloned after only a very limited period of multiplication in the glasshouse, did retain its ability to be nematode transmitted.

A different recombination process also occurs in which the CP and/or 2b gene in RNA2 of some TRV isolates appears to have been derived from PEBV. For this reason, serological analysis is not always successful in discriminating between the different tobraviruses. Attempts to reproduce this recombination by co-inoculating TRV and PEBV to plants in the glasshouse were not successful, suggesting that this may be only a rare occurrence, or may be stimulated by particular environmental conditions. Nevertheless, one report showed that 30% of the TRV isolates recovered from fields in the coastal bulb-growing region in the Netherlands were TRV/PEBV recombinants.

Viral Proteins

RNA1

RNA1 encodes proteins for virus RNA replication and intraplant movement (local and systemic). The 5′ half of RNA1 encodes a large (134–141 kDa) protein that contains methytransferase and helicase domains and is expected to be part of the viral replicase complex.

Readthrough translation of the stop codon of this protein produces a larger protein (194–201 kDa) that contains motifs associated with RNA-dependent RNA polymerase proteins. *In vitro* translation experiments showed that the opal (UGA) translation termination codon of the TRV 134 kDa protein is suppressed by tRNAs that incorporate tryptophan or cysteine at this position. Downstream of the replicase genes is the 1a gene encoding a 29–30 kDa cell-to-cell movement protein. Disruption of this gene prevents accumulation of TRV in inoculated leaves but can be overcome by co-inoculation with tobacco mosaic virus (TMV) or to transgenic plants expressing the TMV 30K movement protein. The C-terminal 1b gene encodes a 12–16 kDa cysteine-rich protein that is involved in seed transmission of PEBV in pea, pathogenicity of TRV and PEBV, and in cultured *Drosophila* cells suppresses RNA silencing. Deletion of the 1b gene prevents systemic movement of TRV in *N. benthamiana*, which can be recovered by co-expression, from TRV RNA2, of other cysteine-rich proteins from PEBV, soil-borne wheat mosaic virus, and barley stripe mosaic virus, as well as by the 2b silencing suppressor protein of cucumber mosaic virus.

RNA2

RNA2 encodes proteins for virus particle formation and transmission by nematodes. Several studies have identified the CP as being involved in the nematode transmission process, and deletion of part of the C-terminal domain of the CP prevented the transmission of PEBV without affecting virus particle formation. Infectious, nematode-transmissible clones of three TRV isolates and one PEBV isolate have been sequenced. In addition to the CP, these RNAs all carry two other genes encoding the 2b and 2c proteins, both of which (for PEBV) have been detected by Western blotting in extracts of virus-infected plants. Mutation studies showed that the 2b gene is necessary for transmission of TRV and PEBV, whereas the 2c protein was only required for transmission of PEBV. This difference may reflect the different species of nematode that transmitted the particular virus isolates used in these studies, rather than a clear mechanistic difference between TRV and PEBV transmission. A further, small open reading frame, encoding a putative 9 kDa protein immediately downstream of and in-frame with the CP gene, is present in PEBV-TpA56 and TRV TpO1. For PEBV, mutation of this gene greatly reduced nematode transmission frequency, although the protein was not detected by Western blotting.

The 2b proteins of the different tobravirus isolates share some amino acid sequence homology with each other, ranging from only 11% identity (TRV Umt1 vs. TRV PaY4) to 99% identity (TRV Umt1 vs. TRV OR2). They also range in size from 238 aa (TRV PaY4) to 354 aa (TRV PpK20). The TRV PaY4 2b protein has been shown to act *in trans*, with nematode transmission of a 2b-mutant virus being complemented by co-infection with wild-type virus. However, this appears to be isolate specific, as the TRV PaY4 2b mutant was complemented only by wild-type TRV PaY4 and not by wild-type TRV PpK20.

The 2b protein may influence nematode transmission by more than one mechanism. In a microscopy study of the roots of *N. benthamiana* plants infected by PEBV, large numbers of particles of the wild-type virus were found in all regions of the root tip, which is where the vector nematodes preferentially feed, whereas virus particles of a mutant deleted for the 2b and 2c genes were present in roots only in much lower numbers. A similar enhancement was found in the efficiency of invasion of roots in *N. benthamiana*, as well as in leaves of *Arabidopsis thaliana*, when TRV carrying the 2b gene was compared with that of TRV lacking the 2b gene.

The 2b protein also physically interacts with the TRV CP (as examined by yeast two-hybrid (Y2H) analysis) and the virus particle (as examined by immunoelectron microscopy). One suggestion is that the 2b protein acts as a bridge to trap virus particles to specific sites of retention on the nematode esophageal cuticle. Aggregates of virus particles have been observed using the electron microscope to collect in this part of the nematode but the co-location of the 2b protein has not yet been demonstrated. It appears that interaction between the CP and 2b protein stabilizes the 2b protein, as in plants infected with wild-type TRV PaY4, carrying both PaY4 CP and PaY42b genes, the 2b protein was detected by Western blotting. However, with a recombinant TRV carrying the TRV PpK20 CP gene and the TRV PaY4 2b gene, the 2b protein could not be detected, possibly because these proteins from two different isolates cannot interact with one another. This is reinforced by results from the Y2H study, where the TRV PpK20 CP and PpK20 2b proteins were found to interact. Similarly, the PaY4 CP and PaY4 2b proteins interacted; however, the PpK20 CP did not interact with the PaY4 2b protein and the PaY4 CP did not interact with the PpK20 2b protein.

Although the 2c protein is involved in transmission of PEBV, little else is known about this protein. Amino acid sequence homologies between the 2c proteins of different TRV isolates range from almost none to over 95% identity. In a Y2H assay, the TRV PpK20 2c protein interacted with the PpK20CP, and removal of the CP C-terminal flexible domain did not affect the interaction.

Tobraviruses as Gene Expression/Silencing Vectors

As RNA1 encodes all the proteins necessary for tobravirus replication and movement, the RNA2 can be modified to carry nonviral sequences without greatly affecting virus infection. Expression vectors have been constructed from all three tobraviruses, in which a duplicate CP promoter sequence is inserted downstream of the native CP gene followed by restriction sites to allow the cloning of other sequences. Together, the tobraviruses can infect a wide range of plant species, often without causing particularly severe symptoms, features which increase their utility as expression vectors.

Plants infected with tobraviruses often undergo a rapid recovery in which infection symptoms and virus levels fall dramatically, although the plants do not become free of virus, most particularly in the meristem regions. However, these plants have developed a strong resistance to further infection by the same virus, most likely by an RNA silencing-based mechanism. Although not well understood, it seems that tobraviruses are potent triggers of RNA silencing but do not encode a strong silencing suppressor protein to counteract this host defense activity. A consequence of this is that when a host plant sequence is inserted into the virus genome, very strong silencing is initiated that targets expression of the host gene itself, a process known as virus-induced gene silencing (VIGS). TRV has become one of the most widely used VIGS vectors for studies of plant gene function, and recent work has shown that PEBV is also a very effective VIGS vector for studies in pea.

Diseases Caused by Tobraviruses

TRV is found in many regions (in Europe, North America, Japan, and Brazil) and has a particularly wide host range, infecting more than 100 species in nature and more than 400 species when tested in the glasshouse, although not all of these infections are systemic. As tobraviruses are transmitted by soil-inhabiting nematodes, infection may be limited in the field to the roots. Weeds may play an important role in the maintenance and spread of tobraviruses, with *Capsella bursa-pastoris*, *Senecio vulgaris*, *Stellaria media*, and *Viola arvensis* being the most commonly found weed hosts of TRV. Virus transmission in seed of these plants was also reported. Many crop plants are infected by TRV, the major diseases being those of potato and ornamental bulbs (narcissus, gladiolus, tulip, lily, and crocus). The symptoms of TRV infection in potato are the formation of arcs and flecks of brown corky tissue in the tuber which is referred to as spraing, and which can make the tuber unfit for sale. Both M-type and NM-type infections can produce spraing symptoms, the biochemical basis for which is not known. It was thought that potato cultivars that did not show spraing symptoms were resistant to TRV; however, recent work has shown that some infections can be symptomless. Nevertheless, even these symptomless infections lead to significant reductions in tuber yield and tuber quality. Inclusion of tubers carrying symptomless infection could have major consequences for the production and distribution of seed potatoes.

PEBV has been reported in several European countries (UK, Netherlands, Italy, Belgium, Sweden) as well as Algeria and Morocco. In the field, it infects mainly legumes, including pea, faba bean, French bean, lupin, and alfalfa. Several weeds and other crop plants may be infected, although often only in the roots.

PepRSV has only been reported from Brazil, where it infects pepper, tomato, and artichoke, as well as local weed species.

See also: Hordeivirus; Nepovirus; Tobamovirus; Viral Suppressors of Gene Silencing.

Further Reading

Harrison BD and Robinson DJ (1978) The Tobraviruses. *Advances in Virus Research* 23: 25–77.

MacFarlane SA (1999) Molecular biology of the tobraviruses. *Journal of General Virology* 80: 2799–2807.

MacFarlane SA and Robinson DJR (2004) Transmission of plant viruses by nematodes. In: Gillespie SH, Smith GL, and Osbourne A (eds.) *SGM Symposium 63: Microbe–Vector Interactions in Vector-Borne Diseases*, pp. 263–285. Cambridge: Cambridge University Press.

Robinson DJ (2005) Tobravirus. In: Fauquet CM, Mayo MA, Maniloff J, Desselberger U, and Ball LA (eds.) *Virus Taxonomy: Eighth Report of the International Committee on Taxonomy of Viruses*, pp. 1015–1019. San Diego, CA: Elsevier Academic Press.

Visser PB, Mathis A, and Linthorst HJM (1999) Tobraviruses. In: Granoff A and Webster R (eds.) *Encyclopedia of Virology*, 2nd edn., pp. 1784–1789. London: Elsevier.

Togaviruses Causing Encephalitis

S Paessler, University of Texas Medical Branch, Galveston, TX, USA
M Pfeffer, Bundeswehr Institute of Microbiology, Munich, Germany

© 2008 Elsevier Ltd. All rights reserved.

Glossary

Arbovirus A virus transmitted to vertebrates by hematophagous (blood-feeding) arthropods.
Cimicid A hemipteran bug of the family *Cimicidae*.
Encephalitis Acute inflammation of the brain.
Enzootic Infection or illness affecting or peculiar to animals of a specific geographic area.

Epidemic disease An illness that spreads rapidly and widely and affects many individuals in an area or a population at the same time.
Epizoodemic An extensive outbreak of a virus in both humans and other vertebrates.
Epizootic Infection or illness in a large number of animals at the same time within a particular region or geographic area, the number of affected animals being in clear excess of that which would be expected for the specific region and period of time.
Etiology The study of causes of diseases or pathologies.
Togavirus A virus of the family *Togaviridae*.

Introduction

In the 1930s, previously unrecognized viruses were isolated from diseased horses in California, Virginia, and New Jersey, and from an infected child in Caracas, Venezuela. These subsequently were named Western equine encephalitis virus (WEEV), Eastern equine encephalitis virus (EEEV), and Venezuelan equine encephalitis virus (VEEV), respectively. Since then, numerous isolations of these viruses have been obtained from infected mosquitoes and from horses, humans, and other vertebrates, mainly birds and rodents. In nature, these viruses are transmitted by hematophagous arthropods.

The virus family *Togaviridae* comprises two genera: *Alphavirus* and *Rubivirus*. Rubiviruses do not usually cause encephalitis and will not be discussed further in this article. Encephalitic alphaviruses are neuroinvasive and cause mild-to-severe neurological symptoms. The alphaviruses WEEV, EEEV, VEEV, and, more rarely, Ross River virus, Chikungunya virus, and Highlands J virus (HJV) can cause encephalitis in equines (see **Figure 1**) or humans.

The prototype alphavirus, Sindbis virus, does not typically cause encephalitis in humans and is encephalitic only under experimental conditions in mice. Therefore, this virus will also be discussed elsewhere. Among the naturally occurring alphaviruses, WEEV, EEEV, and VEEV, have widespread distributions in North, Central, and South America (see **Figure 2**). WEEV is distributed in the US from the mid-Western states of Michigan and Illinois to the West coast and in western Canada, and clinical cases have been reported in 21 states, whereas EEEV is distributed from Texas to Florida along the Gulf coast and from Georgia to New Hampshire along the Atlantic Coast and in eastern Canada, with cases reported in 19 states, including the mid-Western states of Wisconsin, Illinois, and Michigan. Transmission of VEEV occurs predominantly in Central and South America, with the exception of Everglades virus (formerly VEEV subtype II), which is found in Florida, and Bijou Bridge virus (formerly VEEV subtype IIIB), found in cimicid cliff swallow bugs (*Oeciacus vicarius*) in western North America.

Western Equine Encephalitis Virus

This virus was first isolated in 1930 from the brain of an encephalitic horse in California. According to the US Centers for Disease Control and Prevention, 639 confirmed human cases of WEE have occurred in the US from 1964 to 2005. Likely due to changes in irrigation practices and to successful mosquito control programs in the western US, the annual number of cases has been declining, with fewer than ten cases per year since 1988. Infections with WEEV are generally asymptomatic or result in mild disease after an incubation period of 2–7 days with nonspecific symptoms, for example, sudden onset of fever, headache, nausea, vomiting, anorexia, and malaise, although these can be followed by altered mental status, weakness, and signs of meningeal irritation, evidence of more serious infections. Encephalitis occurs in a minority of infected individuals and may lead to neck stiffness, confusion, tonic–clonic seizures, somnolence, coma, and death. Mild-to-severe neurological sequelae can be found in survivors of encephalitis, especially in those <1 year old, at rates between 5% and 30%. In addition, serological studies indicated that the relation of inapparent to apparent infections changes with age. In children <1 year old, the relation is about 1:1, between 1 and 4 years of age it is about 58:1, and in children >14 years old it is >1000:1. The overall case–fatality rate is about 3%. Encephalitis due to WEEV is characterized by vasculitis and focal hemorrhages mainly affecting the basal ganglia and the nucleus of the thalamus. Because small hemorrhages can occur in the white and gray substance throughout the central nervous system (CNS), they can be mistaken for resolved infarcts in elderly patients.

WEEV is maintained in an enzootic cycle between its natural vertebrate hosts, passerine birds, and its most common mosquito vector, *Culex tarsalis*, a species associated with irrigated agriculture and stream drainages in the western US. Transmission to horses and humans is mediated by so-called bridging mosquito vectors, including *Ochlerotatus melanimon* in California, *Aedes dorsalis* in Utah and New Mexico, and *A. campestris* in New Mexico (see **Figure 3**). Depending on the climate, the natural transmission cycle may be maintained throughout the year. In more moderate climate areas, WEEV may persist in yet-unrecognized hosts, or may be reintroduced annually by migratory birds that move great distances. Isolations of WEEV have been documented in the Veracruz region of Mexico and in South America. Genetic analyses of WEEV isolated in South America (Brazil and northern Argentina) revealed a level of nucleotide

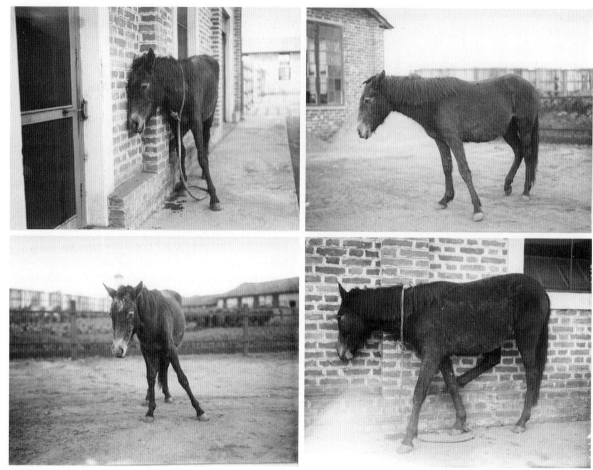

Figure 1 Western equine enceph.alitis in equines.

identity >90% in the E2/6K/E1 coding region when compared to isolates from California, Texas, and as far north as Montana, suggesting a monophyletic nature of the WEEV lineage, with an overall slow evolution. Hahn and colleagues showed that WEEV is a natural recombinant virus, with its capsid amino acid sequence similar to that of EEEV, while the remaining part of the structural polyprotein was more similar to that of Sindbis virus. Further phylogenetic analyses, including the nonstructural genes as well, confirmed that WEEV and the closely related Highlands J, Fort Morgan, and Buggy Creek viruses are recombinant viruses derived from an ancestor formed by an EEEV-like (5′ two-thirds of the genome) and Sindbis virus-like (3′ one-third of the genome) parental virus.

Eastern Equine Encephalitis Virus

The first isolates of EEEV were obtained in 1933 from infected horses in Virginia and New Jersey. The main EEEV transmission cycle occurs between birds and mosquitoes via *Culiseta melanura* mosquitoes. However, the principal arthropod vector for transmission of EEEV to humans or to horses are *Aedes*, *Coquillettidia*, and *Culex* mosquitoes which, unlike *Culiseta melanura*, tend to feed on both birds and mammals. Virus transmission occurs most commonly in and around freshwater hardwood swamps in the Atlantic and Gulf Coast states and in the Great Lakes region. Most cases of EEE have been reported from Florida, Georgia, Massachusetts, and New Jersey. In horses, in some bird species, and in dogs, EEEV can cause severe disease and experimental data suggest that equines can develop viremia upon experimental infection; however, it is believed that horses do not serve as amplifying hosts during epidemics.

According to the US Centers for Disease Control and Prevention, 220 confirmed cases of EEE have occurred in the US from 1964 to 2004. This virus is probably the most virulent of the encephalitic alphaviruses, with a case–fatality rate estimated at >33%. After a 4–10-day incubation period symptoms begin with sudden onset of fever, general muscle pains, and headache of increasing severity. In human cases of encephalitis, fever, headache, vomiting, respiratory symptoms, leucocytosis, hematuria, seizures, and coma may occur. Clinical studies of serologically confirmed and human EEEV infections using

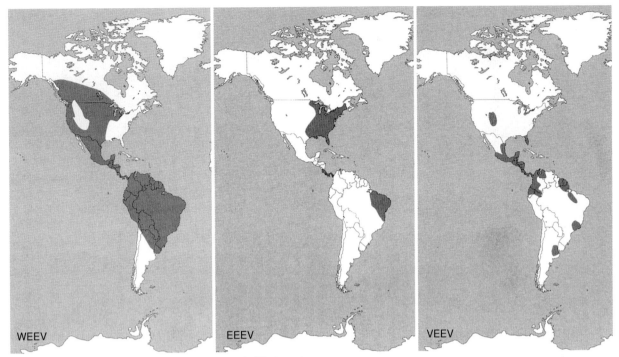

Figure 2 Geographic distribution of equine encephalitic togaviruses.

magnetic resonance imaging and computed tomography have shown changes in the basal ganglia and thalami, suggesting brain edema, ischemia, and hypoperfusion in the early stage of disease. Investigations of gross pathology of fatal human cases revealed brain edema with necrosis, facial or generalized edema, vascular congestion, and hemorrhage in the brain and visceral organs. The predominant micropathological manifestations in the brain include vasculitis, hemorrhage, and encephalitis.

The pathogenesis of EEEV is poorly understood; however, data from experimental infections of mice, guinea pigs, and rhesus monkeys, and from histopathological studies of equine and porcine cases are available. The mouse encephalitis model is well established for several alphaviruses, although it generally lacks the ability to reproduce the vascular component of the disease that is typical for EEE in humans. A hamster model of EEE, used to study acute vasculitis and encephalitis, was described recently. The virus penetrates the brain quickly in animal models and the neuronal phase of the disease develops rapidly. The virus is capable of establishing productive infection in all parts of the mouse or hamster brain. The experimental studies have suggested early infection of periventricular and perivascular neuronal cells in the basal ganglia and hippocampus. It is difficult to explain the early involvement of the basal ganglia and brainstem in experimental infections and to our knowledge no specific anatomic characteristics of these brain regions have been described that might help explain their susceptibility to early infection. In contrast to VEEV (see below), EEEV appears to rapidly invade the brain of infected animals via blood, and the first antigen-positive neuronal cells are located in the basal ganglia and brainstem in the hamster model. Certain findings in animal models, such as the hamster, are similar to early clinical manifestations of EEEV infections in humans. The inflammatory response in the brain is prominent in cases in which animals had survived for at least 5 days, and is produced by macrophages, lymphocytes, and neutrophils. Some of the animal models also display the peripheral pathological changes described in fatal human EEE cases, including congestion and numerous microhemorrhages in the liver, spleen, and lung.

Venezuelan Equine Encephalitis Virus

In 1938, the first isolation of VEEV was obtained from the brain of an encephalitic animal in Venezuela. Like WEEV and EEEV, VEEV is a zoonotic pathogen, transmitted between vector mosquitoes and vertebrate hosts, namely rodents and humans in enzootic cycles, and horses and humans in epidemic or epizootic cycles. Although during epizoodemics essentially any mosquito can be found infected with VEEV, *Ochlerotatus taeniorhynchus* is one of the principal vectors believed to be responsible for the transmission of VEEV during outbreaks, whereas *Culex* (*Melanoconion*) species mosquitoes transmit enzootic strains of VEEV. These viruses are also highly infectious via the aerosol route, and have been responsible for numerous laboratory accidents (>150 cases without an associated perforating injury) and have been developed as a biological weapon in the US and in the former Soviet

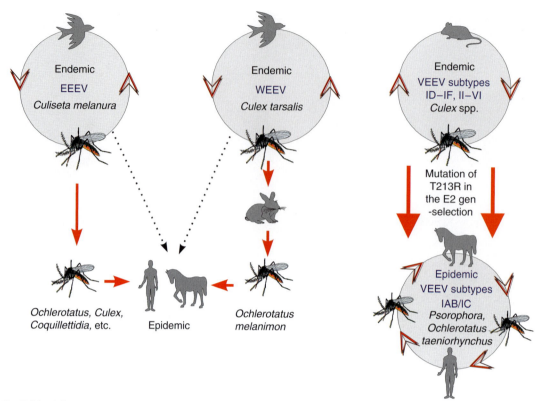

Figure 3 Epidemiology of equine encephalitic togaviruses. Ecological alterations lead to infections of horses and humans with Eastern equine encephalitis virus and Western equine encephalitis virus, but genetic shifts from subtype ID to IC Venezuelan equine encephalitis viruses trigger high viremia in horses, causing them to be virus amplifiers and sustaining epidemics/epizootics.

Union. VEEV infection has also been associated with abortion and fetal death in humans. Equids develop high-titer viremias, which serve as sources of infection for subsequently feeding mosquitoes. In recent years, spillover to humans during equine epizootics has resulted in epidemics of VEEV.

The VEEV antigenic complex, of which VEE virus is the prototype member, is divided into six distinct antigenic subtypes (I–VI). These subtypes correlate with human and equine pathogenicity but major human epidemics and equine epizootics have been associated almost exclusively with subtypes IAB and IC. Enzootic transmission is generally associated with subtypes ID and IE, which are less virulent for horses. In contrast, epizootic subtypes IAB and IC are highly pathogenic for horses, with case–fatality rates of up to 83% reported. The most recent major outbreak occurred in 1995 in Venezuela and Colombia in which 75 000–100 000 human cases occurred, and more than 300 fatal encephalitis cases were recorded. This epidemic was associated with VEEV subtype IC. In 1993, equine disease was associated with VEEV-IE in Mexico and in the period between 1993 and 1995, human cases of VEEV-ID-associated disease occurred in Peru.

Interferons (IFNs) have long been recognized as essential components of the innate immune response to various viral and bacterial infections. It is well established that IFN-α and IFN-β can protect against alphaviral disease and that the virulence correlates with the resistance to IFN-α and IFN-β. Previous studies with VEEV suggested that the IFN-α and IFN-β resistance or sensitivity phenotype correlated with epizootic potential and equine virulence, although others observed little or no difference.

Generally, severe encephalitis in VEEV infection is less common than with EEEV and WEEV infections, although VEEV-associated encephalitis is a more common outcome in children. In adults, VEEV infection usually results in flu-like symptoms and encephalitis is rare. In humans, while the case–fatality rate is low (<1%), neurological diseases, including disorientation, ataxia, mental depression, and convulsions, can be detected in up to 14% of infected individuals, especially children. Neurological sequelae in humans also are common. The predominant pathological findings in fatal human VEE cases include (1) in the CNS: edema, congestion, hemorrhages, vasculitis, meningitis, and encephalitis; (2) in the lungs: interstitial pneumonia, alveolar hemorrhage, congestion, and edema; (3) in lymphoid tissue: follicular necrosis and lymphocyte depletion; and (4) in the liver: diffuse hepatocellular degeneration.

A murine model for VEEV-induced encephalitis and lymphotropism has been established. Experimental studies have demonstrated that the murine model is characterized by biphasic disease, which starts with productive infection of lymphoid tissue and ends in the destruction of the CNS by aggressive viral replication and a 'toxic' neuroinflammatory response, which is uniformly lethal. By the time encephalitis has developed in an infected mouse, infectious virus is usually absent from peripheral organs and blood. However, virus replicates to high titers in the brain and encephalitic mice die 5–7 days after infection. While the classical mouse model is useful to study lethal encephalitis in mice, it does not enable scientists to study VEEV clearance from the brain because naive mice uniformly succumb to CNS infection. To overcome that barrier, and to study the specific anti-VEEV immune response, as well as virus clearance from the brain, researchers use pre-immunized mice that are susceptible to VEEV infection of the CNS but survive the infection. In general, T cells have been shown to play essential roles in the hosts' defense against alphavirus infection, survival, encephalitis, and also in repair of neural damage and homeostasis in the brain. The brain has been proposed to be a distinct immune regulatory tissue, given its status as an immune-privileged site and that T-lymphocyte entry into the CNS under normal homeostatic conditions is thought to be restricted to memory T cells. Upon viral infection in the periphery and the onset of lymphopenic conditions (characteristic of VEEV infection) followed by entry into the brain, a combination of naive T cells and virus-specific memory T cells may migrate to sites of CNS infection. $CD4^+$ T cell effector functions in encephalitis may include a combination of Th1, Th2, and regulatory T cell ($CD25^+$, forkhead family transcription factor, $Foxp3^+$) activities.

$CD8^+$ T cells contribute to both recovery and pathology of encephalitic flavivirus infections, but their role in VEEV-induced encephalitis, immunopathology, and protection is unclear. In general, it has been hypothesized that cell-mediated cytotoxicity is less critical for control of cytopathic viruses such as VEEV, as compared to control of noncytopathic viruses. This general paradigm can be extended to the CNS where elimination of virus from neurons is thought to be via nonlytic mechanisms, due to the poor regenerative capacity of this cell type.

Once the blood–brain barrier is breached by the virus, local antibody production by B cells in the brain plays a role in prevention of viral entry into cells of the CNS and facilitates virus clearance via Fc receptors. Immunohistochemistry studies have shown that B cells are present in the brains of mice that are vaccinated and which survive for 1 month after challenge with VEEV via the intranasal route.

It is believed that VEEV pathogenesis, particularly the development of clinical signs of encephalitis, such as paralysis, depends both on the virus-mediated and immune-mediated neuronal cell death, and that the development of these encephalitic signs depends on two factors, direct damage of neurons and specific characteristics of the neuroinflammatory response. In addition, some studies have indicated that the early neuroinflammatory response in vaccinated animals (rapid adaptive response to infection) is incapable of controlling the initial replication of the challenge VEEV in the brain, which occurs at levels comparable to those measured in naive animals on the first 3 days after intranasal infection. Production of infectious VEEV in the brain of a vaccinated mouse is reduced 7 days after infection and is undetectable 1 month after infection in immunocompetent mice. It is likely that the reduction of VEEV replication in the brain of vaccinated animals on day 7 might impact survival. However, recent studies on genetically modified mice with preexisting anti-VEEV immunity have shown that a high level of replication in the brain over a period of 28 days is not lethal.

Protection from lethal encephalitis in mice appears to be dependent on the presence of $\alpha\beta$ T-cell receptor (TCR)-bearing cells but not on $\gamma\delta$ TCR-bearing cells. The relative degree of immune-mediated inflammation in various TCR-deficient mice does not fully correlate with viral clearance. Remarkably, vaccinated $\gamma\delta$ TCR-deficient mice are protected from lethal viral challenge by intranasal inoculation but VEEV can persist up to 28 days post inoculation. Virus clearance is not affected in vaccinated, immunocompetent mice, or in surviving animals lacking functional IFN-γ receptor. Most notably, $\gamma\delta$ T-cell-deficient mice tolerate (1) high levels of viral replication in the first 7 days after infection (acute phase), which is similar to levels observed in unvaccinated, immunocompetent mice that become paralyzed and succumb to infection; and (2) lower-level virus persistence for at least 28 days that is not observed in immunocompetent mice (asymptomatic, persistent phase). The virus can persist in the brain despite a moderate level of inflammation and cellular infiltration at the site of infection. Such viral persistence has not been observed in other animal models. Observations described in recent studies may provide opportunity in the future to dissect potential intervention points and, perhaps, to add prognostic markers that correlate with protection against VEEV encephalitis in vaccinated humans.

Highlands J Virus

As pointed out earlier, HJV is a member of the recombinant WEEV lineage. Therefore, it is not surprising that this virus initially was listed as a subtype of WEEV but later raised to the taxonomic level of 'virus' based on oligonucleotide and antigenic mapping. HJV is similar epidemiologically to EEEV in that (1) it is transmitted by the same freshwater swamp mosquitoes, *Culiseta melanura*; (2) it shares the same geographical distribution in the eastern parts of North America and as far west as Texas; and (3) its primary vertebrate hosts are passerine birds. Curiously, however, with one recorded exception (an encephalitic horse from Florida in 1964), it has not been linked to equine or human diseases. HJV may cause severe disease in chickens (up to 7% mortality) and young turkeys, and it is responsible for severe losses in egg production in adult turkeys, but has not been detected in the brains of these diseased birds. Thus, the horse case may have had a co-morbidity that went unnoticed. In comparison of VEEV, EEEV, and WEEV, HJV has to be mentioned, but it is not considered an encephalitic alphavirus.

Diagnosis and Treatment

WEEV

IgM antibodies to WEEV usually are detectable at time of onset of the disease in either horses or humans, using 'IgM antibody-capture enzyme-linked immunosorbent assay' (MAC-ELISA) and within a few days after onset by hemagglutination-inhibition and neutralization assays. A fourfold or greater rise or fall in IgG antibody to WEEV in paired acute- and convalescent-phase samples taken at least 10 days apart also indicates infection with WEEV. Depending on the geographical origin of the suspected case or the particular travel history, a neutralization assay may be required to sort out cross-reacting antibodies against VEEV, EEEV, HJV, or other alphaviruses. If direct virus detection is required, virus can be isolated in a variety of vertebrate and mosquito cell lines; however, Vero cells (African Green Monkey kidney cells) are preferentially used. 'Reverse transcription-polymerase chain reaction' (RT-PCR) and real-time RT-PCR protocols for specific amplification of viral RNA are available. Immunohistochemical detection of viral antigen can be performed on brain tissues from dead horses. Clinical parameters usually are within the normal range, but cerebrospinal fluid (CSF) displays elevated levels of protein ($90-110 \text{ mg dl}^{-1}$) and a leukocytosis (up to 500 cm^{-3}, mainly lymphocytes).

EEEV

EEEV infection is diagnosed using serological assays, especially by detection of IgM in serum and CSF in the presence of CNS or febrile disease, and neutralizing antibody testing of acute- and convalescent-phase samples. Virus isolation and detection of viral nucleic acid is possible, as described above for WEEV.

VEEV

Currently, VEEV infection is diagnosed principally by direct detection, for example, nucleic acid or virus isolation from acute-phase serum or spinal fluid or by serological assay, such as detection of VEEV-specific IgM in the CSF using MAC-ELISA or monoclonal antibody-based antigen-capture ELISA. The plaque-reduction neutralization test (PRNT), which like MAC-ELISA is useful in distinguishing VEEV infections from infections with other alphaviruses (see above), cannot be used to identify the serotype. Recently, a VEEV-specific blocking ELISA was described that also identifies serotype-specific antibodies against VEEV in sera of humans, equids, or rodents.

No effective antiviral treatment exists for these encephalitic arboviruses and treatment remains symptomatic.

See also: Rubella Virus; Togaviruses: Equine Encephalitic Viruses.

Further Reading

Bastian FO, Wende RD, Singer DB, and Zeller RS (1975) Eastern equine encephalomyelitis. Histopathologic and ultrastructural changes with isolation of the virus in a human case. *American Journal of Clinical Pathology* 64: 10–13.

Beck CE and Wyckoff RWG (1938) Venezuelan equine encephalomyelitis. *Science* 88: 530.

de la Monte S, Castro F, Bonilla NJ, Gaskin de Urdaneta A, and Hutchins GM (1985) The systemic pathology of Venezuelan equine encephalitis virus infection in humans. *American Journal of Tropical Medicine and Hygiene* 34: 194–202.

Deresiewicz RL, Thaler SJ, Hsu L, and Zamani AA (1997) Clinical and neuroradiographic manifestations of eastern equine encephalitis. *New England Journal of Medicine* 336: 1867–1874.

Ehrenkranz NJ and Ventura AK (1974) Venezuelan equine encephalitis virus infection in man. *Annual Review of Medicine* 25: 9–14.

Femster R (1957) Equine encephalitis in Massachusetts. *New England Journal of Medicine* 257: 701–704.

Garcia-Tamayo J, Carreno G, and Esparza J (1979) Central nervous system alterations as sequelae of Venezuelan equine encephalitis virus infection in rat. *Journal of Pathology* 128: 87–91.

Getting V (1941) Equine encephalomyelitis in Massachusetts. *New England Journal of Medicine* 224: 999–1006.

Griffin DE (2001) Alphaviruses. In: Knipe DM and Howley PM (eds.) *Fields' Virology*, 4th edn., pp. 917–962. New York: Lippincott.

Griffin DE (2003) Immune responses to RNA-virus infections of the CNS. *Nature Reviews Immunology* 3: 493–502.

Paessler S, Aguilar P, Anishchenko M, et al. (2004) The hamster as an animal model for eastern equine encephalitis and its use in studies of virus entrance into the brain. *Journal of Infectious Diseases* 189: 2072–2076.

Paessler S, Ni H, Petrakova O, et al. (2006) Replication and clearance of Venezuelan equine encephalitis virus from the brains of animals vaccinated with chimeric SIN/VEE viruses. *Journal of Virology* 80: 2784–2796.

Philpotts RJ, Brooks TJ, and Cox CS (1980) Laboratory safety for arboviruses and certain other viruses of vertebrates. The subcommittee on arbovirus laboratory safety of the american committee on arthropod-borne viruses. *American Journal of Tropical Medicine and Hygiene* 29: 1359–1381.

Ryzhikov AB, Ryabchikova EI, Sergeev AN, and Tkacheva NV (1995) Spread of Venezuelan equine encephalitis virus in mice olfactory tract. *Archives of Virology* 140: 2243–2254.

Walton TE and Grayson MA (1998) Venezuelan equine encephalomyelitis. In: Monath TP (ed.) *The Arboviruses: Epidemiology and Ecology*, vol. IV, pp. 203–231. Boca Raton, FL: CRC Press.

Weaver SC, Salas R, Rico-Hesse R, et al. (1996) Re-emergence of epidemic Venezuelan equine encephalomyelitis in South America. *Lancet* 348: 436–440.

Togaviruses Causing Rash and Fever

D W Smith, PathWest Laboratory Medicine WA, Nedlands, WA, Australia
J S Mackenzie, Curtin University of Technology, Shenton Park, WA, Australia
M D A Lindsay, Western Australian Department of Health, Mount Claremont, WA, Australia

© 2008 Elsevier Ltd. All rights reserved.

Glossary

Analgesics Medications to relieve pain.
Arbovirus A virus transmitted to vertebrates by hematophagous (blood-feeding) insects and which replicates in the insect, as opposed to 'mechanical' transmission.
Arthritis Acute or chronic inflammation of one or more joints, usually accompanied by pain and stiffness, and sometimes swelling of the joint.
Fascia/fasciitis Fascia are bands or sheaths of connective tissues that support or connect parts of the body. Fasciitis is inflammation of these structures, most commonly the fascia of the sole of the foot and the wrist.
Maculopapular (referring to a rash) A mixture of flat red areas (macules) and small red raised spots (papules).
Nonsteroidal anti-inflammatory agents Medications used to relieve inflammation that are not corticosteroids, aspirin, or acetaminophen. Common ones include ibuprofen, naproxen, and indomethacin.
Sylvatic Involving wild animals.
Synovial Pertaining to the synovium, a thin membrane that lines joints and tendons and secretes synovial fluid.
Tenosynovitis Inflammation of tendons and the synovial membrane surrounding the tendon.
Teratogenic Causing malformations of an embryo or fetus.
Urticarial Characterized by pale or reddened irregular, elevated patches, and severe itching. An urticarial rash is also called 'hives'.

Introduction

Viruses of the family *Togaviridae* (togaviruses) are enveloped, positive-stranded RNA viruses of 60–70 nm diameter, and include viruses of two genera, *Alphavirus* and *Rubivirus*. The latter contains a single species, Rubella virus, a major cause of fever and rash internationally; it is discussed elsewhere in this series. Alphaviruses are distributed worldwide and cause illness in many hundreds of thousands of people each year. Clinical illness ranges from mild febrile illnesses to severe illnesses, such as encephalitis and hemorrhagic fever. One of the most common clinical manifestations of alphavirus infections is fever accompanied by rash and/or arthritis. The alphaviruses causing these illnesses are all closely related and belong to two major and one minor antigenic groups: the Semliki Forest virus (SFV) antigenic group (**Figure 1**), which includes SFV, Chikungunya virus (CHIKV), O'nyong-nyong virus (ONNV), and Ross River virus (RRV); and the Sindbis virus (SINV) virus group (**Figure 2**), comprising SINV and Mayaro virus (MAYV). Barmah Forest virus (BFV) is in its own antigenic complex but is genetically linked to the SFV complex viruses. The diseases they cause are usually named after the virus, though an antigenic subtype of SINV is the cause of Ockelbo disease, Pogosta disease, and Karelian fever, while the disease due to RRV has been called epidemic polyarthritis in the past. All are mosquito-borne viruses, but with differences in their ecology and epidemiology. However, there are significant similarities in the clinical aspects of the diseases caused by these viruses, likely reflecting common pathogenic features.

Clinical Manifestations

Many infected individuals will develop clinical illness, and symptomatic to asymptomatic infection ratios have been estimated at 2:1 to 4:1 for CHIKV, ONNV, and MAYV disease, from 1:80 to 3:1 or higher for RRV infections in Australia, and 1:20 to 1:40 for Ockelbo and Pogosta diseases (**Table 1**).

Infections occur at all ages and in both sexes. The male to female ratio varies in different studies but there are no major gender effects on clinical illness. Infection of children in endemic areas or during epidemics occurs commonly, but clinical disease is less common and usually milder than in adults, and arthritis is rare in children.

Acute Infection

Clinical illness presents mainly as joint pains and muscle pains, accompanied by fever and/or rash in a varying proportion of cases.

Patients typically present with joint pains and muscle pains that are usually preceded by a few days of fever, at least in those who become febrile. Sore throat and

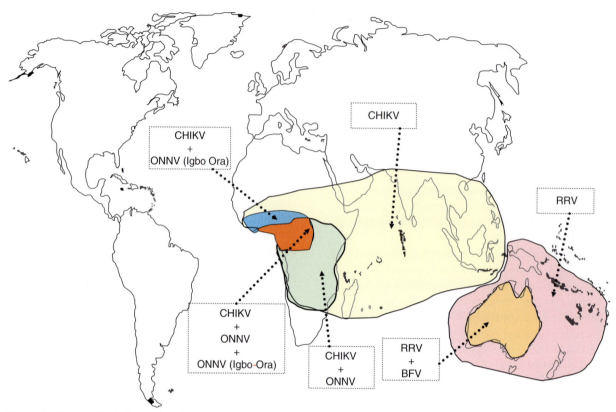

Figure 1 Geographical distribution of the major alphaviruses of the Semliki Forest virus antigenic complex that cause fever, rash, and arthritis in humans.

headache are also commonly reported, and conjunctivitis occurs occasionally. Rash generally follows the other manifestations by a few days, but can sometimes precede them or appear simultaneously. With illness due to RRV and BFV, fatigue is also a prominent clinical feature, and is probably a feature of the other alphaviruses as well. Lymphadenopathy is prominent with some of the alphaviruses, CHIKV causing generalized lymphadenopathy and ONNV causing enlarged posterior cervical lymph nodes.

The pattern of joint involvement is consistent across this group of alphaviruses, commonly involving the ankles, knees, fingers, wrists, elbows, and shoulders. Several joints are involved and the joint involvement is usually symmetric. Other joints including the jaw and the spine may also be affected and back pain is a common complaint. Pain is the most common feature, but most patients also have stiffness of the joints and, less frequently, swelling. The swelling can be due to effusion and/or synovial and soft tissue swelling. During recovery there is a gradual decline in both the number and severity of joint involvement. Inflammation of the fascia of the sole of the foot and the wrist is often reported with RRV and CHIKV infection. Pressure on nerves due to swelling of the fascia may cause tingling in the extremities. This is common with RRV, CHIKV, ONNV, and SINV disease. Muscle pain and tenderness is a prominent part of the clinical illness, usually involving the limbs and shoulders. This can sometimes be more troublesome than the joint pains and it is important to determine whether limb pain is arising from the joints or the muscles.

When rash occurs it is usually maculopapular and most florid on the face, trunk, and upper part of the limbs. However, it can involve the whole body surface, including the palms of the hands, the soles of the feet, and the scalp. Urticarial or vesicular rash occurs in some patients with SINV or BFV disease, but is rare in patients infected with the other viruses. Itchy rash is common with Pogosta disease and ONNV infection. Evidence from RRV and SINV studies suggests that the rashes probably result from presence of virus in the skin and the resulting local immune responses. The RRV rash shows a monocytic infiltrate with presence of RRV antigen within epidermal cells, while the rare purpuric rashes are associated with cytotoxic T-cell responses and capillary leakage.

Gastrointestinal symptoms such as vomiting, abdominal pain, and diarrhea have been reported with RRV and CHIKV illness, particularly in children. Severe illness is rare, other than hemorrhagic disease resembling dengue that can occur following CHIKV infection.

In summary, the majority of infections with the arthritogenic alphaviruses are benign but temporarily debilitating. Fever and rash, if present, usually last less than a week and most patients recover fully within 4 weeks. Joint pains, muscle pains, and lethargy are the slowest to resolve.

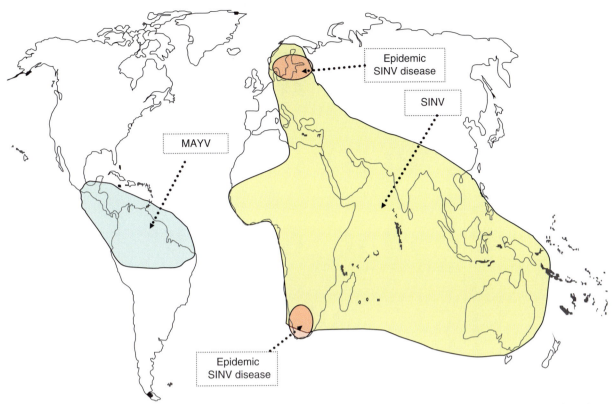

Figure 2 Geographical distribution of the major alphaviruses of the Sindbis virus antigenic complex that cause fever, rash, and arthritis in humans.

Chronic Illness Following Infection

One of the features of alphavirus arthritis is the frequency of prolonged illness, especially persisting joint pains. For SINV, joint pain lasting more than 2 years has been reported for 50% of cases in Finland (i.e., Pogosta disease) and for 25% of cases in Sweden (i.e., Ockelbo disease). Persisting joint pain is also common in Russian cases (i.e., Karelian fever). Similarly, joint pains commonly persist for many months following acute CHIKV infection, and have been reported to follow MAYV infection. Most RRV patients return to full physical activity within 3–6 months, but some suffer from persisting disability due to joint pains, muscle pains, fatigue, and depression. One study found that joint pain persisted for more than 3 months in about 70% of patients. A separate prospective case-controlled study found that at 12 months after onset 90% of patients still had joint pain, 80% had tiredness, 75% had joint stiffness, and 50% had muscle and/or tendon pain.

Infection in Pregnancy

Generally, infection in pregnancy has no special implications for the mother. RRV causes fetal and neonatal death in mice, but there is no evidence of this in humans. During outbreaks in the Pacific Islands possible congenital infection occurred in a small percentage of women, but there was no effect on length of gestation or fetal outcome, and no evidence of congenital malformation. In contrast, during the 2005 Indian Ocean outbreak of CHIKV, there was a 12% rate of transmission to the fetus, and most of these babies had either meningoencephalitis or a coagulation disorder.

Pathogenesis of Alphavirus Arthritis

The virus is introduced subcutaneously by a blood-feeding infected mosquito (**Table 2**) and initial replication probably occurs within subcutaneous tissue and local skeletal muscle. Little is known about the pathogenesis of the systemic manifestations of alphavirus infection, and interest has focused on the joint disease. RRV arthritis induces a predominantly monocytic inflammatory response in synovial fluid in humans. T-cell-derived interferon-γ has been found in joint fluid of RRV-infected humans, and is secreted by RRV-specific T cells of humans and mice. In animal models or *in vitro* a number of alphaviruses including RRV, BFV, CHIKV, and SINV infect synovial monocytes/macrophages and, in the case of RRV, can be found in synovial cells. A combination of release of inflammatory mediators from the infected monocytes/macrophages and the cytotoxic T-cell responses to viral antigens are the likely explanations for the synovial swelling, effusion, and joint pain experienced in acute alphavirus infection.

Table 1 The clinical features of the major alphaviruses causing fever, rash, and arthritis in humans

	Chikungunya virus	O'nyong-nyong virus	Ross River virus	Barmah Forest virus	Sindbis virus (Pogosta)	Sindbis virus (Ockelbo)	Mayaro virus
Incubation period	3–12 days, usually 2–7 days	>8 days	3–21 days, usually 7–9 days	7–9 days	<7 days	<7 days	up to 12 days, usually 6–12 days
Fever	100%	Similar to CHIK	45–55%	50%	45%	40%	100%
Joint pains	80–100%	Similar to CHIK	95–100%	70–85%	95%	95%	50%
Arthritis (Joint stiffness/swelling)	Usually, soft tissue	Similar to CHIK	80–90%	30%	45%	60%	NI[a]
Fatigue	NI[a]	NI[a]	70–80%	80%	NI[a]	NI[a]	NI[a]
Rash	30–50% Maculopapular	30–50% Maculopapular, itchy; some vesicular	50–60% Maculopapular	50–100% Maculopapular; 10% vesicular, some urticarial	90%	95%	30%
Muscle pains	50%	Similar to CHIK	60–90%	70–80%	50%	NI[a]	75%
Tenderness of palms and soles, and/or fasciitis	20–30%	Similar to CHIK	50–60%	NI[a]	NI[a]	NI[a]	NI[a]
Backache	50%	Similar to CHIK	30–60%	NI[a]	NI[a]	NI[a]	NI[a]
Headache	50%	Similar to CHIK	Common	NI[a]	40%	NI[a]	100%
Lymphadenopathy	Generalized, common and prominent	Cervical, common and prominent	Common; generalized in 1–20%	Common; generalized in 5–10%	Not prominent	NI[a]	NI[a]
Severe illness/fatalities	Hemorrhagic disease, encephalitis, death	None reported	Possible meningitis/encephalitis	None reported	None reported	None reported	Hemorrhagic disease
Other	Conjunctivitis reported as common in some outbreaks. Severe congenital infections	Conjunctivitis common	Glomerulonephritis, asymptomatic congenital infections	None reported	None reported	None reported	None reported

[a]NI, no information.

Table 2 Summary of the major ecological characteristics of the arthritogenic alphaviruses

Virus	Reservoir	Major vectors	Geographic distribution
Chikungunya virus	Nonhuman primates, possibly rodents	Africa (rural): *Aedes africanus, Ae. furcifer, Ae. luteocephalus, Ae. taylori*; Asia: *Ae. aegypti, Ae. albopictus*	Africa, Saudi Arabia, SE Asia, Philippines
O'nyong-nyong virus	Unknown	*Anopheles funestus, An. gambiae* (and related species)	Uganda, Kenya, Tanzania, Zaire, Malawi, Mozambique, Senegal, Zambia, Cameroon, and the Central African Republic; Igbo Ora variant in Nigeria, the Central African Republic, Côte d'Ivoire.
Ross River virus	Marsupials, especially macropods	*Culex annulirostris, Ae. vigilax, Ae. camptorhynchus, Ae. notoscriptus, Ae. sagax*	Australia, New Guinea, Irian Jaya, Pacific Islands
Barmah Forest virus	Marsupials, especially macropods	*Cx. annulirostris, Ae. vigilax, Ae. camptorhynchus*	Australia
Mayaro virus	Possibly wild vertebrates	*Haemagogus* spp.	Central America, northern South America, the Amazon Basin, Trinidad
Sindbis virus	Range of wild and domestic birds	Northern Europe: *Ae.* spp., *Culiseta* spp., *Culex* spp.; Africa: *Culex* spp., *Aedes* spp., *Mansonia* spp., *Cx. univittatus* (in South Africa); Australia: *Cx. annulirostris*	Southern and northeastern Africa, Scandinavia, Finland, Russia, Central and Eastern Europe, Asia, Southeast Asia, and Australia

Attempts to isolate RRV from the joint fluid or tissues of patients with persisting joint disease following RRV infection have been unsuccessful, but RRV antigens have been detected within synovial fluid mononuclear cells. RRV RNA has been found within synovial tissue by polymerase chain reaction (PCR) and RRV can persist within macrophage cultures *in vitro* even in the presence of neutralizing antibodies. All this suggests that the chronic arthritis may result from a persistent, nonreplicative infection that induces an ongoing immune response to viral antigens.

There is also evidence that host factors are likely to influence outcomes. RRV joint disease in humans has been associated with HLA-DR7 positivity, which has a possible role in reducing the cytotoxic T-cell responses to the virus.

Diagnosis of Infection

A number of the alphaviruses produce a viremia detectable by culture or nucleic acid amplification during the early stages of infection, though most patients are not seen until after this period. Alphaviruses will grow in mosquito cells such as C6/36, AP-61, or TRA-284 incubated for 3–4 days at 28 °C, followed by blind passage to indicator cells, such as Vero, baby hamster kidney (BHK), or chick embryo cells that are incubated at 37 °C for a few days. The virus can be identified using monoclonal antibodies, by neutralization tests, or by amplification and characterization of viral nucleic acid using virus-specific primers, probes, or product sequencing.

Detection of viral RNA is significantly more sensitive than is culturing and has been used for detection of CHIKV, RRV, and SINV in blood, and occasionally from other clinical material. A variety of assays directed at NSP1 or E2 viral protein sequences have been used either in uniplex or multiplex formats and, more recently, DNA microarrays have been used for virus identification and characterization.

However, the culture and PCR-based methods are expensive, difficult to access, and negative in most cases. Therefore, diagnosis is almost exclusively done using serological tests. Most diagnostic testing uses the enzyme immunoassays (EIAs) or hemagglutination inhibition (HI) tests, with some use of indirect immunofluorescence antibody (IFA) assays. EIAs and IFA tests can be used to specifically detect either immunoglobulin G (IgG) or immunoglobulin M (IgM). HI will detect both IgG and IgM but is relatively insensitive for IgM detection. Neutralizing antibody titers are regarded as the most specific of the tests, but are confined to specialized laboratories.

IgM antibodies nearly always appear within 1 week of onset of illness, followed by a rise in IgG titer. It is characteristic of alphavirus infections that IgM persists for long periods of time. RRV IgM and BFV IgM usually persist for several months and sometimes years, independent of whether the patient has ongoing symptoms. Persisting IgM has also been documented for SINV and CHIKV, and is likely to be common to all the arthritogenic alphaviruses. Therefore detection of IgM does not, by itself, prove recent infection. That is best diagnosed either by seroconversion from IgG negative to IgG positive or by a fourfold or greater rise in rise in IgG titer between acute and convalescent samples tested in parallel.

False-positive IgM results may also occur occasionally and appear to be a problem particularly with EIA tests, but can occur with other tests as well. They are due to either a nonspecific reaction of the assay or to cross-reaction with IgM produced in response to infection with a related virus. For example, false-positive BFV IgM is occasionally seen in patients with genuine RRV IgM, incorrectly suggesting a dual infection. Cross-reacting antibodies are more likely to occur between closely related viruses, such as ONNV and CHIKV, or MAYV and SFV, but also occasionally occur with rubella or with infectious and noninfectious conditions that cause polyclonal B-cell activation.

Therefore, the proper interpretation of alphavirus serology requires a good clinical history of the illness and knowledge of the viruses to which the patient may have been exposed. Nevertheless, standard serological tests performed in the correct clinical context have proved to be very reliable.

Treatment of Alphavirus Infections

There is a very limited literature on the treatment of alphavirus arthritis. Rest and gentle exercise assist many patients, and nonsteroidal anti-inflammatory agents and simple analgesics are commonly used and provide substantial benefit for the musculoskeletal manifestations.

There are currently no specific antiviral agents available for the treatment of alphavirus infections, nor are there likely to be any in the near future. Drugs used for treatment of rheumatoid arthritis help some patients. Hydroxychloroquine has shown a benefit for patients with post-CHIKV arthritis in one small trial, and corticosteroids have been used for acute RRV arthritis, but most practitioners remain reluctant to use these agents until there are more data about long-term benefits and safety.

Barmah Forest Virus

BFV was first isolated from *Culex annulirostris* in southeast Australia in 1974, then from mosquitoes in other areas of eastern and northern Australia. Human cases were not reported until the 1980s in southeastern Australia, Queensland, the far north of Western Australia, and in the Northern Territory. The virus was subsequently found to be endemic in the tropical areas of northern Australia. Since then there has been appearance of human disease in new areas both within the tropical north and in the more southerly temperate areas where RRV is also active. BFV activity in new areas has been marked by human epidemics of varying size, after which it settles into an endemic pattern that is similar to, but not always coincident with, RRV activity. Human infections have now occurred throughout mainland Australia, with seasonality and incidence varying between regions.

BFV has been isolated from many of the same mosquito species that carry RRV. It is assumed that the same marsupial hosts are important in maintenance and amplification of the virus, but the full range of vertebrate hosts remains to be determined. BFV shows genetic homogeneity across Australia, suggesting that the same strain of BFV spreads widely across the country.

BFV disease has been reported in people aged 5–73 years old, although most cases of both BFV disease and RRV disease occur between the ages of 20 and 60 years. Clinically it is very similar to RRV disease, though joint involvement is less common and less severe, while rash is more common and more likely to be vesicular or urticarial.

Chronic illness occurs following BFV arthritis, but there are few data regarding this. Chronicity is less common after BFV infection than following RRV infection and probably occurs in about 10% of patients.

Chikungunya Virus

CHIKV disease was first recognized following an outbreak of arthritis in Tanzania in 1952, and shortly after was isolated from human serum as well as from *Aedes* spp. and *Culex* spp. mosquitoes. The name is a Swahili word meaning 'that which contorts (or bends up)' and refers to the severe joint pains and stiffness associated with this illness.

The virus is widespread across sub-Saharan Africa, Saudi Arabia, the Indian subcontinent, and Southeast Asia, especially the Philippines. Two major genetic lineages exist, of which one is restricted to West Africa. The other is further subdivided into Asian and East African sublineages. While there are some biological differences between these lineages, it is not known whether these have any specific pathogenic implications. There are data to suggest that genetic diversity may be seen among CHIKV strains responsible for individual outbreaks in Africa and in adjacent areas. For example, during the recent Indian Ocean outbreak, the virus was from the East African sublineage, but showed evolution during the course of the outbreak. The Asian sublineage shows a much greater genetic conservation.

In Africa, the virus shows low-level endemic activity in rural areas, where it is believed to be maintained in a cycle involving *Aedes* spp. mosquitoes, nonhuman primates, and, possibly, rodents. The role of nonhuman mammalian hosts in Asia is not known, though seropositive primates have been described. Occurrence of epidemic disease in Africa and Asia is associated with the rainy season and increases in numbers of *Aedes aegypti* and, in Asia, *Ae. albopictus*. Urban outbreaks may continue over 2 or more years, but cases then typically disappear for several years. *Mansonia* spp. and *Culex* spp. have also been implicated as potential vectors in laboratory studies.

Epidemics occur either due to reemergence of the virus in an area of prior activity once sufficient time has passed for a susceptible population to reestablish, or due to movement of the virus into areas without previous activity. They may involve up to half the population. There have been regular occurrences of epidemics in Africa, with the most recent being in Kinshasa in 1999/2000. In Asia a number of outbreaks occurred in the 1960s, followed by a relatively quiescent period in the 1970s, reemergence in Indonesia in 1982, further spread within Malaysia in 1988–89, reemergence in Thailand in 1995, and in Indonesia in 2001–03. The most recent outbreak began in late 2004 in southwestern Indian Ocean islands and subsequently spread over the next couple of years to affect several hundred thousand people in that region, with the largest number of cases on Réunion, the Seychelles, and in several areas of India. Subsequently, small numbers of cases were detected worldwide in travelers returning from epidemic areas.

CHIKV acute illness is more severe than that seen with the other alphaviruses, with fever that is more frequent and higher, worse joint pain, a greater incidence of severe disease, and occasional deaths. While most patients recover within a few days or weeks, chronic joint pains lasting months to years occur in about 10% of patients.

Hemorrhagic manifestations, including petechiae and gingival bleeding due to CHIKV infection, are well described but rare and can lead to a misdiagnosis of dengue. This is complicated by the fact that these two viruses may co-circulate and patients may be simultaneously infected by both. Fatalities in the past have been extremely rare and were reported to occur in the very young or in those with severe hemorrhagic disease, though it is not certain that the fatalities were due to CHIKV. In the recent Indian Ocean outbreak, a number of cases of CHIKV encephalitis and deaths have been reported. The latter have occurred in the elderly or those with co-morbidities.

Mayaro Virus

MAYV was first isolated from a forest worker in Trinidad in 1954, and is now known to be widely distributed in Central America, northern South America, and the Amazon Basin. Transmission is primarily by forest-dwelling *Haemagogus* spp. mosquitoes, though *Mansonia venezuelensis* and *Ae. aegypti* are possible vectors. The virus is probably maintained in a sylvatic cycle between mosquitoes and wild vertebrates. There are two genotypes, one being widespread and the other found only in Brazil.

Occasional outbreaks and sporadic cases occur within the endemic areas following human contact with the forest environment. The illness is very similar to that caused by CHIKV, including hemorrhagic disease in some cases.

Una virus was initially classified as subtype of MAYV, but is now regarded as a separate species and has not been shown to cause human disease.

O'nyong-nyong Virus (Including Igbo-Ora Variant)

ONNV is closely related to CHIKV both virologically and clinically. The disease was first described in Uganda in 1959 and the name is a tribal word for painful joints. It was subsequently isolated from human serum and from mosquitoes. That epidemic eventually involved more than 2 million people in Uganda, Kenya, Tanzania, Zaire, Malawi, Mozambique, Senegal, and Zambia. Spread was thought to have occurred by movement of viremic humans. This virus has also been found in Cameroon and the Central African Republic. A variant called Igbo-Ora was later found in West Africa, in Nigeria, Central African Republic, and Côte d'Ivoire. ONNV disappeared for 35 years before reappearing in Uganda in 1996/1997, with subsequent cases in Kenya. Infections due to the major strain (i.e., not Igbo-Ora) have recently been reported from Côte d'Ivoire and Chad.

Transmission is by *Anopheles funestus* and *An. gambiae*. It is presumed that there is a nonhuman mammalian host to account for maintenance of the virus between epidemics, but it has not yet been identified.

The clinical illness is very similar to CHIKV disease, though fever is usually milder and there is prominent cervical lymphadenopathy. Joint pain persisting for several months has been noted, but the frequency is not known.

Ross River Virus

RRV was named after an area near Townsville in northeastern Australia, where it was first isolated from *Ae. vigilax* in 1966, from mosquitoes collected in 1959. It was isolated from the blood of a febrile child in 1973, then from a case of epidemic polyarthritis in 1979. It has since been found in Papua New Guinea, Irian Jaya, and the Solomon Islands. In 1979–80 an epidemic occurred in various Pacific Islands involving over 50 000 cases, probably originating from a viremic air traveler from Australia.

Three genotypes of RRV have so far been described. Genotype 1 was present in Queensland in northeastern Australia until the mid-1970s, after which it disappeared. Genotype 2 has always been the major type on the east coast of Australia and is now the major current circulating type throughout Australia. Genotype 3 was the dominant type in the southwest of Australia until 1996, since which it has largely been replaced by genotype 2 virus. Pacific Islands isolates belong to genotype 2.

RRV is primarily maintained in a cycle between mosquitoes and vertebrate hosts. Two salt-marsh mosquitoes, *Ae. vigilax* and *Ae. camptorhynchus,* are important in northern and southern coastal areas of Australia, respectively. In some coastal and inland areas, transmission to humans occurs from several freshwater breeding species, especially *Cx. annulirostris* but also *Ae. sagax* and *Ae. normanensis,* as well as *Ae. camptorhynchus* breeding in brackish and freshwater. Other species, such as *Coquillettidia linealis* and *Ae. notoscriptus,* may play significant roles in urban environments.

The major vertebrate hosts for both maintenance and amplification of RRV are the macropod marsupials (kangaroos, wallabies, and euros). Horses and small marsupials, such as possums, may be important in periurban and urban areas. There is circumstantial evidence that RRV may survive for years in desiccation-resistant eggs of some *Aedes* spp.

RRV disease is reported every year in Australia, with between 2000 and 8000 notified cases per annum. In the tropical northern parts of the country most activity is seen during the wet season from December to May. Further south in temperate regions, human infections occur predominantly in late spring, summer, or autumn. In those seasons, rainfall or tidal inundation of coastal breeding sites coupled with warm temperatures leads to an increase in vector numbers. Major outbreaks in these areas occur every 2–4 years. In the arid inland regions of Australia occasional outbreaks occur following heavy rainfall with flooding.

RRV disease has been reported in people ranging in age from less than 1 year old to 88 years old, although most cases occur between the ages of 20 and 60 years. Clinical illness appears to be uncommon in children. The clinical illness is typical of these alphaviruses, though tenosynovitis and fasciitis are more common. Joint pains, muscle pains, and fatigue persist for months or years in 70–90% of patients.

Semliki Forest Virus

SFV was first isolated in 1942 from mosquitoes in Uganda and is distributed across sub-Saharan Africa. Serological surveys show that human infection is common in endemic areas, but it has only been associated with a single reported human outbreak of fever, headache, and joint pains that occurred in the Central African Republic in 1987.

Sindbis Virus (Including Ockelbo Virus)

SINV was first isolated from *Culex univittatus* at Sindbis in Egypt, and subsequently from a wide range of species of mosquitoes and vertebrates in Europe, the Middle East, Africa, India, Asia, the Philippines, and Australia. It was not until 1961 that it was isolated from the blood of a febrile human in Uganda, and then linked to human disease in South Africa in 1963.

SINV is the most widely distributed arbovirus causing human disease. It is found throughout southern and northeastern Africa, Northern, Central, and Eastern Europe, Asia, Southeast Asia, and Australia, but clinical disease is largely restricted to Sweden (Ockelbo disease), Norway (Ockelbo disease), Finland (Pogosta disease), Russia (Karelian fever), and South Africa. Clinical disease has only rarely been reported in Asia. Only isolated cases of illness in humans have been reported from Australia, although serological studies have indicated that human infections occur regularly.

Many species of birds and of other vertebrates have been shown to be capable of being infected with SINV, but disease has been identified only in humans. Mosquitoes feeding on birds appear to be the primary source of human infection. In Sweden the virus has been found primarily in the fieldfare (*Turdus pilaris*), redwing (*T. iliacu*), and song thrush (*T. philomelos*), and in Norway game birds also carry the virus. Transmission to humans is via *Aedes, Culiseta,* and *Culex* spp. in Sweden, Norway, and Finland. In Africa *Culex, Aedes,* and *Mansonia* spp. are important, with *Cx. univittatus* being the primary human vector in South Africa. In Australia, *Cx. annulirostris* is the dominant vector species, but it is also been isolated from *Aedes* spp. No maintenance cycles involving nonhuman vertebrates have been identified as yet in any country.

Two major lineages circulate: the Paleoarctic/Ethiopian and the Oriental/Australian. It is believed that migratory birds are responsible for the spread of the two lineages, though within each there is a genetic heterogeneity consistent with local adaptation and evolution of the virus. Studies in Australia have shown that there is also temporal evolution of the virus within a lineage, presumed to be due to periodic introduction of new variants by migratory birds. A third unique lineage has been identified in the southwest of that country.

The restricted distribution of human epidemics suggests that the Paleoarctic/Ethiopian lineage has a greater ability to cause disease than does the Oriental/Australian lineage. However, the ecology of SINV and the pathogenesis of disease are still poorly understood, so other factors may account for the unusual disease patterns.

Epidemics of Ockelbo disease and Pogosta disease occur during summer and autumn in northern Europe, coincident with the peak mosquito breeding period. Ockelbo occurs each year, while Pogosta disease has shown a consistent periodicity of 7 years since it was first recognized in 1974, the reasons for which are uncertain. Summer–autumn SINV epidemics have also been described in South Africa. Patients usually recover within a few weeks but chronic joint pains are common.

The Future

Alphavirus infections impose a substantial human health and economic burden, without any immediate prospect of antiviral therapy or vaccines. Further research is needed to understand the ecology and epidemiology of these viruses and the pathogenesis of disease.

See also: Arboviruses; Rubella Virus; Togaviruses: Alphaviruses; Togaviruses Causing Encephaitis; Togaviruses: General Features; Togaviruses: Molecular Biology; Togaviruses Not Associated with Human Disease; Vector Transmission of Animal Viruses; West Nile Virus.

Further Reading

Griffin DE (2001) Alphaviruses. In: Knipe DM and Howley PM (eds.) *Fields Virology*, 4th edn., pp. 917–962. Philadelphia, PA: Lippincott Williams and Wilkins.

Harley D, Sleigh A, and Ritchie S (2001) Ross River virus transmission, infection, and disease: A cross-disciplinary review. *Reviews in Clinical Microbiology* 14: 909–932.

Johnson BK (1989) O'nyong-nyong virus disease. In: Monath TP (ed.) *The Arboviruses: Epidemiology and Ecology*, vol. III, pp. 217–223. Boca Raton, FL: CRC Press.

Jupp PG and McIntosh BM (1989) Chikungunya virus disease. In: Monath TP (ed.) *The Arboviruses: Epidemiology and Ecology*, vol. II, pp. 137–157. Boca Raton, FL: CRC Press.

Kurkela S, Manni T, Myllynen J, Vaheri A, and Vapalahti O (2005) Clinical and laboratory manifestations of Sindbis virus infection: Prospective study, Finland, 2002–2003. *Journal of Infectious Diseases* 191: 1820–1829.

Laine M, Luukkainen R, and Toivanen A (2004) Sindbis virus and other alphaviruses as a cause of human arthritic disease. *Journal of Internal Medicine* 256: 457–471.

Mackenzie JS, Broom AK, Hall RA, *et al.* (1998) Arboviruses in the Australian region, 1990 to 1998. *Communicable Diseases Intelligence* 22: 93–100.

Niklasson B (1989) Sindbis and Sindbis-like viruses. In: Monath TP (ed.) *The Arboviruses: Epidemiology and Ecology*, vol. III, pp. 167–176. Boca Raton, FL: CRC Press.

Pinheiro FP and LeDuc JW (1989) Mayaro virus disease. In: Monath TP (ed.) *The Arboviruses: Epidemiology and Ecology*, vol. III, pp. 137–150. Boca Raton, FL: CRC Press.

Powers AM, Aguilar PV, Chandler LJ, *et al.* (2006) Genetic relationships among Mayaro and Una viruses suggest distinct patterns of transmission. *American Journal of Tropical Medicine and Hygiene* 75: 461–469.

Powers AM, Brault AC, Tesh RB, and Weaver SC (2000) Re-emergence of Chikungunya and O'nyong-nyong viruses: Evidence for distinct genetic lineages and distant evolutionary relationships. *Journal of General Virology* 81: 471–479.

Rulli NE, Suhrbier A, Hueston L, *et al.* (2005) Ross River virus: Molecular and cellular aspects of disease pathogenesis. *Pharmacology and Therapeutics* 107: 329–342.

Russell RC (2002) Ross River virus: Ecology and distribution. *Annual Reviews in Entomology* 47: 1–31.

Suhrbier A and Linn ML (2004) Clinical and pathological aspects of arthritis due to Ross River virus and other alphaviruses. *Current Opinions in Rheumatology* 16: 374–379.

Togaviruses Not Associated with Human Disease

L L Coffey, Institut Pasteur, Paris, France

© 2008 Elsevier Ltd. All rights reserved.

Glossary

Cohabitation experiment Fish needle-inoculated with wild-type salmonid alphaviruses (SAVs) are placed in the same tank with naive fish to determine whether SAV can be transmitted from infected to uninfected fish via cohabitation. Alternatively, fish immunized with inactivated SAVs are placed in the same tank as naive fish and then fish experimentally inoculated with wild-type SAVs are introduced to determine whether immunized fish can passively transfer SAV immunity to naive individuals in the same water.

Lice Any of numerous small, flat-bodied, wingless biting or sucking insects of the orders Mallophaga or Anoplura, many of which are external parasites of vertebrates.

Smolt A young salmon at the stage intermediate between the parr and the grilse, when it becomes covered with silvery scales and first migrates from freshwater to the sea.

Introduction

The virus family *Togaviridae* comprises two genera, *Alphavirus* and *Rubivirus*. The latter includes only Rubella virus and will be discussed elsewhere in this encyclopedia. The alphavirus-type species is Sindbis virus which, as do many other alphaviruses, causes fever with rash. There are many medically important alphaviruses, including Eastern equine encephalitis virus, Western equine encephalitis virus, and

Venezuelan equine encephalitis virus; these also will be discussed elsewhere in this encyclopedia.

Most alphaviruses are transmitted by hematophagous arthropods to susceptible terrestrial mammals and, as mentioned, can cause diseases in them ranging from simple fevers to fatal encephalitides. The exceptions are alphaviruses that have not been associated with human rash and fever or encephalitis, or occur only in fish. These and certain other viruses are outlined below and summarized in **Table 1**.

Humans in South America have been shown to have antibody to the mosquito-borne alphaviruses Una virus (UNAV, Semliki Forest antigenic complex) and Trocara virus (not placed in an antigenic complex); these viruses have not been associated with human disease. However, UNAV is closely related to Mayaro virus, a known human pathogen. Antibody to Ndumu virus (NDUV) was found in humans in South Africa in the 1950s and NDUV isolates from *Mansonia* and *Aedes* mosquitoes cause encephalitis in mice; NDUV has not been associated with human disease. Cabassou virus (CABV), a member of the Venezuelan equine encephalitis virus complex, has been isolated from mosquitoes, marsupials, and birds in French Guiana; human disease caused by CABV has not been observed. Whataroa virus (WHAV; Western equine encephalitis virus complex) has been isolated from mosquitoes, birds, and marsupials in New Zealand and from mosquitoes in Australia. Neutralizing antibody to WHAV has been detected in humans in those countries as well; human disease caused by WHAV has not been recorded. Bebaru virus (BEBV), a Semliki Forest complex virus, has been isolated from *Culex* (*Lophoceraomyia*) spp. and *Aedes butleri* mosquitoes in Malaysia and Australia. Neutralizing antibody to BEBV has been detected in Malaysians. Relatively little is known about these six viruses. They have not often been isolated from mosquitoes and their transmission cycles are not defined.

Neutralizing antibodies to Middelburg virus (MIDV), a member of the Middelburg virus complex, have been detected in humans in South Africa, Mozambique, and Angola; no human disease has been associated with MIDV. Neutralizing antibodies to MIDV have been detected in cattle, goats, sheep, and horses in South Africa. However, it is unknown whether livestock participate in the natural transmission of MIDV. Numerous MIDV isolates have been made from *Aedes* and *Mansonia* mosquitoes and virus transmission by *Aedes caballus* has been documented. Experimentally infected *Ae. caballus* successfully transmitted MIDV to lambs starting 4 days post-infection, and the lambs developed viremia and a stiff gait. Despite this, MIDV has not been associated with disease in livestock.

Getah virus (GETV), a member of the Semliki Forest virus complex, was first isolated from mosquitoes in Malaysia and subsequently from mosquitoes in other Asian countries, Australia, and eastern Russia. The first recognition of disease caused by GETV was in racehorses in Japan in 1978. Antibody to this virus has been detected not only in horses but also in humans, pigs, and other mammals in Asia and Australasia. Pathogenicity of GETV is relatively mild and appears to be limited to horses and pigs. Several outbreaks of GETV disease in horses have been reported in Japan and India. Disease is principally characterized by fever, rash, and limb edema and also includes (India) depression, anorexia, and lymphocytopenia. The rash may be caused by immune complex-mediated hypersensitivity. Horses recover completely. Pigs are highly susceptible to GETV infection, with occasional deaths of newborn piglets or fetuses occurring by transplacental infection; adult pigs are less likely to be symptomatic. Because Japanese encephalitis virus, a flavivirus, also causes deaths of fetal pigs, laboratory diagnosis is required to distinguish between infections caused by these two viruses.

Fort Morgan virus (FMV), a Western equine encephalitis complex virus, was first isolated in 1973 from *Oeciacus vicarius* (Hemiptera: Cimicidae) bugs in a nest of cliff swallows occupied by house sparrows (*Passer domesticus*). Subsequently, numerous FMV isolates, plus isolates of a closely related subtype, Buggy Creek virus, were made from *O. vicarius* bugs, house sparrows, and cliff swallows (*Petrochelidon pyrrhonota*); no human disease caused by FMV has been observed. Experimentally infected *O. vicarious* transmitted FMV to nestling sparrows 1 month after infection and for up to 312 days after being held at temperatures simulating hibernation. Mosquitoes of certain *Culex* and *Culiseta* species were refractory to FMV infection, suggesting that transmission is likely mediated only by *O. vicarius*.

Until recently, knowledge of alphaviruses was limited to viruses affecting terrestrial mammals and birds. The recent discovery of alphaviruses in lice parasitizing elephant seals (family Phocidae, genus *Mirounga*) and in fish indicates that alphaviruses occur in both terrestrial and marine environments. It is unknown whether louse-borne Southern elephant seal virus (SESV) causes disease in elephant seals, but outbreaks of sleeping disease and pancreas disease (PD), both caused by salmonid alphaviruses (SAVs), are widespread in farmed fish throughout Western Europe and North America and cause major economic losses to aquaculture industries.

Southern Elephant Seal Virus

SESV was first isolated in 2001 from seal lice, *Lepidophthirus macrorhini*, collected from southern elephant seals (*Mirounga leonina*) on Macquarie Island, 2000 km south of the Australian mainland. Whereas the vector competence of *L. macrorhini* has not been evaluated, Macquarie Island has no mosquitoes and ticks are rare

Table 1 Alphaviruses not associated with human disease or that occur only in fish or have been associated with marine mammals

Virus (abbreviation)	Antigenic complex[a]	Origin	Original source	Distribution	Reservoir	Vector
Aura virus (AURAV)	WEE	Brazil (1959)	Culex (Melanoconion) spp.	Northern South America	Terrestrial mammals?	Mosquitoes
Bebaru virus (BEBV)	SF	Malaysia (1956)	Culex (Lophoceraomyia) spp.	Malaysia, Australia	Not known	Mosquitoes
Cabassou virus (CABV)	VEE	French Guiana (1968)	Culex (Melanoconion) portesi	South and Central America	Not known	Mosquitoes
Fort Morgan virus (FMV)	WEE	Colorado (1973)	Oeciacus vicarius	Western North America	Birds	Oeciacus vicarius (bugs)
Buggy Creek virus (subtype of FMV)	WEE	Oklahoma (1981)	Oeciacus vicarius	Western North America	Birds	Oeciacus vicarius (bugs)
Getah virus (GETV)	SF	Malaysia (1955)	Culex gelidus	Eurasia, Australia	Terrestrial mammals	Mosquitoes
Middelburg virus (MIDV)	Middelburg	South Africa (1957)	Aedes caballus	Africa	Terrestrial mammals?	Mosquitoes
Ndumu virus (NDUV)	Ndumu	South Africa (1959)	Aedes circumluteolus	Africa	Terrestrial mammals	Mosquitoes
Norwegian salmonid alphavirus (NSAV)	Unclassified	Norway (2005)	Salmo salar	Norway	Salmon	Lice?
Salmon pancreatic disease virus (SPDV)	Unclassified	Western Europe (1976)	Salmo salar	Western Europe, United States	Salmon	Lice?
Sleeping disease virus (SDV)	Unclassified	Western Europe (1985)	Oncorhynchus mykiss	Western Europe	Trout	Lice?
Southern elephant seal virus (SESV)	Unclassified	Macquarie Island (2001)	Lepidophthirus macrorhini	Australia	Not known	Lepidophthirus macrorhini (lice)?
Trocara virus (TROV)	Unclassified	Brazil (1984)	Aedes serratus	South America	Not known	Mosquitoes
Una virus (UNAV)	SF	Colombia (1964)	Psorophora ferox	South America, Panama	Terrestrial mammals	Mosquitoes
Whataroa virus (WHAV)	WEE	New Zealand (1962)	Culex pervigilans	New Zealand	Birds, marsupials	Mosquitoes

[a]Antigenic complex abbreviations: SF: Semliki Forest, VEE: Venezuelan equine encephalitis, WEE: Western equine encephalitis. A question mark indicates a postulated host.

on elephant seals, further suggesting that lice are vectors of this virus. SESV has not been isolated from elephant seals, but the prevalence of neutralizing SESV antibody is high in a population of elephant seals on Macquarie Island and is higher in older animals.

Genetic comparisons have revealed that SESV is most closely related to Australasian alphaviruses in the Semliki Forest complex, but antigenic studies showed that SESV is only very distantly related to other members of the Semliki Forest complex. That alphaviruses may be vectored by lice on migrating elephant seals traveling long distances, including between Australasia and South America via Antarctica, suggests that dispersal of some alphaviruses may be by such mechanisms, although seal populations other than those on Macquarie Island have not been sampled for evidence of SESV infection.

Salmonid Alphaviruses

Alphaviruses enzootic in fish include salmon pancreas disease virus (SPDV), sleeping disease virus (SDV), and Norwegian salmonid alphavirus (NSAV). These viruses cause morbidity and mortality in farmed salmon and trout. The SAV shares only ~33% amino acid identity with other alphaviruses, the basis for considering them as quite distinct from other alphaviruses.

Salmon Pancreas Disease Virus

PD has been reported in farmed Atlantic salmon, *Salmo salar*, in Western Europe and the United States since 1976 and causes up to 50% mortality in salmon smolts. Affected smolts develop necrosis of the exocrine pancreas, cardiomyopathy, skeletal myopathy that causes infected fish to swim sluggishly, and inappetance, all of which lead to stunted growth and, sometimes, death. Infected salmon smolts experience a triphasic disease including: (1) a pre-acute stage characterized by pancreatic exocrine necrosis and basophilia of pancreatic acinar cells; (2) an acute phase marked by hemorrhagic pancreatic necrosis and occasional liver and kidney lesions as well as inappetance and inactivity; and (3) a post-acute phase in which the pancreas recovers histopathologically, although survival and weight are lower than in uninfected fish of the same age.

For many years the causative agent of this disease was not known, and attempts to isolate a virus from moribund fish were unsuccessful. The strongest evidence that PD was transmissible came from experimental studies in which PD developed in naive fish injected with spleen homogenates from diseased salmon and from experiments in which naive fish that cohabited in tanks with PD-afflicted fish developed the disease. The etiological agent of PD was identified in 1999 as the alphavirus SPDV on the basis of sequence analysis of viral RNA extracted from tissues of infected farmed salmon. Ectoparasitic sea lice of salmon, *Lepeophtheirus salmonis*, have been postulated as vectors but SPDV-infected lice have not been reported and experimental vector competence studies are lacking. Since PD has been reported only in salmon smolts, natural reservoirs of SPDV presumably are fish. Cohabitation experiments suggest that transmission can occur via direct horizontal spread among salmon in the absence of arthropod vectors, although whether lice were eliminated from experimentally infected salmon is unclear. Further, the stability of SPDV in seawater of various temperatures or evidence of contact transmission among fish or via their excreta has not been evaluated.

As with other alphaviruses, SPDV contains a single-stranded positive-sense RNA genome of approximately 12 kbp. SPDV shares fundamental biological properties with other alphaviruses, although it replicates more slowly in cell cultures than do other alphaviruses and only infects select fish cells. Salmon smolts experimentally infected with SPDV develop hemorrhagic pancreatic necrosis consistent with lesions observed in wild fish. Smolts immunized with inactivated cultured SPDV are protected against cohabitation challenge with wild-type SPDV, indicating that tissue-culture adapted SPDV may have potential as a vaccine candidate to protect against PD.

Sleeping Disease Virus

Outbreaks of SDV have been recognized since 1985 in farmed freshwater rainbow trout, *Oncorhynchus mykiss*, in France, Great Britain, and Norway. Diseased trout lay on their sides at the bottom of their tanks, reminiscent of a sleeping state. Lesions of diseased fish include the sequential development of pancreatic atrophy, cardiac necrosis, and, ultimately, skeletal muscle degeneration. Sleeping disease has been estimated to affect 30% of fish farms in northern France. However, estimates are based on behavioral observations and scant histological studies of moribund fish, and it is difficult to discriminate between pancreatic lesions induced by SDV and those occurring in the similar disease caused by the salmonid birnavirus infectious pancreatic necrosis virus. As with SPDV, mechanisms of SDV transmission are not well understood. One SDV isolate was made from rainbow trout farmed in seawater in close proximity to diseased salmon in Norway, suggesting water-borne transmission. Another possibility is vector-borne transmission via an arthropod or another unrecognized organism that associates with both trout and salmon.

Until 2000, efforts to isolate and characterize SDV proved unsuccessful because the virus does not produce cytopathic effects (CPEs) in most fish cell lines. Isolation of SDV typically requires incubating SDV-infected fish organ homogenates with rainbow trout or Chinook salmon embryo cells at 10 °C for 7–11 days, with CPE generally only becoming evident after several passages. Rainbow trout inoculated intraperitoneally with SDV have detectable viremias within 1 week after inoculation and develop neutralizing antibody to the virus.

Although it exhibits unique ecological characteristics, SDV is presently classified as a strain of SPDV. Genetic analyses of isolates revealed that SDV resembles mammalian alphaviruses with respect to genome organization, conserved nucleotide sequence elements, and amino acid motifs. However, the SDV E3 protein lacks glycosylation. The SDV E1 and E2 glycoproteins diverge genetically from other alphaviruses by 66–70%. This genetic diversity between SDV (and the other SAV) and all other alphaviruses is greater than the degree of diversity between any two mammalian alphaviruses, indicating that the SAVs are the most genetically divergent alphavirus species known. Both SDV and SPDV share only 26% envelope 2 (E2) gene amino acid identity with SESV, a level similar to the identities that SDV and SPDV share with respect to E2 proteins of other alphaviruses. This lack of sequence similarity between marine alphaviruses is evidence against a common aquatic origin of alphaviruses. Comprehensive sequence comparisons using all available SAV sequences with members of other alphavirus complexes are needed if we are to clarify the phylogenetic relationships between SAVs and other alphaviruses, including SESVs.

Sequence comparisons of SDV isolated from geographically distant areas in Europe revealed that strains fall into three genotypes according to location of isolation, but these genetic distinctions are not accompanied by detectable antigenic differences. Even with newly developed monoclonal antibody-based immunoperoxidase detection assays, SDV isolates cannot be differentiated antigenically from one another or from SPDV. Due to difficulties in distinguishing SDV isolates from each other and from SPDV, and to the similar clinical presentations of the three SAVs and infectious pancreatic necrosis virus, definitive diagnosis of SAV is achieved by comparisons of sequenced strains.

Comparative studies revealed that SDV and SPDV share 91% amino acid identity and that SPDV has a 24 nt insertion at the C-terminus of the nsP3 gene that encodes a protein involved in viral RNA synthesis; the phenotypic relevance of this insertion is unknown. Given the rapid and artificial evolution of alphaviruses in cell culture, genomic analyses of SAV that are customarily recovered after extensive fish cell passage should be interpreted with caution. For example, the first published SPDV sequence, derived from virus subjected to repeated passage, contained cell-culture-adapted variant sequences that were not representative of sequences generated from low or unpassaged SPDV isolates.

Norwegian Salmonid Alphaviruses

In 2005, a genetically distinct SAV, NSAV, was identified in diseased salmon and rainbow trout in Norway. NSAV differs genetically from SDV by ~7% and from SPDV by ~9%. Fish infected with NSAV experience disease similar to PD and SD. Unlike certain terrestrial alphaviruses that use small rodents as reservoirs, isolates of NSAV collected from geographically dispersed Norwegian farms raising saltwater fish do not segregate phylogenetically according to physical proximity, suggesting either extensive gene flow within the fish reservoir or a common source of virus. Given that farmed smolts do not migrate in open water, movement of fish by humans for aquaculture is one possible means of NSAV dispersal.

See also: Togaviruses Causing Encephaitis; Togaviruses: Equine Encephalitic Viruses.

Further Reading

Büchen-Osmond C (2003) Taxonomy and classification of viruses. In: Murray PR, Baron EJ, Jorgenson JH, Pfaller MA, and Yolken RH (eds.) *Manual of Clinical Microbiology*, 8th edn., vol. 2, pp. 1217–1226. Washington, DC: ASM Press.

Desvignes L, Quentel C, Lamour F, and Le Ven A (2002) Pathogenesis and immune response in Atlantic salmon (*Salmo salar* L.) parr experimentally infected with salmon pancreas disease virus (SPDV). *Fish and Shellfish Immunology* 12: 77–95.

Fukunaga Y, Kumanomido T, and Kamada M (2000) Getah virus as an equine pathogen. *Veterinary Clinics of North America: Equine Practice* 16: 605–617.

Hodneland K, Bratland A, Christie KE, Endresen C, and Nylund A (2005) New subtype of salmonid alphavirus (SAV), *Togaviridae*, from Atlantic salmon *Salmo salar* and rainbow trout *Oncorhynchus mykiss* in Norway. *Diseases of Aquatic Organisms* 66: 113–120.

Karlsen M, Hodneland K, Endresen C, and Nylund A (2005) Genetic stability within the Norwegian subtype of salmonid alphavirus (family *Togaviridae*). *Archives of Virology* 151: 861–874.

Kokernot RH, McIntosh BM, and Worth CB (1961) Ndumu virus, a hitherto unknown agent, isolated from culicine mosquitoes collected in northern Natal, Union of South Africa. *American Journal of Tropical Medicine and Hygiene* 10: 383–386.

Kumanomido T (2001) Getah virus. In: Service MW (ed.) *The Encyclopedia of Arthropod-Transmitted Infections*, pp. 194–197. Oxon, UK: CABI Publishing.

La Linn M, Gardner J, Warrilow D, et al. (2001) Arbovirus of marine mammals: A new alphavirus isolated from the elephant seal louse, *Lepidophthirus macrorhini*. *Journal of Virology* 75: 4103–4109.

López-Dóriga MV, Smail DA, Smith RJ, et al. (2001) Isolation of salmon pancreas disease virus (SPDV) in cell culture and its ability to protect against infection by the 'wild-type' agent. *Fish and Shellfish Immunology* 11: 505–522.

Munro ALS, Ellis AE, McVicar AH, Mc Lay HA, and Needham EA (1984) An exocrine pancreas disease of farmed Atlantic salmon in Scotland. *Helgoländer Meeresunters* 37: 571–586.

Powers AM, Brault AC, Shirako Y, *et al.* (2001) Evolutionary relationships and systematics of the alphaviruses. *Journal of Virology* 75: 10118–10131.

Tomori O (2001) Middelburg virus. In: Service MW (ed.) *The Encyclopedia of Arthropod-Transmitted Infections*, pp. 339–340. Oxon, UK: CABI Publishing.

Travassos Da Rosa APA, Turell MJ, Watts DM, *et al.* (2001) Trocara virus: A newly recognized alphavirus (*Togaviridae*) isolated from mosquitoes in the Amazon Basin. *American Journal of Tropical Medicine and Hygiene* 64: 93–97.

Villoing S, Béarzotti M, Chilmonczyk S, Castric J, and Brémont M (2000) Rainbow trout sleeping disease virus is an atypical alphavirus. *Journal of Virology* 74: 173–183.

Weston JH, Villoing S, Brémont M, *et al.* (2002) Comparison of two aquatic alphaviruses, salmon pancreas disease virus and sleeping disease virus, by using genome sequence analysis, monoclonal reactivity, and cross-infection. *Journal of Virology* 76: 6155–6163.

Togaviruses: Alphaviruses

A M Powers, Centers for Disease Control and Prevention, Fort Collins, CO, USA

© 2008 Elsevier Ltd. All rights reserved.

Glossary

Alphavirus A virus in the genus *Alphavirus*.
Clade All of the descendants of a common ancestor (also known as a monophyletic group).
Genotype Genetic composition of individuals; used to delineate relatedness of a group of individuals.

Introduction

The alphaviruses are a group of antigenically related arthropod-borne viruses (arboviruses) that were first isolated in the 1930s. The arboviruses were originally divided into one of three serological groups (A, B, or C) based upon results of hemagglutination inhibition (HI) tests as performed by Casals and Brown. Later characterization of the viruses based upon genetic and virion properties led to the inclusion of those viruses clustered in group A in the genus *Alphavirus*.

Alphaviruses can be broadly categorized into three distinct groups based upon the type of illness they produce in humans and/or animals. Disease patterns include: (1) febrile illness associated with a severe and prolonged arthralgia, (2) encephalitis, or (3) no apparent or unknown clinical illness. Any particular alphavirus will fit into only one of the above profiles but additional symptoms such as demyelination may also be present. Additionally, some of the viruses can induce abortion or stillbirth in animals, decreased egg production in avians, or mortality in vertebrates. Further, a small number of alphaviruses, those strains belonging to the Salmon pancreas disease virus (SPDV) species, infect and cause illness only in fish. The alphaviruses have a wide geographic distribution and are present virtually everywhere on the earth. However, patterns of illness do not correlate with geography as viruses causing any given disease syndrome are found under diverse climatic and ecological conditions. Rather, distribution of a particular virus is tied to vertebrate reservoir and invertebrate vector availability.

Taxonomy and Classification

The genus *Alphavirus* is one of two genera within the family *Togaviridae*. While the second genus, *Rubivirus*, contains only a single species (*Rubella virus*), there are 29 recognized species in the genus *Alphavirus* currently divided into at least eight antigenic complexes (**Table 1**).

Evolutionary Relationships

Numerous phylogenetic studies have been performed on members of the genus *Alphavirus* since the advent of sequencing and genetic analyses. Overall, these derived evolutionary relationships closely agree with the antigenic relationships associated with the surface glycoproteins that have been established for members of this group based upon serological techniques. Evolutionary relationships have not been found to vary with the gene(s) or sequences examined with the exception of the members of the western equine encephalitis (WEE) antigenic complex. When viruses in the WEE complex are examined, they are found to be associated with the WEEV genotype when comparing E2 or E1 glycoproteins, while genetic comparisons of other gene regions result in alignment with viruses in the eastern equine encephalitis virus (EEEV) genotype. This unique feature is a result of the fact that most New World WEE complex viruses

Table 1 Alphaviruses[a]

Antigenic complex	Species	Abbreviation	Distribution	GenBank acc. no.
BF	Barmah Forest virus	BFV	Australia	U73745
EEE	Eastern equine encephalitis virus	EEEV	N. America, Mexico, Caribbean, C. America, S. America	U01034
MID	Middelburg virus	MIDV	Africa	AF398374
NDU	Ndumu virus	NDUV	Africa	AF398375
SF	Bebaru virus	BEBV	Malaysia	AF398376
	Chikungunya virus	CHIKV	Africa, Asia	AF192905
	Getah virus	GETV	Asia, Oceania	AF398377
	Mayaro virus	MAYV	S. America, Trinidad	AF398378
	O'nyong nyong virus	ONNV	Africa	M33999
	Ross River virus	RRV	Oceania, Australia	M20162
	Semliki Forest virus	SFV	Africa; Vietnam	X04129
	Una virus	UNAV	S. America, Panama	AF398381
VEE	Venezuelan equine encephalitis virus	VEEV	C., N., S. America	L01442
	Mosso das Pedras virus	MDPV	S. America	AF398382
	Everglades virus	EVEV	Florida (USA)	AF075251
	Mucambo virus	MUCV	S. America, Trinidad	AF398383
	Tonate virus	TONV	Fr. Guiana	AF398384
	Pixuna virus	PIXV	S. America	AF075256
	Cabassou virus	CABV	French Guiana	AF398387
	Rio Negro virus	RNV	Argentina	AF398388
WEE (Sindbis like)	Aura virus	AURAV	S. America	AF126284
	Sindbis virus	SINV	Africa, Eurasia, Australia	V00073
	Whataroa virus	WHAV	New Zealand	AF398394
WEE (recombinants)	Fort Morgan virus	FMV	Western N. America	AF398389
	Highlands J virus	HJV	Eastern N. America	U52586
	Western equine encephalitis virus	WEEV	S. and N. America	J03854
Unclassified	Trocara virus	TROV	S. America	AF252264
	Salmon pancreas disease virus	SPDV	Europe, N. America	AJ012631 AJ238578
	Southern elephant seal virus	SESV	Australia	EMBL DS44746

[a]Classification according to the most recent report of the ICTV; antigenic complexes according to SIRACA.

are descendants of an ancestor that arose by recombination of an EEEV with a Sindbis (SIN)-like virus. This recombination event was a double crossover event that combined the E2/E1 genes of the SINV-like ancestor with the other genes and noncoding regions of an EEEV ancestor.

Phylogenetic analyses have been most useful in the study of the Old World viruses which have been studied in far less detail than the encephalitic alphaviruses that only exist in the western hemisphere. Partial genome sequence analyses have produced conflicting topologies, particularly in the placement of Middelburg virus (MIDV), which has been not been extensively characterized. However, the use of longer sequences in phylogenetic comparisons and evaluations of tree topologies using likelihood tests provides the best estimates of evolutionary relationships given the data available. A phylogeny based upon distance matrix analysis of the E1 gene of representatives of each virus is presented in **Figure 1** demonstrating the monophyletic clustering of viruses within a given antigenic complex. Significantly, the clades generated by all phylogenetic analyses accurately reflect associations of viruses that share medically important characteristics. For example, the encephalitic alphaviruses cluster together and are more closely related to each other than those viruses that express other disease patterns providing evidence of the ancient relationships that exist among the alphaviruses.

Phylogenetic relationships cannot conclusively verify the geographic origin of the alphaviruses as New World and Old World origins are equally parsimonious with each requiring three transoceanic introduction events. Internal branch lengths of distance matrix trees also suggests that the most basal alphaviruses are those that infect

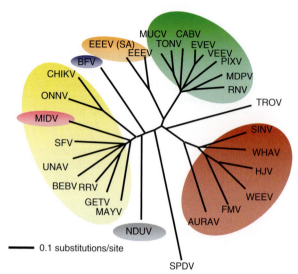

Figure 1 Unrooted phylogenetic tree of all *Alphavirus* species generated from the E1 nucleotide sequences using the F84 algorithm of the neighbor-joining program (SESV is not included because no homologous sequence for this region was available). Virus abbreviations are from **Table 1**. Antigenic complexes are indicated by colored circles.

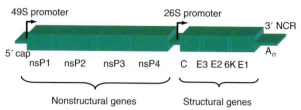

Figure 2 Organization of the alphavirus genome showing genes, promoter elements, and noncoding regions (NCRs).

fish indicating that an ancestral alphavirus adapted to infect fish. Other lineages of alphaviruses evolved as insect-borne viruses as most alphaviral species are recognized to be. Even the southern elephant seal virus (SESV), which has been found to infect lice, supports this hypothesis.

Virion Structure

Alphavirus virions are small (60–70 nm in diameter), icosahedral particles exhibiting $T=4$ symmetry. The single-stranded RNA genome is contained within a capsid core comprised of the virally encoded C protein. This nucleocapsid is then encased in a lipid envelop acquired by budding from the host cell plasma membrane making the virions sensitive to ether and detergents. The envelop membrane also contains the viral E1 and E2 glycoprotein heterodimers that are arranged as trimers on the surface. Binding to cellular receptors occurs via the E2 glycoprotein after which conformational changes in the glycoprotein trimers result following endocytosis.

Genome Organization and Replication

The alphaviruses have a genome consisting of a linear, positive-sense, single-stranded RNA molecule of approximately 11.2–11.8 kbp in size depending upon the species (**Figure 2**). The nonstructural proteins, required for viral replication, are encoded in the 5′ two thirds of the genome, while the structural genes are collinear with the 3′ one third.

The 5′ end of the genome has a 7-methylguanosine cap, while on the 3′ end a polyadenylation tail is present. Just upstream of the poly A tract is a 3′ noncoding region of varying length depending upon the virus. This region contains specific repeat elements that are distinctly associated with each of the different viruses. There are secondary structures associated with this noncoding region that may be involved in replication, host specificity, or virulence patterns.

There are four nonstructural genes, designated nsP1–4, which are necessary for viral replication; each has specific functions during the replication process. The nsP1 protein is required for initiation of synthesis of minus-strand RNA and also functions as a methyltransferase to cap the genomic and subgenomic RNAs during transcription. The second gene, nsP2, encodes a protein that has RNA helicase activity in its N-terminus while the C-terminal domain functions as a proteinase for the alphavirus nonstructural polyprotein. Cleavage of the nonstructural polyprotein by both nsP2 and polyproteins containing nsP2 occurs at three distinct sites and depends upon the composition of the polyprotein. This differential cleavage serves as a regulatory action leading to alternative processing pathways at different times during the infectious process to modulate the rate of replication. For example, early during infection, P123 and P1234 (made by readthrough translation of a stop codon present between the nsP3 and nsP4 genes) are produced by translation of the genomic mRNA. As the infection proceeds, P123 cleaves between nsP1 and nsP2 to produce nsP1 and P23. The P23 polyprotein is autoproteolytic and cleaves itself to generate the individual nsP2 and nsP3 proteins (**Figure 3**). The nsP3 gene encodes a protein consisting of two domains, a widely conserved N-terminal domain and a hypervariable C-terminus, whose functions are not fully understood. The C-terminal region has been shown to tolerate numerous mutations, including large deletions or insertions, and still produce viable viral particles in vertebrate cells. It is possible that some of these nsP3 mutations may play a role in defining cell type and invertebrate vector specificity. The final nonstructural protein, nsP4, encodes the viral RNA-dependent RNA polymerase (RdRp) and contains the characteristic GDD motif associated with this highly conserved element.

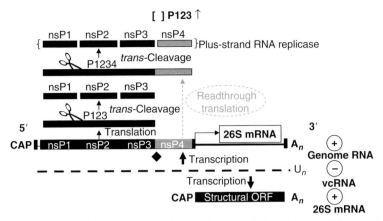

Figure 3 Model for the processing of the alphavirus nonstructural polyprotein during replication. When low levels of P123 are present, *cis*-cleavage of P1234 generates the minus-strand RNA replicase of the virus. This results in primarily negative-sense RNA being generated from the incoming genomic RNA of the virus (upper panel). As the level of the *trans*-acting protease P123 rises in the infected cell, cleavage in *trans* generates other RNA replicase complexes. This results in a shift by the virus from the production of primarily negative-sense RNA to primarily positive-sense RNA. Eventually, replicase complexes capable of producing negative-sense RNA will no longer be present in the infected cell resulting in the complete cessation of negative-sense RNA synthesis (lower panel). The presence of a leaky opal termination codon (depicted as a black diamond) in the virus genome is believed to lead to a more rapid buildup of P123 in the infected cell, and thus a more rapid conversion to the production of positive sense RNA by the virus. Courtesy of Dr. Kevin Myles.

A second polyprotein is generated by translation of a subgenomic mRNA that is generated from an internal promoter immediately downstream of the nonstructural open reading frame. This polyprotein encodes the structural proteins and is processed to produce a capsid protein, the two major envelope surface glycoproteins (E1 and E2) and two small peptides (E3 and 6K) that are not associated with intact virions. E1 and E2 are post-transcriptionally modified in the endoplasmic reticulum and Golgi apparatus before being transported to the plasma membrane where they maintain a close association with each other forming a dimeric spike structure. As virion formation proceeds, the nucleocapsids present in the cytoplasm are moved to the cell membrane. Here, they bind to the surface glycoproteins before budding from the cell. In terms of function, the E2 protein has been found to be an important determinant of antigenicity and cell receptor binding in both the vertebrate host and the insect vector for other alphaviruses. The other major structural protein, E1, has been found to contain domains associated with membrane fusion.

Transmission

The alphaviruses as a whole are transmitted to vertebrate hosts by a broad range of mosquito species. There are some alphaviruses that have non-mosquito invertebrate vectors; Fort Morgan virus (FMV) is transmitted by a cimicid bug (*Oeciacus vicarious*) and SESV is postulated to be transmitted by lice (*Lepidophthirus macrorhini*). Further, the fish alphavirus, salmon pancreas disease virus (SPDV), is not known to have an invertebrate vector. However, for the mosquito-borne alphaviruses, any given virus tends to be transmitted enzootically by only a single or small number of invertebrate species; these very specific host–virus interactions are related to geography and ecological dynamics associated with the mosquito vectors. In contrast to enzootic maintenance, epidemic or epizootic outbreaks may utilize numerous species of mosquitoes but transmission by these species rarely continues once the outbreak subsides. A few of the virus–vector–host interactions have been extensively studied and are described below.

The most well studied alphaviral transmission cycles are those of the equine encephalitides: EEEV, Venezuelan equine encephalitis virus (VEEV), and WEEV. Curiously, all of these viruses, even though they cause encephalitis, exist only in the New World, and are more closely related to each other than other members of the genus *Alphavirus*, utilize extremely different invertebrate vectors in their transmission cycles. EEEV is maintained in North America in a transmission cycle including *Culiseta melanura* mosquitoes and avians. However, this species of mosquito rarely feeds on mammals so epidemic or epizootic transmission to humans or equines typically involves multiple other species of mosquitoes in several genera. The South American cycle of EEEV transmission is not well characterized but isolates of virus have been obtained from *Culex* (*Melanoconion*) mosquitoes implicating them as the likely enzootic vectors. The majority of the viruses within the VEE complex circulate in continuous enzootic habitats between small vertebrate rodent hosts and mosquitoes of the subgenus *Culex* (*Melanoconion*). However, outbreaks of VEEV are a result of transmission by a number of different mammalophilic mosquitoes in the genera *Aedes* and *Psorophora*. Finally, like North American EEEV, WEEV is transmitted in cycles involving avian hosts; however, the predominant mosquito species that vectors this virus is *Culex tarsalis* in North America. WEEV virus may be vertically transmitted by *Aedes* species mosquitoes in temperate climates where year-round maintenance of the virus does not occur.

Of the arthralgic-causing alphaviruses, only a few have well-characterized transmission cycles. Chikungunya virus (CHIKV) (prevalent throughout Africa as well as Southeast Asia and parts of Oceania) appears to be maintained in sylvatic cycles involving wild nonhuman primates and forest-dwelling *Aedes* spp. mosquitoes including *Aedes furcifer*, *Aedes taylori*, *Aedes luteocephalus*, *Aedes africanus*, and *Aedes neoafricanus* in western and central Africa. However, while forest-dwelling mosquito species are primary vectors in west and central Africa, the urban mosquito *Aedes aegypti* has been found to be the most significant vector in Asia with virtually all Asian mosquito isolates coming from this species. Urban or large outbreaks in east Africa are also likely associated with the presence of *Ae. aegypti* mosquitoes. Other common peridomestic species (*Aedes albopictus*, *Aedes vittatus*, and *Anopheles stephensi*) that have been found in abundance in CHIKV endemic areas have also been found to be competent vectors in the laboratory. Both humans and wild nonhuman primates are likely to be the vertebrate reservoirs for CHIKV. The most closely related virus to CHIKV, o'nyong nyong virus (ONNV), is unique among the alphaviruses in that the primary epidemic mosquito vectors are members of the genus *Anopheles*. The virus was isolated from collections of *Anopheles funestus* and *Anopheles gambiae* obtained from epidemic areas and entomological studies have further implicated *An. funestus* as a principal outbreak vector because it was the most abundant species from which isolates of ONNV were obtained during one outbreak. The identity of the vector of ONNV during interepidemic periods remains unknown as does any vertebrate reservoir. Another arthritis-causing alphavirus, Ross River virus (RRV), has been isolated from over 30 different species of mosquitoes; however, the most likely vectors of public health concern are *Aedes vigilax*, *Aedes camptorhynchus*, and *Culex annulirostris* due to their abundance during outbreaks, the number of isolates of virus obtained from these species, and their temporal and spatial association with outbreaks of RRV disease. The vertebrate hosts of RRV are not defined with certainty. Based upon serosurveys, virus isolations, and experimental infection studies, marsupials, macropods, livestock, domestic pets, birds, and humans have all been postulated to play a role either in maintenance or amplification of the virus. Further studies correlating human disease with reservoir status would prove useful in clarifying the role of each of these vertebrates in RRV transmission.

Epidemiology

The epidemiological patterns of the alphaviruses are as diverse as their geography and ecological characteristics. Generally, outbreaks of human or animal illness due to alphavirus infections coincide with peak mosquito seasons in temperate zones while they occur year-round in tropical climates. The magnitude of each outbreak can vary dramatically depending upon whether the outbreak was localized to urban or rural settings. However, in both ecologies, the attack rates can be significant if a virulent alphavirus is the etiological agent. For example, morbidity rates between 30% and 70% are typical during CHIKV epidemics and mortality rates for EEEV outbreaks can reach as high as 70%. The seropositivity rates often correlate with the vector infectivity and transmissibility rates as well as host preference of the mosquito. In general, disease is often more severe in young children and elderly patients and rates of immunity are lower in young children. This indicates that many alphaviruses produce long-lived antibodies that are presumably protective for life. Occupational exposure differences to mosquitoes are likely the cause of any higher incidence rates that may be associated with males compared with females.

See also: Parvoviruses of Arthropods; Togaviruses Causing Encephaitis; Togaviruses Causing Rash and Fever.

Further Reading

Griffin D (2001) Alphaviruses. In: Knipe DM and Howley PM (eds.) *Fields Virology*, 4th edn., vol. 1, pp. 917–962. Philadelphia: Lippincott Williams and Wilkins.

Harley D, Sleigh A, and Ritchie S (2001) Ross River virus transmission, infection, and disease: A cross-disciplinary review. *Clinical Microbiology Reviews* 14: 909–932.

Monath TP and Trent DW (1981) *Togaviral Diseases of Domestic Animals, Comparative Diagnosis of Viral Diseases*, vol. IV, 331–440. New York: Academic Press.

Powers AM, Brault AC, Shirako Y, et al. (2001) Evolutionary relationships and systematics of the alphaviruses. *Journal of Virology* 75: 10118–10131.

Russell RC (2002) Ross River virus: Ecology and distribution. *Annual Reviews of Entomology* 47: 1–31.

Schlesinger S and Schlesinger MJ (2001) *Togaviridae*: The viruses and their replication. In: Knipe DM and Howley PM (eds.) *Fields Virology*, 4th edn., vol. 1, pp. 895–916. Philadelphia: Lippincott Williams and Wilkins.

Strauss JH and Strauss EG (1994) The alphaviruses: Gene expression, replication, and evolution. *Microbiology Reviews* 58: 491–562.

Weaver SC, Frey TK, Huang HV, et al. (2005) Togaviridae. In: Fauquet CM, Mayo MA, Maniloff J, Desselberger U, and Ball LA (eds.) *Virus Taxonomy: Eighth Report of the International Committee on Taxonomy of Viruses*, pp. 999–1008. San Diego, CA: Elsevier Academic Press.

Weaver SC, Powers AM, Brault AC, and Barrett AD (1999) Molecular epidemiological studies of veterinary arboviral encephalitides. *Veterinary Journal* 157: 123–138.

Weston J, Villoing S, Bremont M, et al. (2002) Comparison of two aquatic alphaviruses, salmon pancreas disease virus and sleeping disease virus, by using genome sequence analysis, monoclonal reactivity, and cross-infection. *Journal of Virology* 76: 6155–6163.

Togaviruses: Equine Encephalitic Viruses

D E Griffin, Johns Hopkins Bloomberg School of Public Health, Baltimore, MD, USA

© 2008 Elsevier Ltd. All rights reserved.

Glossary

Arbovirus Virus maintained in an infection cycle involving hematophagous arthropods and vertebrate hosts.
Encephalitis Inflammation of the brain associated with cognitive changes.

History

Nomenclature of the encephalitic alphaviruses recognizes the characteristic geographic distribution of each and the initial identification of the viruses as causes of encephalitis in horses. The first clear record of epidemic equine encephalitis is from 1831, when an outbreak in Massachusetts, USA, resulted in the deaths of 75 horses. Over the next 100 years, several local epidemics of encephalitis in horses occurred along the Atlantic seaboard. The virus of this eastern encephalitis (EEEV) was first isolated from the brain of an affected horse in 1933 by Tenbroek and Merrill. However, 2 years earlier, Meyer, Haring, and Howitt had isolated a virus from central nervous system (CNS) tissues of two horses involved in an epidemic of equine encephalitis in the San Joaquin Valley of California. The EEEV and the western equine encephalitis virus (WEEV) virus were shown to be antigenically distinct. Both diseases occurred in summer epidemics, and, in 1933, Kelsor showed that WEEV could be transmitted by mosquitoes; shortly thereafter, mosquito transmission of EEEV was shown. In 1938, both EEEV and WEEV were isolated from humans with encephalitis occurring in the same regions as the equine cases. In 1936, an epidemic of equine encephalitis occurred in the Goajira region of Venezuela and the virus causing this epidemic, isolated independently by Beck and Wyckoff and Kubes and Rios, was not neutralized by sera from animals immunized with EEEV or WEEV and was designated Venezuelan equine encephalitis virus (VEEV).

Taxonomy and Classification

In 1954, three serologic groups (A, B, and C) of arthropod-borne viruses had been distinguished by Casals and Brown on the basis of cross-reactivity in hemagglutination inhibition (HI) tests. EEEV, WEEV, and VEEV constituted the original group A arboviruses. A second cross-reacting set, including dengue, St. Louis encephalitis, and yellow fever viruses, constituted the group B arboviruses, and the other viruses, mostly from Brazil, were designated group C viruses. As viruses became classified on the fundamental properties of the virion, the group A viruses became the alphaviruses (genus *Alphavirus*) within the family *Togaviridae* of enveloped RNA viruses. HI was used to classify alphaviruses into six broad antigenic complexes; EEE, WEE, and VEE form three of these complexes. The EEE and VEE complexes contain only EEEV and VEEV. The WEE complex includes several viruses in addition to WEEV (e.g., Whataroa, Kyzylagach, Highlands J, Fort Morgan, Aura, and Sindbis), distinguished from each other by complement fixation, neutralization, or other tests.

Viruses within each complex have been subtyped using reactivity with monoclonal antibodies, kinetic HI (EEEV, VEEV), or neutralization (WEEV, VEEV) assays. EEEV and WEEV viruses each have North American and South American subtypes. WEEV and VEEV also can be subdivided into epizootic and enzootic strains. The VEE viruses are differentiated by HI and neutralization tests into subtypes I (VEEV), II (Everglades), III (Mucambo), IV (Pixuna), V (Cabassou), and VI (Rio Negro). Epizootic strains of VEEV are equine virulent and fall within subtype I, which is further subdivided antigenically into IAB and IC (epizootic strains) and ID, IE, and IF (enzootic strains). Sequence information has confirmed the validity of most of these distinctions.

Virion and Genome Structure

Properties of the Virion

Alphavirus virions are 60–65 nm in diameter. The RNA is enclosed in a capsid formed by a single protein arranged as an icosahedron. The nucleocapsid is enclosed in a lipid envelope derived from the host cell plasma membrane and which contains the viral-encoded glycoproteins, E1 and E2. These proteins form heterodimers that are grouped as trimers to form knobs on the virion surface that are arranged with a $T=4$ symmetry. Glycoproteins are arranged with 240 copies of each protein interacting with 240 copies of capsid protein. The virion is sensitive to ether and detergent. Infectivity can be inactivated by heat or acid. The viruses are stable at $-70\,°C$ for long periods.

Properties of the Genome

The 49S genome is a single-stranded, nonsegmented, capped, and polyadenylated message-sense RNA that is infectious. Complete sequence information is available for representative members of all three virus complexes. The genomes contain approximately 11 700 nucleotides and have the nonstructural proteins at the 5'-end and the structural proteins at the 3'-end. Highly conserved regions of genome are the 19 nucleotides at the 3' terminus, important for negative-strand synthesis, and 21 nucleotides at the junction between the nonstructural and structural genes, important for synthesis of the subgenomic RNA. Analysis of codon usage shows an underutilization of the dinucleotide CpG.

Properties of the Viral Proteins

Five potential structural proteins (C, E3, E2, 6K, and E1) are encoded in the 3'-end of the genome. The capsid protein is 259 (EEEV, WEEV) to 275 (VEEV) amino acids long. The N-terminal portion is conserved, basic, and binds the genomic RNA, and the C-terminal portion interacts with the cytoplasmic tail of E2 and with other capsid proteins to form the nucleocapsid. The E3 protein is 59 (VEEV), 60 (WEEV three serologic groups, A, B, and C), or 63 (EEEV) amino acids long, serves as a signal sequence for E2, and, *in vitro*, is shed into the supernatant fluid after cleavage by a furin-like protease in the *trans*-Golgi network. The E2 glycoprotein is a transmembrane protein that is 420 (EEEV) to 423 (WEEV, VEEV) amino acids long and has two (EEEV) or three (WEEV, VEEV) N-linked glycosylation sites. The intracytoplasmic portion has a second stretch of hydrophobic amino acids. The 6K protein is 55 (WEEV, VEEV) or 56 (EEEV) amino acids long and serves as a signal peptide for E1. By analogy to Sindbis virus, small amounts are probably incorporated into the virion. The E1 protein is 439 (WEEV), 441 (EEEV), or 442 (VEEV) amino acids long and has one or two N-linked glycosylation sites. E1 has a short (one or two residue) intracytoplasmic tail and a positionally conserved internal hydrophobic stretch of amino acids in the N-terminal portion of the protein that serves as the fusion peptide for virion entry into the cell. The 5'-end of the genome encodes four nonstructural proteins that function in replication of viral RNA and in production of the genomic and subgenomic RNA.

Replication

For the most part, knowledge of the replication of EEEV, WEEV, and VEEV is based on the extensive studies of Sindbis and Semliki forest viruses. Initial attachment to the cell may involve interaction of the E1 glycoprotein, which has hemagglutination activity, with phospholipids on the cell surface. Efficient subsequent entry involves binding of E2 to cellular proteins followed by receptor-mediated endocytosis. Within the acidified endosomal compartment, the glycoprotein spikes undergo a conformational change that results in fusion with the endosomal membrane and release of the nucleocapsid into the cytoplasm. Acidified endosomes may not be essential for infection of mosquito cells. Once released from the nucleocapsid, the virion RNA serves as mRNA and translation of the nonstructural proteins is initiated. Cytopathic vacuoles form and negative-strand 49S RNA, positive-strand 49S RNA, and positive-strand 26S subgenomic RNA are transcribed. The 26S RNA represents the 3'-portion of the genome from which the structural proteins are translated as a polyprotein. After translation of the capsid protein is complete, it is autocatalytically cleaved from the nascent chain. E2 and E1 are transported into the endoplasmic reticulum, cleaved by cellular proteases, and processed through the Golgi. At the plasma membrane, nucleocapsids align with regions containing the E1–E2 heterodimers and bud from the cell surface. The process may be different in mosquito cells, where intracytoplasmic compartments containing mature virions are seen *in vitro*. However, these structures have not been seen during ultrastructural studies of alphavirus infection in mosquitoes.

Geographic and Seasonal Distribution

The encephalitic alphaviruses are geographically restricted. With few exceptions, alphaviruses are transmitted primarily by mosquitoes and, therefore, transmission is restricted to warm months of the year. EEEV is endemic in eastern Canada and from New Hampshire along the Atlantic seaboard and Gulf Coast to Texas in the United States, in the Caribbean, Central America, and along the north and east coasts of South America. Inland foci exist around the Great Lakes. In the northern part of the range, cases occur between July and October, while in the southern region cases can occur throughout the year. Viruses of the WEE complex are widely distributed throughout the Americas as three closely related but antigenically distinct viruses (WEEV, Highlands J virus, and Fort Morgan virus). Highlands J virus is endemic on the East Coast of the United States in the same areas as EEEV. WEEV is widely distributed in the western plains and valleys of the United States and Canada and is found in South America. Enzootic VEE viruses are perennially active in subtropical and tropical areas of the Americas (subtype II in Florida; ID and IE in Central America; IF and III–VI in South America). Epizootics have appeared in Venezuela, Colombia, Peru, and Ecuador at approximately 10 year intervals.

Host Range and Virus Propagation

The life cycle of the encephalitic alphaviruses involves replication in invertebrate vectors, primarily mosquitoes, and in reservoir vertebrate hosts, primarily birds (EEEV and WEEV) or small mammals (VEEV). In epidemic periods, additional hosts, such as horses and humans, are infected. EEEV causes encephalitis in humans, horses, pigeons, and pheasants. Many birds are susceptible to infection but remain asymptomatic, despite prolonged viremia. The amplifying species for EEEV in North America are wading birds and migratory song birds. In South America, forest-dwelling rodents and marsupials become infected and may provide additional reservoirs. Infection has also been reported in turtles and snakes. In invertebrates, virus is most consistently recovered from *Culiseta melanura* in North America, *Culex taeniopus* in the Caribbean, and *Culex* (*Melanconion*) spp. in South America. Mosquitoes of many species can serve as bridge vectors to mammals.

WEEV may cause encephalitis in horses and humans. The enzootic cycle in North America involves domestic and wild birds and *Culex tarsalis* for WEEV in the western United States and songbirds and *Culiseta melanura* for Highlands J virus in the eastern United States. Serosurveys and virus isolations have demonstrated evidence of natural infection in chickens and other domestic birds, passerine birds, pheasants, rodents, rabbits, ungulates, and snakes.

Epizootic strains of VEEV can cause disease in horses and in humans. The enzootic strains of VEEV can infect horses but these infections are asymptomatic and may protectively immunize horses against epizootic strains. Enzootic strains produce mild disease in humans. A variety of wild birds are susceptible to infection, but small mammals serve as an important reservoir, with efficient transmission of infection by *Culex* (*Melanconion*) spp. mosquitoes.

In addition to the native hosts, a number of laboratory animals are susceptible to infection. All three viruses cause encephalitis in monkeys, mice, rats, guinea pigs, and hamsters. Disease is generally age dependent, so that young animals more often develop fatal infections than do adult animals. Primary isolates of these viruses are often made in newborn mice.

In vitro, the viruses are routinely propagated in cultures of chick embryo fibroblasts, BHK-21 cells, or Vero cells. Most strains will form plaques in these cells and plaque assays provide the usual basis for virus quantitation. Mosquito cell lines support replication, often without cytopathic effect. Many other cell lines (e.g., L-cells, HeLa cells) support replication as well.

Genetics

Alphaviruses show genetic changes by accumulation of point mutations in the genomic RNA, but this occurs at a rate that is much slower than that predicted for other RNA viruses. Recombination is infrequent, but can be demonstrated *in vitro* and occurs at least occasionally in nature because sequence analysis has shown that WEEV is the result of recombination between EEEV and Sindbis-like viruses.

Evolution

The alphaviruses derive from a single unknown protoalphavirus and are part of the alphavirus superfamily of viruses. Viruses in this superfamily have a similar genetic organization and include many RNA plant viruses. In general, amino acids important in secondary structure (e.g., cysteine) have been conserved for the structural glycoproteins E1 and E2, suggesting that the three-dimensional structure is similar for all alphaviruses. Sequence analyses suggest that the encephalitic alphaviruses evolve slowly in nature. The capsid protein and E1 glycoprotein are the most conserved of the structural proteins, whereas the E2 glycoprotein is more divergent. EEEV has evolved independently in North and South America over the last 1000 years and there is currently one group in North America and the Caribbean and three groups in South America, one in the Amazon River basin, one on the coasts of South and Central America and the third in Brazil. North American EEEV isolates are highly conserved, varying by less than 2% over 63 years. South American isolates are evolving more rapidly, perhaps locally within different vector–host relationships. Rates of divergence of WEEV and Highlands J virus of 0.1–0.2% per year have been estimated. It is hypothesized that short transmission seasons and limited host mobility constrain genetic diversity in a geographic region.

Epidemiology

EEEV causes localized outbreaks of equine, pheasant, and human encephalitis in the summers. Cases of equine encephalitis are usually the first indicators of an outbreak. In North America, the primary enzootic cycle is maintained in shaded freshwater swamps where the vector is the ornithophilic swamp mosquito *Culiseta melanura* and the reservoir hosts are migratory passerine songbirds. Young birds are probably most important for virus amplification because they are more susceptible to infection, have a prolonged viremia, and are less defensive toward mosquitoes. Human and equine cases usually occur within 5 mi (8.045 km) of the swamp, with virus being transmitted by epizootic vectors such as *Coquillittidia perturbans*, *Aedes sollicitans*, and, potentially, *Aedes albopictus*. The enzootic vector in the Caribbean is probably *Culex taeniopus* and in South America *Culex (Melanoconion)* spp. Epizootics appear approximately every 5–10 years and are usually associated with heavy rainfall that increases the populations of enzootic and epizootic mosquito vectors. The mechanism of overwintering in northern areas is not known. There is no evidence for overwintering in mosquitoes. Human infections are unusual, with an average of 5–10 cases per year in the United States (**Figure 1**). Serological surveys in the northeastern United States suggest that there are approximately 23 inapparent infections for every case of encephalitis, but children are more susceptible and this declines to only 8:1 for children under 4 years of age.

WEEV is maintained in an endemic cycle involving domestic and passerine birds and *Culex tarsalis*, a mosquito particularly adapted to irrigated agricultural areas. Interseasonal persistence occurs in saltwater marshes, where vertical transmission of WEEV in *Culex tarsalis* has been demonstrated. Occasional isolations have been made from *Culex stigmatosoma*, *Aedes melanimon*, and *Aedes dorsalis*, also competent vectors. Transmission from this enzootic cycle has resulted in small numbers of cases of encephalitis in humans. However, periodically, there have been widespread summer epidemics of equine encephalitis in North America with significant extension into humans. The estimated case to infection ratio is 1:58 in children under 5 years and 1:1150 in adults. Recently, for reasons that are not completely clear, cases of equine and human cases have declined (**Figure 2**). In the eastern United States, Highlands J virus has been isolated from ornithophilic mosquitoes (*Culiseta melanura*, the enzootic vector for EEEV) in freshwater swamp habitats along the Atlantic coast, as well as from birds. Ecologic restriction and limited pathogenicity may explain the paucity of human and equine disease caused by Highlands J virus. The lack of significant human disease during equine outbreaks of WEE in South America may be related to the feeding habits of the vector or to a difference in virulence for humans of the South American strains.

VEEVs are maintained in enzootic cycles by *Culex (Melanoconion)* spp. mosquitoes that live in tropical and subtropical swamps throughout the Americas and breed near aquatic plants. They feed at dawn and dusk on a wide variety of rodents, birds, and other vertebrates. Humans living in these areas have a high prevalence of antibody, but little-recognized disease. Epizootic strains of VEEV arise by mutation from enzootic strains and are isolated only during outbreaks. Epizootics have occurred primarily in Latin America in cattle ranching areas during the rainy season. Formalin-inactivated vaccines containing residual live virus are suspected to be responsible for initiating the 1969–72 outbreak in Central America and which extended to Texas by 1971. During epizootics, horses are an important amplifying species. Virus has been isolated from several species of mosquitoes including *Aedes taeniorhynchus*, *Aedes aegypti*, *Mansonia dubitans*, and *Psorophora confinnis*. The incidence of encephalitis in clinically ill humans is generally less than 5% and the overall mortality less than 1%.

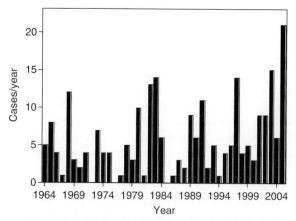

Figure 1 Numbers of human cases of eastern equine encephalitis reported in the United States from 1964 to 2005. Data are from the Centers for Disease Control and Prevention.

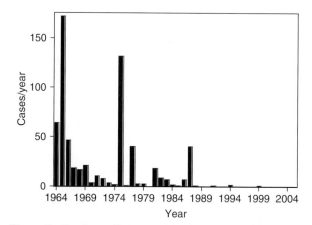

Figure 2 Numbers of human cases of western equine encephalitis reported in the United States from 1964 to 2005. Data are from the Centers for Disease Control and Prevention.

Transmission and Tissue Tropism

The primary mode of alphavirus transmission to birds and mammals is through the bite of an infected mosquito which inoculates virus extravascularly. Mosquitoes become infected by feeding on a viremic host, are able to transmit the virus 4–10 days later (extrinsic incubation), and remain persistently infected. Maintenance of this cycle requires an amplifying host that develops a viremia of sufficient magnitude to infect mosquitoes feeding on it. Other modes of transmission are of occasional importance. EEEV persists in the feather follicles of infected pheasants and secondary transmission among penned pheasants can occur through feather picking and cannibalism. VEEV can be transmitted by the respiratory route between infected horses and to humans in the laboratory. WEEV and VEEV can be transmitted transplacentally.

In mammals, EEEV replicates primarily in muscle and in neurons and glial cells of the CNS, with occasional involvement of liver and lymphatic tissue. Skeletal and myocardial muscle and the CNS of infected birds contain virus. In mosquitoes, there is infection of the midgut, muscle, and salivary glands without involving the nervous system. WEEV replicates primarily in skeletal and cardiac muscle, brown fat, and the choroid plexus, ependyma, and neurons in the CNS of mammals. Little is known of the tissue tropism of WEEV in birds. Epizootic strains of VEEV infect the upper respiratory tract, lymphatic and myeloid tissue, pancreas, liver, and CNS to varying degrees in different mammals. Mosquito infection is initiated in the midgut epithelial cells and spreads through the hemolymph to the salivary glands and flight muscles.

Pathogenicity

EEEV strains vary in their pathogenicity for horses (and presumably for humans), with South American strains being less virulent than North American strains. The molecular basis of these differences have not yet been defined. The North American strains of EEEV are among the most virulent of the alphaviruses, causing severe encephalitis in humans, horses, dogs, pigs, pigeons, emus, quail, and pheasants. The case–fatality rate in humans is 30–50%, up to 90% in horses and 50–70% in pheasants. Laboratory studies indicate a similar neurovirulence of EEEV for monkeys, mice, and hamsters. Hamsters also develop hepatitis and lymphatic organ infection. At 3–4 weeks of age, mice become relatively resistant to peripheral, but not intracranial, inoculation. Birds vary in their susceptibility, with birds of some species developing disease, while birds of many other species show no morbidity or mortality, despite a prolonged viremia. Pheasant deaths are caused by encephalitis while young chickens develop myocarditis. Mosquitoes develop persistent infection of the fat body and salivary glands.

WEEV in the western United States causes epidemics of encephalitis in humans, horses, and emus, but the case–fatality rate of 10% for humans, 20–40% for horses, and 10% for emus is lower than for EEEV in the eastern US. Highlands J virus has been reported to cause rare fatal disease in horses and turkeys. South American strains of WEEV cause fatal encephalitis in horses, but little disease in humans. Epizootic strains are neurovirulent and neuroinvasive in adult mice, whereas enzootic strains are not. Highlands J virus is intermediate in virulence between North and South American strains of WEEV. With increasing age mice become relatively resistant to fatal infection, whereas hamsters remain susceptible to Highlands J virus.

Infection of horses with epizootic strains of VEEV frequently is fatal. Disease is associated with leukopenia and a high-titered, prolonged viremia. In contrast to EEE and WEE, encephalitis is not always apparent and virus is shed in nasal, eye, and mouth secretions as well as in urine and milk. Experimental infection of laboratory animals produces a variety of disease patterns. In guinea pigs and rabbits, VEEV produces necrosis in lymph nodes, spleen, thymus, intestinal, and conjunctival lymphoid tissue, liver, and bone marrow. Hamsters develop encephalitis and pancreatitis in addition to widespread involvement of myeloid and lymphoid tissue. In rats and mice, there is more limited destruction of lymphoid and myeloid tissue and death is usually due to encephalitis. Virus enters the CNS through the olfactory tract and causes neuronal infection and apoptosis. Neonatal mice show extensive replication in many tissues, including brain, myocardium, and pancreas. Comparative studies of the virulent TRD and avirulent TC-83 strains of IA serotype VEEV and construction of recombinant viruses have led to identification of the $5'$-noncoding region and the E2 envelope glycoprotein as important determinants of virulence for mice, whereas changes in nsP3 and E2 determine virulence for horses.

Clinical Features of Infection

EEEV is the most virulent of the encephalitic alphaviruses, causing a high mortality due to encephalitis. Prodromal symptoms of fever, headache, and myalgias are common. The onset of encephalitis tends to be fulminant and is associated with continued fever, increased headache, meningismus, obtundation, and seizures. The overall case–fatality rate is 30% in recent studies, with higher rates in children and the elderly. Recovery is more likely in those individuals that have a long (5–7 day) prodrome and do not develop coma. Sequelae are common, with more than 80% of survivors having significant neurological residua, including paralysis, seizures, and mental retardation. The diagnosis is usually made by detection of antibody in serum or cerebrospinal fluid.

WEEV can cause encephalitis with signs and symptoms similar to those caused by EEEV. There is a 3–5 day prodrome of fever and headache that may progress to irritability, nuchal rigidity, photophobia, and altered mental status. Severe disease, seizures, fatal encephalitis, and significant sequelae are more likely to occur in infants and in young children.

Infection with epizootic strains of VEEV usually causes relatively mild disease in humans. Illness in adults usually is manifested by fever, headache, myalgias, and pharyngitis 2–5 days after exposure. Severe diseases, including fulminant reticuloendothelial infection and encephalitis, may occur in young children. Children recovering from encephalitis may be left with neurological deficits. Fetal abnormalities and spontaneous abortions may occur in infections during pregnancy.

The cerebrospinal fluid during alphavirus encephalitis usually shows a moderate (up to 4000 cells mm^{-3}) pleocytosis with predominance of either polymorphonuclear or mononuclear cells. Protein is usually elevated and the glucose is normal. The electroencephalogram and magnetic resonance imaging scans are usually abnormal, but computed tomographic scans may be normal or indicative only of edema.

Pathology and Histopathology

Initial CNS infection with EEEV in experimental animals is of the capillary endothelial or choroid plexus epithelial cells and spread within the CNS can be cell to cell or through cerebrospinal fluid. The targeted cell within the CNS is the neuron and damage to this cell may be severe and irreversible. Histopathology demonstrates a diffuse meningoencephalitis with widespread neuronal destruction, neuronophagia, gliosis, and perivascular inflammation, with polymorphonuclear and mononuclear leukocytes early, and vasculitis and vessel occlusion late. Hamsters also exhibit necrosis of hepatocytes and lymphatic tissue accompanied by local infiltration of mononuclear leukocytes. Initial infection of mosquitoes is of midgut epithelial cells. Infection is facilitated when virus in the serum is concentrated next to these cells as the infected blood meal clots. Infected midgut epithelial cells subsequently degenerate and slough, and this process may facilitate penetration of the virus into the hemocoel and rapid dissemination of the infection.

Pathological examination of brains from fatal human cases of WEE demonstrates early perivascular extravasation of blood followed by endothelial hyperplasia, perivascular mononuclear and polymorphonuclear inflammation, and parenchymal necrosis. Areas of neuronal degeneration, glial nodules, and demyelination are found. Neonatal mice develop acute inflammation and necrosis in skeletal and smooth muscle, cartilage, and bone marrow. In vertebrate animals with encephalitis, the brain shows multifocal areas of necrosis and widespread lymphocytic infiltration of the leptomeninges and perivascular regions of the brain parenchyma. The heart shows a necrotizing, inflammatory myocarditis. Infiltration of mononuclear leukocytes into areas of lung, liver, and brown fat also occurs.

The pathology of VEE in horses includes cellular depletion of bone marrow, spleen, and lymph nodes, pancreatic necrosis, and, in cases with encephalitis, swelling of vascular endothelial cells, edema, and mononuclear cell cuffing of cerebral vessels and meningitis followed by the later appearance of demyelinating lesions. Small mammals with widespread involvement of the reticuloendothelial system may develop ileal necrosis.

Both innate and adaptive immune responses are induced by infection. Interferon-α/β (IFN-α/β) is produced quickly after infection. Virus-specific IgM is often detectable by enzyme immunoassay or by virus neutralization very early after onset of the disease in both serum and cerebrospinal fluid and provides a means for rapid diagnosis of infection. IgG antibody appears 10–14 days after onset and can be measured by enzyme immunoassay, HI, or neutralization. Many lines of evidence suggest that recovery from infection is primarily dependent on the antibody response. Extensive experimental studies to define the antibody specificity and the mechanisms of recovery and protection have been done using VEEV and Sindbis virus, a member of the WEE complex. Neutralizing and non-neutralizing antibodies are protective. Multiple epitopes on the E1 and E2 glycoproteins induce neutralizing, HI, and protective antibody. Sequencing of monoclonal antibody escape mutants has localized one neutralization domain to E2 residues 180–210 in VEEV. Immunization with peptides has shown that an antibody response to residues 1–25 of the VEEV E2 glycoprotein is also protective.

Cellular immunity has received more limited study, but virus-specific lymphoproliferative and cytotoxic responses have been documented and a mononuclear inflammatory response is common. The importance of cellular immune responses for recovery or for contribution to fatal disease has not been established but IFN-γ can contribute to virus clearance from the CNS. Prior infection with VEEV increases subsequent antibody responses to unrelated antigens. Antithymocyte globulin extends time to death in mice infected with VEEV, suggesting a T-cell-mediated immunopathogenic component to fatal disease.

Prevention and Control

Prevention of infection relies on efforts to control mosquito populations by spraying and reduction of breeding places. Individual use of protective measures such as mosquito repellents and protective clothing are important. Vaccines against EEEV, WEEV, and VEEV are available for horses and against EEEV for birds. Experimental human vaccines against EEEV, WEEV, and VEEV also are available for laboratory workers exposed to these agents. Most of these vaccines consist of formalin-inactivated virus, but TC-83 is a successful live attenuated vaccine against VEEV for use in horses. TC-83 is also used as a vaccine for humans, although side effects are common. No antiviral agents are of proven usefulness in treatment of these infections.

Future

Construction of full-length cDNA alphavirus clones that can be transcribed into infectious RNA provides tremendous potential for understanding the functions of various genes and their importance for replication and virulence in the multiple hosts necessary for maintenance of these viruses in their natural cycles. An understanding of the folding and three-dimensional structures of each of the structural and nonstructural proteins is needed for interpretation of much of the sequence and virulence data that has been acquired. Further information on virulent and avirulent strains and functional analysis of the nonstructural proteins are likely to provide the next level of understanding of virus–host relationships and the conditions that lead to outbreaks.

See also: Central Nervous System Viral Diseases.

Further Reading

Calisher CH (1994) Medically important arboviruses of the United States and Canada. *Clinical Microbiology Reviews* 7: 89–116.

Griffin DE (2007) Alphaviruses. In: Knipe DM and Howley PM (eds.) *Field's Virology*, 5th edn., pp. 1024–1067. Philadelphia: Lippincott Williams and Wilkins.

Weaver SC and Barrett AD (2004) Transmission cycles, host range, evolution, and emergence of arboviral disease. *Nature Reviews Microbiology* 2: 789–801.

Weaver SC, Ferro C, Barrera R, *et al.* (2004) Venezuelan equine encephalitis. *Annual Review of Entomology* 49: 141–174.

Togaviruses: General Features

S C Weaver, University of Texas Medical Branch, Galveston, TX, USA
W B Klimstra and K D Ryman, Louisiana State University Health Sciences Center, Shreveport, LA, USA

© 2008 Elsevier Ltd. All rights reserved.

Glossary

Alphavirus A virus in the genus *Alphavirus*.
Arbovirus A virus transmitted to vertebrates by hematophagous (blood-feeding) insects.
Endemic A disease constantly present in a community in a defined geographic region.
Enzootic A disease constantly present in an animal community in a defined geographic region.
Epidemic Outbreak of human disease above the normal (endemic) incidence.
Epizootic Outbreak of disease in animals above the normal (enzootic) incidence.
Vector An arthropod (mosquito in the case of most alphaviruses) that transmits an arbovirus.
Zoonotic A human disease whose causative agent is maintained in populations of wild animals.

Introduction and History

The *Togaviridae* is a family of enveloped, single-stranded, plus-strand RNA viruses that occur nearly worldwide. The family includes the genus *Alphavirus*, a group of zoonotic viruses maintained primarily in rodents, primates, and birds by mosquito vectors. Human disease occurs when people intrude on enzootic transmission habitats and are bitten by infected mosquitoes, or when the virus emerges to cause epizootics and epidemics. The other genus in the family *Togaviridae* is *Rubivirus*, with Rubella virus, the etiologic agent of 3-day measles (German measles), as its only member.

The first togaviruses to be isolated were the alphaviruses Western equine encephalitis virus (WEEV) and Eastern equine encephalitis virus (EEEV) in 1930 (although EEEV was not described until 1933), followed by Venezuelan equine encephalitis virus (VEEV) in 1938. Serologic tests using hemagglutination inhibition (HI) later indicated that these three viruses were antigenically related, and in 1954 Casals and Brown designated two groups of arboviruses: 'Group A' included WEEV, EEEV, and VEEV, Semliki Forest virus (SFV), and Sindbis virus (SINV), while 'Group B' is now known to include scores of flaviviruses. The HI test also indicated degrees of relatedness within viruses of Group A, with cross-reactions between EEEV and VEEV stronger than reactions between these viruses and WEEV, SFV, or SINV. These interrelationships, which formed the basis for the definitions of the antigenic complexes of alphaviruses, have generally stood the test of time, and are reflected in the phylogenetic relationships described below. Other assays including complement fixation (CF) and neutralization (NT) tests were used in some cases to define relationships on a finer scale. The introduction of the kinetic HI test later allowed further discrimination of certain alphaviruses, such as the antigenic varieties within the VEE complex and between North American and South American EEEV strains. The serological interrelationships among the alphaviruses were later determined to reflect the envelope glycoproteins in the case of HI and NT, and a mixture of structural and nonstructural proteins in the case of CF.

For some time, the *Togaviridae* included four genera: *Alphavirus, Rubivirus, Flavivirus,* and *Pestivirus*. Not until the morphology, RNA genome organizations and coding strategies, as well as the protein and nucleotide sequences were determined for representatives was it recognized that the former two genera differ dramatically from the latter two. The flaviviruses and pestiviruses were then placed into a distinct family, the *Flaviviridae* (now with three genera), and the *Alphavirus* and *Rubivirus* genera were retained in the family *Togaviridae*.

Structure, Systematics, and Evolution

The basis for grouping the alphaviruses and Rubella virus into the *Togaviridae* is that they have very similar genome organizations and coding strategies, as well as primary sequence homology that can be detected in short motifs in the nonstructural protein genes (**Figure 1**). Alphavirus virions have been visualized at high resolution using cryoelectron microscopy. The E1 envelope glycoprotein domain that lies outside of the lipid envelope has been crystallized and its structure solved to atomic resolution. The folds of the alphavirus E1 protein are very similar to those of the flavivirus envelope protein, indicating that these proteins are homologous (share a common ancestral gene) and that the togaviruses and flaviviruses share a distant, common ancestor, at least for the E1/E protein genes which, due to their extensive divergence, cannot be detected from the nucleotide or amino acid sequences.

Alphavirus virions are about 70 nm in diameter; the nucleocapsid, which forms in the cytoplasm of infected cells, includes one molecule of genomic RNA and 240 copies of the capsid protein arranged in hexons and pentons in a $T = 4$ lattice. This icosahedral structure dictates the arrangement of the envelope proteins in the virion. Outside of the plasma membrane-derived lipid envelope, trimeric envelope spikes on the surface of the virion are composed of E1/E2 heterodimers, with the E2 protein most exposed and believed to interact with cellular receptors. The Rubella virion appears as a spherical particle, 60–70 nm in diameter, in electron micrographs. Rubella virus differs from alphaviruses in that the electron translucent zone between the nucleocapsid and envelope proteins is wider and the rubiviruses have a $T = 3$ icosahedral symmetry.

The alphaviruses and rubiviruses are quite distantly related and both share sequence homology with several taxa of plant viruses, some of which are transmitted by insects. These relationships suggest that the ancestor of these viruses was an insect virus that underwent a process called modular evolution that led to several genome rearrangements, including segmentation.

Before the era of molecular genetics, the alphaviruses were grouped into several antigenic complexes based on cross-reactivity in serologic assays including HI and CF (but generally not plaque reduction neutralization; PRNT) (**Table 1**). Within some of these complexes, the VEE complex, for example, differentiation of antigenic subtypes and varieties, some of which have fundamental and critical epidemiological and virulence differences, was accomplished using specialized forms of HI and, in some cases, PRNT. Direct genetic evidence of homology

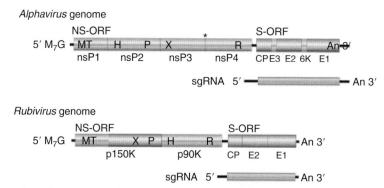

Figure 1 Togavirus genomic coding strategies. Shown are comparative schematic representations of the alphavirus and rubivirus genomic RNAs with untranslated regions represented as solid black lines and open reading frames (ORFs) as open boxes (NS-ORF, nonstructural protein ORF; S-ORF, structural protein ORF). Within each ORF, the coding sequences for the proteins processed from the translation product of the ORF are delineated. The asterisk between nsP3 and nsP4 in the alphavirus NS-ORF indicates the stop codon present in some alphaviruses that must be translationally read through to produce a precursor containing nsP4. Additionally, within the NS-ORFs, the locations of motifs associated with the following activities are indicated: MT, methyltransferase; P, protease; H, helicase; X, unknown function; and R, replicase. The sequences encompassed by the subgenomic RNA (sgRNA) are also shown. Reproduced from Weaver SC, Frey TK, Huang HV, et al. (2005) Togaviridae. In: Fauquet CM, Mayo MA, Maniloff J, Desselberger U, and Ball LA (eds.) *Virus Taxonomy: Eighth Report of the International Committee on Taxonomy of Viruses*, pp. 999–1008. San Diego, CA: Elsevier Academic Press, with permission from Elsevier.

Table 1 Members of the genus *Alphavirus*

Antigenic complex	Virus	Antigenic subtype	Antigenic variety	Clinical syndrome in humans	Distribution
Barmah Forest	Barmah Forest virus (BFV)			Febrile illness, rash, arthritis	Australia
Eastern equine encephalitis (EEE)	Eastern equine encephalitis virus (EEEV)	I–IV		Febrile illness, encephalitis (none recognized in Latin America)	North, Central, South America
Middelburg	Middelburg virus (MIDV)			None recognized	Africa
Ndumu	Ndumu virus (NDUV)			None recognized	Africa
Semliki Forest	Semliki Forest virus (SFV)			Febrile illness	Africa
	Chikungunya virus (CHIKV)			Febrile illness, rash, arthritis	Africa
	O'nyong-nyong virus (ONNV)			Febrile illness, rash, arthritis	Africa
	Getah virus (GETV)			None recognized	Asia
	Bebaru virus (BEBV)			None recognized	Malaysia
	Ross River virus (RRV)	Sagiyama		Febrile illness, rash, arthritis	Australia, Oceania
	Mayaro virus (MAYV)			Febrile illness, rash, arthritis	South and Central America, Trinidad
	Una virus (UNAV)			None recognized	South America
Venezuelan equine encephalitis (VEE)	Venezuelan equine encephalitis virus (VEEV)	I	AB	Febrile illness, encephalitis	North, Central, South America
			C	Febrile illness, encephalitis	South America
			D	Febrile illness, encephalitis	South America, Panama
			E	Febrile illness, encephalitis	Central America, Mexico
	Mosso das Pedras virus (MDPV)		F		
	Everglades virus (EVEV)			Febrile illness, encephalitis	Florida (USA)
	Mucambo virus (MUCV)	III	A	Febrile illness, myalgia	South America, Trinidad
			C	Unknown	Peru
			D	Febrile illness	Peru
	Tonate virus (TONV)	III	B	Febrile illness, encephalitis	Brazil, Colorado (USA)
	Pixuna virus (PIXV)			Febrile illness, myalgia	Brazil
	Cabassou virus (CABV)			None recognized	French Guiana
	Rio Negro virus (RNV)			Febrile illness, myalgia	Argentina
Western equine encephalitis	Sindbis virus (SINV)			Febrile illness, rash, arthritis	Africa, Europe, Asia, Australia
		Babanki		Febrile illness, rash, arthritis	Africa
		Ockelbo		Febrile illness, rash, arthritis	Europe
		Kyzylagach		None recognized	Azerbaijan, China
	Whataroa virus (WHAV)			None recognized	New Zealand
	Aura virus (AURAV)			None recognized	South America
	Western equine encephalitis virus (WEEV)	Several		Febrile illness, encephalitis	Western North, South America
	Highlands J virus (HJV)				Eastern North America
	Fort Morgan virus (FMV)	Buggy Creek		None recognized	Western North America
Trocara	Trocara virus (TROV)				South America
Salmon pancreas disease	Salmon pancreas disease virus (SPDV)			Pancreatic disease (salmon)	Atlantic Ocean and tributaries
	Sleeping disease virus (SDV)			Sleeping disease (trout)	Worldwide
Southern elephant seal virus	Southern elephant seal virus (SESV)			None recognized	Australia

among the alphaviruses, and later between the alphaviruses, rubella virus, and some plant viruses, came first from direct protein sequencing, followed by nucleotide sequencing. Eventually, nucleotide sequencing largely replaced the serological tests for determining relationships among the alphaviruses, although the latter remain useful from an epidemiological standpoint.

Nearly complete phylogenetic trees can now be constructed from homologous E1 envelope glycoprotein gene nucleotide or deduced amino acid sequences (**Figure 2**). These trees largely agree with the original antigenic groupings, with the exception of Middelburg virus, which genetically belongs within the SF complex. The relationships of the New World members of the WEE antigenic complex have more complex relationships when the nonstructural protein genes are analyzed; these viruses group more closely with EEEV than with the Old World SINV-like members of the WEE complex or with the New World member, Aura virus. These relationships reflect an ancient recombination event between a SINV-like ancestor and a virus closely related to an ancestor of EEEV (**Figure 2**). The recombinant ancestor later gave rise to the WEE, Highlands J, and Fort Morgan virus ancestors.

Most estimates for the rate of RNA sequence evolution among alphaviruses are on the order of 10^4 substitutions per nucleotide per year, similar to those of other arboviruses and lower than those of most vertebrate viruses that do not use arthropods as vectors. These low rates may reflect constraints imposed by alternate replication on disparate hosts, as well as by persistent infection of mosquito vectors. However, the timescale for evolution of the togaviruses and alphaviruses is difficult to determine from sequence data because the ability of phylogenetic

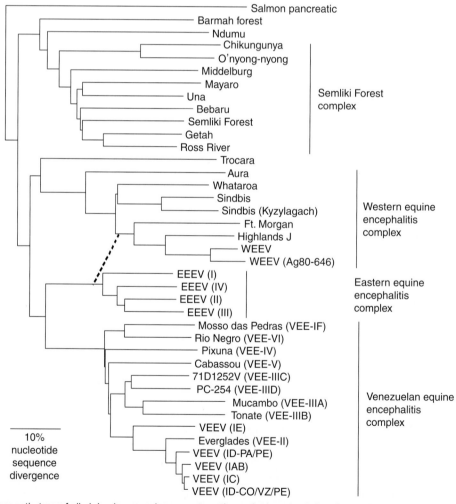

Figure 2 Phylogenetic tree of all alphavirus species except southern elephant seal virus (the homologous sequence region is not available), and selected subtypes and variants, generated from partial E1 envelope glycoprotein gene sequences using the neighbor joining program with the HKY distance formula. Antigenic complexes are shown on the right. The dashed line indicates the ancestral recombination event that led to Highlands J, Fort Morgan, and Western equine encephalitis viruses. Rubella virus cannot be included in this analysis because there is no detectable primary sequence homology with alphavirus structural protein sequences.

methods to compensate for sequential substitutions of the same nucleotides or amino acids is unknown.

Phylogenetic studies of alphaviruses generally point to purifying selection as dominating their evolution, presumably because most mutations are deleterious, especially when efficient replication in widely divergent hosts (mosquitoes and vertebrates) is required. Experimental evolutionary studies employing cell culture model systems also indicate that most adaptive mutations are host cell specific, supporting the hypothesis that the alternating host replication cycle constrains adaptation to new hosts by alphaviruses. The mobility of reservoir hosts appears to influence patterns of alphavirus evolution and their phylogeography. Viruses that use avian hosts, such as EEEV, WEEV, SINV, and Highlands J virus, tend to evolve within a relatively small number of sustained lineages that have very wide geographic distributions. For example, some lineages of WEEV have been sampled from California to Argentina. The efficient dispersal of these viruses by their highly mobile reservoir hosts presumably allows competition among different sympatric lineages to limit virus diversity. In contrast, alphaviruses that use hosts with more limited mobility, such as VEE complex viruses that rely on rodents and other small mammals, tend to exhibit more genetic diversity and lineages confined to smaller, nonoverlapping geographic regions.

Transovarial transmission from infected adult females to their offspring has been documented for a few alphaviruses, but it is not known whether this is important for maintenance of these viruses in nature. Persistent infection of vertebrate hosts also appears to be rare and is not known to be involved in long-term maintenance of alphaviruses. However, mechanisms for the overwintering of alphaviruses in temperate regions, which has been suggested by several genetic studies, remain enigmatic and could include transovarial transmission and/or persistent infection of reservoir hosts.

Compared to the alphaviruses, rubella virus is relatively conserved genetically, with only one serotype described. However, phylogenetic analyses indicate the existence of at least two genotypes, with genotype I isolates predominant in Europe, Japan, and the Western Hemisphere, and the more diverse genotype II isolated in Asia and Europe. There is considerable geographic overlap between these genotypes and subgenotypes, presumably reflecting the efficient dispersal of Rubella virus by infected people.

Diseases

Most of the alphaviruses cause febrile disease in humans and/or domestic animals. However, a few alphaviruses have not been associated with any disease, including Aura, Trocara, Middelburg, Ndumu, Ft. Morgan, and Una viruses. The most severe alphaviral feature, encephalitis, is caused by the New World viruses EEEV, VEEV, and WEEV. In humans, apparent:inapparent infection rates range from very low for EEEV ($c.$ 1:23) to $c.$ 10:1 for VEEV. However, mortality rates in apparent human infections are inversely correlated with these ratios, with $c.$ 50% for EEEV and less than 1% for VEEV. In equids, most infections are apparent, with high mortality rates ($c. \geq 50\%$) for all three viruses. Any of these viruses can also cause disease in a wide range of other domesticated and wild animals, with EEEV documented as the etiologic agent of severe disease in pigs, deer, emus, turkeys, pheasants, cranes, and other birds. Everglades virus is also known to cause neurologic disease in humans, and Highlands J virus has caused documented disease in domestic birds and in a horse.

Several of the Old World alphaviruses including Chikungunya (CHIKV), O'nyong nyong (ONNV), Ross River, and Barmah Forest viruses cause an arthralgic disease syndrome often accompanied by rash. Mayaro virus (MAYV), another member of the SF complex that was probably introduced into South America from the Old World, is the exception to the relationship between hemispheric distributions and disease syndromes. The arthralgic disease caused by these alphaviruses can be highly incapacitating and chronic, with affected persons unable to work or function normally for several months or longer.

Rubella virus is the etiologic agent of rubella, also known as German or 3-day measles, a generally benign infection. Occasionally, more serious complications such as arthritis, thrombocytopenia purpura, and encephalitis can occur. Congenital transmission to the fetus during the first trimester of pregnancy can lead to serious birth defects that comprise congenital rubella syndrome. Congenitally infected children suffer a variety of autoimmune and psychiatric disorders in later life, including a fatal neurodegenerative disease known as progressive rubella panencephalitis. Rubella has been largely controlled through vaccination in most developed countries, although it remains endemic worldwide.

Transmission Cycles

All alphaviruses are zoonotic, and the vast majority have transmission cycles involving mosquito vectors and avian or mammalian reservoir and/or amplification hosts. Exceptions include the fish viruses, sleeping disease virus, and salmon pancreatic disease virus; southern elephant seal virus was isolated from lice and may use these insects as mechanical, rather than biological, vectors, although transmission has not been documented. The mosquito-borne alphaviruses often have relatively narrow vector ranges, relying on one or a few principal mosquito vectors

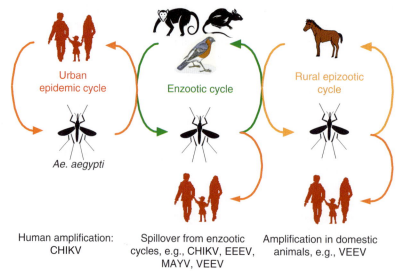

Figure 3 Cartoon showing mechanisms of human infection by alphaviruses. At the center is an enzootic cycle, typically involving avian, rodent, or nonhuman primates as amplification and/or reservoir hosts and mosquito vectors. Humans become infected via direct spillover when they enter enzootic habitats and/or when amplification results in high levels of circulation. Transmission to humans may involve the enzootic vector or bridge vectors with broader host preferences. Right panel: Secondary amplification involving domestic animals can increase circulation around humans, increasing their chance of infection via spillover. In the case of VEEV, mutations that enhance equine viremia are needed for secondary equine amplification. Left panel: CHIKV can use humans for amplification, resulting in urban epidemic cycles and massive outbreaks. Expanded form of virus abbreviations is provided in **Table 1**. Reproduced from Weaver SC (2005) Host range, amplification and arboviral disease emergence. *Archives of Virology* 19(supplement): 33–44, with permission from Springer-Verlag.

(**Figure 3**). Spillover to humans and domesticated animals results in epidemics and epizootics, with the affected animals (including humans) usually not generating sufficient viremia to participate in the transmission cycle (i.e., acting only as dead-end hosts). VEEV temporarily adapts via mutation to utilize equids as highly efficient amplification hosts (with high viremias), resulting in hundreds to several thousands of horses, donkeys, and mules affected during epizootics, and increased spillover to humans that results in epidemics of similar proportions. CHIKV and ONNV cause major epidemics by using humans as amplification hosts and, as vectors, mosquitoes that live in close association with humans, such as *Aedes aegypti, Ae. albopictus* (CHIKV), and *Anopheles* spp. (ONNV). Some alphaviruses, including EEEV, probably use secondary ('bridge') vectors during epidemics and epizootics because the primary mosquito vector (e.g., *Culiseta melanura* for EEEV) has a very narrow host range limited mainly to the animals serving as the reservoir hosts (**Figure 3**).

The infection of mosquito vectors by alphaviruses begins with the ingestion of a blood meal from a viremic vertebrate host. Susceptibility and minimum infectious doses for alphavirus infection of mosquitoes vary dramatically. Following the passage of ingested blood into the posterior midgut of the alimentary tract during feeding, infection of epithelial digestive cells occurs, presumably before the deposition of the peritrophic matrix around the blood meal, which begins with hours of feeding. Recent studies indicate that typical VEEV infections begin with fewer than 100 of the approximately 10 000 epithelial cells becoming infected, and thus the initial midgut infection may represent a bottleneck in alphavirus populations. After replication in the midgut epithelial cells, alphaviruses bud principally from the basal plasma membrane and somehow traverse the basal lamina to enter the hemocoel, or open body cavity of the mosquito. Then, transported by the hemolymph, the virus has access to many internal organs including the salivary glands, where replication leads to virus shedding into the saliva. Upon a subsequent blood meal from a naive host, transmission can occur via saliva deposited extravascularly during probing to locate a blood vessel, or intravascularly during engorgement. The amount of virus transmitted during blood feeding also varies widely, but rarely exceeds 100 infectious units of VEEV, again representing a bottleneck for virus populations. The extrinsic incubation period from an infectious blood meal to transmission by bite can occur in as little as 3 days, depending on the virus, mosquito, and temperature of incubation. However, mosquitoes generally do not re-feed until egg development and oviposition has occurred, which can require a week or more.

The sites of initial alphavirus replication in the vertebrate host are generally known only from studies using subcutaneous or intradermal needle inoculations, rather than by mosquito feeding, and can include Langerhans cells, dendritic cells, dermal macrophages, and fibroblasts in the skin. Because virus is also deposited intravascularly during mosquito feeding, other sites could be important in natural infections. Secondary replication and amplification sites also vary, including lymph nodes, spleen, myocytes, chondrocytes, osteoblasts, and neurons.

Rubella virus is transmitted via the respiratory route among infected humans, with no vector involvement. Transplacental transmission can result in severe complications and sequelae.

Alphaviruses of Greatest Medical and Veterinary Importance

The majority of alphaviruses can cause at least mild disease in humans, and most exceptions are lesser-known viruses that have not been well studied epidemiologically. Recent studies of 'dengue-like' illness in several locations have revealed that several alphaviruses such as VEEV, MAYV, and CHIKV account for a significant number of cases diagnosed clinically as dengue. Proper serological testing or virus detection is needed to accurately determine the cause of many tropical fevers that present with nonspecific 'flu-like' signs and symptoms. Brief descriptions of the most important alphaviruses follow.

Chikungunya Virus

CHIKV is probably the most important alphaviral pathogen worldwide, although it rarely causes fatal disease. CHIKV, named after a local Makonde word meaning 'that which bends up', refers to the characteristic posture assumed by patients suffering severe joint pains. The virus probably occurs in most of sub-Saharan Africa, as well as in India, Southeast Asia, Indonesia, and the Philippines. In Africa, a sylvan transmission cycle between wild primates and arboreal *Aedes* spp. mosquitoes has been characterized, and the virus probably occupies a niche similar to that of the flavivirus, yellow fever virus. In Asia and nearby islands, and recently on islands off the eastern coast of Africa in the Indian Ocean, urban and suburban CHIKV epidemics usually have been associated with *Ae. aegypti* and/or *Ae. albopictus* transmission among human amplification hosts. Explosive epidemics have affected hundreds of thousands to millions of people, and many cases are probably unrecognized because CHIKV infection is difficult to distinguish clinically from dengue. Because CHIKV infection of humans is sporadic, and no reservoir hosts or sylvatic vectors are known from Asia, it is not known whether the virus is zoonotic outside of Africa, where it is believed to have originated. Although CHIKV historically has been regarded as a nonfatal human pathogen, recent fatal cases on islands off the eastern coast of Africa in the Indian Ocean indicate a possible increase in virulence, although improved surveillance is an alternative explanation.

Eastern Equine Encephalitis Virus

In human cases, North American strains of EEEV are the most virulent of the alphaviruses with case–fatality rates generally exceeding 50%. Fortunately, only 220 confirmed cases occurred in the US from 1964 to 2004. A wide variety of domesticated mammals and birds also suffer fatal infections in North America. In South and Central America, EEEV is widespread and equine outbreaks are common, but human EEE is rare, probably due to lower human virulence in the genetically distinct strains that occur there. Attack rates in North America tend to be highest in young children and the elderly.

Transmission cycles of EEEV in North America involve *Culiseta melanura* as the principal enzootic mosquito vector, and passerine birds as reservoir hosts in hardwood swamp habitats. Bridge vectors in the genera *Aedes*, *Coquillettidia*, and *Culex* may be responsible for most transmission to humans and domestic animals. Foci occur in swamps along the eastern seaboard and the coast of the Gulf of Mexico, but have also been documented as far west as Texas and the Dakotas. In South and Central America, the ecology and epidemiology of EEEV are poorly understood. Isolates from mosquitoes suggest that the most important enzootic vectors are members of the subgenus *Culex* (*Melanoconion*), which are also enzootic vectors of most of the VEE complex viruses. Reservoir hosts are not well understood, but may include mammals or birds.

Mayaro Virus

Mayaro virus has been isolated from persons suffering from an arthralgia with rash in Surinam, French Guiana, Colombia, Panama, Brazil, Peru, and Bolivia, and probably occurs in lowland tropical forests throughout much of South America. As are infections with many other alphaviruses, clinical diagnosis of MAYV infection is unlikely and many cases are misdiagnosed as dengue. Sporadic and focal infections usually occur in people living near or working in tropical forests, and forest-dwelling mosquitoes in the genus *Haemagogus* are probably the principal vectors. Wild primates probably serve as reservoir hosts. Mayaro virus activity is often detected during investigations of outbreaks of sylvan yellow fever, probably because both viruses share the same mosquito vectors.

Mayaro virus infection is characterized by a sudden onset of fever, headache, myalgia, chills, and arthralgia, sometimes accompanied by dizziness, photophobia, retro-orbital pain, nausea, vomiting, and diarrhea. Though not known to be fatal, the arthralgia may be severe, may persist for up to 2 months, and usually affects the wrists, ankles, and toes, and less commonly the elbows and knees. Approximately two-thirds of patients develop a fine maculopapular rash on the trunk and extremities.

O'nyong-nyong Virus

ONNV, a close relative of CHIKV, derives its name from the description by the Acholi tribe, meaning 'joint breaker'. The virus was first isolated and characterized in 1959 during an epidemic involving c. 2 million people in Uganda, Kenya, Tanzania, Mozambique, Malawi, and Senegal. Since that epidemic ended in 1962, only a single major epidemic (1996, in the Rakai, Mbarara, and Masaka districts of southwestern Uganda and bordering Bukoba district of northern Tanzania) and a few sporadic reports of virus isolation have been reported. ONNV causes an arthralgia-rash syndrome that can persist for months, and all age groups are affected. Epidemic transmission involves *Anopheles funestus* and *An. gambiae* mosquitoes, which also transmit malaria parasites in Africa. ONNV is the only known alphavirus with *Anopheles* spp. vectors.

Ross River Virus

Ross River virus (RRV) infection has been recognized in Australia since 1928 as associated with epidemic polyarthritis. The disease there may involve up to thousands of people annually, and occurs mainly during the summer and autumn as sporadic cases and small outbreaks, usually among vacationers and other persons living or traveling in rural areas. RRV is maintained in Australia primarily among marsupials and other wild vertebrates, with *Culex annulirostris* and *Aedes vigilax* and other mosquitoes serving as vectors. Human cases and explosive epidemics also have been documented in islands of the South Pacific. Epidemiologic studies implicated *Ae. polynesiensis* as a vector and suggested that humans may have served as amplification hosts.

Following 3–21 days of incubation, disease begins suddenly with headache, malaise, myalgia, and joint pain. Joints may be swollen and tender. Multiple joints are involved; the most commonly affected are the ankles, fingers, knees, and wrists. Pain and loss of function usually last for several weeks, but some patients have persistent or recurrent arthralgia and arthritis for up to a year. About one-half of epidemic polyarthritis patients develop a maculopapular rash, usually lasting 5–10 days; 30–50% also have low-grade fever. Persons 20–50 years of age are most commonly affected, and the incidence is higher in females. Viremia is transient and may precede the onset of arthritis. The skin rash and joint swelling may be due to a local cell-mediated immune response rather than to immune complexes or complement-mediated reaction. The synovial exudate contains no detectable virus and consists almost entirely of mononuclear leukocytes.

Venezuelan Equine Encephalitis Virus

VEE was recognized first in Venezuela in 1936. The etiologic agent, VEEV, has caused explosive equine epizootics and epidemics in many regions of the Americas. The last major epidemic in northern Venezuela and Colombia in 1995 involved approximately 100 000 persons. VEEV strains use two distinct transmission cycles: enzootic, and epidemic or epizootic. The enzootic cycles generally rely on small rodent reservoir hosts and mosquito vectors in the subgenus *Culex* (*Melanoconion*), and occur in tropical forest and swamp habitats from Florida to Argentina. These viruses and subtypes are generally avirulent for equids, but are pathogenic for humans and can cause fatal disease. In contrast, epidemic or epizootic VEEV (subtypes IAB, IC) are virulent for both equids and humans, but have only temporary, unstable transmission cycles involving *Aedes* and *Psorophora* spp. mosquitoes, which transmit the virus among equids circulating high levels of viremia (**Figure 3**). People generally become infected by mosquitoes that previously engorged on viremic equids. VEE outbreaks are believed to occur following mutations of enzootic subtype ID viruses, which result in the transformation to an equid-virulent, IAB or IC subtype.

Following an incubation period of 1–4 days, symptoms of human VEE include fever, lethargy, headache, chills, dizziness, body aches, nausea, vomiting, and/or prostration. Symptoms usually subside after several days, but may recrudesce. Disease occurs in all age groups and both sexes, but severe neurologic signs, including convulsions and seizures, occur primarily in children (4–14% of pediatric cases) and usually late in the illness. Case–fatality rates are estimated at 0.5%.

Western Equine Encephalitis Virus

WEEV was first isolated from a horse brain in the Central Valley of California during a major epizootic in 1930. Like EEEV, WEEV was not established as an etiological agent of human disease until 1938. Later, WEEV was identified nearly throughout the Americas including the eastern US, but more detailed antigenic and genetic studies eventually elucidated other members of the WEE antigenic complex including Highlands J virus in eastern North America and Fort Morgan virus in the central United States. The distribution of WEEV is now believed to include western North America through Central and South America to Argentina. Virus strains from this wide range are remarkably conserved genetically, suggesting efficient transport by birds during migrations.

Antigenically and genetically distinct WEEV subtypes with more limited distributions in South America have also been identified, and some are less virulent in murine models. All of these WEEV-related New World viruses are believed to have descended from an ancient, recombinant alphavirus that obtained its nonstructural and capsid protein genes from an EEEV-like ancestor and its envelope glycoprotein genes from an ancestor of SINV.

In the western US, WEEV is transmitted primarily in agricultural habitats by *Culex tarsalis* mosquitoes among passerine birds, principally sparrows and house finches. A secondary cycle involving *Aedes* spp. mosquitoes and lagomorphs has also been described in central California, and *Aedes* mosquitoes have been implicated as equine epizootic vectors in Argentina. Human infections with WEEV range from inapparent (the vast majority) to a flu-like syndrome to life-threatening encephalitis and meningitis. Symptomatic infection typically includes a sudden onset with fever, headache, nausea, vomiting, anorexia, and malaise, followed by cognitive symptoms, weakness, and meningeal involvement. Young children tend to be affected more severely than adults, and 5–30% suffer permanent neurologic sequelae. The overall human case–fatality rate is about 3%. WEEV also causes encephalitis in equids, with case–fatality rates of 10–50%. Despite a history of producing epidemics with thousands of human cases, WEEV has apparently declined as a human pathogen since the 1970s, probably due in part to irrigation, mosquito control, and culturally reduced human exposure to mosquitoes.

Alphavirus Pathogenesis

Detailed pathogenesis studies of alphavirus infections have been performed for only a few members of the genus. Although they do not normally cause encephalitis in humans or domestic animals, SINV and SFV have served as models for other alphaviruses because they can cause encephalitis in young mice yet require lower levels of biocontainment than the human-virulent members of the genus. In the experimental encephalitis model these viruses cause apoptotic death of neurons after intracerebral inoculation. Studies with these and other aphaviruses described below indicate that the type 1 interferon response is critical for the initial control of infection, and that humoral immunity is most important for clearance of primary infection and for protection against subsequent infections. Cell-mediated immunity has been regarded as less important, and elimination of alphaviruses from the central nervous system is thought to occur via noncytolytic mechanisms.

Among the highly human-pathogenic alphaviruses, VEEV pathogenesis has received the most attention. The murine model produces uniformly fatal encephalitis following all routes of infection. Following subcutaneous infection, dendritic cells are believed to be infected in the dermis, followed by migration to the draining lymph node where high levels of replication result in viremia and lymphodepletion. Also associated with infection is a vigorous innate immune response that may contribute to neuroinvasion and/or febrile illness. VEEV typically invades the central nervous system via olfactory neurons. Encephalitis develops by day 5–7, when the virus has been cleared from the periphery by the adaptive immune response. However, VEEV continues to replicate to high titers in the brain and paralysis is quickly followed by death. Clinical encephalitis probably results from both virus- and immune-mediated neuronal cell death.

The pathogenesis of EEEV is poorly understood, but appears to differ fundamentally from that of VEEV in the murine model; EEEV also produces a biphasic disease, but the initial phase involves replication in fibroblasts, osteoblasts, and skeletal muscle myocytes. Invasion of the central nervous system probably occurs by a vascular route. Hamsters appear to more accurately reproduce the vascular component of EEE, which is prominent in human infections.

Among the alphaviruses that cause rash/arthralgia syndromes, only RRV has received much study as an etiological agent. In the murine model, initial targets of subcutaneous infection are similar to those of EEEV, including bone, joint, and skeletal muscle, with severe associated inflammation involving macrophages, natural killer cells, and CD4+ and CD8+ T lymphocytes. Recent studies with mice deficient in T lymphocytes indicate that the adaptive immune response does not play a critical role in the development of arthritic disease.

Diagnosis and Treatment

Diagnosis of alphavirus infections is rare because clinical signs and symptoms are similar to those caused by many other viral pathogens, including dengue and influenza viruses. Definitive diagnosis generally requires virus isolation from blood taken during the first 2–4 days of illness, or serologic confirmation as described below. Most alphaviruses can be isolated by intracerebral inoculation of newborn mice (usually fatal), and cause cytopathic effects in cells of a variety of mammalian and avian lines. Serodiagnosis can be made by detecting a fourfold or greater change in antibody titers in acute-phase and convalescent-phase serum samples drawn 1–3 weeks apart, using an enzyme-linked immunosorbent assay (ELISA), immunofluorescence, hemagglutination inhibition, CF, or NT. Detection of antiviral immunoglobulin M (IgM) using ELISA has more recently been applied for this purpose and is now the standard assay.

Treatment of most alphavirus infections is symptomatic and supportive. Anticonvulsive therapy may be effective in severe cases of encephalitis, especially in children. Lymphoid depletion caused by VEEV may lead to bacterial infection of the gastrointestinal tract, and antibiotic therapy should be considered in severe cases. Pneumonia secondary to VEE and EEE also is common. Neurologic sequelae occur following severe EEE, VEE, and WEE, including headache, amnesia, anxiety, and motor impairment.

Control of Disease

As is the case for many other arboviruses, control of alphaviral diseases relies upon interruption of transmission by mosquito vectors, and sometimes on vaccination of amplification hosts such as equids in the case of VEE. Mosquito vector control during epidemics and epizootics usually relies upon aerial application of insecticides such as malathion, optimally applied soon after floodwater species emerge from the aquatic immature stages, prior to dispersal, extrinsic infection, and transmission. People present during epidemics should avoid exposure to mosquitoes through limitation of outdoor activities and the use of repellents containing the active ingredient meta-N,N-diethyl toluamide (DEET) or picaridin. Persons entering sylvatic or swamp habitats where enzootic EEEV or VEEV circulate in the Americas, as well as forests of Africa enzootic for CHIKV, should also take these precautions. The diel periodicity of mosquito vectors should also be considered when planning outdoor activities and control measures. For example *Haemagogus* mosquitoes, the principal vectors of MAYV, feed during the daytime in forested neotropical habitats, and bed nets and window screens therefore do not effectively protect people. *Aedes aegypti*, an important urban vector of CHIKV, readily enters and rests in homes, and outdoor applications of adulticides are largely ineffective in controlling epidemics.

See also: Japanese Encephalitis Virus; Rubella Virus; Tick-Borne Encephalitis Viruses; Togaviruses: Alphaviruses; Togaviruses: Equine Encephalitic Viruses; Togaviruses: Molecular Biology; Togaviruses Not Associated with Human Disease.

Further Reading

Calisher CH and Karabatsos N (1988) Arbovirus serogroups: Definition and geographic distribution. In: Monath TP (ed.) *The Arboviruses: Epidemiology and Ecology,* vol. 1, pp. 19–57. Boca Raton, FL: CRC Press.

Casals J and Brown LV (1954) Hemagglutination with arthropod-borne viruses. *Journal of Experimental Medicine* 99(5): 429–449.

Griffin DE (2001) Alphaviruses. In: Knipe DM and Howley (eds.) *Fields Virology,* 4th edn., pp. 917–962. New York: Lippincott Williams and Wilkins.

Powers AM, Brault AC, Shirako Y, *et al.* (2001) Evolutionary relationships and systematics of the alphaviruses. *Journal of Virology* 75(21): 10118–10131.

Strauss JH and Strauss EG (1994) The alphaviruses: Gene expression, replication, and evolution. *Microbiology Reviews* 58(3): 491–562.

Weaver SC (2005) Host range, amplification and arboviral disease emergence. *Archives of Virology* 19(supplement): 33–44.

Weaver SC and Barrett AD (2004) Transmission cycles, host range, evolution and emergence of arboviral disease. *Nature Reviews Microbiology* 2(10): 789–801.

Weaver SC, Frey TK, Huang HV, *et al.* (2005) Togaviridae. In: Fauquet CM, Mayo MA, Maniloff J, Desselberger U, and Ball LA (eds.) *Virus Taxonomy: Eighth Report of the International Committee on Taxonomy of Viruses,* pp. 999–1008. San Diego, CA: Elsevier Academic Press.

Togaviruses: Molecular Biology

K D Ryman and W B Klimstra, Louisiana State University Health Sciences Center at Shreveport, Shreveport, LA, USA
S C Weaver, University of Texas Medical Branch, Galveston, TX, USA

© 2008 Elsevier Ltd. All rights reserved.

Glossary

Apoptosis A cascade of cellular responses to a stimulus (e.g., virus infection) resulting in cell death.
Cytopathic effect Destructive changes in cell morphology, structure, or metabolic processes resulting from virus infection.

Full-length cDNA clone A copy of an RNA virus genome that has been reverse-transcribed into DNA and placed into a subcloning vector. With positive-sense RNA viruses, infectious genomic RNA is generally directly synthesized from bacteriophage transcription promoters located upstream of the virus sequences.

Host cell shut-off Interruption or cessation of host transcription, translation, or other processes that occurs within virus-infected cells.

Interferon A pro-inflammatory cytokine that is produced after virus infection and stimulates antiviral activities in cells expressing the IFNAR receptor complex.

Interferon-stimulated gene (ISG) A gene whose transcription is increased after interferon signaling. Some ISGs are directly antiviral and can inhibit virus transcription or translation.

Positive-sense RNA genome The genomic material of the virus resembles a cellular messenger RNA and the open reading frame can be translated on host ribosomes.

Receptor Structures (e.g., protein, sulfated polysaccharide, phospholipid) found upon the surface of host cells that interact with virus attachment proteins and mediate attachment of virus particles to cell surfaces. These structures may also mediate subsequent events in cell entry such as viral protein rearrangements leading to membrane fusion.

Replicon system A replication-competent but propagation-incompetent form of a cloned virus. In many replicon systems, the replicon genome can be packaged into a virus-like particle by *trans*-expression of structural proteins.

RNA-dependent RNA polymerase (RdRp) The primary RNA synthesis enzyme for RNA viruses which uses viral genomic RNA as a template for transcription/replication.

Introduction

The family *Togaviridae* includes the genera *Alphavirus* and *Rubivirus*. The genus *Alphavirus* is comprised of 29 virus species segregated into five antigenic complexes, most of which are transmitted between vertebrate hosts by mosquito vectors and are capable of replicating in a wide range of hosts including mammals, birds, amphibians, reptiles, and arthropods. In contrast, the sole member of the genus *Rubivirus* is Rubella virus, which is limited to human hosts and is primarily transmitted by respiratory, congenital, or perinatal routes. Together, the viruses are a significant worldwide cause of human and livestock disease. Alphavirus disease ranges from mild to severe febrile illness, to arthritis/arthralgia and fatal encephalitis, while Rubella virus is an important cause of congenital abnormalities and febrile illness frequently accompanied by arthralgia/arthritis in the developing world. At the molecular level, the togaviruses are relatively uncomplicated, small, enveloped virions with single-stranded, positive-sense RNA genomes encoding two polyproteins, one translated from the genomic RNA and the other from a subgenomic mRNA transcribed during replication. However, their replication is very tightly regulated and intimately associated with host cell metabolic processes. Currently, an important aspect of togavirus research is designed to exploit their molecular biology for therapeutic and gene therapy applications. Here, we summarize major aspects of togavirus structure, genome organization, replicative cycle, and interactions with host cells.

Virion Structure and Genome Organization

Togavirus virions are small and enveloped, comprising an icosahedral nucleocapsid composed of 240 capsid (C) protein monomers, cloaked in a lipid envelope studded with membrane-anchored glycoprotein components, E1 and E2. Their tightly packed, regular structure has been revealed by high-resolution reconstruction of the particles by cryoelectron microscopy (**Figure 1**). The envelope glycoproteins of alphaviruses are similarly arranged on the outer surface of particles. E1 and E2 form a stable heterodimer, and three heterodimers interact to form 'spikes' distributed on the virion surface in an icosahedral lattice that mirrors the symmetry of the nucleocapsid. The regularity of the spike distribution is determined by the direct hydrophobic interaction of the E2 glycoprotein tail with a pocket on the surface of the C-protein. E2 projects outward from the virion to form the spikes that interact with host-cell attachment receptors, while the E1 glycoprotein, which mediates fusion with host cell membranes, appears to lie parallel to the lipid envelope. The E2 and E1 proteins are glycosylated at one to four positions, depending on the virus strain, during transit through the host-cell secretory apparatus.

The encapsidated alphavirus genome consists of a nonsegmented, single-stranded, positive-sense RNA molecule of approximately 11–12 kb with a 5′-terminal methylguanylate cap and 3′ polyadenylation, resembling cellular mRNAs (**Figure 2**). The genome is divided into two major regions flanked by the 5′ and 3′ nontranslated regions (NTRs) and divided by an internal NTR: (1) the 5′-terminal two-thirds of the genome encode the four nonstructural proteins (nsPs 1–4); and (2) the 3′ one-third of the genome encodes the three structural proteins (C, precursor E2 [PE2], and precursor E1 [6K/E1]). Overall, the capped and polyadenylated 9.5 kb Rubella virus genome is similarly organized (**Figure 2**), but the 5′-proximal open reading frame (ORF) encodes only two nsPs (P150 and P90) and three structural proteins are encoded by the 3′ ORF (C, E1, and E2).

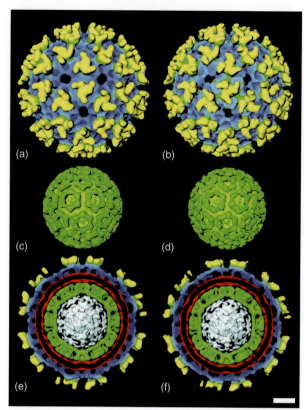

Figure 1 Structure of New World (VEEV) and Old World (SINV) alphaviruses determined by image reconstructions of electron micrographs. Isosurface view along a threefold axis of VEEV (a) and SINV (b) reconstructions showing outer spike trimers (yellow) and envelope skirt region (blue). Isosurface representations of VEEV (c) and SINV (d) nucleocapsids viewed along a threefold-symmetry axis. Cross-sections through VEEV (e) and SINV (f) perpendicular to the threefold axis and in plane with a vertical fivefold axis showing trimers (yellow), skirt region (blue), virus membrane (red), nucleocapsid (green), and RNA genome (white). Scale = 100 Å. Reproduced from Paredes A, Alwell-Warda K, Weaver SC, Chiu W, and Watowich SJ (2007) Venezuelan equine encephalomyelitis virus structure and its divergence from old world alphaviruses. *Journal of Virology* 75: 9532–9537, with permission from American Society for Microbiology.

Infection and Replication Cycle

The togavirus genome encodes very few but, therefore, necessarily multifunctional proteins. The replication of these viruses is very tightly regulated and as little as a single nucleotide change can alter susceptibility to cellular defense mechanisms. The following section outlines molecular aspects of togavirus infection and replication pertinent to the virus–cell interaction. Replication processes of viruses of the family *Togaviridae* have been most extensively studied for the alphaviruses, Sindbis virus (SINV) and Semliki Forest virus (SFV), and extrapolated to the other family members.

Virus Attachment, Entry, and Uncoating

The entry pathways of Rubella virus have undergone only limited study, but there is some evidence that a glycolipid molecule serves as an attachment receptor on some cell types. The nature of the receptor for naturally circulating, 'wild-type' alphaviruses has been examined more extensively, but remains controversial, as is the case with many arthropod-borne viruses. Typically, arthropod-borne viruses are thought to interact either with a single, evolutionarily conserved molecule expressed on most cells, or to utilize different receptors in different hosts and with cells derived from different tissues. The 67 kDa high-affinity laminin receptor (HALR) has been identified as an initial attachment receptor for cell culture-adapted, laboratory strains of SINV and an antigenically related molecule has similarly been identified for Venezuelan equine encephalitis virus (VEEV). In addition, receptor activity for particular cell types has been attributed to several other, as yet uncharacterized, proteins. However, more recent studies have indicated that efficient cell binding by many laboratory strains of alphaviruses is due to interaction with heparan sulfate (HS), a sulfated glycosaminoglycan molecule, and that this phenotype is conferred by positively charged amino acid substitutions in E2 that accompany adaptation to cultured cells. Currently, it is unclear whether or not the 67 kDa HALR or other identified receptor proteins act in concert with HS to bind cell culture-adapted alphavirus strains or if any of these molecules function in the natural replication cycle of wild-type alphaviruses. Moreover, it has become clear that many noncell culture-adapted alphavirus strains bind only weakly to commonly used cell lines, such as those derived from fibroblasts. Recently, DC-SIGN and L-SIGN, C-type lectins, were found to bind carbohydrate modifications on E1 and/or E2 proteins of both HS-binding and non-HS-binding strains of SINV and VEEV, promoting infection. The interaction of the lectins with virion carbohydrates is greatly enhanced by replication of viruses in mosquito cells, most likely due to the retention of high mannose carbohydrate structures, which are processed to complex carbohydrates in mammalian cells. Therefore, C-type lectins may be important attachment receptors for alphaviruses when transmitted by mosquitoes, and receptor utilization could be determined, at least in part, by the cell type in which alphaviruses replicate. Since DC-SIGN and L-SIGN receptors are expressed by a subset of alphavirus-permissive cells, these results further suggest that infection by alphaviruses may be mediated by different receptors on different cell types.

Following receptor interactions, which may initiate uncoating-related conformational changes in E1 and E2, alphavirus and Rubella virus virions enter acidified endosomes via a dynamin-dependent process. The decreasing pH in the endosome results in more dramatic

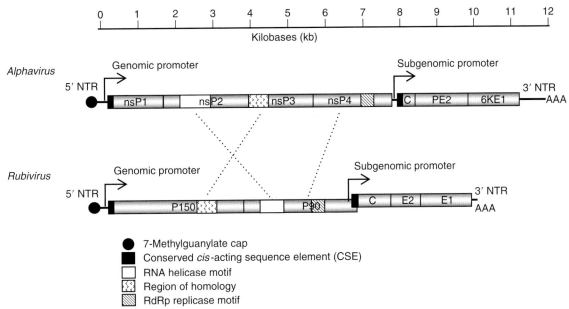

Figure 2 Comparison of *Alphavirus* (SINV) and *Rubivirus* (RUB) genome organization. Regions of nucleotide homology and regions encoding homologous amino acid sequence within the nonstructural protein ORF are shown. ORFs are denoted by boxes and nontranslated regions (NTRs) by lines. Note that for SINV the two ORFs are separated by a NTR region, whereas for RUB the two ORFs overlap in different translational frames. Reproduced from Dominguez G, Wang C-T, and Frey TK (1990) Sequence of the genome RNA of rubella virus: Evidence for genetic rearrangement during togavirus evolution. *Virology* 177: 225–238 with permission from Elsevier.

conformational changes (threshold of ~pH 6.2 with SFV) in the E2/E1 spike leading to formation of E1 homotrimers and exposure of a putative class II fusogenic domain in E1 that is thought to interact with the host cell membrane, promoting fusion of host and virion lipid bilayers. In mammalian cells, SFV membrane fusion is cholesterol-dependent and sphingolipid-dependent, and inhibitors of endosomal acidification block infection. Endosomal fusion leads to a poorly characterized entry of nucleocapsids into the cytoplasm and to their association with ribosomes, most likely directing final uncoating of the genomic RNA and translation initiation.

Genome Translation, Transcription, and Replication

Togavirus genome transcription and replication occur by a strictly positive-sense strategy and are entirely cytoplasmic (**Figure 3**). The alphavirus replicase complex comprises four nsPs encoded in the 5′-terminal ORF, all of which are essential for viral transcription and replication: (1) nsP1 is a guanine-7-methyltransferase and guanylyl-transferase; (2) nsP2 has NTPase, RNA helicase, and RNA triphosphatase activities in its amino-terminal domain, while the carboxy-terminus has highly specific thiol protease activity; (3) nsP3 is a phosphoprotein with unknown function; and (4) nsP4 is the catalytic RNA-dependent RNA polymerase (RdRp). In addition to the final products, some of the processing intermediates have distinct and indispensable functions during the replication process. The 5′-proximal ORF of the Rubella virus genome is also translated as a polyprotein (p200) and cleaved in *cis* by the Rubella virus protease (NS-pro) to form the replication complex comprising P150 and P90. However, the order of its conserved motifs differs from that of the alphavirus nsP polypeptide (**Figure 2**): P150 contains the putative methyltransferase (N-terminal) and protease sequences (C-terminal), while P90 contains the predicted helicase (N-terminal) and RNA polymerase (C-terminal).

Once the nsPs have been translated from the infecting, positive-sense RNA genome, full-length RNAs complementary to the genomic sequence (negative-sense) are synthesized, creating partially double-stranded (ds) RNA replicative intermediates. The RdRp complex then switches to preferential synthesis of positive-sense RNAs, which continues throughout the remainder of the infection replication cycle, while negative-sense RNA transcription ceases entirely. Two positive-sense, capped, and polyadenylated RNAs are synthesized on the negative-sense template: full-length progeny genomes that are packaged into virions and subgenomic (26S) RNAs that are collinear with the 3′ one-third of the genome and encode the structural protein ORF. The 26S RNA accumulates in the cells to 5–20-fold molar excess over genomic RNA and is efficiently translated into a polyprotein that is co-translationally processed to produce the structural proteins.

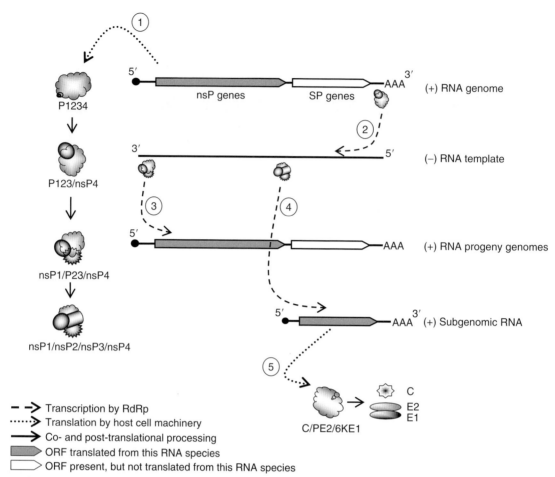

Figure 3 Diagrammatic representation of *Alphavirus* replication in a permissive cell. The replication cycle is depicted as a series of temporally regulated steps: ① Translation and processing of nonstructural polyprotein P1234 or P123; ② Synthesis of complementary, negative-sense RNA by P123/nsP4; ③ Synthesis of progeny genomes by nsP1/P23/nsP4; ④ Synthesis of subgenomic RNAs by nsP1/nsP2/nsP3/nsP4; ⑤Translation and processing of structural polyprotein C/PE2/6KE1. Progeny genomes are then packaged into nucleocapsids and bud at the plasma membrane to release progeny virions.

Alphavirus genome replication and transcription are tightly regulated temporally by sequential processing of the nsPs, which alters replicase composition and RdRp activity. For some alphaviruses (e.g., SFV), the nsPs are translated as P1234, while others (e.g., SINV) translate the nsPs as either P1234 or P123 polyproteins, depending upon read-through of an opal termination codon. The nascent P1234 polyprotein is partially processed by autocatalytic nsP2 protease-mediated cleavage at the 3/4 junction and assembled into the primary P123/nsP4 RdRp complex. This short-lived P123/nsP4 primarily synthesizes genomic, negative-sense RNA replicative intermediates. Membrane-bound P123 is cleaved relatively slowly in *cis* at the 1/2 junction, perhaps to enable formation of protein–protein interactions prior to proteolytic processing, and to form the nsP1–P23–nsP4 replicase that preferentially synthesizes positive-sense RNA. After the aminoterminus of nsP2 has been released, rapid autocatalytic cleavage of P23 produces a mature, stable RdRp complex (nsP1–nsP2–nsP3–nsP4), which transcribes only positive-sense RNA. Evidence suggests that nsP1–P23–nsP4 and the mature replicase may be biased, respectively, toward synthesis of genome-length or subgenomic mRNA, providing further temporal regulation of the replication cycle. Late in infection, when free nsP2 concentrations are high, cleavage at the 2/3 junctions is favored. P1234 polyproteins are rapidly processed into short-lived P12 and P34 precursors, thereby precluding assembly of new replication complexes and terminating negative-strand synthesis. *cis*-Cleavage of the P12 precursor gives rise to nsP1, which targets the plasma membrane, and nsP2. Interestingly, much of nsP2 is transported to the nucleus, suggesting a function beyond the replication and transcription of viral RNAs, as discussed below. P34 is cleaved in *trans* by free nsP2, yielding nsP3, which aggregates in the cytoplasm, and nsP4, which is rapidly degraded by the ubiquitin pathway.

Specificity of togavirus RNA replication by the RdRp replicase is achieved via differential recognition of *cis*-acting conserved sequence elements (CSEs) in the termini of the viral genome and negative-sense genome template. These CSEs are conserved in all alphaviruses, as well as in Rubella virus. A 19-nucleotide (nt) CSE in the 3′ NTR immediately upstream of the poly(A) tail, is the core promoter for synthesis of genome-length, negative-sense RNA replicative intermediates, and probably interacts with the 5′ NTR via translation initiation factors to initiate replication and/or translation. The complement of the 5′ NTR in the minus-strand RNA serves as a promoter for the synthesis of positive-sense genomes. A third cis-acting element that is essential for replication, the 51-nt CSE, is also found near the 5′ end of the genomic RNA within the nsP1 gene. The primary sequence and two-stem–loop secondary structure of this CSE are highly conserved among alphaviruses and serve as replication and translation enhancers. Finally, a 24-nt CSE is found upstream of and including the start of the subgenomic RNA, the complement of which in the negative-strand forms the 26S subgenomic promoter and translation enhancer element.

Togavirus RNA replication occurs on cytoplasmic surfaces of endosome-derived vesicles or cytopathic vacuoles (CPVs). Expression of the nsP polyprotein (particularly nsP1) actively modifies intracellular membranes to create these compartments and to mediate membrane association of the replicase via membrane phosphatidylserine and other anionic phospholipids. The surfaces of CPVs have small invaginations or 'spherules', in which the nsPs and nascent RNA are sequestered, creating a microenvironment for RNA replication, synthesis of structural proteins, and assembly of nucleocapsids. The formation of CPVs in togavirus-infected cells may be linked to stress-induced autophagic mechanisms intended for the targeted degradation of cellular proteins and organelles, but exploited by the virus.

Structural Protein Maturation, Genome Packaging, and Egress of Virions

The subgenomic mRNAs of alphaviruses and of Rubella virus are translated as polyproteins; however, processing of the polyprotein differs between the two. With alphaviruses, the capsid protein is cleaved autoproteolytically from the E2 precursor (known as PE2 or P62) in the cytoplasm and the remaining polyprotein is translocated into the ER lumen, where N-linked oligosaccharide addition occurs. The 6K hydrophobic protein, which is primarily located in the ER membrane, is then cleaved from the polyprotein by the host signalase. With Rubella virus, the capsid-E2 and E2–E1 cleavages are all completed by signalase. Furthermore, Rubella virus lacks the 6K protein and mature E2 is produced in the ER. Alphavirus PE2 and E1 proteins form heterodimers in the ER that are anchored by their transmembrane domains and involve intra- and intermolecular disulfide linkages and associations with host chaperone proteins such as Bip and calnexin/calreticulin. The PE2/E1 heterodimers are routed to the cytoplasmic membrane and PE2 is cleaved into mature E2 by a host furin-like protease as a late event, in a structure between the trans Golgi network and cell surface. The cleaved E3 fragment is lost from some (e.g., SINV), but not all (e.g., SFV) virus particles. PE2-containing virions appear to bud normally from vertebrate cells, but are often defective in conformational rearrangements associated with cell fusion, supporting the hypothesis that the presence of PE2 stabilizes the glycoprotein heterodimer during low pH exposure in the secretory pathway.

Budding of particles occurs through an interaction between C and the cytoplasmic tail of E2; however, it is unclear at which point in the secretory pathway this interaction occurs and whether the interaction is between C monomers, oligomers, or RNA-containing preformed nucleocapsids. With alphaviruses, RNA packaging is directed by an RNA secondary structure in either the nsP1 or nsP2 genes, depending upon the virus, leading to selective packaging of the genome over the subgenome, which is in molar excess. In the accepted model, final budding occurs at the cytoplasmic membrane at sites enriched in E2/E1 heterodimers and is driven by E2 tail-C interactions that force an extrusion of the host lipid bilayer, envelopment of the particle, and release.

Infectious Clone Technology

Knowledge of the genome organization and replication cycle of the togaviruses have enabled the construction of full genome-length cDNA clones from which infectious RNA molecules can be transcribed in vitro and transfected into permissive cells to initiate productive virus replication. This procedure allows easy introduction of specific mutations into the virus genome followed by generations of genetically homogeneous virus populations encoding the mutation(s). Indeed, much of the work described in this article was performed using virus mutants generated from cDNA clones. In addition, 'double-subgenomic promoter viruses' (DP-virus) have been created in which the subgenomic promoter is reiterated immediately upstream of the authentic subgenomic promoter or at the 3′ end of the E1 coding region to drive expression of genes used as reporters of infection or vaccine vectors (e.g., GFP, luciferase, or immunogens from infectious microorganisms) or to test the effect of protein/RNA expression in the context of a togaviral infection (e.g., antiviral or apoptosis-inhibiting proteins, or interfering RNAs). Propagation-defective virus-like particles or 'replicon' systems have also been developed in which the structural protein genes are deleted and replaced with a heterologous protein gene. With alphaviruses, replicon genomes can be packaged to form replicon particles by expressing the structural proteins in trans. Typically, the replicon genome RNA is co-electroporated with two 'helper' RNAs, which are replicated and transcribed by the nsPs provided by the replicon genome and separately encode the capsid and PE2/6K/E1 proteins. Use of these bipartite helper systems minimizes the potential for generation of propagation-competent progeny. In contrast, Rubella virus exhibits a dependence of expression on the nsP ORF from the same nucleic acid as the structural proteins, limiting this approach. Replicon particles can infect a single cell and express the heterologous protein at high levels from the subgenomic promoter but, in the absence of structural protein expression, no progeny virions are produced. Therefore, these vectors allow detailed examination of the interaction of the virus with single cells and can also be used as vaccines with little possibility that adverse effects associated with propagating vial vectors will occur in vaccinated hosts. The results of numerous studies have indicated that togavirus DP-viruses and replicons will likely become important vector systems for the delivery of heterologous gene products to cells *in vitro* and *in vivo*.

Togavirus–Host Cell Interactions

The togavirus–cell interaction can be viewed as a conflict between the cell and the virus: the virus must replicate to propagate itself, generally leading to cytopathic effects (CPEs), while the host cell attempts to suppress virus replication in order to avoid CPE or to save the greater organism from death (in which case the individual cell may be sacrificed). In the following section, these interactions are addressed as a continuum from poorly restrained virus replication leading to rapid CPE on one extreme, to host-cell-circumscribed virus replication and prolonged cell survival on the other.

Cell Viability and CPE

Historically, alphaviruses (e.g., prototypic SINV) have been considered to be strongly cytopathic, causing rapid killing of cultured vertebrate cells that are highly permissive of their replication. In contrast, alphavirus replication in mosquito cells is most often associated with persistent infection and limited cell death. Rubella virus is also cytopathic for vertebrate cells at high infection multiplicities, but can develop persistent infection at low multiplicity and does not infect mosquito cells. Only recently has the complex interaction between the virus infection cycle and cell death begun to be unraveled. Infection of many vertebrate cell lines with alphaviruses or Rubella virus results in morphological changes (e.g., chromatin condensation, nuclear fragmentation, and formation of membrane-enclosed apoptotic bodies), as well as molecular changes (e.g., caspase activation and DNA fragmentation), associated with programmed cell death/apoptosis.

Expression of SINV structural glycoproteins, particularly the transmembrane domains, can cause apoptosis in rat AT-3 prostatic adenocarcinoma cells. Furthermore, mutation from glutamine to histidine at position 55 of the SINV E2 glycoprotein has been associated with increased apoptosis in cultured cells and neurons in mice. However, although accumulation of viral structural proteins accelerates CPE development, apoptosis also occurs in replicon-infected cells in the absence of structural protein synthesis, linked to replication and nsP2 activities. Complicating this picture further is that interaction of high concentrations of UV-inactivated SINV particles with cell surfaces can initiate cell death pathways in cells of some lines, suggesting that viral replication/gene expression is not always required for killing. Finally, evidence also exists for alphavirus-induced necrotic cell death (i.e., not involving active cellular processes) in subpopulations of infected neurons *in vivo*. Therefore, the stimulus for CPE caused by togaviruses may depend upon the virus genotype, dose, the infected cell type, and/or the infection context. At the molecular level, multiple pathways, including mitochondrial cytochrome *c* release, TNF-α death receptor signaling, sphingomyelinase activation and production of ceramide, and redox-stress and inflammatory-stress pathway activation have been associated with togavirus-induced cell death, further supporting the idea that induction of particular mechanisms may be dependent upon multiple virus and host cell factors.

Host-Protein Synthesis Shut-Off

Togavirus replication in highly permissive vertebrate cells devastates cellular macromolecular synthesis, arresting both cellular transcription and translation by independent mechanisms. Transcriptional downregulation of cellular mRNAs and rRNAs can be directly mediated by the mature SINV nsP2 protein even in the absence of viral replication and is critically involved in the production of CPE. This activity is determined by integrity of the nsP2 carboxy-terminal domain, not by helicase or protease activity and can be greatly reduced by point mutations in this region. Moreover, the degree of nsP2-mediated pro-apoptotic stimulus appears to vary considerably between different togaviruses. Nonstructural protein-mediated inhibition of cellular transcription in highly permissive cells, such as fibroblast cell lines, dramatically suppresses the cell's ability to generate an antiviral stress response and is likely beneficial for togavirus replication and dissemination. However, infection with the noncytopathic nsP2 mutants described above restores the cell's response. In contrast, infection of dendritic cells and macrophages, which may be less permissive to infection than fibroblasts or may express a different ensemble of antiviral response mediators, produces a vigorous stress response after virus exposure and large quantities of new mRNAs are synthesized.

Translation of cellular mRNAs is also dramatically inhibited within hours of togavirus infection in highly permissive cells. This process of host-protein synthesis 'shut-off' occurs independent of transcriptional arrest. The degree to which translational shut-off occurs is inextricably linked with the extent of viral RNA replication, such that incomplete translation inhibition is associated with permissivity and mutations in viral replicase complex proteins, particularly nsP2. As a cellular response, translational inhibition is a stress-induced defense mechanism, able to recognize viral infection, amino acid starvation, iron deficiency, and accumulation of misfolded proteins in the endoplasmic reticulum (ER) via the activation of stress kinases. The activity of four distinct stress kinases converges to phosphorylate eukaryotic translation initiation factor (eIF) 2α, inhibiting GTP–eIF2–tRNAi-Met ternary complex formation and globally suppressing translation initiation. However, the expression of specific stress-inducible cellular proteins continues under conditions of phosphorylated eIF2α and generalized shut-off. If the stress response fails to clear the virus and restore homeostasis, the effects of prolonged shut-off become detrimental to the infected cell, leading to CPE and apoptotic cell death. Phosphorylation of eIF2α during togavirus infection is triggered by the presence of dsRNA primarily through activation of the dsRNA-dependent protein kinase (PKR). In some situations, eIF2α phosphorylation appears to be largely responsible for translation inhibition, since the overexpression of a nonphosphorylatable eIF2α mutant abrogates shut-off despite efficient viral replication. However, the arrest of host protein synthesis has also been shown to occur independently of PKR activity and in the apparent absence of eIF2α phosphorylation, by an unknown mechanism.

In response to this cellular defense mechanism, togaviruses have evolved degrees of tolerance to translational shut-off, enabling them to circumvent this block and redirect the cell's inoperative translational apparatus to the synthesis of viral proteins. Indeed, once shut-off has occurred, virtually the only proteins synthesized by the cell are virus-encoded structural proteins, expressed from the subgenomic mRNA. This implies that host protein synthesis shut-off confers an advantage to the virus, allowing usurpation of cellular translation machinery and potentially dampening the host cell's antiviral stress response. Translation enhancer elements located in the 5′ termini of SINV and SFV 26S (and, very likely, Rubella virus) mRNAs, particularly the highly stable RNA hairpin downstream of the AUG start codon, facilitate the continued expression of alphavirus structural proteins during the translation inhibition imposed by eIF2α phosphorylation. SFV infection has been shown to induce the transient formation of stress granules containing cellular TIA-1/R proteins, which sequester cellular mRNAs, but disassemble in proximity to the viral replicase. The temporal correlation between eIF2α phosphorylation, stress granule assembly and localized disassembly, and the transition from cellular to viral protein synthesis suggest that these may be important processes in generalized shut-off of protein synthesis and avoidance of the shut-off by the 26S mRNA. It is likely that the togaviruses also evade and/or antagonize PKR/phospho-eIF2α-independent translation arrest. Notably, the 26S mRNA also has a low requirement for the translation complex scaffolding protein, eIF4G, although integrity of eIF4G is not known to be affected by togavirus replication.

The Interferon-Mediated Antiviral Response

Antiviral activity of interferon-alpha/beta (IFN-α/β) is a critical determinant of the outcome of togavirus infection both *in vitro* and *in vivo*. In mice, the absence of this response results in greatly increased susceptibility to alphavirus infection and disease. In cell culture, IFN-α/β pretreatment profoundly blocks virus replication and protects cells from CPE induced by all togaviruses tested. However, some alphaviruses (e.g., VEEV) appear to be more resistant to the effects of exogenously added IFN-α/β than others (e.g., SINV). All togaviruses tested also stimulate the production of IFN-α/β from cultured cells to varying degrees. Presumably, IFN-α/β induction results from triggering of cellular pattern recognition receptors (PRRs), such as toll-like receptors (TLRs), PKR and/or cytoplasmic RNA helicases such as RIG-I or MDA-5 by 'pathogen-associated molecular patterns' (PAMPs). In addition to IFN-α/β induction, these receptors stimulate a general inflammatory response in infected cells by activating the NF-κB pathway (which is activated after SINV infection). Historically, cytoplasmic dsRNA, produced as a component of the togavirus replicative cycle has been considered the primary PAMP; however, UV-inactivated preparations of some alphaviruses (e.g., SFV) can elicit inflammatory responses in certain cultured cells suggesting multiple pathways of pathogen detection and response. Secreted IFN-α/β signals through its cognate receptor on infected and uninfected cells to upregulate expression of many IFN-stimulated genes (ISGs), producing antiviral proteins, some of which directly inhibit togavirus replicative processes. In less permissive vertebrate cells, induced IFN-α/β can act upon the infected cell to suppress togavirus replication, prevent overt CPE, and promote a persistent infection.

Several studies have evaluated the individual and combined contribution of the two best-characterized IFN-α/β-inducible antiviral pathways, the PKR pathway (described above), and the coupled 2–5A synthetase/RNase L pathway, to the control of alphavirus replication. The latter pathway is composed of the interferon-inducible 2′–5′ oligoadenylate synthetase (OAS) family of dsRNA-dependent enzymes and dormant, cytosolic RNase L. dsRNA-activated 2′–5′ OAS synthesizes 2′–5′-linked oligoadenylates that specifically bind and activate RNase L which then cleaves diverse RNA substrates, thus inhibiting cellular and viral protein synthesis. As described above, constitutively expressed and IFN-α/β-inducible PKR is activated by dsRNA binding to phosphorylate eIF2α, causing a decline in cap-dependent translation of viral and cellular mRNAs. Infection of mice with targeted disruption of the PKR gene, the RNase L gene or both, revealed that PKR, but not RNase L, is involved in the early control of SINV replication *in vivo* and in primary cell cultures. Surprisingly, IFN-α/β-mediated antiviral responses against SINV are largely intact in the absence of PKR and/or RNase L, confirming the existence of 'alternative' IFN-α/β-induced pathway(s) capable of curtailing togavirus replication. Gene transcription and *de novo* protein synthesis are required for this activity. Although the proteins involved are yet to be characterized, one such pathway involves a novel inhibition of cap-dependent translation of infecting viral genomes in the absence of eIF2α phosphorylation. In addition, the interferon-inducible proteins MxA, ISG15, and the zinc-finger antiviral protein (Zap) have some inhibitory effects upon alphavirus replication. Considering the potent inhibitory effects of IFN-α/β toward numerous togaviruses, the expression of additional mechanisms to evade or antagonize the IFN-α/β response is likely critical for togavirus virulence, although convincing evidence of their existence is yet to be presented.

Conclusions and Perspectives

Research into the molecular biology of togaviruses has two primary goals: (1) to understand the relationship of virus interactions with single cells to replication and disease pathogenesis in natural hosts and, subsequently, to identify targets for therapeutic intervention; and (2) to manipulate virus–cell interactions to capitalize upon the potential for use of these viruses as tools for gene therapy and vaccination. In the examples described above, the dissection of virus attachment, entry, translation, replication, effects upon host cells, and host cell responses to infection all have identified vulnerabilities through which virus infection may be curtailed and many laboratories are currently developing antiviral and/or disease-ameliorating strategies. Furthermore, research into virus stimulation of CPE or stress responses has led to an improved understanding of how to, for example, maximize CPE (important for the development of tumor-destroying vectors) or minimize host antiviral activity (important for expression of immunogens by vaccine vectors). Some important issues that remain to be addressed include the full elucidation of the complex relationship of virus replication to development of CPE, the nature of antiviral stress response mediators capable of blocking virus replication, and the mechanisms through which togaviruses antagonize and/or evade these responses.

See also: Togaviruses: Alphaviruses; Togaviruses: General Features; Togaviruses Not Associated with Human Disease.

Further Reading

Dominguez G, Wang C-T, and Frey TK (1990) Sequence of the genome RNA of rubella virus: Evidence for genetic rearrangement during togavirus evolution. *Virology* 177: 225–238.

Griffin DE (2001) In: Fields BN, Knipe DM, Howley PM, *et al.* (eds.) *Fields Virology*, 4th edn., 917pp. Philadelphia: Lippincott.

Li ML and Stollar V (2004) Alphaviruses and apoptosis. *International Reviews of Immunology* 23: 7.

Paredes A, Alwell-Warda K, Weaver SC, Chiu W, and Watowich SJ (2007) Venezuelan equine encephalomyelitis virus structure and its divergence from old world alphaviruses. *Journal of Virology* 75: 9532–9537.

Schlesinger S and Schlesinger MJ (2001) *Togaviridae*: The viruses and their replication. In: Fields BN, Knipe DM, Howley PM, *et al.* (eds.) *Fields Virology*, 4th edn., 895pp. Philadelphia: Lippincott.

Strauss JH and Strauss EG (1994) The alphaviruses: Gene expression, replication, and evolution. *Microbiology Reviews* 58: 491.

Tomato Leaf Curl Viruses from India

S Chakraborty, Jawaharlal Nehru University, New Delhi, India

© 2008 Elsevier Ltd. All rights reserved.

Glossary

Pseudorecombinants New strains of a virus that result from the reassortment of genome nucleic acids during the replication of viruses with divided genomes in mixed infections.

Recombinant A new strain/species of a virus that occurs as a result of the breakage and renewal of covalent links in a nucleic acid chain for rearrangement of nucleic acids in the chain.

Synergism The association of two or more viruses acting at the same time, which enhances symptom severity.

Introduction

Whitefly-transmitted geminiviruses cause epidemics in vegetable, staple, and fiber crops. The diseases are generally associated with local or regional whitefly (*Bemisia tabaci* Gennadius) infestations. They cause enormous economic losses in several crops in the Tropics, which provide ideal conditions for the perpetuation of viruses and the insect vector. Intensive agricultural practices necessitated by the ever-increasing demands of a rapidly growing population and the introduction of new genotypes, cropping pattern, and crops have further aggravated the situation. For example, the continuous cultivation of crops (such as cotton, tomato, pepper, beans, soybean, and melon which are susceptible to the viruses and are attractive hosts for the whiteflies) certainly account for some of the increase in the severity and the vast spread of diseases caused by geminiviruses.

Tomato leaf curl disease (ToLCD) is the most devastating disease of tomato, affecting a large area under cultivation; it can be on the scale of an epidemic. Incidence of this disease was first reported from northern India in 1948. During the 1950s, incidence was reported from central India, followed by occurrence of the disease in the main tomato-growing regions of southern India. Since then, this disease has emerged as a major threat to tomato cultivation, and incidence has increased after introduction of high-yielding tomato varieties during the late 1960s. The disease is ubiquitous with the crop and has been observed in all the tomato-producing areas of the country. Leaf curl disease of tomato is so serious that the ability of small farmers to cultivate tomato in several major production areas, especially during the peak of whitefly infestation season, has been eliminated.

Yield loss varied with the age of plants at the time of infection and varieties being tested. The virus affects all stages of the growth causing 17.6–99.7% loss in yield, depending on the stage of the crop at the time of infection. When plants get infected early, within 20 days after transplanting, they remain stunted and produce few or no fruits and the yield loss may reach up to 100%. However, if the plants are infected at 35 and 50 days after transplanting, the yield loss is reduced to 74.1% and 28.9%, respectively. This indicates that earlier the infection, higher is the yield loss.

Symptoms

Leaf curl disease of tomato is characterized by severe stunting of the plants with upward and downward rolling and crinkling of the leaves. Infected plants exhibit intervenal yellowing, vein clearing, and crinkling and puckering of the leaves, sometimes accompanied by inward rolling of the leaf margins. The disease induces severe stunting, bushy growth, and partial or complete sterility, depending on the stage at which infection has taken place (**Figure 1**).

Figure 1 (a–g) Variation of symptoms induced by begomoviruses infecting tomatoes in India.

Transmission

Under natural conditions, whiteflies transmit the tomato leaf curl viruses (ToLCVs) from infected to healthy plants. Even a single whitefly can transmit the virus. Minimum acquisition access period and inoculation access period of 30 min each is required for successful transmission to occur. Pre-acquisition and pre-inoculation starving of the vector results in higher levels of transmission. ToLCV can persist up to 10 days after acquisition in a single adult whitefly. Females are more efficient transmitters than the males. The virus is also transmitted by dodder (*Cuscuta reflexa* Roxb.). Under artificial conditions, grafting can transmit the disease. It is also known that some of the isolates of ToLCVs are also sap transmissible under laboratory conditions.

Epidemiology

Like other vector-borne diseases affecting crop species, the factors contributing to ToLCD buildup are: (1) availability of virus inoculum and the vector around the fields;

(2) the movement of viruliferous vectors into the freshly sown field; (3) the susceptibility of the variety that builds up vector population and allows establishment of the virus; (4) weather parameters favoring vector population buildup; and (5) vector biotype which can effectively transmit the disease. Interaction of these factors leads to epidemic outbreak. The most important factors responsible for the epidemic are source of inoculum and vector. In India, wherever tomato is grown continuously, there is increase in leaf curl disease incidence. Virus inoculum in the weeds and continuous cropping contribute to rapid disease buildup. Weather parameters, both at macro- and micro-level, affecting the developmental stages of the plant and the life cycle of the vector are important. Tomato is grown in different agroclimatic zones in India, which makes prediction of the outbreak of epidemics more difficult. However, there is definitely a correlation between vector population and disease incidence.

Maximum temperature and rainfall play an important role for spread of the disease in southern India, while minimum temperature and minimum relative humidity influence the whitefly population in the north. ToLCD incidence depends on weather conditions (humidity, rainfall, temperature). It was observed that the disease progresses from February to June, when the dry and hot season with low humidity prevail. The incidence may even reach up to 100% during these months. These types of weather conditions favor whitefly multiplication and disease spread. The whitefly population and ToLCV incidence remain comparatively less during winter and rainy seasons. The tropical climate in southern India allows year-round tomato cropping, which, together with the presence of perennial host plants for both ToLCVs and *B. tabaci*, enables an easy carry-over of ToLCD between growing seasons. Whitefly biotype B can effectively transmit the virus. Overall, factors like long persistence of the virus in the vector, efficient transmission by the biotype, cultivation of tomato throughout the year, and abundance of weed hosts are the contributing factors for the high incidence of the disease under natural conditions.

Host Range

ToLCV strains/species are easily transmitted by grafting and through vector to a wide range of weeds and cultivated crops. Molecular detection using ToLCV-specific primers has led to identification of a range of plant species that can harbor ToLCVs in India. ToLCVs can infect crops such as *Lycopersicon esculentum*, *L. peruvianum*, *L. hirsutum*, *L glandulosum*, *L. pimpinellifolium*, *Capsicum annuum*, *Nicotiana tabacum*, *Vigna unguiculata*, and *Luffa cylindrica*. The viruses perpetuate on many weed hosts, viz. *Acanthospermum hispidum*, *Ageratum conyzoides*, *Blainvella rhomboids*, *Euphorbia hirta*, *Fraveria hirta*, *Parthenium hysterophorus*, *Malvastrum coromandalinum*, and *Croton bonplandianum*.

Genome Organization

Full-length genome of isolates of ToLCVs have been cloned and sequenced. Three types of genomic DNA (DNA-A, DNA-B, and DNA-β) have been found to be associated with begomoviruses causing ToLCD in India.

Apparently, the isolates from southern India have a monopartite genome (DNA-A) associated with DNA-β while both monopartite (DNA-A with DNA-β) and bipartite (DNA-A and DNA-B) begomoviruses have been found to infect tomatoes in northern India. The genome organization of DNA-A (ranges between 2739 and 2759 bp) resembles other begomoviruses having two open reading frames (ORFs) on the viral strand (AV1 and AV2) and four on the complementary strand (AC1, AC2, AC3, and AC4). Genome organization of DNA-B (ranges between 2656 and 2686 bp) and DNA-β (genome size between 1344 and 1376 bp) also resembles other begomoviruses having two ORFs (one on the viral strand, BV1, and the other on the complementary strand, BC1) and one on the complementary strand (βC1), respectively.

Diversity of Tomato-Infecting Begomoviruses

ToLCD appears to be caused by a complex of several viruses based on symptom variations on different indicator hosts. During the 1980s, based on symptoms produced in a particular tomato cultivar, ToLCVs were divided into five groups: (1) severe leaf curl with thickening of veins, (2) severe symptom with enation, (3) screw pattern of leaf arrangement, (4) vein purpling and leaf curl, and (5) exclusively downward curling of leaves. Variability was subsequently also found in the epitope profiles of ToLCVs collected from Karnataka, with groupings suggesting that the tomato crop and some neighboring weed species were hosts to the same ToLCV strains/species. Species status for begomoviruses, however, cannot be conferred based on symptom type or epitope profile.

In accordance with the ICTV *Geminiviridae* study group guidelines, nucleotide sequence identity of DNA-A, and genome organization (see **Table 1**), the following five Indian ToLCV species are to be demarcated: *Tomato leaf curl Bangalore virus*, *Tomato leaf curl Gujarat virus*, *Tomato leaf curl Karnataka virus*, *Tomato leaf curl New Delhi virus*, and *Tomato leaf curl Pune virus*. Except for *Tomato leaf curl Pune virus*, four species have been characterized in detail based on their biological and molecular properties (**Figure 2**).

Table 1 GenBank accession numbers of selected begomoviruses DNA-A, DNA-B and DNA-β sequences used for analysis

Species	Virus name	Accession numbers DNA-A	DNA-B	Abbreviation
DNA-A and DNA-B				
Tomato leaf curl Bangalore virus	Tomato leaf curl Bangalore virus-[Ban1]	Z48182		ToLCBV-[Ban1]
	Tomato leaf curl Bangalore virus-[Kerala]	DQ887537		ToLCBV-[Ker]
	Tomato leaf curl Bangalore virus-[Kolar]	AF428255		ToLCBV-[Kol]
	Tomato leaf curl Bangalore virus-[Ban5]	AF295401		ToLCBV-[Ban5]
	Tomato leaf curl Bangalore virus-[Ban4]	AF165098		ToLCBV-[Ban4]
	Tomato leaf curl Bangalore virus-[Ban AVT1]	AY428770		ToLCBV-[Ban AVT1]
Tomato leaf curl Bangladesh virus	Tomato leaf curl Bangladesh virus	AF188481		ToLCBDV
Tomato leaf curl Gujarat virus	Tomato leaf curl Gujarat virus-[Mirzapur]	AF449999		ToLCGV-[Mir]
	Tomato leaf curl Gujarat virus-[Vadodara]	AF413671		ToLCGV-[Vad]
	Tomato leaf curl Gujarat virus-[Varanasi]	AY190290	AY190291	ToLCGV-[Var]
	Tomato leaf curl Gujarat virus-[Nepal]	AY234383		ToLCGV-[Nepal]
Tomato leaf curl Joydebpur virus	Tomato leaf curl Joydebpur virus	DQ673859		ToLCJoV
Tomato leaf curl Karnataka virus	Tomato leaf curl Karnataka virus-[Bangalore]	U38239		ToLCKV-[Ban]
	Tomato leaf curl Karnataka virus-[Janti]	AY754812		ToLCKV-[Janti]
Tomato leaf curl New Delhi virus	Tomato leaf curl New Delhi virus-[Lucknow]	Y16421	X89653	ToLCNDV-[Luc]
	Tomato leaf curl New Delhi virus-[Mild]	U15016		ToLCNDV-[Mild]
	Tomato leaf curl New Delhi virus-[Severe]	U15015	U15017	ToLCNDV-[Svr]
	Tomato leaf curl New Delhi virus-[Pakistan]	AF448058	AY150305	ToLCNDV-[PK]
	Tomato leaf curl New Delhi virus-[Pakistan-Islamabad]	AF448059	AY150304	ToLCNDV-[PK-IS]
Tomato leaf curl Malaysia virus	Tomato leaf curl Malaysia virus-[India]	DQ629102		ToLCMYV-[IN]
Tomato leaf curl Pune virus	Tomato leaf curl Pune virus	AY754814		ToLCPV
Tomato leaf curl Sri Lanka virus	Tomato leaf curl Sri Lanka virus	AF274349		ToLCSLV
DNA-β				
Tomato leaf curl beta-[Aurangabad]			EF095958	ToLCB-[Aur]
Tomato leaf curl beta-[Bangalore]			AY428768	ToLCB-[Ban]
Tomato leaf curl beta-[Chinthapalli]			AY43855	ToLCB-[Chi]
Tomato leaf curl beta-[Coimbatore]			AY438560	ToLCB-[Coi]
Tomato leaf curl beta-[Jabalpur]			AY230138	ToLCB-[Jab]
Tomato leaf curl beta-[New Delhi]			AJ542490	ToLCNDB
Tomato leaf curl beta-[Pune]			AY838894	ToLCB-[Pune]
Tomato leaf curl beta-[Rajasthan]			AY438558	ToLCB-[Raj]
Tomato leaf curl beta-[Varanasi]			AY438559	ToLCB-[Var]

Tomato Leaf Curl New Delhi Virus

During the mid-1990s, two isolates of this virus were reported from New Delhi and Lucknow. In addition to the severe isolate, a mild isolate was also described from New Delhi. It has a bipartite genome with DNA-A and DNA-B. However, occurrence of DNA-β has also been observed. Infectivity of the cloned DNAs has been demonstrated.

Tomato Leaf Curl Gujarat Virus

Three isolates from Varanasi, Vadodara, and Mirzapur belong to this species. Among them, Varanasi strain has been characterized in detail. DNA-A alone is infectious but DNA-B increases symptom severity. Association of DNA-β has also been observed with this mono-bipartite species under natural conditions. Tomato leaf curl Gujarat virus (ToLCGV) is also sap transmissible to tomato, pepper, N. benthamiana, and N. tabacum. Unexpectedly, ToLCGV-Var DNA-A (AY190290) and DNA-B (AY190291) share a common region (CR) of 155 bp that is only 60% identical which were cognate pair of components that cause severe disease of tomato under field conditions.

Tomato Leaf Curl Bangalore Virus

This virus was reported for the first time from Bangalore, southern India. Several isolates referred to as tomato leaf curl virus Ban1, Ban3, Ban4, Ban5, and Kolar belong

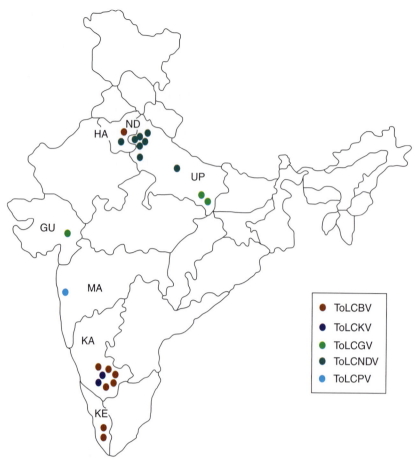

Figure 2 Distribution of five species of tomato-infecting begomoviruses in India. Presence of ToLCVs was located based on availability of full-length DNA-A sequences in GenBank.

to this species. It contains a monopartite DNA-A genome, and a satellite DNA-β molecule was observed to be present with tomato leaf curl Bangalore virus (ToLCBV) infection. Pathogenicity of cloned DNAs has not been demonstrated.

Tomato Leaf Curl Karnataka Virus

This sap-transmissible virus has been isolated from Bangalore. DNA-A alone is infectious and produces typical leaf curl symptoms on tomato. Association of DNA-β has also been observed.

Tomato Leaf Curl Pune Virus

This is the most recently reported new virus from India, whose only DNA sequences are available in GenBank. Tomato leaf curl Pune virus (ToLCPV) contains a DNA-β molecule, which is quite distinct from all other known ToLCVs.

In addition, viruses such as tomato leaf curl Bangladesh virus, tomato leaf curl Malayasia virus, and tomato leaf curl Joydebpur virus are also known to infect tomatoes in India.

Molecular Relationships among ToLCVs

Relationships among the type member of the five species (viz. ToLCBV, ToLCGV, tomato leaf curl Karnataka virus (ToLCKV), tomato leaf curl New Delhi virus (ToLCNDV), and ToLCPV) revealed that there is a great degree of diversity among the Indian tomato-infecting begomoviruses. To examine the diversity of the sequences, phylogenetic trees were generated of the 18 full-length DNA A sequences together with representative sequences present in GenBank (**Figure 3**). The trees constructed using either neighbor joining or most parsimonious methods for full-length sequences were all similar. Members of four different species are closely related and form a well-knitted cluster with the exception of ToLCMYV-[IN]

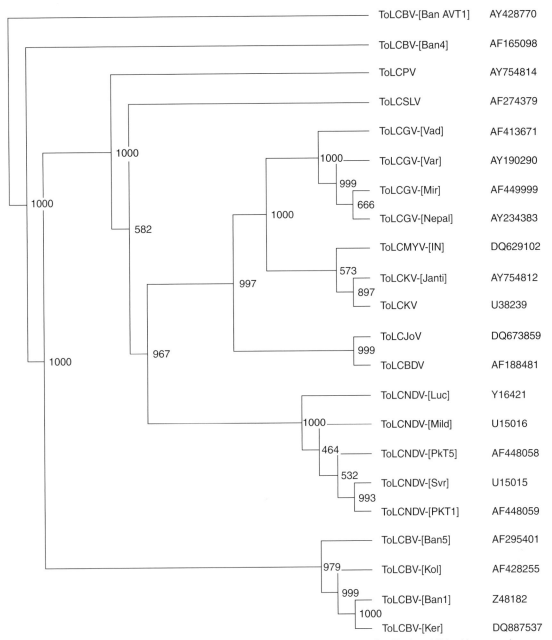

Figure 3 Phylogenetic analyses of begomoviruses causing ToLCD in India with other GenBank-published begomovirus sequences. Numbers at nodes indicate the bootstrap value out of 1000 replicates. The tree was generated with the PHYLIP programs using full-length DNA-A sequence data of Indian ToLCVs and other selected ToLCVs of the SE Asia. The full names of the viruses can be found in **Table 1**.

which was grouped with ToLCKV-Janti and ToLCKV. Full-length DNA-A sequences of Indian ToLCVs formed five different clusters, which generally had 72–82% nucleotide identity between clusters. However, one cluster formed by ToLCKV, belonging to a species known to have arisen through recombination, shared up to 88.3% with ToLCBVs.

Four out of the five sequences typed as ToLCBV shared 94.2–99.5% nucleotide identity with each other. ToLCPV also fell into the ToLCBV group, but had only 88% nucleotide identity to previously published ToLCBVs. ToLCBDVs originated from Bangladesh are distinct and form a different cluster while tomato leaf curl Sri Lanka virus (ToLCSLV) is close to ToLCPV and ToLCBV.

Comparison of DNA-A Sequence

Representatives of five species share varied degree of identity among each other (66–88%). Minimum nucleotide identity (66%) was observed with ToLCGV and

ToLCNDV while maximum identity of 88% was observed between ToLCBV and ToLCPV. Among the five species, ToLCKVs share more than 75% homology with all the species (**Table 2(a)**).

Comparison of DNA-B Sequence

GenBank database revealed the presence of four full-length DNA-B sequences associated with ToLCD in India. DNA-B of ToLCVs is highly conserved among each other in comparison to DNA-A. All the DNA-Bs are closely related (80–97%) (**Table 2(b)**). ToLCNDV, ToLCNDV-[Luc], and ToLCGV-[Var] have the same pairwise comparison profile all along their genomes.

Comparison of DNA-β Sequence

DNA-β molecules associated with ToLCVs share a less degree of homology as compared to DNA-A and DNA-B (**Table 2(c)**), indicating their uniqueness. So far nine full-length sequences have been reported with ToLCD in India and a wide range of 63–96% identity was observed among them. ToLCB-Varanasi share 96% with ToLCB-Aurangabad, followed by ToLCB-Jabalpur having 94% identity with ToLCB-Chinthapalli and by ToLCB-Coimbatore having 93% with ToLCBB. Two DNA-β sequences available from ToLCD-infected samples from Pune share 96% identity with each other.

Comparison of Common Regions

The replication of the B-component by the Rep protein of the A-component is possible because of the existence of the so-called CR, a short stretch of ±200 nt that usually is highly conserved between the two molecules. The CRs of the DNA-A (CRAs) of A-components of several isolates of a particular species are very closely related, with 89–99% identity. The CRAs of different species share 52–77% identity with each other. Sequence identities between the CRs of ToLCGV-[Var] DNA-B and genomic components of ToLCNDV are very high: it is 83% identical to ToLCNDV-[Svr] DNA-A CR and 86% identical to ToLCNDV DNA-B CR, indicating that these sequences could pertain to isolates of the same species. However, the CRs of DNA-A and DNA-B of ToLCGV-[Var] are only 61% identical. The lowest sequence identity (52%) was

Table 2 Percent identity (nucleotide) among begomoviruses causing TLCD in India

(a) DNA-A

	ToLCBV	ToLCGV-Var	ToLCKV	ToLCNDV-Svr	ToLCPV
ToLCBV	–	72	82	72	88
ToLCGV-Var		–	84	66	79
ToLCKV			–	76	83
ToLCNDV-Svr				–	72
ToLCPV					–

(b) DNA-B

	ToLCGV-Var	ToLCNDV-Luc	ToLCNDV 1	ToLCNDV 2
ToLCGV-Var	–	84	85	85
ToLCNDV-Luc		–	80	80
ToLCNDV 1			–	97
ToLCNDV 2				–

(c) DNA-β

	ToLCB-Var	ToLCB-Aur	ToLCB-Chi	ToLCB-Jab	ToLCB-Ban	ToLCB-Coi	ToLCB-Pun	ToLCB-Raj	ToLCB-Nd
ToLCB-Var	–	96	64	65	64	63	63	65	64
ToLCB-Aur		–	64	64	64	63	63	64	65
ToLCB-Chi			–	94	77	77	79	71	70
ToLCB-Jab				–	77	77	78	69	69
ToLCB-Ban					–	93	78	68	68
ToLCB-Coi						–	78	67	68
ToLCB-Pun							–	69	69
ToLCB-Raj								–	80
ToLCB-Nd									–

observed between the CR of ToLCGV-[Var] DNA-A and the CR of ToLCNDV DNA-B. A multiple alignment of the CRs and inter-cistronic regions of the five species causing ToLCD in India revealed that all the isolates of ToLCNDV, ToLCGV, and ToLCKV have similar or very related iteron sequences (GGTGT-XX/X-GGAGT) while ToLCBV and ToLCPV have similar iteron sequences (GGTGG-XX/X-GGTGG) (Table 3). The exception is the ToLCNDV-Mld and ToLCNDV-Luffa isolates whose iteron sequence is GGCGT-CT-GGCGT. The major difference between the CRs of the DNA-A among these viruses is found in the spacer sequence between the two iterons that varies between 2 and 3 nt. It was also observed that a third identical iteron is present at the 5′ end of the CRs. The CRs of DNA-A of ToLCGV-[Var], compared with DNA-B of ToLCGV-[Var], DNA-B of ToLCNDV, and DNA-A of ToLCNDV-Svr, share more than 85% identity in the first 80 nt of their CRs, whereas the region between the TATA box and the hairpin loop is below 50%.

Recombination

Tomato-infecting begomoviruses from India appear to have a great capacity to recombine. The amount of material that viruses may exchange ranges from small fragments of a few nucleotides to very large fragments of 2000 nt or more. When DNA-A was compared, no recombination was observed among ToLCGV-[Var] with other two isolates of the same species, ToLCGV-[Vad] or ToLCGV-[Mir]. Remarkably, a short possible recombination (150 nt) was observed between ToLCGV-[Var] DNA-A and ToLCNDV at the 3′ end of AC1. With ToLCKV, ToLCGV share high nucleotide sequence identity for their IR (approximately the first 100 nt), the 5′ end of AV1 (200 nt), and a long stretch of 1350 nt from the 5′ end of AC3 to the 5′ end of the IR (Figure 4). This indicates that, at these sites, recombination events possibly took place between these two tomato-infecting viruses or with a third unknown virus. Putative recombination sites among isolates of ToLCBV have been identified as AV1, AV2, AC1, and IR of the viral that may account of variability in strains/species.

Pseudo-Recombination and Synergism

Under natural conditions, mixed virus infections in a single plant possess biological and epidemiological implications. For the first time, synergism between two distinct species of begomoviruses infecting tomatoes in India was observed that results in an increase in viral DNA and symptom severity. Also, the occurrence of a more virulent pseudo-recombination between two distinct species may explain the sudden breakdown of resistant tomato cultivars and the development of epidemics in tomato-growing areas in India. Recently, the association of both ToLCGV-[Var] and ToLCNDV-Svr components in a single severely infected tomato plant under natural conditions has been detected. Also, based on coat protein (CP) gene sequences, associations of ToLCGV and ToLCBV, ToLCBV and ToLCKV, and ToLCBV and ToLCNDV have been observed in a single plant. The synergistic role of ToLCGV DNA-A and ToLCNDV-Svr DNA-A, resulting in a much higher level of ToLCNDV-Svr DNA-A in turn, helped to a more efficient replication of the B-components, and particularly the ToLCGV

Table 3 Rep binding sequences of ToLCVs from India

Virus species	DNA-A	DNA-B
ToLCBV	GGTGG-AAT-GGTGG	
ToLCGV	GGTGT-ATT-GAGT	GGTGT-CT-GGTGT
ToLCKV	GGTGT-ACT-GGAGT	
ToLCNDV-Svr	GGTGT-CT-GGAGT	GGTGT-CT-GGTGT
ToLCNDV-Mild	GGCGT-CT-GGCGT	
ToLCPV	GGTGG-AAC-GGTGG	

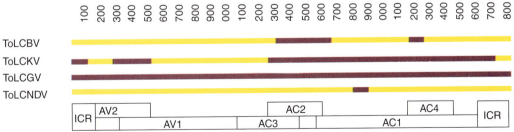

Figure 4 Diagrammatic representation of recombinant fragments between tomato leaf curl Gujarat virus-[Varanasi] and another three biologically characterized ToLCVs originating from India. Each line represents a linearized begomovirus genome (in the sense orientation starting from the origin of replication, in the CR) and each presence of red segments indicates homologous stretches of sequences to ToLCGV-[Var] at the strain level (>89%) within the genome of other Asian geminiviruses. Positions of the regions of the genome are represented at the top of the figure and a representation of a linearized geminivirus genome is represented at the bottom.

DNA-B component, resulting onto greater DNA accumulation in the systemically infected leaves, and consequently more severe symptoms. Since ToLCNDV-Svr and ToLCGV-[Var] belong to the same genus, infect the same hosts, and are transmitted by the same whitefly vector, they are therefore more likely to co-infect the same plants. The synergism between the two viruses will increase the amount of both the viruses in the systemically infected leaves and increase chances of transmission, if need be. As a consequence, doubly infected plants have a considerable potential as sources of inoculum for both viruses, and whiteflies feeding on such plants would, therefore, more easily acquire and transmit both viruses to virus-free plants. This information provides another source of geminivirus biodiversity.

Exchange of genomic components of the members of two distinct species of begomoviruses causing ToLCD in India can form infectious pseudo-recombinants. Transcomplementation between ToLCNDV-Svr DNA-A and ToLCGV DNA-B resulted in a more severe symptom phenotype as compared to wild type. ToLCGV DNA-B can also transcomplement ToLCNDV-Mld (mild isolate) and ToLCKV. The viable nature of these pseudo-recombinants was attributed to the highly conserved nature of AC1 and CR and also to identical iteron sequences. A highly specific interaction between the *Rep* protein of ToLCNDV (mild and severe strains) with their cognate iteron sequences also demonstrated the intimate relationship between these elements and the consequence in terms of DNA accumulation and symptoms. This led to the concept that matching N-Rep and iterons is required for efficient replication and consequently a severe symptomatology, and it also was supportive of the species concept where most of the time, with the exceptions of the recombinant in these regions, these two elements are different between species. Although the perfect match between the N-Rep sequence and the iteron sequences is probably vital in most of the cases, it may not be fatal in that particular one. This suggests that there must be other cooperative factors that are shared between the Rep of ToLCNDV-Mild and ToLCGV-[Var] B component CR, to the point that they compensate a nonmatching iteron interaction.

Replication and Pathogenesis

A highly specific interaction between the *Rep* protein of ToLCNDV (mild and severe strains) and their cognate iteron sequences also demonstrated the intimate relationship between these elements and the consequence in terms of DNA accumulation and symptoms. Two strains of ToLCNDV, viz. severe and mild, share 94% sequence identity on the basis of symptoms on tomato and tobacco. The studies demonstrated that the amino acid at position 10 in Rep protein coupled with a change in the binding site sequence may determine the replication of viral DNA. Change of Asp10 to Asn in Rep protein of the mild strain accompanied by exchange of the 13-mer binding site (making it identical to the severe strain) altered its replication, leading to increased accumulation of viral DNA. In addition, the modified mild strain could replicate heterologous strain DNA-B, indicating that the interaction of Rep protein with its binding site may be essential for replication of viral DNA.

Mutational analysis of ToLCNDV-Svr virion-sense ORFs has been carried out to assign the function. Plants inoculated with infectious DNA which contained deletions in AV2 developed very mild symptoms and accumulated only low levels of both single-stranded (ss-) and double-stranded (ds-) viral DNA, whereas inoculated protoplasts accumulated both ss- and dsDNA to wild-type levels, showing that AV2 is required for efficient viral movement. Mutations in the CP caused a marked decrease in ssDNA accumulation in plants and protoplasts while increasing dsDNA accumulation in protoplasts. The results demonstrated that multiple functions provided by AV2, BV1, and BC1 are essential for viral movement, and that changes in A-component virion-sense mRNA structure or translation affect viral replication.

The role of the movement protein (MP) and nuclear shuttle protein (NSP) in the pathogenicity of ToLCNDV, a bipartite begomovirus, has been elucidated using either potato virus X (PVX) expression vector or by stable transformation of gene constructs under the control of the 35S promoter in *N. tabacum*. No phenotypic changes were observed in any of the three species when the MP was expressed from the PVX vector or constitutively expressed in transgenic plants. Expression of the ToLCNDV NSP from the PVX vector in *N. benthamiana* resulted in leaf curling that is typical of the disease symptoms caused by ToLCNDV in this species. However, expression of NSP from PVX in *N. tabacum* and *L. esculentum* resulted in a hypersensitive response (HR), suggesting that the ToLCVDV NSP is a target of host defense responses in these hosts. The NSP, when expressed as a transgene under the control of the 35S promoter, resulted in necrotic lesions in expanded leaves that initiated from a point and then spread across the leaf. The necrotic response was systemic in all the transgenic plants. N-terminus of NSP is required for the HR. These findings demonstrate that the ToLCNDV NSP is a pathogenicity determinant as well as a target of host defense responses. The necrosis in transgenic tobacco plants is systemic, as it starts from a point on the fully emerged leaf and spreads over the lamina and to other leaves. Thus, ToLCNDV NSP is an avirulence determinant that interacts with the product of a resistance gene encoded by a host defense system, possibly an R gene product, triggering a host defense response involving an HR in *N. tabacum* and tomato.

Management

Until the late 1990s, the main control method employed against ToLCD was intensive use of insecticides targeted at viruliferous immigrant adult *B. tabaci* that spread ToLCVs into and within tomato crops. Identification of sources of resistance is difficult as the leaf curl disease syndrome is caused by different tomato begomoviruses. In the recent past, however, three high-yielding ToLCD-resistant tomato varieties have been developed that can be grown successfully with minimal insecticide use. Artificial screening through whiteflies has been carried out to identify resistant/tolerant varieties from different *Lycopersicon* spp., viz., *L. peruvianum*, *L. peruvianum* f. *glandulosum*, *L. peruvianum* f. *regulare*, *L. esculentum*, *L. chilense*, *L. hirsutum* f. *glabratum*, and *L. pimpinellifolium*. Gene governing resistance has also been mapped in tomato cultivar, H-24 (a derivative from *L. hirsutum* f. *glabratum*) on chromosome 11 against TYLCV and ToLCBV. In order to manage the deployment of this valuable resource, improve the efficacy with which further ToLCV-resistant material is screened, and investigate resistant-genotype/virus interactions, an improved understanding of the diversity and distribution of ToLCVs present will help in developing strategies for ecofriendly management of ToLCD. Agrobacterium-mediated inoculation of cloned DNAs will certainly provide better tool for identification of R genes in tomatoes in the future. Pathogen-derived resistance needs more attention in order to develop broad-spectrum resistance against Indian ToLCVs.

See also: Beta ssDNA Satellites; Potato Viruses; Satellite Nucleic Acids and Viruses; Tomato Yellow Leaf Curl Virus.

Further Reading

Chakraborty S, Pandey PK, Banerjee MK, Kalloo G, and Fauquet CM (2003) *Tomato leaf curl Gujarat virus*, a new begomovirus species causing a severe leaf curl disease of tomato in Varanasi, India. *Phytopathology* 93: 1485–1495.

Chatchawankanphanich O and Maxwell DP (2002) Tomato leaf curl Karnataka virus from Bangalore, India, appears to be recombinant begomovirus. *Phytopathology* 92: 637–645.

Chatterji A, Chatterji U, Beachy RN, and Fauquet CM (2000) Sequence parameters that determine specificity of binding of the replication-associated protein to its cognate in two strains of *Tomato leaf curl virus-New Delhi*. *Virology* 273: 341–350.

Chowda Reddy RV, Colvin J, Muniyappa V, and Seal S (2003) Diversity and distribution of begomoviruses infecting tomatoes in India. *Archives of Virology* 150: 845–867.

Green SK and Kalloo G (1994) Leaf curl and yellowing viruses of pepper and tomato: An overview. *Technical Bulletin No. 21*. Tainan, Republic of China: Asian Vegetable Research and Development Center.

Muniyappa V, Venkatesh HM, Ramappa HK, *et al.* (2000) Tomato leaf curl virus from Bangalore (ToLCV-Ban4): Sequence comparison with Indian ToLCV isolates, detection in plants and insects, and vector relationships. *Archives of Virology* 145: 1583–1598.

Padidam M, Beachy RN, and Fauquet CM (1995) Tomato leaf curl from India has a bipartite genome and coat protein is not essential for infectivity. *Journal of General Virology* 76: 25–35.

Vasudeva RS and Sam Raj J (1948) A leaf curl disease of tomato. *Phytopathology* 38: 364–369.

Tomato Spotted Wilt Virus

H R Pappu, Washington State University, Pullman, WA, USA

© 2008 Elsevier Ltd. All rights reserved.

Taxonomy

The species *Tomato spotted wilt virus* (TSWV) belongs to the genus *Tospovirus* in the family *Bunyaviridae*. Of the more than 300 species of primarily arthropod-borne viruses described in the family *Bunyaviridae*, a small proportion infects plants. Members of other genera in this family are important pathogens of humans and animals. TSWV is considered as the type member of the genus and hence its name formed the basis for coining the genus name. Discovery of a second tospovirus, *Impatiens necrotic spot virus* (INSV), was followed by description of more tospoviruses from several parts of the world. To date, there are more than 14 distinct tospoviruses described. Descriptors for classification of new viruses as tospoviruses include genome organization (**Figure 1**), thrips transmission, host range, and serological and molecular relationships of the nucleoprotein (N) gene. Excellent reviews on various aspects of tospoviruses have been published in the last few years. This review focuses on summarizing recent advances in our understanding of the tospovirus biology, molecular biology, epidemiology, and control.

Biology

TSWV was first described in 1915. The virus has a wide host range; it infects more than 900 plant species that include numerous crops and weeds. The virus is

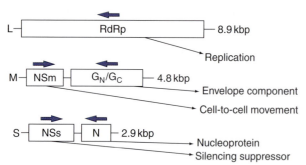

Figure 1 Genome organization of TSWV. Functions of each gene are noted on the right side.

Figure 2 Biological variability of TSWV isolates. Response of *Nicotiana tabacum* to a severe strain (top) and a mild strain (bottom).

mechanically transmissible and is not seed transmitted. In nature, TSWV is transmitted by several species of thrips. Crops that are affected by TSWV include bean, lettuce, peanut, pepper, potato, tobacco, and tomato. Biologically distinct isolates of TSWV exist in nature. Isolates that differ in thrips transmissibility, symptomatology, and symptom severity have been described (**Figure 2**). Variability in the N gene sequence of TSWV isolates suggested geographic delineation in natural virus populations which may be useful for attribution: that is, tracing the potential source or origin of a particular isolate. TSWV causes systemic infection in most of the crops it infects. Infection at early stages of the plant growth causes the most damage that may include severe stunting of the entire plant which often results in death. TSWV epidemics in peanut, pepper, tobacco, and tomato in southeastern United States caused major economic losses and forced shifts in production practices. Losses due to TSWV outbreaks in peanut were estimated at more than US $100 million in Georgia alone in the USA.

Molecular Biology

The morphology and genome structure and organization of tospoviruses share several features with members of other genera in the family *Bunyaviridae*. Particles are pleiomorphic, and 80–120 nm in diameter. The genome of tospoviruses includes three RNAs referred to as large (L), medium (M), and small (S). 5′ and 3′ terminal sequences of the RNAs are conserved. L RNA is in negative sense while M and S RNAs are ambisense in their genome organization. L RNA codes for the RNA-dependent RNA polymerase (RdRp). M RNA encodes precursors for two structural glycoproteins, G_N and G_C, and a nonstructural protein, NSm. The S RNA codes for the nucleocapsid protein (N) and another nonstructural protein, NSs. The three genomic RNAs are tightly linked with the N protein forming ribonucleoproteins (RNPs). These RNPs are encased within a lipid envelope consisting of two virus-encoded glycoproteins, and a host-derived membrane. The genome organization, gene products, and their roles are shown in **Figure 1**. Due to the negative strandedness of the genome, virions contain several molecules of the RdRp to initiate initial rounds of replication of the virion RNAs. Genome expression is facilitated through synthesis of subgenomic RNAs.

Considerable progress has been made in our understanding of the functional roles of various TSWV genes. The 331 kDa RdRp coded by the L RNA serves as a multifunctional, replication-associated protein and is believed to function cooperatively with host-encoded factors. Screening a cDNA library of *Frankliniella occidentalis* using fragments of TSWV RdRp, a putative transcription factor that binds to TSWV RdRp was isolated which was shown to bind to TSWV RNA and enhance TSWV replication. Mammalian cells expressing this putative transcription factor supported TSWV replication. Since TSWV is transmitted by several thrips species it remains to be seen if similar transcription factors exist in other thrips vectors.

Assignment of functions to various TSWV gene products was done using indirect approaches since a reverse genetics system based on an infectious cDNA clone is not yet possible for a negative-sense RNA virus such as TSWV. Using indirect approaches, the functions of the M RNA-encoded glycoproteins and NSm were recently deciphered. The absence of a gene similar to NSm in other genera of the family *Bunyaviridae* suggests that NSm serves a function that facilitates TSWV infection of plants. Virus movement in plants is mediated by a virus-encoded movement protein and a similar function was suggested for NSm and subsequent experimental evidence supported this hypothesis. Recently, the role of NSm in TSWV life cycle was investigated *in vitro* and *in planta*. Virus-encoded movement proteins tend to bind to viral RNAs and facilitate virus movement through plasmodesmata. *In vitro*-expressed NSm interacted with N protein and bound ssRNA in a sequence-nonspecific manner. Members of DnaJ family were found to bind NSm in a yeast-two-hybrid system. Transgenic *Nicotiana tabacum* plants expressing NSm produced symptoms suggestive of TSWV infection. Biochemical analyses of this plant response showed the accumulation of callose, an indicator of the triggering of plant defense response. Constitutive expression of the NSm and its subsequent interference with the plasmodesmata's transport functions could have resulted in abnormal growth pattern seen in these transgenic plants. More direct evidence of the role of NSm in virus movement was obtained by heterologous complementation studies using an infectious cDNA clone of tobacco mosaic virus (TMV) and replacing TMV genes with the NSm gene of TSWV. The NSm gene complemented a movement-deficient mutant of TMV in tobacco plants and facilitated the long-distance movement of the TMV–TSWV NSm hybrid. The NSm protein, when expressed *in planta*, mediated tubule formation in infected protoplasts.

TSWV mutants lacking the envelope were not thrips transmissible, indicating that the determinants for thrips transmission are localized on the glycoprotein-containing envelope. Using two distinct isolates of TSWV that differ in thrips transmission, genetic reassortants (=pseudorecombinants) were generated by coinoculating plants with both isolates. The resulting reassortants were evaluated for their ability to be thrips transmitted and the source of the genomic components of each of these reassortants was determined. Only those reassortants that had the M RNA derived from the thrips-transmissible isolate retained thrips transmissibility confirming the functional contribution of M RNA to thrips transmission (**Figure 3**). Sequence comparisons of the glycoprotein genes of these two isolates showed that a point mutation played a critical role in determining the thrips transmissibility. The roles of G_N and G_C in TSWV–thrips vector interaction are beginning to be elucidated and are discussed under 'Transmission and epidemiology'. Moreover, the M RNA was shown to carry determinants for host adaptation and overcoming host plant resistance as well as pathogen-derived resistance in tomato and tobacco. In the absence of a reverse genetics system, complementation studies

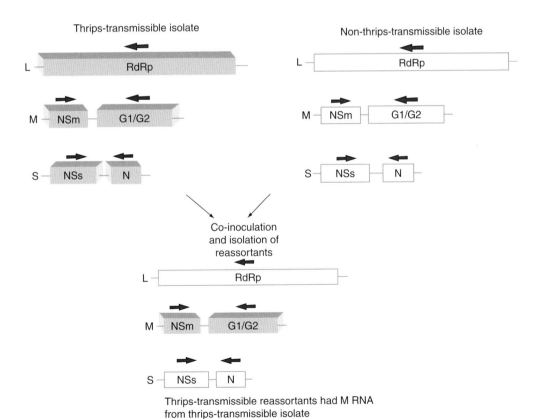

Figure 3 Genetic reassortment as a tool for the functional analysis of tospovirus genes. Localization of thrips transmission determinants on the M RNA of TSWV.

and genetic reassortment tudies such as those mentioned above offer alternate approaches for studying structure–function relationships of various TSWV genes.

The S RNA-encoded proteins, N and NSs, play important roles in the TSWV infection cycle. The N protein, as part of the RNP, serves as the structural protein and may also have some regulatory role in modulating the transcription and replication. Additionally, the involvement of the N protein in particle assembly was suggested based on its interaction with one of the two viral glycoproteins. When co-expressed with G_N and G_C in mammalian BHK21 cells, N protein displayed selective interaction with G_C and the subsequent localization of the N and glycoprotein complex in the Golgi apparatus. The nonstructural protein, NSs, encoded by the S RNA was shown to be a suppressor of RNA silencing.

Diagnosis

TSWV produces a wide range of foliar symptoms. Symptomatology varies depending on the strain, host species, and genotype, and is also influenced by environmental factors such as temperature. In most cases, accurate diagnosis is facilitated by serological or molecular techniques. Purified TSWV is a good immunogen and virus-specific antisera and enzyme-linked immunosorbent assay (ELISA) kits are commercially available. Molecular techniques based on reverse transcription-polymerase chain reaction (RT-PCR) have been developed that can detect the virus in plants and thrips vectors. Sensitive, rapid real-time PCR methods are also available. An NSs-specific monoclonal antibody was produced which was shown to be effective in identifying adults that are capable of transmitting the virus since detection of NSs in thrips is an indication that the virus had multiplied in the vector.

Transmission and Epidemiology

TSWV is transmitted by several species of thrips in a circulative and propagative manner. These include *F. occidentalis, Frankliniella schultzei, Frankliniella intonosa, Frankliniella bispinosa, Frankliniella fusca, Thrips setosus*, and *Thrips tabaci*. The virus is not known to be seed transmitted. Hence, susceptible crops with overlapping production seasons and prevalence of weed hosts and thrips vectors constitute the most important factors for TSWV epidemics. TSWV has to be introduced into a crop by viruliferous thrips and patterns of disease spread mostly suggests primary spread with little secondary spread within the crop. Considerable progress has been made in our understanding of thrips–TSWV interactions, especially in the case of TSWV–*F. occidentalis* association.

For the adult thrips to become a transmitter, the larva has to acquire the virus by feeding on an infected plant. First and second instar larvae are capable of acquiring the virus though the former is more efficient. The virus starts replicating in the larva and survives through the developmental stages. The emerging adult transmits the virus and continues so for life. There is no evidence of transovarial transmission. Thus, the virus has to be acquired by each generation of thrips. Therefore, tospoviruses are capable of replicating in both their host plants and thrips vectors. This intimate biological association between tospoviruses and their thrips vectors had created possibilities for evolution and shifting preferences and specificities between individual tospoviruses and thrips species. New vector species of TSWV began to emerge in the past decade. For example, *F. bispinosa* was reported as a vector of TSWV. For TSWV epidemics to occur, thrips vectors should complete a life cycle on virus-infected host plants. Infected plants that do not support thrips lifecycle can be considered as a dead end in disease epidemiology. The economic impact of TSWV on a number of important crops combined with the fact that thrips-borne inoculum is the most important contributory factor to virus outbreaks resulted in extensive research efforts to understand the basis of TSWV–thrips interactions and the virus and insect factors that contribute to the specificity of virus–vector relations. Of more than 5000 thrips species described, only seven of them are known to transmit TSWV, suggesting complex interactions of recognition, acquisition, replication, and movement at the virus–vector association level. Much is now known about the interaction between TSWV and *F. occidentalis*.

The virus upon acquisition was shown to move through the midgut and subsequently reaches the salivary glands. It is hypothesized that the close proximity of midgut and salivary glands in the thrips larval stage facilitates the virus movement whereas the virus fails to do so as the thrips reaches adult stage. This may explain the inability of adult thrips to transmit the virus if the virus is acquired for the first time in its adult life. The specificity of TSWV and thrips vectors may be due to the presence of a receptor in the vector species which may be absent in nonvector species. A 50 kDa protein and a 94 kDa protein were identified as potential thrips proteins involved in interaction with TSWV. However, cloning the gene for the putative receptor remains to be accomplished. A soluble form of *in vitro*-expressed TSWV G_N protein was used to study its role in recognition by thrips vectors. When thrips were fed with the purified G_N, the protein could be detected in the midgut epithelial cells of the larvae which subsequently resulted in the prevention of TSWV acquisition, suggesting that G_N may be involved in virus recognition by the thrips vector. While G_N potentially mediates the specific recognition between TSWV and its thrips vector, sequence comparisons and biochemical analysis of G_C protein indicated that it facilitates the fusion and subsequent uptake of the virion into vector cells.

Management

TSWV has been an economically important constraint for several crops over the past several decades in several parts of the world. Due to the nonpredictive nature of the outbreaks combined with the lack of forecasting, adoption of preventive measures has not always been practical. The disease cycle has proven to be extremely difficult to break because of the wide and often overlapping host range of both the virus and the thrips vectors.

However, multidisciplinary research in the past decade has led to identification of factors that contribute to TSWV epidemics and development of practical control recommendations which resulted in reduced impact of TSWV in several crops. An integrated management approach must be taken as no single control tactic was found to be effective by itself. The crucial element in the management program is growing virus-resistant or virus-tolerant cultivars. TSWV-resistant cultivars with desirable agronomic traits are now available for peanut, pepper, and tomato. Resistance-breaking strains of TSWV continue to pose a threat, potentially limiting the durability of resistant cultivars. Resistance governed by multigenes should be used where available to make the resistance more durable. Production practices such as growing on plastic mulch or reflective mulch significantly reduced the disease incidence in certain crops. Novel strategies utilizing chemicals such as acibenzolar-*S*-methyl that induce systemic acquired resistance are found to be effective in reducing the incidence of TSWV (**Figure 4**). Thrips management based on selective use of insecticides during the early part of the cropping season combined with resistant cultivars resulted in reduced incidence and increased yield in tomato. Ability to forecast the epidemics by making use of information such as the seasonal dynamics of TSWV transmitters in the vector populations will help growers make appropriate management decisions. Pathogen-mediated resistance due to post-transcriptional gene silencing was shown to be effective in several crops. Expression of single-chain antibodies, TSWV N gene, or various modifications of the N gene were found to be effective in conferring resistance. Genetic engineering can be very useful in situations where natural sources of host plant resistance are not available or difficult to transfer to existing agronomically desirable cultivars. Consumer acceptance and other market-related issues have hampered the efforts to commercialize this technology.

See also: *Tospovirus*; Vector Transmission of Plant Viruses.

Further Reading

Adkins S (2000) Tomato spotted wilt virus – positive steps towards negative success. *Molecular Plant Pathology* 1: 151–157.

Bucher E, Sijen T, De Haan P, Goldbach R, and Prins M (2003) Negative-strand tospoviruses and tenuiviruses carry a gene for a suppressor of gene silencing at analogous genomic positions. *Journal of Virology* 77: 1329–1336.

Culbreath AK, Todd JW, and Brown SL (2003) Epidemiology and management of tomato spotted wilt in peanut. *Annual Review of Phytopathology* 41: 53–75.

Csinos AS, Pappu HR, McPherson RM, and Stephenson MG (2001) Management of *Tomato spotted wilt tospovirus* in flue-cured tobacco with acibenzolar-*S*-methyl and imidacloprid. *Plant Disease* 85: 292–296.

de Medeiros RB, Figueiredo J, Resende RO, and de Avila AC (2005) Expression of a viral polymerase-bound host factor turns human cell lines permissive to a plant- and insect-infecting virus. *Proceedings of the National Academy of Sciences, USA* 102: 1175–1180.

Lewandowski DJ and Adkins S (2005) The tubule-forming NSm protein from tomato spotted wilt virus complements cell-to-cell and long-distance movement of tobacco mosaic virus hybrids. *Virology* 342: 26–37.

Mandal B, Pappu HR, and Csinos AS (2003) Suppressive effect of Actigard on tomato spotted wilt virus. APSnet Image of the week. http://www.aspnet.org/online/Archive/2003/W000022.asp.

Mandal B, Pappu HR, Csinos AS, and Culbreath AK (2005) Response of peanut, pepper, tobacco, and tomato cultivars to two biologically distinct isolates of *Tomato spotted wilt virus*. *Plant Disease* 90: 1150–1155.

Moyer JW (2000) Tospoviruses. In: Hull R (ed.) *Encyclopedia of Microbiology*, vol. 4, pp. 592–597. New York: Academic Press.

Riley DG and Pappu HR (2004) Tactics for management of thrips (Thysanoptera: Thripidae) and tomato spotted wilt virus in tomato. *Journal of Economic Entomology* 97: 1648–1658.

Rinne PL, van den Boogaard R, Mensink MG, *et al.* (2005) Tobacco plants respond to the constitutive expression of the tospovirus

Figure 4 Protective effect of acibenzolar-*S*-methyl against TSWV. Both leaves were mechanically inoculated with TSWV. Leaf on the right was treated with acibenzolar-*S*-methyl prior to inoculation with TSWV. Reproduced from Mandal B, Pappu HR, and Csinos AS (2003) Suppressive effect of Actigard on tomato spotted wilt virus. APSnet Image of the Week, with permission of the American Phytopathological Society.

movement protein NS(M) with a heat-reversible sealing of plasmodesmata that impairs development. *Plant Journal* 43: 688–707.

Sherwood JL, German TL, Moyer JW, Ullman DE, and Whitfield AE (2000) Tomato spotted wilt. In: Maloy OC and Murray TD (eds.) *Encyclopedia of Plant Pathology*, pp. 1030–1031. New York: Wiley.

Sin SH, McNulty BC, Kennedy GG, and Moyer JW (2005) Viral genetic determinants for thrips transmission of tomato spotted wilt virus. *Proceedings of the National Academy of Sciences, USA* 102: 5168–5173.

Soellick T-R, Uhrig JF, Bucher GL, Kellmann J-W, and Schreier PH (2000) The movement protein NSm of tomato spotted wilt tospovirus (TSWV): RNA binding, interaction with TSWV N protein, and identification of interacting plant proteins. *Proceedings of the National Academy of Sciences, USA* 97: 2373–2378.

Takeda A, Sugiyama K, Nagano H, *et al.* (2002) Identification of a novel RNA silencing suppressor, NSs protein of tomato spotted wilt virus. *FEBS Letters* 532: 75–79.

Whitfield AE, Ullman DE, and German TL (2005) Tospovirus–thrips interactions. *Annual Review of Phytopathology* 43: 459–489.

Tomato Yellow Leaf Curl Virus

H Czosnek, The Hebrew University of Jerusalem, Rehovot, Israel

© 2008 Elsevier Ltd. All rights reserved.

Glossary

Agroinoculation An alternative route for viral infection; the viral genome is cloned, usually as a head-to-head dimer, in the T-DNA of *Agrobacterium tumefaciens* and is delivered to plants by inoculation.

Introgression Incorporation of chromosomal segments of wild tomato species in the domesticated tomato by crosses and selection; introgression lines are used to localize genes to the tomato chromosomes.

Rolling circle Mechanism of replication followed by many viral DNA and by begomoviruses in particular.

Viruliferous whitefly Insect that has acquired virus from an infected plant and is ready to infect other host plants.

Whitefly Insect that belongs to the order *Homoptera*; they cause damage to plants by feeding and by vectoring plant viruses.

Introduction

In the late 1950s the tomato cultures in the Jordan valley of Israel were unexpectedly affected by a disease of unknown etiology. The disease was accompanied by large populations of whiteflies. The suspicion that the whiteflies were the vector of a viral disease was confirmed following controlled transmission experiments in the laboratory. The virus was named tomato yellow leaf curl virus (TYLCV). The virus was isolated and its genome sequenced in the late 1980s.

TYLCV is a member of the genus *Begomovirus* of the family *Geminiviridae*, which includes viruses transmitted by the whitefly *Bemisia tabaci*. Begomoviruses have a genome either split between two circular single-stranded (ss) DNA molecules of approximately 2700 nt each named DNA A and DNA B (bipartite) or with a single genomic DNA A-like molecule (monopartite). TYLCV is monopartite. The relationships between the virus, the vector, and the host tomato plant have been the object of many studies.

From the early 1960s tomato cultures have been under the constant threat of TYLCV-like begomoviruses worldwide. TYLCV has quickly spread to the Middle East, Central Asia, North and West Africa, Southeast Europe, the Caribbean islands, Southeast USA, and Mexico. TYLCV-related begomoviruses have been identified in Italy, the Maghreb and Western Africa, and the Arabian Peninsula. Breeding programs for resistance have started in the mid-1970s and several commercial varieties with adequate resistance have been released. Several loci tightly linked to TYLCV resistance have been assigned to the small arm of tomato chromosome 6. A variety of strategies have been devised based on the pathogen-derived resistance concept, which involves the expression of functional as well as dysfunctional viral genes. RNA-mediated virus resistance based on antisense RNA and post-translational gene silencing was efficient but was highly sequence dependent.

Virus Structure

Like all geminiviruses, TYLCV has a characteristic particle of twinned morphology of approximately 20×30 nm in size (**Figure 1(a)**). The virus capsid (total m.w. 3330 kDa) consists of two joined, incomplete icosahedra, with a $T=1$ surface lattice containing a total of 22 capsomeres each containing five units of a 260-amino-acid coat protein (30.3 kDa). TYLCV has a single 2787 nt (total m.w. 980 kDa) covalently closed-genomic circular ssDNA.

Figure 1 Virions, host tomato plants and whitefly vector. (a) Viral particles with twinned morphology of approximately 20 × 30 nm in size. (b) Left – infected tomato plant with typical symptoms; right – noninfected tomato plant. (c) Adult *B. tabaci* whiteflies, approximately 2 mm in size.

Host Range

The domesticated tomato *Solanum esculentum* (formerly *Lycopersicon esculentum*) is the primary host of TYLCV. Most of the wild tomato species such as *S. chilense*, *S. habrochaites* (formerly *L. hirsutum*), *S. peruvianum*, and *S. pimpinellifolium* include accessions that are symptomless carriers and are used as genitors in breeding programs for TYLCV resistance. Several cultivated plants (bean (*Phaseolus vulgaris*), petunia (*Petunia hybrida*), and lisianthus (*Eustoma grandiflorum*)) are hosts of TYLCV and present severe symptoms upon whitefly mediated inoculation. Weeds, such as *Datura stramonium* and *Cynanchum acutum*, present distinct symptoms, whereas others, such as *Malva parviflora*, are symptomless carriers. Plants used to rear whiteflies, such as cotton (*Gossypium hirsutum*) and eggplant (*Solanum melongena*), are immune to the virus. Experimental hosts of the virus include jimsonweed (*Datura stramonium*). Some plants recalcitrant to whitefly mediated inoculation, such as *Nicotiana benthamiana* and *N. tabacum*, may be infected by TYLCV DNA clones using agroinoculation.

Transmission to Tomato

In nature TYLCV is transmitted exclusively by the whitefly *B. tabaci* in a persistent manner (**Figure 1**). Symptoms included growth arrest leaflets cupped inward with yellow margins, and barely produced fruits (**Figure 1**). *B. tabaci* thrives in commercial fields of cotton, tomato, eggplant, and pepper, in tropical and subtropical countries. It also infests the greenhouses in these regions as well as those in temperate countries. Whiteflies may cause dramatic damages due to feeding and the transmission of begomoviruses. *B. tabaci* comprises many types (or biotypes) that can be distinguished by their plant host range, fertility, as well as with molecular markers (especially from the mitochondrial genome). Among the many biotypes, B and Q are extremely efficient vectors of TYLCV.

Transmission of TYLCV to tomato by *B. tabaci* B biotype has been studied thoroughly. The virus is transmitted to tomato plants after vector-feeding on infected tomato plants or alternative hosts. The incidence of the disease is directly correlated with the pressure of the whitefly population. One to three viruliferous insects are able to infect a tomato plant. The efficient acquisition access period (AAP) is 15–30 min, the latent period is 8–24 h, and the efficient inoculation access period (IAP) is at least 15 min. Female *B. tabaci* are more efficient vectors than males. The ability of the whiteflies to transmit TYLCV to tomato test plants steadily decreases with age, from 100% to 10–20% during their adult life time.

Symptoms develop on inoculated seedlings 2–3 weeks after insect first inoculation feeding. In the field inoculation can occur immediately after transplantation. Infected seedlings remain stunted and do not yield fruits. Infection at a later stage affects vegetative growth and fruit setting. Disease incidence increases rapidly and in severely affected regions results in yield reduction of up to 100%.

The viral DNA replicates in the nuclei of infected cells and is mostly phloem limited. Apart from whiteflies, TYLCV can be transmitted by grafting and agroinoculation. It is not mechanically transmitted and it is not propagated by seeds.d

Genome Organization and Expression

A schematic drawing of the TYLCV genome is shown in **Figure 2**. It replicates according to a rolling circle mechanism. The viral genome encodes two large open reading frames (ORFs) on the viral strand (V1 and V2), and four on the complementary strand (C1–C4). A 313 nt long intergenic region (IR) contains a 29 nt long stem–loop structure with the conserved nanonucleotide TAATATTAC which is the origin of replication (Ori) of the virus. The IR also contains the promoters of the V1, V2, C1, and C4 genes.

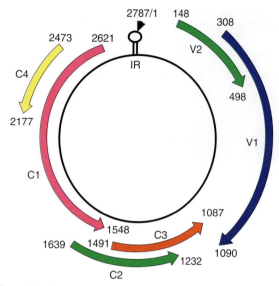

Figure 2 Genome organization of tomato yellow leaf curl virus. The single-stranded virion DNA comprises 2787 nt. Open reading frames (ORFs) of virion-sense and complementary-sense strand polarity are designated (V) and (C), respectively. ORFs are represented by an arrow; numbers indicate first and last nucleotide of each ORF. V1 encodes the capsid protein (CP), V2 a movement protein, C1 the replication initiator protein (Rep), C2 a transcriptional activator protein (TrAP), C3 a replication enhancer protein (REn), and C4 a symptom and movement determinant. IR: intergenic region. The conserved inverted repeat flanking the conserved nanonucleotide sequence TAATATTAC is symbolized by a stem–loop; an arrow head indicates the cleaving position of Rep in the TAATATT/AC loop; A at the cutting site (/) is nucleotide number one, by definition.

- V1. The V1 ORF encodes the capsid protein (CP). The CP is a multifunctional protein: The N- and C-termini of CP monomers interact to form the capsid (one viral genomic molecule is encapsidated in each geminate particle). The CP has a nuclear localization signal (NLS) and it is able to shuttle the viral genomic DNA in and out of the nucleus. It is essential for infectivity, and it is the only viral protein recognized by the insect vector.
- V2. The V2 ORF has properties analogous to those of a movement protein and in some cases mutants affect symptom expression.
- C1. The C1 ORF encodes the replication initiator protein (Rep). It is the only viral protein necessary for replication. The functional protein is an oligomer. The Rep protein recognizes the Ori located in the IR and specifically cleaves the nanonucleotide TAATATTAC between nucleotides 7 and 8. Together with plant host polymerase(s) it initiates viral DNA replication according to the rolling-circle model. Short inverted repeats located in the IR upstream the stem–loop structure are recognized by the Rep protein and also involved in the replication process. Rep also interacts with another viral protein, the viral replication enhancer protein (REn), encoded by ORF C3 and with a variety of cellular proteins such as a retinoblastoma-like protein, an interaction that may enhance plant cell replication and thereby viral multiplication.
- C2. The ORF C2 encodes a transcriptional activator protein (TrAp). Both ORFs C2 and C3 are transcribed from a promoter located in the 3′ end of the C1 ORF. TrAp enhances transcription of the viral strand promoter. It is able to bind to viral genomic DNA and possesses an NLS. In addition, the C2 gene product acts as silencing suppressors.
- C3. The ORF C3 encodes the replication enhancer protein (REn). This protein interacts with itself to form oligomers. REn interacts with Rep and with cell-cycle associated host proteins to increase the amount viral DNA (genomic and double-stranded) in the infected plant.
- C4. The ORF C4 encodes a protein not essential for infectivity but contributes to the spread of the virus in the plant and induction of symptoms. The protein may also act as a silencing suppressor, but this has not been definitively proved.

Geographical Distribution

The name TYLCV was coined in the early 1960s to describe a virus transmitted by the whitefly *B. tabaci* that affected tomato cultures in Israel. Early diagnosis of TYLCV was essentially based on symptom observation, although symptoms vary greatly as function of soil, growth conditions, and climate. Serology has been of limited use because whitefly transmitted geminiviruses share many epitopes. The analysis of DNA sequences has become the tool of choice, allowing one to accurately identify the virus and to evaluate its relationship with other TYLCV isolates. TYLCV has been reported in the mid- and late 1970s in Cyprus, Jordan, and Lebanon. It has been identified in Egypt and Turkey in the early 1980s. The virus has spread to Turkey, Iran and the Asian republics of the former USSR, and to Saudi Arabia and Yemen during the mid–late 1990s. Two TYLCV isolates closely related to the Middle Eastern virus have been described in Japan in the late 1990s. In China, TYLCV isolate TYLCVs has been identified in the southwest province of Guangxi. In the early 1990s two virus isolates belonging to a new species related to the Middle Eastern TYLCV, named tomato yellow leaf curl Sardinia virus, have been identified in Sardinia (TYLCSV-Sar) and in Sicily (TYLCSV-Sic), Italy. The Sardinian isolate TYLCSV-Sar has been discovered in Spain in the early 1990s. By he mid-1990s the Middle Eastern TYLCV was found in Portugal and in Spain; in the latter country, it tended to displace the previously established Italian virus.

Recombinants between the two viruses have been found. TYLCV has invaded North Africa, possibly from Spain and Italy. In Morocco, both the Italian and the Middle Eastern strain were discovered in the early 2000s. In Tunisia the Italian virus was identified in the early 2000s. TYLCV appeared in Southern France in 1999. In 2000, the Middle Eastern strain of TYLCV was identified in Crete, Attiki, and Southern Peloponnesus in Greece. In East Africa, TYLCV was present in Sudan as early as the late 1970s. In Tanzania a virus distinct from the Mediterranean isolates has been sequenced. In the Réunion Island the virus was detected in the late 1990s. The Middle Eastern strain of TYLCV has appeared in the Western Hemisphere in the mid-1990s in the Caribbean Islands, first in the Dominican Republic, then Cuba, Jamaica, Puerto Rico, and the Bahamas. From there, the virus has reached the USA, identified first in Virginia in the late 1990s, then in Florida, Georgia, Louisiana, North Carolina, and Mississippi. It seems that TYLCV is now found in several regions of Mexico.

Hence during the last decade TYLCV has spread extremely fast and constitutes a major limiting factor to tomato cultivation, worldwide. In many regions, the invading Mediterranean strain of TYLCV coexists with local TYLCV strains, but in several cases it has almost replaced them, as in Southern Spain. The Sardinian isolate of TYLCSV has been recently detected in Jordan and in Israel, indicating that the expansion of TYLCSV is not unidirectional, from the Eastern Mediterranean region to other parts of the world as thought not long ago.

Taxonomy and Classification

TYLCV is the name of the virus originally isolated in Israel. Sequence comparisons of TYLCV-like viruses analyzed worldwide have revealed that the name TYLCV was used mistakenly to name a complex of closely as well as distantly related begomoviruses affecting tomato. Viruses with nucleotide sequence homology of more than 89% have been considered to be strains of the same species while viruses with homologies of less than 89% have been considered as belonging to different virus species. Accordingly, the begomoviruses affecting tomato have been classified into several species: all the TYLCV isolates known today have a monopartite genome.

The members of the TYLCV species, as well as members of species closely related to TYLCV updated as of December 2006, are listed in **Table 1**; the acronym and the GenBank accession number of the DNA sequence are provided.

This classification is rendered even more complicated by the recent discovery that recombination between members of different species of geminiviruses happens relatively frequently. For example, a naturally occurring recombination has been recently found in the Almeria region, Spain, between TYLCV and TYLCSV. The recombinant virus, coined tomato yellow leaf curl Malaga virus (TYLCMalV-[ES], AF271234), probably occurred because the two virus species have coexisted in the tomato plants grown in the greenhouse.

Additional begomoviruses that infect tomato cultures have not been assigned to the TYLCV and are discussed elsewhere in this encyclopedia. These viruses clearly differ from the various TYLCV isolates in the symptoms they induce on tomato, in their host range, and in their nucleotide sequence.

Virus–Vector Relationship

Path of the Virus in the Insect

Like all begomoviruses TYLCV follows a definite path in its *B. tabaci* vector. Whiteflies feed when their stylets reach the virus-rich phloem of infected leaves. TYLCV can be detected in the insect head approximately 10 min after feeding started. From the esophagus the virus reaches the midgut where it can be detected approximately 30 min after it was first found in the head. At this point, some of the viruses might be excreted through the hindgut. The crossing of TYLCV from the midgut to the hemolymph is surprisingly fast. The virus reaches the hemolymph 30 min after it was first detected in the midgut, 90 min after the beginning of the AAP. The crossing of the gut is likely to be an active process involving specialized unknown receptors. To avoid degradation in the hemolymph, geminiviruses interact with a GroEL-like chaperonin produced by the insect endosymbiotic bacteria and excreted in the hemolymph. TYLCV can be detected in the salivary glands approximately 5.5 h after it was first detected in the haemolymph, 7 h after the beginning of the AAP, approximately 1 h before the insects are able to infect tomato plants. Once the virus reaches the salivary gland, crossing several cell walls that may constitute selective receptor-mediated barriers, it is almost immediately excreted into the salivary pump and from there into the plant together with the saliva. The capsid (and the coat protein) is the only begomoviral structure recognized by the would-be receptors and by the endosymbiotic chaperonin.

Retention of TYLCV in the Insect Vector

The virus can be detected in every stage of vector development. Following a 48 h AAP on TYLCV-infected tomato, viral DNA remains associated with *B. tabaci* for the entire adult life of the insect while infectivity decreases with time, but not entirely. Various TYLCV isolates present sometimes different parameters of interactions with their whitefly host. For example, the maximum retention period

Table 1 List of 59 virus isolates (written in black) used in this article belonging to 15 different strains (written in red) and six different species of the *Tomato yellow leaf curl virus* cluster (written in green). The accession numbers of the complete A component sequence is indicated in the second column and the abbreviation of the name of the virus isolates and strains is indicated in the third column

Tomato leaf curl Sudan virus		
Tomato leaf curl Sudan virus – Gezira		ToLCSDV-Gez
Tomato leaf curl Sudan virus – Gezira [Sudan:Gezira:1996]	AY044137	ToLCSDV-Gez[SD:Gez:96]
Tomato leaf curl Sudan virus – Shambat		ToLCSDV-Sha
Tomato leaf curl Sudan virus – Shambat [Sudan:Shambat:1996]	AY044139	ToLCSDV-Sha[SD:Sha:96]
Tomato leaf curl Sudan virus – Yemen		ToLCSDV-YE
Tomato leaf curl Sudan virus – Yemen [Yemen:Tihamah:2006]	EF110890	ToLCSDV-YE[YE:Tih:06]
Tomato yellow leaf curl Axarquia virus		
Tomato yellow leaf curl Axarquia virus – [Spain:Algarrobo:2000]	AY227892	TYLCAxV-[ES:Alg:00]
Tomato yellow leaf curl Malaga virus		
Tomato yellow leaf curl Malaga virus – [Spain:421:1999]	AF271234	TYLCMalV-[ES:421:99]
Tomato yellow leaf curl Mali virus		
Tomato yellow leaf curl Mali virus – Ethiopia		TYLCMLV-ET
Tomato yellow leaf curl Mali virus – Ethiopia [Ethiopia:Melkassa:2005]	DQ358913	TYLCMLV-ET[ET:Mel:05]
Tomato yellow leaf curl Mali virus – Mali		TYLCMLV-ML
Tomato yellow leaf curl Mali virus – Mali [Mali]	AY502934	TYLCMLV-ML[ML]
Tomato yellow leaf curl Sardinia virus		
Tomato yellow leaf curl Sardinia virus – Italy		TYLCSV-IT
Tomato yellow leaf curl Sardinia virus – Italy [Italy:Sardinia:1988]	X61153	TYLCSV-IT[IT:Sar:88]
Tomato yellow leaf curl Sardinia virus – Sicily		TYLCSV-Sic
Tomato yellow leaf curl Sardinia virus – Sicily [Israel:Henryk:2005]	DQ845787	TYLCSV-Sic[IL:Hen:05]
Tomato yellow leaf curl Sardinia virus – Sicily [Italy:Sicily]	Z28390	TYLCSV-Sic[IT:Sic]
Tomato yellow leaf curl Sardinia virus – Sicily [Tunisia:Bkalta 3:2002]	AY736854	TYLCSV-Sic[TN:Bk3:02]
Tomato yellow leaf curl Sardinia virus – Spain		TYLCSV-ES
Tomato yellow leaf curl Sardinia virus – Spain [Spain:Almeria 2:1992]	L27708	TYLCSV-ES[ES:Alm2:92]
Tomato yellow leaf curl Sardinia virus – Spain [Spain:Canary]	AJ519675	TYLCSV-ES[ES:Can]
Tomato yellow leaf curl Sardinia virus – Spain [Spain:Murcia 1:1992]	Z25751	TYLCSV-ES[ES:Mur1:92]
Tomato yellow leaf curl Sardinia virus – Spain[Morocco:Agadir:2002]	AY702650	TYLCSV-ES[MA:Aga:02]
Tomato yellow leaf curl virus		
Tomato yellow leaf curl virus – Gezira		TYLCV-Gez
Tomato yellow leaf curl virus – Gezira [Sudan:1996]	AY044138	TYLCV-Gez[SD:96]
Tomato yellow leaf curl virus – Iran		TYLCV-IR
Tomato yellow leaf curl virus – Iran [Iran]	AJ132711	TYLCV-IR[IR]
Tomato yellow leaf curl virus – Israel		TYLCV-IL
Tomato yellow leaf curl virus – Israel [Australia:Brisbane:2006]	1934[a]	TYLCV-IL[AU:Bri:06]
Tomato yellow leaf curl virus – Israel [China:Shangai 2:2005]	AM282874	TYLCV-IL[CN:SH2:05]
Tomato yellow leaf curl virus – Israel [Cuba]	AJ223505	TYLCV-IL[CU]
Tomato yellow leaf curl virus – Israel [Dominican Republic]	AF024715	TYLCV-IL[DO]
Tomato yellow leaf curl virus – Israel [Egypt:Ismaelia]	AY594174	TYLCV-IL[EG:Ism]
Tomato yellow leaf curl virus – Israel [Egypt:Nobaria:1991]	EF107520	TYLCV-IL[EG:Nob:91]
Tomato yellow leaf curl virus – Israel [Israel:Rehovot:1986]	X15656	TYLCV-IL[IL:Reo:86]
Tomato yellow leaf curl virus – Israel [Italy:Sicily:2004]	DQ144621	TYLCV-IL[IT:Sic:04]
Tomato yellow leaf curl virus – Israel [Japan:Haruno:2005]	AB192966	TYLCV-IL[JR:TosH:05]
Tomato yellow leaf curl virus – Israel [Japan:Misumi:Stellaria]	AB116631	TYLCV-IL[JR:Mis:Ste]
Tomato yellow leaf curl virus – Israel [Japan:Miyazaki]	AB116629	TYLCV-IL[JR:Miy]
Tomato yellow leaf curl virus – Israel [Japan:Omura:Eustoma]	AB116630	TYLCV-IL[JR:Omu:Eus]
Tomato yellow leaf curl virus – Israel [Japan:Omura]	AB110217	TYLCV-IL[JR:Omu]
Tomato yellow leaf curl virus – Israel [Japan:Tosa:2005]	AB192965	TYLCV-IL[JR:Tos:05]
Tomato yellow leaf curl virus – Israel [Jordan:Tomato:2005]	EF054893	TYLCV-IL[JO:Tom:05]
Tomato yellow leaf curl virus – Israel [Lebanon:Tomato:2005]	EF051116	TYLCV-IL[LB:Tom:05]
Tomato yellow leaf curl virus – Israel [Mexico:Culiacan:2005]	DQ631892	TYLCV-IL[MX:Cul:05]
Tomato yellow leaf curl virus – Israel [Morocco:Berkane:2005]	EF060196	TYLCV-IL[MO:Ber:05]
Tomato yellow leaf curl virus – Israel [Puerto Rico:2001]	AY134494	TYLCV-IL[PR:01]
Tomato yellow leaf curl virus – Israel [Spain:Almeria:Pepper:1999]	AJ489258	TYLCV-IL[ES:Alm:Pep:99]
Tomato yellow leaf curl virus – Israel [Tunisia:2005]	EF101929	TYLCV-IL[TN:05]
Tomato yellow leaf curl virus – Israel [Turkey:Mersin:2004]	AK812277	TYLCV-IL[TR:Mer:04]
Tomato yellow leaf curl virus – Israel [US:Florida]	AY530931	TYLCV-IL[US:Flo]
Tomato yellow leaf curl virus – Mild		TYLCV-Mld
Tomato yellow leaf curl virus – Mild [Israel;1993]	X76319	TYLCV-Mld[IL;93]
Tomato yellow leaf curl virus – Mild [Japan:Aichi]	AB014347	TYLCV-Mld[JR:Aic]

Continued

Table 1 Continued

Tomato yellow leaf curl virus – Mild [Japan:Atumi]	AB116633	TYLCV-Mld[JR:Atu]
Tomato yellow leaf curl virus – Mild [Japan:Daito]	AB116635	TYLCV-Mld[JR:Dai]
Tomato yellow leaf curl virus – Mild [Japan:Kisozaki]	AB116634	TYLCV-Mld[JR:Kis]
Tomato yellow leaf curl virus – Mild [Japan:Osuka]	AB116636	TYLCV-Mld[JR:Osu]
Tomato yellow leaf curl virus – Mild [Japan:Shimizu]	AB110218	TYLCV-Mld[JR:Shi]
Tomato yellow leaf curl virus – Mild [Japan:Shizuoka]	AB014346	TYLCV-Mld[JR:Shz]
Tomato yellow leaf curl virus – Mild [Japan:Yaizu]	AB116632	TYLCV-Mld[JR:Yai]
Tomato yellow leaf curl virus – Mild [Jordan:Cucumber:2005]	EF158044	TYLCV-Mld[JO:Cuc:03]
Tomato yellow leaf curl virus – Mild [Jordan:Homra:2003]	AY594175	TYLCV-Mld[JO:Hom03]
Tomato yellow leaf curl virus – Mild [Jordan:Tomato:2005]	EF054894	TYLCV-Mld[JO:Tom:03]
Tomato yellow leaf curl virus – Mild [Lebanon;LBA44:05]	EF185318	TYLCV-Mld[ILB;LBA44:05]
Tomato yellow leaf curl virus – Mild [Portugal:2:1995]	AF105975	TYLCV-Mld[PT:2:95]
Tomato yellow leaf curl virus – Mild [Reunion:2002]	AJ865337	TYLCV-Mld[RE:02]
Tomato yellow leaf curl virus – Mild [Spain:72:1997]	AF071228	TYLCV-Mld[ES:72:97]
Tomato yellow leaf curl virus – Mild [Spain:Almeria:1999]	AJ519441	TYLCV-Mld[ES:Alm:99]

[a]Australian accession.
Species names are in italic fonts and strain and isolate names are in roman fonts.

of isolates of TYLCSV (TYLCSV-[IT:Sar]) was 8 days from the end of the AAP compared to 5 weeks for TYLCV.

Both TYLCV and TYLCSV can be detected in the eggs and in adult progeny of viruliferous whiteflies in low incidence. However while the TYLCV progeny was able to transmit the virus to tomato plants, the TYLCSV progeny did not. Although the question of the possible replication of TYLCV in its insect vector remains controversial, transcripts of viral genes V1, V2, and C3 accumulated in *B. tabaci* long after transfer of viruliferous whiteflies on nonhost cotton plants. In contrast to TYLCV, transcripts of the bipartite begomovirus ToMoV did not accumulate in the whiteflies.

Deleterious Effects of TYLCV on Life Expectancy and Fertility of the Whitefly Vector

Three-day-old insects raised on eggplants (a TYLCV nonhost) following a 48 h AAP on TYLCV-infected tomato plants showed a reduction of 17–23% in their life expectancy compared to insects that have not acquired the virus, and to a 40–50% decrease in the mean number of eggs laid. These deleterious effects, in addition to the invasion of the reproductive system, suggest that TYLCV has some features reminiscent of an insect pathogen.

Sexual Transmission

TYLCV can be transmitted among whiteflies in a sex-dependent manner, in the absence of any other source of virus. TYLCV was transmitted from viruliferous males to females and from viruliferous females to males, but not among insects of the same sex. The recipient insects were able to efficiently inoculate tomato test plants. In the recipient insects TYLCV was first detected in the hemolymph, but not in the head as in the case of acquisition from infected plants. Therefore the virus follows, at least in part, the circulative pathway associated with acquisition from infected plants. Insect to insect virus transmission was instrumental in increasing the number of whiteflies capable of infecting tomato test plants in a whitefly population. Accordingly, a plant virus can be sexually transmitted among its insect vector.

Breeding Tomato for Resistance to TYLCV

Incorporation of Resistance Trait from Wild Tomato Species

The domesticated tomato *S. esculentum* is extremely susceptible to TYLCV. Breeding tomato for resistance started in the early 1960s and is going on since then. It has been initially observed that some accessions of wild tomato species exhibited resistance to the virus and could be crossed with the domesticated tomato. The first successful breeding program was initiated with *S. peruvianum*. Although this type of resistance was controlled by three to five recessive genes, a first hybrid with acceptable resistance was released in 1988. Advanced breeding lines and commercial varieties containing this type of resistance were released subsequently. Resistance genes from several different accessions of the wild tomato species *S. chilense* have been incorporated into breeding programs in the early 1990s. For each *S. chilense* accession, resistance is determined by a different semidominant gene and several minor genes. The wild tomato species *S. habrochaites* has been used in the late 1990s to produce breeding lines in which resistance is under the control of a major dominant locus.

After more than 25 years of effort, the best resistant cultivars and breeding lines give far higher yields upon

Figure 3 TYLCV-infected tomato fields planted with susceptible and resistant tomato genotypes. Note that the susceptible plants are stunted, and they will not produce fruits; for comparison, the resistant plants will remain symptomless, or have mild symptoms, and will yield.

infection than those of susceptible cultivars; disease symptoms are absent or mild but all of them contain various amounts of virus (**Figure 3**). It is interesting to note that in many cases (especially when the resistance originates from certain accessions of *S. chilense* and of *S. habrochaites*), the tomato lines are resistant to TYLCV and also to a number of begomoviruses in very different parts of the world, some of them monopartite and others bipartite. For example, line Lh902 that originated from *S. habrochaites* is resistant to TYLCV in the Mediterranean region as well as to a wide range of bipartite geminiviruses in Central America.

Genetically Engineered Tomato for Resistance

A variety of strategies have been devised based on the expression of functional as well as dysfunctional viral genes such as coat protein and Rep protein. A tomato interspecific hybrid expressing the TYLCV coat protein gene under the control of the 35S CaMV promoter responded upon whitefly mediated TYLCV inoculation showed expression of delayed disease symptoms and recovery from disease. Tomato inbred lines have been transformed with a 3′ truncated Rep gene (two-fifths of the gene) and the virus intergenic region from the Florida isolate of TYLCV (TYLCV-[US:Flo]). Resistance was achieved upon whitefly mediated inoculation in field conditions. However, these plants were susceptible to the bipartite begomovirus ToMoV from Florida. Similarly, tomato plants transformed with the Rep gene of TYLCSV (TYLCSV-[IT-Sar]) truncated in its 3′ end (leaving 210 amino acids out of 257) exhibited resistance to TYLCSV-[IT:Sar] but were susceptible to the related strain from Spain TYLCSV-ES[ES]. The viral Rep gene altered in the NTP-binding site crucial for virus replication has been used as *trans*-dominant proteins to block virus replication.

RNA-mediated virus resistance was highly sequence dependent. Expression of antisense sequence of the TYLCV-[IT-Sar] Rep gene in transgenic *N. benthamiana* resulted in repression of nearly all virus replication. The incorporation of a ribozyme structure into the antisense construct did not increase the efficiency of the system. Resistant tomato plants have been developed which exploit the mechanism of post-transcriptional gene silencing via double-stranded RNA sequences. Noncoding conserved regions from the genomes of TYLCV, TYLCV-mild (TYLCV-Mld[IL]), and from the TYLCV strains from Egypt (TYLCV-[EG:Ism]), and TYLCSV from Sardinia (TYLCSV-[IT:Sar]) and Spain (-TYLCSV-ES[ES]) were used to design a construct that can trigger broad resistance against the different TYLCVs. A high level of resistance was obtained when plants were inoculated with TYLCV-[EG:Ism], TYLCV-Mld[IL], and TYLCSV-ES[ES]. A positive correlation between resistance and the accumulation of virus-specific siRNAs was observed in silenced plants.

Localization of Loci Linked to Resistance to TYLCV in Tomato for Possible Use in Marker-Assisted Breeding

Mapping genes in tomato has been facilitated by the development of saturated maps based on DNA polymorphism (RFLP, AFLP, SSR, SCAR, etc.). A TYLCV-resistance gene originating from *S. chilense* accession LA 1969, *Ty-1*, has been mapped to tomato chromosome 6 using RFLP markers. Three additional loci linked to resistance against TYLCV have been mapped to chromosome 6 using resistant tomato inbreds derived from *S. chilense* accessions LA1932, LA2779, and LA1938. Another resistant gene originating from *S. pimpinellifolium* has been mapped, using RAPD markers, also to chromosome 6 but to a locus different from *Ty-1*. It can be noted that a locus conferring resistance to a nonrelated TYLCV isolate belonging to the species tomato leaf curl Taiwan virus from Taiwan, originating from *L. hirsutum* LA1777, has been located in chromosome 11 using RFLP markers.

See also: African Cassava Mosaic Disease; Bean Golden Mosaic Virus; Beet Curly Top Virus; Emerging Geminiviruses; Tomato Leaf Curl Viruses from India; Satellite Nucleic Acids and Viruses.

Further Reading

Czosnek H (ed.) (2007) *Tomato Yellow Leaf Curl Virus Disease: Management, Molecular Biology, Breeding for Resistance.* Dordrecht: Springer.

Lapidot M and Friedmann M (2002) Breeding for resistance to whitefly-transmitted geminiviruses. *Annals of Applied Biology* 140: 109–127.

Nakhla MK and Maxwell DP (1998) Epidemiology and management of tomato yellow leaf curl disease. In: Hadidi A, Khetarpal RK, and Koganezawa H (eds.) *Plant Virus Disease Control*, pp. 565–583. St. Paul, MN: APS Press.

Picó B, Diez MJ, and Nuez F (1996) Viral diseases causing the greatest economic losses to tomato crop. Part II: The tomato yellow leaf curl virus – A review. *Scientia Horticulturae* 67: 151–196.

Tombusviruses

S A Lommel and T L Sit, North Carolina State University, Raleigh, NC, USA

© 2008 Elsevier Ltd. All rights reserved.

Glossary

Modular evolution Evolution of viral genomes involving the combination of common genes/gene families.

Movement protein A plant viral protein that potentiates viral cell-to-cell movement through plasmodesmata.

Origin of assembly Unique RNA sequence and structure that specifically binds CP to initiate capsid assembly.

Polycistronic The characteristic of a given RNA containing more than one gene (open reading frame).

Quasiequivalence Capsid protein subunits arranged so that they are in somewhat equivalent environments with respect to their adjacent subunits.

(Silencing) suppressor Virally encoded product that inhibits a stage in the host RNA silencing pathway.

Subgenomic RNAs Less-than-full-length RNAs that are produced during replication, usually to express internal open reading frames.

Introduction

The *Tombusviridae* is a relatively large and diverse family of single-stranded, positive-sense, RNA plant viruses with common morphological, structural, molecular, and genetic features. Due to their small size and extremely high virus titer in experimental hosts, viruses in the family are particularly well characterized in terms of virion structure, replication, gene expression strategies, local and systemic movement, suppression of host gene silencing and associated satellite viruses, and defective interfering RNAs.

The family is constituted based on a unifying phylogenetic and biological feature. The RNA-dependent RNA polymerase of tombusviruses is highly conserved in terms of sequence identity, genomic structure, and gene expression and function. Biologically, tombusviruses share the property of being primarily soil transmitted, often without a biological vector, and accumulate to high levels in the roots of infected plants. Beyond these constants the family is remarkably diverse in biology, pathology, host range, and genome organization.

Many of the viruses now comprising the family *Tombusviridae* have been studied for a number of decades. Viruses like tomato bushy stunt virus (TBSV), carnation ringspot virus (CRSV), carnation mottle virus (CarMV), and cymbidium ringspot virus (CymRSV) were first described and virions purified and characterized in the 1940s. The taxonomic and phylogenetic relatedness of many of the viruses now comprising this family was not resolved until the mid-1980s, when genomes of these viruses were first cloned and sequenced.

Taxonomy, Phylogeny, and Evolution

The family *Tombusviridae* of plant viruses is composed of the genera *Aureusvirus, Avenavirus, Carmovirus, Dianthovirus, Machlomovirus, Necrovirus, Panicovirus,* and *Tombusvirus*, with more than 43 species and 15 tentative species recognized. Several genera are represented by many species whereas three genera are monotypic. The type species of the genus *Aureusvirus* is *Pothos latent virus*. This genus is quite similar to the genus *Tombusvirus* but is distinguished by having significantly different sized movement and silencing suppressor proteins. The type species of the monotypic genus *Avenavirus* is *Oat chlorotic stunt virus*. This species constitutes a separate genus because the genome organization is intermediate between those of the genera *Carmovirus* and *Tombusvirus*. Furthermore its capsid protein (CP) is significantly larger than those found in other genera whose CPs have a protruding (P) domain. *Carnation mottle virus* is the type species of the genus *Carmovirus*. This genus is distinguished by having two small proteins associated with virus movement and a CP with a P domain. The genus *Dianthovirus*, of which

Carnation ringspot virus is the type species, has the most dramatic taxonomic distinction: its genome is split into two segments. *Maize chlorotic mottle virus* is the type and monotypic species of the genus *Machlomovirus*. This genus is structurally quite similar to the genus *Panicovirus* but contains an additional open reading frame (ORF) at the 5′ end of its genome which nearly completely overlaps the polymerase. The genus *Necrovirus* is represented by *Tobacco necrosis virus A*. The genome organization and expression strategy are quite similar to those in the carmoviruses, but the necroviruses have the phylogenetically distinct CPs lacking a P domain. The genus *Panicovirus* is also represented by a single species, *Panicum mosaic virus*. This virus is distinguished by having carmovirus-like movement proteins (MPs), a CP without a P domain, as well as several accessory genes (**Figure 1**).

This family serves as an excellent example supporting the concept of modular evolution of viruses. The polymerase is the sole shared module binding the family. Within the limitations of the viral polymerase the family has taken great liberties with the arrangement and expression of the various genes on its polycistronic RNA. While it is true that all members of the family have a $T=3$ icosahedral virion, different genera achieve this structure using two phylogenetically distinct CP modules. At this time, it appears that this family has acquired at least three phylogenetically different cell-to-cell movement modules. The sources of the virus suppressor of host gene silencing appear to be equally diverse. In addition, various species within a particular genus have acquired additional unique accessory modules, possibly for transmission by fungal vectors.

The minimal viral polymerases of the *Tombusviridae* lack any identifiable helicase motif, and do not have a nucleotide triphosphate binding motif. They do, however, have the canonical glycine–aspartate–aspartate (GDD) motif found in most RNA polymerases. All *Tombusviridae* polymerases are similarly expressed from an interrupted ORF by one of two translational regulatory mechanisms: terminator readthrough or −1 ribosomal frameshifting (**Figure 1**). The *Tombusviridae* polymerase is also phylogenetically related to that in the genus *Luteovirus* and, to a lesser extent, *Enamovirus* but not with the genus *Polerovirus* in the family *Luteoviridae*. Polymerases of tombusviruses belong to the Sindbis-like superfamily of RNA polymerases, but they are often categorized in a separate subgroup because of their reduced size and missing motifs.

The virions of tombusviruses are morphologically similar. They all form icosahedral $T=3$ symmetry particles approximately 30–38 nm in diameter. Virions are formed from 180 copies of a single CP subunit ranging from 25 to 48 kDa in size. Virions parse into two morphological subclasses the first of which have rough or granular surfaces and the second smooth surfaces. The smooth particles, such as those found in the genera *Machlomovirus*, *Necrovirus*, and *Panicovirus*, have smaller CPs and lack the P domain found in the larger CPs (**Figure 2**).

There are also common elements in the viral genome structures of tombusviruses. They are all small, single-stranded, positive-sense RNA viruses, encoding four to six ORFs on a single RNA molecule, with the exception of the dianthoviruses, which have a split or bipartite genome. In addition, members of the *Tombusviridae* all rely on the generation of one or more subgenomic RNAs (sgRNAs) to express genes from the polycistronic viral genome.

Virion Structure

Although the underlying size and structure of the virions are similar, the members can be broadly subdivided into two groups based on the presence or absence of a P domain at the c-terminus of the CP. Genera containing P domains (*Aureusvirus*, *Avenavirus*, *Carmovirus*, *Dianthovirus*, and *Tombusvirus*) produce virions that are 32–38 nm in diameter and display a granular surface. These CP subunits range in size from 37 to 48 kDa. The X-ray crystal structures of TBSV and turnip crinkle virus (TCV; *Carmovirus*) revealed the discrete organization of each CP subunit. Generally, the CP subunit can be divided into four domains: the RNA-binding (R) domain is located at the N-terminus followed by the arm (a), shell (S), and P domains (66, 35, 67, and 110 amino acids, respectively, in TBSV). The R domain contains many basic residues and is found in the interior of the virion. Cryoelectron microscopy reconstructions have further revealed the presence of internal ordered cages of RNA intertwined with CP residues from the R domains beneath the virion surface. This internal scaffold may play a role in directing specific packaging of viral RNA and formation of the icosahedral virions. The virion structure is primarily formed by the globular S domains (stabilized by a pair of Ca^{2+} per subunit) which are composed of two sets of four-stranded antiparallel β-sheets. The S domain is also the most highly conserved region of the CP subunit. There is a flexible hinge of five residues between the S and P domains that allows the CP subunit to adopt different configurations by varying the angle between the domains. This feature of the CP subunit overcomes the structural constraints imposed by the icosahedral morphology. P domains (containing antiparallel β-sheet structures in a jellyroll conformation, with one six-stranded β-sheet and one four-stranded β-sheet) of adjacent CP subunits dimerize to produce 90 projections leading to the granular surface texture. Genera lacking the P domain (*Machlomovirus*, *Necrovirus*, and *Panicovirus*) produce virions that are 30–32 nm in diameter with a smooth surface similar to viruses in the genus *Sobemovirus*. These shorter CP subunits range from 25 to 29 kDa in size.

Tombusviruses 147

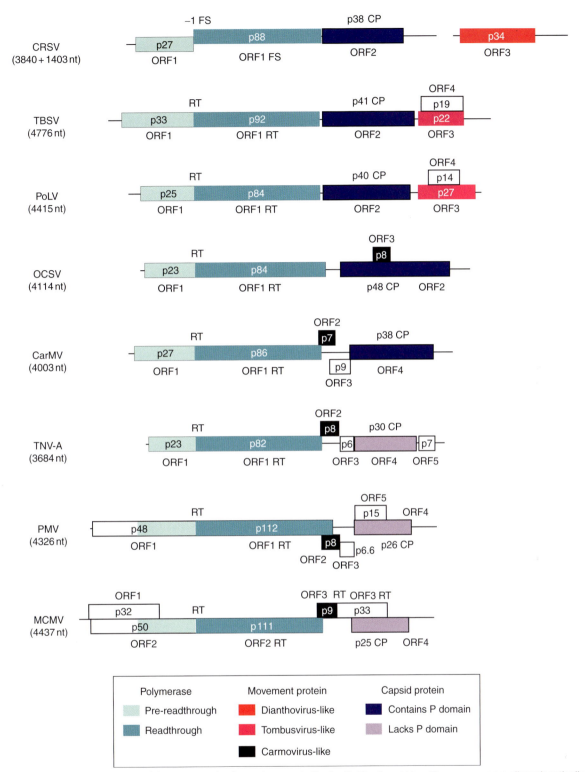

Figure 1 Genomic organization of the type species for each genus in the family *Tombusviridae*. The genomes are aligned vertically at the site of the polymerase readthrough event. Boxes represent known and predicted ORFs with the sizes of the respective proteins (or readthrough products) indicated, for example, p27 for 27 kDa protein. Similarly, colored boxes represent proteins with extensive sequence conservation and function. RT, translational readthrough of termination codon; –1 FS, -1 ribosomal frameshifting event; CP, capsid protein; CRSV, carnation ringspot virus (genus *Dianthovirus*); TBSV, tomato bushy stunt virus (genus *Tombusvirus*); PoLV, pothos latent virus (genus *Aureusvirus*); OCSV, oat chlorotic stunt virus (genus *Avenavirus*); CarMV, carnation mottle virus (genus *Carmovirus*); TNV-A, tobacco necrosis virus A (genus *Necrovirus*); PMV, panicum mosaic virus (genus *Panicavirus*); MCMV, maize chlorotic mottle virus (genus *Machlomovirus*).

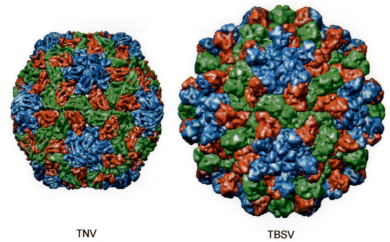

Figure 2 Virion images of tombusviruses based on X-ray crystal structures. Left image is tobacco necrosis virus (TNV) at 2.25 Å. The capsid protein of this virus does not contain a protruding domain and is representative of species in the genera *Machlomovirus*, *Panicovirus*, and *Necrovirus*. Right image is TBSV at 2.9 Å. The capsid protein of this virus contains a protruding domain and is representative of species in the genera *Avenavirus*, *Aureusvirus*, *Carmovirus*, *Dianthovirus*, and *Tombusvirus*. For both images, the individual subunits are colored according to their various conformations: A (blue), B (red), and C (green). From Shepherd CM, Borelli IA, Lander G, et al. (2006) VIPERdb: A relational database for structural virology. *Nucleic Acids Research* 34 (Database Issue): D386–D389, VIPERdb.

Due to the quasi-equivalent nature of the CP subunits when arranged in an icosahedron, each subunit can take on one of three distinct conformations, termed A, B, and C (respectively, blue, red, and green in **Figure 2**). The A and B conformations differ in the angle between the S and P domains and their arrangement on the virion surface (A: fivefold axes, B: threefold axes). The C conformation differs from A and B in that their R/a domains intertwine to form an ordered internal structure around threefold axes of symmetry termed the β-annulus. The three different CP subunit conformations pack together as either AB or CC dimers in the virion particle. For TCV, virion assembly initiates with three CP dimers and viral RNA followed by formation of the virion shell according to structural constraints imposed by CP/RNA interactions. In TCV, the origin of assembly RNA sequence (which specifically initiates the interaction with CP subunits) is contained within a 186 nt region at the 3′ end of the CP ORF. Assembly studies with TCV also showed that only RNAs equal to or smaller than 4.35 kbp in size could be packaged, suggesting very strict size limitations for icosahedral virions.

Virions of tombusviruses have an M_r of $\sim 8.2–8.9 \times 10^6$ and produce a single, well-defined band upon centrifugation with sedimentation coefficients ranging from 118 to 140$S_{20,w}$. Virion densities range from 1.34 to 1.36 g cm^{-3} in CsCl gradients. Virions are stable at acidic pH, but expand above pH 7 and in the presence of ethylenediaminetetraacetic acid (EDTA). Virions are resistant to elevated temperatures although thermal inactivation usually occurs above 80 °C. Due to the lack of a lipid membrane, virions are insensitive to organic solvents and nonionic detergents.

Most genera produce virions containing a single molecule of positive-sense, linear single-stranded RNA (ssRNA) ranging in size from 3.7 to 4.8 kb depending on the genus. The genomic RNA constitutes about 17% of the particle weight. The exception is the genus *Dianthovirus* where virions contain two genomic RNAs: a large RNA-1 of $\sim$3.9 kb and a smaller RNA-2 of 1.5 kb. The genomic RNAs lack a 5′ cap structure and the 3′ ends are not polyadenylated nor do they form tRNA-like structures. Satellite RNAs (satRNAs) of nonviral origin are associated with several genera. Satellite viruses (which code for their own CP) are also associated with several genera but these are readily distinguishable due to their significantly smaller dimensions. Both small RNAs modify the symptoms of their respective helper viruses.

Genome Organization and Replication Strategy

The dimensions of the icosahedral virions constrain the genome sizes of the various members of the family *Tombusviridae*. This forces the members to adopt several strategies for expression of their genetically compact, polycistronic genomes. Their unifying genomic feature is the presence of an interrupted ORF for the virally encoded RNA-dependent RNA polymerase at the 5′ end of the genome. The interruption is manifested as either an amber termination codon which may be read through (most members) or a -1 ribosomal frameshifting signal (genus *Dianthovirus* only). In either case, this leads to the direct translation of two ORFs with identical amino termini from the genomic RNA (**Figure 1**). These polymerase subunits share a high degree of sequence similarity within the family. Notably, they do not contain a helicase-type motif and the genomes do not encode a helicase-like ORF. ORFs downstream of the polymerase are expressed via 3′ co-terminal sgRNAs with the resident ORFs produced with an assortment of translational strategies such as ribosomal scanning and frameshifting. Aside from the single CP ORF mentioned previously, the genomes of *Tombusviridae* also encode one or more MPs that are involved in cell-to-cell spread of the virus. These range in size from the 34–35 kDa version found in the genus *Dianthovirus* to the 22 kDa product found in the genus *Tombusvirus* and down to the 6–9 kDa pair found in the genus *Carmovirus*. In most cases, the MPs can potentiate the local movement of unencapsidated viral RNA but systemic infection generally requires encapsidation. The exception is for members of the genus *Tombusvirus* which can infect systemically without CP albeit at a reduced rate. The other prominent protein product with a clearly defined function is the genus *Tombusvirus* p19 protein which is a potent suppressor of post-transcriptional gene silencing (PTGS). p19 binds double-stranded short interfering RNAs (siRNAs) that are generated by a Dicer-like RNase as part of the host RNA-silencing response to viral infection. This binding prevents incorporation of the siRNAs into the RNA-induced silencing complex (RISC) to prevent further cleavage of viral RNAs. Although other members of the family may not encode a unique suppressor protein, it is known that other proteins can have suppressor activity aside from their primary functions. This is the case for TCV where the CP is the viral suppressor of PTGS. Various additional ORFs are also encoded with undefined functions. One other method utilized by members of the family *Tombusviridae* for gene expression is genome segmentation which is only employed by the genus *Dianthovirus* where RNA-2 is monocistronic and encodes the MP.

Since the genomic RNAs are positive sense, they are directly infectious as unencapsidated RNAs. However, the lack of a 5′ cap structure and a 3′ poly(A) tail precludes the normal circularization procedure employed by mRNAs for translation. Instead, a cap-independent mechanism involving a long-distance RNA–RNA interaction between sequence elements in the 5′ and 3′ noncoding regions has been demonstrated in TBSV and tobacco necrosis virus-A (TNV-A) and proposed in the case of red clover necrotic mosaic virus (RCNMV; genus *Dianthovirus*) RNA-1. The direct translation readthrough strategy ensures that the longer ORF (containing the catalytic site of the polymerase) is present in only 5–10% of the translation products since readthrough is inefficient. Translation of the virally encoded polymerase subunits is followed by localization to and proliferation of cellular membranes. The location of viral replication varies dependent on the particular virus and is specified by the pre-readthrough portion of the polymerase. In all cases studied to date, the polymerase anchors on membranes, causing proliferations. The advent of yeast-based replication systems for TBSV, cucumber necrosis virus (CNV; genus *Tombusvirus*), and CymRSV has provided valuable insights into the host components utilized by the viral replication complex.

Once the polymerase has been assembled, it binds to the 3′ terminus to initiate minus-strand synthesis. If full-length copies are generated, these serve as templates for synthesis of progeny genomic RNAs. This synthesis is highly asymmetric with positive-stranded RNA accumulating at much higher levels than negative-stranded RNA. Occasionally, positive-strand synthesis will be terminated prematurely which leads to the formation of templates for plus-strand sgRNA synthesis. This premature termination is controlled by long-distance RNA–RNA interactions which occur in *cis* (for instance, in TBSV) or in *trans* (for instance, RCNMV). Generally, sgRNAs are produced later in the infection cycle and behave as monocistronic mRNAs. However, the sgRNAs also demonstrate various translational strategies such as ribosomal scanning in the case of p19 expression which is nested within the p22 ORF in TBSV. In this case, the p22 start codon is suboptimal and occasionally read through by ribosomes before translation at the optimal p19 start codon.

During replication, defective interfering (DI) RNAs (which are not packaged into virions) arise in viruses belonging to some genera. These DI RNAs are derived from the viral genome and are replicated to very high titers which lead to their interfering with viral genomic replication levels. DI RNAs have been used extensively to delimit critical sequence elements required for RNA replication.

Transmission, Host Range, and Epidemiology

The natural host range of a given species is relatively narrow; however, the experimental host range tends to be broad. Members can infect either monocotyledonous or dicotyledonous plants, but no species can infect both. Natural infections can be limited to or are often concentrated in the root system. Many species induce a necrosis symptom in the foliar parts of the plant. All species are readily transmitted by mechanical inoculation and by host vegetative propagation and some may be transmitted by contact and through seeds. A number of species are readily detected in the soil, surface waters, rivers, lakes, and even the ocean. Transmission by the chytrid fungi in the genus *Olpidium* and beetles have also been reported for members of several genera. Most, if not all, members can be transmitted through the soil either dependent on, or independent of, a biological vector. This rather unusual soil and water mode of transmission is unique to this plant virus family and is based on the unusually robust constitution of the virion.

Geographical distribution of particular species varies from wide to restricted. Most species occur in temperate regions although legume-infecting carmoviruses and a tentative member of the genus *Dianthovirus* have been recorded from tropical areas.

In nature, viruses in the family *Tombusviridae* can be spread in a variety of ways, including seed and pollen transmission, mechanical transmission, vegetative propagation and grafting, growth in infected soil, and by vectors such as fungi, thrips, and beetles. As mentioned above, one hallmark of members of the family *Tombusviridae* is the amount of virus present in the roots of infected plant tissue, and the ease of transmission through the soil, whether that transmission is dependent on a biological vector or not. For viruses that can be transmitted by fungi, there is a considerable amount of specificity required between the virus and the fungal vector. For example, CNV (*Tombusvirus*), melon necrotic spot virus (*Carmovirus*), and cucumber leaf spot virus (*Aureusvirus*) are transmitted by the root-inhabiting chytrid fungus *Olpidium bornovanus*, whereas TNV (*Necrovirus*) is transmitted exclusively by *Olpidium brassicae*.

Pathogenesis

For the species in which it has been studied, the virus titers and pathogenic effects are at least as high, if not higher, in the root system of the host as in the other parts of the plant. This is consistent with the robust nature of the virions and that viruses in the family are transmitted through the soil.

Many tombusviruses generate DI RNAs that can affect pathogenesis. DI RNAs are essentially deletion mutants of the viral genome which are generated by errors in replication such as rearrangement or recombination. Several species such as TBSV and TCV produce DI RNAs during viral infection. For TBSV the presence of DI RNAs results in attenuation of symptoms, whereas for TCV the presence of DI RNAs intensifies symptoms. Many species, including both TBSV and TCV, are also known to harbor satRNAs. These small RNAs are generally not derived from viral sequences, but they depend on the helper virus for replication and packaging. As with DI RNAs, they can attenuate TBSV or intensify symptoms (TCV).

The origin of nonhomologous satRNAs is unclear at present, but like DI RNAs, satRNAs can be generated from recombination events with other satRNAs, the viral genome, DI RNAs, or host sequences. Both DI RNAs and satRNAs have been useful molecular tools for studying viral replication and recombination. Since both satRNAs and DI RNAs are dependent on the host or parent virus for replication, they must have retained any sequence or structural signals required for recognition by the viral polymerase, and therefore can help determine exactly what those signals are.

Recombination can facilitate viral evolution as well as repair of viral genomes, and so it is an important factor in the virus life cycle and it can have profound effects on pathogenesis. In the family *Tombusviridae*, recombination in the genera *Carmovirus* and *Tombusvirus* has been shown to repair damaged or deleted 3′ ends of virus-associated RNAs, as well as generate new satRNAs or DI RNAs. Recombination can occur between both homologous and nonhomologous sequences, probably by a copy-choice mechanism in which the viral replicase jumps from one template to another during the replication process.

Plants infected with tombusviruses display a distinctive cytopathology, observed by electron microscopy as dense-staining features. These features are known as multivesicular bodies, and are believed to be sites of viral replication as well as accumulation of exceedingly high concentrations of progeny virions. To date, cell biology studies have shown that species can replicate on, remodel, and proliferate membranes of peroxisomes, mitochondria, and the cortical ER. There appears to be no correlation between a specific organelle membrane used for replication and taxa.

See also: Assembly of Viruses: Enveloped Particles; Assembly of Viruses: Nonenveloped Particles; *Carmovirus*; Defective-Interfering Viruses; Luteoviruses; Machlomovirus; *Necrovirus*; Satellite Nucleic Acids and Viruses; Togaviruses: Alphaviruses; Vector Transmission of Plant Viruses; Viral Suppressors of Gene Silencing.

Further Reading

Fauquet CM, Mayo MA, Maniloff J, Desselberger U, and Ball LA (eds.) (2005) *Virus Taxonomy: Eighth Report of the International Committee on Taxonomy of Viruses.* San Diego, CA: Elsevier Academic Press.

Kneller EL, Rakotondrafara AM, and Miller WA (2006) Cap-independent translation of plant viral RNAs. *Virus Research* 119: 63–75.

Martelli GP, Gallitelli D, and Russo M (1988) Tombusviruses. In: Koenig R (ed.) *The Plant Viruses, Polyhedral Virions with Monopartite RNA Genomes,* vol. 3, pp. 13–72. New York: Plenum.

Miller WA and White KA (2006) Long-distance RNA–RNA interactions in plant virus gene expression and replication. *Annual Review of Phytopathology* 44: 447–467.

Morris TJ and Carrington JC (1988) Carnation mottle virus and viruses with similar properties. In: Koenig R (ed.) *The Plant Viruses, Polyhedral Virions with Monopartite RNA Genomes,* vol. 3, pp. 73–112. New York: Plenum.

Nagy PD and Pogany J (2006) Yeast as a model host to dissect functions of viral and host factors in tombusvirus replication. *Virology* 344: 211–220.

Nagy PD and Simon AE (1997) New insights into the mechanisms of RNA recombination. *Virology* 235: 1–9.

Robbins MA, Reade RD, and Rochon DM (1997) A cucumber necrosis virus variant deficient in fungal transmissibility contains an altered shell protein domain. *Virology* 234: 138–146.

Rochon D, Kakani K, Robbins M, and Reade R (2004) Molecular aspects of plant virus transmission by olpidium and plasmodiophorid vectors. *Annual Review of Phytopathology* 42: 211–241.

Rossman MG and Johnson JE (1989) Icosahedral RNA virus structure. *Annual Review of Biochemistry* 58: 533–573.

Russo M, Burgyan J, and Martelli GP (1994) Molecular biology of Tombusviridae. *Advances in Virus Research* 44: 381–428.

Scholthof HB (2006) The tombusvirus-encoded P19: From irrelevance to elegance. *Nature Reviews Microbiology* 4: 405–411.

Shepherd CM, Borelli IA, Lander G, *et al.* (2006) VIPERdb: A relational database for structural virology. *Nucleic Acids Research* 34 (Database Issue): D386–D389.

White KA and Nagy PD (2004) Advances in the molecular biology of tombusviruses: Gene expression, genome replication, and recombination. *Progress in Nucleic Acid Research and Molecular Biology* 78: 187–226.

Torovirus

A E Hoet, The Ohio State University, Columbus, OH, USA
M C Horzinek, Utrecht University, Utrecht, The Netherlands

© 2008 Elsevier Ltd. All rights reserved.

Glossary

Colostrum-deprived (CD) calves Conventional calves that did not receive antibody-rich first milk or colostrum during the first 6–12 h of life, therefore lacking of passive immunity from the mother (calves are hypogammaglobulinemic individuals when they born).

Gnotobiotic (Gn) calves Animals that are obtained and maintained in germ-free environments, in which the composition of any associated microbial flora, if present, is fully defined.

History

Equine torovirus (EToV), originally referred to as Berne virus, was accidentally isolated in equine kidney cells in 1972 from a rectal swab taken from a horse with diarrhea. EToV is the only torovirus that has been propagated in cell culture, in lines of equine dermis or embryonic mule skin cells, where it causes a cytopathic effect that results in cell lysis. While Berne virus was not neutralized by antisera against known equine viruses, serologic cross-reactions were observed in neutralization tests and enzyme-linked immunosorbent assay (ELISA) using sera from calves that had been experimentally infected with morphologically similar particles, then named 'Breda' viruses.

The first Breda virus, now called bovine torovirus (BToV), was discovered in 1979 during an investigation in a dairy herd in Breda (Iowa), in which severe neonatal calf diarrhea had been a problem for three consecutive years. After this initial report, other strains of BToV were identified in beef calves from Ohio, and in a colostrum-deprived (CD) calf from Iowa. Despite repeated attempts, BToV cannot be adapted to grow in cell or tissue cultures and must be passaged in gnotobiotic (Gn) calves, which has hampered its biochemical, biophysical, and molecular characterization. Most of the studies on the pathogenesis and pathology of torovirus infections have been done in BToV-infected Gn and CD calves, as well as in limited field studies; in contrast, most of the biochemistry and morphogenesis data on toroviruses are based on EToV studies.

In 1984, torovirus-like (TVL) particles were detected in the feces from human patients with gastroenteritis by using electron microscopy (EM). Since then, reports of human toroviruses (HToVs) in children and adults with acute diarrhea have appeared in several countries. TVL particles have also been detected in fecal samples from pigs, and named porcine torovirus (PToV). Proof that the observed structures were not artifacts was obtained when toroviral RNA sequences were found in the feces of piglets and in stools from humans with diarrhea. In recent years, TVL particles have also been detected in turkeys and have been associated with a 'stunting syndrome'.

Taxonomy and Classification

Toroviruses are single-stranded, positive-polarity RNA viruses with a peplomer-bearing envelope. The term *torus* (Latin) refers to the circular convex molding in the form of a doughnut that some columns or pilaster have at their bases; indeed, it was the unique biconcave disk and C-shape of the virion in the extracellular environment that suggested this naming. Since 1992, the genus *Torovirus* has been included with the genus *Coronavirus* and the newly recognized genus *Bafinivirus* in the family *Coronaviridae*, based on similarities in genomic organization and replication strategies. Toro- and coronaviruses are also ancestrally related: their polymerase and envelope genes diverged from those of a common predecessor. Because of their inclusion in the family *Coronaviridae*, the nomenclature for coronavirus genes, mRNAs, and structural proteins have also been applied to toroviruses. However, the lack of sequence homology in the structural genes and the absence of antigenic relatedness with coronaviruses justify their taxonomic position as a separate genus. The International Committee on Taxonomy of Viruses (ICTV) presently recognizes four species in the genus *Torovirus*: *Equine torovirus*, *Bovine torovirus*, *Porcine torovirus*, and *Human torovirus*.

The families *Coronaviridae* and *Arteriviridae*, as well as the new family *Roniviridae*, are the constituents of the order *Nidovirales*, the second order in animal virology (after the order *Mononegavirales*). This assignment is based on their similar basic genomic organization and common replication strategy: the synthesis of a 3′ co-terminal nested set of subgenomic mRNAs, and the possession of two open reading frames (ORFs) connected by a frameshift site to express a replicase directly from the genomic RNA. This nested set of mRNAs was the foundation for the name of the order *Nidovirales* (from Latin *nidus*, 'the nest').

Virion Properties

Torovirus particles possess a nucleocapsid with helical symmetry coiled into a hollow tube (diameter 23 nm, average length 104 nm, periodicity 4.5 nm). Extracellular, negatively stained torovirus virions are generally observed as kidney- or C-shaped particles (105–140 nm × 12–40 nm). They can also be seen as spherical or oval particles (89 ± 7 nm × 75 ± 9 nm) or rod-shaped virions (35 nm × 170 nm), depending on the different orientations of the virions with respect to the electron beam. A graphic representation of a torovirion is shown in **Figure 1**. A tightly fitting envelope, 11 nm thick, surrounds the virion structure bearing prominent drumstick-shaped peplomers (17–24 nm), and a fringe of shorter spikes (8–10 nm), which represent the spike and the hemagglutinin-esterase (HE) proteins, respectively. Intracellularly, toroviruses are observed as elongated tubules with rounded ends (rod-shaped virions, 35–42 nm × 80–105 nm), located in the cytoplasm of infected intestinal cells.

Genome

The torovirus genome consists of a single-stranded, polyadenylated RNA of positive (messenger) polarity, which is about 28.5 kbp in length. Recently, the BToV genome has been completely sequenced, and it has been shown to comprise 28 475 nt and contain six ORFs (see **Figure 1**), each of which encodes a known protein. ORF1a and ORF1b are the most 5′ proximal reading frames and constitute the replicase (RNA-dependent RNA polymerase) gene, which is expressed as a large precursor protein directly from the genomic RNA by a ribosomal frameshift mechanism, similar to other nidoviruses. The large product of these ORFs is apparently involved in the synthesis of a negative-strand RNA and the onset of genomic and subgenomic RNA synthesis. The other four ORFs correspond to structural protein genes and are expressed by the production of a 3′

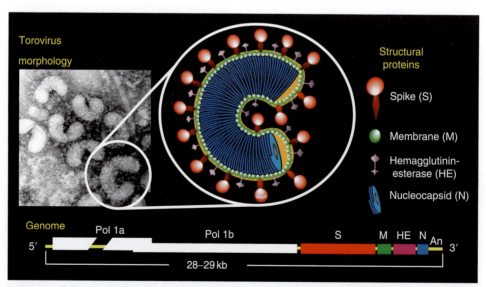

Figure 1 Typical extracellular morphology of a torovirus as visualized by immune electron microscopy (IEM), with a graphic representation of torovirus virion and its structural proteins and genome. Modified with permission from Hoet AE (in press) Torovirus pathogenesis and immune responses. In: Perlman S, Snijder EJ, and Gallagher TM (eds.) *The Nidoviruses*, copyright 2008 by the ASM Press.

co-terminal nested set of four mRNAs. The BToV ORF2 (4752 nt), ORF3 (702 nt), ORF4 (1248–1251 nt), and ORF5 (504 nt) encode the spike (S), membrane (M), HE, and nucleocapsid (N) proteins, respectively.

Proteins

Proteins with molecular weights of 20, 22, 37, and 80–100 kDa have been identified in EToV virions. Detergent treatment of virions releases the 22, 37, and 80–100 kDa proteins, which indicates their association with the envelope. Similar polypeptides of 20, 37, 85, and 105 kDa have been identified in purified BToV by means of surface radioiodination. The nucleocapsid (N) protein (167 aa) is the most abundant polypeptide found in the virion (80–84%) with a predicted molecular weight of 18.3–19.2 kDa. Blotting experiments performed with EToV have identified this internal protein as the only RNA-binding polypeptide in the virion. The M glycoprotein (233 aa, 26.5 kDa) is the second most abundant protein (~13%) and is probably associated with the envelope. Computer analysis has revealed characteristics of a class III membrane protein lacking a cleaved signal sequence, but containing three successive transmembrane α-helices in the N-terminal half. The M protein accumulates in intracellular membranes, predominantly those of the endoplasmic reticulum and is believed to play a role in assembly, maturation, and nucleocapsid recognition during the budding process.

The heterogeneous, N-glycosylated, 80–100 kDa protein is recognized by both neutralizing and hemagglutination-inhibiting monoclonal antibodies and is therefore identified as the spike (S) protein projecting from the virion surface. The spike (S) gene encodes an apoprotein (1581 aa) with a molecular weight of about 178 kDa. The deduced amino acid sequence contains domains typical of type I membrane glycoproteins: an N-terminal signal sequence, a putative C-terminal transmembrane anchor, and a cytoplasmic tail. The HE is a class I membrane N-glycosylated protein (416–417 aa) of 65 kDa that is also located in the BToV envelope. It has an N-terminal signal sequence, a C-terminal transmembrane domain, several N-glycosylation sites, and a putative 'FGDS' motif which displays acetylesterase activity specific for N-acetyl-9-O-acetylneuraminic acid. The HE protein could be an additional receptor-binding protein (next to the spike protein) but without a function in viral entry. EToV virions lack an HE protein, only possessing a partial sequence in ORF4.

Physical Properties

Buoyant densities of 1.16, 1.17, 1.18, and 1.14 g ml^{-1} have been reported for the virions of EToV, BToV serotype 1, BToV serotype 2, and HToV, respectively. EToV is remarkably stable in the environment, and relatively resistant to phospholipase C, trypsin, chymotrypsin, and even deoxycholate; however, Triton X-100 and organic solvents destroy its infectivity. EToV is stable within a wide range of pH, being inactivated only below pH 2.5 or above pH 10.3. BToV1 appears to be less stable than both EToV and BToV2 as changes in its sedimentation behavior and density have been observed after prolonged storage at −70 °C. The infectivity of a fecal preparation containing BToV1 has been reported to be lost completely after 3 weeks at 4 °C, whereas EToV in cell-free supernatant remained stable for 92 days. Storage of toroviruses at −20 to −70 °C helps to preserve infectivity. However, even at these temperatures the viruses will deteriorate, though at a slower rate. Repeated cycles of freezing and thawing of purified BToV2 results in loss of peplomers and disintegration of virions.

Replication and Morphogenesis

Because it can be grown in cell culture, the morphogenesis of EToV has been most extensively studied. Ultrastructural and immunofluorescence (IF) studies on intestinal cells from BToV-infected calves have shown similarities with EToV morphogenesis.

Attachment, Entry, and Uncoating

Attachment to the apical surface of enterocytes is apparently mediated through the spike proteins, but the HE proteins may be involved as well. Entry or penetration of BToV into enterocytes is apparently by receptor-mediated endocytosis. Lysosomal degradation in vesicles containing the virus is probably responsible for uncoating and subsequent release of the BToV RNA. The site at which this occurs has not yet been determined.

EToV replication occurs in the cytoplasm. Preformed tubular capsids bud through membranes of the Golgi stack and of the endoplasmic reticulum. A host cell nuclear function seems to be required since ultraviolet (UV) pre-irradiation of cells, actinomycin D, and α-amanitin have been reported to reduce virus yields. The replication cycle takes around 10–12 h to complete.

Transcription

In EToV-infected cells, five virus-specific, polyadenylated mRNAs are found with sizes of >20.0, 7.5, 2.1, 1.4, and 0.8 kbp. Northern (RNA) blot hybridizations with restriction fragments from cDNA clones have shown that the five EToV mRNAs form a 3′ co-terminal nested set. Sequence analysis has revealed the presence of four complete ORFs with initiation codons coinciding with the

5' ends of EToV RNAs 2–5, respectively; RNA 5 is contiguous on the consensus sequence. EToV RNAs 1–3 are transcribed independently, as has been shown by UV transcription mapping. The genes for M, HE, and N are preceded by short noncoding 'intergenic' regions, containing a transcription-regulating element (TRE) conforming to the consensus sequence 5' (C)ACN$_{3-4}$CUUUAGA 3'. A copy of this sequence is also present at the extreme 5' terminus of the genome. In contrast, the S gene overlaps with the replicase gene and the N-terminal 28 residues of S are in fact encoded by an internal (−1) reading frame within ORF1b; moreover, there is no TRE. The production of this 3' co-terminal nested set of mRNAs characterizes the toroviruses and justifies their inclusion into the order *Nidovirales*.

Translation

No RNA-dependent RNA polymerase is found in torovirus virions. The torovirus replicase is probably translated as soon as the RNA is liberated. Translation yields two large polyproteins, from which by proteolytic cleavage the various subunits of the viral replicase/transcriptase are derived, as well as accessory proteins of as yet unknown function. Downstream of ORF1b, there are the genes for the structural proteins S, M, HE, and N (as ordered from 5' to 3'); these are translated from four subgenomic mRNAs, numbered 2–5 (with the genomic RNA as RNA1).

Post-Translational Processing

The N-glycosylated S protein is derived from the processing of a 200 kDa precursor present in infected cells, but not in virions. Eighteen potential N-glycosylation sites, two heptad repeat domains, and a possible 'trypsin-like' cleavage site exist in the spike protein amino acid sequence. The mature S protein consists of two subunits and their electrophoretic mobility upon endoglycosidase F treatment suggests that the predicted cleavage site is functional *in vivo*. The heptad repeat domains are probably involved in the generation of an intrachain coiled-coil secondary structure; similar interchain interactions can play a role in the formation of the observed S protein dimers. The intra- and interchain coiled-coil interactions may stabilize the stalk of the torovirus peplomers.

Assembly, Budding, Egress, and Maturation

About 10 h after infection, EToV particles are observed within parts of the unaltered Golgi apparatus, and extracellularly. At that time, tubular structures of variable length, diameter, and electron density appear in the cytoplasm, and also in the nucleus of infected cells, probably representing preformed nucleocapsids. It is unknown whether the accumulation of nucleocapsids in the nucleus reflects a nuclear phase in the replication of EToV or some sort of defective assembly. Viruses predominantly bud into the lumen of Golgi cisternae. The preformed nucleocapsid tubules approach the Golgi membrane with one of the rounded ends oriented toward the membrane and attach to it laterally. During budding, the nucleocapsid is apparently stabilized, leading to a higher electron density and a constant diameter (23 nm). Release into the intestinal lumen is probably through reverse pinocytosis. Virus maturation apparently occurs intracellularly during the egress process, where the virus nucleocapsid appearance changes from a straight rod (intracellular) into a torus shape (extracellular). The characteristic torus morphology of BToV is only observed in extracellular viral particles or in vacuoles near the cell surface, and never in the cytoplasm.

Geographic Distribution

In cattle, toroviruses have been detected by ELISA, reverse-transcriptase polymerase chain reaction (RT-PCR), and/or immune electron microscopy (IEM) in Austria, Belgium, Canada, Costa Rica, France, Germany, Great Britain, Hungary, Japan, Netherlands, New Zealand, South Africa, and the USA. BToV-seropositive cattle have also been reported in Belgium, France, Germany, India, Switzerland, United Kingdom, and the USA, with seroprevalence ranging between 55% and 94.6%. Most adult horses tested in Switzerland possess neutralizing antibodies to EToV. HToV appears to occur in Canada, Brazil, France, Great Britain, India, the Netherlands, and the USA.

Host Range

Neutralizing antibodies to EToV have been found in sera from horses, cattle, goats, sheep, pigs, rabbits, and feral mice, but not in humans or in carnivores. The host range of BToV appears to be restricted to cattle; however, seropositive reactions to BToV have also been detected in several ungulate species.

TVL particles have been observed in the feces of children and adults with diarrhea. Interestingly, these can be aggregated after the addition of anti-BToV calf sera, can be detected by using a BToV antigen capture ELISA, and the hemagglutination of rat erythrocytes by TVL particles can be inhibited by BToV antisera. These observations indicate antigenic cross-reactivity between HToV and BToV, and may point to a zoonotic connection. TLVs have also been seen in fecal samples of cats and dogs but neither serologic nor molecular identification has been obtained.

Genetic Relationships

There is limited information available on genetic variation amongst toroviruses. The complete genome sequence has been reported only for BToV. However, partial genome sequences have been obtained and used to chart the genetic diversity. In general, there is little divergence (20–40%) among known genotypes of EToV, BToV, PToV, and HToV in the (S, M, HE, and N) genes. Phylogenetic analyses have shown that all BToV strains are closely related, whether they are of European or American origin. Also, all PToV variants form a distinct genetic cluster. However, BToV and PToV sequences are sufficiently different to be assigned as different genotypes. HToVs show a high degree of similarity to New World BToVs (83%), and less with European strains of BToV (73%) and PToV (56%).

Antigenic Properties

In addition to the typical torovirus morphology, BToV, EToV, HToV, PToV, and Lyon-4 virus (Breda-like BToV detected in France) share common antigens. Currently, only one strain of EToV has been isolated, and all attempts to obtain a second equine isolate have been fruitless. Two strains of BToV have been reported in addition to the original isolate described by Gerald Woode and colleagues. One of the strains was detected in feces from a 5-month-old diarrheal calf in Ohio; the second (Iowa strain) was recovered from a 2-day-old experimental animal. On the basis of their cross-reactivity in ELISA, IEM, and hemagglutination/hemagglutination inhibition (HA/HI) assays using rat or mouse erythrocytes, the three isolates were assigned to two serotypes: BToV1, represented by the Iowa 1 isolate, and BToV2, comprising the Ohio and the second Iowa isolate.

Antigenic cross-reactivity has been demonstrated by ELISA, IEM, HI, and immunoblotting between BToV and HToV, indicating a close relationship. Several authors reported that HToV particles detected in humans with persistent diarrhea, and morphologically similar to BToV, could be agglutinated by BToV antiserum; stronger reactions are observed when BToV-2 antisera are used. Hyperimmune sera to BToV have also been used to detect toroviruses in humans by ELISA and IEM.

Epidemiology

Several epidemiological studies have demonstrated a high seroprevalence of BToV antibodies in several different groups of cattle, indicating that the virus may circulate with high frequency in these populations. BToV have also been detected by ELISA and/or RT-PCR in cases of gastroenteritis in cattle. Up to 44% of the BToV-positive samples from these cases did not contain other major enteric pathogens. Calves up to 4 months of age are highly susceptible to diarrhea induced by BToV, especially those below 3 weeks of age. The virus has also been recognized in 5–6-month-old beef calves arriving from sales barns. Intermittent BToV shedding can occur in young calves during the first 10 months of life. Older calves and adult animals can also shed BToV at different ages, perhaps by intermittent subclinical infections or by contracting new BToV infections. Levels of maternal BToV-specific antibodies circulating in the calf influence the clinical outcome of the infection; a seronegative neonatal calf is about seven times more likely to develop diarrhea than a seropositive calf.

Transmission and Tissue Tropism

It has been suggested that the transmission of BToV is via the oral/nasal route by direct contact with contaminated feces or nasopharyngeal secretions. Oral inoculation of calves with BToV has been shown to induce diarrhea with virus shed in feces under experimental conditions. The nasal route is another possible pathway for entry as BToV antigen and viral RNA have been detected in the nasal secretions of feedlot calves. Additionally, diarrhea has been induced after intranasal inoculation in Gn and CD calves. For bovine coronavirus (BCoV), respiratory tract infections have been reported to occur prior to enteric infections, indicating the possible importance of this route of transmission in the spread and pathogenesis of this distantly related group of enteric nidoviruses. It is possible that BToV, like BCoV, could initially replicate in nasal epithelial cells, and thereby amplify before being swallowed and infecting the intestinal tract. This hypothesis for BToV pathogenesis should be further studied.

BToV has a tissue tropism for enterocytes located from the lower half of the villi extending into the crypts, affecting the caudal portion of the small intestine (midjejunum through ileum) and the large intestine. Infection of other types of cells and organs by BToV has not been reported.

Pathogenesis and Clinical Features of Infection

All BToV strains are pathogenic, causing mild to profuse diarrhea in experimentally and naturally infected young calves. Twenty-four to seventy-two hours post exposure, the first clinical signs are observed (mild fever, depression, weakness, and anorexia), followed by a greenish-yellow to bright yellow watery diarrhea that lasts for 3–5 days. Calves may develop severe dehydration and die. Fecal virus shedding begins 24–72 h post infection, coinciding with the onset of diarrhea, and lasts for 2–6 days. Shedding peaks around 3–4 days post infection. Mixed infections with other enteric viruses, such as rotaviruses or astroviruses, result in a more severe watery diarrhea than is induced by either virus alone. In CD calves with a normal intestinal flora, diarrhea is generally more severe than in Gn calves. Sporadic and recurrent shedding of BToV can last for up to 4 months. BToV has also been detected in nasal samples. Further studies are needed to analyze BToV replication and shedding in the respiratory tract as well as its role in respiratory pathologies.

EToV seroconversion occurs in horses between 10 and 12 months of age, without the appearance of symptoms. Experimentally infected animals (intravenous route) have been reported to seroconvert without clinical signs. Oral infection experiments in horses have not been reported to date.

Several studies have shown an association of HToV infection with diarrhea in children. In one study, 35% of children with enteritis shed HToVs in their feces, but only 14.5% of the asymptomatic controls shed the virus (statistically significant difference – odds ratio 3.1). Affected children showed watery diarrhea, vomiting three to four times daily, and dehydration as a consequence; neither fever, nor the presence of other enteric pathogens, was recorded. Recently, fecal excretion of HToV has been associated with nosocomial infections in infants with necrotizing enterocolitis (NEC). Immunocompromised children appear to be highly susceptible to disease following HToV infection.

Pathology and Histopathology

The target organs of BToV in calves are the lower half or two-thirds of the small intestine and the entire large intestine, particularly the spiral colon. There is little macroscopic evidence of the infection. Histological examination shows villous atrophy and epithelial desquamation in randomly scattered areas from the mid-jejunum to the lower small intestine, as well as areas of necrosis in the large intestine. As shown by IF, both crypt and villus epithelial cells contain antigen. The watery diarrhea is probably a result of the loss of reabsorptive capacity of the colonic mucosa, combined with malabsorption in the small intestine. The germinal centers of Peyer's patches are depleted of lymphocytes and may occasionally show fresh hemorrhage. The dome epithelial cells, including the M-cells, display the same cytopathic changes as seen in the absorptive cells of the villi. Virions are found in cells of both the small and large intestine, and between enterocytes at the basal and lateral plasma membranes. In macrophages of the lamina propria, virions in various stages of degradation are found.

Antigen is detected as early as 48 h after infection in epithelial cells of the lower half of the villi and crypts of the affected areas, as well as in dome epithelium. Fluorescence is cytoplasmic (although a few nuclei may be faintly stained) and generally most pronounced in the intestines with the least tissue damage. The mid-jejunum is infected first and the infection eventually reaches the large intestine. Diagnosis by IF should be performed preferentially on sections of the large intestine from calves killed after the onset of diarrhea (i.e., several days after infection of the epithelium).

Immune Response

Up to the age of 4 months, all calves in a sentinel experiment regularly excreted BToV in their feces. They showed early serum IgM responses despite the presence of IgG1 isotype maternal antibodies, but no IgA seroconversion. Antibody titers then decreased below detection; persistent IgG1 titers developed in only a few animals. After introduction into the dairy herd at 10 months of age, all calves developed diarrhea and shed virus. Seroconversion for all antibody isotypes was observed at this stage, indicating lack of mucosal memory. In contrast, coronavirus infection in the presence of maternal antibodies leads to isotype switch and a memory response.

Prevention and Control

There are no specific preventive measures for this virus; however, general hygiene, biosecurity practices, and the intake of adequate, protective amounts of colostrum can be used to prevent BToV infections. There are no reports on the effects of disinfection or heat sterilization on toroviruses.

Acknowledgments

The authors would like to thank R. de Groot, Utrecht, The Netherlands, for the use of unpublished information from his torovirus review article in press.

See also: Coronaviruses: General Features; Coronaviruses: Molecular Biology.

Further Reading

Cavanagh D and Horzinek MC (1993) Genus *Torovirus* assigned to the *Coronaviridae*. *Archives of Virology* 128: 395.
De Groot RJ (in press) Molecular biology and evolution of toroviruses.
Draker R, Roper RL, Petric M, and Tellier R (2006) The complete sequence of the bovine torovirus genome. *Virus Research* 115: 56–68.
Hoet AE (in press) Torovirus pathogenesis and immune responses. In: Perlman S, Snijder EJ, and Gallagher TM (eds.) *The Nidoviruses*. Washington, DC: ASM Press.

Hoet AE, Cho KO, Chang KO, *et al.* (2002) Enteric and nasal shedding of bovine torovirus (Breda virus) in feedlot cattle. *American Journal of Veterinary Research* 63: 342–348.

Hoet AE and Saif LJ (2004) Bovine torovirus (Breda virus) revisited. *Animal Health Research Reviews* 5: 157–171.

Hoet AE, Smiley J, Thomas C, *et al.* (2003) Association of enteric shedding of bovine torovirus (Breda virus) and other enteropathogens with diarrhea in veal calves. *American Journal of Veterinary Research* 64: 485–490.

Horzinek MC (1999) Molecular evolution of corona- and toroviruses. *Advances in Experimental Medicine and Biology* 473: 61–72.

Koopmans M and Horzinek MC (1994) Toroviruses of animals and humans: A review. *Advance Virus Research* 43: 233.

Lodha A, de Silva N, Petric M, and Moore AM (2005) Human torovirus: A new virus associated with neonatal necrotizing enterocolitis. *Acta Paediatrica* 94: 1085–1088.

Smits SL, Lavazza A, Matiz K, *et al.* (2003) Phylogenetic and evolutionary relationships among torovirus field variants: Evidence for multiple intertypic recombination events. *Journal of Virology* 77: 9567–9577.

Snijder E and Horzinek MC (1995) The molecular biology of toroviruses. In: Siddell S and ter Meulen V (eds.) *Coronaviruses*, pp. 219–238. New York: Plenum Press.

Vries AAF, de Horzinek MC, Rottier PJM, *et al.* (1997) The genome organization of the *Nidovirales*: Similarities and differences between arteri-, toro-, and coronaviruses. *Seminars in Virology* 8: 33.

Tospovirus

M Tsompana and J W Moyer, North Carolina State University, Raleigh, NC, USA

© 2008 Published by Elsevier Ltd.

Glossary

Ambisense genome Viral RNA genome with open reading frames in both the viral- and viral complementary (vc) sense on the same genome segment.

Envelope Membrane-like structure that packages genome segments.

IGR The intergenic region is the untranslated, A-U rich region found between the two open reading frames on the S and M RNA segments.

Negative sense genome Viral RNA genome that codes for proteins in the vc sense. Transcription of vc mRNA is required for translation of viral proteins.

Nucleocapsid Viral RNA encapsidated in the nucleoprotein.

Protoplast Plant cell lacking its cell wall.

RNP Ribonucleoprotein complex consisting of the viral RNA genome segment, nucleoprotein, and a small number of polymerase molecules.

Virion Quasispherical structure containing the viral genome and bounded by a membrane-like envelope.

History

Diseases now known to be caused by tomato spotted wilt virus (TSWV) were first reported in 1915 and were shown to be of viral etiology by 1930. This taxon of plant viruses was categorized as a monotypic virus group consisting of a single virus (TSWV) until the report of impatiens necrotic spot virus (INSV) in 1991. Thus, most of the characteristics which define the genus *Tospovirus* were obtained through investigation of TSWV even after the discovery of additional viruses in the genus (**Table 1**). Biological investigations beginning in the 1940s revealed a virus that had an unusually large host range and occurred in nature as a complex mixture of phenotypic isolates. However, it was one of the least stable viruses and most difficult plant viruses to mechanically transmit. Although the enveloped virions were observed in the 1960s, molecular characterization and elucidation of the genome organization were not completed until the early 1990s. The virus was shown to be vectored by thrips in the 1930s and later transmitted in a persistent manner. Thrips were demonstrated to be a host for replication of the virus and that replication was required for transmission in the early 1990s. Later it was recognized that limited, localized replication may occur in thrips that does not result in the thrips becoming viruliferous. Advances in gene function and cellular biology have been limited due to the absence of a robust *in vitro* plant or thrips cell culture system, and lack of an efficient reverse genetics system. However, limited progress has been made utilizing gene expression systems and classical viral genetics.

Taxonomy and Classification

Tospoviruses constitute the only genus of plant-infecting viruses in the family *Bunyaviridae*; however, these viruses share many molecular characteristics typical of other members of this virus family. They have an enveloped virion containing the viral genome which is distributed among three RNA segments that replicate in a manner consistent with that of other negative strand viruses. All three segments have highly conserved, complementary

Table 1 List of *Tospovirus* species[a,b,c]

Tospovirus species	Abbreviation
Groundnut bud necrosis virus (Peanut bud necrosis virus)	GBNV
Groundnut ringspot virus	GRSV
Groundnut yellow spot virus (Peanut yellow spot virus)	GYSV
Impatiens necrotic spot virus	INSV
Tomato chlorotic spot virus	TCSV
Tomato spotted wilt virus	TSWV
Watermelon silver mottle virus	WSMoV
Zucchini lethal chlorosis virus	ZLCV
Tentative *Tospovirus* species	
Capsicum chlorosis virus (Gloxinia tospovirus) (Thailand tomato tospovirus)	CACV
Chrysanthemum stem necrosis virus	CSNV
Iris yellow spot virus	IYSV
Groundnut chlorotic fan-spot virus	GCFSV
Physalis severe mottle virus	PhySMV
Watermelon bud necrosis virus	WBNV

[a]http://www.ncbi.nlm.nih.gov/ICTVdb.
[b]Whitfield AE, Ullman DE, and German TL (2005) Tospovirus–thrips interactions. *Annual Review of Phytopathology* 43: 459–489.
[c]Synonyms are indicated inside parentheses.

termini resulting in a pan-handle structure and genes with functions similar to those of viruses in other genera are located in similar locations on the genome. However, the genome organization is distinct from the other genera. The small (S) and middle (M) segments each encode two genes in opposite or ambisense polarity.

Classification of a *Tospovirus* population as a distinct species (virus) is based upon the similarity of sequence between the nucleocapsid genes of the respective viruses. This is in contrast to the system used to differentiate viruses in other genera which traditionally relied on serological neutralization of infectivity or other biological properties (hemagluttination) mediated by the glycoproteins. Tospovirus isolates with greater than 90% nucleotide similarity in the nucleocapsid gene are classified as isolates of the same species (virus). Serologically related isolates with 80–90% sequence identity are subjectively classified as strains or as distinct species depending on other criteria. Isolates with less than 80% identity are classified as distinct species.

Geographic Distribution

TSWV, the type member of the tospoviruses is found worldwide in temperate regions in association with its thrips vector. The wide host-range of TSWV and its thrips vector is consistent with the geographic distribution. Other tospoviruses have more well-defined distribution. For example, GBNV, WBNV, and WSMoV, that are transmitted by *Thrips palmi*, a thrips species found only in the subtropics are only known to occur in Southeast Asia. Another anomaly is INSV. While INSV is reported to occur around the world, it is almost entirely limited to greenhouse-grown floral crops.

Host-Range and Virus Propagation

TSWV has one of the most diverse host-ranges of any plant-infecting virus. The virus infects over 925 plant species belonging to 70 botanical families, both monocots and dicots. In addition, TSWV infects approximately ten thrips species. Important economic plants susceptible to TSWV include tomato, potato, tobacco, peanut, pepper, lettuce, papaya, and chrysanthemum. Other tospoviruses (e.g., IYSV) have much narrower host-ranges and thus the broad host-range of TSWV is not characteristic of the genus. These viruses can be transmitted mechanically or by their thrips vector, but are not transmitted transovarially, by plant seeds or pollen. Purified RNA preparations are not infectious. There are no robust plant or insect culture systems for tospoviruses. However, plant and insect protoplasts have been successfully inoculated.

Virion Properties

Tospovirus virions are quasispherical, enveloped particles 80–120 nm in diameter (**Figure 1**). Two viral coded glycoproteins, G_N and G_C, are embedded in the viral envelope and form surface projections 5–10 nm long. Ribonucleoprotein (RNP) particles consisting of the viral RNA encapsidated in the nucleoprotein (nucleocapsid), and a small number of polymerase molecules are contained within the envelope. Nucleocapsids are pseudocircular due to noncovalent bonding of the complementary RNA termini. Intact virions as well as carefully prepared RNPs retrieved from sucrose or $CsSO_4$ gradients are infectious. There are several reports that TSWV and INSV isolates, while infectious, are defective for virion formation.

Genome Properties

Tospoviruses have a single-stranded, tripartite RNA genome with segments designated as L, M, and S in order of decreasing size (**Figure 2**). The termini of each of the RNA segments consist of an eight nucleotide sequence (5′ AGAGCAAU 3′) that is strictly conserved among all tospoviruses. The remaining untranslated region at the termini also has a high degree of complementarity. Base pairing at the termini between the inverted complementary sequences supports a pan-handle structure that most likely serves as a promoter for replication.

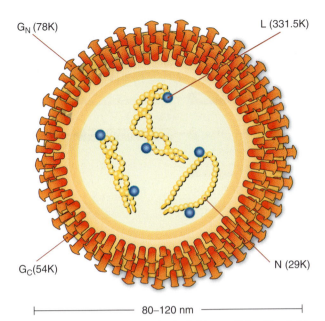

Figure 1 Tospovirus quasispherical virion particles. The S, M, and L RNA genomic segments are encapsidated by the nucleoprotein, are in association with L protein molecules, and form pan-handle structures due to the complementarity of their 5′ and 3′ ends. The glycoproteins G_N and G_C are embedded within the viral envelope.

The L RNA is 8.9 kbp and codes for the L or RdRp protein in the viral complementary (vc) sense (**Figure 2**). The M and S RNAs are in ambisense orientation. The M RNA is 4.8 kbp and codes in the viral sense for the nonstructural protein NSm and for the G_N/G_C precursor glycoprotein in the vc sense. The S RNA is 2.9 kbp and codes in the viral sense for the nonstructural protein NSs and the nucleocapsid protein in the vc sense (**Figure 2**).

TSWV M and S RNA IGRs have variable lengths, are A–U rich, and are the most hypervariable regions of the genome. The 5′ and 3′ ends of the IGRs are conserved, separated by variable sequences, deletions, and insertions. In addition, highly conserved sequences are embedded within the S RNA IGR. A 33 nucleotide (nt) duplication occurring in the S RNA IGR of some isolates has been correlated with loss of competitiveness in mixed infections of isolates with and without the duplication. A 31 nt conserved sequence, with significantly higher GC-content compared to the remaining S RNA IGR, has also been found in some TSWV isolates. The IGRs of the M and S segments have high inclination for base pairing thought to be involved in initiation and termination of transcription. There is speculation that the termination of transcription is dictated by a conserved nucleotide sequence (CAACUUUGG) in the center of the S and M RNA IGR or that it is due to a secondary structure highly stabilized in the 31 nt region referred to above.

Full length molecules of the M and S RNAs are found in infected tissue and purified virions in both the viral and vc sense (approximate ratio of 10:1), consistent with ambisense segments from other viruses. Defective interfering RNAs (DIs) associated with attenuated symptom expression and increased replication rate are also frequently observed. DIs from the L ORF in TSWV infected tissue are the result of and associated with attenuated infectivity. Deletions, frameshift, and nonsense mutations in the G_N/G_C ORF have been shown to interfere with thrips transmissibility and virion assembly. Recently, frameshift and nonsense mutations with unknown effect have also been identified in the N ORF. The formation of DIs is favored by repeated mechanical passage in certain plant hosts, high inoculum concentration, and low temperatures. Available evidence supports the hypothesis that secondary structure rather than sequence is the primary determinant of the site of deletion. There is also a high frequency of DIs that maintain the original reading frame resulting in translation of truncated proteins whose existence was confirmed in nucleocapsid preparations.

Protein Properties

The 331.5 kDa L protein encoded by the L RNA has been identified as the putative RdRp, through sequence homology with other members of the *Bunyaviridae* and identification of sequence motifs characteristic of polymerases. RdRp activity has been associated with detergent-disrupted TSWV virion preparations. The 33.6 kDa NSm protein encoded by the M RNA has been shown to induce tubule structures in plant protoplasts and insect (*Spodoptera* and *Trichoplusia*) cells. Induction of tubules in plants, ability to change the size exclusion limit of plasmodesmata, an early expression profile and complementation of cell-to-cell and systemic movement in a movement-defective tobacco mosaic virus vector is evidence that NSm is the TSWV movement protein and that it supports long-distance movement of viral RNAs. In thrips, NSm does not aggregate into tubules, indicating that this protein might not have any function in the vector's life cycle. It is also known that NSm specifically interacts with the N protein, the At-4/1 intra- and intercellular trafficking plant protein and binds single-stranded RNA in a sequence-nonspecific manner. An NSm homolog is absent in the animal infecting *Bunyaviridae*. The 127.4 kDa G_N/G_C precursor glycoprotein also coded by the M RNA contains a signal sequence that allows its translation on the endoplasmic reticulum. Proteolytic cleavage of the polyprotein does not require other viral proteins. The M_rs of G_N and G_C is 78 kDa and 54 kDa respectively. Evidence for the involvement of the glycoproteins in thrips transmission is provided by: (1) their interaction with proteins of the thrips vector, (2) their association with the insect midgut during acquisition, (3) the loss of thrips transmissibility of envelope-deficient mutants, (4) the presence of a

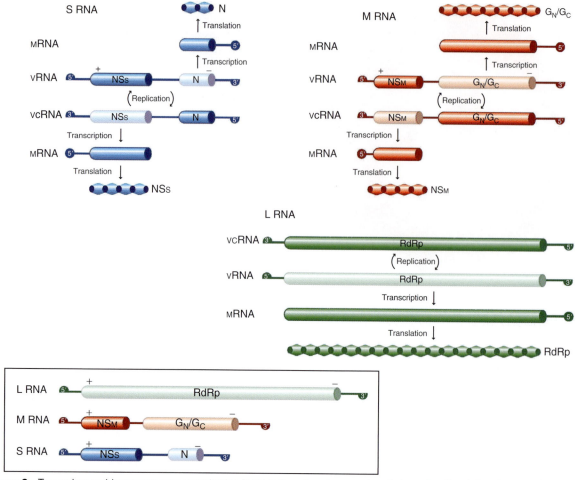

Figure 2 Tospovirus ambisense genome organization (inset in figure) and expression strategy. Positive (+) and negative (−) sense ORFs are dark and light shaded tubes respectively. Proteins from the S and M segments are translated from subgenomic mRNAs which are capped with 10–20 nt of non-viral orgain at the 5' end.

glycoprotein sequence motif that is characteristic for cellular attachment domains and (5) the observation that only reassortants with the M RNA of a thrips-transmissible isolate rescue thrips transmissibility. Specifically, G_N is involved in virus binding and/or entry in thrips midgut cells, whereas G_C is a possible fusion protein playing a significant role in pH-dependent virus entry. It is also believed that these proteins are implicated in virion assembly. The NSs protein encoded on the S RNA is 52.4 kDa and accumulates to high levels as loose aggregates or paracrystalline arrays of filaments. NSs has RNA silencing suppressor activity, affects symptom expression in TSWV-infected plants, and is not present in the mature virus particle. The N protein, also encoded by the S RNA, ranges in size from 29 to 3 kDa depending on the virus. This protein encapsidates the viral RNA segments, is highly abundant, and is the predominant protein detected in serological assays. A 'head-to-tail' interaction of the nucleoprotein N terminus (aa 1–39) with the C terminus (aa 233–248) results in multimerization.

Replication

Replication of viral RNA and assembly of virions occurs in the cytoplasm of both plant and insect cells. Tospovirus replication, however, has mainly been described based on plant infection. Upon entry into the plant cell, the virus loses its membrane and releases infectious nucleocapsids into the cytoplasm. In thrips cells, infection by tospoviruses is accommodated by binding of the viral surface glycoproteins to a host cell receptor(s) (possibly a 50 kDa and/or a 94 kDa protein). This is followed by release of infectious nucleocapsids into the cytoplasm, through fusion between the viral and thrips membranes possibly initiated by low pH. Depending on the concentration of N protein, the viral RNA is either transcribed or replicated. At low N concentrations, the polymerase transcribes mRNAs that are translated into the virus proteins. Translation of proteins from the S and M ambisense RNAs occurs from subgenomic mRNAs (**Figure 2**). The S and M subgenomic mRNAs are capped at the 5′ terminus with 10–20 nucleotides of nonviral origin indicating that tospoviruses utilize a cap-snatching mechanism to regulate transcription. Leader sequences of alfalfa mosaic virus have also been detected as caps of TSWV mRNA in mixed infections of the two viruses. The TSWV transcriptase has a reported preference for caps with multiple base complementarity with the viral template. Upon increase of N protein concentration, the polymerase switches its mode to replication with the viral RNA serving as the template. Replicated viral RNAs form RNPs that can presumably associate with the NSm protein for movement through plasmodesmata to adjacent plant cells through tubular structures. Alternatively, RNPs form new virions by associating with the glycoproteins and budding through the Golgi membranes. Virions are initially double-membraned, but soon coalesce and form groups of virions with a single membrane surrounded by another membrane.

Pathogenicity and Cytopathology

Tospoviruses are noted for the severity of the diseases they cause in plants. Symptoms are highly variable, depending on the virus, the virus isolate, the host plant, time of the year, and environment, and are thus of little diagnostic value. Chlorosis, necrosis, ring or line patterns, mottling, silvering, and stunting often appear on inoculated and systemically infected leaves. Systemic invasion of plants is frequently nonuniform. Stems and petioles may exhibit necrotic lesions. Observed symptoms often mimic disease and injury caused by other biotic and abiotic stresses. Infection of younger plants results in severe stunting and high mortality rates. TSWV has been shown to affect more severely *Datura*, *Nicotiana*, and *Physalis* plants under a specific temperature regime (daytime, $29 \pm 2\,°C$, nighttime, $24 \pm 3\,°C$). The effect of tospovirus infection on thrips has been controversial, due to the confounded effects of plant host, virus, and environment on the insect vector and the genetic variability of thrips and virus populations. TSWV infection of *F. occidentalis* provided evidence that thrips exhibit an immune response to the virus. Recent work with TSWV-infected *F. fusca* reared on infected foliage indicated a direct effect of the virus on thrips resulting in reduced fitness. The same study showed that the plant infection status and the TSWV isolate have also an effect on the insect, explaining the variable results obtained from independent studies of virus pathogenicity on the insect vector.

Tospoviruses induce characteristic cytopathic structures that are host and virus-isolate dependent. In addition to virions, inclusions of viroplasms consisting of the NSs or N protein may be abundant in the cytoplasm. NSs may aggregate in loose bundles (e.g., TSWV) or in highly ordered paracrystalline arrays (e.g., INSV). Excess N protein occurs in granular electron dense masses. NSs and nucleocapsid protein inclusions have been observed in infected plant and insect cells. NSm protein induces tubule structures in plant protoplasts and insect cells.

Transmission and Epidemiology

Tospoviruses are transmitted from plant to plant by at least ten thrips species in the genera *Frankliniella*, *Scirtothrips*, and *Thrips*. Among the more common vectors are *Frankliniella occidentalis*, *F. fusca*, *F. schultzei*, *F. intonsa*, *F. bispinosa*, *Thrips palmi*, *T. setosus*, and *T. tabaci*. Thrips feed on the cytoplasm of plant cells. The contents of infected cells are ingested and the virus is transported along the lumen of the digestive tract to the midgut, the primary binding and entry site into the insect cells. The brush border of the midgut lumen is the first membrane barrier that the virus encounters. The virus replicates in the midgut and crosses the basement membrane into the visceral muscle cells. The virus subsequently enters the primary salivary glands. It has been hypothesized that the virus moves from the midgut to the salivary glands through infection of ligament-like structures, or when there is direct contact between membranes of the visceral muscles and the primary salivary glands during the larval stages of development. A less plausible hypothesis is that the virus infects the salivary glands after entry and circulation in the hemocoel. Viral inoculum is introduced into plants in the insect saliva coincident with feeding on the plant by adult thrips.

The process of successful acquisition occurs only by larvae and acquisition rates decrease as larvae develop, affecting adult vector competency. Vector competency is also determined by the thrips' feeding preference on a particular host, the uniformity of distribution of virus in plant cells, the rate of virus replication in the midgut, and the extent of virus migration from the midgut to the visceral muscle cells and the salivary glands. In some instances the virus can be acquired by adult thrips and infects midgut cells, but is unable to spread further possibly due to the formation of an age-dependent midgut barrier (e.g., basal lamina). Research has shown the existence of thrips transmitters with detectable levels of virus, nontransmitters with detectable virus, and nontransmitters with no detectable virus, supporting multiple sites for vector specificity between tospoviruses and thrips. Evidence for replication of the virus in the insect vector is based on the accumulation of NSs and the visualization of viral inclusions in midgut epithelial cells, muscle cells, and the salivary glands. Although the virus is maintained transtadially throughout the life of the insect, there is no evidence for transovarial transmission. Thus, each generation of thrips must acquire the virus during the larval stages.

The primary dispersal of tospoviruses is by adult thrips and dissemination of infected somatic tissue in vegetatively propagated crops. These viruses are thought to move long distances in thrips carried by wind currents. They may also survive in commercial agricultural systems in weeds that serve as a bridge between crops. Infected summer weeds (e.g., in NC *I. purpurea, I. hederacea, M. verticillata, A. palmeri, C. obtusifolia, R. scabra, Ambrosia artemisiifolia* L., *Polygonum pensylvanicum* L., *and Chenopodium album* L.) are the principal source for spread of TSWV to winter annual weeds, from which the virus is spread to susceptible crops in spring. Secondary spread within a crop can only occur in crops that concomitantly support virus infection and reproduction of the vector as only the larval stage can acquire the virus for transmission. Transmission through plant seed and pollen has not been conclusively demonstrated. The emergence of these viruses as serious pathogens in crops has been attributed to the increased prevalence of *F. occidentalis* as an agricultural pest on a worldwide basis.

Genetics and Evolution

The knowledge base for genetics and evolution of tospoviruses has been derived almost exclusively from TSWV. TSWV has a characteristic ability to adapt to new or resistant hosts and to lose phenotypic characters following repeated passage in experimental hosts, especially *Nicotiana benthamiana*. The virus occurs in plants as a heterogeneous mutant population with one or two predominant haplotypes and 9–21 rare haplotypes. Recent research shows that natural TSWV variants evolve in nature through recombination, random genetic drift, and mutation. Intergenomic recombination is important for the genesis and evolution of ancestral TSWV lineages. Genetic drift during thrips transmission and mutation concurrent with virus population growth, shape the genetic architecture of the most recently evolved lineages. The existence of single viral strains as mutant populations and recombination in ancestral viral lineages arm TSWV with a unique genetic reservoir for causing disease and spreading in epidemic proportions in nature. Additional research at the species level supports a distinct TSWV geographical structure and the occurrence of species-wide population expansions. TSWV is also known to use reassortment of genome segments to adapt to resistant hosts under specific laboratory conditions. The determinants of adaptation to resistance in tomato and pepper have been mapped to the M and S RNAs respectively. Little is known about the thrips–tospovirus coevolution and the genetic diversity of the thrips vector itself. The altered status of *Thrips tabaci* as a TSWV vector is one of the very few likely examples of coevolution between tospoviruses and their insect vector.

Detection and Diagnosis

Tospoviruses have certain unique biological properties that are useful for diagnosis. These viruses can be mechanically transmitted by gently rubbing inoculum on plants dusted with carborundum. *Nicotiana glutinosa* L., *Chenopodium quinoa* Wild., and garden petunia give characteristic lesions that progress as spots or concentric zones, and sometimes as lethal necrosis. Tospoviruses can also be identified by electron microscopy of leaf-dip preparations on thin sections of infected plants. Additional techniques for identification are based on the enzyme-linked immunosorbent assay (ELISA) using polyclonal and monoclonal antibodies, and detection of viral-specific nucleic acids using ribo- and cDNA-probes. The reverse transcription-polymerase chain reaction (RT-PCR) is the most powerful and commonly used technique for detecting small amounts of tospovirus RNA. Real time RT-PCR has been successfully used to detect and quantify TSWV in leaf soak and total RNA extracts from infected plants and thrips. RT-PCR with degenerate primers can detect five distinct tospovirus species. Tissue selection and sampling strategy are critical factors in TSWV diagnosis and detection regardless of the technique. Because, TSWV titer varies throughout the plant and does not spread uniformly throughout plants that are 'systemic' hosts, sampling strategies should be validated in each situation.

Prevention and Control

Tospoviruses cause significant economic losses annually, due to suppressed growth, yield, and reduced quality. These viruses can be partially managed in well-defined cropping systems such as glasshouses by obtaining uninfected plant propagules, implementing a preventative thrips control program in high risk areas, together with constant monitoring of production areas for thrips and infected plants. However, these strategies are costly and require intensive management. Control in field crops is problematic due to the array of external sources of inoculum. Vector control is generally ineffective against the introduction of virus from external sources, due to thrips' high fecundity, ability to develop insecticide resistance, and to infest many TSWV-susceptible crops. Some measure of control can be achieved using thrips-proof mesh tunnels in the field and reflective mulches. Cultural practices such as utilization of virus-tested planting stock, careful selection of planting dates, removal of cull piles and weeds, rotation with nonsusceptible crops, prevention of planting TSWV-susceptible crops adjacent to each other, reduced in-field cultivation to avoid movement of thrips from infected sources, can reduce the spread of tospoviruses. In peanuts, higher plant density, planting from early until late May and application of selected insecticides have reduced the incidence of TSWV. In flue-cured tobacco, early-season treatment with activators of plant defenses and insecticides have also significantly reduced TSWV incidence.

Deployment of resistant cultivars has provided benefits in only three of the crops infected by TSWV. Although little is known about the benefits of host resistance against most of the tospoviruses, TSWV defeated nearly every resistance gene deployed against it in many crops. Single-gene resistance is available for TSWV in a limited number of tomato (*Sw-5*) and pepper (*tsw*) cultivars. Naturally occurring, resistance-breaking isolates of TSWV have been recovered from pepper and tomato cultivars containing their respective resistance genes. A co-dominant cleaved amplified polymorphic sequence (CAPS) marker has been developed for TSWV marker-assisted selection in pepper. 'Field' resistance has been reported for some peanut varieties. Progress has been made in understanding the genetic basis of the ability of TSWV to overcome single gene resistance by mapping determinants to specific segments of the TSWV genome and characterizing the selection process. Pathogen-derived resistance utilizing the N and NSm genes has been effective in some greenhouse and field tests; however, isolates have been obtained that overcome nucleocapsid mediated resistance. Best suppression of TSWV epidemics has been achieved with the integrated use of moderately resistant cultivars, chemical, and cultural practices. The impact of these viruses on agricultural production is in large part due to the absence of durable forms of resistance in the affected crops or other highly effective control measures.

Future Perspectives

The last decade has been characterized by exciting developments in our understanding of tospovirus molecular biology, evolution, and virus–host relationships. Further progress in understanding replication and gene function requires the development of efficient reverse genetics, plant and insect culture systems for tospoviruses. In addition, effective management of these viruses will depend on a deeper understanding of thrips' genetic diversity, virus–thrips coevolution and the changes in viral and thrips population dynamics upon exertion of specific selective forces. Such understanding can be acquired only through integrated research at the interface of virology, entomology, and ecology.

See also: Tomato Spotted Wilt Virus.

Further Reading

Adkins S (2000) Tomato spotted wilt virus-positive steps towards negative success. *Molecular Plant Pathology* 1: 151–157.

Best RJ (1968) Tomato spotted wilt virus. In: Smith KM and Lauffer MA (eds.) *Advances in Virus Research*, pp. 65–145. New York: Academic Press.

de Avila AC, de Haan P, Kormelink R, *et al.* (1993) Classification of tospoviruses based on phylogeny of nucleoprotein gene sequences. *Journal of General Virology* 74: 153–159.

German TL, Ullman DE, and Moyer JW (1992) Tospoviruses: Diagnosis, molecular biology, phylogeny, and vector relationships. *Annual Review of Phytopathology* 30: 315–348.

Goldbach R and Peters D (1994) Possible causes of the emergence of tospovirus diseases. *Seminars in Virology* 5: 113–120.

Prins M and Goldbach R (1998) The emerging problem of tospovirus infection and nonconventional methods of control. *Trends in Microbiology* 6: 31–35.

Sin SH, McNulty BC, Kennedy GG, and Moyer JW (2005) Viral genetic determinants for thrips transmission of tomato spotted wilt virus. *Proceedings of the National Academy of Sciences, USA* 102: 5168–5173.

Tsompana M, Abad J, Purugganan M, and Moyer JW (2005) The molecular population genetics of the tomato spotted wilt virus (TSWV) genome. *Molecular Ecology* 14: 53–66.

Ullman DE, Meideros R, Campbell LR, Whitfield AE, and Sherwood JL (2002) Thrips as vectors of tospoviruses. In: Plumb R (ed.) *Advances in Botanical Research*, pp. 113–140. London: Elsevier.

Ullman DE, Sherwood JL, and German TL (1997) Thrips as vectors of plant pathogens. In: Lewis TL (ed.) *Thrips as Crop Pests*, pp. 539–565. London: CAB International.

Whitfield AE, Ullman DE, and German TL (2005) Tospovirus–thrips interactions. *Annual Review of Phytopathology* 43: 459–489.

Totiviruses

S A Ghabrial, University of Kentucky, Lexington, KY, USA

© 2008 Elsevier Ltd. All rights reserved.

Glossary

Mycoviruses Viruses that infect and multiply in fungi.

Pseudoknot A secondary structure in viral mRNA that slows movement of the ribosome and may cause a frameshift that allows entry to an alternative reading frame during translation.

Ribosomal frameshifting Ribosomes switching reading frame on an mRNA, in response to the presence of a slippery site and/or a pseudoknot, to synthesize a protein or a polyprotein from two overlapping reading frames.

Introduction

The discovery of the killer phenomenon in the 1960s in yeast (*Saccharomyces cerevisiae*) and in the smut fungus (*Ustilago maydis*) eventually led to the discovery of the isometric double-stranded (ds) RNA mycoviruses with nonsegmented genomes, currently classified in the genus *Totivirus* (family *Totiviridae*). Killer strains of yeast or smut secrete a protein toxin to which they are immune, but which is lethal to sensitive cells. The precursor to the killer toxin is encoded by a satellite dsRNA, which is dependent on a helper virus with nonsegmented dsRNA genome (totivirus) for encapsidation and replication. Unlike the helper totiviruses associated with the yeast and smut killer systems, other members of the family *Totiviridae*, including the viruses that infect filamentous fungi (members of the newly proposed genus *Victorivirus*) and those that infect parasitic protozoa (members of the genera *Giardiavirus* and *Leishmaniavirus*), are not known to be associated with killer phenotypes. Purified preparations of some of these viruses, however, can contain dsRNA species suspected of being satellite or defective dsRNAs.

The isometric dsRNA totiviruses that infect fungi and protozoa are unique among dsRNA viruses in that their genomes are undivided, whereas the genomes of all other dsRNA viruses are segmented. The yeast and smut totiviruses and associated killer systems, as well as the totiviruses infecting parasitic protozoa, are discussed elsewhere in this encyclopedia. In addition to examining the similarities and differences among members of the family *Totiviridae*, this article will focus on the totiviruses that infect filamentous fungi, a group of viruses that are phylogenetically more closely related to each other than to other totiviruses and utilize a different strategy to express their genomes.

Taxonomy and Classification

The family *Totiviridae* encompasses a broad range of viruses characterized by isometric particles, ~40 nm in diameter, that contain a monosegmented dsRNA genome coding for a capsid protein (CP) and an RNA-dependent RNA polymerase (RdRp). At present, three genera are recognized: *Totivirus*, *Giardiavirus*, and *Leishmaniavirus*.

Viruses currently placed in the genus *Totivirus* infect yeast, smut fungi, or filamentous fungi, whereas those in the latter two genera infect parasitic protozoa. Two distinct RdRp expression strategies have been reported for species in the family *Totiviridae*: those that express RdRp as a fusion protein (CP–RdRp) by ribosomal frameshifting, such as saccharomyces cerevisiae virus L-A and the viruses that infect parasitic protozoa; and those that synthesize RdRp as a separate nonfused protein by an internal initiation mechanism, as shown for Hv190SV and proposed for all of the other totiviruses that infect filamentous fungi. Although *Helminthosporium victoriae 190S virus* (Hv190SV) is the only species recognized by the International Committee on Taxonomy of Viruses (ICTV) as a member of the genus *Totivirus*, the complete nucleotide sequences of several tentative totiviruses that infect filamentous fungi have recently been reported (**Table 1**). Sequence and phylogenetic analyses have demonstrated that these viruses have many properties in common with Hv190SV and that they are more closely related to each other than to the viruses infecting the yeast and smut fungi. A new genus (genus *Victorivirus*) was recently proposed to accommodate Hv190SV and the related viruses infecting filamentous fungi, with Hv190SV as the type species. The name 'Victorivirus' is derived from the specific epithet of *H. victoriae*, the host of the proposed type species.

In addition to fungal and protozoal hosts, totiviruses may also have crustacean hosts. A nonsegmented dsRNA virus with isometric particles, designated penaeid shrimp infectious myonecrosis virus (IMNV), was recently isolated from diseased penaeid shrimp and tentatively assigned to the family *Totiviridae*. IMNV is the causal agent of the shrimp myonecrosis disease characterized by necrosis of skeletal muscle, particularly in the distal abdominal segments and tail fan. Phylogenetic analysis based on the RdRp region of viruses in the totivirus family suggests that giardia lamblia virus (GLV; genus *Giardiavirus*) is the closest relative to IMNV. Interestingly, both IMNV and GLV, unlike all other members of the family *Totiviridae*, are infectious as purified virions. IMNV is presently unclassified and it has yet to be determined whether it is a member of a novel genus in the family *Totiviridae* or whether IMNV is a member of a novel family of dsRNA viruses that infect invertebrate hosts. The complete nucleotide sequences of several members and tentative members of the family *Totiviridae* have been published and the GenBank accession numbers are listed in **Table 1**.

Virion Properties

The buoyant densities in CsCl of virions of members of the totivirus family range from 1.36 to 1.43 g cm^{-3}, and the sedimentation coefficients of these virions range from

Table 1 List of virus members and tentative members in the family *Totiviridae*, length of their genomes, size of encoded gene products and GenBank accession numbers

Virus[a]	Abbreviation	Genome length (nt)	Size (kDa) CP	RdRp/CP–RdRp[b]	GenBank accession no.
Genus: *Totivirus*					
Saccharomyces cerevisiae virus L-A	ScV-L-A	4579	76	171[b]	NC_003745
Saccharomyces cerevisiae virus L-BC	ScV-L-BC	4615	78	176[b]	NC_001641
Ustilago maydis virus H1	UmV-H1	6099	81	201[b]	NC_003823
Genus: *Victorivirus*					
Helminthosporium victoriae 190S virus	Hv190SV	5179	81	91	NC_003607
Chalara elegans RNA virus 1	CeRV1	5310	81	96	NC_005883
Coniothyrium minitans RNA virus	CmRV	4975	81	93	NC_007523
Epichloe festucae virus 1	EfV-1	5109	80	90	AM261427
Gremmeniella abietina RNA virus L	GaRV-L	5133	81	90	NC_003876
Helicobasidium mompa totivirus 1-17	HmTV-1-17	5207	83	93	NC_005074
Magnaporthe grisea virus 1	MgV-1	5359	77	91	NC_006367
Sphaeropsis sapinea RNA virus 1	SsRV-1	5163	89	92	NC_001963
Sphaeropsis sapinea RNA virus 2	SsRV-2	5202	83	91	NC_001964
Genus: *Giadiavirus*					
Giardia lamblia virus	GLV	6277	98	210[b]	NC_003555
Trichomonas vaginalis virus 1	TVV-1	4648	74	160[b]	U57898
Trichomonas vaginalis virus 2	TVV-2	4674	79	162[b]	NC_003873
Trichomonas vaginalis virus 3	TVV-3	4844	79	156[b]	NC_004034
Genus: *Leishmaniavirus*					
Leishmania RNA virus 1-1	LRV1-1	5284	82	98	NC_002063
Leishmania RNA virus 1-4	LRV1-4	5283	83	99	NC_003601
Leishmania RNA virus 2-1	LRV2-1	5241	78	88	NC_002064
Unclassified					
Penaeid shrimp myonecrosis virus	IMNV	7560	99	196[b]	NC_007915
Eimeria brunetti RNA virus 1	EbRV	5358	83	98	NC_002701

[a]The names of the ICTV-recognized virus species are written in italics.
[b]RdRp is expressed or proposed to be expressed as a CP–RdRp fusion protein.

160S to 190S (S_{20w}, in Svedberg units). Particles lacking nucleic acid sediment with apparent sedimentation coefficients of 90–105S. Isolates of ScV-L-A and UmV-H1 may have additional components, containing satellite or defective dsRNAs, with different sedimentation coefficients and buoyant densities. Virion-associated RdRp activity can be detected in all totiviruses examined to date. Protein kinase activity is associated with Hv190SV virions; capsids contain phosphorylated forms of CP.

Virion Structure and Composition

The totiviruses have isometric particles, approximately 40 nm in diameter, with icosahedral symmetry (**Figure 1**). The capsids are single-shelled and encompass a single major polypeptide. The capsids consist of 120 CP subunits of molecular mass in the range of 76–98 kDa. The capsid structures of three members of the family *Totiviridae* have been determined, at least one (the yeast L-A virus) at near atomic resolution using X-ray crystallography, and the other two (UmV-H1 and Hv190SV) at moderate resolutions (~1.4 nm) using cryo-transmission electron microscopy combined with three-dimensional image reconstruction. In all cases, the capsids of the fungal totiviruses are made up of 60 asymmetric CP dimers arranged in a '$T=2$' layer. Compared to the yeast L-A capsid, the Hv190SV capsid shows relatively smoother outer surfaces. The quaternary organization of the Hv190SV particle, however, is remarkably similar to the yeast L-A and the cores of the larger dsRNA viruses of plants and animals: the A-subunits cluster around the fivefold axis and B-subunits around the threefold axis.

The ScV-L-A CP removes the 5′ cap structure of host mRNA and covalently attaches it to the histidine residue at position 154. The decapping activity is required for efficient translation of viral RNA. The published yeast L-A capsid structure reveals a trench at the active site of decapping. The decapping activity has yet to be demonstrated for any other totivirus CPs.

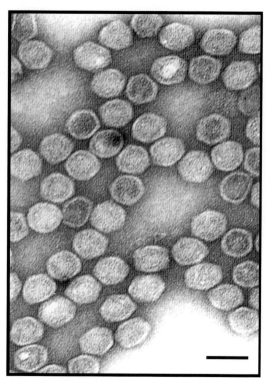

Figure 1 Negative contrast electron micrograph of particles of an isolate of *Helminthosporium victoriae 190SV*, the type species of the newly proposed genus *Victorivirus*. Scale = 50 nm.

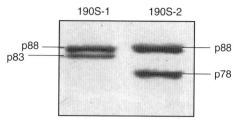

Figure 2 Sodium dodecyl sulfate-polyacrylamide gel electrophoresis (SDS-PAGE) analysis of Hv190SV sedimenting components 190S-1 and 190S-2. Purified preparations of Hv190SV contain two types of particles, 190S-1 and 190S-2, which differ slightly in sedimentation rates. The 190S-1 capsids contain p88 and p83, occurring in approximately equimolar amounts, and the 190S-2 capsids comprise similar amounts of p88 and p78. The capsid proteins p88 and p83 are phosphorylated, whereas p78 is nonphosphorylated.

Although a single gene encodes the capsid of Hv190SV, like other totiviruses, the Hv190SV capsid comprises two closely related major CPs, either p88 and p83 or p88 and p78. The capsids of all other totiviruses so far characterized appear to contain only a single major CP. Interestingly, HmTV-1-17, a totivirus infecting a filamentous fungus, has similar capsid heterogeneity to that of Hv190SV. It would be of interest to determine whether capsid heterogeneity and post-translational modification (phosphorylation and proteolytic processing) of the primary CP is a common feature of totiviruses infecting filamentous fungi (members of the tentative genus *Victorivirus*). Purified Hv190S virion preparations contain two types of particles, 190S-1 and 190S-2, which differ slightly in sedimentation rates (190S-1 is resolved as a shoulder on the slightly faster sedimenting component 190S-2) and capsid composition. The 190S-1 and 190S-2 virions are believed to represent different stages in the virus life cycle. The 190S-1 capsids contain p88 and p83, occurring in approximately equimolar amounts, and the 190S-2 capsids comprise similar amounts of p88 and p78. p88 and p83 are phosphoproteins, whereas p78 is nonphosphorylated (**Figure 2**). Totivirus virions encapsidate a single molecule of dsRNA, 4.6–6.3 kbp in size. Some totiviruses may additionally contain satellite dsRNAs or defective dsRNAs, which are encapsidated separately in capsids encoded by the totivirus genome.

Genome Organization and Expression

In general, the genome organization of the totiviruses infecting fungi and protozoa are similar: each virus genome contains two large open reading frames (ORFs); the 5′ proximal ORF encodes a CP and the 3′ ORF encodes an RdRp (see the genome organization of Hv190SV as an example; **Figure 3**). Except for LRV2-1 and UmV-H1, the RdRp ORF overlaps the CP ORF and is in the −1 frame (ScV-L-A, ScV-L-BC, Hv190SV, TVV-2, and GLV) or in the +1 frame (LRV1-1 and LRV1-4) with respect to the CP ORF. The RdRp ORF of LRV2-1 does not overlap the CP ORF, and is separated from it by a stop codon. The Umv-H1 genome contains only a single ORF that encodes a polyprotein that is predicted to be autocatalytically processed by a viral papain-like protease to generate the CP and RdRp proteins.

The totiviruses express their RdRps either as CP–RdRp (gag-pol-like) fusion proteins or as separate nonfused proteins. Expression of RdRp as a CP–RdRp fusion protein via −1 ribosomal frameshifting has been well documented for ScV-L-A. Virion-associated CP–RdRp has been detected as a minor protein in the capsids of ScV-L-A, ScV-L-BC, TVV, and GLV. Although CP–RdRp fusion proteins were neither detected *in vivo* nor associated with virions of LRV1-1, LRV1-4, or LRV2-1, expression of RdRp as a fusion protein by +1 ribosomal frameshifting or ribosomal hopping (LRV2-1) has been proposed.

The overlap region between ORF1 and ORF2 of ScV-L-A (130 nt), LRV1-1 and LRV1-4 (71 nt), and GLV (122 nt) contains the structures necessary for ribosomal frameshifting including a slippery site and a pseudoknot structure to promote fusion of ORF1 and ORF2 *in vivo*. Although the overlap region in TVV is short (14 nt), it contains a potential ribosomal slippage heptamer.

The overlap regions in the dsRNA genomes of Hv190SV and Hv190S-like viruses (members of the

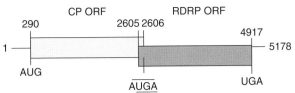

Figure 3 Genome organization of *Helminthosporium victoriae 190S virus*, the type species of the newly proposed genus *Victorivirus*. The dsRNA genome encompasses two large overlapping open reading frames (ORFs) with the 5' ORF encoding a capsid protein (CP) and the 3' ORF encoding an RNA-dependent RNA polymerase (RdRp). Note that the termination codon of the CP ORF overlaps the initiation codon of the RdRp ORF in the tetranucleotide sequence AUGA. Adapted from Ghabrial SA and Patterson JL (1999) *Encyclopedia of Virology*, 2nd edn., pp. 1808–1812. New York: Academic Press, with permission from Elsevier.

newly proposed genus *Victorivirus*), on the other hand, are of the AUGA type where the initiation codon of the RdRp ORF overlaps the termination codon of the CP ORF (**Figure 3**), suggesting that expression of RdRp occurs by a mechanism different from translational frameshifting. For Hv190SV, the stop codon of the CP ORF (nucleotide position 2606-UGA-2608) overlaps with the start codon (nucleotide position 2605-AUG-2607) for the RdRp ORF in the sequence AUGA (**Figure 3**). The complete nucleotide sequences of Hv190SV and eight other putative members of the family *Totiviridae* infecting filamentous fungi (members of the genus *Victorivirus*) have been reported and the sequences deposited in the GenBank, with Hv190SV being biochemically and molecularly best characterized of these (**Table 1**).

The 5' end of the plus strand of all totiviruses dsRNAs examined to date is uncapped and the 3' end is not polyadenylated. The 5' end of Hv190SV dsRNA is uncapped and highly structured and contains a relatively long (289 nucleotides) 5' leader with two minicistrons. These structural features of the 5' untranslated region (UTR) of Hv190SV dsRNA suggest that the CP-encoding ORF1 (with its AUG present in suboptimal context according to the Kozak criteria) is translated via a cap-independent mechanism. The 5'-UTR of the leishmaniavirus LRV-1-1 functions as an internal ribosome entry site (IRES). Translation of the uncapped giardiavirus GLV mRNA in *Giardia lamblia* is initiated on a unique IRES element that contains sequences from a part of the 5'-UTR and a portion of the capsid coding region.

The UGA codon at position 2606–2608 of the Hv190SV genomic plus strand was verified by site-directed mutagenesis to be the authentic stop codon for ORF1 (**Figure 3**). The RdRp-encoding downstream ORF2 of Hv190SV is in a −1 frame with respect to ORF1 (**Figure 3**) and is expressed via an internal initiation mechanism (a coupled termination–initiation mechanism is proposed). The Hv190SV RdRp is detectable as a separate, virion-associated component, consistent with its independent translation from ORF2. The tetranucleotide AUGA overlap region, or a very similar structure, is characteristic of the overlap region of all putative members of the totivirus family that infect filamentous fungi (victoriviruses).

The initial report on the molecular characterization of IMNV, a monosegmented dsRNA virus infecting penaeid shrimp and tentatively assigned to the family *Totiviridae*, concluded that the viral genome encompasses two non-overlapping ORFs with ORF1 encoding a polyprotein comprised of a putative RNA-binding protein and a CP. The coding region of the RNA-binding protein was located in the first half of ORF1 and contained a dsRNA-binding motif in the first 60 amino acids. The second half of ORF1 encoded a CP, as determined by amino acid sequencing, with a molecular mass of 99 kDa. ORF2 encoded a putative RdRp with the eight conserved motifs characteristic of totiviruses. Phylogenetic analysis based on the RdRp clustered IMNV with GLV, the type species of the genus *Giardiavirus* in the family *Totiviridae*. Important novel features of the genome organization of IMNV, however, were most recently uncovered that has significant bearing on how the viral proteins are expressed. These features include two encoded '2A-like' motifs, which are likely involved in ORF1 polyprotein 'cleavage', a 199 nt overlap between ORF1 and ORF2, and the presence a 'slippery heptamer' motif and predicted RNA pseudoknot in the region of ORF1–ORF2 overlap. The latter features probably allow ORF2 to be translated as a fusion with ORF1 by '−1' ribosomal frameshifting. Although the generation of CP as a polyprotein and the potential involvement of encoded '2A-like' peptides (GDVESNPGP and GDVEENPGP) in processing of the polyprotein to release the major CP represent novel features not shared by totiviruses, the potential expression of RdRp as a CP–RdRp fusion protein via ribosomal frameshifting is a common strategy utilized by totiviruses for expressing their RdRps. Experimental evidence for the presence of the putative CP–RdRp fusion protein in purified virions or in infected tissues, however, is lacking.

Virus Replication Cycle

Limited information is available on the replication cycle of totiviruses and has mainly been derived from *in vitro* studies of virion-associated RNA polymerase activity and the isolation of particles representing various stages in the replication cycle. In *in vitro* reactions, the RNA polymerase activity associated with virions of the fungal totiviruses ScV-L-A, UmV-H1, and Hv190SV, isolated from lag-phase cultures, catalyzes end-to-end transcription of dsRNA, by a conservative mechanism, to produce mRNA for CP, which is released from the particles. Purified ScV-L-A virions, isolated from log-phase cells, contain a less-dense class of particles, which package only plus-strand RNA. In *in vitro* reactions, these particles exhibit a replicase activity that catalyzes the synthesis of minus-strand RNA to form dsRNA. The resultant mature particles, which attain the same density as that of the dsRNA-containing virions isolated from the cells, are capable of synthesizing and releasing plus-strand RNA.

A proposed life cycle of Hv190SV is depicted in **Figure 4**. Host-encoded protein kinase and protease have been shown to be involved in post-translational modification of CP. Phosphorylation and proteolytic processing are proposed to play a role in the virus life cycle; phosphorylation of CP may be necessary for its interaction with viral nucleic acid and/or phosphorylation may regulate dsRNA transcription/replication. Proteolytic processing

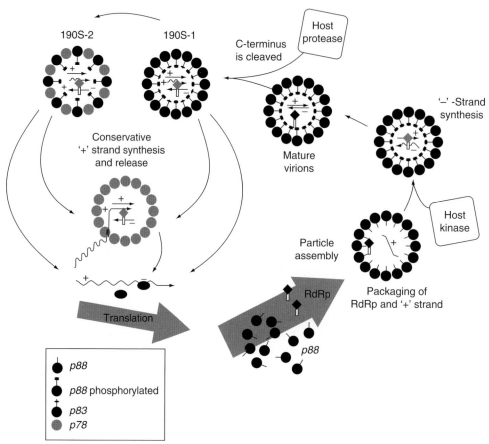

Figure 4 Life cycle of Hv190SV. Mature virions contain a single dsRNA molecule and their capsids are composed entirely or primarily of the capsid protein (CP) p88. Virions representing different stages of the virus life cycle can be purified from the infected fungal host *Helminthosporium victoriae* including the well-characterized 190S-1 and 190S-2 virions. These two types of virions differ in sedimentation rate, phosphorylation state, and CP composition; 190S-1 capsids contain p88 and p83, whereas the 190S-2 capsids contain p88 and p78 (p88 is the primary translation product of the CP gene; p83 and p78 represent post-translational proteolytic processing products of p88 at its C-terminus). p88 and p83 are phosphorylated, whereas p78 is nonphosphorylated. The virions with phosphorylated CPs (p88+p83) have significantly higher transcriptase activity *in vitro* than those containing the nonphosphorylated p78. Transcription occurs conservatively and the newly synthesized plus-strand RNA is released from the virions. Phosphorylation of CP is catalyzed by a host kinase, and is proposed to play a regulatory role in transcription/replication. A host-encoded protease catalyzes the proteolytic processing of phosphorylated p88; this occurs in two steps, leading first to p83 (the generation of the 190S-1 virions) and then to p78 (190S-2 virions). The conversion of p88-p83-p78 is proposed to play a role in the release of the plus-strand RNA transcripts from virions. The released plus-strand RNA is the RNA that is translated into CP and RNA-dependent RNA polymerase (RdRp) and packaged in capsids assembled from the primary translation product p88. It is not known whether p88 is phosphorylated before or after assembly. Synthesis of minus-strand RNA occurs on the plus-strand RNA template inside the virion; phosphorylation may be involved in turning on the replicase activity. Adapted from Ghabrial SA and Patterson JL (1999) *Encyclopedia of Virology*, 2nd edn., pp. 1808–1812. New York: Academic Press, with permission from Elsevier.

and cleavage of a C-terminal peptide, which leads to dephosphorylation and the conversion of p88 to p78, may play a role in the release of the plus-strand RNA transcripts from virions (**Figure 4**).

Biological Properties

There are no known natural vectors for the transmission of the fungal totiviruses. They are transmitted intracellularly during cell division, sporogenesis, and cell fusion. Although the yeast totiviruses are effectively transmitted via ascospores, the totiviruses infecting the ascomycetous filamentous fungi are essentially eliminated during ascospore formation. The leishmaniaviruses are not infectious as purified virions and are propagated during cell division. The giardiavirus GLV and the unclassified IMNV, on the other hand, are infectious as purified virions. Successful transfection of the protozoa *Giardia lamblia* has also been accomplished via electroporation with plus-strand RNA transcribed *in vitro* from GLV dsRNA. Whereas IMNV causes a myonecrosis disease in its crustacean host, GLV is associated with latent infection of its flagellated protozoan human parasite *G. lamblia*. GLV is released into the medium without lysing the host cells and the extruded virus can infect many virus-free isolates of the protozoan host.

There are no known experimental host ranges for the fungal totiviruses because of the lack of suitable infectivity assays. As a consequence of their intracellular modes of transmission, the natural host ranges of fungal totiviruses are limited to individuals within the same or closely related vegetative compatibility groups. Furthermore, mixed infections with two or more unrelated viruses are common, probably as a consequence of the ways by which fungal viruses are transmitted in nature. Dual infection of yeast with ScV-L-A and ScV-L-BC and the filamentous fungus *Sphaeropsis sapinea* with SsRV-1 and SsRV-2 are examples of mixed infections involving totiviruses. Dual infection of *H. victoriae* with Hv190SV and the chrysovirus Hv145SV is an example of a mixed infection involving two unrelated viruses belonging to two different families, a totivirus and a chrysovirus.

With the exception of IMNV, an unclassified virus tentatively assigned to the family *Totiviridae*, which causes a myonecrosis disease in shrimp, totiviruses are generally associated with symptomless infections of their hosts. A possible exception to this rule is Hv190SV, since mixed infections with Hv190SV and Hv145SV are associated with a debilitating disease of the fungal host. The disease phenotype and the two viruses are transmitted via hyphal anastomosis and diseased isolates are characterized by reduced growth, excessive sectoring, aerial mycelial collapse, and generalized lysis (**Figure 5**). The disease phenotype of *H. victoriae*, the causal agent of Victoria blight of oats, is of special interest not only because diseased isolates are hypovirulent, but also because of the probable viral etiology. The disease phenotype was also transmitted by incubating fusing protoplasts from virus-free fungal isolates with purified virions containing both the Hv190S and 145S viruses. The frequency of infection and stability of the newly diseased colonies, however, were very low, and verification of transmission was based on virus detection by immune electron microscopy. A systematic molecular approach based on DNA-mediated transformation of a susceptible virus-free *H. victoriae* isolate with full-length cDNA clones of viral dsRNAs is currently being pursued to verify the viral etiology of the disease phenotype and elucidate the roles of the individual viruses in disease development.

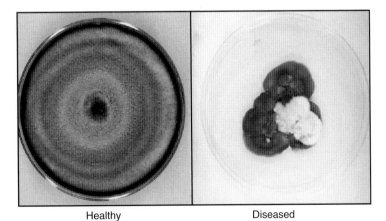

Figure 5 Colony morphology of virus-free (healthy), virus-infected (diseased) isolates of *Helminthosporium victoriae*. The colonies were grown for a week on potato dextrose agar medium at room temperature. The healthy colony shows uniform mycelial growth that extends to the entire plate, whereas the diseased colony is stunted and shows excessive sectoring.

A transformation vector for *H. victoriae* based on the hygromycin B resistance marker was constructed and used to transform an *H. victoriae* virus-free isolate with a full-length cDNA clone of Hv190SV dsRNA. The Hv190SV cDNA was inserted downstream of a *Cochliobolus heterostrophus GPD1* promoter and upstream of an *Aspergillus nidulans trpC* terminator signal. The hygromycin-resistant transformants expressed the Hv190SV CP, as determined by Western-blotting analysis. Transformation of a normal virus-free fungal isolate with a full-length cDNA of Hv190SV dsRNA conferred a disease phenotype. Symptom severity varied among the transformants from symptomless to severely stunted and highly sectored (**Figure 6**). Symptom severity correlated well with the level of viral capsid accumulation. Integration of the viral genome in the host DNA was verified by Southern blot hybridization analysis and expression of the viral dicistronic mRNA was demonstrated by northern analysis. Like natural infection, the primary translation product of the *CP* gene (p88) was phosphorylated and proteolytically processed to generate p83 and p78. The CP assembled into virus-like particles indistinguishable in appearance from empty capsids normally associated with virus infection. The empty capsids accumulated to significantly higher levels than in natural infections. Although the RdRp was expressed and packaged, no dsRNA was detected inside the virus-like particles or in the total RNA isolated from mycelium. Despite inability to launch dsRNA replication from the integrated viral cDNA, the demonstration that the ectopic 190SV cDNA copies were transcribed and translated and that the resultant transformants exhibited a disease phenotype provides convincing evidence for a viral etiology for the disease of *H. victoriae*. The role of the Hv145SV in disease development, however, has yet to be elucidated.

Virus–Host Relationships

The yeast killer system, comprised of a helper totivirus (ScV-L-A) and associated satellite dsRNA (M-dsRNA), is one of the very few known examples where virus infection is beneficial to the host. The ability to produce killer toxins by immune yeast strains confers an ecological advantage over sensitive strains. The use of killer strains in the brewing industry provides protection against contamination with adventitious sensitive strains. Totiviruses maintain only the genes that are essential for their survival (RdRp and CP), but make efficient use of host proteins. The host cells have evolved to support only a defined level of virus replication, beyond which virus infection may become pathogenic. Because of amenability to genetic studies, the yeast–virus system has provided significant information on the host genes required to prevent viral cytopathology. A system of six chromosomal genes, designated superkiller (or SKI) SKI2, SKI3, SKI4, SKI6, SKI7, and SKI8, negatively control the copy number of the totivirus ScV-L-A and its satellite M-dsRNAs. The only crucial function of these genes is to block virus multiplication. Mutations in any of these

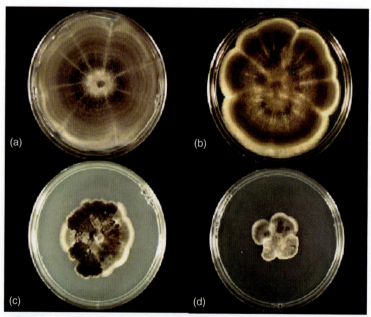

Figure 6 Colony morphology of a virus-free *Helminthosporium victoriae* isolate transformed with a full-length cDNA clone of the totivirus Hv190SV dsRNA. Colonies transformed with vector alone (a) or with recombinant vector containing full-length cDNA to Hv190SV dsRNA (b–d) were grown for a week at room temperature. Note that colonies (b)–(d) show increasingly more severe symptoms (reduced growth and sectoring).

SKI genes lead to the development of the superkiller phenotype as a consequence of the increased copy number of M-dsRNA. The SKI genes affect primarily the initiation of translation rather than the stability of mRNA, and are thus part of a cellular system that specifically blocks translation of nonpolyadenylated mRNAs (like the plus-strand transcripts of totiviruses). About 30 chromosomal genes, termed MAK genes (for maintenance of killer), are required for stable replication of the satellite M-dsRNA. Only three of these MAK genes are necessary for the helper virus (ScV-L-A) multiplication. Mutants defective in any of 20 MAK genes show a decreased level of free 60S ribosomal subunits. Since the *mak* mutations affecting 60S subunit levels are suppressed by *ski* mutations, and since the latter act by blocking translation of nonpolyadenylated mRNAs, the level of 60S ribosomal subunits is critical for translation of nonpolyadenylated mRNAs.

The Hv190S totivirus that infects the plant pathogenic fungus *H. victoriae* utilizes host-encoded proteins (a protein kinase and a protease) for post-translational modification of its CP. Phosphorylation and proteolytic processing of CP may play a role in regulating transcription and the release of plus-strand transcripts from virions (**Figure 4**).

The *H. victoriae*-virus system is well characterized and provides a useful model system for studies on virus–host interactions in a plant pathogenic fungus. A major attribute of this system is the fact that the virus-infected *H. victoriae* isolates exhibit a disease phenotype (**Figure 5**), which is rare among fungal viruses. Modulation of fungal gene expression and alteration of phenotypic traits as a consequence of mycovirus infection are little understood, with the exception of the chestnut blight fungus–hypovirus system. It was previously demonstrated that the fungal gene *Hv-p68* is upregulated as a result of virus infection and proposed that upregulation of this gene might play a role in virus pathogenesis (**Figure 7**). Hv-p68, a novel alcohol oxidase/RNA-binding protein, belongs to the large family of FAD-dependent GMC oxidoreductases with 67–70% sequence identity to the alcohol oxidases of methylotrophic yeasts. Hv-p68, however, shows only limited methanol-oxidizing activity and its expression is not induced in cultures supplemented with methanol as the sole carbon source (**Figure 7**). The natural substrate for Hv-p68 is not known, but the structurally similar alcohol oxidases are known to oxidize primary alcohols irreversibly to toxic aldehydes. Overexpression of Hv-p68 and putative accumulation of toxic intermediates was proposed as a possible mechanism underlying the disease phenotype of virus-infected *H. victoriae* isolates. Overexpression of Hv-p68 in virus-free fungal isolates, however, resulted in a significant increase in colony growth and did not induce a disease phenotype. Thus, overexpression of Hv-p68 *per se* is not sufficient to induce the disease phenotype in the absence of virus infection. If the function of *Hv-p68* is similar to that of

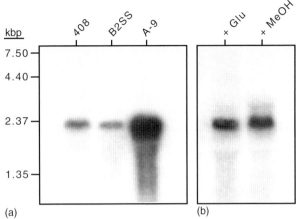

Figure 7 Northern analysis of Hv-p68 mRNA transcript levels. (a) Total RNA (25 μg) isolated from the virus-free isolates 408 and B-2ss and the virus-infected isolate A-9 were electrophoresed on a formaldehyde/agarose gel, blotted, and hybridized under high stringency conditions with a radiolabeled probe for Hv-p68. Hv-p68 mRNA, 2.25 kbp in size, was detected in RNA samples from all three *H. victoriae* isolates. The level of Hv-p68 transcript in cultures of the virus-infected isolate A-9, however, was at least 15-fold higher than that for the virus-free isolates. (b) Total RNA (30 μg) isolated from cultures of the virus-free isolate B-2ss grown for 3 days in minimal medium supplemented with either glucose (Glu) or methanol (MeOH). Similar amounts of Hv-p68 mRNA were detected in total RNA isolated from fungal cultures supplemented with either glucose or methanol, as a carbon source. Reproduced from Soldevila AI and Ghabrial SA (2001) A novel alcohol oxidase/RNA-binding protein with affinity for mycovirus double-stranded RNA from the filamentous fungus *Helminthosporium* (*Cochliobolus*) *victoriae*. *Journal of Biological Chemistry* 276: 4652–4661, with permission from the American Society for Biochemistry and Molecular Biology, Inc.

the homologous alcohol oxidases of methylotrophic yeasts or other filamentous fungi, that is, irreversible oxidation of primary alcohols into aldehydes, then overexpression of Hv-p68 could lead to an accumulation of toxic aldehydes and development of disease phenotype. The finding that colonies overexpressing the Hv-p68 protein did not exhibit the disease phenotype and grew more rapidly than the nontransformed wild-type suggests that accumulation of toxic aldehydes did not occur, and that such aldehydes were probably assimilated into carbohydrates via the xylulose monophosphate pathway.

Hv-p68, which co-purifies with viral dsRNA (mainly that of Hv145SV), is a multifunctional protein with alcohol oxidase, protein kinase, and RNA-binding activities. Hv-p68 is detectable as a minor component of Hv190SV capsid. Recent evidence strongly suggests that Hvp68 is the cellular protein kinase responsible for phosphorylation of the capsid proteins. The RNA-binding activity of Hv-p68 has been demonstrated by gel mobility shift and northwestern blot analysis. Furthermore, the RNA-binding domain of Hv-p68 was mapped to the N-terminal region that contains the ADP-binding domain. Because dsRNA-binding proteins are known to sequester dsRNA and suppress antiviral host defense mechanisms, it is feasible that overexpression of the dsRNA-binding protein Hv-p68 may lead to the induction of the disease phenotype by suppressing host defense. This idea is consistent with the finding that overexpression of Hv-p68 led to enhancement in the accumulation of Hv145S dsRNA. The role of Hv145SV in the development of the disease phenotype, however, is not yet clear. Furthermore, it is curious that Hvp68 co-purifies with Hv145S dsRNA since the latter is predicted to be confined to the viral capsids where virus replication takes place. Recent results, however, suggest that a significant proportion of the Hv145V dsRNA does not appear to be encapsidated. Considering the multifunctional nature of the Hv-p68 protein, additional studies are needed to determine whether or not Hv-p68 upregulation has a role in viral pathogenesis.

Evolutionary Relationships among Totiviruses

Sequence comparison analysis of the predicted amino acid sequences of totivirus RdRps indicated that they share significant sequence similarity and characteristically contain eight conserved motifs. This sequence similarity was common to all totiviruses so far characterized including the totiviruses that infect the yeast, smut, and filamentous fungi, as well as those infecting parasitic protozoa. As indicated earlier, the viruses infecting filamentous fungi (members of the newly proposed genus *Victorivirus*) express their RdRp separate from the CP by an internal initiation mechanism,

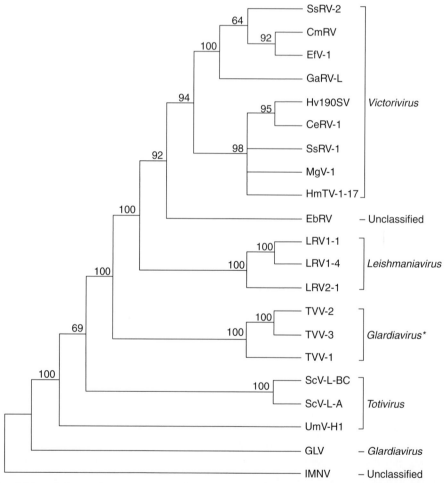

Figure 8 Neighbor-joining phylogenetic tree constructed based on the RdRp conserved motifs and flanking sequences. The RdRp sequences were derived from aligned deduced amino acid sequences of members of the family *Totiviridae* using the program CLUSTAL X. Motifs 1–8 and the sequences between the motifs, as previously designated by Ghabrial SA in 1998, were used. See **Table 1** for virus name abbreviations and GenBank accession numbers. The phylogenetic tree was generated using the program PAUP*. Bootstrap numbers out of 1000 replicates are indicated at the nodes. The tree was rooted with the RdRp of the penaeid shrimp infectious myonecrosis virus (IMNV), an unclassified virus tentatively assigned to the family *Totiviridae*, which was included as an outgroup.

whereas the other members of the family *Totiviridae* express their RdRps as fusion proteins (CP–RdRp), mainly via a ribosomal frameshifting mechanism. Phylogenetic analysis based on multiple alignments of amino acid sequences of totivirus RdRp conserved motifs (**Figure 8**) reflects these differences as the viruses infecting filamentous fungi (victoriviruses) are most closely related to each other and form a distinct large well-supported cluster (bootstrap value of 94%). Likewise, phylogenetic trees based on the CP sequences showed similar topology to those based on RdRp sequences (**Figure 9**). The fact that independent alignments of CP and RdRp sequences give similar phylogenetic relationships (**Figures 8** and **9**) further supports the creation of a new genus to accommodate the viruses infecting filamentous fungi.

The leishmaniaviruses LRV1, LRV2, and LRV4 (percent sequence identities of 46%) are most closely related to each other and form a discrete cluster with 100% bootstrap value. This is also true for the two yeast viruses ScV-L-A and ScV-LBC (RdRp identity of 32%), members of the genus *Totivirus*. UmV-H1, the third member in the genus *Totivirus* forms a phylogenetic clade distinct from the cluster of the two yeast viruses. This may reflect the difference in RdRp expression strategy between these viruses since UmV-H1 expresses its RdRp via the generation of a polyprotein followed by proteolytic processing to release the viral protein, whereas the yeast viruses express their RdRps as CP–RdRp fusion proteins via a −1 ribosomal frameshifting. It is of interest that Hv190SV and related viruses (victoriviruses) are phylogenetically more closely to the leishmaniaviruses than to the yeast viruses (**Figures 8** and **9**). It was previously hypothesized that leishmaniaviruses and the fungal totiviruses are of old origin having existed in a single-cell-type progenitor

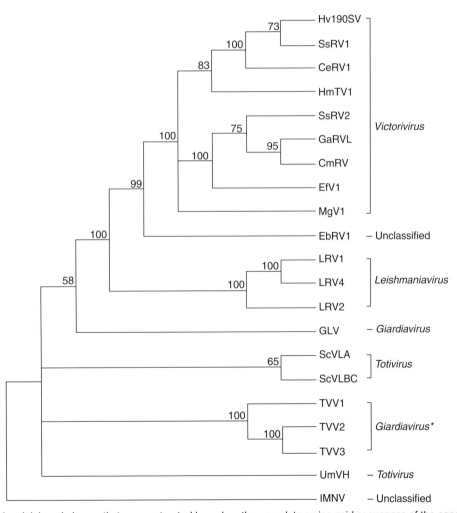

Figure 9 Neighbor-joining phylogenetic tree constructed based on the complete amino acid sequences of the capsid proteins of viruses in the family *Totiviridae*. The CP multiple sequence alignment was performed with the program CLUSTAL X and the phylogenetic tree was generated using the program PAUP*. See **Table 1** for virus name abbreviations and GenBank accessions numbers. Bootstrap numbers out of 1000 replicates are indicated at the nodes. The tree was rooted with the CP of the penaeid shrimp infectious myonecrosis virus (IMNV), an unclassified virus tentatively assigned to the family *Totiviridae*, which was included as an outgroup.

prior to the divergence of fungi and protozoa. The finding that the unclassified Eimeria brunetti RNA virus 1 (EbRV-1), which infects an apicomplexan parasitic protozoa, is more closely related to members of the new genus *Victorivirus* than to protozoal viruses is consistent with this idea. EbRV-1 may represent an intermediate progenitor in the evolution of 'victoriviruses'.

It is worth noting that the classification of TVV1, TVV2, and TVV4 as tentative members of the genus *Giardiavirus* is not supported by phylogenetic analyses based on the totivirus CPs or RdRps since they do not cluster with GLV, the type species of the genus *Giardiavirus*. The trichomonas vaginalis viruses are most closely related to each other as they form a distinct phylogenetic cluster with 100% bootstrap support (**Figures 8** and **9**). It would probably be justifiable to create a new genus in the family totiviridae to accommodate these viruses.

Taxonomic considerations of totiviruses may benefit greatly from elucidating the capsid structure of representatives of the genera *Leishmaniavirus* and *Giardiavirus* (including the tentative members TVVs). Furthermore, resolving the capsid structure of the unclassified viruses EbRV-1 and IMNV may prove useful for their taxonomic placement. Information on the genome organization and RdRp expression strategy of EbRV-1 is lacking and such information would be crucial in determining whether a novel genus in the family *Totiviridae* should be created for EbRV-1. IMNV has some features that distinguishes it from totiviruses including that it expresses its CP as a polyprotein that undergoes autoproteolytic processing. It is likely that IMNV would be classified as member of a novel family of dsRNA viruses that infect invertebrate hosts.

See also: Fungal Viruses; Giardiaviruses; Leishmaniaviruses; Ustilago Maydis Viruses; Viral Killer Toxins; Yeast L-A Virus.

Further Reading

Castón JR, Luque D, Trus BI, *et al.* (2006) Three-dimensional structure and stoichiometry of *Helminthosporium victoriae190S* totivirus. *Virology* 347: 323–332.

Cheng J, Jiang D, Fu Y, Li G, Peng Y, and Ghabrial SA (2003) Molecular characterization of a dsRNA totivirus infecting the sclerotial parasite *Coniothyrium minitans*. *Virus Research* 93: 41–50.

Ghabrial SA (1998) Origin, adaptation and evolutionary pathways of fungal viruses. *Virus Genes* 16: 119–131.

Ghabrial SA and Patterson JL (1999) *Encyclopedia of Virology*, 2nd edn., pp. 1808–1812. New York: Academic Press.

Ghabrial SA, Soldevila AI, and Havens WM (2002) Molecular genetics of the viruses infecting the plant pathogenic fungus *Helminthosporium victoriae*. In: Tavantzis S (ed.) *Molecular Biology of Double-Stranded RNA: Concepts and Applications in Agriculture, Forestry and Medicine*, pp. 213–236. Boca Raton, FL: CRC Press.

Huang S and Ghabrial SA (1996) Organization and expression of the double-stranded RNA genome of *Helminthosporium victoriae* 190S virus, a totivirus infecting a plant pathogenic filamentous fungus. *Proceedings of the National Academy of Sciences, USA* 93: 12541–12546.

Icho T and Wickner RB (1989) The double-stranded RNA genome of yeast virus L-A encodes its own putative RNA polymerase by fusing two open reading frames. *Journal of Biological Chemistry* 264: 6716–6723.

Nibert ML (2007) '2A-like' and 'shifty heptamer' motifs in penaeid shrimp infectious myonecrosis virus, a monosegmented double-stranded RNA virus. *Journal of General Virology* 88: 1315–1318.

Nomura K, Osaki H, Iwanami T, Matsumoto N, and Ohtsu Y (2003) Cloning and characterization of a totivirus double-stranded RNA from the plant pathogenic fungus, *Helicobasidium mompa* Tanaka. *Virus Genes* 23: 219–226.

Park Y, James D, and Punja ZK (2005) Co-infection by two distinct totivirus-like double-stranded RNA elements in *Chalara elegans* (*Thielaviopsis basicola*). *Virus Research* 109: 71–85.

Preisig O, Wingfield BD, and Wingfield MJ (1998) Coinfection of a fungal pathogen by two distinct double-stranded RNA viruses. *Virology* 252: 399–406.

Romo M, Leuchtmann A, and Zabalgogeazcoa I (2007) A totivirus infecting the mutualistic fungal endophyte *Epichloë festucae*. *Virus Research* 124: 38–43.

Soldevila A and Ghabrial SA (2000) Expression of the totivirus *Helminthosporium victoriae*190S virus RNA-dependent RNA polymerase from its downstream open reading frame in dicistronic constructs. *Journal of Virology* 74: 997–1003.

Soldevila AI and Ghabrial SA (2001) A novel alcohol oxidase/RNA-binding protein with affinity for mycovirus double-stranded RNA from the filamentous fungus *Helminthosporium* (*Cochliobolus*) *victoriae*. *Journal of Biological Chemistry* 276: 4652–4661.

Tuomivirta TT and Hantula J (2005) Three unrelated viruses occur in a single isolate of *Gremmeniella abietina* var. *abietina* type A. *Virus Research* 110: 31–39.

Wickner RB, Wang CC, and Patterson JL (2005) Totiviridae. In: Fauquet CM, Mayo MA, Maniloff J, Desselberger U, and Ball LA (eds.) *Virus Taxonomy: Eighth Report of the International Committee on Taxonomy of Viruses*, pp. 581–590. San Diego, CA: Elsevier Academic Press.

Transcriptional Regulation in Bacteriophage

R A Weisberg, D M Hinton, and S Adhya, National Institutes of Health, Bethesda, MD, USA

Published by Elsevier Ltd.

Introduction

Most bacteriophages regulate their own gene expression, many regulate host gene expression, and a few even regulate genes of other phages. Regulation enables temperate phages to establish and maintain lysogeny; it allows both temperate and virulent phages to carry out developmental programs in which groups of genes are expressed in an ordered, temporal sequence as their products are needed during the lytic growth cycle, and it helps phages to commandeer the transcriptional resources of their hosts. In this article we describe the biological roles and mechanisms of regulation employed by several extensively studied model phages of *Escherichia coli*: temperate phage λ and virulent phages T4, T7, and N4. We emphasize transcriptional regulation, although we present examples of regulated RNA and protein stability as well. Investigations of these regulatory strategies have not only elucidated much about phage biology, but have also yielded many insights into regulatory mechanisms that are used by the host. We also describe several other strategies found in some less extensively studied phages.

Transcript Initiation and Elongation by Host RNA Polymerase

When a phage chromosome enters a cell, it encounters a host RNA polymerase (RNAP) that has specific requirements for promoter recognition. Prokaryotic cellular RNAPs consist of a large, multisubunit core, consisting of two copies of the α subunit, and one each of the β, β′, and ω subunits. Core has the ability to synthesize RNA, while a sixth subunit, σ-factor, is needed to identify and bind to promoters, and to initiate RNA synthesis at specific start sites. Primary σ-factors are used during exponential growth for the expression of housekeeping genes while alternate σ factors are needed to transcribe bacterial genes required under certain growth conditions or at times of stress. Many phages direct the synthesis of transcription factors that interact with, modify, or replace host σ. Another class of phage-encoded factors targets core subunits, and these factors can regulate either transcript initiation or transcript elongation.

Hundreds of prokaryotic σ factors have been identified, and they share up to four regions of similar sequence and function. Primary σ factors, like σ^{70} of *E. coli*, have three well-characterized regions with the potential to recognize and bind promoters. Residues in region 2 interact with a −10 DNA element (positions −12 to −7), residues in region 3 contact an extended −10 (TGn) motif (positions −15 to −13), and residues in region 4 contact a −35 element (positions −35 to −29) (**Figure 1**; the negative numbers indicate the number of base pairs upstream of the transcription start point). Only two of these three contacts are necessary for good promoter activity, and the majority of host promoters are −10/−35 promoters that use the DNA binding of regions 2 and 4. The −10/−35 promoters also require an interaction between residues in region 4 and a structure in core, called the β-flap, which is required to position region 4 correctly for its contact with the DNA.

RNAP also contacts host promoters through interactions between the C-terminal domain (CTD) of the α-subunits of polymerase core and A–T-rich sequences located between −40 and −60 (UP elements). Although RNAP contains two α subunits, they are not equivalent; one is bound to β and the other is bound to β′. Each α-CTD can interact specifically with a promoter proximal UP element, centered at position −41, or a promoter distal UP element, centered at position −52. In addition, the α-CTD domains can interact nonspecifically with DNA located in the −40 to −60 region.

Initiation of transcription from a −10/−35 promoter by host RNAP is a multistep process. After binding to the promoter through the interactions of σ^{70} with the −10 and −35 elements, the enzyme distorts a region around the −10 element, a step called isomerization, and becomes

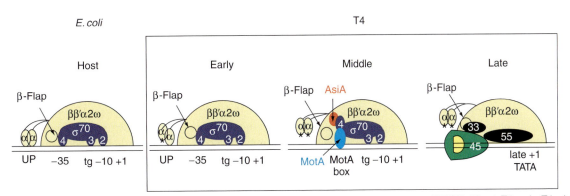

Figure 1 Changes to (*E. coli*) RNAP during T4 infection. Panels from left to right present cartoons of RNAP at host, T4 early, T4 middle, and T4 late promoters. Core polymerase (β, β′, α₂, and ω) is in tan. CTDs of α-subunits, which can interact with upstream promoter DNA (UP), and the β-flap are indicated. Region 2, 3, and 4 of σ^{70}, which recognize a −10 element, TGn, and a −35 element, respectively, are in dark blue. T4-catalyzed ADP-ribosylation of an α-CTD, which prevents the interaction of α-CTD with DNA, is denoted with a star. Recognition of a T4 middle promoter, which contains a σ^{70} −10 element and a MotA box element, requires σ^{70}-containing RNAP, the T4 activator, MotA, and the T4 coactivator, AsiA. Recognition of a T4 late promoter, which has a TATA element at −10, requires core polymerase, a T4 σ-factor composed of gp33 and gp55, and the T4 activator gp45. See text for details.

competent to initiate synthesis of an RNA chain that is complementary to the template strand of the DNA and that grows in a 5′ to 3′ direction. As the chain grows, its association with the template and enzyme is stabilized, principally through the formation of 8–9 bp of RNA:DNA hybrid adjacent to the 3′-end. Upstream of this point, the nascent transcript dissociates from the template strand of the DNA, and an additional 5–6 nt fill an exit channel before the chain emerges from the enzyme. During this process, the newborn elongation complex (EC) releases its grip on the promoter and begins translocating along the template at a rate of about 50 bp^{-1}s. Occupancy of the exit channel by RNA weakens binding of σ^{70} to the core and facilitates σ^{70} release from the EC. However, the kinetics of σ^{70} release are controversial, and there is evidence that it can remain associated with the EC even after considerable translocation.

Transcript elongation stops when the EC reaches a sequence that encodes a transcription terminator, at which point RNAP dissociates from the template and the RNA. Two classes of terminators are known in *E. coli*: intrinsic and Rho-dependent. Intrinsic terminators are nascent transcripts that can form a stable, base-paired stem–loop or hairpin that is followed by a U-rich stretch. Termination occurs within the U-rich stretch, typically 6–7 nt from the base of the hairpin. Although the nascent terminator transcript is sufficient for termination, the efficiency can be altered by proteins. Rho protein, which is necessary for termination at the second class of terminator, binds to nascent RNA upstream of the termination point. Once bound, it can use ATP to translocate in a 5′ to 3′ direction toward the EC. It is believed that termination occurs when Rho catches up to a paused EC.

Phage T4

T4 is the best-characterized member of the T4-type phages of the family *Myoviridae*, consisting of phages distinguished by a contractile tail. T4 has a 169 kbp linear chromosome, which is terminally redundant and circularly permuted. Consequently, the T4 genome is shown as a circle (**Figure 2**). Injection of the DNA into the host initiates a pattern of T4 gene expression that results in a burst of phage about 20 min after infection. The production of phage proteins is regulated primarily at the level of temporally controlled transcription through the synthesis of early, middle, and late RNAs.

T4 Early Transcription

Productive T4 infection requires exponentially growing cells. Consequently, polymerase containing σ^{70} is the major species present when T4 infects. A top priority for T4 is to rapidly commit the host polymerase to its agenda.

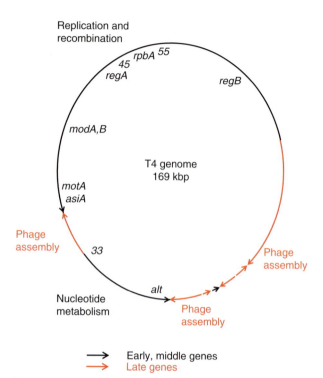

Figure 2 Map of bacteriophage T4. The 169 kbp linear T4 DNA is represented as a circle because it is circularly permuted and terminally redundant. Black regions contain early/middle genes, which include genes whose products are needed for DNA replication and recombination and nucleotide metabolism. Red regions contain late genes, whose products are primarily involved in phage assembly. The positions of specific genes referred to in the text are also shown.

It accomplishes this task through multiple strategies. Immediately after infection, T4 programs strong transcription from its early promoters, which are recognized by σ^{70} containing host RNAP. These promoters are needed for expressing early genes, many of which encode proteins that are involved in quickly moving the host resources to the phage, such as nucleases that degrade the host DNA and proteins that will be needed for middle gene expression. T4 early promoters have the recognition features of the σ^{70}-dependent $-10/-35$ promoters, but these promoters also have the extended -10 sequence for σ^{70} region 3 and UP elements for α-CTD contact. Consequently, the 40 or so early promoters are extremely strong, and they compete very efficiently with the more than 600 host promoters for productive transcription initiation by the available polymerase.

To increase further the advantage of early promoters over host promoters, T4 encodes a protein, Alt, that chemically modifies host RNAP by ADP-ribosylation. This modification specifically increases the activity of T4 early promoters relative to those of the host. Alt is actually present within the phage head, and it is injected into the host along with phage DNA. Thus, the Alt modification begins immediately after infection. Although Alt-dependent ADP-ribosylation affects a minor fraction of many *E. coli* proteins, it primarily modifies a specific residue (Arg265) on one of the α-subunits. It is thought that this modification occurs on the same α on every polymerase, but which α-subunit is unknown. Arg265 is located within the CTD of the α-protein, and normally this residue directly contacts UP element DNA. However, ADP-ribosylation of Arg265 eliminates this DNA interaction. Sequence differences between T4 early promoters and host promoters in the −40 to −60 region (the α-CTD binding sites) may explain how this asymmetric ADP-ribosylation favors T4 early promoters.

T4 transcription gains another advantage over the host early after infection because T4 DNA itself is modified, containing glucosylated, hydroxymethyl cytosines. This modification protects the phage genome from multiple nucleases that are encoded by T4 early genes and digest unmodified host DNA within the first few minutes of infection. In addition, T4 encodes an early protein (Alc) that specifically terminates transcription on unmodified host DNA.

T4 Middle Transcription

T4 middle gene expression begins about 1 min after infection. Middle genes include genes that encode T4 replication proteins, transfer RNAs, and proteins that will be needed later to switch from middle to late promoter recognition. At the start of middle transcription, two early gene products, the ADP-ribosylating enzymes ModA and ModB, complete the ADP-ribosylation of Arg 265, resulting in the modification of both α-subunits. This prevents any contact between the α-CTD domains and the −40 to −60 sequences, thereby further decreasing initiation from host promoters. In addition, ADP-ribosylation of the second α-subunit now lessens the activity of T4 early promoters. Thus, the action of ModA/ModB should contribute to the switch from T4 early to T4 middle transcription. Early gene expression is also decreased by the action of another early gene product, RegB endoribonuclease, which cleaves within the sequence GGAG present in the ribosome binding site of many early transcripts. Although RegB is present even after late gene expression has begun, and some middle and late T4 transcripts contain the RegB recognition sequence, only cleavage of early RNA is observed. In particular, the mRNA of the T4 MotA protein, which is required for middle promoter activation (see below), is an RegB substrate. Evidence suggests that the secondary structure of the mRNA is important for RegB recognition, but the exact nature of RegB specificity is not yet understood.

T4 middle RNA is synthesized using two separate strategies. First, early and middle genes are located together with almost all middle genes downstream of early genes (**Figure 2**). Consequently, transcription from T4 early promoters produces middle RNA as these early transcripts extend into middle genes. Although there has been speculation that this extension involves an active T4 antitermination process, perhaps like that of λ (below), definitive evidence for such a mechanism has not been found and the control of this extension is not yet understood.

Second, T4 activates the initiation of transcription at more than 30 middle promoters. These promoters contain the σ^{70} −10 element and are dependent on RNAP containing σ^{70} for transcription. However, they lack the σ^{70} −35 element and have instead a different consensus sequence, the 'MotA box', centered at −30 (**Figure 1**). Activation of middle promoters occurs by a process called σ appropriation and requires two T4-encoded proteins, MotA and AsiA. MotA is a transcription activator that interacts both with the MotA box and with the far C-terminus of σ^{70}, a region that contains some of the residues that normally interact with the β-flap. AsiA is a small protein that binds tightly to σ^{70} region 4. When MotA is present, AsiA coactivates transcription from T4 middle promoters. By itself AsiA also inhibits transcription from σ^{70}-dependent promoters that require an interaction with the −35 element.

AsiA binds to multiple residues in σ^{70} region 4 and dramatically changes its conformation. Some of these residues normally interact with the −35 element, and some with the β-flap, but the structure that results from AsiA binding no longer interacts with either. Thus, AsiA binding inhibits transcription from σ^{70}-dependent promoters that require these interactions. Furthermore, by removing the interaction of σ^{70} with the −35 sequences, AsiA helps MotA interact with the MotA box, centered at −30. This is because the 9 bp MotA box includes sequences that would normally interact at least indirectly with σ^{70} region 4. In addition, AsiA frees the far C-terminus of σ^{70} to interact with MotA, since the structural changes prevent this portion of σ^{70} from interacting with the β-flap.

Interestingly, AsiA binds rapidly to free σ^{70}, but poorly, if at all, to σ^{70} that is present in polymerase. Because there is an excess of core relative to σ^{70} in *E. coli*, very little free σ^{70} is available at any given time. Consequently, the binding of AsiA to σ^{70} must occur when σ^{70} is released from core, which occurs in this case after initiation at early promoters. Thus, transcription from early promoters directly promotes the switch to middle transcription, helping to coordinate the start of middle transcription with the vitality of early transcription.

T4 Late Transcription

Late genes primarily encode proteins needed to form virions and to process replicated T4 DNA. The switch from middle to late transcription begins about 6 min after infection. The synthesis of some early and middle proteins decreases because translation of their mRNA is repressed by the T4-encoded middle protein, RegA, which competes with ribosomes for binding to the translational initiation sites of some middle mRNAs. RegA binds to an AU-rich motif but the precise requirements for binding are unknown. In addition, middle RNA synthesis decreases because the previous ADP-ribosylation of α-CTD and the binding of a T4 middle protein, RpbA, to core polymerase favor transcription from late promoters, described below, over transcription from middle promoters.

Transcription from T4 late promoters is accomplished by host core polymerase in combination with three phage-encoded middle proteins: gp33, gp55, and gp45 (**Figure 1**). Gp55 shares sequence similarity with σ region 2, and like region 2, gp55 recognizes and contacts a specific DNA element, in this case TATAAATA, centered at position −10 (**Figure 1**). Gp55 interacts with gp45 and the β′-subunit of RNAP. Gp33 interacts with gp45 and the β-flap of RNAP. Gp45 forms a ring that encircles T4 DNA and can move along it in either direction. Interaction of RNAP-bound gp33 with DNA-bound gp45 connects RNAP to DNA upstream of the gp55 DNA contact. Thus, gp33 and gp55 together can be considered a T4 σ-factor for late promoters. Interestingly, gp45 is also the DNA polymerase clamp protein, a processivity factor that is needed to keep T4 DNA polymerase from dissociating from DNA during replication. As a consequence, transcription from late promoters requires active DNA replication, because gp45 is loaded onto the DNA through its role as the DNA polymerase clamp. This connection of late transcription to the replication of phage DNA serves to coordinate the expression of late genes, whose functions are primarily DNA packaging and capsid assembly, with the amount of phage DNA. T4 late transcription continues until 20 min after infection, when the fully formed phage with the packaged phage genomic DNA are released through lysis of the cell.

Phages T7 and T3

T7 and T3 belong to the family *Podoviridae*, comprising phages with short tails. Their virions respectively contain about 39.9 and 38.7 kbp of double-stranded nonpermuted DNA. A short segment is directly repeated at each end ('terminal redundancy'). The chromosomal arrangement of genes with identical functions is similar in the two phages, and to several other sequenced phages of the family, in accord with the modular evolution of phages. We will describe the transcriptional regulation of T7, because it is the best-studied member of the entire group, mentioning T3 when appropriate. The genetic map and the transcription pattern of T7 are shown in **Figure 3**. T7 encodes 56 open reading frames (ORFs) most of whose functions are known. The ORFs include three cases of programmed translational frameshifting.

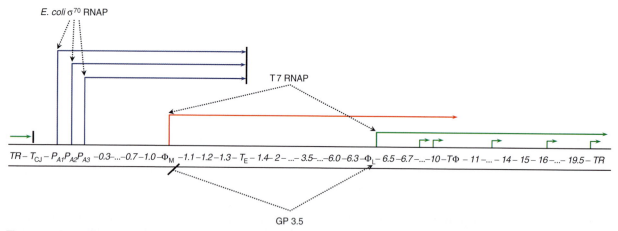

Figure 3 Phage T7 promoters, transcripts, repressors, and activators. Only genes and sites mentioned in the text are shown, and the diagram shows the gene order as it appears in packaged DNA. *Escherichia coli* RNAP transcribes the early promoters, P_{A1}, P_{A2}, and P_{A3} (blue arrows), and T7 RNAP transcribes the middle and late promoters, Φ_M and Φ_L (red and green arrows, respectively). Five other T7 RNAP-dependent promoters located downstream of Φ_L are also shown. T_E, T_Φ, and T_{CJ} are the terminators discussed in the text. TR means terminal repeat. The transcript from Φ_L (green arrow) continues through the left terminal repeat and ends at T_{CJ} because T7 chromosome ends are joined by a single copy of the terminal repeat during DNA replication (see text). Activation and repression of Φ_L and Φ_M by Gp3.5 are indicated by dotted lines ending in an arrow and bar, respectively. A bar across the transcript arrow heads signify terminated transcripts.

Temporal Expression of Genes

Studies of T7 protein synthesis after infection of the host show three temporal stages of gene expression. Early or class I genes (genes *0.3* to *1.3*), middle or class II genes (*1.4* to *6.3*) and late or class III genes (*6.5* to *19.5*) with some overlap, as described later. All T7 genes are transcribed from the same strand, from left to right. The genes in each class are physically contiguous, and the three groups, I, II and III, are also positioned from left to right in the phage chromosome in that order.

Early genes are transcribed by host σ^{70}-RNAP, and their products modify the intracellular environment so as to facilitate phage development. Only gene *1.0* in this group is essential for phage growth; it encodes a single polypeptide RNA polymerase (T7 RNAP) that specifically transcribes the class II and III genes. The class II gene products are involved in phage-specific DNA metabolic reactions, whereas the class III genes encode DNA packaging, virion assembly, and host cell lysis proteins.

DNA Entry and Coupled Transcription

Transcription of T7 and T3 phage DNA, unlike that of some other phages, starts before the complete genome is injected into the host. After adsorption of the phage, the products of phage genes *14*, *15*, and *16* are released from the virion and form a tunnel from the tail tip across the outer and inner membranes of the host for DNA entry into the cytosol. However, only 850 bp from the left end of the phage chromosome (**Figure 3**), which include the major early promoters P_{A1}, P_{A2}, and P_{A3}, enter the host. Transcription from these promoters by host RNAP must begin before additional DNA can enter. The reason for the barrier to further DNA entry is not known, but it is DNA sequence independent. Mutations that allow DNA entry in the absence of transcription map in the middle of gene *16*, suggest that the middle segment of Gp16 is at least partly responsible for the blockade. Although transcription from the three early promoters in the 850 bp region is sufficient, T7 RNAP-mediated transcription takes over in completing the DNA-entry process in normal infection.

Early Transcription

As mentioned earlier, T7 transcription begins immediately after infection at the strong adjacent P_{A1}, P_{A2}, and P_{A3} promoters, although several weak promoters, as well as some RNAP binding sites, which do not initiate transcription, are distributed in the class I region (not shown in **Figure 3**). The importance, if any, of these other promoters and the RNAP binding sites is not known. Transcription from the early promoters continues after complete genome entry and terminates at different sites with incomplete efficiency within the $\sim$8000 bp long polycistronic transcription unit or at a Rho-independent terminator (T_E) located at the end of the transcription unit encompassing genes *0.3–1.3*. T7 early transcription terminates at T_E both *in vivo* and *in vitro*, but T3 early transcription terminates at T_E only *in vivo*, suggesting the involvement of a termination factor in the latter phage.

Middle Transcription

Transcription of the class II genes (from *1.4* to *6.3*) by T7 RNAP begins about 4–6 min and continues until about 15 min after infection. These genes are transcribed from several T7 RNAP specific promoters, of which the major one, Φ_M, is shown in **Figure 3**. Note that three class I genes – *1.1*, *1.2*, and *1.3* – are also transcribed by T7 RNAP because they are located downstream of the middle promoter Φ_M but upstream of the early terminator T_E (**Figure 3**). T7 but not host RNAP ignores the T_E terminator when transcribing genes *1.1* to *1.3*. Transcription from Φ_M terminates weakly at the T7 RNAP terminator, T_Φ, which is located between genes *10* and *11*. Specific transcription from Φ_M can be demonstrated *in vitro*, but only on templates that lack the class III promoters, indicating that a mechanism exists for high level transcription from Φ_M *in vivo*. It has been suggested that Φ_M has much-reduced affinity for T7 RNAP compared to that of class III promoters, but the middle promoter is stimulated *in vivo* by an unusually high local concentration of T7 RNAP, which is encoded by an adjacent gene.

Late Transcription

There are several T7 RNAP promoters that direct expression of late genes (**Figure 3**). The first late promoter, Φ_L, directs transcription of all genes located to the right of gene *6.3*, from gene *6.5* to the last gene, *19.5*. The late genes are transcribed from 8 min after infection until cell lysis occurs. Compared to the strength of Φ_M, Φ_L is intrinsically strong. Note that class III genes *6.5* through *10* are transcribed from both Φ_M and Φ_L, presumably because these gene products are needed in larger amounts. Although T7 RNAP initiating at Φ_L also terminates at T_Φ with reduced efficiency, all late transcription terminates at site T_{CJ}. T_{CJ} is downstream of Φ_L in phage DNA that is actively replicating and transcribed within infected cells, but the two sites are separated by the ends of the packaged T7 chromosome (**Figure 3**). This is because replication produces concatemers – repeating units of nonredundant T7 sequence that are separated by the terminal repeat – and packaging produces the chromosome ends from the concatemers. T_{CJ} has the sequence 5′-ATCTGTT with no secondary structure potential. It acts as a transcription pause site *in vitro*.

T7 RNA Polymerase

T7 RNAP is a single subunit enzyme of 98 000 Da. It recognizes a 19 bp sequence (consensus: TAATAC-GACTCACTATAGG) centered on position −8 from the start point of transcription (+1), which is the first G of the final GG sequence. There are 17 identified middle and late promoters that are used exclusively by T7 and not by host RNAP. T7 RNAP does not recognize host promoters and many of the host terminators, whether the latter are Rho-dependent or Rho-independent. T7 RNAP can utilize both T7 and T3 promoters but T3 RNAP is somewhat specific. The open complexes made by T7 RNAP distort at least 10 bp of double-stranded promoter DNA and are unstable until short RNA oligomers are made. Both T3 and T7 RNAP initiate transcription with 5′-GG. Significantly, the T7 enzyme catalyzes RNA synthesis 5–10 times faster than the host RNAP. As mentioned, besides Φ_M and Φ_L, there are several other promoters distributed among class II and III genes that are transcribed by T7 RNAP, and transcription from these promoters increases the products of certain genes. The 3′-ends of the RNA made by the phage enzyme without the aid of any other factors at T_Φ are also different from that by the host enzyme, $G(U)_6G$-3′ for T7 versus $G/C(U)_{6-7}$-3′ for the host. As mentioned below, the termination efficiency is modulated by the phage protein Gp3.5. T3, but not T7, RNAP needs the *E. coli* protein DnaB helicase for initiation, elongation, and termination of replication *in vivo*. DnaB is essential for both phage DNA replication and RNA synthesis in high salt *in vitro*. The significance of the sharing of a protein between transcription and replication is not known.

Regulation of Transcription

T3 and T7 regulate gene expression mainly at the level of transcription. We mentioned that the gene products made from early transcripts create a favorable environment for phage growth and produce T7 RNAP. At this point, transcription of class II genes begins. As soon as the synthesis of middle proteins reaches an optimal level, both early and middle transcription are turned off, and the transcription machinery is directed exclusively to expression of late genes. The temporal control and relative amounts of transcription of different genes are achieved in several ways. First, temporal control is facilitated by positioning of the three groups of genes, I, II, and III, in the order from left to right; and second, mechanisms exist to switch from early to middle and from middle to late transcription (below). In addition, some gene products are needed in large amounts for a sustained period. The corresponding genes are located in areas of overlapping transcription and possess additional, strategically located promoters in the phage chromosome. For example, as noted earlier, genes *1.1* through *1.3* are transcribed as part of both early and middle transcription units. Next, differential levels of transcription of some class I genes are achieved by 'polarity', that is, expression of promoter proximal genes in higher amounts than promoter distal ones. Polarity is caused by Rho-dependent transcription termination signals in the early transcription unit. Finally, the mode of entry of T7 DNA into the host is also an important element in temporal control of transcription.

There are three components of the early to middle switch, and all are due to the products of early transcription. First, gene *0.7* product inactivates host RNAP. Gp0.7 is a seryl–threonyl protein kinase that phosphorylates the β- and/or β′-subunits of host RNAP. Phosphorylation decreases early transcription to about 25%. The kinase inactivates itself in a timely fashion after it is no longer needed. Second, T7 RNAP, the gene *1.0* product, accumulates. This enzyme recognizes the middle and late promoters. The host RNAP activity is also inhibited by the third factor, the middle gene *2.0* product. Gp2.0 binds to the host RNAP in a 1:1 stoichiometry. Although Gp2.0 is essential for phage growth, Gp0.7 is not. Much needs to be learned about Gp2.0 action.

The switch from middle to late transcription requires the product of middle gene *3.5*. This protein binds to T7 RNAP, forming a complex that inhibits transcription strongly from Φ_M but weakly from the late promoter, Φ_L. Biochemical experiments have shown that Gp3.5 inhibits the transition from the initiating to the elongating complex. The Φ_M–T7 RNAP complexes are much weaker than the corresponding late promoter complexes. Gp3.5 destabilizes the former complexes more readily than the latter ones. Interestingly, Gp3.5 has three additional functions. It is a lysozyme, which is required for release of progeny phage after infection, it stimulates DNA replication, and, indirectly, it enhances termination at T_{CJ}. Ordinarily, T_{CJ} is a pause site for T7 RNAP, and Gp3.5 increases pausing at this site.

RNA Processing

The long polycistronic RNA molecules made at early, middle, and late times after phage infection are cut at specific sites by the host enzyme, RNase III, to yield discrete, relatively stable mRNAs. The processing sites in early T7 RNA have been studied in more detail. RNase III cleaves at specific sites within hairpin structures that are located between coding sequences. Although some of the RNA molecules that result from RNase III cleavage are translated more efficiently than the uncut RNA, RNase III cleavage is not essential for phage growth.

Phage N4

The virulent phage N4 uses transcription strategies that differ significantly from those of T7 or T4. Early genes are located at the left end of the N4 chromosome, in the first 10 kbp of the ~70 kbp, linear, double-stranded DNA. Transcription of the early genes requires a phage-coded, virion-encapsulated polymerase, vRNAP, which is injected into the host along with the phage DNA. This polymerase is distantly related to the T7 family of RNA polymerases (above). N4 early promoter sequences have a conserved motif and a sequence that can form a hairpin with a stem of 5–7 nt and a loop of 3 nt in single-stranded DNA. *In vivo* transcription by vRNAP requires the activity *E. coli* gyrase and *E. coli* SSB (*Eco*SSB), the host single-stranded DNA binding protein that is normally a member of the host DNA replication machinery. It is thought that gyrase-catalyzed, negative supercoiling of the DNA leads to local melting and extrusion of the hairpin, which then provides the features needed for polymerase promoter recognition. *Eco*SSB melts the complementary strand hairpin while the template strand hairpin, which is resistant to SSB melting, remains available for vRNAP recognition. In addition, SSB prevents annealing of the RNA product to the template DNA. Thus, *Eco*SSB acts as an architectural transcription factor as well as a recycling factor that makes the template DNA available for multiple rounds of early transcription.

Middle N4 genes encode the N4 replication proteins and are located within the left half of the N4 genome. Middle genes are transcribed by a second phage-encoded RNA polymerase, N4 RNAPII, composed of the N4 early products p4 and p7. This polymerase is also a member of the T7 polymerase family. N4 middle promoters contain an AT-rich sequence at the transcription start site. By itself, N4 RNAPII is inactive on double-stranded DNA and has very limited activity on single-stranded DNA. Active transcription requires another N4 protein, gp2, which interacts specifically with single-stranded DNA and with N4 RNAP II. It is thought that an as yet unidentified protein binds to the promoter element while gp2 stabilizes a single-stranded region at the promoter start and brings N4 RNAPII to the transcription start site. This system differs significantly from the other well-characterized phage transcription systems and may provide a good model system for investigating mitochondrial transcription, which is also carried out by a member of the T7 polymerase family.

Transcription of N4 late genes, which are located in the right half of the phage genome, requires *E. coli* RNA polymerase. Late promoters have regions of limited similarity to σ^{70} DNA elements and when present in linear DNA, they are weak promoters for σ^{70} polymerase. Transcription *in vivo* and with linear templates *in vitro* requires the N4 SSB protein. Surprisingly, the single-stranded DNA-binding function of SSB is not needed for this activation. Rather a transactivating domain of SSB that interacts with the β'-subunit of *E. coli* RNA polymerase core is required. The mechanism of this activation is not known, but is thought to occur at a step after initial binding of RNA polymerase to the promoter. Thus, this late N4 system should also provide insight into a mode of transcriptional activation that has not been previously characterized.

Phage λ and the Regulation of Lysogeny and Lysis

Lambda is the founding member of a large family of temperate bacteriophages that are related by common gene organization and limited sequence similarity. Although λ belongs to the *family Siphoviridae* by virtue of its long, noncontractile tail, several close relatives are not members of this group. Lambda chromosomes isolated from virions consist of about 49 kbp of linear, nonpermuted, mostly double-stranded DNA. Complementary single-stranded extensions of 12 bases at each end enable the chromosome to circularize by end-joining after it is transferred from the virion into an infected cell. Transcription of all phage genes, both early and late, is catalyzed by host RNAP. Infection with λ, unlike infection with virulent phages, does not always lead to cell lysis and liberation of progeny phage. Instead, infected cells frequently survive and give rise to lysogens – cells that contain a quiescent, heritable copy of the λ chromosome called prophage (**Figure 4**). The prophage expresses few genes and replicates in synchrony with the host, so that each daughter cell also contains a prophage. The 'decision' between lytic growth and lysogen formation after infection is controlled at many levels (below). Indeed, λ, like other temperate species, has evolved sophisticated mechanisms to ensure that a substantial fraction but not all infected cells survive and become lysogenic, and that intermediate responses, for example, survival without inheritance of a prophage (abortive lysogeny), are rare.

Once a lysogen has formed, loss of the prophage, or 'curing', is infrequent. Lambda and its relatives insert their chromosomes into specific sites in the bacterial chromosome during the establishment of lysogeny and remain there during subsequent cell division. Although site-specific prophage insertion is widespread, it is not a general feature of lysogeny; some temperate phages that are not closely related to λ use other strategies that also ensure prophage retention. For example, phage Mu, inserts its chromosome nonspecifically, while phage P1 does not insert at all, but instead exists in lysogens as a single-copy plasmid whose replication is coupled to that of the host chromosome. Lysogeny is not a dead end; the prophage can switch into an

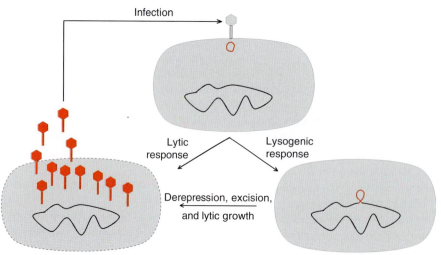

Figure 4 The λ life cycle. After infection, the linear phage chromosome enters the cell and the single-stranded ends cohere to form a ring (in red, top panel). When the infected cell enters the lytic cycle, early and late phage genes are fully expressed, phage DNA replication ensues, the resulting replicas are packaged into phage heads, and the cell lyses, releasing infectious particles into the environment (left bottom panel). When the infected cell enters the lysogenic cycle, the phage expresses mainly early genes to a limited extent. Among these genes are *cI* and *int*. CI prevents further phage gene expression, and Int catalyzes insertion of the phage into the bacterial chromosome (lower right panel). The resulting lysogen passes a copy of the inserted, repressed prophage to its descendents. When repression fails, the prophage excises and enters the lytic cycle.

active state in which many virus genes are expressed, the prophage is excised from the bacterial chromosome, and normal lytic growth ensues (**Figure 4**). This transition, known as the genetic (or, more correctly, epigenetic) switch, is normally infrequent but can be induced experimentally by treatment of cells with DNA damaging agents. Such treatment causes a signal cascade that leads to autoproteolytic cleavage of repressor, a phage protein that prevents most phage gene expression (below). Experimental induction of lysogens by DNA damage is not universal; lysogens of some temperate phages, such as P2, are not inducible.

The Establishment and Maintenance of Lysogeny

Immediately after infection, host RNAP initiates transcription at the λ early promoters, P_L and P_R (**Figure 5**). P_R directs the transcription of the *cII* gene, whose product is the central effector of the lysogenic response. CII protein promotes lysogeny by activating the coordinate initiation of transcription at three λ promoters: P_{RE}, P_{Int}, and P_{aQ} (**Figure 5**). P_{RE}, the promoter for repressor establishment, directs the transcription of the *cI* gene, and P_{RE} activation thus provides a burst of CI protein synthesis shortly after infection. CI, or repressor, prevents transcription of nearly all λ genes in an established lysogen, and, indeed, is required for the maintenance of lysogeny. CI blocks the initiation of early transcripts by binding to operator sites that overlap P_L and P_R (below). It blocks late gene transcription indirectly, by preventing the

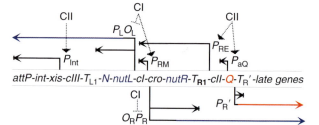

Figure 5 Phage λ promoters, transcripts, repressors, activators, and antiterminators. Only genes and sites mentioned in the text are shown. The top, light, horizontal line represents a DNA strand written with 5′ to 3′ chemical polarity, and the parallel bottom line represents the complementary strand written 3′ to 5′. Activation or repression of individual promoters by CI and CII proteins is indicated by dotted lines terminated by an arrowhead or bar, respectively. Transcripts and their orientation are indicated by heavy solid lines with arrowheads, with each transcript adjacent to the transcribed strand. A bar across an arrowhead indicates a terminated transcript. T_{L1} and T_{R1} are the first of several transcription terminators in the P_L and P_R operons, respectively, and T_R' is the first terminator in the P_R', or late operon. N suppresses terminators by binding to ECs after they transcribe the *nutL* and *nutR* sites (see text). This is indicated by a color change from black to blue in the transcript. Q suppresses termination at T_R' and downstream terminators by binding to the EC shortly after initiation at P_R' (see text). This is indicated by a color change from black to red in the transcript.

synthesis of Q, an early protein that activates transcription of late genes (**Figure 5**). CI is not only a repressor, it is also an activator. It activates the initiation of transcription at P_{RM}, the promoter for repressor maintenance, a second promoter for *cI* transcription. Thus, once *cI* transcription has initiated from P_{RE}, it is self-sustaining. This is essential to maintain lysogeny because CI prevents continued transcription of *cII*.

P_{Int}, the second CII-activated promoter, directs the transcription of the *int* gene. Int protein or integrase is a topoisomerase that binds to specific sequences within the phage and host 'attachment sites'. These two sites differ from each other in the number, type, and disposition of Int binding sites. The phage attachment site also contains sequences that specifically bind several molecules of IHF, a host-encoded DNA bending protein that promotes the formation of a recombinogenic structure that can capture the bacterial attachment site. Int then uses its topoisomerase activity to catalyze breakage, exchange, and rejoining of the four DNA strands at specific positions within the two sites. This inserts the prophage into the bacterial chromosome and ensures its inheritance by the progeny of the infected cell. Reversal of insertion, or excision, is rare as long as CI continues to repress transcription (below).

Finally, P_{aQ}, the third CII-activated promoter, directs the synthesis of an antisense RNA that inhibits the production of Q, a phage protein that promotes transcription of late genes. Inhibition of Q synthesis increases the frequency of lysogeny after infection by delaying the production of late phage proteins that would kill the cell. Once lysogeny is established, repressor prevents further transcription of Q, as noted earlier.

The rate of CII accumulation after infection is controlled by the action of several phage and host gene products, a few of which are described in more detail below. Initially, λ Cro protein and, eventually, CI inhibit initiation of the *cII* transcript at P_R. Lambda N protein promotes transcript elongation through *cII* by suppressing T_{R1}, a rho-dependent transcription terminator (**Figure 5**). Host RNase III in concert with the antisense OOP RNA (not shown in **Figure 5**) cleaves the *cII* transcript. Finally, host FtsH protease degrades CII, but λ CIII protein inhibits its degradation. The net effect of these multiple layers of control is probably to make the lysis–lysogeny decision responsive to internal and environmental cues. A complete quantitative description of the kinetic details of the response is cherished goal of systems biologists.

Dimers of CII and dimers, tetramers, and octamers of CI bind to specific sites that overlap the promoters they regulate. Both use a helix–turn–helix DNA binding motif to recognize their sites. CII binds specifically to sequences that straddle the −35 element of the relevant promoters, and binding is thought to activate transcription initiation through contacts between the tetramer and σ^{70} RNAP. These contacts enhance binding of RNAP to the promoter and isomerization of the bound complex. CI dimers bind to repeated tandem operator sites that flank and interpenetrate each promoter: O_{L1}, O_{L2}, and O_{L3} for P_L,
and O_{R1}, O_{R2}, and O_{R3} for P_R. One dimer binds to each operator, and neighboring dimers can interact cooperatively with each other as well as with dimers bound near the other promoter as shown in **Figure 6**. Cooperative interactions strengthen operator binding and make it more sensitive to changes in CI concentration, thus sharpening the transition between full repression and full derepression and decreasing the probability of biologically unproductive intermediate states. Repressor dimers bound to O_{L1} and O_{L2} occlude binding of RNAP to P_L, and, similarly, binding to O_{R1} and O_{R2} occludes P_R. Repressor bound at O_{R2} has an additional function; it contacts the σ-subunit of RNAP bound to P_{RM}, promoting isomerization of the bound enzyme and subsequent *cI* transcription. O_{R3} binds repressor relatively weakly and is not required for repression

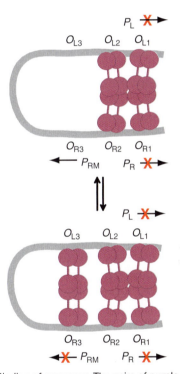

Figure 6 Binding of repressor. The pairs of purple dumbbells represent repressor dimers bound to subsites of O_L and O_R. One end of each dumbbell represents the N-terminal portion of the protein, which contains a DNA-binding domain. The other end represents the C-terminal portion of the protein, which contains regions that can participate in cooperative interactions with similar regions in neighboring bound dimers to form tetramers, and with a tetramer bound to a distant operator to form an octamer. At low concentrations of repressor, the cooperative interactions and the intrinsic affinities of the binding sites lead to occupancy of subsites 1 and 2 of each operator. In this configuration, promoters P_L and P_R are repressed and promoter P_{RM} is activated (top of figure). At higher concentrations of repressor, all six subsites are occupied and all three promoters are repressed (bottom of figure). Repression is indicated by a red 'X' over the arrows representing transcripts.

of P_R. This operator is occupied only at relatively high repressor concentration, and occupancy prevents a further increase in concentration by repressing P_{RM}. Thus, *cI* transcription from P_{RM} is both positively and negatively autoregulated.

Prophage Excision

If the level of repressor in a lysogen drops below the level needed to inhibit transcription, the prophage expresses genes needed for excision and lytic growth. Two λ proteins, Int and Xis, are needed for prophage excision. Int, as noted earlier, catalyzes breakage and rejoining of DNA strands. Xis is a directionality factor; it is required for excision and inhibits insertion. Analogous directionality factors are produced by many other temperate phages. Xis molecules bind specifically and cooperatively to sequences within the attachment sites and help to recruit Int to an adjacent binding sequence. This favors recombination between the two attachment sites that flank the prophage and inhibits recombination between the phage and host attachment sites. Interestingly, when Xis concentration is limiting, the host-encoded Fis protein can alleviate the deficiency. It is likely that Fis and IHF, the other host-encoded protein needed for recombination, help to couple the efficiencies of insertion and excision to the physiological state of the host.

Although Int and Xis are encoded by adjacent genes, the protein levels are differentially controlled so that the efficiency and direction of recombination are coupled to the appropriate stages of the phage life cycle. In infected cells destined for lysogeny, *int* but not *xis* is transcribed from CII-activated P_{Int} (**Figure 5**). Xis, which is made from the P_L transcript, is unstable, so it inhibits insertion only briefly when transcription from P_L is repressed after infection. After prophage induction, both *int* and *xis* are transcribed from P_L, and their products promote rapid prophage excision. Excision is usually followed by phage DNA replication and the expression of late genes. Although excision is not a prerequisite for the subsequent events, it is necessary for synthesis of concatemers of λ DNA, a substrate for packaging phage chromosomes into virions.

Lytic Growth

Lambda lytic growth is broadly similar to that of virulent phages, although the regulatory strategies differ considerably. The Cro, N, and Q proteins regulate transcription mainly during lytic growth. None of them is produced in lysogens because CI directly represses their syntheses. Cro is a transcriptional repressor, similar to CI in binding specificity and structure. However, subtle differences in binding specificity and the absence of cooperative interactions weaken Cro repression of P_L and P_R, and adapt it to serve as a negative regulator of early gene transcription during lytic growth. Cro is essential for lytic growth because it represses transcription of *cII*, thus enabling a fraction of infected cells to enter the lytic pathway. Both N and Q are 'antiterminators'; they act during transcript elongation to suppress termination. N enhances expression of early, and Q of late genes (**Figure 5**). Nearly all λ genes are located downstream of terminators and therefore require antitermination for full expression. *Cro* and *N* are exceptions, and, indeed, they are normally expressed before any other λ proteins are produced.

What is the role of terminators in λ biology? Their role in the expression of early genes is unclear, since termination is rapidly and efficiently suppressed as N accumulates after infection. Possibly the delay in early gene expression until the concentration of N reaches an effective level influences the lysis–lysogeny decision. Indeed, it has been shown that the efficiency of translation of the N message is increased by RNase III cleavage, and that the intracellular concentration of RNase III varies with the growth phase of the culture. Alternatively, or in addition, termination adds to the effect of repressor in silencing early gene expression in lysogens. In any event, early gene terminators are likely to confer some evolutionary advantage because several are conserved in phages related to λ. By contrast, the role of termination in controlling late gene expression is clear because P_R', the promoter for these genes, is constitutive. Thus, termination prevents transcription of late genes in lysogens and premature transcription of late genes during lytic growth.

Antitermination by N and Q is limited to λ transcription because both proteins bind specifically to phage nucleic acid before they act. N binds nascent transcripts of the *nutL* and *nutR* sites, which are located downstream of P_L and P_R, respectively (**Figure 5**). RNA binding alters N structure and facilitates its transfer to the nearby elongation complex, where it acts to suppress termination. Several host proteins enhance N-dependent antitermination: NusA, NusB, NusE/RpsL, and NusG. Some of these proteins bind *nut* RNA, some bind the elongation complex, and one, NusA, binds N. All of these proteins have roles in host biology that are independent of N, and their abundance and activity might link the efficiency of antitermination to as yet unknown aspects of host physiology. Nevertheless, N is clearly the central actor since some level of antitermination has been observed *in vitro* in the absence of all of the Nus factors and the *nut* site when N is present in excess. Despite considerable effort, the location of the N binding site on the EC and the nature of the antiterminating modification are still unknown. The modification is stable *in vitro* in the presence of NusA and persists *in vivo* for considerable time and distance after it has been established. Persistence is important for λ physiology since important terminators, notably those preceding the *Q* and *xis* genes, are located more than 5 kbp from the *nut* sites.

Q binds to a site in the nontranscribed region of the λ late promoter, P_R' (**Figure 5**). It is transferred from this site to the EC after synthesis of a 16 or 17 nt transcript. At these points the EC pauses because the σ^{70}-subunit, which has not yet been released, binds to a transcribed DNA sequence that resembles the extended −10 regions of certain genuine *E. coli* promoters. Sigma-70 interacts with Q after its transfer to the EC, but the locations of any additional contacts are unknown. Q modification of the EC decreases the duration of the 16/17 pause and suppresses downstream terminators.

Q and N suppress nearly all terminators that have been tested, both intrinsic and rho-dependent, and also increase the average rate of elongation. The mechanisms of suppression and, indeed, of termination are controversial. It is widely accepted that formation of the hairpin stem of intrinsic terminators disrupts the upstream (5′) segment of the RNA:DNA hybrid, which consists mainly of relatively weak rU:dA base pairs. It is not clear how hairpin formation is linked to hybrid disruption, and what the subsequent steps in termination consist of. Among the mechanisms that have been considered to explain N and Q action are inhibition of hairpin formation, strengthening of the weak RNA:DNA hybrid, and increase of elongation rate so that the EC escapes from the termination zone.

It is interesting that the termination/antitermination mode of gene control is not general among temperate phage. For example, it has not been found in phages Mu or P2, which regulate transcription exclusively through activators and repressors of initiation. However, several interesting variants of the termination/antitermination strategy have been discovered in other phages. In lysogens of the mycobacterial phages L5 and Bxb1, the repressor arrests the progress of the EC by binding to numerous 'stoperator' sites in the phage chromosome and physically blocking further translocation. The repressor of coliphage P4 also prevents transcript elongation, but by a quite different mechanism. P4 repressor is an RNA that activates termination by pairing with phage transcripts upstream of terminators. It has been suggested that the resulting RNA:RNA duplexes alter the secondary structure of the transcripts in such a way as to favor termination. Finally, HK022, an *E. coli* phage related to λ, uses antitermination to express its early genes but lacks a protein analogous to N. Instead, the HK022 P_L and P_R operons encode short, structured RNA segments that bind to the EC and suppress downstream terminator sites as efficiently as N does for λ. These RNAs act exclusively *in cis*, thus limiting antitermination to phage transcripts.

Conclusions

Both temperate and virulent phages alter the transcriptional and translational machinery of their hosts to suit their own ends. We have described phage-encoded mechanisms that change promoter utilization, transcript elongation rate, and transcript termination efficiency. These changes can be modest or drastic, ranging from chemical modification of a host RNAP subunit to synthesis of a new and structurally dissimilar polymerase. The selective advantage conferred by some of these changes is poorly understood and may be an adaptation to evolutionary pressures that are not apparent in laboratory conditions. An example that we have not previously described is the existence of phage-encoded transfer RNAs that duplicate or augment existing host activities. Among the differences between the regulatory strategies used by temperate and by virulent phages is the lethal character of some mechanisms used by the latter group. Most striking among these is degradation of host DNA, which can be seen soon after infection by T4 and T7. Another is inactivation of host RNAP. Temperate phages can also kill their hosts, but this typically occurs after the phage enters the lytic growth pathway. The essence of the temperate lifestyle is peaceful coexistence of the prophage and the host. A prophage that conferred a selective disadvantage on its lysogenic host would probably not last long in nature. If there is a consistent difference between the regulatory strategies of the two phage types, it is the ability of temperate species to prevent the expression of genes that harm the host. In contrast, the essence of being a lytic phage is to invade, take over, and destroy as efficiently as possible. Nevertheless, the extent of this difference should not be exaggerated. Some virulent phages have close relatives that are known only as prophage-like components of bacterial chromosomes, and some temperate phages have close relatives that have lost elements required for repressing gene expression and are, at first view, indistinguishable from virulent species. In any event, the tactics of both the temperate and virulent phages have taught researchers many elegant mechanisms for gene expression and control.

Acknowledgment

The authors are grateful to R. Bonocora for designing Figure 1 and to Rodney King and Bill Studier for comments. This work was supported by the Intramural Research Program of the NIH, National Institute of Child Health and Human Development, National Institute of Diabetes and Digestive and Kidney Diseases, and the National Cancer Institute.

See also: Genome Packaging in Bacterial Viruses; History of Virology: Bacteriophages; Icosahedral ssDNA Bacterial Viruses.

Further Reading

Calendar R (2006) *The Bacteriophages*. New York: Oxford University Press.

Greive SJ and von Hippel PH (2005) Thinking quantitatively about transcriptional regulation. *Nature Reviews Molecular Cell Biology* 6(3): 221–232.

Miller ES, Kutter E, Mosig G, Arisaka F, Kunisawa T, and Ruger W (2003) Bacteriophage T4 genome. *Microbiology and Molecular Biology Reviews* 67(1): 86–156.

Oppenheim AB, Kobiler O, Stavans J, Court DL, and Adhya S (2005) Switches in bacteriophage lambda development. *Annual Review of Genetics* 39: 409–429.

Ptashne M (2004) *A Genetic Switch. Phage Lambda Revisited*. Cold Spring Harbor, NY: Cold Spring Harbor Press.

Transmissible Spongiform Encephalopathies

E D Belay and L B Schonberger, Centers for Disease Control and Prevention, Atlanta, GA, USA

© 2008 Elsevier Ltd. All rights reserved.

Glossary

Alleles Mutually exclusive forms of the same gene on homologous chromosomes.

Codon A sequence of three nucleotide bases in a gene that specifies an amino acid to be incorporated into a protein during its synthesis.

Encephalopathy Brain disease resulting from infectious agents such as prions.

Genetic polymorphism The regular occurrence in a population of two or more components of a gene with greater frequency than can be explained by recurrent mutation.

Mutation A permanent inheritable change in the genetic material of a host.

Prion Small, proteinaceous infectious particles that resist inactivation by procedures that affect nucleic acids; they are believed to be the causative agents of scrapie and other spongiform encephalopathies of animals and humans.

Prion protein A normal cellular version of the infectious prions encoded by the prion protein gene; its cellular function is not well understood.

Introduction

Prion diseases, also known as transmissible spongiform encephalopathies, constitute a group of fatal, human, and animal subacute neurodenerative diseases caused by an unconventional agent. Strong evidence indicates that their etiology and pathogenesis involve modification of a host-encoded normal cellular protein known as the prion protein (PrP^C). A unique hallmark of this group of diseases includes their sporadic occurrence without any apparent environmental source of infection or genetic association with prion protein gene mutations or as a result of disease transmission from infected humans and animals. Neuropathologically, most prion diseases manifest with widespread neuronal loss, spongiform lesions, and astrogliosis. Characteristically, no signs of inflammation are detected in pathologic samples. However, the presence of an abnormal, pathogenic prion protein, often called a scrapie prion protein (PrP^{Sc}) after the first recognized prion disease, scrapie, is demonstrable in the brain and often in other tissues of humans and animals affected by prion diseases. The incubation period of acquired forms of prion diseases is usually measured in years and sometimes in decades.

Etiologic Agent of Prion Diseases

Studies that characterized the causative agents of prion diseases initially focused on understanding the agents responsible for scrapie in sheep. This disease is the most common prion disease of animals and was first reported in the 1730s in England. Before the 1980s, prion diseases were widely believed to be caused by 'slow viruses' despite the fact that no viral particles or disease-specific nucleic acids were identified in association with scrapie transmission in laboratory animals. The scrapie agent could not be grown in cell culture, a feature that hampered research to characterize the agent. The successful transmission of scrapie to laboratory mice in 1961 contributed to overcoming this hurdle and greatly facilitated efforts to understand the nature of the agent. Quantitation of infectivitiy in a sample, however, still was tedious and required a year to complete.

Two distinctive properties of the scrapie agent led to the suspicion that the agent was devoid of nucleic acids and, thus, may not be a virus but primarily composed of a protein. These properties included (1) resistance of the scrapie agent to procedures, such as treatment with ultraviolet light and ionizing radiation, that normally inactivate other microorganisms, including viruses and (2) the reduction of the infectivity of the scrapie agent by procedures that denature or degrade proteins. The concept that the scrapie agent might replicate in the absence of nucleic acids or might just be a protein was postulated as early as the 1960s by Alper and colleagues, Pattison and Jones, and Griffith. In 1982, Prusiner and colleagues described the successful purification of a hydrophobic protein, the presence of which was required for scrapie transmission in laboratory animals. Prusiner introduced the term 'prion' to describe this protein by borrowing and mixing the first few letters from the descriptive phrase 'proteinaceous infectious' particle. Since then, additional evidence has accumulated indicating that prions may be acting alone in causing prion diseases. However, the critical steps in the production, propagation, and pathogenesis of this infectious protein remain unclear. As a result, the study of prions has become an important, relatively new area of biomedical investigation. Some critics of the prion hypothesis still believe that nucleic acids undetected by current methods may play a crucial role in the pathogenesis of prion diseases.

Prions appear to be composed largely or entirely of the abnormal protein designated as PrP^{Sc}. This protein is an abnormal conformer of a host-encoded cellular protein, PrP^C, which in humans is encoded by the prion protein gene located on the short arm of chromosome 20. PrP^C is a structural component of cell membranes of neurons and other tissues. Its normal function is poorly understood. However, it appears to be involved in supporting neuronal synaptic activity, copper binding, and neuroprotective functions by interacting with other cell-surface proteins.

The underlying pathophysiologic mechanism in the occurrence of prion diseases involves the biochemical conversion of PrP^C into the pathogenic PrP^{Sc}. This conversion occurs by a poorly defined post-translational autocatalytic process, possibly requiring the aid of cofactors such as proteins or nucleic acids. The initial instigating PrP^{Sc} molecule may originate from exogenous sources or within the brain from somatic or germline prion protein gene mutations. Knockout mice devoid of the prion protein gene are resistant to scrapie infection, indicating that the production of PrP^C is required for the generation and propagation of PrP^{Sc}. During its conversion, PrP^{Sc} acquires more β-sheet structure that renders it resistant to proteolytic enzymes, conventional disinfectants, and standard sterilization methods. A higher proportion of the tertiary structure of PrP^C, on the other hand, is composed of α-helices which make it more sensitive to denaturation by proteinase-K treatment. Removal of the neuroprotective functions of PrP^C as more of it becomes converted to the pathogenic PrP^{Sc} and accumulation of PrP^{Sc} in neurons have been suggested as major contributory factors in the underlying pathogenesis of prion diseases and widespread neuronal death.

Prion Diseases of Animals

Prion diseases of animals include scrapie in sheep, goats, and moufflon; bovine spongiform encephalopathy (BSE) in cattle; feline spongiform encephalopathy in domestic and zoo cats; ungulate spongiform encephalopathy in exotic zoo ruminants; chronic wasting disease (CWD) in deer, elk, and moose; and transmissible mink encephalopathy (TME) in mink (**Table 1**). Epidemiologic evidence indicates that feline and ungulate spongiform encephalopathies were caused by the BSE agent possibly transmitted via consumption of BSE-contaminated feed. Although strong evidence is lacking, speculations have persisted that scrapie in sheep may have been the original source of prion diseases in other animals, such as BSE and CWD.

TME occurred in outbreaks among ranched mink primarily in the US but also in Canada, Finland, Germany, and Russia. The last known outbreak of TME occurred in 1985 in Wisconsin. Nonepizootic cases of TME have not been reported.

Scrapie

Although scrapie was recognized as a distinct clinical entity of sheep over 250 years ago, many aspects of the disease including its natural origin in flocks and the precise means by which it usually spreads remain uncertain. Experimentally, the disease was first transmitted by intraocular inoculation of scrapie brain extracts. In the 1940s, more than 1500 sheep developed scrapie from receipt of a vaccine against louping ill virus that contained scrapie-contaminated lymphoid tissue.

Scrapie transmission may occur by different postulated mechanisms. A commonly cited source of transmission, for example, is the placenta and amniotic fluid of scrapie-infected ewes. These tissues are known to harbor the infectious agent and can cause scrapie when fed to sheep. They may contaminate pastures and barns that, in turn, may remain potentially infectious for years. Another possible source of spread is feces because prion replication occurs in gut lymphoid tissues after oral inoculation in sheep and goats. The importance of oral transmission is supported by experimental studies that detected prions in sheep tonsils examined early during the incubation period. Other poorly defined scrapie transmission mechanisms include (1) the vertical transfer of the scrapie agent and (2) the possible chance occurrence of scrapie caused by hypothesized rare, spontaneous changes in the animal's cellular prion protein.

Scrapie occurs endemically in many countries, including Europe and North America. Australia and New Zealand have sizable sheep populations but are generally recognized as free of the disease. To protect their 'scrapie free' status, these countries have established extensive

Table 1 Animal and human prion diseases

Type of prion disease	Affected host	Year first described or identified
Animal prion diseases		
Scrapie	Sheep and goats	1730s
Bovine spongiform encephalopathy	Cattle	1986
Chronic wasting disease	Deer, elk, and moose	1967
Transmissible mink encephalopathy[a]	Mink	1947
Feline spongiform encephalopathy[b]	Domestic and wild cats	1990
Ungulate spongiform encepahlopathy[b]	Exotic ruminants (e.g., kudu, nyala)	1986
Human prion diseases		
Kuru		1950s
Sporadic CJD		1920s
Iatrogenic CJD		1974[c]
Variant CJD		1996
Familial CJD		1924
Gerstmann–Sträussler–Scheinker syndrome		1936
Fatal familial insomnia		1986

[a]The last known outbreak of transmissible mink encephalopathy occurred in 1981 in Wisconsin.
[b]The known feline and ungulate spongiform encephalopathies are believed to have resulted from BSE transmission.
[c]The first report of iatrogenic CJD was in a recipient of cornea obtained from a CJD decedent; human pituitary growth hormone-associated CJD was first reported in 1985 and dura mater graft-associated CJD in 1987.
CJD, Creutzfeldt–Jakob disease.

safeguards to prevent the introduction of scrapie into their herds from imported animals.

The breed of sheep and polymorphisms of the prion protein gene can greatly influence susceptibility to scrapie. Experimental transmissions with scrapie-infected tissues, for example, have confirmed the differing susceptibility to scrapie of different breeds of sheep. Other studies of Suffolk sheep in the US indicated that susceptibility to scrapie was highly correlated with a polymorphism in the prion protein gene at codon 171 (glycine or arginine); the presence of arginine confers resistance to the disease.

Chronic Wasting Disease

Mule deer, white-tailed deer, and Rocky Mountain elk are the major known natural hosts for CWD. In 2005, a hunter-killed moose was confirmed with CWD in Colorado, suggesting that this member of the deer family too is a natural host. The disease was first identified as a fatal wasting syndrome of captive mule deer in the late 1960s in research facilities in Colorado. Subsequently, it was identified in mule deer in a research facility in Wyoming and then in captive elk in facilities in both Colorado and Wyoming. In 1977, CWD was first recognized as a spongiform encephalopathy.

In the early 1980s, the disease was recognized in free-ranging elk. By the mid-1990s, CWD was regarded as endemic in deer and elk in a contiguous area in northeastern Colorado and southeastern Wyoming. The known geographic extent of CWD has increased dramatically in the US since 1996. As of mid-2006, CWD in free-ranging cervids had been reported in 74 counties of 11 states, including Colorado, Kansas, Illinois, Nebraska, New Mexico, New York, South Dakota, Utah, West Virginia, Wisconsin, and Wyoming. In Canada, since 2001, CWD has been identified in wild deer and elk in Saskatchewan and since 2005 in several mule deer in Alberta.

Among cervids, CWD is most likely transmitted by direct animal contact or indirectly by exposure to a contaminated environment. Unlike other prion diseases, CWD is highly communicable in its natural host, particularly among captive cervids. In addition to becoming emaciated, animals with CWD characteristically develop polydipsia, polyuria, increased salivation, and difficulty in swallowing. These characteristics probably contribute to residual contamination of the environment that appears to be a very important source of transmission.

CWD does not appear to occur naturally outside the cervid family. It has been transmitted experimentally by intracerebral injection to laboratory mice, ferrets, mink, squirrel monkeys, goats, and cattle. However, experimental studies have not demonstrated transmission of CWD to cattle after oral challenge or after having cattle reside with infected deer herds.

No human cases of prion disease with strong evidence of a link with CWD have been identified despite several epidemiologic investigations of suspected cases. In addition, transgenic mice experiments indicate the existence of a significant species barrier against CWD transmission to humans. Nevertheless, concerns remain that this species barrier may not provide complete protection to humans against CWD and that the level and frequency of human exposures to CWD may increase with the spread of CWD in North America.

As a precaution, efforts to reduce human exposure to the CWD agent are generally recommended. Meat from depopulated CWD-infected captive cervids has not been allowed to enter the human food or animal feed supply. To minimize their risk of CWD exposure, hunters are encouraged to consult with their state wildlife agencies to identify areas where CWD occurs and to continue to follow advice provided by public health and wildlife agencies.

Bovine Spongiform Encephalopathy

Prion diseases became a focus of worldwide attention after a large outbreak of BSE in cattle emerged in the UK and spread to other countries. This attention increased dramatically when evidence accumulated indicating that the BSE agent was responsible for the occurrence of variant Creutzfeldt–Jakob disease (CJD) in humans. By far the largest number of BSE cases was reported from the UK, followed by other European countries.

Although BSE was first recognized in the UK in 1986; earlier cases probably occurred since the early 1980s. The number of UK BSE cases increased rapidly in the second half of the 1980s and early 1990s, peaked in 1992 with 37 280 confirmed cases, and has markedly declined since then (**Figure 1**). In 2005, the number of confirmed UK BSE cases had dropped to 225.

Clinically, the signs of BSE include neurologic dysfunction, including unsteady gait with falling and abnormal responses to touch and sound. In some animals, the onset of BSE can be insidious and subtle and may be difficult to recognize. During the early phase of the UK BSE outbreak, the public media introduced the popular term 'mad cow' disease to describe the strange disease causing fearful and aggressive behavior in some of the cattle infected with BSE.

Although the original source of the BSE outbreak is unknown, the two most accepted hypotheses are cross-species transmission of scrapie from sheep and the spontaneous occurrence of BSE in cattle. Data on the latter hypothesis may become available with increased detection and monitoring rates of atypical, as well as typical cases of BSE. According to the first hypothesis, transmission of scrapie to cattle occurred because of the practice of feeding cattle protein derived from rendered animal carcasses including those of scrapie-infected sheep. In the past, cattle feed rendering in the UK involved several treatment steps, including exposure of the feed to prolonged heating in the presence of a hydrocarbon solvent. Some researchers have suggested that omission of these steps in the late 1970s and early 1980s in the UK contributed to the emergence of BSE by allowing scrapie infectivity to survive the rendering process. Regardless of the origin of BSE, the epidemiologic evidence indicates that feeding cattle rendered BSE-infected carcasses greatly amplified the BSE outbreak. Several other factors may have contributed to the emergence of BSE in the UK, including a relatively high rate of endemic scrapie, a high population ratio of sheep to cattle, and the inclusion of rendered meat and bone meal at high rates in cattle feed.

Since the BSE outbreak was first detected, an estimated >2 million cattle have been infected with BSE in the UK. Approximately half of these BSE-infected cattle would have been slaughtered for human consumption, potentially exposing millions of UK residents. Beginning in 1988, UK animal and public health authorities implemented several protective measures to prevent further exposure of animals and humans to BSE-infected cattle products. The implementation of these measures, particularly animal feed bans, led to the dramatic decline in the UK BSE outbreak.

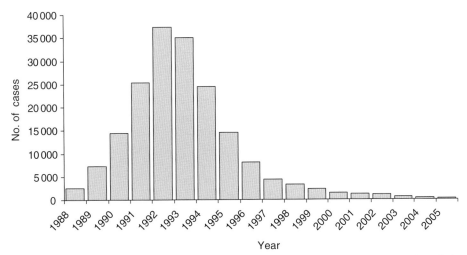

Figure 1 Bovine spongiform encephalopathy cases reported in the UK. BSE cases shown are by year of restriction: 442 BSE cases identified before 1988 are not included.

BSE was reported for the first time outside the UK in Ireland in 1989 and in Portugal and Switzerland in 1990. By August 2006, the number of countries that reported one or more BSE cases in native cattle increased to 25, including 21 countries in Europe. The four countries outside Europe that reported BSE cases are Canada, Israel, Japan, and the US. The BSE outbreak appears to be declining in most European countries, although some cases continue to occur.

In North America, BSE was first detected in 1993 in a cow that had been imported into Canada from the UK. Rendered cohorts of this cow may have been responsible for the 11 BSE cases subsequently identified during 2003–06 among cattle born in Canada; one of these cases was identified in Washington State but was later traced to a farm in Canada. At least six of these 11 BSE cases were born after the 1997 ruminant feed ban which was instituted to prevent BSE transmission among cattle. Because of the continued occurrence of new BSE infections after the 1997 ruminant feed ban, a specified risk material ban was recently instituted in Canada to further reduce cattle exposure to BSE by excluding potentially infectious nervous tissues from all animal feed. In 2005 and 2006, respectively, BSE was confirmed in an approximately 12-year-old cow born and raised in Texas and 10-year-old cow from Alabama. The source of BSE infection for these two cows remains unknown.

Because cattle carcasses were included in the production of animal feed, potential transmission of BSE to other animals was considered during the early phase of the BSE outbreak in the UK. BSE-like diseases were identified in zoo animals (ungulate spongiform encephalopathy) beginning in the late 1980s and in domestic cats (feline spongiform encephalopathy) beginning in 1990, indicating the potential for the BSE agent to cross the species barrier and spread to other animals. This development led to the establishment of enhanced CJD surveillance in the UK to monitor the possible transmission of BSE to humans.

Prion Diseases of Humans

Prion diseases of humans include kuru, CJD, variant CJD, Gerstmann–Sträussler–Scheinker syndrome (GSS), and fatal familial insomnia (FFI) (**Table 1**). Kuru is a fatal ataxic disease that was first described in the 1950s among the Fore tribe of the highlands of Papua New Guinea. In 1959, W. J. Hadlow described the similarity of the neuropathology of this disease with that of scrapie. In 1963, Gajdusek and colleagues successfully transmitted kuru by inoculating kuru brain tissue intracerebrally into chimpanzees, making this disease the first human prion disease to be successfully transmitted to experimental animals.

Kuru is the first epidemic human prion disease to be investigated. Since its investigation began in 1956, over 2700 cases have been documented. Strong epidemiologic evidence suggests that kuru spread among the Fore people by ritualistic endocannibalism. After this practice ended in the late 1950s, no children born after 1959 developed the disease and the number of new cases dramatically declined. Likely incubation periods of seven male cases recently reported by Collinge and colleagues ranged from 39 to 56 years and the incubation periods may have been up to 7 years longer. These male cases and four female cases had kuru from July 1996 to June 2004. The majority of these 11 kuru patients were reported to be heterozygous at the polymorphic codon 129 of the prion protein gene, a genotype associated with extended incubation periods and resistance to prion disease.

Creutzfeldt–Jakob Disease

CJD is the most common form of prion disease in humans. It was first recognized in the 1920s and bears the name of two German neurologists, Creutzfeldt and Jakob, who separately reported patients with rapidly progressive neurodegenerative diseases. Two of the patients initially reported by Jakob had the typical neuropathological features that we now recognize as CJD.

CJD is usually characterized by the onset of dementia, ataxia, or behavioral abnormalities. Later in the course of the illness, CJD patients commonly develop dysarthria, movement disorders such as myoclonus and tremors, and akinetic mutism. The presence of a characteristic electroencephalogram (EEG) finding of triphasic, periodic sharp wave complexes can be demonstrated with multiple testing in about 75% of patients. Elevated levels of 14-3-3 proteins in the cerebrospinal fluid (CSF) can also be demonstrated in most CJD patients. Elevated CSF 14-3-3 is a marker for rapid neuronal death and in the appropriate clinical context can often help in making a premortem diagnosis of CJD. This test is nonspecific, however, and may be elevated in other neurologic conditions that result in rapid neuronal death. More recently, magnetic resonant imaging (MRI) findings showing high intensity in the basal ganglia and cortical regions of the brain have been correlated with a CJD diagnosis. The median age of CJD patients at the time of death is 68 years with approximately 70% of cases occurring between 55 and 75 years of age. Typically, the disease progresses rapidly over a period of several weeks. Over 50% of the patients die within 6 months and 80% within 1 year of disease onset. A definitive diagnosis of CJD can only be made by histopathologic or immunodiagnostic testing of brain tissues obtained at autopsy or biopsy. Histopathologic examination of brain tissue demonstrates the hallmark triad of spongiform lesions, neuronal loss, and astrogliosis. Immunodiagnostic assays, such as immunohistochemistry and

Western blot testing, that show the presence of PrP^{Sc} confirm the CJD diagnosis.

Historically, three different forms of CJD have been reported: sporadic, iatrogenic, and familial CJD. Sporadic CJD occurs in the absence of outbreaks with no known environmental source of infection. Decades of research has not identified a specific source of infection for sporadic CJD that accounts for about 85% of patients. Spontaneous generation of the pathogenic prions was hypothesized as a cause for sporadic CJD, possibly resulting from random somatic mutations or errors during prion protein gene expression. Iatrogenic CJD, on the other hand, is associated with transmission of the CJD agent via medical interventions such as administration of contaminated human pituitary hormones and the use of contaminated dura mater grafts and neurosurgical equipment. Familial CJD has been associated with the presence of inheritable prion protein gene mutations. Beginning in the mid-1990s, the emergence of a variant form of CJD, linked with BSE transmission to humans, was reported. This newly emerged disease differs from other forms of CJD by the young age of affected patients and its clinical and pathologic features (**Table 2**).

Sporadic CJD is a heterogeneous disorder that can be further subdivided into five different subtypes based on the Western blot characteristics of protease-resistant fragment of PrP^{Sc} and the polymorphism at codon 129 of the host prion protein gene. These different subtypes correlate with characteristic clinical and neuropathologic phenotypes. The most common subtype is associated with a 21 kDa PrP^{Sc} fragment, designated type 1, and the presence of methionine at the polymorphic codon 129 of the prion protein gene.

Variant Creutzfeldt–Jakob Disease

In 1996, the identification of a cluster of young patients with a prion disease was reported in the UK as part of the CJD surveillance system that was established in response to concerns about the potential spread of BSE to humans. Because the patients' age and their clinical progression and neuropathologic profile were different from other endemic CJD patients, the term 'new variant CJD' was initially used to describe this emerging prion disease in humans. This term was later shortened to variant CJD. Since 1996, the number of variant CJD cases increased and strong scientific evidence indicated that variant CJD resulted from BSE transmission to humans. As of June 2007, a total of 202 variant CJD patients was reported worldwide, including 165 patients from the UK, 21 from France, four from Ireland, three from the US, two each from Netherlands and Portugal, and one each from Canada, Italy, Japan, Saudi Arabia, and Spain. At least seven of the non-UK variant CJD patients (two each from the US and Ireland, and one each from Canada, France, and Japan) were believed to have acquired variant CJD during their past residence or visit in the UK.

Variant CJD can be distinguished from the more common classic CJD by the clinical and laboratory findings (**Table 2**). The median age at death of variant CJD patients is 40 years younger than sporadic CJD patients (28 and 68 years, respectively), but their median illness duration is longer (14 and <6 months, respectively). Unlike classic CJD patients, variant CJD patients predominantly have psychiatric manifestations at onset with delayed appearance of frank neurologic signs and the typical 'pulvinar sign' on the MRI. The diagnostic EEG

Table 2 Clinical and pathologic characteristics distinguishing variant Creutzfeldt-Jakob disease (variant CJD) from classic CJD

Characteristic	Variant CJD	Classic CJD
Median age (range) at death (years)	28 (14–74)	68 (23–97)[a]
Median duration of illness (months)	13–14	4–5
Clinical presentation	Prominent psychiatric/behavioral symptoms, painful sensory symptoms, delayed neurologic signs	Dementia, early neurologic signs
Periodic sharp waves on electroencephalogram	Almost always absent	Often present
'Pulvinar sign' on magnetic resonance imaging[b]	Present in >75% of cases	Very rare or absent
Presence of 'florid plaques' on neuropathologic Sample	Present in great numbers	Rare or absent
Immunohistochemical analysis of brain tissue	Marked accumulation of PrP-res[c]	Variable accumulation
Presence of agent in lymphoid tissue	Readily detected	Not readily detected
Increased glycoform ratio on Western-blot analysis of PrP-res	Present	Not present
Genotype at codon 129 of prion protein	Methionine/valine[d]	Polymorphic

[a]US CJD surveillance data 1979–2001.
[b]Symmetrical high signal in the posterior thalamus relative to that of other deep and cortical gray matter.
[c]Protease-resistant prion protein.
[d]A patient with preclinical vCJD related to blood-borne transmission was heterozygous for methionine and valine.

finding that is common in classic CJD patients is very rare in patients with variant CJD. For unknown reasons, all variant CJD patients tested to date had methionine homozygosity in the prion protein gene at codon 129, which is polymorphic for methionine or valine. This homozygosity is present in approximately 35–40% of the general UK population. Similar to classic CJD, a definitive diagnosis of variant CJD requires laboratory testing of brain tissues. In addition to the spongiform lesion, neuronal loss, and astrogliosis typical of most prion diseases, the neuropathology in variant CJD patients is characterized by the presence of numerous 'florid plaques', consisting of amyloid deposits surrounded by a halo of spongiform lesions.

Studies in the UK have indicated the probable secondary person-to-person transmission of the variant CJD agent in three patients by blood collected up to 3.5 years before variant CJD onset in the donors. Because a large proportion of the UK population has potentially been exposed to the BSE agent, concerns still exist about additional secondary spread of the agent via blood products and possibly via contaminated surgical instruments.

Prion Diseases of Humans Associated with Genetic Mutations

One of the intriguing properties of prion diseases in humans is the fact that they can be both infectious and inheritable. The inherited or genetic forms of prion diseases are associated with insertion, deletion, or point mutations of the open reading frame of the prion protein gene. At least 24 different point mutations of the prion protein gene have been described in association with human prion diseases. These genetic prion diseases have widely varying clinical and neuropathologic manifestations. Historically, genetic forms of prion diseases, in part based on their phenotypical expression, were classified as familial CJD, GSS, and FFI. Beginning in 1989, many types of insertion mutations associated with markedly heterogeneous phenotypes have been reported in familial clusters.

In addition to influencing susceptibility to variant CJD, the polymorphism at codon 129 markedly influences the clinicopathologic phenotype of several inherited prion diseases. The most striking example of this influence is the phenotype associated with codon 178 mutation that substitutes aspartic acid with asparagine. Patients who have this mutation in combination with methionine on the mutant allele at codon 129 present with the FFI phenotype, whereas patients who have valine at codon 129 of the mutant allele present with the familial CJD phenotype. The codon 129 polymorphism may also influence the age at onset and duration of illness in some prion diseases.

Familial CJD

Patients with familial CJD generally have clinicopathologic phenotype similar to nongenetic forms of CJD. The disease has a dominant inheritance pattern and over half of affected family members carrying the mutation eventually die of CJD. Familial CJD has been reported among many family clusters from Canada, Europe, Japan, Israel, the US, and several Latin American countries. It is most frequently associated with a mutation substituting glutamic acid with lysine at codon 200 of the prion protein gene. Arguably, familial CJD associated with codon 200 mutation is the most common inheritable form of prion disease in humans. The largest familial cluster was reported among Jews of Libyan and Tunisian origin. About 14 other less-frequent mutations associated with familial CJD have been reported from many countries.

Gerstmann–Sträussler–Scheinker syndrome

The term GSS is used to describe a heterogeneous group of inherited human prion diseases that are characterized by a long duration of illness (median: ~5 years) and the presence of numerous PrP-amyloid plaques, primarily in the cerebellum. GSS carries the name of three physicians who in 1936 reported the disorder among patients spanning many generations of an Austrian family. The disease in this family was later shown to be associated with a mutation at codon 102 of the prion protein gene and may have been first identified as early as 1912.

At least 13 different types of prion protein gene mutations or a combination of mutations in at least 56 kindred or families have been reported in association with the GSS phenotype. Familial clusters with the GSS phenotype have been reported from Canada, Europe, Japan, Israel, Mexico, and the US. Many of the GSS mutations are associated with a greater degree of variability in the disease phenotype than other inherited forms of prion diseases. The most frequent GSS mutation results in leucine for proline substitution at codon 102 and is coupled with methionine at the polymorphic codon 129 of the mutant allele. Patients with this mutation commonly manifest with ataxia, dysarthria, movement disorders, and possibly dementia and akinetic mutism. The illness can last for up to 6 years in some patients with the GSS 102 mutation. In other forms of GSS, an illness duration exceeding 20 years has been reported.

Fatal familial insomnia

FFI is a human prion disease with predominant involvement of the thalamus, resulting in a clinical phenotype characterized often by intractable insomnia and autonomic nervous system dysfunction, including abnormalities in temperature regulation, increased heart rate, and hypertension. The neuropathologic lesions are more severe in the thalamus than other regions of the brain. FFI is primarily associated with a mutation at codon 178 of the prion protein gene resulting in a substitution of aspartic acid with asparagine in combination with methionine at the polymorphic codon 129 of the mutant allele. Occasionally, sporadic FFI cases with no apparent mutation in the prion protein gene have been reported. FFI has been identified in Australia, Canada, Japan, the US, and several European countries.

Disclaimer

The findings and conclusions in this report are those of the authors and do not necessarily represent the views of the funding agency.

See also: Prions of Yeast and Fungi.

Further Reading

Belay ED, Maddox RA, Williams ES, Miller MW, Gambetti P, and Schonberger L (2004) Chronic wasting disease and potential transmission to humans. *Emerging Infectious Diseases* 10: 977–984.

Belay ED and Schonberger LB (2005) The public health impact of prion diseases. *Annual Review of Public Health* 26: 191–212.

Prusiner SB (1982) Novel proteinaceous infectious particles cause scrapie. *Science* 216: 136–144.

Prusiner SB (2003) *Prion Biology and Diseases,* 2nd edn., New York: Cold Spring Harbor Laboratory Press.

Tumor Viruses: Human

R Grassmann and B Fleckenstein, University of Erlangen – Nürnberg, Erlangen, Germany
H Pfister, University of Köln, Cologne, Germany

© 2008 Elsevier Ltd. All rights reserved.

Glossary

RDA (representational difference analysis) PCR-based technique for the identification and cloning of DNA sequences (e.g., virus DNA) present in a particular cell but not in a matched reference cell.

Introduction

Viruses are responsible for about 15–20% of all cancers worldwide. Human tumor viruses constitute a heterogeneous group of viruses, which are causally linked to the development of malignant diseases in humans. Conventionally, the term is confined to those viruses that are likely to cause cancer by malignant conversion of infected cells. Accordingly, the human tumor viruses include the Epstein–Barr virus (EBV), the Kaposi's sarcoma-associated herpesvirus (KSHV), the high-risk human papillomaviruses (HPV), the hepatitis B virus (HBV), the hepatitis C virus (HCV), and the human T-cell leukemia virus (HTLV-1). Other viruses, such as HIV, support tumor development and growth indirectly, for example, by inhibiting the immune response to the tumor but do not infect the progenitors of malignant cells. Besides their systematic differences, most tumor viruses share several of the following features:

- establishment of chronic or long-term persistent infection;
- presence of viral gene functions, which interfere with cellular growth control, apoptosis control, DNA repair, or genomic stability; and
- capacity to transform cells in culture and/or to be oncogenic in experimental animal systems.

These features are not unique to tumor viruses but also shared by other human viruses, which are not yet clearly demonstrated but were suspected to cause human malignancies. These include some subtypes of human adenoviruses, the human polyomaviruses (BK, JC), many subtypes of papillomaviruses, and HTLV-2.

Virally induced tumors contain mostly viral genomes or parts of viral genomes, frequently integrated into the cellular genomes. As a hallmark of their infection prior to malignant conversion, many virally induced tumors are clonal in respect to integrated viral sequences. All human malignancies caused by tumor viruses are rare consequences of the infection and develop after long-term viral persistence. This indicates the necessity of rare secondary events, which are crucial to viral oncogenesis. Among them could be accumulation of genetic damage including cellular mutations induced by viral or nonviral factors, including physical and chemical carcinogens.

EBV-Associated Lymphoproliferative Malignancies

EBV, a member of the *Herpesviridae*, was the first virus to be linked to the oncogenesis of a human malignant disease. In 1964 it was identified by Epstein, Achong, and Barr in a B-cell line derived from an African Burkitt's lymphoma. Besides Burkitt's lymphoma, there are two other histologically and clinically distinct types of EBV-associated B-cell lymphoma, Hodgkin's disease and the lymphomas of immunosuppressed individuals. These three EBV-associated lymphoid malignancies differ in the patterns of the viral latent-gene expression and seem to be derived from cells at different positions in the B-cell differentiation pathway. Besides B-lymphoid malignancies, the virus was also found associated with some rare types of T-cell and natural killer (NK) cell lymphomas.

Lymphocyte Transformation *In Vitro*

EBV is capable of immortalizing human primary B lymphocytes, which resemble phenotypically activated B lymphocytes and are capable of proliferating permanently in culture. Cell lines usually contain EBV genomes as nonintegrated covalently closed circular double-stranded DNA in various copy numbers. Due to a block in structural gene expression, the transformed cells usually do not synthesize progeny viruses; instead, they express a series of latency-associated genes, which are also active in various combinations in human malignancies associated with EBV. These genes include five EBV-associated nuclear antigens (EBNA1, 2, 3A–C), two latent membrane proteins (LMP1, 2), and several noncoding RNAs. Most of these genes have important functions in the viral latent persistence. EBNA2 and LMP1 are essential for the *in vitro* transformation of B cells, as has been confirmed by using recombinant forms of EBV that lack individual latent genes. These studies have also highlighted a crucial role for EBNA1, EBNA-LP, EBNA3A, and EBNA3C in the transformation process. The main transforming protein of EBV is LMP1. It fulfills the criteria of a classic oncogene (e.g., rodent-fibroblast transformation). Although it lacks any homology to cellular proteins, LMP1 functionally mimics co-stimulatory receptors of the tumor necrosis factor (TNF) superfamily. Independent of a TNFR ligand, it exerts its pleiotropic effects, including the induction of cell-surface adhesion molecules and activation antigens, and the upregulation of anti-apoptotic proteins. These signals account for both the growth- and survival-stimulating functions of LMP1.

Lymphomas in Immunosuppressed Individuals

Individuals who are compromised in their cellular immunity (T-cell immune reaction) are at risk for the development of EBV-positive B-cell lymphoproliferative diseases. These include the immunoblastic lymphomas in patients with genetic immunodeficiencies, in AIDS patients, and the post-transplantation lymphomas (PTLs) in patients under immunosuppressive therapy after organ transplantation. PTLs are polyclonal or monoclonal lesions, which mostly arise within the first year of allografting, when immunosuppression is most severe. Almost all of these early onset tumors are EBV positive. The growth-stimulating EBV latent genes including EBNA2 and LMP1 are expressed; it suggests that the PTL consists of virus-transformed cells, which closely resemble *in vitro*-transformed lymphocytes that grow out in the absence of effective T-cell surveillance.

Hodgkin's Lymphoma

EBV-induced infectious mononucleosis was recognized as a risk factor for the development of Hodgkin's lymphoma. This malignant tumor is characterized by the predominance of a nonmalignant infiltrate which vastly outnumbers the malignant cells (Hodgkin cells, Reed–Sternberg cells: HRS cells). All HRS cells within a tumor are part of the same clone and are probably derived from crippled germinal-center cells that have been rescued from the germinal-center reaction.

The infiltrate distinguishes the subtypes, the nodular sclerosing (NS), mixed cellularity (MC), and rarer lymphocyte-depleted (LD), which are to different extents associated with EBV. Approximately 40% of Hodgkin's lymphoma in the developed world is associated with EBV and between 50% and 90% of MC and LD subtypes. The EBV genome in the tumor is clonal and present in every HRS cell which is evidence for a causal role in the pathogenesis of Hodgkin's lymphoma. The HRS cells express a particular subset of latent-cycle proteins – EBNA1, LMP1, and LMP2. A plausible pathogenetic role for the virus could be the prevention of apoptosis in cells that have undergone a defective germinal center reaction. In particular, LMP1 by stimulating co-stimulatory (TNFR) pathways is capable of replacing a T-cell signal that prevents apoptosis. Whether EBV continues to contribute to the malignant phenotype at the time of tumor presentation is not known to date.

Burkitt's Lymphoma

Burkitts lymphoma is a worldwide distributed B-cell lymphoma with defined subtypes. Whereas only 15–25% of the sporadic form, which prevails in Europe and Northern America, is EBV associated, the endemic form is nearly to 100% positive for EBV. Endemic Burkitts lymphoma is a frequent childhood tumor in the humid lowlands of eastern and central Africa. It presents in children around 8 years of age as a unilateral swelling of the jaw. Due to early metastasis the disease is usually multifocal at diagnosis.

Burkitt's lymphoma and derived cell lines (from both EBV-positive and EBV-negative cases) are now confirmed to be of germinal-centroblast origin. In this stage of B-cell development, the germinal-center reaction of the B cell (including somatic hypermutations of the V chains and isotype switching) is almost finished and cells are starting to enter the memory compartment. Common denominators of all BL forms are chromosomal translocations, which result in increased expression of the cMYC protein. The contribution of EBV to the pathogenesis of Burkitt's lymphoma remains still unclear. The most compelling evidence of EBV's involvement in endemic BL is the high frequency (98%) of tumors carrying viral DNA and the presence of clonal EBV in all of the tumor cells. The viral gene expression is limited to mainly EBNA1. The transforming LMP1 protein is absent and even seems to be incompatible with the growth of BL cells *in vivo*. Thus, the virus might have an initiating role in which growth-transforming B-cell infections establish a pool of target cells that are at risk of a subsequent *MYC* translocation. Epidemiological evidence indicates that co-infection with malaria and HIV is an important cofactor, which increases tumor incidence, possibly by chronic stimulation of the B-cell system.

Lymphoid Malignancies with T-Cell or NK-Cell Characteristics

EBV is also associated with three types of T-cell malignancies. In particular, the nasal T-cell lymphoma has a high rate of association. It is an extranodal lymphoma of the angiocentric type, which primarily develops in the nosal caveola. The tumor is relatively frequently found in Southeast Asia. Frequently nasal EBV-associated lymphomas resemble the phenotype of NK cells (CD3−/CD56+). The EBV-gene expression includes EBNA1, LMP1, and LMP2 in various extents.

EBV-Associated Carcinomas

The infection of epithelia by EBV eventually results in malignant transformation and the development of carcinomas. Such EBV-associated cancers are the anaplastic nasopharyngeal carcinoma (NPC), a subset of gastric adenocarcinomas and certain salivary gland carcinomas.

Nasopharyngeal Carcinoma

NPCs are highly malignant neoplasias, which mostly occur in adults between the ages of 20 and 50 years. The prognosis is poor; most frequently, NPC presents with early metastasis into cervical lymph nodes and the skull. The anaplastic (undifferentiated) form of NPC (aNPC) shows the most consistent worldwide association with EBV (virtually 100% worldwide). The anaplastic type of the NPC is recognized as a separate clinical entity. It differs from other types of NPC by its low grade of differentiation of the tumor cells and a characteristic tendency to extended lymphocyte infiltrations. The aNPC is particularly common in areas of China and Southeast Asia, reaching a peak incidence of around 20–30 cases per 100 000. An etiologic role of the virus in tumor development is supported by following characteristic features of the tumor cells: (1) virtually all aNPC contain EBV-DNA, and (2) the viral episomes in the tumors are monoclonal. From this observation one can deduce that the infection event has occurred prior to tumorigenesis. This assumption is supported by the presence of monoclonal EBV-genomes in the noninvasive progenitor lesions, the *in-situ*-NPC. (3) Besides EBNA1, the tumors express LMP2 and partly (40%) the viral oncoprotein LMP1. The lack of LMP1 expression in some tumors however seems to be a secondary late event since it has been reported that the premalignant lesions of NPC all express LMP1. Thus, the presence of the growth-signal-mediating LMP1 and LMP2 is additional evidence for the causal role of the virus in tumorigenesis.

Other Carcinomas

EBV is present in a high proportion (>90%) of lymphoepithelioma-like gastric carcinomas, which morphologically closely resemble NPC. About 5–25% of gastric adenocarcinomas are also associated with EBV. These tumors display a restricted pattern of EBV latent-gene expression, including EBNA1 and LMP2A. A possible role of EBV in the pathogenesis of gastric carcinomas seems to be confined to late tumorigenesis, which is suggested by the absence of EBV infection in premalignant gastric lesions. A subset of salivary gland carcinomas have also been found to be EBV-positive including the expression of latent genes like EBNA1 and LMP1 and 2 in parts of the tumors. Recently, EBV has also been detected in some carcinomas of the breast and liver.

Malignant Diseases Related to the KSHV

By using a polymerase chain reaction (PCR)-based technique for the selective amplification of unknown sequences ('representational difference analysis') Chang and co-workers identified in 1994 a new human herpesvirus in the tissue of Kaposi's sarcoma (KS). The new virus was designated as human herpesvirus type 8 (HHV8) or KSHV. It is highly associated with KS but also found in several rare lymphoproliferative conditions, including Castleman's disease and body cavity-based lymphoma (synonymous: primary effusion lymphoma).

Kaposi's Sarcoma

The tumor was first described in 1872 by Moritz Kaposi as 'multiple idiopathic pigmented sarcoma of the skin'. It is a multifocal, proliferative lesion of spindle-shaped cells with slit-like vascular spaces in skin and mucous membranes of the oral cavity, gastrointestinal tract, and pleura. The tumor cells, termed KS spindle cells, are likely of endothelial origin. KSs can be grouped into four clinical subtypes: (1) classic, (2) endemic or African, (3) transplantation-associated or immunosuppressive therapy-associated, and (4) epidemic or HIV/AIDS-associated. Based on the following evidence, a causal role of KSHV in the pathogenesis of KS is now widely accepted: (1) KSHV genomes are regularly detected in all subtypes of KS; (2) KSHV is present in the endothelial and spindle cells, the neoplastic component of the tumor; and (3) epidemiological and prospective cohort studies in HIV-infected individuals show high correlation of KSHV infections with a later development of KS. Viral gene expression in the latently infected KS cells is likely to support the growth of tumor cells. Among the viral proteins synthesized are a viral homolog of a D-type cyclin (vCYC), which is capable of stimulating the cell cycle in the G1 phase, and a viral homolog of FLICE-inhibitory protein (vFLIP), which is a potent suppressor of extrinsic apoptosis. Furthermore, latently infected KS spindle cells produce the 'latency-associated nuclear antigen' (LANA-1) that is required for the replication of circular viral episomes and the activation of a wide range of cellular genes.

B-Lymphoid Malignancies

'Multicentric Castleman's disease' (MCD) is a lymphoproliferative condition, characterized by enhanced B-cell proliferation and vascular proliferation in expanded germinal centers. In these lesions KSHV can be detected. In a proportion of B cells surrounding the follicular centers of MCD, several viral proteins were found to be expressed. These include LANA-1, the viral interleukin-6 (vIL-6), and the several proteins with homology to interferon regulatory factors (vIRF-1/K9, K10.5/LANA-2, K10). The signal transduction cascades induced by these viral proteins might be relevant for the pathogenesis of this disease. The second B-lymphoid malignancy associated with KSHV is the 'primary effusion lymphoma' (synonymous: body cavity-based lymphoma). This rare lymphoma develops mostly in AIDS patients in the spaces of pleura, pericard, or peritoneum.

Papillomaviruses as Major Cause of Human Cancers

There is nowadays sufficient evidence that certain, so-called 'high-risk' types of HPV are carcinogenic to humans in the cervix (HPV16, 18, 31, 33, 35, 39, 45, 51, 52, 56, 58, 59, and 66). In developing countries, carcinoma of the cervix uteri is the most frequent type of female cancer. Even in industrialized countries, in spite of extensive preventive screening, this cancer is among the most frequent female malignancies. HPV DNA can be found in virtually all cervical cancers, and the early genes E6 and E7 are usually expressed. The most prevalent type in epidermoid carcinomas is HPV16. HPV18 may be preferentially associated with adenocarcinomas. Other high-risk types have been detected in a few cases of squamous cell carcinoma each. HPV DNA was also demonstrated in more than 50% of the less-prevalent carcinomas of the vulva and penis (basaloid and warty tumors), the vagina, and the anus; HPV16 is again the most frequent type, followed by HPV18. HPV6 and HPV11 with a lower carcinogenic potential were detected in verrucous carcinomas of vulva, penis, and anus.

HPV16 and other members of the high-risk papillomavirus group immortalize primary human keratinocytes and induce resistance to differentiation stimuli. Histological abnormalities can be observed in stratifying keratinocyte cultures that resemble those in precancerous, intraepithelial lesions *in vivo*. The cells are not tumorigenic in nude mice initially, but quickly change to an aneuploid karyotype, which is in keeping with frequently occurring abnormal mitoses in HPV16-positive lesions. At higher passage level, malignant clones reproducibly arise, which indicates that HPV infection is sufficient to induce cancer cells in combination with additional spontaneous or virus-induced modifications. The viral genes E6 and E7 are required to trigger these effects. They encode proteins that inactivate tumor suppressors and modulate cell-cycle regulation, DNA repair, and apoptotic processes, for example, by interacting with the cellular proteins p53, p105-RB (the retinoblastoma protein), p21, p27, Bak, and PDZ domain proteins. E6 activates the catalytic subunit of the telomerase as an important step in cell immortalization. E6 and E7 induce chromosomal instability, mitotic defects, and aneuploidy, which will finally contribute to tumor progression. The E6 and E7 proteins of low-risk viruses display much lower affinities to the cellular proteins, in parallel with a lower or nondetectable transforming potential *in vitro*.

Much attention has been paid to the possible role of viral DNA integration in tumor progression. HPV18 DNA appears integrated into the cellular genome in almost all cervical cancers, and HPV16 DNA in about two-thirds of the cases. This is in contrast with benign and premalignant lesions, where the viral DNA usually persists extrachromosomally. There is no evidence for a specific integration site, but HPV DNA has been repeatedly detected in the vicinity of the *myc* proto-oncogene in combination with an overexpression of the cellular gene. Probably, more important is the upregulation of viral oncogene expression following critical integration events, which result in a clonal selection of the affected cells.

The persistence of viral DNA and the continual expression of transforming genes in advanced cancers suggest that HPV functions are also involved in the maintenance of the malignant phenotype. An experimental suppression of E6 and E7 expression inhibited the proliferation of HPV-positive cervical cancer cell lines and reduced the cloning efficiency in semisolid medium, thus indicating that the viral proteins are still modulating the growth of malignant cells.

The genital tract HPVs are also responsible for many HPV infections at extragenital mucosal sites such as the oral cavity, the oropharynx, and most notably the larynx. However, cancers arising in this field harbor HPV DNA less frequently than genital tumors (oral cavity and larynx on average about 25%, oropharynx 35%, preferentially carcinoma of the tonsil). The reason for the striking difference between the genital and aerodigestive tracts is not known. Either the etiology of many oral and laryngeal cancers is unrelated to HPV, or the relevant HPV types are not yet characterized or the viral DNA is no longer necessary for cancer cells and is finally lost.

Different HPV types induce various proliferative skin lesions that are benign, like plantar, common and flat warts. An association between HPV and skin cancer becomes obvious in epidermodysplasia verruciformis (EV). EV patients are infected with a subgroup of HPVs, which induce characteristic persisting macular lesions disseminated over the body. Many EV patients develop squamous cell skin carcinomas, mainly at sun-exposed sites, which suggest a co-carcinogenic effect of ultraviolet light. The DNA of HPV5 or 8 persists extrachromosomally in high copy number in more than 90% of the cancers. HPV14, 17, 20, or 47 were occasionally detected. The prevalence of specific HPVs is in striking contrast with the plurality of HPV in benign lesions and has been interpreted as reflecting a higher oncogenic potential of these types.

The carcinogenic potential of HPV8 could be clearly demonstrated in transgenic mice with the early genes of HPV8 under control of the keratin 14 promoter, which regularly develops papillomas with moderate and severe dysplasia and carcinomas in 6% of animals without any further treatment with physical or chemical carcinogens. E6 turned out to be the major oncogene in the mouse skin, necessary and sufficient for carcinogenesis. In contrast with genital HPV, no complex formation could be detected between HPV8 E6 and the cellular p53 protein, which suggests different strategies of transformation. Human keratinocytes transduced with recombinant retroviruses expressing HPV8 E7 invaded the dermis when tested in organotypic skin cultures.

A high prevalence of HPV DNA in squamous and basal cell carcinomas of the skin, particularly of immunosuppressed but also of immunocompetent patients, has been demonstrated by highly sensitive PCRs. Evidence is accumulating for many novel HPV types related to EV HPVs and cutaneous types. That there is a strong association between genital HPV16 and rare squamous cell neoplasms from the finger is remarkable. Individual skin tumors were frequently noted to be infected by several HPVs. No single HPV type predominates in skin cancers of non-EV patients, so far as is known. HPV DNA persists at very low concentrations in many skin cancers, usually at less than one genome copy per cancer cell. The relevance of these findings to the pathogenesis of cutaneous cancer remains to be determined. The possibilities discussed above for carcinomas of the aerodigestive tract are also valid for skin carcinomas.

Chronic Hepatitis B and C in Hepatocarcinogenesis

Hepatocellular carcinoma (HCC) is among the most common fatal malignancies in humans worldwide. An association with HBV from the Hepadnavirus family was suggested by the geographical coincidence of a high incidence of HCC in Southeast Asia and equatorial Africa with high rates of chronic HBV infections, frequently contracted congenitally. Prospective studies demonstrated about a 100-fold increased risk of hepatoma among carriers of the HBV surface antigen (HBsAg). Integrated HBV DNA can be detected in a large proportion of the tumors from high-risk areas and in hepatoma-derived cell lines. HBV is the first human tumor virus against which vaccination programs have been initiated on a broad basis. First signs of a decrease in the incidence of hepatoma in populations vaccinated in the 1970s substantiate the viral role in cancer development.

Liver cancer usually develops only after several decades of chronic HBV-induced hepatitis and may thus be triggered by accumulating genetic damage due to inflammation and continuous cell regeneration. A specific contribution of HBV might be expected from *cis* effects following integration of viral DNA, but except for a few case reports no consistent evidence has been obtained for the activation of particular proto-oncogenes. A transactivation of transcription may be more relevant; this can be achieved by the viral X protein, the large surface protein, and a truncated preS$_2$/S protein. The viral preS$_2$/S gene, which normally encodes a surface protein, appears frequently disrupted in cancers as a consequence of DNA integration and then gives rise to the transactivator. All HBV transactivators exert pleiotropic effects via the protein kinase C/raf-controlled signal pathway, finally activating transcription factors such as AP-1 and NF-κB and proliferation. The analysis of viral integration patterns and functional assays suggest that at least one transactivator may function in most hepatomas.

Mutations in the p53 tumor suppressor gene occur in about 30% of human hepatomas. They are observed more often in countries with dietary contamination by mutagenic aflatoxin and seem to be a late event in liver carcinogenesis. The X protein was shown to interact with elements of the DNA repair system, which may increase the mutation rate of p53.

Two types of large HBsAg with deletions at the preS$_1$ and preS$_2$ regions were detected in 60% of HCC patients. The preS mutant proteins can initiate endoplasmic reticulum stress to induce oxidative DNA damage and genomic instability. Liver cancer arose not only in mice transgenic for the X gene but also in mice transgenic for preS mutants.

More recently, seroepidemiological evidence was obtained for a correlation between HCV infections and hepatoma. Antibodies against HCV were detected in between 13% and over 80% of liver cancer patients around the world. Over 60% of acute hepatitis C becomes chronic and may progress to cirrhosis and HCC. Latency periods between primary infection and cancer are usually measured in decades, but in some cases the intervals are rather short (5–10 years). The cumulative prevalence of hepatoma in cirrhotic HCV-infected patients is over 50%, indicating that HCV substantially increases the risk of HCC. HCV is related to flaviviruses and pestiviruses and is the first human tumor-related virus with an RNA genome and no DNA intermediate during replication. The role of the virus in carcinogenesis is not yet clear. Liver injury during chronic hepatitis may be responsible for malignant conversion, but there is also some evidence that HCV is more directly involved. HCV appears to persist and replicate in hepatocytes during malignant transformation. Viral proteins interact with many host-cell factors and affect cell signaling, transcription, translation, proliferation, and apoptosis. The HCV capsid core protein and the nonstructural proteins NS3 (a serine proteinase), NS4B, and NS5A revealed transformation potential in tissue culture. Both HCV core and NS5A target the Wnt-β-catenin pathway, which appears crucial in human HCC. As discussed for HBV, HCV also induces endoplasmic reticulum and oxidative stress.

The Role of HTLV-1 in the Oncogenesis of Adult T-Cell Leukemia/Lymphoma

In 1977 adult T-cell leukemia/lymphoma, a malignancy of CD4-positive T lymphocytes, was first recognized as a distinct clinical entity by Uchiyama, Takatsuki, and colleagues. Three years later, in 1980, Poeisz, Gallo, and co-workers isolated the first human retrovirus, HTLV-1 from the lymphoma-subtype of ATLL. HTLV-1, but not the closely related HTLV-2 is highly associated with ATLL. Besides ATLL, HTLV also is linked to the pathogenesis of a chronic neurodegenerative disorder, called HTLV-associated myelopathy/tropical spastic paraparesis.

Transformation of T Cells

HTLV-1-infected T lymphocytes derived from leukemic and nonleukemic patients, in contrast to normal T cells, regularly give rise to permanent cultures, which are capable of expressing all viral proteins. The virus *in vitro* has immortalizing capacity: primary human T lymphocytes can be transformed to permanent growth in tissue culture, which phenotypically resembles ATLL cells. A nonstructural regulatory protein of the virus, Tax, is capable of mediating the transformation. Besides regulating viral transcription, this multifunctional protein interferes with cellular control of survival, proliferation, and genomic stability. It combines many features characteristic of oncogenes, among them the ability to immortalize primary human T cells to permanent growth. These cells closely resemble HTLV-1-transformed and patient-derived T cells. Tax is also capable of inducing malignant growth in animal models. In HTLV-1 Tax transgenic mice it induces leukemia which is similar to the clinical pattern of ATLL.

Adult T-Cell Leukemia/Lymphoma

About 1–3% of all HTLV-infected individuals develop ATLL, mostly after several decades of asymptomatic viral persistence. Typically, the disease manifests at the age of 40–60 years. Frequent symptoms are cutaneous lesions, pulmonary complications, hepatosplenomegaly, and hypercalcemia, which at least in part can be attributed to the capacity of the malignant cells to infiltrate organs. Generally, ATLL is categorized into four forms: (1) acute, (2) chronic, (3) smoldering, and (4) lymphoma-type, which are all associated with HTLV-1. The smoldering and chronic ATLL are less-severe forms of the disease, which may convert to the aggressive acute ATLL. The lymphoma-type is characterized by the presence of extensive lymphadenopathy in the absence of blood or bone marrow involvement. Acute ATLL is the most frequent form (more than 55%) and until now barely curable and mostly fatal. Leukemia cells exhibit an unusual and characteristic morphology with lobulated nuclei. Regarding the surface phenotype they resemble T helper cells, from which they are probably derived (CD4), expressing high amounts of the interleukin 2 receptor (IL2Rα). Viral gene expression including the Tax gene is rather low and the role of the protein for the maintenance of the malignant proliferation unclear. A causative role of HTLV-1 in ATLL is now generally accepted because of the following evidence: (1) the geographic correspondence of ATLL and HTLV-1; (2) the almost 100% association of HTLV-1 and ATLL; (3) the clonal integration of the provirus in

ATLL cells, which indicates infection prior to malignant transformation; and (4) the oncogenic capacity of HTLV-1 and its oncoprotein Tax in animal models.

General Conclusions and Future Perspectives

In most cases, long latency periods of many years or several decades elapse between primary infection by tumor viruses and first symptoms of cancer. All human tumor viruses are widespread in the world population. They contribute to malignant disease, mainly by initiation of oncogenesis, but regular infection does not immediately result in cancer. Thus, all human tumor viruses are important or necessary risk factors for particular cancers, but require additional events to induce malignant disease. This implies that in principle, all virally induced cancers can be prevented by protective immunization. As examples, vaccination against HBV and recently, against high-risk HPV are powerful prophylactic means to fight the associated cancers. The low manifestation rates of virus-induced malignancies also imply that the mere proof of an infection with a tumor virus is of limited value for the management of patients and cancer prevention. Specific diagnostic tests have to be designed that evaluate parameters of the viral infection more closely related to malignant conversion. In many but not all cases, continuous viral expression is detectable in malignant tumors, which raises the prospect of virus-specific pharmacological interference for adjuvant cancer therapy or cancer immunotherapy. For instance, the neoplastic phenotype of HPV-positive genital carcinoma cells seems to be affected by viral functions and may thus be a promising therapeutical target. Even in cases, in which primary infection and initial growth transformation apparently lead, through tumor progression, to a constitutive form of proliferation where viral gene products are not necessary for growth, virus may be used to target the tumor. For instance, stimulating specifically the immune response against viral antigens may be an appropriate future strategy.

See also: Epstein–Barr Virus: General Features; Epstein–Barr Virus: Molecular Biology; Hepatitis B Virus: General Features; Hepatitis B Virus: Molecular Biology; Hepatitis C Virus; Human Immunodeficiency Viruses: Antiretroviral agents; Human Immunodeficiency Viruses: Molecular Biology; Papillomaviruses: Molecular Biology of Human Viruses; Retroviral Oncogenes.

Further Reading

Cougot D, Neuveut C, and Buendia MA (2005) HBV-induced carcinogenesis. *Journal of Clinical Virology* 34(supplement 1): S75.

Grassmann R, Aboud M, and Jeang KT (2005) Molecular mechanisms of cellular transformation by HTLV-1 Tax. *Oncogene* 24(39): 5976–5985.

IARC Working Group (2007) *IARC Monographs on the Evaluation of Carcinogenic Risks to Humans, Vol. 90, Human papillomaviruses.* Lyon, France: International Agency for Research on Cancer.

Levrero M (2006) Viral hepatitis and liver cancer: The case of hepatitis C. *Oncogene* 25: 3834–3847.

Neipel F and Fleckenstein B (2005) Human herpesvirus-8. In: ter Meulen V and Mahy B (eds.) *Topley and Wilsons, Microbiology and Microbial Infections: Virology*, pp. 541–558. London: Arnold.

Pfister H (2003) Human papillomavirus and skin cancer. *Journal of the National Cancer Institute* (Monograph) 31: 52.

Proietti FA, Carneiro-Proietti AB, Catalan-Soares BC, and Murphy EL (2005) Global epidemiology of HTLV-I infection and associated diseases. *Oncogene* 24(39): 6058–6068.

Schulz TF (2006) The pleiotropic effects of Kaposi's sarcoma herpesvirus. *Journal of Pathology* 208(2): 187–198.

Thorley-Lawson DA (2005) EBV the prototypical human tumor virus – just how bad is it? *Journal of Allergy and Clinical Immunology* 116(2): 251–261.

Young LS and Rickinson AB (2004) Epstein–Barr virus: 40 years on. *Nature Reviews Cancer* 10: 757–768.

zur Hausen H (2006) *Infections Causing Human Cancer.* Weinheim, Germany: Wiley-VCH Verlag.

Tymoviruses

A-L Haenni, Institut Jacques Monod, Paris, France
T W Dreher, Oregon State University, Corvallis, OR, USA

© 2008 Elsevier Ltd. All rights reserved.

Glossary

Icosahedron A solid having 20 faces and 12 vertices.
Open reading frame Region of a genome that can be translated into a protein.
Phylogeny The complete evolutionary history of group of animals, plants, bacteria, viruses, etc.
Pseudoknot RNA structure formed by base-pairing between nucleotides within the loop subtending a stem and nucleotides outside of this loop.

sgRNA Subgenomic RNA, derived from a viral RNA genome during replication, serves as mRNA for the expression of genes that are translationally silent in the genomic RNA.
Vein-clearing Yellowing along the leaf veins.

The Family and Its Distinguishing Features

The members of the family *Tymoviridae* are presented in **Table 1**. Like the family itself, the genus *Tymovirus* derives its name from the type species, *Turnip yellow mosaic virus*. Turnip yellow mosaic virus (TYMV) was first isolated in 1946 and is by far the most intensively studied member of the family. Indeed, TYMV is one of the best-characterized plant viruses. The family was recently created in recognition of the close relationships between the genera *Tymovirus*, *Marafivirus*, and the founding member of newly created genus *Maculavirus*, grapevine fleck virus (GFkV). Poinsettia mosaic virus (PnMV) is currently a family member unassigned to a genus. Complete genome sequences are available for the type species of each of the three genera of the *Tymoviridae* and for PnMV.

The members of the family *Tymoviridae* are characterized by their icosahedral, nonenveloped, ~29 nm virions, that can readily be visualized by negative-staining electron microscopy (EM). Infections produce a characteristic

Table 1 Members of the family *Tymoviridae*

Species name	Virus abbreviation	Complete sequence accession no.
Genus *Tymovirus*		
Andean potato latent virus	APLV	
Belladonna mottle virus	BeMV	
Cacao yellow mosaic virus	CYMV	
Calopogonium yellow vein virus	CalYVV	
Chayote mosaic virus	ChMV	AF195000
Clitoria yellow vein virus	CYVV	
Desmodium yellow mottle virus	DYMoV	
Dulcamara mottle virus	DuMV	AY789137
Eggplant mosaic virus	EMV	J04374
Erysimum latent virus	ErLV	AF098523
Kennedya yellow mosaic virus	KYMV	D00637
Melon rugose mosaic virus	MRMV	
Okra mosaic virus	OkMV	EF554577
Ononis yellow mosaic virus	OYMV	J04375
Passion fruit yellow mosaic virus	PFYMV	
Peanut yellow mosaic virus	PeYMV	
Petunia vein banding virus	PetVBV	
Physalis mottle virus	PhyMV	Y16104
Plantago mottle virus	PlMoV	AY751779
Scrophularia mottle virus	SrMV	AY751777
Turnip yellow mosaic virus	TYMV	J04373, X16378, X07441
Voandzeia necrotic mosaic virus	VNMV	
Wild cucumber mosaic virus	WCMV	
Tentative members		
Anagyris vein yellowing virus	AVYV	AY751780
Nemesia ring necrosis virus	NeRNV	AY751778
Genus *Marafivirus*		
Bermuda grass etched-line virus	BELV	
Maize rayado fino virus	MRFV	AF265566
Oat blue dwarf virus	OBDV	U87832
Citrus sudden death-associated virus	CSDaV	AY884005, DQ185573
Tentative members		
Grapevine asteroid mosaic-associated virus	GAMaV	
Grapevine rupestris vein feathering virus	GRVFV	AY706994
Genus *Maculavirus*		
Grapevine fleck virus	GFkV	AJ309022
Tentative member		
Grapevine red globe virus	GRGV	
Unassigned virus		
Poinsettia mosaic virus	PnMV	AJ271595

mixture of filled, infectious virions and empty or near-empty capsids. Regular surface features, representing prominent peaks formed by pentamers and hexamers of coat protein (CP) molecules, are evident by EM (**Figure 1**).

The genomes of members of the *Tymoviridae* are composed of a single positive-stranded RNA generally 6.0–6.5 kb long, although the genome of GFkV is 7.5 kb long. Subgenomic (sg) RNAs of 1 kb or less are associated with CP expression. All *Tymoviridae* genomes have a distinctive skewed nucleotide composition that is rich in C residues (32–50%). The unifying characteristic of the genome design of all *Tymoviridae* members is the presence of a long open reading frame (ORF) that covers most of the genome and encodes the replication polyprotein with identifiable domains: methyltransferase, papain-like proteinase, helicase, and RNA-dependent RNA polymerase (RdRp) in order N- to C-terminal (**Figure 2**). The CP ORF is situated downstream of the polyprotein ORF, to which it is fused in the marafivirus and PnMV genomes. Close familial relationships are easily discerned from alignments of the sequences of the RdRp and CP genes.

Most viruses in the family have narrow host ranges. Many of the tymoviruses have been isolated from noncrop hosts and have thus far not presented major disease threats to crops. Marafiviruses are associated with significant crop losses, perhaps resulting from their more effective transmission by flying insects.

Tymoviruses

Properties and Distinguishing Characteristics

Although virtually all information about tymoviruses has been derived from studies on TYMV, it is considered to be generally applicable to the other members of the genus. Virions of tymoviruses contain genomic RNAs 6.0–6.7 kb long and produce a single 3′ collinear sgRNA less than 1 kb in length encoding the ~20 kDa CP. Their genomes include a distinctive 16-nt-long 'tymobox' sequence, and the genomes of most species possess a 3′ tRNA-like structure (TLS) that can be efficiently esterified with valine. The genomes encode three ORFs (**Figure 2**), two of which almost completely overlap: the overlapping protein (OP) and replication protein (RP) ORFs, which begin at AUG codons separated by only 4 nt. Tymoviruses replicate in all major tissues of their host plants, and accumulate to high levels (more than $0.1\,\mathrm{mg\,g^{-1}}$ leaf tissue). Both filled and empty or near-empty particles accumulate. They are readily transmitted mechanically under laboratory conditions, and are spread over limited distances by beetle vectors in nature. Infection produces mosaic symptoms and a distinctive cytopathy that is evident upon EM observation by the appearance of small vesicles on the surface of chloroplasts, together with vacuolation and clumping of chloroplasts.

Capsid Structure

Virions of tymoviruses are highly stable ~29 nm, $T=3$ icosahedra formed by 180 copies of the single CP, arranged as 12 pentamers and 20 hexamers. These groupings form the vertices with fivefold and sixfold symmetry that constitute the surface peaks visible upon high-resolution EM observation (**Figure 1**). Intersubunit stabilization is provided primarily by hydrophobic protein–protein contacts, allowing the formation of stable shells that appear to be devoid of RNA. These empty or near-empty capsids can account for about one-third of the particles present in infected tissues and are readily identifiable by internal staining in negative-contrast EM (**Figure 1**). They

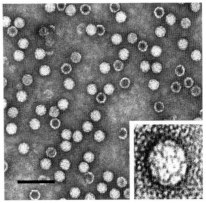

Figure 1 Tymovirus particle structure. A cartoon of the arrangement of subunits into pentamers and hexamers to build the $T=3$ icosahedron is shown at left. A negative-contrast electron micrograph of virions and 'empty' particles of belladonna mosaic tymovirus is shown at right, with the high-magnification inset showing the prominent surface structure. Scale = 100 nm. Image courtesy Dr. D.-E. Lesemann. Reproduced from Mayo MA, Dreher TW, and Haenni A-L (2000) Genus *Tymovirus* In: van Regenmortal MHV, Fauquet CM, Bishop DHL, et al. (eds.) *Seventh Report of the International Committee on Taxonomy of Viruses*. New York: Academic Press, with permission from Elsevier.

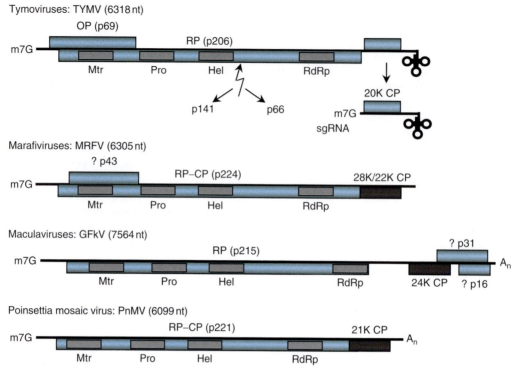

Figure 2 Genomes of the type members of the genera of the family *Tymoviridae* and of PnMV. All genomes have a 5′-m7GpppG cap, but the 3′ terminal structures vary. TYMV has a valine-specific tRNA-like structure (cloverleaf), GFkV and PnMV have a poly(A) tail, while MRFV has neither. The known or predicted (?) expressed ORFs are indicated with the molecular weight (K, kDa) of the predicted protein. The RPs all possess methyltransferase (Mtr), papain-like proteinase (Pro), helicase (Hel), and RdRp domains. The TYMV RP is cleaved as indicated, and similar cleavages are expected for the other viruses. The TYMV OP is the viral movement/RNAi suppressor protein. The TYMV CP is expressed from a sgRNA, and all other CPs are probably also expressed from sgRNAs (not shown). The 28 kDa MRFV CP (whose true size is likely to be closer to 25 kDa) may be produced by proteolytic cleavage from the RP–CP fusion protein, and an analogous event may occur with PnMV. Expression of the MRFV p43 and GFkV p31 and p16 has not yet been validated.

sediment as the 'top component' at 45–55S in CsCl density gradients, and readily separate from the 'bottom component', the infectious 110–120S virions that contain the genomic RNA. Minor components of intermediate density contain a range of subgenomic-size RNAs that can be translated to yield CP. Roughly equimolar amounts of genomic and sgRNAs are encapsidated, but the precise disposition of sgRNA in the various particles has not been determined. Like many other viral CPs, tymoviral CPs fold into eight-stranded, β-barrel, jelly-roll structures. However, unlike some other CPs, there is no positively charged domain at the N terminus for interaction with RNA and charge neutralization during packaging. Charge neutralization is thought to be provided by polyamine molecules (mostly spermidine) associated with the RNA. The crystal structures of empty and infectious particles have been determined for TYMV and physalis mottle virus. Consistent with the dominance of protein–protein interactions in particle stabilization, empty and infectious particles have very similar structures.

Packaging signals in tymoviral RNAs have yet to be identified, though RNA recruitment may involve conserved hairpins in the 5′ untranslated region (UTR) and interaction of C-rich segments of the RNA with CP at low pH. Localized areas of low pH (5–6) have been postulated to arise at the surfaces of photosynthesizing chloroplasts. Tymoviral replication occurs in characteristic vesicles that form at the chloroplast surfaces (see below), and capsid formation is most active near these vesicles, suggesting a possible coupling between replication and encapsidation during infection. Despite much effort, there has been no success in developing a cell-free packaging system. These efforts have focused on the possible role of 'artificial top component' (ATC) capsids as decapsidation and encapsidation intermediates. ATCs are protein shells devoid of RNA that can be made from infectious virions by treatments such as freeze–thawing and exposure to high pH and pressure. They are similar in structure to infectious virions, but lack a capsomere of six CP molecules that is ejected during the treatment, allowing RNA escape.

Genome Organization

Tymoviral genomic RNAs possess a 5′-m7GpppG cap and terminate at the 3′ end in -CC(A) except for dulcamara mottle virus (DuMV) that is in most cases part of a TLS. The typical tymoviral TLS is just over 80 nt long, has a distinctive pseudoknot close to the 3′ terminus, and is a close structural mimic of cellular tRNA^{Val}. A valine-specific anticodon is present, and the 3′ terminus can be specifically aminoacylated with valine. The valylated RNAs of some, though not all tymoviruses, can form a tight complex *in vitro* with the GTP-bound form of translation elongation factor eEF1A. Most molecules of encapsidated RNA lack the terminal A residue in the -CC(A) end, which is thought to be added by the host tRNA-specific CCA-nucleotidyltransferase at the beginning of infection. The main roles of the TYMV TLS are thought to be (1) as a 3′ enhancer of translation initiation, (2) as a regulator of the onset of RNA replication by modulating access by the polymerase to the 3′ end, and (3) in maintaining an intact-CCA 3′ end. A minority of tymoviral genomic RNAs have 3′ UTRs that lack the typical valine-specific TLS (DuMV, erysimum latent virus, and nemesia ring necrosis virus, NeRNV).

Tymoviral RNAs act as cap-dependent messenger RNAs. Of the three ORFs, two (OP and RP) are expressed directly from the genomic RNA, whereas the third (CP) is expressed from the 5′-capped sgRNA (**Figure 2**). Although the sequence contexts around the OP and RP initiation codons vary, 4 nt always separates these closely spaced AUGs. The RP ORF covers most of the viral genome, and encodes the ~200 kDa RP, the only viral protein essential for supporting viral RNA replication in protoplasts. This ORF encodes discrete domains (**Figure 2**) that are discernable by virtue of sequence relationships to similar domains encoded by a wide variety of positive-stranded RNA viral genomes. Particularly the RdRp domain is well conserved and appropriate for phylogenetic comparisons. Closely related, yet distinct, tymoviral genomes have RdRp domains with about 70% nucleotide sequence identity. The RP is translated as a precursor protein that subsequently undergoes maturation cleavage catalyzed by the papain-like proteinase domain. The single known cleavage separates the RdRp domain from the remainder of the RP; thus, for TYMV, the precursor p206 is processed to yield p141 and p66.

The OP ORF almost entirely overlaps the RP ORF. Its length and the sequence of the encoded protein are highly variable. OPs range between 49 and 82 kDa, and have only 25–40% amino acid sequence identity between the most closely related pairs of viruses. OP expression is needed for establishment of infection and for spread of the virus in plants. The OP is a suppressor of the host RNA silencing antiviral response and is also believed to be the viral movement protein. It is a substrate for ubiquitin-dependent degradation by the proteasome. Long-distance movement of the virus in plants requires expression of the CP.

The CP ORF initiates close to the end of the RP ORF, sometimes even a little upstream. CP sequences are variably conserved among tymoviruses, with 36–86% amino acid sequence identity between the most closely related pairs of viruses.

Tymoviral genomes possess a highly conserved sequence, the 16 nt tymobox (–GAGUCUGAAUUGCUUC–) with small variations in some viruses, especially wild cucumber mosaic virus, just upstream of the CP ORF. The tymobox and its associated 'initiation box' sequence CAA(U/G) positioned 8 or 9 nt downstream of the tymobox are believed to serve as the core elements of the sgRNA promoter. The tymobox overlaps with the 3′ end of the RP ORF, resulting in the presence of the tripeptide – ELL – near the C terminus of all RPs.

Replication Cycle

The ultrastructural changes that reflect viral replication activity and that result in the distinctive pathology of the chloroplasts have been particularly well described for TYMV. As is typical of positive-stranded RNA viruses, the replication cycle is completed entirely in the cytoplasm, and RNA replication occurs in association with membranes, specifically the chloroplast outer membrane. Vesicles 50–80 nm in diameter form as invaginations of the two outer membranes of chloroplasts. They form before the appearance of virions, which are initially found close to vesicle clusters and later throughout the cytoplasm; in some cases, empty capsids accumulate in nuclei. The RdRp has been localized to zones on chloroplasts that are rich in vesicles, and this localization depends on protein–protein interaction between the proteinase and RdRp domains of the mature RPs p141 and p66, respectively.

RNA replication occurs via the production of full-length minus strands, whose synthesis in TYMV is directed by the 3′ terminal –CCA serving as promoter and initiation site. No other minus-strand promoter elements have been identified in the 3′ UTR. As mentioned above, the tymobox appears to function with the initiation box as the core of the promoter directing sgRNA synthesis by internal initiation on the minus strand. The sgRNA of TYMV has a m7GppppA 5′ terminus.

Studies on the functions of viral proteins and of *cis*-acting sequences in the genomic RNA have been greatly facilitated by the use of 'infectious clones', that is, molecularly cloned cDNA versions of the genome from which infectious RNA can be derived.

Infection and Transmission

Tymoviruses cause chlorotic mosaic, vein-clearing, and mottling symptoms, generally without strong stunting. Host ranges are mostly narrow, and to date, no tymoviruses infecting monocot plants have been isolated. Tymoviruses are transmitted over limited distances by chrysomelid beetles, and some are weakly seed-transmissible. They can also be transmitted mechanically.

Phylogenetic Relationships and Species Demarcation

In the past, serology using antisera raised against intact virus was the main criterion in classifying tymoviruses and in distinguishing between species. This is no longer the most convenient approach, however, and it has in fact been shown that distinct viruses whose genomic and CP sequences have identities less than 80% and 90%, respectively, can appear to be serologically identical. This seems to be due to similar or identical dominant epitopes within otherwise distinct CPs.

The study of relationships among tymoviruses and to other viruses will undoubtedly in the future rely on the interpretation of sequences derived from the genomic RNA. The phylogenetic trees based on the sequences of the CP and the RdRp (**Figure 3**) show similar relationships, both among the tymoviruses and to the other members of the family *Tymoviridae*. This suggests that recombination between these coding regions has not been a strong evolutionary force among the *Tymoviridae*. However, the genome of NeRNV, which has a tobamoviral-type TLS capable of aminoacylation with histidine, indicates that recombination can occur and shape tymoviral genome evolution. The complete genome sequences are available for 13 tymoviruses or tentative tymoviruses (**Table 1**).

Alignments based on the RP protein sequences indicate that the capilloviruses, carlaviruses, trichoviruses, and potexviruses, all members of the *Flexiviridae*, are the next most closely related virus groups. More distant sequence relationships indicate that members of the family *Tymoviridae* belong to the alpha-like virus group.

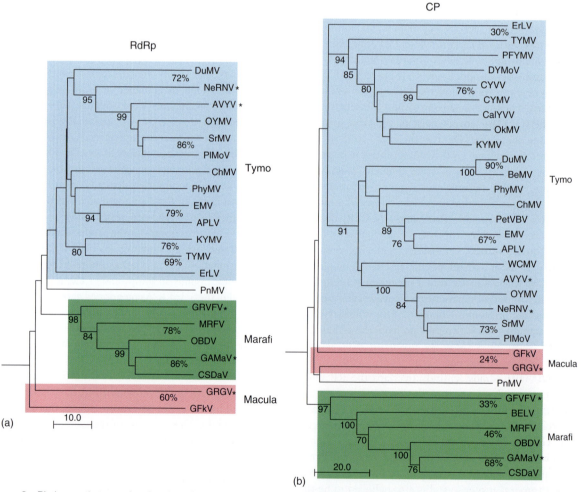

Figure 3 Phylogenetic trees showing the relationships between members of the family *Tymoviridae* based on neighbor-joining amino acid alignments of (a) the core RdRp domain and (b) the CP sequences. Bootstrap values (%) are indicated at well-supported nodes, and percent identities taken from pairwise alignments are indicated between selected sequences. Branch lengths reflect the number of amino acid differences. Viruses from each of the three genera are separately shaded as marked. Viruses with tentative taxonomic status are marked with an asterisk.

Marafiviruses

Properties and Distinguishing Characteristics

Three current members of the genus *Marafivirus* are viruses principally infecting monocots (grasses), while another infects citrus and the two tentative marafiviruses were isolated from grapevine. Plant host ranges are narrow. The genomes are 6.3–6.8 kb long, with a single ORF encoding an RP/CP fusion protein covering most of the genome. Two forms of the same CP (∼25 and ∼21 kDa) are found in virions. A single 3′-collinear sgRNA (<1 kb) encoding the smaller form of the CP is produced during infection. A slightly modified form of the tymobox sequence, termed 'marafibox', is present in the genome near the junction between the RdRp- and capsid-coding sequences. Marafiviruses are phloem-limited and not mechanically transmissible. The grass-infecting marafiviruses are transmitted by cicadellid leafhoppers in a persistent-propagative manner, involving virus replication in the insect.

A distinctive feature of the marafiviruses is the presence of two CPs in the virions, a major CP ∼21 kDa form and a minor N-terminally extended 22–28 kDa form. The latter is thought to arise by proteolytic release from the polyprotein, whereas the former is expressed from the sgRNA. The production of this sgRNA is probably under the control of the marafibox, a 16 nt sequence (–CA(A/G)GGUGA AUUGCUUC–) very closely related to the tymobox, and the –CA(A/U)-initiation box located 8 nt downstream. The roles of the two forms of CP are unclear.

Virion Structure, Genome Organization, Replication Cycle, and Phylogenetic Relationships

Marafiviruses produce virions and empty capsids that are similar to those produced by tymoviruses when observed by negative-staining EM. However, due to the phloem-limited replication of marafiviruses, the yield of virus is low, limiting the usefulness of EM for viral identification. Complete genome sequences are available for four confirmed or tentative marafiviruses (**Table 1**).

Marafivirus genomic RNAs appear to be 5′-capped but they lack the 3′-TLS that is distinctive of tymoviruses. Apart from maize rayado fino virus (MRFV), marafivirus genomes possess a 3′ poly(A) tail. Almost the entire genome of marafiviruses is devoted to encoding a large 224–240 kDa polyprotein, which includes replication-associated and CP domains (**Figure 2**). The relationships between the CP and RdRp domains of the marafiviruses and other members of the family *Tymoviridae* are indicated in **Figure 3**. The polyprotein is believed to be proteolytically processed by the viral proteinase at two sites, between the helicase and RdRp domains as in tymoviruses, and between the RdRp and CP domains. Although the only experimentally mapped polyprotein cleavage site in a member of the family *Tymoviridae* is that occurring in TYMV, amino acid sequence alignments have discerned probable cleavage sites occurring immediately downstream of Gly-Gly or Gly-Ala dipeptides at the helicase–RdRp and RdRp–CP junctions in marafivirus polyproteins.

At the present state of relative paucity in marafivirus genome sequences, the taxonomic implications of evident variation in genome organization are uncertain. In addition to lacking a poly(A) tail, MRFV is the only marafivirus possessing a significant overlapping ORF, encoding a putative 43 kDa protein, whose expression *in vivo* has not been verified. Related but strongly interrupted coding sequences have been discerned overlapping the methyltransferase domains of other marafiviral genomes and, based on sequence similarities to tymoviral OP, it has been postulated that these sequences represent variously degenerate versions of the tymoviral OP. The loss of OP during evolution may explain the phloem limitation of marafiviruses. Citrus sudden death-associated virus (CSDaV) is the only marafivirus with an additional ORF near the 3′ end of the genome. This ORF almost completely overlaps the CP ORF and could encode a 16 kDa protein.

The three grass-infecting marafiviruses, MRFV, OBDV, and Bermuda grass etched-line virus, are serologically related, but are not serologically cross-reactive with tymoviruses or more distantly related viruses. Despite the indications of genome variability mentioned above, the marafiviruses do form a phylogenetic group with interspecies relationships similarly close to those between the various tymoviruses (**Figure 3**). Sequence relationships among the RP and CP ORFs clearly link the marafiviruses to the other members of the *Tymoviridae*. Like the tymoviral CPs, marafivirus CPs lack a cluster of basic amino acids at the N terminus.

Infection and Transmission

The best-studied marafiviruses are MRFV and OBDV. They cause stunting and chlorotic leaf spots or streaks, general leaf discoloration, or enations on leaf and stem veins. Cytological symptoms are restricted to the phloem and adjacent parenchyma cells, with clearly discernible hyperplasia and hypertrophy. These viruses are not mechanically transmissible or transmissible through seed, but are vectored in nature by leafhoppers. Significantly, MRFV and OBDV are maintained for long periods and through moults in leafhopper vectors, and transmission can occur from viruliferous vectors over a period of several weeks. OBDV replicates in the aster leafhopper (*Macrosteles fascifrons*), and the transmission behavior of MRFV is also consistent with replication in the cicadellid leafhopper (*Dalbulus maidis*). MRFV is commonly associated and co-transmitted with maize stunt spiroplasma or maize bushy stunt mycoplasma in causing economically significant outbreaks of corn stunt syndrome. OBDV can also be co-transmitted with a mycoplasma. The presence of virus in the leafhopper vectors is not associated with detectable symptoms or loss of reproductive fitness.

The most economically significant marafivirus is probably CSDaV. This virus is believed to be the causative agent of the citrus sudden death syndrome, which has had devastating effects on orange trees in Brazil. Preliminary evidence suggests transmission by aphids. The grapevine-infecting tentative marafiviruses, grapevine asteroid mosaic-associated virus and grapevine rupestris vein-feathering virus, have no known insect vector.

Maculaviruses

Properties and Distinguishing Characteristics

Recent sequencing of the complete genome of GFkV supported the creation of the new genus *Maculavirus*. The distinguishing characteristics were the length of the GFkV genome, the absence of a tymobox/marafibox, the lack of a fused RP/CP ORF as present in marafivirus genomes, and the lack of phylogenetic clustering with either tymoviruses or marafiviruses. Grapevine red globe virus (GRGV) is a tentative member of the genus. Both viruses infect only *Vitis* species (grapevine and relatives). In common with marafiviruses, these maculaviruses are restricted to the phloem and are not sap transmissible. No insect vectors have been identified. As for all other members of the family *Tymoviridae*, maculaviruses produce ~29 nm particles with prominent surface features observable by EM. Top component and infectious bottom component particles are produced. GFkV infections are cytologically distinct from marafivirus infections through the presence of severely modified mitochondria ('multivesiculate bodies').

The genome of GFkV (7564 nt) is the longest among the members of the family *Tymoviridae*, and has the highest cytosine content (49.8%). The arrangement of RP and CP ORFs is similar to that of tymoviruses (**Figure 2**). The RP ORF encodes a 215 kDa protein with replication-associated domains that present sequence relatedness to the tymoviral and marafiviral RPs. The amino acid sequence of the RdRp domain shows the closest relationship to sister genera (59% and 67% identities with the RdRps of a tymovirus and marafivirus, respectively).

The 24.5 kDa GFkV CP has 23–31% sequence identity with CPs of tymoviruses and marafiviruses, and like other CPs of members of the family *Tymoviridae* it lacks a cluster of basic amino acids at the N terminus. The GFkV genome encodes two additional proteins from ORFs that partially overlap the CP ORF in the 3' region of the genome. No functions for these putative proteins are predicted from sequence comparisons. Close relationship between GFkV and GRGV is rather weakly supported by sequence comparisons (**Figure 3**).

The GFkV genome is probably capped at the 5' end and has a poly(A) tail at the 3' end, in common with most marafiviruses and PnMV. It is the only known member of the *Tymoviridae* to lack a tymobox- or marafibox-like sequence. Nevertheless, two or more sgRNAs, ~1.3 and ~1.0 kb in length, are present in infected tissues. These RNAs appear to be packaged in both top- and bottom-component particles, in the latter case apparently together with the genomic RNA.

Poinsettia Mosaic Virus

Properties and Distinguishing Characteristics

PnMV possesses properties characteristic of both the tymoviruses and marafiviruses. The ~28 nm virus particles have the EM structure typical of the *Tymoviridae* and separate into typical top components containing the sgRNA and infectious bottom components. Like tymoviruses, PnMV is produced in high yield during infection, is not phloem-limited, and can be mechanically transmitted in the laboratory. Replication is associated with chloroplast cytopathy, though not with the appearance of the replication-associated vesicles typical of tymovirus infection. The properties and coding arrangement of the 6.1 kb RNA are more closely aligned with those of the marafiviruses (**Figure 2**). The genomic RNA is polyadenylated, contains a marafibox-related putative sgRNA promoter, and possesses a single long ORF that encodes an RP/CP fusion protein (p221). No OP ORF is present. The 21–24 kDa PnMV CP is presumably expressed from the 0.65 kbp sgRNA; the involvement of potential cleavage of the RP/CP fusion protein to produce a second CP variant is not known. Although PnMV sequences are more closely related to the marafiviruses than tymoviruses, the relationships appear to be insufficiently close to warrant inclusion in the genus *Marafivirus* (**Figure 3**).

In commercial poinsettia cultivation, PnMV is transmitted by vegetative propagation; the virus is not transmissible via seed or pollen. The natural host range is restricted to *Euphorbia* sp., especially poinsettia, *E. pulcherrima*. Symptoms appear seasonally, varying from inapparent to light mottling. PnMV is often associated with poinsettia cryptic virus (family *Partitiviridae*) and with a phytoplasma that is responsible for the desirable free-branching phenotype.

See also: Flexiviruses.

Further Reading

Bradel BG, Preil W, and Jeske H (2000) Sequence analysis and genome organisation of poinsettia mosaic virus (PnMV) reveal closer relationship to marafiviruses than to tymoviruses. *Virology* 271: 289–297.

Canady MA, Larson SB, Day J, and McPherson A (1996) Crystal structure of turnip yellow mosaic virus. *Nature Structure Biology* 3: 771–781.

Ding SW, Howe J, Keese P, *et al.* (1990) The tymobox, a sequence shared by most tymoviruses: Its use in molecular studies of tymoviruses. *Nucleic Acids Research* 18: 1181–1187.

Dreher TW (2004) *Turnip yellow mosaic virus*: Transfer RNA mimicry, chloroplasts and a C-rich genome. *Molecular Plant Pathology* 5: 367–375.

Edwards MC, Zhang Z, and Weiland JJ (1997) Oat blue dwarf marafivirus resembles the tymoviruses in sequence, genome organization, and expression strategy. *Virology* 232: 217–229.

Francki RIB, Milne RG, and Hatta T (eds.) (1985) *Atlas of Plant Viruses, Vol. I.* Boca Raton, FL: CRC Press.

Hirth L and Givord L (1988) Tymoviruses. In: Koenig R (ed.) *The Plant Viruses*, pp. 163–212. New York: Plenum.

Jakubiec A, Notaise J, Tournier V, *et al.* (2004) Assembly of *Turnip yellow mosaic virus* replication complexes: Interaction between the proteinase and polymerase domains of the replication proteins. *Journal of Virology* 78: 7945–7957.

Kadaré G, Rozanov M, and Haenni A-L (1995) Expression of the turnip yellow mosaic virus proteinase in *Escherichia coli* and determination of the cleavage site within the 206 kDa protein. *Journal of General Virology* 76: 2853–2857.

Koenig R, Pleij CW, Lesemann DE, Loss S, and Vetten HJ (2005) Molecular characterization of isolates of anagyris vein yellowing virus, plantago mottle virus and scrophularia mottle virus: Comparison of various approaches for tymovirus classification. *Archives of Virology* 150: 2325–2338.

Maccheroni W, Alegria MC, Greggio CC, *et al.* (2005) Identification and genomic characterization of a new virus (*Tymoviridae* family) associated with citrus sudden death disease. *Journal of Virology* 79: 3028–3037.

Martelli GP, Sabanadzovic S, Abou-Ghanem Sabanadzovic N, Edwards MC, and Dreher T (2002) The family *Tymoviridae*. *Archives of Virology* 147: 1837–1846.

Mayo MA, Dreher TW, and Haenni A-L (2000) Genus *Tymovirus* In: van Regenmortal MHV, Fauquet CM, Bishop DHL, *et al.* (eds.) *Seventh Report of the International Committee on Taxonomy of Viruses.* New York: Academic Press

Morch MD, Boyer JC, and Haenni A-L (1988) Overlapping open reading frames revealed by complete nucleotide sequencing of turnip yellow mosaic virus genomic RNA. *Nucleic Acids Research* 16: 6157–6173.

Sabanadzovic S, Ghanem-Sabanadzovic NA, Saldarelli P, and Martelli GP (2001) Complete nucleotide sequence and genome organization of grapevine fleck virus. *Journal of General Virology* 82: 2009–2015.

Weiland JJ and Dreher TW (1989) Infectious TYMV RNA from cloned cDNA: Effects *in vitro* and *in vivo* of point substitutions in the initiation codons of two extensively overlapping ORFs. *Nucleic Acids Research* 17: 4675–4687.

Umbravirus

M Taliansky, Scottish Crop Research Institute, Dundee, UK
E Ryabov, University of Warwick, Warwick, UK

© 2008 Elsevier Ltd. All rights reserved.

Glossary

Fibrillarin An abundant nucleolar protein that participates in the processing and modification of rRNAs and a methyltransferase.

Nucleolus A prominent subnuclear domain and the site of transcription of ribosomal RNA (rRNA), processing of the pre-rRNAs, and biogenesis of pre-ribosomal particles, and also participates in other aspects of RNA processing and cell function.

Phloem Plant vascular system used for rapid long-distance transport of assimilates and macromolecules.

Plasmodesmata Plant-unique intercellular membranous channels that span plant cell walls linking the cytoplasm of adjacent cells.

Introduction

Umbraviruses differ from other plant viruses in that they do not encode a coat protein (CP) and, thus no conventional virus particles are formed in infected plants. The name of the genus *Umbravirus* is derived from the Latin *umbra*, which means a shadow or an uninvited guest who comes with an invited one. This name reflects the way in which umbraviruses depend for survival in nature on an assistor virus, which is always a member of the family *Luteoviridae*. For transmission between plants, the CP of the luteovirus forms aphid-transmissible hybrid virus particles encapsidating umbraviral RNA. In nature, each umbravirus is associated with one particular luteovirus, although in experiments transcapsidation is not needed for umbravirus accumulation within infected plants because functions such as protection and movement of the virus RNA do not require the presence of the luteovirus or its CP. Moreover, under experimental conditions, mechanical transmission of umbraviruses can take place without the aid of an assistor virus. This implies that umbraviruses encode some product(s) that functionally compensate for the lack of a CP (see below).

Currently the genus *Umbravirus* includes seven distinct virus species: *Carrot mottle virus* (CMoV), *Carrot mottle mimic virus* (CMoMV), *Groundnut rosette virus* (GRV), *Lettuce speckles mottle virus* (LSMV), *Pea enation mosaic virus-2* (PEMV-2), *Tobacco mottle virus* (TMoV), and *Tobacco bushy top virus* (TBTV). These viruses together with the tentative members of the genus *Umbravirus* are listed in **Table 1**. Some of these viruses have been known since the early days of plant virology. The first to be described was TMoV, reported from Zimbabwe and Malawi in 1945. The most economically important umbravirus is GRV, which is endemic throughout sub-Saharan Africa. Sporadic, unpredictable outbreaks of groundnut rosette disease (actually caused by a satellite RNA of GRV) cause severe crop losses. Pea enation mosaic disease outbreaks are also sporadic and localized, but losses of nearly 90% in peas and up to 50% in field beans have been reported. One or both of the carrot-infecting umbraviruses probably occurs worldwide wherever carrots are grown, but they are uncommon in commercial crops because the insecticides used to control carrot fly also control the aphid vectors of CMoV and CMoMV. LSMV is reported only from California, USA.

Virus Properties

Although no virions are formed in plants infected with umbraviruses unaccompanied by the assistor virus, the infectivity of CMoV and GRV in buffer extracts of infected leaves is surprisingly stable: infectivity remains for several hours at room temperature and is resistant to treatment with ribonuclease. Infectivity is, however, sensitive to a treatment with organic solvents suggesting that lipid-containing structures are involved in the protection of umbravirus RNA.

Table 1 Virus members of the genus *Umbravirus*

Virus	Abbreviation	Sequence	Accession no.	Assistor virus
Carrot mottle mimic virus	(CMoMV)	Complete genomic RNA	U57305	Carrot red leaf virus (CaRLV)
Carrot mottle virus	(CMoV)			Carrot red leaf virus (CaRLV)
Groundnut rosette virus	(GRV)	Complete genomic RNA	Z69910	Groundnut rosette assistor virus (GRAV)
		Satellite RNAs	Z29702–Z29711	
Lettuce speckles mottle virus	(LSMV)			Beet western yellows virus (BWYV)
Pea enation mosaic virus-2	(PEMV-2)	Complete genomic RNA	U03563	Pea enation mosaic virus-1 (PEMV-1)
		Satellite RNA	U03564	
Tobacco bushy top virus	(TBTV)	Complete genomic RNA	AF402620	Unknown
Tobacco mottle virus	(TMoV)	Partial genomic RNA	AY007231	Tobacco vein distorting virus (TVDV)
Tentative members in the genus				
Sunflower crinkle virus	(SuCV)			Unknown
Sunflower yellow blotch virus	(SuYBV)			Unknown
Tobacco yellow vein virus	(TYVV)			Tobacco yellow vein assistor virus (TYVAV)

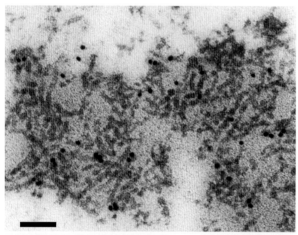

Figure 1 Electron microphotograph of a section of a *Nicotiana benthamiana* cell expressing GRV ORF3 protein. The section shows a complex of filamentous RNP particles embedded in an electron-dense matrix. The section was labeled by *in situ* hybridization with an RNA probe specific for viral RNA. Scale = 100 μm.

Indeed, in plants infected with a number of umbraviruses, including CMoV, enveloped structures ~50 nm in diameter were observed. However, an infective fraction from GRV-infected tissue contained complexes with a buoyant density of 1.34–1.45 g cm^{-3} consisting of filamentous ribonucleoprotein (RNP) particles, composed of the umbraviral ORF3 protein (see below) and virus RNA, embedded in a matrix (**Figure 1**). The relationship between the enveloped and filamentous structures is unclear.

Genome Organization and Expression

The genome of umbraviruses consists of a single linear segment of positive-sense, single-stranded RNA (ssRNA) (**Figure 2**). The complete genomic RNA sequences of CMoMV, GRV, PEMV-2, and TBTV are comprised of 4201, 4019, 4253, and 4152 nt, respectively. There is no polyadenylation at their 3' termini and there is no information about modifications of their 5' termini. At the 5' end, a very short noncoding sequence precedes ORF1, which encodes a putative 31–37 kDa protein. ORF2, which overlaps the 3' end of ORF1, potentially encodes a 63–65 kDa protein, but lacks an AUG initiation codon. It is likely that ORF1 and ORF2 are translated as a single 94–98 kDa protein by a −1 translational frameshifting mechanism; the sequence associated with frameshifting events in several animal and plant viruses is found in the region at the 3' end of the ORF1. The predicted amino acid sequence of the ORF2-encoded product has similarities with the sequences of RNA-dependent RNA polymerases (RdRps) of viruses in the families *Tombusviridae* and *Luteoviridae* and contains all eight conserved motifs of RdRp of positive-strand RNA viruses. A short untranslated stretch of nucleotides separates ORF2 from ORF3 and ORF4, both of which almost completely overlap in different reading frames and encode 26–29 kDa products. The ORF4 contains motifs characteristic of the cell-to-cell movement proteins (MPs) of plant viruses, in particular those of cucumoviruses. The ORF3 products from different umbraviruses possess up to 50% homology to each other but show no

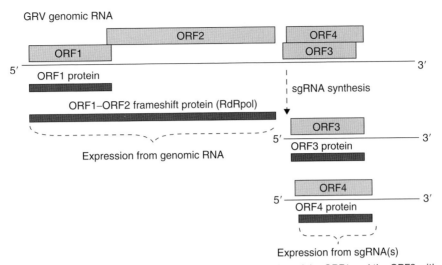

Figure 2 Organization and expression strategy of the GRV genome. Translation of the ORF1 and the ORF2 with a frameshift event results in production of the ORF1–ORF2 frameshift protein; subgenomic RNA(s) (sgRNA(s)) synthesis is required for expression of the ORF3 and ORF4. Solid lines represent RNA molecules; gray boxes represent open reading frames (ORFs); black boxes represent translation products.

significant similarity to any other viral or nonviral product. The ORF3 and OFR4 products are likely to be translated from subgenomic RNA(s) – 3′ terminal RNAs of the appropriate size were detected in GRV-infected plant tissue. It should be pointed out that the sequenced genomes of umbraviruses do not encode a potential CP(s). The essential role of the ORF3–ORF4 block in movement of umbraviruses within plants has been experimentally demonstrated and is discussed below.

Replication

Replication of umbravirus RNA presumably involves the ORF1/ORF2-encoded RdRp. Leaves of plants infected with umbraviruses contain abundant double-stranded RNA (dsRNA) including a major species of about 4.4–4.8 kbp corresponding in size to that expected for a double-stranded form of the viral genomic RNA which may be an umbravirus RNA replication intermediate. No other details of the replication mechanism have been elucidated.

Satellite RNA

Satellite RNAs are associated with some umbraviruses. In the case of GRV, satellite RNA is found in all naturally occurring isolates, and is primarily responsible for the symptoms of groundnut rosette disease. GRV satellite RNA is an ssRNA of about 900 nt which relies on GRV for its replication and, more unusually, is also required for the groundnut rosette assistor virus (GRAV)-dependent aphid transmission of GRV. Thus, unlike most virus satellite RNAs, it is essential for the biological survival (though not the replication) of its helper virus. The role of the satellite RNA in the transmission process is to mediate transcapsidation of GRV RNA by GRAV protein to form stable aphid-transmissible hybrid virus particles. Although different GRV satellite RNA variants contain up to five potential ORFs, none of the ORFs is essential for any of the functions and biological properties that have been ascribed to GRV satellites. In contrast, the satellite RNA that is associated with some isolates of PEMV-2 is not required for transcapsidation of PEMV-2 RNA by the CP of its assistor virus PEMV-1 or for aphid transmission of the hybrid particles, and other umbraviruses, such as CMoV, do not have satellite RNAs, yet are transcapsidated by their assistors and thus transmitted by aphids. The reasons for these differences have not been explained.

Cell-to-Cell Movement Function

The highly conserved ORF4-encoded proteins of umbraviruses exhibit significant sequence similarity with cell-to-cell MPs of other plant viruses, in particular the 3a proteins of cucumoviruses. Therefore, it has been suggested that this protein is involved in cell-to-cell trafficking of umbravirus RNA through plasmodesmata. This suggestion has been confirmed by a number of genetic, cytological, and biochemical approaches. By using the gene replacement strategy, it was demonstrated that that the GRV ORF4 protein could functionally replace the MPs of unrelated viruses, Potato virus X (PVX) (all the products encoded by the triple gene block and the CP) and cucumber mosaic virus (CMV) (the 3a MP and the CP). Localization of the GRV ORF4 protein has much in common with localization of MPs of other plant virus groups. The green fluorescent protein (GFP)-tagged GRV ORF4 protein targeted to

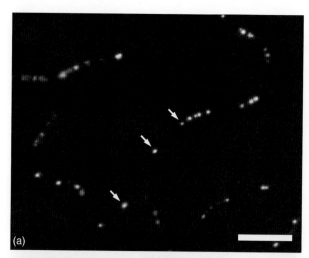
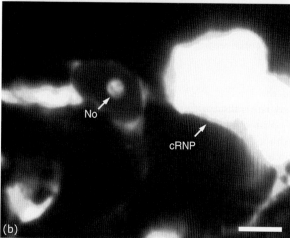

Figure 3 Confocal laser scanning images showing the localization of (a) GRV ORF4 protein fused to GFP to plasmodesmata (shown by arrows) and (b) GRV ORF3 protein fused to GFP to the nucleolus (No) and cytoplasmic inclusions containing ORF3 protein RNP particles (cRNP). Scale = 25 μm (a), 10 μm (b).

plasmodesmata (**Figure 3**(a)). Also, this protein formed extended tubular structures on the surface of protoplasts infected either with GRV or with the heterologous virus expressing the GFP-tagged GRV ORF4. The GRV ORF4 protein binds to RNA *in vitro* in a noncooperative manner which may make the viral RNA of the RNP complex accessible for translation and replication.

Phloem-Dependent Long-Distance Movement Function

One of the striking features of umbraviruses is their ability to move long distance within the plant without having a conventional CP. Involvement of CPs in long-distance movement of viral infection has been shown for a number of plant viruses, including tobacco mosaic virus (TMV) which utilizes the CP exclusively for its long-distance but not cell-to-cell movement function. By using the gene replacement strategy it was demonstrated that the ORF3 proteins of GRV, PEMV-2, and ToMV were able to functionally substitute for the TMV CP in the long-distance movement process. The hybrid TMV mutants expressing the umbraviral ORF3 proteins were able to move rapidly through the phloem. It was also shown that specific mutations in the ORF3 protein of PEMV-2 abolish the ability of this virus to move long distance without affecting its cell-to-cell spread within inoculated leaves.

It should be noted that the mechanisms of the long-distance movement facilitated by umbraviral ORF3 proteins is different from those mediated by suppressors of RNA silencing such as the 2b protein of CMV and the HC-Pro protein of potyviruses. Rather than supporting virus infection by suppressing the host RNA-mediated response, the ORF3 protein seems to protect RNA by binding to viral RNA. It was shown that the ORF3 proteins of GRV, PEMV-2, and ToMV increase the stability of viral RNA and protect it from RNases. Immunogold-labeling and *in situ* hybridization experiments showed that the GRV ORF3 protein accumulated in cytoplasmic granules consisting of filamentous RNP particles composed of the ORF3 protein and viral RNA (**Figure 1**). It has been suggested that these particles may be a form in which viral RNA moves long-distance through the phloem. Also the ORF3-containing RNP complex may protect viral RNA from nucleases and the RNA silencing machinery.

Involvement of the Nucleolus in Umbravirus Systemic Infection

In addition to cytoplasmic granules, the GRV ORF3 protein was also found in nuclei, predominantly targeting nucleoli (**Figure 3**(b)). Sequence analysis of the ORF3 proteins of umbraviruses revealed the presence of both a nuclear localization signal (NLS) and a nuclear export signal (NES), the functional roles of which in nuclear import and export of the GRV ORF3 has been confirmed experimentally by genetic analysis. Functional analysis of ORF3 protein mutants revealed a correlation between the ORF3 protein nucleolar localization and its ability to form RNP particles and transport viral RNA long distances. It was also shown that the ORF3 protein interacts with a nucleolar protein, fibrillarin, redistributing it from the nucleolus to the cytoplasm, and that such an interaction is absolutely essential for umbravirus long-distance movement through the phloem.

Interaction with Assistor Virus

Although umbraviruses accumulate and spread very efficiently within infected plants, they depend on the assistance of luteoviruses for their survival in nature, as they require encapsidation by luteoviral CP for horizontal transmission by aphids. In turn, umbraviruses facilitate the movement of phloem-limited luteoviruses to and from the phloem, as well as the cell-to-cell movement of luteoviruses between mesophyll and epidermal cells. It was shown that the ability to promote cell-to-cell movement of luteoviruses is a unique feature of the umbraviral ORF4 MP. Moreover, the ORF4 MP of GRV can facilitate cell-to-cell movement of *Potato leafroll virus* even when it is expressed from heterologous PVX or CMV genomes. In most instances, the luteovirus partner does not depend on the umbravirus infection for its survival in nature. A complex consisting of PEMV-1 (the genus *Enamovirus*, family *Luteoviridae*) and PEMV-2 is a notable exception. Unlike other members of the family *Luteoviridae*, PEMV-1 on its own lacks the ability to move, even through the phloem; both long-distance and cell-to-cell movement functions of PEMV-1 are provided by the umbraviral component of the complex, PEMV-2. Such a strong mutual dependence and adaptation between umbraviruses and luteoviruses suggest a long co-evolution, which has resulted in establishing a range of interactions from facultative coexistence (GRV/GRAV) to complete dependence (PEMV-1/PEMV-2).

Similarity with Other Taxa

Amino acid sequence comparisons showed that the putative RdRp encoded by CMoMV, GRV, PEMV-2, and TBTV belong to the so-called supergroup 2 of RNA polymerases, as do those of viruses in the genera *Carmovirus*, *Necrovirus*, *Machlomovirus*, and *Tombusvirus*. Since these enzymes are the only universally conserved proteins of positive-strand RNA viruses, the genus *Umbravirus* might be considered to be in or close to the family *Tombusviridae*.

Host Range

Individual umbraviruses are confined in nature to one or two host plant species. For example, groundnut is the only known natural host of GRV and the entire rosette disease complex (GRV, its satellite RNA and GRAV). Experimental host ranges of umbraviruses are broader but still restricted. They usually induce a leaf mottle and/or mosaic in infected plants.

Transmission

In nature, umbraviruses are transmitted by aphids, but only from plants that are infected also with an assistor luteovirus. The mechanism of the dependent transmission is the encapsidation of the umbraviral RNA by the CP provided by the assistor virus. Hence, the transmission of the dependent umbravirus occurs in the same persistent (circulative, nonpropagative) manner as that of the assistor-luteovirus.

Prevention and Control

For the avoidance of pea enation mosaic disease, it is recommended that pea or faba bean crops should be sited away from alfalfa and clover fields. Early sowing and close spacing can reduce groundnut rosette disease incidence, probably by inhibiting the landing response of the vector. This approach is also effective against tobacco rosette disease.

However, the best control approach is to use resistant cultivars if they are available. Resistance to GRV controlled by two independent recessive genes has been found and groundnut lines possessing this resistance have been developed.

A possibility for the future is the deployment of transgenic forms of resistance. Strategies for engineering transgenic resistance against umbraviruses include the transformation of plants with translatable or nontranslatable sequences from the umbraviruses themselves or their satellite RNAs.

See also: Luteoviruses; Plant Virus Vectors (Gene Expression Systems).

Further Reading

Demler SA, Borkhsenious ON, Rucker DG, and de Zoeten GA (1994) Assessment of the autonomy of replicative and structural functions encoded by the luteo-phase of pea enation mosaic virus. *Journal of General Virology* 75: 997–1007.

Kim SH, Ryabov EV, Brown JW, and Taliansky M (2004) Involvement of the nucleolus in plant virus systemic infection. *Biochemical Society Transactions* 32: 557–560.

Nurkiyanova KN, Ryabov EV, Kalinina NO, et al. (2001) Umbravirus-encoded movement protein induces tubule formation on the surface of protoplasts and binds RNA incompletely and noncooperative. *Journal of General Virology* 82: 2579–2588.

Ryabov EV, Robinson DJ, and Taliansky ME (1999) A plant virus-encoded protein facilitates long distance movement of heterologous viral RNA. *Proceedings of the National Academy of Sciences, USA* 96: 1212–1217.

Ryabov EV, Robinson DJ, and Taliansky M (2001) Umbravirus-encoded proteins that both stabilize heterologous viral RNA *in vivo* and mediate its systemic movement in some plant species. *Virology* 288: 391–400.

Taliansky M, Roberts IM, Kalinina N, et al. (2003) An umbraviral protein, involved in long-distance RNA movement, binds viral RNA and forms unique, protective ribonucleoprotein complexes. *Journal of Virology* 77: 3031–3040.

Taliansky ME and Robinson DJ (2003) Molecular biology of umbraviruses: Phantom warriors. *Journal of General Virology* 84: 1951–1960.

Ustilago Maydis Viruses

J Bruenn, State University of New York, Buffalo, NY, USA

© 2008 Elsevier Ltd. All rights reserved.

Glossary

Kex2p, Kexin, furin A membrane-bound proteinase in the Golgi that processes some secreted proteins in many eukaryotes.

Ustilago maydis Smut fungus. A plant pathogen infecting maize, wheat, oats, and barley that can adopt either yeast or mycelial form.

Introduction

The *Ustilago maydis* viruses belong to the genus *Totivirus* within the family *Totiviridae*. They are a related group of viruses that, like many fungal and protozoan viruses, exist as permanent passengers in their host cells, to which they are not deleterious. These are double-stranded RNA (dsRNA) viruses with similar structure and life cycle, derived from a very ancient ancestor but still possessing recognizable sequence similarities in their common protein, the viral RNA-dependent RNA polymerase (RdRp). They are not naturally infectious and are passed from cell to cell by meiosis or mitosis, although they will, inefficiently, infect protoplasts. In some cases, they provide a selective advantage to the host, encoding cellular toxins that kill cells of the same or related species that lack the virus. These killer toxins are apparently derived from cellular genes that have been co-opted by resident viruses. In some cases (as discussed below) cellular ancestors of these toxins still exist. When multiple species of genomic and satellite dsRNA exist in the fungal and protozoan dsRNA viruses, they are separately encapsidated. This is no disadvantage to viruses that do not depend on an infectious cycle for propagation.

Taxonomy

The fungal and protozoan dsRNA viruses generally fall into three groups: the totiviruses, the chrysoviruses, and the partitiviruses, all of which are closely related as judged by the sequence of their RdRps (**Figure 1**). In this scheme, only the giardiaviruses (which are presently classified as a separate genus within the family *Totiviridae*) are, rather inappropriately, marooned between the chrysoviruses and the partitiviruses. However, the *Ustilago maydis* viruses are placed closest to the *Saccharomyces* viruses, with which they have much in common. The three different *Ustilago maydis* viruses (UmVP1H1, UmVP6H1, and UmVP1H2) for which sequence information exists are more closely related to each other than to any other fungal viruses. This is now a common phenomenon among the fungal viruses, in which several similar dsRNA viruses may share the same host cell: ScV-L1 and ScV-La (*Saccharomyces cerevisiae*), DdV1 and DdV2 (*Discula destructiva*), and GaV-L1 and GaV-L2 (*Gremmeniella abietina*). However, there are cases in which two unrelated dsRNA viruses occupy the same cell (HvV145S and HvV190S; SsV1 and SsV2). Aside from the similarities in RdRp sequence, the *Ustilago maydis* and *Saccharomyces* viruses share a number of other characteristics.

Genome Structure

The totiviruses are dsRNA viruses with a single defining property: a single segment of dsRNA encodes all necessary

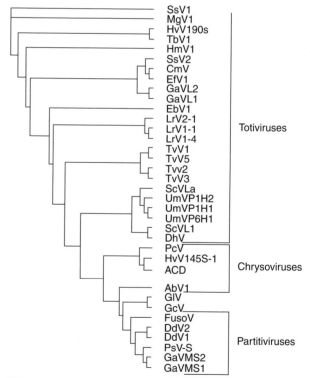

Figure 1 Cladogram of fungal and protozoan dsRNA virus RdRps. For clarity, statistics have been omitted, but each branch is supported by at least 500/1000 bootstraps. This is a Treeview rendering of a Clustal X alignment of complete RdRp sequences. Only a few of the known partitiviruses are shown due to space limitations.

viral functions. All RNA viruses of eukaryotes have the same problem of expression: eukaryotic ribosomes generally initiate only in one place on an mRNA. Encoding more than one protein on a single RNA requires some method of obviating this limitation. The totiviruses have adopted a number of strategies for this purpose: programmed ribosomal frameshifting (ScV, GLV, LRV, TVV), stop and go translation, using overlapping termination and initiation codons (HvV190S, CmRV or CmV), internal initiation (SsV or SsRV), and proteolytic processing (UmV).

The UmV dsRNAs fall into three categories: H (heavy), M (medium), and L (light). The H segments that have been partially or completely sequenced code for viral capsid and RdRp polypeptides. The M segments encode secreted polypeptides, some of which are toxins that kill susceptible cells (killer toxins). The L segments have no significant open reading frames, and are exact copies of the 3' portions of M plus strands.

The one UmV H segment entirely sequenced (P1H1) has a single open reading frame encoding a polypeptide of 1820 amino acids with an N-terminal capsid and a C-terminal RdRp sequence. There is a putative papain-like protease between the capsid and RdRp domains, so that generation of the capsid polypeptide appears to occur by endoproteolytic cleavage. The proportion of cap and cap-pol fusion protein in ScV is quite critical for capsid assembly; too much of either results in loss of the virus. If this is the case in UmV, the proportions must be controlled by the efficiency of endoproteolytic cleavage; with some low but finite probability, the protease must fail, generating a fusion protein containing the RdRp domain. Alternatively, the RdRp domain might be packaged as a single domain, as it must be in the partitiviruses. In either case, this is a rather sloppy way of doing things, since a large proportion of the C-terminal domain must be discarded as useless. However, it is no more wasteful than the processing of the poliovirus polyprotein.

There are three main isolates of *Ustilago maydis* whose viral dsRNAs have been characterized – P1, P4, and P6, named after the killer toxins that some viral segments encode (see further ahead). Some isolates have only three viral dsRNAs (H1, M1, L1), while others may have as many as seven segments. A summary of the known coding regions of these dsRNAs is shown in **Table 1**. In all cases, only the viral plus strand has an open reading frame, and the predicted protein products have been demonstrated by *in vitro* protein synthesis, and/or by isolation *in vivo*. Presumably all totivirus mRNAs, like their genomic plus strands, are uncapped and lack polyA. This has only been directly demonstrated for ScV.

The *Ustilago maydis* Killer Toxins

The KP1, KP4, and KP6 toxins are encoded by P1M2, P4M2, and P6M1. The KP4 toxin is the best characterized of these. The KP4 toxin is synthesized as a 127-amino-acid preprotoxin from which the amino-terminal 22 amino acids are removed by signal peptidase during secretion. The resulting toxin is an unglycosylated peptide of 105 amino acids and 11.1 kDa. The KP4 toxin is the only killer toxin known to be not processed by Kex2p. Its three-dimensional structure is known (**Figure 2**). It is a monomer with five disulfides stabilizing a structure with seven beta-strands and three alpha-helices, consisting of a single beta-sheet with five antiparallel strands and two antiparallel alpha-helices at about a 45° angle to the beta-strands. Its structure is somewhat similar to that of the SMK toxin, even though the latter consists of a heterodimer. The SMK toxin is encoded by the nuclear genome of *Pichia farinosa*, rather than by a viral dsRNA, and it has essentially no primary structure similarity to the KP4 toxin. SMK is synthesized as a preprotoxin which is proteolytically processed to produce the heterodimer, so we can view the KP4 toxin as a preprotoxin that has lost

Table 1 Protein products of the UmV dsRNAs

dsRNA	Length (bp)	Product	Length (aa)	Cleavage	Function
P1H1	6099	cap-pol	1820	None	RdRp
		cap	?	Endoprotease	capsid
P1M1	1504	prepropp	256		prepropp
		P1M1 alpha	103	sp, Kex2p	?
		P1M1 beta	83	Kex2p	?
P1M2	1034	KP1 pptoxin	291		pptoxin
		KP1 alpha	120	sp, Kex2p	?
		KP1 beta	117	Kex2p	KP1 toxin
P4M2	1006	KP4 pptoxin	128		pptoxin
		KP4 toxin	105	sp	KP4 toxin
P6M2	1234	KP6 pptoxin	219		pptoxin
		KP6 alpha	79	sp, Kex2p	KP6 alpha
		KP6 beta	81	Kex2p	KP6 beta

RdRp, RNA-dependent RNA polymerase; Prepropp, prepropolypeptide; sp, signal peptidase; Kex2p, Kex2p secretory proteinase; pptoxin, preprotoxin; ?, unknown.

its Kex2p sites (or SMK as a preprotoxin that has gained Kex2p sites).

The latter interpretation seems more likely, because there are other fungal proteins encoded by nuclear genomes that are highly similar to the KP4 toxin (**Figure 3**). The two proteins shown aligned with the KP4 toxin are from *Aspergillus oryzae* (BAE57860) and from *Gibberella zeae* (XP_380238). These are both predicted proteins without known function. However, both are predicted (by the SignalP 3.0 Server) to be secreted proteins, and their mature peptides are predicted to begin with the same amino-terminal amino acid sequence as the KP4 toxin, LGINCR, or with an extra amino-terminal K (in the *A. oryzae* peptide). None of these proteins has predicted Kex2p cleavage sites (xyKR, xyRR, or xyPR, where x is hydrophobic and y is anything), and the KP4 toxin is known to be processed only by signal peptidase, alone among fungal killer toxins. Five of the ten cysteines in the KP4 toxin are conserved in both of these proteins, which share 25% and 36% sequence identity with the KP4 toxin, and a sixth cysteine may also be shared. The *A. oryzae* peptide has eight total cysteines and the *G. zeae* peptide 9. Hence each would have to be missing at least one of the five disulfides present in the KP4 toxin, and if they shared the KP4 structure would have to be stabilized by other interstrand interactions. K42 is known to be critical for KP4 toxin function, but this residue is not in a region conserved in the three proteins (**Figure 3**). The SMK toxin, with a structure somewhat similar to that of KP4, has only four disulfides. The striking similarity of these three proteins, all secreted, all lacking any N-linked sites for glycosylation, all processed only by signal peptidase, implies that the KP4 toxin was captured by the virus from a cellular transcript, since the *G. zeae* peptide, the *A. oryzae* peptide, and SMK are all encoded by nuclear genes. This evolutionary process is more obvious with the ScV k1 and k2 toxins, whose viral genes preserve the polyA they derived from the 3' end of cellular messengers as internal portions of the viral plus strands.

The KP4 toxin is the only one of the UmV killer toxins about which we have some knowledge of mode of action. The KP4 toxin interferes with the function of calcium channels both in susceptible fungal cells and in mammalian cells. Calcium transport and signaling in fungi are now known to be necessary for many cellular processes and to involve the products of dozens of genes. The fungal calcium channel consists of a typical alpha-subunit (encoded by CCH1 in *S. cerevisiae*) and at least one more polypeptide (encoded by MID1 in *S. cerevisiae*). Presumably, the target for KP4 is located in the cell wall or cell membrane, since its effects are reversible, but it need not be the calcium channel itself. The SMK toxin is known to target a cell membrane ATPase in *S. cerevisiae*, which is susceptible to the SMK toxin but not to KP4. KP4 seems to act by interfering with calcium-regulated signal transduction pathways, thereby preventing cell growth and division. Resistance to KP4, which is encoded by a nuclear gene, might involve an alteration of any component of these signal transduction pathways.

The KP6 toxin is composed of two polypeptides, KP6-alpha and-beta, which are processed from a single precursor protein (a prepropolypeptide) of 219 amino acids encoded on P6M2. The preprotoxin is cleaved by

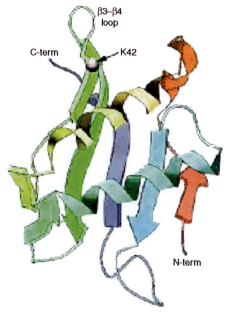

Figure 2 Structure of the KP4 toxin shown as a ribbon rendering. Figure courtesy of Tom Smith.

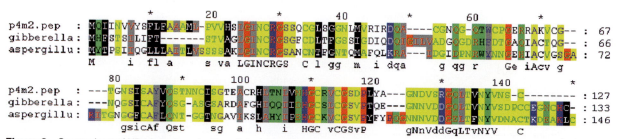

Figure 3 Comparison of KP4 (p4m2.pep) with cellular homologs. This is a Gendoc rendering of a Clustal X alignment showing physical–chemical similarities (color) and similarity (small letters in consensus line) or identity (capital letters in consensus line).

signal peptidase after residue 19; by Kex2p cleavages after residues 27, 108, and 138; and by Kex1p cleavage after residue 137 to yield two polypeptides, alpha of 79 amino acids (8.6 kDa) and beta of 81 amino acids (9.1 kDa). Alpha and beta are not necessarily found as a multimer and can be added separately to susceptible cells to reconstitute killer activity; both are necessary for activity. Unlike the KP4 toxin, KP6 has irreversible effects on susceptible cells, actually killing cells rather than inhibiting growth. The structure of KP6-alpha is known (**Figure 4**). The monomer is a four-stranded antiparallel beta-sheet with two alpha-helices on one side, a small alpha-helix at the N-terminus, and a separate beta-strand, with four disulfide bonds, involving all eight cysteines in the polypeptide. In the crystal, the polypeptide forms trimers held together by salt bridges, and the trimers sit on each other forming an hour-glass shape, suggestive of an ion pore. However, expression of the KP6-beta peptide alone in susceptible cells is lethal, suggesting a mode of action more like that of colicins or diphtheria toxin, in which one of the toxin polypeptides is introduced into the cell by the action of one or more other toxin polypeptides and acts lethally within the cell. KP6-beta also has six cysteines, suggesting a structure with three disulfides. The two monomer nature of the KP6 toxin is reminiscent of the *S. cerevisiae* k1 toxin, which also has two monomers derived from a single preproprolypeptide by signal peptidase, Kex2p, and Kex1p processing. However, in the k1 toxin, the alpha and beta monomers are covalently bound by intermolecular disulfides, which are absent in KP6. The toxin is properly processed and secreted from *S. cerevisiae*, although there are some differences; none of the *Ustilago maydis* killer toxins (or other secreted proteins, apparently) are glycosylated, and processing of mutant precursors is different in yeast and *Ustilago maydis*. The KP6-alpha polypeptide has one possible site for N-linked glycosylation, which is used in yeast but not in *Ustilago maydis*. Unlike the KP4 toxin, the KP6 toxin polypeptides have no significant similarity to any proteins currently in the databases. Resistance to the KP6 toxin maps to the nuclear genome, as with KP4.

The KP1 toxin is encoded by P1M2 and translated as a preprotoxin of 291 amino acids, subsequently cleaved by signal peptidase and Kex2p to produce mature polypeptides alpha (120 amino acids) and beta (117 amino acids). Only beta is required for toxin activity. KP1-beta, like the other *Ustilago maydis* killer polypeptides, has an unusual number of cysteines (six) which, if all were involved in disulfides as in KP6-alpha and in KP4, would indicate three disulfides. P1M2 is different from the toxin encoding dsRNAs in UmVP4 and UmVP6 (P4M2 and P6M1). The 3' noncoding regions of the plus strands of P4M2 and P6M1 are entirely homologous to the L segments in UmVP4 and UmVP6, respectively, while the L segment in UmVP1 is derived from P4M1. P4M1 does encode a prepropolypeptide, whose processed polypeptides are secreted but have no detectable relation to the KP1 toxin (**Table 1**). Nothing is presently known of the mechanism of action of the KP1 toxin, and none of the UmVP1M1 or UmvP1M2 polypeptides have any homologs in the databases. Resistance to the KP1 toxin may be nuclear or cytoplasmic, hinting that some immunity may result from expression of one or more viral dsRNAs, as with the *S. cerevisiae* k1 toxin.

Replication and Transcription

Like all dsRNA viruses, UmV has a capsid-associated transcriptase that makes the viral plus strands. However, little is known of the nature of the transcripts or the mechanism of transcription. One peculiarity of this system is the presence of dsRNAs (L, 354–355 bp) entirely homologous to the 3' ends of the plus strands of their cognate dsRNAs (M). Presumably, like all viral dsRNAs, these are the result of replication of plus strands within nascent viral particles. It is not clear if the source of the L plus strands is internal transcription initiation or processing of M plus strands. Normal-sized L plus strands (with exactly the predicted 5' ends) are produced either by the viral transcriptase or by RNAP II transcription of cDNA clones in *Ustilago maydis* or heterologous systems, implying that a cleavage event must be responsible for production of the L segments.

This cleavage must be either an inherent property of the cognate M plus strand or cleavage by a ubiquitous nuclease activity recognizing a universal feature of the M plus strands. No ribozyme activity can be demonstrated in the cognate M plus strands *in vitro*. However,

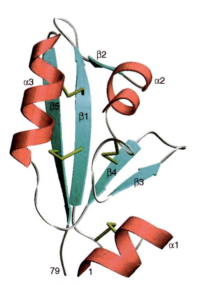

Figure 4 Structure of KP6-alpha shown as a ribbon rendering. Courtesy of the American Society for Biochemistry and Molecular Biology.

all the cognate M plus strands have a peculiar predicted secondary structure, in which the predicted L plus strand includes a very long predicted hairpin stem (23 perfect base pairs in P4M2) immediately following the predicted cleavage site (which is highly conserved). These should make ideal substrates for the recently described cellular machinery (dicer) responsible for siRNA production, which is ubiquitous in eukaryotes.

The function of the L segments is obscure: production of toxin proceeds perfectly well from cDNA constructs lacking the L encoding region, in either homologous or heterologous systems, and the P1M2 segment (encoding the KP1 toxin) has no cognate L segment. The L segments must include a packaging signal, but it is not clear what prevents its inclusion in an M plus strand sequence lacking the 3′ L sequence, a situation that does exist with P1M2. In the closely related virus ScV, a sequence as small as 18 bases serves as the packaging signal.

Transcription in dsRNA viruses may be either semiconservative (resulting in displacement of the parental plus strand) or conservative (*de novo* synthesis of plus strands). Transcription in reovirus and ScV is conservative and in bacteriophage phi6 is semiconservative. An interesting difference in the structure of the viral RdRp may be responsible for this. The phi6 RdRp has an insertion of some 20 amino acids within the N-terminal F motif, an insertion that is absent in the reovirus RdRp (and absent in all the other known RdRp and reverse transcriptase structures). None of the totivirus, chrysovirus, or partitivirus RdRps has a phi6-like insertion; most may therefore be conservative transcriptases. However, UmVP1M2 has been reported to be semiconservatively transcribed.

Viral Structure

Like other fungal viruses, UmV has a small icosohedral capsid of about 41 nm in diameter, with a sedimentation coefficient of 172S and a buoyant density in CsCl of 1.42 g ml^{-1}. All dsRNA viruses so far characterized in detail (bluetongue virus, or BTV; reovirus, or Reo; and ScV, the *S. cerevisiae* virus) are icosohedral with an inner (or complete) capsid structure of 120 copies of a capsid protein. This is a highly unusual arrangement in viruses, which generally behave according to the Caspar and Klug rules for assembly of icosohedral viruses, which call for the simplest capsids to consist of 60 copies of a single polypeptide ($T = 1$) or 180 copies of a single polypeptide ($T = 3$). However, the dsRNA viruses do not really violate this stricture, since they have 60 copies of a capsid polypeptide in one configuration and 60 copies in another, so that, in effect, they have 60 copies of an asymmetric dimer, in which the two monomers share portions with identical shape and portions with different shape. The *Ustilago maydis* viruses obey this generalization, although, like all the fungal and protozoan viruses, they have a single capsid shell, not two, as in BTV and Reo. One variation of this arrangement exists in the *Penicillium chrysogenum* virus (PcV), in which the two monomers are actually part of a single polypeptide chain, the two halves of which probably share regions of identical structure.

This unique capsid structure must play an important role in properly arranging the viral polymerase within the particle to allow transcription and replication of the genomic dsRNA. In ScV and UmV, this is accomplished by making the polymerase as the carboxy-terminal portion of a fusion protein in which the N-terminal is the capsid polypeptide. In ScV, two copies of this fusion protein are incorporated into each viral particle. Each viral particle has one copy of a single species of viral dsRNA. In UmV and ScV (as well as the other totiviruses), this single species encodes both capsid polypeptide and RdRp. When additional dsRNAs are present (e.g., toxin-encoding species), these are also separately encapsidated, and are considered satellite dsRNAs: they are completely dependent on the capsid/RdRp-encoding dsRNA for their transcription, replication, and packaging. Satellite dsRNAs may be present in more than one copy per virion. This is a considerably simpler strategy than pursued by the reoviruses, orbiviruses, and cystoviruses, all of which package multiple species of dsRNA in each viral particle and as many as 12 copies of the viral RdRp in each.

Genetic Engineering

The production of killer toxins by a plant pathogen (*Ustilago maydis*) has excited some interest in using the toxin to genetically engineer resistance to the fungus in its host crop plants. Since it is possible to produce both KP6 and KP4 toxins in heterologous systems using cDNA expression vectors, transgenic maize (KP6 and KP4) and wheat (KP4) plants have been constructed that express the toxins. These plants do secrete active killer toxins and exhibit some resistance to toxin-sensitive species of *Ustilago*. Since the toxins are harmless to animals when eaten, this may provide a new method of preventing the periodic smut infestations that occasionally decimate susceptible grain crops.

See also: Viral Killer Toxins; Yeast L-A Virus.

Further Reading

Bruenn J (2002) The double-stranded RNA viruses of *Ustilago maydis* and their killer toxins. In: Tavantzis SM (ed.) *dsRNA Genetic Elements. Concepts and Applications in Agriculture, Forestry, and Medicine*, pp. 109–124. Boca Raton, FL: CRC Press.

Bruenn J (2003) A structural and primary sequence comparison of the viral RNA dependent RNA polymerases. *Nucleic Acids Research* 31: 1821–1829.

Bruenn J (2004) The *Ustilago maydis* killer toxins. In: Schmitt MJ and Schaffrath R (eds.) *Microbial Protein Toxins*, pp. 157–174. Heidelberg: Springer.

Cheng RH, Caston JR, Wang G-J, *et al.* (1994) Fungal virus capsids: Cytoplasmic compartments for the replication of double-stranded RNA formed as icosahedral shells of asymmetric Gag dimers. *Journal of Molecular Biology* 244: 255–258.

Gage MJ, Rane SG, Hockerman GH, and Smith TJ (2002) The virally encoded fungal toxin KP4 specifically blocks L-type voltage-gated calcium channels. *Molecular Pharmacology* 61: 936–944.

Gu F, Khimani A, Rane S, Flurkey WH, Bozarth RF, and Smith TJ (1995) Structure and function of a virally encoded fungal toxin from *Ustilago maydis*: A fungal and mammalian Ca^{2+} channel inhibitor. *Structure* 3: 805–814.

Jiang D and Ghabrial SA (2004) Molecular characterization of Penicillium chrysogenum virus: reconsideration of the taxonomy of the genus Chrysovirus. *Journal of General Virology* 85: 2111–2121.

Kang J, Wu J, Bruenn JA, and Park C (2001) The H1 double-stranded RNA genome of *Ustilago maydis* virus-H1 encodes a polyprotein that contains structural motifs for capsid polypeptide, papain-like protease, and RNA-dependent RNA polymerase. *Virus Research* 76: 183–189.

Li N, Erman M, Pangborn W, Duax WL, *et al.* (1999) Structure of *Ustilago maydis* killer toxin KP6 α-subunit: A multimeric assembly with a central pore. *Journal of Biological Chemistry* 274: 20425–20431.

Naitow H, Tang J, Canady M, Wickner RB, and Johnson JE (2002) L-A virus at 3.4 A resolution reveals particle architecture and mRNA decapping mechanism. *Nature, Structural Biology* 9: 725–728.

Soldevila AI and Ghabrial SA (2000) Expression of the Totivirus Helminthosporium victoriae 190S virus RNA-dependent RNA polymerase from its downstream open reading frame in dicistronic constructs. *Journal of Virology* 74: 997–1003.

Wickner RB (1996) Double-stranded RNA viruses of *Saccharomyces cerevisiae*. *Microbiological Reviews* 60: 250–265.

Vaccine Production in Plants

E P Rybicki, University of Cape Town, Cape Town, South Africa

© 2008 Published by Elsevier Ltd.

Glossary

Immunogenicity Ability of a substance to stimulate the acquired immune response in a recipient.
Pharming The production of pharmaceutically relevant substances in plants.
Vaccine Substance which elicits an immune reaction which protects against infection by a natural pathogen.

Introduction

The concept of plant-produced vaccines, and in particular viral vaccines for animals and humans, is both recent – the first vaccine-relevant protein was made in plants only in 1989 – and controversial, with dire misgivings still being expressed concerning a number of issues. The central planks of the justification for plant-produced pharmaceuticals in general, and for the production of vaccines in particular, have been that their production in plants is both safe and potentially very cheap. A strong additional argument has been that the use of food plants will allow edible vaccines to be locally produced where they are needed most – in the developing world. Thus, the kind of argument that has developed is that plant production of especially viral vaccines allows the possibility of producing large amounts of high-grade vaccines for very low cost (1) for diseases where there is a high burden of preventable disease, (2) where existing vaccines are expensive, (3) where there are no vaccines for 'orphan diseases', and (4) where therapeutic vaccination could be very effective in treating disease.

The need for low-cost vaccines is borne out by the fact that at least 3 million people die every year of vaccine-preventable diseases, yet it takes many years on average for any new vaccine's price to come down to levels (<US$ 1 per dose) where it can be incorporated into the extended programme of immunization (EPI) bundle.

As with any maturing technology, however, problems have come to light along with proofs of concept, and the current view of the field is very different to the early one. Thus, concerns over genetic 'contamination' of food crops have largely blunted the prospect of edible viral vaccines delivered via unprocessed food plants; regulatory concerns – both for production and registration – have tempered the optimism of early years concerning widespread and rapid adoption of the technology; production via transgenic plants has, on the whole, proved to be less reliable and lower-yielding than was hoped; immunogenicity of model antigens delivered orally has often proved to be disappointingly low; and, most importantly, it has been realised that costs of production are far less important in determining the final price of a vaccine than was generally assumed.

However, there have been notable successes as well: a number of animal model systems have been thoroughly tested, and protection against disease has been obtained; production systems have to some extent been worked out, and the first products have been licenced; transient expression systems are proving to have great potential for even large-scale production; and, most significantly, large pharmaceutical companies are increasingly interested in the prospects of making high-value products in plants.

The increasing exposure of the concept to public view has also brought its share of problems: it was after all the public outcry over 'contamination' of food plants with vaccine protein genes that led to the involvement of regulatory agencies in the issue, and a moratorium on the use of food crops for vaccine production. Moreover, public acceptance of human vaccines made in plants is by no means assured, and the anti-genetically modified (anti-GM) lobby has already made an issue of the means of production. Together with these sobering developments has come the realization that high-value products not meant for use in humans are probably the first best use of the advantages of plant production of biopharmaceuticals.

For example, it is probably a far better idea to use plants to produce proteins and other macromolecules as reagents or components of diagnostic kits than it is to produce far more regulated vaccine components – as well as being a much quicker route to financial gain.

While the initially determinedly rosy picture of very cheap, edible vaccines against orphan diseases and developing country scourges has not been realized, there is now an altogether more realistic consensus view on how to proceed with the application of this still very promising technology to human and animal health. This article covers the historical development of the concept of plant-produced viral vaccines and relevant expression systems, as well as central issues affecting production, purification, and application of such vaccines.

History

The first publication on the production of vaccine-relevant proteins in plants was in 1989, when A Hiatt and colleagues at the Scripps Institute in La Jolla produced mouse-derived monoclonal antibodies (MAbs) in transgenic tobacco (see **Table 1**). This was soon followed in 1992 by the first of a series of papers – in this case involving hepatitis B virus (HBV) surface antigen (HBsAg) – by the group of Charles J. Arntzen and Hugh Mason on the production of human viral vaccine candidate antigens in transgenic plants. While they did not at the time hold out the promise of edible vaccines, they did conclude that plants did produce antigens that were equivalent to those produced in more conventional systems, and they held promise as low-cost vaccine production systems. Concurrent with these advances was a rapid and parallel improvement in the understanding of mucosal immunity and the fact that the mucosa-associated lymphoid tissue (MALT) – and, in particular, the gut-associated lymphoid tissue (GALT) – was probably the premier 'organ' of the entire adaptive immune response. It was perhaps inevitable that these two should come together in the concept of 'edible vaccines', where vaccine antigens would be introduced into people and animals in food, given that most human and animal viruses gain entry to their hosts via mucosal surfaces.

The first proof of concept for edible plant-produced vaccines was performed in the mid-1990s using a bacterial protein – *Escherichia coli* heat-labile enterotoxin (LT-B) – but this was soon followed by work on HBsAg, secretory antibodies, Norwalk virus capsid protein, and a *Vibrio cholerae* enterotoxin subunit vaccine (CTB). These studies laid important ground in terms of demonstrating conclusively that plant-produced vaccine antigens were effectively homologous to those produced by more conventional routes, and that oral dosing or gavage of experimental animals elicited significant immune responses, in the absence of replication of the agent. However, the first demonstration that a plant-produced viral antigen was protective against disease came from an animal model system and the production of the foot-and-mouth disease virus (FMDV) structural protein VP1: mice that were parenterally immunized with a crude extract of transgenic *Arabidopsis thaliana* plants were protected against viral challenge. Another development along these lines in 1999 was the proof that injection of rabbits with rabbit hemorrhagic

Table 1 Landmarks in plant-produced vaccine research

Year	Landmark
1989	Expression of immunoglobulin genes in plants
1992	Expression of hepatitis B surface antigen in transgenic plants
1995	Oral immunization with a plant-produced recombinant bacterial antigen
1996	Plant expression of Norwalk virus capsid protein and its oral immunogenicity
1998	Use of a recombinant plant monoclonal secretory antibody in humans
1998	LT-B gene in potatoes protects mice against LT
1999	Oral immunization with potato-produced VP60 protein protects against hemorrhagic disease virus
1999	Protection against FMDV by vaccination with foliar extracts of rTMV-infected plants expressing FMDV VP1
2000	Human responses to Norwalk virus vaccine delivered in transgenic potatoes
2001	Oral immunogenicity of oral HPV L1 co-administered with *E. coli* enterotoxin mutant R192G or CpG DNA
2002	Plant-derived HPV 16 E7 oncoprotein induces immune response and specific tumor protection in mice
2002	Boosting of a DNA measles immunization with an oral plant-derived measles virus vaccine
2003	Production of HPV-16 VLPs in transgenic potato and tobacco and oral immunogenicity
2003	Expression of HPV-16 VLPs in transgenic tobacco and parenteral immunogenicity
2003	Oral immunogenicity of HPV-11 VLPs expressed in potato
2003	Oral immunization with rotavirus VP7 in transgenic potatoes induces high titers of mucosal neutralizing IgA
2004	Conformational analysis of hepatitis B surface antigen fusions in an *Agrobacterium*-mediated transient expression system
2004	Oral TGEV vaccine in maize seed boosts lactogenic immunity in swine
2004	Mucosally targeted HIV gp41-derived fusion protein with CTB elicits transcytosis-blocking Abs
2005	Magnifection – TMV-based transient agroinfection platform
2005	Immunogenicity as a boost in humans of potatoes containing HBsAg
2005	Development of a plant-based vaccine for measles
2006	Proof of concept of a plant-produced papillomavirus vaccine in a rabbit model
2006	A plant-derived edible rotavirus subunit vaccine is stable for over 50 generations in transgenic plants

disease virus (RHDV) VP60, produced in transgenic plants, could protect them against virus challenge.

Another important development in the mid-1990s was the application of transient plant virus-based expression systems to the expression of vaccines. The first experimental vaccines produced were chimeras of tobacco mosaic virus (TMV) capsid protein fused to malarial peptides; by the late 1990s, a HIV-1 gp41 peptide fused to cowpea mosaic virus capsids had been tested in mice both parenterally and orally. However, it was again an animal model which provided the best proof of concept, and again with FMDV VP1: production of whole VP1 in plants infected with recombinant TMV and parenteral vaccination of mice with leaf extracts allowed successful protection against live challenge with FMDV. This work provided some of the first evidence that full-length foreign proteins could successfully be produced using a plant virus, in amounts that were sufficient to allow immunization using only crude extracts. However, the technology was limited in that the choice of appropriate vectors was limited, as was the choice of host plants, and vectors were often unstable and did not express large proteins very well.

By the late 1990s, it was apparent that many of the early fears voiced by vaccinologists and immunologists about oral delivery of vaccines 'tolerizing' the recipients to the vaccine – and thereby abolishing an immune response to it – were essentially groundless. The basis for a good response to an orally delivered vaccine was that it should be particulate, and that some form of adjuvant may be required. This was borne out by work with human papillomavirus (HPV) virus-like particles produced in a baculovirus expression system, where the number of particles necessary for a strong systemic humoral and mucosal response to orally introduced VLPs in mice could be reduced by an order of magnitude by co-administration with CpG DNA or mutant *Escherichia coli* LT-B (LT R192G). This work, together with reports on oral immunization with respiratory syncytial virus (RSV) F protein and HBsAg, showed that strong Th1-type responses could be elicited by such treatment. The state of the art in mucosal immunity at the time can be summarized as follows: (1) the mucosal immune system consisted of different compartments (such as gut, respiratory, genital mucosa) which intercommunicated, so that, for example, intranasal immunization gave significant immunity at the vaginal mucosa; (2) oral vaccination elicited antigen-specific mucosal secretory IgA as well as systemic immunity, and CD8+ T-cell responses could be elicited. While co-administration of adjuvants such as LT-B, CTB, CpG DNA, and saponins considerably improved mucosal responses and allowed lower dosages, fusion of nonpolymerizing polypeptides to self-assembling adjuvant entities such as LT-B and CTB also allowed significant improvements in immunogenicity.

By 2000, the first human trial results of Norwalk virus capsid protein made in potato had shown that the vaccine was both tolerated and immunogenic, with almost all volunteers developing a significant immune response. Shortly afterward, the Arntzen group showed that a plant material containing HBsAg given orally was an effective prime for a parenteral boost with HBsAg. In 2002, the first successful combination of DNA prime–plant-produced protein boost vaccination in mice with a measles envelope glycoprotein vaccine showed that plant-produced vaccines could be used in very effective combination with DNA vaccines. However, yields of virus vaccine proteins in transgenic plants are low, and oral immunization even with concentrated extracts is not very successful. In contrast, production of antibodies in plants has been more successful and many different kinds of immunoglobulins with various specificities have been produced.

Plant tissues used for production of viral vaccines include leaf and stem tissues of tobaccos of various varieties, *Arabidopsis*, alfalfa, spinach, and potatoes; of aquatic weeds such as *Lemna* spp. (duckweed); seeds of tobaccos, beans, and maize; fruits like tomatoes and strawberries; carrots; single-cell cultures of *Chlorella* and *Chlamydomonas*; suspension cell cultures of tobacco and other plants; hairy root cultures derived from various plants via *Agrobacterium rhizogenes*; and transformed chloroplasts of a variety of plant species. Experience gained from these various systems showed that there was no way of predicting whether or not a given DNA sequence would express protein at a reasonable level, or that the protein would be stable. Although chloroplast expression sometimes gave high yields, this system was not suitable for glycosylated proteins. Single-cell cultures offered few advantages over conventional fermentation/or cell culture techniques, and it seemed that accumulation of proteins in seeds was preferable to expression in green tissue, because of easier purification and higher accumulation levels.

A major setback to the development of plant-produced vaccines occurred in 2002 when soybean and maize harvests in two states in the USA were found to contain maize seeds from 'volunteer' plants engineered to express transmissible gastroenteritis virus (TGEV) capsid protein. The company involved – ProdiGene – had proprietary technology allowing high levels of accumulation of foreign proteins in especially maize seeds, and had potential oral vaccine products including a hepatitis B vaccine, an LT-B vaccine to treat *E. coli* infections in humans, as well as other products under development with large commercial partners. ProdiGene was fined and forced to clean up the seed by the US Department of Agriculture, which has since issued guidelines to prevent a recurrence. The fallout from these incidents has led to an effective moratorium on the development of 'pharmed' products in food crops including vaccine pharming. Another limitation became obvious in 2005 when the Mason–Arntzen group at Arizona State University released results of a small human trial of a potato-based HBsAg vaccine. In this trial, volunteers previously immunized parenterally with a conventional vaccine ate doses of potato containing ~850 µg of antigen and 63% of them developed increased serum anti-HBsAg titers. However, the immunogenicity was still disappointingly low, despite the ingestion of relatively large doses of antigen.

Growing public unease over the use of transgenic plants for biopharmaceutical production, with the attendant risks of 'contamination' of crop plant gene pools, has prompted the development of other technologies for producing such materials in plants. One such technology is the *Agrobacterium tumefaciens*-based transient expression system: this allows very high levels of expression without the uncertainties inherent in the regeneration and propagation of transgenic plants, and was being used very successfully to produce antibody molecules by 2004. Its main advantage is that the simultaneous expression of a large number of constructs can very rapidly be investigated. Indeed, the Mason group used transient 'agroexpression' very productively in 2004 to explore the conformation and chimeric fusion properties of HBsAg.

By 2005, another major advance in expression technology was announced, which combined the advantages of agroinfection and of viral vectors: this was the 'magnifection' technology of ICON Genetics, which was based on the introduction of recombinant DNA into plants by agroinfection and the subsequent amplification of gene expression by the recombinational reconstitution of a self-replicating TMV-based RNA genome.

Recently, serious reservations have been expressed by researchers and companies about the regulatory acceptance of plant-produced vaccines and other therapeutics, for human and even for animal use. The observed reluctance of large companies to buy into the technology, or to fund it, is probably due to the lack of appropriate regulatory structures and some conservatism in regulatory agencies concerning plant products. However, there is no reason to believe that existing frameworks and regulations could not be adapted to handle plant-produced vaccines: this was demonstrated in 2006 when Dow AgroSciences registered a plant-produced vaccine against Newcastle disease virus (NDV), which affects poultry. This was the first plant-based vaccine registered in the US, and the first to be licenced by the USDA's Animal and Plant Health Inspection Service. The HN envelope glycoprotein of NDV was produced in suspension-cultured tobacco cells, thus avoiding all the production and regulatory problems associated with whole plant production. In the same year, the Centre for Genetic Engineering and Biotechnology (CIGB) in Cuba registered a tobacco-produced MAb for use in purification of their commercially produced HBV vaccine: this development obviates the need for MAb production in ascitic fluid from mice, and considerably reduces the cost of manufacture of a vaccine which is distributed worldwide.

The introduction in mid-2006 of Merck's yeast-produced Gardasil anti-human papillomavirus (HPV) L1 major capsid protein VLP vaccine was hailed as a major development in cancer prevention, and the vaccine is predicted to be a 'blockbuster'. However, at a cost of US$ 360 per course of three doses, it will not make inroads into limiting cervical cancer incidence in developing countries. In 2003, three groups published simultaneous accounts of plant production of HPV L1 proteins. While the antigens were correctly folded and assembled, the yields were low and oral and parenteral immunogenicities were also low: however, these problems were circumvented for HPV-16 L1 in early 2007 by a combination of codon optimization and intracellular targeting, and excellent parenteral immunogenicity was achieved.

Obstacles to Vaccine Pharming

The following obstacles impede the development and potential application of plant-based vaccines, especially in developing countries:

1. increasingly stringent regulatory barriers in developed countries which deter commercial development of new vaccines generally, and vaccines depending on new technology in particular;
2. declining commercial interest in producing vaccines due to negligible return on vaccines needed in developing countries (people who do need vaccines cannot afford them and have to rely on governments and/or aid agencies for making them available);
3. earlier expectations from the technology have not been fulfilled;
4. many developing countries have little or no infrastructure for the approval of experimental drugs and vaccines, or for GM crops;
5. there is widespread and profound suspicion in developing countries of 'GM products', usually fueled by developed country activists;
6. the 'guinea pig' argument: why should developing country populations be used to test products not used in developed countries, where existing but expensive 'best practice' products are already available? and
7. production of vaccine antigens in plants may not significantly decrease the cost of the final vaccines, given that downstream processing and other costs remain the same.

Most of these obstacles could possibly be overcome: countries could pass relevant legislation; regulatory bodies could adopt suitable guidelines; publicity/and information campaigns could change public perceptions, both in the developed as well as in the developing world.

The 'guinea pig' argument could be countered by testing products in developed countries at the same time as in developing countries.

However, a more serious obstacle is that conventional pharma industry is not engaging with plant production technology primarily because it is seen as technology in search of a product, when the technology already available is adequate to the task. Moreover, it is a serious misconception that using plants to produce vaccines would lower prices significantly, as cost of materials and processing is only a minor component of total retail vaccine cost, that is, about 24%. Thus, even a significantly lower cost of vaccine material would not change the cost of the product much, especially as processing costs would remain the same however cheaply one could produce raw vaccine ingredients.

The Way Forward

Vaccine pharming has moved on toward maturity in an encouraging way in recent years, as the dogma of 'cheap oral vaccines' has to some extent shifted to an acceptance that processed and possibly injectable products may be preferred to raw or crudely processed oral vaccines. Indeed, as new needle-free developments in vaccine delivery such as oral and nasal and transdermal delivery are becoming increasingly popular, plant-made vaccines may find a niche, especially in products that need less processing than those intended for parenteral delivery.

It is possible that practitioners will come to accept that success will not be achieved by trying to bring cheap vaccines and therapies for poor people, but by producing generics. For example, it may be a better idea to concentrate on HPV vaccines, which currently are very expensive and have only a few remaining years of patent protection, rather than on HBV vaccines, which are out of patent, are produced conventionally by several high-volume producers, and currently sell for low enough prices to be included in EPI packages.

Another important area to focus on is animal vaccines: these are necessarily very low cost for meat animals such as poultry, sheep, and cattle, and animal model systems have shown very satisfactorily that viral and other plant-produced vaccines could be very useful. The development times are shorter and regulatory hurdles are lower than for similar human vaccines, and edible or minimally processed oral vaccines are probably far more feasible for animals than humans. Major targets could be the already-tested FMDV vaccine for cattle, sheep, pigs, and even wild animals; Newcastle disease and other chicken and poultry viruses; H5N1 and other influenza viruses in domestic and wild fowl; and African horsesickness virus which is currently threatening to emerge into Europe.

It is possible that expression of viral antigens in, for example, maize seed could allow very cheap production of orally dosable vaccines which could be stored for years in a dried form for very little expense, to be pulled out when and if the need for emergency barrier immunization occurred.

It is important for the development of plant-made vaccines that several large companies are interested in the technology, and have done a lot of thorough investigation of its potential. It is especially significant that the first product – Dow's Newcastle disease vaccine – has run the regulatory course and was accepted in 2006.

What is probably needed for the successful future application of pharmed products, especially in developing countries, is a highly effective set of plant-derived products – vaccines and therapeutics – which have been approved by regulatory bodies such as the US Food and Drug Administration (FDA) and European Medicines Agency (EMEA). This will lead to public acceptance and a more welcoming stance by industry, both of which are essential if the technology is to survive.

See also: Plant Virus Vectors (Gene Expression Systems); Vaccine Strategies; Vector Transmission of Plant Viruses.

Further Reading

Daniell H (2006) Production of biopharmaceuticals and vaccines in plants via the chloroplast genome. *Biotechnology Journal* 1: 1071–1079.

Gerber S, Lane C, Brown DM, *et al.* (2001) Human papillomavirus virus-like particles are efficient oral immunogens when coadministered with *Escherichia coli* heat-labile enterotoxin mutant R192G or CpG DNA. *Journal of Virology* 75: 4752–4760.

Gleba Y, Klimyuk V, and Marillonnet S (2005) Magnifection – a new platform for expressing recombinant vaccines in plants. *Vaccine* 23: 2042–2048.

Kirk DD and Webb SR (2005) The next 15 years: Taking plant-made vaccines beyond proof of concept. *Immunology and Cell Biology* 83: 248–256.

Mason HS, Lam DM, and Arntzen CJ (1992) Expression of hepatitis B surface antigen in transgenic plants. *Proceedings of the National Academy of Sciences, USA* 89: 11745–11749.

Thanavala Y, Huang Z, and Mason HS (2006) Plant-derived vaccines: A look back at the highlights and a view to the challenges on the road ahead. *Expert Review of Vaccines* 5: 249–260.

Voinnet O, Rivas S, Mestre P, and Baulcombe D (2003) An enhanced transient expression system in plants based on suppression of gene silencing by the p19 protein of tomato bushy stunt virus. *Plant Journal* 33: 949–956.

Vaccine Safety

C J Clements, The Macfarlane Burnet Institute for Medical Research and Public Health Ltd., Melbourne, VIC, Australia
G Lawrence, The Children's Hospital at Westmead, Westmead, NSW, Australia and University of Sydney, Westmead, NSW, Australia

© 2008 Elsevier Ltd. All rights reserved.

Glossary

Adjuvant A chemical additive to vaccines that improves the presentation of the antigen to the immune system of the recipient, thereby increasing the effectiveness of the vaccine. The commonest adjuvant is an aluminum salt.

Adventitious agent A contaminating virus or other infectious organism.

Adverse event following immunization Any medical incident that follows immunization and is believed to be caused by the immunization.

Anaphylaxis A hypersensitivity reaction characterized by sudden onset and rapid progression of signs and symptoms in at least two organ systems. It results in release of histamine from mast cells and may follow administration of drugs and vaccines. The severity of the reaction varies from mild symptoms of flushing to severe bronchospasm, collapse, and death. It is treatable with adrenaline.

Auto-disable syringe A form of disposable syringe that automatically locks after one use, preventing its reuse. It is the syringe design of choice for immunization.

Causality assessment A scientific method for assessing the relationship between an adverse event and the receipt of a dose of vaccine.

Clinical trial A research activity that involves the administration of a test regime to humans to evaluate its efficacy and safety. There are classically four phases in clinical trials assessing vaccines.

Guillain–Barré syndrome A polyradiculoneuritis resulting in various degrees of motor and sensory disturbance. The condition frequently follows an acute infection or administration of vaccine.

High-titer measles vaccine A measles vaccine that contains at least one log greater ($>\log 10^4$) numbers of live viral particles than a standard vaccine preparation ($\log 10^3$).

Hypotonic hyporesponsive episode (HHE) The sudden onset of limpness, reduced responsiveness, and change in skin color in an infant following the administration of a vaccine. Typically (though not exclusively), this may follow administration of pertussis-containing vaccines.

Immunization The process of presenting an antigen to the immune system with the purpose of inducing an immune response. Immunization is synonymous with vaccination.

Injection equipment Any material needed to introduce a vaccine into the body of the recipient. Classically refers to needles and syringes, but new technologies for administering vaccines will widen its usage.

Neurovirulence test One of several tests that is required of live viral vaccines before release to ensure that the vaccine virus does not negatively affect the central nervous system.

Presentation How a vaccine is presented for use (e.g., as a liquid, freeze-dried powder, one-dose vial, 10-dose vial, etc.).

Preservative A chemical (most frequently thiomersal/thimerosal) that is included in a liquid vaccine preparation that inhibits the growth of organisms that inadvertently contaminate the vaccine.

Program error A mistake made by vaccination staff that occurs during administration of a vaccine.

Protective immunity Immunity that protects against actual disease, not simply evoking an immune response.

Regulatory authority An institution/laboratory that takes responsibility for quality control of vaccines used in a country.

Safety box A tough cardboard box into which can be placed used injection equipment. It is generally then transported to an incinerator or can be burnt on the spot, thus safely disposing of medical waste that might otherwise be a hazard.

Stabilizer Proteins or chemicals that are added to a liquid vaccine to maintain it in a stable or unchanging state.

SV40 Simian virus 40 is a virus found in certain monkeys and cell cultures derived from monkey tissues.

Vaccine diluent A liquid (usually sterile water or saline) that is used to reconstitute freeze-dried vaccine.

Vaccine safety The discipline of ensuring that vaccines are manufactured, transported, and administered with minimum risk to the recipient, vaccinator, or general public.

Introduction

Vaccines are arguably the most effective weapon in the modern fight against infectious diseases. They avert millions of deaths and cases of disease every year. But there is no such thing as a 'perfect' vaccine that protects everyone who receives it and is entirely safe for everyone. The scientist attempts to create a vaccine that protects nearly everyone, is extremely effective, and is as safe as humanly possible. That still leaves some vaccine recipients who are not fully protected; and worse, a very small number who suffer negative consequences from the vaccine itself.

There are many aspects that can be considered in the design phase which influence the ultimate safety and public acceptability of a vaccine (**Figure 1**). Vaccine safety can be categorized into aspects related to the manufacture of the vaccine, delivery of the vaccine to the recipient through an immunization programme, and individual and population responses to the vaccine. This article describes the efforts taken to ensure that a vaccine is manufactured, tested, and administered as safely as possible, and identifies how vaccine safety is measured and monitored after a vaccine is licensed for use in the population.

Vaccines are designed to evoke a protective response from the immune system as if it were the actual infectious disease. While a vaccine is designed to maximize the body's reaction and minimize any pathological effects, effective vaccines (i.e., capable of inducing protective immunity) may also produce some undesirable side effects. Clinical trials determine whether side effects are tolerable or not, and detect certain adverse events before the vaccine reaches general distribution. Most adverse reactions are mild and clear up quickly.

Design factors can help minimize programmatic errors in the administration of the vaccine by reducing the chance of human error. The most common types of programmatic errors are unsafe or contaminated injections. It is sobering to realize that up to one-third of injections (not just for vaccination) are not sterile. New technologies have been introduced in the last decade that make injections much safer, including an auto-disable syringe that can only be used once and must be discarded because the plunger mechanism locks after use. This device helps minimize transmission of diseases that result from the reuse of contaminated injection equipment.

Vaccine adverse reactions can be due to the vaccine itself, or to immunological factors in the individual such

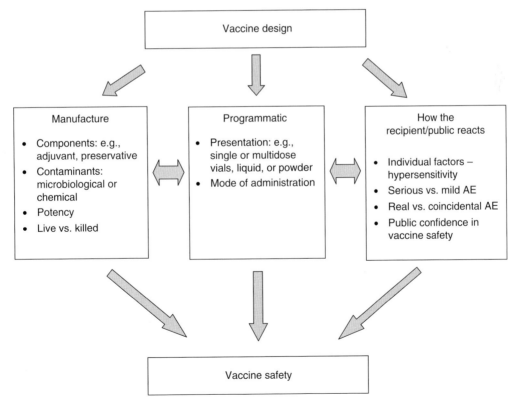

Figure 1 Components of vaccine safety.

as anaphylaxis or hypersensitivity reactions. However, the majority of adverse events thought to be related to the administration of a vaccine are actually not due to the vaccine itself – many are simply coincidental events. For example, many children have a cough around the time of vaccination, but this is independent of the vaccine, not caused by it. An appropriate causality assessment, using internationally consistent methods, is necessary to fully determine whether a particular vaccine causes a specific reaction. Even when scientists have been able to show that a particular event is unrelated to vaccine administration, the public may still perceive there to be a link.

Vaccine Design

Manufacture

A vaccine consists of many parts, only one of which is the antigen by which it is known. Other components may include an adjuvant, a preservative, antibiotics, stabilizers, and certain ingredients that remain from the manufacturing process. There may be components not stated on the information sheet that are classified as proprietary that the manufacturers are not obliged to declare. The safety of the vaccine can be affected by each and all of its components.

How a vaccine is manufactured will also determine its content. For instance, thiomersal may be used in the manufacture of pertussis vaccines to maintain sterility during the manufacturing process. While the bulk of the chemical is removed, the final product will still contain minute amounts of thiomersal. Influenza vaccines are grown on chicken eggs – trace amounts of egg proteins not removed in the purification process can cause severe allergic reactions in susceptible individuals.

Safeguards in vaccine manufacture now make it highly unlikely that currently available vaccines are contaminated with infectious agents that cause disease. Manufacturers supplying the international market are required to submit vaccines to extensive tests to detect known potential pathogens. As well, they must test cell lines used in the production of vaccines for the presence of a variety of infectious agents. Once manufactured, vaccines must undergo rigorous clinical trials for safety and effectiveness before being licensed by regulatory authorities.

Procedures designed to detect and prevent the presence of adventitious agents in vaccines include characterization of the cell substrates used for vaccine production as well as testing the products by specified methods. For example, manufacturers are required to test their cell lines used for the production of poliovirus vaccines (generally monkey kidney cells) for a variety of infectious agents such as tuberculosis, SV40, herpes viruses (including simian cytomegalovirus), Coxsackie virus, and lymphocytic choriomeningitis virus. Cell tissue culture tests must show an absence of cytopathogenic effects. Adult and suckling animals are also used to screen for viable microbial agents. A neurovirulence safety test is conducted for the live oral poliovirus vaccine.

The advent of bovine spongiform encephalitis (BSE) in the 1990s generated concern that the etiological agent might be contaminating vaccines through the use of calf serum during manufacture. Industry has taken steps to ensure that the risk to vaccines from BSE is virtually zero.

New technologies can also give rise to new concerns during manufacture. For instance, some modern vaccines are grown on chick embryo fibroblasts. These vaccines may test positive by a polymerase chain reaction assay for avian leukocytosis virus, a retrovirus not unlike the HIV/AIDS virus. Because the assay is so sensitive, it detects nonreplicating fragments of the virus that have no pathological significance. The cell substrates used for measles/mumps/rubella (MMR) vaccine are derived from flocks free of avian leukosis virus. MMR vaccine is also tested extensively for adventitious viral activity. As technology improves, regulatory authorities are likely to require the application of new methods to screen for infectious agents in vaccines.

Potency

The potency (or viral titer) of a vaccine can influence both its safety and its ability to induce protective immunity. When developing new rotavirus vaccines, the safety and efficacy of low-, medium-, and high-viral-titer formulations of human–bovine reassortant vaccines were assessed in randomized controlled clinical trials of infants. The medium-potency formulation was selected for use in phase III trials as it was found to be more efficacious than the high- and low-titer formulations and had a slightly lower, but not clinically significant, incidence of fever. In another example, a high-titer formulation of measles vaccine was used in some African countries in 1989–92 to protect infants from as early as 6 months of age. Unexpected results showed that the survival of infants who received the high-titer vaccine before 9 months of age was substantially lower than those who received the standard titer vaccine at 9 months of age. The high-titer vaccine was withdrawn from use.

Live or killed vaccine

There are three main types of viral vaccines currently in widespread use: live attenuated virus (e.g., MMR, yellow fever, varicella vaccines), killed virus (e.g., influenza vaccine), and recombinant protein (e.g., hepatitis B vaccine). Safety issues differ to some extent for the three types of vaccine. For all, the risk of an adverse event is highest in the immediate post-vaccination period due to reactions to vaccine components and to programmatic errors. Live-attenuated viral vaccines must undergo viral replication in the recipient to produce sufficient antigen to generate a protective immune response. The vaccine recipient may experience a mild form of the disease, particularly at the time of peak viral load. Febrile seizure in young children 5–14 days after receiving MMR vaccine is a known adverse event. There is also the potential risk of reversion to virulence if the vaccine strain of virus undergoes genetic mutations either *in vitro* during manufacture or in the vaccinee, that can cause significant illness. Killed vaccines usually contain trace amounts of the inactivating agent (often formaldehyde) that can cause reactions in some individuals. There is also the potential risk of incomplete inactivation with the effect that the vaccine could contain trace amounts of live virus capable of replication in the host to cause disease. An atypical variant of measles infection has been reported in some recipients of killed measles vaccine who were at risk of developing severe delayed hypersensitivity reaction following exposure to wild measles virus. As well as containing no viral genetic material, recombinant protein vaccines have the advantage of presenting fewer antigens to the immune system, thus minimizing the risk of hypersensitivity reaction. However, they are often produced in yeast cells, traces of which may remain after purification and cause allergic reactions in some individuals.

Programmatic Factors, Presentation, and Administration

A 'programmatic error' is caused by an error or errors in the handling or administration of a vaccine. The error is usually the fault of a person rather than the fault of the vaccine or other technology. It can generally be prevented through staff training and an adequate supply of safe injection equipment although vaccine design factors can help minimize the chance of human error.

A programme error may lead to a cluster of events. Improper immunization practice can result in abscesses or other blood-borne infections. Freeze-dried vaccines (such as measles-containing vaccines and yellow fever) are presented as a dry powder in a vial that needs reconstituting with liquid. Errors can occur in reconstituting the powder, such as the use of inappropriate or contaminated liquid. Once a vaccine is reconstituted, it is open to the environment and makes an ideal culture medium for organisms such as *Staphylococcus*. Fatal toxic shock syndrome has been reported following administration of a multidose vial of reconstituted measles vaccine that had become contaminated. In such circumstances, a number of infants immunized from the same vial may die within a short time of injection. To avoid this catastrophe, field staff are required to discard reconstituted vaccine after 6 h, or at the end of the vaccination session, whichever is sooner.

There are options for the way a vaccine is presented that can affect its safety in the field. To reduce cost, vaccines are often presented in multidose vials. It is common in developing countries for diphtheria–tetanus–pertussis (DTP) and measles vaccines to be in 10-dose vials, while BCG is often in 20-dose vials. The downside of multidose vials is that they have the potential for contamination (see above). In contrast, single-dose presentations are unlikely to become contaminated as the vial is discarded once the single dose has been used. Prepackaged mono-dose vials reduce the need for human handling (and hence reduce the risk of contamination) but are more costly per unit. Industrialized countries tend to favor mono-dose presentations because of their increased safety and convenience.

There are a number of different modes of administration of vaccines, and these affect safety in different ways. Aerosol presentations are attractive because they do not need a needle or syringe, thus reducing the risk of contamination. However, administering an inhaled vaccine to a 9-month-old infant is not without its practical problems. Nasal delivery bypasses the intestinal barrier, but recipients with purulent rhinitis (a common situation in developing countries) may interfere with successful seroconversion. Edible vaccines and vaccines presented using skin patches are also attractive for their apparent freedom from these safety concerns.

The injectable vaccine is by far the commonest presentation in the world today. Injections are one of the most common healthcare procedures. Five to ten percent of some 16 billion injections administered worldwide each year are given for immunization. Unsafe injections or unsafe practices in relation to immunization are not only responsible for cases of hepatitis B and C, HIV, and other serious and potentially lethal side effects suffered by vaccine recipients, but may also pose an occupational hazard to health providers and an environmental hazard to communities. Furthermore, unsafe injection practices can seriously impede the progress made by immunization programmes, and have a substantial effect on global immunization coverage. Fortunately, a number of technologies are at hand that help make injections safer.

The reuse of standard single-use disposable syringes and needles becomes unsafe when they are not disposed of after use but are reused multiple times. The auto-disable syringe, which is now widely available at low cost, presents the lowest risk of person-to-person transmission of blood-borne pathogens (such as hepatitis B or HIV) because it cannot be reused. The auto-disable syringe is the equipment of choice for administering vaccines, both in routine immunization and mass campaigns. In 1999, the World Health Organization (WHO) and other United Nations organizations issued a joint statement promoting the use of auto-disable syringes in immunization services. 'Safety boxes', puncture-proof containers – for the collection and disposal of used disposable and auto-disable syringes, needles, and other injection materials – reduce the risk from contaminated needles and syringes. The safe disposal of used needles and syringes is a critical component of any vaccination program if infection is to be prevented. Poor management of injection-related waste exposes patients, healthcare workers, waste handlers, and, indeed, the wider community, to infections, toxic effects, and injuries.

Program errors may occur when a new vaccine is integrated into an existing immunization programme that has important (and sometimes subtle) differences in its presentation, preparation, or administration. Errors such as the omission of a component, injection of oral vaccines, and the use of incorrect diluent to reconstitute freeze-dried vaccines are examples of this type of errors. While thermostabilty of a vaccine is not strictly an issue of vaccine safety, failure to maintain the cold chain is a significant component of programmatic error. This can lead to vaccine failure and the need to revaccinate populations, thus re-exposing them to the risk, albeit small, of vaccine adverse reactions.

How the Recipient/Public Reacts

While the recipient of a dose of vaccine will hopefully react by developing immunity to the challenging antigen, it is common for there to be various other less desirable reactions following vaccine administration. The WHO describes an adverse vaccine reaction as either 'common' or 'rare'. Common reactions are generally mild, settle without treatment, and have no long-term consequences. More serious reactions are very rare – usually of a fairly predictable (albeit extremely low) frequency.

Rates of common, mild vaccine reactions

The purpose of a vaccine is to induce immunity by causing the recipient's immune system to react to the vaccine. It is not surprising that vaccination commonly results in certain mild side effects (**Table 1**). Local reaction, fever, and systemic symptoms can result as part of the normal immune response. In addition, some vaccine components (e.g., aluminum adjuvant, antibiotics) can lead to reactions. A successful vaccine reduces these reactions to a minimum while inducing maximum immunity.

Rates of rare, more severe vaccine reactions

Most of the rare vaccine reactions (e.g., seizures, thrombocytopenia, hypotonic hyporesponsive episodes, persistent inconsolable screaming) are self-limiting and do not lead to long-term problems (**Table 2**). Anaphylaxis, while potentially fatal, is treatable without leaving any long-term effects.

Public perceptions of vaccine safety

While vaccine manufacture and administration have been getting progressively safer over the last 30 years, the perception of the public has been changing. Certain reports of adverse events following immunization (AEFIs) published in the medical literature over the past few years have resulted in controversy. While generating provocative hypotheses, the studies on which these reports are based have generally not fulfilled the criteria needed to draw conclusions about vaccine safety with any degree of certainty. Yet these reports have had a major influence on public debate and opinion making. Sadly, health professionals have not always handled public announcements about such matters very well and the debate has often spilt over into the political arena and policymaking. At times, mishandling of public information has resulted in a reduced public acceptance of a vaccine. Even when scientific evaluation has been correct, the public has often been swayed by other factors. This again underlines the importance of a correct assessment of causality.

Until very recently, communication with the public by health professionals and vaccine manufacturers has often been on a paternalistic basis: "We know what is best for you." The level of education of the general public has risen, helped by the Internet; freedom-of-information laws in many countries have allowed the public to access what was previously privileged information. As a result, members of the public are often well informed on a given public health issue such as vaccine safety.

Table 1 Summary of common minor viral vaccine reactions

Vaccine	Local reaction (pain, swelling, redness)	Fever	Irritability, malaise, and nonspecific symptoms
Hepatitis B	Adults up to 30% Children up to 5%	1–6%	
Measles/MMR	Up to 10%	Up to 5%	Up to 5%
Oral polio (OPV)	None	Less than 1%	Less than 1%[a]
Influenza	>10%	Up to 10%	Up to 10%

[a]Diarrhoea, headache, and/or muscle pains.
Note: the rates due to the vaccine administration will be lower as these symptoms occur independently as part of normal childhood.
Source: Supplementary information on vaccine safety. Part 2: Background rates of adverse events following immunization. WHO/V&B/00.36. http://www.who.int/vaccines-documents/DocsPDF00/www562.pdf.

Table 2 Summary of rare, serious viral vaccine reactions, onset interval, and rates

Vaccine	Reaction	Onset interval	Rate per million doses
Hepatitis B recombinant vaccine	Anaphylaxis	0–1 h	1–2
	Guillain–Barré syndrome (plasma derived)	1–6 weeks	5
Measles/MMR[a]	Febrile seizures	5–12 days	333
	Thrombocytopenia (low platelets)	15–35 days	33
	Anaphylaxis	0–1 h	1–50
Oral polio (OPV)	Vaccine-associated paralytic poliomyelitis (VAPP)	4–30 days	1.4–3.4[b]
Japanese encephalitis	Serious allergic reaction		10–1000
	Neurological event		1–2.3
Yellow fever	Post-vaccination encephalitis	7–21 days	500–4000 in infants less than 6 months[c]
	Allergic reaction/anaphylaxis	0–1 h	5–20
	Catastrophic system collapse resembling naturally acquired infection	2–7 days	2–32 (higher rates in the elderly or history of thymoma)[d]

[a]Reactions (except anaphylaxis) do not occur if already immune (~90% of those receiving a second dose); children over 6 years unlikely to have febrile seizures.
[b]VAPP risk is higher for first dose (1 per 1.4–3.4 million doses) compared to 1 per 5.9 million for subsequent doses and 1 in 6.7 million doses for contacts.
[c]Isolated cases with no denominator make it difficult to assess the rate in older children and adults, but it is extremely rare (less than four cases per 8 million doses).
[d]Studies have reported >10-fold higher risk of this rare adverse event among elderly vaccine recipients. There is a suggestion that reduced thymus gland activity (through age, immunosuppression, or surgical removal) increases the risk.
Source: Supplementary information on vaccine safety. Part 2: Background rates of adverse events following immunization. WHO/V&B/00.36. http://www.who.int/vaccines-documents/DocsPDF00/www562.pdf.

Discussions about vaccine safety now must contend with this new era. It is no longer sufficient for professionals to be factually correct – they must present information in ways that are perceived to be correct. A key component of this is representing themselves in ways that generate the confidence and trust of the public. In Jordan, a vaccine scare followed hard on the heels of a scare about water contamination, with major consequences. The public lost confidence in the truthfulness of government spokespersons, even though their claims that the vaccines in question were safe later proved to be correct. A similar situation occurred in the United Kingdom when there was a large decline in MMR immunization rates after publication of a now-disproved hypothesis that MMR vaccine caused autism. One of the factors attributed to this was loss of public confidence in the government and health spokespersons because of the way the BSE situation had been handled some years earlier.

Monitoring and Assessment of Vaccine Safety

The clearest and most reliable way to determine whether an adverse event is causally related to vaccination is by comparing rates of the event in a vaccinated and nonvaccinated group in a randomized clinical trial. Such trials, however, are unlikely to be large enough to assess very rare events, and postmarketing surveillance systems are required to identify events potentially related to vaccination. Postmarketing surveillance capability is improving; more countries now have monitoring systems, and more importance is attached to the reporting of suspected links between vaccination and adverse events. These systems have been successful in bringing to light serious adverse events after vaccines have been approved for use by regulatory authorities. A recent example is intussusception after administration of reassortant rhesus rotavirus vaccine.

Evaluating Adverse Events – Causality Assessment

Clinical events such as anaphylaxis or rash may occur subsequent to vaccination. But are they 'caused' by vaccination? It would be highly likely that collapse within minutes of receiving a dose of vaccine was due to the vaccine. But a rash 3 weeks later may not be – more likely it was coincidental. It is the area between these two scenarios that needs clarification.

Assessment of whether a given vaccine causes a particular adverse reaction varies from the casual observation to the carefully controlled study. The public frequently forms a decision about a vaccine's safety based on the information available to them – often a report based on unscientific observations or analyses that fail to stand the scrutiny of rigorous scientific investigation. Submitting a study to a scientific process rather than to partially informed opinion is crucial in determining whether a vaccine actually causes a given reaction. If undertaken carelessly or without scientific rigor, the study results will be inconclusive at best, may result in the inappropriate withdrawal of a valuable vaccine from use, or at worst may result in the exposure of a population to a dangerous vaccine.

When the Global Advisory Committee on Vaccine Safety (GACVS) undertakes causality assessment, the criteria it looks for in arriving at conclusions include consistency of observations, strength of the association, specificity, temporal relation, and biological plausibility. Clearly, not all these criteria need to be present, and neither does each carry equal weight for a causal relationship between an adverse event and the vaccine to be determined. Biological plausibility is a less robust criterion than the others described.

An association between vaccine administration and an adverse event is most likely to be considered strong when the evidence is based on well-conducted human studies that demonstrate a clear association in a study specifically designed to test the hypothesis of such association. Such studies may be randomized controlled clinical trials, cohort studies, case-control studies, or controlled case-series analyses. Case reports by themselves, however numerous and complete, do not fulfill the requirements for testing hypotheses.

Surveillance of AEFIs

Vaccines are generally administered to healthy individuals, so any report of vaccine administration causing harm needs to be taken seriously. Although the vaccines in general used in national immunization programmes are extremely safe and effective, adverse events can occur following vaccine administration and no vaccine is perfectly safe. In addition to the vaccines themselves, the process of immunization is a potential source of adverse events. An AEFI is any adverse event that follows immunization that is believed to be caused by the immunization. Using this terminology allows description and analysis of the event without prejudging causality. Immunization will tend to be blamed for any event that happens after immunization. AEFIs can be classified into five categories:

- vaccine reaction: a reaction due to the inherent properties of the vaccine;
- programme error: an error in the immunization process;
- coincidental event: unrelated to the immunization, but has a temporal association;
- anxiety-related: a reaction can arise from the pain of the injection rather than the vaccine; and
- unknown: in some cases the cause of the AEFI remains unknown.

It is important to investigate a report of an AEFI to determine into which of the above categories it fits. The

Table 3 Benefits of monitoring AEFIs

Improves vaccine safety by creating awareness among health professionals of the risks of vaccination
Identifies urgent problems that need investigation and action
Improves capacity to respond to AEFI reports
Improves the quality of the vaccination programme by monitoring performance and increasing staff confidence
Stimulates other aspects of programme monitoring, e.g., case reporting
Public perceives that health authorities understand and monitor vaccine safety, building public confidence in immunization
Detects signals for potential follow-up and research
Estimates rates for serious AEFIs
   For comparison between products
   To determine risks and benefits of immunization
   To validate prelicensure data
Identifies programmatic errors and batch problems

response to an AEFI will vary depending on the cause. However, many countries have ineffective reporting systems or no system at all. Currently, only 35% of 192 countries, and only 25% of 165 nonindustrialized countries, have an adequately functioning system for monitoring adverse events following immunization. The benefits of good surveillance of AEFIs are shown in **Table 3**.

Benefits of monitoring AEFIs

As vaccine-preventable infectious diseases continue to decline, people have become increasingly concerned about the risks associated with vaccines. Furthermore, technological advances and continuously increasing knowledge about vaccines have led to investigations focused on the safety of existing vaccines which have sometimes created a climate of concern. Allegations regarding vaccine-related adverse events that are not rapidly and effectively dealt with can undermine confidence in a vaccine and ultimately have dramatic consequences for immunization coverage and disease incidence.

Not only is public awareness of vaccine safety issues rising, there are increasing opportunities for AEFIs to occur due to various factors. For instance, an increasing number of vaccine doses is now given in campaigns than ever before (**Table 4**). Polio virus is targeted for eradication and measles elimination is being implemented in many parts of the world. As a result, vast numbers of doses of vaccine are being given over short time intervals. Thus, even if the AEFI rates stay the same, the actual number of cases is likely to be much higher than would be expected over the same time period, simply because so many doses have been administered.

Classical Safety Events

SV40

Between 1959 and 1963, both inactivated and live attenuated poliovirus vaccines were inadvertently contaminated

Table 4 Possible effect of immunization campaigns on vaccine safety

A real rise in program errors due to pressure on staff
  Increased workload on staff can increase human error
  Extra or new staff may not be fully trained in administration technique
Because large numbers of doses are administered there may be a rise in the absolute number of AEFIs even though rates may not change
Campaigns can generate popular opposition and rumors about vaccine safety
Rate of AEFIs prior to campaign may be unknown, so that any report of an AEFI appears to be due to the campaign

with simian virus 40 (SV40), a monkey virus known to be oncogenic for newborn hamsters. SV40 was found to be present in monkey kidney tissue used for propagation of the poliovirus during manufacture. From 1961, manufacturers were required by control authorities to test for SV40 using infectivity assays and different cell cultures, and lots positive for SV40 were not released. A few years later, the source of monkeys used for production was changed to species that do not harbor SV40. There is no evidence that polio vaccine administered since the early 1960s has contained SV40, and large epidemiological studies have not identified an elevated cancer risk in persons who received SV40-contaminated vaccines, but fragments of SV40 DNA have recently been identified in certain human tumors.

Swine flu

Influenza vaccine has been the source of a number of issues related to safety over the years. Mass immunization against swine-like influenza was carried out in 1976–77 using the A/New Jersey swine-influenza vaccine. Initially it appeared that this vaccine generated an increased risk of acquiring vaccine-related Guillain–Barré syndrome (GBS). In contrast, the 1978–79 influenza vaccine was not associated with a statistically significant excess risk of GBS. Although the original Centers for Disease Control study of the relation between A/New Jersey/8/76 (swine flu) vaccine and GBS demonstrated a statistical association and suggested a causal relation between the two events, controversy has persisted. At least one recent review of the data seemed to confirm the link. Another review suggested that there was no increased risk of acquiring GBS associated with the influenza vaccines administered during these seasons and that the causative 'trigger agent' in the A/New Jersey (swine) influenza vaccine administered in 1976 has not been present in subsequent influenza vaccine preparations.

There was no increase in the risk of vaccine-associated GBS from 1992–93 to 1993–94. A study by Lasky suggested only slightly more than one additional case of GBS per million persons vaccinated against influenza. More recently, an oculorespiratory syndrome has been described following influenza vaccine.

Recent Safety Events

A number of events in recent years has provoked concern that certain vaccines might be unsafe. Most of these concerns have proved to be unfounded, and some remain unproven or incompletely understood.

Thiomersal

Thiomersal is a mercury-containing preservative used in certain liquid vaccines such as DTP and hepatitis B vaccines. It came into the public gaze in 1999 as the result of a study being undertaken to estimate how much mercury was being administered to children in North America through vaccines. It transpired that the amount exceeded the allowed maximum according to environmental health permitted limits. Scientists assumed that ethyl mercury in thiomersal would have the same toxicology as methyl mercury, a chemical about which a great deal was known. Over the ensuing years, evidence was accumulated that showed it actually had a different metabolic pathway in the human body compared with methyl mercury. Both were metabolized in the liver and excreted in the gut, but methyl mercury was reabsorbed and accumulated in the body. Ethyl mercury, on the other hand, was found to be passed in the stool and lost from the body, thus avoiding a cumulative effect. Overwhelming scientific evidence indicates no clear link between neurodevelopmental disorders or autism and the receipt of vaccines containing thiomersal. One set of authors clearly flies in the face of the weight of scientific evidence, yet support a widely held public belief that thiomersal (in the quantities present in vaccines) is bad for children. Live viral and live bacterial vaccines do not contain thiomersal, largely because the preservative would kill the vaccine organism, rendering the vaccine useless.

Hepatitis B vaccine and multiple sclerosis

Reports of multiple sclerosis developing after hepatitis B vaccination have led to the concern that this vaccine might be a cause of multiple sclerosis in previously healthy subjects. A number of reliable studies have been consistent in showing that there is no association between hepatitis B vaccination and the development of multiple sclerosis.

MMR autism and bowel disease

Reports have proposed a link between the administration of the MMR vaccine and juvenile autism, Crohn's disease, and other forms of inflammatory bowel disease. The weight of scientific evidence strongly favors the conclusion that there is no direct association between measles virus, measles vaccine, autism, or inflammatory bowel disease.

Intranasal influenza vaccine and Bell's palsy

After the introduction of an inactivated intranasal influenza vaccine that was used only in Switzerland, 46 cases of Bell's palsy were reported. The relative risk of Bell's palsy in one study was found to be 19 times the risk in the controls, corresponding to 13 excess cases per 10 000 vaccinees within 1–91 days after vaccination. As a result, this vaccine is no longer in clinical use.

Yellow fever vaccine

Until recently, yellow fever vaccine was considered to be very safe. It has been given to many millions of people over more than 60 years to protect against yellow fever virus infection, which can be fatal in 5% or more of cases. From 1996 onward, a previously unrecognized adverse event of multiorgan failure and death was detected through passive surveillance in the United States, Australia, and other countries. Subsequent studies found that the risk of this rare adverse event was approximately 3 per 1 million doses, and vaccine recipients aged 65 years and older had a 10-fold higher risk than younger vaccinees, particularly following the first dose of the vaccine.

HIV Infection

The era of the HIV epidemic has had an impact on vaccine safety in a number of ways.

- It may be transmitted by injection (not only during vaccination). It could thus be transmitted by incorrect injection practices during routine immunization to a small number of infants attending a clinic at the same time. And it could theoretically be transmitted during mass campaigns where many more children might be at potential risk.
- Staff may be infected because of needle-stick injury or other improper handling of infected body fluids.
- It may alter the immune response of the recipient, either making him/her vulnerable to some negative effect of the vaccine, or diminishing the effect the vaccine has on protecting the infant. Some vaccines are not as effective in more immune-suppressed individuals, and severely immune-depressed infants must avoid vaccines such as BCG as they are at risk of disseminated BCG, which may be lethal.
- The course of the disease might be altered. The activation of CD4+ T lymphocytes following immunization could potentially augment HIV replication and result in accelerated progression to disease. Several, but not all, investigators have described increased HIV RNA plasma levels lasting several days following immunization.

On the positive side, the HIV epidemic has alerted health workers to the potential dangers of contamination during vaccine administration and has resulted in the safer use of syringes and proper disposal of injection equipment.

Conclusions

The design and manufacture of vaccines are of such high standards today that they have become extremely safe. But it will never be possible to say that everything is known about the subject and that vaccines are totally safe. Nor can the unpredictable human element in administration ever be totally excluded from the equation. But new methods of safety testing will emerge as each new threat is discovered. And new technologies will complement existing safety measures. However, communicating to vaccine recipients or their parents that the vaccine they are about to receive is neither totally effective nor totally safe makes health professionals uncomfortable. But the reality is that vaccines are many orders of magnitude safer than contracting the disease. Through constant vigilance, we can anticipate that vaccines will remain extremely safe – as safe as humanly possible. Any lesser alternative is unacceptable.

See also: AIDS: Vaccine Development; Antigenicity and Immunogenicity of Viral Proteins; DNA Vaccines; Immune Response to viruses: Antibody-Mediated Immunity; Neutralization of Infectivity; Vaccine Strategies; Yellow Fever Virus.

Further Reading

Afzal MA, Minor PD, and Schild GC (2000) Clinical safety of measles, mumps and rubella vaccine. *Bulletin of the World Health Organization* 78(2): 199–204.

Bellaby P (2003) Communication and miscommunication of risk: Understanding UK parents' attitudes to combined MMR vaccination. *British Medical Journal* 327: 725–728.

Causality assessment of adverse events following immunization(2001) *Weekly Epidemiology Record, 23 March,* 12, 76: 85–92. http://www.who.int/vaccine_safety/reports/wer7612.pdf(accessed February 2007).

Clements CJ and McIntyre PB (2006) When science is not enough – A risk/benefit profile of thiomersal-containing vaccines. *Expert Opinion on Drug Safety* 5(1): 17–29.

Cutts FT, Clements CJ, and Bennett JV (1997) Alternative routes of measles immunization: A review. *Biologicals* 25: 323–338.

Dellepiane N, Griffiths E, and Milstein JB (2000) New challenges in assuring vaccine quality. *Bulletin of the World Health Organization* 78: 2155–2162.

Halsey N (2003) Vaccine safety: Real and perceived issues. In: Bloom BR and Lambert P-H (eds.) *The Vaccine Book,* pp. 371–385. Amsterdam: Academic Press.

Kharabsheh S, Al Otoum H, Clements CJ, et al. (2001) Mass psychogenic illness following Td vaccine in Jordan. *Bulletin of the World Health Organization* 79(8): 764–770.

Madsen KM, Lauritsen MB, Pedersen CB, et al. (2003) Thimerosal and the occurrence of autism: Negative ecological evidence from Danish population-based data. *Pediatrics* 112(3, Part 1): 604–606.

Moss WJ, Clements CJ, and Halsey NA (2003) Immunization of children at risk for infection with human immunodeficiency virus. *Bulletin of the World Health Organization* 81(1): 61–70.

Safranek TJ, Lawrence DN, Kurland LT, et al. (1991) Reassessment of the association between Guillain–Barré syndrome and receipt of swine influenza vaccine in 1976–1977: Results of a two-state study. Expert Neurology Group. *American Journal of Epidemiology* 133(9): 940–951.

Shah K and Nathanson N (1976) Human exposure to SV40: Review and comment. *American Journal of Epidemiology* 103: 1–12.

Simonsen L, Kane A, Lloyd J, Zaffran M, and Kane M (1999) Unsafe injections in the developing world and transmission of bloodborne pathogens: A review. *Bulletin of the World Health Organization* 77(10): 789–800.

Supplementary information on vaccine safety. Part 2: Background rates of adverse rates of adverse events following immunization. WHO/V&B/00.36. http://www.who.int/vaccines-documents/DocsPDF00/www562.pdf(accessed February 2007).

Thompson NP, Montgomery SM, Pounder RE, and Wakefield AJ (1995) Is measles vaccination a risk factor for inflammatory bowel disease? *Lancet* 345: 1071–1074.

Relevant Websites

http://www.cioms.ch – Council for International Organizations of Medical Sciences (CIOMS) – WHO Working Group on Vaccine Pharmacovigilance.

http://www.brightoncollaboration.org – The Brighton Collaboration.

http://www.who.int – The WHO Expert Committee on Biological Standardization; WHO Programme for International Drug Monitoring; WHO's Global Advisory Committee on Vaccine Safety (GACVS).

http://www.who-umc.org – Uppsala Monitoring Centre (UMC), A WHO Collaborating Center.

Vaccine Strategies

I Kusters and J W Almond, sanofi pasteur, Lyon, France

© 2008 Elsevier Ltd. All rights reserved.

Introduction

To develop a new viral vaccine, get it licensed, and bring it to market is a lengthy, complex, and very expensive task. It requires a detailed knowledge of all aspects of the virus, especially its structure, epidemiology, pathology, and immunobiology, and demands a close collaboration between fundamental scientists, regulatory authorities, and industrial scientists and engineers. Completely new vaccines against human viruses appear infrequently on the market and the cost and complexity of their development has escalated with time, mainly due to the increased regulatory pressures to have highly defined products and to ensure complete clinical safety and high efficacy. Over the past 20 years or so, six new human virus vaccines have been developed for licensure in major markets; these are hepatitis B, Japanese encephalitis, hepatitis A, varicella (recently with a zoster formulation), and, over the past year, rotavirus, and human papilloma virus (HPV). The introduction of HPV vaccines is a technological triumph that offers to dramatically reduce the incidence of cervical cancer worldwide in the long term. The newest rotavirus vaccines seem, so far, to be free from the complications that led to the withdrawal of the earlier Rotashield vaccine in 1999 and promise to be very effective in reducing the burden of rotavirus diarrhea worldwide. In spite of this impressive progress, however, virus vaccine development has not accelerated over recent decades, and there remains a significant list of human virus diseases of widespread prevalence for which there are no vaccines available. These include human immunodeficiency virus (HIV), hepatitis C, hepatitis E, Eptein–Barr virus, herpes simplex virus (HSV), cytomegalovirus (CMV), respiratory syncytial virus (RSV), and parainfluenza viruses (PIV) (see **Table 1**).

In the veterinary world, the list of successes is significantly more impressive, and both the time cost of developing a new vaccine can be substantially reduced by the ability to reach proof of concept through direct challenge experiments, often with the wild-type virus. In this arena, the requirements are often very different to those for human vaccines, especially for vaccines of production animals where ease of administration and cost per dose are paramount. Recognizing that many approaches and outstanding fundamental challenges are similar for human and veterinary virus vaccines, this article focuses principally on the former with occasional reference to veterinary vaccines where they illustrate particular concepts.

Fueled by the new technologies of genomics, proteomics, and molecular immunology, the past 20 years have seen an impressive increase in our knowledge of all aspects of virology, providing insights to guide new vaccine concepts. The biological properties of viruses influencing choice of strategy include pathogenesis, serotype diversity,

Table 1 Human viruses causing disease with important medical need for which no vaccines are available

Adenoviruses	Human immunodeficiency virus (HIV)
Chikungunya	Human metapneumovirus (hMPV)
Cytomegalovirus (CMV)	
Dengue	Norwalk virus
Enterovirus 71 (EV71)	Parainfluenza virus (PIV)
Epstein–Barr virus (EBV)	Parvovirus B19
Hantavirus	Respiratory syncytial virus (RSV)
Hepatitis C (HCV)	Rhinovirus
Hepatitis E (HEV)	SARS
Herpes simplex virus (HSV)	West Nile virus

antigenic variation, immune evasion mechanisms, latency, and route of transmission. New vaccine candidates have been described for a good number of the viruses listed in **Table 1**, and although many of these are in an early, pre-proof-of-concept stage, some are substantially developed and offer realistic prospects of licensure over the next decade. The most promising of these, based on pre-clinical and clinical results obtained so far, are dengue, hepatitis E, and HSV and CMV. Significant challenges remain however, and really promising candidate vaccines against pathogens such as HIV, hepatitis C, and infant RSV remain elusive. This article focuses on the strategies available to develop new viral vaccines and discusses some of the challenges posed by the more difficult targets.

Types of Vaccines

There are multiple possible approaches to the development of a viral vaccine that can be generally described as follows:

1. Killed whole or split virus vaccines. This approach requires that the virus can be grown to high titer in cell culture or other scalable medium such as hens eggs; that the virus can be successfully and completely inactivated using an agent such as formaldehyde or B-propiolactone without destroying immunogenicity; that, from an industrialization perspective, the immunogenic dose is low to modest with respect to virus yield (in the 10 μg range) and that the killed whole or split particle elicits protective immunity. This approach has had excellent successes in the form of vaccines such as inactivated polio vaccine (IPV), hepatitis A (HAV) vaccine, and influenza.
2. Subunits or single proteins prepared by recombinant DNA methods and fermentation processes in cell culture. This approach may work well when a single protein can provide immunity and where the expression system allows appropriate folding and processing of the viral protein.
3. Live-attenuated vaccines. There are several approaches possible.
    a. Use of a closely related animal virus that is not well adapted for efficient and widespread replication in humans and therefore does not cause disease, but nevertheless provokes an immune response that protects against the corresponding human virus. The best known example of this is the use of vaccinia virus to vaccinate against smallpox, but a similar approach has been used for rotaviruses, with genetic reassortment to confer appropriate antigenicity.
    b. Development of an empirically attenuated human virus by multiple passages in tissue culture, typically of nonhuman origin, and/or passage in animals. Attenuation is usually achieved by the accumulation of a number of mutations that affect the efficient functioning at normal body temperature of various genes or gene products, thereby reducing virulence. Replication competence, however, is maintained at a sufficient level to stimulate a protective immune response. Although there are some drawbacks to this approach from a safety perspective (e.g., a risk of reversion), this method has provided a bedrock of vaccinology over many decades and has worked well for viruses such as polio – oral polio vaccine (OPV), mumps, measles, rubella, and yellow fever.
    c. Live-attenuated vaccines prepared by knowledge-based manipulation of the viral genome. There are several examples of candidate vaccines in this category including HSV and influenza.
4. Vectored or chimeric virus approaches. This is where an existing virus vaccine can be modified genetically to carry genes encoding antigens from a foreign virus. The chimeric vaccine should retain the attenuation and growth characteristics of the parent vaccine strain but stimulate immunity against the foreign virus.
5. Naked DNA. This is where a DNA encoding viral antigens plus appropriate expression control sequences is administered directly to the recipient. Expression of the DNA leads to an immune response against the antigens encoded.

Principles of Vaccine Development and Examples

When developing a new vaccine, the choice of approach is made very much on a case-by-case basis, and for a given virus is driven by knowledge of its biology, structure, antigenic diversity, and pathogenesis. High importance should be given to what type of immunity arises as a result of natural infection and whether the pathogen can cause persistent and/or repeated infections in a single host. Experiments in animal models may also allow the dissection of the immune response to identify correlates of protection. The use of primates in particular can be useful if the disease produced is similar to that observed in humans. However, many viruses are highly host specific and may have evolved strategies to evade immune responses that may also be host specific (such as recruitment of downregulators of complement fixation). Care must therefore be taken as results in animal models may not be entirely reproducible in the natural host.

Killed Vaccines

Evidence that circulating antibodies are sufficient to provide immunity may come, for example, from the observation that the disease is modified or exacerbated in immune deficiencies such as hypo-gammaglobulinaemia and/or that passive immune globulin can protect against infection and disease. The latter observation was made for several viruses prior to vaccine development, including hepatitis A, suggesting that the key to vaccine development in these cases would be the stimulation of a strong humoral immune response. Such responses can often be adequately provided by killed vaccines, and so this was an obvious choice of approach. The strategy will also be influenced by the successes and failures with closely related viruses (either human or animal) that have similarities in epidemiology, pathogenesis, and mode of transmission. Thus, for HAV, the successful paradigm of the inactivated polio virus vaccine from the same virus family (*Picornaviridae*) provided further confidence that a killed whole virus particle approach would be effective. Indeed, inactivated HAV vaccines were developed successfully on this basis by several companies in the early to mid-1990s. Current HAV vaccines are prepared by propagating the virus in an approved cell substrate, human fibroblasts or human diploid cell culture MRC-5, purification, inactivation using formalin, and adjuvanted with aluminum hydroxide. The vaccines are given parenterally, as a two-dose series, 6–18 months apart.

A further example of a killed vaccine is rabies vaccine. Several WHO-recommended inactivated rabies vaccines are available currently. They are all similar in being whole virus, used after inactivation by β-propiolactone, and purification and concentration by ultracentrifugation and/or ultrafiltration. The vaccines are required to have a protective potency defined as >2.5 international units (IU). The potency is determined by the National Institute of Health (NIH) potency test which is based on assays using intra-cerebral challenge of previously immunized mice. The main differences between the rabies vaccines currently available lies in the cell culture used for production. The cell cultures used for the WHO-recommended vaccines are MRC-5 cells for purified human diploid cell vaccine (HDCV), Vero cells for purified Vero cell culture rabies vaccine (PVRV), and primary duck or chicken embryo fibroblasts for purified duck embryo vaccine (PDEV) and purified chick embryo culture vaccine (PCECV), respectively.

When a new virus emerges that poses a severe threat to human health such as the human coronavirus which caused severe acute respiratory syndrome (SARS) in 2002, it is necessary to start working immediately to develop a vaccine. The huge challenge in such a scenario is time. Generally, to develop a vaccine from basic research through animal studies, clinical lot development, analytical test development, clinical trials, industrial scale-up, and licensure takes 8–12 years. These timelines can be compressed in case of extreme urgency, but this compression is not unlimited. Although experience on existing vaccines can be exploited, all processes and procedures need to be evaluated, validated, and implemented. Working with a BMBL Section III Laboratory Biosafety Level 3 (BSL 3) agent, such as human SARS coronavirus, is not exceptional for vaccine manufacturers, but many precautions need to be taken in regards to equipment and laboratory practices, as little was known about the SARS virus in the early stages. The choice of the vaccine approach was indeed influenced by the time factor. If the virus grows well in cell culture, an inactivated viral vaccine is usually the option of choice as it is the fastest to accomplish. Fortunately, in case of the human SARS coronavirus, the virus did grow well on Vero cells, was efficiently inactivated either by formol or β-propiolactone, and appeared to be very immunogenic in several animal models as well as in human beings. After 2003, no more human SARS cases were observed and the development of such vaccines has generally been put on hold for the present.

Recombinant Protein Vaccines

As discussed above, a further simple strategy, particularly when only antibodies are required, is to use recombinant DNA methods to express a single surface structural protein of the virus in a host–vector system such as *Escherichia coli* or one of several yeast species. This approach may be adopted when the virus cannot be propagated efficiently in culture, making a killed vaccine approach impossible, when the inactivation process may diminish immunogenicity, or when a focused immune response against a specific protein is required. This approach has proved very successful for hepatitis B, where the vaccine is composed of particulate complexes of the virus surface glycoprotein HBsAg produced in yeast. These particles mimic virus-like particles produced during natural infection and induce a highly protective and long-lasting immune response. This vaccine has been on the market since 1992 as a three-dose series of injection, each containing 5–20 μg HBsAg. A new version of this vaccine has recently been licensed with a formulation containing 20 μg HBsAg, adjuvanted with monophosphoryl lipid A (MPL) and alum, which reduces the incidence of nonresponders compared to a population vaccinated with the licensed vaccine.

The recently introduced human papilloma virus vaccines have also been developed using the recombinant protein approach (see below). But as with killed vaccine approaches, one size does not fit all, and for many viruses the approach of recombinant protein expression has not proved successful for a variety of reasons. For example, many viruses have a complex structure that cannot easily be reproduced in foreign hosts at high yield, particularly when the final structure is formed from several conformationally interdependent proteins. Incorrectly folded or immature proteins may not elicit functional, protective antibodies. Second, for some viruses, immune responses to several proteins together may be necessary to provide complete protection against disease. Third, recombinant proteins administered conventionally are generally poor at providing cellular responses of a Th1 profile that may be necessary for protection against viruses such as HIV, hepatitis C virus (HCV), and members of the herpes family. Consequently, for many viruses, live-attenuated approaches or more complex production systems and/or methods of delivery are required as discussed below.

Live-Attenuated Vaccines

Live-attenuated vaccines are used to prevent diseases such as yellow fever, polio, mumps, measles, and rubella. They are based on viral strains that have lost their virulence, but are still capable of replicating sufficiently well to provoke a protective immune response. They cause infection but without inducing the clinical manifestations, eliciting a humoral as well as cellular immune response. Historically, the attenuation was obtained by passage in animals. The first demonstration of attenuation of a virus in cell culture was that of the yellow fever virus by Lloyd and Theiler. The attenuation resulted from a prolonged passage on cultures of chick embryo tissue. Another example is that of the development of the oral polio vaccines by Albert Sabin in the 1950s. Wild-type strains of each of the three serotypes were passaged in monkey testicular tissue both *in vitro* and *in vivo*, while testing in monkeys was performed by intracerebral inoculation at various stages of passage. Eventually, strains were selected that were unable to induce paralysis in animals. The number of passages and cloning steps required to achieve the desired level of attenuation varied between the serotypes. These vaccines are still routinely used in many countries of the world and have been the principal tool with which the WHO has pursued its campaign of global polio eradication.

The advent of genome sequencing and recombinant DNA techniques in the 1980s allowed the key mutations conferring attenuation, empirically introduced by Sabin's passages, to be identified. In addition to temperature sensitivity mutations affecting protein structure, all three attenuated strains had in common, point mutations in the $5'$ noncoding regions which affected the stability of RNA secondary structure believed to be important for interaction with host factors and for internal entry of ribosomes.

Rotavirus

As discussed above, live-attenuated vaccines may also be based on a closely related animal virus. The rotavirus vaccine licensed by Merck in 2006 is based on the bovine WC3 rotavirus. The original monovalent bovine strain is naturally attenuated for human beings, but does not induce protective immunity. To improve the effectiveness of the strain for human use, reassortant strains were prepared which contained genes encoding capsid proteins from the most common human serotypes on a background of the bovine strain. The present vaccine contains five single gene reassortants, each containing a gene for a capsid protein from human serotypes G1, G2, G3, G4, and P1A (**Figure 1**). A three-dose regimen with $2.0–2.8 \times 10^6$ infectious units per reassortant, administered orally beginning at age 6–12 weeks with a 4–10 week interval between doses, provides 70% protection against both mild and severe rotavirus diarrhea.

Nowadays, for some viruses, attenuated viral vaccines can be designed on a more rational basis, by specifically targeting virulence factor functions that may not be essential for virus replication, especially in cell culture, but are necessary *in vivo* to counter host innate defense mechanisms. An example here is the NS1 gene of influenza A viruses. This protein is able to downregulate interferon production in the virus-infected cell and some deletion mutants of NS1 lack this function and are therefore much more easily controlled by the host interferon response and are thereby attenuated. The augmented interferon (IFN) response provoked by such viruses may also have the advantage of providing stronger immune stimulation resulting in increased immune responses. So far, such strains have only been tested in animal models but they offer promise as future influenza vaccine strains.

A drawback of many live-attenuated vaccines is that they require a cold chain from point of production to point of use and this may pose logistical difficulty, especially in developing countries. Also, the safety of live-attenuated viral vaccines is under constant scrutiny because of the risk, albeit small, that the mutations conferring attenuation will revert to wild type, allowing the virus to become virulent again. This is the reason why some live-attenuated vaccines are not recommended for immunosuppressed patients.

Viral Vectors

If neither the killed nor the attenuated vaccine approach is appropriate or feasible, one can consider the use of a viral vector. In this case, an attenuated virus is used as a backbone carrying immunogenic proteins of the virus of interest. In general, the viral proteins chosen are the membrane and/or envelope proteins as these proteins are presented on the outside of the virus particle and recognized by the immune system. An example of a viral vector is the yellow virus vaccine strain 17D. This vaccine strain was developed in the 1930s, since which time over 400 million people have been immunized with this vaccine. The strategy here is to use the 17D vaccine as a vector to deliver the two structural proteins, the premembrane PRE-M and envelope proteins from closely related

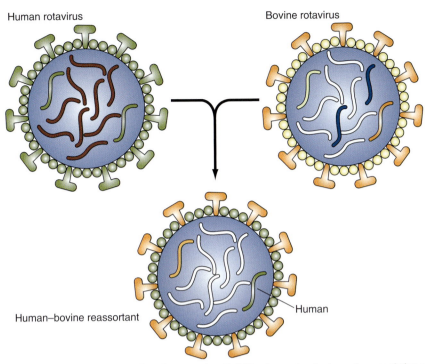

Figure 1 Rotavirus reassortant to generate oral live virus vaccine. RotaTeq is a polyvalent vaccine consisting of five human-bovine reassortants: four G serotypes (G1, G2, G3, G4) representing 80% of the G strains circulating worldwide, and one P serotype representing >75% of the P strains circulating worldwide. Reprinted by permission from Macmillan Publishers Ltd: *Nature Medicine* (Buckland BC (2005) The process development challenge for a new vaccine. *Nature Medicine* 11: S16–S19.), copyright (2005).

flaviviruses. The resulting chimeric virus needs to be viable and to replicate efficiently in an acceptable cell substrate for vaccine production. The chimerivax dengue virus approach (by Acambis in collaboration with sanofi pasteur) has been developed using the PRE-M and envelope genes of wild-type clinical isolates. The technology involved is illustrated in **Figure 2**. The yellow fever virus genome has been cloned as an infectious cDNA. This infectious cDNA is manipulated to remove the PRE-M and E-Genes of yellow fever virus and exchange them for the coat protein genes of each of the dengue virus serotypes. Thus four individual chimerivax cDNAs are constructed. Transcribing these cDNAs to RNA provides infectious RNA with which to transfect cells in culture. Thus, the resulting virus is a heterologous virus containing the immunizing antigens of dengue virus with the replicative engine of the yellow fever 17D vaccine. Chimeric dengue viruses are expected to mimic the biological properties of yellow fever 17D which has the excellent characteristics of providing minimal reactogenicity and lifelong immunity. The candidate vaccines have been characterized in preclinical models, neurovirulence in mice, viremia and immunogenicity in monkeys, and shown to have desirable characteristics. Moreover, when the four constructs are mixed and administered to monkeys, seroconversion against all four dengue serotypes appears to occur simultaneously in most infected animals. Moreover, the antibodies generated seem to be functional in that they neutralize dengue viruses in plaque reduction tests. The chimerivax dengue viruses grow well in culture and are well suited to industrial scaleup. In human volunteers, the chimeric viruses are safe and well tolerated and elicit specific immune responses against the different dengue serotypes. These strains therefore provide excellent candidates for further development.

DNA

Since the early 1990s, there has been considerable interest in the possible use of naked DNA as a vaccine delivery method. Naked DNA has the advantage that it can be taken up by cells and express the viral protein encoded. Depending on the conditions, this expression can be mid- to long term, thereby providing a substantial stimulation of the immune response. DNA vaccinology has apparently worked well in mice, but, so far, results in humans have been mainly disappointing requiring milligram amounts of DNA. Delivery of the DNA on colloidal gold, however, seems to offer a better prospect of success, as reported recently for hepatitis B virus and influenza. Regulatory issues concerning the use of DNA vaccines and its possible insertion integration into chromosomal genes are potential drawbacks to this type of approach, especially for use in prophylactic vaccination in infants. Further work is needed on safety issues before it can be seriously considered as a means to vaccinate populations.

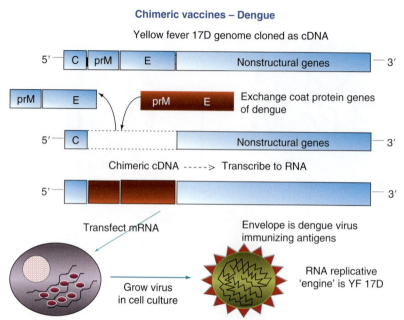

Figure 2 A DNA copy of the genome of yellow fever virus (blue) is manipulated to replace the prM and E genes by those of a related flavivirus such as Dengue (red). Transfection of mRNA transcribed from the resulting cDNA produces a 'chimeric virus' in cell culture.

The Challenge of Antigenic Variation

Antigenic Variation

Antigenic variation is displayed by a number of important pathogenic viruses and poses a particular problem for vaccine developers. The variation may be manifest in different ways depending on the virus' natural biology. Thus, for some viruses such as foot-and-mouth disease virus (FMDV), rhinoviruses, and HPV, multiple antigenic variants or serotypes co-circulate, sometimes with particular geographical patterns or ecological niches. Individual strains may show antigenic drift, presumably generating new serotypes over time. The rate of drift and the generation of new serotypes are not well understood for these viruses but may involve genetic recombination as well as cumulative mutational change. Other viruses may show a different pattern of antigenic variation: for example, influenza A viruses circulate as a limited number of subtypes (currently two in humans, H1 and H3), and each of these accumulate antigenic changes over several seasons (antigenic drift), escaping the most recently generated population-based immunity as they evolve. Occasionally, a new subtype may emerge (antigenic shift) either through genetic reassortment of a human strain with an avian strain, or possibly through direct evolution from an avian strain. Generally new subtypes displace existing subtypes as was the case when H2 emerged to displace H1 in 1957 and when H3 displaced H2 in 1968. For reasons that are not clear, this displacement did not occur when H1 reemerged in 1976 and since then there have been two influenza A subtypes co-circulating in the human population. This type of 'longitudinal' antigenic variation is clearly different from that of the multiple serotype viruses discussed above, and is generally more tractable in terms of vaccine development. Yet a different pattern is observed with HIV and to some extent HCV, where a limited number of genetic clades may contain very many different antigenic variants and where longitudinal variation to escape recently generated immune responses occurs within a single persistently infected individual. This type of natural biology generates a plethora of antigenic variants that can co-circulate. Providing immunological protection against all of these is an immense challenge for vaccine developers.

Multiple Serotype Vaccines

So what are the strategies available to develop vaccines against antigenically variable viruses? Most straightforwardly, one can generate simple killed vaccines against the currently circulating strains as discussed above and use these where the virus is prevalent. This strategy has had some success in the case of FMDV, where the geographical range of the virus may be (at least partially) restricted by regulations on movement of susceptible farm animals, and the vaccines are cheap and quick to prepare. However, even though simple killed vaccines have been shown to work for individual serotypes of rhinoviruses, it is difficult to imagine that this approach would be effective for this virus where, presumably because of widespread human contact and international travel, there seems to be a freer global circulation of multiple and unpredictable serotypes each winter. Moreover, for most people, rhinovirus infections are relatively trivial and therefore the balance of medical need versus industrial feasibility/commercial attractiveness of preparing vaccines in advance against multiple strains does not favor such a strategy. It would perhaps be a different matter if it were possible to design a simple rhinovirus vaccine that provided cross-protection against all serotypes.

The case of HPV is more manageable because of the small number of serotypes associated with severe disease. Thus, in this case, a significant impact on disease can be made by vaccinating against just a few of the many different genotypes. Both currently available vaccines contain HPV types 16 and 18 to vaccinate against cervical cancer, and the Merck vaccine contains, in addition, HPV types 6 and 11 to vaccinate against genital warts (condylomata accuminata). The absence of efficient cell culture systems for papillomaviruses has required the development of eukaryotic expression systems to produce virus-like particles composed of the L1 capsid protein, which are highly immunogenic. Future, second-generation vaccines will likely incorporate one or more of the additional highly or moderately oncogenic serotypes, such as 31, 33, and 35.

Influenza Annual Vaccination

For influenza, the limited number of circulating subtypes in any particular season makes it possible to adopt a strategy of annual vaccination. Thus, current influenza vaccines are trivalent, containing strains of H1 and H3 of influenza A and an influenza B strain. The strategy of annual vaccination is not without risk and requires a high level of international cooperation on disease surveillance and strain isolation, construction of high-yielding seed viruses, preparation of reagents for formulation, and industrial production. Following strain selection and recommendation by WHO, vaccine production has to occur over a very tight time schedule to ensure that vaccine is ready for the following winter (**Figure 3**). Occasionally problems arise such as a mismatch between the selection of a particular vaccine strain and the virus that eventually circulates during the following winter. This may compromise the effectiveness of that particular component of the vaccine. Other potential problems include less than optimal growth of the high growth reassortant seed at the industrial scale, leading to less vaccine being produced, and the lateness of seeds or reagents impacting on prompt delivery of vaccine for the flu season. The vast majority of influenza vaccine used currently is partially purified, killed whole or split virus prepared in embryonated hen's eggs, formulated to 15 μg of hemagglutinin (HA) of each strain, and provides generally good protection that correlates well with the induction of virus neutralizing or hemagglutination-inhibiting antibodies. However, vaccines prepared using different technologies are arriving on the market or are in advanced stages of development. These include live-attenuated (cold-adapted) strains, licensed by MedImmune in the USA in 2003, influenza-recombinant surface protein (hemagglutinin) produced in a baculovirus expression system from Protein Sciences, and inactivated

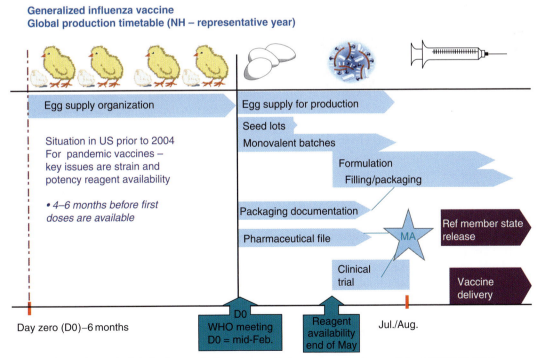

Figure 3 Approximate time schedule for the production of annual influenza vaccine from embryonated hens eggs.

virus vaccines prepared in cell cultures such as Vero, Madin–Darby canine kidney (MDCK), or PerC-6. This exploration of alternative technologies in recent years has been fueled by several criteria, perhaps the most important of which is greater and more flexible scaleup capability. The recent concerns over the possible emergence of an H5 pandemic has focused attention on the present worldwide limits in capacity, most notably in a situation of 'surge' demand, and governments have responded by providing the industry with incentives to increase capacity and diversify methods of production. A further response to this concern has been 'dose sparing' clinical studies on influenza vaccines adjuvanted with alum and other, proprietary adjuvants. These studies, using an H5N1 strain, suggest that it may be possible to reduce the vaccine dose from 2×90 μg HA nonadjuvanted, to 2×30 μg HA adjuvanted with aluminum hydroxide or even as low as 3.75 μg HA adjuvanted with new proprietary adjuvants. Moreover, there have been renewed suggestions that it may be possible to develop vaccines with a broader and perhaps multiseasonal protective effect by stimulating cellular immune responses, particularly against nucleoprotein (NP), a claim currently made for the live-attenuated approach, and even a universal flu vaccine, for example, based on the well-conserved M2 virus surface protein. Animal challenge experiments, especially using multiple arrays of the M2 protein, have been encouraging to date and suggest it may be possible to provoke a much stronger response against M2 than that induced by natural infection. Whether such a response will provide solid protection in humans however remains to be established.

HIV Approaches

The vast array of antigenic variants of HIV and HCV renders the approaches discussed above extremely difficult for these types of viruses. For HIV in particular, many strategies have been tried, so far without significant success. Early on in the AIDS pandemic the focus was on the use of simple recombinant surface glycoproteins, aimed at inducing neutralizing antibodies in the hope that even limited protection against homologous or closely related strains would provide a proof of concept that could be built upon. Unfortunately, this type of approach was not successful and a large phase III clinical trial carried out by Vaxgen using *E. coli*, which produced gp120, was not able to provide convincing evidence of protection even against strains closely related to that present in the vaccine formulation. Subsequently, there has been considerable effort on the induction of cellular responses, especially CD8 cytotoxic T lymphocytes, and more recently on a balanced cellular response to include CD4 effector mechanisms. The objective here is to provide the means for the immune system to launch an immediate attack on the first cells to become infected following exposure to the virus. Ideally, such an attack would prevent the primary viremia by eliminating the virus before it becomes established in the body. However, more realistically, there is evidence from primate studies that strong preexisting T-cell responses can control the primary viremia and reduce the viral set point during the asymptomatic phase. A low viral set point is associated with slow or no progression to AIDS. In addition, the HIV evades immune surveillance by actively downregulating the major histocompatibility complex I (MHCI) molecules on the surface of infected cells by the HIV nef protein. Vaccination strategies to produce cellular responses have mainly used vectors such as the vaccinia virus strains MVA and NYVAC, canarypox, adenoviruses, or naked DNA, either with multiple doses of a single type of construct or in heterologous prime-boost strategies. Antigens delivered have ranged from substantial regions of gag-pol-env of HIV to multiple copies in a 'string of beads' format of defined T-cell epitopes presented by common human leukocyte antigen (HLA) haplotypes. In general, these studies have not delivered T-cell responses of sufficient magnitude to be strongly encouraging, although one such strategy based on canarypox delivery of gag-pol-env antigens, followed by boosting with a recombinant env protein, has progressed to a clinical phase III study. Although many commentators have expressed doubt about whether this approach will show efficacy, it may generate useful information on the role of cellular responses in controlling HIV loads.

Most recently, HIV vaccine efforts have again turned to the induction of neutralizing antibodies, this time aimed specifically at epitopes that have been defined by studying unusual but highly informative broadly neutralizing monoclonal antibodies. The fact that such antibodies exists is highly encouraging from a vaccine perspective. The concepts here are based on the notion that certain conserved but crucial regions of gp120 or gp41 are naturally poorly immunogenic, either because they are relatively hidden in the conformationally folded protein or are shielded by strongly immunogenic noncritical domains or by glycosylation, or because they are only transiently exposed during the structural rearrangements that accompany cell binding and virus penetration. It is argued that because antibodies against these regions are neutralizing they will be protective if they can be generated prophylactically with sufficient avidity and at sufficient titer. This 'cryptic epitope' idea has been discussed for several viruses over many years and is akin to that mentioned above for influenza M2, in that the objective is to generate a far stronger response against a particular antigen or epitope than that resulting from natural infection. Such antibodies, once induced, will need to have the kinetic properties necessary to effectively neutralize the virus *in vivo*. So far there are no examples among virus vaccines that prove this concept. For HIV gp160, the particular construction, presentation, and formulation of molecules able to raise high-titer antibodies against these conserved regions (many of which are imprecisely defined) are far from obvious. Nevertheless, the induction of prophylactic immune responses of this type is certainly worth detailed investigating in detail, given the magnitude of the HIV problem. The challenge of developing a vaccine against HIV however remains immense.

Conclusions and Perspectives

The easy viral vaccine targets of significant medical importance have been done. The viruses against which we do not have vaccines today are either of regionalized or sporadic importance medically and the incentives to develop them have not been sufficiently large or they are viruses that pose significant challenges in terms of their biological characteristics. Thus, HIV and HCV pose challenges because of their antigenic variation and the fact that natural immune responses are unable to protect and/or eliminate the virus. RSV poses challenges because of the immunopotentiation of pathogenesis that, for infants, must be avoided at all costs.

Nevertheless, there are grounds for optimism. A new generation of adjuvants, making it possible to selectively orientate immune responses toward Th1 or Th2 as necessary, promises the possibility of being able to 'improve on nature' in terms of immune response provoked by the viral antigens. New developments in vectors and virus 'chimeras' offer promise for vaccines such as dengue and perhaps RSV and parainfluenza viruses and targeted modification of immunomodulatory genes may offer prospects of new vaccines against herpes family viruses. Fundamental studies on virus pathogenesis, epidemiology, and immunobiology are greatly aided by new technologies such as genomics and proteomics and it is likely that the improved understanding will increase the technical and scientific feasibility of developing new viral vaccines in the years ahead.

See also: AIDS: Vaccine Development; Antigenic Variation; Antigenicity and Immunogenicity of Viral Proteins; DNA Vaccines; Immune Response to viruses: Antibody-Mediated Immunity; Neutralization of Infectivity; Vaccine Production in Plants; Vaccine Safety.

Further Reading

Buckland BC (2005) The process development challenge for a new vaccine. *Nature Medicine* 11: S16–S19.
Garber DA, Silvestri G, and Feinberg MB (2004) Prospects for an AIDS vaccine; Three big questions, no easy answers. *Lancet* 4: 379–413.
Koelle D and Corey L (2003) Recent progress in herpes simplex virus immunobiology and vaccine research. *Clinical Microbiological Reviews* 16(1): 69–113.
Plotkin SA, Rupprecht CE, and Koprowski H (2005) Rabies vaccine. In: Plotkin SA and Orenstein WA (eds.) *Vaccines,* 4th edn., pp. 1011–1038. Philadelphia: Saunders.

Vaccinia Virus

G L Smith, Imperial College London, London, UK

© 2008 Elsevier Ltd. All rights reserved.

The Origin of Vaccinia Virus

The origin and natural host of vaccinia virus (VACV) remain unknown and have been discussed in detail by Baxby. Although cowpox virus (CPXV) is believed to have been used by Jenner in 1796 for vaccination against smallpox, Downie demonstrated in 1939 that the smallpox vaccine strains in use in the twentieth century were biologically distinct from CPXV. They were called VACV after *vacca,* Latin for cow. Analysis of the genomes of many orthopoxviruses (OPVs) by restriction endonuclease digestion and sequencing confirmed that VACV is a distinct OPV species. Genome sequence analyses showed that VACV was not formed directly from CPXV and variola virus (VARV) by recombination or indirectly from either virus by passage and mutation.

So where did VACV come from? The most probable explanation is that VACV is an OPV that is either no longer endemic or present in a host species in which it is not recognized. Horsepox virus (HSPV) has been proposed, and several anecdotes are consistent with this theory. First, Jenner and other early vaccinators took vaccine from horses when the supply of cowpox (a rare disease) was short. Second, the VACV interferon gamma (IFN-γ)-binding protein binds equine IFN-γ (and IFN-γ from several other species). Third, the sequence of an OPV that caused an epidemic in horses in Mongolia is more closely related to VACV strains than to other OPVs. However, none of these observations proves that VACV has an equine origin. It is possible that VACV in horses is a zoonosis and that its natural host is elsewhere, rather as CPXV and monkeypox virus (MPXV) are misnomers since their natural reservoirs are different types of rodent.

Currently, VACV-like viruses are circulating among buffaloes in northern India and cattle in South America, but it remains unclear whether these represent infections of animals with the smallpox vaccine that was transferred from man or whether they are natural reservoirs of VACV. VACV can infect many species and has a relatively broad host range, like CPXV and unlike VARV and camelpox virus (CMLV).

If one assumes that Jenner used CPXV for smallpox vaccination in 1796 (and this is uncertain because there is no virus isolate to study), sometime between 1796 and 1939 VACV replaced CPXV as the smallpox vaccine. It is not known when this happened, but anecdotal evidence suggests that it had occurred by the middle of the nineteenth century. First, the smallpox vaccine that was taken from England to the USA in 1856 and that became the New York City Board of Health vaccine is VACV and not CPXV. Second, the histological descriptions of cells infected by the smallpox vaccines in the late nineteenth century mentioned the cytoplasmic B-type inclusion bodies that are characteristic of OPV infections and made by both VACV and CPXV, but failed to report the much more obvious A-type inclusion bodies made by CPXV and not VACV. The failure to report A-type inclusions is consistent with the vaccine used at that time not being CPXV.

VACV: The Smallpox Vaccine

Widespread vaccination with VACV led to the control of smallpox in many countries by the 1970s, its removal from the Indian subcontinent by 1975, and its eradication by 1979. The last naturally occurring case was in Somalia in 1977, but the last person to die of smallpox was in Birmingham, UK, in 1978, following a laboratory escape of the virus. The eradication of smallpox remains the greatest triumph of the World Health Organisation (WHO) and a fine example of the benefit of cooperation among nations. Smallpox and its eradication are described beautifully in a monumental book published by the WHO in 1988.

Although VACV was used to achieve the global eradication of smallpox, it was a vaccine with an imperfect safety record. Several types of complication were noted. Generalized vaccinia was a systemic skin infection that usually resolved without sequelae. Progressive vaccinia was a grave condition that occurred in people with T-cell immunodeficiency. In these cases, the virus spread from the vaccination site without restriction and was often fatal. Eczema vaccinatum was a skin condition associated with infection of eczematous patients and usually resolved safely. There were also unpredictable neurological complications, such as encephalitis and encephalopathy, that were sometimes fatal. Finally, there were cases of myocarditis, and these have been evident again during recent vaccinations in the USA. All types of complication were more frequent in primary vaccinees than in those vaccinated previously. Large-scale studies of the complications of smallpox vaccination in the USA during the late 1960s recorded nine deaths after 14.2 million vaccinations. Without careful screening of vaccinees, it is probable that widespread vaccinations today with the old vaccines would induce a greater number of complications because of more widespread eczema and increased immunodeficiency due to infection with human immunodeficiency virus.

Attenuated VACV Strains

The complications of smallpox vaccination led investigators to develop safer candidate vaccines for smallpox. These were of two types: virus preparations from which infectivity had been removed by chemical or physical means, and virus strains that had diminished virulence. Trials with killed virus preparations failed to induce protection in models of OPV infection because these vaccines were composed largely of intracellular mature virus (IMV) and so lacked the antigens present in the outer envelope of the extracellular enveloped virus (EEV) (see the section titled 'Virion structure'). These antigens are important for the induction of protective immunity against OPV. The attenuated virus vaccines show more promise and several were produced, including modified virus Ankara (MVA) in southern Germany, LC16m8 in Japan, and NYVAC in USA. However, although these viruses are certainly less reactogenic in man, their ability to induce protective immunity is uncertain because they were never used in populations that were exposed to smallpox. They might induce protection against smallpox or they might not. While their utility as smallpox vaccines is uncertain, they do show considerable promise as recombinant vaccines for the delivery of antigens from other pathogens against which protective immunity is needed. In addition, these attenuated virus vaccines might be used for primary smallpox vaccination, followed by vaccination with the traditional vaccines known to protect against smallpox, in order to reduce adverse reactions. Indeed, this strategy was used with the MVA strain in Germany.

Taxonomy

Vaccinia virus is the most intensively studied species in the genus *Orthopoxvirus* of the family *Poxviridae*, and several VACV strains have been sequenced. The first strain sequenced was Copenhagen, which has been a reference for comparison with the sequences of additional VACV strains and other OPVs. Other commonly used, sequenced VACV strains are Western Reserve (WR), Lister (and derivatives LC16m0 and LC16m8), and MVA. All VACV strains cluster closely together phylogenetically, as do all other OPV species except for CPXV, where two strains (GRI-90 and Brighton Red) are more divergent and might eventually be considered as separate species. HSPV appears to be the OPV most closely related to VACV.

Genome Structure

The VACV genome is a linear, double-stranded DNA (dsDNA) molecule of between 185 and 200 kbp with covalently closed termini that are linked by hairpin loops. There are inverted terminal repeats (ITRs) that vary in length between different VACV strains. These contain blocks of tandem repeats and a few genes that, consequently, are diploid. VACV genes are tightly packed so that there is little noncoding DNA, and mRNAs are not spliced. Overall, the genome may be divided broadly into three regions. The central region (about 100 kbp) encodes proteins that are conserved between VACV strains, other OPVs, and, indeed, all chordopoxviruses. Most (89) of these proteins are essential for some aspect of virus replication. In contrast, the left and right terminal regions encode proteins that are more variable between different VACV strains and other OPVs and are nonessential for virus replication. Instead, these genes affect virus host range, virulence, and interactions with the host immune system. Like other OPVs, the VACV genome is rich (67%) in adenine and thymine (A + T) residues.

Virion Structure

The structure of the VACV particle has been studied extensively and different models have been presented. There is agreement that there are two morphologically distinct forms of virus produced from each infected cell, and these have been called IMV and EEV. The terms mature virus (MV) and extracellular virus (EV) have been used by some authors recently. Each virion is brick-shaped, has dimensions of 250 nm × 350 nm, and is surrounded by one or more lipid membranes. The formation of IMV and EEV is considered later (see the section titled 'Assembly'). Studies in the early 1960s proposed that the IMV particle is surrounded by a single membrane, which is synthesized *de novo* in the cytoplasm and is not linked to cellular membranes. Subsequently, this view was challenged and it was proposed that the IMV form was surrounded by a double layer of membrane derived from and continuous with preexisting cellular membranes between the endoplasmic reticulum and the Golgi stack. Current evidence favors a single membrane covered by a proteinaceous layer. Inside this, there is a dumbbell-shaped virus core with lateral bodies in the concavities of the core. The core contains pores in its outer layer and is surrounded by a palisade of protein spikes. Within the core, the virus genome is packaged with structural proteins and enzymes that are needed to transcribe virus genes upon infection.

The EEV form comprises an IMV that is surrounded by an additional lipid membrane derived from the *trans*-Golgi network or early endosomes. The extra lipid envelope is fragile and contains antigens that are absent from the IMV particle.

Replication Cycle

Entry

The entry of VACV starts with the binding of either an IMV or EEV to the cell surface. The cell receptors for these virions have not been identified, but are different. The entry of IMV is considered first. IMV particles can bind glycosaminoglycans (GAGs) or laminins on the cell surface. Although several virus proteins (H3, D8, and A27) were reported to bind GAGs, these interactions seem nonessential because virions lacking these proteins individually retain infectivity. After binding to the cell, the IMV particle enters either by fusion at the cell surface in a pH-independent manner (**Figure 1**) or by endocytosis followed by fusion with the vesicle membrane. Unlike other enveloped viruses, which usually have a single protein mediating fusion of the virus and cellular membranes, the VACV fusion machine consists of a complex of at least nine proteins that are all essential for entry.

The entry of EEV is more complicated, owing to the need to remove both lipid envelopes before the core can be released into the cytoplasm. This is achieved by shedding the EEV outer envelope via a nonfusogenic mechanism that is triggered by interaction of EEV surface proteins with GAGs on the cell surface (**Figure 2**). This enables the IMV particle within the EEV envelope to bind to the cell surface and enter in the same way as a free IMV particle. Once a virus core is released into the cell, it is transported on microtubules deeper into the cell, and virus factories develop near the nucleus.

Transcription

Within minutes of infection, VACV mRNAs are synthesized by the virus DNA-dependent RNA polymerase, which is packaged in virions together with associated transcription factors and enzymes to cap and polyadenylate the mRNAs. Virus mRNAs are extruded from the core, presumably through the pores in the core wall. Early proteins include enzymes for synthesis of nucleoside triphosphate precursors and for DNA replication, and factors that block the innate immune response to infection. After early proteins have been expressed, the virus core is

246 Vaccinia Virus

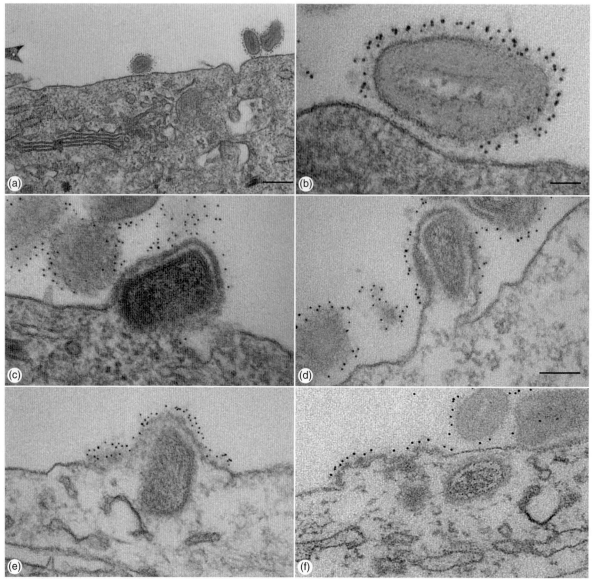

Figure 1 Electron micrographs showing the entry of VACV IMV by fusion at the cell surface. Purified VACV strain Western Reserve (WR) IMV (325 p.f.u./cell) was spinoculated onto PtK$_2$ cells to study the binding and entry of IMV, and samples were either retained on ice or incubated at 37 °C for 10 min. The IMV surface antigen D8 was labeled with D8-specific monoclonal antibody (mAb) for immunoelectron microscopy. Images show IMV particles bound to cells at 4 °C (a) and 37 °C (b), the IMV membrane fusing with the plasma membrane (c and d), the core entering into the cytosol (e), and the core leaving the site of entry (f). Scale = 200 nm (a), 50 nm (b), and 100 nm (d). Panels (c)–(f) are all shown at the same magnification. Reproduced from Carter GC, Law M, Hollinshead M, and Smith GL (2005) Entry of the vaccinia virus intracellular mature virion and its interactions with glycosaminoglycans. *Journal of General Virology* 86: 1279–1290, with permission from the Society for General Microbiology.

uncoated further and the genome is released for DNA replication. Once virus DNA replication has commenced, the pattern of transcription changes and intermediate genes are transcribed. Several intermediate genes encode late transcription factors needed for expression of late genes. Late genes encode the structural proteins that make up new virus particles, enzymes that may be packaged into these virions, and additional factors that influence the host response to infection. Host protein synthesis is shut off shortly after VACV infection, and this is mediated by the D10 protein, which removes the methylated cap structure of mRNAs. Virus and host mRNAs are both cleaved, and, since the former are more abundant, virus protein synthesis predominates.

DNA Replication

Virus DNA replication takes place in cytoplasmic factories from which most cellular organelles are excluded. The virus genome encodes many of the factors and

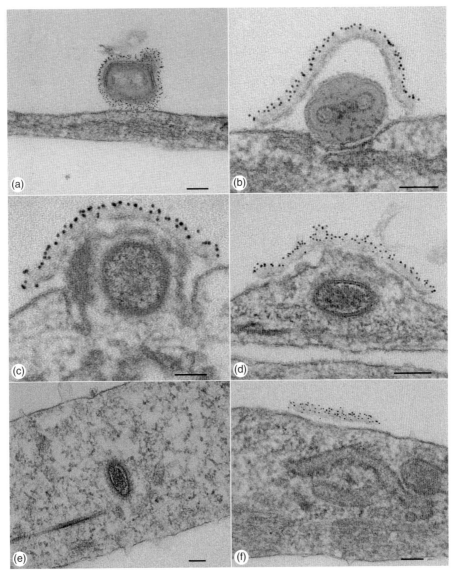

Figure 2 Electron micrographs showing the entry of EEV into cells by ligand-induced nonfusogenic dissolution of the EEV outer envelope followed by fusion of the IMV with the plasma membrane as in **Figure 1**. Fresh EEV of VACV strain WR was spinoculated onto PtK2 cells at 4 °C (a) and then incubated at 37 °C for 10 min (b–f). The EEV surface was labeled by rat anti-B5 monoclonal antibody (mAb) 19C2 (10), then by rabbit anti-rat IgG followed by 6 nm protein A–gold conjugate, and the samples were processed for electron microscopy. Scale = 100 nm (for all panels). Reproduced from Law M, Carter GC, Roberts KL, Hollinshead M, and Smith GL (2006) Ligand-induced and nonfusogenic dissolution of a viral membrane. *Proceedings of the National Academy of Sciences, USA* 103: 5989–5994, with permission from the National Academy of Sciences, USA.

enzymes needed for DNA replication, such as a DNA polymerase, nucleoside phosphohydrolase, a single-stranded DNA-binding protein, a processivity factor, and a protein kinase. In addition, the virus encodes a topoisomerase and DNA ligase, although the latter is (surprisingly) not essential for virus replication. DNA replication starts by the introduction of a nick into one DNA strand near the terminal hairpin loop. This enables the loop to unfold and serve as a template for DNA synthesis. After elongation to the end of the unfolded hairpin, the nascent and parent DNA strands separate, refold to create the hairpin, and thereby enable DNA replication to continue down the dsDNA genome by strand displacement. On reaching the distal end of the genome, the polymerase elongates round the hairpin and along the opposite DNA strand to produce concatemeric DNA molecules of greater than unit length. These are resolved into unit-length monomers by introduction of specific nicks near the terminal hairpins followed by re-ligation. Up to 10 000 copies of the virus genome may be produced from each infected cell, but only a fraction of these may be packaged into new virus particles.

Assembly

Virus assembly occurs in cytoplasmic factories in which the first visible structures are crescent shaped and are composed of virus protein and host lipid (**Figure 3**). The available evidence favors a model in which these structures contain a single lipid bilayer covered by a protein layer. The surface layer contains the D13 protein, which is essential for morphogenesis. Viruses lacking the D13 protein, or treated with the drug rifampicin, fail to form crescents. The D13 protein forms trimers that assemble into a lattice, or sheet, which appears to stabilize the formation of a nascent membrane. The crescents then grow into complete ovals or spheres into which the virus genome is packaged before the immature virion is sealed. Proteolytic cleavage of some of the capsid proteins and condensation of the virus core lead to the formation of an infectious IMV.

Most IMV particles remain within the cell until it lyses, but some are transported on microtubules to near the microtubule organizing center where they are wrapped by a double membrane (derived from either the *trans*-Golgi network or endosomes) that has been modified by the inclusion of several virus proteins (mostly glycosylated). The wrapping process produces a virion with three membranes that is called intracellular enveloped virus (IEV) (or wrapped virus (WV), using the nomenclature adopted recently by some authors), which is transported on microtubules to the cell surface. Upon reaching the cell periphery, the virus outer membrane fuses with the plasma membrane to externalize the virion by exocytosis. This surface virion is called cell-associated enveloped virus (CEV) and is surrounded by two membranes. CEV induces the formation of actin tails from beneath the plasma membrane where the CEV sits, and growth of these tails propels the virion away from the cell and into surrounding cells. VACV mutants unable to synthesize actin tails form small plaques, indicating the importance of this mechanism for cell-to-cell spread of virus. Alternatively, virions exposed on the cell surface may be released as EEV.

EEV are important for long-range spread of virus infection in cell culture and *in vivo*. The ability to spread in this way is conveniently studied by the comet assay, in which virus plaques are allowed to form on cell monolayers in liquid culture medium. Under these conditions, EEV particles spread from the original plaque in a unidirectional manner (caused by convection currents), giving rise to a comet-shaped plaque in which the primary plaque represents the head and secondary plaques form the tail. Antibodies directed against the EEV surface, but not IMV, inhibit comet formation. EEV particles not only are important for virus spread *in vivo* but are also important targets for antibodies that protect against OPV-induced disease.

The release of EEV from the cell surface or the retention of CEV is influenced by not only the cell type and virus proteins, primarily the A34 protein, but also the B5 and A33 EEV-specific envelope glycoproteins.

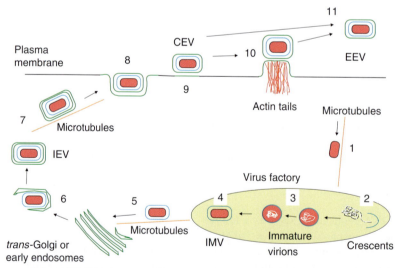

Figure 3 Schematic presentation of VACV morphogenesis. 1. After VACV entry, the virus core is transported on microtubules to a site near the nucleus where a virus factory develops. 2. A virus crescent forms and becomes associated with virus DNA associated with core proteins. 3. The virus membrane is sealed to form immature virions, which undergo condensation associated with proteolytic cleavage of several capsid proteins to form IMV (4). 5. IMV is transported on microtubules. 6. IMV is wrapped by modified membranes derived from the *trans*-Golgi network or endosomes to form intracellular enveloped virus (IEV). 7. IEV is transported on microtubules to the cell periphery. 8. IEV fuses with the plasma membrane to expose a cell-associated enveloped virus (CEV) on the cell surface by exocytosis. 9. CEV on the cell surface may induce the formation of actin tails (10) or dissociate from the cell to form EEV. 11. EEV may be formed by release of CEV from the cell surface before or after formation of actin tails. Modified from Smith GL, Vanderplasschen A, and Law M (2002) The formation and function of extracellular enveloped vaccinia virus. *Journal of General Virology* 83: 2915–2931, with permission from the Society for General Microbiology.

Virulence

VACV virulence and the mechanisms by which infection leads to disease (pathogenesis) have been studied in several animal models. The validity of any single model is uncertain because the natural host of VACV is unknown, but mice have been used most often and have been infected by dermal scarification or intranasal, intradermal, intraperitoneal, intracranial, or intravenous routes. The intranasal or intradermal routes are considered more physiologically relevant and have been used widely. Following intradermal infection, the virus causes a localized mild disease characterized by the formation of a lesion that heals in about 2 weeks. Virus does not spread from the site of infection and animals remain healthy. The intranasal model causes a more systemic illness in which the animals lose weight and the virus may spread to other tissues. The outcome of infection is influenced by the dose and strain of virus administered.

Many VACV proteins have been reported to affect the outcome of infection in one model or another. These proteins may be grouped into three types. The first group comprises enzymes (such as thymidine kinase, thymidylate kinase, and ribonucleotide reductase) that synthesize nucleoside precursors, or enzymes involved in DNA metabolism (e.g., DNA ligase). The second group comprises proteins that affect virus transport, release, and cell-to-cell spread. These include intracellular proteins F12 and A36, and the EEV proteins F13, A33, A34, and B5. The third and largest group comprises proteins that interfere with the host response to infection, which have been called immunomodulators.

VACV immunomodulators are numerous and target many aspects of the host immune system. They may be subdivided into those proteins that act within the infected cell and those that are secreted and act extracellularly. Intracellular proteins inhibit signaling pathways, such as those leading from Toll-like receptors (TLRs) or receptors for interleukins (ILs), tumor necrosis factor (TNF), or IFNs to the production of pro-inflammatory cytokines or IFNs. Such proteins include A46, A52, K1, M2, and N1. Other intracellular proteins (B13, F1, and N1) may block apoptosis, and yet others (K3 and E3) block the action of IFN-induced antiviral proteins. One intracellular protein contributes to virulence by synthesizing steroid hormones (A44). Extracellular proteins bind to complement factors (VCP), type I and type II IFNs (B8 and B18), cytokines IL-1β, IL-18, and TNF-α (B15, C12, and CrmC), and CC chemokines (vCKBP, B7, and A41). These VACV proteins function either by preventing soluble factors binding to their natural receptors on cells, or by preventing the establishment of chemokine concentration gradients, thereby diminishing the host response to infection. Collectively, these immunomodulatory proteins provide a substantial defense against the innate immune system. The contribution of individual proteins to virus virulence has been studied by using mutants lacking specific genes in several mouse models of infection. Loss of specific genes can either diminish, enhance, or have no effect on virulence, and the outcome can vary depending on the model of infection used.

The degree of virus attenuation derived from loss of proteins involved in virus transport and spread is in general much greater than that derived from removal of immunomodulators or enzymes synthesizing nucleoside triphosphate precursors.

Host Range

VACV infects a very broad range of cultured cells and only a few cell types, such as Chinese hamster ovary (CHO) cells, are nonpermissive. VACV also infects a broad range of host species, rather like CPXV and unlike the narrow host range of VARV and CMLV.

VACV Expression Vectors

VACV was developed as an expression vector in 1982. Since that time, this expression system has been refined to synthesize proteins in significant quantities in mammalian cells, and is employed widely as a laboratory tool by immunologists, virologists, and cell and molecular biologists. Recombinant VACV strains have been proposed as live vaccines for infectious diseases and cancer, and as oncolytic tools to target human cancers. The widespread use of VACV-based therapies has been hampered by their imperfect safety record. Therefore, attenuated strains are being utilized and engineered to improve immunogenicity by removal or modification of immunomodulatory proteins.

Antiviral Drugs

During the last 5 years, there has been a concerted effort to develop drugs against OPVs due to concerns about the possible use of VARV in bioterrorism. Given the highly conserved nature of OPV replication, the targets for these drugs are mostly conserved between OPVs and work against both VARV and VACV. Two particularly encouraging drugs are cidofovir (and acylated derivatives thereof) and ST-246. The former is a licensed drug for treating human cytomegalovirus infection and the latter is a new compound that targets morphogenesis by blocking formation of IEV particles by inhibiting wrapping of IMVs.

Immune Response to Vaccination

VACV infection in man and animals induces a robust antibody and T-cell response. It is clear that both are important in protection against VACV-induced disease, and a T-cell response is essential for recovery from infection. The latter was evident historically from the life-threatening infections caused by vaccination of children suffering from T-cell immunodeficiency. The exact correlates of protection against smallpox are unknown, not least because smallpox was eradicated before much of modern immunology was developed. However, recent vaccination programs to defend against the perceived threat of bioterrorism with VARV have prompted detailed studies of the immune responses induced by VACV strains that protected against smallpox. These studies showed that both antibody and T-cell responses are very long-lived in humans (many decades) and that antibody responses were induced against a wide range of antigens. Several proteins on the surface of IMV are targets for neutralizing antibodies (including A27, D8, and H3), whereas the only antigen on the EEV surface that is a target for neutralizing antibody, in the absence of complement, is B5. B5 is therefore an important component of vaccines against OPV diseases, including smallpox. It is notable that one of the attenuated vaccines proposed as a safer vaccine against smallpox (LC16m8) lacks the B5 antigen.

Future Perspectives

VACV remains the only vaccine to have eradicated a human disease, smallpox. Despite the eradication of this disease 30 years ago, there remain many good reasons to continue to study VACV today. The ongoing development of this virus as a vaccine for infectious disease and cancer remains central, and its wide range of immunomodulatory proteins for interference with the innate immune system make it an excellent model for studying interactions between viruses and their host.

Acknowledgment

The author is a Wellcome Principal Research Fellow.

See also: Cowpox Virus; Poxviruses; Smallpox and Monkeypox Viruses.

Further Reading

Baxby D (1981) *Jenner's Smallpox Vaccine: The Riddle of the Origin of Vaccinia Virus.* London: Heinemann.

Broyles SS (2003) Vaccinia virus transcription. *Journal of General Virology* 84: 2293–2303.

Carter GC, Law M, Hollinshead M, and Smith GL (2005) Entry of the vaccinia virus intracellular mature virion and its interactions with glycosaminoglycans. *Journal of General Virology* 86: 1279–1290.

Condit RC, Moussatche N, and Traktman P (2006) In a nutshell: Structure and assembly of the vaccinia virion. *Advances in Virus Research* 66: 31–124.

Fenner F, Anderson DA, Arita I, Jezek Z, and Ladnyi ID (1988) *Smallpox and Its Eradication.* Geneva: World Health Organization.

Fenner F, Wittek R, and Dumbell KR (1989) *The Orthopoxviruses.* London: Academic Press.

Lane JM, Ruben FL, Neff JM, and Millar JD (1969) Complications of smallpox vaccination, 1968: National surveillance in the United States. *New England Journal of Medicine* 281: 1201–1208.

Law M, Carter GC, Roberts KL, Hollinshead M, and Smith GL (2006) Ligand-induced and nonfusogenic dissolution of a viral membrane. *Proceedings of the National Academy of Sciences, USA* 103: 5989–5994.

Moss B (1996) Genetically engineered poxviruses for recombinant gene expression, vaccination, and safety. *Proceedings of the National Academy of Sciences, USA* 93: 11341–11348.

Moss B (2006) Poxvirus entry and membrane fusion. *Virology* 344: 48–54.

Moss B (2007) *Poxviridae*: The viruses and their replication. In: Knipe DM (ed.) *Fields Virology,* 5th edn; vol. 2, pp. 2905–2946. Philadelphia, PA: Lippincott Williams and Wilkins.

Seet BT, Johnston JB, Brunetti CR, *et al.* (2003) Poxviruses and immune evasion. *Annual Review of Immunology* 21: 377–423.

Smith GL (2007) Genus *Orthopoxvirus*: Vaccinia virus. In: Mercer AA, Schmidt A, and Weber O (eds.) *Poxviruses,* pp. 1–45. Berlin: Springer.

Smith GL, Vanderplasschen A, and Law M (2002) The formation and function of extracellular enveloped vaccinia virus. *Journal of General Virology* 83: 2915–2931.

Relevant Website

http://www.poxvirus.org – Poxvirus Bioinformatics Resource Center.

Varicella-Zoster Virus: General Features

J I Cohen, National Institutes of Health, Bethesda, MD, USA

Published by Elsevier Ltd.

Glossary

Acantholysis Breakdown of a cell layer in the epidermis by separation of individual epidermal keratinocytes from their neighbors.

Dysesthesia Distortion of any sense, especially touch.

History

Descriptions of vesicular rashes characteristic of chickenpox (varicella) date back to the ninth century. In 1875, Steiner showed that chickenpox was an infectious agent by transmitting the disease from chickenpox vesicle fluid to previously uninfected people. Shingles (zoster) has been recognized since ancient times. In 1909, Von Bokay suggested that chickenpox and shingles were related infections, an idea that was confirmed experimentally in the 1920s and 1930s when children inoculated with fluid from zoster vesicles were shown to contract chickenpox. In 1943, Garland suggested that zoster was due to reactivation of varicella virus that remains latent in sensory nerve ganglia, and, in 1954, Hope–Simpson reaffirmed this hypothesis.

The viral agents of varicella and zoster were first cultivated by Weller in 1952 and shown on morphologic, cytopathic, and serologic criteria to be identical. In 1984, Straus and colleagues showed that viruses isolated during sequential episodes of chickenpox and zoster from the same patient had identical restriction endonuclease patterns, proving the concept of prolonged latent carriage of the virus. In 1983, Gilden showed that varicella-zoster virus (VZV) DNA is latent in human sensory ganglia and, Hyman and colleagues showed that VZV RNA is present in human trigeminal ganglia. In 1974, Takahashi developed a live attenuated VZV vaccine for prevention of varicella, and, in 2005, Oxman and colleagues showed that a high potency formulation of this vaccine is effective in reducing rates of zoster and postherpetic neuralgia.

Taxonomy and Classification

VZV (species *Human herpesvirus 3*) is a member of genus *Varicellovirus* in subfamuly *Alphaherpesvirus* of the family *Herpesviridae*. Other alphaherpesviruses that infect humans include herpes simplex viruses 1 and 2 (HSV-1 and HSV-2), and, rarely, B virus (a macaque herpesvirus). All of these agents exhibit relatively short replicative cycles, destroy the infected cell, and establish latent infection in sensory ganglia.

Simian varicella virus (SVV) is the most closely related, well-characterized virus to VZV. Natural infection with SVV can cause a varicella-like illness in Old World monkeys; the virus establishes latency in trigeminal ganglia, and can spontaneously reactivate to cause a rash that can transmit the virus to naive animals. The complete sequence of SVV has been determined and the virus shares nearly an identical set of genes with VZV. SVV is not known to infect humans.

Geographic and Seasonal Distribution

Varicella and zoster infections occur worldwide. Over 90% of varicella occurs during childhood in industrialized countries located in the temperate zone, but infection is commonly delayed until adulthood in tropical regions. Zoster may occur less frequently in tropical areas, because of later acquisition of primary infection. Varicella infection is epidemic each winter and spring, while zoster occurs throughout the year, without a seasonal preference.

Host Range and Virus Propagation

The reservoir for VZV is limited to humans. The virus inherently grows poorly in nonhuman animals or cell lines. Myers and colleagues, however, developed a guinea pig animal model of VZV infection by adapting the virus for growth in guinea pig embryo cells *in vitro*. Inoculation of animals results in a self-limited viremic infection and the emergence of both humoral and cellular immunity. Latent VZV DNA has been demonstrated in dorsal root and trigeminal ganglia. Rats, cotton rats, and mice inoculated with VZV develop latent infection of dorsal root ganglia. Inoculation of VZV into fetal thymus–liver implants in severe combined immune deficiency (SCID) mice results in virus replication in T cells; inoculation of virus into subcutaneous fetal skin implants reproduces many of the histopathologic features of varicella. Intravenous inoculation of SCID mice with VZV-infected human T cells results in virus infection of human skin xenografts.

An alternative, but less ideal, animal model involves the common marmoset (*Callithrix jacchus*). VZV replicates in the lungs with a mild pneumonia and a subsequent humoral immune response. Inoculation of chimpanzees with VZV results in a transient rash containing viral DNA and evokes a modest humoral immune response. None of these animal models have, as yet, reproduced the disease pattern seen in humans, namely a vesicular rash and spontaneous reactivation from latency.

VZV is usually cultured in human fetal diploid lung cells in clinical laboratories. The virus has been cultivated in numerous other human cells including melanoma cells, primary human thyroid cells, astrocytes, Schwann cells, and neurons, and can be grown in some simian cells including primary African green monkey kidney cells and Vero cells, and in guinea pig embryo fibroblasts.

VZV is extremely cell associated. The titer of virus released into the cell culture supernatant is very low, and preparation of cell-free virus, by sonication or freeze–thawing cells, usually results in a marked drop in viral titer. Therefore, virus propagation is usually performed by passage of infected cells onto uninfected cell monolayers. VZV is detected by its cytopathic effect with refractile rounded cells that gradually detach from the monolayer, or by staining with fluorescein-labeled antibody.

Genetics

Several markers can be used to distinguish different strains of the virus. These include temperature sensitivity, plaque size, antiviral sensitivity, and restriction endonuclease cleavage patterns. The molecular basis for most of these strain differences is unknown, but viruses that are resistant to acyclovir usually have mutations in their thymidine kinase gene; other resistant strains have mutations in the DNA polymerase gene.

The genome of the prototypical laboratory strain VZV Dumas consists of 124 884 bp. The identification of many viral genes was made by analogy to HSV-1 genes with similar sequences and by genetic-complementation studies in which cell lines expressing selected VZV proteins were used to support the growth of HSV-1 mutants. Cosmids and bacterial artificial chromosomes (BACs) derived from VZV have been developed to allow targeted deletion, insertion, or site-directed point mutations in individual viral gene products. Using these systems, 22 viral genes have been shown to be dispensable, and 5 genes have been shown to be required, for virus replication *in vitro*. Recombinant VZV containing Epstein–Barr virus (EBV), hepatitis B virus, HIV, or HSV genes have been constructed.

Evolution

Comparison of the nucleotide and predicted amino acid sequences of VZV with HSV-1 and HSV-2 indicates that these viruses originated from a common ancestor. They share similar gene arrangements and only five genes of VZV do not appear to have HSV counterparts.

VZV is more distantly related to all other human herpesviruses, but many of the nonstructural proteins involved in viral replication have conserved elements and activities. Comparison of VZV, for example, with EBV shows that the majority of VZV genes are homologous with EBV. Three large blocks of genes are conserved, although rearranged within the two genomes.

VZV isolates have been classified into three different clades, corresponding to Japanese, European, and mosaic genotypes (combination of European and Japanese). Other investigators have proposed two clades: a Singapore/Japanese and a North American/European genotype.

Serologic Relationships and Variability

There is only one serotype of VZV. Antibodies detected by the complement-fixation test and virus-specific immunoglobulin M (IgM) antibodies decline rapidly after convalescence from varicella. Other, more sensitive serologic tests, including immune adherence hemagglutination (IAHA), fluorescence antibody to membrane antigen (FAMA), and enzyme-linked immunosorbent assay (ELISA), recognize antibodies that persist for life. VZV-specific antibodies are boosted by both recrudescent infection (zoster) and exposure to others with varicella.

Variability of VZV strains has been shown primarily by differences in restriction endonuclease patterns. Passage of individual strains *in vitro* eventually results in minor changes in restriction endonuclease patterns, predominantly through deletion or reiteration of small repeated elements scattered throughout the genome. Other than these sites, the genome sequence is remarkably stable. For example, the sequence of the thymidine kinase gene has been determined for several epidemiologically unrelated wild-type and acyclovir-resistant strains and found to possess >99% nucleotide and amino acid sequence identity.

Epidemiology

Varicella may occur after exposure of susceptible persons to chickenpox or herpes zoster. Over 95% of primary infections result in symptomatic chickenpox. Over 90% of individuals in temperate countries are infected with VZV before age 15.

Zoster is due to reactivation of VZV in patients who have had prior chickenpox; some of these patients may not recall the primary infection. Zoster is not clearly related to exposure to chickenpox or to other cases of zoster. About 10–20% of individuals ultimately develop herpes zoster – the risk of which rises steadily with age (**Figure 1**). Severely immunocompromised patients, such as those with the acquired immunodeficiency syndrome (AIDS), have a particularly high incidence of zoster. Recurrent zoster is uncommon; less than 4% of patients experience a second episode. Asymptomatic viremia has been detected in bone marrow transplant recipients and has been followed by recovery of cell-mediated immunity.

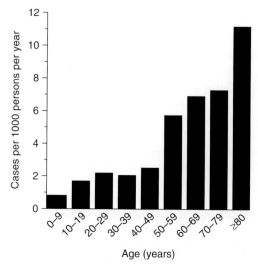

Figure 1 Incidence of herpes zoster. Reproduced from Kost RG and Straus SE (1996) Postherpetic neuralgia – pathogenesis, treatment, and prevention. *New England Journal of Medicine* 335: 32–42, with permission from Massachusetts Medical Society.

Transmission and Tissue Tropism

VZV is transmitted by the respiratory route. VZV has been detected by polymerase chain reaction (PCR) in room air from patients with varicella or zoster. Intimate, rather than casual, contact is important for transmission. Chickenpox is highly contagious; about 60–90% of susceptible household contacts become infected. Herpes zoster is less contagious than chickenpox. Only 20–30% of susceptible contacts develop varicella. Patients with varicella are infectious from two days before the onset of the rash until all the lesions have crusted.

Primary infection with VZV results in viral replication in the upper respiratory tract and oropharynx with lesions present on the respiratory mucosa. T cells subsequently becomes infected and transmit virus to the skin. Virus has been detected by PCR in the oropharynx and virus has been cultured from the blood early during varicella. Virus infection of epidermal cells in the skin is thought to be initially limited by interferon; later T cells trafficking in the skin may become infected and disseminate the virus to various organs throughout the body including the nervous system. During latency, VZV DNA can be detected in ganglia of the cranial nerves (e.g., trigeminal ganglia), dorsal roots (e.g., thoracic and trigeminal ganglia), and autonomic nervous system (e.g., celiac and vagus nerve) by PCR. Using *in situ* hybridization, VZV has been detected predominantly, if not exclusively, in neurons. During latency, only 6 of the 70 known viral genes are expressed.

Zoster is due to reactivation of virus in the sensory ganglia. The factors leading to its reactivation are not known, but are associated with neural injury and cellular immune impairment. Reactivated virus spreads down the sensory nerve to the skin, where the resulting vesicles are typically confined to a single dermatome. Viremia and subsequent cutaneous or visceral dissemination of lesions may occur in zoster, especially in immunocompromised patients.

Pathogenicity

Passage of wild-type VZV in cell culture by Takahashi in 1974 led to attenuation of the virus and changes in its temperature sensitivity and infectivity for certain cell lines. The resulting Oka vaccine strain has nucleotide polymorphisms at 31 sites. Comparison of the complete nucleotide sequence of the Oka vaccine strain with its parental virus shows 42 nucleotide and 20 amino acid differences. Multiple nucleotide changes in several genes are responsible for attenuation of the vaccine for growth in skin. The Oka vaccine strain can be distinguished from wild-type strains by differences in restriction endonuclease patterns.

Clinical Features of Infection

The incubation period for chickenpox is 2 weeks, with a range of 10–21 days. The disease begins with fever and malaise, followed by a generalized vesicular rash (**Figure 2(b)**). Lesions tend to appear first on the head and trunk and then spread to the extremities. New lesions usually follow viremic waves for 3–5 days and, in the normal host, most lesions are crusted and healed by 2 weeks. Lesions in different stages coexist in an individual. The disease is usually self-limited in the normal host.

Complications of varicella are more common in neonates, children with malnutrition, immunocompromised patients (e.g., malignancy or immunosuppressive therapy), pregnant women, and older adults. These complications include bacterial superinfection of the skin, pneumonia, hepatitis, encephalitis, thrombocytopenia, and purpura fulminans. Reye syndrome occurs in rare children who take aspirin to treat varicella fevers. Prior to the widespread use of the varicella vaccine in the USA, there were about 3–4 million cases of varicella each year with about 100 deaths; however, the number of cases has dropped sharply and death from varicella in the USA is very rare.

Zoster usually presents with pain and dysesthesias 1–4 days before the onset of the vesicular rash. The rash is usually painful and confined to a single dermatome (**Figure 2(c)**), but may involve several adjacent dermatomes. Fever and malaise often accompany the rash. Vesicles often are pustular by day 4 and become crusted by day 10 in the normal host.

Postherpetic neuralgia (PHN), manifested by pain lasting for weeks to several years in the area of the initial rash, is the most common and disconcerting complication of zoster in the normal host. Less common complications include encephalitis, myelitis, the Ramsay–Hunt syndrome (lesions in the ear canal, with auditory and facial nerve involvement), ophthalmoplegia, facial weakness, and pneumonitis. Immunocompromised patients with zoster are more likely to develop disseminated disease (**Figure 2(d)**) with neurologic, ocular, or visceral involvement. Patients with AIDS have a high frequency of zoster and may develop recurrent or chronic disease with verrucous, hyperkeratotic skin lesions.

Pathology and Histology

Varicella lesions are readily recognized in the skin and mucous membranes. However, similar lesions also occur in the mucosa of the respiratory and gastrointestinal tracts, liver, spleen, and any tissue, and remain unrecognized except in severe cases. With severe disease, there is inflammatory infiltration of the small vessels of most organs. Zoster causes inflammation and necrosis of the

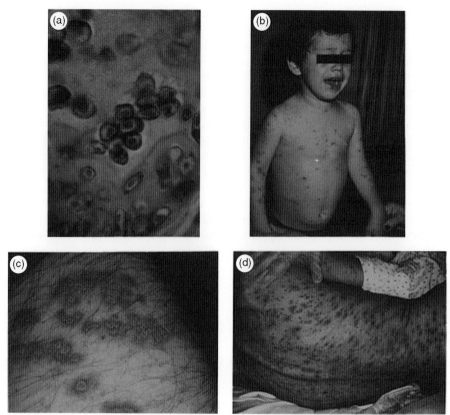

Figure 2 Histopathology and clinical findings of varicella-zoster virus infections. (a) Eosinophilic intranuclear inclusions from a skin biopsy in a patient with herpes zoster (original magnification ×400). (b) Chickenpox in a child. (c) Localized zoster in an adult. (d) Disseminated zoster in a patient with chronic lymphocytic leukemia. Reproduced from Straus SE, Ostrove JM, Inchauspe G, et al. (1988) Varicella-zoster virus infections: Biology, natural history, treatment, and prevention. *Annals of Internal Medicine* 108: 221–237, with permission from American College of Physicians.

sensory ganglia and its nerves, and skin lesions that are histopathologically identical to those seen with varicella.

Cutaneous lesions due to VZV begin with infection of capillary endothelial cells followed by direct spread to epidermal epithelial cells. The epidermis becomes edematous with acantholysis and vesicle formation. Mononuclear cells infiltrate the small vessels of the dermis. Initially, vesicles contain clear fluid with cell-free virus, but later the vesicles become cloudy and contain neutrophils, macrophages, interferon, and other cellular and humoral components of the inflammatory response pathways. Subsequently, the vesicles dry leaving a crust that heals usually without scarring.

Cells infected with VZV show eosinophilic intranuclear inclusions with multinucleate giant cell formation (**Figure 2(a)**). These changes are not specific for VZV, as they are seen with HSV and cytomegalovirus infections.

Immune Response

Infection with VZV elicits both a humoral and cellular immune response. The ability of VZV immune globulin (VZIG) to attenuate or prevent infection in exposed children (see the next section) indicates that virus-specific antibody is important in protection from primary infection. The presence of VZV-specific immunoglobulin G (IgG) does not correlate, however, with protection from zoster. Antibody to VZV is often present by the time the rash of varicella first appears. Virus-specific IgM, IgG, and IgA are present within 5 days of symptomatic disease; however, only IgG persists for life. Antibodies to viral glycoproteins gE, gB, gH, and the immediate-early 62 protein (IE62) have been detected during acute infection, and the titers of antibodies to these proteins are boosted during recurrent infection. The mere presence of antibody to VZV glycoproteins in children with leukemia who had received live varicella vaccine is not adequate to prevent breakthrough varicella or zoster.

Cellular immune responses to VZV are more important in recovery from acute varicella infection and for prevention of, and recovery from, zoster. The level of cellular immunity correlates with disease severity during acute varicella. Cytotoxic T cells that lyse virus-infected cells are present by 2–3 days after the onset of the rash of varicella. Cell-mediated immunity, as measured by lymphocyte proliferative response, is directed against cells expressing glycoproteins gE, gB, gH, gI, and gC, the IE4, IE62, and IE63 proteins, and other viral proteins. Interferon is present in VZV vesicles.

Most varicella infections result in lifelong immunity to reinfection. Second episodes of varicella are rare; these individuals tend to have reduced humoral and cellular immunity to VZV at the time of the second infection. Zoster is associated with a reduction in cellular immunity to VZV that, in the normal host, is partially restored in response to this recurrent infection. Recurrent zoster is uncommon, except in severely immune-deficient patients, such as those with AIDS.

Prevention and Control

Prevention of varicella can be achieved by restricting exposure or by resorting to either immunoglobulin prophylaxis or vaccination with live, attenuated virus. If given within 4 days of exposure to the virus, VZIG prevents or attenuates varicella in seronegative persons. The preparation has no effect in modifying zoster. VZIG is recommended for individuals (1) with recent, close contact to patients with varicella or zoster, (2) who are susceptible to varicella, and (3) who fall in a high-risk category. The latter include premature or certain newborn infants, pregnant women, and patients with congenital or acquired cellular immunodeficiencies. Supplies of VZIG are currently very limited in the USA and VariZIG, prepared from high-titer human immune serum, has replaced VZIG.

The live, attenuated varicella vaccine (Oka strain) was licensed in the USA in 1995 and is recommended for vaccination of healthy children and adults. Most children develop adequate humoral and cellular immunity to varicella after a single dose of vaccine; an additional dose enhances the degree of immunity and is recommended for susceptible children and adults. A rash may follow vaccination. It is usually mild, but can be severe if the vaccine is given to patients experiencing periods of profound cellular immune impairment. The live vaccine virus establishes neural latency and can reactivate. Thus, zoster has been reported in vaccinees, especially those who are immunocompromised, but the rate appears to be no higher than that following natural infection. Vaccination may be used for postexposure prophylaxis.

A high-titer formulation of the Oka vaccine virus was licensed in the USA in 2006. It reduced the frequency of zoster and PHN by 51% and 67%, respectively, in healthy persons ≥60 years old. The most common side effects were mild injection-site reactions; no significant severe adverse reactions were attributed to the vaccine.

Patients with varicella or zoster should be isolated from susceptible persons until all lesions have crusted. This is particularly important for hospital workers and immune-deficient patients.

Acyclovir and leukocyte interferon have been used in the treatment of varicella and zoster in immunocompromised patients. Interferon proved to be an inadequate and impractical therapy. Acyclovir is the current treatment of choice for selected infections. It results in a shorter duration of symptoms and decreased visceral dissemination of varicella or zoster in the immunocompromised host. Acyclovir also prevents spread of trigeminal zoster to the eye and modestly shortens the duration of varicella and zoster symptoms in the normal host. Analogs of acyclovir, such as famciclovir and valaciclovir, result in higher levels of antiviral activity than oral acyclovir and have been licensed for oral therapy of zoster in the USA. Acyclovir-resistant strains of VZV have been reported in patients with AIDS; these infections are best treated with foscarnet. Corticosteroids, when used early during zoster, reduce acute pain. Herpes zoster, particularly in elderly patients, may lead to prolonged and severe PHN. Treatment of PHN is difficult and often unsatisfactory, but many patients experience improvement with gabapentin, pregabalin, or tricyclic antidepressant drugs like amitriptyline.

Future Perspectives

Widespread vaccination of children with the attenuated, live varicella vaccine has reduced the incidence and severity of varicella. Use of a high-titer live varicella vaccine was shown to lower the incidence of zoster and postherpetic neuralgia in the elderly. Further research is needed into the mechanisms of VZV latency and reactivation as well as identification of which viral proteins are critical for protection from varicella and zoster. Since the live varicella vaccine can reactivate to cause zoster and since breakthrough cases of varicella continue to occur in vaccinated persons, further knowledge of VZV latency and immunity should lead to safer and more effective vaccines against varicella and herpes zoster.

See also: Herpes Simplex Viruses: General Features; Herpes Simplex Viruses: Molecular Biology; Herpesviruses: Discovery; Herpesviruses: Latency; Varicella-Zoster Virus: Molecular Biology.

Further Reading

Cohen JI, Brunell PA, Straus SE, and Krause PR (1999) Recent advances in varicella-zoster virus infection. *Annals of Internal Medicine* 130: 922–932.

Cohen JI, Straus SE, and Arvin AM (2007) Varicella-zoster virus: Replication, pathogenesis, and management. In: Knipe DM, Howley PM, Griffin DE, et al. (eds.) *Fields Virology,* 5th edn, pp. 2773–2818. Philadelphia, PA: Lippincott Williams and Wilkins.

Gershon A and Arvin A (2000) *Varicella-Zoster Virus.* Cambridge: Cambridge University Press.

Gilden DH, Cohrs RJ, and Mahalingham R (2003) Clinical and molecular pathogenesis of varicella-zoster virus infection. *Viral Immunology* 16: 243–258.

Hambleton S and Gershon AA (2005) Preventing varicella-zoster disease. *Clinical Microbiology Reviews* 18: 70–80.

Heininger U and Seward JF (2006) Varicella. *Lancet* 368: 1365–1376.

Kimberlin D and Whitley R (2007) Varicella-zoster vaccine for the prevention of zoster. *New England Journal of Medicine* 356: 1338–1343.

Kost RG and Straus SE (1996) Postherpetic neuralgia – pathogenesis, treatment, and prevention. *New England Journal of Medicine* 335: 32–42.

Ku CC, Besser J, Abendroth A, Grose C, and Arvin AM (2005) Varicella-zoster virus pathogenesis and immunobiology: New concepts emerging from investigations with the SCIDhu mouse model. *Journal of Virology* 79: 2651–2658.

Mitchell BM, Bloom DC, Cohrs RJ, Gilden DH, and Kennedy PG (2003) Herpes simplex virus and varicella zoster latency in ganglia. *Journal of Neurovirology* 9: 194–204.

Oxman MN, Levin MJ, Johnson GR, et al. (2005) A vaccine to prevent herpes zoster and postherpetic neuralgia in older adults. *New England Journal of Medicine* 352: 2271–2284.

Straus SE, Ostrove JM, Inchauspe G, et al. (1988) Varicella-zoster virus infections: Biology, natural history, treatment, and prevention. *Annals of Internal Medicine* 108: 221–237.

Varicella-Zoster Virus: Molecular Biology

W T Ruyechan and J Hay, The State University of New York, Buffalo, NY, USA

© 2008 Elsevier Ltd. All rights reserved.

Virion and Genome Structure

The Virion

The virion of varicella-zoster virus (VZV), like those of all herpesviruses, is enveloped and approximately spherical, with a diameter of 100–300 nm. A minimum of 30 proteins ranging in size from ~250 to 17 kDa have been identified as components of the virion. The viral double-stranded DNA genome is contained in an icosahedral nucleocapsid 100 nm in diameter that has 5:3:2 axial symmetry and is embedded within a structure known as the tegument. The tegument is composed of virus-encoded proteins, several of which are involved in expression of the viral genome (see below). The thickness of the tegument can vary, even within a single virion. The lipid envelope of the virus, which is derived from cellular membranes, surrounds the tegument and contains a minimum of eight viral glycoproteins.

The Genome

The VZV genome is a linear, double-stranded DNA molecule of $c.$ 125 kbp with a $G+C$ content of 46%. The genome consists of two covalently joined segments, the unique long (U_L) and unique short (U_S) regions, each bounded by a terminal and internal inverted repeat (TR_L/IR_L (88 bp) and TR_S/IR_S (7.3 kbp) respectively). The repeats allow structural rearrangements of the unique regions to occur during viral DNA replication and, as a result, encapsidated genomes exist as a mixture of four isomers. Two isomers are found predominantly in viral particles and are termed P (prototype) and I_S (inverted U_S). Two other isomers, which contain U_L inverted with respect to U_S in the P and I_S configurations, are designated I_L and I_{SL}. These represent 2–5% of the virion DNA. A small amount of full-length closed circular DNA (representing 0.1–5% of the total) has also been isolated from virions. The function of these molecules in relation to the viral life cycle remains unknown.

Analysis of the complete sequence of the VZV Dumas strain revealed the protein-coding potential of a total of 71 open reading frames (ORFs) that were named sequentially according to location in the genome (ORF1–ORF71, with ORF33.5 overlapping ORF33 and ORF62–ORF64 in IR_S duplicated as ORF69–ORF71 in TR_S). Two other genes (ORFS/L and ORF9A) were added subsequently. The genome map thus contains 71 unique genes, the first of which (ORFS/L) is located at the left end of the genome and the last (ORF68) at the right end of U_S, with ORF62–ORF64 in IR_S duplicated as ORF69–ORF71 in TR_S. The VZV ORFs encode proteins involved in regulation of viral gene expression, viral DNA replication, and regulation of nucleotide pools, and structural proteins including those required for DNA packaging and capsid assembly, as well as the viral glycoproteins (**Table 1**).

Understanding of the roles of the VZV gene products was long hampered owing to the highly cell-associated nature of the virus, which precluded plaque purification, and to the lack of a good animal model of VZV infection. The development of a cosmid transfection system and, more recently, a VZV bacterial artificial chromosome (BAC), coupled with a severe combined immunodeficiency (SCID) mouse/human (hu) xenograft model of pathogenesis, has resulted in considerable progress in the delineation of VZV gene function.

Productive Infection

Productive VZV infection begins with fusion of the virus envelope with the membrane of a susceptible cell. This occurs via interactions between one or more VZV

Table 1 Functions of VZV genes

Gene	HSV	Pred. mol. wt. (Da)	Properties	Function
1	***[a]	12 103	Hydrophobic C-terminus	Unknown
2	***	25 983		Unknown
3	UL55	10 149		Unknown
4	UL54	51 540	Hydrophilic dimer	(IE) Transcription regulation
5	UL53	38 575	Hydrophobic/membrane	Glycoprotein K
6	UL52	122 541		DNA primase[b]
7	UL51	28 245		Tegument protein
8	UL50	44 816		dUTPase
9	UL49	32 845	Interacts with IE62 in cytoplasm/tegument component	Tegument/trafficking
9A	UL49.5	9800	Present in cell membrane	Glycoprotein N
10	UL48	46 573		Tegument/transactivator
11	UL47	91 825	Hydrophilic/acid N-terminus	Tegument[b]
12	UL46	74 269		Tegument
13	***	34 531		Thymidylate synthase
14	UL44	61 350	N-terminal repeats	Glycoprotein C
15	UL43	44 522	Hydrophobic	Membrane?
16	UL42	46 078	DNA-binding, polymerase processivity	DNA polymerase processivity subunit[b]
17	UL41	51 365		Virion host shutoff/tegument
18	UL40	35 395		Ribonucleotide reductase (small subunit)
19	UL39	86 823		Ribonucleotide reductase (large subunit)
20	UL38	53 969		Capsid
21	UL37	115 774		Tegument[b]
22	UL36	306 325		Tegument
23	UL35	24 416	Hydrophilic/STQ rich	Capsid/hexon tip[b]
24	UL34	30 451	Hydrophobic C-terminus	Phosphoprotein
25	UL33	17 460	Hydrophilic N-terminus	DNA packaging[b]
26	UL32	65 692		DNA packaging[b]
27	UL31	38 234	Hydrophilic/basic N-terminus	Nuclear matrix associated[b]
28	UL30	130 041	Hydrophilic	DNA polymerase catalytic subunit
29	UL29	132 133	Hydrophilic	Major ssDNA binding protein
30	UL28	86 968		DNA packaging[b]
31	UL27	98 026		Glycoprotein B
32	***	15 980	Hydrophilic/acidic	Unknown
33	UL26	66 043		Protease/capsid maturation[b]
33.5	UL26.5	34 000		Capsid assembly[b]
34	UL25	65 182		DNA packaging/capsid[b]
35	UL24	28 973	Basic/nuclear	Important for replication in skin and T cells
36	UL23	93 646	Hydrophilic	Pyrimidine deoxynucleoside kinase
37	UL22	93 646	Hydrophobic	Glycoprotein H
38	UL21	60 395		Tegument[b]
39	UL20	27 078	Hydrophobic/membrane	Virion egress[b]
40	UL19	154 971		Major capsid protein
41	UL18	34 387		Capsid
42/45	UL15	82 752	Spliced gene product	DNA packaging/DNA terminase?[b]
43	UL17	73 905		DNA packaging/tegument[b]
44	UL16	40 243		Tegument[b]
46	UL14	22 544		Tegument[b]
47	UL13	54 347		Protein kinase/tegument
48	UL12	61 268		Deoxyribonuclease
49	UL11	8907	Hydrophilic	Virion component[b]
50	UL10	48 669	Hydrphobic	Glycoprotein M
51	UL9	94 370		DNA origin binding protein/helicase
52	UL8	86 343		Helicase/primase complex accessory protein[b]
53	UL7	37 417		Unknown
54	UL6	86 776		Virion component[b]
55	UL5	98 844		DNA helicase[b]
56	UL4	27 166	S- and T-rich	Unknown
57	***	8079	Hydrophilic/basic	Tegument/virion egress[b]

Continued

Table 1 Continued

Gene	HSV	Pred. mol. wt. (Da)	Properties	Function
58	UL3	25 093	Hydrophilic/basic	Unknown
59	UL2	34 375		Uracil-DNA glycosylase
60	UL1	17 616	Acidic	Glycoprotein L
61	IE110	50 913	Zinc ring finger	Transcriptional regulator
62/71	IE175	139 989	Dimer/phosphoprotein	(IE) major transcriptional activator/tegument
63/70	US1	30 494	Hydrophilic	(IE) transcriptional regulator/tegument
64/69	US10	19 868		Virion component
65	US9	11 436	Hydrophobic	Tegument[b]
66	US3	43 677		Protein kinase
67	US7	39 362	Required for growth in skin and T cells	Glycoprotein I
68	US8	69 953	Required for virus growth	Glycoprotein E
S/L	***	24 640	Spans junction of the genome termini	Plaque morphology/cell adhesion

[a]Asterisks indicate genes that do not have herpes simplex virus (HSV) counterparts.
[b]Function inferred from data with HSV.
The predicted molecular weights are based on the sequence of the Dumas strain of VZV.

glycoproteins and as yet unidentified receptors on the cell surface. Upon entry and uncoating, the capsid is translocated to the nuclear membrane and the viral genome enters the nucleus, presumably through a nuclear pore. Viral tegument proteins also enter the nucleus, including several that regulate viral gene expression. Transcription and viral DNA replication are then believed to occur in a regulated temporal cascade, as is typical of all herpesviruses. Both viral and cellular proteins are involved in these processes. Progeny viral DNA molecules in the form of concatamers are presumably generated via a rolling-circle mechanism, and then cleaved and packaged into preassembled capsids. Tegument assembly is believed to occur at least in part in the nucleus. The infectious virion is believed to be formed following the de-envelopment/re-envelopment model currently extant for herpesvirus morphogenesis (see below). The virus grows slowly in tissue culture and is highly cell associated, and purified virions exhibit low levels of infectivity. The cell-associated nature of the virus has also led to speculation that, during infection of specific cell types, infectious spread occurs primarily via cell-to-cell fusion, whereas for other cell types intact cell-released virions are required.

Genome Replication

Replication of the VZV genome has not been subject to extensive biochemical analysis, primarily due to the difficulty in obtaining sufficient numbers of synchronously infected cells. Consideration of the relative proportions of the isomeric genome forms and the frequency of novel junctions in DNA isolated from virions and from infected cells have led to the following model. Upon entry into the nucleus, the viral genome circularizes and undergoes a limited number of rounds of bidirectional replication. This initiates at one or both copies of the origin of DNA replication (ori$_S$), which is located in TR$_S$ and IR$_S$. The minimal ori$_S$ sequence consists of a series of 16 AT repeats and an upstream consensus binding sequence (5′-CGTTCGCACTT-3′) for the origin-binding protein encoded by ORF51. During this phase, inversion of U$_L$ and U$_S$ is believed to occur via intramolecular recombination mediated by the inverted repeats. Replication then shifts to a rolling-circle mechanism, generating head-to-tail concatamers, and this phase accounts for the bulk of viral DNA synthesis. Finally, the newly synthesized DNA is cleaved into unit-length molecules and packaged into preassembled capsids in the nucleus. The unequal proportions of isomeric genome forms within particles are postulated to result from preferential cleavage of concatemers at the novel junction between TR$_L$ and TR$_S$ generated by the initial circularization event, rather than at the normal junction between IR$_L$ and IR$_S$.

The full complement of viral and cellular proteins required for complete replication of the VZV genome has not yet been enumerated. However, orthologs of the seven genes required for origin-dependent replication of herpes simplex virus (HSV-1) DNA have been identified. These include the proteins encoded by ORF6, ORF16, ORF28, ORF29, ORF51, ORF52, and ORF55. ORF6, ORF55, and ORF52 encode the primase, helicase, and an accessory factor, respectively, which are believed to form a heterotrimeric complex by analogy with HSV-1. ORF28 encodes the catalytic subunit of the viral DNA polymerase and is the target of the antiviral drugs acyclovir and valacyclovir. ORF16 encodes the polymerase processivity factor that interacts directly with the catalytic subunit. The ORF29 protein is the major single-strand DNA-binding protein (SSB) and, as indicated above, the ORF51 protein binds to ori$_S$ in a site-specific manner and also contains a helicase activity. Of these seven factors, only the DNA polymerase, the origin-binding protein, and SSB have been characterized at a biochemical level.

Overview of Gene Expression

Upon entry into the cell nucleus, the VZV genome is transcribed by the host cell's general transcription apparatus, including RNA polymerase II (RNA Pol II). By analogy with other alphaherpesviruses, the ORFs are believed to be expressed in three distinct kinetic classes: immediate early (IE), early (E), and late (L). However, clear evidence for this temporal cascade has been difficult to obtain owing to the low titers and cell-associated nature of VZV grown in tissue culture. Only three proteins have been shown to be expressed under IE conditions. These are IE4 (encoded by ORF4), IE62 (ORF62 and its counterpart in TRs, ORF71), and IE63 (ORF63 and its counterpart, ORF70) the latter two of which are expressed from ORF62 and ORF63 in IR$_S$ (and their duplicates in TR$_S$, ORF70 and ORF71).

VZV promoters have been shown in general to be essentially quiescent in the absence of virus-encoded transcriptional activators. Several VZV-encoded proteins have been shown in transient transfection assays to be capable of transactivating or transrepressing specific viral promoters, promoters from heterologous virus families, and cellular promoters. These include the IE62, IE4, IE63, ORF10, and ORF61 proteins. However, IE62 is considered to be the primary viral transactivator during the lytic phase of VZV replication. It is known to activate expression of genes from all three putative kinetic classes of VZV genes and can act independently of all other viral factors. Efficient activation of VZV promoters by IE62 usually occurs via synergistic interactions between this viral protein and ubiquitous cellular transcription factors whose consensus binding sites are found within VZV promoters. Predominant among these factors are specificity factor-1 (Sp1) and upstream factor (USF), which are predicted to be involved in regulating the expression of over half of the viral genome. When other viral factors are present, they have been shown to be capable of modulating the level of IE62 activation, even in some cases where the promoter in question appeared responsive only to IE62.

Promoter Structure

Results from analysis of a small number of VZV promoters, and predictions of the positions of putative promoters, indicate they appear to be relatively typical RNA Pol II promoters with TATA-like elements and binding sites for ubiquitous cellular transcription factors upstream of the TATA elements. Due to the compact nature of the VZV genome, many promoters are predicted to lie completely or partially within the coding sequences of genes immediately upstream. VZV promoters are also characterized by the fact that the TATA elements required or predicted to be positioned correctly for expression are frequently atypical and therefore less likely to be bound with high affinity by the TATA-binding protein (TBP). Thus far, no TATA-less VZV promoters have been identified and no promoters showing dependence on the presence of an initiator element (INR) at or just downstream of the start site of transcription are known. This most likely reflects the fact that only a few promoters have been analyzed in detail at the functional level, rather than any specific absence of these alternatives.

Potential VZV promoters can be separated into three general categories: A, B, and C type. This classification is preliminary as it is based on a meager amount of functional data. A-type promoters contain only a TATA element. B-type promoters contain a TATA element and one or more functional upstream binding sites for ubiquitous cellular transcription factors with which IE62 can interact physically or functionally. Examples of such promoters are that for ORF67 (encoding a glycoprotein, gI) and the regulatory element controlling divergent expression of ORF28 (DNA polymerase) and ORF29 (SSB). C-type promoters contain a TATA element, binding sites for factors with which IE62 can interact, and binding sites for factors that function to activate the promoter independently of IE62. The only known example of this type of promoter is that driving expression of the gene (ORF62) that encodes IE62 itself. A fourth class of promoter, at this time hypothetical, is one that is activated in the absence of IE62 by cellular or other viral proteins.

IE Gene Expression

Examination of the putative promoter regions of the three IE genes reveals, as noted above, that only the promoter for ORF62 (and its duplicate, ORF71), which encode IE62, contains binding sites for the cellular factor Oct-1. The presence of the Oct-1 recognition sequence and the existence of ORF10, the ortholog of HSV-1 VP16, indicate that IE62 can be expressed via the same mechanism as the HSV-1 IE genes, whereas IE4 and IE63 might be expressed via mechanisms possibly novel to VZV. By analogy with the HSV-1 IE expression mechanism, ORF10 would interact with Oct-1 bound to the promoter and the cellular transcription factor host cell factor-1 (HCF-1) to activate IE62 expression (**Figure 1**). This does appear to be one mechanism by which IE62 is expressed, since the existence of Oct-1/ORF10/HCF-1 complexes interacting with the IE62 promoter has been demonstrated, and depletion of HCF-1 by small interfering RNA (SiRNA) significantly reduces ORF10-mediated transactivation of the IE62 core promoter.

IE62 has also been shown to regulate expression of its own promoter and, unlike HSV-1 ICP4, is capable of positive as well as negative regulation. An alternative pathway for initial expression of the three IE genes involves the direct action of IE62 that enters the infected

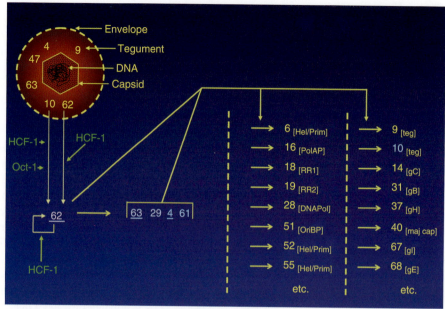

Figure 1 Schematic of the VZV virion and data on lytic phase gene expression. Numbers refer to the ORFs within the VZV genome. The IE genes are underlined. VZV genes whose protein products are known to be directly involved in viral gene expression are shown in blue. Cellular factors involved in IE gene expression are shown in green. Cellular factors such as Sp1 and USF known to be involved in non-IE gene expression are not shown. The vertical dashed lines delineate ORFs whose gene products are primarily involved in DNA replication and nucleotide metabolism or in the structure of the virion.

cell as part of the virion tegument. This pathway involves the cellular factor HCF-1, which has been shown to interact functionally with IE62. IE62 and HCF-1 cooperate in activating the IE62 promoter, resulting in a positive feedback loop allowing for efficient IE62 expression and IE62-mediated activation of the IE4 and IE63 promoters. This mechanism appears to be unique among the alphaherpesviruses to VZV.

Non-IE Gene Expression

Expression of the majority of VZV genes homologous to HSV-1 E and L genes is believed to be mediated primarily through the action of IE62 in conjunction with other viral proteins (the IE4, IE63, ORF61, and ORF29 proteins) and cellular factors. Information regarding expression of these genes is limited owing to the small number of promoters that have been analyzed. It is currently believed that IE62 is capable of transactivating the great majority of VZV promoters in the absence of other viral gene products. However, this rule is not hard and fast since the responsiveness to IE62 of promoters expressing structural proteins such as gI appears to be much less than that observed with IE promoters and the promoters controlling expression of DNA polymerase and SSB. Based on their ability to interact directly with IE62 and the frequent presence of their binding sites in VZV promoters, the ubiquitous cellular factors USF and Sp1 appear to play a significant role in IE62-mediated expression of VZV genes. They are believed to stabilize TBP binding, and both have been shown to interact directly with IE62. The roles of these transcription factors in IE62-mediated activation of specific VZV promoters have been validated in the SCID/hu mouse model of VZV pathogenesis.

These results represent only the beginnings of our understanding of the full spectrum of VZV gene regulation. As indicated above, IE4 and the ORF61 protein can also activate some non-IE promoters in the absence of IE62. Moreover, in the case of low levels of IE62 in the infected cell, expression of the ORF47 kinase and IE4 remains at essentially wild-type levels, in contrast to other known IE62-responsive genes, thus suggesting that there may be alternative non-IE62-mediated mechanisms for expression of some VZV genes. Other cellular transcription factors are also likely to be involved in VZV gene expression, including those in the CREB family (AP-1 and ATF-2). Finally, it is not known whether HCF-1 is required for activation of all IE62-responsive promoters, or whether this factor is important only for IE gene expression.

Protein Translation

Little is known about the details of translation in VZV-infected cells. It is assumed that mRNAs are translated in a fashion similar to that in HSV-1- and other alphaherpesvirus-infected cells, and that translation occurs on both free and membrane-bound polysomes.

Following translation, viral proteins are efficiently transported to specific cellular organelles (e.g., the nucleus and *trans*-Golgi network), where they perform their roles in viral replication and particle assembly. It has been shown recently that IE63 increases expression of the cellular factor EF-1α, which is involved in the initiation of translation. Thus, VZV infection may increase the efficiency of translation.

Post-Translational Processing

The most extensively studied post-translational processes involving VZV proteins are glycosylation and phosphorylation. Glycosylation is mediated by cellular genes, whereas phosphorylation involves the kinases encoded by ORF47 and ORF66, as well as cellular kinases. The VZV glycoproteins are all N-linked glycosylated and the majority show additional modifications such as O-linked glycosylation, sulfation, and phosphorylation. For example, gE contains both O-linked and N-linked glycans and is heavily sulfated and phosphorylated, with phosphorylation occurring at the level of both the polypeptide chain and glycosyl residues. The phosphorylated amino acids are serine and threonine but not tyrosine, and the enzymes involved are probably cellular casein kinases I and II. The enzyme responsible for phosphorylation of the complex oligosaccharides has not been identified, and roles for the specific phosphorylated or sulfated residues are unknown.

Viral proteins that are modified primarily by phosphorylation include the IE62, ORF61, and IE63 proteins. Phosphorylation of IE62 is known to involve the ORF66 and ORF47 kinases, as well as cellular kinases such as casein kinases I and II and protein kinase C. The function of the majority of these phosphorylation events is unknown, but phosphorylation by the ORF66 kinase plays a role in intracellular localization of IE62. Specifically, upon phosphorylation by the ORF66 kinase, IE62 is excluded from the infected cell nucleus, resulting in a diminution of viral gene expression and incorporation of IE62 into the virion particle. The ORF47 and ORF66 kinases also play roles in viral cell tropism and pathogenesis, since viruses lacking these functions are deficient for growth in human T cells and primary skin cells.

Assembly, Release, and Cytopathology

Models for assembly and release of VZV virions are derived from extensive electron microscopic studies and from analysis of the synthesis and trafficking of VZV glycoproteins in infected cells. However, no consensus model exists. Fully assembled capsids containing unit-length genomes representative of all four possible isomeric forms appear to be released from the nucleus, having acquired a portion of the tegument and a lipid coat from the nuclear lamella. These particles are then de-enveloped and bud into the Golgi apparatus, where additional tegument proteins are acquired. The mature viral envelope derived from Golgi membranes and containing the fully processed viral glycoproteins is then added. The particles are either released from the cell by fusion of virus-containing vesicles with the plasma membrane or of vesicles containing enveloped particles with lysosomes, in which case complete or partial degradation of the virions may take place. The involvement of lysosomes is proposed to rest on their mannose-6-phosphate receptors, which are capable of binding this sugar present on the surface of VZV particles. This process does not occur in the skin of the human host since the cells in the superficial layers of the skin lack lysosomal vesicles. This may explain why virus isolated from vesicle fluid is more infectious than that obtained in tissue culture and why it exhibits a higher and more stable titer.

In tissue culture, cells infected with VZV initially show increased refractility followed by rounding and swelling. Multinucleated cells are observed late in infection, although the extent of syncytium formation is cell type dependent and usually is less than that seen with syncytial HSV-1 strains. The infected cells eventually detach from the support surface. The nuclei of infected cells are larger than those of uninfected cells, with marginally located chromatin and peripherally located nucleoli.

Gene Expression during Latency

There is evidence that, during VZV latency, a small number of lytic genes are expressed at both the RNA and protein levels. These include ORF4, ORF21, ORF29, ORF62, and ORF66. These proteins play unknown roles during latency, and are made even more intriguing in that they are localized primarily in the cytoplasm rather than the nucleus. The relative frequencies of detection of expression of these genes varies considerably, with the two most frequently detected being the products of ORF63 (IE63) and ORF29 (SSB). Therefore, the promoters regulating expression of these two genes appear to be the best candidates in which to find clues relating to the mechanisms of latent gene expression.

The putative IE63 promoter borders on ori_S, which is present in TR_S and IR_S. Little work has been published concerning the promoter-regulating expression of IE63. A putative TATA element has been identified by computer analysis 28 nt upstream from the transcriptional start site, as have several potential binding sites for Sp1 and CAAT. None of these elements has yet been authenticated by site-specific mutation.

More information is available concerning expression of ORF29. The promoter lies within an intergenic region situated between the divergently oriented ORF28 and ORF29 protein-coding regions, and shares a required USF site with the ORF28 promoter. ORF28 and ORF29 can be expressed either coordinately or independently, and the observed expression of only ORF29 during latency may involve neuron-specific cellular factors that recognize sequences within the ORF29 promoter and lead to its activation.

An alternative regulatory mechanism could involve variations in the chromatin structures of the promoter regions of the latently expressed genes. Like the HSV-1 genome, the VZV genome is episomal during latency and is likely to be bound by histones. This raises the possibility that, during latency, the promoters and coding regions of the putative latency-associated genes, particularly ORF63 and ORF29, could be histone-free (euchromatic) and non-repressed. How and why these specific regulatory elements could be singled out remains to be established.

HCF-1 may also play a role in reactivation of VZV from latency. In contrast to the situation in other tissues, HCF-1 is not detected in the nuclei of sensory ganglia but rather in the cytoplasm. However, experimental conditions that induce reactivation of HSV-1 result in rapid nuclear localization of HCF-1. Thus, redistribution of HCF-1 may trigger expression of IE62, ultimately resulting in the production of infectious virus.

Subjects for Future Investigation

Molecular Mechanism of Vaccine Attenuation

The live-attenuated vaccine for prevention of varicella has been in use in Japan for over two decades and was approved for use in the USA in 1995, where coverage is now approaching 95% of the susceptible pediatric population. This vaccine has also recently been shown to be efficacious for the prevention of zoster and for mitigation of postherpetic neuralgia. This being said, the vaccine represents a mixture of strains and retains the capacity to establish a latent infection and reactivate. Moreover, the molecular mechanism of attenuation remains unknown. A large number of mutations are present in vaccine virus IE62, but it is clear that the cause of attenuation extends beyond IE62 and is multifactorial. An important area of future study would involve definition of the mechanism of attenuation and production of a better characterized, and therefore safer, second-generation vaccine.

Molecular Basis of Post-Herpetic Neuralgia

One of the problems facing the current aging population in the USA and the rest of the world is the morbidity associated with post-herpetic neuralgia resulting from episodes of zoster. Nothing is known about the roles that viral genes play in this process. Investigation of the effects of specific VZV proteins on the expression or elaboration of neuropeptides and neuronal processes will be important in elucidating the molecular mechanisms responsible for post-herpetic neuralgia. This may aid the development of drugs specific for amelioration of this syndrome. Particular focus would be appropriate on the IE63 and ORF29 proteins, which have been detected in human ganglia.

VZV and Cellular Processes

There is now a considerable and continuously expanding body of information concerning the interaction of various VZV gene products with host cell components. Furthermore, recently reported gene array analyses indicate that the expression of numerous cellular genes is affected during infection. However, little information is available on the specific VZV factors involved in these changes and in what ways these changes are important for infection and pathogenesis. VZV-induced alteration of host cell function is known to differ to some extent from that observed with HSV-1, since VZV infection, unlike HSV-1 infection, inhibits the cellular nuclear factor kappa B (NFκB) pathway. Examination of these questions at the molecular level in the context of the various cell types infected by VZV will reveal important information, which will not only provide insights into VZV infection but also into fundamental cellular processes.

See also: DNA Vaccines; Herpes Simplex Viruses: General Features; Herpes Simplex Viruses: Molecular Biology; Herpesviruses: General Features; Herpesviruses: Latency; Persistent and Latent Viral Infection; Pseudorabies Virus; Varicella-Zoster Virus: General Features.

Further Reading

Cohen JI and Straus SE (2001) Varicella zoster virus and its replication. In: Fields BM, Knipe DM, and Howley PM (eds.) *Virology*, pp. 2707–2730. Philadelphia, PA: Lippincott-Raven.

Gomi Y, Sunamachi H, Mori Y, Nagaike K, Takahashi M, and Yamanishi K (2002) Comparison of the complete DNA sequences of the varicella vaccine and its parental virus. *Journal of Virology* 76: 11447–11459.

Grose C, Maresova L, Medighesi G, Scott GK, and Thomas G (2006) Endocytosis of varicella zoster virus glycoproteins: Virion envelopment and egress. In: Sandri-Goldin RM (ed.) *Alpha Herpesviruses: Molecular and Cellular Biology*, pp. 157–174. Norfolk, UK: Caister Academic Press.

Kenyon TK, Homan E, Storlie J, Ikoma M, and Grose C (2003) Comparison of varicella zoster virus ORF47 protein kinase and casein kinase II and their substrates. *Journal of Medical Virology* 70: 95–102.

Kinchington PR and Cohen JI (2000) Viral proteins. In: Arvin AM and Gershon AA (eds.) *Varicella Zoster Virus: Virology and Clinical Management*, pp. 74–104. Cambridge: Cambridge University Press.

Ku C-C, Besser J, Abendroth A, Grose C, and Arvin AM (2005) Varicella-zoster virus pathogenesis and immunobiology: New concepts emerging from investigations with the SCIDhu mouse model. *Journal of Virology* 79: 2651–2658.

Mitchell BM, Bloom DC, Cohrs RJ, Gilden DH, and Kennedy PGH (2003) Herpes simplex virus-1 and varicella-zoster virus latency in ganglia. *Journal of Neurovirology* 9: 194–204.

Nagaike K, Mori Y, Gomi Y, et al. (2004) Cloning of the varicella-zoster virus genome as an infectious bacterial artificial chromosome in *Escherichia coli*. *Vaccine* 22: 4069–4074.

Narayanan A, Nogueira ML, Ruyechan WT, and Kristie TM (2005) Combinatorial transcription of the HSV and VZV IE genes is strictly determined by the cellular coactivator HCF-1. *Journal of Biological Chemistry* 280: 1369–1375.

Oxman MN, Levin MJ, Johnson GR, et al. (2005) A vaccine to prevent herpes zoster and postherpetic neuralgia in older adults. *New England Journal of Medicine* 325: 2271–2284.

Ruyechan WT (2006) Varicella zoster virus transcriptional regulation and the roles of VZV IE proteins. In: Sandri-Goldin RM (ed.) *Alpha Herpesviruses: Molecular and Cellular Biology*, pp. 1–20. Norfolk, UK: Caister Academic Press.

Ruyechan WT and Hay J (2000) DNA Replication. In: Arvin AM and Gershon AA (eds.) *Varicella Zoster Virus: Virology and Clinical Management*, pp. 51–73. Cambridge: Cambridge University Press.

Silverstein S and Straus SE (2000) Pathogenesis of latency and reactivation. In: Arvin AM and Gershon AA (eds.) *Varicella Zoster Virus: Virology and Clinical Management*, pp. 123–141. Cambridge: Cambridge University Press.

Varicosavirus

T Sasaya, National Agricultural Research Center, Ibaraki, Japan

© 2008 Elsevier Ltd. All rights reserved.

Glossary

Big-vein disease of lettuce A lettuce disease that is characterized by abnormal dilation or enlargement of a vein or artery in a lettuce leaf, decreasing market value and reducing the proportion of marketable lettuce plants. The name 'big-vein disease' refers to the appearance of the major symptoms.

L protein A main subunit of the polymerase complex, responsible for most of the functions required for transcription and replication: RNA-dependent RNA polymerase, mRNA 5′ capping, 3′ poly(A) synthesis, and protein kinase activities.

Olpidium brassicae An obligate parasitic root-infecting fungus and a member of the *Chytridiomycetes* that seldom appears to be deleterious to the host plants but acts as vector of plant viruses.

Stop-and-start model A model for RNA synthesis of nonsegmented negative-sense RNA viruses that the viral polymerase complex transcribes a gene, polyadenylates and releases mRNA, and reinitiates transcription of the next gene under the direction of the gene-junction sequence.

History

Big-vein disease of lettuce (*Lactuca sativa* L.) was first reported in 1934. The causal agent was postulated to be a root-infecting virus, but the virus particles remained unidentified for more than half a century. In 1983, a rod-shaped virus was found in field-grown lettuce showing big-vein symptoms. Because the rod-shaped virus is always associated with big-vein-affected lettuce, it was regarded as the agent that induced big-vein disease, although this virus alone had not been shown to actually induce big-vein symptoms in lettuce. The rod-shaped virus was named *Lettuce big-vein virus* and was classified as the type species of the genus *Varicosavirus*, which had been established by the International Committee on the Taxonomy of Viruses (ICTV) in 2000.

Since 2000, however, a second virus assigned to the genus *Ophiovirus* and named mirafiori lettuce virus (MiLV; also referred to as mirafiori lettuce big-vein virus), has been consistently isolated by various researchers from field-grown lettuce showing big-vein symptoms. Mechanical inoculation of partially purified MiLV preparations and transmission experiments using MiLV- or lettuce big-vein associated virus (LBVaV)-carrying fungal vectors independently (and conclusively) showed that MiLV, but not LBVaV, is the causal agent of big-vein disease. Therefore, lettuce big-vein virus was renamed lettuce big-vein associated virus by ICTV in 2005. However, the etiology of big-vein disease is still not fully understood, because big-vein affected plants serologically negative for MiLV but positive for LBVaV were reported in a field experiment in Italy. A transgenic line carrying the coat protein (CP) gene of LBVaV in antisense orientation showed resistance both to LBVaV infection and to MiLV infection, and was thus resistant to big-vein disease.

Stunt disease of tobacco (*Nicotiana tabacum* L.) reported only from Japan was first recorded in 1931 as a seed-bed disorder. Stunt disease seriously damaged tobacco leaf production in Japan until the 1960s, when the disease was found to be transmitted by an obligate parasitic soil-borne fungus, and tobacco growers paid more attention to tobacco seed-bed sanitation. Because tobacco with stunt symptoms invariably has a varicosavirus and inoculation of tobacco seedlings with partially purified virion preparation produces typical tobacco stunting, the virus is regarded as the causal agent of tobacco stunt disease. Because of marked differences in the host reaction between the varicosavirus and LBVaV, the virus was initially considered to be a distinct species, rather than a strain of LBVaV, and named tobacco stunt virus (TStV). LBVaV and TStV are serologically indistinguishable and a recent comparison of nucleotide and amino acid sequences for LBVaV and TStV confirmed that TStV is a strain, a tobacco strain, of LBVaV.

In 2000, ICTV listed camellia yellow mottle virus and freesia leaf necrosis virus as tentative varicosaviruses. However, in 2005 ICTV dropped the two viruses from the genus *Varicosavirus* due to the lack of reliable information on their biological and virion properties. LBVaV is thus the only confirmed virus species of the genus *Varicosavirus*.

LBVaV was initially believed to have a divided genome consisting of two components of double-stranded (ds) RNA, approximately 7.0 and 6.5 kbp in size. Once LBVaV purification was established and the complete nucleotide sequence of LBVaV determined, a precise reinvestigation of LBVaV genome components showed that LBVaV is not a dsRNA virus but a single-stranded RNA virus with a bipartite genome. Negative-sense and positive-sense RNAs are separately encapsidated in the virions, with the negative-sense RNA being predominant.

Taxonomy and Classification

LBVaV particles are fragile, nonenveloped rods about 18 nm in diameter with a dominant particle length of 360 nm. Virus particles have a central canal about 3 nm in diameter and an obvious helix with a pitch of about 5 nm, resembling the inner striated nucleocapsid core of the rhabdovirus. Virus particles are very unstable *in vitro*, so the helix of particles, especially those in purified preparations, tends to loosen and come partially uncoiled, even if fixed with glutaraldehyde.

Criteria for membership in the genus are nonenveloped rod-shaped particles, a two-segmented negative-sense RNA genome, and transmission through moist soil by motile zoospores of an obligate parasitic soil-borne fungus. The fungal vector is reported to be *Olpidium brassicae*, but the taxonomy and nomenclature of the fungus needs to be reevaluated.

Genetics and Evolution

Nucleotide sequence and genome-mapping studies of LBVaV showed that the LBVaV genome consists of 12 878 nucleotides (nt) and is divided into two segments (**Figure 1**). The larger LBVaV genome segment (LBVaV RNA1) contains antisense information for a small gene and a large gene that encodes a large protein (L protein)

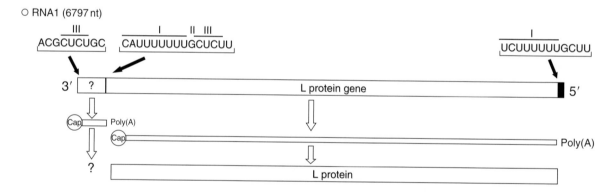

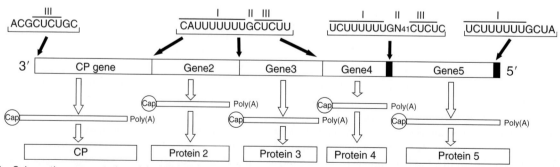

Figure 1 Schematic representation of the LBVaV genome and expression of the consecutive transcription of monocistronic mRNAs. Gene-junction regions, including a gene-end sequence (I), an intergenic sequence (II), and a gene-start sequence (III), are shown above negative-sense genomic RNAs.

of 2040 amino acids. The smaller LBVaV genome segment (LBVaV RNA2) contains antisense information for five major genes, which have coding capacities of 397, 333, 290, 164, and 368 amino acids, with the gene at the 3' end of the LBVaV RNA2 encoding the CP. FASTA and BLAST analyses of the LBVaV CP and L protein revealed the amino acid sequence similarities to the nucleocapsid (N) proteins and the L proteins of rhabdoviruses, respectively. However, the translated sequences from the other four LBVaV genes showed no striking similarity to other known viral sequences.

Even though LBVaV is a two-segmented negative-sense RNA virus with a fragile nonenveloped rod shape, LBVaV shares several features with a rhabdovirus, which has a linear nonsegmented negative-sense RNA genome and large enveloped particles with a prominent fringe of peplomers.

L Protein Encoded on RNA1

As with other negative-sense RNA viruses, the L protein of LBVaV is positively charged and contains polymerase and RNA binding domains. Alignment of the LBVaV L protein sequence with those of several other negative-sense RNA viruses shows the conservation of functional motifs of an RNA-dependent RNA polymerase (**Figure 2**). The L protein of LBVaV contains a conserved premotif A, which presumably plays an important role in RNA template binding and positioning, and four motifs that correspond to motifs A, B, C, and D, thought to comprise the palm and finger regions of the polymerase active site in negative-sense RNA viruses. A putative ATP binding motif, $Kx_{17}GxGxG$, proposed to be involved in ATP binding associated with polyadenylation or protein kinase activity, is present at position 1643–1665, and LBVaV actually transcribes 3' polyadenylated mRNAs.

The L protein is the only region sufficiently conserved to establish evolutionary relationships among negative-sense RNA viruses. Even though the LBVaV genome is segmented, most of the conserved motifs previously identified in the L proteins of nonsegmented negative-sense RNA viruses are conserved in the L protein of LBVaV (**Figure 2**), particularly the conserved sequence GDN in motif C, which likely constitutes the evolutionary and functional equivalent of the sequence GDD in polymerases of positive-sense RNA viruses. This is in striking contrast to the segmented genome of negative-sense RNA viruses whose polymerase has the conserved sequence SDD. Amino acids G at position 662 and W at position 671 in motif B, specific to the nonsegmented viruses, are also conserved in the L protein of LBVaV. Tetrapeptide E(F/Y)xS, located downstream from motif D, which is specifically conserved in polymerases of segmented negative-sense RNA viruses, is not found in the L protein of LBVaV.

Phylogenetic data and pair-wise amino acid sequence comparisons of the partial LBVaV L protein with those of nonsegmented negative-sense RNA viruses also prove that LBVaV is more closely related to viruses of the family

```
              Premotif A                     Motif A
LBVaV  518  KEREIKVAARMYSLMTERMRYYFVLTEGL <39> NINIDFSKWNTNMR <54>
LNYV   490  KEREMNPTPRMFALMSHLMRVYVVITESM <48> CMSLDFEKWNGHMR <55>
NCMV   544  KEREMNPVARMFALMTLKMRSYVVITENM <42> CINMDFEKWNLNMR <56>
SYNV   567  KEREMKTKARFFSLMSYKLRMYVTSTEEL <41> SMNIDFSKWNONMR <54>
RYSV   537  KERELKIMARFFALLSFKMRLYFTATEEL <41> VINMDFVKWNQOMR <55>
VSIV   530  KERELKLAGRFFSLMSWKLREYFVITEYL <42> ANHIDYEKWNNHQR <56>
RABV   543  KERELKIEGRFFALMSWNLRLYFVITEKL <42> AFHLDYEKWNNHQR <58>
BEFV   556  KERELKEEGRFFSLMSYELRDYFVSTEYL <42> ANNIDYEKWNNYQR <56>
IHNV   494  KEMELKIKGRGFGLMTFMPRLLQVLRESI <40> NKSLDINKFCTSQR <67>
TSWV  1286  KMQRTKT-DREIYLMSMKVKMMLYFIEHT <49> FLSADQSKWSASGL <69>
RSV   1416  KNQHGG--LREIYVLNIFERIMQKTVEDF <41> STSDDASKWNQGHY <75>

              Motif B                        Motif C           Motif D
LBVaV  MCYRGHLGGFEGLRQKGWTVATVCLL <12> LMGQGDNQII <64> LDGRQLPQWYKKT
LNYV   YSFTGHKGGQEGLRQKGWTIFTVVCL <12> IMGMGDNQVL <64> LKGVPLSMDLKKI
NCMV   KSYEGHIRGFEGLRQKGWTVFTVVLI <12> LMGQGDNQVL <64> YKGVPLCSSLKRI
SYNV   WSRTGDESGKEGLRQKGWTITTVCDI <12> LIGGGDNQVL <63> YSGVPLRGRLKVI
RYSV   VCWIDDGAGKEGIRQKAWTIMTVCDI <12> LVGGGDNQVL <64> YKGVPLRSPLKQV
VSIV   VCWQGQEGGLEGLRQKGWTILNLLVI <12> VLAQGDNQVI <63> FRGVIRGLETKRW
RABV   TCWNGQDGGLEGLRQKGWSLVSLLMI <12> VLAQGDNQVL <63> FRGNILVPESKRW
BEFV   VCWEGQKGGLEGLRQKGWSILNYLMI <12> ILAQGDNQTI <63> IEGTIKGLPTKRW
IHNV   GVFSGLKGGIEGLCQYVWTICLLLRV <12> ILAQGDNVII <62> HCPQHLTLAIKKA
TSWV   NTYPVSMNWLQGNLNYLSSVYHSCAM <18> WIVHSDDNAT <32> KSFCITLNPKKSY
RSV    SYIETETGMMQGILHYTSSLFHAIFL <31> NMESSDDSSF <31> GTYLGIYKSPKST
```

Figure 2 Comparison of L protein motifs conserved among LBVaV, eight rhabdoviruses (lettuce necrotic yellows virus (LNYV), northern cereal mosaic virus (NCMV), sonchus yellow net virus (SYNV), rice yellow stunt virus (RYSV), vesicular stomatitis Indiana virus (VSIV), rabies virus (RABV), bovine ephemeral fever virus (BEFV), and infectious hematopoietic necrosis virus (IHNV)), and tomato spotted wilt virus (TSWV) and rice stripe virus (RSV), two segmented negative-sense RNA viruses of plants. Numbers at the beginning of lines indicate the position of the first displayed amino acid. Numbers within brackets indicate the numbers of amino acids not represented in the figure. Conserved residues recognized previously for L proteins of negative-sense RNA viruses are shown in bold letters and residues characteristic of the division between nonsegmented and segmented negative-sense RNA viruses are underlined.

Rhabdoviridae than to those of the family *Bornaviridae*, *Filoviridae*, and *Paramyxoviridae*. Phylogenetic analysis of the partial L proteins between LBVaV and rhabdoviruses shows that the L protein of LBVaV clusters with viruses of the family *Rhabdoviridae*, and that the L protein of LBVaV is more distantly related to those of plant rhabdoviruses (genera *Cytorhabdovirus* and *Nucleorhabdovirus*) than they are to each other. On the other hand, the L protein of LBVaV is more closely related to those of plant rhabdoviruses than those of other rhabdoviruses (**Figure 3**).

Proteins Encoded on RNA2

The first gene at the 3′ end of genomic RNA encodes the CP which shows low but significant similarity to the N protein of rhabdoviruses. The other genes encode proteins of unknown functions, while the second and fifth encoded proteins (proteins 2 and 5) have features slightly similar to corresponding proteins of rhabdoviruses.

Gene Bank Database Homology searches using the amino acid sequence of the LBVaV CP consistently retrieve N proteins of rhabdoviruses. Phylogenetic analysis of the LBVaV CP and N proteins of rhabdoviruses shows that the LBVaV clusters with viruses of the family *Rhabdoviridae*, and that LBVaV CP is most closely related to the N protein of infectious hematopoietic necrosis virus (IHNV) a fish rhabdovirus (genus *Novirhabdovirus*), and to a lesser extent with plant rhabdoviruses. In a PSI-BLAST search against the ProDom protein domain database, the carboxy-terminal part of the LBVaV CP shows similarity to two domains; one is the homologous domain among N proteins of plant rhabdoviruses and another is the homologous domain among those of fish rhabdoviruses.

The LBVaV protein 2 is encoded by the second gene that corresponds in position to a gene coding for the phosphoprotein (P protein) of rhabdoviruses, which interacts with the N protein and the L protein and constitutes the active polymerase complex. However, the 333 amino acid protein encoded by the second gene of RNA2 has no significant amino acid sequence relatedness to P proteins of rhabdoviruses. The overall structure of the LBVaV protein 2 and P proteins, assessed by amino acid composition and by relative hydropathicity distribution, is similar. Coiled-coil regions of P proteins of rhabdoviruses have been used to predict the regions for their oligomerization and interaction. Analysis using the COILS2 program predicts that the LBVaV protein 2 contains three possible coiled-coil regions at the amino-terminal (aa 29–48), central (aa 122–135), and carboxy-terminal (aa 239–253) domains that are similar in position, to those in the P protein of sonchus yellow net virus (SYNV). In site-directed mutagenesis experiments using the bacterially expressed P protein of vesicular stomatitis Indiana virus (VSIV), detailed analysis of phosphorylation sites for cellular casein kinase II in the P protein confirmed that five of ten potential phosphorylation sites of the P protein, which may regulate transcription and replication, are required to form the active polymerase complex. The LBVaV protein 2 also has ten potential phosphorylation sites, and the five essential phosphorylation sites for the P protein of VSIV are conserved in positions similar to those of the LBVaV protein 2. A predominantly basic 20-amino acid-long region near the carboxy terminus of the P protein of VSIV, which may act as a site for binding to the L protein, appears to be highly conserved in the carboxy-terminal region of the LBVaV protein 2.

The LBVaV protein 5, which corresponds in position to a gene coding for the glycoprotein (G protein) of rhabdoviruses, also has little direct relatedness to the envelope-associated glycoproteins of rhabdoviruses. The LBVaV protein 5 does not contain any canonical sites for N-linked glycosylation, and does not contain the N terminal signal peptide or carboxy-terminal hydrophobic transmembrane anchor domains, which are common among G proteins of rhabdoviruses. Conservation of cysteine residues is another common feature of G proteins of rhabdoviruses. Alignment using the Clustal W program for G proteins with plant rhabdoviruses shows that 11 cysteine residues are situated at conserved positions. In the case of LBVaV, six of ten cysteine residues are located at similar conserved positions in G proteins of plant rhabdoviruses. The LBVaV protein 5 may thus result from the degeneration of G proteins of plant

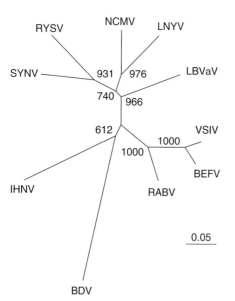

Figure 3 Computer-generated tree illustrating phylogenetic relationships between LBVaV and selected rhabdoviruses on the basis of aligned amino acid sequences of the partial L protein between premotif A and motif D. The tree is constructed using neighbor joining, and numbers at nodes are confidence values for 1000 bootstrap replicates; bootstrap values below 500 are not shown. Borna disease virus (BDV) is used as an out-group. See **Figure 2** for abbreviations of virus names.

rhabdoviruses. The six cysteine residues in the LBVaV protein 5 that may play an important role in activities, such as forming the secondary structure of G proteins of plant rhabdoviruses appear to have been stringently conserved among the LBVaV protein 5 and G proteins of plant rhabdoviruses. The other five cysteine residues, however, appear to have evolved extensively to accommodate diverse host requirements and vector adaptations.

Transcription Termination/Initiation Strategy

From the LBVaV genome, capped and polyadenylated monocistronic mRNAs are transcribed. In the gene-junction regions of LBVaV, transcription termination/polyadenylation and initiation signal sequences comparable to those of rhabdoviruses especially cytorhabdoviruses are recognized (**Figure 4**) suggesting that LBVaV and rhabdoviruses may use a similar mechanism of transcription, a stop-and-start model, to differentially express individual genes from a contiguous virus genome.

Gene-junction regions of LBVaV, and likewise rhabdoviruses, are grouped into three elements: a gene-end sequence (element I) at the 3′ end of each gene on the genomic template, which is required for transcription termination/polyadenylation; a short nontranscribed intergenic sequence (element II), which is not represented in mRNAs and separates each gene; and a gene-start sequence (element III) located at the beginning of each subsequent gene, which plays an important role in transcription initiation. Gene-end sequences of rhabdoviruses consist of an AU-rich region, a cytidylate, and a poly(U) tract and the poly(U) tract is thought to be a template for the poly(A) tail of mRNA due to reiterative transcription or slippage. The gene-end sequences of LBVaV are either 3′-UAUNCAUUUUUUU-5′ (type A) or 3′-AUAAUCUUUUUU-5′ (type B), reminiscent of those observed in rhabdoviruses. When the gene-end sequence of LBVaV is type A, transcription of the next gene is initiated immediately after a single nucleotide G, while for type B, the two genes (genes 5 and L) were located at the 5′ most ends of each RNA (negative-sense) or transcription is reinitiated after a long 42 nt intergenic sequence located between genes 4 and 5, suggesting that the two variations of the gene-end sequences may influence signaling initiation of downstream mRNA synthesis as reported for VSIV in which the gene-end sequence plays an important role both in transcription termination/polyadenylation and in reinitiating transcription of downstream genes.

The short nontranscribed intergenic sequence in most rhabdoviruses generally consists of a few nucleotides. Although intergenic sequences of rhabdoviruses are highly conserved in sequence and length in the genera *Vesiculovirus* and *Nucleorhabdovirus* but not in the genera *Lyssavirus*, *Ephemerovirus*, and *Cytorhabdovirus*, the first residue of the intergenic sequence is G. Site-directed mutagenesis experiments in the intergenic sequence using the infectious cDNA of VSIV indicated that the G residue is required for efficient termination of the upstream transcript. In the case of LBVaV, the intergenic sequence, except between genes 4 and 5, is highly conserved, consisting of a single nucleotide, G. Intergenic sequences of fish rhabdoviruses in the genus *Novirhabdovirus* also consist of a single nucleotide, G or A.

In contrast to most rhabdoviruses, including nucleorhabdoviruses whose gene-start sequence in gene-junction regions is 3′-UUGU-5′, the conserved gene-start sequence of LBVaV is 3′-CUCU-5′. Sequence 3′-CUCU-5′ is also conserved in gene-junction regions of lettuce necrotic yellows virus (LNYV) and northern cereal mosaic virus (NCMV) in the genus *Cytorhabdovirus*, suggesting that sequence 3′-CUCU-5′ may be important in acting as a transcription initiation signal for both LBVaV and cytorhabdoviruses, which replicate in the cytoplasm of plant cells.

Host Range and Geographic Distribution

LBVaV is extremely unstable *in vitro* making a mechanical transmission difficult, time-consuming, and inefficient. Since LBVaV inocula are frequently contaminated with MiLV and specific detection of LBVaV and MiLV has become feasible only very recently, available information on the symptomatology and experimental host range of LBVaV needs to be verified. We recently confirmed through fungal vector transmission experiment, however, that LBVaV systemically infected seven species in four families, and the infected plants including lettuce did not show any symptoms. In contrast, the experimental host range of the tobacco strain of LBVaV encompasses 41 species in nine dicotyledonous families

		I	II	III
LBVaV	TypeA	UNCAUUUUUU	G(N)$_x$	CUCU
	TypeB	AAUCUUUUUU		
LNYV		AUUCUUU	G(N)$_x$	CUCU
NCMV		AUUCUUUU	GA	CUCU
SYNV		AUUCUUUU	GG	UUGW
RYSV		AUUAUUUU	GGG	UUGU
VSIV		AUACUUUUUU	GA	UUGU
RABV		GYACUUUUUU	G(N)$_x$	UUGU
IHNV		URUCUUUUUU	G/A	CCRW

Figure 4 Comparison of gene-junction regions between LBVaV and selected rhabdoviruses. Gene-junction regions are separated into three elements: element I is a gene-end sequence that constitutes the poly(U) tract at the 3′ end of each gene on negative-sense genomic RNAs and plays the role of a transcription termination/polyadenylation signal; element II is a short intergenic sequence that is not transcribed during mRNA synthesis; and element III is a gene-start sequence that constitutes the initiation site for transcription of each mRNA. Bold type in viral sequences indicates consensus nucleotides, R and W indicate A/G and A/U, and (N)x corresponds to a variable number of nucleotides. See **Figure 2** for abbreviations of virus names.

following mechanical inoculation and 35 plant species in 13 families after fungal vector transmission. The tobacco strain does not infect lettuce, but induces vein clearing, vein necrosis, mottling, and stunting in many *Nicotiana* species.

LBVaV is widespread in cool to temperate regions and several parts of subtropical regions in Europe, Asia, and North and South America, but the tobacco strain probably occurs only in Japan.

Transmission and Vectors

In nature, LBVaV is transmitted through motile zoospores of an obligate parasitic soil-inhabiting fungus. LBVaV is very stable in soil within long-lived, resting spores of the fungus for 20 years or more. It remains unclear whether the virus replicates inside fungal spores.

The natural vector for LBVaV is reported to be *Olpidium brassicae* (Wor.) Dang. *O. brassicae* was first found in a cabbage (*Brassica oleracea*) root in 1878 and the species name of the fungus is derived from the genus name of cabbage. Many strains of *O. brassicae* differing significantly in host specificity, morphology, and vector capacity have been recognized. In particular, a crucifer strain, which readily multiplies in the roots of crucifer plants such as cabbage and mustard, is differentiated from a noncrucifer strain that can not multiply in the roots of crucifer plants. The crucifer strain is host-specific and the noncrucifer strain highly plurivorous. Morphologically, zoosporangia and exit tubes of the crucifer strain differ slightly from those of the noncrucifer strain. The noncrucifer strain is generally a vector of MiLV and tobacco necrosis virus but the noncrucifer strain fails to transmit these viruses. Furthermore, the crucifer strain shows sexual reproductivity, although the noncrucifer strain does not reproduce sexually. Zoospores from a single-sporangium isolate of the crucifer strain do not develop a resting spore, but after mating between different mating (sexual) types, a resting spore forms. In contrast, zoospores from the noncrucifer strain develop a resting spore without mating. Based on these differences, the crucifer and noncrucifer strains have been divided into two distinct species, but this proposal has long been ignored. Recent rDNA-ITS analysis of *O. brassicae* strongly supported the notion that crucifer and noncrucifer strains are separate species rather than strains of the same *Olpidium* species, and a new species name, *O. virulentus*, has been proposed for the noncrucifer strain. The noncrucifer strain was confirmed to be a vector of LBVaV, whereas there is no reliable evidence that the crucifer strain (*O. brassicae*) is capable of transmitting LBVaV.

See also: Ophiovirus; Plant Rhabdoviruses; Rabies Virus; Vector Transmission of Plant Viruses.

Further Reading

Banerjee AK, Barik S, and De BP (1991) Gene expression of nonsegmented negative strand RNA viruses. *Pharmacology and Therapeutics* 51: 47.

Das T, Gupta AK, Sims PW, Gelfand CA, Jentoft JE, and Banerjee AK (1995) Role of cellular casein kinase II in the function of the phosphoprotein (P) subunit of RNA polymerase of vesicular stomatitis virus. *Journal of Biological Chemistry* 270: 24100–24107.

Einer-Jensen K, Krogh TN, Roepstorff P, and Lorenzen N (1998) Characterization of intramolecular disulfide bonds and secondary modifications of the glycoprotein from viral hemorrhagic septicemia virus, a fish rhabdovirus. *Journal of Virology* 72: 10189–10196.

Lot H, Campbell RN, Souche S, Milne RG, and Roggero P (2002) Transmission by Olpidium brassicae of *Mirafiori lettuce virus* and *Lettuce big-vein virus*, and their roles in lettuce big-vein etiology. *Phytopathology* 92: 288–293.

Rose JK and Whitt MA (2001) *Rhabdoviridae*: The viruses and their replication. In: Knipe M and Howley PM (eds.) *Field Virology*, 4th edn., pp. 1221–1224. New York: Raven Press.

Sasaya T, Ishikawa K, and Koganezawa H (2002) The nucleotide sequence of RNA1 of *Lettuce big-vein virus*, genus *Varicosavirus*, reveals its relation to nonsegmented negative-strand RNA viruses. *Virology* 297: 289.

Sasaya T and Koganezawa H (2006) Molecular analysis and virus transmission tests place *Olpidium virulentus*, a vector of *Mirafiori lettuce big-vein virus* and tobacco stunt virus, as a distinct species rather than a strain of *Olpidium brassicae*. *Journal of General Plant Pathology* 72: 20–25.

Sasaya T, Kusaba S, Ishikawa K, and Koganezawa H (2004) Nucleotide sequence of RNA2 of *Lettuce big-vein virus* and evidence for a possible transcription termination/initiation strategy similar to that of rhabdoviruses. *Journal of General Virology* 85: 2709–2717.

Tordo N, Haan PD, Goldbach R, and Poch O (1992) Evolution of negative-stranded RNA genomes. *Seminars in Virology* 3: 341–357.

Vector Transmission of Animal Viruses

W K Reisen, University of California, Davis, CA, USA

© 2008 Elsevier Ltd. All rights reserved.

Glossary

Anthroponosis Virus transmitted among human hosts.
Arbovirus Virus transmitted by blood-feeding arthropods among vertebrates.
Diapause Arthropod hibernation.
Gonotrophic cycle Cycle of blood feeding and egg deposition.
Transmission Distribution of virus usually by the bite of an infectious vector.

Vector Animal (usually an arthropod) that distributes viruses among hosts.
Vector competence Ability of vector to become infected with and transmit a virus.
Vectorial capacity Mathematical measure of case distribution or the force of transmission.
Viremia The presence of viruses within the vertebrate host peripheral circulatory system.
Zoonosis Virus transmitted mainly among animal hosts with incidental transmission to humans.

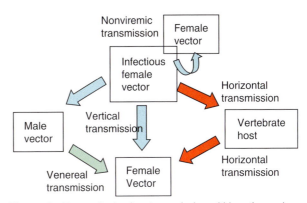

Figure 1 Types of arbovirus transmission within arthropod vectors and between arthropod vectors and vertebrate hosts. Vertical transmission can be trans-stadial, as in ticks, or transgenerational, as in mosquitoes.

Vector Transmission

This article describes how vectors (usually arthropods and especially insects) distribute or transmit viruses among vertebrate hosts. Although occasional instances of mechanical transmission by contaminated mouthparts of the feeding arthropod have been documented and nonviremic transmission among blood-feeding vectors has been demonstrated in the laboratory and occurs in nature, virus transmission generally is biological and propagative (**Figure 1**). This type of transmission demonstrates that the virus infects and multiplies within the tissues of both arthropod and vertebrate hosts before transmission can occur. Viruses transmitted by arthropods have been lumped into a loose assemblage termed 'arboviruses' that includes viruses within the families *Bunyaviridae, Flaviviridae, Rhabdoviridae*, and *Togaviridae*. Transmission rates and therefore the numbers of cases typically remain at low levels, unless there is a disruption in system balance such as a climate anomaly, virus mutation to a virulent form, or an ecological disturbance, or large-scale host movement that enables virus amplification to epidemic levels.

Horizontal Transmission

Arthropod vectors, such as mosquitoes, feed on vertebrates to obtain blood which is used as a dietary supplement to stimulate and support egg development. If the vertebrate host is circulating a sufficient quantity of virus within its peripheral blood (i.e., is viremic) and if the arthropod is susceptible to infection, then the arthropod may become infected soon after blood feeding. After the blood meal is digested and the eggs are laid, completing the gonotrophic cycle, the surviving arthropods will then seek a subsequent blood meal to initiate the maturation of the next batch of eggs. The duration of the gonotrophic cycle essentially delineates the frequency of vector–host contact and therefore the number of opportunities for transmission by an arthropod during its lifetime. Horizontal arbovirus transmission occurs after the arthropod becomes infectious and is capable of expectorating virus when it feeds to obtain the next and subsequent blood meals. All arthropods expectorate salivary secretions while blood feeding, and this provides a mechanism to deliver the virus to the host thus completing horizontal transmission. After becoming infectious, most arthropods continue transmitting virus throughout the remainder of their lifetime. West Nile virus (WNV) is an example of a virus maintained by horizontal transmission from bird to mosquito to bird.

Vertical Transmission

Some arboviruses acquired by the arthropod host are transmitted vertically to the next blood-feeding life stage. In ticks that feed once per life stage, transmission is trans-stadial (across life stages) from the larvae or nymphs that acquire the virus, molt, and then pass the virus to nymphs and/or adults that transmit their acquired infection to vertebrate hosts during the next blood meal. Tick-borne encephalitis virus (TBEV) is an example of a virus that is vertically maintained and transmitted within populations of the tick vectors, *Ixodes persulcatus* or *Ixodes ricinus*. In Diptera, acquisition of virus by the host-seeking female may be followed by passage of virus through the egg, larval, and pupal stages to adults in the next generation when transmission to vertebrate hosts is facilitated by infectious females. LaCrosse virus (LACV) is an example of a virus that is maintained vertically by efficient vertical transmission by its primary mosquito host, *Aedes triseriatus*, but is also transmitted horizontally to chipmunk reservoir hosts and occasionally to humans. In contrast, vertical passage of WNV occurs occasionally, but still may be important in virus amplification and persistence. In most arthropods vertical transmission is not 100% efficient, and therefore for a virus to be maintained, amplification by horizontal transmission is necessary to increase the number of infected individuals within the vector population.

Venereal Transmission

Although male mosquitoes and other arthropods may become infected vertically, they usually do not blood-feed and therefore generally are 'dead end' hosts for the viruses that infect them. However, in some instances infected males are able to transmit virus venereally to uninfected females during mating. These females may become infective and then transmit virus horizontally as well as vertically. LACV is also an example of a virus that may be transmitted venereally and then horizontally and vertically by infected females. In contrast, all male tick life stages require a blood meal for molting and therefore can become infected orally. In some instances these males, orally infected as immatures, can transmit virus venereally.

Nonviremic Transmission

When multiple vectors feed in close proximity, virus may be transmitted from infectious to uninfected arthropods without the host developing a viremia. This method of transmission initially was described for TBEV among *Ixodes* ticks that often attach near to one another on the host and require 4–5 days to blood-feed. Recently, however, nonviremic transmission has also been demonstrated for *Culicoides* flies as well as mosquitoes in laboratory experiments.

Types of Transmission Cycles

Epidemiologists have classified virus transmission cycles by the types of vertebrate hosts, with zoonoses maintained by transmission among animal hosts with occasional tangential involvement of humans, and anthroponoses maintained by direct transmission among humans (**Figure 2**).

Zoonoses

Many arboviruses are zoonoses that are maintained and amplified in transmission cycles involving one or more vectors and a variety of animal hosts. When amplification progresses to elevated levels, transmission may spill over to tangentially include domestic animals and/or humans. St. Louis encephalitis virus is an example of a zoonosis that is maintained and amplified by horizontal transmission by mosquitoes among birds, but spills over to tangentially infect humans (**Figure 2**). In this instance, birds serve as maintenance reservoir and/or amplification hosts, because they are susceptible and capable of producing viremias of sufficient titer to infect blood-feeding mosquitoes. Humans are 'dead end' hosts, because they do not produce sufficient viremias to infect additional mosquitoes. Vectors that participate in the basic zoonotic cycle are termed maintenance vectors, whereas those that carry the virus from the enzootic cycle (among animals) to domestic animals or humans are termed 'bridge' vectors. Bridge vectors may not be able to maintain transmission in the absence of more susceptible and effective maintenance vectors. Although in many zoonoses humans are 'dead end hosts', humans still become infected and may develop serious disease.

Anthroponoses

In some circumstances, humans produce a viremia sufficiently elevated to infect vectors and maintain a vector–human transmission cycle independent of animal hosts. Dengue viruses (DENVs) are examples of anthroponotic viruses maintained by human–mosquito–human transmission (**Figure 2**). Although there have been reports of horizontal enzootic and vertical transmission of DENV, these events rarely have been associated with or have led

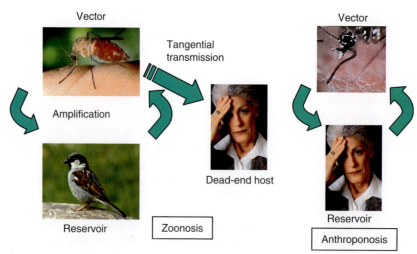

Figure 2 Comparison between a bird-maintained zoonotic virus (such as the St. Louis encephalitis virus) and an anthroponotic virus (such as dengue virus). Activities to decrease or interrupt virus amplification and thereby prevent human cases.

to major epidemics (transmission to or among humans). The distinction between zoonoses and anthroponoses becomes blurred with viruses such as yellow fever virus, in which there is a distinct enzootic primate–mosquito–primate maintenance cycle in jungle habitats and then an epidemic human–mosquito–human cycle in urban habitats.

Factors Enabling Transmission

The efficiency of arbovirus transmission depends upon genetic factors that encode for susceptibility to infection and for biological traits that enable vector survival and repeated susceptible host contact. These properties may be summarized, respectively, as vector competence and vectorial capacity.

Vector Competence

The innate ability of an arthropod to acquire and transmit an arbovirus is termed vector competence. Infection occurs when the ingested virus binds to receptor sites in the lumen of the midgut of the vector and then enters the midgut cells. Susceptibility to infection typically relates to the number of receptor sites and to the affinity of the infecting virus to bind with these sites. Many arthropods lack suitable binding sites, do not get infected, and have what is termed a midgut infection barrier. Others allow midgut infection and perhaps low-level virus replication, but the infection remains midgut-limited and does not disseminate to infect the salivary glands. Post-transcriptional gene silencing by RNA interference (RNAi) is a natural antiviral response in mosquitoes. As a countermeasure, some arboviruses evade RNAi or produce RNAi suppressors that limit antiviral defense. These viruses replicate to high levels and escape the midgut; however, not all infect the salivary glands or they infect the salivary glands but are not transmitted. These salivary gland barriers may be similar to midgut infection and escape barriers and are referred to as salivary gland infection and escape barriers. Still other less-known vector attributes may downregulate infections after or during dissemination, also limiting transmission. In these vectors, transmission peaks soon after infection, but then decreases over time, so that older vectors have lower transmission rates than younger vectors. In general, however, competent vectors frequently become infected and efficiently transmit co-evolved viruses. The risk of infection and transmission by competent vectors increases as a function of vector–host contact which may be measured by the vector's reproductive age.

Vectorial Capacity

In general, vector competence defines the ability of a vector to become infected with and transmit an arbovirus, usually under laboratory conditions. By comparison, vectorial capacity (C) is an epidemiological measure of the force of transmission or the daily case dissemination rate, and combines ecological factors relating to the frequency of reservoir host contact (ma, daily vector density in relation to host abundance, squared to account for refeeding), daily vector survival (P), the duration of the extrinsic incubation period (e), and vector competence (V):

$$C = \frac{ma^2 V(P^e)}{-\ln P}$$

Although originally developed to describe the rate of malaria transmission, this formula has been used to describe arbovirus transmission dynamics or to compare the importance of different vector species. For a zoonosis such as WNV, ma^2 becomes bites per bird per night. Frequently, measures of vector abundance (m, originally the man-biting rate) are combined with measures of vector host selection (a, the anthropophagic index or frequency of feeding on humans) to estimate bites per host per day. For zoonoses these terms must be changed to reflect the maintenance host species and the frequency with which they are fed upon by maintenance vector species. Vector blood-meal host-selection patterns determine the vertebrate hosts most frequently infected, and their susceptibility and viremia response then determines the effectiveness of transmission. In general, vectors with a narrow host range that includes susceptible host species will more effectively and rapidly amplify viruses than those with a broad host range that includes both susceptible and nonsusceptible host species. The latter vector species may be important as bridge vectors; for example, *A. aegypti* is a very effective vector of DENV because it feeds almost entirely on humans, whereas *A. albopictus* is less efficient because of its broad host range.

Daily survival (P) typically is measured by examination of host-seeking female reproductive age or from mark–release–recapture studies. In the current formula, P is considered to be constant throughout the life of the vector; however, some studies indicate that mortality rates may actually decrease as a function of vector age, thereby increasing vectorial capacity. The extrinsic incubation period (e) is discussed in detail below, but essentially is the time from infection to transmission by the vector. Vector competence (V) is the same as defined above and measures the ability of the vector to become infected and transmit virus.

Extrinsic Incubation Period

Early in arthropod-transmitted viral epidemics, there are distinct time periods between episodes of new cases that delineate the 'extrinsic incubation period' or virus development outside of the vertebrate or human host. In reality, this measure of transmission dynamics is related to the time from virus acquisition by the vector until the transmission leads to new cases. Because arthropod body temperature approximates ambient conditions and because virus replication is both temperature limited and dependent, the duration of the extrinsic incubation period changes with latitude and as a function of climate variation, being shortest during summer. Degree–day models can be generated in the laboratory by incubating infected females at different temperatures and measuring the time until they are capable of transmission. Transmission is temperature limited, because viruses have threshold below which they will not replicate. This lower threshold delineates viral distribution in time and space. Theoretically, viruses with a lower (i.e., cool) temperature threshold are expected to have a wider geographic distribution and can be transmitted more effectively at northern latitudes than those with a higher (warm) threshold. Above this lower limit, the rate of viral replication within the vector is dependent upon the temperature of the vector environment and typically increases as an exponential function of ambient temperature. Vector behavior may modify the temperature environment in comparison with ambient conditions, especially during cool periods of the year or day. Species that rest in houses, for example, have a more ameliorated temperature than those that rest outdoors. Nocturnal species that rest in burrows during the day and come out to seek blood-meal hosts at night always enter a warmer environment and therefore their body temperatures may slightly exceed ambient conditions. Both examples of insect behavioral change of body temperature can alter virus replication, transmission rates, and therefore virus amplification.

Phases of Transmission

Transmission dynamics vary over time and space and may be grouped into three phases: introduction or maintenance, amplification or epidemic, and subsidence.

Maintenance

Seasonal changes in vector abundance or the autumnal cessation of blood-feeding behavior limit or interrupt arbovirus maintenance transmission. In general, viruses either persist within host populations or become regionally extinct and require reintroduction. At southern latitudes with sufficient vector abundance and warm temperatures, viruses may be transmitted continually during winter. In southern areas with alternating dry and wet seasons, vector abundance changes related to precipitation also dictate the extent and timing of the transmission season. At northern latitudes mechanisms for local maintenance may include persistent infections within either vector or vertebrate host populations that are acquired during fall and then relapse during the following spring. Alternatively, viruses may become regionally extinct and are reintroduced the following spring. Unfortunately, data supporting most overwintering mechanisms are fairly fragmentary despite years of investigation. Recent studies of the molecular genetics of multiple viral isolations made over time and space show patterns of change indicating mutation or extinction, followed by the introduction and persistence of new genotypes. Often, rather small changes in the viral genome can result in marked changes in virulence that affect fitness and therefore amplification. The changes enhancing virulence may be associated with increased host and geographical range. WNV is an example of a virus that has changed genetically, extended its distribution markedly, and has caused serious epidemics throughout northern latitudes in Europe and North America.

Amplification

After maintenance or introduction at low levels, arboviruses undergo seasonal amplification (or increase in the numbers of infected hosts) during favorable weather periods. The rate of amplification is dependent upon vector competence and vectorial capacity factors, but may be limited by vertebrate host immunity and climate. Acquired immunity or depopulation of the primary vertebrate host population during the previous season or seasons may affect transmission during the subsequent season, because a large percentage of infected vectors transmit virus to immune hosts that do not become infective to other vectors. In short-lived vertebrate hosts, such as house sparrows, population turnover rates are high and elevated herd immunity levels short-lived. However, elevated immunity rates in long-lived hosts, such as humans, may limit infection to younger, nonimmune cohorts. A second factor that may constrain transmission is the effect of climate on vector abundance. Arthropod populations tend to vary markedly in response to climate variation. During years when climate is unfavorable, abundance may remain low and below thresholds necessary to effectively amplify virus. During these years maintenance transmission remains at very low levels or virus may disappear during the current and subsequent seasons. In contrast, rapid vector population increases and longer survivorship during favorable weather periods frequently are accompanied by concurrent increases in virus amplification to epidemic levels.

Subsidence

Epizootic and/or epidemic-level transmission may be relatively short-lived and generally is followed by subsidence. Typically, the amplification curve attains asymptotic epidemic levels and then begins to subside due to decreasing numbers of susceptible hosts, seasonal changes in day length or climate, and/or focused intervention measures. In general, the more intensive and extensive the epidemic transmission, the faster and more widespread the seroconversion or depopulation rates within the primary host population, and the more rapid the focal subsidence. During years with low-level transmission, sufficient naive hosts may remain and provide a source for viral amplification during the following season. Eventually, the numbers of susceptible hosts decrease, slowing the force of transmission and decreasing the numbers of new cases.

Seasonal changes in day length and weather can produce comparable dramatic decreases in epidemic transmission without exhausting the number of available hosts which will remain available for amplification during the following season. At temperate latitudes, the onset of cold weather and shortening photoperiod may induce diapause in the primary vector(s), terminate blood feeding, and interrupt transmission, leading to rapid subsidence. At lower latitudes, comparable seasonality of transmission can be induced by variation in rainfall. Amplification the following season may be dependent upon the intensity of transmission during the end of the previous season and the amount of virus present in the overwintering host populations.

Risk Factors

The risk of infection and disease in vertebrates is related directly to the degree of exposure to host-seeking infectious arthropods and to the vertebrate hosts' immune response to infection.

Residence

Undoubtedly, place of residence markedly affects the risk of infection. Mostly due to the ecology of the arthropod vectors and vertebrate hosts, most arboviruses tend to be focal in their distribution and associated with particular landscapes. Humans often become infected by zoonotic viruses when housing areas are constructed within transmission foci or when human alteration of the landscape creates environments that bring vector and maintenance host together. Rural housing within or adjacent to wooded areas supporting the mosquito *A. triseriatus* has been found to be a critical risk factor for LACV infection in children. In contrast, urban *Culex* populations produced in municipal wastewater systems interfacing with periurban communally roosting American crow populations have led the spatial aggregation of WNV infection risk in humans.

Climate

Climate variation affects temperature and precipitation patterns, mosquito abundance and survival, and therefore arbovirus transmission patterns. These changes often vary at different temporal scales. At temperate latitudes seasonal changes in temperature and photoperiod drive transmission cycles, whereas at tropical latitudes seasonal changes may relate more to wet and dry seasons. The magnitude of annual temperature and precipitation changes may have decadal or shorter cycles based on changes in sea surface temperature, such as the El Niño/southern oscillation change in the Pacific sea surface temperature that markedly alters precipitation and temperature patterns in parts of the Americas. These cycles alter storm tracks that affect mosquito and avian abundance, the intensity and frequency of rainfall events, and groundwater depth.

Age

In the absence of acquired immunity, different viruses seem to exhibit a predilection to cause disease in different host age classes. This may be related to differential biting rates by vector mosquitoes on different host age groups or due to age-related differences in innate immunity. Host-seeking mosquitoes seem to feed more readily on adult than on nestling birds of the same species. With regard to illness, for example, western equine encephalomyelitis virus (WEEV) tends to cause more frequent and severe disease in infants (<1 year), whereas SLEV causes more severe illness in the elderly (>65 years). For endemic viruses, the prevalence of previous infection increases as a function of age, and therefore older cohorts tend to be protected by a greater prevalence of acquired immunity than younger cohorts. This age-related pattern of acquired immunity can change the apparent-to-inapparent ratio of clinical disease for endemic viruses.

Occupation

Different occupations place workers in different environments at different times, and therefore vary their risk for mosquito and arbovirus exposure. For example, farm workers who pick and pack vegetables outdoors at night in southern California were found to have an 11% seroprevalence rate against SLEV, whereas persons residing within Los Angeles were found to have a lower 1.7% exposure rate. In contrast, persons sitting at home in late afternoon in Thailand have a high attack rate by *A. aegypti* and a high exposure rate to dengue viruses, compared to persons working outdoors from morning to night in agricultural professions.

Socioeconomic Status

Historically, socioeconomic status has been related closely to the distribution of cases during urban epidemics. Homes and municipal drainage systems frequently are not well maintained in low-income neighborhoods, and this has shown to be related to the distribution of human cases. In the tropics, lack of a municipal water system and domestic

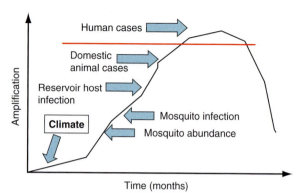

Figure 3 Surveillance indicators used to measure enzootic virus amplification and forecast the risk of human infection. Climate variation can provide the earliest indication of pending risk and may drive the rate of viral amplification.

water storage has been linked strongly to *A. aegypti* abundance and to disease risk.

Forecasting Risk

The cascade of different surveillance measures becoming positive over time as virus amplifies can be used to forecast the risk of human infection, especially for endemic zoonoses. **Figure 3** shows the approximate relative times when surveillance measures might become positive along the enzootic amplification curve for a virus such as WEEV. Usually, human cases are the last factor to be detected and tangential transmission occurs only after considerable WEEV amplification in the mosquito–bird cycle. These surveillance data can be used to focus on mosquito control.

See also: Animal Rhabdoviruses; Epidemiology of Human and Animal Viral Diseases.

Further Reading

Black WC, IV and Moore CG (2006) Population biology as a tool for studying vector-borne diseases. In: Beaty BJ and Marquardt WC (eds.) *The Biology of Disease Vectors*, pp. 393–416. Niwot, CO: University Press of Colorado.

Edman JD (2000) Arthropod transmission of vertebrate parasites. In: Eldridge BF and Edman JD (eds.) *Medical Entomology*, pp. 151–164. Dordrecht, The Netherlands: Kluwer.

Eldridge BF (2000) The epidemiology of arthropodborne diseases. In: Eldridge BF and Edman JD (eds.) *Medical Entomology*, pp. 165–186. Dordrecht, The Netherlands: Kluwer.

Hardy JL, Houk EJ, Kramer LD, and Reeves WC (1983) Intrinsic factors affecting vector competence of mosquitoes for arboviruses. *Annual Review of Entomology* 28: 229–262.

Reisen WK (2002) Epidemiology of vector-borne diseases. In: Mullen GR and Durden LA (eds.) *Medical and Veterinary Entomology*, pp. 16–27. New York, NY: Academic Press.

Vector Transmission of Plant Viruses

S Blanc, INRA–CIRAD–AgroM, Montpellier, France

© 2008 Elsevier Ltd. All rights reserved.

Glossary

Fitness The relative ability of an individual (or population) to survive and reproduce in a given environment.

Helper component (HC)-transcomplementation An HC encoded by a viral genome X mediates the vector transmission of a virus particle containing a viral genome Y.

Horizontal transmission The transmission of a virus, parasite, or other pathogen from one individual to another within the same generation, as opposed to vertical transmission.

Pierce-sucking insects Insects adapted to sap or blood feeding, with the mouthparts transformed into long chitin needles that can pierce and penetrate tissues and allow pumping up their content.

Quasispecies Ensemble of mutant viral genomes constituting a viral population.

Vector Organism acquiring a pathogen on an infected host and inoculating it in a new healthy one.

Vertical transmission The transmission of a pathogen from the parent(s) to the offspring, usually through the germline.

Introduction

Viruses are intracellular parasites diverting the host cellular machinery for their own replication and offspring particles production. As such, they most often negatively affect the hosting cells, sometimes even killing them, and hence repeatedly and unavoidably face the problem of moving on and colonizing new healthy and potent 'territories'. Within a single host, viruses can both diffuse from cell to cell and be transported on longer distances by the vascular system. While animal viruses use membrane fusions (if enveloped) or membrane receptors to penetrate healthy cells, plant virus entry during the host invasion is always resulting from a passage through 'tunnels' traversing the cell wall, called plasmodesmata, and ensuring a cytoplasm continuity between adjacent cells. Any viral population can grow this way only until the physical limits of the host are reached. Then, a critical passage in the 'outside world' separating two compatible hosts has to be successfully achieved. Because animals are motile and often come in contact, some associated viruses can directly access either blood or permissive tissues of a healthy host and operate a cell entry resembling that involved during invasion of single hosts. However, a most frequently adopted strategy relies on additional organisms, capable of sampling the virus population within an infected host, transporting, disseminating, and efficiently inoculating infectious forms of this virus within host population. Such organisms are designated vectors, giving rise to the term vector transmission. Vector transmission is found frequently in animal viruses and, presumably due to stable hosts and to the need of covering considerable distances between them, has been adopted by the vast majority of plant viruses. Each virus species is submitted to different ecological conditions; hence, an impressive complexity of host–virus–vector interactions has been unraveled over a century of research efforts. The object of this chapter is to synthesize the knowledge available at present in the field of vector transmission of viruses, with a special emphasis on plant viruses, where a great diversity of strategies have been discovered and documented. Indeed, the numerous patterns of vector transmission described for plant viruses include all those reported in animal viruses and many more.

Plant Virus Vectors

Any organism that is creating a break into the cell wall, either for penetrating a plant or simply for feeding on it, and that is capable of covering the distance between two separated plants, can possibly be used as a vector by viruses, for traveling through space and time. Vectors have been described in groups of organisms as diverse as parasitic fungi, nematodes, mites, and most importantly insects (**Table 1**). The pattern of virus uptake, preservation, transport, dissemination, and inoculation can be very

Table 1 Vectors and mode of transmission in families of plant viruses

Family[a]	Vector	Mode of vector-transmission[b]
Bromoviridae genus Alfamovirus	Aphids	Noncirculative capsid strategy
Bromoviridae genus Cucumovirus	Aphids	Noncirculative capsid strategy
Bromoviridae genus Ilarvirus	Thrips	?
Bromoviridae genus Oleávirus	?	?
Bromoviridae genus Bromovirus	Beetle	?
Bunyaviridae	Thrips, planthopper	Circulative propagative
Caulimoviridae	Aphid, mealybug, leafhopper	Noncirculative helper strategy
Circoviridae	Aphid	Circulative nonpropagative
Closteroviridae	Aphid, whitefly, mealybug	Noncirculative
Comoviridae genus Comovirus	Beetle	?
Comoviridae genus Fabavirus	Aphid	Noncirculative
Comoviridae genus Nepovirus	Nematode	Noncirculative capsid strategy
Geminiviridae	Leafhopper, whitefly	Circulative nonpropagative[c]
Luteoviridae	Aphid	Circulative nonpropagative
Partiviridae	?	?
Potyviridae genus Potyvirus	Aphid	Noncirculative helper strategy
Potyviridae genus Ipomovirus	Whitefly	Noncirculative
Potyviridae genus Macluravirus	Aphid	Noncirculative
Potyviridae genus Rymovirus	Mite	Noncirculative
Potyviridae genus Tritimovirus	Mite	Noncirculative
Potyviridae genus Bymovirus	Fungus	Circulative
Reoviridae	Planthopper, leafhopper	Circulative propagative
Rhabdoviridae	Leafhopper, aphid	Circulative propagative
Sequiviridae	Aphid, leafhopper	Noncirculative helper strategy
Tombusviridae	Fungus	Noncirculative

[a]The families are broken down to the genus level when they contain genera with totally different vectors and mode of transmission.
[b]The helper or capsid strategies (see **Table 2**) are mentioned when experimentally demonstrated for at least one of the member species. When no complement is added to either circulative or noncirculative, it reflects the lack of further information.
[c]For at least one member species (*Tomato yellow leaf curl virus*, TYLCV), replication within the vector is still being debated.
The noncirculative viruses, or assimilated as discussed in the text, are in blue. The circulative viruses, or assimilated as described in the text, are in green.

different, due to the specific biology of all three (plant, virus, and vector) partners. However, viruses transmitted by 'pierce-sucking' insects are quantitatively predominant, and the classification established for their various modes of transmission is widely used as a reference for comparison with others. For this reason, hemipteran insect transmission will be described first in details and succinctly compared later on with that by other types of vectors.

Transmission of Plant Viruses by Insects

History of the Classification of the Different Modalities of Transmission

The transmission of plant viruses has been investigated for over a century with the most common vectors being sap-feeding insects with pierce-sucking mouth parts, particularly aphids, and also whiteflies, leafhoppers, planthoppers, and mealybugs. Pioneer studies have demonstrated the complexity and diversity of the interactions between plant viruses and their insect vectors. Even as late as the 1950s, scientists, using the tools at hand, were merely measuring quantitative traits such as the time required for virus acquisition on infected plants and the time during which the virus remained infectious within the vector. Three categories were then defined: (1) the nonpersistent viruses, acquired within seconds and retained only a few minutes by their vectors; (2) the semipersistent viruses, acquired within minutes to hours and retained during several hours; and (3) the persistent viruses that require minutes to hours for acquisition, and can be retained for very long periods, often until the death of the vector. It is important to note that, though the classification and terminology have changed in the last decades, these categories are still used by a number of authors, and thus often encountered in the literature.

In an early study on 'nonpersistent' viruses, the transmission of potato virus Y was abolished by chemical (formaldehyde) or ultraviolet (UV) treatments of the extremity of the stylets of live viruliferous aphids, demonstrating that infectious virus particles were retained there. It was first believed that the transmission of 'nonpersistent' viruses could be assimilated to mechanical transmission, stemming from nonspecific contamination of the stylets, the vector acting simply as a 'flying needle'. Consistent with this was the repeated demonstration that 'nonpersistent' viruses are lost upon moulting of the viruliferous vectors. Later on, the hypothesis of virus uptake during sap ingestion and inoculation during putative regurgitation led to a change from vectors as flying needles to vectors as 'flying syringes', the virus–vector relationship still being considered as nonspecific (**Figure 1**). It is interesting that, while in plant viruses recent data unequivocally convinced the scientific community that the situation is much more complex, likely involving specific receptors in vectors for specific virus species (see the next section), in animal viruses very few studies are available at present and this mode of transmission is still referred to as 'mechanical vector transmission'. The prime conclusion from these experiments is that the virus–vector association occurs externally, on the cuticle lining the food or salivary canal in the insect stylets. Because semipersistent viruses are also lost upon vector moulting, their association with the vector was also proposed to be external, likely in the stylets, though a

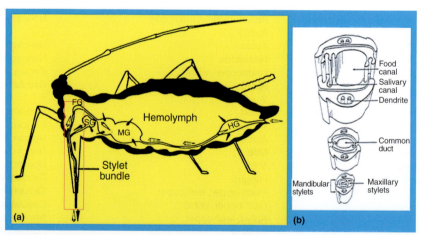

Figure 1 Different routes of plant viruses in their aphid vectors. (a) The white arrows represent the ingestion of circulative viruses, whereas the black arrows materialize their cycle within the aphid body, and inoculation in a new host plant. The red square area indicates the region of the anterior feeding system, where noncirculative viruses are retained in their vectors. (b) Cross sections of the stylet bundle illustrating the inner architecture of maxillary stylets which defines interlocking structures, food canal and salivary canal, fused at the distal extremity into a single common duct, where most noncirculative viruses are thought to be retained (see text). Adapted from Taylor CE and Robertson WM (1974) Electron microscopy evidence for the association of tobacco severe etch virus with the Maxillae in Myzuspersical (Sulz.). *Journal of Phytopathology* 80: 257–266.

possible location 'upstream', on the cuticle lining the anterior gut of the insect, was also proposed in some cases.

In sharp contrast, many persistent viruses were observed within the vector body by electron microscopy, in various organs and tissues, indicating an internal association with the vectors. Such viruses were shown to pass through the gut epithelium into the hemolymph and join the salivary glands to be ejected together with saliva (**Figure 1**). A latent period of hours to days after acquisition, during which the virus cannot be efficiently inoculated, is consistent with the time needed for completing this cycle within the vector body. Moreover, microinjection of purified persistent viruses within the insect hemolymph subsequently resulted in efficient transmission to new healthy plants, proving that virus within the vector body can get out and be inoculated to host plants.

Altogether, these results prompted a revision of the classification of the modes of transmission in the late 1970s, based on qualitative criteria and still valid today (**Table 2**). The non- and semipersistent viruses were grouped in a new category designated 'noncirculative viruses', and the persistent viruses were named 'circulative viruses'. While circulative animal viruses (arboviruses) in fact infect their vectors where they efficiently replicate, some circulative plant viruses can seemingly operate their cycle in the vector body without any cell infection and replication. Hence, the category 'circulative' has been broken down into the two subcategories: 'propagative' and 'nonpropagative'. The various families and/or genera of plant viruses and their associated vectors and modes of transmission are listed in **Table 1**.

During the last decades, the implementation of molecular and cellular biology has provided invaluable tools for studying the molecular mechanisms of virus–vector interaction. The data currently available for each category are summarized in the following subsections.

Circulative Transmission

Logically, circulative viruses are ingested by vectors, while feeding on infected plants. Some viruses are limited to phloem tissues, which the insect vector can reach within minutes to hours depending on the species, which explains the long feeding period required for their acquisition. As schematized in **Figure 1(a)**, the viruses cross the mid- or hindgut epithelium, are released into the hemolymph, and can then adopt various pathways to traverse the salivary glands, and be released in their lumen, wherefrom they will be inoculated upon salivation into healthy hosts. During this basic cycle, the virus encounters and must overcome diverse cellular barriers, where the existence of specific virus–vector interaction has long been established experimentally, though specific receptors have not been identified so far.

Propagative transmission

Propagative transmission of plant viruses is the homolog of that of arboviruses in vertebrates. Members of the virus families *Rhabdoviridae*, *Reoviridae*, *Bunyaviridae*, and the genus *Marafivirus* are transmitted this way. In compatible virus–vector associations, after infecting the gut epithelium, virus particles are released in the hemocoel cavity and colonize various organs and tissues of the vector, including, ultimately, the salivary glands. The viruses can either diffuse in the hemolymph and concomitantly infect different organs, or follow a constant pattern of spread from organ to organ, as demonstrated for rhabdoviruses which move in (and spread from) the central

Table 2 Different modes of plant virus transmission by insects with pierce-sucking mouth parts

	Circulative		Noncirculative	
Transmission modes[a]	Propagative	Nonpropagative	Capsid strategy	Helper strategy
Acquisition time[b]	Minutes to hours	Minutes to hours	Seconds to hours	Seconds to hours
Retention time[c]	Days to months	Days to months	Minutes to hours	Minutes to hours
Inoculation time[d]	Minutes to hours	Minutes to hours	Seconds to minutes	Seconds to minutes
Association with vectors[e]	Internal	Internal	External	External
Replication in vectors	Yes	No	No	No
Requirement of an HC[f]	No	No	No	Yes

[a]These modes of transmission were established and are widely accepted for virus transmission by pierce-sucking insects. As discussed in the text, they sometimes also apply to other types of vector.
[b]The length of time required for a vector to efficiently acquire virus particles upon feeding on an infected plant.
[c]The length of time during which the virus remains infectious within its vector, after acquisition.
[d]The length of time required for a vector to efficiently inoculate infectious virus particles to a new healthy plant.
[e]Internal means that the virus enters the inner body of its vector, passing through cellular barriers. External means that the virus binds the cuticle of the vector and never passes through cellular barriers.
[f]A helper component (HC) is involved in cases where the virus particles do not directly recognize vectors, acting as a molecular bridge between the two.

nervous system. Within the vector, all these cases are very similar to genuine infection of an insect host; hence, it is difficult to decide whether insects are proper vectors of propagative viruses, or should rather be considered as alternative hosts. Apart from the genus *Marafivirus*, all propagative viruses are in families that comprise viruses that infect animals, suggesting that they might have evolved from insect viruses by secondarily acquiring the capacity to replicate in plants.

Nonpropagative transmission

This particular association between insect vectors and plant viruses is reasonably well understood only for members of some plant virus species in the family *Luteoviridae*. Such viruses have developed original mechanisms of viral transport, both when passing through gut and salivary gland barriers, and when traveling into the hemocoel cavity.

The cycle of luteoviruses within their vector body involves specific ligand-receptor-like recognition at the cell entry of both the gut epithelium and the salivary glands. While viral ligands are known to be structural proteins of the coat and extension thereof, very little is known about the putative counterpart receptors on the cell membranes of the vectors. Despite this lack of full understanding of the molecular process, many electron microscopy and molecular studies have determined in detail the route of luteovirus particles within the vector, across cellular layers (**Figure 2**). Once the virus reaches either the apical membrane of the gut epithelium, or the basal membrane of the accessory salivary gland cells, and attaches to the specific receptors, it provokes invagination of the plasmalemma, forming small coated virus-containing vesicles. Soon after budding, the coated vesicles deliver the virus particles to a larger uncoated membrane endosomal compartment. Interestingly, luteoviruses mostly escape the route of degradation of internalized material ending in lysosomes. Instead, the virus particles become concentrated in the endosomes, and *de novo* elongated uncoated vesicles are repacked, transporting the viruses to the basal or apical membrane, in gut and accessory salivary gland cells, respectively. The elongated vesicles, which contain rows of virions (**Figure 2**), finally fuse with plasma membranes and release the virus either into the hemocoel cavity or into the lumen of the salivary ducts. Despite these extensive searches of luteovirus particles in their insect vectors, by several independent groups of scientists, no particles have been observed in any organ other than the gut or the accessory salivary glands. Furthermore, classical monitoring of the viral titers within the vector, for assessing viral replication, failed to demonstrate an increase over time. Consequently, it is generally accepted that virus particles, either included in membrane vesicles or suspended in the hemolymph, never come in contact with

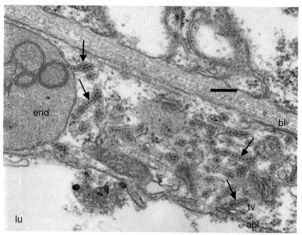

Figure 2 Transcytosis of cucurbit aphid-born yellows virus (family *Luteoviridae*) in hindgut cell of the aphid vector *Myzus persicae*. Luteovirus particles present in the gut lumen (lu) are internalized from the apical plasmalemma (apl) and transported to the basal lamina (bl) in a complex pattern involving different vesicular structures, described in the text. A network of uncoated tubular vesicles is visible (tv indicated by arrows), sometimes connected to the endosome (end). The bar represents 100 nm. The photograph is provided by Catherine Reinbold and Véronique Brault (INRA, Colmar, France). Reproduced from Blanc S (2007) Virus transmission – getting out and in. In: Waigmann E and Heinlein M (eds.) *Viral Transport in Plants*, pp. 1–28. Berlin: Springer, with kind permission of Springer Science and Business Media.

the cell cytoplasm, thus precluding any possibility of viral replication.

Virus transfer into the hemolymph is believed to occur by passive diffusion. However, the possible impact of the insect immune system at this step of the virus life cycle is often discussed. A study demonstrated that a major protein of the aphid hemolymph, the symbionin, was mandatory for efficient luteovirus transmission. The symbionin is a homolog of the *Escherichia coli* chaperone GroEL, secreted in aphids by endosymbiotic bacteria of the genus *Buchnera*. Eliminating symbiotic bacteria, and thus symbionin, by antibiotic treatments significantly reduced the aphid efficiency as a vector. Consistently, direct evidence of a physical interaction between symbionin and luteovirus particles has been detected in several viral species, and virus mutants deficient in symbionin binding are poorly transmitted. Two hypotheses can explain the positive action of symbionin and are still debated, since no direct proof could be experimentally obtained: (1) it exhibits protective properties, masking the virus to the immune system and maintaining its integrity during transfer through the hostile hemolymph environment, or (2) its putative chaperon activity ensures correct folding facilitating transfer into the salivary glands.

A similar phenomenon was later demonstrated for other circulative nonpropagative viruses (in the family *Geminiviridae*) by other vectors (whiteflies), suggesting that symbionin participation may be a general phenomenon in this mode of transmission.

Noncirculative Transmission of Plant Viruses

This mode of transmission is by far the most frequently encountered in plant viruses and concerns over 50% of the viral species described to date. Noncirculative viruses do not enter the body of their vectors. They simply attach to receptor sites located externally on the cuticle lining the anterior part of the digestive tract, most often the alimentary/salivary canals within the stylets or the foregut region, and wait until the vector has moved to another plant, where they contrive to be released to initiate a new infection. It has often been proposed that difference between non- and semipersistent viruses, in acquisition and retention time within the vector, was due to a differential location of the binding sites, the former being retained in the stylets and the latter higher up in the foregut. This distinction, however, is not experimentally supported at present, and these two categories should be considered with caution (they are no longer included in **Table 2**). Because noncirculative viruses are adsorbed on the vector cuticle during sap ingestion, and released in many cases during salivation, the favored but still hypothetical location for the retention sites is at the distal extremity of the maxillary stylets, where a single common salivary/food canal is serving for both sap uptake and saliva ejection (**Figure 1(b)**). The best-studied cases of noncirculative transmission are those of viruses in the genera *Cucumovirus*, *Potyvirus*, and *Caulimovirus*, transmitted by various aphid species. Mutagenesis studies of the viral proteins interacting with the vector's mouthparts have clearly indicated that some single-amino-acid substitutions can specifically abolish the virus transmissibility by certain vector species but not by others. Such results, obtained both with cucumber mosaic virus and cauliflower mosaic virus, demonstrate a very specific virus–vector recognition, and strongly suggest the existence of a receptor in the insect mouthparts. Unfortunately, here again and despite the quantitative importance of noncirculative transmission in plant viruses, the viral ligands are well characterized (see below), but the receptors on the insect cuticle stand as the major black box in this field of research, their chemical nature and even their precise location remaining largely hypothetical.

The viral protein motifs directly involved in the attachment to vector putative receptors have been characterized in a number of cases. The frequent occurrence of both transmissible and nontransmissible isolates in the same virus species has greatly facilitated the identification of viral gene regions involved in vector transmission, by simple sequencing of viral genomes, and reverse genetic approaches have been successfully used for definitive confirmation. These investigations clearly revealed two distinct viral strategies for controlling the molecular mechanisms of virus–vector association (**Figure 3**), which represent the currently preferred subcategories in noncirculative transmission (**Table 2**).

The helper strategy

The development of artificial feeding of insect vectors, through stretched parafilm membranes, made it possible to assess the transmission of purified virus particles. The primary striking result was that, in most cases, purified virions are not transmissible, suggesting the requirement of an additional component that is probably eliminated during purification. Sequential feeding on plants infected with transmissible and nontransmissible isolates of such virus species showed that the former produced a compound that could be acquired by the vector, and subsequently mediated the transmission of the nontransmissible isolate. This compound, first described in potyviruses and caulimoviruses, was designated helper component (HC). In both cases, the HC was later purified and demonstrated to be a viral nonstructural protein produced upon plant cell infection, named P2 in caulimoviruses and HC-Pro in potyviruses. P2 and HC-Pro are responsible for specific recognition of the vector. They can be acquired on their own by the aphid and bind to putative receptors within its mouthparts. They exhibit a separated domain, which specifically binds homologous virus particles, thus creating a molecular bridge between virus and vector (**Figure 3**). It is important to realize that HCs can be acquired on their own by the vector, subsequently scavenging virions from homologous species acquired in other locations of the same plant, or even in other plants. This phenomenon extending the assistance between related viral genomes during vector transmission has been termed HC-transcomplementation (**Figure 3**), and may have some important implications in the population genetics and evolution of a virus species (discussed below).

Capsid Strategy

However, for some viruses transmitted by aphids, particularly in the genera *Cucumovirus*, *Alfamovirus*, and *Carlavirus*, purified viral particles are readily acquired and transmitted by the vector. This indicated unequivocally that the coat protein of members of the species must be capable of direct attachment to vector receptors (**Figure 3**). An experiment confirming this conclusion was the demonstration that tobacco mosaic virus, which is not transmitted by any vector, can be transmitted by aphids when its RNA genome is encapsidated *in vitro* into the coat protein of an aphid-transmissible cucumovirus. Recently, the amino acid positions in the coat protein of cucumber mosaic virus, involved in binding to the putative receptors of aphids, have been identified. Their substitution differentially affects the efficiency of transmission by distinct

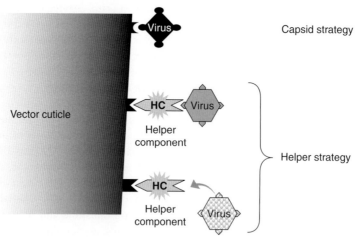

Figure 3 Two molecular strategies for virus–vector interaction in noncirculative transmission of plant viruses. Both strategies allow the retention of virus particles in the vector mouth parts or foregut on putative receptors located at the surface of the cuticular lining. In the capsid strategy, a motif of the coat protein is able to directly bind to the vector's receptor. In the helper strategy, virus–vector binding is mediated by a virus-encoded nonstructural protein (HC), which creates a reversible 'molecular bridge' between the two. HC can be acquired alone, prior to the virion, and thereby allows HC-transcomplementation. In this case, an HC encoded by a genome X (for instance that encapsidated in the gray virion) can subsequently assist in the transmission of a genome Y of the same population, encapsidated in the dotted virion. This possible sequential acquisition of HC and virion is symbolized by the arrow. Reproduced from Froissart R, Michalakis Y, and Blanc S (2002) Helper component-transcomplementation in the vector transmission of plant viruses. *Phytopathology* 92: 576–579, with permission from The American Phytopathological Society.

aphid species, definitely confirming that no HC is required for virus/vector interaction.

Transmission by Beetles

The only cases of insect transmission diverting clearly from one of the patterns detailed above is that described in insect with 'biting-chewing' mouthparts, the best documented examples being beetles, though other insects have a similar feeding behavior.

In contrast to insect with 'pierce-sucking' mouthparts, the beetle feeding is damaging tissues and killing cells, implying an obligate translocation of deposited viruses toward adjacent live cells where they can initiate infection. Beetles have long been reported to acquire and retain viruses from numerous virus species, usually those with highly stable virus particles, and to release them together with the regurgitation of disrupted plant material, lubricating the mouthparts during feeding on new host plants. Surprisingly, however, only some rare species are actually successfully transmitted. A high amount of RNase activity has been found in the beetle regurgitation liquid, and shown to block infection by non-beetle-transmissible viruses. Hence, those viral species that are efficiently transmitted are likely capable of translocation in the vascular system, and/or transfer to unwounded cells, away from the RNase activity. Some virus particles have been detected within the hemolymph of their beetle vector, only seconds after acquisition on infected plants. No correlation between the presence of virions in the hemolymph and success of transmission has yet been established, precluding a possible comparison with the circulative or noncirculative transmission described above.

Transmission by Noninsect Vectors

Although quantitatively less important, noninsect vectors have been identified in several plant virus species. Some are arthropod mites with a morphology distantly related to that of insects, others are totally unrelated organisms such as nematodes or even fungi. However, in many instances, a comparison with the mode of transmission defined in **Table 2** remains possible, demonstrating that this classification can be used as a general basis applicable to all cases of vector transmission.

Transmission by Mites

The transmission of plant viruses by mites has been far less studied, and the molecular and cellular mechanisms of virus–vector relationships remain poorly understood. Several different viral species have been studied and the results suggest the existence of different types of interactions with their respective vectors. This variety is illustrated by two examples: (1) the presence of massive amounts of bromegrass mosaic virus (BMV, family *Bromoviridae*) particles within cells of the gut epithelium of the insect *Eriophyes tulipae* seems to indicate that the virus replicates within the vector, resembling circulative-propagative transmission; (2) wheat streak mosaic virus (WSMV, family *Potyviridae*) appears to be noncirculative. WSMV was recently demonstrated to interact with its mite vectors through an HC-Pro homolog, thus likely suggesting a helper strategy. Further investigations on additional viral species are required to evaluate the diversity of the modes of transmission found with mite vectors, but it seems likely that they will be closely related to those described for insect vectors in **Table 2**.

Transmission by Nematodes

The transmission by nematode vectors is particularly interesting. Such vectors are moving very slowly in the soil (one or a few meters per year), and cannot disseminate the transmitted virus over long distances. Instead, they retain virus particles, for several months or even years, and usually transmit them in the same location but from plants of a growing season n to plants of the growing season $n + 1$. Consequently, nematode transmission is resulting in a time travel for the virus, rather than a space travel.

Nematodes feed by piercing root cells with protractile stylets, constituting the anterior part of their feeding apparatus, and by ingesting the cell content together with viruses when host plants are infected. In all cases described in the literature, the virus particles are adsorbed on the cuticle lining the foregut and are presumably released from there when inoculated in a new host plant. Since viruses are lost upon moulting of the nematode, the virus–vector association is external and this mode of transmission can be assimilated to the noncirculative transmission. Interestingly, the two subcategories of noncirculative transmission also appear to be represented in nematode transmission. Indeed, among the viruses belonging to two viral genera using nematodes as vectors, *Nepovirus* and *Tobravirus*, viruses in the former genus have been shown to directly attach to the retention sites through their coat protein (capsid strategy), whereas in the latter genus, viruses produce a nonstructural protein which allows transcomplementation (see the helper strategy in **Figure 3**). As in other virus–vector associations, the viral proteins involved in vector recognition were identified, but the putative receptor in the nematodes remains unknown.

Transmission by Fungi

Parasitic fungi are coming in contact with their host plants under the form of motile zoospores, which can digest the root cell wall and penetrate into the cytoplasm, from where they will colonize the whole plant. Two different patterns of virus transmission exist. Some viruses, for instance in the genus *Bymovirus*, are present within the fungus cytoplasm early during formation of the future zoospores in infected plant cells. They will remain inside the zoospore until its cytoplasm is injected in the next host cell. In other cases, for instance in cucumber necrotic virus (genus *Tombusvirus*), the best-studied example of a fungus-transmitted virus, virions are specifically retained at the surface of the zoospore envelope, and inoculated into the plant upon cell wall digestion and fungal penetration. In this case, the receptors of the vector were partially characterized. They have been shown to be distributed at the surface of the zoospore of *Olpidium bornovanus*, and their chemical nature was identified as a glycoprotein.

Altogether, the transmission of plant viruses by fungi can tentatively be regarded as homologous to the categories noncirculative, for those viruses associated to the external coat of the zoospore, and circulative for those internalized in its cytoplasm. Unfortunately, not enough information is available to definitely decide whether viral replication is effective within the fungus, and hence whether both propagative and nonpropagative subcategories are present.

Transmission and Evolution

One aspect of vector transmission that is rarely considered in the literature is its evident impact on the population genetics and evolution of viruses. Indeed, while vector transmission is always regarded as an excellent means of maintenance and spread within host populations, it also implies that the virus is able to solve certain problems. First of all, one must bear in mind that a virus population rapidly accumulates very numerous mutations and develops a swarm of genome variants generally designated 'quasispecies'. In such a genetic context, the virus sample collected, transported, and inoculated by the vector will be highly variable. This sample, which gives rise to a new population after each round of vector transmission, will obviously condition the composition of the viral lineage, and thus its evolution. In this regard, two key parameters must be considered:

1. The distribution of viral variants within the host plant determines what is actually accessible for sampling by the vector. If variants are homogeneously distributed within the whole plant, several can be acquired together by the vector. In contrast, if they are physically isolated in different cells, organs, or tissues, only one or few will be collected.
2. The pattern of the virus acquisition by the vector can dramatically change the composition of the transmitted viral sample. When a given vector collects the virus all at once in a single location, the diversity of genome variants transmitted tends to be restricted. On the other hand, collecting the viral sample bits by bits upon successive probing in different locations of the infected host will increase the genetic diversity of the pool of viral genomes transmitted.

It has been repeatedly demonstrated experimentally that successive genetic bottlenecks imposed in viral lineages are drastically decreasing the mean fitness of RNA virus populations, sometimes even driving them toward extinction. One can then easily conceive that viruses will develop strategies to avoid or compensate for genetic bottlenecks during their life cycle. It is highly probable that different strategies of vector transmission found in plant viruses have different impacts on the fitness and evolution of viruses, through the different pools of genome variants that are collected and inoculated. For instance, in the helper strategy, it has been suggested that HC-transcomplementation (**Figure 3**) allows the vector to sample viruses in several steps and in several locations of the host, thus opening the genetic bottleneck, and perhaps attenuating its detrimental effect.

Future Prospects

From a mechanistic standpoint, future advances will mainly concern the characterization of the vector receptors that are used by viruses to ensure successful transmission. Indeed, none of these molecules have been identified so far, in any vector species and category of transmission. Apart from the example of the fungus-transmitted CNV described above, not even their chemical nature and precise location have been established.

From an evolutionary standpoint, two aspects have been mostly neglected in spite of their undeniable importance: (1) the size of the virus sample transmitted by a given vector remains largely unknown in most cases, (2) the distribution of the genome variants, constituting the swarm of mutants of the viral quasispecies, within different cells, organs, and tissues of the host plant is largely unresolved, precluding any evaluation of what genetic diversity from a given viral population is actually accessible to the vectors.

See also: Vector Transmission of Animal Viruses.

Further Reading

Blanc S (2007) Virus transmission – getting out and in. In: Waigmann E and Heinlein M (eds.) *Viral Transport in Plants*, pp. 1–28. Berlin: Springer.

Froissart R, Michalakis Y, and Blanc S (2002) Helper component-transcomplementation in the vector transmission of plant viruses. *Phytopathology* 92: 576–579.

Gray S and Gildow FE (2003) Luteovirus–aphid interactions. *Annual Review of Phytopathology* 41: 539–566.

Ng JC and Falk BW (2006) Virus–vector interactions mediating non-persistent and semi-persistent transmission of plant viruses. *Annual Review of Phytopatholy* 44: 183–212.

Power AG (2000) Insect transmission of plant viruses: A constraint on virus variability. *Current Opinion in Plant Biology* 3(4): 336–340.

Rochon D, Kakani K, Robbins M, and Reade R (2004) Molecular aspects of plant virus transmission by olpidium and plasmodiophorid vectors. *Annual Review of Phytopathology* 42: 211–241.

Taylor CE and Robertson WM (1974) Electron microscopy evidence for the association of tobacco severe etch virus with the Maxillae in Myzuspersical (Sulz.). *Journal of Phytopathology* 80: 257–266.

Whitfield AE, Ullman DE, and German TL (2005) Tospovirus–thrips interactions. *Annual Review of Phytopathology* 43: 459–489.

Vegetable Viruses

P Caciagli, Istituto di Virologia Vegetale – CNR, Turin, Italy

© 2008 Elsevier Ltd. All rights reserved.

Glossary

Incubation The time period from the initial infection (inoculation) to the manifestation of the disease (symptoms). For virus diseases of vegetables, incubation can vary between 2 and 40 days, most commonly between 1 and 3 weeks.

Meristem A tissue in plants consisting of undifferentiated cells, found in zones of the plant where growth (cell division) can take place (meristematic dome). The term 'meristem tip' is used to denote the meristem dome together with 1–2 primordial leaves and measuring between 0.1 and 0.5 cm in height.

Roguing The practice of removing diseased (or abnormal) plants from a crop.

Vector An organism capable of transmitting a pathogen from one plant host to another by establishing specific relationships with the pathogen itself.

Vegetables

The plant species/subspecies cultivated as vegetables are close to 200 and belong to more than 30 botanical families, representing, with fruits, about 35% of the total food derived from plants (including cereals, vegetable oils, and sugar crops).

Viruses

According to different estimates, 3–5% of overall vegetable production is lost to virus infections, but losses can be occasionally very high, and are normally higher where pest control is lower, for example, in developing countries.

Keeping in mind that viruses of potato, sweet potato, cassava, and legumes are not considered in this article, as they are dealt with elsewhere in this encyclopedia, a total of about 200 viruses (excluding cryptic viruses, satellites, and viroids) have been reported as naturally infecting 67 of the 129 vegetables considered here. These viruses belong to 30 genera in 15 families and to nine genera unassigned to families; eight viruses are in three families, but have not been assigned to genera, and three more viruses have not been assigned to any taxa (**Table 1**).

Table 1 Taxonomic assignment of vegetable viruses, their numbers per genus, and their vectors

Genome type	Family	Genus	No. of virus species	Vectors
Single-stranded DNA	Geminiviridae	Begomovirus	35	Whiteflies
		Curtovirus	2	Leafhoppers
		Mastrevirus	1	Leafhoppers
	Nanoviridae	Babuvirus	1	Aphids
Double-stranded DNA RT	Caulimoviridae	Badnavirus	4	Mealybugs
		Caulimovirus	1	Aphids
Double-stranded RNA	Reoviridae	Fijivirus	1	Planthoppers
Single-stranded, negative-sense RNA	Rhabdoviridae	Cytorhabdovirus	2	Aphids
		Nucleorhabdovirus	2	Aphids
			1	Leafhoppers
		Unassigned in the family	2	Aphids
			1	Leafhoppers
			2	Not known
	Bunyaviridae	Tospovirus	3	Thrips
	Unassigned	Ophiovirus	2	Fungi
		Varicosavirus	1	Fungi
Single-stranded, positive-sense RNA	Sequiviridae	Sequivirus	2	Aphids
	Comoviridae	Comovirus	3	Beetles
		Fabavirus	1	Aphids
		Nepovirus	11	Nematodes
	Potyviridae	Ipomovirus	1	Whiteflies
		Potyvirus	40	Aphids
		Unassigned in the family	1	Aphids
	Luteoviridae	Polerovirus	3	Aphids
		Unassigned	2	Aphids
	Tombusviridae	Aureusvirus	1	Fungi
		Carmovirus	3	Fungi
		Necrovirus	1	Fungi
		Tombusvirus	5	Fungi
	Bromoviridae	Alfamovirus	1	Aphids
		Cucumovirus	2	Aphids
		Ilarvirus	5	(Thrips)
	Tymoviridae	Tymovirus	4	Beetles
	Closteroviridae	Closterovirus	3	Aphids
		Crinivirus	5	Whiteflies
	Flexiviridae	Allexivirus	6	Mites
		Carlavirus	8	Aphids
		Potexvirus	9	None known
	Unassigned	Benyvirus	1	Plasmodiophorids
		Ourmiavirus	2	None known
		Pomovirus	1	Plasmodiophorids
		Sobemovirus	1	None known
		Tobamovirus	7	None known
		Tobravirus	2	Nematodes
		Umbravirus	3	Aphids
	Unassigned family	Unassigned genus	1	Plasmodiophorids
			2	Not known

As tomatoes, watermelons, cabbages, onions, cucumbers, and eggplants make up almost 50% of the total vegetable production (legumes, potatoes, sweet potatoes, and cassava excluded), it is no wonder that more than 90 viruses have been found in these crops, with about 50 viruses in tomatoes, the most cultivated vegetable of all (about 13% of the total) (**Table 2**).

Ecology and Epidemiology

Factors normally considered in epidemiological studies are hosts, pathogens, and environment. To these, we can add vectors and human action.

Host

Among possible reactions of a plant to virus attack, only susceptibility and tolerance (plant systemically infected with very mild or no visible effects) favor the spread of viruses. It is evident, from both the number of viruses isolated and the number of vegetables found naturally infected, that most vegetables escape virus infections, for a number of different reasons, including resistance, possibly as a consequence of centuries of human selection for good-looking products. In fact, only 38 vegetables are infected, under natural conditions, by more than one virus, and only 25 by more than two viruses.

Pathogen

The main virus features that affect epidemiology are variability, host range, and transmission pathways. The higher the variability of a virus, the higher is its chance to adapt to new hosts and environments. Most vegetable viruses (as most plant viruses) have a single-stranded RNA (ssRNA) genome, the genome most prone to errors during replication, as it lacks the proofreading mechanisms that are present in DNA replication. Although it has some drawbacks, this feature ensures a high level of variability to viruses with a single genome component. Viruses with a divided genome can also count on natural reassortment of multiple-component genomes to increase their variability. This can strongly influence the host range of the virus. Out of 76 viruses infecting more than one vegetable crop, only 12 naturally infect more than five crops, with eight of them having an ssRNA divided genome (viz., alfalfa mosaic virus, arabis mosaic virus, cucumber mosaic virus, lettuce infectious yellows virus, radish mosaic virus, squash mosaic virus, tobacco rattle virus, and tomato black ring virus); six of these have vegetable hosts in more than one botanic family. Viruses of vegetables with a divided genome are only 30% of the total. The analysis of the host-range done for this work does not include all cultivated plants and, furthermore, the host range of viruses is not usually restricted to cultivated plants, but nevertheless the number of crops naturally infected is a good indicator of the host-range width. The host range is also strictly interconnected to the ways viruses are transmitted. Viruses, even those infecting centenary trees, need to be able to pass from one susceptible host to another. The means by which viruses move from host to host can be summarized as (1) vectors (flagellate protists: plasmodiophorids, fungi, nematodes, and arthropods); (2) seed and pollen; (3) vegetative propagation; and (4) mechanical transmission.

1. Vectors of the viruses infecting vegetables are shown in **Table 1** (column 5). Transmission by vectors involves all epidemiological factors: the virus, the vector, the host plant, the environment, in both the biotic and abiotic aspects, and man-action, in the form of movement of vectors, viruses, and host plants around the world.

2. Out of the 134 viruses that infect vegetables and have been studied for seed transmission, only 33 are seed transmitted in some plants. However, only 15 (listed

Table 2 The top vegetables in the world (excluding potato, sweet potato, cassava, and legumes), their production, and the number of viruses found in each crop in natural infection

Vegetable	Family	Production[a]	Viruses[b]
Tomato	Solanaceae	126.1	53
Watermelon	Cucurbitaceae	97.1	14
Cabbages	Brassicaceae	66.7	2
Onion	Alliaceae	62.1	4
Yam	Dioscoreaceae	46.7	3
Cucumber	Cucurbitaceae	41.6	20
Plantain banana	Musaceae	33.0	2
Eggplant	Solanaceae	30.4	8
Pepper	Solanaceae	27.2	28
Carrot (and turnip)	Apiaceae	23.8	13
Lettuce and chicory	Asteraceae	21.9	21
Squash, pumpkin	Cucurbitaceae	20.4	3
Cauliflower and broccoli	Brassicaceae	17.7	3
Garlic	Alliaceae	14.2	9
Spinach	Chenopodiaceae	12.8	10
Taro	Araceae	11.3	3
Asparagus	Asparagaceae	6.6	6
Okra	Malvaceae	5.3	2
Leek	Alliaceae	1.8	3
Artichoke	Asteraceae	1.4	14
Total		881	193[c]

[a]Data, expressed in millions of tons, refer to year 2004; from data made available by the Food and Agriculture Organization, United Nations.
[b]Number of viruses detected from natural infections; the same virus may infect more than one vegetable.
[c]Total of different virus species detected in vegetables.

in **Table 3**) are seed-transmitted in vegetables. Nevertheless, seed transmission is of paramount importance, as it can start virus infection in a newly planted crop. All viruses known to be seed-transmitted in vegetables have single-stranded, positive-sense RNA genome.

3. Viruses that are vegetatively propagated are listed in **Table 4** together with the number of viruses reported from natural infection in each crop. With the exception of artichoke, the number of virus detected is linked to the distribution of crops in terms of both the total production and cultivation areas outside the center of origin.

4. Mechanical transmission is particularly important for viruses (for instance, tobamo- and potexviruses) that reach high concentration in their vegetable hosts, particularly in the epidermis and hairs, and maintain infectivity for a long time in the environment. The risk is particularly high for crops frequently manipulated during the growing season, as man and his tools can act as carriers. Mechanical transmission due to repeated cuts for harvesting are known for zucchini yellow mosaic virus in zucchini squash, and for radish mosaic virus in rocket (eruca).

The case of tobacco mosaic virus (TMV) is worth a special mention, as an example of a 'seed-mechanical' transmission: TMV can contaminate the tegument of a tomato seed and from here infect the seedling during its growth. A special case is also the pollen-mediated transmission of tobacco streak virus and pelargonium zonate spot virus. These viruses can be mechanically inoculated by thrips that rub contaminated pollen on the leaf surface while moving on a host plant.

Environment

The environment has an influence on virus spread through both biotic and abiotic factors. Among biotic factors, the most important is wild vegetation that can act as reservoir for viruses, and fauna that may include potential vectors. Among abiotic factors, soil has an influence on vectors of soil-borne viruses. Light, sandy soils favor the spread of nematode-borne viruses, while heavy, water-retaining soils favor the spread of plasmodiophorid- and fungal-borne viruses. Rain and humidity, wind, and, most of all, temperature has a strong influence on virus epidemiology, because of their effects on arthropod vectors. High humidity and moderate temperatures favor multiplication of aphids, while strong winds and rains impair their long-distance movements. Higher temperatures are more suited for whiteflies and leafhoppers. So, it is not by chance that whitefly-borne viruses are more common in areas with subtropical climate, and

Table 3 Viruses transmitted through vegetable seeds

Family	Virus (genus)	Vegetable	Species	Family
Comoviridae	Squash mosaic virus (Comovirus)	Watermelon	Citrullus lanatus (C. vulgaris)	Cucurbitaceae
		Squash, pumpkin	Cucurbita maxima	Cucurbitaceae
	Chicory yellow mottle virus (Nepovirus)	Parsley	Petroselinum crispum subsp. tuberosum	Apiaceae
		Chicory	Cichorium intybus	Asteraceae
	Beet ringspotvirus (Nepovirus)	Mangolds (beet roots, fodder beet)	Beta vulgaris	Chenopodiaceae
		Swiss chard	Beta vulgaris var. cicla	Chenopodiaceae
	Tobacco ringspot virus (Nepovirus)	Lettuce	Lactuca sativa	Asteraceae
	Tomato black ring virus (Nepovirus)	Celery	Apium graveolens	Apiaceae
		Lettuce	Lactuca sativa	Asteraceae
		Tomato	Solanum lycopersicon	Solanaceae
Potyviridae	Celery latent virus (Potyvirus?)	Celery	Apium graveolens	Apiaceae
	Lettuce mosaic virus (Potyvirus)	Lettuce	Lactuca sativa	Asteraceae
Tombusviridae	Tomato bushy stunt virus (Tombusvirus)	Pepper	Capsicum annuum	Solanaceae
		Tomato	Solanum lycopersicon	Solanaceae
	Cucumber leaf spot virus (Aureusvirus)	Cucumber	Cucumis sativus	Cucurbitaceae
Bromoviridae	Alfalfa mosaicvirus (Alfamovirus)	Pepper	Capsicum annuum	Solanaceae
	Spinach latent virus (Ilarvirus)	Spinach	Spinacia oleracea	Chenopodiaceae
Unassigned to family	Cucumber green mottle mosaic virus (Tobamovirus)	Squash, pumpkin	Cucurbita maxima	Cucurbitaceae
		Watermelon	Citrullus lanatus (C. vulgaris)	Cucurbitaceae
	Pepper mild mottle virus[a] (Tobamovirus)	Pepper	Capsicum annuum	Solanaceae
	Tobacco mosaic virus[a] (Tobamovirus)	Pepper	Capsicum annuum	Solanaceae
		Tomato	Solanum lycopersicon	Solanaceae
	Tomato mosaic virus[a] (Tobamovirus)	Tomato	Solanum lycopersicon	Solanaceae

[a]Transmission of tobamoviruses is not a true seed transmission as a result of embryo infection, but a 'seed-mechanical' transmission; TMV can contaminate the tegument of a seed and from here infect the seedling during its growth.

Table 4 Vegetatively propagated vegetables from which viruses have been reported, and the number of viruses reported in natural infections

Species	Common name	Family	No. of viruses reported
Cynara scolimus	Artichoke	Asteraceae	14
Allium sativum	Garlic	Alliaceae	9
Asparagus officinalis	Asparagus	Asparagaceae	6
Allium ascalonicum	Shallot	Alliaceae	5
Colocasia esculenta or C. antiquorum	Taro	Araceae	4
Ullucus tuberosus	Olluco	Basellaceae	4
Canna edulis	Achira	Cannaceae	3
Dioscorea spp.	Yam	Dioscoreaceae	3
Arracacia xanthorrhiza	Arracacha	Apiaceae	2
Musa spp.	Plantain banana	Musaceae	2
Tropaeolum tuberosum	Mashua	Tropaeolaceae	2
Rorippa (Nasturtium) officinale or R. officinale × R. microphyllum	Water cress	Brassicaceae	1
Bambuseae (tribe)	Bamboo shoot	Poaceae	1

that the global warming is accompanied by an increase of their importance and diffusion areas.

Control Measures

Control of virus and virus diseases in vegetables, as in most crops, is indirect and based upon a number of strategies for preventing virus infection. Seeds should be as virus-free as possible, using certified seeds whenever possible, disinfecting seeds of crops susceptible to tobamoviruses by a number of simple, but effective chemical treatments, when certified seed is not available. Multiplication material of vegetatively propagated species (**Table 4**) need to be selected and virus-free vegetative stock prepared and maintained. For freeing plant material from viruses one can use (1) heating of dormant plant parts (thermotherapy) and growing small shoot tips removed immediately after heat treatment; (2) meristem tip culture, particularly efficient for eliminating viruses from plants of the family Araceae, for example, may be preceded by or associated with heat treatment (very effective for garlic, for example); (3) chemical treatment with antiviral substances, like ribavirin, either alone or with DHT; usually disappointing, it may work in association with the other two methods. Seedlings should be protected before and soon after transplanting as plants are usually very susceptible at this stage. After transplanting, it is important to prevent virus spread from outside and within the crop. Contamination from the soil can be avoided by removing debris from previous crop (particularly for tobamo-, tombus-, and carmoviruses) or by crop rotation, alternating crops susceptible to different viruses; crop rotation may help also in controlling vectors of limited mobility (nematodes, fungi, plasmodiophorids). Crops can be protected against airborne vectors by physical means, either excluding the vectors (nets and covers of various types, wind-breakers) or repelling them (mulching with repellent effect, e.g., UV-reflective silver plastic), or by chemical treatments. Biologic control of vectors is not very efficient, particularly against introduction of viruses into the crop, as efficient vectors can be occasional visitors of the crop rather than regular colonizers. Control of weeds and volunteer plants from the previous crop is also very important in reducing early infections. When a vegetable is grown on a wide territory crop after crop, a 'crop-free period' can be very efficient in preventing virus infections (e.g., carrot virus Y in carrots in southern Australia). As for avoiding the introduction of viruses in areas (islands, countries, continents) where they are not present yet, most agriculturally advanced countries set quarantine rules aimed at excluding specific viruses (or their vectors) or limiting their further spread; setting and applying quarantine measures can be quite complex, but it may be worth particularly for viruses transmitted through seed or in dormant vegetative parts. In order to control the spread of viruses within the crops, one needs to reduce the movement of viruses. For vector-borne viruses, a number of chemical treatments are available for controlling the different vectors. Hygiene of operator and implements (by washing with a solution of 3% trisodium orthophosphate, for example) is particularly important for viruses readily mechanically transmissible (tobamo-, potex-, carmoviruses, but also comoviruses) in crops that need frequent tending. Roguing can be effective only if virus incidence is still low (e.g., less than 10%), the incubation is short, and the movement of the virus from plant to plant, by whatever means, is slow. When roguing plants hosting vectors, care should be taken not to disperse them while removing plants. A further possibility for protecting a vegetable crop from virus damage is artificial infection of the plants with a strain of the virus causing only mild disease symptoms (cross-protection) or associating a satellite to a virus to the same aim (e.g., cross-protection is used for protecting cucurbits from damages caused by zucchini yellow mosaic virus and tomatoes from tomato mosaic virus; co-infection with a satellite can be used to reduce damages by cucumber mosaic virus in pepper and tomato). Both methods have a number of objections, based on ecological and practical considerations, so they are not recommended as general practice, but still can reduce virus damages at least for some time in some crops. Another method of choice to avoid virus infection of vegetables is resistance, to vectors in few instances (aphids, whiteflies, thrips) and to viruses in many cases. For analysis of resistance to viruses, either natural or engineered, readers are referred to specific articles of this encyclopedia. Table 5 lists some of the crops, and their viruses, for which natural resistance is commercially available. Transgenic lines of vegetables have been developed that resist to very damaging viruses, but they are not listed here because, for a number of reasons, transgenic seeds are not available to the majority of growers yet.

A Few Special Cases

A few viruses infecting vegetables and the symptoms they cause in their hosts are hereafter described, chosen as examples among the many vegetable virus diseases. For details on the structure and properties of the viruses mentioned, the reader is referred to the specific articles of this encyclopedia.

Cucumber Mosaic Virus

Cucumber mosaic virus (CMV) is a positive-sense ssRNA virus with a tripartite genome encapsidated in isometric particles measuring about 30 nm in diameter. CMV has a very wide host range (more than 800 species in 70 botanic families). Natural infections have been reported from at least 10 vegetable species and in a large number of weeds, the most important virus reservoir. In many of these weeds CMV is not associated to any symptoms and is efficiently seed-transmitted. The virus is transmitted by several aphid species in a nonpersistent manner: infectivity is retained by vectors for a few hours and is lost with molt. Symptoms of CMV in vegetables like tomato plants infected at an early stage are chlorotic local lesions between the secondary leaf vein. Necrosis, first in the form of brown leaf areas, and later as brown lines along petioles and stems, proceeds toward the bottom of the plant, which can die in a few weeks after infection. Fruits, if present, are misshaped, sun-burnt, and necrotic. Plants infected at a mature stage normally show growth reduction, bushy appearance, and leaf deformation (shoestring-like leaves). Fruits have reduced size and do not ripen. Depending on a number of factors, including the presence of satellite RNAs, plants may have normal-looking leaves but show necrosis of fruits (losses are quite reduced in this case) or show typical, but stronger symptoms (in this case, losses can be very high). Symptoms of CMV infection in 'pepper' plants depend on the virus strain and on the plant age at the moment of infection. In young pepper plants CMV causes leaf yellowing and narrowing, and necrotic symptoms, on both the foliage and the fruit. If older plants are infected, CMV symptoms can be mild and appear on lower, yellowing leaves as green ring spots or oak-leaf patterns, but also as a mild mosaic, with a general dull appearance. Cucurbits infected by CMV, particularly zucchini, squash, and melons, show strong mosaic

Table 5 Vegetables and viruses for which natural resistance of different degrees has been incorporated in commercially available cultivars

Species	Common names	No. of viruses reported	Genetic (natural) resistance available to
Lactuca sativa	Lettuce	20	Lettuce mosaic virus (Potyvirus)
Brassica oleracea var. capitata	Cabbage	2	Turnip mosaic virus (Potyvirus)
Spinacia oleracea	Spinach	10	Cucumber mosaic virus (Cucumovirus)
Cucumis sativus	Cucumber	20	Cucumber mosaic virus (Cucumovirus), papaya ringspot virus (Potyvirus), watermelon mosaic virus (Potyvirus), zucchini yellow mosaic virus (Potyvirus)
Cucurbita maxima	Squash, pumpkin	3	Cucumber mosaic virus (Cucumovirus), papaya ringspot virus (Potyvirus), watermelon mosaic virus (Potyvirus), zucchini yellow mosaic virus (Potyvirus)
Cucurbita pepo	Zucchini, marrows, pumpkin, squash	14	Cucumber mosaic virus (Cucumovirus), papaya ringspot virus (Potyvirus), watermelon mosaic virus (Potyvirus), zucchini yellow mosaic virus (Potyvirus)
Capsicum annuum	Pepper	27	Cucumber mosaic virus (Cucumovirus), pepper mottle virus (Potyvirus), potato virus Y (Potyvirus), tobacco etch virus (Potyvirus), tobacco mild green mosaic virus (Tobamovirus), tobacco mosaic virus (Tobamovirus), tomato spotted wilt virus (Tospovirus)
Solanum lycopersicon	Tomato	53	Tobacco mosaic virus (Tobamovirus), tomato mosaic virus (Tobamovirus), tomato spotted wilt virus (Tospovirus)
Solanum melongena	Eggplant (aubergine)	8	Cucumber mosaic virus (Cucumovirus), tomato mosaic virus (Tobamovirus)

symptoms and leaf narrowing, and distortion (**Figure 1**). Fruits are pitted and misshapen, unmarketable. To worsen things, mixed infections with other viruses are common. Celery and parsley infected by CMV show leaf yellowing and necrosis. 'Lettuce' infected by CMV shows intense mosaic, vein chlorosis and, frequently, vein browning and necrosis when temperature drops below 13 °C. CMV infection of spinach causes a variety of symptoms including stunting, yellowing, and mottling of the older leaves and malformation of the younger leaves.

Garlic Viruses

Because of exclusive vegetative propagation, by cloves or, in some cultivars, by inflorescence bulbils, garlic is particularly prone to accumulate complex virus infections. The most known virus-associated syndrome in garlic is mosaic. Symptoms include various degrees of leaf discolorations, most pronounced in the younger leaves, chlorotic mottling, striping, and streaking. Plants are also stunted compared to healthy looking plants under the same conditions. Most probably, two potyviruses are responsible for mosaic in garlic: one is a garlic strain of onion yellow dwarf virus (OYDV-G), the other a specific strain of leek yellow stripe virus (LYSV-G). Garlic clones, first freed from OYDV and LYSV and then mechanically inoculated with either virus, have been shown to suffer yield reduction from 17% to 60% depending on the virus–cultivar combinations. Yield losses are worse in case of mixed infections or chronic infections (vegetatively propagated material originally infected). Two aphid-borne carlaviruses, garlic common latent virus (GarCLV) and shallot latent virus (SLV), also known as garlic latent virus (GarLV), have also been identified as parts of the garlic virus complex. A number of allexiviruses (genus *Allexivirus*; family *Flexiviridae*) have been identified in recent years in garlic. Although none of them has been associated with any particular symptom, some of them cause yield effects, particularly in mixed infections with OYDV and LYSV. Garlic dwarf virus (GDV) (family *Reoviridae*; genus *Fijivirus*), originally isolated in France, causes well-defined symptoms of dwarfing and leaf thickening.

Tobamoviruses

Tobamoviruses are possibly the most-studied viruses of plants. Tobamoviruses have rod-shaped particles 300 × 18 nm, each containing a single molecule of positive-sense, ssRNA. Particles are very stable. No biological vector is known, but because of high stability and high concentration in plant tissues, tobamoviruses are readily transmitted mechanically during crop tending. They are also seed-transmitted but, as mentioned before, in a manner very different from true seed transmission. Tobacco mosaic virus (TMV) (from which the name of the genus is derived) and tomato mosaic virus (ToMV) are the most known, but other viruses, pepper mild mottle virus (PMMoV) and tobacco mild green mosaic virus (TMGMV), infect vegetables in the family Solanaceae and can be differentiated by biological and serological tests. TMV, ToMV, and PMMoV commonly infect peppers, causing chlorotic mosaic, leaf distortion, sometimes systemic necrosis, and defoliation, depending on the usual factors: plant cultivar and age, virus strain, light intensity, and temperature. Fruits are disfigured, with discolored or necrotic areas. PMMoV usually causes milder symptoms on leaves but is more severe on fruits. TMGMV has been found occasionally in pepper. Tomato plants are mostly infected by ToMV, but occasionally also by TMV. The most characteristic symptoms are mottled areas of light and dark green on the leaves. Plants infected at an early stage of growth are yellowish and stunted. Leaves may also be malformed, narrowed, although not as much as with CMV, or showing enations (outgrowths) on the lower leaf lamina. High temperature can mask leaf symptoms. Fruits can be from almost normal to misshapen and be reduced in size and number, showing uneven ripening, corky or necrotic rings, internal browning. TMV and ToMV can infect eggplant occasionally, causing mild symptoms on both leaves and fruits.

Tomato Spotted Wilt Virus

Tomato spotted wilt virus (TSWV) is the type member of the genus *Tospovirus* in the family *Bunyaviridae*. Virus particles are roughly isometric, 80–90 nm in diameter, enveloped in a lipoprotein membrane, contain three single-stranded linear RNAs, one of negative and two of ambisense polarity, associated with a nucleoprotein to form the nucleocapsid. TSWV is transmitted by thrips in a circulative, propagative manner. It can infect hundreds of species in 70 botanical families. The first symptoms of TSWV infection in tomato are chlorotic

Figure 1 Symptoms of cucumber mosaic virus infection on zucchini leaves. Courtesy of V. Lisa.

spots, 3–4 mm in diameter, on apical leaves, rapidly turning to bronze and dark-brown (necrosis). If infected when young, the plant will eventually die. If infected later, dark brown streaks also appear on stems and leaf petioles; growing tips are greatly stunted, and usually severely affected with systemic necrosis. Fruits, mostly reduced in size, will display characteristic symptoms – immature fruit have mottled, light green rings with raised centers that develop to orange and red discoloration patterns on mature fruits (**Figure 2**). TSWV-infected pepper shows symptoms similar to those described for tomatoes. Plants infected at a very early stage are usually severely stunted and yield no fruit. Plants infected later show chlorotic or necrotic rings on leaves and stems. Some cultivars react with flower and leaf drop. Fruits develop necrotic or discolored spots or rings. Lettuce and endive infected when young turn yellow, collapse, and die. Older plants develop marginal wilting, yellowing, and necrosis of the leaves. Typical is the bending of midribs as a consequence of symptoms appearing only on one half of the leaf. The plant may also look twisted. Endive appears to be more susceptible than lettuce. Chicory infected by TSWV has a stunted, bushy, yellowish look, with chlorotic necrotic spots along the midribs and secondary veins. Eggplants and artichokes can also be naturally infected by TSWV, but only occasionally.

Turnip Mosaic Virus

Turnip mosaic virus (TuMV) is a typical member of the genus *Potyvirus* in the family *Potyviridae*. It has nonenveloped, filamentous virions, usually flexuous, with a modal length of 720 nm, containing a monopartite, positive-sense ssRNA genome. It is transmitted by several species of aphids in a nonpersistent manner. It has a relatively wide host range; naturally, infections occur mainly in the family Brassicaceae and in few weeds outside this family. All *Brassica* species are susceptible to TuMV. The virus also infects lettuce, watercress, radish, and rocket, and it is present all over the world. Young leaves of infected cauliflower plants show chlorotic ringspots followed, as the leaf ages, by yellow or brownish spots surrounded by circular or irregular necrotic rings. Generally, these occur in the vicinity of the leaf veins. Leaf blight appears in infected sections when many lesions coalesce. Outer leaves of infected cabbages may show ringspot pattern or appear more uniformly necrotic, but, as the leaves age, a chlorotic color almost replaces the normal green tissue. In white cabbage, the occurrence of internal necrotic lesions, sunken and coalescent, has been correlated with the presence of field symptoms of turnip mosaic virus. This postharvest disorder may result in the loss of large quantities of stored cabbage. Symptom expression is temperature sensitive, with most pronounced symptoms at temperatures ranging from 22 to 30 °C, and is very much masked below 17 °C. Other *Brassica* species, like turnip, broccoli, and Chinese cabbage, react to TuMV infection with different degrees of mosaic, necrosis, leaf distortion, and reduced growth.

Tomato Yellow Leaf Curl Virus and Related Begomoviruses

Tomato yellow leaf curl virus (TYLCV) is a member of the genus *Begomovirus* in the family *Geminiviridae*. It has geminate particles measuring 20×30 nm and encapsidating a monopartite genome of circular, ssDNA 2.8 kbp in size. The virus is transmitted by the whitefly *Bemisia tabaci* in a circulative manner, but it is not mechanically transmissible. Seed transmission has not been reported. TYLCV has a relatively narrow natural host range. Infected tomato plants are stunted, with branches and petioles tending to assume an erect position. Leaflets of the infected plants are smaller than those of healthy ones and upward curled with margins more or less yellow (**Figure 3**). Flowers have a normal appearance, but fruit production is strongly reduced. TYLCV was first described in Israel and it is now known to be present in the Mediterranean basin, in Egypt and Sudan, in Iran, and in the Caribbean basin. A number of monopartite begomoviruses, transmitted by *B. tabaci*, inducing very similar symptoms on tomato plants, with some differences in the host range and significant differences in their sequences, have been described in recent years and classified as distinct species of the genus *Begomovirus* (e.g., *Tomato yellow leaf curl Sardinia virus*). Similar symptoms on tomato plants are induced by other begomoviruses, named in the same way, but are actually bipartite viruses (e.g., tomato yellow leaf curl China, -Iran, -Thailand virus); some of these viruses may be mechanically transmissible with difficulties. A few other begomoviruses, named tomato leaf curl someplace virus, where someplace is the name of the area (state, country) of origin, may induce some yellowing of tomato leaves, despite the name. Molecular diagnostics are therefore the only tools for identification of begomoviruses inducing yellow leaf curl symptoms in tomato plants.

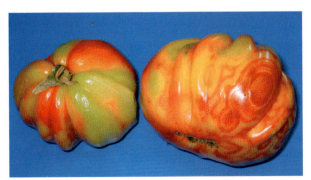

Figure 2 Symptoms of tomato spotted wilt virus infection on tomato fruits. Courtesy of the late P. Roggero.

Figure 3 Leaves of a tomato plant infected by tomato yellow leaf curl virus (right) compared to leaves of a healthy plant (left).

Figure 4 Symptoms of zucchini yellow mosaic virus infection on zucchini fruits. Courtesy of the late P. Roggero.

Zucchini Yellow Mosaic Virus

Zucchini yellow mosaic virus (ZYMV) is a typical member of the genus *Potyvirus* (family *Potyviridae*) (see TuMV). It is transmitted by several species of aphids in a nonpersistent manner and its natural host range is essentially limited to members of the family Cucurbitaceae. ZYMV is a 'recent' virus – it was first detected in 1973 in southern Europe, the first epidemics were reported in 1979–80, and in 1996 it was already present practically in all the areas where cucurbits are grown. We still lack an acceptable explanation for the rapid worldwide diffusion of the virus. Early symptoms in infected zucchini appear as vein clearing of fine leaf veins, followed by a general yellowing of the leaf with dark green areas. Later on, the mosaic becomes stronger, and vein banding appears. Leaves can be deformed to shoestring appearance, but the most affected are fruits, that are severely malformed and often develop longitudinal cracks (**Figure 4**). Similarly, squash and pumpkin plants develop knobby areas on the fruits resulting in prominent deformations. In watermelons, as in melons, ZYMV induces yellow mosaic, severe malformations of leaves and fruits, that often display radial or longitudinal cracks. On cucumber, symptoms of ZYMV infection are less severe than on the other cucurbit crops; mosaic of variable intensity, followed by dark green vein banding and discoloration of fruits.

See also: *Alphacryptovirus* and *Betacryptovirus*; Cucumber Mosaic Virus; Diagnostic Techniques: Microarrays; Diagnostic Techniques: Plant Viruses; Flexiviruses; Plant Resistance to Viruses: Engineered Resistance; Plant Resistance to Viruses: Geminiviruses; Plant Resistance to Viruses: Natural Resistance Associated with Dominant Genes; Plant Resistance to Viruses: Natural Resistance Associated with Recessive Genes; Plant Virus Diseases: Economic Aspects; Potyviruses; *Tobamovirus*; Tomato Spotted Wilt Virus; Tomato Yellow Leaf Curl Virus; *Tospovirus*; Vector Transmission of Plant Viruses.

Further Reading

Conti M, Gallitelli D, Lisa V, et al. (1996) *I principali Virus delle Piante Ortive.* Milan: Bayer.

Jones JB, Jones JP, Stall RE, and Zitter TA (eds.) (1991) *Compendium of Tomato Diseases.* St. Paul, MN: The American Phytopathological Society.

Pernezny K, Roberts PD, Murphy JF, and Goldberg NP (eds.) (2003) *Compendium of Pepper Diseases.* St. Paul, MN: The American Phytopathological Society.

Ryder EJ (1999) *Lettuce, Endive and Chicory.* Cambridge: CABI Publishing.

Thresh JM (ed.) (2006) *Advances in Virus Research 67: Plant Virus Epidemiology.* San Diego, CA: Academic Press.

Zitter AT, Hopkins DL, and Thomas CE (eds.) (1996) *Compendium of Cucurbit Diseases.* St. Paul, MN: The American Phytopathological Society.

Relevant Website

http://vegetablemdonline.ppath.cornell.edu – Cornell Plant Pathology Vegetable Disease Web Page.

Vesicular Stomatitis Virus

S P J Whelan, Harvard Medical School, Boston, MA, USA

© 2008 Elsevier Ltd. All rights reserved.

Historical Perspective

Vesicular stomatitis viruses (VSVs) are transmitted naturally by arthropods to a broad range of animal species. A clinically significant acute disease is manifest in domesticated animals, notably cattle, horses, and pigs, and it is characterized by fever and the appearance of vesicular lesions in the mouth, tongue, udder teats, and hoof coronary bands. Symptoms are therefore similar to those following infection with the apthovirus, foot-and-mouth disease virus (FMDV) and, consequently, rapid diagnosis is important in livestock. VSV infection was first described in the USA in 1916, following an epidemic in cattle and horses. However, reports from 1862 describe a clinically similar disease in army horses during the American Civil War. Today the virus is distributed throughout the Americas and is enzootic in Central America. In Panama, estimates suggest that up to 29% of the human population has been exposed to the virus as judged by the presence of neutralizing antibodies. Infection of humans can result in a mild febrile illness, but is generally asymptomatic. VSVs have been described rarely outside the Western Hemisphere.

Taxonomy and Classification

VSVs have a nonsegmented negative-sense RNA genome and are assigned to the order *Mononegavirales*. Within this order, they are assigned to the family *Rhabdoviridae*, based upon their characteristic 'bullet' shape. Within the rhabdovirus family, they are further assigned to the genus *Vesiculovirus*. A list of the nine currently recognized species assigned to the genus is provided in **Table 1**, along with their geographic distribution and sources of the virus in nature. The *Eighth Report of the International Committee on the Taxonomy of Viruses* lists an additional 19 members tentatively assigned to this genus. The type species is *Vesicular stomatitis Indiana virus*. Vesicular stomatitis Indiana virus (VSIV) has been widely studied in laboratories as a prototype of all the *Mononegavirales* and will be the primary focus of this article. Much is known about the replication and molecular biology of VSIV. VSIV was isolated following an outbreak of a vesicular disease of cattle in Richmond, Indiana, in 1925. The infectious agent was maintained by serial passage in animals and eventually became the Indiana serotype of VSV. *Vesicular stomatitis New Jersey virus* is also classified as a species in the genus *Vesiculovirus*. It was isolated following an outbreak in cattle in 1926 and is serologically and genetically distinct from VSIV. Vesicular stomatitis Alagoas virus (VSAV) was isolated from domesticated animals in Alagoas Brazil during an outbreak of VSV.

Structure of VSV

Particles

A schematic of VSIV is shown in **Figure 1** along with an electron micrograph showing virus particles. The particles appear bullet-shaped and are approximately 180 nm long and 70 nm in diameter. The particles comprise 74% protein, 20% lipid, 3% RNA, and 3% carbohydrate. The virus possesses a lipid envelope that is decorated with trimeric spikes of the 67 kDa attachment glycoprotein (G). The interior of the particle contains a ribonucleoprotein (RNP) core of the genomic RNA complexed with the viral nucleocapsid (N) protein. This N-RNA core is associated with the RNA-dependent RNA polymerase (RdRp), the viral components of which are a 241 kDa large (L) protein and a tetramer of a 29 kDa accessory

Table 1 Geographic distribution and sources of natural isolation of viruses representing the nine recognized species in the genus *Vesiculovirus*

Virus	Geographic distribution	Source
Carajas virus (CJSV)	Brazil	Phlebotomine sandflies
Chandipura virus (CHPV)	India, Nigeria	Mammals, sandflies
Cocal virus (COCV)	Argentina, Brazil, Trinidad	Mammals, mosquitoes, mites
Isfahan virus (ISFV)	Iran, Turkmenistan	Sandflies, ticks
Maraba virus (MARAV)	Brazil	Phlebotomine sandflies
Piry virus (PIRYV)	Brazil	Mammals
Vesicular stomatitis Alagoas virus (VSAV)	Brazil, Columbia	Mammals, sandflies
Vesicular stomatitis Indiana virus (VSIV)	Americas	Mammals, mosquitoes, sandflies
Vesicular stomatitis New Jersey virus (VSNJV)	Americas	Mammals, mosquitoes, midges, blackflies, houseflies

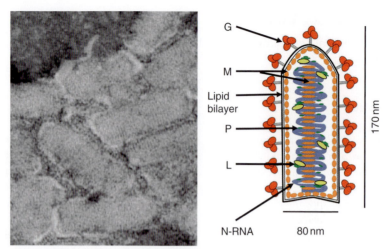

Figure 1 Electron micrograph of VSIV particles and a schematic illustration of the virion. At left, a negative-stained image of a group of virus particles is shown. At right, a schematic illustration of the virion is shown along with the dimensions of a viral particle. N-RNA, nucleocapsid protein coated RNA; P, phosphoprotein; M, matrix protein; G, attachment glycoprotein; L, large polymerase subunit. Kindly provided by David Cureton, Harvard Medical School.

phosphoprotein (P). Together, these form the internal helical nucleocapsid. Another major structural component of the virus particles is the 26 kDa matrix (M) protein, which is located below the membrane and associated with the nucleocapsid. The approximate composition of the particle is one molecule of RNA, 1200 copies of N, 500 copies of P, 1800 copies of M, 1200 copies of G, and 50 copies of L. The role of these proteins in viral replication is described in more detail below.

Genome

VSIV has a nonsegmented, negative-sense RNA genome of 11 161 nt. The genome comprises a 50 nt 3′ leader region (le), five genes that encode in order the N, P, M, G, and L proteins, and a 59 nt 5′ trailer region (**Figure 2**). The 5′ and 3′ ends of the genome are not modified and contain a 5′ triphosphate and 3′ hydroxyl. A key feature of all mononegaviruses is that the RNA genome is not found naked within infected cells. Instead, it is present as a ribonucleoprotein complex, in which it is completely covered by the viral nucleocapsid protein. This N-RNA template, associated with the viral L and P proteins, comprises the transcription-competent core of the virus, and delivery of this complex is required to initiate the infectious cycle. Consequently, in contrast to positive-sense RNA viruses, the naked RNA is not infectious.

Viral Replication Cycle

The replication cycle (**Figure 3**) should be considered a continuum of events. However, it is convenient to divide the cycle into the following three stages.

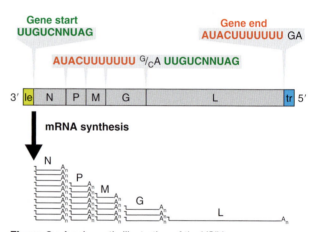

Figure 2 A schematic illustration of the VSIV genome highlighting the polymerase regulatory elements. The genome is represented 3′–5′ as a series of boxes, comprising le = leader region, N = nucleocapsid, P = phospho, M = matrix, G = glyco, L = large polymerase, and tr = trailer region. The conserved cis-acting elements that regulate polymerase activity during mRNA synthesis are shown for emphasis. Colors: green, the conserved residues of the gene-start sequence; red, the conserved residues of the gene-end sequence; and black, the nontranscribed residues of the gene junction. The leader and trailer regions also contain key elements that regulate polymerase activity and serve as promoters as described in the text.

Attachment, Entry, and Uncoating

Attachment of VSV to host cells is mediated by the glycoprotein which binds to the surface of cells. Given that VSV G can mediate infection of almost all cells in culture, either the receptor for VSV must be widely distributed, or the virus may be able to utilize multiple surface molecules for attachment. Phosphatidylserine (PS) was long thought to be the receptor for VSV but

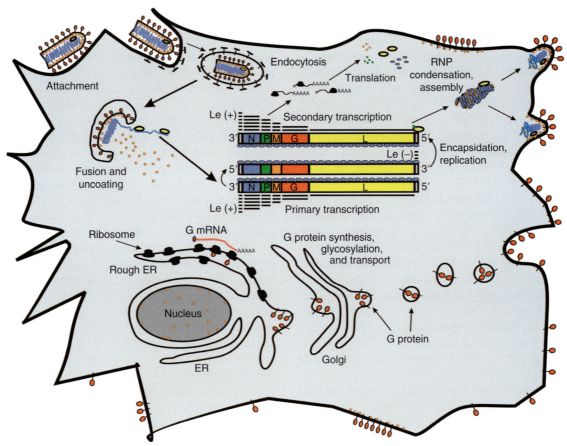

Figure 3 A schematic illustration of the replication cycle of VSV. The replication cycle (described in the text) is depicted showing attachment of virus to the cell, internalization, release of the viral core into the cytoplasm, primary viral mRNA synthesis, mRNA translation, genomic replication, secondary viral mRNA synthesis, assembly, and budding of infectious particles. Kindly provided by David Cureton, Harvard Medical School.

recent studies have questioned this finding. Importantly, the PS binding site in VSV G is internal to the trimer in its prefusion form. Following attachment to the cell, the virus is internalized via clathrin-dependent endocytosis and delivered to an early endosome. The pH threshold necessary to trigger the conformational alterations in VSIV G that promote fusion of the viral and cellular membranes is approximately 6.2. This pH is reached in the early endosome, and viral and cellular membranes fuse to release the transcription-competent RNP core into the cytoplasm of the cell. Recent work has called into question this conventional view of the VSV entry pathway and has posited a new model for viral entry. In this model, fusion and RNP release are spatially and temporally separated. Specifically, the low pH encountered during endocytic transport triggers fusion and the delivery of the RNP core into an intralumenal vesicle within an endosomal carrier vesicle. A subsequent (cell-mediated) fusion event is then required to fuse the membrane of the intralumenal vesicle with the limiting membrane of the cell and deliver the RNP into the cytoplasm to initiate the infectious process. Irrespective of the precise route by which virus enters the cell, the end result is the delivery of the transcription-competent core into the cytoplasm. This is accompanied by the release of M protein, which can migrate to the nucleus where it plays a role in inhibiting host gene expression.

Gene Expression

Following the delivery of the transcription-competent core into the cytoplasm, RNA synthesis can begin. The replication cycle of VSV is entirely cytoplasmic, occurring efficiently in enucleated cells. The demonstration that purified VSV particles contain a functional RdRp that is active *in vitro* has led to major advances in our understanding of viral gene expression. During RNA synthesis, the polymerase uses the encapsidated genomic RNA as template in two distinct reactions: (1) transcription of five mRNAs that encode the N, P, M, G, and L proteins; and (2) replication to yield full-length antigenomic, and then genomic RNA strands. Our current understanding of gene expression is summarized as follows. In response to a specific promoter element that is provided by sequences within the 3′ leader region and the conserved residues of the first (N) gene-start (3′-UUGUCNNUAG-5′) sequence, the RdRp initiates mRNA synthesis. Synthesis commences at the N gene-start sequence and generates an mRNA that is capped, methylated, and polyadenylated, each of these reactions occurring co-transcriptionally. Termination at the end of the N gene is achieved by the polymerase recognizing a highly conserved sequence element referred to as the gene-end 3′-AUACUUUUUUG/C-5′. This sequence signals the polymerase to stutter on the U_7 tract to generate the polyA tail, and leads to termination of mRNA synthesis. Termination at the end of the N gene is essential for the polymerase to be able to transcribe the P gene. A poorly understood event that is localized to a short region at the N–P gene junction results in the synthesis of approximately 30% less P mRNA than N mRNA. This sequential and polar synthesis of the viral mRNAs continues through the entire genome and provides a gradient of viral mRNA synthesis such that $N > P > M > G > L$. These products of mRNA synthesis are illustrated in **Figure 2**.

Among the notable steps of mRNA synthesis are the unusual mechanisms by which the 5′ and 3′ ends of the RNA are formed. Each stage of mRNA cap formation is distinct from those employed in other systems. Specifically, the 5′ end of the pppApApCpApG mRNA is capped by an unusual ribonucleotidyltransferase activity that transfers the monophosphate RNA onto GDP derived from GTP to form the G*ppp*ApApCpApG mRNA cap structure. In contrast, all other capping reactions are catalyzed by an RNA guanylyltransferase that transfers GMP onto the 5′ end of a diphosphate RNA through a reaction that involves a covalent enzyme GMP intermediate. The VSV L protein is responsible for this novel reaction, which appears to involve a covalent intermediate between L and the viral mRNA. The resulting G*ppp*ApApCpApG cap structure is then methylated at guanine N-7 and ribose 2′-O positions to yield 7^mG*ppp*A^{m}pApCpApG. These activities are also provided by the L protein, and again differ from conventional mRNA cap methylation reactions. For VSIV, the two enzymatic activities have been shown to share a single binding site for the methyl donor, S-adenosyl-L-methionine (SAM). In contrast, other cap methylation reactions are normally executed by two distinct proteins with separate binding sites for the methyl donor SAM. Formation of the 3′ end of the RNA is also unusual. Specifically, polyadenylation occurs in a pseudo-templated fashion in which the polymerase complex reiteratively transcribes the conserved U tract present at the end of each VSV gene.

The viral mRNAs are efficiently translated by the host translation machinery, but how they compete with cellular mRNAs for translation is not well understood. Viral protein synthesis is essential for replication of the genomic RNA. Ongoing translation provides a continuous supply of soluble N protein that drives the encapsidation of the nascent RNA chain. This process is intimately linked to genomic RNA replication which first results in the production of a full-length encapsidated complementary antigenome RNA. This antigenome can then serve as template to produce more progeny genomic RNAs for use as templates for further mRNA synthesis in a process referred to as secondary transcription.

Precisely how the different polymerase activities are regulated in infected cells remains poorly understood. Two functionally distinct pools of polymerase have been purified from cells. One initiates internally at the N gene-start sequence and functions as the viral transcriptase. A second complex initiates at the 3′ end of the genome and functions as the viral replicase. These complexes are reported to differ in their composition such that the transcriptase comprises the viral P and L proteins, together with several cellular proteins including translation elongation factor-1α, heat shock protein 60, and the host cell RNA guanylyltransferase. In contrast, the replicase is reported to comprise the viral N, P, and L proteins.

In addition to the species of RNA described above, two short leader RNAs are generated during RNA synthesis: a 47 nt Le+ from the 3′ end of the genomic RNA and a 45 nt Le− from the 3′ end of the antigenomic RNA. The function of these RNAs is poorly understood, although a role for the Le+ in the shutoff of host gene expression has been described. A long-standing model for the regulation of RNA synthesis in VSV postulates that polymerase initiates all RNA synthesis at position 1 of the genome, and during synthesis of Le+ a crucial regulatory decision is made to either terminate leader and initiate mRNA synthesis at the N gene start, or alternatively to read through the leader–N gene junction and synthesize the full-length antigenome. The obligatory requirement for protein synthesis to provide a source of N protein to encapsidate the nascent RNA during genome replication led to the suggestion that N protein availability switches polymerase activity from transcriptase to replicase.

In recent years, there has been an accumulation of evidence that conflicts with this model. Specifically, a VSV mutant containing a single amino acid change in the template-associated N protein produces an excess of N mRNA over Le+ *in vitro*, suggesting that polymerase can synthesize N independently of Le+. In another series of experiments, recombinant VSVs, containing a 60 nt gene inserted between the leader region and the N gene, were employed. The recombinant viruses were examined to determine the effect of altering the potential number of ultraviolet (UV)-induced dimers between adjacent uracil residues. Such dimers block progression of the polymerase. These experiments showed that changing the UV sensitivity of the Le+ had no effect on the sensitivity of the 60 nt mRNA in infected cells, suggesting that polymerase can initiate synthesis internally at the first gene start. In addition, two separate pools of polymerase can be isolated from infected cells, one that initiates internally at the N gene start and the second that initiates at the 3′ end of the genome. These findings support the hypothesis that mRNA synthesis can initiate independently of leader synthesis, and show that polymerase function is not simply switched by N protein levels. However, viral gene expression is controlled, it results in the exponential amplification of the input genomic RNA, yielding progeny genomes that can be assembled into infectious particles, the next phase of the viral replication cycle.

Assembly and Budding

Assembly of infectious virus particles involves many critical interactions. Our current understanding of this intricate process is that the matrix (M) protein complexes with the RNP core and represses transcription of viral mRNAs. This condensed RNP complex acquires a lipid envelope that has been modified by the insertion of an externally oriented glycoprotein. Details of how the RNP is transported to the site of budding and how the matrix interacts with and condenses the RNP are poorly understood. The M protein is only found associated with RNPs at sites of viral budding, and how genomic RNPs (rather than antigenomic RNPs) are specifically selected for budding is unclear. The M protein contains two 'late domains' (PTAP and PPPY motifs) that appear to be critical in the late phase of the assembly-release pathway. Amino acid substitutions in the PPPY motif result in the accumulation of bullet-shaped virions that are stalled at a late stage of virus budding. This motif, and similar motifs in proteins from other enveloped viruses, target virus for budding through interaction with components of the endosomal sorting complex required for transport or ESCRT pathway. The release of infectious particles completes the replication cycle and provides progeny virions for infection of the next cell.

The kinetics of the viral replication cycle are rapid. In mammalian cells in culture, one infectious particle can produce 10 000 infectious progeny within 8 h. While much remains to be explored about the biology of VSV, it is one of the best understood animal viruses. Studies on VSV should continue to prove informative in understanding how enveloped viruses enter and bud from their host cells, and how nonsegmented negative-strand RNA viruses (mononegaviruses) express their genetic information.

Functions of the Viral Proteins

Nucleocapsid protein

The N protein coats the viral genomic RNA and the positive-sense antigenomic replicative intermediate to form ribonucleoprotein (RNP) complexes. The interaction with the N protein renders the RNA resistant to cleavage by ribonucleases. It is in this form that the RNA is presented to the polymerase to serve as template. The N protein comprises 422 amino acids and has a molecular weight of approximately 48 kDa. The crystal structure of a complex of 10 molecules of N protein bound to 90 nt of RNA has been solved, revealing that N has a bilobed structure with RNA bound between the lobes. The structure indicates that N protein must be either transiently displaced or substantially remodeled during copying of the RNA genome by the viral polymerase.

Phosphoprotein

P is a multiply phosphorylated acidic protein of 265 amino acids that functions as an essential polymerase cofactor and plays an additional role in maintaining N protein in a soluble form necessary for RNA encapsidation. Sequence analysis has identified three domains of P. An acidic N-terminal domain (domain I) of 150 amino acids contains phosphorylation sites at Ser 60, Thr 62, and Ser 64. Phosphorylation of these residues by the host casein kinase II leads to the oligomerization of P protein and together with the large polymerase protein L, assembly of the polymerase complex. Domain I is separated from domain II by a highly variable hinge region comprising residues 150–210. Domain II (residues 210–244) contains additional phosphorylation sites at Ser 226, Ser 227, and Ser 233 that appear to be important for RNA replication. Domain III is basic and comprises the C-terminal 21 amino acids. A crystal structure of a fragment comprising amino acids 107–177 of the P protein has been solved, providing evidence that P protein functions as a tetramer. In addition to P, two proteins (C and C′) are produced from the P gene. These proteins are small (55 and 65 amino acids, respectively), are highly basic, and have not been detected in virus particles. Recombinant viruses that are unable to produce C and C′ replicate normally in cultured mammalian cells. The role of these proteins is unclear but may be important for infections in insects and/or mammalian hosts *in vivo*.

Matrix protein

The matrix (M) protein is the major structural component of virus particles. M is a small 229-amino-acid multifunctional protein. It condenses viral RNPs and drives budding of virus particles from the host plasma membrane. In addition, M downregulates host gene expression by directly interacting with Rae 1, thus inhibiting nuclear transport. The crystal structure of amino acids 48–229 of the matrix protein has been solved and reveals regions of M that may be required for membrane association. Two N-terminal truncations of M are generated in infected cells by initiation of translation at methionines 33 and 51. A recombinant virus engineered to ablate the expression of these two alternate forms of M shows a modest reduction in cytopathic effect in culture cells.

Glycoprotein

The attachment glycoprotein (G) is a trimer present on the surface of the virion. It is responsible for attachment of virus to cells and promotes fusion of the viral and cellular membranes during endocytosis. The protein is synthesized as a 511-amino-acid polypeptide that is co-translationally inserted into membranes in the endoplasmic reticulum (ER). An N-terminal 16-amino-acid signal sequence is cleaved from the nascent polypeptide upon insertion. A 20-amino-acid hydrophobic sequence acts as the membrane anchor and serves as a stop-transfer signal during ER translocation. The remaining 29 amino acids of G remain as a cytoplasmic tail. Co- and post-translational modifications result in N-glycosylation of two asparagine residues in the ectodomain, and palmitoylation of a cysteine in the cytoplasmic domain. The role of palmitoylation in infection is unclear as some VSV strains lack this modification. However, glycosylation permits association of G with calnexin, an ER-resident chaperone, and mutations that prevent glycosylation result in G protein aggregation. Mature G protein is selectively transported to the basolateral surface of polarized epithelial cells. At the plasma membrane, VSV G clusters in microdomains that are the sites of virus assembly and budding. The determination of the crystal structure of a fragment of the G protein in both pre- and postfusion conformations has provided new insight into the entry process. The structures show that each G protein monomer contains two hydrophobic fusion loops that insert into the target cell membrane. Substantial rearrangements of the G protein that are driven by the low pH environment encountered in the endocytic pathway drive fusion of the viral and cellular membranes.

Large polymerase protein

The large polymerase protein (L) comprises 2109 amino acids. This 241 kDa multifunctional protein is responsible for template binding, ribonucleotide polymerization, and the co-transcriptional modification of the 5′ and 3′ terminus of the viral mRNAs so that they are capped and methylated. To date, structures are not available for any portion of the L protein. Amino acid sequence alignments of the L genes of representative members of the families *Rhabdoviridae*, *Paramyxoviridae*, and *Filoviridae* have led to the identification of six regions of sequence conservation (CRI–CRVI), separated by regions of no or low sequence homology. These conserved regions are thought to represent the functional domains of L protein. CRIII contains clearly identifiable motifs found in all polymerases and in which alterations to a universally conserved aspartic acid residue eliminate polymerase activity in reconstructed RNA synthesis assays. CRVI functions as a messenger RNA cap-modifying enzyme. Amino acid substitutions to this region disrupt mRNA cap methylation at both the guanine N-7 and ribose 2′-O positions of the cap structure. The regions of L involved in other modifications have not yet been assigned. As described above, L protein has been reported to associate with a number of host proteins, including translation elongation factor-1α, heat shock protein 60, and the host RNA guanylyltransferase. However, the functional significance of these interactions is not yet certain.

Host Range and Transmission

In nature, VSVs can infect a broad range of animals including mammals and invertebrates. Overt disease is typically only seen in cattle, horses, and swine, but there is evidence of natural infection in a range of wild ruminants, ungulates, carnivores, marsupials, and rodents. Infection can be transmitted directly from animal to animal, but this requires an abrasion or a means of introducing the virus below the skin. Humans can also be infected by contact with vesicular lesions or the saliva of infected animals, resulting in an influenza-like illness. Aerosol transmission has also been reported in humans. Insect vectors appear to be important in natural transmission of the virus. VSVs have been isolated on more than 40 occasions from hematophagous insects including mosquitoes, biting midges (*Culicoides* spp.), phlebotomine sand flies (*Lutzomyia* spp.), and black flies (*Simulium* spp.). In addition, VSV has been experimentally transmitted to mice from infected mosquitoes and between insects through co-feeding. Susceptible insect hosts are also capable of transovarial transmission and these infected progeny have been shown to infect mammalian hosts. Recovery of vesicular stomatitis New Jersey virus (VSNJV) from nonbiting flies suggests that mechanical transmission may occur. There is also evidence that grasshoppers (*Melanoplus sanguinipes*) can be infected experimentally, and cattle that ingest infected grasshoppers can develop disease.

Consistent with the ability to infect a broad range of experimental animals, VSV replicates efficiently in a variety of cell lines of vertebrate and invertebrate origin. The Syrian hamster kidney cell line, BHK-21, is commonly used to generate viral stocks of high titer (10^9–10^{10} pfu ml^{-1}). VSV can also infect and replicate to high titers in cultured cells derived from insects, reptiles, and fish, and replication can occur in several insect cell lines. Infection of mammalian, avian, and some insect cells results in a cytopathic effect (CPE). In fibroblasts, this is seen as an increase of membrane blebbing followed by extensive cell rounding. The M protein is responsible for these changes in infected cells. A noncytopathic infection occurs in some insect cells including lines derived from *Aedes aegypti*, *Aedes albopictus* (C6/36), and *Drosophila melanogaster*. Experimentally, the virus has been shown to express its genetic information in spheroblasts derived from the yeast, *Saccharamoyces cerevisae*, and to produce infectious virions in embryonic cells derived from the nematode *Caenorhabditis elegans*. Thus, the host requirements for viral replication are provided by an extremely broad range of eukaryotes.

Evolution

Several principles of evolutionary biology have been explored using VSV as a model system. The viral polymerase has an error rate of 10^{-3} to 10^{-5} per nucleotide per round of replication. Given the genome size is 11 161 nt, each time the genome is copied, on average, at least one nucleotide change is introduced. These mutations are typically deleterious, others will be neutral, and some may offer a selective advantage. Consequently, this high mutation rate contributes to VSV's ability to rapidly evolve and adapt to environmental changes. The high mutation rate also begs the question as to why VSV does not mutate itself out of existence. Rather than existing as a single defined sequence, the virus exists as a swarm of mutant sequences around a consensus sequence or a viral quasispecies. Critically, it is this quasispecies that is the biologically relevant target for selection.

Experiments in which the diversity of the viral population is artificially restricted by performing serial plaque-to-plaque transfers (thus restricting the infectious virus population to 1 on each transfer), rapidly result in fitness losses. Such an accumulation of deleterious mutations in populations of asexual organisms lacking compensatory mechanisms such as sex or recombination was predicted by Muller in a concept referred to as Muller's ratchet. This principle has been nicely illustrated in studies of VSV, which is not known to undergo homologous recombination. Studies with VSV have also illustrated other fundamental principles of population genetics. Work by the laboratory of John Holland demonstrated that VSVs of equal fitness could coexist for several generations, but one population would then rapidly outgrow the other. This demonstrated that populations accumulate mutations and, infrequently, mutations provide a selective advantage which leads to its outgrowth. Importantly, the mutation rates in both populations were the same, reflecting an intrinsic property of the viral polymerase. This is referred to as the Red Queen effect. This is a hypothesis proposed by the evolutionary biologist Leigh van Valen in reference to Lewis Carroll's book *Through the Looking Glass*, in which the Red Queen comments to Alice that "it takes all the running you can do, to keep in the same place."

Despite the intrinsically high polymerase error rate, VSVs isolated in nature can show a remarkable genetic stability. For example, sequence analysis of viruses from different hosts and different years has shown that those isolated from multiple hosts within the same region over a period of decades were more closely related than viruses isolated in the same year, but from a different ecological zone. This suggests that ecological factors are a major driving force in the evolution of VSV. Consistent with this, viruses from similar ecological zones 800 km apart were found to be more closely related than viruses from different ecological zones 25 km apart. VSV evolution is unusual in this apparent lack of a 'molecular clock' in which the genome sequence does not show a clear relationship with year of isolation.

The evolutionary origins of VSVs are uncertain but the conservation of gene order and sequence similarity to other members of the *Mononegavirales* indicate they diverged from a common ancestor. Although the number of genes differ, their order is maintained across the *Mononegavirales* as 3′ N–P–M–G–L 5′. The significance of this gene order has been elegantly tested in engineered VSIV recombinants by shuffling the order of the three central genes (P, M, and G), and by moving the N and G genes. Remarkably, all gene orders yield infectious virus, and the replication kinetics of several recombinant viruses in which the P, M, and G genes were shuffled were similar to those of the wild-type parent virus. Consistent with its critical role in driving RNA encapsidation during genome replication, moving the N gene from its promoter proximal position diminished viral replication. These experiments suggest that the conserved gene order may reflect an ancestral gene order that has remained frozen because of the lack of a mechanism for homologous recombination in the *Mononegavirales*.

Geographic and Seasonal Distribution

VSVs are typically restricted to the Americas. Within this region, VSIV and VSNJV are the two most commonly isolated serotypes and have the broadest geographic distribution ranging from Peru to Canada. Vesicular stomatitis Alagoas virus (VSAV) is the major serotype that is isolated in Brazil. Both VSIV and VSNJV are seen in the United States although most outbreaks are associated with VSNJV. In the US, the last outbreak was from 2004 to 2006. In this outbreak, viral infection was reported in Texas, New Mexico, and Colorado in 2004, and the outbreak spread to include Arizona, Utah, Wyoming, Nebraska, Montana, and Idaho in 2005, with a small number of cases reported in Wyoming in 2006. Overall some 1000 animals were affected, representing the most significant outbreak of disease in the US for almost 10 years. Vesicular stomatitis is endemic in many Latin American countries. In 2002, VSNJV and VSIV were reported in Panama, Costa Rica, Nicaragua, El Salvador, and Mexico, and VSNJV was reported in Honduras, Guatemala, and Belize. In 2004, VSNJV and VSIV were isolated in Columbia, Venezuela, Ecuador, and Peru, and VSIV in Brazil and Bolivia. Although infection is reported throughout the Americas, the highest incidence usually is in tropical regions with more sporadic outbreaks in temperate regions. In temperate regions, outbreaks typically peak in late summer and end by the first frost. In tropical regions, peak incidence typically coincides with the end of the rainy season. These peaks correlate with high insect population levels and, consistent with the role of insect vectors in disease spread, outbreaks tend to spread along waterways. Although, on occasion, VSV has been reported in parts of Europe and Africa, such reports are rare and are probably linked to importation of infected animals.

Pathogenesis and Pathology

Following exposure of animals to VSV, the incubation period is usually 2–4 days. Prior to the development of lesions, animals can show signs of depression, lameness, a fever of up to 41.5 °C, and excessive salivation. In infected cattle, swine and horses, pink to white papules appear in the mouth, lips, gums, nose, teats, and feet. In these regions, the epithelium separates from the basal layer forming a vesicle that fills with clear yellowish fluid that contains very high titers of infectious virus. These vesicles combine and readily rupture. In some cases, lesions appear at a secondary site relative to the initial point of inoculation. The development of secondary lesions is suggestive of a viremia, but virus has not been isolated from the blood of experimentally inoculated natural hosts. Infection results in significant losses for the livestock industry as weight loss can be significant (up to 135 kg in beef cattle), and dairy cattle usually cease milk production. In the absence of secondary infections, the vesicular lesions typically heal in 1–2 weeks and the animals start to regain weight. Viral RNA has been isolated from animals several months after infection although infectious virus has not been recovered.

Immune Response

Our understanding of the immune response to VSV in natural hosts is limited. However, in experimental animals, notably mice, the immune response to VSV has been characterized. Interferon plays an important role in the resistance of older mice to VSV infection and, consistent with this, adult mice that are unable to produce interferon succumb to infection. Different strains of VSV induce different levels of interferon, but there is no clear correlation between levels of interferon *in vitro* and pathogenesis in animals. Neutralizing antibody has also been shown to play a role in the defense against VSV infection in experimentally infected swine and mice. The target of these neutralizing antibodies is the viral attachment protein G, and adoptive transfer experiments have shown that they are sufficient to protect young mice against infection. In endemic areas, susceptible hosts often have detectable neutralizing antibody responses to VSV. However, these responses are not protective, perhaps reflecting the fact that viral replication is largely confined to the epithelium. Experimental infection of mice with VSV also induces a strong cytotoxic T-lymphocyte (CTL) response. The N protein is the predominant antigen recognized and the resulting CTLs are not serotype specific. However, the functional significance of these CTLs in protection against infection is uncertain as mice incapable of mounting such a response survive VSV infection.

Diagnosis, Prevention, and Control

The similarity of symptoms between VSV and FMDV make rapid diagnosis important, especially in countries that are free of FMDV. Diagnosis is achieved typically by analysis of samples of fluid from vesicular lesions using reverse transcription-polymerase chain reaction (RT-PCR), enzyme-linked immunosorbent assay (ELISA), or growth of virus in cell culture. VSV is listed as a notifiable disease by the World Organisation for Animal Health (OIE) and a positive diagnosis results in quarantine of the affected area. To control spread of disease, farm equipment is disinfected using a 2% bleach solution to interrupt animal-to-animal spread, and insect control measures can be introduced. Vaccines against VSIV and VSNJV have been used to control outbreaks in Central and South America. Although vaccination is not routinely practiced throughout the Americas, experiments demonstrate a substantive decrease in clinical infections in vaccinated animals.

Future Perspectives

VSVs will continue to serve as an important prototype of the nonsegmented negative-strand RNA viruses. The advantages of VSV as a model for these viruses are: (1) its relative safety, in that the virus is not a significant human pathogen; (2) abundant viral replication in a broad range of cultured cells yielding up to 10 000 pfu $cell^{-1}$ 8–12 h post inoculation of mammalian cells; (3) a robust reverse genetic system which permits the generation of helper-dependent viruses; (4) crystal structures of the N-RNA template, a portion of the P and M proteins, and the pre- and postfusion forms of the G protein; and (5) an *in vitro* system for viral mRNA synthesis using purified recombinant polymerase. Capitalizing on these advantages will continue to provide new mechanistic insights into the molecular details of viral entry, gene expression, and assembly. Among the key questions to address are understanding: (1) how and where the viral RNP core enters a cell; (2) how and where the N-RNA serves as template for gene expression; (3) how the large multifunctional polymerase serves to initiate, cap, methylate, polyadenylate, and terminate mRNA synthesis in response to specific sequence elements; (4) how the activities of the polymerase are controlled between mRNA synthesis and genome replication; and (5) how the nucleocapsid templates are selected for assembly and budding from the cell. In addition to the fundamental questions regarding the molecular biology of these viruses, the future holds promise for unraveling the significance of the different host species and vectors in the biology of infection. Genomic approaches will likely yield clues as to how VSV interacts with host cells of insect and mammalian origin, and how these cells respond to infection. Many questions remain to be answered regarding the ecology of the virus including the maintenance of the virus, its transmission, and its relative genetic stability in nature. Such studies are not only warranted in understanding VSV as a model virus, but also for the prospects of using VSV as an oncolytic virus and a live-attenuated vaccine vector for human disease. These potential applications should only add to the urgency with which studies on this prototypic virus are pursued.

See also: Chandipura Virus; Rabies Virus; Quasispecies.

Further Reading

Eigen M (1993) Viral quasispecies. *Scientific American* 269: 42–49.
Gaudier M, Gaudin Y, and Knossow M (2002) Crystal structure of the vesicular stomatitis virus matrix protein. *EMBO Journal* 21: 2886–2892.
Green TJ, Zhang X, Wertz GW, and Luo M (2006) Structure of the vesicular stomatitis virus nucleoprotein-RNA complex. *Science* 313: 357–360.
Lyles DS and Rupprecht CE (2006) In: Knipe DM, Howley PM, Griffin DE, Lamb RA, and Martin MA (eds.) *Fields Virology,* 5th edn., p. 1364. Philadelphia: Lippincott Williams and Wilkins.
Novella IS (2003) Contributions of vesicular stomatitis virus to the understanding of RNA virus evolution. *Current Opinion in Microbiology* 6: 399–405.
Ogino T and Banerjee AK (2007) Unconventional mechanism of mRNA capping by the RNA-dependent RNA polymerase of vesicular stomatitis virus. *Molecular Cell* 25: 85–97.
Roche S, Bressanelli S, Rey FA, and Gaudin Y (2006) Crystal structure of the low-pH form of the vesicular stomatitis virus glycoprotein G. *Science* 313: 187–191.
Roche S, Rey FA, Gaudin Y, and Bressanelli S (2007) Structure of the prefusion form of the vesicular stomatitis virus glycoprotein G. *Science* 315: 843–848.
Rodriguez LL, Fitch WM, and Nichol ST (1996) Ecological factors rather than temporal factors dominate the evolution of vesicular stomatitis virus. *Proceedings of the National Academy of Sciences, USA* 93: 13030–13035.
Wertz GW, Perepelista VP, and Ball LA (1998) Gene rearrangement attenuates expression and lethality of a nonsegmented negative strand RNA virus. *Proceedings of the National Academy of Sciences, USA* 95: 3501.
Whelan SPJ, Barr JN, and Wertz GW (2004) Transcription and replication of non-segmented negative-strand RNA viruses. *Current Topics in Microbiology and Immunology* 283: 61–119.

Viral Killer Toxins

M J Schmitt, University of the Saarland, Saarbrücken, Germany

© 2008 Elsevier Ltd. All rights reserved.

Glossary

ER-associated degradation (ERAD) A cellular quality control mechanism ensuring efficient removal of misfolded and/or unassembled proteins from the endoplasmic reticulum (ER) lumen and their subsequent elimination by the cytoplasmic ubiquitin–proteasome system.
Heterokaryon Coexistence of two or more genetically different nuclei in a common cytoplasm.
Importins A family of proteins that transport macromolecules into the eukaryotic nucleus.
Killer virus system A killer virus system consists of a helper totivirus and a satellite double-stranded (ds) RNA encoding the unprocessed precursor of a secreted protein toxin that also gives functional immunity.
Preprotoxin Toxin precursor in the cytoplasm of a killer yeast which is post-translationally imported into the secretory pathway for toxin processing, maturation, and secretion.
Satellite dsRNA An encapsidated dsRNA that is dependent on a helper virus for encapsidation and replication.
Spheroplasts A yeast cell whose cell wall has been enzymatically removed.

Introduction

Although double-stranded (ds) RNA viruses had previously been identified in filamentous fungi, dsRNA viruses in yeast were discovered as determinants of the killer phenomenon in *Saccharomyces cerevisiae* and are now known to be associated with the presence of cytoplasmically inherited members of the totivirus family, which are frequently found in different yeast genera. Among these, the killers of *S. cerevisiae*, *Zygosaccharomyces bailii*, *Hanseniaspora uvarum*, and *Ustilago maydis* – the latter being the cause of corn smut – are best characterized. Characteristic for all killers is the secretion of protein toxins that are lethal to sensitive strains of different species and genera. Cell killing is usually achieved in a receptor-mediated process, requiring initial toxin binding to components of the outer yeast cell surface (such as β-1,6-D-glucans, α-1,3-mannoproteins, or chitin) and subsequent toxin transfer to a secondary plasma membrane receptor. Depending on the toxin, final lethality can be caused by plasma membrane damage, G1 or S phase cell-cycle arrest, and/or by rapid inhibition of DNA synthesis.

In the yeasts *S. cerevisiae*, *Z. bailii*, and *H. uvarum* as well as in the maize smut fungus *U. maydis*, the killer phenotype is cytoplasmically inherited and caused by an infection with dsRNA viruses of the family *Totiviridae* that are widely distributed among yeast and higher fungi. Since the majority of fungal mycoviruses are noninfectious and symptomless in the corresponding host, they are often classified as cryptic viruses or virus-like particles (VLPs). All known fungal viruses spread vertically by cell–cell mating and/or heterokaryon formation. In *S. cerevisiae*, diploids formed by mating of a killer with a sensitive strain are likewise killers, as are all haploid progeny of subsequent meiosis. In contrast, virus-free strains are usually sensitive nonkillers, while those containing ScV-L-A and a toxin-encoding M-dsRNA are killers (see below). Sensitive strains survive mating with killers, and cytoplasmic mixing of the multiple M-dsRNA copies during zygosis accounts for the inheritance pattern during meiosis. Extracellular spread of virions is generally hampered by the rigid yeast and fungal cell-wall barrier, and fungal viruses have adopted a strategy of transmission via mating and hyphal fusion (which occurs frequently in nature) making an extracellular route of spread dispensable. While some of these viruses can be associated with adverse phenotypic effects on the fungus (like La France disease in *Agaricus bisporus*, plaque formation in *Penicillium*, and hypovirulence in *Cryphonectria*), dsRNA viruses and their associated satellite dsRNAs in *S. cerevisiae*, *U. maydis*, *H. uvarum*, and *Z. bailii* are responsible for a killer phenotype which is based on the secretion of a polypeptide toxin (killer toxin) that is lethal to a variety of sensitive yeast and fungal strains. With the exception of toxin-secreting strains of *Z. bailii*, killer toxin production is usually associated with specific immunity, protecting the corresponding killer yeast against its own viral toxin.

dsRNA Viruses and Killer Phenotype Expression in *S. cerevisiae*

On the basis of killing profiles and the lack of cross-immunity, three major killer types (K1, K2, K28) have so far been identified in *S. cerevisiae*. Each of them produces a specific killer toxin and a self-protecting immunity component. Killer phenotype expression correlates with the presence of two types of dsRNA species stably persisting in the cytoplasm of the infected host: the genomic dsRNA of the helper virus, ScV-L-A, and one of three toxin-coding satellite dsRNAs that have been recognized so far (ScV-M1, ScV-M2, or ScV-M28). The ScV-L-A and M-dsRNAs are separately encapsidated into capsids encoded by ScV-L-A dsRNA and are present in high copy number in the yeast cell cytoplasm. *In vivo*, ScV-L-A does not confer a phenotype nor does it lead to host-cell lysis or slower cell growth. While the killer phenotype can be transmitted to sensitive yeast cell spheroplasts (harboring ScV-L-A) by transfection with an ScV-M preparation (either during mating or cotransformation with a dsDNA plasmid), extracellular transmission occurs rarely, if at all, in nature. The survival strategy adopted by these dsRNA viruses appears to be a balanced host interaction resulting in stable maintenance, little if any growth disadvantage, and vertical transmission. Mechanisms of exiting and entering the host cell through its tough and rigid cell wall are rendered unnecessary by relatively efficient horizontal transmission during the frequent zygosis events in yeast. Acquisition of a toxin-encoding M satellite dsRNA provides positive selection for both this dsRNA and ScV-L-A, since virus-free segregants are killed.

As summarized in **Table 1**, the linear dsRNA genome of ScV-L-A contains two open reading frames (ORFs) on its plus-strand RNA: ORF1 encodes the major capsid protein Gag necessary for encapsidation and viral particle structure, the second gene (ORF2) represents the RNA-dependent RNA polymerase Pol which is *in vivo* expressed as a Gag–Pol fusion protein by a −1 ribosomal frameshift event. In contrast to L-A, each M-dsRNA genome contains a single ORF coding for a preprotoxin (pptox) representing the unprocessed precursor of the mature and secreted toxin that also gives functional immunity. Since each toxin-coding ScV-M dsRNA depends on the coexistance of ScV-L-A for stable maintenance and replication, the killer viruses resemble classical satellites of ScV-L-A. Although the presence of all three M-dsRNAs with different killer specificities in a single cell is excluded at the replicative level of the M genomes, this limitation can be by-passed by introducing cDNA copies of the K2 and K28 pptox genes into a

Table 1 Members of the family *Totiviridae* involved in killer phenotype expression

Totivirus/satellite dsRNA	Virus host	Function of virus	dsRNA (kb)	Encoded protein(s)
ScV-L-A	*Saccharomyces cerevisiae*	Helper virus	L_A (4.6)	Gag, major capsid protein; Gag-Pol, RDRP[a]
ScV-M1	*Saccharomyces cerevisiae*	Satellite dsRNA	M_1 (1.6)	K1 preprotoxin
ScV-M2	*Saccharomyces cerevisiae*	Satellite dsRNA	M_2 (1.5)	K2 preprotoxin
ScV-M28	*Saccharomyces cerevisiae*	Satellite dsRNA	M_{28} (1.8)	K28 preprotoxin
UmV-H	*Ustilago maydis*	Helper virus	H (6.1)	Gag, major capsid protein; Gag-Pol, RDRP[a]
UmV-P1	*Ustilago maydis*	Satellite dsRNA	M_{P1} (1.4)	KP1 preprotoxin
UmV-P4	*Ustilago maydis*	Satellite dsRNA	M_{P4} (1.0)	KP4 toxin
UmV-P6	*Ustilago maydis*	Satellite dsRNA	M_{P6} (1.2)	KP6 preprotoxin
HuV-L	*Hanseniaspora uvarum*	Helper virus	L_A (4.6)	Gag, major capsid protein; Gag-Pol, RDRP[a]
HuV-M	*Hanseniaspora uvarum*	Satellite dsRNA	M_{Hu} (1.0)	KT470 toxin precursor
ZbV-L	*Zygosaccharomyces bailii*	Helper virus	L_{Zb} (4.6)	Gag, major capsid protein; Gag-Pol, RDRP[a]
ZbV-M	*Zygosaccharomyces bailii*	Satellite dsRNA	M_{Zb} (2.1)	prepro-zygocin

[a]RDRP, RNA-dependent RNA polymerase.
Adapted from Schmitt MJ and Breinig F (2002) The viral killer system in yeast: From molecular biology to application. *FEMS Microbiology Reviews* 26: 257–276, with permission from Blackwell Publishing.

natural K1 killer, resulting in stable triple-killers producing three different virus toxins at a time and simultaneously expressing triple-toxin immunity.

As typical for many dsRNA viruses, the replication cycles of ScV-L-A and M-dsRNAs depend on the presence of specific packaging signals at or near the 3′ end of the viral single-stranded RNA (ssRNA) transcript. These RNA regions function as viral binding sites (VBSs) and consist of a stem–loop structure whose stem is interrupted by an unpaired protruding A residue (**Figure 1**). For replication of L-A, two sequence elements are essential: an internal replication enhancer (IRE), which is essentially indistinguishable from the VBS, and a small stem–loop structure 5 bp from the 3′-terminus (3′-TRE). While L-A contains just a single VBS element, the toxin-coding transcripts of M1, M28, and M_{Zb} (encoding the *Z. bailii* viral toxin zygocin) each have two such VBS domains. *In vivo*, these VBS elements are *cis*-active sequences that are recognized by Gag–Pol and subsequently packaged into new viral particles. The replicase reaction on the (+)ssRNA template takes place *in viro* (i.e., within the viral capsid) and requires a correct 3′-end sequence and structure. Within the intact and mature virion, conservative transcription of the plus-strand from the dsRNA template requires recognition of its very 5′-terminal sequence by Gag–Pol. In M28 (as in other dsRNAs), the plus-strand initiates with 5′-GAAAA(A); since there are little additional conserved sequences immediately downstream, this terminal recognition element (5′-TRE) may be all that is necessary for transcription initiation (**Figure 1**).

Viral Replication Cycle

Yeast ScV-L-A virions are noninfectious icosahedral particles 39 nm in diameter that show certain similarities to mammalian reoviruses and rotaviruses. Each L-A virus consists of a single copy of the 4.6 kb L-A dsRNA genome which is encapsidated by 60 asymmetric dimers of the 76 kDa coat protein Gag and two copies of the 171 kDa Gag–Pol fusion protein. During conservative replication, the single-stranded plus-strand RNA (L-A(+)ssRNA) is transcribed within the viral particle (*in viro*) and subsequently extruded into the cytoplasm where it serves as (1) messenger RNA for translation into the viral proteins Gag and Gag–Pol and (2) RNA template which is packaged into new viral particles. Once this coat assembly is completed, Gag–Pol functions as replicase, synthesizes a minus-strand, and generates the dsRNA genome of the mature virus. The replication cycle of the toxin-coding M satellite dsRNA resembles that of L-A with the exception that each M virion can accept two copies of the smaller M-dsRNA genome before the ssRNA transcript is extruded into the cytoplasm; in analogy to certain DNA bacteriophages, this phenomenon has been named 'headful packaging'.

Viral Preprotoxin Processing and Toxin Maturation

In totivirus infected killer yeasts, the toxin-encoding M(+)ssRNA transcript is translated on free cytosolic ribosomes into a pptox precursor which is post-translationally imported into the secretory pathway for further processing, maturation, and toxin secretion. Interestingly, intracellular pptox processing in killer strains of either *S. cerevisiae* (K1, K2, K28), *Z. bailii* (zygocin), or *U. maydis* (KP4) is mechanistically conserved, resulting in the secretion of a biologically active monomeric or α/β heterodimeric virus toxin (**Figure 2**). In case of K28 killer strains containing ScV-M28, pptox is processed into a heterodimer whose α- and β-subunit is covalently linked by a single disulfide bond. Since the unprocessed toxin precursor resembles a secretory protein, it contains a hydrophobic

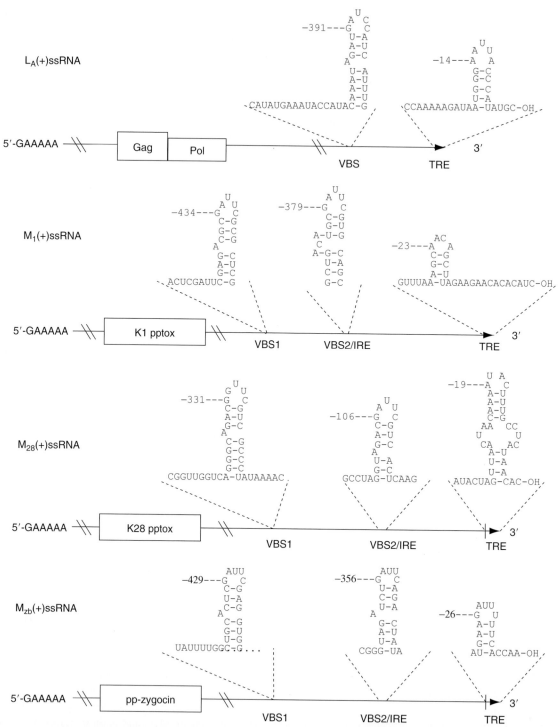

Figure 1 Comparison of the 5'- and 3'-ssRNA sequences in ScV-L-A and in the toxin-coding transcripts of the satellite ScV-M1, ScV-M28, and ZbV-M dsRNAs. Potential cis-active sequences within the 3'-termini are indicated (VBS, viral binding site; IRE, internal replication enhancer; TRE, 3'-terminal recognition element). Numbers are shown for distance from the 3' end of each (+)ssRNA.

signal sequence for pptox import into the ER lumen followed by the toxin subunits α (10.5 kDa) and β (11.0 kDa) which are separated from each other by a potentially N-glycosylated γ-sequence (**Figure 2**). During passage through the yeast secretory pathway the K28 toxin precursor is enzymatically processed to the biologically active heterodimer in a way that is highly homologous to prohormone conversion in mammalian cells. In a late Golgi compartment, the N-glycosylated γ-sequence is removed by the action of the furin-like endopeptidase Kex2p, the

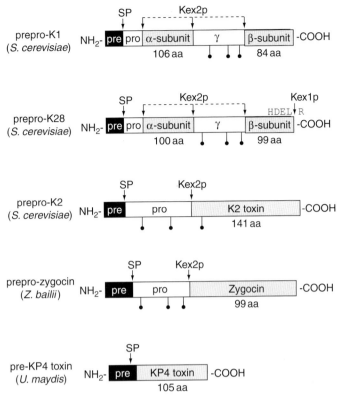

Figure 2 Analogy of toxin precursor processing in totivirus-infected killers of S. cerevisiae, Z. bailii, and U. maydis. Schematic outline of the precursor proteins (preprotoxins) encoding either a heterodimeric α/β toxin (K1, K28) or a monomeric virus toxin (K2, zygocin, KP4). Signal peptidase (SP) cleavage and Kex2p endopeptidase processing sites are indicated. Potential N-glycosylation sites are indicated with black circles (aa, amino acids).

C-terminus of β is trimmed by carboxypeptidase Kex1p cleavage, and the biologically active protein is secreted as a 21 kDa α/β heterodimer whose β-C-terminus carries a four-amino-acid epitope (HDEL) that represents a classical ER retention signal. Since this signal is initially masked by a terminal arginine residue (HDEL*R*), ER retention of the toxin precursor is prevented until the protoxin enters a late Golgi compartment where Kex1p cleavage uncovers the HDEL signal of the toxin (**Figure 3**).

Endocytosis and Intracellular Transport of the K28 Virus Toxin

In contrast to most virally encoded yeast killer toxins, which do not enter their host but rather kill a sensitive cell by disrupting plasma membrane function, K28 is taken up by receptor-mediated endocytosis and subsequently targeted to the secretory pathway. Once it has reached an early endosomal compartment, the toxin travels the secretion pathway in reverse and translocates into the cytosol. Responsible and essential for this retrograde transport is a short-amino-acid motif at the carboxyterminus of the K28 β-subunit (HDEL) which functions as ER targeting/retention signal that is normally present at the C-terminus of resident chaperones in the ER lumen such as Kar2p/BiP and Pdi1p, the protein disulfide isomerase. In yeast and higher eukaryotes, H/KDEL-carrying proteins are recognized by a membrane-bound H/KDEL receptor ensuring efficient recycling of H/KDEL-proteins from an early Golgi compartment back to the ER. In case of the K28 virus toxin, this sequence allows retrograde transport from early endosomes (via Golgi) to the ER from where the toxin enters the cytoplasm and transduces its lethal signal into the nucleus. Endocytotic uptake and retrograde transport is a common strategy realized in certain prototypes of bacterial toxins such as *Pseudomonas* exotoxin A, *Escherichia coli* heat-labile toxin (HLT), or even Shiga toxin. All these protein toxins are family members of microbial A/B toxins, which are usually internalized by receptor-mediated endocytosis, followed by reverse secretion via Golgi and ER. Interestingly, many of these toxins contain putative ER retention signals at their C-termini, and H/KDEL-dependent mechanisms have, therefore, been postulated to be of major importance for toxin entry into mammalian cells. In this respect, a major difference between the yeast K28 virus toxin and bacterial A/B toxins is that K28 itself

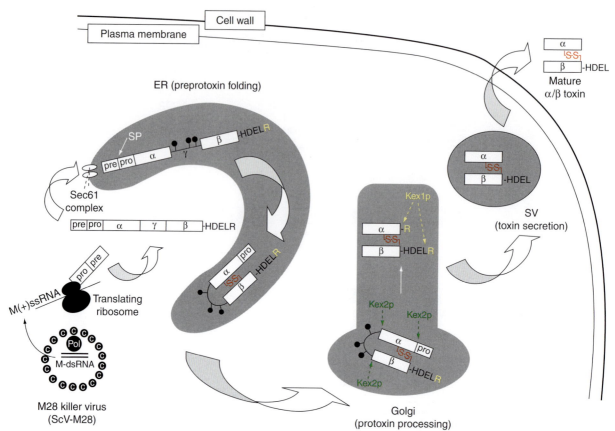

Figure 3 Preprotoxin processing and toxin secretion in an ScV-M28-infected killer yeast. After *in vivo* translation of the preprotoxin coding killer virus transcript, the toxin precursor is post-translationally imported into the endoplasmic reticulum (ER) through the Sec61 complex. Signal peptidase (SP) cleavage in the ER lumen removes the N-terminal signal sequence (pre-region) and protoxin folding is mediated by lumenal ER chaperones. The intervening γ-sequence is N-glycosylated and a single disulfide bond between α and β is generated. In a late Golgi compartment, Kex2p endopeptidase cleaves the pro-region, removes the γ-sequence, and carboxypeptidase Kex1p cleavage trimms the C-termini of both subunits, leading to the secretion of mature α/β toxin whose β-C-terminal HDEL motif is uncovered and thus accessible for interaction with the HDEL receptor of the target cell. SV, secretory vesicle. Adapted from Schmitt MJ and Breinig F (2002) The viral killer system in yeast: From molecular biology to application. *FEMS Microbiology Reviews* 26: 257–276, with permission from Blackwell Publishing.

is produced and secreted by a eukaryotic (yeast) cell, and therefore the C-terminal ER-targeting signal in the toxin precursor is initially masked by a terminal arginine residue which ensures successful pptox passage through the early secretory pathway. Once the toxin has reached a late Golgi compartment, Kex1p cleavage removes the β-C-terminal arginine residue and thereby uncovers the ER targeting signal of the virus toxin. To ensure proper access of the ER targeting signal to the K/HDEL receptor of the sensitive target cell, many A/B toxins (including the yeast K28 virus toxin) contain a unique disulfide bond at or near the C-terminus. Consequently, mutant toxin variants with altered inter- and/or intra-subunit disulfide bonding are nontoxic *in vivo* due to the incapability to reach their intracellular target. Thus, disulfide bond formation in microbial and viral A/B toxins is of major importance to ensure interaction competence of the toxins with the K/HDEL-receptor of the target cell.

ER Exit and Nuclear Entry of the K28 Virus Toxin

During host-cell penetration, K28 retrotranslocates from the ER into the cytosol and dissociates into its subunit components. The β-subunit is subsequently polyubiquitinated and proteasomally degraded while the cytotoxic α-subunit enters the nucleus and causes cell death (**Figure 4**). ER exit of the α/β heterodimeric toxin is mediated by the Sec61 complex, termed translocon, which functions as major transport channel in the ER membrane of yeast and higher eukaryotes. In yeast, each translocon resembles a core heterotrimeric complex consisting of the transmembrane protein Sec61p and the two smaller subunits Sbh1p and Sss1p. Besides being the major channel for co- and post-translational protein import into the ER, Sec61p is also involved in the export and removal of malfolded and/or misassembled proteins from the secretory pathway to initiate their proteasomal

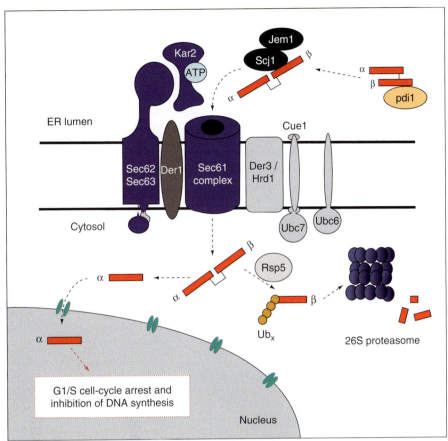

Figure 4 Retrotranslocation of the K28 virus toxin from the ER and its lethal effect in the nucleus. After endocytotic uptake and retrograde transport via Golgi and ER, the toxin is gated through the Sec61p export channel by the help of lumenal ER chaperones such as Pdi1p, Kar2p, Jem1p, and Scj1p. Within the cytosol, β is ubiquitinated and proteasomally degraded while α enters the nucleus and causes cell death. The α-toxin interacts with essential host proteins involved in eukaryotic cell-cycle control and causes cell death through G1/S cell-cycle arrest and inhibition of DNA synthesis. Cellular components of the ER quality control system ERAD (such as Cue1p, Ubc7p, Der3p/Hrd1p, Ubc6p, and Der1p) are not involved in ER-to-cytosol export of the K28 virus toxin (ERAD components are shown in gray color). Reproduced from Leis S, Spindler J, Reiter J, Breinig F, and Schmitt MJ (2005) *Saccharmyces cerevisiae* K28 toxin – A secreted virus toxin of the A/B family of protein toxins. In: Schmitt MJ and Schaffrath R (eds.) *Topics in Current Genetics 11: Microbial Protein Toxins*, pp. 111–132. Berlin, Heidelberg: Springer, with kind permission of Springer Science and Buisness Media.

degradation in the cytosol. In addition to its central function in protein quality control in the ER, Sec61 is also responsible for ER retrotranslocation of certain plant, microbial and viral A/B toxins such as ricin, cholera toxin, *Pseudomonas* exotoxin A, and the yeast K28 virus toxin. In contrast to microbial and plant A/B toxins, however, retrotranslocation of K28 from the ER lumen is independent of ubiquitination and proteasome activity and classical components normally involved in ER-associated protein degradation (ERAD) are not required for ER exit of this virus toxin. In K28 intoxicated cells, toxin translocation competence in the ER strongly depends on the activity of lumenal Hsp70 chaperones (such as Kar2p/BiP and Pdi1p), Hsp40 cochaperones (such as Scj1p and Jem1p), and additionally requires proper maintenance of calcium homeostasis in the ER (**Figure 4**). So far it is not known what cellular component within or near the ER membrane is responsible for toxin exit, but K28 might be a fruitful tool to identify a novel transport pathway for protein transport across the ER membrane.

K28 Affects DNA Synthesis, Cell-Cycle Progression, and Induces Apoptosis

Although the cytotoxic α-component of K28 (10.5 kDa) can enter the nucleus by passive diffusion, extension of α by a classical nuclear localization sequence (NLS) significantly enhances its *in vivo* toxicity due to faster and more efficient nuclear import mediated by α/β importins of the host. Within the nucleus, α interacts with host proteins of essential function in eukaryotic cell cycle control and initiation of DNA synthesis. Thus, as the virus toxin targets evolutionary highly conserved proteins with basic and central function, toxin resistance mechanisms based on mutations in essential chromosomal host genes hardly occur *in vivo*, indicating that the toxin has developed a very efficient strategy to penetrate and kill its target cell. Most interestingly, while high toxin concentrations (10 pmol or higher) cause necrotic cell killing via cell-cycle arrest and inhibition of DNA synthesis, treatment with low doses of viral killer toxins (<1 pmol) results in an apoptotic host-cell response triggered by the accumulation of reactive oxygen species, ROS (**Figure 5**). Since toxin concentration is usually low in the natural environment of a killer yeast, toxin-induced apoptosis is probably an important prerequisite for efficient cell killing. Furthermore, since apoptosis is also important in the pathogenesis of virus infections in mammals, it is not surprising that toxin-encoding yeast killer viruses (as has been shown for ScV-M1 and ScV-M2) can also induce a programmed suicide pathway in noninfected yeast. Although viral killer toxins were shown to be primarily responsible for this phenomenon, yeast killer viruses are not solely responsible for triggering a cell death pathway in yeast.

Lethality of Membrane Damaging Viral Killer Toxins

Yeast viral killer toxins kill sensitive cells in a receptor-mediated fashion by interacting with receptors at the level of the cell wall and the cytoplasmic membrane (**Figure 5**). The initial step involves rapid toxin binding to a primary receptor R1 which is localized within the mannoprotein or β-1,6-glucan fraction of the cell wall. In the second step the toxin translocates to the plasma membrane and interacts with a secondary receptor R2 (**Figure 5**). To date, only the membrane receptor for the K1 virus toxin has been identified: Kre1p, an O-glycosylated cell surface protein initially GPI-anchored to the plasma membrane and involved in β-1,6-glucan biosynthesis and K1 cell wall receptor assembly. Once bound to the plasma membrane, ionophoric virus toxins (such as K1 and zygocin) disrupt cytoplasmic membrane function by forming cation-selective ion channels, while K28 enters the cell and acts in the nucleus: DNA

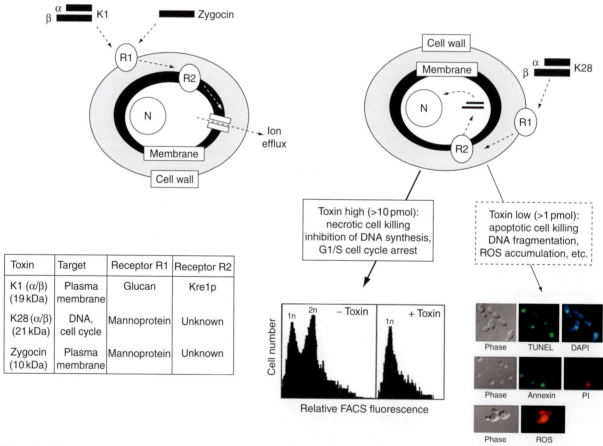

Figure 5 Receptor-mediated toxicity of the viral killer toxins K1, zygocin, and K28. Killing of a sensitive yeast is envisaged in a two-step process involving initial toxin binding to receptors within the cell wall (R1) and the cytoplasmic membrane (R2). After interaction with the plasma membrane, ionophoric toxins such as K1 and zygocin disrupt cytoplasmic membrane function, while K28 enters the cell by endocytosis and diffuses into the nucleus to cause cell death (note that the cell surface receptors R1 and R2 are different in all three toxins; see also table inset). At high toxin doses (>10 pmol) sensitive cells arrest in the cell cycle with pre-replicated DNA (1n; left panel), while cells treated with K28 in low concentrations (<1 pmol) respond with apoptosis as shown by typical apoptotic markers such as chromosomal DNA fragmentation (TUNEL-positive cells), accumulation of reactive oxygen species (ROS), and phosphatidylserin exposure at the external surface of the plasma membrane detected by annexin-V staining (right panel).

synthesis is rapidly inhibited and cells arrest at the G1/S boundary of the cell cycle (**Figure 5**).

Ion channel formation in yeast membranes induced by the K1 virus toxin was initially reported using patch-clamping techniques as a result from direct toxin action. However, this observation is inconsistent with the complete resistance seen in immune yeast cell spheroplasts, and so far receptor-independent channels have not been observed, neither in yeast membranes nor in *Xenopus laevis* oocytes. Similar to K1 (and probably K2 as well), zygocin represents a membrane-damaging virus toxin which is produced and secreted by ZbV-M-infected killer strains of the osmotolerant spoilage yeast *Z. bailii*. Zygocin itself is a monomeric, nonglycosylated protein toxin with an unusual broad killing spectrum, being equally active against phytopathogenic as well as human pathogenic yeasts including *Candida albicans*, *C. glabrata*, *C. tropicalis*, and *Sporothrix schenkii*. Since even filamentous fungi such as *Fusarium oxysporum* and *Colletotrichum graminicola* are effectively killed by the toxin, zygocin represents a virus toxin with significant antimycotic potential. Similar to K1 but significantly more efficient, zygocin disrupts plasma membrane integrity and causes rapid cell killing. Its ionophoric mode of action has been reinforced by *in silico* sequence analysis, identifying a stretch of potential α-helical conformation that forms an amphipathic structure characteristic for membrane-disturbing antimicrobial peptides such as alamethicin, melittin, and dermaseptin. In addition, this feature is accompanied by a transmembrane helix at the C-terminus of zygocin which is predicted to favor a membrane permeabilizing potential, not by activating native ion channels but rather by establishing pores by itself after toxin oligomerization. It is therefore assumed that the hydrophobic part in zygocins' amphipathic α-helix is responsible for toxin binding to the target cell. The postulated model of zygocin action resembles that of human α-defensins. In analogy to alamethicin, toxicity of zygocin is probably mediated by incorporation of its transmembrane helix into the plasma membrane, a process solely driven by the natural transmembrane potential of the energized yeast and fungal plasma membrane. Thus, zygocins' mode of action portrays the lethal mechanism of antimicrobial peptides that are produced by virtually all higher eukaryotes. Mechanisms of resistance against antimicrobial peptides are rare and often limited to changes in the composition of the cytoplasmic membrane. In major contrast to mammalian cells, the outer leaflet of yeast and fungal membranes is enriched in negatively charged lipids. Due to the cationic net charge of antimicrobial peptides (including zygocin), an affinity to these lipids facilitates toxin adsorption to the target membrane. Consistent with that, deletion of chromosomal genes whose gene products affect plasma membrane lipid composition (such as *PDR16* and *PDR17*) causes a dramatic decrease in zygocin sensitivity because toxin binding to the plasma membrane is largely prevented in the genetic background of a yeast *pdr16/17* mutant. In contrast to K1, a zygocin-specific membrane receptor is not required for its *in vivo* toxicity as the physicochemical properties of zygocin allow efficient plasma membrane interaction independent of any membrane receptor or docking protein.

Self-Protection in Killer Virus-Infected Yeast – Toxin Immunity

For many decades it was unknown how a killer virus-infected yeast protects itself against its own secreted toxin. In killer yeast, functional immunity is essential for survival since the toxins often target and inhibit central eukaryotic cell functions. This is in major contrast to bacterial toxins such as cholera toxin and Shiga toxins which selectively kill eukaryotes, thus making immunity dispensable in a prokaryotic host. Recently, the mechanism of protecting immunity against the K28 virus toxin has been elucidated. It is now known that immune cells take up external toxin (either produced by itself or by other K28 killers) and translocate it back to the cytosol where the reinternalized α/β toxin rapidly forms a complex with the pptox precursor that has not yet been imported into the ER. Within this complex, the K28 heterodimer is selectively ubiquitinated and proteasomally degraded while the pptox moiety of the complex is in part released to be either imported into the ER (to give active virus toxin) or to complex a newly internalized K28 heterodimer. In this process, the amount of cytosolic ubiquitin is critical for immunity and overexpression of mutant ubiquitin (blocked in polyubiquitin chain formation) results in a significant decrease in toxin secretion and a suicidal phenotype based on nonfunctional immunity. Alternatively, decreasing cytosolic ubiquitin causes an increase in toxin secretion, while immunity is not impaired as sufficient pptox is available for K28 complex formation. This simple and highly efficient mechanism ensures that a toxin-producing killer yeast is fully protected against the lethal action of its own toxin. In contrast to K28 immunity, the precise mechanism(s) of self-protection against the ionophoric virus toxins K1 and zygocin is still obscure and remains largely unknown.

See also: Fungal Viruses; Totiviruses; Ustilago Maydis Viruses; Yeast L-A Virus.

Further Reading

Breinig F, Sendzik T, Eisfeld K, and Schmitt MJ (2006) Dissecting toxin immunity in virus-infected killer yeast uncovers an intrinsic strategy of self-protection. *Proceedings of the National Academy of Sciences, USA* 103: 3810–3815.

Bruenn JA (2002) The double-stranded RNA viruses of *Ustilago maydis* and their killer toxins. In: Tavantzis SM (ed.) *dsRNA Genetic Elements: Concepts and Applications in Agriculture, Forestry, and Medicine*, pp. 109–124. Boca Raton, FL: CRC Press.

Bruenn JA (2005) The *Ustilago maydis* killer toxins. In: Schmitt MJ and Schaffrath R (eds.) *Microbial Protein Toxins*, pp. 157–174. Berlin, Heidelberg: Springer.

El-Sherbeini M and Bostian KA (1987) Viruses in fungi: Infection of yeast with the K1 and K2 killer virus. *Proceedings of the National Academy of Sciences, USA* 84: 4293–4297.

Ivanovska J and Hardwick JM (2005) Viruses activate a genetically conserved cell death pathway in a unicellular organism. *Journal of Cell Biology* 170: 391–399.

Leis S, Spindler J, Reiter J, Breinig F, and Schmitt MJ (2005) S. cerevisiae K28 toxin – a secreted virus toxin of the A/B family of protein toxins. In: Schmitt MJ and Schaffrath R (eds.) *Microbial Protein Toxins*, pp. 111–132. Berlin, Heidelberg: Springer.

Reiter J, Herker E, Madeo F, and Schmitt MJ (2005) Viral killer toxins induce caspase-mediated apoptosis in yeast. *Journal of Cell Biology* 168: 353–358.

Schmitt MJ and Breinig F (2002) The viral killer system in yeast: From molecular biology to application. *FEMS Microbiology Reviews* 26: 257–276.

Schmitt MJ and Breinig F (2006) Yeast viral killer toxins: Lethality and self-protection. *Nature Reviews Microbiology* 4: 212–221.

Schmitt MJ and Neuhausen F (1994) Killer toxin-secreting double-stranded RNA mycoviruses in the yeasts *Hanseniaspora uvarum* and *Zygosaccharomyces bailii*. *Journal of Virology* 68: 1765–1772.

Weiler F and Schmitt MJ (2005) Zygocin – a monomeric protein toxin secreted by virus-infected *Zygosaccharomyces bailii*. In: Schmitt MJ and Schaffrath R (eds.) *Microbial Protein Toxins*, pp. 175–187. Berlin Heidelberg: Springer.

Wickner RB (1996) Double-stranded RNA viruses of *Saccharomyces cerevisiae*. *Microbiology Reviews* 60: 250–265.

Viral Membranes

J Lenard, University of Medicine and Dentistry of New Jersey (UMDNJ), Piscataway, NJ, USA

© 2008 Elsevier Ltd. All rights reserved.

Introduction

Viruses of many kinds possess lipids as integral components of their structure. Lipid-containing, or enveloped, viruses include: *Baculo-, Bunya-, Corona-, Filo-, Herpes-, Lenti-, Orthomyxo-, Paramyxo-, Pox-, Retro-, Rhabdo-,* and *Togaviridae*. Despite the great diversity of these viruses in regard to structure, replicative strategy, host range and pathogenicity, the function of the lipids in each case is the same – to form a membrane surrounding the encapsidated viral RNA or DNA genome. The lipids of the viral envelope form a continuous bilayer that functions as a permeability barrier protecting the viral nucleocapsid from the external milieu. Embedded in this bilayer are numerous copies of a limited number of virally encoded transmembrane proteins (often just one or two) that are required for virus entry into a potential host cell. These proteins mediate two essential functions: attachment of the virion to the cell surface; and fusion of the viral envelope with a cell membrane, resulting in accession of the viral nucleocapsid containing the genome to the cellular cytoplasm.

The membrane is acquired during viral assembly within an infected cell. This generally occurs by the budding of the previously assembled viral nucleocapsid through a cell membrane, incorporating host cell lipids and viral membrane proteins, but excluding virtually all cell membrane proteins. The particular cell membrane through which viral budding occurs is characteristic for each virus. Many viruses bud through the plasma membrane, but bunyaviruses bud through the Golgi apparatus, coronaviruses chiefly through the endoplasmic reticulum, and herpesviruses through the nuclear membrane. Poxviruses, which are among the largest and most complex animal viruses, are unique in acquiring two discrete membranes through a series of interactions with different cellular organelles, in a process that is not well understood.

Viral membranes have been extensively studied for many years, and for many reasons. The budded viral envelopes themselves represent isolated, readily purifiable subdomains of discrete cellular membrane bilayers. The envelope proteins that mediate entry of viruses into infectible cells are regarded as prototypes of membrane fusion machines. The initial binding of virions to potential host cells is mediated by prototypical ligand–receptor interactions in which viral proteins (ligands) bind to cell surface proteins (receptors). These properties of viral envelopes represent potential targets for antiviral therapy; they also provide models for important cellular phenomena.

Viral Bilayers

For many non-pathogenic viruses that can be readily purified in sufficient quantities in the laboratory, the bilayer arrangement of lipids in the viral envelope has been directly demonstrated using physical methods. It is assumed that the lipids in all viral envelopes are similarly arranged in a bilayer, based on those observations, and on the fact that the host cell membranes from which the viruses bud all contain bilayers, and also because it is the only physically reasonable arrangement of lipids that could supply the required protection from environmental stresses. Intact virions are impermeant to proteases and other enzymes.

Indeed, virions can swell and shrink in response to changes in osmolarity, showing that viral envelopes are impermeant to small molecules and ions as well as large proteins, and thus must consist of intact bilayers that completely surround the encapsidated viral genome.

The lipid composition of various viruses grown under different conditions, and in different cell types, has been studied in detail. These have shown that wide variations in lipid composition are tolerated in many viruses; they have provided no evidence that any substantial fraction of envelope lipids is bound to viral envelope proteins specifically, or exists in a nonbilayer conformation.

Recent studies have shown that many viruses bud from specialized regions of the plasma membrane known as 'rafts'. Rafts are regions of the plasma membrane characterized by high concentrations of sphingolipids (sphingomyelin and sphingoglycolipids) and cholesterol. These components participate in the formation of separate, partially miscible phases, distinguishable from the more fluid phase(s) which are relatively enriched in unsaturated phospholipids. Raft and non-raft phases co-exist as contiguous bilayers, but diffusion is relatively restricted within raft phases, and exchange of molecules between phases occurs more slowly than diffusion within a single phase. Although the existence of local heterogeneity in lipid and protein structure has been demonstrated in cell membranes using a variety of physical and detergent extraction techniques, a precise definition of rafts has not been achieved. There is little agreement regarding their size, or whether they are nucleated by lipids or by proteins. Despite these uncertainties, certain membrane constituents are clearly concentrated in specialized regions rich in sphingolipids and cholesterol, and these have been identified as the sites of important cellular signaling and transport functions, and of assembly and budding for many enveloped viruses. Detailed lipid analyses of purified budded virions have confirmed their origin in rafts, and have in turn helped to define the lipid composition of virus-associated rafts.

Viral Membrane Proteins

The proteins of viral membranes, like those of other membranes, may be classified as either integral or peripheral. Integral proteins are those that cross the membrane bilayer at least once, and thus cannot be solubilized without disrupting the bilayer, that is, without using detergents. Peripheral proteins are also membrane associated, but they do not cross the membrane and they can be removed from it by treatment with aqueous salts, high pH or chaotropic agents, which leave intact the hydrophobic interactions that stabilize the bilayer.

Most integral membrane proteins of enveloped viruses span the bilayer only once, although exceptions exist. Each transmembrane (anchoring) domain is a sequence of 18–27 predominantly hydrophobic amino acid residues. Because transmembrane sequences are inherently insoluble in water, integral membrane proteins require detergents for extraction from the bilayer and solubilization. When detergents and lipids are both removed from purified viral proteins, they tend to aggregate into rosettes, forming a kind of protein micelle, with the transmembrane sequences clustered together at their centers in order to maximize hydrophobic interactions and minimize contact with water. Purified viral membrane proteins can be reinserted into lipid bilayers of defined composition by mixing the detergent-solubilized protein with lipids, then removing the detergent by dialysis. These reconstituted viral membranes ('virosomes') often possess the native receptor binding and fusion activities.

The functions of the major integral viral membrane proteins are: first, to attach the virus to the uninfected host cell; and second, to effect penetration of the genome into the host cell cytoplasm through membrane fusion of the viral envelope with a host cell membrane. As much as 90% of each viral receptor binding/fusion protein is external to the viral membrane and thus accessible to removal and/or degradation by added proteases. Viral membrane proteins are often morphologically identifiable in electron micrographs as 'spikes' on the outer surface of membrane particles. Under favorable conditions, nearly the entire external domain may be rendered soluble and recovered intact and correctly folded after proteolytic removal from the virion, facilitating crystallization and structural analysis. These domains possess oligosaccharide side chains identical to those of cellular integral proteins in structure and attachment sites, and often possess disulfide bonds as well, reflecting the viral proteins' normal procession through the cell's endoplasmic reticulum–Golgi system (**Figure 3**).

While these proteins constitute the major fraction (>95%) of viral integral membrane proteins, there is an additional class of small integral proteins that oligomerize within the bilayer to form channels that facilitate the transport of ions or small molecules. These proteins have been called 'viroporins'. They include the M2 protein of influenza, the 6K protein of alphaviruses, and Vpu of HIV-1. They are thought to function in various ways to facilitate the assembly and release of new viral particles from the infected cell.

Peripheral membrane proteins are attached to the viral membrane by a combination of hydrophobic and electrostatic interactions. They may penetrate the bilayer to some extent, but they do not cross it as integral proteins do. Viral peripheral membrane proteins include M1 of influenza, M of paramyxoviruses, and MA of retroviruses. All enveloped viruses except the togaviruses encode an M-like peripheral protein that functions to bring together the envelope and nucleocapsid components during viral assembly.

Cellular Virus Receptors: Virus Membranes as Ligands

The first step in infection, attachment of the virus to the outer surface of the host cell, is performed by specific membrane proteins of enveloped viruses. One or more unique cellular 'receptors' are recognized by each species or strain of virus. The presence of a specific cellular receptor is often the major factor determining the susceptibility of a particular species to infection; it also determines the infectibility of different tissues or cells within infected individuals.

Different viruses may bind to any of a large number of different cell surface proteins, carbohydrates, or lipids. Binding serves several purposes. Most generally, it attaches the virus to the uninfected cell, maintaining proximity, and increasing effective viral concentration on the cell surface. More specifically, interaction of viral spikes with specific cell surface proteins may initiate conformational changes that activate the viral proteins for fusion. Binding to certain cell surface proteins may also promote endocytosis of the virus, by any of several cellular pathways. Endocytosis introduces the viral envelope into the lower pH of the endosome, which is required for activation of some viral fusion proteins. Activation of certain cell surface receptors by virus binding may also initiate specific signaling cascades within the cell, which may be useful to the virus during subsequent steps of infection.

Each enveloped virus exhibits unique binding specificities of its membrane proteins with particular cell surface features, resulting in a unique combination of these effects. For example, the HIV-1 recognition protein gp120 exhibits a near total specificity for binding to the CD4 receptor on immune cells. Further activation of the virus' fusion protein gp40 occurs by interaction with a co-receptor, either the chemokine receptor CCR5 or CXCR4. In contrast, orthomyxo- and many paramyxoviruses have much broader specificity. Their recognition proteins (HA or HN, respectively) bind to sialic acid residues attached to various cell surface proteins or lipids. Different strains show preference for sialic acid in different covalent linkages. Rhabdoviruses such as vesicular stomatitis or rabies virus are still less specific, binding indiscriminately to negative charge clusters, whether created by lipids, proteins, oligosaccharides, or surface-bound polyanions. This nonspecific binding property helps to account for the broad host range of these viruses, although some have also been reported to bind specifically to acetylcholine receptors, which may explain their neurotropism.

Viral Fusion

Fusion of the viral envelope with a cell membrane is facilitated by integral viral membrane proteins. The best studied viruses (HIV-1, orthomyxo-, paramyxo-, retro-, toga-, and rhabdoviruses) each possess a single fusogenic glycoprotein, but herpes- and poxviruses may possess several that work together. Virosomes consisting only of a purified viral fusion protein reconstituted into a lipid bilayer vesicle fuse readily with protein-free lipid bilayers, suggesting that the fusion proteins can act on host cell lipids and do not require participation by host cell proteins.

Because of their ready availability and ease of purification, viral fusion proteins have served as prototypes for understanding biological fusion reactions. Several general principles have emerged, which have been found to apply to at least one major class of cellular fusion reactions (those mediated by proteins called SNAREs) as well as all characterized viral fusions.

First, both viral and cellular fusion proteins act directly on the lipid bilayer to facilitate rearrangements identical to those that occur during protein-free lipid bilayer fusion. Fusion is thought to occur through a series of steps, constituting the so-called 'stalk–pore' pathway (**Figure 1**). The two closely apposed bilayers (**Figure 1(a)**) dimple

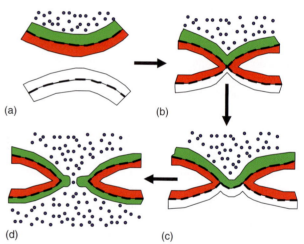

Figure 1 The stalk–pore mechanism of membrane bilayer fusion. (a) Initial pre-fusion state. The two leaflets of the cell membrane bilayer are colored green and red. The cytoplasm is indicated by blue dots. The viral membrane is colorless. (b) The stalk structure. In this state, continuity is established between the outer leaflets of the two membrane bilayers, allowing lipid mixing. (c) The fusion diaphragm constitutes a single bilayer separating the two aqueous compartments, comprising the inner leaflets of the cellular and viral membrane bilayers. The outer leaflets have already fused, and their lipids have mixed. (d) The fusion pore arises from rearrangement of the limited-area fusion diaphragm, perhaps facilitated by fusion proteins. Expansion of the fusion pore allows complete mixing of aqueous compartments and completes fusion. Fusion can occur even between protein-free lipid bilayers under certain conditions, but fusion proteins increase efficiency, probably by acting at each step in this pathway. Reproduced from Chernomordik LV and Kozlov MM (2005) Membrane hemifusion: Crossing a chasm in two leaps. *Cell* 123: 375–382.

towards each other to form the 'stalk' (**Figure 1(b)**). This thins to form a 'hemifusion diaphragm' (**Figure 1(c)**), which now separates the two aqueous compartments by a single membrane bilayer in place of the two that separated them in the pre-fusion state. The hemifusion bilayer consists of the inner leaflets of the two original bilayers, the outer leaflets having already fused with consequent mixing of their lipid components. The hemifusion diaphragm can then rearrange to form a 'fusion pore' (**Figure 1(d)**), which must stabilize and widen to allow aqueous mixing, and thus complete the fusion reaction.

Precisely how viral membrane proteins promote fusion via the stalk–pore mechanism remains a subject of active research. Several properties of viral fusion proteins are known to be essential for activity, however. All viral and cellular fusion proteins are oligomeric, usually trimeric for virus fusion proteins. In most cases, several (probably 5–7) trimers must act cooperatively in order to complete the fusion reaction. One attractive idea is that the several fusion protein trimers encircle a limited area of bilayer, into which the fusion proteins can then transfer the energy released by their ensuing conformational transitions (**Figure 2**) in order to effect the lipid rearrangements required for fusion. Viral fusion proteins might potentiate any or all of the fusion steps: initial stalk formation, hemifusion diaphragm formation from the stalk, fusion pore formation from the hemifusion diaphragm, expansion of the initial fusion pore to complete fusion.

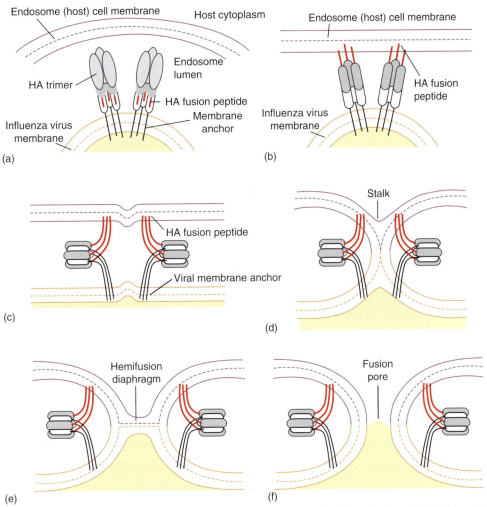

Figure 2 Proposed stalk–pore mechanism for membrane fusion mediated by a class I virus fusion protein, influenza hemagglutinin, HA. (a) The influenza virus has been internalized into the host cell endosome by receptor-mediated endocytosis (not shown). (b) Acidification of the endosome causes a conformational change in HA, which causes the fusion peptide (red) to become exposed; it subsequently inserts into the endosomal membrane. (c) Further refolding and clustering of HA trimers leads to bending of the two membranes toward each other. The resulting stalk (d) rearranges into the hemifusion diaphragm (e), where the virus interior is separated from the cytoplasm by a single bilayer, composed of the internal leaflets of the original membranes. This eventually rearranges, resulting in the formation of the fusion pore (f), which initially flickers and then dilates to complete the fusion reaction (not shown). Intra-leaflet lipid mixing, illustrated in **Figure 1**, is not shown. From Cross KJ, Burleigh LM, and Steinhauer DA (2001) Mechanism of cell entry by influenza virus. *Expert Reviews in Molecular Medicine* 6: 1–18, Cambridge University Press.

Within these general principles two distinct classes of viral fusion proteins have been recognized, possessing radically different architecture. Type I viruses include orthomyxo-, paramyxo-, retro-, and coronaviruses. Type II viruses include toga- and flaviviruses. Despite their different structures, these two classes of proteins facilitate the same lipid modifications during fusion.

All virus fusion proteins must remain inactive during biosynthesis and assembly so as to prevent premature, indiscriminate, counterproductive fusion within the infected cell. Both type I and type II fusion proteins are activated by a two-step process potentiated by interactions with a host cell. The inactive type I precursor protein is first cleaved at a specific site by limited proteolysis during assembly, generating a metastable, active form (**Figure 2(a)**). This then undergoes a conformational change mediated either by interaction with a specific cellular co-receptor (HIV-1), or by interaction with viral recognition proteins (paramyxoviruses) or by the low pH inside an endosome (orthomyxoviruses). Type II viral fusion proteins acquire their active conformation by an incompletely understood rearrangement of viral envelope proteins to form fusion protein trimers followed by interactions with specific lipids, notably cholesterol.

Activated fusion proteins of either type possess the following three structural features, which are required for complete fusion (**Figure 2**):

1. *Transmembrane domain.* This is the helical hydrophobic sequence (around 20 residues) that defines the fusion protein as an integral protein, and fixes it irreversibly in the viral bilayer. The transmembrane domain is inserted into the membrane bilayer during synthesis on membrane-bound ribosomes and appears not to rearrange during subsequent processing, activation or fusion. Type I and type II fusion proteins possess similar transmembrane domains. Not all transmembrane domains can participate in fusion reactions; a certain amount of conformational flexibility is required.

The transmembrane domain is required to complete the fusion reaction. Constructs in which the external, fusogenic domain of the influenza HA protein was attached only to the outer leaflet of the viral bilayer by a covalent bond with a lipid were capable of inducing hemifusion only, but were unable to complete the reaction.

2. *Fusion peptide.* This is a second relatively hydrophobic sequence that inserts into the cell membrane bilayer, and thus serves to bind the virus and cell membranes together. Exposure of the fusion peptide, enabling it to penetrate the target cell membrane, is an essential aspect of fusion protein activation, and requires a conformational change from a precursor form (**Figure 2(b)**). In type I proteins, this results from proteolytic activation; the influenza fusion peptides, for example, comprise the newly created N termini. Type II fusion peptides are located in a protruding loop of the protein structure, which is exposed by a poorly understood interaction with lipids, notably cholesterol. As with the transmembrane domain, a certain amount of conformational flexibility is required in the fusion peptide.

3. *A rigid, oligomeric rod-like structure connecting the fusion peptide with the transmembrane domain.* This consists of a helical coiled-coil in type I proteins (**Figure 2(b)**), and an arrangement of β-sheet domains in type II proteins. Once the fusion peptide has inserted into the target membrane, fusion is completed by the rearrangement of this metastable structure to its lowest free energy form (**Figures 2(d)–2(e)**). In order to assume this form, the rigid oligomer folds back upon itself, forming a 'hairpin' (**Figures 2(d)–2(e)**), thus dragging the cell membrane, tethered by the fusion peptide, toward the viral membrane, tethered by the transmembrane domain. The free energy released by this rearrangement is transferred to the lipid bilayers, providing the energy necessary to complete the fusion reaction. In the final fused product, the fusion peptide and the transmembrane domain are adjacent to each other in the same membrane, held in proximity by the fully stable, rigid hairpin (**Figure 2(f)**).

Membrane Synthesis and Viral Assembly

Viruses generally use the housekeeping mechanisms already operating in the infected cell in order to make maximal use of their limited genomes. Hence, viral membrane protein synthesis is carried out on host cell membrane-bound ribosomes, which inserts them into the endoplasmic reticulum membrane in the correct orientation (**Figure 3**). There they are glycosylated by the host cell machinery and assembled into appropriate oligomers, as directed by their own primary amino acid sequence. Most viral glycoproteins are then passed on to the Golgi by cellular mechanisms, where they are further glycosylated by host cell enzymes. For this reason, the envelope proteins of vesicular stomatitis virus (VSV), influenza, and a few other enveloped viruses have provided valuable tools to study the glycosylation and transport of membrane proteins through the cellular endoplasmic reticulum–Golgi–plasma membrane system. Because host cell protein synthesis is often inhibited by infection with these viruses (by a variety of cytopathic mechanisms), large amounts of a single viral membrane protein are produced and correctly processed in infected cells, without competition by cellular proteins.

Likewise, viral proteins are targeted to specific cellular locations by cellular processes. The viral proteins display the same amino acid sequence 'addresses' as host cell proteins, which are recognized by the host cell glycosylation and transport machinery. For example, the single VSV glycoprotein, named G, is glycosylated in the endoplasmic

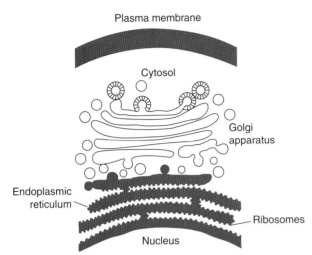

Figure 3 The endoplasmic reticulum–Golgi–plasma membrane system of a cell. All viral and cellular integral membrane proteins are synthesized by ribosomes bound to the endoplasmic reticulum membrane. Proteins destined for the plasma membrane are transported first to the proximal region of the Golgi (the *cis* face), then sequentially through the Golgi cisternae, to the *trans* face and out to the plasma membrane. In polarized cells, targeting to the apical or basolateral surface of the cell occurs from the *trans* face of the Golgi. Assembly and budding of different enveloped viruses occurs at characteristic points within this membrane system.

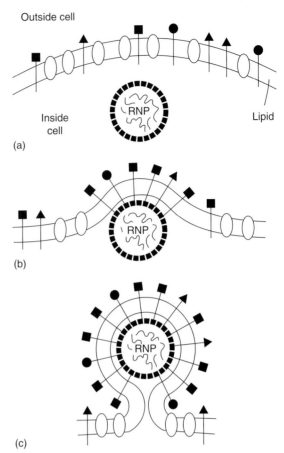

Figure 4 One kind of virus budding. Viral glycoproteins, inserted into the cellular membrane at the endoplasmic reticulum and processed through the Golgi to the plasma membrane (see **Figure 3**), associate with the assembled viral nucleocapsid. The direct association pictured here is characteristic of togaviruses. For other viruses, possessing helical nucleocapsids, the association is mediated by a peripheral membrane protein. Cellular membrane proteins are excluded from the envelope of the mature virion. This may occur during assembly, as pictures, or by prior formation of a viral membrane patch (or raft), before the nucleocapsid arrives at the membrane.

reticulum, the oligosaccharide is modified in the Golgi, and the mature protein is targeted to the basolateral plasma membrane of polarized cells after entry into the late Golgi. The influenza HA protein, on the other hand, is glycosylated and delivered to the apical plasma membrane of the same polarized cells after passage through the intracellular membrane system. The retention of coronavirus glycoproteins by the endoplasmic reticulum, and of bunyavirus glycoproteins by the Golgi, reflects the operation of the same cellular mechanisms that retain resident cellular proteins in these organelles. The localization of viral membrane proteins in turn, determines the location of viral assembly and budding.

The budding process consists of the wrapping of a viral glycoprotein-enriched piece of membrane around the previously assembled nucleocapsid, which contains the viral genome (**Figure 4**). Remarkably, the completed viral envelope contains viral proteins and host cell lipids, with host cell membrane proteins being almost completely excluded. There is much less discrimination, however, between different viral proteins than between viral and cellular proteins, since viral envelope proteins of one kind can assemble with nucleocapsids of another, resulting in the formation of pseudotype virions. Pseudotypes have proven useful in redirecting specific viral genomes to alternate host cells since membrane attachment proteins are major determinants of host cell specificity (see above). It has been suggested that the discrimination between viral and cellular membrane proteins may arise from the exclusion of cellular proteins from virus-associated raft-like lipid phases.

Different viruses have been described, that bud at every stage in the endoplasmic reticulum–Golgi–plasma membrane pathway (**Figure 3**). While paramyxo-, orthomyxo-, rhabdo-, and togaviruses (and many others) generally bud from the plasma membrane, they have also been shown to bud intracellularly under certain conditions. Some retroviruses assemble at the plasma membrane, while others do not; this has provided a classical basis for distinguishing between different types of retroviruses. Other viruses normally bud intracellularly, from the endoplasmic reticulum or Golgi apparatus, for example, coronaviruses and bunyaviruses respectively, but these have also been observed to bud further down the pathway. In these cases, the nucleocapsid assembles on the cytoplasmic face of the membrane, and then buds into the intracellular organelle. The newly formed virion may then be secreted out of the cell through the normal secretory pathway, although this does not always occur efficiently.

In all enveloped viruses except togaviruses, budding is mediated by a peripheral membrane protein, usually called M or MA, which links the glycoprotein-containing patch of lipids with the viral nucleocapsid, containing the viral genome. The M proteins interact specifically with nucleocapsids of their own viral species, but they do not always interact specifically with the corresponding viral glycoproteins. Instead, they may concentrate on the cytoplasmic side of the raft-like lipid phases that accumulate various viral glycoproteins. This could provide the structural basis for the formation of pseudotype virions, and might explain why many viruses contain widely varying ratios of glycoproteins to M or nucleocapsid proteins.

In contrast, togaviruses, which lack any M protein, possess an icosahedral nucleocapsid, which interacts directly with the cytoplasmic domain of the viral membrane protein. Completed virions contain an equal number of nucleocapsid and membrane protein molecules. Both are in a similar geometric arrangement, mediated by specific protein–protein interactions between them.

As described above, the lipids of the viral membrane are taken from the host cell membrane during budding. No new lipids are specifically synthesized in response to viral infection. Alterations in cellular lipid metabolism have been reported to result from some viral infections in cultured cells, but these are probably secondary to other cytopathic effects; there is no indication that they play an important role in the progress of infection.

See also: Baculoviruses: Molecular Biology of Nucleopolyhedroviruses; Bunyaviruses: General Features; Coronaviruses: General Features; Filoviruses; Herpesviruses: General Features; Orthomyxoviruses: Molecular Biology; Reticuloendotheliosis Viruses; Togaviruses: General Features.

Further Reading

Alberts B, Johnson A, Lewis J, Raff M, Roberts K, and Walter P (2002) *Molecular Biology of the Cell,* 4th edn., chs. 10–13. New York: Garland Science (Taylor and Francis Group).

Briggs JAG, Wilk T, and Fuller SD (2003) Do lipid rafts mediate virus assembly and pseudotyping? *Journal of General Virology* 84: 757–768.

Chernomordik LV and Kozlov MM (2005) Membrane hemifusion: Crossing a chasm in two leaps. *Cell* 123: 375–382.

Cross KJ, Burleigh LM, and Steinhauer DA (2001) Mechanism of cell entry by influenza virus. *Expert Reviews in Molecular Medicine* 6: 1–18, Cambridge University Press.

Gonzalez ME and Carrasco L (2003) Viroporins. *FEBS Letters* 552: 28–34.

Harrison SC (2006) Principles of virus structure. In: Knipe DM, Howley PM, Griffin DE, *et al.* (eds.) *Fields Virology,* 5th edn, pp. 59–98. Philadelphia: Lippincott Williams and Wilkins.

Jardetzky TS and Lamb RA (2004) A class act. *Nature* 427: 307–308.

Kielian M and Rey FA (2006) Virus membrane fusion proteins: More than one way to make a hairpin. *Nature Reviews Microbiology* 4: 67–76.

Simons K and Vaz WLC (2004) Model systems, lipid rafts and cell membranes. *Annual Review of Biophysics and Biomolecular Structure* 33: 269–295.

Smith AE and Helenius A (2004) How viruses enter animal cells. *Science* 304: 237–242.

Viral Pathogenesis

N Nathanson, University of Pennsylvania, Philadelphia, PA, USA

© 2008 Elsevier Ltd. All rights reserved.

Introduction

Viral pathogenesis deals with the interaction between a virus and its host. Included within the scope of pathogenesis are the stepwise progression of infection from virus entry through dissemination to shedding, the defensive responses of the host, and the mechanisms of virus clearance or persistence. Pathogenesis also encompasses the disease processes that result from infection, variations in viral pathogenicity, and the genetic basis of host resistance to infection or disease. A subject this broad cannot be treated in a single entry, and this article focuses on the dissemination of viruses and their pathogenicity.

Sequential Steps in Viral Infection

One of the cardinal differences between viral infection of a simple cell culture and infection of an animal host is the structural complexity of the multicellular organism. The virus must overcome a number of barriers to accomplish the stepwise infection of the host, beginning with entry, followed by dissemination, localization in a few target tissues, shedding, and transmission to other hosts.

Entry

Some viruses (such as papillomaviruses, poxviruses, and herpes simplex viruses) will replicate in cells of the skin or mucous membranes. Following infection of skin, viruses may spread further by passage of intact virions or virus-infected macrophages and dendritic cells to the regional lymph nodes. A considerable number of viruses cause sexually transmitted infection, which is usually acquired by either by penetration through breaks in the skin or by direct invasion of the superficial epithelium of the mucous membranes. Contaminated blood, semen, or secretions may introduce hepatitis B virus directly into the circulation via breaks in skin and membranes, while human immunodeficiency virus (HIV) may be trapped on the surface of dendritic cells (Langerhans cells) in the epidermis and then transported to draining lymph nodes. Although the skin is a formidable barrier, mechanical injection that breaches the barrier is a 'natural' route of entry for many viruses. More than 500 individual arboviruses are maintained in nature by a cycle that involves a vector and a vertebrate host. When the infected vector takes a blood meal, virus contained in the salivary gland is injected. Also, virus may be transmitted accidentally by contaminated needles, injection of a virus-contaminated therapeutic, transfusion with virus-contaminated blood or blood products, or tattooing.

The oropharynx and gastrointestinal tract are important portals of entry for many viruses, and enteric viruses may invade the host at a variety of sites from the oral cavity to the colon. Some viruses produce localized infections that remain confined to the gastrointestinal tract whereas others disseminate to produce systemic infection. Most enteric viruses, such as rotaviruses, infect the epithelium of the small or large intestine, but some enteroviruses, such as poliovirus, replicate in the lymphoid tissue of the gut and nasopharynx. There are numerous barriers to infection via the enteric route. First, much of the ingested inoculum will remain trapped in the luminal contents and never reach the wall of the gut. Second, the lumen constitutes a hostile environment because of the acidity of the stomach, the alkalinity of the small intestine, the digestive enzymes found in saliva and pancreatic secretions, and the lipolytic action of bile. Third, the mucus that lines the intestinal epithelium presents a physical barrier protecting the intestinal surface. Fourth, phagocytic scavenger cells and secreted antibodies in the lumen can reduce the titer of infectious virus. Viruses that are successful in using the enteric portal tend to be resistant to acid pH, proteolytic attack, and bile, and some may actually exploit the hostile environment to enhance their infectivity. The importance of anal intercourse as a risk factor for hepatitis B and HIV infection has led to the recognition that some viruses can gain entry through the lower gastrointestinal tract, but the exact mechanism of HIV infection through the anocolonic portal remains to be determined.

Respiratory infection may be initiated either by virus contained in aerosols that are inhaled by the recipient host, or by virus that is contained in nasopharyngeal fluids and is transmitted by hand-to-hand contact. Aerosolized droplets are deposited at different levels in the respiratory tract depending upon their size: those over $10\,\mu m$ diameter are deposited in the nose, those $5-10\,\mu m$ in the airways, and those $<5\,\mu m$ in the alveoli. Once deposited, the virus must bypass several effective barriers, including phagocytic cells in the respiratory lumen, a covering layer of mucus, and ciliated epithelial cells that clear the respiratory tree of foreign particles. The temperature of the respiratory tract varies from $33\,°C$ in the nasal passages to $37\,°C$ (core body temperature) in the alveoli. Viruses (such as rhinoviruses) that can replicate well at $33\,°C$ but not at $37\,°C$ are limited to the upper respiratory tract and, conversely, viruses that replicate well at $37\,°C$ but not at $33\,°C$ (such as influenza virus) mainly infect the lower respiratory tract. Most respiratory viruses initiate infection by replicating in epithelial cells lining the alveoli or the respiratory tree, but some viruses will also replicate in phagocytic cells located either in the respiratory lumen or in subepithelial tissues.

Spread

Once infection is established at the site of entry, most viruses will spread locally by cell-to-cell transmission of infection. Some viruses remain localized near their site of entry while others disseminate widely. Viruses that are released only at the apical surface of infected cells tend to remain localized, while those that are released at the basal surface tend to disseminate. Usually, the first step in dissemination of the infection is the transport of virus via the efferent lymphatic drainage from the initial site of infection to the regional lymph nodes, either as free virions or in virus-infected phagocytic cells.

The single most important route of dissemination is the circulation, which can potentially carry viruses to any site in the body. There are several sources of viremia. Virus may enter with the efferent lymphatic flow, or may be shed from infected endothelial cells or circulating mononuclear leukocytes. Viruses can circulate either free in the plasma phase of the blood ('plasma' viremia) or associated with formed elements ('cell-associated' viremia), and these two types of viremia have quite different characteristics. An 'active' viremia is caused by the active replication of virus in the host, which occurs after a lag period required for tissue replication and shedding into the circulation. Termination of viremia is often quite abrupt and coincides with the appearance of neutralizing antibody in the serum.

Plasma viremia reflects a dynamic process in which virus continually enters the circulation and is removed. The rate of turnover of virus within the plasma compartment is best expressed as transit time, the average duration of a virion in the blood compartment. Typically, transit times range from 5 to 60 min and tend to decrease as the size of the virions increases. Circulating viruses are removed by the phagocytic cells of the reticuloendothelial system, principally in the liver (Kupffer cells), and to a lesser extent in the lung, spleen, and lymph nodes. Once the host has developed circulating antibody, plasma virus is very rapidly neutralized in the circulation and transit time is reduced by several fold. Plasma viremias are usually short-lived (about a week), but there are notable exceptions.

A number of viruses replicate in cells that are found in the circulation, particularly monocytes, B or T lymphocytes, or (rarely) erythrocytes, and produce a 'cell-associated' viremia. Cell-associated viremia may be of short duration, as in the case of ectromelia and other poxviruses. However, in many instances, cell-associated virus titers are low, and viremia persists for the life of the host rather than terminating when neutralizing antibody appears.

There are several ways in which a virus might cross the vascular wall, but the precise mechanism for penetrating the blood–tissue barrier is unknown in most instances. Potential routes into tissues include passage between endothelial cells at sites where there are no tight junctions, translocation of virus-containing endosomes through endothelial cells, replication within endothelial cells, and tissue invasion of virus-infected lymphocytes or monocytes.

Some viruses can disseminate by spreading through the axons of peripheral nerves. Although less important than viremia, neural spread plays an essential role for certain viruses (such as rabies viruses and several herpesviruses) while other viruses (such as poliovirus and reovirus) can utilize both mechanisms of spread.

Localization or Tropism

One of the salient features that distinguishes viruses is their localization or tropism within the animal host. The names of individual viral diseases often reflects the organs or tissues that are involved; thus, smallpox, poliomyelitis, and hepatitis each have their characteristic features. Tropism is regulated both by viral dissemination and cellular susceptibility. Disease localization may not correspond to the distribution of infection since it reflects both the spread of the virus and the host response to infection.

The localization of a virus is determined at several phases of infection, including the portal of entry, systemic spread by viremia or the neural route, and the invasion of local organs or tissues.

Most viruses are quite selective in the cell types that are infected *in vivo* and this selectivity plays a significant role in localization. One major determinant of cellular susceptibility is the presence of viral receptors, namely, molecules on the cellular surface that act as receptors for the specific virus, usually by binding a protein on the virion surface (the viral attachment protein). For instance, poliovirus can only replicate in primate tissues, because the poliovirus receptor is expressed only by primate cells. Transgenic mice expressing the the poliovirus receptor can be infected and develop typical paralytic poliomyelitis after intracerebral injection.

Cellular susceptibility also may be determined at post-entry steps in the replication cycle. An example is Newcastle disease virus, a paramyxovirus of birds. Virulent isolates of Newcastle disease virus encode a fusion protein that is readily cleaved by furin, a proteolytic enzyme present in the Golgi apparatus, so that the virion is activated during maturation prior to reaching the cell surface, and before budding of nascent virions. This makes it possible for virulent strains of the virus to infect many avian cell types, thereby increasing its tissue host range, and causing systemic infections that are often lethal. Avirulent strains of Newcastle disease virus cannot be cleaved by furin and depend upon extracellular proteases, which restricts the spread of the virus beyond the respiratory tract.

For oncogenic viruses, the tumors may represent a distant echo of virus localization. Epstein–Barr virus (EBV) infects epithelial cells and B cells. Latently infected B cells are immortalized, and, rarely, these B cells undergo further transformation into lymphomas by virtue of additional genetic events such as chromosomal transposition (t8:14).

Shedding and Transmission

Acute viral infections are characterized by brief periods (days to a few weeks) of intensive virus shedding into respiratory aerosols, feces, urine, or other bodily secretions or fluids. Persistent viruses are often shed at relatively low titers, but this may be adequate for transmission over the prolonged duration of infection (months to years). Although many viruses replicate in the skin, relatively few are spread from skin lesions. Several viruses are shed in the semen, including hepatitis B and human immunodeficiency virus. A number of viruses are excreted in colostrum and milk, including cytomegalovirus, mumps, rubella, and HIV. Blood is also an important source of transmitted virus, particularly for those viruses that produce persistent plasma or cell-associated viremias, such as hepatitis B, C, and D viruses, HIV, HTLV-I, HTLV-II, and cytomegalovirus, and which can be transmitted by contaminated needles and transfusions. A number of viruses have been isolated from the urine but viruria is probably not important for transmission of most viruses.

Once shed, there are a several means by which a virus is transmitted from host to host in a propagated chain of infection. Probably the most important mechanism is contamination of the hands of the infected transmitter from feces, oral fluids, or respiratory secretions expelled during coughing or sneezing. The virus is then passed by hand-to-hand contact leading to oral, gastrointestinal, or respiratory infection. A second common route is inhalation of aerosolized virus. A third significant mechanism involves direct person-to-person contact (oral–oral, genital–genital, oral–genital, or skin–skin). Finally, indirect person-to-person transmission can occur via blood or contaminated needles. Common source transmissions are less frequent but can produce dramatic outbreaks, via contaminated water, food products, or biologicals such as blood products and vaccines.

Viral Virulence

Virulence refers to the ability of a virus to cause illness or death in an infected host, relative to other isolates or variants of the same agent. The study of virulence variants can provide important insights into pathogenesis, and carries the potential for development of attenuated live virus variants that can be used as vaccines. One approach to virulence is to map the genetic determinants that underlie the virulence phenotype; the other approach is a study of pathogenesis, which describes the differences in infections with viral variants of different virulence. Measures of virulence may be quantitative, or qualitative, since variants of a single virus may differ markedly in their cellular tropism, their mode of dissemination in the infected host, or the disease phenotype that they produce.

Viral Genetic Determinants of Virulence

Viral virulence is encoded in the viral genome and expressed through structural proteins, nonstructural proteins, or noncoding sequences. A large body of information about virus variants has established several principles.

1. The use of mutants has made it possible to identify the role of individual genes and proteins as determinants of the biological behavior of many viral variants.
2. There is no 'master' gene or protein that determines virulence, and attenuation may be associated with changes in any of the structural or nonstructural proteins, or in the noncoding regions of the genome.
3. The virulence phenotype can be altered by very small changes in the genome, if they occur at 'critical sites'. At such sites, a single point mutation leading to the substitution of a specific amino acid or base is often sufficient to alter virulence. For most viruses with small genomes (<20 kb), only a few discrete 'critical sites' have been discovered, usually fewer than 10 per genome.
4. It is possible to create variants with attenuating mutations at several critical sites and these may be more attenuated than single point mutants, and less prone to reversion.
5. Attenuating mutations are often host range mutations that affect replication in some cells but not in others.
6. Although many attenuating mutations have been sequenced, relatively few have been characterized at a biochemical or structural level as to their mechanism of action.

Genetic determinants of viral virulence and attenuation may be illustrated with examples from a few selected virus groups. The attenuation of the vaccine strains of poliovirus is due to critical sites, both in the $5'$ nontranslated region (NTR) upstream from the long open reading frame and in selected structural and nonstructural proteins. Each of the three strains of oral polio vaccine (OPV) carry attenuating point mutations in the $5'$ NTR (at positions 480, 481, or 472, in types 1, 2, and 3, respectively). Attenuating mutations at these sites are associated with reduced neurovirulence in monkeys and reduced ability to replicate in the central nervous system of transgenic mice bearing the poliovirus receptor. Each of the OPV strains carries at least one mutation that is associated with an alteration in a viral structural or nonstructural protein and confers temperature sensitivity and reduced neurovirulence, and these mutations involve different proteins for the three OPV strains. It has been suggested that mutations in the structural proteins produce attenuation via structural transitions that occur either during virion uncoating or assembly.

Bunyaviruses are negative stranded RNA viruses with a trisegmented genome. The large (L) RNA segment encodes the viral polymerase; the middle (M) RNA segment encodes two glycoproteins (G1 and G2) and a nonstructural protein, Ns_m; the small (S) segment encodes the nucleoprotein (N) and a nonstructural protein, Ns_s. Attenuation can involve either the ability of the virus to replicate in the central nervous system (neurovirulence) or its ability to cause viremia and reach the central nervous system (neuroinvasiveness). Reduced neuroinvasiveness of one bunyavirus virus strain is associated with reduced ability to replicate in striated muscle and consequent low viremogenicity. Attenuation maps to the M RNA segment encoding the viral glycoproteins suggesting that there is an alteration in the infection of myocytes (but not of neurons). Another attenuated bunyavirus showed a striking reduction in its ability to replicate in the central nervous system of adult mice, and attenuation maps to the L RNA segment. The attenuated virus replicates poorly in C1300 NA neuroblastoma and other murine cell lines, and presumably has a mutant polymerase that restricts viral replication in neurons and certain other cell types.

Reoviruses are double-stranded 10-segmented RNA viruses. Most of the virulence determinants represent qualitative differences between type 1 Lang and type 3 Dearing in tissue and organ tropism rather than quantitative differences in disease severity

1. Reovirus type 1 Lang disseminates through the blood and causes an ependymitis in the brain while type 3 Dearing disseminates through the neural route and causes a neuronotropic encephalitis in the brain. The S1 segment encodes the σ1 protein, which is the viral attachment protein, and is the major determinant of dissemination and brain cell tropism.
2. The M1 segment encoding the μ2 protein influences replication in cardiac myocytes and myocarditis.
3. The M2 segment encoding the μ1 protein affects the protease sensitivity of the virion, and quantitative neurovirulence.
4. The L1 segment encoding λ3 protein influences replication in cardiac myocytes and myocarditis.
5. The L2 segment encoding λ2 protein influences replication levels in the intestine, titers of shed virus, and horizontal transmission between mice.

A new class of virus-encoded proteins has been recognized that contribute to the virulence of viruses by mimicking normal cellular proteins. This group of 'cell-derived' genes has been identified primarily within the genomes of large DNA viruses that have a greater capacity to maintain accessory genes than do viruses with small genomes. 'Virokines' are secreted from virus-infected cells and mimic cytokines, thereby perturbing normal host responses. 'Viroceptors' resemble cellular receptors for cytokines (including antibodies or complement components) that are thereby diverted from their normal cellular targets. Some virus-encoded proteins interfere with antigen presentation and immune induction, while others prevent apoptosis, or interrupt intracellular signaling initiated by cytokines or interferons.

Pathogenic Mechanisms of Virulence

There are many sequential steps in a viral infection and a difference in the comparative replication of two virus variants at any one of these steps can account for their relative degrees of virulence. Attenuated variants usually will replicate or spread less briskly in one or several tissues associated with the pathogenic process. A reduction in viremia can be an important mechanism for the reduction in virulence. For instance, the Mahoney strain of type 1 poliovirus causes a much higher viremia and paralytic rate than several other poliovirus strains that are less viremogenic. Different variants of a single virus can exhibit differences in their tropism for organs, tissues, and cell types, which confers a multidimensional character upon virulence. The three attenuated strains in the oral poliovirus vaccine provide an example. Both attenuated and wild-type polioviruses are able to replicate in the gastrointestinal tract to similar levels, judging from the shedding of virus in the feces, but the attenuated strains show a reduction of about 10 000-fold (relative to the wild-type strains) in their ability to replicate in the spinal cord after intraspinal injection. In this instance, virulence for the target organ (the spinal cord), rather than viremogenicity, appears to be the most important determinant of virulence.

Host Determinants of Susceptibility and Resistance

Studies with inbred animals have documented that the outcome of infections often varies in different strains of mice, and genetic analyses have been conducted to determine the form of inheritance of susceptibility, and to identify the responsible gene. Several generalizations may be made:

1. Most genetic loci control susceptibility to a specific family of viruses and not to all viruses.
2. Susceptibility can often be mapped to a single autosomal locus, but multiple loci have been identified in some instances.
3. Susceptibility may be either dominant or recessive.
4. Where loci have been mapped, they usually are distant from the major histocompatibility locus, and in most such instances, the mechanism of susceptibility is not immunological.
5. The exact mechanism of susceptibility has not been well defined in most instances, although there are some examples that map to the major histocompatibility complex (MHC), implying an immunological explanation.

It is difficult to investigate genetic determinants of host susceptibility in outbred populations of animals or humans. However, recent studies of HIV-1 have provided substantial evidence of a number of human genetic determinants of susceptibility to infection or to the rate of progression to overt AIDS. The most striking of these is a polymorphism in the gene that encodes the chemokine receptor, CCR5, which is the major co-receptor for HIV-1. Some individuals have a deletion in the gene (the Δ32 deletion) such that no CCR5 protein is expressed. Those who are homozygous for the Δ32 mutation are at very low risk of HIV infection, and those who are heterozygous for the Δ32 deletion can be infected but have prolonged incubation periods to onset of AIDS by comparison with CCR5+ homozygous individuals.

The host response to a viral infection may also be influenced by a variety of physiological variables such as age, sex, stress, and pregnancy, all of which have been shown to influence the outcome of infection with some viruses. In general, very young animals are more susceptible than adult hosts to acute viral infections, and are more likely to undergo severe or fatal illness. The innate high susceptibility of infants born of nonimmune mothers is revealed only in those rare instances where a virus has disappeared from an isolated community so that all age groups lack immune protection. A classical example is measles, which can be a devastating disease with a mortality of up to 25% in nonimmune infants. Less commonly, advanced age can also be associated with increased susceptibility to viral infection. St. Louis encephalitis and West Nile encephalitis are examples of this phenomenon.

See also: Innate Immunity: Defeating; Innate Immunity: Introduction; Persistent and Latent Viral Infection.

Further Reading

Barber DL, Wherry EJ, Casopust D, et al. (2006) Restoring function in exhausted CD8 T cells during chronic viral infection. *Nature* 439: 682–687.

Martin MP and Carrington M (2005) Immunogenetics of viral infections. *Current Opinion in Immunology* 17: 510–516.

Nathanson N, Ahmed R, Brinton MA, et al. (eds.) (2007) *Viral Pathogenesis and Immunity,* 2nd edn. London: Academic Press.

Nathanson N and Tyler K (2005) The pathogenesis of viral infections. In: Mahy BWJ and ter Meulen V (eds.) *Topley and Wilson's Microbiology,* pp. 236–269. London: Hodder Arnold.

Racaniello VR (2006) One hundred years of poliovirus pathogenesis. *Virology* 344: 9–16.

Tumpey TM, Basler CF, Aguilar PV, et al. (2005) Characterization of the reconstructed 1918 Spanish influenza pandemic virus. *Science* 310: 77–80.

Viral Receptors

D J Evans, University of Warwick, Coventry, UK

© 2008 Elsevier Ltd. All rights reserved.

Introduction

The majority of viruses have an extracellular phase in the life cycle, which is a necessary part of virus transmission and dissemination. To initiate the replication cycle for the production of progeny virions, the virus must first enter a cell. Cell entry is mediated by the specific interaction of the virus with molecules on the cell surface – the receptor – resulting in attachment, a process that leads to internalization of either part or all of the virus particle, or the genome. A cell that expresses the cognate cell-surface receptor for a virus is termed susceptible. Not all susceptible cells to which the virus binds are capable of supporting infection and replication, perhaps due to the absence of necessary intracellular components. Cells that support virus binding, entry, replication, and progeny virion release are termed permissive.

With certain notable exceptions, such as viruses of fungi that exhibit no extracellular stage in the life cycle or plant viruses introduced mechanically by insect vectors, the virus receptor therefore forms the primary determinant of virus tropism. The identification and characterization of the receptor therefore provides important insights into the early stages of the virus replication cycle, and has been a major research focus of molecular virologists worldwide.

A Definition of Virus Receptor and Co-Receptors

Our enhanced understanding of virus–receptor interactions and the early events in cell entry have demonstrated that this process is, in many cases, considerably more complicated than originally thought, involving multiple cell-surface or intracellular molecules with different roles in the entry events. A classical definition of a virus receptor would be the cell-surface molecule that mediates virus attachment and cell entry. Our current understanding of these processes demonstrates that many viruses use different molecules to mediate attachment and post-attachment entry events. For these reasons the term co-receptor is often used to indicate an accessory molecule implicated in cell entry. Both receptors and co-receptors will be considered within the scope of this article. Rather than aim to provide a complete catalog of viruses and receptors, the focus of this article is to use key examples to illustrate specific types of virus–receptor interactions and introduce the general concepts involved in receptor identification and characterization.

The Identification of Virus Receptors

Since a working definition of a receptor is a specific cell-surface molecule involved in virus attachment, the identification of the receptor requires a means of detecting virus binding, and a requirement to demonstrate the specificity of the process.

Direct detection of a virus bound to the cell requires some form of labeling – whether metabolic labeling with a radioisotope or labeling with a conjugated fluorescent marker – and subsequent purification of the virus particle, often by density gradient centrifugation. Indirect methods of detecting virus can also be used, for example, the hemagglutination phenotype demonstrated by sialic acid binding viruses such as influenza A virus, or by detection of a bound virus with antibodies to the virus particle.

The specificity of the observed binding event typically requires the inhibition of the process using a defined block. The routine availability of monoclonal antibodies has hugely facilitated this approach to receptor identification, though alternate methods such as the use of enzymatic pretreatment of cells with heparinase or neuraminidase (NA) have also been used to demonstrate interactions with nonprotein cell-surface molecules.

Finally, in cases where the receptor is a single protein molecule, the specificity of the virus interaction can be further demonstrated by transformation of a nonsusceptible cell with DNA encoding the protein. Following expression at the cell surface, the acquisition of a virus binding phenotype is tested, perhaps combined with the demonstration of specificity by the inhibition with monoclonal antibodies. For example, the molecular cloning of the poliovirus receptor (PVR) involved expression of the protein in murine fibroblasts, which conferred susceptibility to an already permissive cell type.

Typically, a combination of the approaches outlined above is necessary to unequivocally demonstrate the identity of a virus receptor. Multiple confirmatory approaches are required as each technique has inherent uncertainties. For example, monoclonal antibodies may cross-react with closely related membrane proteins, or transformation of DNA may induce receptor expression, rather than encode the receptor *per se*. Similar techniques, and the associated caveats, are subsequently used in the initial stages of characterization of the virus–receptor interaction, for example, the receptor protein domain(s) to which the virus attaches, or the cell or tissue specificity of attachment.

Specific Examples of Virus Receptors

The examples used below illustrate a range of different types of virus–receptor interactions – those involving a single type of target molecule on the cell surface, either proteins or carbohydrate moieties, and those in which multiple receptors and co-receptors have been implicated. Due to their pivotal role in determining tissue and host tropism, and hence influence on pathogenesis, particular effort has been made in identifying receptors for important human and animal virus pathogens. However, the broad concepts illustrated below apply equally to the cell-surface proteins of bacteria used by bacteriophages – where the exquisite specificity of the interaction is used in classification and typing of *Salmonella*.

Sialic Acid

Influenza A is an enveloped virus that causes epidemic and pandemic respiratory tract infections in humans, but is naturally an enteric pathogen of aquatic birds. Both of the external virus glycoproteins are involved in interaction with the cellular receptor, sialic acid, the first identified receptor for a human virus. The trimeric hemagglutinin (HA) molecule, consisting of a stalk and globular head, carries a conserved cleft or pit at the membrane-distal end into which the sialic acid binds. Sialic acids are a family of negatively charged monosaccharides derived from neuraminic acid; those used as virus receptors are usually attached to a penultimate galactose by either an $(\alpha 2,3)$- or $(\alpha 2,6)$-linkage. The structure of an HA–sialyllactose complex has been determined at atomic resolution and provides insight into the specificity of binding, which in human strains involves $(\alpha 2,6)$-linked sialic acid in contrast to the binding of avian strains which exclusively interact with $(\alpha 2,3)$-linked sialic acid. Interaction of influenza with sialic acid is a low-affinity process, attachment being facilitated by multimeric binding to many HA molecules. The tetrameric NA virus glycoprotein has enzymatic activity that hydrolyses the glycosidic linkage of the receptor sialic acid, so enabling the release of progeny virus from the cell surface. NA-inhibitors have been developed as successful antiviral therapies.

Although influenza A is the best-characterized sialic acid-binding virus, there are many other examples, including human parainfluenza virus (hPIV) types 1–3 (and the murine hPIV analog Sendai virus), picornaviruses (rhinovirus type 67, enterovirus type 70), polyomaviruses (JC and BK virus), type 3 reoviruses, group C rotaviruses and adeno-associated virus. As with influenza A virus, predominant bias to either $(\alpha 2,6)$- or $(\alpha 2,3)$-linked sialic acid has been detected in each of these examples.

Heparan Sulfate and Other Glycosaminoglycans

Glycosaminoglycans (GAGs) (including heparan, chondroitin, and dermatan sulfate) are negatively charged sulfated sugars which are widely implicated in attachment of a range of viruses as diverse as herpes simplex virus (HSV), enteroviruses, flaviviruses, and respiratory syncytial virus. However, despite the isolation of clinical virus isolates with GAG-binding capacity, definitive evidence that GAGs have a role in the cell infection process *in vivo* is yet to be obtained. The relevance of GAGs has been further questioned by the observation that foot and mouth disease virus (FMDV), a picornavirus that binds $\alpha_v\beta_3$ integrins *in vivo*, can readily adapt to use GAGs in cell culture – this process is discussed further below.

GAG binding is routinely demonstrated by inhibiting virus binding to the cell surface, or virus infection, with soluble analogs such as heparin sulfate, by heparinase pretreatment of cells, or by analyzing binding to cells deficient in GAG synthesis. The interaction of FMDV and heparin has also been determined crystallographically and shown to involve a region of the capsid distinct from the surface-exposed loop that binds the *in vivo* receptor, $\alpha_v\beta_3$ integrin. GAG-binding domains are usually linear or conformational regions of the virus surface that are rich in basic residues, and adaptation to GAG binding often involves the acquisition of basic (lysine or arginine) amino acid substitutions.

Most viruses shown to bind GAGs are also known to interact with at least one additional cell surface receptor. Glycoprotein C (gC) of HSV-1 and HSV-2 binds HS, and the gD interaction with certain forms of GAGs appears necessary for postbinding prefusion events in cell infection. However, gD also interacts with HVEM (herpes virus entry mediator, a member of the tumor necrosis factor receptor family) and nectins 1 and 2, suggesting that HSV may either exhibit considerable redundancy in cell-surface interactions or, interestingly, that the virus carries a number of binding activities tailored to particular cell or tissue types. Similarly, GAG-binding echoviruses appear to also require the primary receptor for entry, decay accelerating factor (DAF) (Williams and Evans, unpublished data).

The cell–receptor interactions of human immunodeficiency virus (HIV) have received particular attention. In addition to the primary receptor, CD4, and the chemokine co-receptors (see below), syndecans with GAG chains have been implicated in macrophage infection by HIV. Similarly, infection of CD4-negative microvascular endothelial cells, a probable route via which the virus gains access to the central nervous system (CNS), has been shown to be a lipid raft-independent process involving cell-surface proteoglycans. Again, as with HSV, the complexity of interaction with multiple cell receptors, and the *in vivo* function, is yet to be satisfactorily dissected.

Histo-Group Antigens as Receptors

The wide range of nonprotein molecules used for attachment by viruses also includes the histo-blood group antigens (HBGAs), which serve as receptors for the noroviruses, positive-strand RNA viruses that cause acute gastroenteritis. HBGAs are complex carbohydrates present both as free oligosaccharides and attached to the surface of mucosal epithelia. The HBGA phenotypes of humans are polymorphic and complex, and are associated with multiple gene families. This was consistent with the demonstration that norovirus infection was influenced by both prior acquired immunity and the genetic background – patients with blood group O were significantly more susceptible to infection than those with blood group B. Subsequent binding studies with norovirus virus-like particles (VLPs) – which retain the receptor-binding capacity of the virus which cannot be cultured *in vitro* – and specific antibodies to blood-group antigens, soluble blood-group antigens or glycosidases, have confirmed a role for these molecules in virus attachment.

Immunoglobulin Superfamily Proteins

Our understanding of the range of cell-surface proteins that function as virus receptors is influenced by the methods utilized to identify the protein. Since these methods are predominantly geared to isolating single-gene traits, for example, cDNA library screening, it is unsurprising that the majority of proteins definitively identified as receptors are single proteins, rather than one component of a multiprotein complex. Perhaps reflecting their relative abundance at the cell surface, proteins belonging to the immunoglobulin superfamily are implicated in the attachment of many viruses.

Poliovirus, the causative agent of paralytic poliomyelitis, binds the membrane-distal Ig-like domain of CD155, the PVR. Receptor binding initiates a series of irreversible conformational changes in the virus particle, which results in the externalization of capsid proteins usually hidden in the mature infectious particle, and the delivery of the virus genome to the cell cytoplasm. Data are accumulating to suggest that this process involves formation of a pore traversed by the genome, though other models exist. That PVR is sufficient to mediate these processes has been elegantly demonstrated by several methods; PVR expression in murine cells renders them fully permissive for poliovirus infection, and is the basis for the generation of PVR-transgenic mouse lines for neurovirulence studies, the conformational changes observed at the cell surface can be recapitulated using poliovirus and soluble PVR (sPVR) *in vitro*, and – most recently – the bound virus–receptor complex has been visualized by crystallography and, using cryo-EM, studied at the surface of PVR-loaded membrane vesicles.

Human rhinoviruses are very closely related to members of the enterovirus genus to which poliovirus belongs. The rhinoviruses can be divided into two predominant groups based upon their receptor usage; of these, the major receptor group bind to the membrane-distal Ig-like domain of intracellular adhesion molecule type 1 (ICAM-1). Like poliovirus, receptor binding initiates conformational changes in the rhinovirus particle that are considered essential for infection. A further similarity is the region of the capsid involved in the receptor interaction. In both poliovirus and the major group rhinoviruses, the receptor binds to a canyon or cleft surrounding the fivefold axis of icosahedral particle symmetry. Two decades ago, Rossmann proposed that this canyon is too narrow to allow the binding of virus neutralizing antibodies, thereby allowing the virus to retain a relatively invariant receptor-binding domain in the presence of immunoselection. More recent crystallographic studies have demonstrated that some neutralizing antibodies penetrate deep into the rhinovirus canyon. This suggests that the canyon has less to do with evasion of immune surveillance, but has instead evolved in response to receptor binding requirements and to accommodate the structural changes that occur post-binding.

The Ig-like protein that has received some of the most intense scrutiny is CD4, the primary receptor for HIV. The identification of CD4 as a receptor for HIV went some way to explain the tropism and pathogenesis of the virus; CD4+ T cells were often depleted in HIV-infected individuals. Antibodies to CD4 blocked virus attachment, fusion and infection, and expression of CD4 in murine cells permitted virus binding. Detailed analysis using monoclonal antibodies specific for each Ig-like domain of CD4, and subsequent site-directed mutagenesis of the receptor, demonstrated that the attachment protein of HIV (the virus surface glycoprotein gp120) binds to a surface-exposed loop of the most membrane-distal of the four Ig-like domains of CD4. Gp120 carries the conserved CD4 attachment site which is protected, though not completely hidden, from interaction with neutralizing antibodies by the diversity, conformational flexibility, and heterogeneous glycosylation of the surrounding protein. One characteristic exhibited by HIV-infected long-term nonprogressors is the presence of a potent neutralizing antibody response to the CD4 binding site – because the virus must retain a largely invariant receptor-binding domain, these antibodies are active against divergent virus isolates, and significant effort has been expended in generating vaccine candidates capable of inducing a similar immune response. Other therapeutic approaches based upon the conserved interaction of gp120 and CD4 have been investigated. Although soluble CD4 receptors, lacking the transmembrane domains, block HIV infection *in vitro*, they have been largely ineffective in clinical trials. These disappointing results have been attributed to the need to maintain high serum concentrations of a protein with a short half-life, and the ability of the virus to spread directly from cell to cell by fusion, rather than by extracellular virus. Similarly, approaches involving toxins fused to soluble CD4 that are used to target gp120 expressed at the surface of infected cells have only been promising in *in vitro* assays.

Complement Control Proteins as Virus Receptors

A number of viruses have been shown to bind to complement control proteins, also termed the regulators of complement activity (RCA) family proteins. These proteins, characterized by a two-disulfide protein fold termed a short consensus repeat (SCR), are involved in protecting autologous cells from complement-mediated lysis.

Epstein–Barr virus (EBV) attachment to B lymphocytes occurs by the interaction of the virus gp350/220 envelope glycoprotein with the cell-surface C3dg receptor designated CR2 (CD21). CR2 was identified as the receptor following the observed similarity between gp350/220 sequences and C3dg, the complement protein target for CR2 activity. Further supporting evidence included the demonstration that multimerized peptides of the proposed gp350/220 or C3dg binding domains inhibited attachment, and the receptor identity was confirmed by expression studies in CR2-deficient target cells. CR2 possesses 16 extracellular SCR domains, the two most membrane-distal being the site of gp350/220 binding, as demonstrated by the construction of murine-human CR2 chimeric receptors, and using SCR-specific monoclonal antibodies in blocking studies.

Laboratory-adapted measles virus binds another RCA family protein member, CD46, a four-SCR domain protein that acts as a cofactor in the proteolytic inactivation (and hence cell protection) of C3b/C4b. Antisera to CD46 blocks virus binding and infection, and transfection of rodent cells with CD46 renders them susceptible to infection by the Edmonston and Hallé strains of measles virus. Further studies demonstrated that clinical isolates of measles cultured in human B-cell lines exhibited a cell tropism inconsistent with the use of CD46 as the receptor – further details are provided in the section on laboratory adaptation.

DAF, a 55 kDa RCA-family protein, is the receptor for a significant number of human enteroviruses, including the coxsackie B viruses and many echoviruses. In contrast to many of the other examples presented here, where binding is to the membrane-distal domain of the receptor, the majority of the known DAF-binding enteroviruses interacts with the third SCR, together with the second and/or fourth SCR domains. Recent completion of the atomic structure of DAF shows it to be a rod-shaped protein, with the four SCR domains projecting in a broadly linear fashion from the top of a heavily glycosylated Ser/Thr-rich stalk, which distances the complement-control functions from the cell membrane. Unsurprisingly, considering the receptor structure, interaction of DAF with echoviruses does not involve the canyon on the virus surface, but instead occurs across the twofold axis of symmetry, with the receptor interface occupying a surface-exposed region of the capsid. In the case of the coxsackieviruses, DAF binding serves both to attach the virus to the cell surface and initiate a cascade of signaling events which results in the transport of the virus to the co-receptor for infection – CAR, the coxsackie and adenovirus receptor, which is located at tight junctions and not normally accessible to the virus.

Co-Receptors for Infection

The distinction between a receptor and a co-receptor for infection is largely semantic. As our understanding of virus binding and entry improves, it is clear that this is often a multistage process involving a cascade of interactions between the virus and cell-surface proteins. The initial interaction is usually defined as being to the receptor, with subsequent interactions being mediated by co-receptors. This order can be reversed if the first interaction is a relatively nonspecific event – for example, the binding of HSV to HS – in which case co-receptor binding may be considered to temporally precede interaction with a more specific molecule that perhaps defines the cell or tissue tropism of the virus. Therefore, the discrimination between nonspecific attachment molecules, receptors, and co-receptors – in which all are sometimes required for infection – is open to interpretation. What is clearly important for many viruses is the requirement for a range of interactions between the virus and the cell surface. Indeed, as our understanding of virus infection improves, the number of examples in which a single receptor is implicated – such as poliovirus – is dwindling, and may represent the exception, rather than the rule.

Although the coxsackie B viruses bind DAF at the cell surface, infection also requires CAR, an Ig-superfamily protein with two extracellular domains, which interacts with the canyon of coxsackie B viruses or the fiber of adenovirus types 2 and 5. Like the poliovirus–PVR interaction, binding of coxsackie B viruses to CAR – at the cell surface or in solution – triggers the irreversible conformational changes associated with particle uncoating. However, CAR is naturally located in tight junctions where it mediates homotypic cell adhesion, and is therefore not immediately accessible for virus attachment at the apical cell surface. Recent elegant studies have demonstrated that virus binding to DAF at the latter site results in Abl-kinase mediated signaling events which cause actin rearrangements that allow virus movement to the tight junctions where uncoating occurs. DAF binding also activates Fyn-kinase, which results in the phosphorylation of caveolin and subsequent virus entry in caveolin vesicles. The coxsackie B viruses therefore exploit the signaling capacity of their low-affinity receptor (DAF) to traffic to the high-affinity, uncoating receptor (CAR) and thereby cross the epithelial barrier.

Like poliovirus and rhinovirus, binding of HIV to CD4 initiates conformational changes in the trimeric gp120 that are a prerequisite for the subsequent fusion events. However, unlike the two picornavirus examples, HIV must also interact with a co-receptor as a necessary precursor to the fusion of viral and cellular membranes. This requirement for a co-receptor explains why HIV can bind to human CD4 expressed on murine cells – which lack the co-receptor – without undergoing membrane fusion. A mouse-CD4 cell line was transformed with a human cDNA library and used to functionally screen for additional proteins which would enable gp120-mediated membrane fusion. This, or analogous, screens were used to define the co-receptor for HIV as one of a related series of receptors for chemotactic cytokines such as RANTES, MIP-1α and MIP-1β. The chemokine receptors identified, CXCR4 or CCR5, belong to the seven-transmembrane-domain G-protein-coupled receptor superfamily. Co-receptor specificity appears to be determined by the third hypervariable domain (V3) of gp120 – a target for potently neutralizing antibodies *in vitro* – with an additional conserved domain of gp120 contributing to generic features of the chemokine receptor attachment. The critical conserved residues of the latter domain remain hidden from immune surveillance until exposed by the structural changes in gp120 that occur as a consequence of CD4 binding. The importance of the chemokine co-receptor in HIV infection is supported by the observed genetic resistance to HIV of individuals with a deletion and frameshift within CCR5. Similarly, co-receptor identification provided an explanation for the observed suppression of HIV infection by the chemotactic cytokines released by CD8+ T cells. Finally, the identification of chemokine receptors as critical co-receptors of HIV infection has also contributed to our understanding of virus tropism and evolution *in vivo*. CCR5-tropic strains of HIV are predominant for transmission and during the asymptomatic phase of virus infection. CCR5 is expressed at high levels in activated/memory cells and in gut-associated lymphoid tissue, the primary site of HIV replication. As infection progresses, CXCR4-tropic viruses evolve, able to exploit the wider tissue expression of the CXCR4 receptor. The, perhaps logical, assumption that this broadening of cell and tissue tropism is the cause of disease progression is contradicted by the failure to observe a switch to CXCR4-tropic viruses in a significant proportion of patients with acquired immune deficiency syndrome (AIDS).

Laboratory Adaptation and Receptor Usage by Clinical Virus Isolates

Due to the possession of error-prone replication strategies, many viruses exhibit an exquisite ability to adapt to the environment in which they are grown – for example, by the generation of mutations that confer resistance to neutralizing antibodies. The same variation can confound the correct identification of virus receptors in laboratory studies, due to the adaptation to alternate receptors for cell attachment and entry.

The example of heparan sulfate binding by FMDV has already been used to illustrate this process. HS-adapted FMDV isolates are apathogenic in animals, presumably reflecting the evolutionary adaptability of the virus to an environment in which receptor usage is not a rate-limiting process in infection, transmission, or dissemination. The virus interacts with the *in vivo* receptor ($\alpha_v\beta_3$ integrin) using a conserved Arg-Gly-Asp amino acid triplet (RGD motif) on a surface-exposed loop of the VP1 capsid protein. Laboratory-adapted HS-binding FMDV strains can dispense with $\alpha_v\beta_3$ integrin binding altogether, as demonstrated by the use of reverse genetics to mutate the RGD motif to another sequence. Being acid-labile, it is likely that uncoating of FMDV is mediated by entry of the virus–receptor complex into an endosome. This is supported by the further demonstration that it is possible to engineer anti-FMDV antibodies onto the cell surface to which the virus is bound, and via which infection can occur in the absence of either HS or integrin binding.

CD46, like most other complement-control proteins, is expressed ubiquitously on serum-exposed cells. Clinical isolates of measles virus, isolated in either B95a cells or human B-cell lines, do not infect all CD46-expressing cell lines, suggesting that they may use an alternate receptor. Subsequent functional expression cloning studies resulted in the isolation of a cDNA encoding human SLAM (signaling lymphocyte-activation molecule; also known as CDW150) that conferred susceptibility to B95a-cell cultured measles virus. SLAM is an Ig-superfamily protein expressed on a range of immune system cells (macrophages, dendritic cells, some B and T cells) that, when liganded to SLAM on an adjacent cell, transduces a signal resulting in T_H2 cytokine (IL13 and 4) production. All measles virus isolates tested can infect cells via SLAM, though some – including the lab-adapted Edmonston strain – can also use CD46. However, studies suggest that CD46 binding is irrelevant *in vivo* and arises upon culture by the substitution of as few as a single amino acid in the HA protein of the virus.

Summary

Virus receptors are key components of the early events involved in cell infection. The examples outlined above should illustrate that the definition of receptors acting solely as attachment molecules is overly simplistic. In addition to attachment, receptors also actively contribute to entry by initiating conformational changes in the virus that lead to uncoating. In addition, they provide mechanisms for internalization – which increasingly appear to involve signaling events that occur upon receptor binding – in which the virus may subvert a natural receptor cycling process. Furthermore, the identification of virus receptors contributes significantly to our understanding of host, tissue, and cell tropism, and helps explain aspects of virus pathogenesis. The selected bibliography provides further examples illustrating aspects of virus receptor identification and function.

See also: Cytokines and Chemokines; Foot and Mouth Disease Viruses; Influenza; Measles Virus; Viral Pathogenesis.

Further Reading

Belnap DM, McDermott BMJ, Filman DJ, et al. (2000) Three-dimensional structure of poliovirus receptor bound to poliovirus. *Proceedings of the National Academy of Sciences, USA* 97: 73–78.

Bergelson JM, Cunningham JA, Droguett G, et al. (1997) Isolation of a common receptor for coxsackie B viruses and adenoviruses 2 and 5. *Science* 275: 1320–1323.

Coyne CB and Bergelson JM (2006) Virus-induced Abl and Fyn kinase signals permit coxsackievirus entry through epithelial tight junctions. *Cell* 124: 119–131.

Evans D and Almond J (1998) Cell receptors for picornaviruses as determinants of cell tropism and pathogenesis. *Trends in Microbiology* 6: 198–202.

Hogle JM (2002) Poliovirus cell entry: Common structural themes in viral cell entry pathways. *Annual Review of Microbiology* 56: 677–702.

Lindahl G, Sjobring U, and Johnsson E (2000) Human complement regulators: A major target for pathogenic microorganisms. *Current Opinion in Immunology* 12: 44–51.

Lusso P (2006) HIV and the chemokine system: 10 years later. *EMBO Journal* 25: 447–456.

Maddon PJ, Dalgleish AG, McDougal JS, Clapham PR, Weiss RA, and Axel R (1986) The T4 gene encodes the AIDS virus receptor and is expressed in the immune system and the brain. *Cell* 47: 333–348.

Mendelsohn CL, Wimmer E, and Racaniello VR (1989) Cellular receptor for poliovirus: Molecular cloning, nucleotide sequence, and expression of a new member of the immunoglobulin superfamily. *Cell* 56: 855–865.

Olofsson S and Bergstrom T (2005) Glycoconjugate glycans as viral receptors. *Annals of Medicine* 37: 154–172.

Pettigrew DM, Williams DT, Kerrigan D, Evans DJ, Lea SM, and Bhella D (2006) Structural and functional insights into the interaction of echoviruses and decay-accelerating factor. *Journal of Biological Chemistry* 281: 5169–5177.

Rossmann MG (1989) The canyon hypothesis. Hiding the host cell receptor attachment site on a viral surface from immune surveillance. *Journal of Biological Chemistry* 264: 14587–14590.

Tan M and Jiang X (2005) Norovirus and its histo-blood group antigen receptors: An answer to a historical puzzle. *Trends in Microbiology* 13: 285–293.

Ward T, Pipkin PA, Clarkson NA, Stone DM, Minor PD, and Almond JW (1994) Decay accelerating factor (CD55) identified as the receptor for echovirus 7 using CELICS, a rapid immuno-focal cloning method. *EMBO Journal* 13: 5070–5074.

Yanagi Y, Takeda M, and Ohno S (2006) Measles virus: Cellular receptors, tropism and pathogenesis. *Journal of General Virology* 87: 2767–2779.

Viral Suppressors of Gene Silencing

J Verchot-Lubicz, Oklahoma State University, Stillwater, OK, USA
J P Carr, University of Cambridge, Cambridge, UK

© 2008 Elsevier Ltd. All rights reserved.

Glossary

Dicer, or Dicer-like (DCL) enzymes RNAse III or RNAse-III-like enzymes responsible for digesting the noncoding regions of mRNAs to produce 21–24 nt single-strand RNAs known as miRNAs and siRNAs.

Green fluorescent protein (GFP) This is derived from jellyfish and fluorescence green. Excitation wavelength is 488 nm and emission is above 520 nm. Fusions involving GFP are often used to study protein subcellular targeting or distribution in tissues.

MicroRNAs (miRNAs) Single-strand RNAs that are 21–24 nt in length are found in eukaryotes and arise from noncoding regions of transcripts. These are produced by nucleolytic processing by DICER, and RNAse-III-like enzyme. These are crucial components of the RNAi pathway.

RNA interference (RNAi) Similar to post-transcriptional gene silencing. More specifically, cellular or synthetic small RNA molecules can target homologous mRNA for degradation thereby preventing gene expression.

RNA silencing or post-transcriptional gene silencing (PTGS) Mechanism regulating gene expression by regulating RNA accumulation after transcription. Mechanism involves RNA degradation machinery to shut off gene expression.

Short interfering RNAs (siRNAs) Double-strand RNAs that are 21–24 nt in length which are generated by DICER or Dicer-like enzymes. SiRNAs can spread systemically in *C. elegans* and may cause silencing in distal organs. Some single-strand RNAs are made double-strand by RNA-dependent RNA polymerases. These double-stranded products are then cleaved by DICER.

Transcriptional gene silencing (TGS) Silencing of genes in the nucleus. A small RNA molecule triggers *de novo* DNA methylation thereby blocking transcription. Small RNA typically is homologous to the target gene.

Virus-induced gene silencing (VIGS) Viral RNAs can trigger for PTGS similar to small RNAs. Several plant viruses have been engineered as vectors for use in experiments shutting off gene expression by PTGS. Fragments of genes, antisense RNAs, small RNAs can be introduced into the viral vector and silencing is induced upon inoculation with the recombinant virus.

Introduction

RNA silencing, also known as post-transcriptional gene silencing (PTGS) or RNA interference (RNAi) is a mechanism regulating gene expression in a wide range of eukaryotes. RNA silencing is a mechanism in which small RNAs block gene expression by targeting homologous mRNAs without impacting nuclear DNA. Andrew Hamilton and David Baulcombe first showed that short antisense RNAs of 20–25 nt which share homology with target mRNAs are produced in silenced tissues. Since 1999 additional short RNAs have been identified, including microRNAs and siRNAs, which target homologous RNA sequences either for sequence specific degradation or, in some instances, function to repress mRNA translation.

PTGS is induced by double-stranded RNAs in most eukaryotic systems. Since most RNA viruses form double-stranded RNA replication intermediates, replicating viruses often trigger PTGS, which subsequently degrades all homologous RNAs within the cell. This has led researchers to suggest that PTGS may have originally evolved as an antiviral defense mechanism. The ability of PTGS to target viral RNAs for degradation was demonstrated in the early 1990s when transgenic plants expressing untranslatable transcripts of the viral coat protein or replicase gene were found to be resistant to infection by the homologous virus while remaining susceptible to unrelated viruses. Virus resistance was also reported in experiments using transgenic plants which failed to accumulate detectable levels of the transgenically expressed coat protein RNA.

The laboratory of William Dougherty provided the first reports of transgenic plants recovering from virus infection. A set of transgenic tobacco plants expressing a nontranslatable tobacco etch virus (TEV) coat protein RNA were initially susceptible to virus infection but then recovered and became highly resistant to secondary inoculation. Viral RNA was undetectable and transgene RNAs showed lower steady-state level accumulation in recovered leaves indicating that an RNA degradation mechanism was triggered. Contemporary studies by Rob Goldbach's laboratory showed that nontranslatable transcripts for the tomato spotted wilt virus N protein-protected transgenic plants from infection by the virus. The idea that viruses trigger a cellular antiviral defense pathway which degrades homologous RNAs was further supported by experiments in the laboratory of

David Baulcombe showing that recovery can occur during infection of nontransgenic tobacco plants. *Nicotiana clevelandii* inoculated with the nepovirus tomato black ring virus (TBRV) strain W22 initially showed clear virus symptoms but later recovered. Plants were resistant to secondary inoculation with the same W22 strain but were susceptible to inoculation with the heterologous nepovirus, tobacco ringspot virus (TRSV), and showed partial protection to secondary infection with the TBRV strain BUK. Insertion of a fragment of the TBRV genome into the potato virus X (PVX) genome was sufficient to block infection of PVX in recovered plants indicating that plants contain an inducible sequence-specific degradation mechanism that may be a component of the plant immune response to RNA virus infection.

This same PTGS mechanism also provides cross-protection against secondary virus infection seen in the TBRV and PVX experiments described above and in experiments showing that plants recovered from infection with tobacco rattle virus containing GFP (TRV-GFP) were resistant to secondary infection with PVX-GFP. Technology based on PTGS has been developed to engineer virus resistance in transgenic plants.

Viral-induced gene silencing (VIGS) in transgenic plants can result in methylation of the cognate transgene or nuclear gene. It has been suggested that *de novo* DNA methylation triggered by transcriptional gene silencing (TGS) or PTGS is driven by siRNAs imported into the nucleus, which recruit DNA methyltransferases to similar target sequences. Thus, RNA viruses, such as PVX or tobacco rattle virus (TRV), have been engineered as vectors used for knocking out host gene expression. These are termed VIGS vectors. Entire genes or fragments of genes can be inserted into the viral vector and following inoculation, induce silencing of the cognate endogenous gene. One of the earliest examples of VIGS targeting an endogenous gene was insertion of the phytoene desaturase gene (PDS) into PVX. PDS is involved in carotenoid biosynthesis and affects plants' susceptibility to photobleaching. Virus-induced PDS silencing causes leaves to lose all green color and bleach white under normal lighting conditions. The most popular research tools to study TGS and PTGS are transgenic plants expressing the green fluorescent protein (GFP) and PVX or TRV vectors containing fragments of the GFP coding sequence. GFP expression can be monitored using a UV lamp and over time GFP expression disappears throughout the entire plant. In 1997, Olivier Voinnet and David Baulcombe first reported GFP-transgene silencing by an RNA virus and also demonstrated the silencing signal can spread systemically to suppress gene expression in distal regions. Thus, the use of VIGS to suppress expression of endogenous genes has become an important tool for analysis of gene function.

DNA viruses can also be used as gene silencing vector. The chalcone synthase gene, which is involved in flower pigmentation in *Petunia hybrida* flowers, was inserted into tobacco yellow dwarf virus genome and flower pigmentation was completely altered in virus infected petunia. Cabbage leaf curl virus (CabLCV), a member of the genus *Begomovorus* of the family *Geminiviridae*, has been engineered to express any endogenous targeted gene of *Arabidopsis* and proved to be an efficient tool for scientists studying gene expression in this model plant. Similarly, the African cassava mosaic virus (ACMV), another begomovirus, modified to be a silencing vector, was demonstrated to be able to silence a variety of endogenous genes in cassava, thereby providing a useful tool to breeders for that crop.

Viral-Derived Short Interfering RNAs

Small RNAs of approximately 21–24 nt are found in all eukaryotes and belong to two general classes: microRNAs (miRNAs) and short interfering RNAs (siRNAs). miRNAs are 21–24 nt ssRNAs and arise from nonprotein coding regions of transcripts which are nucleolytically processed by an RNAse-III-like enzyme called DICER in animals and *Caenorhabditis elegans*, or DICER-like (DCL) in *Arabidopsis*. siRNAs similar to miRNAs range in size from 21 to 24 nt, but are dsRNAs derived from longer double-stranded RNA (dsRNA), including RNAs containing inverted repeats or replicative forms of RNA viruses.

Animals and *C. elegans* encode only a single DICER while *Arabidopsis* encodes four DCL proteins. While studies in *Drosophila* and plants show that DICER plays a role in antiviral defense, currently there is no direct evidence that RNA silencing acts as a natural antiviral defense mechanism in vertebrates. Dicer 2 mutants in *Drosophila* are hypersusceptible to virus infection. All four DCL proteins in *Arabidopsis* are involved in generating siRNAs from DNA and RNA viruses. DCL1 and DCL4 proteins produce 21 nt siRNAs, DCL2 produces 22 nt siRNAs, and DCL3 produces 24 nt siRNAs. While reports show all four DCL proteins contribute to the production of siRNAs from geminiviruses (CaLCuV) and pararetroviruses (cauliflower mosaic virus; CaMV) in plants, DCL2 and DCL4 generate siRNAs for defense against RNA viruses.

Cellular RNA-dependent RNA polymerases (RDR) also contribute to the formation of siRNAs. Single-stranded RNAs are made double stranded by cellular RDRs and then are cleaved by DICER to produce siRNAs. *HEN1* encodes a methyltransferase which acts alongside the DCL proteins to methylate the 3′-terminal nt protecting siRNAs from degradation. siRNAs then guide sequence-specific RNA-induced silencing complexes (RISCs) to target sequences for degradation. The RISC is comprised of several proteins including *ARGONAUTE* (*AGO*), which bind siRNAs or target sequences. RDR proteins use the siRNAs as primers for synthesis of dsRNAs from target viral RNAs. *Arabidopsis* encodes three *RDR* genes named *RDR1*, *RDR2*, and *RDR6*. *RDR1* is salicylic acid inducible. *Arabidopsis* plants showing defects in RDR1 show increase susceptibility to TMV and TRV. *RDR6* is also known as *SDE1* or *SGS2* and is required for transgene-generated short RNAs. *SDE5* is a factor recently identified which acts with *RDR6* to generate dsRNAs. *SDE5* may function in the nuclear transport of dsRNAs produced by *RDR6*. Mutant plants deficient in *RDR6* show increased susceptibility to cucumber mosaic virus (CMV) but not to other viruses.

Research in plants and in *C. elegans* first showed that a systemic silencing signal which spreads to distal organs is likely to be an siRNA or dsRNA. Research has shown that 21 nt siRNAs can spread 10–15 cells in tobacco and *Arabidopsis* plants. For more extensive movement, additional rounds of signal amplification are needed to produce a second generation of siRNAs. A model of transitive RNA silencing was proposed by Voinnet and colleagues in 2003 in which dsRNAs are synthesized *de novo* by RDR6, which are then cleaved by DICER to generate the second generation of 21 nt siRNAs. Cycles of siRNA propagation and movement into neighboring tissues lead to general silencing throughout surrounding tissues. Systemic silencing spread was proposed to rely on vascular transport of longer 25 nt siRNAs.

RNAi-Based Antiviral Therapies

Viral siRNAs accumulate to significant levels in plants and insects but have not been characterized in human cells infected with any RNA viruses. However, synthesized siRNAs have been shown to block replication and accumulation of a wide range of animal RNA and DNA viruses in cell cultures and vertebrate systems. There are examples of siRNAs targeted to specific sequences in the genomes of viruses including poliovirus (PV), foot and mouth disease virus (FMDV), hepatitis virus A (HVA), influenza virus, SARS-COV, HIV, and hepatitis virus B (HVB) which reduce virus titer and inhibit replication in cell cultures and in mice. Researchers using synthesized RNAs targeting different viral RNA sequences reported this strategy to be a successful form of antiviral therapy. For monopartite RNA viruses such as PV, FMDV, HVA, short RNAs targeting conserved sequences corresponding to genes encoding structural proteins or the viral replicase have been successful.

One of the important issues in developing RNAi therapeutics has been the pressure on target sequences to mutate which causes the virus to escape the suppressive activity of the siRNA molecule. To address this concern researchers have relied on bioinformatics tools and GenBank database to search entries of virus sequences to identify highly conserved nucleic acid elements ranging from 21 to 25 nt in length. Comparisons among virus isolates have been crucial for determining the most conserved regions. Among viruses, which have high mutation rates, therapies combining synthesized short RNA molecules targeting several conserved sequences have been most effective at reducing the occurrence of escape viruses.

The polyomavirus Simian virus 40 (SV40) as well as members of the family *Herpesviridae* including Epstein–Barr virus (EBV), herpes simplex virus 1 (HSV-1), Kaposi's sarcoma-associated virus (KSHV), and were found to encode miRNAs. At least 23 miRNAs have been identified in EBV-infected lymphocytes, many of which map to the BART gene. All BART gene miRNAs accumulate mainly during latency suggesting that they likely play a role in this stage of infection. KSHV is also associated with lymphomas and 12 miRNAs have been identified in latent infected cells. For KSHV a different set of miRNAs are seen during lytic infection suggesting that specific miRNAs are expressed during different stages of the viral life cycle. Viral-encoded miRNAs function to regulate viral gene expression and to downregulate host transcription. Since many of the herpesviral miRNAs associate with latency, they likely play a role in enabling the virus to evade the host immune system for many years. Further research is needed to find out if viral miRNAs affect tumorogenesis and if RNAi technology can be used to alter the onset of cancer.

Viral Suppressors of RNA Silencing Counter Cellular Defenses to Promote Infection

Plant and insect viruses encode silencing suppressor proteins which inhibit one or more steps in the miRNA or siRNA degradation pathway, thus countering the antiviral defense machinery. Many silencing suppressors bind dsRNA, siRNAs, and miRNAs. The potyvirus HC-Pro blocks the RISC from acting on target RNAs (**Figure 1**). HC-Pro was the first identified suppressor of RNA silencing in plants and was discovered by researchers studying viral synergistic diseases. Synergism is a phenomenon in which one virus shows increased titer and symptom induction due to the presence of a second but unrelated virus. Many, but not all, examples of synergy involve co-infection in which one of the viral partners is a potyvirus. Building on the early studies of PVX/PVY synergy, Vicki Vance, James Carrington, and their colleagues carried out further investigations of PVX/PVY synergy at the molecular level. They examined virus-specific RNA production in doubly infected plants and used PVX-derived vectors and transgenic plants to express segments of PVY and other potyviral genomes. Transgenic plants expressing the potyviral P1/HC-Pro sequence developed the same synergistic response when inoculated with unrelated

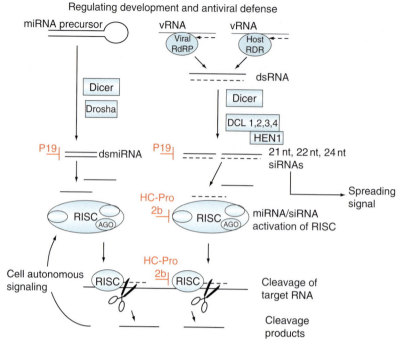

Figure 1 Model for RNA silencing updated from Himber C, Dunoyer P, Moissiard G, Ritzenthaler C, and Voinnet O (2003) Transitivity-dependent and -independent cell-to-cell movement of RNA silencing. *EMBO Journal* 22(17): 4523–4533; Matzke MA and Matzke AJ (2004) Planting the seeds of a new paradigm. *PLoS Biology* 2: E133; Blevins T, Rajeswaran R, Shivaprasad PV, et al. (2006) Four plant Dicers mediate viral small RNA biogenesis and DNA virus induced silencing. *Nucleic Acids Research* 34(21): 6233–6246. Double-stranded RNAs (dsRNAs) are produced either by hairpin interactions or by viral and host RNA-dependent RNA polymerases which synthesize complementary RNAs using viral RNA (vRNA) or mRNA templates. Dicer or Dicer-like proteins cleave the dsRNAs to smaller miRNAs or siRNAs which then guide the RISC complex to the target RNAs. Target RNAs are cleaved into short RNAs which can interact with the RISC complex to continue the cycle of cell autonomous RNA degradation. Some siRNAs are diverted prior to the RISC complex and move cell-to-cell and long distance to perpetuate silencing in surrounding and distal tissues. Examples of viral silencing suppressors are identified in red which block silencing by acting on the siRNAs or RISC complex.

viruses such as PVX, TMV, and CMV. P1/HC-Pro silencing suppression was demonstrated by crossing the transgenic plants with transgenic plants that were silencing for the GUS gene. The progeny lines showed restored GUS expression indicating that P1/HC-Pro suppressed transgene silencing. In a second set of experiments the GFP transgene was silencing by PVX vector containing GFP. When P1/HC-Pro was added, GFP expression was restored. Contemporaneously with this work, it was shown that CMV is also able to counteract RNA silencing and that this is mediated by the CMV 2b protein.

A reversal of silencing assay was developed for identification of viral silencing suppressor proteins. This assay uses the GFP expressing *Nicotiana benthamiana* line 16c which is susceptible to GFP silencing by infiltrating leaves with a suspension of *Agrobacterium* carrying the GFP gene. This induces silencing of GFP throughout the entire plant. GFP expression is restored to silenced plants following infection with PVX containing a gene for a silencing suppressor. Variations on this approach include grafting of transgenic plants that are silenced for reporter gene expression to scions that are transgenic for the same reporter gene and for the candidate silencing suppressor protein. This approach was used to demonstrate that the CMV 2b protein inhibits systemic silencing in plants. Using this approach a wide number of viral silencing suppressor proteins have been identified (**Table 1**). Remarkably, some viruses encode more than one silencing suppressor. For example, citrus tristeza virus encodes at least three proteins with silencing suppressor activity, which can confound attempts by the host at generating resistance to the virus.

In general, silencing suppressor proteins inhibit production of siRNAs. Many silencing suppressor proteins bind siRNAs preventing their incorporation into the RISC. The tombusvirus p19 is one example whose crystal structure was recently described. *In vitro* assays show that p19 binds 21 nt siRNAs. The potyvirus P1/HC-Pro acts to block the RISC reducing siRNA accumulation while enhancing miRNA accumulation. The cucumovirus 2b inhibits the spread of the silencing signal but also accumulates in the nucleus where it interferes with silencing-induced DNA methylation.

Several viral silencing suppressor proteins have cross-kingdom activity, that is, they are able to suppress RNA silencing in both insect and plant cells. In a seminal

Table 1 Viral silencing suppressors (April 2007)

Host	Genome	Virus genus	Virus name(s)	Suppressor(s)	Other name/function(s)
Plant	DNA	Begomovirus	Tomato leaf curl virus	C2	Transcriptional activator
			Tomato yellow leaf curl virus	C2	Adenosine kinase inhibitor
			African cassava mosaic virus	AC2, AC4	
			Mungbean yellow mosaic virus	AC2	
			Tomato golden mosaic virus	AL2	
		Curtovirus	Beet curly top virus	L2	
	(+) ssRNA	Aureusvirus	Pothos latent virus	P14	
		Benyvirus	Beet necrotic yellow vein virus	P14	Regulate RNA2, CP accumulation
				P31	Enhance P14 activity in roots
		Carmovirus	Turnip crinkle virus	P38	CP
			Hibiscus chlorotic ringspot virus	CP	
		Closterovirus	Beet yellows virus	P21	
			Beet yellow stunt virus	P22	
			Citrus tristeza virus	P20	
				P23	
				CP	
			Grapevine leafroll-associated virus-2	P24	
		Crinivirus	Sweet potato chlorotic stunt virus	P22	
				RNAse 3	
		Comovirus	Cowpea mosaic virus	Small CP	
		Cucumovirus	Cucumber mosaic virus	2b	Systemic movement virulence
			Tomato aspermy virus	2b	
		Furovirus	Soil-borne wheat mosaic virus	19K	
		Hordeivirus	Barley stripe mosaic virus	γb	
			Poa semilatent virus		
		Ipomovirus	Cucumber vein yellowing virus	P1	Protease
		Pecluvirus	Peanut clump virus	P15	
		Polerovirus	Beet western yellows virus		
			Cucurbit aphid-borne yellows virus		
			Potato leafroll virus		
		Potexvirus	Potato virus X	TGBp1	Intercellular movement
		Potyvirus	Potato virus Y	HC-Pro	Systemic movement, protease, transmission
			Tobacco etch virus		
			Turnip mosaic virus		
			Zucchini yellow mosaic virus		
		Sobemovirus	Rice yellow mottle virus	P1	Systemic movement
			Cocksfoot mottle virus		
		Tobamovirus	Tobacco mosaic virus	126K	Replicase component
			Tomato mosaic virus		
		Tobravirus	Tobacco rattle virus	16K	Seed transmission
		Tombusvirus	Artichoke mottled crinkle virus	P19	Virulence, movement
			Carnation Italian ringspot virus		
			Cymbidium ringspot virus		
			Tomato bushy stunt virus		
		Tymovirus	Turnip yellow mosaic virus	P69	Virulence, movement
		Vitivirus	Grapevine virus A	P10	
	(−) ssRNA	Tenuivirus	Rice hoja blanca virus	NS3	
		Tospovirus	Tomato spotted wilt virus	NS$_S$	
	dsRNA	Phytoreovirus	Rice dwarf virus	Pns10	
	Viroid	Pospiviroid	Potato spindle tuber viroid	RNA secondary structure	
		Avsunviroid	Avocado sunblotch viroid		
Fungus	dsRNA	Hypovirus	Cryphonectria hypovirus 1-EP713	p29	Protease

Continued

Table 1 Continued

Host	Genome	Virus genus	Virus name(s)	Suppressor(s)	Other name/function(s)
Animal	DNA	Adenovirus	Adenovirus	VA1 RNA	
		Poxvirus	Vaccinia	E3L	Interferon antagonist
	(+) ssRNA	Flavivirus	Hepatitis C virus	Core protein	
		Nodavirus	Flock house virus	B2	
			Nodamura virus		
			Striped jack nervous necrosis virus		
			Greasy grouper nervous necrosis virus		
		Picorna-like	Cricket paralysis virus	N-terminal domain of non-structural protein	
	(−) ssRNA	Orthomyxovirus	Influenza A, B, C viruses	NS1	Interferon antagonist
		Orthobunyavirus	La Crosse virus	NS$_S$	
	Retrovirus	Lentivirus	Human immunodeficiency virus-1	Tat	
		Spumavirus	Primate foamy virus type 1	Tas	
	dsRNA	Orthoreovirus		σ3	Outer shell protein
	Viroid-like	Deltavirus	Hepatitis delta virus	RNA secondary structure	

Updated from tables collated by Bucher E, Lohuis D, van Poppel PM, Geerts-Dimitriadou C, Goldbach R, and Prins M (2006) Multiple virus resistance at a high frequency using a single transgene construct. *Journal of General Virology* 87: 3697–3701; Li F and Ding SW (2006) Virus counterdefense: Diverse strategies for evading the RNA-silencing immunity. *Annual Review of Microbiology* 60: 503–531; Palukaitis and MacFarlane (2006); and Silhavy D and Burgyan J (2004) Effects and side-effects of viral RNA silencing suppressors on short RNAs. *Trends in Plant Science* 9: 76–83.

study, the group of S.W. Ding used both transgenic plants harboring silenced reporter genes and cultured insect cells to demonstrate that the B2 protein of flock house virus (FHV), an insect-infecting virus, was a silencing suppressor. Since then, a variety of insect and vertebrate-infecting, as well as fungus-infecting, viruses have been shown to encode proteins with RNA silencing activity under various assay conditions (**Table 1**). However, the results obtained by assaying viral proteins for silencing suppression in nonhost systems need to be interpreted with caution. For example, the influenza virus NS1 protein, which inhibits the interferon system in human cells, also inhibits RNA silencing in plants and in drosophila cells but not in HeLa cell cultures.

Recently, Deleris and colleagues created lines of *Arabidopsis thaliana* plants carrying single or multiple mutant alleles of the *DCL* genes. These plants offer an additional and less ambiguous method for identifying silencing suppressors, at least for plant viruses. This is illustrated by experiments using a turnip crinkle virus (TCV) derivative in which the coat protein gene, which also functions as a silencing suppressor, was replaced by a *GFP* gene. Compared to wild-type TCV, this modified virus is compromised in movement and symptom induction. However, symptom induction and systemic movement were restored in double *dcl2–dcl4* mutant plants; to the same extent seen in plants constitutively expressing a transgene encoding TCV coat protein.

Viroids are infectious small circular RNAs which do not encode a protein and rely on host DNA-dependent RNA polymerases for replication. Biao Ding's laboratory detected small 21 nt RNAs that are seen in PSTVd-infected plants and are active in RISC-mediated cleavage of target RNAs containing the GFP coding sequence fused to a homologous RNA segment, but the small RNAs do not impact PSTVd accumulation. While further experiments found no indication that PSTVd RNA suppresses silencing, experiments showed that PSTVd secondary structure blocks RISC-mediated cleavage. Thus, instead of suppressing RNA silencing, secondary structure of the PSTVd genome provides protection against degradation by the silencing machinery.

Silencing Suppressors Affect Plant Development

Many plant viral suppressors of RNA silencing have the ability to cause disease by altering the normal course of plant development. This was revealed by analysis of mutant viruses. For example, naturally occurring mutations in the gene for the TMV 126 kDa replicase protein result in a masked (symptomless) phenotype which relates to decreased viral silencing suppression activity and rate of systemic movement. Site-specific mutations introduced into the tobacco vein mottling virus (a potyvirus) HC-Pro gene also altered symptom expression. More drastic forms of mutagenesis, such as complete deletion of the 2b protein gene, created CMV strains that did not induce symptoms.

Transgenic plants expressing known or candidate viral silencing suppressor genes have been used to characterize the effects of the protein on plant gene expression, metabolism, and development, in the absence of virus infection. In most cases, stable expression of RNA silencing suppressors such as potyviral P1/HC-Pro proteins, tombusviral P19 proteins, beet yellows virus P21 protein, TCV CP and ACMV or SLCMV AC4 proteins, strongly disrupted plant development. This was most often seen as stunting of the plants and deformation of the stem, petioles, leaves, and flowers.

In the case of the CMV 2b protein, the strength of the altered phenotype in 2b-transgenic *Arabidopsis* plants corresponded with the severity of the symptoms induced by the strain from which the gene had been obtained. Transgenic expression of a 2b protein from a severe strain strongly inhibited growth of primary roots while both mild and severe strain 2b proteins enhanced the elongation of lateral roots. Overall, the effects of severe strain 2b proteins expressed as transgenes mimicked the phenotypes resulting from mutations in genes regulating the auxin response pathway and in *ago1* mutant plants.

Viral Suppression of the miRNA Pathway

It has been assumed that silencing suppressors alter plant development because they target elements that are common to, or highly similar between, the antiviral siRNA pathway and miRNA-directed regulation of host mRNA accumulation and translation (**Figure 1**). Since miRNAs negatively regulate the mRNA levels of several factors with roles in development (e.g., scarecrow-like factors, auxin response factors etc.), silencing suppressors which have the ability to interfere with components of small RNA-directed pathways can trigger or inhibit aspects of plant development.

Northern analysis has been used to study changes in the levels of miRNAs and target mRNAs accumulation in transgenic *Arabidopsis* plants expressing viral silencing suppressors. This method of analysis has provided evidence that some viral silencing suppressors disrupt of miRNA-regulated gene expression. The technique has also been used to monitor of levels of longer dsRNA species, miRNA precursor transcripts, miRNA duplexes, as well as the mRNA targets and their breakdown products (**Figure 1**). In combination with *in vivo* or *in vitro* studies of RNA or protein binding by silencing suppressor proteins, comparisons of steady-state levels of these various RNA species in nontransgenic versus transgenic plants expressing the silencing suppressors have helped reveal the target(s) of silencing suppressors within the miRNA pathway.

Is interference with miRNA-regulated gene expression entirely due to a case of mistaken identity, in which silencing suppressors accidentally or incidentally inhibit common targets or similar steps within the miRNA and siRNA-directed pathways? In some cases, for example where the mode of action of a silencing suppressor is to bind dsRNAs in a relatively nonsize selective manner, such as the TCV CP or the aureusvirus P14 protein, there would be a greater potential for cross-inhibition between the two pathways. In cases where the silencing suppressor acts by size-selective RNA binding, or can bind selectively to protein components of the silencing pathways (e.g., binding only to specific members of the AGO or DCL protein families), there is the possibility that silencing suppressors discriminate between components of the two RNA silencing pathways (**Figure 1**). Speculatively, this ability to discriminate suggests the possibility that natural selection operates on the genes for silencing suppressor to produce factors that minimize damage to host plants, or produce developmental changes that in some way favor the replication, spread, or transmission of the virus.

A recent breakthrough in animal virus research revealed that HIV-1 infection suppresses the microRNA pathway in a manner that promotes HIV-1 infection, but it is not evident whether this phenomenon depends on a viral silencing suppressor protein. Knockdown of *Dicer* and *Drosha* in HIV-1 infected cells showed that these two RNAse III enzymes contribute to suppression of HIV-1 infection. Microarray experiments identified miRNAs that were upregulated or induced only in HIV-1 infected cells. miR-17/92 is a polycistronic miRNA cluster which is downregulated during HIV-1 replication. miR-17/92 cluster includes miR17–5p and miR20a which target the histone acetylase PCAF, a cofactor for Tat in HIV-1. Thus, the miRNAs do not target HIV-1 but cellular factors necessary for HIV-1 gene expression. Thus, HIV-1 suppression of miR-17/92 cluster ensures a necessary supply of PCAF for virus replication.

See also: Plant Resistance to Viruses: Natural Resistance Associated with Recessive Genes; Plant Resistance to Viruses: Engineered Resistance; Plant Resistance to Viruses: Natural Resistance Associated with Dominant Genes; Virus Induced Gene Silencing (VIGS).

Further Reading

Blevins T, Rajeswaran R, Shivaprasad PV, et al. (2006) Four plant Dicers mediate viral small RNA biogenesis and DNA virus induced silencing. Nucleic Acids Research 34(21): 6233–6246.

Bucher E, Lohuis D, van Poppel PM, Geerts-Dimitriadou C, Goldbach R, and Prins M (2006) Multiple virus resistance at a high frequency using a single transgene construct. Journal of General Virology 87: 3697–3701.

Cai X, Schafer A, Lu S, et al. (2006) Epstein–Barr virus microRNAs are evolutionarily conserved and differentially expressed. PLoS Pathogens 2(3): e23.

Deleris A, Gallego-Bartolome J, Bao JS, Kasschau KD, Carrington JC, and Voinnet O (2006) Hierarchical action and inhibition of plant Dicer-like proteins in anti-viral defense. *Science* 313(5783): 68–71.

Galiana-Arnoux D, Dostert C, Schneemann A, Hoffmann JA, and Imler JL (2006) Essential function *in vivo* for Dicer-2 in host defense against RNA viruses in drosophila. *Nature Immunology* 7(6): 590–597.

Gitlin L, Stone JK, and Andino R (2005) Poliovirus escape from RNA interference: Short interfering RNA-target recognition and implications for therapeutic approaches. *Journal of Virology* 79(2): 1027–1035.

Hamilton A, Voinnet O, Chappell L, and Baulcombe D (2002) Two classes of short interfering RNA in RNA silencing. *EMBO Journal* 21(17): 4671–4679.

Hamilton AJ and Baulcombe DC (1999) A species of small antisense RNA in posttranscriptional gene silencing in plants. *Science* 286(5441): 950–952.

Himber C, Dunoyer P, Moissiard G, Ritzenthaler C, and Voinnet O (2003) Transitivity-dependent and -independent cell-to-cell movement of RNA silencing. *EMBO Journal* 22(17): 4523–4533.

Li F and Ding SW (2006) Virus counterdefense: Diverse strategies for evading the RNA-silencing immunity. *Annual Review of Microbiology* 60: 503–531.

Li HW, Li WX, and Ding SW (2002) Induction and suppression of RNA silencing by an animal virus. *Science* 296(5571): 1319–1321.

Marathe R, Anandalakshmi R, Smith TH, Pruss GJ, and Vance VB (2000) RNA viruses as inducers, suppressors and targets of post-transcriptional gene silencing. *Plant Molecular Biology* 43(2–3): 295–306.

Matzke MA and Matzke AJ (2004) Planting the seeds of a new paradigm. *PLOS Biology* 2: E133.

Pfeffer S, Sewer A, Lagos-Quintana M, *et al.* (2005) Identification of microRNAs of the herpesvirus family. *Nature Methods* 2(4): 269–276.

Pfeffer S and Voinnet O (2006) Viruses, microRNAs and cancer. *Oncogene* 25(46): 6211–6219.

Segers GC, van Wezel R, Zhang XM, Hong YG, and Nuss DL (2006) Hypovirus papain-like protease p29 suppresses RNA silencing in the natural fungal host and in a heterologous plant system. *Eukaryotic Cell* 5(6): 896–904.

Silhavy D and Burgyan J (2004) Effects and side-effects of viral RNA silencing suppressors on short RNAs. *Trends in Plant Science* 9: 76–83.

Viroids

R Flores, Instituto de Biología Molecular y Celular de Plantas (UPV-CSIC), Valencia, Spain
R A Owens, Beltsville Agricultural Research Center, Beltsville, MD, USA

© 2008 Elsevier Ltd. All rights reserved.

Glossary

Catalytic RNA RNA molecules that are able to catalyze, in a protein-free medium, specific reactions involving the formation or breakage of covalent bonds. In nature, these reactions are usually transesterifications (self-cleavage and ligation) affecting the catalytic RNA itself.

Hammerhead structure The conserved secondary/tertiary structure shared by the smallest class of natural ribozymes. Most have been found in one or both strands of certain viroid and viroid-like satellite RNAs where they mediate self-cleavage of multimeric intermediates arising from replication through a rolling-circle mechanism.

Ribozyme RNA motif responsible for the catalytic activity of certain RNA molecules. In nature, they are found embedded within catalytic RNAs.

Introduction

Viroids are the smallest known agents of infectious disease – small (246–401 nt), highly structured, circular, single-stranded RNAs that lack detectable messenger RNA activity. While viruses have been described as 'obligate parasites of the cell's translational system' and supply some or most of the genetic information required for their replication, viroids can be regarded as 'obligate parasites of the cell's transcriptional machinery'. Thus far, viroids are known to infect only plants.

The first viroid disease to be studied by plant pathologists was potato spindle tuber. In 1923, its infectious nature and ability to spread in the field led Schultz and Folsom to group potato spindle tuber disease with several other 'degeneration diseases' of potatoes. Nearly 50 years were to elapse before Diener's demonstration in 1971 that the molecular properties of its causal agent, potato spindle tuber viroid (PSTVd), were fundamentally different than those of conventional plant viruses.

Genome Structure

Efforts to understand how viroids replicate and cause disease without the assistance of any viroid-encoded polypeptides have prompted detailed analysis of their structure. Viroids possess rather unusual properties for single-stranded RNAs (e.g., a pronounced resistance to digestion by ribonuclease and a highly cooperative thermal denaturation profile), leading to an early realization that they might have an unusual higher-order structure.

To date, the complete sequences of 29 distinct viroid species plus a large number of sequence variants have been determined (**Table 1**). All are single-stranded circular RNAs containing 246–401 unmodified nucleotides. Theoretical calculations and physicochemical studies indicate that PSTVd and related viroids assume a highly base-paired, rod-like conformation *in vitro* (**Figure 1**). Pairwise sequence comparisons suggest that the series of short double helices and small internal loops that comprise this so-called 'native' structure are organized into five domains whose boundaries are defined by sharp differences in sequence similarity.

The 'central domain' is the most highly conserved viroid domain and contains the site where multimeric PSTVd RNAs are cleaved and ligated to form circular progeny. The 'pathogenicity domain' contains one or more structural elements which modulate symptom expression, and the relatively small 'variable domain' exhibits the greatest sequence variability between otherwise closely related viroids. The two 'terminal domains' appear to play an important role in viroid replication and evolution. Although these five domains were first identified in PSTVd, apple scar skin viroid (ASSVd) and related viroids also contain a similar domain arrangement. Certain viroids such as *Columnea* latent viroid (CLVd), Australian grapevine viroid (AGVd), and *Coleus blumei* viroid 2 (CbVd 2) appear to be 'mosaic molecules' formed by exchange of domains between two or more viroids infecting the same cell. RNA rearrangement/recombination can also occur within individual domains, leading, in coconut cadang-cadang (CCCVd) and citrus exocortis (CEVd) viroids, to duplications of the right terminal domain plus part of the variable domain. This domain model is not shared by avocado sunblotch (ASBVd) and related viroids.

Much less is known about viroid tertiary structure, especially *in vivo* where these molecules almost certainly accumulate as ribonucleoprotein particles. UV-induced

Table 1 Classification of viroids of known nucleotide sequence

Family[a]	Genus[a]	Name	Abbreviation	Nucleotides[b]
Pospiviroidae	Pospiviroid	Chrysanthemum stunt	CSVd	354–356
		Citrus exocortis	CEVd	368–375 (463–467)
		Columnea latent	CLVd	370–373
		Iresine	IrVd	370
		Mexican papita	MPVd	359–360
		Potato spindle tuber	PSTVd	356–361 (341)
		Tomato apical stunt	TASVd	360–363
		Tomato chlorotic dwarf	TCDVd	360
		Tomato planta macho	TPMVd	359–360
	Cocadviroid	Citrus viroid IV	CVd-IV	284
		Coconut cadang-cadang	CCCVd	246–247 (287–301)
		Coconut tinangaja	CTiVd	254
		Hop latent	HLVd	256
	Hostuviroid	Hop stunt[c]	HSVd	294–303
	Apscaviroid	Apple dimple fruit	ADFVd	306,307
		Apple scar skin[d]	ASSVd	329–334
		Australian grapevine	AGVd	369
		Citrus bent leaf	CBLVd	315,318
		Citrus dwarfing	CDVd	294,297
		Grapevine yellow speckle 1	GVYSVd 1	366–368
		Grapevine yellow speckle 2[e]	GYSVd 2	363
		Pear blister canker	PBCVd	315,316
	Coleviroid	*Coleus blumei* 1	CbVd 1	248–251
		Coleus blumei 2	CbVd 2	301,302
		Coleus blumei 3	CbVd 3	361–364
Avsunviroidae	Avsunviroid	Avocado sunblotch	ASBVd	246–251
	Pelamoviroid	Chrysanthemum chlorotic mottle	CChMVd	398–401
		Peach latent mosaic	PLMVd	335–351
	Elaviroid	Eggplant latent	ELVd	332–335

[a]Classification follows scheme proposed by Flores *et al.* (see VIII Report of the International Committee on Taxonomy of Viruses) with minor modifications. The nucleotide sequences of blueberry mosaic, burdock stunt, *Nicotiana glutinosa* stunt, pigeon pea mosaic mottle, and tomato bunchy top viroids are currently unknown; consequently, these viroids have not been assigned to specific genera. Whether apple fruit crinkle and citrus viroid original source should be considered variants or new viroid species of genus *Apscaviroid* is pending.
[b]Sizes of variants containing insertions or deletions arising *in vivo* are shown in parentheses.
[c]Includes cucumber pale fruit, citrus cachexia, peach dapple, and plum dapple viroids.
[d]Includes pear rusty skin and dapple apple viroids.
[e]Formerly termed grapevine viroid 1B.

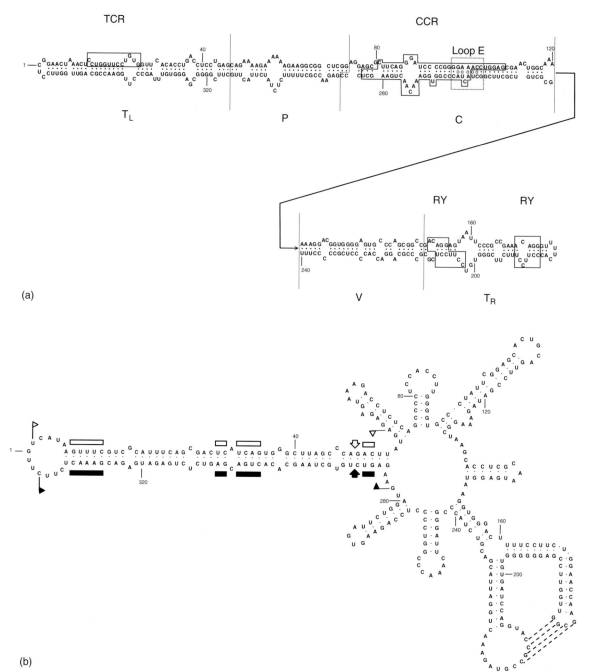

Figure 1 (a) The rod-like secondary structure of PSTVd (intermediate strain) showing the five domains characteristic of members of the family *Pospiviroidae:* the terminal left (T_L), pathogenicity (P), central (C), variable (V), and terminal right (T_R). The central conserved region (CCR) is located within the C domain and contains a UV-sensitive loop E motif with noncanonical base pairs (denoted by circles). The T_L domains of genera *Pospiviroid* and *Apscaviroid* contain a terminal conserved region (TCR), while those of genera *Hostuviroid* and *Cocadviroid* contain a terminal conserved hairpin (not shown). The T_R may also contain 1–2 copies of a protein-binding RY motif. (b) The branched secondary structure of PLMVd (reference variant). Plus and minus self-cleavage domains are indicated by flags, nucleotides conserved in most natural hammerhead structures by bars, and the self-cleavage sites by arrows. Black and white symbols refer to plus and minus polarities, respectively. Nucleotides involved in a pseudoknot are indicated by broken lines. Redrawn with modifications from Gross HJ, Domdey H, Lossow C, *et al*. (1978) Nucleotide sequence and secondary structure of potato spindle tuber viroid. *Nature* 273: 203–208; Hernández C and Flores R (1992) Plus and minus RNAs of peach latent mosaic viroid self-cleave *in vitro* through hammerhead structures. *Proceedings of the National Academy of Sciences, USA* 89: 3711–3715; Bussière F, Ouellet J, Côté F, Lévesque D, and Perreault JP (2000) Mapping in solution shows the peach latent mosaic viroid to possess a new pseudoknot in a complex, branched secondary structure. *Journal of Virology* 74: 2647–2654.

cross-linking of two nucleotides within a loop E motif in the central domain of PSTVd provided the first definitive evidence for such tertiary interactions. Similar UV-sensitive structural elements have also been discovered in a number of other RNAs including 5S eukaryotic rRNA, adenovirus VAI RNA, and the viroid-like domain of the hepatitis delta virus genome. Loop E forms during the conversion of multimeric PSTVd RNAs into monomers. The ability of ASBVd-related RNAs to undergo spontaneous self-cleavage mediated by hammerhead ribozymes as well as the presence of pseudoknots critical for infectivity in some other members of the *Avsunviroidae* (**Figure 1**) provide additional evidence for the functional importance of viroid tertiary structure.

Classification

Based upon differences in the structural and functional properties of their genomes, viroids species are assigned to one of two taxonomic families (see **Table 1**). Members of the family *Pospiviroidae* (type member PSTVd) have a rod-like secondary structure that contains five structural–functional domains and several conserved motifs. Most members of the family *Avsunviroidae* (type member ASBVd), in contrast, appear to adopt a branched conformation, and multimeric RNAs of all family members behave as catalytic RNAs and undergo spontaneous self-cleavage (**Figure 1**). Differences in their sites of replication also support this classification scheme; that is, PSTVd and ASBVd replicate in the nucleus and the chloroplast, respectively, and the same appears to occur for other members of each family. Each family is subdivided into genera according to certain demarcating criteria. Groups of sequence variants that show >90% sequence identity in pairwise comparisons and share some common biological property are arbitrarily defined as viroid species. *In vivo*, each viroid species is actually a 'quasispecies', that is, a collection of closely related sequences subject to a continuous process of variation, competition, and selection. There is phylogenetic evidence for an evolutionary link between viroids and other viroid-like subviral RNAs (**Figure 2**).

Host Range and Transmission

All viroids are mechanically transmissible, and most are naturally transmitted from plant to plant by man and his tools. Individual viroids vary greatly in their ability to infect different plant species. PSTVd can replicate in about 160 primarily solanaceous hosts, while only two members of the family Lauraceae are known to support ASBVd replication. HSVd has a particularly wide host range that includes herbaceous species as well as woody perennials. Many natural hosts are either vegetatively propagated or crops that are subjected to repeated grafting or pruning operations. PSTVd, ASBVd, and CbVd1 are vertically transmitted through pollen and/or true seed, but the significance of this mode of transmission in the natural spread of disease is unclear. PSTVd can be encapsidated by the coat protein of potato leafroll virus (PLRV, a polerovirus) as well as velvet tobacco mottle virus (VTMoV, a sobemovirus), and epidemiological surveys suggest that PLRV facilitates viroid spread under field conditions.

Commonly used techniques for the experimental transmission of viroids include the standard leaf abrasion methods developed for conventional viruses, 'razor slashing' methods in which phloem tissue in the stem or petiole is inoculated via cuts made with a razor blade previously dipped into the inoculum, and, in the case of CCCVd, high-pressure injection into folded apical leaves. Viroids can also be transmitted by either plant transformation or 'agroinoculation' during which a modified *Agrobacterium tumefaciens* Ti plasmid is used to introduce full-length viroid-complementary DNA into the potential host cell. Either technique can overcome the marked resistance of some hosts to mechanical inoculation. Identification of the molecular mechanism(s) that determine viroid host range remains an important research goal.

Symptomatology

Viroids and conventional plant viruses induce a very similar range of macroscopic symptoms. Symptom expression is usually optimal at the same relatively high temperatures (30–33 °C) that promote viroid replication. Stunting and leaf epinasty (a downward curling of the leaf lamina resulting from unbalanced growth within the various cell layers) are considered the classic symptoms of viroid infection. Other commonly observed symptoms include vein clearing, veinal discoloration or necrosis, and the appearance of localized chlorotic/necrotic spots or mottling in the foliage. Symptoms may also be expressed in flowers and bark, and fruits or tubers from viroid-infected plants may be abnormally shaped or discolored. Viroid infection of certain citrus rootstock/scion combinations may result in tree dwarfing (**Figure 3**). Viroid infections are often latent and rarely kill the host.

Viroid infections are also accompanied by a number of cytopathic effects – chloroplast and cell wall abnormalities, the formation of membranous structures in the cytoplasm, and the accumulation of electron-dense deposits in both chloroplasts and cytoplasm. Metabolic changes include dramatic alterations in growth regulator levels.

336 Viroids

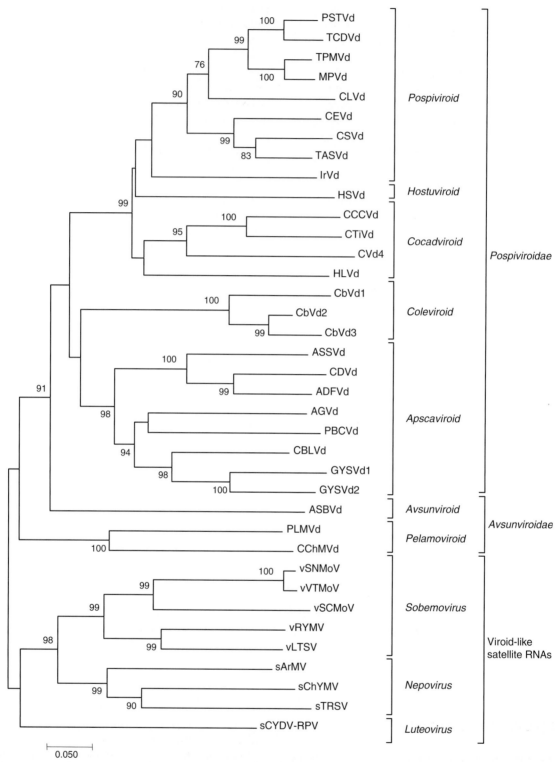

Figure 2 Neighbor-joining phylogenetic tree obtained from an alignment manually adjusted to take into account local similarities, insertions/deletions, and duplications/rearrangements described in the literature for viroid and viroid-like satellite RNAs. Bootstrap values were based on 1000 random replicates (only values >70% are shown). Viroid abbreviations are those used in **Table 1**. Viroid-like satellite RNAs: lucerne transient streak virus (sLTSV); rice yellow mottle virus (sRYMV); subterranean clover mottle virus (sSCMoV); *Solanum nodiflorum* mottle virus (sSNMoV)); velvet tobacco mottle virus (sVTMoV); tobacco ringspot virus (sTRSV); *Arabis* mosaic virus (sArMV); chicory yellow mottle virus (sChYMV); cereal yellow dwarf virus-RPV (sCYDV-RPV). Adapted from Elena SF, Dopazo J, de la Peña M, Flores R, Diener TO, and Moya A (2001) Phylogenetic analysis of viroid and viroid-like satellite RNAs from plants: A reassessment. *Journal of Molecular Evolution* 53: 155–159.

Figure 3 (a) Symptoms of PSTVd infection in Rutgers tomato approximately 4 weeks after inoculation of cotyledons with PSTVd strains causing mild, intermediate, and severe symptoms. (b) Symptoms of ASBVd infection in avocado fruits and leaves. (c) Viroid-induced dwarfing of citrus growing on susceptible rootstocks: All trees in the block were graft-inoculated with CDVd shortly after transfer to the field; only one tree (right foreground) escaped infection. Note the difference in height.

Geographic Distribution

Although PSTVd, HSVd, CEVd, and ASBVd are widely distributed throughout the world, other viroids have never been detected outside the areas where they were first reported. Several factors may contribute to this variation in distribution pattern. Among the crops most affected by viroid diseases are a number of valuable woody perennials such as grapes, citrus, various pome and stone fruits, and hops. Propagation and distribution of improved cultivars is highly commercialized, with the result that many cultivars are now grown worldwide. The international exchange of plant germplasm also continues to increase at a rapid rate. In both instances, the large number of latent (asymptomatic) hosts facilitates viroid spread.

Epidemiology and Control

Viroid diseases pose a potential threat to agriculture, and several are of considerable economic importance. Ready transmission of PSTVd by vegetative propagation, foliar contact, and true seed or pollen continues to pose a serious threat to potato germplasm collections and breeding programs. Coconut cadang-cadang has killed over 30 million palms in the Philippines since it was first recognized in the early 1930s. While many viroids were first detected in ornamental or crop plants, most viroid diseases are thought to result from chance transfer from endemically infected wild species to susceptible cultivars. Several lines of circumstantial evidence are consistent with this hypothesis:

1. The experimental host ranges of several viroids include many wild species, and these wild species often tolerate viroid replication without the appearance of recognizable disease symptoms.
2. Although co-evolution of host and pathogen is often accompanied by appearance of gene-for-gene vertical resistance, no useful sources of resistance to PSTVd infection have been identified in the cultivated potato.
3. Viroids and/or viroid-related RNAs closely related to TPMVd and CCCVd have been detected in weeds and other wild vegetation growing near fields containing viroid-infected plants.

Growers and plant pathologists are unlikely to have simply overlooked diseases with symptoms as severe as those of chrysanthemum stunt or cucumber pale fruit, two diseases first reported after World War II. Large-scale monoculture of genetically identical crops and the commercial propagation/distribution of many cultivars are two comparatively modern developments which would facilitate the development of serious disease problems following the chance transfer of viroids from wild hosts to cultivated plants. Viroid diseases may also arise by transfer between cultivated crop species. For example, pears provide a latent reservoir for ASSVd; likewise, while HSVd infections of grapes are often symptomless, this viroid causes severe disease in hops. In both instances, the two crops are often grown in close proximity.

Because no useful sources of natural resistance to viroid disease are known, diagnostic tests continue to play a key role in efforts to control viroid diseases. Since viroids lack a protein capsid, the antibody-based techniques used to detect many plant viruses are not applicable. Tests based upon their unique molecular properties have largely supplanted biological assays for viroid detection. Problems with viroid bioassays include the length of time required for completion (weeks to years) and difficulties in detecting mild or latent strains. Several rapid (1–2 day) protocols involving polyacrylamide gel electrophoresis (PAGE) under denaturing conditions take advantage of the circular nature of viroids. Using these protocols, nanogram amounts of viroid can be unambiguously detected without the use of radioactive isotopes. In recent years, diagnostic procedures based upon nucleic acid hybridization or the polymerase chain reaction (PCR) are being widely used. The simplest methods involve the hybridization of a nonradioactively labeled viroid-complementary DNA or RNA probe to viroid samples that have been bound to a solid support followed by colorimetric or chemiluminescent detection of the resulting DNA–RNA or RNA–RNA hybrids. Such conventional 'dot blot' assays can detect picogram amounts of viroids using clarified plant sap or tissue prints rather than purified nucleic acid as the viroid source. PCR-based protocols are finding increasing acceptance in those cases where either this level of sensitivity is inadequate or a number of closely related viroids are present in the same sample.

Molecular Biology

Although devoid of messenger RNA activity, viroids replicate autonomously and cause disease in a wide variety of plants. Much has been learned about the molecular biology of viroids and viroid–host interaction over the past 25 years, but the precise nature of the molecular signals involved remains elusive. A series of questions first posed by Diener summarizes the many gaps in our current understanding of the biological properties of these unusual molecules:

1. What molecular signals do viroids possess (and cellular RNAs evidently lack) that induce certain DNA-dependent RNA polymerases to accept them as templates for the synthesis of complementary RNA molecules?
2. What are the molecular mechanisms responsible for viroid replication? Are these mechanisms operative in uninfected cells? If so, what are their functions?
3. How do viroids induce disease? In the absence of viroid-specified proteins, disease must arise from direct interaction(s) of viroids (or viroid-derived RNA molecules) with host-cell constituents. Infections by PSTVd and ASBVd induce RNA silencing (see below).
4. What determines viroid host range? Are viroids restricted to higher plants, or do they have counterparts in animals?
5. How did viroids originate?

Replication

A variety of multimeric plus- and minus-strand RNAs have been detected by nucleic acid hybridization in viroid-infected tissues. Based on their analysis, viroid replication has been proposed to proceed via a 'rolling circle' mechanism that involves reiterative transcription of the incoming plus circular RNA to produce a minus-strand RNA template. ASBVd and related viroids utilize a symmetric replication cycle in which the multimeric minus strand is cleaved to unit-length molecules and circularized before serving as template for the synthesis of multimeric plus strands. PSTVd and related viroids utilize an asymmetric cycle in which the multimeric minus strand is directly transcribed into multimeric plus strands. In both cases, the multimeric plus strands are cleaved to unit-length molecules and circularized.

A diversity of host-encoded enzymes have been implicated in viroid replication. Low concentrations of α-amanitin specifically inhibit the synthesis of both PSTVd plus and minus strands in nuclei isolated from infected tomato, strongly suggesting the involvement of DNA-dependent RNA polymerase II, transcribing an RNA template, in the replication of PSTVd and related viroids. In nuclear extracts, transcription of the PSTV plus strand by RNA polymerase II starts in the left terminal loop; furthermore, incubation of active replication complexes containing CEVd with a monoclonal antibody directed against the carboxy-terminal domain of RNA polymerase II results in the immunoprecipitation of both CEVd plus- and minus-strand RNAs. Mature PSTVd plus strands accumulate in the nucleolus and the nucleoplasm, while *in situ* hybridization indicates that minus-strand RNAs are confined to the nucleoplasm. The identity of the polymerase(s) responsible for replication of members of the family *Avsunviroidae* in the chloroplast is less certain. ASBVd synthesis is resistant to tagetitoxin, strongly indicating the involvement of a nuclear-encoded chloroplastic RNA polymerase. Initiation sites for both ASBVd plus- and minus-strand synthesis have been mapped to the AU-rich terminal loops of their respective native structures.

In vitro evidence indicates that specific cleavage of multimeric PSTVd plus-strand RNAs requires (1) rearrangement of the conserved central region to form a branched structure containing a GNRA tetraloop and (2) the action of one or more host-encoded nucleases. Other less-efficient processing sites can also be used *in vivo*. Plus- and minus-strand RNAs of ASBVd and related viroids, in contrast, undergo spontaneous self-cleavage through hammerhead ribozymes to form linear monomers (**Figure 4**). Addition of certain chloroplast proteins acting as RNA chaperones facilitates this hammerhead ribozyme-mediated self-cleavage reaction. The final step in viroid replication is the ligation of linear monomers to form mature circular progeny. Plant cells are known to contain RNA ligase activities which can act upon the 5′ hydroxyl and 2′,3′ cyclic phosphate termini formed by either cleavage pathway.

Movement

Upon entering a potential host cell, viroids must move to either the nucleus (*Pospiviroidae*) or chloroplast (*Avsunviroidae*) before beginning replication. Available data suggest that PSTVd enters the nucleus as a ribonucleoprotein complex formed by the interaction of cellular proteins with specific viroid sequence or structural motifs. VirP1, a bromodomain-containing protein isolated from tomato, has a nuclear localization signal and binds to the terminal right domain of PSTVd. Proteins such as TFIIIA and ribosomal protein L5 that bind to the loop E motif may also be involved in viroid transport into the nucleus. How ASBVd or other members of the family *Avsunviroidae* enter and exit the chloroplast is currently unknown.

To establish a systemic infection, viroids leave the initially infected cell – moving first from cell to cell and then long distances through the host vasculature. Upon injection into symplasmically isolated guard cell in a mature tomato leaf, fluorescently labeled PSTVd RNA does not move. Injection into interconnected mesophyll cells, in contrast, is followed by rapid cell-to-cell movement through the plasmodesmata. Long-distance movement of viroids, like that of nearly all plant viruses, occurs in the phloem where it follows the typical source-to-sink pattern of photoassimilate transport. Viroid movement in the phloem almost certainly requires formation of a ribonucleoprotein complex, possibly involving a dimeric lectin known as phloem protein 2 (PP2), the most abundant protein in phloem exudate. Movement of PSTVd in the phloem appears to be sustained by replication in supporting cells and is tightly regulated by developmental and cellular factors. For example, *in situ* hybridization reveals the presence of PSTVd in vascular tissues underlying the shoot apical meristem of infected tomato, but entry into the shoot apical meristem itself appears to be blocked. Another important control point for PSTVd trafficking is the bundle sheath–mesophyll boundary in the leaf. By disrupting normal pattern of viroid movement, it may be possible to create a plant that is resistant/immune to viroid infection.

Pathogenicity

Sequence comparisons of naturally occurring PSTVd and CEVd variants as well as infectivity studies with chimeric viroids, constructed by exchanging the pathogenicity domains of mild and severe strains of CEVd, have clearly shown that the pathogenicity domain in the family *Pospiviroidae* contains important determinants of symptom expression. Symptom expression is also affected by the

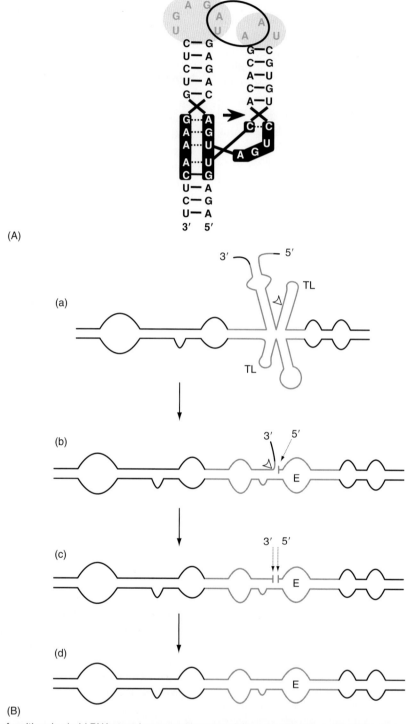

Figure 4 Cleavage of multimeric viroid RNAs requires rearrangement of the native structure. (A) During transcription, the strands of both polarities of members of the family Avsunviroidae can fold into hammerhead structures (here illustrated for the hammerhead of the PLMVd plus RNA) and self-cleave accordingly. Nucleotides conserved in most natural hammerhead structures are on a black background, and the self-cleavage site is denoted by an arrow. A circle delimits the presumed tertiary interaction between terminal loops enhancing the catalytic activity. Watson-Crick and noncanonical base pairs are represented by continuous and discontinuous lines, respectively. After self-cleavage, the RNA adopts a new conformation favoring ligation (or self-ligation). (B) Processing of a longer-than-unit-length plus PSTVd RNA transcript in a potato nuclear extract. The central conserved region of the substrate for the first cleavage reaction (a) contains a tetraloop (denoted TL). After dissociation of the 5′ segment from the cleavage site, the new 5′ end refolds and is stabilized by formation of a UV-sensitive loop E (b), while the 3′ end partially base-pairs with the lower strand. Single-stranded nucleotides at the 3′ end are then cleaved between positions 95 and 96 (c), and ligation of the 5′ and 3′ termini (d) results in formation of mature circular progeny. From Baumstark T, Schröder ARW, and Riesner D (1997) Viroid processing: Switch from cleavage to ligation is driven by a change from a tetraloop to a loop E conformation. *EMBO Journal* 16: 599–610.

rate of viroid replication, and sequence changes in the variable domain have been shown to regulate progeny titers in infected plants. Studies with TASVd revealed the presence of a third pathogenicity determinant in the left terminal loop. Also, a single U/A change position 257 in the central domain of PSTVd results in the appearance of severe stunting and a 'flat top' phenotype. In the family *Avsunviroidae*, determinants of pathogenicity have been mapped to either a tetraloop capping a hairpin stem in chrysanthemum chlorotic mottle viroid (CChMVd) or an insertion that folds into a hairpin also capped by a tetraloop in peach latent mosaic viroid (PLMVd).

The ability of novel viroid chimeras to replicate and move normally from cell to cell implies certain basic similarities between their structures *in vitro* and *in vivo* but provides no information about the nature of the molecular interactions responsible for symptom development. Until recently, it was widely assumed that the mature viroid RNA was the direct pathogenic effector. Just like viruses, however, viroid replication is also accompanied by the production of a variety of small (21–26 nt) RNA molecules. The role of these small interfering RNAs (siRNAs) in viroid pathogenicity is not yet clear, but the inverse relationship between accumulation levels of the mature viroid RNAs and the corresponding siRNAs for members of the family *Avsunviroidae* suggest that the latter may regulate the titer of the former. Also, recovery of tomato plants from the symptoms of severe PSTVd infections is preceded by the accumulation of PSTVd-specific siRNA.

Viroid infections are accompanied by quantitative changes in a variety of host-encoded proteins. Certain of these are 'pathogenesis-related' proteins whose synthesis or activation is part of a general host reaction to biotic or abiotic stress, but others appear to be more specific. In tobacco, PSTVd infection results in the preferential phosphorylation of a host-encoded 68 kDa protein that is immunologically related to an interferon-inducible, dsRNA-dependent mammalian protein kinase of similar size. The human kinase is differentially activated by PSTVd strains of varying pathogenicity *in vitro*, while infection of tomato by intermediate or severe strains of PSTVd induces the synthesis of PKV, a dual-specificity, serine/threonine protein kinase. Broad changes in host gene expression following PSTVd infection have been detected by complementary DNA macroarray analysis.

Host Range

Possibly as a result of its involvement in the cleavage/ligation of progeny RNA, nucleotides in the central domain of PSTVd and related viroids appear to play an important role in determining host range. For example, a single nucleotide substitution in the loop E motif results in a dramatic increase in the rate of PSTVd replication in tobacco. The biological properties of CLVd also suggest that this domain contains one or more host-range determinants. CLVd appears to be a natural mosaic of sequences present in other viroids; phylogenetic analysis (see **Figure 2**) suggests that it can be considered to be a PSTVd-related viroid whose conserved central domain has been replaced by that of HSVd. Like HSVd (but not PSTVd or related viroids), CLVd can replicate and cause disease in cucumber.

Origin and Evolution

Much of the early speculation about viroid origin involved their possible origin as 'escaped introns' (i.e., descent from normal host RNAs). More recently, however, viroids have been proposed to represent 'living fossils' of a precellular RNA world that assumed an intracellular mode of existence sometime after the evolution of cellular organisms. The presence of ribozymes in members of the *Avsunviroidae* strongly supports this view.

The inherent stability of viroids and viroid-like satellite RNAs (structurally similar to viroids but functionally dependent on helper viruses) which arises from their small size and circularity would have enhanced the probability of their survival in primitive, error-prone RNA self-replicating systems and assured their complete replication without the need for initiation or termination signals. Most viroids (but not satellite RNAs or random sequences of the same base composition) also display structural periodicities with repeat units of 12, 60, or 80 nt. The high error rate of prebiotic replication systems may have favored the evolution of polyploid genomes, and the mechanism of viroid replication (i.e., rolling-circle transcription of a circular template) provides an effective means of genome duplication.

Viroids and viroid-like satellite RNAs all possess efficient mechanisms for the precise cleavage of their oligomeric replication intermediates to form monomeric progeny. PSTVd and related viroids appear to require proteinaceous host factor(s) for cleavage, but others (members of the family *Avsunviroidae* and viroid-like satellite RNAs) contain ribozymes far smaller and simpler than those derived from introns. Thus, ASBVd and the other self-cleaving viroids may represent an evolutionary link between viroids and viroid-like satellite RNAs. No viroid is known to code for protein, a fact that is consistent with the possibility that viroids are phylogenetically older than introns.

Phylogenetic evidence for an evolutionary link between viroids and other viroid-like subviral RNAs has been presented by Elena *et al.* (see **Figure 2**). Among several subviral RNAs possibly related to viroids is carnation small viroid-like RNA, a 275 nt circular molecule with self-cleaving hammerhead structures in both its plus and minus strands that has a DNA counterpart. This novel retroviroid-like element shares certain features with both viroids and a small RNA transcript from newt.

See also: Hepatitis Delta Virus; Origin of Viruses; Plant Resistance to Viruses: Natural Resistance Associated with Dominant Genes; Quasispecies; Ribozymes; Satellite Nucleic Acids and Viruses.

Further Reading

Baumstark T, Schröder ARW, and Riesner D (1997) Viroid processing: Switch from cleavage to ligation is driven by a change from a tetraloop to a loop E conformation. *EMBO Journal* 16: 599–610.

Bussière F, Ouellet J, Côté F, Lévesque D, and Perreault JP (2000) Mapping in solution shows the peach latent mosaic viroid to possess a new pseudoknot in a complex, branched secondary structure. *Journal of Virology* 74: 2647–2654.

Diener TO (1979) *Viroids and Viroid Diseases.* New York: Wiley-Interscience.

Ding B, Itaya A, and Zhong X (2005) Viroid trafficking: A small RNA makes a big move. *Current Opinion in Plant Biology* 8: 606–612.

Elena SF, Dopazo J, de la Peña M, Flores R, Diener TO, and Moya A (2001) Phylogenetic analysis of viroid and viroid-like satellite RNAs from plants: A reassessment. *Journal of Molecular Evolution* 53: 155–159.

Flores R, Hernandez C, Martinez de Alba AE, Daros JA, and DiSerio F (2005) Viroids and viroid–host interactions. *Annual Review of Phytopathology* 43: 117–139.

Flores R, Randles JW, Owens RA, Bar-Joseph M, and Diener TO (2005) Viroids. In: Fauquet CM, Mayo MA, Maniloff J, Desselberger U, and Ball AL (eds.) *Virus Taxonomy: Eighth Report of the International Committee on Taxonomy of Viruses*, pp. 1145–1159. San Diego, CA: Elsevier Academic Press.

Gross HJ, Domdey H, Lossow C, et al. (1978) Nucleotide sequence and secondary structure of potato spindle tuber viroid. *Nature* 273: 203–208.

Hadidi A, Flores R, Randles JW, and Semancik JS (eds.) (2003) *Viroids.* Collingwood: CSIRO Publishing.

Hernández C and Flores R (1992) Plus and minus RNAs of peach latent mosaic viroid self-cleave *in vitro* through hammerhead structures. *Proceedings of the National Academy of Sciences, USA* 89: 3711–3715.

Hull R (2002) *Matthews' Plant Virology*, 4th edn. New York: Academic Press.

Virus Classification by Pairwise Sequence Comparison (PASC)

Y Bao, Y Kapustin, and T Tatusova, National Institutes of Health, Bethesda, MD, USA

© 2008 Elsevier Ltd. All rights reserved.

Glossary

Demarcation A mapping of ranges of pairwise distances into taxonomic categories.

Introduction

Virus classification is very important for virus research. It is also an extremely difficult task for many virus families. Traditionally, virus classification relied on properties such as virion morphology, genome organization, replication mechanism, serology, natural host range, mode of transmission, and pathogenicity. Yet viruses sharing the above properties can reveal tremendous differences at the genome level. For example, classification of many phages is currently based on presence, structure, and length of a tail, and this approach has been shown not to correlate with genomic information, leading to a very difficult situation and hundreds of unclassified phages.

Molecular virus classification based on virus sequences has been used increasingly in recent years, thanks to the growing number of viral sequences available in the public sequence databases. The most commonly used sequence comparison methods include multiple sequence alignment and phylogenetic analysis. Another molecular classification method that has drawn more and more attention from virologists is pairwise sequence comparison (PASC). In this article, we briefly describe various sequence comparison methods, introduce the PASC tool, and compare it with other methods.

Sequence Comparison Methods

A universal approach to compare biological sequences, in a sense of producing meaningful results at various levels of divergence, is in the realm of sequence alignment. An alignment is an arrangement of residues of two or more sequences in a way that reveals their possible relatedness, with space characters inserted into the sequences to indicate single-residue insertions and deletions. A variety of algorithms and programs are available to suit a wide range of problems requiring sequence alignments as parts of their solutions. Depending on the specifics of a problem, different types of algorithms or their combinations may work best. Most alignment algorithms can be broadly categorized by the scope of their application on sequences (local vs. global), or by the number of sequences involved (pairwise vs. multiple).

Each pairwise alignment can be viewed as an array of per-residue operations transforming one sequence to the other. These operations are substitutions (called matches and mismatches in nucleotide alignments), insertions, and deletions. A generalization of this concept to multiple alignments is possible. Alignments are scored using a scoring scheme appropriate for a biological context. The widely used affine scheme assigns substitution scores to substitutions and a penalty to each space, and an additional penalty to each gap defined as a maximal consecutive run of spaces. Given a set of sequences, an alignment is called optimal if it has the maximal score over all possible alignments. Optimal alignments are not necessarily unique; two or more alignments can be tied with the same score.

Local algorithms are capable of detecting similarities between arbitrary parts of sequences. Applications involving local alignments are numerous, including search for orthologous genes or conserved protein domains. Alignments produced with local algorithms are tractable to mathematical analysis, which allowed the building of tools that evaluate statistical significance of the alignments. The algorithm for computing optimal local alignments is known as Smith–Waterman and it runs in time proportional to the product of the sequences' lengths. Since this is too slow for large-scale searches, many heuristic methods (with BLAST being the most popular) have been developed, allowing matching typical queries against gigabase-sized archives of sequence in a matter of seconds.

Fast as they are, algorithms like BLAST are not suitable for all applications. Although they are capable of picking out segments of high similarity, no segment of input sequences is guaranteed to be a part of the resulting alignments, and some segments may belong to more than one individual alignment. Additional post-processing steps are often required in order to produce consistent sets of local alignments. This complicates the use of local alignments in applications involving uniform computing of identities.

When sequences in the set are expected to align end-to-end, global alignment algorithms are applicable. The strict algorithm for computing an optimal global alignment is known as Needleman–Wunsch. With the running time estimate being the same as in the Smith–Waterman, global alignments allow straightforward evaluation of identities as they provide unambiguous mapping for every residue. An important end-space free variant is used when one of the sequences is expected to align in the interior of another. Global algorithms are normally not suitable for applications aiming to capture rearrangement events.

Multiple sequence alignments can technically be viewed as a generalization of the pairwise case. However, they often serve different goals, such as a detection of weak and/or dispersed similarities over a set of sequences known to share a common function or structure. Computing an optimal multiple sequence alignment is a computationally costly task, and most implementations use various heuristics to approximate the alignment in a reasonable time. Note that even when the optimal multiple alignment is available, pairwise alignments inferred from it are not guaranteed to be optimal. There are many multiple sequence alignment tools available: CLUSTALW, DIALIGN, MAFFT, MUSCLE, PROBCONS, ProDA, and T-COFFEE, etc.

Phylogenetic analysis is probably the most frequently used molecular virus classification tool. A phylogeny or evolutionary tree is a mathematical structure which is used to model the historical relationships between groups of organisms or sequences. Main types of methods used to construct phylogenetic trees include distance-based methods (such as neighbor-joining), parsimony, maximum likelihood, and other probabilistic inference techniques. The most common distance-based methods utilize multiple sequence alignments to estimate the evolutionary distance between each pair of sequences and reconstruct the tree from the distances. Either protein sequences or DNA sequences can be used.

Phylogenetic analysis was used in the vast majority of virus families described in the Eighth Report of the International Committee on the Taxonomy of Viruses (ICTV) to support their classification. It has also been applied to the classification of a large group of distantly related viruses. For example, phages consist of many different families. Therefore, the conventional phylogenetic analysis that uses genomic sequences or individual protein sequences would not work for the classification of phages as a whole group. A 'phage proteomic tree' was developed to classify phages by the overall similarities of all protein sequences present in the phage genomes.

Although phylogenetic analysis is well established as a tool for virus classification, it is usually computationally intensive, and requires expertise to perform the analysis and to interpret the results. A more robust method is preferred so that researchers without an advanced computer system and advanced knowledge about the phylogenetic analysis can also use it. Also, as discussed below, despite the fact that some of the sequence alignment methods (such as BLAST) are very fast and easy to carry out, their results will not reveal the taxonomic relationships between the two viruses. A method that can place a new virus in the appropriated taxonomic position is desired. PASC is a good combination of the two methods.

The Principle of PASC

In the PASC system, pairwise global alignment is performed on complete genomes or particular protein sequences for each viral family, and their percentage of identity is calculated. The number of virus pairs at each percentage is plotted. The distribution of the identities is not evenly spread, but rather clustered into groups of peaks for viruses at strain, species, genus, and subfamily levels. The percentage range of each peak serves as a good reference for taxonomic classification based on sequence similarities. This method has been applied to polioviruses using the protein and nucleotide sequences of the VP1 gene, as well as the whole genome sequence, coronaviruses using the protein sequences of the polymerase and helicase, potyviruses using the protein and nucleotide sequences of the complete ORF and the coat protein gene, geminiviruses using the complete sequences of DNA-A, flexiviruses using the protein and nucleotide sequences of the three major viral gene products (replication protein, triple gene block and coat protein), papillomaviruses using the nucleotide sequences of the L1 gene, and poxviruses using the protein sequences of the DNA polymerase. There is an increasing interest to expand PASC to other virus families.

In order to apply PASC to a larger number of virus families and be used by a wide range of virologists, the following should be considered:

1. The same algorithm and sequence dataset should be used to determine the demarcations and to place new viruses in the right taxonomic position.
2. The sequence dataset used for demarcation determination and the virus taxonomy database should be updated frequently to reflect the most recent status.
3. The algorithm should be robust and fast enough for large sequence sets.
4. The system should be readily accessible to researchers worldwide and easy to use.

PASC Implementation at NCBI

The National Center for Biotechnology Information (NCBI) has developed a web-based PASC system that meets all of the criteria mentioned above. In this implementation, complete viral genomes are organized into groups corresponding to broad taxonomic entities such as family or floating genus. Within each group, alignments are pre-computed and stored in a database for each pair of the genomes. The alignments are used to evaluate identities, defined as the ratio of matching residues over the total alignment length. The alignments are computed using the pairwise global algorithm with the affine scoring scheme assigning one to matches and minus one to mismatches and nonterminal spaces and gaps. Since the genomes vary in lengths, terminal spaces are not penalized during the alignment computing but taken into account when computing the identity.

PASC interface is built around a histogram of pairwise identities. The primary feature of the interface is the comparison of an external sequence such as a newly sequenced viral genome, with genomes in a user-selected group. After the sequence is submitted, PASC will start computing the alignments, or extracting them from the database if the query is a member of the group. At the end of the process, a user is presented with a list of closest matches. Matches can be selected to visualize their positions on the identity distribution chart.

The PASC system at NCBI not only reproduced results for virus families for which PASC had been applied to, but also generated data with well-separated identity distributions that can be used as taxonomy demarcations for other virus families such as *Caliciviridae*, *Flaviviridae*, and *Togaviridae* among others.

Applications of PASC

PASC can be used to define taxonomy demarcations for many viral families. Two examples are shown here. In the first example, 45 complete genomes from the family *Luteoviridae* were used to construct the distribution of the pairwise identities among the genomes (**Figure 1**). Pairs located at 88% and up are all from different strains in the same species; pairs between 53% and 85% are mostly from different species in the same genus; and pairs located at 52% and lower are all from different genera. These percentages can therefore serve as demarcations for classification of luteoviruses. In the second example, the pairwise identities of 38 RNA-dependent RNA polymerase (RdRp) gene segment of viruses in the family *Reoviridae* were calculated and plotted (**Figure 2**). Similar to the above mentioned example, 81% and up, between 59% and 76%, and 57% and below can be used as boundaries for strains, species, and genera of reoviruses based on the identities between their polymerase genes. It should be noted that the determination of demarcation using PASC is not always straightforward. For some virus families, phenotypic characteristics of the viruses are required to be taken into account.

PASC can place newly sequenced viruses into the correct taxonomy group. For example, a genomic sequence of a luteovirus (GenBank accession number AY956384) appeared recently in the international sequence databases with the name chickpea chlorotic stunt virus, which is not an official ICTV species name. When this sequence was tested with the luteoviruses in the PASC system, it was found that the virus with the highest sequence similarity to it is cucurbit aphid-borne yellows virus in the genus *Polerovirus*. The similarity is 61.7%, which is in the demarcation of different species in the same genus. It can thus be suggested that this virus is a member of a new species in the genus *Polerovirus*.

Finally, PASC can identify possible questionable classifications in the existing groups when the peaks on the graph are very well separated. **Table 1** lists the pairs of the RdRp segment of reoviruses whose identities are between 54% and 54.5% in **Figure 2**. From the description above, this region represents pairs of viruses from different genera.

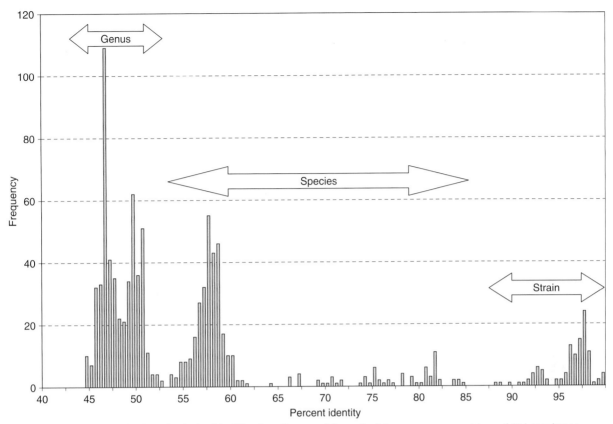

Figure 1 Frequency distribution of pairwise identities from the complete nucleotide sequence comparison of 45 luteoviruses.

This is true for most of the pairs in the table. However, the group also includes a pair containing St. Croix river virus and palyam virus, which both currently belong to the genus *Orbivirus*. Further investigation using PASC revealed that palyam virus has identities higher than 59% with many other viruses in the family, and therefore is indeed an orbivirus. The highest identity of St. Croix river virus with other viruses is only about 54%, which is lower than the demarcation for species in the same genus. This suggests that the species *St. Croix river virus* should probably be placed in a new genus in the *Reoviridae* family.

Advantages of PASC Compared to Other Methods

Unlike other classification methods based on the phenotypic properties of viruses, PASC is a quantitative tool. For those virus families that are suitable for PASC analyses, demarcations can be easily determined and new viruses can be clearly placed into the correct taxonomy. However, there are times when PASC alone cannot give a definite classification, and other viral properties have to be considered.

Compared with another quantitative approach, phylogenetic analysis, PASC is less computationally intensive and can be easily updated with new sequence data. In addition, PASC results are relatively easier to interpret, which can be potentially done by a computer program without human intervention. It would therefore be possible to set up an automatic system for high throughput classification.

Many researchers use BLAST to search sequence databases to find best matches for viral sequences of interest. Although BLAST is readily available and easy to run, it is not the best tool for virus classification. This is because BLAST is a local alignment program, and, as discussed above, it may not take highly variable regions into account when calculating identities. Even when an output from BLAST covers the whole sequence as a single alignment, information about the taxonomy relationship between the query virus and the virus closest to it is not immediately available. For example, an identity of 75% could be within the range of the same species in one viral family, but in the range of different species in another family. On the contrary, PASC suggests explicitly whether the query virus is in the same species as some existing viruses, or if it should be assigned within a new species or genus in the family.

The total number of virus sequences in the GenBank/EMBL/DDBJ databases is more than 4 times now than 5 years ago (from about 109 000 to about 446 000 in

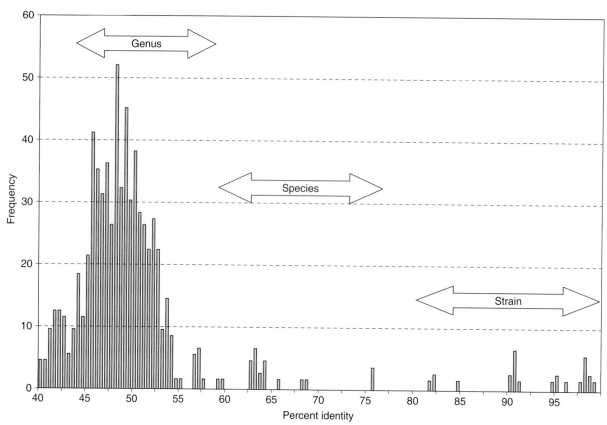

Figure 2 Frequency distribution of pairwise identities from the nucleotide sequence comparison of the RdRp segments of 38 reoviruses.

Table 1 Pairs of the RdRp segment of the reoviruses with identities between 54% and 54.5%

Identity	Same genus?	Same species?	Genome 1[a]	Genome 2
0.544895	No	No	20279540 Coltivirus|Eyach virus	37514915 Mycoreovirus|Mycoreovirus 3
0.542353	No	No	8574569 Seadornavirus|Banna virus	24286507 Cypovirus|Cypovirus 1|Dendrolimus punctatus cypovirus 1
0.541812	No	No	8574569 Seadornavirus|Banna virus	32470626 Orthoreovirus|Mammalian orthoreovirus|Mammalian orthoreovirus 3
0.541744	No	No	8574569 Seadornavirus|Banna virus	25808995 Orbivirus|Bluetongue virus|Bluetongue virus 2|Corsican bluetongue virus
0.541257	Yes	No	50253405 Orbivirus|St. Croix River virus	50261332 Orbivirus|Palyam virus
0.541096	No	No	22960700 Coltivirus|Colorado tick fever virus	32349409 Mycoreovirus|Mycoreovirus 1|Cryphonectria parasitica mycoreovirus-1 (9B21)
0.540999	No	No	8574569 Seadornavirus|Banna virus	14993633 Cypovirus|Cypovirus 1
0.54089	No	No	14993610 Cypovirus|Cypovirus 14	20177438 Fijivirus|Nilaparvata lugens reovirus

[a]The numbers correspond to sequences in GenBank. The taxonomy lineages from the genus level of the viruses are also shown.

October 2006). It is possible to test the PASC system on many virus families now. New sequencing technology makes it possible to generate large amounts of virus sequences from environmental samples without the need to isolate and purify virus particles. In such cases, a molecular based method is the only way to classify the viral sequences, and PASC can be very useful for this purpose.

Limitations of PASC

Although PASC has been applied successfully to several virus families, the approach has some limitations.

First of all, PASC is not suitable for virus families whose current classification is largely based on virus morphologies, such as phages in the families *Siphoviridae* and *Podoviridae*. The whole genome PASC may not work well for virus families with highly diverse sequences. This includes viruses with low overall sequence similarities or large differences in genome sizes and organization. For example, in the family *Herpesviridae*, the percentage of identity between different species in the genus *Varicellovirus* ranges from 39% (between cercopithecine herpesvirus 9 and suid herpesvirus 1) to 83% (between bovine herpesvirus 1 and bovine herpesvirus 5), while the percentage of identity between a virus in the genus *Simplexvirus* and one in the genus *Varicellovirus* could be as high as 54% (between cercopithecine herpesvirus 2 and bovine herpesvirus 5). The overlap of such identities makes it impossible to determine the species and genus demarcations for herpesviruses. In the family *Poxviridae*, the largest genome (canarypox virus) is almost 3 times as big as the smallest one (bovine papular stomatitis virus). The huge differences in the genomes sizes will introduce large artifacts when calculating the identities. In such cases, single gene or a cluster of genes needs to be used in PASC instead of whole genomes. The polymerase protein sequences of poxviruses have been used to perform PASC and a good result was obtained. However, not every single gene can be used for PASC. The genes must be present in all viruses of a family and be very conserved. They need to be tested extensively and accepted by the research community as a useful taxonomic criterion before being applied to the PASC system.

Second, for those families whose PASC were constructed with whole genomes, the query sequences to be tested on PASC to determine their taxonomic positions have to be complete genome sequences as well in order to get an accurate prediction. Although a percentage of identity can be obtained when a partial genome is used, the value will be smaller than what it really should be if a complete genome were used because of the way percentage of identity is calculated in this system. In addition, this value obtained with a partial sequence may not reflect the real taxonomic position of this virus, if, for example, recombination is frequent and important for this virus. This will reduce the number of sequences that can be tested by PASC.

Last, it is almost impossible to get identical PASC results when different methods are used. As mentioned above, there are many pairwise sequence alignment programs available. Even within a single program, variation of parameters can affect alignments. After an alignment is obtained, there can be various ways to calculate the distance. For demarcations computed using different types of distances, it may be difficult to choose a rational reason to privilege one demarcation threshold over another. Moreover, once a demarcation is adopted and a researcher uses a different definition to measure the distance between a new virus and an existing one, the comparison with the histogram will sometimes be misleading. These issues can be overcome by using a centralized PASC system where the alignment identities of new viruses are computed using the same algorithm and the same parameter set that were used to compute identities within the family and to create the demarcation.

Conclusion

PASC is a molecular classification tool for many virus families. It calculates the pairwise identities of virus sequences within a virus family and displays their distributions, and can help determine the demarcations at strains, species, genera, and subfamilies level. PASC has many advantages over conventional virus classification methods. The tool has been successfully applied to many virus families, although it may not work well for virus families with highly diverse sequences. The PASC tool at NCBI established distributions of identity for a number of virus families. A new virus sequence can be tested with this system within a few minutes to suggest the taxonomic position of the virus in these families. This system eliminates the potential discrepancies in the results caused by different algorithms and/or different data used by the virology community. Data in the system can be updated automatically to reflect changes in virus taxonomy and additions of new virus sequences to the public database. The web interface of the tool makes it easy to navigate and perform analyses.

See also: Taxonomy, Classification and Nomenclature of Viruses; Virus Species; Virus Databases.

Further Reading

Edgar RC and Batzoglou S (2006) Multiple sequence alignment. *Current Opinion in Structural Biology* 16: 368–373.

Fauquet CM, Mayo MA, Maniloff J, Desselberger U, and Ball LA (eds.) (2005) *Virus Taxonomy, Classification and Nomenclature of Viruses: Eighth Report of the International Committee on the Taxonomy of Viruses.* London: Academic Press.

Felsenstein J (2004) *Inferring Phylogeny.* Sunderland, MA: Sinauer Associates.

Gusfield D (1997) *Algorithms on Strings, Trees, and Sequences: Computer Science and Computational Biology.* Cambridge: Cambridge University Press.

Page RD and Holmes EC (1998) *Molecular Evolution: A Phylogenetic Approach.* Oxford: Blackwell Science.

van Regenmortel MHV (2007) Virus species and virus identification: Past and Current Controversies. *Infection, Genetics and Evolution* 7: 133–144.

van Regenmortel MHV, Bishop DH, Fauquet CM, *et al.* (1997) Guidelines to the demarcation of virus species. *Archives of Virology* 142: 1505–1518.

Relevant Website

http://www.ncbi.nlm.nih.gov – NCBI, PASC.

Virus Databases

E J Lefkowitz and M R Odom, University of Alabama at Birmingham, Birmingham, AL, USA
C Upton, University of Victoria, Victoria, BC, Canada

© 2008 Elsevier Ltd. All rights reserved.

Glossary

Annotation The process of adding descriptive information to a data set and the individual data elements that comprise that data set. May include both computational analyses and manual curation.

Curation The process of manually adding descriptive information or explanatory material to a data set usually based on a review of the existing literature.

Database A collection of information, usually stored electronically on a computer.

Database, hierarchical (XML) Extensible markup language. A formally defined hierarchical organization of data.

Database management system (DBMS) A database software package designed to operate on a particular computer platform and operating system.

Database schema A map of the database structure, including all of the fields, tables, and relationships that comprise the framework for storage of information within that database.

Database, relational A database structure that provides the ability to define formal relationships between all data elements stored within the database.

Ontology A controlled vocabulary used to formally describe (for our purposes) a biological process or entity.

SQL Structured Query Language. A computer language used to query relational databases for information.

Introduction

In 1955, Niu and Frankel-Conrat published the C-terminal amino acid sequence of tobacco mosaic virus capsid protein. The complete 158-amino-acid sequence of this protein was published in 1960. The first completely sequenced viral genome published was that of bacteriophage MS2 in 1976 (GenBank accession number V00642). Sanger used DNA from bacteriophage phiX174 (J02482) in developing the dideoxy sequencing method, while the first animal viral genome, SV40 (J02400), was sequenced using the Maxam and Gilbert method and published in 1978. Viruses therefore played a pivotal role in the development of modern-day sequencing methods, and viral sequence information (both protein and nucleotide) formed a substantial subset of the earliest available biological databases. In 1965, Margaret O. Dayhoff published the first publicly available database of biological sequence information. This *Atlas of Protein Sequence and Structure* was available only in printed form and contained the sequences of approximately 50 proteins. Establishment of a database of nucleic acid sequences began in 1979 through the efforts of Walter Goad at the US Department of Energy's Los Alamos National Laboratory (LANL) and separately at the European Molecular Biology Laboratories (EMBL) in the early 1980s. In 1982, the LANL database received funding from the National Institutes of Health (NIH) and was christened GenBank. In December of 1981, the Los Alamos Sequence Library contained 263 sequences of which 50 were from eukaryotic viruses and 12 were from bacteriophages. By its tenth release in 1983, GenBank contained 1865 sequences (1 827 214 nucleotides) of which 449 (457 721 nucleotides) were viral. In August of 2006, GenBank (release 154) contained approximately 59 000 000 records, including 367 000 viral sequences.

The number of available sequences has increased exponentially as sequencing technology has improved. In addition, other high-throughput technologies have been developed in recent years, such as those for gene expression and proteomic studies. All of these technologies generate enormous new data sets at ever-increasing rates. The challenge, therefore, has been to provide computational systems that support the storage, retrieval, analysis, and display of this information so that the research scientist can take advantage of this wealth of resources to ask and answer questions relevant to his or her work. Every article in this encyclopedia contains knowledge that has been derived in part from the analysis of large data sets. The ability to effectively and efficiently utilize these data sets is directly dependent on the databases that have been developed to support storage of this information. Fortunately, the continual development and improvement of data-intensive biological technologies has been matched by the development and improvement of computational technologies. This work, which includes both the development and utilization of databases as well as tools for storage and analysis of biological information, forms a very important part of the bioinformatics field. This article provides an overview of database structure and how that structure supports the storage of biological information. The different types of data associated with the analysis of viruses are discussed, followed by a review of some of the various online databases that store general biological information as well as virus-specific information.

Databases

Definition

A database is simply a collection of information, including the means to store, manipulate, retrieve, and share that information. For many of us, lab notebook fulfilled our initial need for a 'database'. However, this information storage vehicle did not prove to be an ideal place to archive our data. Backups were difficult, and retrieval more so. The advent of computers – especially the desktop computer – provided a new solution to the problem of data storage. Though initially this innovation took the form of spreadsheets and electronic notebooks, the subsequent development of both personal and large-scale database systems provided a much more robust solution to the problems of data storage, retrieval, and manipulation. The computer program supplying this functionality is called a 'database management system' (DBMS). Such systems provide at least four things: (1) the necessary computer code to guide a user through the process of database design; (2) a computer language that can be used to insert, manipulate, and query the data; (3) tools that allow the data to be exported in a variety of formats for sharing and distribution; and (4) the administrative functions necessary to ensure data integrity, security, and backup. However, regardless of the sophistication and diverse functions available in a typical modern DBMS, it is still up to the user to provide the proper context for data storage. The database must be properly designed to ensure that it supports the structure of the data being stored and also supports the types of queries and manipulations necessary to fully understand and efficiently analyze the properties of the data.

Data

The development of a database begins with a description of the data to be stored, all of the parameters associated with the data, and frequently a diagram of the format that will be used. The format used to store the data is called the database schema. The schema provides a detailed picture of the internal format of the database that includes specific containers to store each individual piece of data. While databases can store data in any number of different formats, the design of the particular schema used for a project is dependent on the data and the needs and expertise of the individuals creating, maintaining, and using the database. As an example, we will explore some of the possible formats for storing viral sequence data and provide examples of the database schema that could be used for such a project.

Figure 1(a) provides an example of a GenBank sequence record that is familiar to most biologists. These records are provided in a 'flat file' format in which all of the information associated with this particular sequence is provided in a human-readable form and in which all of the information is connected in some manner to the original sequence. In this format, the relationships between each piece of information and every other piece of information are only implicitly defined, that is, each line starts with a label that describes the information in the rest of the line, but it is up to the investigator reading the record to make all of the proper connections between each of the data fields (lines). The proper connections are not explicitly defined in this record. As trained scientists, we are able to read the record in **Figure 1(a)** and discern that this particular amino acid sequence is derived from a strain of Ebola virus that was studied by a group in Germany, and that this sequence codes for a protein that functions as the virus RNA polymerase. The format of this record was carefully designed to allow us, or a computer, to pull out each individual type of information. However as trained scientists, we already understand the proper connections between the different information fields in this file. The computer does not. Therefore, to analyze the data using a computer, a custom software program must be written to provide access to the data.

Extensible markup language (XML) is another widely used format for storing database information. **Figure 1(b)** shows an example of part of the XML record for the Ebola virus polymerase protein. In this format, each data field can be many lines long; the start and end of a data record contained within a particular field are indicated by tags made of a label between two brackets ('<label>...</label>'). Unlike the lines in the GenBank record in **Figure 1(a)**, a field in an XML record can be placed inside of another, defining a structure and a relationship between them. For example, the *TSeq_orgname* is placed inside of the *TSeq* record to show that this

GenBank record

```
LOCUS       NP_066251                2212 aa            linear   VRL 30-MAR-2006
DEFINITION  polymerase [Zaire ebolavirus].
ACCESSION   NP_066251
VERSION     NP_066251.1  GI:10313999
DBSOURCE    REFSEQ: accession NC_002549.1
KEYWORDS    .
SOURCE      Zaire ebolavirus (ZEBOV)
  ORGANISM  Zaire ebolavirus
            Viruses; ssRNA negative-strand viruses; Mononegavirales;
            Filoviridae; Ebola-like viruses.
REFERENCE   1  (residues 1 to 2212)
  AUTHORS   Volchkov,V.E., Volchkova,V.A., Chepurnov,A.A., Blinov,V.M.,
            Dolnik,O., Netesov,S.V. and Feldmann,H.
  TITLE     Characterization of the L gene and 5' trailer region of Ebola virus
  JOURNAL   J. Gen. Virol. 80 (Pt 2), 355-362 (1999)
   PUBMED   10073695
...
REFERENCE   8  (residues 1 to 2212)
  AUTHORS   Volchkov,V.E.
  TITLE     Direct Submission
  JOURNAL   Submitted (20-AUG-1998) Institute of Virology, Philipps-University
            Marburg, Robert-Koch-Str. 17, Marburg 35037, Germany
COMMENT     PROVISIONAL REFSEQ: This record has not yet been subject to final
            NCBI review. The reference sequence was derived from AAD14589.
            Method: conceptual translation.
FEATURES             Location/Qualifiers
     source          1..2212
                     /organism="Zaire ebolavirus"
                     /strain="Mayinga"
                     /db_xref="taxon:186538"
     Protein         1..2212
                     /product="polymerase"
                     /function="synthesis of viral RNAs; transcriptional RNA
                     editing"
                     /note="L"
                     /calculated_mol_wt=252595
     CDS             1..2212
                     /gene="L"
                     /locus_tag="ZEBOVgp7"
                     /coded_by="NC_002549.1:11581..18219"
                     /citation=[1]
                     /db_xref="GeneID:911824"
ORIGIN
        1 matqhtqypd arlsspivld qcdlvtracg lyssyslnpq lrncklpkhi yrlkydvtvt
       61 kflsdvpvat lpidfivpvl lkalsgngfc pveprcqqfl deiikytmqd alflkyylkn
...
     2101 ndynqqrqsr tqtyhfirta kgritklvnd ylkfflivqa lkhngtwqae fkklpelisv
     2161 cnrfyhirdc nceerflvqt lylhrmqdse vklierltgl lslfpdglyr fd
//
```
(a)

XML record

```
<?xml version="1.0"?>
<!DOCTYPE TSeqSet PUBLIC "NCBI_TSeq.dtd">
<TSeqSet>
<TSeq>
  <TSeq_seqtype value="protein"/>
  <TSeq_gi>10313999</TSeq_gi>
  <TSeq_accver>NP_066251.1</TSeq_accver>
  <TSeq_taxid>186538</TSeq_taxid>
  <TSeq_orgname>Zaire ebolavirus</TSeq_orgname>
  <TSeq_defline>polymerase [Zaire ebolavirus]</TSeq_defline>
  <TSeq_length>2212</TSeq_length>
  <TSeq_sequence>
MATQHTQYPDARLSSPIVLDQCDLVTRACGLYSSYSLNPQLRNCKLPKHIYRLKYDVTVTKFLSDVPVAT
LPIDFIVPVLLKALSGNGFCPVEPRCQQFLDEIIKYTMQDALFLKYYLKNVGAQEDCVDEHFQEKILSSI
...
QLQIQRSPYWLSHLTQYADCELHLSYIRLGFPSLEKVLYHRYNLVDSKRGPLVSITQHLAHLRAEIRELT
NDYNQQRQSRTQTYHFIRTAKGRITKLVNDYLKFFLIVQALKHNGTWQAEFKKLPELISVCNRFYHIRDC
NCEERFLVQTLYLHRMQDSEVKLIERLTGLLSLFPDGLYRFD
  </TSeq_sequence>
</TSeq>
</TSeqSet>
```
(b)

Figure 1 Data formats. Examples of two different formats for organizing sequence data are shown. (a) An example of a GenBank flat file sequence record. (b) Part of the same record using a hierarchical XML format.

organism name applies only to that sequence record. If the file contained multiple sequences, each *TSeq* field would have its own *TSeq_orgname* subfield, and the relationship between them would be very clear. This self-describing hierarchical structure makes XML very powerful for expressing many types of data that are hard to express in a single table, such as that used in a spreadsheet. However, in order to find any piece of information in the XML file, a user (with an appropriate search program) needs to traverse the whole file in order to pull out the particular items of data that are of interest. Therefore, while an XML file may be an excellent format for defining and exchanging data, it is often not the best vehicle for efficiently storing and querying that data. That is still the realm of the relational database.

'Relational database management systems' (RDBMSs) are designed to do two things extremely well: (1) store and update structured data with high integrity, and (2) provide powerful tools to search, summarize, and analyze the data. The format used for storing the data is to divide it into several tables, each of which is equivalent to a single spreadsheet. The relationships between the data in the tables are then defined, and the RDBMS ensures that all data follow the rules laid out by this design. This set of tables and relationships is called the schema. An example diagram of a relational database schema is provided in **Figure 2**. This Viral Genome Database (VGD) schema is an idealized version of a database used to store viral genome sequences, their associated gene sequences, and associated descriptive and analytical information. Each box in **Figure 2** represents a single object or concept, such as a genome, gene, or virus, about which we want to store data and is contained in a single table in the RDBMS. The names listed in the box are the columns of that table, which hold the various types of data about the object. The 'gene' table therefore contains columns holding data such as the name of the gene, its coding strand, and a description of its function. The RDBMS is

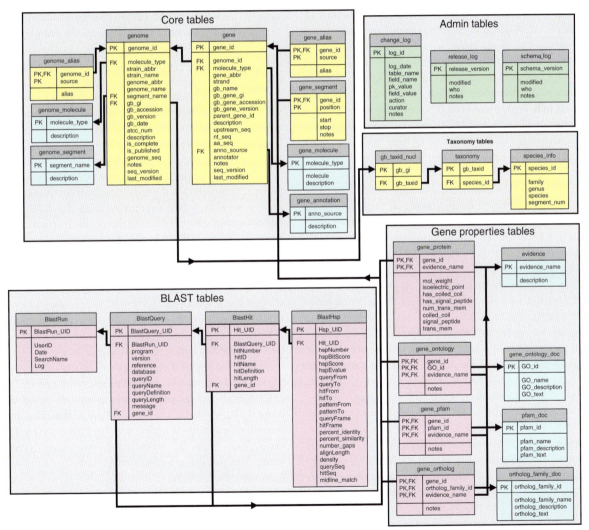

Figure 2 VGD Database schema. Underlying structure of a Viral Genome Sequence Database (VGD). Data tables are grouped according to the type of information contained in each set of tables. Each table contains a set of fields that hold a particular type of data. Lines and arrows display the relationships between fields as defined by the foreign key (FK) and primary key (PK) that connect two tables. (Each arrow points to the table containing the primary key.) Tables are color-coded according to the source of the information they contain: yellow, data obtained from the original GenBank sequence record and the ICTV Eighth Report; pink, data obtained from automated annotation or manual curation; blue, controlled vocabularies to ensure data consistency; green, administrative data.

able to enforce a series of rules for tables that are linked by defining relationships that ensure data integrity and accuracy. These relationships are defined by a foreign key in one table that links to corresponding data in another table defined by a primary key. In this example, the RDMS can check that every gene in the 'gene' table refers to an existing genome in the 'genome' table, by ensuring that each of these tables contains a matching 'genome_id'. Since any one genome can code for many genes, many genes may contain the same 'genome_id'. This defines what is called a one-to-many relationship between the 'genome' and 'gene' tables. All of these relationships are identified in **Figure 2** by arrows connecting the tables.

Because viruses have evolved a variety of alternative coding strategies such as splicing and RNA editing; it is necessary to design the database so that these processes can be formally described. The 'gene_segment' table specifies the genomic location of the nucleotides that code for each gene. If a gene is coded in the traditional manner – one ORF, one protein – then that gene would have one record in the 'gene_segment' table. However, as described above, if a gene is translated from a spliced transcript, it would be represented in the 'gene_segment' table by two or more records, each of which specifies the location of a single exon. If an RNA transcript is edited by stuttering of the polymerase at a particular run of nucleotides, resulting in the addition of one or more nontemplated nucleotides, then that gene will also have at least two records in the 'gene_segment' table. In this case, the second 'gene_segment' record may overlap the last base of the first record for that gene. In this manner, an extra, nontemplated base becomes part of the final gene transcript. Other more complex coding schemes can also be identified using this, or similar, database structures.

The tables in **Figure 2** are grouped according to the type of information they contain. Though the database itself does not formally group tables in this manner, database schema diagrams are created to benefit database designers and users by enhancing their ability to understand the structure of the database. These diagrams make it easier to both populate the database with data and query the database for information. The core tables hold basic biological information about each viral strain and its genomic sequence (or sequences if the virus contains segmented genomes) as well as the genes coded for by each genome. The taxonomy tables provide the taxonomic classification of each virus. Taxonomic designations are taken directly from the *Eighth Report of the International Committee on Taxonomy of Viruses* (ICTV). The 'gene properties' tables provide information related to the properties of each gene in the database. Gene properties may be generated from computational analyses such as calculations of molecular weight and isoelectric point (pI) that are derived from the amino acid sequence. Gene properties may also be derived from a manual curation process in which an investigator might identify, for example, functional attributes of a sequence based on evidence provided from a literature search. Assignment of 'gene ontology' terms (see below) is another example of information provided during manual curation. The BLAST tables store the results of similarity searches of every gene and genome in the VGD searched against a variety of sequence databases using the National Center for Biotechnology Information (NCBI) BLAST program. Examples of search databases might include the complete GenBank nonredundant protein database and/or a database comprised of all the protein sequences in the VGD itself. While most of us store our BLAST search results as files on our desktop computers, it is useful to store this information within the database to provide rapid access to similarity results for comparative purposes; to use these results to assign genes to orthologous families of related sequences; and to use these results in applications that analyze data in the database and, for example, display the results of an analysis between two or more types of viruses showing shared sets of common genes. Finally, the 'admin' tables provide information on each new data release, an archive of old data records that have been subsequently updated, and a log detailing updates to the database schema itself.

It is useful for database designers, managers, and data submitters to understand the types of information that each table contains and the source of that information. Therefore, the database schema provided in **Figure 2** is color-coded according to the type and source of information each table provides. Yellow tables contain basic biological data obtained either directly from the GenBank record or from other sources such as the ICTV. Pink tables contain data obtained as the result of either computational analyses (BLAST searches, calculations of molecular weight, functional motif similarities, etc.) or from manual curation. Blue tables provide a controlled vocabulary that is used to populate fields in other tables. This ensures that a descriptive term used to describe some property of a virus has been approved for use by a human curator, is spelled correctly, and when multiple terms or aliases exist for the same descriptor, the same one is always chosen.

While the use of a controlled vocabulary may appear trivial, in fact, misuse of terms, or even misspellings, can result in severe problems in computer-based databases. The computer does not know that the terms 'negative-sense RNA virus' and 'negative-strand RNA virus' may both be referring to the same type of virus. The provision and use of a controlled vocabulary increases the likelihood that these terms will be used properly, and ensures that the fields containing these terms will be easily comparable. For example, the 'genome_molecule' table contains the following permissible values for 'molecule_type': 'ambisense ssRNA', 'dsRNA', 'negative-sense ssRNA', 'positive-sense ssRNA', 'ssDNA', and 'dsDNA'. A particular viral genome must then have one of these values entered into the 'molecule_type' field of the 'genome' table, since this field is a foreign key to the 'molecule_type' primary key of the 'genome_molecule' table. Entering 'double-stranded DNA' would not be permissible.

Annotation

Raw data obtained directly from high-throughput analytical techniques such as automated sequencing, protein interaction, or microarray experiments contain little-to-no information as to the content or meaning. The process of adding value to the raw data to increase the knowledge content is known as annotation and curation. As an example, the results of a microarray experiment may provide an indication that individual genes are up- or downregulated under certain experimental conditions. By annotating the properties of those genes, we are able to see that certain sets of genes showing coordinated regulation are a part of common biological pathways. An important pattern then emerges that was not discernable solely by inspection of the original data. The annotation process consists of a semiautomated analysis of the information content of the data and provides a variety of descriptive features that aid the process of assigning meaning to the data. The investigator is then able to use this analytical information to more closely inspect the data during a manual curation process that might support the reconstruction of gene expression or protein interaction pathways, or allow for the inference of functional attributes of each identified gene. All of this curated information can then be stored back in the database and associated with each particular gene.

For each piece of information associated with a gene (or other biological entity) during the process of annotation and curation, it is always important to provide the evidence used to support each assignment. This evidence may be described in a Standard Operating Procedure (SOP) document which, much like an experimental protocol, details the annotation process and includes a description of the computer algorithms, programs, and analysis pipelines that were used to compile that information. Each piece of information annotated by the use of this pipeline might then be coded, for example, 'IEA: Inferred from Electronic Annotation'. For information obtained from the literature during manual curation, the literature reference from which the information was obtained should always be provided along with a code that describes the source of the information. Some of the possible evidence codes include 'IDA: Inferred from Direct Assay', 'IGI: Inferred from Genetic Interaction', 'IMP: Inferred from Mutant Phenotype', or 'ISS: Inferred from Sequence or Structural Similarity'. These evidence codes are taken from a list provided by the Gene Ontology (GO) Consortium (see below) and as such represent a controlled vocabulary that any data curator can use and that will be understood by anyone familiar with the GO database. This controlled evidence vocabulary is stored in the 'evidence' table, and each record in every one of the gene properties tables is assigned an evidence code noting the source of the annotation/curation data.

As indicated above, the use of controlled vocabularies (ontologies) to describe the attributes of biological data is extremely important. It is only through the use of these controlled vocabularies that a consistent, documented approach can be taken during the annotation/curation process. And while there may be instances where creating your own ontology may be necessary, the use of already available, community-developed ontologies ensures that the ontological descriptions assigned to your database will be understood by anyone familiar with the public ontology. Use of these public ontologies also ensures that they support comparative analyses with other available databases that also make use of the same ontological descriptions. The GO Consortium provides one of the most extensive and widely used controlled vocabularies available for biological systems. GO describes biological systems in terms of their biological processes, cellular components, and molecular functions. The GO effort is community-driven, and any scientist can participate in the development and refinement of the GO vocabulary. Currently, GO contains a number of terms specific to viral processes, but these tend to be oriented toward particular viral families, and may not necessarily be the same terms used by investigators in other areas of virology. Therefore it is important that work continues in the virus community to expand the availability and use of GO terms relevant to all viruses. GO is not intended to cover all things biological. Therefore, other ontologies exist and are actively being developed to support the description of many other biological processes and entities. For example, GO does not describe disease-related processes or mutants; it does not cover protein structure or protein interactions; and it does not cover evolutionary processes. A complementary effort is under way to better organize existing ontologies, and to provide tools and mechanisms to develop and catalog new ontologies. This work is being undertaken by the National Center for Biomedical Ontologies, located at Stanford University, with participants worldwide.

Access (Searching for Information)

The most comprehensive, well-designed database is useless if no method has been provided to access that database, or if access is difficult due to a poorly designed application. Therefore, providing a search interface that meets the needs of intended users is critical to fully realizing the potential of any effort at developing a comprehensive database. Access can be provided using a number of different methods ranging from direct query of the database using the relatively standardized 'structured query language' (SQL), to customized applications designed to provide the ability to ask sophisticated questions regarding the data contained in the database and mine the data for meaningful patterns. Web pages may be designed to provide simple-to-use forms to access and query data stored in an RDBMS.

Using the VGD schema as a data source, one example of an SQL query might be to find the gene_id and name of all the proteins in the database that have a molecular weight between 20 000 and 30 000, and also have at least one transmembrane region.

Many database providers also provide users with the ability to download copies of the database so that these users may analyze the data using their own set of analytical tools.

Output (Utilizing Information)

When a user queries a database using any of the available access methods, the results of that query are generally provided in the form of a table where columns represent fields in the database and the rows represent the data from individual database records. Tabular output can be easily imported into spreadsheet applications, sorted, manipulated, and reformatted for use in other applications. But while extremely flexible, tabular output is not always the best format to use to fully understand the underlying data and the biological implications. Therefore, many applications that connect to databases provide a variety of visualization tools that display the data graphically, showing patterns in the data that may be difficult to discern using text-based output. An example of one such visual display is provided in **Figure 3** and shows conservation of synteny between the genes of two different poxvirus species. The information used to generate this figure comes directly from the data provided in the VGD. Every gene in the two viruses (in this case crocodilepox virus and molluscum contagiosum virus) has been compared to every other gene using the BLAST search program. The results of this search are stored in the BLAST tables of the VGD. In addition, the location of each gene within its respective genomic sequence is stored in the 'gene_segment' table. This information, once extracted from the database server, is initially text but it is then submitted to a program running on the server that reformats the data and creates a graph. In this manner, it is much easier to visualize the series of points formed along a diagonal when there are a series of similar genes with similar genomic locations present in each of the two viruses. These data sets may contain gene synteny patterns that display deletion, insertion, or recombination events during the course of viral evolution. These patterns can be difficult to detect with text-based tables, but are easy to discern using visual displays of the data.

Information provided to a user as the result of a database query may contain data derived from a combination of sources, and displayed using both visual and textual

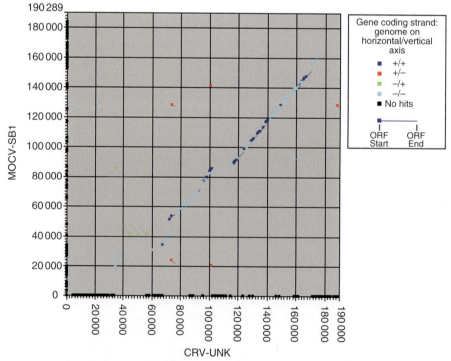

Figure 3 Gene synteny plot. A comparison of gene sequence and genomic position conservation between crocodilepox virus (horizontal axis) and molluscum contagiosum virus (vertical axis). All predicted proteins encoded by each virus were compared to each other using BLASTP. Each pair of proteins showing some measure of similarity, as determined by a BLAST expect (E) value <0.00001, was plotted according to the location of each gene on each respective genome. The color of the points reflects the identity of the coding strand of each gene. Black points along either of the two axes represent proteins unique to that genome.

feedback. **Figure 4** shows the web-based output of a query designed to display information related to a particular virus gene. The top of this web page displays the location of the gene on the genome visually showing surrounding genes on a partial map of the viral genome. Basic gene information such as genome coordinates, gene name, and the nucleotide and amino acid sequence are also provided. This information was originally obtained from the original GenBank record and then stored in the VGD database. Data added as the result of an automated annotation pipeline are also displayed. This includes calculated values for molecular weight and pI; amino acid composition; functional motifs; BLAST similarity searches; and predicted protein structural properties such as transmembrane domains, coiled-coil regions, and signal sequences. Finally, information obtained from a manual curation of the gene through an extensive literature search is also displayed. Curated information includes a mini review of gene function; experimentally determined gene properties such as molecular weight, pI, and protein structure; alternative names and aliases used in the literature; assignment of ontological terms describing gene function; the availability of reagents such as antibodies and clones; and also, as available, information on the functional effects of mutations. All of the information to construct the web page for this gene is directly provided as the result of a single database query. (The tables storing the manually curated gene information are not shown in **Figure 2**.) Obviously, compiling the data and entering it into the database required a substantial amount of effort, both computationally and manually; however, the information is now much more available and useful to the research scientist.

Errors

No discussion of databases would be complete without considering errors. As in any other scientific endeavor, the data we generate, the knowledge we derive from the data, and the inferences we make as a result of the analysis of the data are all subject to error. These errors can be introduced at many points in the analytical chain. The original data may be faulty: using sequence data as one example, nucleotides in a DNA sequence may have been misread or miscalled, or someone may even have mistyped the sequence. The database may have been poorly designed; a field in a table designed to hold sequence information may have been set to hold only 2000 characters, whereas the sequences imported into that field may be longer than 2000 nucleotides. The sequences would have then been automatically truncated to 2000 characters, resulting in the loss of data. The curator may have mistyped an Enzyme Commission (EC) number for an RNA polymerase, or may have incorrectly assigned a genomic sequence to the wrong taxonomic classification. Or even more insidious, the curator may have been using annotations provided by other groups that had justified their own annotations on the basis of matches to annotations provided by yet another group. Such chains of evidence may extend far back, and the chance of propagating an early error increases with time. Such error propagation can be widespread indeed, affecting the work of multiple sequencing centers and database creators and providers. This is especially true given the dependencies of genomic sequence annotations on previously published annotations. The possible sources of errors are numerous, and it is the responsibility of both the database provider and the user to be aware of, and on the lookout for, errors. The database provider can, with careful database and

Figure 4 Gene record. An example of a gene record derived from data in the Viral Genome Database. The record contains a map showing the gene's location; basic descriptive information; analytical data; manually curated descriptive information; links to protein sequence analyses and BLAST similarity data; and the gene sequence itself.

application design, apply error-checking routines to many aspects of the data storage and analysis pipeline. The code can check for truncated sequences, interrupted open reading frames, and nonsense data, as well as data annotations that do not match a provided controlled vocabulary. But the user should always approach any database or the output of any application with a little healthy skepticism. The user is the final arbiter of the accuracy of the information, and it is their responsibility to look out for inconsistent or erroneous results that may indicate either a random or systemic error at some point in the process of data collection and analysis.

Virus Databases

It is not feasible to provide a comprehensive and current list of all available databases that contain virus-related information or information of use to virus researchers. New databases appear on a regular basis; existing databases either disappear or become stagnant and outdated; or databases may change focus and domains of interest. Any resource published in book format attempting to provide an up-to-date list would be out-of-date on the day of publication. Even web-based lists of database resources quickly become out-of-date due to the rapidity with which available resources change, and the difficulty and extensive effort required to keep an online list current and inclusive. Therefore, our approach in this article is to provide an overview of the types of data that are obtainable from available biological databases, and to list some of the more important database resources that have been available for extended periods of time and, importantly, remain current through a process of continual updating and refinement. We should also emphasize that the use of web-based search tools such as Google, various web logs (Blogs), and news groups, can provide some of the best means of locating existing and newly available web-based information sources. Information contained in databases can be used to address a wide variety of problems. A sampling of the areas of research facilitated by virus databases includes

- taxonomy and classification;
- host range, distribution, and ecology;
- evolutionary biology;
- pathogenesis;
- host–pathogen interaction;
- epidemiology;
- disease surveillance;
- detection;
- prevention;
- prophylaxis;
- diagnosis; and
- treatment.

Addressing these problems involves mining the data in an appropriate database in order to detect patterns that allow certain associations, generalizations, cause–effect relationships, or structure–function relationships to be discerned. **Table 1** provides a list of some of the more useful and stable database resources of possible interest to virus researchers. Below, we expand on some of this information and provide a brief discussion concerning the sources and intended uses of these data sets.

Data

Major repositories of biological information

The two major, overarching collections of biological databases are at the NCBI, supported by the National Library of Medicine at the NIH, and the EMBL, part of the European Bioinformatics Institute. These large data repositories try to be all-inclusive, acting as the primary source of publicly available molecular biological data for the scientific community. In fact, most journals require that, prior to publication, investigators submit their original sequence data to one of these repositories. In addition to sequence data, NCBI and EMBL (along with many other data repositories) include a large variety of other data types, such as that obtained from gene-expression experiments and studies investigating biological structures. Journals are also extending the requirement for data deposition to some of these other data types. Note that while much of the data available from these repositories is raw data obtained directly as the result of experimental investigation in the laboratory, a variety of 'value-added' secondary databases are also available that take primary data records and manipulate or annotate them in some fashion in order to derive additional useful information.

When an investigator is unsure about the existence or source of some biological data, the NCBI and EMBL websites should serve as the starting point for locating such information. The NCBI Entrez Search Engine provides a powerful interface to access all information contained in the various NCBI databases, including all available sequence records. A search engine such as Google might also be used if NCBI and EMBL fail to locate the desired information. Of course PubMed, the repository of literature citations maintained at NCBI, also represents a major reference site for locating biological information. Finally, the journal *Nucleic Acids Research* (NAR) publishes an annual 'database' issue and an annual 'web server' issue that are excellent references for finding new biological databases and websites. And while the most recent NAR database or web server issue may contain articles on a variety of new and interesting databases and websites, be sure to also look at issues from previous years. Older issues contain articles on many existing sites that may not necessarily be represented in the latest

Table 1 Virus databases and other related information

Information resource name	Sponsor	URL	Description
Indices of databases			
All the Virology on the WWW	David M. Sander; Tulane University	http://www.virology.net/	Collection of virology information on the Internet
Nucleic Acids Research 2006 Database Issue	Journal: Nucleic Acids Research	http://nar.oxfordjournals.org/content/vol34/suppl_1/index.dtl	Issue describing molecular biology databases
Nucleic Acids Research 2006 Web Server Issue	Journal: Nucleic Acids Research	http://nar.oxfordjournals.org/content/vol34/suppl_2/index.dtl	Issue describing molecular biology websites
General sites			
NCBI	National Center for Biotechnology Information (NCBI), National Library of Medicine, National Institutes of Health (NIH)	http://www.ncbi.nlm.nih.gov/	Resource for molecular biology information
PubMed	National Library of Medicine, NIH	http://www.pubmed.gov/	Biomedical literature database
GenBank Sequence Database	NCBI	http://www.ncbi.nlm.nih.gov/Genbank/	Annotated collection of publicly available nucleotide sequences
NCBI Trace Archive	NCBI	http://www.ncbi.nlm.nih.gov/Traces/	NCBI repository of sequence trace files
NCBI Assembly Archive	NCBI	http://www.ncbi.nlm.nih.gov/Traces/assembly/	NCBI repository of sequence assembly files
EMBL Sequence Database	European Molecular Biology Laboratory (EMBL)	http://www.ebi.ac.uk/embl/	Annotated collection of publicly available nucleotide sequences
DDBJ Sequence Database	DNA Data Bank of Japan	http://www.ddbj.nig.ac.jp/	Annotated collection of publicly available nucleotide sequences
UniProt (Universal Protein Resource)	EMBL; Protein Information Resource, Georgetown University; Swiss Institute of Bioinformatics	http://www.ebi.ac.uk/uniprot/	Protein sequence repository
Taxonomy			
ICTVnet	International Committee on Taxonomy of Viruses (ICTV)	http://www.danforthcenter.org/iltab/ictvnet/	ICTV official website
ICTVdb	ICTV	http://www.ncbi.nlm.nih.gov/ICTVdb	ICTV database of virus taxonomy
NCBI Taxonomy Browser	NCBI	http://www.ncbi.nlm.nih.gov/Taxonomy/	Taxonomy-based retrieval of sequence data
Virus-oriented			
NCBI Viral Genomes Resource	NCBI	http://www.ncbi.nlm.nih.gov/genomes/VIRUSES/viruses.html	NCBI resources for the study of viruses
ASV	American Society for Virology (ASV)	http://www.asv.org/	Society to promote the exchange of information and stimulate discussion and collaboration among scientists active in all aspects of virology
IUMS-Virology	International Union of Microbiological Societies	http://www.iums.org/divisions/divisions-virology.html	Society to promote the study of microbiological sciences internationally

Continued

Table 1 Continued

Information resource name	Sponsor	URL	Description
Viruses: From Structure to Biology	ASV; Washington University School of Medicine	http://medicine.wustl.edu/~virology/	Historical overview of virus research
BRCs (Bioinformatics Resource Centers for Biodefense and Emerging or Re-Emerging Infectious Diseases)	National Institute of Allergy and Infectious Diseases (NIAID), NIH	http://www.brc-central.org/	NIAID/NIH-funded Bioinformatics Resource Centers providing web-based genomics resources to the scientific community on viruses in category A, B, and C priority pathogens
VBRC (Viral Bioinformatics Resource Center)	University of Alabama at Birmingham; NIAID	http://www.vbrc.org/	Bioinformatics resources directed at *Arenaviridae*, *Bunyaviridae*, *Flaviviridae*, *Filoviridae*, *Paramyxoviridae*, *Poxviridae*, and *Togaviridae*
Viral Bioinformatics-Canada	University of Victoria, Canada	http://www.virology.ca/	Viral genomics resources
PATRIC (PathoSystems Resource Integration Center)	Virginia Bioinformatics Institute; NIAID	http://patric.vbi.vt.edu/	Bioinformatics resources directed at *Caliciviridae*, *Coronaviridae*, hepatitis A virus, hepatitis E virus, *Lyssaviridae*
ViroOligo	Oklahoma State University	http://viroligo.okstate.edu/	Virus-specific oligonucleotides for PCR and hybridization
VirGen	Bioinformatics Centre, University of Pune, India	http://bioinfo.ernet.in/virgen/	Annotated and curated database for complete viral genome sequences
VIDE (Virus Identification Data Exchange)	Australian Centre for International Agricultural Research; University of Idaho	http://image.fs.uidaho.edu/vide/	Online descriptions and lists of plant viruses
RNAs and Proteins of dsRNA Viruses	Institute for Animal Health, UK Biotechnology and Biological Sciences Research Council	http://www.iah.bbsrc.ac.uk/dsRNA_virus_proteins/	Databases for the study of *Reoviridae*, *Cystoviridae*, *Birnaviridae*, *Totiviridae*
euHCVdb (European HCV Database)	Institute for the Biology and Chemistry of Proteins, CNRS and Lyon University	http://euhcvdb.ibcp.fr/	European hepatitis C virus database
HCV Database Project	Los Alamos National Laboratories; NIAID	http://hcv.lanl.gov/	HCV sequence and immunology database
Hepatitis Virus Database	National Institute of Genetics (Japan)	http://s2as02.genes.nig.ac.jp/	Database for the study of hepatitis B, C, and E
HERVd (Human Endogenous Retrovirus Database)	Institute of Molecular Genetics, Academy of Sciences of The Czech Republic	http://herv.img.cas.cz/	Human endogenous retrovirus database
HIV Drug Resistance Database	Stanford University	http://hivdb.stanford.edu/	HIV drug resistance database
HIV Sequence Database	Los Alamos National Laboratories; NIAID	http://www.hiv.lanl.gov/	HIV sequence, resistance, immunology, and vaccine trials databases
HIV-1, Human Protein Interaction Database	Southern Research Institute, Birmingham, Alabama; NCBI	http://www.ncbi.nlm.nih.gov/RefSeq/HIVInteractions/	Interactions of HIV-1 proteins with those of the host cell

BioHealthBase (Influenza)	University of Texas Southwestern Medical Center; NIAID	http://www.biohealthbase.org/	Bioinformatics resources directed at influenza
Influenza Sequence Database	Los Alamos National Laboratories	http://www.flu.lanl.gov/	Influenza sequence database
Influenza Virus Resource	NCBI	http://www.ncbi.nlm.nih.gov/genomes/FLU/	Influenza sequence database
IVDB (Influenza Virus Database)	Beijing Genomics Institute	http://influenza.genomics.org.cn/	Integrated information resource and analysis platform for influenza virus
Influenza Genome Sequencing Project	NIAID	http://www.niaid.nih.gov/dmid/genomes/mscs/influenza.htm	NIAID influenza sequencing project
Poxvirus Bioinformatics Resource Center	University of Alabama at Birmingham; NIAID	http://www.poxvirus.org/	Poxvirus genomic sequences, gene annotations, and analysis
SARS-CoV RNA SSS Database	Center for Modern Biology, School of Life Sciences, Yunnan University, Kunming, China	http://www.liuweibo.com/sarsdb/	SARS Secondary Structural Sequence database
ACLAME (A Classification of Genetic Mobile Elements)	Department of Macromolecular Conformation and Bioinformatics, Free University of Brussels, Belgium	http://aclame.ulb.ac.be/	Database of mobile genetic elements
Subviral RNA Database	University of Ottawa	http://subviral.med.uottawa.ca/	Viroids and viroid-like RNAs
Genomes of the T4-like phages	Department of Biochemistry, Tulane University	http://phage.bioc.tulane.edu/	Sequences of T4-like bacteriophages
DPVweb (Descriptions of plant viruses)	Association of Applied Biologists (UK)	http://www.dpvweb.net/	Plant virus database
ILTAB (International Laboratory for Tropical Agricultural Biotechnology)	Danforth Plant Sciences Center, St. Louis, Missouri	http://www.danforthcenter.org/iltab/	*Caulimoviridae*, *Geminiviridae*, and *Potyviridae* database
Functional motifs			
InterPro	European Bioinformatics Institute (EBI)	http://www.ebi.ac.uk/interpro/	Database of protein families, domains, and functional sites
CDD (Conserved Domain Database)	NCBI	http://www.ncbi.nlm.nih.gov/Structure/cdd/cdd.shtml	Conserved Domain Database and search service
PFAM (Protein families)	Washington University in St. Louis	http://pfam.wustl.edu/	Database of protein families and Hidden Markov models
PRINTS (Protein fingerprints)	University of Manchester, England	http://www.bioinf.man.ac.uk/dbbrowser/PRINTS/	Conserved motifs used to characterize a protein family
PROSITE	Swiss Institute of Bioinformatics	http://ca.expasy.org/prosite/	Database of protein families and domains
VIDA (Virus database of homologous protein families)	Wohl Virion Centre, University College London and Genome Informatics Research Lab, Universitat Pompeu Fabra, Barcelona	http://www.biochem.ucl.ac.uk/bsm/virus_database/	Homologous protein families from complete and partial virus genomes
VOCs (Viral Orthologous Clusters)	University of Victoria, Canada	http://athena.bioc.uvic.ca/workbench.php?tool=vocs	Viral orthologous clusters database

Continued

Table 1 Continued

Information resource name	Sponsor	URL	Description
VOGs (Viral Clusters of Orthologous Groups)	NCBI	http://www.ncbi.nlm.nih.gov/genomes/VIRUSES/vog.html	Viral clusters of orthologous groups database
Structural			
PDB (Protein Data Bank)	Research Collaboratory for Structural Bioinformatics	http://www.rcsb.org/	Three-dimensional structure database
Macromolecular Structure Database	EBI	http://www.ebi.ac.uk/msd/	Macromolecular structure database
Dali	Institute of Biotechnology, University of Helsinki, Finland	http://ekhidna.biocenter.helsinki.fi/dali/	Database of 3D structure comparisons
The Big Picture Book of Viruses	David M. Sander; Tulane University	http://www.virology.net/Big_Virology/	Catalog of virus pictures on the Internet
VIPERdb (Virus particle explorer)	Scripps Research Institute Structural Biology	http://viperdb.scripps.edu/	Icosahedral virus capsid structures
Virus World	Institute for Molecular Virology, University of Wisconsin-Madison	http://virology.wisc.edu/virusworld/	Images of virus capsids from X-ray crystallography and cryo-EM data
Pathway			
Reactome	Cold Spring Harbor Laboratory, EBI, Gene Ontology Consortium	http://www.reactome.org/	Curated knowledgebase of biological pathways
BioCyc	SRI International, Menlo Park, California	http://www.biocyc.org/	Collection of databases describing genome and metabolic pathways
KEGG (Kyoto Encyclopedia of Genes and Genomes)	Kyoto University and the Human Genome Center at University of Tokyo	http://www.genome.jp/kegg/	Database of biological pathways and genes
Other			
GO (Gene Ontology Database)	Gene Ontology Consortium	http://www.geneontology.org/	Development of controlled vocabularies (ontologies) that describe gene products in terms of their associated biological processes, cellular components, and molecular functions
National Center for Biomedical Ontology	Stanford University	http://bioontology.org/	Development of innovative technologies and methods that allow scientists to create, disseminate, and manage biomedical ontologies
GEO (Gene Expression Omnibus)	NCBI	http://www.ncbi.nlm.nih.gov/geo/	A gene expression/molecular abundance data repository

IEDB (Immune Epitope Database and Analysis Resource)	La Jolla Institute for Allergy and Immunology	http://www.immuneepitope.org/	Dissemination of immune epitope information
IntAct (Protein interactions)	EBI	http://www.ebi.ac.uk/intact/	Database and analysis system for protein interactions
Analytical tools			
ExPASy	Swiss Institute of Bioinformatics (SIB)	http://ca.expasy.org/	Tools for the analysis of protein sequences, structures, etc.
PASC (Pairwise Sequence Comparison)	NCBI	http://www.ncbi.nlm.nih.gov/sutils/pasc/	Analysis of pairwise identity distribution within viral families
NCBI Viral Genotyping Tool	NCBI	http://www.ncbi.nlm.nih.gov/projects/genotyping/	Identification of viral sequence genotypes
Medically oriented			
AIDSinfo	US Department of Health and Human Services	http://aidsinfo.nih.gov/	Information on HIV/AIDS treatment, prevention, and research
AMA Infectious diseases	American Medical Association	http://www.ama-assn.org/ama/pub/category/1797.html	Online information resources
CDC NCID (National Center for Infectious Diseases)	US Centers for Disease Control and Prevention (CDC)	http://www.cdc.gov/ncidod/diseases/	Centers for Disease Control and Prevention (USA) disease index
CIPHI Fact Sheets	Canadian Institute of Public Health Inspectors	http://action.web.ca/home/ciphiont/readingroom.shtml	Information pages for public health, including viral diseases
CSEI (Center for the Study of Emerging Infections)	Institute for BioSecurity, St Louis University, School of Public Health	http://www.emerginginfections.slu.edu/	Information on emerging infections
Diseases & Conditions	New York State Department of Health	http://www.health.state.ny.us/diseases/	General public information
ENIVD (European Network for Diagnostics of "Imported" Viral Diseases)	European Commission	http://www.enivd.org/	European Network for Diagnostics of 'Imported' Viral Diseases
EPR (Epidemic and Pandemic Alert and Response)	World Health Organization (WHO)	http://www.who.int/csr/disease/en/	WHO integrated global alert and response system for epidemics and other public health emergencies
European Centre for Disease Prevention and Control	The European Union	http://www.ecdc.eu.int/	European Centre for Disease Prevention and Control
MedScape	WebMD	http://www.medscape.com/	Information and educational tools for specialists
Merck Manual of Diagnosis and Therapy, Infectious Diseases	Merck & Co., Inc.	http://www.merck.com/mrkshared/mmanual/section13/sec13.jsp	Online reference of infectious diseases
MMWR (Morbidity and Mortality Weekly Report)	CDC	http://www.cdc.gov/mmwr/	CDC Morbidity and Mortality Weekly Report
NIH Vaccine Research Center	Dale and Betty Bumpers' Vaccine Research Center, NIAID, NIH	http://www.vrc.nih.gov/	NIH vaccine research

journal publication, but are nevertheless still available and current.

There are several websites that serve to provide general virus-specific information and links of use to virus researchers. One of these is the NCBI Viral Genomes Project, which provides an overview of all virus-related NCBI resources including taxonomy, sequence, and reference information. Links to other sources of viral data are provided, as well as a number of analytical tools that have been developed to support viral taxonomic classification and sequence clustering. Another useful site is the All the Virology on the WWW website. This site provides numerous links to other virus-specific websites, databases, information, news, and analytical resources. It is updated on a regular basis and is therefore as current as any site of this scope can be.

Taxonomy and classification

One of the strengths of storing information within a database is that information derived from different sources or different data sets can be compared so that important common and distinguishing features can be recognized. Such comparative analyses are greatly aided by having a rigorous classification scheme for the information being studied. The International Union of Microbiological Societies has designated the International Committee on Taxonomy of Viruses (ICTV) as the official body that determines taxonomic classifications for viruses. Through a series of subcommittees and associated study groups, scientists with expertise on each viral species participate in the establishment of new taxonomic groups, assignment of new isolates to existing or newly established taxonomic groups, and reassessment of existing assignments as additional research data become available. The ICTV uses more than 2600 individual characteristics for classification, though sequence homology has gained increasing importance over the years as one of the major classifiers of taxonomic position. Currently, as described in its Eighth Report, the ICTV recognizes 3 orders, 73 families, 287 genera, and 1950 species of viruses. The ICTV officially classifies viral isolates only to the species level. Divisions within species, such as clades, subgroups, strains, isolates, types, etc., are left to others. The ICTV classifications are available in book form as well as from an online database. This database, the ICTVdb, contains the complete taxonomic hierarchy, and assigns each known viral isolate to its appropriate place in that hierarchy. Descriptive information on each viral species is also available. The NCBI also provides a web-based taxonomy browser for access to taxonomically specified sets of sequence records. NCBI's viral taxonomy is not completely congruent with that of ICTV, but efforts have been under way to ensure congruency with the official ICTV classification.

Nucleotide sequence data

The primary repositories of existing sequence information come from the three organizations that comprise the International Nucleotide Sequence Database Collaboration. These three sites are GenBank (maintained at NCBI), EMBL, and the DNA Data Bank of Japan (DDBJ). Because all sequence information submitted to any one of these entities is shared with the others, a researcher need query only one of these sites to get the most up-to-date set of available sequences. GenBank stores all publicly available nucleotide sequences for all organisms, as well as viruses. This includes whole-genome sequences as well as partial-genome and individual coding sequences. Sequences are also available from large-scale sequencing projects, such as those from shotgun sequencing of environmental samples (including viruses), and high-throughput low- and high-coverage genomic sequencing projects. NCBI provides separate database divisions for access to these sequence datasets. The sequence provided in each GenBank record is the distillation of the raw data generated by (in most cases these days) automated sequencing machines. The trace files and base calls provided by the sequencers are then assembled into a collection of contiguous sequences (contigs) until the final sequence has been assembled. In recognition of the fact that there is useful information contained in these trace files and sequence assemblies (especially if one would like to look for possible sequencing errors or polymorphisms), NCBI now provides separate Trace File and Assembly Archives for GenBank sequences when the laboratory responsible for generating the sequence submits these files. Currently, the only viruses represented in these archives are influenza A, chlorella, and a few bacteriophages.

An important caveat in using data obtained from GenBank or other sources is that no sequence data can be considered to be 100% accurate. Furthermore, the annotation associated with the sequence, as provided in the GenBank record, may also contain inaccuracies or be out-of-date. GenBank records are provided and maintained by the group originally submitting the sequence to GenBank. GenBank may review these records for obvious errors and formatting mistakes (such as the lack of an open reading frame where one is indicated), but given the large numbers of sequences being submitted, it is impossible to verify all of the information in these records. In addition, the submitter of a sequence essentially 'owns' that sequence record and is thus responsible for all updates and corrections. NCBI generally will not change any of the information in the GenBank record unless the sequence submitter provides the changes. In some cases, sequence annotations will be updated and expanded, but many, if not most, records never change following their initial submission. (These facts emphasize the responsibility that submitters of sequence data have to ensure the accuracy of their original submission and to update their sequence data and annotations as necessary.) Therefore, the user of the information has the responsibility to ensure, to the extent possible, its accuracy is sufficient to support any conclusions derived from that information. In recognition of these problems, NCBI established the Reference Sequence (RefSeq) database project, which attempts to provide reference sequences for genomes, genes, mRNAs, proteins, and RNA sequences that can be used, in NCBI's words, as "a stable reference for gene characterization, mutation analysis, expression studies, and polymorphism discovery". RefSeq records are manually curated by NCBI staff, and therefore should provide more current (and hopefully more accurate) sequence annotations to support the needs of the research community. For viruses, RefSeq provides a complete genomic sequence and annotation for one representative isolate of each viral species. NCBI solicits members of the research community to participate as advisors for each viral family represented in RefSeq, in an effort to ensure the accuracy of the RefSeq effort.

Protein sequence data

In addition to the nucleotide sequence databases mentioned above, UniProt provides a general, all-inclusive protein sequence database that adds value through annotation and analysis of all the available protein sequences. UniProt represents a collaborative effort of three groups that previously maintained separate protein databases (PIR, SwissProt, and TrEMBL). These groups, the National Biomedical Research Foundation at Georgetown University, the Swiss Institute of Bioinformatics, and the European Bioinformatics Institute, formed a consortium in 2002 to merge each of their individual databases into one comprehensive database, UniProt. UniProt data can be queried by searching for similarity to a query sequence, or by identifying useful records based on the text annotations. Sequences are also grouped into clusters based on sequence similarity. Similarity of a query sequence to a particular cluster may be useful in assigning functional characteristics to sequences of unknown function. NCBI also provides a protein sequence database (with corresponding RefSeq records) consisting of all protein-coding sequences that have been annotated within all GenBank nucleotide sequence records.

Virus-specific sequence databases

The above-mentioned sequence databases are not limited to viral data, but rather store sequence information for all biological organisms. In many cases, access to nonviral sequences is necessary for comparative purposes, or to study virus–host interactions. But it is frequently easier to use virus-specific databases when they exist, to provide a more focused view of the data that may simplify many of the analyses of interest. **Table 1** lists many of these virus-specific sites. Sites of note include the NIH-supported Bioinformatics Resource Centers for Biodefense and Emerging and Reemerging Infectious Diseases (BRCs). The BRCs concentrate on providing databases, annotations, and analytical resources on NIH priority pathogens, a list that includes many viruses. In addition, the LANL has developed a variety of viral databases and analytical resources including databases focusing on HIV and influenza. For plant virologists, the Descriptions of Plant Viruses (DPV) website contains a comprehensive database of sequence and other information on plant viruses.

Structural information

The three-dimensional structures for quite a few viral proteins and virion particles have been determined. These structures are available in the primary database for experimentally determined structures, the Protein Data Bank (PDB). The PDB currently contains the structures for more than 650 viral proteins and viral protein complexes out of 38 000 total structures. Several virus-specific structure databases also exist. These include the VIPERdb database of icosahedral viral capsid structures, which provides analytical and visualization tools for the study of viral capsid structures; Virus World at the Institute for Molecular Virology at the University of Wisconsin, which contains a variety of structural images of viruses; and the Big Picture Book of Viruses, which provides a catalog of images of viruses, along with descriptive information.

Functional motifs and orthologous clusters

Ultimately, the biology of viruses is determined by genomic sequence (with a little help from the host and the environment). Nucleotide sequences may be structural, functional, regulatory, or protein coding. Protein sequences may be structural, functional, and/or regulatory, as well. Patterns specified in nucleotide or amino acid sequences can be identified and associated with many of these biological roles. Both general and virus-specific databases exist that map these roles to specific sequence motifs. Most also provide tools that allow investigators to search their own sequences for the presence of particular patterns or motifs characteristic of function. General databases include the NCBI Conserved Domain Database; the Pfam (protein family) database of multiple sequence alignments and hidden Markov models; and the PROSITE database of protein families and domains. Each of these databases and associated search algorithms differ in how they detect a particular search motif or define a particular protein family. It can therefore be useful to employ multiple databases and search methods when analyzing a new sequence (though in many cases they will each detect a similar set of putative functional motifs). InterPro is a database of protein families, domains, and functional sites that combines many other existing motif databases. InterPro provides a search tool, InterProScan, which is able to utilize several different search algorithms dependent on the database to be searched. It allows users to choose which of the available databases and search tools to use when analyzing their own sequences of interest. A comprehensive report is provided that not only summarizes the results of the search, but also provides a comprehensive annotation derived from similarities to known functional domains. All of the above databases define functional attributes based on similarities in amino acid sequence. These amino acid similarities can be used to classify proteins into functional families. Placing proteins into common functional families is also frequently performed by grouping the proteins into orthologous families based on the overall similarity of their amino acid sequence as determined by pairwise BLAST comparisons. Two virus-specific databases of orthologous gene families are the Viral Clusters of Orthologous Groups database (VOGs) at NCBI, and the Viral Orthologous Clusters database (VOCs) at the Viral Bioinformatics Resource Center and Viral Bioinformatics, Canada.

Other information

Many other types of useful information, both general and virus-specific, have been collected into databases that are available to researchers. These include databases of gene-expression experiments (NCBI Gene Expression Omnibus – GEO); protein–protein interaction databases, such as the NCBI HIV Protein-Interaction Database; The Immune Epitope Database and Analysis Resource (IEDB) at the La Jolla Institute for Allergy and Immunology; and databases and resources for defining and visualizing biological pathways, such as metabolic, regulatory, and signaling pathways. These pathway databases include Reactome at the Cold Spring Harbor Laboratory, New York; BioCyc at SRI International, Menlo Park, California; and the Kyoto Encyclopedia of Genes and Genomes (KEGG) at Kyoto University in Japan.

Analytical Tools

As indicated above, the information contained in a database is useless unless there is some way to retrieve that information from the database. In addition, having access to all of the information in every existing database would be meaningless unless tools are available that allow one to process and understand the data contained within those databases. Therefore, a discussion of virus databases would not be complete without at least a passing reference to the tools that are available for analysis. To populate a database such as the VGD with sequence and analytical information, and to utilize this information for subsequent analyses, requires a variety of analytical tools including programs for

- sequence record reformatting,
- database import and export,
- sequence similarity comparison,
- gene prediction and identification,
- detection of functional motifs,
- comparative analysis,
- multiple sequence alignment,
- phylogenetic inference,
- structural prediction, and
- visualization.

Sources for some of these tools have already been mentioned, and many other tools are available from the same websites that provide many of the databases listed in **Table 1**. The goal of all of these sites that make available data and analytical tools is to provide – or enable the discovery of – knowledge, rather than simply providing access to data. Only in this manner can the ultimate goal of biological understanding be fully realized.

See also: Evolution of Viruses; Phylogeny of Viruses; Taxonomy, Classification and Nomenclature of Viruses; Virus Classification by Pairwise Sequence Comparison (PASC).

Further Reading

Ashburner M, Ball CA, Blake JA, et al. (2000) Gene ontology: Tool for the unification of biology. The Gene Ontology Consortium. *Nature Genetics* 25(1): 25–29.

Bao Y, Federhen S, Leipe D, et al. (2004) National Center for Biotechnology Information Viral Genomes Project. *Journal of Virology* 78(14): 7291–7298.

Fauquet CM, Mayo MA, Maniloff J, Desselberger U, and Ball LA (2005) *Virus Taxonomy: Classification and Nomenclature of Viruses. Eighth Report of the International Committee on Taxonomy of Viruses.* San Diego, CA: Elsevier Academic Press.

Galperin MY (2006) The Molecular Biology Database Collection: 2006 update. *Nucleic Acids Research* 34(database issue): D3–D5.

Joshi-Tope G, Gillespie M, Vastrik I, et al. (2005) Reactome: A knowledgebase of biological pathways. *Nucleic Acids Research* 33(database issue): D428–D432.

Kellam P and Alba MM (2002) Virus bioinformatics: Databases and recent applications. *Applied Bioinformatics* 1(1): 37–42.

Korber B, LaBute M, and Yusim K (2006) Immunoinformatics comes of age. *PLoS Computational Biology* 2(6): e71.

Kuiken C, Korber B, and Shafer RW (2003) HIV sequence databases. *AIDS Reviews* 5(1): 52–61.

Kuiken C, Mizokami M, Deleage G, et al. (2006) Hepatitis C databases, principles and utility to researchers. *Hepatology* 43(5): 1157–1165.

Lefkowitz EJ, Upton C, Changayil SS, et al. (2005) Poxvirus bioinformatics resource center: A comprehensive *Poxviridae* informational and analytical resource. *Nucleic Acids Research* 33(database issue): D311–D336.

Lesk AM (2005) *Database Annotation in Molecular Biology.* Chichester, UK: Wiley.

Lindler LE, Lebeda FJ, and Korch G (2005) *Biological Weapons Defense: Infectious Diseases and Counterbioterrorism.* Totowa, NJ: Humana Press.

Natarajan P, Lander GC, Shepherd CM, et al. (2005) Exploring icosahedral virus structures with VIPER. *Nature Reviews Microbiology* 3(10): 809–817.

Rubin DL, Lewis SE, Mungall CJ, et al. (2006) National Center for Biomedical Ontology: Advancing biomedicine through structured organization of scientific knowledge. *Omics* 10(2): 185–198.

Yusim K, Richardson R, Tao N, et al. (2005) Los Alamos Hepatitis C Immunology Database. *Applied Bioinformatics* 4(4): 217–225.

Virus Entry to Bacterial Cells

M M Poranen and A Domanska, University of Helsinki, Helsinki, Finland

© 2008 Elsevier Ltd. All rights reserved.

Glossary

Adsorption Initial interaction between a virus particle and a cellular receptor molecule.
Bacteriophage (or phage) A virus that infects bacteria.
Capsid The protective protein coat of a virus particle.
Cell envelope Plasma membrane and cellular structures located outside the plasma membrane.
Cell wall Cellular structures located outside the plasma membrane.
Lipopolysaccharide A unique glycolipid of the outer membrane of Gram-negative bacteria.
Peptidoglycan A polymer consisting of long glycan chains cross-linked via peptide bridges and forming homogenous layer outside the plasma membrane of eubacteria component of the cell wall.
Receptor A specific molecule or molecular assembly exposed on the surface of a cell to which a virus entering the cell attaches.
Receptor-binding protein A virion protein responsible for the interaction of a virion with a specific cellular receptor molecule.
Vertex Fivefold symmetry position of the icosahedra; one icosahedral particle has 12 fivefold symmetry positions.
Viral envelope An outer lipid–protein bilayer of a virus.
Virion A virus particle, the extracellular form of a virus.

Introduction

Viruses are intracellular parasites that are dependent on the metabolic apparatus of the cell. Unlike other parasitic self-replicating systems, like plasmids and viroids, viruses possess an extracellular phase that allows spread from one infected cell or organism to another. Consequently, viruses have to have means to infect new host cells. The entry into a suitable host cell is a key event for the viral reproduction and survival.

Host Cell Barriers

The nature of the host cell wall has a great influence on the viral entry strategy. Gram-positive bacteria have a single internal lipid bilayer and a thick cell wall made of peptidoglycan while Gram-negative cells are covered by an internal membrane, a thin layer of peptidoglycan, and an outer membrane. In addition, bacterial cells may secrete polysaccharides that make a protective extracellular capsule on the surface of the cell.

The relatively strong and inert cell wall of eubacteria efficiently restricts the passage of macromolecules. In addition, bacterial cells do not have endocytic-like uptake systems, which are commonly utilized by eukaryotic viruses to gain access into the host cell. These features of the host strongly influence the mechanism employed by bacterial viruses to gain access into the host cytoplasm. In fact, the capsids of most bacteriophages are never internalized into the host cell; only the viral genome with some necessary protein factors is delivered across the host envelope.

Virions as a Genome Delivery Devise

Virions represent an extracellular form of the virus. It is a vehicle, which allows the virus to resist the harsh environment outside the cell. In addition to its protective nature, the main task of the virion is to recognize the host and to deliver the viral genome with necessary accessory factors to the new host cell. The mechanism of the genome delivery is typically reflected in the structure of the capsid (see below).

The nature of the viral genome influences the mechanisms of virus entry. Viruses that have genomes which cannot be expressed using the enzymatic apparatuses of the host, need to also bring viral polymerases into the host cell. This applies to all viruses having dsRNA or negative-sense ssRNA genomes. Regardless of the type of the genome, many viruses deliver some accessory protein factors inside the host. These are required in the early stage of the infection, prior to the viral genome expression, either to complete the entry process or for successful genome replication and expression.

Host Recognition

Host cell recognition by viruses is a highly specific process. Basically, all bacteriophages have, on their exterior surface, some protein that binds to a receptor molecule exposed on the surface of a susceptible cell. The receptor-binding proteins are often localized in the vertices of the icosahedral virions or the tips of the helical virions. In tailed bacteriophages the initial recognition is carried out by the fibers that are connected to the distal end of the tail (**Figure 1(a)**). The specific recognition leads to irreversible structural rearrangements in the virion components; the viral receptor-binding complexes overcome an energy barrier and fold into a minimal energy state. These conformational changes lead to more tight attachment and eventually trigger the entry process. The rigid structure of the virion is destabilized so that the genome delivery can be accomplished. The icosahedral capsids of bacteriophages seldom disassemble completely; only the structures at the vertices become labile making openings for genome release.

Many bacteriophages, including icosahedral ssRNA bacteriophages (e.g., MS2, Qβ), filamentous ssDNA bacteriophages (e.g., Ff, as M13, fd, and fl), and enveloped dsRNA phage (φ6), utilize a bacterial pilus as their primary receptor. The pilus enables efficient capture of the virion at a distance from the cell surface, and the retraction of the pilus translocates the bound phages to the host envelope (**Figure 1(b)**). This allows phages to get access to the cell surface regardless of the polysaccharide capsule, which could restrict the easy access to the cell surface. Some phages (e.g., K1-5, K5, K1E), however, may also use the polysaccharide capsule as the initial site of recognition and binding. Typically, these phages have enzymatic activities associated with the virion for capsule degradation.

Other receptor sites for bacteriophages are lipopolysaccharides, various cell envelope and flagellar proteins, as well as cell-wall carbohydrates. The tailed phages infecting Gram-negative bacteria use either the lipopolysaccharide moieties (e.g., T2, T4, T7) or envelope proteins, such as porins and transporters (e.g., PP01, T1, T5, λ) as their receptors, while phages infecting Gram-positive bacteria typically attach to the cell-wall teichoic acids. Membrane proteins or peptidoglycan moieties rarely are used as attachment sites for bacteriophages infecting Gram-positive bacteria.

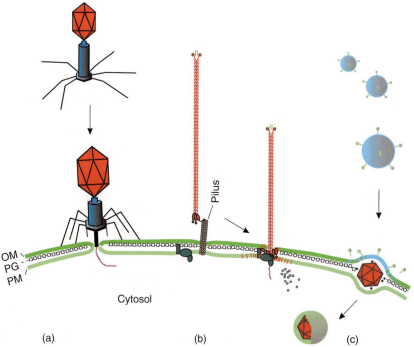

Figure 1 Schematic presentation of the main entry strategies utilized by bacterial viruses. OM, outer membrane; PG, peptidoglycan; PM, plasma membrane. (a) Genome delivery through an icosahedral vertex; given example is a phage with long-contractile tail from the family *Myoviridae* (e.g., T4). Primary interaction between the phage and the host cell is mediated by the tail fibers (black). Contraction of the tail sheath (blue) facilitates the penetration of the cell envelope. The viral genome (purple) is released from the head (red) of the phage virion through the tail tube (black) into the host cytosol. The protein capsid remains outside the cell. (b) Dissociation of filamentous phage capsid at the cell envelope. The receptor binding protein located at the tip of the helical virion interacts with the bacterial pilus. Pilus retraction brings the virion on the cell surface enabling interaction with the co-receptor molecule (green). The viral genome (purple) is released into the host cytosol as the virion proteins are inserted into the plasma membrane. (c) Penetration of the nucleoprotein complex of an enveloped dsRNA virion. The viral spike proteins (green) on the virion surface mediate the interaction between the host and the virion. Fusion between the host outer membrane and phage envelope (blue) takes place leading to the mixing of host and viral lipids. The nucleoprotein assembly (nucleocapsid; red) released into the bacterial periplasm penetrates the peptidoglycan network with the aid of a lytic enzyme (black) located on the nucleocapsid surface. Subsequently, the host plasma membrane is penetrated via an endocytic-like process.

Cell-Wall Penetration

The virions of many bacteriophages contain peptidoglycan-hydrolyzing enzymes. These are specialized proteins that locally and temporarily disrupt the peptidoglycan network, thus allowing the penetration of the cell wall. Similar enzymes are also involved in the release of progeny phages from the host cell at the end of the infection cycle. However, when assembled into virions, they likely have an active and crucial role in the entry process.

The lytic activity of virion-associated peptidoglycan-hydrolyzing enzymes is stringently controlled to ensure localized openings in the cell wall. This allows maintaining cell integrity, which is essential for successful production of progeny viruses during the infection cycle. For this reason such enzymes are tightly incorporated into the virus particle. The peptidoglycan-hydrolyzing enzymes of tailed bacterial viruses are often associated with the tail structures or, in the case of some short-tailed phages (e.g., T7), with internal head proteins that are ejected at the beginning of the infection cycle. Icosahedral dsDNA phages with internal membrane and no tail (e.g., PRD1, Bam35) as well as the enveloped dsRNA phage φ6 also possess peptidoglycan-hydrolyzing enzymes in their virions (see below; **Figure 1(c)**).

The most common cell-wall-degrading enzymes found in bacteriophage virions are lysozymes and lytic transglycosylases. Both lysozymes and lytic transglycosylases cleave the same glycosidic bond between N-acetylmuramic acid and N-acetylglucosamine units of the glycan strands. The end product of lysozyme is a disaccharide with reducing end of N-acetylmuramic acid, whereas that of lytic transglycosylase is a disaccharide containing a ring structure (anhydromuramic acid).

Not all phages have specialized enzymes in their virions for the penetration of the host cell wall during entry. These phages rely on preexisting channels within the host envelope. This applies to bacterial viruses with ssDNA and ssRNA genomes. For example, infection of filamentous ssDNA phages absolutely depends on a protein complex assembled in the bacterial plasma membrane spanning the peptidoglycan layer (**Figure1(b)**).

Genome Delivery Mechanisms of Phages

In general, entry strategies of bacterial viruses fall into three main categories: (1) genome delivery through an icosahedral vertex, (2) virion dissociation at the cell envelope, and (3) virion penetration via membrane fusion and endocytic-like event (**Figure 1**). All known tailed dsDNA phages (e.g., T4, T5, T7), dsDNA phages with an internal membrane (tectiviruses), as well as icosahedral ssDNA and ssRNA phages deliver their genome through a genome delivery apparatus located at one of the capsid vertices (**Figure 1(a)**). Capsids of filamentous phages and membrane-containing phage PM2 disassemble completely at the cell envelope (**Figure 1(b)**), whereas the enveloped dsRNA bacterial viruses utilize a unique membrane fusion-type uncoating at the outer membrane and virus subparticle internalization through the plasma membrane (**Figure 1(c)**).

Icosahedral Tailed dsDNA Bacterial Viruses

The tailed dsDNA bacteriophages use the tail structure as a genome delivery devise (**Figure 1(a)**). The tail is a protein tube of various length and complexity attached at one of the vertices of the icosahedral head. During the entry process the viral genome enclosed within the head travels through the tail into the host cell. Despite the numerous types of tailed bacterial viruses, they all possess one of the following three types of tails: long contractile, long noncontractile, or short noncontractile tail. The type of tail serves as the main criterion for classification of tailed bacteriophages and also influences the genome delivery mechanism.

The contractile tail which is common for members of the family *Myoviridae* is a complex, long tube assembly surrounded by a sheath that is terminated by a base plate typically decorated with terminal tail fibers (**Figure 1(a)**). The tail fibers act as sensor to detect the host bacterium. Interaction between the tail fibers and their surface receptors leads to conformational changes in the base plate and subsequent irreversible adsorption of the virus on bacterial cells. This triggers contraction of the tail sheath, which then drags the base plate along the tail tube making the tube protrude from the base plate (**Figure 1(a)**). Subsequently, the tail tube penetrates the cell envelope and phage DNA is released into the cell through a transmembrane channel made from viral proteins, host proteins, or a combination of both. The force required for penetration is associated with the contraction of the tail sheath. The peptidoglycan-hydrolyzing activity present in the tail tube facilitates the penetration of the peptidoglycan layer. Some viral proteins enter the cell with the phage DNA in order to protect it from host exonuclease activities and are implicated in the early stages of phage genome transcription.

The bacteriophages of the family *Syphoviridae* have long noncontractile tails. The noncontractile tail is composed of a relatively flexible tube ending with a tip (or straight fiber) and few tail fibers attached to the tube. In contrast to myoviruses, tails of syphoviruses do not possess a contractile sheath and therefore do not contract during DNA ejection. Nevertheless, the long noncontractile tail undergoes considerable conformational changes, which occur after primary interaction with the host cell. These conformational changes mostly concern the distal part of the tail (straight fiber). The straight fiber penetrates the cell envelope and concomitantly shrinks in length while it increases in diameter. This signals the DNA to be released from the head of the virus. The straight fiber of the tail provides the channel for DNA penetration through the host cell envelope. Alternatively, viral DNA travels through a pre-existing pore within the host envelope, as many bacteriophages with noncontractile tails adsorb to channel forming protein complexes such as porins and transporters.

Members of the family *Podoviridae* have short noncontractile tails decorated with tail fibers. Tail fibers are responsible for host recognition and binding. Most of these viruses bind to the lipopolysaccharide of the outer membrane of Gram-negative bacteria. In contrast to the long tails of myo- and syphoviruses, the tails of podoviruses are too short to span the host cell envelope. After tail fiber attachment to lipopolysaccharide, several proteins are ejected from the virion. In the intact virion these proteins either form the internal core structure of the head or are located in the tail. Proteins ejected from the virion are involved in the formation of the channel across the cell envelope, thus extending the short tail and allowing the phage genome to travel from the virion into the cytoplasm.

Icosahedral dsDNA Bacterial Viruses with an Internal Membrane

The internal membrane of membrane containing dsDNA bacteriophages plays an important role during the entry process. According to the genome, the icosahedral membrane containing dsDNA phages fall into two families, *Tectiviridae* with the type member PRD1 and *Corticoviridae* with the only member PM2. The genome of tectiviruses is a linear dsDNA molecule with terminal proteins attached to the 5'-end, whereas the PM2 genome is a highly supercoiled circular dsDNA molecule. Obviously, these viruses use different strategies to deliver their genomes into the host cell.

Icosahedral dsDNA bacterial viruses with an internal membrane from the family *Tectiviridae* (e.g., PRD1) attach to the host cell receptor via a spike protein. The receptor structure for tectiviruses infecting Gram-negative cells is encoded by a multiple drug resistance conjugative plasmid. Specific interactions between the phage and the host induce dissociation of the spike complex from one of the capsid vertices and subsequent formation of a pore in the viral capsid. The internal membrane is transformed into a tubular tail-like structure, which protrudes from the capsid through the pore and penetrates the outer layers of the host cell. The rigid protein capsid stays intact outside the cell. Two lytic enzymes are associated with the virion (membrane) and function during the peptidoglycan penetration. A current model suggests that the viral integral membrane fuses with the host plasma membrane, thus connecting the cytoplasm with the interior of the viral membrane leading to delivery of the viral genome into the host cytoplasm. The members of the family *Tectiviridae* that infect Gram-positive bacteria (e.g., Bam35) attach to the host cell peptidoglycan structure. The internal membrane of these viruses also transforms into a tubular structure suggesting a similar mechanism for DNA entry into the host cell. The membrane tube of tectiviruses forms a channel for DNA translocation and thus is a counterpart to the tails of dsDNA bacteriophages discussed above.

Bacteriophage PM2 (*Corticoviridae*) infects Gram-negative marine bacteria from the genus *Pseudoalteromonas*. It has a circular dsDNA genome. The internal membrane of PM2 does not transform into a tubular structure in contrast to the internal membranes of tectiviruses. Instead, the capsid of the virus completely dissociates on the cell surface and the lipid core faces the host's outer membrane. The viral membrane most likely fuses with the bacterial outer membrane delivering the highly supercoiled dsDNA genome into the periplasm, from where it travels to the host cytosol.

Pleomorphic dsDNA Bacterial Viruses

The quasi-spherical virions of pleomorphic dsDNA bacterial viruses (*Plasmaviridae*) are nucleoprotein complexes within a lipid-protein membrane. The only classified member of the family *Plasmaviridae* is Acholeplasma phage L2. Acholeplasma are wall-less bacteria; thus, the phage adsorbs directly on the plasma membrane. It is assumed that adsorption leads to fusion of viral and host cell membranes resulting in entry of the nucleoprotein complexes into the cell.

Icosahedral ssDNA Bacterial Viruses

The icosahedral ssDNA bacterial viruses of the family *Microviridae* infect either free-living enterobacteria (e.g., φX174, G4, and α3) or obligate intracellular parasites lacking a cell wall, like chlamydia (e.g., Chp2) and spiroplasma (SpV4).

Icosahedral ssDNA bacterial viruses infecting free-living enterobacteria contain large spike complexes at the vertices of the virion. The spike complex is composed of five major spike proteins (G in φX174) and one minor spike protein (H in φX174) located at the spike apex. These spikes are responsible for the phage binding to the host cell, as well as for ejection of the ssDNA genome into the host cytoplasm. Bacteriophages infecting free-living enterobacteria adsorb to the lipopolysaccharide structures of the cell's outer membrane. Although microviruses are tail-less, they may follow an entry pathway similar to that of tailed phages (e.g., T4). Initially, φX174-like viruses adsorb reversibly to the cells. This is analogous to the interaction of the tail fibers of tailed phages with the host cell. After reversible adsorption, which allows detection of the host bacterium, follows the irreversible adsorption. Virus binding to a suitable lipopolysaccharide induces conformational changes in the spike proteins and leads to DNA release from the viral capsid. Similarly to the tail of tailed phages, the major spike proteins constitute a channel through which the ssDNA travels into the host cytosol. The minor spike protein penetrates through the host plasma membrane along with the viral DNA. The viral capsid lacking DNA and one of 12 minor spike proteins remains outside the cell.

Large spike complexes are not present in the virions of phages that infect parasitic bacteria. Instead, they have elaborate viral coat protein protrusions on the threefold axis of symmetry not seen in the φX174-like viruses. These protrusions are likely responsible for the receptor recognition. Bacteriophages infecting parasites mostly use protein receptors present in the plasma membrane of wall-less bacteria.

Filamentous ssDNA Bacterial Viruses

Filamentous ssDNA bacterial viruses (family *Inoviridae*) mostly infect Gram-negative bacteria. Usually, they utilize the sites of bacterial envelope where outer and inner membranes are in close proximity. The protein complexes functioning as co-receptors are preferentially localized at these sites of the cell envelope.

The receptor-binding protein of filamentous bacteriophages is a minor coat protein (pIII in Ff phages) located at one end of the extended phage particle (**Figure 1(b)**). At the beginning of the infection process this protein binds to the tip of the bacterial pilus. Bacterial pilus is a complex assembly immobilized in the bacterial plasma membrane that spans the periplasmic space and protrudes through the outer membrane (**Figure 1(b)**). Pili are able to retract when they bind to the solid surface and are used for bacterial motility. Upon phage adsorption the pilus rapidly retracts bringing the phage particle through the outer membrane close to its co-receptor, a protein complex (TolQRA for Ff phages) located in the inner membrane and periplasm of the bacterial cell envelope (**Figure 1(b)**). Virus adsorption to the primary receptor (pilus) uncovers the co-receptor-binding domain in the receptor-binding protein of the virion. This allows the interaction between the receptor-binding protein of the virus and the co-receptor to proceed, thereby anchoring the virion to the bacterial plasma membrane. This triggers insertion of adjacent major coat proteins of the filamentous phage particle into the bacterial plasma membrane leading to uncoating of the virion and ssDNA translocation into the host cytoplasm (**Figure 1(b)**). The uncoating of the virion and genome translocation both require the functional co-receptor complex, which is involved in the formation of the channel for DNA traveling through the peptidoglycan and the plasma membrane.

Icosahedral Enveloped dsRNA Bacterial Viruses

The dsRNA bacteriophages (family *Cystoviridae*) have a unique entry mechanism among bacteriophages. As other dsRNA viruses, bacteriophages with a dsRNA genome deliver their genome within a large protein capsid, which protects the enclosed dsRNA genome throughout the infection cycle. Internalization of large nucleoprotein complexes into bacterial cells is rare. The protein capsid of dsRNA bacterial viruses, however, penetrates the host plasma membrane via a mechanism similar to endocytic entry of animal viruses (**Figure 1(c)**).

The virion of dsRNA bacteriophages contains three structural layers that sequentially assist in the penetration of the host outer membrane, the peptidoglycan layer, and the plasma membrane. The spikes protruding from the virion surface are involved in the receptor recognition and binding (**Figure 1(c)**). The dsRNA bacteriophages adsorb either to a bacterial pilus (e.g., φ6), which then retracts, or to polysaccharide (e.g., φ13). After phage adsorption at the outer membrane, the viral envelope fuses with the host membrane, thus uncoating the virion and placing the resulting viral nucleoprotein complex (or nucleocapsid) into the periplasm (**Figure 1(c)**). The fusion between the viral envelope and bacterial outer membrane is driven by the transmembrane proteins of the viral envelope whose fusogenic properties are activated after virus adsorption. The lytic enzyme of the virus located between the envelope and nucleocapsid locally digests the host cell-wall peptidoglycan, thereby allowing the nucleocapsid to reach the plasma membrane (**Figure 1(c)**). Eventually, the nucleocapsid penetrates the plasma membrane via an endocytic-like route. This involves the formation of the plasma membrane curvature at the contact site resulting in an intracellular vesicle, which then pinches off from the plasma membrane (**Figure 1(c)**). The transcriptionally active viral particle is released from the vesicle into the host cytosol.

Icosahedral ssRNA Bacterial Viruses

All of the known icosahedral ssRNA phages (e.g., MS2, Qβ, GA, family *Leviviridae*) are pilus specific and utilize bacterial pili as attachment site. The virions of icosahedral ssRNA phages contain a single attachment protein (or maturation protein), which carries out the initial interaction with the host pilus. The attachment protein is covalently linked to both ends of the genomic ssRNA. In addition, this protein is partially exposed on the capsid surface likely forming one of the capsid vertices. Virus adsorption initiates the cleavage of the attachment protein into two fragments and later dissociation of those fragments from the viral capsid. As the pilus retracts into the cell, the attachment protein fragments together with the genomic ssRNA are pulled inside the host cell. The empty capsid remains outside the cell.

Bacteriophages with ssRNA genomes do not possess the peptidoglycan-hydrolyzing enzymes present in many other bacterial viruses which facilitate the peptidoglycan penetration during the entry process. Instead, ssRNA phages rely on the host cell pilus assembly to reach the cytoplasm.

Energetics

The viral genome transport across the host membrane is not a passive process. The force required for cell envelope penetration comes from different sources. As discussed earlier, phages with long contractile tails gain energy from the contraction of the tail sheath. This powerful process allows viral DNA to pass the host membrane in a very short time (e.g., 30 s for T4). Another type of energy comes from the pressure inside the viral capsid, which often is very high: In the capsids of dsDNA bacteriophages, the pressure may reach 50 atm ($\sim$5 MPa)! This internal pressure of the virion, built up during DNA packaging, likely facilitates DNA ejection from the capsid during the initial stages of genome delivery.

The viral genome translocation may also be dependent on different host cell activities. The genome of T7 bacteriophage is pulled into the cytoplasm first by the host RNA polymerase and then by the RNA polymerase encoded from the phage genome. The retraction of bacterial pili and the transfer of the virus particles along the pili also depend on the cellular energy. Also, in many cases viral genome translocation requires membrane potential, which might provide energy to move DNA or RNA, stabilize a transmembrane channel, or regulate the genome translocation process.

See also: Icosahedral Enveloped dsRNA Bacterial Viruses; Icosahedral ssRNA Bacterial Viruses; Icosahedral Tailed dsDNA Bacterial Viruses.

Further Reading

Bennett NJ and Rakonjac J (2006) Unlocking of the filamentous bacteriophage virion during infection is mediated by the C domain of pIII. *Journal of Molecular Biology* 356: 266–273.

Fane BA, Brentlinger KL, Burch AD, et al. (2006) ϕX174 et al., the Microviridae. In: Calendar R (ed.) *The Bacteriophages,* 2nd edn., pp. 129–145. New York: Oxford University Press.

Grahn AM, Daugelavicius R, and Bamford DH (2002) Sequential model of phage PRD1 DNA delivery: Active involvement of the viral membrane. *Molecular Microbiology* 46: 1199–1209.

Kivela HM, Daugelavicius R, Hankkio RH, Bamford JKH, and Bamford DH (2004) Penetration of membrane-containing double-stranded-DNA bacteriophage PM2 into *Pseudoalteromonas* hosts. *Journal of Bacteriology* 186: 5342–5354.

Letellier L, Boulanger P, de Frutos M, and Jacquot P (2003) Channeling phage DNA through membranes: From *in vivo* to *in vitro*. *Research in Microbiology* 154: 283–287.

Ponchon L, Mangenot S, Boulanger P, and Letellier L (2005) Encapsidation and transfer of phage DNA into host cells: From *in vivo* to single particles studies. *Biochimica et Biophysica Acta* 1724: 255–261.

Poranen MM, Daugelavicius R, and Bamford DH (2002) Common principles in viral entry. *Annual Review of Microbiology* 56: 521–538.

Rydman PS and Bamford DH (2002) Phage enzymes digest peptidoglycan to deliver DNA. *ASM News* 68: 330–335.

Van Duin J and Tsareva N (2006) Single-stranded RNA phages. In: Calendar R (ed.) *The Bacteriophages,* 2nd edn., pp. 175–196. New York: Oxford University Press.

Virus Evolution: Bacterial Viruses

R W Hendrix, University of Pittsburgh, Pittsburgh, PA, USA

© 2008 Elsevier Ltd. All rights reserved.

Glossary

Chromosome The physical DNA or RNA molecule present in the virion and containing the information of the genome. The chromosome can differ from the genome by, for example, having a terminal repetition of part of the genomic sequence or being circularly permuted relative to other chromosomes in the population.

Genome The totality of the genetic complement of a virus. The genome is a conceptual object usually expressed as a sequence of nucleotides.

Homologous Having common ancestry. Homology of two sequences is often inferred on the basis of a high percentage of identity between the sequences, but homology itself is either present or not and is not expressed as a percentage.

Novel joint A novel juxtaposition of sequences in a genome resulting from a nonhomologous recombination event.

Bacteriophage Evolution

Bacterial viruses ('bacteriophages' or 'phages' for short) have probably been evolving for 4 billion years or more, but it is only in recent years that we have come to a relatively detailed view of the genetic mechanisms that underlie phage evolution. Phages do not leave fossils in the conventional sense, but the sequences of phage genomes contain more detailed information about how the phages have evolved than any conventional fossil has. Because of the importance of sequence, our understanding of phage evolution has increased dramatically as more phage sequences have become available over the past few years. Even more important than the individual sequences, however, is the fact that we have multiple genome sequences that we can compare to each other. It is in such comparisons of different genomes that we see the unmistakable signatures of past evolutionary events, which would be completely invisible in an examination of a single genome alone.

Most of the work on phage evolution has addressed evolution of the double-stranded DNA (dsDNA) tailed phages, the members of the order *Caudovirales*, and most of the discussion here deals with that group. Toward the end of the article, we will consider evolution of other groups of phages as well as recent evidence suggesting deep evolutionary connections between some of the bacteriophage groups and viruses that infect members of the Eukarya and Archaea.

The Tailed Phage Population

Nothing in phage evolution makes sense except in the light of what we have learned about the nature of the global population of tailed phages. The most remarkable fact is that the size of the population is almost incomprehensibly large – a current estimate is that there are about 10^{31} individual phage particles on the planet. To give a feeling for the magnitude of this number, if these phages were laid end to end, they would reach into space a distance of ~200 million light years. This is apparently a very dynamic population; ecological studies of marine viruses suggest that the entire population turns over every few days. To replenish the population would then require about 10^{24} productive infections per second on a global scale. Each such infection is an opportunity for evolutionary change, either by mutation during replication or by recombination with the DNA in the cell, which will almost always include DNA of other phages in the form of resident prophages. At a lower frequency, cells may be co-infected by two different phages, affording a different opportunity for genetic exchange. As described below, the evolutionary events that are inferred from genome sequence comparisons include extremely large numbers of improbable events, and the only way these events could have given rise to the existing phage population is if they have had an extremely large number of opportunities to occur.

Evolutionary Mechanisms in the dsDNA Tailed Phages

The Nature of the Genomes

The genomes of the dsDNA tailed phages use DNA very efficiently, with typically about 95% of the sequence devoted to encoding proteins. Spaces between genes can often be identified as containing expression signals such as transcription promoters or terminators. Genes are typically arranged in groups that are transcribed together. The specific types of genes in a genome, their numbers, and how they are arranged along the DNA are all highly variable among phages. Genome size is also variable, with the smallest known phages in this group having genomes of about 19 kbp and about 30 genes and the largest consisting genomes of 500 kbp and nearly 700 genes.

Types of Evolutionary Changes

Comparisons of genome sequences between phages reveal the sorts of changes in the genome sequence that in aggregate constitute phage evolution. The first of these is point mutation, in which one base pair is substituted for another. The number of point mutational differences seen between two homologous sequences varies from none to so many that our ability to detect any residual similarity of the sequence, often measured at the more sensitive level of the encoded amino acid sequence, has vanished. The amount of difference between two sequences due to point mutation is a function of how long it has been since they diverged from a common ancestral sequence, because such differences accumulate progressively over time. In practice, however, it is an unreliable measure of time for a number of reasons: we do not know the mutational rates for phages in a natural setting, we do not know that those rates have been constant over evolutionary time, and different genes accumulate changes at different rates because any mutations that are detrimental to the function of the encoded protein will be lost from the population by natural selection, and different proteins have different tolerances for mutational changes in their amino acid sequences.

The other major type of evolutionary change to a genome is DNA recombination. Particularly important is 'illegitimate' or 'nonhomologous' recombination, in which two different DNA sequences are joined together to form an association of sequences that did not exist before, known as a 'novel joint'. Nonhomologous recombination can join parts of two or more genomes into one new genome; it can also cause deletion or inversion of a sequence within a genome by mediating recombination between two sequences in the genome. Nonhomologous recombination can in principle – and, it appears, also in practice – produce virtually any novel joint that can be imagined. The results of laboratory experiments have shown that homologous recombination happens many orders of magnitude more frequently than nonhomologous recombination. Unlike nonhomologous recombination, homologous recombination does not leave a novel joint that can be detected in analysis of the sequence, but it has the potential to reassort any flanking novel joints, bringing them and their associated sequences together in new combinations and providing a mechanism for them to move rapidly through the population.

Genome Comparisons

There is typically nothing in a single phage genome sequence, viewed in isolation, that reveals the presence of changes in the phage's evolutionary past, of either mutational or recombinational origin. When multiple genomes are compared, a wealth of differences corresponding to such changes is revealed. **Figure 1** illustrates a comparison for a group of four rather similar genomes of phages infecting enteric bacterial hosts. The physical gene map shown is that of *Escherichia coli* phage HK97, and the histograms indicate the locations of sequences in the other phages that match HK97. It is apparent that the genome sequences match each other in a patchwork fashion, and the parts of the sequence that match the HK97 sequence are different for the three phages. In other words, the genomes are genetic mosaics with respect to each other. In a pairwise comparison of genomes, the transitions between where sequences match and where they do not are abrupt, even when they are examined at the level of the nucleotide sequence. This implies that there have been nonhomologous recombination events in the ancestry of the phages, creating novel joints in the resulting recombinant phages. These novel joints are detected when they are compared with a sequence that did not suffer that particular recombination event. In the comparisons shown in **Figure 1**, there is evidence for about 75 ancestral nonhomologous recombination events, occurring in either the ancestry of HK97 or in that of one of the three phages being compared. It is worth noting that this probably underestimates the number of such events, because any novel joints that lie in the common ancestry of all these phages would not have been detected.

A striking feature of the novel joints in the sequence revealed by such genome sequence comparisons is that they fall predominantly at gene boundaries. While some are precisely at the gene boundaries, some also fall in the middle of spaces between genes or even a few codons into the upstream or downstream gene's coding region. More rarely, we can see novel joints in the interior of a gene's coding region. In the case of some such genes for which the encoded proteins are well characterized, the novel joints in the DNA sequence fall at a position corresponding to a domain boundary of the protein. Another feature of the genome comparisons that extends across all the groups of phages examined to date is that there are clusters of genes that are never, or rarely, separated by nonhomologous recombination. These are typically genes whose protein products are known to function together, such as the genes encoding the structural proteins of the head.

A frequent consequence of nonhomologous recombination among viruses is the transfer of genetic material from the genomes of one viral lineage into the genomes of a different viral lineage, a process known as 'horizontal transfer' or 'lateral transfer' of genes. This is the process that gives rise to the mosaicism in the genomes discussed above. It also means that different parts of a genome may, and most often do, have different evolutionary histories. This last fact leads to an interesting and still unresolved difficulty for viral taxonomists. That is, attempts to represent the relationships among viruses by a hierarchical taxonomy, as is conventionally done, necessarily fail to capture the multiple different sets of hierarchical relationships displayed by the individual exchanging genetic modules, deriving from their different evolutionary histories. The viruses, of course, are unaware of human attempts to classify them and so are unaffected by the resulting controversies.

Evolutionary Mechanisms

When the mosaicism of phage genomes was first seen in DNA heteroduplex mapping experiments in the late 1960s, it was proposed that there might be special sites in the DNA, possibly at gene boundaries, that served as points of high-frequency nonhomologous recombination, either through short stretches of homology or through a site-specific recombination mechanism. This view was formulated as the 'modular evolution' model of phage evolution, in which it was proposed that exchange between genomes took place repeatedly at these special sites, leading to the observed mosaic relationship between genomes.

With the current availability of many more genomes and data at the level of nucleotide sequence, it has become clear that, while the mosaic results of phage evolution are much as described nearly 40 years ago, the mechanisms by which that state is achieved are fundamentally different than what had been proposed. That is, the observations are best explained by a model of rampant nonhomologous recombination among phage genomic sequences, with the sites of recombination being distributed, to a first approximation, randomly across the sequence with no regard for gene boundaries or other features of the sequence. Most recombinants produced this way will be nonfunctional

Virus Evolution: Bacterial Viruses 373

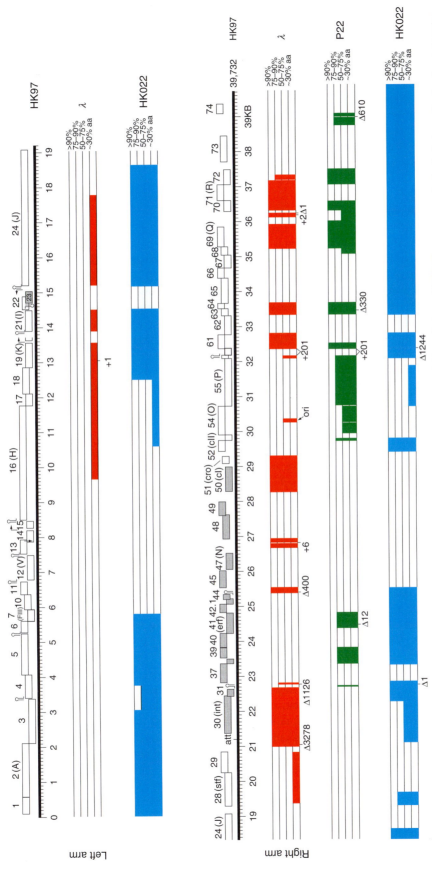

Figure 1 The horizontal black line, scaled in kilobase pairs, represents the genome sequence of *Escherichia coli* phage HK97. The genes are indicated above the line by rectangles, with open rectangles indicating genes transcribed rightward and shaded rectangles genes transcribed leftward. The colored histograms indicate the areas of sequence similarity between HK97 and the three phages indicated to the right of each histogram: *E. coli* phages lambda and HK022 and *Salmonella enterica* phage P22. Adapted from Juhala RJ, Ford ME, Duda RL, Youlton A, Hatfull GF, and Hendrix RW (2000) Genomic sequences of bacteriophages HK97 and HK022: Pervasive genetic mosaicism in the lambdoid bacteriophages. *Journal of Molecular Biology* 299: 27–52.

because, for example, they have glued together parts of two encoded functional proteins into a nonfunctional chimera. The tiny fraction of such recombinants that are fully functional and competitive in the natural environment will be those that have recombination joints where they do no harm, like at gene boundaries; the rest will be rapidly eliminated from the population by natural selection. We would also expect that two recombining genomes would not typically be lined up in register, and this would most often lead to nonfunctional recombinants. There are a few examples of genomes with short quasi-duplications that can be explained by a slightly out of register recombination event, supporting this supposition.

Nonhomologous recombination also has the potential to bring novel genes into a genome and to mediate large-scale reorganizations of genomes, as seen for example in *E. coli* phage N15 which has head and tail structural genes homologous to those of phage lambda and, in the other half of the genome, genes from a very different, apparently plasmid-like, source. Another example is the *Shigella* phage SfV, which has head genes in the phage HK97 sequence family and tail genes like those of phage Mu.

The entire process described above can be viewed as a classical Darwinian scenario in which tremendous diversity is generated in the population by a combination of point mutations and homologous and nonhomologous recombination, and the stringent sieve of natural selection acts to eliminate all but the fittest recombinants. The survivors have largely, though not entirely, lost the appearance of having been generated by what is essentially a random mutational process.

Evolution of Other Phage Types

Although most of the genome comparisons that have led to our current understanding of phage evolution have been carried out with the tailed phages, there has been some work done with other groups. The most extensive has been done with members of the families *Microviridae* and the *Inoviridae*, phages with small, circular single-stranded DNA (ssDNA) chromosomes, typified by the well-studied *E. coli* phages φX174 and M13. For both of these groups there is evidence for horizontal exchange of sequences, as in the tailed phages, but it is quantitatively much less prominent than what is seen in the tailed phages. There are several differences between the tailed phages and these two groups of smaller phages, some or all of which may contribute to the differences in evolutionary outcomes seen between them. These include 10- to 100-fold differences in the sizes of the genomes and the fact that the members of the *Microviridae* and *Inoviridae* do not encode recombination functions, among others.

Another study looked at evidence for evolutionary change among members of the family *Cystoviridae*, a group of phages with a double-stranded RNA (dsRNA) genome, divided into three physical segments. Comparisons among environmentally derived genome sequences showed no evidence for recombinational exchange within the segments but clear indications of reassortment of the segments. Thus it appears that evolution in this group of viruses achieves the same general outcome as evolution of the tailed phages, namely horizontal exchange of genetic modules, but by a very different mechanism. Analogous horizontal exchange by reassortment of the segments of a segmented genome has been extensively characterized in the influenza viruses, a group of animal viruses with a single-stranded RNA (ssRNA) genome of eight segments.

Deep Evolutionary Connections

Although there are some genes that are found in all tailed phages – for example, genes encoding the major capsid protein – these are very diverse in sequence. This diversity is sufficiently great that it is not possible, on the basis of their sequences, to group them into a single sequence family, at either the DNA or the protein sequence levels. Thus the subsets of capsid protein sequences that share demonstrable sequence similarity may represent independent nonhomologous lineages of capsid proteins, or alternatively they may all be part of the same lineage but have diverged in sequence to the point that in many cases no surviving sequence similarity can be demonstrated. However, recent structural work on the polypeptide folds of capsid proteins appears to be extending our analytical reach farther back in evolutionary time and giving a resolution to this question. High-resolution X-ray structures of the capsid proteins of phages HK97 and T4 demonstrate that they have a common fold which might indicate common ancestry despite the absence of any surviving sequence similarity. Cryoelectron microscopic structures of phages representing two more capsid protein sequence groups, P22 and φ29, make a less rigorous but still convincing case that these capsid proteins also share in a common polypeptide fold and so are most likely part of the same lineage. Remarkably, a similar experiment on (the animal virus) herpes simplex virus argues that the shell-forming domain of its major capsid protein also shares the tailed phage capsid protein fold and so may be part of the same lineage.

A similar and more completely documented case for a virus lineage extending across the cellular domains of the hosts they infect can be made for a group that includes bacteriophage PRD1 (a member of the family *Tectiviridae*), archaeal virus STIV, and eukaryotic viruses adenovirus, PBCV1 (*Phycodnaviridae*), and mimivirus. This is again based on shared capsid protein folds, in this case an unusual double 'jellyroll' fold. Finally, there is compelling structural similarity between the cystoviruses, the group of phages with segmented dsRNA chromosomes, and a characteristic double protein shell discussed above, and the reoviruses of plant and animal hosts.

The simplest interpretation of these observations, though not the only possible interpretation, is that there were already viruses resembling these different types of contemporary viruses prior to the divergence of cellular life into the three current domains. The different types of viruses in this view would have co-evolved with their hosts as the hosts divided into domains and descended to the present time, with in the end nothing to indicate their common origins except the shared structure of their capsid proteins. Whatever the truth of this matter, an important caveat to such an interpretation is that the evidence for common viral ancestries across host domains is at present based primarily on conserved properties of capsid proteins, and so any conclusions about common lineages, strictly speaking, only apply to capsid protein lineages. However, despite such reservations, the data make an intriguing though still tentative case that viruses resembling at least three groups of contemporary viruses had already evolved into forms something like those of contemporary viruses before the three domains of cellular life had begun to separate, ~3.5 billion years ago.

See also: Evolution of Viruses; Origin of Viruses; Phylogeny of Viruses; Quasispecies.

Further Reading

Brüssow H and Hendrix RW (2002) Phage genomics: Small is beautiful. *Cell* 108: 13–16.

Casjens SR (2005) Comparative genomics and evolution of the tailed-bacteriophages. *Current Opinion in Microbiology* 8: 451–458.

Fuhrman JA (1999) Marine viruses and their biogeochemical and ecological effects. *Nature* 399: 541–548.

Hatfull GF, Pedulla ML, Jacobs-Sera D, *et al.* (2006) Exploring the mycobacteriophage metaproteome: Phage genomics as an educational platform. *PLoS Genetics* 2: e92. http://genetics.plosjournals.org/perlserv/?request=get-document&doi=10.1371%2Fjournal.pgen.0020092 (accessed June 2007).

Juhala RJ, Ford ME, Duda RL, Youlton A, Hatfull GF, and Hendrix RW (2000) Genomic sequences of bacteriophages HK97 and HK022: Pervasive genetic mosaicism in the lambdoid bacteriophages. *Journal of Molecular Biology* 299: 27–52.

Nolan JM, Petrov V, Bertrand C, Krisch HM, and Karam JD (2006) Genetic diversity among five T4-like bacteriophages. *Virology Journal* 3: 30–44.

Suttle CA (2005) Viruses in the sea. *Nature* 437: 356–361.

Virus-Induced Gene Silencing (VIGS)

M S Padmanabhan and S P Dinesh-Kumar, Yale University, New Haven, CT, USA

© 2008 Elsevier Ltd. All rights reserved.

Glossary

Functional genomics The study of genes with respect to the role they play within biological processes.

High-throughput screening A method to efficiently test a large number of putative genes to identify candidates that may regulate a specific biological process.

Hypersensitive response A defense response in plants that initiates cell death to restrict the growth of pathogen.

Knockdown Reduction in the expression of a gene.

Knockout Complete inhibition of gene expression.

Reverse genetics Approaches used to define the function of a gene or sequence of DNA within the context of the organism.

T-DNA Transfer DNA from *Agrobacterium tumefaciens*, a vector routinely used for transforming plants.

Introduction

Virus-induced gene silencing (VIGS) is an RNA silencing-based technique used for the targeted downregulation of a host gene through the use of a recombinant virus. The term VIGS was originally coined to describe the phenomenon of recovery of a host plant from viral infection. Today, its usage is predominantly in reference to a tool for turning down host gene expression, especially in plants. In principle, a plant gene of interest can be silenced by infecting the plant with a viral vector that has been modified to express a nucleic acid sequence homologous to the host gene. Viral infection and synthesis of double-stranded RNA (dsRNA) viral replication intermediates initiate the innate RNA silencing or post-transcriptional gene silencing (PTGS) pathway, ultimately leading to the degradation of host transcript. The potency, specificity, and speed of PTGS have thus been harnessed to create an efficient gene knockout system. The ability of VIGS to rapidly initiate silencing and generate mutant phenotypes without the use of laborious transgenic approaches has made it the tool of choice for characterizing genes, especially in plant species where conventional analytical techniques have had limited success.

PTGS and the Principle behind VIGS

The term virus-induced gene silencing was first used by A. van Kammen to describe the process by which some plants were able to recover from viral infection, coincident with the loss of viral RNA from the infected tissue. The recovery process was soon understood to be associated with the plant inherent RNA-silencing mechanism which is an evolutionarily conserved antiviral system. Most plant viruses are RNA viruses and produce dsRNA intermediates during replication. In contrast to host messenger RNA, the viral dsRNA is made up of relatively long stretches of complementary RNA strands. The host identifies these as foreign and triggers the RNA-silencing pathway. The plant recovery is thus the successful outcome of the actions of this surveillance system.

The phenomenon of RNA silencing, initially observed in plants where it was called PTGS, has since been known to operate in almost all eukaryotic species including fungi, worms, flies, and mammals. In fungi, the process is called quelling while in animal systems it is known as RNAi (RNA interference). In plants and other eukaryotes, the mechanism of RNA silencing of viruses is highly conserved. The dsRNA which is generally synthesized by a viral polymerase or in some cases a host-encoded RNA-dependent RNA polymerase is recognized and cleaved by Dicer, a ribonuclease III (RNase III)-like enzyme. This results in the production of 21–24 nt duplex molecules called short-interfering RNAs (siRNA's). In an ATP-dependent manner, the siRNA's are denatured and one strand is exclusively incorporated into a multi-subunit nuclease complex called the RNA-induced silencing complex (RISC). Within the RISC, siRNA serves as a guide to recognize and base pair with the homologous RNA (in this case the viral RNA), eventually leading to degradation of the target RNA. This occurs via an RNase H-like cleavage mechanism directed by the RISC. One of the key factors that make PTGS such a potent system of host defense is that once initiated, it can maintain the silencing effect by relaying a diffusible silencing signal throughout the plant. As the virus moves within the host tissue, those cells that have received the silencing signal are primed to recognize the viral RNA and initiate its degradation.

PTGS and its efficiency against viral genomes can be exploited to create recombinant viral vectors that serve as tools for gene silencing. Viruses carrying segments of host genes, when used to infect plants, would produce siRNA's specific to the host mRNA. RISC-mediated degradation of target host mRNAs would ultimately lead to downregulation of gene expression. The infected plant would thus have a phenotype similar to a loss-of-function mutant of the gene of interest. The efficiency of viruses in initiating silencing was first shown in plants infected with a recombinant tobacco mosaic virus (TMV) carrying a fragment of the plant gene *phytoene desaturase* (*PDS*), a regulator of carotenoid biosynthesis. Infected *Nicotiana benthamiana* plants showed degradation of the host PDS mRNA and resultant alterations in the pigment synthesis pathway led to significant photobleaching in the leaves. The potential of this gene knockout system was immediately recognized and has since been applied to more than ten plant species. The *PDS* gene now serves as the conventional gene used for testing the efficacy of a virus in inducing silencing (**Figure 1**).

The Development of VIGS as a Tool for Functional Genomics

Several VIGS vectors have been developed over the past 8 years (**Table 1**). The choice of VIGS vector plays a key role in efficient silencing and there are many factors to be considered when choosing the virus to be used for VIGS. Ideally, the virus must produce little or no symptoms during infection, thereby facilitating easy visualization and interpretation of the mutant phenotype. It must induce persistent silencing. Viruses with strong silencing suppressors are avoided since they can interfere with the establishment of silencing. It is advantageous to have infectious cDNA clones of the virus for cloning purposes and the virus must retain infectivity after insertion of foreign DNA. The virus should also show uniform spread, infect most cell types including the meristem, and preferably show a broad host range.

RNA Virus-Based Vectors

TMV, an RNA virus belonging to the genus *Tobamovirus*, was the earliest viral vector to be used for VIGS but the severity of symptoms in susceptible tissue and inability to invade the meristem limited its application. A PVX (potato virus X)-based vector produced milder infection symptoms but displayed a narrower host range and did not infect the growing points of infected plants. Tobacco rattle virus (TRV) was shown to overcome all these disadvantages and currently is the most widely used VIGS vector, especially in solanaceous hosts. It can infect all cell types including the meristem and so can be used to investigate the role of genes in early developmental processes. The silencing is more persistent when compared to other viral vectors and it is documented to infect a wide range of hosts.

TRV is a plus-sense RNA virus with a bipartite genome. RNA-1 encodes the viral replicase and movement

Figure 1 Silencing of the *PDS* gene (phytoene desaturase) in *N. benthamiana* using TRV-based VIGS system. *Nicotiana benthamiana* plants infected with the empty vector TRV-VIGS do not have a noticeable phenotype while TRV-*NbPDS*-infected plants show photobleaching, a hallmark of PDS silencing.

Table 1 Selected list of plant viruses used as VIGS vectors for gene silencing

Virus	Genus	Host	Method of infection
Tobacco mosaic virus	*Tobamovirus*	*N. benthamiana*	Inoculation of infectious viral RNA
Potato virus X	*Potexvirus*	*N. benthamiana*, potato	Inoculation of infectious viral RNA; agro-inoculation
Tobacco rattle virus	*Tobravirus*	*N. benthamiana*, *Arabidopsis thaliana*, tomato, pepper, petunia, potato, poppy	Agro-inoculation
Pea early browning virus	*Tobravirus*	Pea	Agro-inoculation
Satellite Tobacco mosaic virus (with helper TMV)	*Tobamovirus*	Tobacco	Inoculation of infectious viral RNA
Bean pod mottle virus	*Comovirus*	Soybean	Inoculation of infectious viral RNA
Tomato golden mosaic virus (TGMV)	*Geminivirus*	*N. benthamiana*	Microprojectile bombardment
Cabbage leaf curl virus (CaLCuV)	*Geminivirus*	*A. thaliana*	Microprojectile bombardment
African cassava mosaic virus (ACMV)	*Geminivirus*	Cassava	Microprojectile bombardment
Barley stripe mosaic virus	*Hordeivirus*	Barley, wheat	Inoculation of infectious viral RNA
Fescue strain of Brome mosaic virus (F-BMV)	*Bromovirus*	Rice, maize, barley	Inoculation of infectious viral RNA

proteins, while RNA-2 contains genes for the coat protein and two nonstructural proteins, one of which is needed for vector transmission. Since the latter two genes are dispensable for viral replication and spread in plants, RNA-2 could be modified to replace the nonstructural genes with a multiple cloning site for insertion of host-derived gene sequences. The host sequence to be inserted into the viral vector could theoretically be as small as 23 nt (the size of an siRNA) but in practice the ideal size range is between 300 and 1500 bp.

There are many ways of introducing the virus into the plant system and these include rubbing the plants with infectious viral transcripts or biolistic delivery. However, the most effective means appear to be incorporation of the viral cDNA into *Agrobacterium tumefaciens*-based plant transformation vectors (T-DNA vectors) under the control of constitutive promoters followed by infiltration of the *Agrobacterium* cultures into the host. This ensures not just ease in the inoculation procedure but also efficient transformation and high levels of viral RNA production within plant cells. For the TRV system, VIGS can be induced by simply mixing individual *Agrobacterium* cultures containing RNA-1 and modified RNA-2 and then co-infiltrating into young plants using a needle-less syringe. Alternatively, the cultures can also be vacuum-infiltrated or sprayed onto the plant.

The appearance of VIGS-related phenotype depends on a number of factors including the plant species used, the age of the plant, and environmental factors like temperature and humidity. Under optimal conditions in N. benthamiana plants, TRV-induced silencing phenotypes can be visualized as early as 1 week. TRV is also one of the few viruses that have been modified into a highly efficient cloning and expression system for use in large-scale functional genomics screens. One of the most useful features of TRV vectors is their ability to induce VIGS in a number of solanaceous hosts like N. benthamiana, tomato, potato, pepper, petunia, poppy (Eudicot species), and the model system Arabidopsis thaliana (family Brassicaceae).

A number of other viruses have been adopted for VIGS in species where TRV has not been efficacious (Table 1). For tobacco plants, researchers have developed the satellite virus-induced silencing system (SVISS), which is a two-component system that uses a modified satellite TMV (that acts as the inducer of silencing) and a helper TMV-U2 virus (that promotes viral replication). Most of the VIGS vectors thus far described have been designed for dicotyledonous species though a majority of the economically important food crops such as rice, maize, wheat, and barley are monocots. The RNAi machinery has been shown to function in monocots and hence research in recent years has focused on finding ideal monocot VIGS vectors. The hordeivirus barley stripe mosaic virus (BSMV) was the first virus adapted for VIGS in barley and subsequently applied to wheat. A specific strain of brome mosaic virus isolated from tall fescue (F-BMV) was cloned, characterized, and modified for use as a VIGS vector. F-BMV-based vectors were shown to effectively silence genes from rice, barley, and some cultivars of maize. The successful adaptation of VIGS technology in monocots will, in the future, aid in the functional characterization of genes in these otherwise recalcitrant systems.

DNA Virus-Based Vectors

DNA viruses belonging to the family Geminiviridae like tobacco golden mosaic virus (TGMV) and cabbage leaf curl virus (CaLCuV) have been successfully employed for silencing and serve as good alternatives to RNA viruses. CaLCuV was also the first VIGS vector to be used to knockdown gene expression in Arabidopsis. Geminiviruses are highly conserved in their genetic sequence and so, in theory, these viruses can be quickly adapted for VIGS based on information from one virus. Other beneficial properties of geminiviruses are that they are genetically more stable than RNA-based vectors, can invade meristematic tissue, and infect a wide range of plants including economically important crops. In fact, begomoviruses like African cassava mosaic virus (ACMV) and pepper huasteco yellow vein virus (PHYVV) have been successfully employed as VIGS vectors in cassava and pepper. As a caveat, it must be noted that most of the geminiviruses used so far appear to induce some symptoms of infection and thus care must be taken when interpreting VIGS phenotypes.

Applications of VIGS in Aiding Plant Gene Characterization

There have been numerous examples validating VIGS as an effective method for loss-of-function studies in plant systems. The preeminent use of this technique has been in studying plant–pathogen interactions, especially in deciphering plant defense pathways. Resistance against pathogens in many plants is dictated by the presence of specific R genes and each gene exclusively initiates defense against one race of a given pathogen. Numerous studies have used VIGS to study genes that are activated or regulated in R gene pathways. These include N gene-mediated resistance to TMV, Rx1-mediated resistance to PVX, Mla13-mediated resistance to powdery mildew, Cf-4 regulated resistance to Cladosporium fulvum, and Pto-mediated resistance to Pseudomonas syringae in tomato. In some of these studies, cDNA libraries inserted into VIGS vectors were screened to identify genes that were essential for the R gene-activated hypersensitive response (HR) which is a hallmark of plant defense. These studies in parallel identified common components in disease resistance like the chaperone protein HSP90, mitogen-activated protein (MAP) kinase pathway proteins, regulators of the protein degradation pathway, and TGA and WRKY transcription factors. The use of VIGS has also aided in the identification of novel pathways that play a role in defense or pathogenesis. For instance, autophagy, an evolutionarily conserved programmed cell death response, was found to be important for limiting cell death induced during viral infection. Similarly, P58^{IPK}, a plant ortholog of the inhibitor of mammalian dsRNA-activated protein kinase (PKR), was shown to be recruited by TMV to maintain infectivity. While the function of PKR in defense has been well characterized in animal systems, this was the first study showing a role for this protein in plant viral pathogenesis.

Meristem invasion by viruses like TRV make them ideal for use in silencing genes that may be important for cell proliferation, tissue differentiation, and flower development. Studies on floral genes have mainly been limited to model systems like Arabidopsis and petunia, but with the advent of VIGS it is now possible to test floral homeotic genes in a wide range of hosts and analyze conservation or divergence of gene functions between species. As a case study, in solanaceous plants, the ortholog of Arabidopsis and Antirrhinum floral homeotic gene AP3/DEFICIENS was silenced and found to produce a phenotype similar to that seen in Arabidopsis. The TRV system was also used to initiate VIGS in tomato fruits showing its potential in studying fruit development. An *N. benthamiana* cystein protease, calpain, was shown to play a key role in cell differentiation and organogenesis using the TRV-VIGS system. Similarly, the role of proliferating cell nuclear antigen (PCNA) was investigated with geminivirus-derived vectors. The primary advantage of using VIGS for such studies is not just that it expands the study of plant growth and differentiation into nonhost systems, but also that it overcomes problems of embryo lethality or sterility which would be unavoidable if true mutants of these genes were being studied.

Among its other applications, VIGS has been adopted for analyzing biosynthetic pathways (sterol synthesis and pigment synthesis), characterizing genes involved in stress response or hormone response, organelle biogenesis, and to look at transport of proteins within the cell (especially nuclear transport mediated by different importins).

Advantages of Using VIGS as a Tool for Functional Genomics

In plants, the conventional methodology for gene function analysis has been to use insertional mutagenesis to shut down expression of the gene of interest and study the consequence of its loss to the plant. The T-DNA plasmid of *Agrobacterium*, which is capable of inserting into the plant genome, has for a long time been one of the tools of choice for disrupting plant genes. An alternative approach has been to use transposons. While being powerful tools for reverse genetics, they came with a number of limitations. Laborious and time-consuming screens had to be carried out to search for T-DNA insertions or transposon tagged lines. Loss of genes essential in the early developmental stages led to embryo lethality while functionally redundant genes rarely showed a discernible phenotype when knocked out and so these genes could not be identified in screens. These techniques were also not amenable for genomic-scale studies because of the difficulty of achieving genome saturation and problems with multiple insertions. The advent of RNA-silencing technology meant that it was possible to disrupt the function of a specific gene by the introduction of gene-specific self-complementary RNA, also known as hairpin RNA (hpRNA). This process still required plant transformation and screening to identify silenced plants. Many of the constraints imposed by these methodologies have been overcome with the discovery of VIGS and its subsequent use a tool.

The biggest advantage with this technology is the short time needed to go from gene sequence to functional characterization of its knockdown. While screening for a potential T-DNA knockout would take at least a few months, with VIGS it can be scaled down to 2 weeks. Thus relationships between genes and phenotypes/functions can be established quickly. This factor alone has catapulted VIGS into one of the most widely accepted and popular tools for gene analysis. VIGS is not as labor intensive as the other techniques since it does not involve transformation of plant tissue. Application of the virus and induction of VIGS are easy, especially after incorporation of viral sequences into intermediate plant transformation vectors. VIGS can, in general, be induced at any stage during a plant's life cycle – thus it is still possible to characterize genes that may otherwise be lethal when knocked out during early stages of development. One can overcome redundancy by carefully choosing the gene sequence to be inserted into the viral vector so as to maximize (or minimize) the silencing of related genes. Many VIGS vectors can be used on multiple host species and this can prove useful for testing gene functions against different genetic backgrounds. These vectors have accelerated the characterization of genes in many nonmodel hosts like *N. benthamiana*, petunias, potato, soybean, and even poplars where conventional techniques achieved very limited success. Finally, as has been shown with TRV and PVX, VIGS vectors can be adapted for use in high-throughput functional genomic screens.

Limitations of VIGS

It must be noted that, like all other techniques, VIGS also has its limitations. Above all, it can only induce a transient silencing response. In almost all cases, we observe an eventual recovery of the host from the viral infection and subsequent replenishment of gene transcript levels. Due to the nature of the silencing phenomenon, even during the peak silencing stages, one cannot be assured of complete knockout of transcript levels. In spite of using weak or attenuated viruses, it is not possible to avoid some of the host gene expression changes that are associated with the viral infection and these factors must be taken into consideration while interpreting the results. Ideally, therefore, VIGS should be used in conjunction with or validated by other techniques of gene function analysis. The use of recombinant viruses also requires greater precautionary measures and care must be taken to avoid accidental transmission of the modified or infectious virus.

Conclusions

With the sequencing of numerous genomes completed or near completion, we have in our hands a vast amount of genetic data that is to be deciphered, genes that need to be characterized, and pathways that are to be elucidated. The need of the hour is for powerful functional analysis tools and in the last few years VIGS has proven to be the breakthrough technology that aids in rapid and robust gene characterization. Plant gene analysis has traditionally been carried out in few model systems whose genomes have been sequenced and are also easily amenable to transformation techniques. VIGS is one of the few tools that can be applied on a broad spectrum of plants, since most plants are susceptible to viral infections. VIGS will therefore play a crucial role in promoting gene analysis in many nonmodel systems including those that have duplicated genomes or do not have a sequenced genome. A plethora of new viruses are being added to the list of VIGS vectors and it is exciting to note the addition of vectors for monocots like rice, wheat, and maize. Small grain cereals are among the world's most important food crops but gene discovery and annotation in these systems have been carried out on only a small scale mainly due to the lack of tools for reverse genetics. The adaptation of VIGS in monocots will therefore play an important role in gene analysis and annotation in these economically important crops.

Within the animal kingdom, RNA-silencing technology has been harnessed to carry out genomewide silencing screens. In human cells, *Drosophila*, and *Caenorhabditis elegans*, dsRNA or hpRNA's designed to systematically target a majority of predicted genes have been screened to identify novel players in cell growth, proliferation, and cell death. We now have all the tools necessary to carry out similar whole genome functional discovery studies in plants. The successful application of VIGS in two completed plant genome sequences, *Arabidopsis* and rice, opens up the possibility of carrying out systematic loss-of-function studies in these systems. Many plant genomes have been partially annotated in the form of expressed sequenced tags (EST) libraries, which serve as inventories of expressed genes. These EST libraries can also serve as ideal resources for use in VIGS-mediated high-throughput screening. In the postgenomics era, VIGS will prove to be a technique that will help accelerate the conversion of genomic data to functionally relevant information and contribute to our understanding of the molecular processes occurring within plants.

See also: Plant Antiviral Defense: Gene Silencing Pathway.

Further Reading

Baulcombe DC (1999) Fast forward genetics based on virus-induced gene silencing. *Current Opinion in Plant Biology* 2: 109–113.

Burch-Smith T, Miller J, and Dinesh-Kumar SP (2003) PTGS Approaches to large-scale functional genomics in plants. In: Hannon G (ed.) *RNAi: A Guide to Gene Silencing*, p. 243. New York: Cold Spring Harbor Laboratory Press.

Burch-Smith TM, Anderson JC, Martin GB, and Dinesh-Kumar SP (2004) Applications and advantages of virus-induced gene silencing for gene function studies in plants. *Plant Journal* 39: 734–746.

Carrillo-Tripp J, Shimada-Beltran H, and Rivera-Bustamante R (2006) Use of geminiviral vectors for functional genomics. *Current Opinion in Plant Biology* 9: 209–215.

Dinesh-Kumar SP, Anandalakshmi R, Marathe R, Schiff M, and Liu Y (2003) Virus-induced gene silencing. *Methods in Molecular Biology* 236: 287–294.

Ding XS, Rao CS, and Nelson RS (2007) Analysis of gene function in rice through virus-induced gene silencing. *Methods in Molecular Biology* 354: 145–160.

Lu R, Martin-Hernandez AM, Peart JR, Malcuit I, and Baulcombe DC (2003) Virus-induced gene silencing in plants. *Methods* 30: 296–303.

Robertson D (2004) VIGS vectors for gene silencing: Many targets, many tools. *Annual Review of Plant Biology* 55: 495–519.

Virus Particle Structure: Nonenveloped Viruses

J A Speir and **J E Johnson,** The Scripps Research Institute, La Jolla, CA, USA

© 2008 Elsevier Ltd. All rights reserved.

Introduction

Nonenveloped viruses provide model systems for high-resolution structural study of whole virus capsids (and their components) and the principles of large-scale nucleoprotein tertiary and quaternary interactions. In many cases, they are relatively small, stable, and highly symmetric particles that can be purified as homogeneous samples in large quantities. These properties make them suitable for the techniques of X-ray crystallography, X-ray fiber diffraction, and cryoelectron microscopy (cryo-EM) combined with image reconstruction. Enveloped virus particles generally have greater flexibility and asymmetry, two features that inhibit application of high-resolution techniques.

The first crystal structures of intact virus capsids published in the late 1970s and early 1980s are examples of the simplest forms of replicating organisms. They were small, spherical, positive-sense single-stranded (ss) RNA plant viruses about 170–350 Å in diameter with $T=1$ or $T=3$ symmetry, which were easy to purify in gram quantities. By the mid- to late 1980s, the structures of small ssRNA animal viruses (poliovirus and the common cold virus, rhinovirus) and the intact, 3000 Å-long ssRNA helical tobacco mosaic virus (TMV) were also completed. The range of virus types, sizes, and complexity addressed by X-ray crystallography has grown significantly with advances in X-ray technology and computing. Detailed polypeptide and nucleic acid models have now been fitted into data between 4–1.8 Å resolution. Crystal structures of 800 Å diameter virus capsids with $T = 25$ icosahedral symmetry have now been determined and show quaternary polypeptide interactions as intricate and varied as their biology.

Spherical particles larger than 800 Å in dimension have been studied by cryo-EM providing detailed images of some of the most complex structures ever visualized at high resolution. This technique is also readily applied to smaller viruses like those described above. Typical cryo-EM studies produce 15–30 Å resolution image reconstructions, but under ideal conditions sub-nanometer resolution can be obtained. The upper diameter limit for applying the cryo-EM technique to viruses is yet to be hit; however, with increased sample thickness (e.g., particle diameter), comes an increase in data collection and analysis restrictions. Currently, the two largest spherical virus structures determined with cryo-EM are the 2200 Å diameter PpV01 algae virus, that has an icosahedral protein shell with $T = 219$ quasi-symmetry, and the 5000 Å diameter mimivirus, found in amoebae, that has an icosahedral protein shell with estimated $T = 1179$ quasi-symmetry. While the PpV01 structure was determined at 30 Å resolution, the thickness of ice needed to embed mimivirus (7500 Å for shell + attached fibers) limited the image reconstruction to 75 Å resolution, which is the reason for the estimated T number. Thus, cryo-EM can provide detailed images of symmetrical virus particles of almost any size and is an increasingly valuable tool in virus structural studies.

In cases where capsid size or other factors prohibit obtaining near-atomic resolution data, structures of the individual proteins or their assembly products can be determined by nuclear magnetic resonance (NMR) and/or X-ray crystallography. Capsid subunit folds, nucleic acid-binding sites, and some details of quaternary structure can be gathered in this way. Exciting leaps in understanding virus capsid assembly, interactions, and dynamics have come from combining individual structures with cryo-EM particle images to build complete pseudo-atomic models for the larger, more-complex virus capsids. The combined approach generates discoveries that could not have been obtained from either method alone. This article highlights some of the important features of the more than 200 high-resolution nonenveloped virus structures determined to date.

The Building Blocks

The β-Barrel Fold

The earliest spherical virus capsid structures, both of animal and plant viruses, revealed a strikingly common fold for the structural subunits that continues to be found in new virus structures: the antiparallel β-barrel (**Figure 1**). It has been found mainly in icosahedral particles, but is also known to form the bacilliform particles of alfalfa mosaic virus (AMV). The barrel is formed by two back-to-back antiparallel β-sheets with a jelly-roll topology defining a protein shell approximately 30 Å thick. The β-barrels position in the capsid shells in a variety of orientations with the insertions between β-strands having a great range of sizes, giving each virus the ability to evolve a distinct structure, function, and antigenic identity. Indeed, the structure of nudaurelia capensis omega virus (NωV), an insect tetravirus, revealed an entire 133-residue immunoglobin-like domain inserted between strands E and F that forms a major portion of the particle surface. Just as prevalent, the very N- and C-termini of the capsid proteins are often extended polypeptides that do not form part of the barrel. Those portions of the extended termini close to the barrel can have ordered structure to various degrees depending on the local (quasi-)symmetry environment (see **Figure 1**), and tend to engage in varied types of interactions, such as binding one or more nearby subunits and/or nucleic acid, that play critical roles in particle assembly, stability, and nucleic acid packaging.

Other Folds

While the β-barrel remains a common element in a wide range of virus structures, other capsid subunit folds have been discovered in both enveloped (not shown) and nonenveloped particles (**Figure 2**). The nonenveloped subunits also form icosahedral shells as well as helical rods. The structure determination of the $T = 3$, ssRNA bacteriophage MS2 in 1990 revealed the first new fold for an icosahedral virus capsid subunit. It consists of a five-stranded β-sheet that faces the inside of the particle, with a hairpin loop and two helices facing the exterior. The only structural similarity to the particles formed by β-barrels was their assembly into a $T = 3$ quasi-icosahedral shell. Just 10 years later, the head (capsid) structure of the $T = 7$, dsDNA tailed bacteriophage HK97 showed another new capsid subunit fold, and the same fold has also been found in bacteriophage T4 capsids. Other than the high probability that a capsid subunit with a β-barrel is going to form an icosahedral particle, no other general themes can be derived from the variety of subunit folds to predict particle size, quasi-symmetry, or host specificity.

Completely Symmetric Capsids

T = 1 Icosahedral Particles

The simplest of the spherical particles, the $T = 1$ capsids, have only one protein subunit in each of the 60 positions related by icosahedral symmetry. This has been the symmetry found in three of the four known encapsulated satellite viruses, one of which is the 180 Å diameter,

Virus Particle Structure: Nonenveloped Viruses

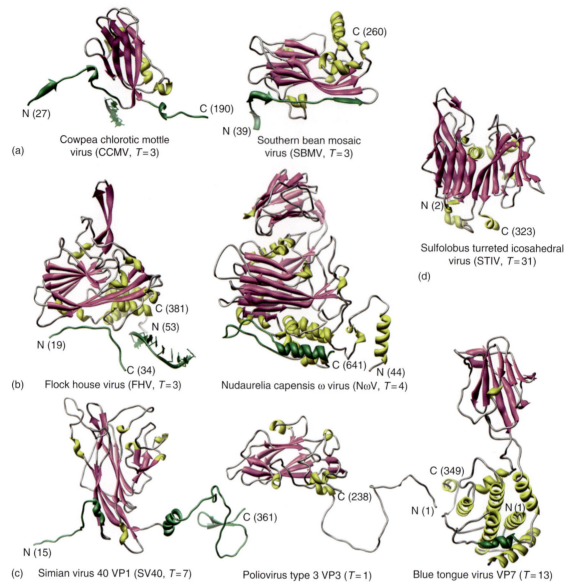

Figure 1 Structure of (a) plant, (b) insect, (c) vertebrate, and (d) archaea virus protein subunits that assemble into icosahedral shells. The name of the virus appears below the corresponding protein subunit along with the capsid triangulation number. The N- and C-termini are labeled with the residue numbers in brackets. The N-termini (and sometimes the C-termini) usually consist of highly charged residues associated with the interior of the virus particle and/or the packaged nucleic acid. As such, they do not follow icosahedral symmetry and are regularly disordered (invisible) in the final structures. Many virus subunit structures determined to near atomic resolution have the β-barrel fold and/or insertions with nearly all β-structure (colored magenta). Multiple copies (from 180 to 900) of the single subunit shown for each virus, except for that of poliovirus, form the entire icosahedral protein shell. Assembly of icosahedral virus particles with more than 60 subunits requires quasi-symmetric interactions often involving subtle to extensive differences in structure at the subunit N- and C-termini. The subunit regions involved in quasi-symmetric interactions critical to virion structure and assembly are colored green (only a single variation is shown for each virus; not yet known for STIV). The 'switch' in structure between identical subunits is a response to differences in the local chemical environment, defined by the number of subunits forming the icosahedral shell, in order to maintain similar bonding between neighboring subunits. The structural variations include the presence or absence of highly ordered RNA structure (green stick models) in FHV and CCMV. Poliovirus utilizes multiple copies of two additional subunits highly similar to VP3 to form a complete virion. Thus, there is no quasi-symmetry in poliovirus (note the absence of any green highlights) since neighboring subunits are different proteins.

ssRNA satellite tobacco mosaic virus (STMV) particle (**Figures 3(a)** and **3(b)**). The small, compact STMV capsid subunits (159 a.a.) have the β-barrel fold with the β-strands running roughly tangential to the particle curvature with the B–C, D–E, F–G, and H–I turns pointing toward the fivefold axes. This is a common subunit orientation and will be simply called 'tangential' going forward. The turns between strands are all short (no insertions),

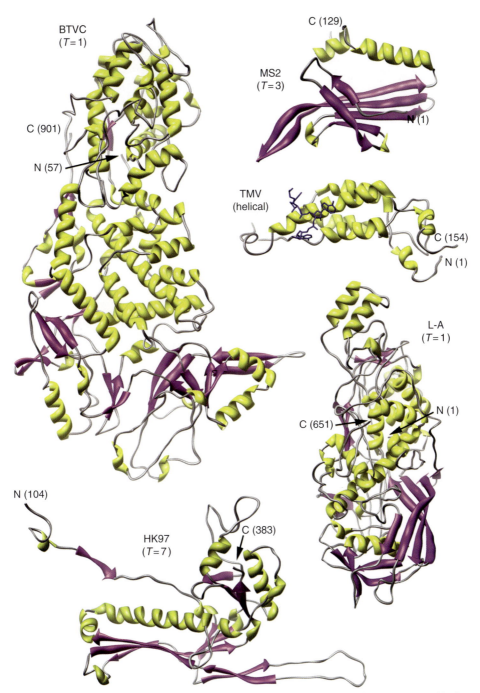

Figure 2 Structure of capsid subunits with folds other than the β-barrel. The name of the virus appears next to the corresponding protein subunit along with the capsid triangulation number (if applicable). The N- and C-termini are labeled with the residue numbers in brackets. With the exception of MS2, helical structure is dominant, with some β-sheet formations mixed into the folds. The subunit of the helical TMV rod is mostly composed of a four antiparallel α-helix bundle that forms the core of the rods, but has a small β-sheet away from the axis that forms the rod surface. The three bound nucleotides per TMV subunit are shown as a blue stick model (see the section titled 'Helical particles'). Both the BTVC (inner shell) and L–A subunits form $T=1$ particles with two copies of the subunit in the icosahedral asymmetric unit (discussed in text), but do not share any detectable structural homology. The BTVC inner shell subunit is the largest capsid subunit determined to near-atomic resolution.

with only an extended N-terminus jutting 60 Å from the barrel that helps form both the threefold and twofold symmetry related interactions (trimers and dimers) via hydrogen bonds and extension of antiparallel β-sheets.

The strongest inter-subunit contacts occur at the capsid subunit dimers, where a large number of both hydrophobic and hydrophilic interactions form across a large buried surface between the subunits. The interior surface of the

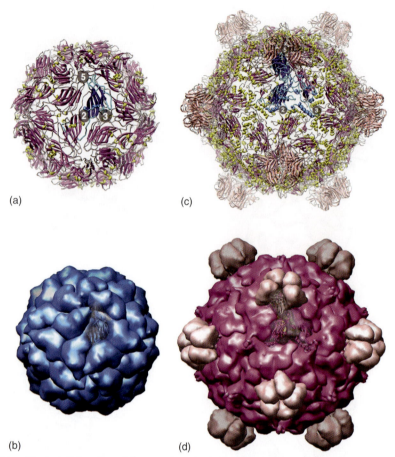

Figure 3 Structures of two $T=1$ particles viewed down an icosahedral twofold rotation axis (approximately to scale). STMV is displayed in (a) and (b), and ΦX174 in (c) and (d). The top images show all 60 capsid subunits rendered as ribbon drawings (β-strands – magenta or pink; helices – yellow), with the icosahedral asymmetric unit (IAU) shown all in blue. The fivefold, threefold, and twofold icosahedral axes are labeled with semitransparent gray circles containing the appropriate rotation number in white. The images are depth-queued such that features further away fade into the white background. The back of the particles are purposely faded (invisible) to make a clearer representation. The bottom images are surface representations of the same structures above, with the IAU shown as a transparent surface to show the ribbon drawing of the capsid subunits. Note the more compact look of STMV due to the close approach of complementary surfaces on the β-barrels vs. the more spread-out look of ΦX174 due to its use of extended loops to create a contiguous capsid. The exterior of the ΦX174 shell includes the G-protein (pink subunits), which surrounds hydrophilic channels at the fivefold axes, and has a β-barrel fold that is oriented nearly parallel to the fivefold axes such that they form 12 broad spikes on the particle surface. This will be called the 'axial' orientation. This increases the particle dimensions at these axes by 60 Å but does not contribute to the enclosed volume as they sit atop an essentially complete protein shell. (b, d) Courtesy of the Viperdb web site.

STMV capsid dimers binds a large section of ordered double-helical RNA. Two additional symmetry-related subunits interact with the RNA via their extended N-termini. The protein–RNA interactions are split between direct side-chain bonds to the ribose sugars, and water-mediated hydrogen bonds by residues of both the local subunit dimer β-barrels and N-termini. Altogether the ordered RNA accounts for about 59% of the packaged nucleic acid. Although STMV is one of the simplest icosahedral viruses, its features demonstrate many of the basic protein–protein and protein–RNA bonding schemes that have been observed in larger and more complex particles.

While an advantage of the $T=1$ capsid is its equivalent subunit environments, its major limitation is the small particle volume. To create larger shells, the subunits either have to be larger or there have to be more of them. The former is seen in the crystal structure of the ssDNA bacteriophage ΦX174 capsid (**Figures 3(c)** and **3(d)**). Its 426 a.a. F proteins form a 340 Å diameter $T=1$ icosahedral particle, which has nearly fourfold greater volume compared to the satellite virus capsids. Each subunit has the β-barrel fold, but it accounts for only one-third of the F protein. The remainder of the capsid subunit is composed of long loops inserted between the β-strands that make up the majority of inter-subunit interactions and capsid surface. ΦX174 has additional capsid proteins, but the contiguous protein shell is made up of only the F protein. While a larger subunit can create a viable amount of enclosed volume, native $T=1$ particles are uncommon. This may be due to inefficiencies in dealing with the larger amount of genome dedicated to the major capsid protein and the more complex structural requirements of a single large subunit to build a stable and functional particle.

Helical Particles

The only high-resolution structures of helical viruses are from three members of the genus *Tobamovirus*, whose type member is TMV. The tobamoviruses form rod-shaped particles 3000 Å long and 180 Å in diameter, with a central hole about 40 Å across. The particles are built from over 2100 identical capsid protein subunits that are 158 a.a. in size, and arranged in a right-handed helix with 49 subunits in every three turns and a pitch of 23 Å. The subunit has a narrow, elongated structure composed of a four-helix bundle on one end, and a small β-sheet on the other (**Figure 2**). The subunits sit perpendicular to the particle axis, with the helix end of the subunit nearest the center and the β-sheet end forming the outer surface of the rod. Neighboring subunits stack upon one another in the particle helix such that two helices of each protein interact with two others from the neighboring subunit, forming another four-helix bundle to build the rod.

The structure of the TMV particle determined in the late 1980s revealed the first details of virus capsid protein interactions with nucleic acid. A long loop inserted between two of the subunit helices forms the interior channel of the particle that includes a continuous RNA binding site at the edge of the helices. The entire RNA genome is well ordered in the TMV structure by conforming to the helical symmetry of the coat protein stacking. There are three nucleotides bound to each protein subunit (**Figure 2**) over the entire length of the rod, which adds subunits during assembly until the entire genome is encapsidated. The RNA is bound at its phosphate groups by direct, calcium-mediated, and probably water-mediated interactions with nearby arginine and aspartic acid residues, and at its sugars and bases by hydrophobic complementary, polar interactions, and a few hydrogen bonds from local arginine and aspartic acid residues. This mode of nucleic acid binding showed that, since the viral genome sequences do not share the symmetry of their enclosing particles, interactions with the coat protein are complementary to the nucleic acid, but non-base-specific to ensure that the entire genome is encapsidated. The same general principles continue to be found in the ordered protein–nucleic acid interactions observed in icosahedral particle structures.

Quasi-Symmetric Capsids

T = 3 Icosahedral Particles

Icosahedral particles built from multiples of 60 identical capsid subunits exist in great numbers, and are arranged with many variations on the quasi-symmetry first predicted by Caspar and Klug in 1962. In one of the closest matches to their predictions, 90 coat protein dimers (180 total coat proteins) of the ssRNA plant cowpea chlorotic mottle virus (CCMV) assemble with nearly ideal $T = 3$ quasi-symmetry into the shape of a 280 Å diameter truncated icosahedron, which is defined by planar pentamer (formed by A subunits) and hexamer (formed by B and C subunits) morphological units (capsomers) at the fivefold and threefold (quasi-sixfold) axes, respectively (**Figure 4**, top). The 190 a.a. capsid subunit folds into a small β-barrel with short loops between strands, but has both extended N- and C-termini (**Figure 4**, bottom left). All 180 subunits have the axial orientation with the N-termini extending toward the center of the pentamers and hexamers, and the C-termini extending outward into neighboring capsomers by reaching across the twofold axes. The very ends of the C-termini are clamped in between the β-barrel and N-terminal extension of the twofold related subunit, forming the largest number of interactions between capsid subunits and tying the capsomers together. This occurs at both the twofold and quasi-twofold axes with high fidelity. The dimer interactions also define the dihedral angles between the planar capsomers; the angles between hexamer–hexamer (twofold) and pentamer–hexamer (quasi-twofold) interfaces only differ by 4° (**Figure 4**, bottom right), producing a smooth particle curvature. A triangular unit of A, B, and C subunits defines the icosahedral asymmetric unit (IAU), and also a high-fidelity quasi-threefold axis. Calcium and RNA binding sites are preserved at all three interfaces (calcium site exterior, RNA site interior, and under the calcium site) with nearly equivalent positions of the coordinating side chains.

The pentamer and hexamer units show more differentiation. Besides subtle contact changes between the β-barrels, the N-termini of the hexamer subunits form a nearly prefect sixfold symmetric parallel β-tube about the threefold axes in an impressive display of quasi-sixfold interactions. The N-termini of the pentamer subunits extend a bit toward the fivefold axes but form no structure due to disorder after residue 42. Interestingly, deletion of the β-hexamer in one study did not prevent hexamers from forming, but did destabilize the particle. Other than the differences seen at the capsomer centers, the remainder of the interactions between the A, B, and C subunits are subtly changed to accommodate the differences in chemical environments, rather than switched from one form to another, and may explain the highly pleomorphic properties of the CCMV capsid subunit.

T = 4 Icosahedral Particles

The ssRNA insect tetraviruses remain the only known viruses that have a nonenveloped, $T = 4$ capsid. The crystal structure of the 400 Å diameter NωV capsid in

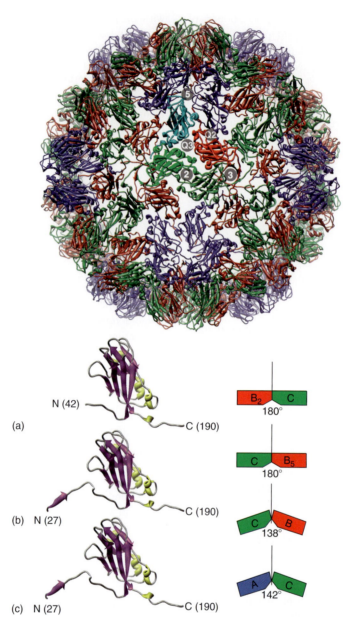

Figure 4 Structure of the $T=3$ CCMV particle viewed down an icosahedral twofold axis. Icosahedral and quasi-symmetry (Q) axes are labeled as in **Figure 3**. The quasi-sixfold axes are not labeled as they coincide with the icosahedral threefold axes. (top) The entire CCMV particle with A-subunits colored blue, B colored red, and C colored green. An IAU is shown above the twofold axis in brighter colors and larger ribbons. The calcium sites between quasi-threefold related subunits are strongly implicated in CCMV particle expansion, which occurs in the absence of divalent metal ions at high pH. The CCMV crystal structure shows that under these conditions the charged residues of the calcium-binding site would cause a repulsion between quasi-threefold related subunits, and a cryo-EM structure of the swollen form of CCMV confirmed the opening of large holes at the quasi-threefolds. Indeed, swollen CCMV breaks apart into subunit dimers and RNA in high salt concentrations, and under various solution conditions, can reassemble into a number of structures including $T=1$, $T=3$, and $T=7$ particles, long tubes, multi-shelled particles and tubes, and laminar plates. It is one of the most pleomorphic viral subunits ever studied, but amazingly only makes $T=3$ particles in plant infections. (Bottom left) Individual structures of the proteins in the A, B, and C subunit environments colored as in **Figure 2**. Note they are closely identical except for the lack of extended N-terminal structure in A, and slightly varying directions taken by the C-termini away from the β-barrels (not readily visible in these views). (Bottom right) Diagrammatic representation of the flat and bent subunit/capsomer interfaces. Individual blocks represent the subunits as viewed tangentially to the particle curvature. Colors and letters designate the subunit type with the dihedral angle between them given underneath.

1996 showed the 644 a.a. capsid subunit has three domains: an exterior Ig-like fold, a central β-barrel in tangential orientation, and an interior helix bundle (**Figure 5**, right side). The helix bundle is created by the N- and C-termini of the polypeptide before and after the β-barrel fold, and the Ig-like domain is an insert between the E and F β-strands. All 240 of the NωV capsid subunits undergo autoproteolysis at residue 571 after assembly,

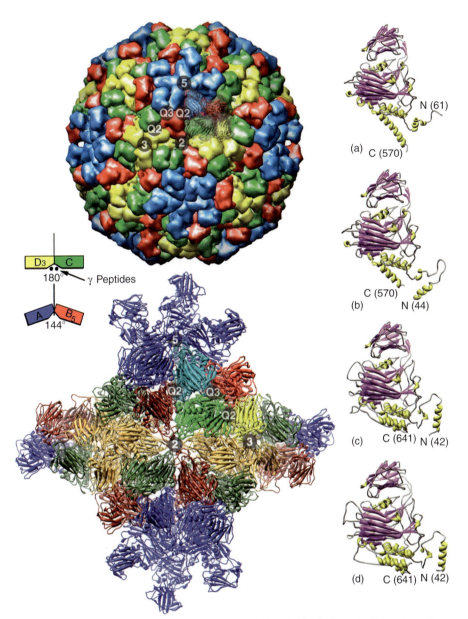

Figure 5 Structure of the $T=4$ NωV particle viewed down an icosahedral twofold axis. Icosahedral and quasi-symmetry (Q) axes are labeled as in **Figure 3**. The low-fidelity quasi-sixfold axes are not labeled as they coincide with the icosahedral twofold axes. (Right) Individual structures of the proteins in the A, B, C, and D subunit environments colored as in **Figure 2**. While the basic folds and domains are similar, they differ considerably at both the N- and C-termini, in loop structure at one end of the barrel, and slightly in position and orientation of the Ig domains. In particular, the C- and D-subunits have additional ordered polypeptide in the helical domain that dictates switching between bent and flat contacts in the capsid architecture (see below and text). (Left) Surface representation of the entire $T=4$ particle (as in **Figure 1**) and a more detailed ribbon representation of the area between the four fivefold axes surrounding the center twofold axis. A diagrammatic representation of the subunit/capsomer interfaces is also shown as in **Figure 3**. Complete pentamers are only shown on the top and bottom of the ribbon image. The A-, B-, and C-subunits are colored as in **Figure 3**, and the D-subunits are colored yellow. An asymmetric unit is shown above and to the right of the twofold axis in brighter colors and larger ribbons (also shown in surface representation above as ribbons). The $T=4$ arrangement can be thought of as two sets of triangles: one set is formed by the quasi-threefold related A-, B-, and C-subunits, and the other is formed by the icosahedral threefold related D-subunits. The dihedral angles of the interfaces between two ABC units (144°), and between ABC and DDD units (180°; both are quasi-twofold related contacts) creates flat triangles with fivefold axes at the corners, and bending between them along the lines from fivefold to twofold axes. Surface image courtesy of the Viperdb web site.

leaving the cleaved portion of the C-terminus, called the γ-peptide, associated with the capsid. The fold and tangential orientation of the β-barrel together with the autoproteolysis revealed a strong relationship between tetraviruses and the $T=3$ insect nodaviruses. Indeed, the folds and catalytic sites superimpose with little variation, and a γ-peptide also exists in the nodaviruses. Unlike in CCMV, the quasi-symmetry of these insect particles is obviously controlled by a molecular switch (**Figure 5**, left side). In NωV, a segment of the γ-peptide (residues 608–641) is only ordered in the C and D subunits, and functions as a wedge between the ABC and DDD units to prevent curvature and create a flat contact. The interface between two ABC units is bent partly due to the lack of ordered γ-peptide. A similar situation exists in the nodaviruses, except that the ordered polypeptide between twofold related C-subunits in the $T=3$ particle comes from the A-subunit N-terminus, and ordered duplex RNA also fills the interface. The structural information from the tetravirus group has defined important relationships and some common themes in the biology of the insect viruses.

T = 7 Icosahedral Particles

All three $T=7$ capsid structures determined at near atomic resolution are dsDNA viruses with unique features: two mammalian polyomaviruses, SV40 and polyoma virus, and the head of the λ-like tailed bacteriophage HK97 from the family *Siphoviridae*. The 500 Å diameter SV40 capsid is constructed of 72 pentamers arranged with $T=7d$ quasi-symmetry, not 12 pentamers and 60 hexamers as predicted by quasi-equivalence theory. Thus, 60 of the SV40 pentamers are actually in hexavalent environments as they are located in positions that are occupied by hexamers in an ideal $T=7$ lattice. As the valency of a true $T=7$ lattice is present, it was a surprise to discover it could be achieved without the expected number of identical subunits or domains (i.e., five subunits instead of six). The pentamers contain five copies of the 364 a.a. VP1 protein, which have the axial β-barrel orientation and extended C-terminal arms. Interlocking secondary structures of the G-strands and DE loops with clockwise neighbors form the intimately associated pentamers. The C-arms of all five subunits in each pentamer reach outward to one of three specific types of neighboring pentamers depending on the local environment (threefold and two types of twofold clusters), and form another β-strand between the N-terminus and BIDG sheet of the neighboring subunit (a similar clamp to that of CCMV). This ties the neighboring pentamers together in a general way determined by the position and valency of each pentamer. Remarkably, the VP1 proteins and clamp structures in the pentamer cores are structurally identical, and only the direction and ordered structure of the C-terminal arms are changing as they switch contacts between the nonequivalent pentamers.

By comparison, the 650 Å diameter HK97 capsid is a true $T=7l$ icosahedron, composed of 420 subunits arranged in 12 concave pentamers and 60 planar hexamers. Strikingly, the 385 a.a. protein subunit has a new virus fold (**Figure 6**). Each subunit forms two covalent isopeptide bonds (cross-links) with neighboring subunits on opposite sides of the protein to create topologically linked protein rings arranged with icosahedral symmetry, also a new find in protein structure studies. Notably, the cross-links are formed between subunits that encircle the pentamers and hexamers (one from each neighboring capsomer), not between the subunits forming them. This defines how their complex interweaving occurs at the particle threefold and quasi-threefold axes (**Figure 6**). The fact that this is the final mature state of the HK97 head particle, and that the cross-links are not yet formed in particles at the early to mid-stages of the capsid maturation process, has led to informative studies on capsid dynamics using HK97 as a model system. The final step of HK97 maturation involves formation of the final 60 cross-links in the transition from expansion intermediate IV (EI-IV) to the mature capsid (head II). The pentons appear to be dynamic, oscillating up and down relative to the contiguous capsid shell, until cross-linked. See the article 'Principles of Virus Structure' for animations of the entire transition from the prohead state through head II.

T = 13 Icosahedral Particles

The 700 Å diameter dsRNA blue tongue virus core (BTVC) particle remains the largest molecular structure ever determined by X-ray crystallography to near-atomic resolution. The core particle has two shells: an inner $T=1$ particle, composed of 120 large (901 a.a.) VP3 proteins (two in each icosahedral asymmetric unit), which packages a few RNA polymerases and 10 strands of dsRNA, and an outer $T=13l$ particle composed of 780 VP7 proteins (349 a.a.), for a total of 900 capsid proteins and a molecular weight of over 54 000 000 Da. The VP7 subunits form a trimer in solution. The trimer structure was determined by X-ray crystallography and combined with cryo-EM images of BTVC to determine the entire BTVC crystal structure. The VP7 subunit has two domains, an outer β-barrel and an inner helix bundle (**Figure 1**), both of which have extensive contacts in the trimer. In an interesting twist, the β-barrel is actually an insert between two helices in the linear VP7 sequence, and the helix bundle is responsible for the majority of interactions between trimers in the assembled shell. The VP7 trimers are postulated to 'crystallize' or assemble on the inner shell's surface into the 12 pentamers and 120 hexamers, which follow quasi-symmetry with high fidelity having very little differences in their intercapsomer interactions. In contrast, the inner $T=1$ shell breaks from quasi-equivalence theory by using large-scale flexing of the two huge VP3 proteins to generate similar shapes of interactions in the absence of similar contacts, and build a nearly featureless hydrophobic surface for interaction with VP7. Together, these shells present a dazzling array

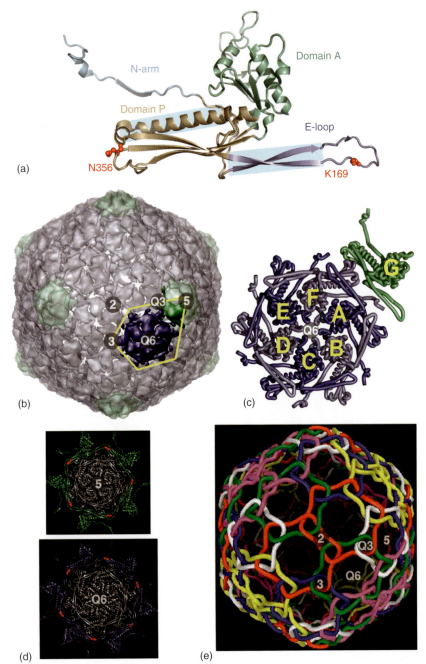

Figure 6 Structure of the $T=7$ HK97 particle. Icosahedral and quasi-symmetry (Q) axes are labeled as in **Figure 3**. (a) HK97 subunit fold. The small HK97 subunits have mixed α/β structure, forming two domains in the shape of an L that have a continuous hydrophobic core, and placing the two residues involved in cross-linking other subunits (red) on opposite sides of the protein. (b) Surface representation of the entire HK97 capsid viewed down an icosahedral twofold axis, with pentamers colored separately from hexamers. The IAU is outlined in yellow. Unlike the $T=3$ and $T=4$ structures, none of the hexamers sit on icosahedral axes. (c) Ribbon diagram of the seven subunits in the IAU showing how they align such that the P-domains and E-loops overlap to define the sides of the local capsomers, but place their tips (with the cross-linking residues) outside the capsomer to be accessible to neighboring subunits. (d) The two types of cross-linked subunit 'circles' or 'rings' in HK97. The surrounding cross-linked subunits (one from each neighboring capsomer) are shown in green (5-circle) or blue (Q6-circle), with the cross-link shown in red. (e) Diagrammatic representation of the interlocking cross-links in HK97. The 5- or Q6-circles, now with subunit and cross-links just represented by tubes, are colored separately showing how they form molecular chainmail by interweaving at the icosahedral and quasi-threefold axes. This molecular chainmail stabilizes the unusually thin 18-Å-wide protein shell (only half the width usually formed by capsids composed of β-barrel subunits), which can be described as a protein balloon. The capsids are also stabilized by other extensive inter-subunit interactions with up to nine other coat proteins. The N-arm, which is not involved in the cross-links, extends 67 Å away from the subunit to contact subunits in neighboring capsomers. (a–d) Courtesy of Dr. Lu Gan. (e) Courtesy of Gabe Lander.

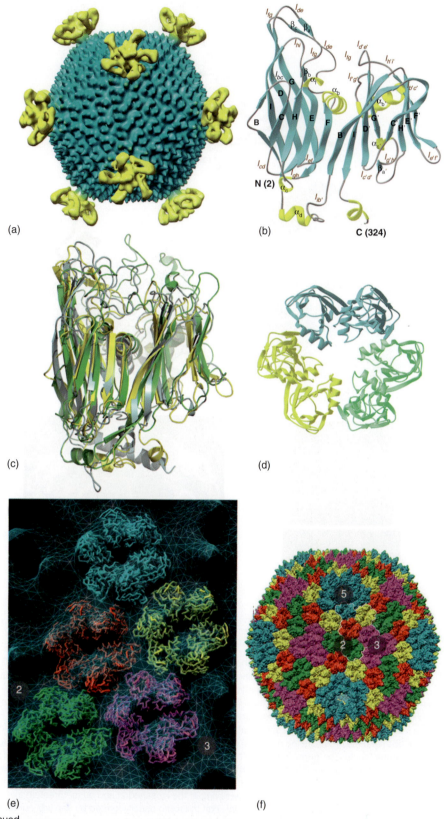

Figure 7 Continued

of protein interactions that go from near perfect conformity with quasi-equivalent theory to defining a new category of viral subunit interactions in the $T=1$ particle.

$T=25$ and Larger Icosahedral Particles: The Double-Barrel Subunit

Virus capsids having to assemble with $T=25$ quasi-symmetry or higher have avoided the enormous challenge of accurately placing 1500 or more individual proteins in a shell by reducing the problem to just a few related hexon locations. The crystal structure of the hexon from the dsDNA, $T=25$ adenovirus capsid revealed it is a trimer of three identical 967 a.a. proteins. Each protein has two axially oriented β-barrel domains. In the trimer, the barrels are almost precisely related by a pseudo-sixfold axis such that each forms a corner of the hexon base. The hexon is the building block for the particle and is extremely stable due to interpenetrating polypeptides in the trimer. Thus, a stable trimer satisfies the hexavalent lattice positions in $T=25$ adenovirus as the stable pentamers do in $T=7$ SV40. Combined with cryo-EM images of the capsid, these structures revealed that there are only four different hexon-binding sites in adenovirus (pentons have a different subunit), and that much larger hexon facets can be constructed in this manner. In a recent series of groundbreaking discoveries that surprised virus researchers, a combination of both cryo-EM and crystallography methods revealed that the hexon structures of large dsDNA viruses in all domains of life have similar 'double-barrel trimers' forming pseudo-hexagonal shapes. These include adenovirus (eukarya), $T=25$ PRD1 (bacteria), $T=169$ paramecium bursaria chlorella 1 virus (PBCV-1, green algae), and, most recently, $T=31$ sulfolobus turreted icosahedral virus (STIV, archaea) (**Figure 7**). When the double-barrel subunits from all four viruses are superimposed, over 46% of their polypeptides are within 2.2 Å r.m.s.d. of each other. For the first time, these apparently complicated and unrelated viruses infecting quite different hosts have been solidly linked by their structures, supporting the hypothesis that at least one virus lineage originated before life separated into three domains.

Bacilliform Particles

The ssRNA plant alfalfa mosaic virus (AMV), a member of the family *Bromoviridae* that includes CCMV, forms four types of bacilliform particles composed of one of the four genomic RNAs and a single 220 a.a. capsid subunit. The particles are 180 Å in diameter and range in length from 300 to 570 Å. The spherical ends have $T=1$ symmetry, while the cylindrical bodies have a repeating sixfold symmetry with large holes at the hexamer centers. After a mild trypsin treatment and removal of RNA, the capsid subunit readily forms $T=1$ particles that crystallize and their structure has been determined. The $T=1$ particle is 220 Å in diameter with large holes 35 Å across at the fivefold axes. The subunit is a simple β-barrel in the axial orientation with extended termini at the base. The C-termini reach across the twofold axes to hook between the N-terminus and β-barrel of the symmetry-related subunit in a strikingly similar fashion to the structure of the CCMV particle. Thus, this structure of the AMV endcaps helps demonstrate how the highly pleomorphic capsid proteins of bromoviruses can assemble into several particle forms via strong but flexible dimer contacts depending on solution conditions.

Psuedo-Symmetric Capsids

$P=3$ Icosahedral Particles

Instead of having three copies of a single protein in slightly different chemical environments, a number of viruses have evolved with two or three different proteins in each environment that have unique primary structures but similar folds. The coat protein arrangements are similar to that proposed for $T=3$ lattices such that they are given the $P=3$ (pseudo $T=3$) designation. Unlike the single protein in T capsid arrangements, the different

Figure 7 Structure of the $T=31$ STIV capsid and major coat protein (MCP). (a) The cryo-EM image reconstruction of the entire STIV capsid viewed down an icosahedral twofold axis. The shell formed by 900 copies (300 trimers) of the MCP is shown in blue, and the 200 Å tall turrets that compose the fivefold axes are shown in yellow. (b) Crystal structure of the MCP, with β-strands colored cyan and helices yellow. Like other large dsDNA viruses, the protein folds into two covalently linked β-barrels, one shown on the left and the other on the right. Interestingly, the large capsid subunit of the plant comoviruses is the only one outside the large dsDNA virus families with the double-barrel fold. (c) Superposition of the major coat proteins from STIV (yellow), PRD1 (cyan), and PBCV-1 (green) in approximately the same orientation as (b), showing the closely similar double-barrel folds. (d) Model of the STIV MCP trimer, based on that seen in the other dsDNA viruses. Each protein is colored separately, and each barrel makes the corner of one of the five unique hexon units seen in the assembled $T=31$ virus. (e) Top view of the STIV asymmetric unit from the cryo-EM structure (shown as a blue lattice) with Cα traces of the MCP trimer fitted in the hexon positions. Icosahedral axes are labeled as in previous figures, with the blue hexon just below the fivefold axis. The highly correlated fit between the atomic model and cryo-EM reconstruction reveals possible inter-subunit interactions, quasi-symmetry, and how the main protein shell is constructed, shown in (f). (f) Pseudo-atomic model of the complete MCP shell using the same color scheme as in (e), with the asymmetric unit outlined in white. Note the modularity of the hexons, which theoretically can be exploited to form capsids of extremely large size such as the 5000 Å diameter mimivirus. Courtesy of Dr. Reza Khayat et al., as published in Khayat R, Tang L, Larson ET, et al. (2005) Structure of an archaeal virus capsid protein reveals a common ancestry to eukaryotic and bacterial viruses. *Proceedings of the National Academy of Sciences, USA* 102: 18944–18949.

capsid proteins in P arrangements are free to mutate, function, and form interactions independent of one another but more of the genome must be used to encode multiple capsid proteins. Two well-studied families with $P = 3$ capsids are the ssRNA animal picornaviruses and plant comoviruses, which have a common genome organization. The structures of several human rhinovirus (HRV) serotypes (common cold virus) are 320 Å diameter particles with the VP1, VP2, and VP3 capsid subunits in a $T = 3$-like IAU (**Figure 8**). These subunits all have the β-barrel fold oriented tangentially and similar overall structure except for some loops and extended protein termini. The exterior of the particle has a deep cleft circling the fivefold axes (viral canyon) that binds the ICAM receptor on the surface of cells and which is also bound by neutralizing antibodies.

The capsid of the plant bean-pod mottle virus (BPMV) is also 320 Å in diameter, but has only two subunits, small and large. Both fold into β-barrel domains: one in the small subunit and two that remain covalently attached in the large subunit. The subunits assemble into a $T = 3$-like shell such that the small subunit is equivalent to VP1, and the large subunit to VP2 and VP3 in the HRV capsids.

The BPMV structure had the first ordered protein–RNA interactions seen in an icoshedral virus. Seven ribonucleotides in a single RNA strand were visible inside the capsid shell near the icosahedral threefold axis in a shallow pocket formed by nine different segments from both domains of the large capsid protein. A few specific interactions and a larger number of nonbonded contacts bind the ordered RNA segment.

Other Icosahedral Particles with Uncommon Architectures

While shell arrangements other than T lattices are possible with multiple single or unique capsid proteins in the IAU, these are so far rare. The inner $T = 1$ shell of the BTVC is described above in the $T = 13$ virus section. A very similar $T = 1$ shell has been described for the 440 Å diameter dsRNA yeast L–A virus capsid structure, which also has two copies of the single 680 a.a. in the IAU. This arrangement is emerging as a common structure among the inner protein shells of dsRNA viruses that transcribe mRNA inside the particle and extrude the nascent strands through holes in the shell.

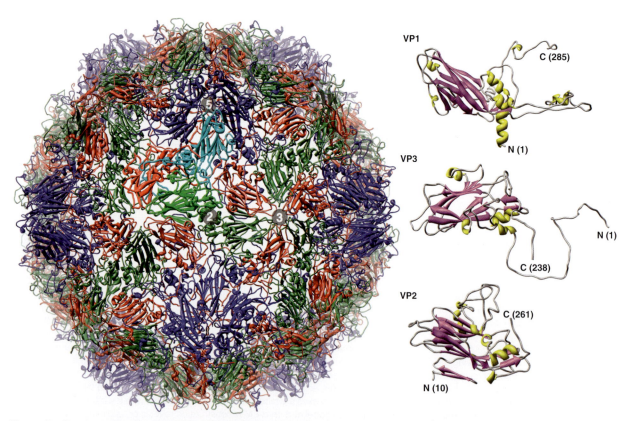

Figure 8 Structure of the $P = 3$ HRV-16 capsid viewed down an icosahedral twofold axis. Icosahedral axes are labeled as in **Figure 3**, but note there are is no quasi-symmetry since the proteins have unique primary structures. The VP1 proteins are colored blue, VP2 green, and VP3 red. An asymmetric unit is shown above and to the left of the twofold axis in brighter colors and larger ribbons (note the color and asymmetric unit definitions are a bit different than in $T = 3$ particles). Structures of the three different proteins are shown to the right and colored as in **Figure 2**. Note the slight tilt of the VP1 subunits that helps to create the canyon surrounding the pentamers.

Concluding Remarks

A great deal of our early knowledge of virus structure came from the nonenveloped viruses, and they continue to provide valuable data in understanding the virus life cycle. The important role of viruses in disease makes it imperative we understand the chemistry and biology of their capsids in ever greater detail. While some overall themes emerge, such as the general nature of capsid protein–protein interactions, protein–RNA interactions, and protein shell organizations, their remarkable variation and complexity defies attempts to form a comprehensive understanding of how they assemble and function. Recent structural evidence pointing to the existence of a primordial virus may lead to important new insights as virus structural studies address even more complex particles at an accelerated rate.

See also: Capsid Assembly: Bacterial Virus Structure and Assembly; Cryo-Electron Microscopy; Electron Microscopy of Viruses; Tobacco Mosaic Virus; Virus Particle Structure: Principles.

Further Reading

Burnett RM (2006) More barrels from the viral tree of life. *Proceedings of the National Academy of Sciences, USA* 103: 3–4.

Chiu W and Johnson JE (eds.) (2003) *Advances in Protein Chemistry, Vol. 64: Virus Structure.* San Diego: Academic Press.

Freddolino PL, Arkhipov AS, Larson SB, McPherson A, and Schulten K (2006) Molecular dynamics simulations of the complete satellite tobacco mosaic virus. *Structure* 14: 437–449.

Khayat R, Tang L, Larson ET, *et al.* (2005) Structure of an archaeal virus capsid protein reveals a common ancestry to eukaryotic and bacterial viruses. *Proceedings of the National Academy of Sciences, USA* 102: 18944–18949.

Rossmann MG, Morais MC, Leiman PG, and Zhang W (2005) Combining X-ray crystallography and electron microscopy. *Structure* 13: 355–362.

Roy P, Maramorosch K, and Shatkin AJ (eds.) (2005) *Advances in Virus Research, Vol. 64: Virus Structure and Assembly.* San Diego: Academic Press.

Relevant Website

http://viperdb.scripps.edu/ – Viperdb Web Site (A Database of Virus Structures with Many Visualization and Analysis Tools).

Virus Particle Structure: Principles

J E Johnson and J A Speir, The Scripps Research Institute, La Jolla, CA, USA

© 2008 Elsevier Ltd. All rights reserved.

Introduction

The virion is a nucleoprotein particle designed to move the viral genome between susceptible cells of a host and between susceptible hosts. An important limitation on the size of the viral genome is its container, the protein capsid. The virion has a variety of functions during the virus life cycle (**Table 1**); however, the principles dictating its architecture result from the need to provide a container of maximum size derived from a minimum amount of genetic information. The universal strategy evolved for the packaging of viral nucleic acid employs multiple copies of one or more protein subunit types arranged about the genome. In most cases the subunits are arranged with well-defined symmetry, but there are examples where assembly intermediates and sometimes mature particles are not globally symmetric. The assembly of subunits into nucleoprotein particles is, in many cases, a spontaneous process that results in a minimum free energy structure under intracellular conditions. The two broad classes of symmetric virions are helical rods and spherical particles.

Helical Symmetry

The nucleoprotein helix can, in principle, package a genome of any size. Extensive studies of tobacco mosaic virus (TMV) show that protein subunits will continue to add to the extending rod as long as there is exposed RNA. Protein transitions required to form the TMV helix from various aggregates of subunits are now understood at the atomic level. It is clear that subunits forming the helix display significant polymorphism in the course of assembly; however, excluding the two ends of the rod, all subunits are in identical environments in the mature helical virion. This is the ideal protein context for a minimum free energy structure. In spite of these packaging and structural attributes the helical virion must be deficient in functional requirements that are common for animal viruses because they are found only among plant, bacterial, and archeal viruses. Even among plant viruses, only 7 of the 25 recognized groups are helical. The large majority of all viruses are roughly spherical in shape.

Icosahedral Symmetry

The architectural principles for constructing a 'spherical' virus were first articulated by Crick and Watson in 1956. They suggested that identical subunits were probably distributed with the symmetry of Platonic polyhedra (the tetrahedron, 12 equivalent positions; the octahedron, 24 equivalent positions; or the icosahedron, 60 equivalent positions). Subunits distributed with the symmetry of the icosahedron (**Figure 1(a)**) provide the maximum sized particle, for a given sized subunit, in which all copies of a subunit lie in identical positions. The repeated interaction of chemically complementary surfaces at the subunit interfaces leads naturally to such a symmetric particle. The 'instructions' required for assembly are contained in the tertiary structure of the subunit (**Figure 1(b)**). The actual assembly of the protein capsids is a remarkably accurate process. The use of subunits for the construction of organized complexes places strict control on the process and will naturally eliminate defective units. The reversible formation of noncovalent bonds between properly folded subunits leads naturally to error-free assembly and a minimum free energy structure.

Table 1 Functions of the virus capsid in simple RNA viruses

Assembly	Subunits must assemble to form a protective shell for the RNA
Package	Subunits specifically package the viral RNA
Binding to receptors	The capsid may actively participate in virus infection processes binding to receptors and mediating cell entry (animal) and disassembly
Transport	Virion transport within the host (plant)
Mutation	Capsid protein mutation to avoid the immune system
RNA replication	Some capsid proteins function as a primer for viral RNA replication

Virus Structures

Crystallographic studies of nearly 90 unique, icosahedral viruses have demonstrated that there are a limited

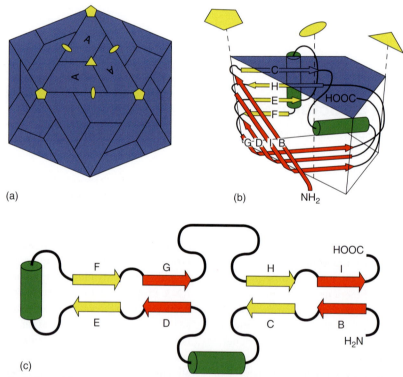

Figure 1 (a) The icosahedral capsid contains 60 identical copies of the protein subunit (blue) labeled A. These are related by fivefold (yellow pentagons at vertices), threefold (yellow triangles in faces), and twofold (yellow ellipses at edges) symmetry elements. For a given sized subunit this point group symmetry generates the largest possible assembly (60 subunits) in which every protein lies in an identical environment. (b) A schematic representation of the subunit building block found in many RNA and some DNA viral structures. Such subunits have complementary interfacial surfaces which, when they repeatedly interact, lead to the symmetry of the icosahedron. The tertiary structure of the subunit is an eight-stranded β-barrel with the topology of the jellyroll (see c-, β-strand and helix coloring is identical to b). Subunit sizes generally range between 20 and 40 kDa with variation among different viruses occurring at the N- and C-termini and in the size of insertions between strands of the β-sheet. These insertions generally do not occur at the narrow end of the wedge (B–C, H–I, D–E, and F–G turns). (c) The topology of viral β-barrel showing the connections between strands of the sheets (represented by yellow or red arrows) and positions of the insertions between strands. The green cylinders represent helices that are usually conserved. The C–D, E–F, and G–H loops often contain large insertions.

number of folds utilized in forming viral capsids. By far the most common fold is the eight-stranded antiparallel β-sandwich shown schematically in **Figure 1(b)**. The details of all the folds observed in nonenveloped virus capsids and their distribution are discussed elsewhere in this encyclopedia.

Early ideas explaining spherical virus architecture were extended on the basis of physical studies of small spherical RNA plant viruses. The large yields and ease of preparation made them ideal subjects for investigations requiring substantial quantities of material. Protein subunits forming virus capsids of this type are usually 20–40 kDa. An example of a virus consistent with the Crick and Watson hypothesis is satellite tobacco necrosis virus (STNV) which is formed from 60 identical 25 kDa subunits. The particle's outer radius is 80 Å and the radius of the internal cavity is 60 Å providing a volume of 9×10^5 Å^3 for packaging RNA. A single hydrated ribonucleotide in a virion will occupy on average roughly 600–700 Å^3. Thus, the STNV volume is adequate to package a genome of only 1200–1300 nt. STNV is a satellite virus and the packaged genome codes for only the coat protein. The 'helper virus', tobacco necrosis virus, supplies proteins required for RNA replication. Most simple ribovirus genomes contain coding capacity for at least two proteins; roughly, 1200 nt for the capsid protein and 2500 nt for an RNA-directed RNA polymerase. The inner radius required to package such a minimal genome is 90 Å. Consistent with this requirement were experimental studies showing that the vast majority of simple spherical viruses had outer radii of at least 125 Å, which corresponds to inner radii of roughly 100 Å for a typical shell thickness for a 400 residue subunit. Such particles had to be formed from more than 60 subunits, yet X-ray diffraction patterns of crystalline tomato bushy stunt virus (TBSV) and turnip yellow mosaic virus (TYMV) were consistent with icosahedral symmetry.

Quasi-Equivalent Virus Capsids

Although a number of investigators developed hypotheses explaining the apparent inconsistent observations, Caspar and Klug, in 1962, derived a general method for the construction of icosahedral capsids that contained multiples of 60 subunits. The method systematically enumerates all possible quasi-equivalent structures and is similar to that derived by Buckmeister Fuller to construct geodesic domes. The quasi-equivalent theory of Caspar and Klug explained the distribution of morphological units (features identifiable at low resolution by electron microscopy often corresponding to hexamer, pentamer, trimer, or dimer aggregates of the subunits) on all structures observed to date, but the results from high-resolution crystallographic studies have shown some remarkable inconsistencies with the microscopic principles upon which the theory is based.

Quasi-equivalence is best visualized graphically. Formally, subunits forming quasi-equivalent structures must be capable of assembling into both hexamers (which are conceptually viewed as planar) and pentamers (which are convex because one subunit has been removed from the planar hexamer and yet similar (quasi-equivalent) contacts are maintained). Caspar and Klug proposed that, logically, atomic interactions of fivefold-related and sixfold-related subunits could be closely similar, as only the dihedral angle between subunits would change. They argued that this difference could be accommodated by allowed variations in noncovalent bond angles between residues stabilizing the oligomers. This prediction was correct for some virus capsids, as illustrated below for cowpea chlorotic mottle virus (CCMV), but in many viruses there is an explicit molecular switch that changes the oligomeric state. If subunits assembled as all hexamers the result would be a sheet of hexamers and a closed shell could not form (**Figure 2(a)**). The rules of quasi-equivalence described a systematic procedure for inserting pentamers into the hexagonal net in such a way as to form a closed shell with exact icosahedral symmetry. **Figure 2** illustrates this principle and the selection rules for inserting pentamers. **Figure 3** illustrates how the morphogenesis of such an assembly may occur using the crystallographic structure of CCMV as inspiration. CCMV was the first virus structure determined that agreed in detail with the predictions of Caspar and Klug.

The quasi-equivalence theory has been universally successful in describing surface morphology of spherical viruses observed in the electron microscope and, prior to the first high-resolution crystallographic structure of a virus, it was assumed that the underlying assumptions of Caspar and Klug were essentially correct. The structure of TBSV determined at 2.9 Å resolution revealed an unexpected variation from the concept of quasi-equivalence which was defined as "any small, non random, variation in a regular bonding pattern that leads to a more stable structure than does strictly equivalent bonding." Unlike CCMV, the structure of TBSV showed that differences occurring between pentamer interactions and hexamer interactions were not small variations in bonding patterns, but almost totally different bonding patterns. **Figure 4** shows diagrammatically the subunit interactions in the shell of TBSV, SBMV, BBV, and TCV. These high-resolution structures revealed that the mathematical concept of quasi-equivalence predicted surface lattices accurately, but not for the reasons expected. Bonding contacts between quasi-threefold-related subunits are maintained with little deviation from exact symmetry while quasi-twofold contacts and icosahedral twofold contacts (which are predicted to be very similar) are quite different. The hexamer quasi-symmetry is better described as a trimer of dimers in the TBSV and related structures. Unlike the conceptual model and the CCMV capsid, the particle curvature in TBSV results from both pentamers and hexamers.

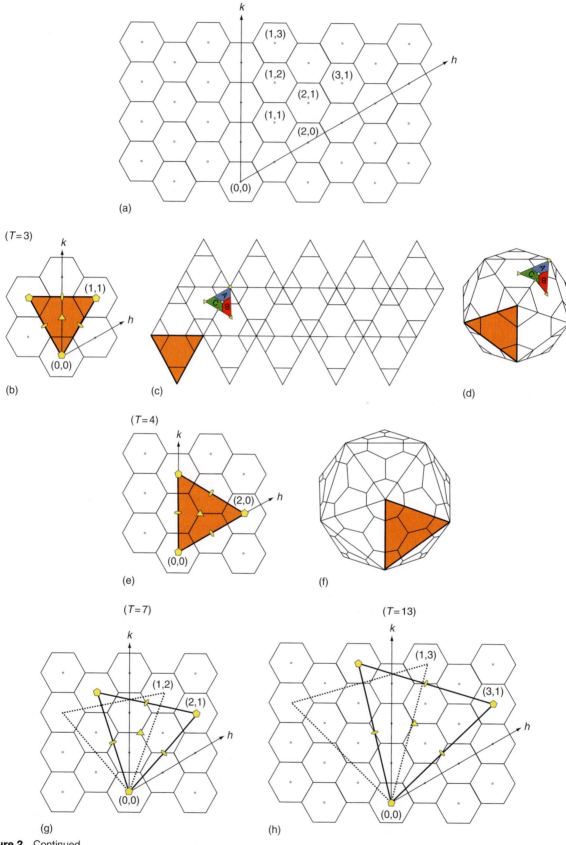

Figure 2 Continued

The high-resolution $T=3$ structures showed that the overall features of the quasi-equivalent theory were correct, but the underlying concepts of quasi-equivalent bonding had to be revised. The first low-resolution structure (22.5 Å) of a $T=7$ virus required an even greater conceptual adjustment to the underlying principles of quasi-equivalence. Rayment, Baker, and Caspar reported, in 1982, that the polyomavirus capsid contained 72 capsomers as previously reported from electron microscopy studies, but that all the capsomers were pentamers of protein subunits even when they were located at hexavalent lattice points. The $T=7$ surface lattice predicts 12 pentamers and 60 hexamers; thus, the prediction of the number and position of the morphological units is correct, but the fine structure of the morphological units is incorrect. Although the result was highly controversial when first reported, additional electron microscopy studies and the 3.5 Å resolution X-ray structure of polyomavirus have fully confirmed the all-pentamer structure. This result clearly shows the limits of theory in predicting virus structure and indicates that further understanding of capsid structure will come only from experimental studies. The structure of the polyomavirus and its relatives illustrates an important concept when considering surface lattice formation. The feature of greatest importance is not the symmetry of the morphological unit positioned on the hexamer sites, but only its ability to accommodate six neighbors (it is a hexavalent position). Normally morphological units with sixfold symmetry accomplish this, but here, rather acrobatic molecular switching has permitted a pentamer of subunits to accommodate six neighbors. Although the all pentamer capsid has been observed for the $T=7$ structure of papilloma viruses, cauliflower mosaic virus appears to have the hexamer/pentamer distribution predicted by quasi-equivalent theory as do the $T=7$ capsids of the lambda-like bacteriophage, HK97. A substantial number of complex virus structures have been determined by cryo-electron microscopy and the surface lattices agree well with the predictions of quasi-equivalence. Thus there is considerable confidence in the lattice assignments, but the capsomer structure and therefore number of subunits must be carefully confirmed.

Figure 2 Geometric principles for generating icosahedral quasi-equivalent surface lattices. These four constructions show the relation between icosahedral symmetry axes and quasi-equivalent symmetry axes. The latter are symmetry elements that hold only in a local environment. (a) It is assumed in quasi-equivalence theory that hexamers and pentamers (a hexamer contains six units and a pentamer contains five units) can be interchanged at a particular position in the surface lattice. Hexamers are initially considered planar (an array of hexamers forms a flat sheet as shown) and pentamers are considered convex, introducing curvature in the sheet of hexamers when they are inserted. Inserting 12 pentamers at appropriate positions in the hexamer net generates the closed icosahedral shell, composed of hexamers and pentamers. The positions at which hexamers are replaced by pentamers are defined by the indices h and k measured along the labeled axes. The values of (h,k) used in the following examples are labeled. To construct a model of a particular quasi-equivalent lattice, one face of an icosahedron (equilateral triangles colored orange in (b–f)) is generated in the hexagonal net. The origin (0,0) is replaced with a pentamer, and the (h,k) hexamer is replaced by a pentamer. The third replaced hexamer is identified by threefold symmetry (i.e., complete the equilateral triangle). Each quasi-equivalent lattice is identified by a number $T = h^2 + hk + k^2$ where h and k are the indices used above. T indicates the number of quasi-equivalent units in the icosahedral asymmetric unit. For the purpose of these constructions it is convenient to choose the icosahedral asymmetric unit as 1/3 of an icosahedral face defined by the triangle connecting a threefold axis to two adjacent fivefold axes. Other asymmetric units can be chosen such as the triangle connecting two adjacent threefold axes and an adjacent fivefold axis (see (c) and **Figure 4**). The total number of units in the particle is $60T$, given the symmetry of the icosahedron. The number of pentamers must be 12 and the number of hexamers is $(60T - 60)/6 = 10(T - 1)$. (b) One face of the icosahedron for a $T = 3$ surface lattice is identified by the orange triangle with the bold outline. The yellow symmetry labels are the same as those defined in **Figure 1**. The hexamer replaced has coordinates $h = 1, k = 1$. The icosahedral asymmetric unit is 1/3 of this face and it contains three quasi-equivalent units (two units from the hexamer coincident with the threefold axis and one unit from the pentamer). (c) Arranging 20 identical faces of the icosahedron as shown can generate the three-dimensional model of the quasi-equivalent lattice. Three quasi-equivalent units labeled A (blue), B (red), and C (green) are shown. These correspond to the three quasi-equivalent units defined in **Figures 3** and **4** rather than the alternative definition used in (a) and (b). (d) The folded icosahedron is shown with hexamers and pentamers outlined. The orange face represents the triangle originally generated from the hexagonal net. The $T = 3$ surface lattice represented in this construction has the appearance of a soccer ball. The trapezoids labeled A, B, and C identify quasi-equivalent units in one icosahedral asymmetric unit of the rhombic triacontahedron discussed in **Figure 4**. (e) An example of a $T = 4$ icosahedral face ($h = 2, k = 0$). In this case the hexamers are coincident with icosahedral twofold axes. (f) A folded $T = 4$ icosahedron with the orange face corresponding to the face outlined in the hexagonal net. Note that folding the $T = 4$ lattice has required that the hexamers have the curvature of the icosahedral edges. (g) A single icosahedral face generated from the hexagonal net for a $T = 7$ lattice. Note that there are two different $T = 7$ lattices ($h = 2, k = 1$ in bold outline; and $h = 1, k = 2$ in dashed outline). These lattices are the mirror images of each other. To fully define such a lattice, the arrangement of hexamers and pentamers must be established as well as the enantiomorph of the lattice. (h) A single icosahedral face for a $T=13$ lattice is shown. The two enantiomorphs of the quasi-equivalent lattice ($h = 3, k = 1$ – bold; and $h = 1, k = 3$ – dashed) are outlined. The procedure for generating quasi-equivalent models described here does not exactly correspond to the one described by Caspar and Klug in 1962. Caspar and Klug distinguish between different icosadeltahedra by a number $P = h^2 + hk + k^2$ where h and k are integers that contain no common factors but 1. The deltahedra are triangulated to different degrees described by an integer f that can take on any value. In their definition $T = Pf^2$. The description in this figure has no restrictions on common factors between h and k; thus, $T = h^2 + hk + k^2$ for all positive integers. The final models are identical to those described by Caspar and Klug.

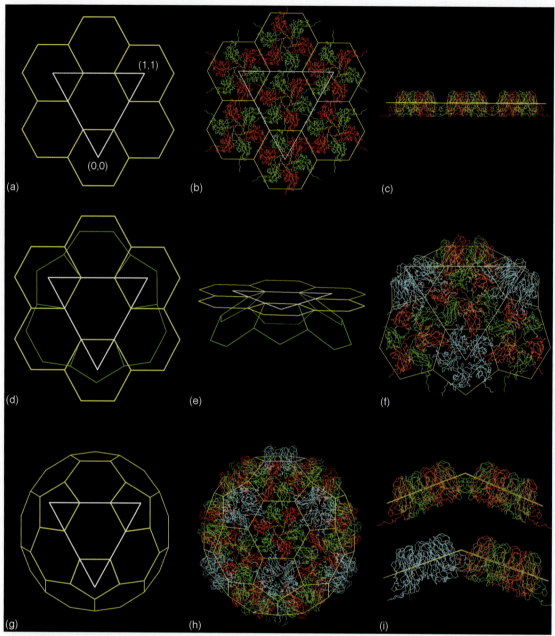

Figure 3 Molecular graphics construction of a $T = 3$ quasi-equivalent icosahedron. (a) Hexagonal sheet overlaid with the triangular coordinates (white) for a theoretical $T = 3$ quasi-equivalent icosahedron ($h = 1$, $k = 1$, see **Figure 2(b)**). The sheet has true sixfold rotational symmetry about axes passing through the hexamer centers, which are normal to the sheet. (b) Copies of the hexamer coordinates from the CCMV X-ray structure (colored by asymmetric unit position, see **Figures 2(c)–2(d)** and **4**) can be positioned in the sheet by simple translations. (c) A side view of the modeled sheet demonstrates its planarity. (d) Hexamers at the corners of the white ($h = 1$, $k = 1$) triangle become pentamers. The planar sheet (yellow model) takes on curvature to maintain contacts between the polygons (green model). (e) The magnitude of the pentamer-induced curvature is displayed in the side view of the partial polyhedron. (f) Coordinates of the CCMV X-ray structure fit this construction without any manipulation. (g) A completed $T = 3$ icosahedral model. The 12 pentamers generate curvature that closes the structure. This cage (a truncated icosahedron) accurately describes the geometric morphology of CCMV (h) which is composed of modular, planar, pentamers (12) and hexamers (20). Angular pentamer–hexamer and hexamer–hexamer interfaces (i) stabilize curvature in the absence of convex pentamers used to construct the soccer ball of **Figure 3(d)** (see also **Figure 4**).

Picorna-Like Virus Capsids

A number of viral capsids are constructed with pseudo $T = 3$ symmetry. These structures contain β-barrel subunits (**Figure 1(b)**) in the quasi-equivalent environments formed in $T = 3$ structures, but each of the three β-barrels in the asymmetric unit has a unique amino acid sequence. Rather than 180 identical subunits, the $P = 3$ particles contain 60 copies each of three different subunits (**Figure 5**). These structures do not require quasi-equivalent bonding

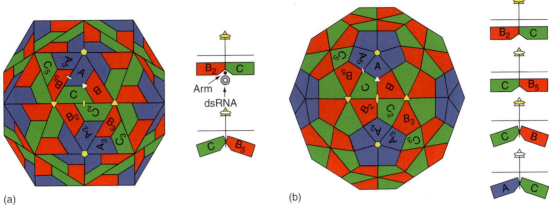

Figure 4 Although quasi-equivalence theory can predict, on geometrical principles, the organization of hexamers and pentamers in a viral capsid, the detailed arrangement of subunits can only be established empirically. High-resolution X-ray structures of $T=3$ plant and insect viruses show that the particles are organized like the icosahedral rhombic triacontahedron or truncated icosahedron. A convenient definition of the icosahedral asymmetric unit for both geometrical shapes is the wedge defined by icosahedral threefold axes left and right of the particle center and an icosahedral fivefold axis at the top. The icosahedral asymmetric unit contains three subunits labeled A (blue), B (red), and C (green) (see **Figures 2(c)–2(d)**). The asymmetric unit polygons represent chemically identical protein subunits that occupy slightly different geometrical (chemical) environments as indicated by differences in their coloring. Polygons with subscripts are related to A, B, and C by icosahedral symmetry (i.e., A to A_5 by fivefold rotation). The shapes of the $T=3$ soccer ball model in **Figure 2(d)**, truncated icosahedron in **Figure 3**, and rhombic triacontahedron are all different; however, the quasi-symmetric axes are in the same positions relative to the icosahedral symmetry axes for all three models. Quasi-threefold and quasi-twofold axes are represented by the white symbols. The quasi-sixfold axes are coincident with the icosahedral threefold axes in $T=3$ particles as shown in **Figures 2(b)–2(d)** and **3**. (a) The rhombic triacontahedron is constructed by placing rhombic faces perpendicular to icosahedral twofold symmetry axes (yellow ellipse). Thus, the A, B, and C polygons are co-planar within each asymmetric unit. The shape of the subunit in $T=3$ plant and insect viruses is nearly identical to the shape of the subunit in the $T=1$ virus and they pack in a very similar fashion. The $T=1$ subunits in one face (**Figure 1(a)**) are related by an icosahedral threefold axis while the $T=3$ subunits in one face are related by a quasi-threefold axis. The dihedral angle between subunits C and B_5 (juxtaposed across quasi-twofold axes) is 144° and is referred to as a bent contact (bottom right image), while the dihedral angle between subunits C and B_2 (juxtaposed across icosahedral twofold axes) is 180° and is referred to as a flat contact (top-right image). Two dramatically different contacts between subunits with identical amino acid sequences are generated by the insertion of an extra polypeptide from the N-terminal portion of the A or C subunits into the groove formed at the flat contact. This polypeptide is called an 'arm'. The flat contact can also be upheld by insertion of nucleic acid structure into the same groove. Neither N-terminal arms nor nucleic acid structure has been observed in the groove across the quasi-twofold axis; thus, C and B_5 are in direct contact as in, for example, the X-ray structure of flock house virus (FHV). (b) A truncated icosahedron achieves curvature at different interfaces compared to the rhombic triacontahedron. Interactions between B_2–C and between C–B_5 polygons are both defined by 180° dihedral angles (side view at top right), whereas bends similar in magnitude occur within the asymmetric unit at the B–C and C–A polygon interfaces (138° and 142°, respectively; side view at bottom right). This creates the planar pentamer and hexamer morphological units characteristic of the truncated icosahedron and the CCMV X-ray structure (**Figure 3(h)**).

because each unique interface will have different amino acids interacting, rather than the same subunits forming different contacts. The animal picornaviruses have capsids of this type. Animal virus capsids undergo rapid mutation which avoids recognition by the circulating immune system. Capsids composed of three subunit types could mutate in one subunit without affecting the other two. This would be less likely to affect assembly or other functions of the particle in $P=3$ shells than it would in $T=3$ shells. At least one plant virus group displays $P=3$ shells, the comoviruses. An interesting variation occurs in these capsids when compared with the picornaviruses. Two of the domains forming the shell are contained in a single polypeptide chain. This phenomenon is readily understood in the context of the synthesis of the subunits in picorna and comoviruses. In both cases the proteins are synthesized as a polyprotein that is subsequently cleaved into functional proteins by a virally encoded protease. Clearly one of the cleavage sites in picornaviruses is missing in the comoviruses, resulting in these two domains being still a 'polyprotein'. Tobacco ringspot virus (TRV) is an example where all three subunits are linked in a single polypeptide chain like 'beads on a string'. It is interesting to note that the X-ray structure of TRV showed that the linkage of subunits in the capsid is the same as that predicted for the precursor polyprotein of picornaviruses based on the their X-ray structures, supporting the divergent evolution of the $P=3$ capsids from plant viruses to animal viruses.

Virus Particle Maturation

While $T=3$ and $P=3$ animal viruses undergo maturation processes that confer infectivity, there is little change in their particle morphology during this process.

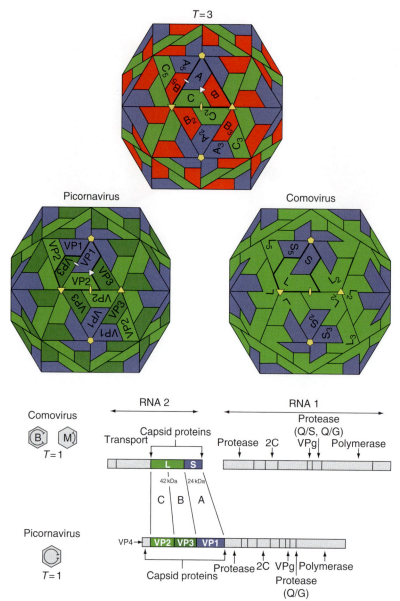

Figure 5 A comparison of $T = 3$, picornavirus, and comovirus capsids. In each case, one trapezoid represents a β-barrel and the icosahedral asymmetric units are outlined in bold. The icosahedral asymmetric unit of the $T = 3$ shell contains three identical subunits labeled A, B, and C (see **Figure 4**). The asymmetric unit of the picornavirus capsid contains three β-barrels, but each has a characteristic amino acid sequence labeled VP1, VP2, and VP3. The comovirus capsid is similar to the picornavirus capsid except that two of the β-barrels (corresponding to the green VP2 and VP3 units) are covalently linked to form a single polypeptide, the large protein subunit (L), while the small protein subunit (S) corresponds to VP1 (note the similar color shading). The individual subunits of the comovirus and picornavirus capsids are in identical geometrical (chemical) environments (e.g., VP1 and S are always pentamers) technically making these $T = 1$ capsids, but they are often referred to as pseudo $T = 3$, or $P = 3$ capsids. Comoviruses and picornaviruses have a similar gene order, and the nonstructural 2C and polymerase genes display significant sequence homology. The relationship between the capsid subunit positions in these viruses and their location in the genes is indicated by color-coding and the labels A, B, and C in the gene diagram.

In contrast, bacteriophages, such as HK97, display dramatic subunit reorganization during maturation. (The head (capsid) of the HK97 bacteriophage first assembles as a metastable prohead (prohead I and II), and then undergoes a complex maturation process that involves large-scale macromolecular transitions that expand the protein shell. There are three well-defined expansion intermediates (EI-I/II, EI-III, EI-IV) that end with the EI-IV state, which is the final step before head maturation is complete. The crosslinks between the 420 subunits are formed during these steps, except for the last 60 situated around the 12 pentamers. Formation of the final 60 crosslinks is the final step that creates head II, the mature capsid.) The X-ray structure of mature HK97 showed that each subunit in the $T = 7l$ capsid makes contact with nine other subunits, displaying molecular promiscuity at an unprecedented level. These interactions, together with covalent cross-links between subunits create an exceptionally stable particle. It is clear, however, that such a structure cannot be achieved in a single step. It is too complicated. The solution is to assemble in stages. The initial assembly product is called a procapsid. This is formed by three gene products: 415 copies of the capsid protein, 12 copies of the portal subunit, and about 60 copies of a virally encoded protease that is inside the 500 Å diameter particle. This is an equilibrium assembly that is similar to the assembly of simpler viruses. Subunits are confined to discrete volumes and contact only immediate neighbors, again, as observed with simple viruses. Completion of assembly signals protease activation, and 103 residues are cut from each capsid protein and the protease auto-digests, leaving an empty particle that is a substrate for DNA packaging enzymes. When the terminase complex, also virally encoded, attaches to the double-stranded DNA viral genome and the portal, DNA packaging ensues and after ~10% of the DNA is packaged, the particle undergoes an extraordinary reorganization involving 40° rotations of subunits and 40 Å translations, totally remodeling the contacts resulting in the highly interlocked structure in the mature virion. Maturation of this type is entirely programmed into the original tertiary and quaternary structures and is an excellent example of a molecular machine.

Concluding Remarks

This overview of virus structure and assembly illustrates the remarkable level of sophistication associated with one or a few gene products acting in concert. The requirement for maximum utility from minimal genetic information has led to the evolution of proteins with exceptionally high functional density that are studied to develop better understanding of virus-associated disease and as accessible paradigms for cellular function.

See also: Capsid Assembly: Bacterial Virus Structure and Assembly; Tobacco Mosaic Virus; Virus Particle Structure: Nonenveloped Viruses.

Further Reading

Caspar DLD (1980) Movement and self-control in protein assemblies. Quasi-equivalence revisited. *Biophysics Journal* 32: 103–138.
Caspar DLD and Klug A (1962) Physical principles in the construction of regular viruses. *Cold Spring Harbor Symposia on Quantitative Biology* 27: 1–24.
Crick FHC and Watson JD (1956) Structure of small viruses. *Nature* 177: 473–475.
Johnson JE and Speir JA (1997) Quasi-equivalent viruses: A paradigm for protein assemblies. *Journal of Molecular Biology* 269: 665–675.
Natarajan P, Lander GC, Shepherd CM, Reddy VS, Brooks CL, III, and Johnson JE (2005) Exploring icosahedral virus structure with VIPER. *Nature Reviews Microbiology* 3: 809–817.
Speir JA, Bothner B, Qu C, Willits DA, Young MJ, and Johnson JE (2006) Enhanced local symmetry interactions globally stabilize a mutant virus capsid that maintains infectivity and capsid dynamics. *Journal of Virology* 80: 3582–3591.
Tang L, Johnson KN, Ball LA, Lin T, Yeager M, and Johnson JE (2001) The structure of pariacoto virus reveals a dodecahedral cage of duplex RNA. *Nature Structural Biology* 8: 77–83.
Zandi R, Reguera D, Bruinsma RF, Gelbart WM, Rudnick J, and Reiss H (2001) Origin of icosahedral symmetry in viruses. *Proceedings of the National Academy of Sciences, USA* 101: 15556–15560.
Zlotnick A (2005) Theoretical aspects of virus capsid assembly. *Journal of Molecular Recognition* 18: 479–490.

Virus Species

M H V Van Regenmortel, CNRS, Illkirch, France

© 2008 Elsevier Ltd. All rights reserved.

Glossary

Ecological niche The ecological niche of a virus refers to certain biological properties of the virus, such as host range, tissue tropism in the host, and type of vector, rather than a geographical space or a particular environment. In the absence of the virus, its ecological niche property is also absent and the notion of a vacant niche is thus meaningless.

Polythetic class A polythetic class consists of members that exhibit overall similarity and have a large number of common properties. However, the members of a polythetic class do not all share a common character that constitutes a single defining property of the class. In contrast, the members of a monothetic class, such as those of the order *Mononegavirales*, possess such a single defining property, that is, the presence of a negative-sense, single-stranded RNA (ssRNA) genome.

Introduction

The question of what is a virus species is related to the general problem of how the world of viruses should be partitioned and ordered for achieving a coherent scheme of distinct and easily recognizable viral entities. In view of the variability of viruses arising from the error-prone process of genomic replication, it is often difficult to decide whether a newly encountered virus is the same as one seen previously, for this requires that we give an answer to the vexing question: "How different must two virus isolates be in order to be considered different viruses, rather than the same virus?" Virologists usually have no difficulty in distinguishing pathogenic variants of a virus, while still recognizing these variants as the same kind of virus. Even without realizing it, they pass judgment on the significance of the observed differences between individual virus isolates and, if the extent of difference appears small enough, they will consider the variant to be the same virus. In taxonomic parlance, they will say that the variant belongs to the same virus species.

'Species' is the universally accepted term for the lowest taxonomic clustering of living organisms. Although viruses are not organisms, they are considered to be biological entities because they possess some of the properties of biological agents, such as having a genome and being able to adapt to particular hosts and changing environments. The classification system of viruses therefore uses the same hierarchical ranks of order, family, genus, and species used in all biological classifications.

Although the species category is the most fundamental one in any biological classification, it took many years before virologists started to assign viruses to separate species. Virus genera and families were readily accepted but there was considerable opposition to the idea that the species concept could be applied to viruses. One reason for this reluctance was the lack of agreement among biologists about what a species actually is. It is in fact remarkable that a century and a half after the appearance of Charles Darwin's *The Origin of Species*, there is still no general agreement about what constitutes a plant, animal, or microbial species. In different branches of biology, no less than 22 different species concepts have been applied.

Distinguishing between Abstract Classes and Real Objects

A basic requirement for clear thinking is the ability to distinguish between real, tangible objects, such as viruses, and the abstract classes used in taxonomy, such as families and species, which are mental constructs that exist only in the minds of people who think about them. Although a taxonomic class is defined by properties possessed by concrete objects, it is only an abstract thought that has no real existence outside of human minds. For this reason, viral species, genera, and families cannot cause diseases and cannot be purified, sequenced, or seen through electron microscopes. Only the viruses themselves can be studied in this manner. Classes used in biological classification have a hierarchical structure. A class, such as a particular species, can belong only to one higher rank immediately above it such as a particular genus and that genus, in turn, can only belong to one family. The relationship known as 'class membership' is the relationship between an abstract class or thought and the members of that class which are real, concrete objects located in space and time. Class membership is thus able to bridge two different logical categories, the abstract and the concrete. It is important to be able to establish a certain link between these two logical categories, as it is impossible for an object such as a virus to be 'part' of an abstraction, such as a species. Similarly, an abstract thought cannot be 'part of' a concrete object. The correct statement is to say that the viruses that are handled by virologists are members of certain abstract constructs, such as species or genera. These constructs are invented by taxonomists for the purpose of introducing some order to the bewildering variety of viruses.

A universal class, also known as an Aristotelian class, is defined by properties that are constant and immutable. This allows members of such a class to be recognized with absolute certainty, because one or more property is necessarily present in every member of the class. Virus families, for instance, are universal classes because they consist of members, all of which share a number of defining properties that are both necessary and sufficient for class membership. Allocating a virus to a family is thus an easy task, because a few structural or chemical attributes will suffice to allocate the virus to a particular family. For instance, all the members of the family *Adenoviridae* are nonenveloped viruses that have an icosahedral particle and double-stranded DNA, with projecting fibers at the vertices of the protein shell.

Unfortunately, not all properties of members of classes correspond to unambiguous and stable properties such as the presence or absence of a DNA genome or of a particular particle morphology. Many qualitative properties of concrete objects are inherently vague and do not possess precise borderlines. For instance, in the description of viruses, properties used for recognizing members of species, such as the degree of genome sequence similarity between virus isolates or the nature of the symptoms induced by a virus in its host, whether mild or severe, tend to be inherently imprecise and fuzzy. As a result the species classes that can be conceptualized on the basis of such properties will themselves be fuzzy and membership in the class will then be a matter of convention or stipulation. This explains some of the difficulties one encounters when dealing with the species level in any biological classification.

When fuzziness is accepted as an unavoidable characteristic of species, it becomes possible to describe species in terms of continuums with hazy boundaries. In a similar way, colors can be distinguished conceptually in spite of the continuous nature of the spectrum of electromagnetic waves, and mountain peaks are given names in spite of the absence of sharp boundaries in geological rock formations.

Species in Biology

The traditional view of species is that they correspond to groups of similar organisms that can breed among themselves and produce fertile offspring. The classical definition of so-called 'biological species' states that "species are groups of interbreeding natural populations which are reproductively isolated from other such groups." The reproductive isolation often simply reflects a geographic isolation or a behavioral incompatibility. This definition is only applicable to organisms that reproduce sexually and it has limited value in the plant kingdom because of the high frequency of hybridization between different species of plants. Even in animals, the criterion of reproductive isolation does not always hold, as shown by the ability of dogs, wolves, and coyotes to interbreed although they are members of different species of the same genus.

In order to make the definition of biological species applicable to asexual organisms, it was later modified as follows: "a species is a reproductive community of populations, reproductively isolated from others, that occupies a specific niche in nature." Some authors reject the view that asexual organisms can form biological species but most biologists disagree with that view, as this would render the species concept inapplicable to a large portion of the biological realm.

Another species concept is that of evolutionary species, which has been defined as "a single lineage of ancestor–descendant populations which maintains its identity from other such lineages and which has its own evolutionary tendencies and historical fate." Such a concept does not provide any guidance on how far back in time a species can be traced. Life on Earth is a biological and historical continuum and it is as difficult to demarcate boundaries in time that would separate individual evolutionary species as it is to define clear-cut breeding discontinuities that would separate different biological species. These difficulties led to the view that species are similar to fuzzy sets with unclear boundaries and that it is impossible to draw sharp boundaries between them, as is done with universal classes such as genera and families. In practice this means that it is not possible to rely on a single defining property for differentiating between two species. This led to the proposal that species are so-called 'polythetic classes' defined by a combination of properties, each of which might occur outside any given class yet could be absent in a member of the particular class (**Figure 1**).

The members of a polythetic class exhibit overall similarity and have a large number of characters in common. However, in contrast to the members of a monothetic or universal class, they need not all share a common character that could be used as a defining property of the class. A polythetic class is actually a cluster concept based on the concept of family resemblance introduced in philosophy by Ludwig Wittgenstein.

What Is a Virus Species?

In the past, many virologists were opposed to the introduction of species in virus classification because they assumed that the only legitimate species concept was that of biological species defined by sexual reproduction, gene pools, and reproductive isolation. Such a concept is obviously not applicable to viruses that are replicated as clones. Another reason for the reluctance to adopt the species category in virus classification was the absence of a satisfactory definition of a virus species. Various definitions of virus species had been proposed over the years but none gained general acceptance. One definition, for instance, stated that "virus species are strains whose properties are so similar that there seems little value in giving them separate names." Such a definition is not very helpful as it simply replaces undefined species by undefined strains and suggests that attributing names to viruses is the same activity as constructing a taxonomy.

In 1991, the International Committee on Taxonomy of Viruses (ICTV), which is the body established by the International Union of Microbiological Societies to make decisions on matters of virus classification and

Figure 1 Schematic representation of the distribution of properties (1 to 5) in five members of a polythetic class. Each member possesses several of these properties (4 out of 5) but no single property is present in all the members of the class. The missing property in each case is represented by the gray section.

nomenclature, endorsed the following definition which was proposed by the author: "a virus species is a polythetic class of viruses that constitute a replicating lineage and occupy a particular ecological niche." This definition is not only based on phenetic similarities among members of species but also stresses the internal cohesion present in biological lineages that share a common biotic niche. Another important aspect of the definition is that a virus species is defined as a polythetic class rather than as a traditional universal class (see **Figure 1**). The ICTV has been using the concept of polythetic class for the creation of separate species taxa.

The total DNA or RNA sequence found in a virion corresponds to the viral genome and is part of the viral phenotype because it corresponds to a portion of the virion's chemical structure. The phenome of the virus corresponds to its observable physical properties, including the morphology and molecular constitution of the virion, as well as the biochemical activities and relational properties of the virus. The phenotype has a temporal dimension as it changes over time, for instance during replication.

A classification based on genome sequences is actually a phenotypic classification with the characters being molecular, rather than morphological or relational. Although putative phylogenetic relationships can be inferred from sequences, they remain tentative and depend on many tacit assumptions. There is little justification for the common assumption that when species are demarcated on the basis of a proposed phylogeny, this produces a classification that is necessarily more correct or useful than a classification based on other phenotypic characters.

Defining Species as a Taxonomic Class Is Not the Same As Demarcating Individual Viral Species

As a definition is the explanation of the meaning of a word or concept, it is possible to give a definition of the concept species. In contrast, it is not possible to give a definition of an object such as a particular virus, as objects can only be described by enumerating their properties and cannot be 'defined'. When viruses or any other object are given proper names, their names do not have a meaning that can be captured in a definition. When we define the name of an abstract species class by means of defining properties, we state the conditions for that name to apply to any given member of the species.

Two types of species definitions must be distinguished. One definition concerns the taxonomic species category corresponding to the lowest level in a hierarchical system of classification. The definition of virus species accepted by the ICTV is a definition of that type and it was introduced because many virologists thought that the species concept was not applicable to viruses. However, such a definition of the category species should not be confused with the definitions or demarcations of each of the 1950 species classes listed in the Eighth ICTV Report published in 2005. Confusing the two meanings of species, that is, as a particular class which has certain viruses as members and as a taxonomic category, respectively, would amount to confuse the concept of chemical element (the class of all elements) with the class corresponding to a single element, such as gold defined by its atomic number 79.

It must be emphasized that once the taxonomic category of virus species had been accepted by virologists, this did not do away with the difficult task of demarcating and defining the hundreds of individual species classes that had to be created on the basis of different combinations of properties for each polythetic species class. In order to demarcate individual species, it was necessary to rely on properties that were not present in all the members of the genus to which the species belongs, because such properties obviously would not permit individual species to be differentiated. Characteristics such as virion morphology, genome organization, method of replication, and number and size of viral proteins are shared by all the members of a genus and are thus not useful for distinguishing individual species in a genus. The properties that are useful for discriminating between individual species within a genus are the natural host range, cell and tissue tropism, pathogenicity, mode of transmission, certain physicochemical and antigenic properties of virions, and small differences in genome sequence. Unfortunately, these properties can be altered by only one or a few mutations and they may therefore vary in different members of the same species. This is one of the reasons species demarcation is not an easy task and often requires drawing boundaries across a continuous range of genomic and phenotypic variability. Taxonomic decisions at the level of species are very much a matter of opinion and adjudication rather than of logical necessity. There is indeed no precise degree of genome difference that could be used as a cutoff point to differentiate between two species nor is there a simple quantitative relationship between the extent of genome similarity and the similarity in phenotypic and biological characteristics of a virus.

There is also a strong subjective element in virus classification, as the practical need to distinguish between individual viruses is not the same in all areas of virology. From a human perspective, not all infected hosts are equally relevant. Human pathogens or pathogens that infect animals and plants of economic importance tend to be studied more intensively than the very many viruses that infect insects or marine organisms. As a result fine distinctions based on minor differences in host range or pathogenicity may be emphasized in the case of viruses of particular interest to humans and the criteria used for differentiating individual species may thus depend on the type of host that is infected. There is, of course, no reason the criteria used for distinguishing between species should be the same in all the virus genera and families, because the need to make certain distinctions are not the same in all fields of virology.

Virus Identification

In order to identify a virus as a member of a species, use is made of so-called diagnostic properties that allow the virus to be recognized as being similar to other viruses. This, of course, means that the species had to be created and defined by taxonomists beforehand. Only when the properties of many members of an established species are compared, is it possible to discover which diagnostic property or set of properties will discriminate between the members of that species and other species, thereby allowing viruses to be identified. A single diagnostic character may sometimes suffice for virus identification but this character should not be confused with the collective set of combined properties which has been used initially to demarcate and define the species as a polythetic class. Species are not defined by means of one or more diagnostic properties used for virus identification but are defined by taxonomists, who stipulate which covariant sets of shared properties have to be present in most members of the species. It is the frequent, combined occurrence of these properties in individual members of a species that allows one to predict many of the properties of a newly discovered virus once it has been identified as a member of a particular species. If a species could be defined monothetically instead of polythetically, that is, by a single diagnostic character and nothing else, the identification of a virus as a member of a species would not be very informative. A single diagnostic character, such as a nucleotide motif or the reactivity with a monoclonal antibody, may suffice to identify a virus but such a property should not be mistaken for a single defining characteristic of the species.

This apparent contradiction between the need for many characters to define and delineate a species and the fact that a single property may suffice to identify a virus disappears when it is realized that defining an abstraction, such as a species, is a task different from identifying a concrete object, such as a virus. Providing a definition of an abstract taxonomic class and identifying concrete objects are not equivalent tasks and they abide by different logical rules.

Species Names and How to Write Them

After the category species was accepted as the lowest taxonomic class to be used in viral taxonomy, the ICTV decided in 1998 that the existing English common names of viruses were to become the official species names and that, to denote the difference, species names should be written in italics with a capital initial letter. This typography, which is similar to that used for other taxa such as virus genera and families, makes it possible to distinguish virus species officially recognized by the ICTV and written in italics from other viral entities such as tentative species and virus strains written in Roman characters.

When ICTV introduced italicized virus species names, it did not intend that the existing names of viruses written in Roman characters should be abandoned. In their writings, virologists need to refer continuously to the viruses they study and because these viruses are not taxonomic abstractions, their names should be written in Roman characters.

In most cases it is necessary only once in a scientific paper to draw attention to the taxonomic position of the virus and this can be done by mentioning that the virus under study, for example, measles virus, is a member of the species *Measles virus*, genus *Morbillivirus*, and family *Paramyxoviridae*. In publications written in languages other than English, the common names of the virus remain those used in that language. It will suffice, for instance, to say once: "le virus de la rougeole est un membre de l'espèce *Measles virus*."

It is correct to say that a virus, a strain, or an isolate of the species *Cucumber mosaic virus* has been sequenced but incorrect to say the species itself has been sequenced. Unfortunately, some authors confuse viruses with species and write that the species *Cucumber mosaic virus* (instead of the virus cucumber mosaic virus) has been isolated from a tobacco plant, is transmitted by an aphid vector, is the causal agent of a mosaic disease, and has been sequenced. This is of course incorrect because virus species are man-made taxonomic classes and do not have hosts, vectors, or sequences. For the same reason, the genus *Cucumovirus* or the family *Paramyxoviridae* cannot be isolated, visualized in an electron microscope, or sequenced.

At a time when English is more widely understood than Latin, the use of italicized English instead of italicized Latin for the names of virus species is more practical and is in line with the emergence of English as the modern language of international scientific communication. Most virologists are opposed to the Latinization of species names and welcomed the introduction of well-known English species names. They certainly preferred this to the creation of an entirely new set of Latin names for the 1950 different virus species that are recognized at present.

Binomial Names for Virus Species

A problem with the current way of naming virus species is that the name of the species (for instance *Measles virus*) differs only in typography from the name of the virus (measles virus). One way to facilitate the distinction between virus names and species names would be to change the current names of virus species into nonlatinized binomial names. Such a system, which has been advocated by plant virologists for many years, consists in replacing the word '*virus*' appearing at the end of the existing species name by the genus name which also ends in '*-virus*'. *Measles virus* then becomes *Measles morbillivirus*, *Hepatitis A virus* becomes *Hepatitis A hepatovirus*, and *Tobacco mosaic virus* becomes *Tobacco mosaic tobamovirus*. The advantage of such a system, which could be implemented without problems for about 98% of all virus species names, is that inclusion of the genus name in the species name provides additional information about the properties of the virus. A changeover to binomial species names would not affect the common names of viruses in English or other languages since names such as measles virus or 'virus de la rougeole' would remain the same. The ICTV is currently debating the possibility of introducing binomial species names and some decision is likely to be made in the near future.

Given that the common names of viruses are used repeatedly in scientific texts there is a need for abbreviating them and the ICTV has published several lists of recommended acronyms for virus names. Since the names of virus species are used only very seldom in publications, there is no need to abbreviate them. If binomial names of virus species were introduced in the future, the abbreviations of common names of viruses will of course not be affected.

See also: Nature of Viruses; Phylogeny of Viruses; Quasispecies; Taxonomy, Classification and Nomenclature of Viruses.

Further Reading

Calisher CH and Mahy BWJ (2003) Taxonomy: Get it right or leave it alone. *American Journal of Tropical Medicine and Hygiene* 68: 505–506.

Claridge MF, Duwah HA, and Wilson MR (eds.) (1997) *Species: The Units of Biodiversity.* London: Chapman and Hall.

Fauquet CM, Mayo MA, Maniloff J, Desselberger U, and Ball LA (eds.) (2005) *Virus Taxonomy, Classification and Nomenclature of Viruses: Eighth Report of the International Committee on Taxonomy of Viruses,* San Diego, CA: Elsevier Academic Press.

Lewontin RC (1992) Genotype and phenotype. In: Keller EF and Lloyd EA (eds.) *Keywords in Evolutionary Biology*, pp. 137–144. Cambridge, MA: Harvard University Press.

Pigliucci M (2003) Species as family resemblance concepts: The (dis-)solution of the species problem? *Bioessays* 25: 596–602.

Van Regenmortel MHV (1990) Virus species, a much overlooked but essential concept in virus classification. *Intervirology* 31: 241–254.

Van Regenmortel MHV (2003) Viruses are real, virus species are man-made taxonomic constructions. *Archives of Virology* 148: 2481–2488.

Van Regenmortel MHV (2006) Virologists, taxonomy and the demands of logic. *Archives of Virology* 151: 1251–1255.

Van Regenmortel MHV (2007) Virus species and virus identification: Past and current controversies. *Infection, Genetics and Evolution* 7: 133–144.

Van Regenmortel MHV and Mahy BWJ (2004) Emerging issues in virus taxonomy. *Emerging Infectious Diseases* 10: 8–13.

Viruses and Bioterrorism

R F Meyer and S A Morse, Centers for Disease Control and Prevention, Atlanta, GA, USA

Published by Elsevier Ltd.

Introduction

Man has known that biological organisms and toxins were useful as weapons of war long before the germ theory of disease was understood. However, as the twentieth century came to a close, the perceived difficulties in production, weaponization, and deployment of these biological weapons as well as a belief that moral restraints would preclude the use of these weapons gave many a false sense of security. Recently, a number of events have served to focus attention on the threat of terrorism and the potential for the use of biological, chemical, or nuclear weapons against the military, civilian populations, or agriculture for the purpose of causing illness, death, or economic loss. This possibility became a reality in October 2001 when someone sent spores of *Bacillus anthracis* to media companies in New York City and Boca Raton, Florida, resulting in five deaths, considerable panic throughout the United States and other countries, and raised the awareness of our vulnerability.

There are more than 1400 species of infectious organisms that are known to be pathogenic for humans; many additional organisms are capable of causing disease in animals or plants. Realistically, only a few of these infectious agents pose serious problems or are capable of affecting human, animal, or plant health on a large scale. Even fewer of these agents are viruses. Viruses that could be used as weapons against humans, animals, or plants generally possess traits including ease of production and dissemination, transmissibility, environmental stability, and high morbidity and mortality rates.

Definitions

The use of biological agents is often characterized by the manner in which they are used. For the purposes of this article 'biological warfare' is defined as a special type of warfare conducted by a government against a target; 'bioterrorism' is defined as the threat or use of a biological agent (or toxin) against humans, animals, or plants by individuals or groups motivated by political, religious, ecological, or other ideological objectives. Furthermore, terrorists can be distinguished from other types of criminals by their motivation and objective; however, criminals may also be driven by psychological pathologies and may use biological agents. When criminals use biological agents for murder, extortion, or revenge, it is called a 'biocrime'.

Historical Perspective

The use of viral agents for biological warfare has a long history, which predates their recognition and isolation by culture. Their early use is consistent with what, at the time, was known about infectious diseases, particularly smallpox. In the sixteenth century, the Spanish explorer, Francisco Pizarro, presented the indigenous peoples of South America with variola-contaminated clothing, which resulted in widespread epidemics of smallpox. During the French and Indian War (1745–67), Sir Jeffrey Amherst, commander of the British forces in North America, suggested the deliberate use of smallpox to 'reduce' Native American tribes hostile to the British. Captain Ecuyer (one of Amherst's subordinates), fearing an attack on Ft. Pitt from Native Americans, acquired two variola-contaminated blankets and a handkerchief from a smallpox hospital and, in a gesture of good will, distributed them to the Native Americans. As a result, several outbreaks of smallpox occurred in various tribes in the Ohio River valley. In 1775, during the Revolutionary War, the British attempted to spread smallpox among the Continental forces by inoculating (variolation) civilians fleeing Boston. In the South, there is evidence that the British were going to distribute slaves who had escaped during hostilities, and were sick with smallpox, back to the rebel plantations in order to spread the disease.

The use of viruses other than Variola major is a more recent phenomenon and reflects an increased knowledge of how to grow and stabilize viruses for delivery purposes. Allegations have been made by the government of Cuba that the CIA was responsible for the massive outbreaks of swine fever in 1971 and dengue fever in 1980 that ravaged the country. However, subsequent investigations have failed to find substantive proof of CIA involvement in these outbreaks. The Aum Shinrikyo, a religious cult responsible for the 1995 release of sarin gas in the Tokyo subway system, was also involved in biological warfare activity and sent a team of 40 people to Zaire to acquire Ebola virus. Fortunately, they were unsuccessful in this endeavor. In 1997, unknown farmers in New Zealand deliberately and illegally introduced rabbit hemorrhagic disease virus (a calicivirus) onto the south island as an animal control tool to kill feral rabbits.

Over the past two decades, the human immunodeficiency virus (HIV) has been involved in a number of biocrimes. This most likely reflects the availability of HIV-contaminated blood as a source of this virus. For example, in 1990, Graham Farlow, an asymptomatic HIV-positive inmate at a prison in New South Wales, Australia, injected a guard with HIV-contaminated blood. The guard became infected with HIV; Farlow subsequently died of AIDS. In 1992, Brian T. Stewart, a phlebotomist at a St. Louis, MO hospital, injected his 11-month-old son with HIV-contaminated blood during a fight over payment of child support. In 1993, Iwan E. injected his former girlfriend with 2.5 ml of HIV-contaminated blood after she broke up with him. In 1994, Dr. Richard J. Schmidt, a married Louisiana gastroenterologist, injected a former lover with HIV-contaminated blood. Molecular typing of the HIV strains demonstrated that she contracted the same strain of HIV as found in one of Dr. Schmidt's patients. In perhaps the most famous case, Dr. David Acer, a Florida dentist infected with HIV, transmitted the disease to six of his patients between 1987 and 1990. The intentional infection of these patients is a possibility although there is no direct evidence. In spite of these incidents, HIV has not been included on lists of threat agents for public health bioterrorism preparedness. However, some contend that HIV has great weapon potential if the goal is to destabilize a society.

Viruses have also been involved in suspected incidents or hoaxes. In 1999, an article appeared suggesting that the CIA was investigating whether Iraq was responsible for causing the outbreak of West Nile fever in the New York City area. The story relied heavily on a previous story written by an Iraqi defector, claiming that Saddam Hussein planned to use West Nile virus strain SV 1417 to mount an attack. The investigation indicated that there was no known evidence of bioterrorism involved in the spread of West Nile virus. A fictional 'virus' was also involved in one of the largest bioterrorism hoaxes in 2000. According to e-mail messages widely circulated on the Internet, an organization known as the Klingerman Foundation was mailing blue envelopes containing sponges contaminated with a fictional pathogen called the 'Klingerman virus'. According to the e-mail alert, 23 people had been infected with the virus, including 7 who died.

Viruses as Bioweapons

Advances in viral culture and virus stabilization made during the second half of the twentieth century facilitated large-scale production of viral agents for aerosol dissemination. A report for the United Nations on chemical and biological weapons and the effects of their possible use gave estimates on the numbers of casualties produced by a hypothetical biological attack (**Table 1**). Three viruses (Rift Valley fever virus, Tick-borne encephalitis virus, and Venezuelan Equine Encephalomyelitis (VEE) virus) were evaluated in a scenario in which 50 kg of the agent was released by aircraft along a 2 km line upwind of a population center of 500 000. The viral agents produced fewer casualties and impacted a smaller area when compared with the bacterial agents used in this hypothetical model. Of note, smallpox was apparently not evaluated because it had not yet been eradicated and level of vaccine-induced immunity in the population was high.

Viral agents were part of the biological weapons arsenal of both the Soviet Union and the United States (**Table 2**). VEE virus was stockpiled by both countries as an incapacitating agent; Variola major and Marburg viruses were stockpiled as lethal agents by the Soviet Union. The Soviet Union reportedly conducted a live field test of Variola major virus on Vozrozhdeniye Island in the Aral Sea in the 1970s, in which 400 g of the virus was released into the atmosphere by explosion. Unfortunately, a laboratory technician who was collecting plankton samples from an oceanographic research vessel 15 km from the island became infected. It was reported that after returning home to Aralsk, she transmitted the infection to several people including children. All those infected died. A number of other viruses that infect humans (e.g., Ebola virus, Lassa fever virus, enterovirus 70) or livestock (e.g., foot and mouth disease virus, rinderpest, Newcastle disease virus) have also been studied for their offensive capabilities or for the development of medical and veterinary countermeasures.

Today, with the increased level of concern, a number of viruses have been cited as possible weapons for use against humans or animals (**Table 2**). The requirements for an ideal biological warfare agent include availability, ease of production, stability after production, a susceptible population, absence of specific treatment, ability to incapacitate or kill the host, appropriate particle size in aerosol so that the virus can be carried long distances by prevailing winds and inhaled deeply into the lungs of unsuspecting victims, ability to be disseminated via food or water, and the availability of a vaccine to protect certain groups. Other factors such as the economic and psychological impact of an attack on animal agriculture with a viral agent must also be considered.

Variola major is considered to be the major viral threat agent for humans. Thus, considerable effort has been expended toward preparing the public health and medical communities for the possibility that this agent will be employed by a terrorist. Variola major is considered to be an ideal terrorist weapon because it is highly transmissible by the aerosol route from infected to susceptible persons; the civilian populations of most countries contain a high proportion of susceptible persons; the disease is associated with a high morbidity and about 30% mortality; initially, the diagnosis of a disease that has not been seen for almost 30 years would be difficult; and, other than the vaccine, which may be effective in the first few days post infection, there is no proven treatment available.

Alphaviruses (**Table 2**) are also of concern because they can be produced in large amounts in inexpensive and unsophisticated systems; they are relatively stable and highly infectious for humans as aerosols, and strains are available that produce incapacitating (e.g., VEE) or lethal infections (EEE case fatality rates range from 50–75%). Furthermore, the existence of multiple serotypes of VEE and EEE viruses, as well as the inherent difficulties of inducing efficient mucosal immunity, make defensive vaccine development difficult.

The filoviruses and arenaviruses that cause hemorrhagic fever have also been considered as agents that might be used by terrorists because of their high virulence and capacity for causing fear and anxiety. The filoviruses, Ebola and Marburg, can also be highly infectious by the airborne route. Humans are generally susceptible to infection with these viruses with fatality rates greater than 80%, and infection can be transmitted between humans through direct contact with virus-containing body fluids. There are five species of arenaviruses (Lassa fever, Junin, Machupo, Guanarito, and Sabia) that can cause viral hemorrhagic fevers with a case fatality rate of about 20%. Large quantities of these viruses can be produced by propagation in cell culture. Infection occurs via the respiratory pathway suggesting that dissemination via aerosol might be used by a terrorist. Human to human transmission has also been reported with aerosol transmission the most likely route for at least some of the secondary cases. The filoviruses and arenaviruses discussed above are BSL-4

Table 1 Estimates of casualties produced by hypothetical biological attack

Agent	Downwind Reach (km)	Dead	Incapacitated
Rift Valley fever	1	400	35 000
Tick-borne encephalitis	1	9500	35 000
VEE	1	200	19 800
Francisella tularensis	>20	30 000	125 000
Bacillus anthracis	>20	95 000	125 000

Note. These estimates are based on the following scenario: release of 50 kg of agent by aircraft along a 2 km line upwind of a population center of 500 000.

Table 2 Classification of viral agents that are considered to be of concern for bioterrorism and biowarfare and those that have been weaponized or studied for offensive or defensive purposes as part of former or current national biological weapons programs

Nucleic acid	Family	Genus	Species
Negative-sense single-stranded RNA	Arenaviridae	Arenaviruses	Lassa fever[a,b]
			Junin[a,b]
			Machupo[a,b]
			Sabia
			Guanarito
	Bunyaviridae	Phlebovirus	Rift Valley fever[b]
		Nairovirus	Crimean-Congo HF
		Hantavirus	Hantaan and related viruses[b]
			Sin Nombre
	Orthomyxoviridae	Influenzaviruses	Influenza A[b]
	Filoviridae	Filovirus	Ebola[a]
			Marburg[c]
	Paramyxoviridae	Henipavirus	Nipah virus
		Morbillivirus	Rinderpest[a,b,d,e,f]
		Avulavirus	Newcastle disease virus[b]
Positive-sense single-stranded RNA	Flaviviridae	Flavivirus	Yellow fever[a,b,d]
			Dengue[b]
			Tick-borne encephalitis virus[a]
			Japanese encephalitis virus[a]
			Omsk hemorrhagic fever virus
	Togaviridae	Alphavirus	Venezuelan equine encephalomyelitis virus[c,g]
			Eastern equine encephalomyelitis virus[b]
			Western equine encephalomyelitis virus[b]
			Chikungunya virus[b]
	Picornaviridae	Enterovirus	Enterovirus 70[h]
		Hepatovirus	Hepatitis A virus
		Apthovirus	Foot and mouth disease virus[f,i]
Double-stranded DNA	Poxviridae	Orthopoxvirus	Variola major[c,b,j]
			Camelpox[h]
	Asfarviridae	Asfivirus	African swine fever virus[a]

[a] Studied by the Soviet Union BW program.
[b] Studied by the U.S. BW program.
[c] Weaponized by the Soviet Union BW program.
[d] Studied by the Canada BW program.
[e] Studied by the France BW program.
[f] Studied by the Germany BW program.
[g] Weaponized by the U.S. BW program.
[h] Studied by the Iraq BW program.
[i] Studied by the Iran BW program.
[j] Studied by the North Korea BW program.

agents and diagnostic capacities for infections caused by these viruses are limited.

Impact of Biotechnology

Because the nucleic acid of many viruses, including some that are currently not threats, can be manipulated in the laboratory, the potential for genetic engineering remains a serious threat. Biotechnology, which has had a tremendous impact on the development of medicines, vaccines, and in the technologies needed to counter the threat of naturally occurring disease, can also be used to modify viruses with unintended consequences or even for the development of novel biological agents. Several examples involving viruses are presented below.

Mousepox Virus

An Australian research group was investigating virally vectored immunocontraceptive vaccines based on ectromelia virus, the causative agent of the disease termed mousepox. They created a recombinant virus, which expressed the mouse cytokine IL-4 in order to enhance the antibody-mediated response to other recombinant antigens carried on the virus vector. Instead, the ectromelia virus vector expressing IL-4 altered the host's immune response to this virus resulting in lethal infections in normally genetically resistant mice (e.g., C57BL/6). Additionally, this virus also caused lethal infections in mice previously immunized against infection with ectromelia virus. The creation of this 'supermousepox' virus led to speculation that similar genetic engineering could be performed on Variola major leading to a biological weapon that would be effective against an immunized population.

Pandemic Influenza

The influenza pandemic of 1918–19, which followed World War I, was uniquely severe, causing an estimated 20–40 million deaths globally. This pandemic happened before the advent of viral culture and very little was known about the virus until the discovery of the polymerase chain reaction (PCR). Recently, the complete coding sequences of all eight viral RNA segments has been determined by using reverse transcription-PCR (RT-PCR) to amplify the viral RNA sequences from formalin-fixed and frozen tissue samples from individuals who died during this pandemic in an effort to shed light on both the reasons for its extraordinary virulence and evolutionary origin. More recently, researchers reconstructed the 1918 Spanish influenza pandemic virus using reverse genetics and observed that this reconstructed virus exhibited exceptional virulence in the model systems examined and that the 1918 hemaglutinin and polymerase genes were essential for optimal virulence.

Synthetic Genomes

A full-length poliovirus complementary DNA (cDNA) (c. 7500 bp) has been synthesized in the laboratory by assembling oligonucleotides of plus- and minus-strand polarity. The synthetic poliovirus cDNA was transcribed by RNA polymerase into viral RNA, which translated and replicated in a cytoplasmic extract of uninfected HeLa S3 cells, resulting in the *de novo* synthesis of infectious poliovirus. The publication of this research raised concerns that more complicated viruses (e.g., Variola major or Ebola) could be synthesized from scratch based on publicly available sequences, or that viruses could be created that do not exist in the wild.

Recognition, Response, and Deterrence

An effective defense requires a comprehensive approach that includes: prevention of access to viral stocks; improved means of detecting deliberately induced disease outbreaks; rapid medical recognition of specific syndromes (e.g., hemorrhagic fever syndrome); rapid laboratory identification of viruses in patient specimens; prevention of person–person transmission; reliable decontamination procedures; development of effective vaccines; and development of effective antiviral therapy.

Rapid and accurate detection of biological threat agents is the basis of an effective public health response to bioterrorism. In order to address this issue, CDC in collaboration with other partners established a national network of laboratories called the Laboratory Response Network (LRN), which was provided with the tools to accomplish this mission. Rapid assays utilizing advanced molecular and immunological technologies for detection of agents such as variola virus, as well as emerging public health threats such as SARS coronavirus and H5N1 influenza virus, were distributed to member laboratories. Equipment, training, and proficiency testing are elements of the LRN and contribute to a uniform operational plan. The importance of high-quality standardized testing for detection of these agents is exemplified by the rapid need for medical countermeasures to protect or treat civilian populations. Accurate laboratory analysis is a major element in the decision process for deployment of the Federal Government's Strategic National Stockpile (SNS) of medical countermeasures.

As part of the effort to deter biological terrorism and strengthen the law enforcement response to such an act, the US recently established a microbial forensic laboratory known as the National Bioforensics Analysis Center that operates in partnership with the Federal Bureau of Investigation. Scientists are already developing methods for the forensic investigation of incidents involving viruses.

Summary

For the terrorist, the use of a viral agent would pose a challenge due to problems associated with acquisition, cultivation, and dissemination. The target for an attack with a viral agent can range from humans to animals and plants. Therefore, agricultural targets are also a major concern. Nature has provided many challenges to combating viral diseases. Viral agents are much more prone to genetic variation and mutation, and can be manipulated or created in the laboratory to take on desired characteristics. Differentiating between natural and intentional viral disease outbreaks can be challenging. Unlike bacterial diseases, many of which are treatable, there are fewer medical countermeasures to employ when dealing with viral infections. Laboratory diagnostic methods and reagents must continuously be refined to account for genetic changes and variants. Thus, the challenge of developing bioterrorism countermeasures is significant. Fortunately, this effort contributes to combating natural disease events more effectively, which has global benefits.

See also: AIDS: Disease Manifestation; AIDS: Global Epidemiology; AIDS: Vaccine Development.

Further Reading

Bray M (2003) Defense against filoviruses used as biological weapons. *Antiviral Research* 57: 53–60.

Breeze RG, Budowle B, and Schutzer SE (eds.) (2005) *Microbial Forensics*. Burlington, MA: Academic Press.

Carus WS (2002) *Bioterrorism and Biocrimes. The Illicit Use of Biological Agents since 1900*. Amsterdam, The Netherlands: Fredonia Books.

Cello J, Paul AV, and Wimmer E (2002) Chemical synthesis of poliovirus cDNA: Generation of infectious virus in the absence of natural template. *Science* 297: 1016–1018.

Charrel RN and de Lamballerie X (2003) Arena viruses other than Lassa virus. *Antiviral Research* 57: 89–100.

Enserink M (2002) Did bioweapons test cause a deadly smallpox outbreak? *Science* 296: 2116–2117.

Henderson DA, Inglesby TV, Gartlett JG, et al. (1999) Smallpox as a biological weapon. *JAMA* 281: 2127–2137.

Hopkins DR (1983) *Princes and Peasants. Smallpox in History*. Chicago, IL: University of Chicago Press.

Jackson RJ, Ramsay AJ, Christensen SD, Beaton S, Hall DF, and Ramshaw IA (2001) Expression of mouse interleukin-4 by a recombinant ectromelia virus suppresses cytolytic lymphocyte responses and overcomes genetic resistance to mousepox. *Journal of Virology* 75: 1205–1210.

Lanciotti RS, Roehrig JT, Duebel V, et al. (1999) Origin of the West Nile virus responsible for an outbreak of encephalitis in the northeastern United States. *Science* 286: 2333–2337.

Metzker ML, Mindell DP, Liu X, Ptak RG, Gibbs RA, and Hillis DM (2002) Molecular evidence of HIV-1 transmission in a criminal case. *Proceedings of the National Academy of Science, USA* 99: 14292–14297.

Rotz LD, Khan AS, Lillibridge SR, Ostroff SM, and Hughes JM (2002) Public health assessment of potential biological terrorism agents. *Emerging Infectious Diseases* 8: 225–229.

Sidwell RW and Smee DF (2003) Viruses of the Bunya- and Togaviridae families: Potential as bioterrorism agents and means of control. *Antiviral Research* 57: 101–111.

Tumpey TM, Basler CF, Aguilar PV, et al. (2005) Characterization of the reconstructed 1918 Spanish influenza pandemic virus. *Science* 310: 77–80.

World Health Organization(1970) *Health Aspects of Chemical and Biological Weapons*. Geneva: WHO.

Viruses Infecting Euryarchaea

K Porter, B E Russ, A N Thorburn, and M L Dyall-Smith, The University of Melbourne, Parkville, VIC, Australia

© 2008 Elsevier Ltd. All rights reserved.

Glossary

Alkaliphilic Having a requirement for an environment with a high pH.

Burst size The number of infectious virus particles released per cell.

Carrier state Persistent infection of a host cell by a virus, with the surviving host persistently carrying and continually producing virus without entering a lysogenic state.

Circular permutation A change in the sequence of the linear DNA termini that does not alter the relative sequence (e.g., circular permutation of ABCDEFGH could generate BCDEFGHA, CDEFGHAB, etc).

Concatamer Two or more DNA molecules that are linked together to form a long, linear DNA molecule.

Cured A host cell that was once a lysogen, but no longer carries viral DNA in any form.

Halophilic Having a requirement for an environment with a high salt concentration.

Headful packaging The mechanism of packaging viral DNA based on the size of the virus head, rather than the length of the viral genome.

Hyperthermophile Having a requirement for an environment with a high temperature ($>80\,°C$).

Insertion sequences Repetitive sequences of DNA that can move from one site to another within the viral DNA.

Integrase/recombinase An enzyme which can integrate viral DNA into the genome of its host cell.

Lysogen A host cell that has been infected by a virus that remains dormant, despite the presence of viral DNA.

Lytic virus A virus that is able to infect a host, replicate, and subsequently leave the host cell by rupturing (lysing) the host cell.

Methanogenic Having the ability to produce methane.

Monovalent A virus that has a host range limited to one species.

Prophage A virus that is dormant within the host cell.

Protein-primed replication Replication of DNA via the interaction of the DNA polymerase with specific proteins, rather than DNA or RNA primers.

Temperate virus A virus that is able to infect a host, but remain dormant within the host cell.

Terminal redundancy Linear DNA with the same sequence at each end.

Transduction The transfer of host DNA from one host cell to another by a virus.

Transfection The introduction of pure viral genomic DNA into a host cell, producing viable virus.

Introduction

In comparison with the viruses infecting Bacteria and Eukarya, our understanding of viruses that infect the third domain of life, the Archaea, is still in the early stages. The first archaeal virus was accidentally discovered in 1974, 3 years before the recognition of Archaea as a separate domain. Twenty-two years later, a total of only ~35 archaeal viruses and virus-like particles (VLPs) had been reported, the vast majority of which displayed head-and-tail morphologies and possessed linear double-stranded DNA (dsDNA) genomes. By 2006, at least 63 viruses and VLPs had been described, which infect members of the two major archaeal kingdoms, Crenarchaeota, including the extremely thermophilic sulfur-metabolizers, and Euryarchaeota, including the anaerobic methanogens, the extreme halophiles, and some hyperthermophiles. All have dsDNA genomes and while the total number of archaeal viruses is still miniscule compared with the >5100 published bacteriophages, they display a wide diversity of morphologies and characteristics. This article focuses on virus representatives infecting the members of the kingdom Euryarchaeota, a group for which the diversity of known representatives has blossomed in recent years.

Early Euryarchaeal Viruses and VLPs

In the early days of euryarchaeal virus research (1974–93), the range of virus and VLP representatives of the Euryarchaeota was limited. Attempts to isolate these particles were rare and many of the published representatives were discovered accidentally, for example, as contaminants in 'pure' preparations of archaeal flagella. In hindsight, the inability to culture the environmentally dominant Archaea restricted the isolation of viruses to those that infected easily grown laboratory strains, for example, *Halobacterium* for the haloarchaea and *Methanothermobacter* and *Methanobrevibacter* for the methanogens. Except for one oblate VLP with a circular dsDNA genome observed in cultures of *Methanococcus voltae* strain A3, all representatives possessed head-and-tail morphologies with complex protein profiles and genomes of linear dsDNA, similar to the 'classical' bacteriophages of the families *Myoviridae* (head-and-tail viruses with contractile tails) and *Siphoviridae* (head-and-tail viruses with noncontractile tails).

Like their hosts, the viruses of the extremely halophilic Archaea generally required high levels of NaCl or $MgSO_4$ for stability, although halobacterium salinarum viruses Hh-1 and ΦN were relatively stable in low salt, with the latter virus retaining infectivity even after prolonged incubation in distilled water. Many of the extremely halophilic viruses (or haloviruses) were lytic, with latent periods ranging from 6 to 17 h and burst sizes between 60 and 1300 PFU/cell. These latent periods and burst sizes appeared to be regulated by external salt concentrations, a phenomenon proposed to be significant for virus replication during changing salinity conditions in the environment. Some haloviruses such as halobacterium salinarum virus ΦH were truly temperate, while others such as halobacterium salinarum virus S45 persistently infected their host cells.

The viruses of the methanogens were lytic (by definition, no infectivity could be shown for VLP A3), with latent periods ranging from 4 to 9 h and burst sizes between 6 and 20 PFU/cell. However, some controversy still exists regarding the virus–host relationship for ψM1 and relatives (see below).

In general, the characterization of the early euryarchaeal virus and VLP isolates did not progress to the molecular level and it is likely that most of the early isolates are now lost. The exceptions were ΦH, which is one of the most thoroughly studied archaeal viruses, and methanothermobacter marburgensis virus ψM1, and relatives ψM2 and ψM100 (see below). ΦH was isolated after spontaneous lysis of its host culture and was able to infect a range of *Hbt. salinarum* strains. The virion morphologically resembled members of the family *Myoviridae* and contained a linear dsDNA genome of *c.* 59 kbp, which was packaged by a headful mechanism. The genome was partially sequenced and was shown to be transcribed in early, middle, and late phases. Although the virus particles contained linear DNA, ΦH was temperate and circular viral genomes (plasmids) could be isolated from infected cells. Genomic DNA was able to be transfected into *Hbt. salinarium* and *Haloferax volcanii* cells, producing viable virus particles. This was the first demonstration of transfection in Archaea, and allowed the optimization of plasmid transformation. However, ΦH was highly unstable, due to the activity of insertion sequences and inversion of the 'L-fragment', which could act as an autonomous plasmid. For this reason, work on ΦH was discontinued in the late 1990s.

Current Euryarchaeal Viruses and VLPs

The published viruses and VLPs infecting members of the kingdom Euryarchaeota in the readily accessible literature are listed in **Table 1**. This table excludes nine uncharacterized head-and-tail viruses with *c.* 50 nm diameter heads, isolated in 1977 from the Great Salt Lake in Utah, USA, on uncharacterized *Halobacterium* isolates. The remainder of this article focuses on detailing the advances in research for euryarchaeal viruses and VLPs that have been studied since 2001.

ψM1, ψM2, and ψM100; Viruses of the Methanogenic Archaea

A monovalent virus of *M. marburgensis*, ψM1 was isolated several times from an anaerobic sludge-bed reactor in

Table 1 Viruses and virus-like particles of the kingdom Euryarchaeota

Virus	Publication date	Isolating host	Isolated from	Infection pathway	Genome size (kbp)	GC content (%)	Particle size (nm)	Number of proteins; size (kDa)	Lipids
Order *Caudovirales*, family *Myoviridae*									
Viruses with an isometric head, contractile tail, and linear, double-stranded DNA genome									
ΦCh1	1997	*Natrialba magadii* DSM 3394	*Natrialba magadii*	Lytic, temperate	58.5[a,b] AF440695	61.9	70 (head); 130 (tail)	9; 15–80	nd[c]
ΦH	1982	*Halobacterium salinarum* ATCC 29341	*Halobacterium salinarum*	Temperate	59[d]	65	64 (head); 170 (tail)	>3; 22–80	nd
HF1	1993	*Haloferax lucentense* NCIMB 13854	Cheetham, VIC, Australia	Lytic, carrier state	75.7[b] AY190604	55.8	58 (head); 94 (tail)	>4; 20–55	nd
HF2	1993	*Halorubrum coriense* ACAM 3911	Cheetham, VIC, Australia	Lytic, carrier state	77.7[b] AF222060	55.8	58 (head); 94 (tail)	>4; 20–55	nd
Hs1	1974	*Halobacterium salinarum* strain 1	*Halobacterium salinarum*	Lytic, carrier state	nd	nd	50 (head); 120 (tail)	nd	nd
Ja.1	1975	*Halobacterium salinarum* NRC 34001	The Salt Ponds of Yallahs, Jamaica	Lytic	230	nd	90 (head); 150 (tail)	nd	nd
S41	1998	*Halobacterium salinarum* NRC 34001	Little Salt Pond in Yallahs, Jamaica	Lytic, carrier state	nd	nd	89 (head); 141 (tail)	nd	nd
S50.2	1998	*Halobacterium salinarum* NRC 34001	Little Salt Pond in Yallahs, Jamaica	Lytic, carrier state	nd	nd	63 (head); 78 (tail)	nd	nd
S4100	1998	*Halobacterium salinarum* NRC 34001	Little Salt Pond in Yallahs, Jamaica	Lytic, carrier state	nd	nd	56 (head); 85 (tail)	nd	nd
S5100	1990	*Halobacterium salinarum* NRC 34001	Little Salt Pond in Yallahs, Jamaica	Lytic, carrier state	nd	nd	65 (head); 76 (tail)	nd	nd
Order *Caudovirales*, family *Siphoviridae*									
Viruses with an isometric head, noncontractile tail, and linear, double-stranded DNA genome									
ΦF1	1993	*Methanobacterium thermoformicicum* FF1	Anaerobic sludge-bed reactor	Lytic	85	nd	70 (head); 160 (tail)	nd	nd
ΦF3	1993	*Methanobacterium thermoformicicum* FF1	Anaerobic sludge-bed reactor	Lytic	36	nd	55 (head); 230 (tail)	nd	nd
ΦN	1988	*Halobacterium salinarum* NRL/JW	*Halobacterium salinarum*	nd	56[e]	70	55 (head); 85 (tail)	>1; 53	nd
ψM1	1989	*Methanothermobacter marburgensis* DSM 2133	Experimental anaerobic sludge digester	Lytic, possible integration	26.8[b] AF065411, AF065412	46.3	55 (head); 210 (tail)	3; 10–35	nd
ψM2	1989	*Methanothermobacter marburgensis* DSM 2133	Spontaneous deletion mutant of ψM1	Lytic, possible integration	26.1[b] AF065411	46.3	55 (head); 210 (tail)	3; 10–35	nd
ψM100	2001	*Methanothermobacter wolfeii*	Defective provirus in host chromosome	Defective provirus	28.8[b] AF301375	45.4	nd	nd	nd
B10	1982	*Halobacterium* sp. B10	nd	Lytic	nd	nd	nd	nd	nd
Hh-1	1982	*Halobacterium salinarum* ATCC 29341	Fermented anchovy sauce	No lysis during release, carrier state	37.2	67.05	60 (head); 100 (tail)	>1; 35	nd

Continued

Table 1 Continued

Virus	Publication date	Isolating host	Isolated from	Infection pathway	Genome size (kbp)	GC content (%)	Particle size (nm)	Number of proteins; size (kDa)	Lipids
Hh-3	1982	*Halobacterium salinarum* ATCC 29341	Fermented anchovy sauce	Lytic, carrier state	29.4	62.15	75 (head); 50 (tail)	lt; 1; 47	nd
PG	1986	*Methanobrevibacter smithii* strain G	Rumen fluid samples	Lytic	50	nd	nd	nd	nd
PMS1	1986	*Methanobrevibacter smithii*	nd	Lytic	35	nd	nd	nd	nd
S45	1984	*Halobacterium salinarum* NRC 34001	The Salt Ponds of Yallahs, Jamaica	No lysis, persistent infection	nd	nd	40 (head); 70 (tail)	nd	nd
VTA	1999	*Methanococcus voltae* PS	*Methanococcus voltae*	nd	4.4	nd	40 (head); 61 (tail)	nd	nd

Genus *Salterprovirus*
Viruses that are spindle-shaped, with a tail and a linear, double-stranded DNA genome, with terminal-bound proteins.

Virus	Publication date	Isolating host	Isolated from	Infection pathway	Genome size (kbp)	GC content (%)	Particle size (nm)	Number of proteins; size (kDa)	Lipids
His1	1998	*Haloarcula hispanica* ATCC 36930	Avalon saltern, VIC, Australia	Lytic, carrier state	14.9[b] AF191796	40	74 × 44 (body); 7 (tail)	nd	Possible
His2	2006	*Haloarcula hispanica* ATCC 36930	Pink Lakes, VIC, Australia	Lytic, carrier state	16.1[b] AF191797	40	67 × 44 (body); nd (tail)	>4; 21–62	Possible

Unclassified
Viruses that are spherical, with a linear, double-stranded DNA genome

Virus	Publication date	Isolating host	Isolated from	Infection pathway	Genome size (kbp)	GC content (%)	Particle size (nm)	Number of proteins; size (kDa)	Lipids
SH1	2005	*Haloarcula hispanica* ATCC 36930	Serpentine Lake, WA, Australia	Lytic, carrier state	30.9[b] AY950802	68.4	70	15; 4–185	Yes

Unclassified
Virus-like particles that are polyhedral

Virus	Publication date	Isolating host	Isolated from	Infection pathway	Genome size (kbp)	GC content (%)	Particle size (nm)	Number of proteins; size (kDa)	Lipids
CWP	1988	*Pyrococcus woesei*	nd	nd	No nucleic acid	nd	nd	3	nd

Unclassified
Virus-like particles that are oblate, with a circular, double-stranded DNA genome

Virus	Publication date	Isolating host	Isolated from	Infection pathway	Genome size (kbp)	GC content (%)	Particle size (nm)	Number of proteins; size (kDa)	Lipids
VLPA3	1989	*Methanococcus voltae* A3	*Methanococcus voltae* A3	Temperate	23	nd	52 × 70	1 (and 3 minor); 13	nd

Unclassified
Virus-like particles that are lemon-shaped, with a tail, and a circular, double-stranded DNA genome

Virus	Publication date	Isolating host	Isolated from	Infection pathway	Genome size (kbp)	GC content (%)	Particle size (nm)	Number of proteins; size (kDa)	Lipids
VLP PAV1	2003	'*Pyrococcus abysii*' GE23	'*Pyrococcus abysii*'	Nonlytic, persistent infection	17.5[b]	nd	120 × 80 (body); 15 (tail)	>3; 6–36	Possible

[a]Genome completely sequenced.
[b]Also packages host derived RNA. Some genome adenine residues are methylated.
[c]nd, not determined.
[d]Some sequence data available: 405323, 405325, AH004327, S63992, S63933, S63994, X00805, X80161, X80162, X80163, X80164, X52504.
[e]All genome cytosine residues are methylated.

1989. The particles resemble bacteriophages of the family *Siphoviridae*, with a head diameter of *c.* 55 nm and a tail of *c.* 210 nm × 10 nm, composed of individual segments and an enlarged terminal segment (**Figure 1**). The ψM1 virion contains a 26.8 kbp genome and three major structural proteins. The proteins appear to be encoded by two open reading frames (ORFs), designated 13 and 18, whose products are post-translationally modified to produce the three proteins. To form the larger protein, the products of the ORFs are covalently cross-linked. The smaller proteins are then formed by processing of this protein at the N- and C-termini. The product of ORF 13 is similar to proteins found in Gram-positive Bacteria and bacteriophages; for example, it shows 28% identity over 330 amino acids to prophage PBSX protein XkdG in *Bacillus subtilis*.

The linear dsDNA genome of ψM1 is terminally redundant and circularly permuted, and packaged via a headful packing mechanism. Upon passage in laboratory cultures a spontaneous ψM1 deletion mutant designated ψM2, which lacked a 0.7 kbp segment (DR1), was found to become dominant. The 26.1 kbp ψM2 genome (AF065411) (**Figure 2**) and the DR1 ψM1 segment (AF065412) were sequenced. The ψM2 genome has a guanine–cytosine (GC) content of 46.3% and encodes 31 ORFs, seven of which have been assigned putative functions, again similar to those in Gram-positive Bacteria and bacteriophages. This suggests some gene exchange between bacterial and archaeal viruses, or a common ancestor of the bacterial and archaeal viruses. DR1 encodes ORF A, a duplication of ψM2 ORF 27. Although this element affects the stability of ψM1, the insertion and deletion of DR1 does not appear to interrupt any downstream ORFs.

Under laboratory conditions, ψM1 is a lytic virus, with no evidence of lysogen formation. The virus produces a pseudomurein endoisopeptidase, a lytic enzyme that cleaves host pseudomurein, encoded by *peiP* (ORF 28). Unexpectedly, ORF 29 encodes a putative site-specific integrase/recombinase, suggesting that ψM1 and ψM2 are actually temperate. This theory is supported by the absence of a viral DNA polymerase, which is usually found in lytic viruses, and the capacity of ψM1 particles to mediate transduction of resistance and biosynthesis markers, an ability as yet unobserved in other archaeal viruses. In addition, ψM1-resistant cultures of *Methanothermobacter wolfeii* spontaneously lyse and carry a chromosomal prophage, known as ψM100, which is homologous to ψM1 and ψM2. ψM100 appears to be a defective virus, as no VLPs are observed in autolysates but pseudomurein endoisopeptidase PeiW, encoded by ψM100 ORF 28, is present. The 28.8 kbp ψM100 sequence (AF301375) (**Figure 2**) has a GC content of 45.4%. It contains a 2.8 kbp fragment, IRFa, which has a GC content of 33.4% and apparently originates from a source not homologous to ψM1 and ψM2. The remaining 26.0 kbp of the ψM100 genome is 70.8% identical to ψM2. The lytic/temperate relationship between ψM1-like viruses and *M. marburgensis* remains unresolved.

Viruses of the Extremely Halophilic Archaea

Head-and-Tail Viruses HF1 and HF2

In 1993, a deliberate search for haloviruses infecting a wider range of haloarchaea resulted in the isolation of HF1 and HF2 from Cheetham Saltworks in Victoria, Australia. Both viruses are lytic, but can enter unstable carrier states in laboratory cultures. They have mutually exclusive host ranges, HF1 infecting a wide range of hosts from the genera *Haloarcula*, *Halobacterium*, and *Haloferax*, while HF2 strictly infects *Halorubrum* species. HF2 is sensitive to chloroform exposure, but HF1 is relatively insensitive to this solvent. In aspects other than host range and chloroform sensitivity, HF1 and HF2 are very similar. Their particles have identical morphologies, resembling bacteriophages of the family *Myoviridae*, with heads of *c.* 58 nm in diameter and tails of *c.* 94 nm in length (**Figure 3**). The structural proteins of HF1 and HF2 are very similar, with four major and several minor proteins. They also share similarly sized genomes of linear dsDNA.

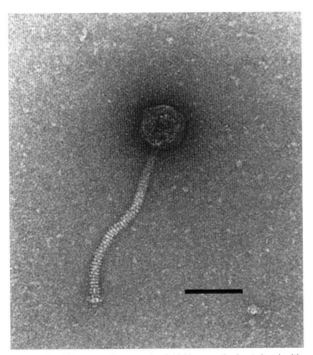

Figure 1 Electron micrograph of ψM1, negatively stained with uranyl acetate. Scale = 70 nm. Adapted from *Archives of Microbiology*, 152, 1989, p. 106, Characterization of ψM1, a virulent phage of *Methanobacterium thermoautotrophicum* Marburg, Meile L, Jenal U, Studer D, Jordan M, and Leisinger T, figure 1, © Springer-Verlag 1989, with kind permission from Springer Science and Business Media.

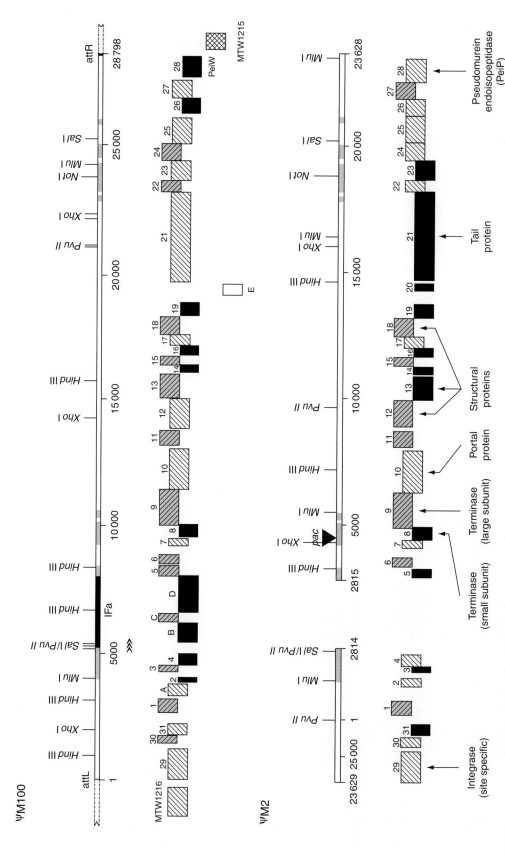

Figure 2 Schematic representation of ψM100 with its flanking regions and comparison to that of ψM2. The ORFs are represented by numbered boxes and their vertical placement indicates the gene location in one of the six possible reading frames. Homologous ORFs carry the same numbers, and the assigned functions for some ORFs of ψM2 are reported. MTW1215 and MTW1216 represent the two ORFs encoded by the chromosome sequences flanking ψM100. The experimentally determined pac locus for ψM2 is shown as a solid black triangle. The IFa fragment (black bar) and the DNA regions with at least 200 bp with >98% identity in ψM100 and ψM2 (gray bars) are indicated. Three arrowheads represent three contiguous copies of a direct repeat of 125 nt in the vicinity of the left end of the IFa fragment. Adapted from Luo Y, Pfister P, Leisinger T, and Wasserfallen A (2001) The genome of archaeal prophage ψM100 encodes the lytic enzyme responsible for autolysis of *Methanothermobacter wolfeii*. *Journal of Bacteriology* 183: 5789, with permission from American Society for Microbiology.

The genome of HF2 (AF222060) replicates via concatamer formation. It is 77.7 kbp in length (**Figure 4**), has a GC content of 55.8%, and contains 121 closely spaced ORFs. Approximately 12% of ORFs show similarity to known sequences. The HF2 ORFs are transcribed as operons, with distinct early, middle, and late transcripts. The ORFs are likely to be strategically organized into three groups: (1) genes involved in establishing infection; (2) genes involved in DNA synthesis, modification, and replication; (3) and genes involved in virus assembly and release. There are long intergenic repeats throughout the genome, which appear to be linked to transcription regulation. The genomic arrangement suggests that such viruses mediate genetic exchange across a range of hosts, resulting in mosaic virus genomes (see below).

The HF1 genome (AY190604) is 75.9 kbp in length (**Figure 4**), has a GC content of 55.8%, and encodes 117 ORFs, of which 13% show similarity to known sequences (excluding those of HF2). The first 48 kbp of the HF1 genome is identical to that of HF2, apart from one silent base change, and the remaining sequence is 87% identical. In the latter region, HF2 contains two unique ORFs, two pairs of ORFs that are joined in HF1, and several ORFs with less than 70% amino acid identity to those in HF1. Divergence in this region of the genome probably explains the wide differences in viral host range and chloroform sensitivities of the two viruses. The path of sequence divergence also indicates that a recent recombination event exchanged the end of one virus for that of a third, unknown but related, virus.

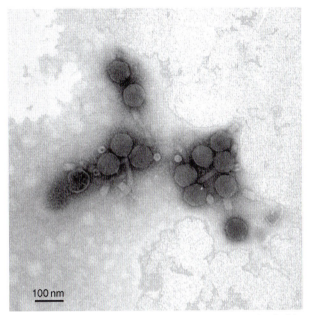

Figure 3 Electron micrograph of HF2, negatively stained with uranyl acetate.

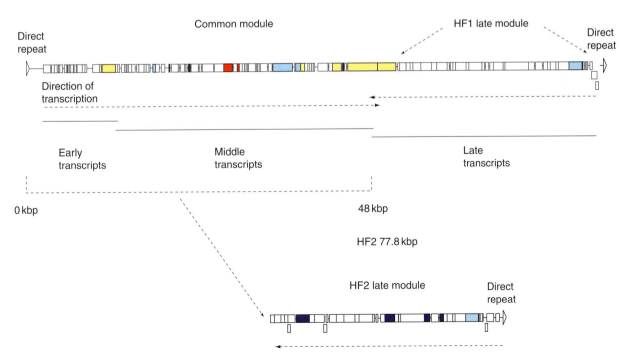

Figure 4 Closely related haloviruses HF1 (75.898 kbp) and HF2 (77.670 kbp) have genomes that are identical from 0 to 48 kbp, after which there is significantly lower homology. Direction of transcription is indicated below the genes in the diagram, as is the pattern of gene expression. Genes are drawn to scale. Genes highlighted in red are involved in transcription; dark blue – virion structure; yellow–nucleic acid modification; light blue – replication and repair.

The Haloalkaliphilic Virus, ΦCh1

Although most extremely halophilic Archaea live at a neutral pH, there is a group that is not only extremely halophilic, but is also alkaliphilic. In 1997, a haloalkaliphilic virus, ΦCh1, was isolated after spontaneous lysis of a culture of the haloalkaliphilic archaeon *Natrialba magadii*. ΦCh1 is monovalent, producing plaques on lawns of ΦCh1-cured *Nab. magadii* cells. The virus is also temperate, with lysogens containing genomic ΦCh1 DNA integrated into the *Nab. magadii* chromosome. Virus integration is potentially mediated by putative site-specific recombinases, Int1 and Int2, encoded by ORFs 35 (*int1*, which has been shown to mediate inversion and excision reactions, resulting in small, circular molecules) and 45 (*int2*).

ΦCh1 particles resemble the family *Myoviridae*, with a head of *c.* 70 nm in diameter and a contractile tail *c.* 130 nm × 20 nm (**Figure 5**). The ΦCh1 virion contains at least nine structural proteins, designated A to I in decreasing size order. Proteins A and H are encoded by ORF 19 and protein E is encoded by ORF 11 (gene *E*). Gene *E* is transcribed late in cell infection and protein E is found to be associated with the host membrane, suggesting a role in DNA packaging or assembly of coat proteins during viral exit.

ΦCh1 virus particles contain both linear dsDNA and host-derived RNA species. Genomic ΦCh1 DNA is modified at adenine residues within various sequences, including 5′-GATC-3′. Although DNA methylation is a common viral mechanism for avoiding host-mediated restriction, it was unexpected in ΦCh1, as *Nab. magadii* does not have methylated DNA. An N^6-adenine methyltransferase, encoded by ORF 94 of ΦCh1, is transcribed and expressed late in cell infection. These factors suggest that ΦCh1 may infect other hosts.

The ΦCh1 genome has a circular replicative form and is circularly permuted and terminally redundant, suggesting a headful packing mechanism. The 58.5 kbp genome (AF440695) has a GC content of 61.9% and is predicted to contain 98 ORFs (**Figure 6**). In a similar manner to head-and-tail bacteriophages, the ORFs appear to be organized into three transcriptional units encoding (1) structural and morphogenesis proteins; (2) replication, stabilization, and gene regulation proteins; and (3) proteins of mostly unknown functions. Around half of the ORFs show similarity to sequences in the databases, although most of these are proteins of unknown function. The genome sequence of ΦCh1 shows strong similarity to the incomplete genome of ΦH, which is surprising given that the two viruses were isolated from distinctly different environments and hosts. This unexpected relationship between supposedly diverse viruses suggests that similar viruses may be widely distributed throughout various hypersaline environments across the planet.

His1 and His2: Spindle-Shaped Haloviruses of the Genus *Salterprovirus*

From 1974 to 1998, all of the viruses described for the kingdom Euryarchaeota were of head-and-tail morphology. Nevertheless, in waters such as the Dead Sea, the most abundant VLPs observed by direct electron microscopy were spindle shaped. In a study of a Spanish saltern, spindle-shaped VLPs were found to increase in abundance with increasing salinity. In 1998, a spindle-shaped halovirus, His1, was reported. Isolated from the Avalon saltern in Victoria, Australia, on a lawn of *Haloarcula hispanica*, the His1 virion is *c.* 74 nm × 44 nm, with a short 7 nm tail (**Figure 7(a)**). His1 resembles crenarchaeal virus SSV1 and due to the morphological similarities and comparable genome sizes of these two viruses, His1 was initially, and mistakenly, placed in the *Fuselloviridae* family of archaeal viruses (type species, SSV1).

In 2006, a second spindle-shaped virus named His2 was described. It was isolated from the Pink Lakes in Victoria, Australia, and also plaques on *Har. hispanica*. By negative-stain electron microscopy, His2 particles are

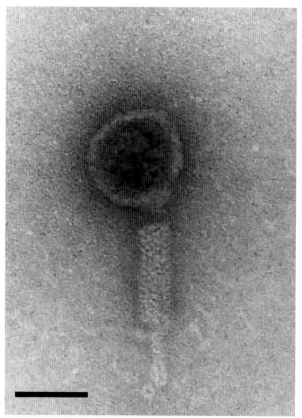

Figure 5 Electron micrograph of ΦCh1, negatively stained with phosphotungstate. A contracted tail and the anchor structure between head and tail are visible. Scale = 50 nm. Adapted from Witte A, Baranyi U, Klein R, *et al.* (1997) Characterization of *Natronobacterium magadii* phage ΦCh1, a unique archaeal phage containing DNA and RNA. *Molecular Microbiology* 23: 605, with permission from Blackwell Science Ltd.

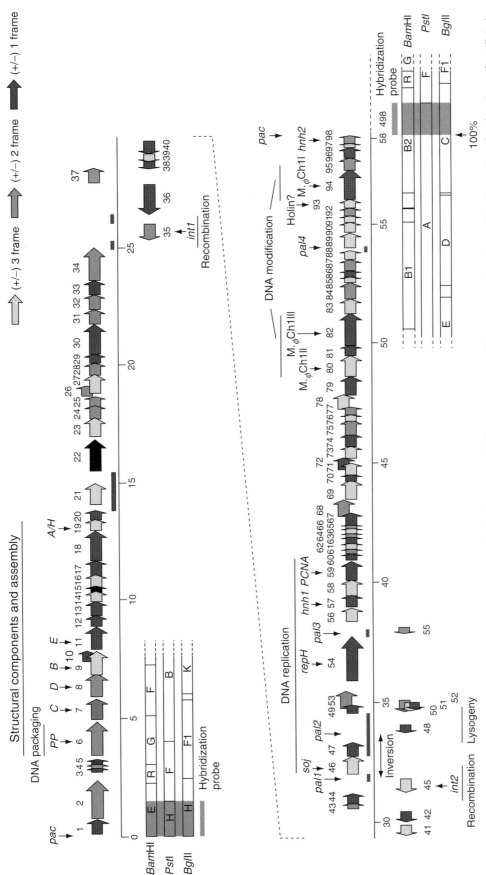

Figure 6 Linear representation of the 58 498 bp ΦCh1 genome, starting at the *pac* site. ORFs are represented by arrows and numbered. Putative and verified functions of predicted gene products are indicated. The shades of the arrows indicate the frames of the ORFs: light gray, third reading frame; mid-gray, second reading frame; and dark gray, first reading frame for both orientations. A partial restriction map showing the region embedding the *pac* site and the fragments that are created during the cut at *pac* is given beneath the genome map. Both maps are correlated with each other. The binding region for the hybridization probe that was used to map the *pac* site is shaded. Low G+C regions are indicated by black bars, and the segment that is inverted in pΦHL with respect to ΦCh1 is indicated by an arrow. The putative binding sites for site-specific recombinases, *pal1* to *pal4*, are also indicated. Adapted from Klein R, Baranyi U, Rössler N, Greineder B, Scholz H, and Witte A (2002) *Natrialba magadii* virus ΦCh1: First complete nucleotide sequence and functional organization of a virus infecting a haloalkaliphilic archaeon. *Molecular Microbiology* 45: 852, with permission from Blackwell Science Ltd.

more flexible than His1 particles, with the virions being c. 67 nm × 44 nm, with a short tail (**Figure 7(b)**). The His2 virion contains at least four structural proteins, designated viral protein (VP) 1 (62 kDa) to VP 4 (21 kDa). To date, the ORFs encoding VPs 2 and 3 have not been determined. VP 1 is encoded by ORF 29 and predicted to be exported to the cell surface and glycosylated. ORF 29, and the downstream ORFs 30, 31, and 33, have closely similar sequences in at least four species of extremely halophilic Archaea, although the segments of similarity do not appear to be part of provirus genomes. The functions of these archaeal homologs have not yet been determined.

Although both monovalent haloviruses are virulent, they appear to enter a carrier state and have the capacity to exit cells without causing cell lysis. His1 and His2 are relatively stable in the pH range 3–9; however, His1 is more resistant than His2 to exposure to both increased temperature and lowered ionic environments. Both His1 and His2 are sensitive to chloroform and have low buoyant densities and flexible virus particles. These factors suggest that the virions contain lipids.

The genomes of haloviruses His1 (AF191796) and His2 (AF191797) are linear dsDNA of 14.5 and 16.1 kbp, respectively (**Figure 8**). The genomes have a GC content of c. 40%, about 23% lower than *Har. hispanica*, suggesting that the haloviruses are not well equipped to replicate in *Haloarcula*. Moreover, no tRNA genes have been detected in the genomes of these viruses.

The His1 and His2 genomes each contain 35 ORFs and show almost no sequence similarity to one another or to known sequences. The notable exceptions are the putative DNA polymerases of the two viruses, which are weakly similar to one another at the nucleotide level and show 42% identity at the protein level. The DNA polymerases appear to be members of family B DNA polymerases that use the protein-priming mechanism. The His1 and His2 genomes have inverted terminal repeat (ITR) sequences at each end (105 and 525 nt, respectively), and the 5′ end of each is linked to a terminal protein (TP), which is essential for DNA replication. These characteristics are typical of systems that replicate DNA via protein priming.

At the molecular level, His1 and His2 are distantly related to one another; however, they do not appear to be related to fuselloviruses at all, with distinctly different genome structures, replication strategies, virus–host relationships, and genome sequences. Consequently, haloviruses His1 and His2 have been placed in an independent virus genus, *Salterprovirus*.

Spherical Halovirus SH1

In addition to the spindle-shaped and head-and-tail VLPs observed in hypersaline waters such as the Dead Sea, isometric particles were also present in high numbers. These could have been classical head-and-tail VLPs, which had lost their tails through the negative-stain treatment, which tends to disrupt halophilic virus particles. However, in 2003, a spherical halovirus named SH1 that resembled these isometric particles was first noted. It was isolated from Serpentine Lake, Rottnest Island, Western Australia, on *Har. hispanica*. The SH1 virion is c. 70 nm in diameter and displays an outer capsid layer with a compact core particle of c. 50 nm in diameter (**Figure 9**). Morphologically, it resembles viruses such as the bacteriophage PRD1, human adenovirus, and archaeal virus STIV, which all share common architecture.

SH1 is virulent, infecting both *Har. hispanica* and an uncharacterized isolate of the genus *Halorubrum*. The virion is stable to exposure to temperatures of 50 °C and in a pH range of 6–9, but is sensitive to exposure to both chloroform and lowered ionic environments. SH1 particles contain a lipid layer and 15 structural proteins,

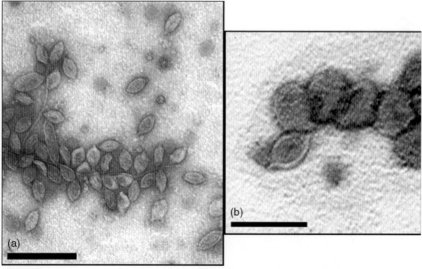

Figure 7 Electron micrograph of (a) His1 and (b) His2, negatively stained with uranyl acetate. Scale = 200 nm (a); 100 nm (b).

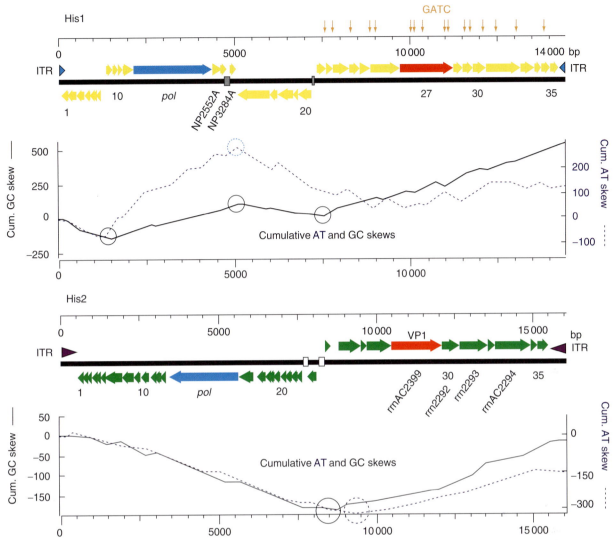

Figure 8 Diagram of the His1 and His2 genomes. The major features are shown, including the predicted ORFs (yellow and green arrows), putative DNA polymerases (blue arrows), VP1 and its His1 homolog (red arrows), terminal inverted repeat sequences (purple arrowheads), and regions of nucleotide sequence repeats (gray boxes). Scale bars, in bp, are shown above each genome diagram. ORFs are numbered (left to right), and some have been named (*pol*, DNA polymerase; VP1, major capsid protein) or the accession numbers of closely related sequences given. The template strand for each ORF is indicated by the direction of arrows and their position above or below the black line. The vertical arrows about the His1 genome scale bar show positions of the sequence GATC. Plots of cumulative GC skew (solid black line) and AT skew (dashed blue line) are shown below each genome diagram, with inflection points indicated by circles. Adapted from Bath C, Cukalac T, Porter K, and Dyall-Smith ML (2006) His1 and His2 are distantly related, spindle-shaped haloviruses belonging to the novel virus group, *Salterprovirus*, *Virology* 350: 233, with permission from Elsevier.

designated VP 1 to VP 15 in decreasing size order. Details on the locations and possible functions of the SH1 virion proteins are available.

The genome of SH1 (AY950802) is linear dsDNA of 30.9 kbp (**Figure 10**), with a GC content of 68.4%. The viral genome contains 56 ORFs, which show very little sequence similarity to known sequences. It contains 309 bp ITRs. The molecular features of SH1 indicate that it is a member of a novel virus group. Based on the evidence from electron microscopy surveys, haloviruses similar to SH1 are likely to be widespread and dominant members of hypersaline systems.

PAV1, A VLP of the Hyperthermophilic Order Thermococcales

No virus has ever been reported to infect the hyperthermophilic Euryarchaeota. However, two major groups of VLPs, rod-shaped particles and spindle-shaped particles, have been observed in enrichment cultures of samples obtained from deep-sea hydrothermal vents. In 2003, such spindle-shaped VLPs were discovered in supernatant of a '*Pyrococcus abyssii*' culture and were designated '*P. abysii*' virus 1 (PAV1). PAV1 is continuously released, but does not cause lysis of host cells and cannot be

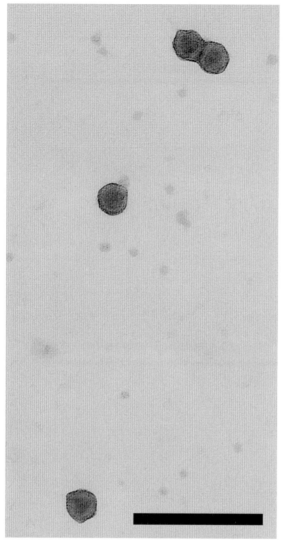

Figure 9 Electron micrograph of SH1, stained with uranyl acetate. Scale = 200 nm.

induced to infectivity using ultraviolet or γ irradiation, mitomycin C, or heat or pressure shock. No viral genomes integrated into the host chromosome have been detected.

PAV1 particles are flexible and similar in appearance to both the fuselloviruses and the salterproviruses, being *c.* 120 nm × 80 nm, with a tail of *c.* 15 nm and tail fibers (**Figure 11**). The virion is composed of at least three structural proteins, of *c.* 6, 13, and 36 kDa. PAV1 is sensitive to exposure to chloroform, Triton X-100, proteinase K, and sodium dodecyl sulfate. Combined with its low buoyant density and flexible particles, this suggests that the particles contain lipid.

The genome of PAV1 is circular dsDNA of 17.5 kbp. It contains 24 putative ORFs, which show no similarity to sequences in the databases. A detailed description of the PAV1 genome should be published soon.

Conclusion

At first glance, it may appear that the virus representatives of the kingdom Euryarchaeota resemble the more thoroughly studied bacterial and eukaryal viruses, or the viruses of the kingdom Crenarchaeota. However, on a molecular level, the euryarchaeal viruses are clearly distinct and radically different from any other known viruses. Many of the representatives, such as PAV1 and SH1, show almost no similarity to any sequences in the databases and even the head-and-tail representatives, which appear to resemble the 'classical' bacteriophages in both morphology and genome structure, encode many previously unobserved genes.

Fortunately, it appears that we are now in a good position to culture euryarchaeal viruses. Whether all the dominant morphological forms of euryarchaeal viruses have been isolated is still uncertain; however, the current

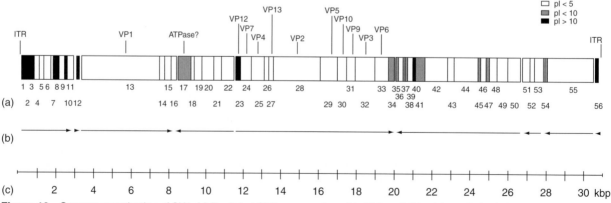

Figure 10 Genome organization of SH1. (a) Predicted ORFs are numbered 1–56 from left to right and shaded according to the calculated isoelectric point of the predicted gene products. Structural virion components (VP 1 to VP 14) determined by protein chemistry methods are marked. (b) Direction of transcription is depicted by arrows. (c) Scale bar. Adapted from Bamford DH, Ravantti JJ, Rönnholm G, *et al*. (2005) Constituents of SH1, a novel, lipid-containing virus infecting the halophilic euryarchaeon *Haloarcula hispanica*. *Journal of Virology* 79: 9102, with permission from American Society for Microbiology.

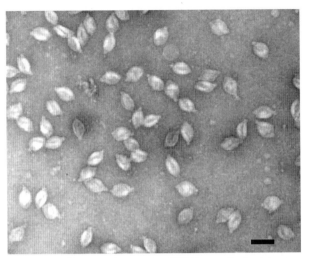

Figure 11 Electron micrograph of PAV1, negatively stained with 2% uranyl acetate. Scale = 100 nm. Adapted from Geslin C, Le Romancer M, Erauso G, Gaillard M, Perrot G, and Prieur D (2003) PAV1, the first virus-like particle isolated from a hyperthermophilic euryarchaeote, "*Pyrococcus abyssi*". *Journal of Bacteriology* 185: 3890, with permission from American Society for Microbiology.

cultivation techniques appear to have isolated all of the dominant morphological forms of haloviruses (head-and-tail, spindle shaped, and spherical) that have been observed in hypersaline waters. The current representatives suggest that there may be relatively few dominant virus families infecting the kingdom Euryarchaeota, and that these virus families may be widespread. This view may change radically when the metagenomic studies of archaeal environments, such as those performed by the Venter Institute program, begin to bear fruit. In any case, the major goal now is to develop genetic systems to analyze the viruses already isolated, in order to explore and understand their genetic properties, ecological significance, and evolutionary impact.

See also: Crenarchaeal Viruses: Morphotypes and Genomes; Fuselloviruses of Archaea.

Further Reading

Bamford DH, Ravantti JJ, Rönnholm G, et al. (2005) Constituents of SH1, a novel, lipid-containing virus infecting the halophilic euryarchaeon *Haloarcula hispanica*. *Journal of Virology* 79: 9097–9107.

Bath C, Cukalac T, Porter K, and Dyall-Smith ML (2006) His1 and His2 are distantly related, spindle-shaped haloviruses belonging to the novel virus group, *Salterprovirus*. *Virology* 350: 228–239.

Bath C and Dyall-Smith ML (1998) His1, an archaeal virus of the. *Fuselloviridae* family that infects *Haloarcula hispanica*. *Journal of Virology* 72: 9392–9395.

Dyall-Smith M, Tang S-L, and Bath C (2003) Haloarchaeal viruses: How diverse are they? *Research in Microbiology* 154: 309–313.

Fauquet CM, Mayo MA, Maniloff J, Desselberger U, and Ball LA (eds.) (2005) *Virus Taxonomy: Eighth Report of the International Committee on Taxonomy of Viruses*. San Diego, CA: Elsevier Academic Press.

Geslin C, Le Romancer M, Erauso G, Gaillard M, Perrot G, and Prieur D (2003) PAV1, the first virus-like particle isolated from a hyperthermophilic euryarchaeote, "*Pyrococcus abyssi*". *Journal of Bacteriology* 185: 3888–3894.

Klein R, Baranyi U, Rössler N, Greineder B, Scholz H, and Witte A (2002) *Natrialba magadii* virus ΦCh1: First complete nucleotide sequence and functional organization of a virus infecting a haloalkaliphilic archaeon. *Molecular Microbiology* 45: 851–863.

Luo Y, Pfister P, Leisinger T, and Wasserfallen A (2001) The genome of archaeal prophage ψM100 encodes the lytic enzyme responsible for autolysis of *Methanothermobacter wolfeii*. *Journal of Bacteriology* 183: 5788–5792.

Meile L, Jenal U, Studer D, Jordan M, and Leisinger T (1989) Characterization of ψM1, a virulent phage of *Methanobacterium thermoautotrophicum* Marburg. *Archives of Microbiology* 152: 105–110.

Pfister P, Wasserfallen A, Stettler R, and Leisinger T (1998) Molecular analysis of. *Methanobacterium* phage ψM2. *Molecular Microbiology* 30: 233–244.

Porter K, Kukkaro P, Bamford JKH, et al. (2005) SH1: A novel, spherical halovirus isolated from an Australian hypersaline lake. *Virology* 335: 22–33.

Rössler N, Klein R, Scholz H, and Witte A (2004) Inversion within the haloalkaliphilic virus ΦCh1 DNA results in differential expression of structural proteins. *Molecular Microbiology* 52: 413–426.

Tang S-L, Nuttall S, and Dyall-Smith M (2004) Haloviruses HF1 and HF2: Evidence for a recent and large recombination event. *Journal of Bacteriology* 186: 2810–2817.

Tang S-L, Nuttall S, Ngui K, Fisher C, Lopez P, and Dyall-Smith M (2002) HF2: A double-stranded DNA tailed haloarchaeal virus with a mosaic genome. *Molecular Microbiology* 44: 283–296.

Witte A, Baranyi U, Klein R, et al. (1997) Characterization of *Natronobacterium magadii* phage √Ch1, a unique archaeal phage containing DNA and RNA. *Molecular Microbiology* 23: 603–616.

Visna-Maedi Viruses

B A Blacklaws, University of Cambridge, Cambridge, UK

© 2008 Elsevier Ltd. All rights reserved.

Glossary

Dyspnea Shortness of breath.

Endocytosis Where a particle is enveloped by the cell membrane and internalized into a vacuole, an endosome.

Germinal center Area where B cells undergo proliferation after encountering their specific antigen with helper T lymphocytes.

Glomerulonephritis Inflammation of the glomeruli.

Hyperplasia Increase in the number of cells.

Leukoencephalitis Inflammation of the brain caused by leukocytes.
Lymphoid follicle Organized lymphocyte clusters.
Pneumonitis Inflammation of the lung.

Introduction

Visna-maedi virus (VMV) is a retrovirus in the genus *Lentivirus*. Other lentiviruses are human (HIV), simian (SIV), feline (FIV), and bovine (BIV) immunodeficiency viruses, caprine arthritis encephalitis virus (CAEV), and equine infectious anemia virus (EIAV). VMV causes a persistent infection of sheep leading to pneumonitis, demyelinating leukoencephalitis, mastitis, and arthritis, eventually killing the host. VMV originally came to prominence during an epidemic of lung and a separate wasting disease in sheep in Iceland from the 1930s to the 1950s. It was isolated from cases of interstitial pneumonitis (*maedi* = shortness of breath in Icelandic) and demyelinating leukoencephalomyelitis (causes *visna* = wasting in Icelandic) and was the first lentivirus isolated in 1957. Study of the Icelandic epidemic, with scrapie and ovine pulmonary adenomatosis (caused by another retrovirus, Jaagsiekte sheep retrovirus), allowed Björn Sigurdsson to introduce the concept of slow viral infections in 1954. These infections are typified by long incubation periods with pathology developing slowly and progressively, eventually leading to death of the host. The derivation of the genus name, *lenti* (Latin for slow), comes from this.

Clinical lung disease typical of maedi, and in many cases now known to be caused by VMV viruses, has been described since the early 1900s in other countries. Maedi is therefore also known as ovine progressive pneumonia or Montana sheep disease (USA), Graaf Reinet disease (South Africa), zwoegerziekte (the Netherlands), and la bouhite (France). VMV (and/or CAEV) is present in most countries worldwide apart from Iceland (eradicated during the 1940s and the 1950s), Australia, and New Zealand. Where infections are present, economic losses due to loss of milk production, failure to thrive, early culling, and death of animals are seen.

The major target cells for VMV are macrophages and dendritic cells. These are also common target cells for all the other lentiviruses. Aspects of lentivirus infection such as early infection and sites of persistence in accessory cells of the immune system may therefore be studied with VMV without the complication of lymphocyte infection.

Small Ruminant Lentiviruses

Historically, when a lentivirus was isolated from sheep, it was classified as VMV and from goats as CAEV. Although it was known that experimentally VMV could infect goats and CAEV infect sheep, it was not until recently that the natural interspecies transmission of the viruses was shown by molecular epidemiology. Sequencing data from geographically distant countries and from incidences of interspecies transmission into previously uninfected populations shows that VMV and CAEV are able to infect either species as well as other wild small ruminant species. The viruses have therefore now been grouped together as the small ruminant lentiviruses (SRLVs) although the terms VMV and CAEV are also in common use. This article focuses on studies using 'classical' VMV strains in sheep but also draws on data from CAEV and goats.

Visna-Maedi Virus

Virion and Genome

The virion is enveloped with a spherical or coffin-shaped core like other lentiviruses. The virion packages a diploid RNA genome of 9189–9256 bp. Several strains and molecular clones have been fully sequenced. It is a complex retrovirus with three structural genes, *gag, pol,* and *env*, and three auxilliary genes, *vif, tat* (or *vpr*), and *rev* (**Figure 1**). The functions of the different gene products have been deduced by direct study of VMV and CAEV and also by comparison to HIV and other lentiviruses (**Table 1**).

Replication Cycle

Replication of the virus occurs in a manner similar to that of other retroviruses. Virus binds directly to the cellular receptor or via Fc receptors if coated in antibody. The virion core is released into the cytoplasm by fusion of the envelope with the plasma membrane or, if the particle has been taken up by endocytosis, with the endosomal membrane. The RNA genome is reverse-transcribed to a double-stranded DNA (dsDNA) intermediate (provirus). The process of DNA replication duplicates the RNA termini forming the long terminal repeats (LTRs) of the provirus which contain the virus promoter, mRNA start site, and polyadenylation signals. The provirus integrates into the host cell genome with no apparent preferred site for integration. It is this integrated provirus that serves as template for viral mRNA and genome production. VMV has a complex mRNA expression pattern. *In vitro*, during productive replication of fibroblast cells, there is early synthesis and transport of *tat/vpr* and *rev* doubly spliced mRNA to the cytoplasm. Once REV is expressed, late synthesis and transport of *gag, pol, env,* and *vif* single or unspliced mRNAs or genomic RNA are mediated. This allows expression of the viral structural proteins, virion assembly (by budding through the plasma membrane or into vacuoles), and thus productive replication.

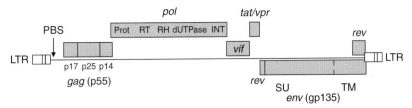

Figure 1 Proviral map of VMV. The provirus structure of VMV is shown with the position of open reading frames indicated. The proviral long terminal repeats (LTRs) contain the U3, R, and U5 regions (in this order) made during reverse transcription. The viral RNA genome contains R/U5 and U3/R at the 5′ and 3′ termini, respectively, where R is a short direct repeat. The primer binding site (PBS) for initiation of reverse transcription is indicated (binds tRNA lysine). The open reading frames are *gag*, *pol*, *vif*, *tat/vpr*, *env*, and *rev*, which contains two exons.

Table 1 Gene products of VMV

ORF	mRNA	Products	Function

gag	Full-length unspliced	Precursor Gag p55 cleaved to:
		  Matrix p17
		  Capsid p25
		  Nucleoprotein p14
			Major core proteins of virion

pol	Full-length unspliced	Precursor Gag/Pol polyprotein made by ribosomal frameshift, cleaved to:	
		Protease	Aspartic protease cleaves Gag and Pol
		Reverse transcriptase	Synthesizes proviral cDNA
		RNase H	Degrades genomic RNA to allow second-strand cDNA synthesis
		dUTPase	Reduces dUTP incorporation and so A-to-G mutation in second-strand synthesis. Mutants attenuated *in vitro* and *in vivo*
		Integrase	Integrates provirus into host genome
env	Singly spliced	Precursor Env gp135 cleaved to:	
		Surface subunit, SU gp110	Binds to cellular receptor
		Transmembrane subunit, TM gp41	Fuses viral envelope to cellular membrane (plasma or endosomal) to release virion core into cytoplasm
vif	Singly spliced	VIF	Assumed to be like HIV-1 VIF: inhibits incorporation of cytidine deaminases into virion and so stops new cDNA degradation or G-to-A mutations. Mutants attenuated *in vitro* and *in vivo*
tat/vpr	Doubly spliced	TAT/VPR	Historically called TAT but little evidence to support TAT function, more similar to VPR: LTR has no TAR element, high promoter activity without TAT, and TAT increases promoter activity by two- to threefold; cellular location nuclear; not secreted; packaged in virion; arrests cells in G2 phase of cell cycle. Mutants viable *in vivo* and *in vitro*
rev	Doubly spliced	REV	Interacts with rev-responsive element (RRE) in *env* coding region to allow export of unspliced and singly spliced mRNA species from the nucleus

Target Cells

Monocyte/macrophages and dendritic cells are the major target cell type infected *in vivo* although other cell types have also been shown to either harbor viral nucleic acid or express viral protein, for example, lung, gut, and mammary gland epithelial cells. The frequency of infected cells *in vivo* is very low with often less than 1% of the macrophage population being infected. There is no productive infection of lymphocytes unlike the primate lentiviruses. The range of cells that VMV infect *in vitro* is wider than macrophages with fibroblasts, an important resource for growing virus in the laboratory, and epithelial, endothelial, and smooth muscle cells all reported to replicate virus.

Replication in the monocyte/macrophage lineage is tightly linked to the maturation stage of the macrophage. Three replication states have been suggested: latent, restricted, and productive. Latent is where cells contain provirus but do not express viral products. In productive infection *in vitro* and *in vivo*, cells express approximately 5000 copies of viral RNA per cell, contain viral proteins, and produce virions. *In vivo* however, most infected cells express low quantities (approximately 5–200 copies) of viral RNA but no protein. This is called restricted replication and is seen in immature macrophages, for example, blood monocytes. It is only when macrophages mature in tissues that productive viral replication is found. Thus macrophages have been called 'Trojan horses' in SRLV infection in that they silently deliver the virus to tissues where virus replication can occur.

The cellular receptors for SRLV have not been identified although VMV strains K1514 and EV1 use a receptor(s) widely expressed on different cells and species. For K1514 this has been mapped to sheep chromosome 3p and the syntenic region of human chromosome 2 (2p25 > q13), and for EV1 to mouse chromosomes 2 and/or 4. If a one-component receptor is assumed, this excludes ovine major histocompatibility complex (MHC) class II (previously identified as a possible receptor by Env binding) and many of the cellular proteins used as co-receptors by other lentiviruses, for example, CXCR4 and CCR5. Studies using viral binding to cellular blots have identified possible candidate receptor proteins that have not yet been fully identified although a membrane proteoglycan is implicated. CAEV strains 63 or Cork may use a receptor which is much more restricted in expression in different species and although it has not been identified, is known, from interference and inhibition assays, to be different from the receptor for VMV strain K1514.

The cellular tropism of VMV is also affected by the ability of the virus to express the provirus. Different promoter activities are the consequence of LTR sequence and its ability to bind different transcription factors. Unlike the primate lentiviruses, SRLV LTRs have no NF-κB binding sites, but interact with AP-1 (only K1514), AP-4, and the AML/PEBP2/CBF family of transcription factors. Nothing is known about the changes that occur in maturing macrophages to allow viral expression although replication is known to be dependent on activation of the extracellular signal-related kinase/p38 mitogen-activated protein (ERK/MAP) kinase pathways.

Transmission

There are two important routes of transmission for VMV. The first is respiratory by aerosolized lung exudates between animals that are closely housed and/or in contact for long periods of time. The second is oral by colostrum/milk from infected dams to offspring.

As VMV is usually cell associated *in vivo* (there are few reports of cell-free virus), it has commonly been thought that the infective moiety in lung exudates and colostrum/milk was virus-infected macrophages or epithelial cells. However, recent data suggest that cell-free virus may be important in respiratory transmission. The presence of cell-free virus in the lining fluid of lungs with severe lesions has been shown and the lower lung is a more efficient point of infection than the trachea. Aerosolized particles small enough to reach the lower lung could contain free virus but infected cells would be too large. Once in the lower lung, a variety of possible target cells are present: macrophages (both alveolar and interstitial), dendritic cells, and lung epithelial cells, all of which may allow virus replication.

It is less clear in studies looking at oral transmission of VMV to lambs using colostrum/milk whether cell-free virus or virus-infected cells are important. Induction of lactation corresponds to virus expression in mammary tissue and colostrum contains cell-free virus as well as infected macrophages and epithelial cells. In a transmission study in lambs infected naturally by colostrum, ileal epithelial cells at the tips of the villi were important in early viral replication with virus antigen expression also seen in mononuclear cells in the lamina propria (macrophages), Peyer's patches, and mesenteric lymph nodes (macrophages and dendritic cells). Cell-free virus in colostrum may infect epithelial cells, but these cells could also transcytose viral particles. Similarly infected cells, especially macrophages in colostrum/milk, could transcytose into the lamina propria or beyond to infect the lamb. The appearance of infected mononuclear cells in the lamina propria within 10 h suggests that this route of infection also occurs. Therefore both cell-free and cell-associated virus could infect lambs via colostrum.

Immunity

Acute Infection Cannulation Model

Immune responses to VMV are slow to be detected in blood after natural infection. Seroconversion may take 6 months to 2 years to occur and sporadic T-cell reactivities in peripheral blood lymphocytes are often reported. However, when infection is studied in an acute infection cannulation model (**Figure 2**), immune responses are detected much earlier. The induction of antibody specific for Gag p25 can be shown within 4 days by enzyme-linked immunosorbent assay (ELISA) and neutralizing antibody (presumably specific for Env) within 10 days in efferent lymph. T-cell reactivity (by proliferation to Gag antigen and VMV-specific cytotoxic T-lymphocyte precursors (pCTLs)) is detected within 7–15 days. A low percentage of sheep also show directly active cytotoxic T lymphocytes (CTLs) in efferent lymphocytes. Therefore the kinetics of the immune response to VMV are within normal ranges seen to other viruses. However, it takes

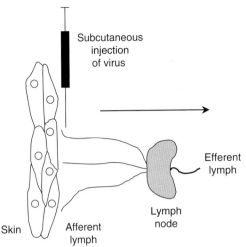

Figure 2 Lymphatic drainage diagram. The acute infection cannulation model for VMV infection of sheep uses subcutaneous injection of virus. The cellular drainage from the site of infection may be collected by cannulation of the lymphatic vessels draining the area. The lymph node directly draining the site of infection may also be biopsied. Lymphatic drainage flows from the tissue through the afferent lymph to the lymph node and then to the efferent lymph (indicated by an arrow). Virus in lymph cells (detectable by 1–4 days) or plasma (none detected) may be quantified, cells carrying virus identified (macrophages and dendritic cells), and immune reactivity monitored (see text). Even using this model, infected cells are at very low frequency, for example, 1 in 10^5–10^6 cells, although increased levels of infected cells rise to a peak by approximately 10–14 days and then decrease to low persistent levels.

much more time for this immune reactivity to be detected systemically. The slow replication rate of VMV and low frequencies of virus-infected cells may be responsible for this weak systemic immune response.

Antibody Responses

In blood, the antibody response is first seen to Gag antigens, in particular capsid antigen p25 (usually 1 month after experimental infection). The anti-Env response develops slowly, but as disease develops it eventually predominates. Other viral proteins also induce antibody responses. Neutralizing antibody responses are targeted to Env and develop very slowly (usually 2–3 months after experimental infection). Virus neutralization causes inhibition of viral binding to the cell (using fibroblasts as target cells) or inhibition of viral reverse transcription and/or integration (using macrophages as target cells). At least five epitopes have been defined in Env by monoclonal antibody reactivity and one immunodominant region is found in transmembrane (TM) Env. There are regions of hypervariability in Env sequences linked to antibody reactivity and escape (see below). VMV-specific antibodies are also detectable in colostrum and milk, cerebrospinal fluid (CSF), and synovial fluid.

Good IgM and IgG1 responses to VMV are induced but in natural infection, sheep do not make an IgG2-isotype response to viral proteins. Persistently infected sheep are able to make an IgG2 response to viral proteins immunized in adjuvant, and it is therefore thought that there may be abnormalities in the antigen-presenting function or T-cell help to B cells during natural infection. This isotype restriction is not seen in goats infected with CAEV; indeed, the predominant isotype induced is an indicator of whether pathology may develop, with strong IgG1 responses to Env SU associated with development of arthritis. The antiviral antibodies induced precipitate viral antigen in agar gel immunodiffusion tests and activate complement but the lack of an IgG2 response is associated with the lack of measurable antiviral antibody-dependent cellular cytotoxicity (ADCC) reactivity in antibody from infected sheep.

T-Lymphocyte Responses

Peripheral blood T-lymphocyte (PBL) proliferative responses to virus antigens are often intermittent with reactivity induced by 1–6 weeks post infection or later, depending on route and amount of virus challenge. Proliferating cells respond to Gag and Env although other viral antigens have not been tested. The response to Gag is mediated by CD4+ T cells. Proliferative T-cell responses to VMV have been detected in CSF by 3 weeks post infection and this reactivity coincides with the initiation of lesion development. Directly active CD8+ CTLs are seen in a low proportion of infected sheep (<20%) but all sheep develop pCTLs (CD8+ T lymphocytes) in their PBL by 3 months after experimental infection. These are induced by day 12 in lymph nodes draining the site of infection and are detectable in lymph nodes and efferent lymph of persistently infected sheep. Infected macrophages can present antigen via MHC class II to stimulate CD4+ T lymphocytes and MHC class I to act as targets for CD8+ CTL as well as to stimulate effector CTL function. Infected macrophages in lesions may therefore present antigen to lymphocytes.

Protective Immune Responses

Protective immune mechanisms have not been defined against VMV. After experimental infection, viral-infected cell frequencies peak, usually 1–2 weeks post infection, and then drop to levels maintained during persistent infection, suggesting that immune mechanisms are induced which control the level of infection. However, experiments to define protective T-lymphocyte subset(s) by *in vivo* antibody depletion of lymphocytes have not been successful at defining protective cell types. Indeed the experiments showed that CD4+ T lymphocytes are required for efficient infection of sheep by

VMV although this lymphocyte subset is not infected. In contrast to the immunodeficiency viruses, no role was detected for CD8+ T cells in control of virus replication. However, the low levels of VMV replication and the return of CD8+ T lymphocytes to the peripheral circulation within 14 days of the start of depletion may mean that sufficient CTL function remained to control the infection. Further work therefore needs to be carried out to investigate protective immune mechanisms against VMV.

Immune Abnormalities

There is no gross immunodeficiency after infection by SRLV; however, there are conflicting reports of immune abnormalities in SRLV-infected animals including changes in total immunoglobulin levels, CD4/CD8 T-cell ratios, decreased suppressor cell activity, and reduced delayed-type hypersensitivity (DTH) responses in skin and increased responses to certain bacterial infections which may be traced to macrophage dysfunction. The specific clinical state of animals is hard to normalize between studies and this may lead to the differing results.

Pathogenesis

Viral Entry

Cell-free virus or virus-infected cells enter the body where they may infect dendritic cells or macrophages in the primary infection site. These cells carry virus to the lymph node where infection is established and then macrophages leaving the lymph node disseminate virus throughout the body leading to a persistent infection. An important tissue for persistent infection of the animal is the bone marrow. Small foci of cells showing restricted replication of VMV have been identified which also express macrophage markers. These are probably myeloid stem cells although no study has defined these well. Infected stem cells may act as a continual source of infected monocyte/macrophages and/or dendritic cells which constantly seed the periphery and tissues. Where the cells mature and express viral antigen, the immune response to the infected cells then causes pathology. Recent molecular data question this theory as very few bone marrow samples from CAEV-infected goats assayed by polymerase chain reaction (PCR) for both proviral DNA and RNA were positive for virus. However, the levels of infection in bone marrow have not been defined well and may be below the detection limit of the assays used. Further work must be done to clarify this point. If bone marrow is not a constant source of infected cells, the virus must be maintained in infected tissues, perhaps lymph nodes and spleen. *In vivo* VMV may be isolated from many tissues including lung, mammary gland, joints, CNS, lymph nodes, spleen, and bone marrow. However, Kupffer cells of the liver are not permissive for virus infection and liver is not involved in pathology of the infection.

Clinical Signs

Infection by VMV does not cause obvious clinical signs for many months and sometimes years (2–5 years). Once clinical signs occur there is a progressive increase in the severity of the clinical signs with time. The diseases caused by SRLV are all typified by chronic active inflammatory processes. Within tissues, there is lymphocyte infiltration and proliferation, and the formation of organized lymphoid follicles and germinal centers. Lesions consist mostly of lymphocytes, macrophages, and plasma cells, often with CD8+ T lymphocytes predominating. This is overlaid with tissue-specific pathology (see **Table 2** for affected organs). Rarely, more acute disease may be seen in younger animals, usually leukoencephalitis, arthritis, or pneumonitis with high mortality.

Viral Cytopathic Effect or Immunopathology?

What causes the gradual buildup of pathology is not clear but it has two contributing factors: the effect of virus infection on infected cell and tissue function and the immune response to the virus. Little is known about the direct effect of virus on tissues although peptide analogs of the *tat/vpr* gene product are neurotoxic in rats and mice transgenic for the *tat/vpr* gene show lymphoproliferative disorders. Indeed, the number of virus-infected cells in lesions is low, so direct viral damage may be minimal.

There is a major immunopathological component to lesion formation. Early studies using immunosuppression reduced the number of CNS lesions seen in sheep infected intracerebrally, although the extent of infection within the CNS was the same, suggesting that direct viral toxicity was not a mediator of pathology. Similarly immunization with viral antigen before or after infection or superinfection with virus often increases the level of pathology seen. There is a positive correlation between virus load and pathology, and it is thought the immune response to viral antigen drives lesion formation. This is difficult to comprehend in early lesions where little virus antigen is detectable but once started, a positive-feedback loop of inflammation and increased virus expression drives progressive damage.

Both lymphocytes and macrophages taken from tissues with pathology show an activated phenotype: for example, they upregulate MHC class I and II, and adhesion molecule expression. However, there is variability in findings with functional assays and cytokine expression as to whether cells are activated or inhibited. Alveolar macrophages from lungs with pneumonitis show increased production of granulocyte monocyte colony stimulating

Table 2 Diseases caused by SRLV

Clinical sign (name)	Age of animal infected	Pathology	Infected cells
Dyspnea/pneumonia (maedi)	Adult sheep, 2–5 years old Goats, 1–6 months old	Interstitial pneumonitis with perivascular infiltrates; lymphofollicular hyperplasia (some with germinal centers) around vessels, bronchi, and bronchioles; smooth muscle hyperplasia near terminal bronchioles and alveolar ducts	Type I and II pneumocytes, interstitial and alveolar macrophages, endothelial cells and fibroblast-like cells Cell-free virus in lung lining fluid
Wasting and ascending paralysis (visna)	Adult sheep, 2–5 years old Goats 1–6 months old	Periventricular encephalitis (white matter) with perivascular cuffing and infiltration of the parenchyma by mononuclear cells; focal demyelination in brain and spinal cord; meningitis; areas of necrosis; raised cell numbers in CSF; intrathecal antibody production	Interstitial macrophages, microglial cells, rare endothelial, choroidal epithelial cells and fibroblasts of choroid plexus, astrocytes Cell-free virus in CSF rare
Enlarged, hardened udder ('hard bag')	Adult ewes and does	Indurative mastitis with lymphoid hyperplasia, and plasma cells in glandular interstitium, occasional infiltrates into ductal walls and lumens; periductal proliferation and follicle formation; gradual loss of epithelium; diffuse fibrosis; increased numbers of cells and cellular debris in lumen causing increased somatic cell count in milk	Epithelial cells, macrophages, endothelial cells and fibroblast-like cells infected Cell-free virus in colostrum/milk
Swollen joints, lameness ('big knee')	Goats, 1–2 years old Rare in sheep	Chronic arthritis with hyperplastic synovial membrane; subsynovial mononuclear cell infiltrates (follicle formation with germinal centers); increased synovial fluid with raised numbers of lymphocytes (some macrophages and synovial cells) and IgG1 levels (produced *in situ*); angiogenesis in villi and subintima; areas of cellular necrosis; frequent mineralization; increase in fibrous connective tissue around synovial membranes	Synovial membrane cells, macrophages, fibroblasts, endothelial cells. Cell-free virus in synovial fluid
Other affected organs			
Lymphadenopathy		Cortical hyperplasia (increased T cells); germinal center expansions and increased B-lymphocyte areas	Macrophages (and dendritic cells)
Kidney		Glomerulonephritis and interstitial nephritis; medulla and corticomedullary junction lesions; proliferation of mesangial and kidney endothelial cells; lymphatic clusters and follicles	Tubular epithelial cells, macrophages
Third eyelid		Lymphoproliferative inflammation	Macrophages, and glandular, ductal and surface epithelia
Skin		Decreased DTH responses	

factor (GM-CSF), interleukin (IL)-8, and fibronectin but not IL-6, IL-10, or transforming growth factor (TGF) β. Phenotypic changes seen *in vivo* may be caused by immune activation during the inflammatory reaction; however, some studies on infected macrophages *in vitro* suggest direct viral effects on cytokine expression and function. VMV infection of macrophages causes increased expression of IL-8 mRNA but decreased phagocytic and chemotactic activity. MHC class I and II, and LFA-1 and -3 expression are unaffected by infection. CAEV-infected macrophages show a decreased response to bacterial products: inducible nitric oxide synthase (iNOS), tumor necrosis factor (TNF)-α, IL-1β, IL-6, and IL-12p40 induction are all affected. Infection of cells has also been shown to lead to cell death via apoptosis. Infection of macrophages has a complex effect on function which is also altered by the presence of other cells. Co-culture of VMV-infected macrophages with autologous lymphocytes induces the secretion of lentivirus-induced interferon (lentiferon). This is probably a mixture of interferon (IFN)-α and IFN-γ. Effects of lentiferon include inhibition of proliferation and maturation of monocytes and of virus assembly in infected cells, thus decreasing virus replication, but it also increases MHC class II expression on macrophages which may contribute to lymphoproliferative responses.

The cytokines important in upregulating versus inhibiting VMV expression are poorly understood, partly due to the lack of purified ovine cytokines. Only GM-CSF has been shown to increase VMV transcription and antigen production in macrophages. GM-CSF and TNF-α treatment of cells increases CAEV LTR promoter activity and so may cause increased virus replication. However, there is mixed evidence as to the role that type I IFN plays against VMV infection: pretreatment of choroid plexus cells with a source of ovine type I IFN does not inhibit VMV replication, while pretreatment of cells with human IFN-α does block replication. *In vivo* treatment of lambs early in infection with IFN-τ reduces virus replication and disease development. Although not a cytokine, NO is also known to inhibit virus replication in macrophages.

Autoimmunity

The lesions caused by SRLV in the CNS and joints are very similar to those seen in multiple sclerosis and rheumatoid arthritis of humans. These have links to autoimmunity, and indeed autoimmunity would explain the chronic development of pathology seen in these infections. However, there is very little evidence for this as a mechanism of lesion formation with VMV. There is no response to myelin basic protein (lymphocytes or antibody) or lipid antigen (antibody) in sheep with visna although elevated levels of autoimmune antibody (specific for rheumatoid factor, single-stranded DNA (ssDNA), and cardiolipin) in serum have been detected in infected sheep.

Role of Lymphocytes

CD4+ T lymphocytes are important in promoting infection and lesion formation. At the time of acute infection, CD4+ T lymphocytes are necessary for efficient establishment of infection and later lymphocytes are necessary for lesion development. Therefore, immune responses to unrelated antigens, whether secondary infections or experimental immunizations or vaccinations, may cause increased inflammatory reactions that accelerate lesion development.

Role of the Host and Virus Genetics

Breed resistance to infection and disease development has been documented in sheep and goats. In sheep, the Icelandic breed was particularly susceptible to VMV infection and development of visna, while the Karakul flock, from which infected animals introduced the disease to Iceland, never showed signs of disease. Border Leicester rams are resistant to visna but develop arthritis. This and agricultural practice (e.g., are animals milked?) may explain, in part, the geographical distribution of clinical signs of VMV infection; for example, Iceland commonly saw pneumonitis and encephalomyelitis, UK usually notes pneumonitis, and Spain has documented pneumonitis and mastitis. With interspecies transmission of SRLV, it is not clear whether strains more closely resembling VMV in goats will cause increased incidences of pneumonic disease or whether it is goat genetics which determines joints as a major site of lesion development. In USA, the VMV strains isolated are more closely related to CAEV than European VMV strains and yet pneumonia is still commonly seen in sheep. In goats with CAEV, where only $\sim$30% of animals develop arthritis, linkages to certain MHC class I alleles have been associated with development of disease.

Strains of virus that can be differentiated phenotypically *in vitro* can cause different patterns of lesion development. Usually strains growing to high titer and causing cytopathic effects in macrophages are the most virulent *in vivo*. Strains selected for increased neurovirulence grow better in sheep choroid plexus cells than those that cause pneumonia. This is linked to LTR sequence variation. Therefore, both host and viral genetics are important elements in disease progression.

Immune Evasion

SRLVs persist in the host despite an immune response that includes antibody, CD4+, and CD8+ T lymphocytes. Several mechanisms of viral immune evasion have been noted that may allow this to occur. First, integration into the host genome allows latent infection and persistence in cells without expression of antigen. Similarly, gene expression is tightly regulated with no expression of antigen in immature macrophages. Thus these cells are invisible to the immune system. Second, low-titer neutralizing antibody responses producing low-affinity antibody mean that virus may be able to dissociate from antibody and infect cells. Third, neutralization escape mutants have been shown to emerge during infection. Mutants arise by point mutation in Env, one of the first documented incidences of antigenic drift in viruses. Later, the neutralizing antibody response may broaden to include these mutants. It is thought these mutants have a selective advantage and help the virus to persist in the host. Whether escape mutants are important in VMV pathogenesis is debatable as the original parental virus may be isolated late after infection, showing it has not been cleared from animals and replaced by the mutants. Similarly, neutralization mutants arise when antibody is not present, and their presence is unrelated to severity of pathology. Fourth, the ability of VMV to cause cell fusion means that cell-free virus is not required (and very rarely seen *in vivo*) as virus may pass from cell to cell directly and so is not exposed to antibody. Fifth, antibody-mediated enhancement of infection of macrophages has been shown in which antibody-coated virus is taken up into macrophages via Fc receptors, thus increasing the effective viral dose. Sixth, there is no ADCC response to the virus. Seventh, although technically not immune evasion, the virus uses the CD4+ T-cell response to help replication, either by activation of macrophages or attraction of macrophages to the relevant tissue.

Control

Because of its insidious nature, control of VMV infection is difficult. Due to the economic losses and animal welfare problems that the disease causes, control programs are in place in many countries. These are based on serological tests, usually ELISAs (often to capsid p25 and TM Env) or agar gel immunodiffusion tests. However, because of the length of time taken for induction of a detectable antibody response in blood, and the possibility of fluctuations in the response, testing regimes involve repeated tests to maintain accreditation of SRLV-free status. PCR has not been used routinely for VMV diagnosis due to the strain variation that is present within SRLV and the low provirus load in blood of persistently infected animals.

Several options are open to try to derive SRLV-free flocks. Iceland eradicated the disease in the 1940s and the 1950s by culling all infected flocks and restocking from SRLV-clean animals. Others have used removal of serologically positive animals (and their progeny) from the flock after each test round until serologically free status is achieved. In the Netherlands, there was a successful policy of re-deriving flocks from 'snatched' lambs. These are lambs taken at birth from their mothers before they suckle infected colostrum. The lambs are then hand-reared on either heat-inactivated or bovine colostrum. Once VMV-free flocks are achieved, careful control of importation of animals into flocks/countries is necessary to maintain this status, as it is importation of infected asymptomatic animals that then live in close contact with the host flock that often leads to spread of infection into previously free areas, for example, Iceland (1933) and Finland (1981). With proof now of interspecies transmission, SRLV-free status will only be maintained if both sheep and goats are kept separate or both species kept SRLV free.

Treatment and Vaccination

At present, there is no treatment against infection although certain anti-retroviral drugs do have activity against VMV. There have now been several vaccination studies using inactivated virus preparations, live-attenuated viruses, and subunit vaccines administered by a variety of vectors, for example, protein, plasmid, and vaccinia virus, using a variety of routes and different adjuvants or cytokine immunomodulators. None has produced sterilizing immunity and some have increased the levels of pathology seen. Induction of neutralizing antibody by immunization with recombinant Env has proven problematic using baculo- and vaccinia virus-expressed antigen in sheep. The best result has been with bacterial-expressed Env gp70 but only low titers of antibody were seen using this immunogen. Some vaccines have reduced virus load early in infection but pathology has still developed, although in some cases this is less severe than control animals. The most promising result has come from CAEV in goats using DNA prime and then protein boost with Env in Freund's incomplete adjuvant. There was decreased viral replication in lymph node and synovium, and reduced development of severe arthritis for more than 18 months using this vaccine.

The possibility of inducing antibody that enhances the ability of virus to infect macrophages, the induction of CD4+ T lymphocytes responses that may increase the efficiency of the primary infection, as well as increasing the pathological response are all aspects of the immune response to VMV that needs much further study before a successful vaccine will be released to the field.

Conclusions

VMV has many features in common with the other lentiviruses including infection of macrophages and dendritic cells. There is now an understanding of the type of lesion VMV induces in many tissues. However, there is still much work to be done with VMV to answer questions on sites of persistence of the virus, triggers of lesion development, protective immune responses, and vaccine control of the infection.

See also: Equine Infectious Anemia Virus; Human Immunodeficiency Viruses: Molecular Biology; Human Immunodeficiency Viruses: Pathogenesis; Immune Response to viruses: Antibody-Mediated Immunity; Simian Immunodeficiency Virus: Animal Models of Disease; Simian Immunodeficiency Virus: General Features; Viral Pathogenesis.

Further Reading

Blacklaws B, Bird P, and McConnell I (1995) Early events in infection of lymphoid tissue by a lentivirus, maedi-visna. *Trends in Microbiology* 3: 434–440.

Blacklaws BA, Berriatua E, Torsteinsdottir S, *et al.* (2004) Transmission of small ruminant lentiviruses. *Veterinary Microbiology* 101: 199–208.

de Andres D, Klein D, Watt NJ, *et al.* (2005) Diagnostic tests for small ruminant lentiviruses. *Veterinary Microbiology* 107: 49–62.

Eriksson K, McInnes E, Ryan S, *et al.* (1999) CD4(+) T-cells are required for the establishment of maedi-visna virus infection in macrophages but not dendritic cells *in vivo*. *Virology* 258: 355–364.

Eriksson K, McInnes E, Ryan S, *et al.* (1999) *In vivo* depletion of CD8+ cells does not affect primary maedi visna virus infection in sheep. *Veterinary Immunology and Immunopathology* 70: 173–187.

Gendelman HE, Narayan O, Kennedy-Stoskopf S, *et al.* (1986) Tropism of sheep lentiviruses for monocytes: Susceptibility to infection and virus gene expression increase during maturation of monocytes to macrophages. *Journal of Virology* 58: 67–74.

Gendelman HE, Narayan O, Molineaux S, *et al.* (1985) Slow, persistent replication of lentiviruses: Role of tissue macrophages and macrophage precursors in bone marrow. *Proceedings of the National Academy of Sciences, USA* 82: 7086–7090.

Georgsson G, Pálsson PA, and Pétursson G (1990) Some comparative aspects of visna and AIDS. In: Racz P, Haase AT, and Gluckman JC (eds.) *Modern Pathology of AIDS and Other Retroviral Infections*, pp. 82–98. Basel: Karger.

Haase AT (1986) Pathogenesis of lentivirus infections. *Nature* 322: 130–136.

Peterhans E, Greenland T, Badiola J, *et al.* (2004) Routes of transmission and consequences of small ruminant lentiviruses (SRLVs) infection and eradication schemes. *Veterinary Research* 35: 257–274.

Ravazzolo AP, Nenci C, Vogt HR, *et al.* (2006) Viral load, organ distribution, histopathological lesions, and cytokine mRNA expression in goats infected with a molecular clone of the caprine arthritis encephalitis virus. *Virology* 350: 116–127.

Ryan S, Tiley L, McConnell I, and Blacklaws B (2000) Infection of dendritic cells by the maedi-visna lentivirus. *Journal of Virology* 74: 10096–10103.

Sigurdsson B (1954) Mædi, a slow progressive pneumonia of sheep: An epizoological and a pathological study. *British Veterinary Journal* 110: 255–270.

Sigurdsson B (1954) Rida: A chronic encephalitis of sheep. With general remarks on infections which develop slowly and some of their special characteristics. *British Veterinary Journal* 110: 341–354.

Thormar H (2005) Maedi-visna virus and its relationship to human immunodeficiency virus. *AIDS Reviews* 7: 233–245.

Watermelon Mosaic Virus and Zucchini Yellow Mosaic Virus

H Lecoq and C Desbiez, Institut National de la Recherche Agronomique (INRA), Station de Pathologie Végétale, Montfavet, France

© 2008 Elsevier Ltd. All rights reserved.

Glossary

Mild-strain cross-protection A plant systemically infected by a mild virus strain will not develop additional symptoms when inoculated by a severe strain of the same virus. Most often, the severe strain does not multiply in the cross-protected plant.

Filiformy A leaf deformation symptom in which the leaf blade is drastically reduced but not the veins, giving a shoestrings aspect to the leaf.

Transmission propensity A measure of vector importance quantifying the natural ability of a species to inoculate a plant with a virus under conditions that allow vectors to move and feed freely.

History and Taxonomy

Watermelon Mosaic Virus

Mosaic diseases of cucurbit crops were first reported in the 1920s and the early literature contains a diversity of names for viruses or virus diseases that were only partially characterized. In 1940, Milbrath reported a severe watermelon (*Citrullus lanatus*) mosaic disease in California, but the nomenclature of the causal agent remained controversial until 1979. In 1965, Webb and Scott compared ten isolates from southern USA of what was then called the watermelon mosaic virus (WMV) complex. Based on cross-protection experiments, serological relationships and host range reactions they divided them into two groups: WMV1 and WMV2. WMV1 and WMV2 were considered as different viruses, also different from a watermelon mosaic isolate reported from South Africa in 1960 by van Regenmortel. Further work by Milne and Grogan in 1969 increased the confusion by concluding that WMV1 and WMV2 were strains of the same virus. In 1979, Purcifull and Hiebert definitively clarified the situation by demonstrating that WMV1 and WMV2 were indeed serologically distinct entities, and that WMV1 was closely related to papaya ringspot virus (PRSV). Today, WMV1 is considered as the W strain of PRSV, while WMV2 is referred to as watermelon mosaic virus (WMV). In addition, the same authors showed that a watermelon mosaic virus isolate from Morocco was a third serological entity. This isolate is considered as a distinct virus, Moroccan watermelon mosaic virus (MWMV), to which probably also belongs the South African isolate.

So, from the initial watermelon mosaic virus complex emerged three different virus species: *Watermelon mosaic virus*, *Papaya ringspot virus*, and *Moroccan watermelon mosaic virus*.

Zucchini Yellow Mosaic Virus

An apparently new cucurbit virus was isolated in 1973 from a zucchini squash (*Cucurbita pepo*) plant in Northern Italy, and Lisa *et al.* described this virus as belonging to a new potyvirus species, *Zucchini yellow mosaic virus*, in 1981. In 1979, many melon (*Cucumis melo*) crops were devastated in southwestern France by an apparently new virus disease, whose causal agent was tentatively named muskmelon yellow stunt virus; very rapidly, it appeared that it was a strain of zucchini yellow mosaic virus (ZYMV). Within a few years, ZYMV was reported in many countries on the five continents, and in this regard, ZYMV appears as a typical example of an emerging plant virus.

Classification

Based on particle morphology, aphid transmissibility, serological relationships, ability to induce pinwheel cytoplasmic inclusions in host cells, genome organization, and nucleotide sequences, WMV and ZYMV were identified as members of the genus *Potyvirus*, family *Potyviridae*. Several other potyviruses have been shown to infect cucurbit crops including PRSV-W, MWMV, zucchini yellow fleck virus (ZYFV), melon vein banding mosaic virus (MVBMV), telfairia mosaic virus (TeMV), turnip mosaic virus (TuMV), clover yellow vein virus (ClYVV), and bean yellow mosaic virus (BYMV). However, these viruses have either only limited geographic distribution or minor economical incidence.

Molecular analyses based on the coat protein (CP) coding sequence revealed that cucurbit-infecting potyviruses belong to several 'clusters' of closely related species: ZYMV and WMV belong to a cluster that also contains mostly legume-infecting potyviruses, whereas PRSV, MWMV, and ZYFV are grouped in a 'PRSV-like' cluster containing mostly cucurbit-restricted viruses (**Figure 1**).

Symptoms

Watermelon Mosaic Virus

WMV induces a diversity of symptoms according to the isolate and the host cultivar. On leaves, symptoms are mosaics, vein banding, more or less severe leaf deformations, and filiformy. Some isolates induce discoloration and deformation on fruits of zucchini squash susceptible cultivars while other isolates do not affect fruit and yield quality. Mosaic and discoloration are also observed on leaves and fruits of some melon cultivars (**Figure 2**).

Zucchini Yellow Mosaic Virus

Since its first detection, ZYMV was recognized as a virus causing extremely severe symptoms leading to complete yield losses in the case of early contamination. This severity of symptoms was probably an important factor for the rapid identification of ZYMV soon after its first outbreaks, in many countries. In melon, leaf symptoms include vein clearing, yellow mosaic, leaf deformation, occasionally with blisters and enations. There is often a severe plant stunting. Some ZYMV isolates can induce a rapid and complete wilt in cultivars possessing the *Fn* gene. On fruits, a diversity of symptoms are observed: external mosaic or necrotic cracks, internal marbling, and hardening of the

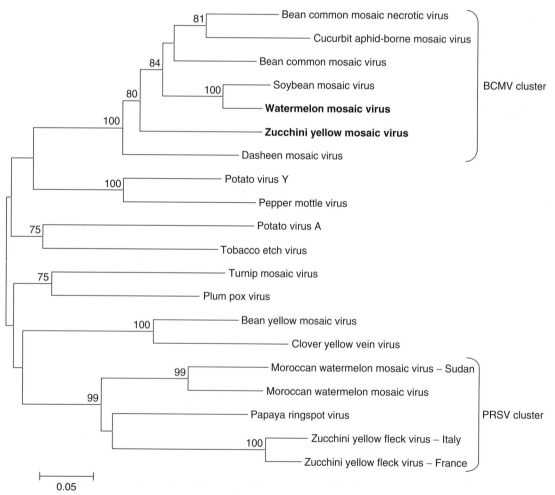

Figure 1 Relationships based on CP amino acid sequences, of cucurbit-infecting potyviruses. Branch lengths indicate the molecular divergence of viruses (the scale bar represents 0.05 mutations per residue). Figures at some nodes represent bootstrap values (in %), indicating the robustness of each node. Only values above 75% are indicated.

Figure 2 Mosaic symptoms on a leaf and fruit of a melon plant infected by a moderately severe WMV isolate.

Figure 3 Severe mosaic and deformations on leaves and fruits of a zucchini squash plant infected by ZYMV. Reproduced from Astier S, Albouy J, Maury Y, and Lecoq H (2001) *Principes de virologie végétale: Génome, pouvoir pathogène, écologie.* Paris: INRA Editions, with permission from QUAE.

flesh. Seeds are occasionally severely deformed and have poor germination rates. In zucchini squash, symptoms are very severe on leaves with mosaic, yellowing, leaf distortion and sometimes very severe filiformy. Fruits are generally severely misshaped with prominent knobs and are of course unmarketable (**Figure 3**). In cucumber (*Cucumis sativus*) and watermelon, mosaic and deformations are generally observed on leaves and fruits.

Synergism

Synergism has been observed between WMV or ZYMV and other viruses infecting cucurbits which can be expressed either by increase in virus multiplication rates or by more severe symptoms. Significant increase in cucumber mosaic virus (CMV) multiplication rate was observed in cucumber plants co-infected by ZYMV. CMV could also partially overcome a resistance in cucumber when in mixed infection with ZYMV. In grafted cucumbers, double infection by CMV and ZYMV induces a severe and rapid wilting reaction which is not observed with infections by CMV or ZYMV alone. When WMV or ZYMV are in mixed infection with the polerovirus, cucurbit aphid-borne yellows virus (CABYV), CABYV multiplication rate and symptom intensity are increased. As for other potyviruses, these synergetic effects could be related to the strong potyvirus gene silencing suppressor.

Geographic Distribution

Both WMV and ZYMV are now widely distributed in the major cucurbit production areas worldwide. Their geographic distributions are broadly overlapping and frequent mixed infections are observed in the fields.

For unknown reasons, WMV appears to be rare or absent in cucurbits in subtropical or tropical areas. For instance, in exhaustive surveys conducted in Nepal, Sudan, and French West Indies, no WMV was detected, although ZYMV was relatively abundant. In Florida, WMV is frequent in northern and central counties but is not detected in southern counties, whereas ZYMV can be found throughout the state. This cannot be related to a lack of potential WMV vectors or reservoirs because both are abundant in tropical or subtropical areas.

ZYMV is present worldwide in almost all countries where cucurbits are grown, under temperate, Mediterranean, subtropical, and tropical climatic conditions. It affects highly mechanized cropping systems (such a glasshouse crop production in Northern Europe) as well as more traditional agroecosystems (such as flood-irrigated crops on the Nile banks). ZYMV has been reported in very remote areas including semi-desertic regions or islands.

Host Range

WMV has a relatively wide experimental host range for a potyvirus. It infects over 170 species in 26 mono- or dicotyledonous families. Besides cucurbits, WMV causes mosaic diseases in legumes (pea, broad bean) and orchids (vanilla, *Habenaria radiata*) and infects many weeds that can serve as alternative hosts. Generally, naturally infected weeds do not present evident symptoms of viral infection.

ZYMV has a relatively narrow host range. In natural conditions, it infects mostly cultivated or wild cucurbits but also a few flower species (*Delphinium, Althea*) or weeds.

Diagnostic Method

The confusion that was prevalent in the early descriptions of watermelon mosaic diseases was mainly due to the convergence of symptoms caused by WMV, PRSV, and MWMV in cucurbits and to the lack of proper diagnostic tools. Symptoms and virus particle morphology were clearly insufficient to differentiate these viruses. The production of specific polyclonal antisera, and the development of simple serological tests, such as the gel double-diffusion test in agar containing sodium dodecyl sulfate and sodium azide (SDS-ID), brought a major contribution to the proper diagnosis of cucurbit potyviruses. In particular, this method contributed to the rapid and unequivocal identification of ZYMV in several countries soon after its first observation. Now, double antibody sandwich enzyme-linked immunosorbent assay (DAS-ELISA) is generalized and commercial kits are available for WMV and ZYMV. Recently, dipstick serological tests based on the lateral flow technique have been developed that allow an easy and rapid diagnosis of ZYMV in the fields.

Monoclonal antibodies (MAbs) have been produced for ZYMV and WMV. They proved to be very useful to study the serological variability and to differentiate ZYMV and WMV subgroups. They have also been used successfully to analyze virus interactions, and in particular cross-protection efficiency and specificity.

Many partial nucleotide sequences are now available for ZYMV and WMV, particularly in the CP coding region, which allowed the development of specific primers for each virus. However, in routine testing, DAS-ELISA seems to be more reliable than RT-PCR.

Vector Relationships

WMV is transmitted by at least 35 aphid species in 19 genera. Fewer aphid species were tested for their ability to transmit ZYMV, and 11 were identified as ZYMV vectors. *Aphis craccivora*, *Aphis gossypii*, *Macrosiphum euphorbiae*, and *Myzus persicae* are efficient WMV and ZYMV vectors. Some aphid species were shown to be poor or nonvectors of WMV and ZYMV, which suggests some level of specificity in the virus–vector interaction.

WMV and ZYMV are transmitted on the nonpersistent mode: they are acquired and transmitted during very short probes (a few seconds to minutes), and their retention period in the vector is relatively short (a few hours).

WMV and ZYMV as typical potyviruses require the presence of a virus-encoded helper component (HC-Pro) protein for transmission. HC-Pro from WMV and ZYMV are interchangeable and both mediate efficiently the transmission of purified virions of both species.

Several ZYMV isolates that have lost aphid transmissibility have been characterized, and a unique feature for this virus is that single amino acid mutants have been identified in the three domains important for transmission. ZYMV-NAT has an A to T substitution in the DAG motif in the CP, ZYMV-PAT a T to A substitution in the PTK motif of HC-Pro, and ZYMV-R1A a K to E substitution in the KLSC motif of HC-Pro. These mutants led to the identification of an interaction between the HC-Pro and CP through their PTK and DAG domains.

The nontransmissible isolate ZYMV-NAT (having the DTG motif in the CP) could be transmitted by aphids from plants infected concomitantly by a transmissible isolate of PRSV. This occurred through heteroencapsidation, a phenomenon by which ZYMV RNA is completely or partially encapsidated by PRSV CP, which is functional in aphid transmission. An aphid nontransmissible isolate deficient for HC-Pro can also be transmitted by aphids when in mixed infection with an isolate that has a functional HC-Pro. The transmissible isolate provides its functional HC-Pro to mediate the transmission of the deficient isolate. These two mechanisms can contribute to the maintenance, in natural conditions, of variants which have lost their vector transmissibility.

An interesting interaction has been observed between ZYMV and *Aphis gossypii*, an aphid vector colonizing cucurbit crops. *Aphis gossypii* lives longer and produces more offspring on ZYMV-infected than on noninfected plants. In addition, more alatae are produced on infected plants, which may stimulate the spread of ZYMV. These phenomena might be related to the observed changes in phloem exudates' composition (free amino acids, sugars) in virus-infected plants.

Epidemiology

Virus Sources

WMV has not been described as seed borne in cucurbits or other crops, but may be transmitted through vegetative propagation in vanilla. ZYMV seed transmission remains controversial. It has been reported in squash but other studies repeatedly failed to observe seed transmission. This is an important issue, since ZYMV seed transmission would be the simplest explanation for the rapid ZYMV dissemination throughout the world in the 1980s. If ZYMV seed transmission does occur, it is at a low rate and may be only for some strains or hosts. Another possible way for long-distance dissemination of ZYMV is through the globalization of vegetable production and trade. It has been shown that ZYMV-infected fruits imported from Central America into Europe could be very efficient virus sources for aphids.

Virus sources from which epidemics will initiate could be overwintering weeds or crops infected during the previous crop season. In tropical and subtropical regions, cucurbit crops or weeds grow all year round, and viruses could easily move from an old infected crop or cucurbit weed to a young planting. In more temperate regions, non-cucurbit weeds were found to be efficient reservoirs for WMV but not for ZYMV. Winter protected-crops could contribute to ZYMV overwintering in Mediterranean regions and residential gardens were found to be important sources of ZYMV in California.

Efficient Vectors

Many potential aphid vector species have been identified for WMV and ZYMV. A study was conducted to compare the vector capacity of two aphid species, one colonizing cucurbits (*A. gossypii*), the other not (*A. craccivora*). Two parameters were used. Transmission efficiency was measured in the laboratory with single aphids exposed in sequence to an infected plant and then to four healthy plants. Transmission propensity was measured by arena tests (more representative of natural conditions) in which aphids could move between plants and feed without interference. It was shown that *A. craccivora* had both a higher efficiency and propensity to disseminate ZYMV than *A. gossypii*. This highlights the importance of noncolonizing transient vector species in the epidemiology of ZYMV.

Pattern of Spread

The same general pattern is observed for WMV and ZYMV. The first contaminations occur generally shortly after planting, depending upon the availability of virus sources in the environment. The secondary virus spread then occurs when aphid flights, particularly of noncolonizing species, spread the virus from the primary infection foci to the rest of the crop. The diseased plants are often distributed in large patches that rapidly extend and join each other, leading to the complete contamination of the crop within a few weeks. Simultaneously, the infected crop will serve as a source of virus to contaminate weeds or nearby young plantings. Epidemic development curves have an overall S-shape, generally fitting well with the logistic model.

Control Methods

Prophylactic Measures

Prophylactic measures are intended to prevent or limit the contact of viruliferous aphids with cultivated plants. They are not specific for a particular virus and are generally efficient for all aphid-borne viruses. These include careful weeding near plantings and avoiding overlapping crops in the same area to reduce virus and aphid sources near new plantings. Plastic mulches have a repelling action on aphids and significantly delay WMV and ZYMV spread. However, they confer only a temporary protection that is limited to the early stages of the crop, because their efficiency decreases when plant growth covers their surface. Row covers of different types (unwoven, perforated plastics, etc.) can also be used; they physically prevent winged aphids from reaching the plants, but they must be removed to allow insect pollination necessary for cucurbits. Both methods have a major drawback: they require a lot of plastic material that farmers must dispose in an environmentally sound way after the crop cycle. Insecticides applications have generally been found inefficient in limiting WMV and ZYMV spread. This is to relate to the large number of winged aphids that land on the plants and to the rapidity of the transmission process. Oil applications can delay virus spread when inoculum pressure is moderate. When applicable, a 1 month crop-free period has been shown to be efficient in limiting ZYMV virus spread.

Cross-Protection

Although the mechanism of cross-protection has not been fully elucidated (it probably relies on the gene silencing machinery), this method has been developed at a commercial level to protect cucurbit crops against ZYMV. The principle is simple (**Figure 4**): when a mild virus isolate (i.e., that has no significant impact on commercial yield) is inoculated to young seedlings, it protects the plant from subsequent contaminations by severe isolates of the same virus. The mild strain ZYMV-WK is a natural variant of a severe aphid nontransmissible isolate. Although efficient against most ZYMV isolates, ZYMV-WK does not protect against very divergent isolates such as those from Réunion Island, indicating some specificity in the protection. A single amino acid change (R to I) in the FRNK conserved domain of the HC-Pro is responsible for symptom attenuation of ZYMV-WK. A complete technological package (mild strain production, quality control protocols, inoculation machines) has been developed to implement commercially ZYMV cross-protection. Mild WMV isolates have been reported that could also have a potential for cross-protection.

Resistant Cultivars

The use of virus-resistant cultivars is probably the easiest and cheapest way to control plant viral diseases at the farmer's level. Breeding for resistance still relies mainly upon searching for resistance characters in germplasm collections and introgression of the resistance gene(s) into commercially acceptable cultivars. Considerable efforts have been made to look for resistance to WMV and ZYMV in genetic resources, and some WMV- or ZYMV-resistant commercial cultivars are now available. Some resistance genes confer complete and durable resistance (such as the *zym* gene in cucumber), while others confer only partial resistances or may be overcome by virus evolution (such as the *Zym* gene in melon). An

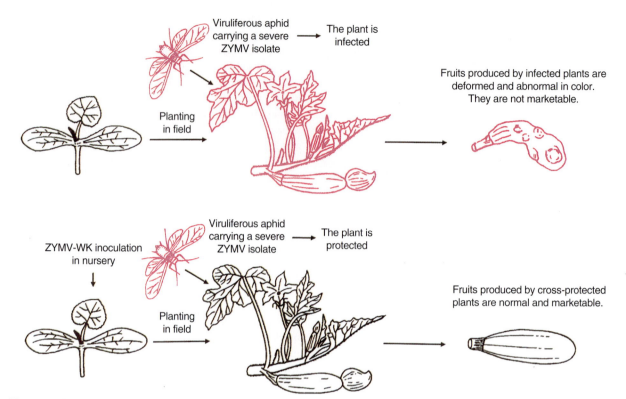

Figure 4 Representation of mild strain cross-protection as applied with ZYMV-WK in zucchini squash. Reproduced from Astier S, Albouy J, Maury Y, and Lecoq H (2001) *Principes de virologie végétale: Génome, pouvoir pathogène, écologie*. Paris: INRA Editions, with permission from QUAE.

interesting situation is observed for ZYMV resistance in squash. Although the resistance level was high in the original accession of *Cucurbita moschata* in which the resistance was identified, when transferred through interspecific crosses to zucchini squash (*C. pepo*), the resistance phenotype was different: ZYMV multiplies but the plants present only very mild symptoms (a phenomenon called tolerance). However, tolerance appears not to be stable since aggressive variants of the virus (i.e., causing severe symptoms in tolerant plants) may emerge in these plants. A single amino acid change in the P3 gene is sufficient to confer this aggressive phenotype. However, the aggressive variants are counter-selected when in competition with common ZYMV isolates in susceptible cucurbits. This genetic load associated with aggressiveness could be a factor that will make the tolerance durable. In melon, a resistance to WMV and ZYMV transmission by *A. gossypii* was found to be governed by the single dominant gene *Vat*. This gene is present in many commercial cultivars, but confers only limited protection in the fields, probably because WMV and ZYMV are also transmitted by many other aphid species in natural conditions.

In the last two decades, attempts were made to obtain WMV and ZYMV transgenic resistant plants using the pathogen-derived resistance approach. Different constructs were tested to obtain resistant WMV or ZYMV plants and the best results were obtained with the full-length CP gene and with ribozymes. Freedom II, a transgenic squash hybrid containing the WMV and ZYMV CP genes, was released in the USA in 1995, as the first virus-resistant transgenic crop to be commercially cultivated in the world. It proved to have a very efficient resistance to WMV and ZYMV in field conditions. Similar biotech crops are presently grown mainly in southeastern USA, particularly during late summer and fall when WMV and ZYMV inoculum pressure is high.

Variability

Only limited biological variability has been reported for WMV. This concerns mainly differences in symptom intensity, host range, or aphid transmissibility. In contrast, from its first description, ZYMV appeared to have a very important biological diversity. When collections of field isolates were compared, an important variability was revealed in host range and symptoms on susceptible hosts, with isolates producing mild or atypical mosaic symptoms, necrosis, or wilting reactions. When these isolates were inoculated to melon or squash varieties possessing resistance genes, different pathotypes could be differentiated. Important variability has also been observed in aphid transmissibility and several aphid nontransmissible or poorly transmissible isolates have been described.

Limited serological variability was observed in WMV and ZYMV when polyclonal antibodies were used. However, the development of monoclonal antibodies against WMV and ZYMV allowed the characterization of serotypes closely correlated to the molecular variability.

The molecular variability of ZYMV and WMV was assessed mostly on the CP region, particularly the N-terminal part of the CP that is known for potyviruses to be highly variable and is frequently used for molecular studies. More than 200 partial CP sequences are now available for ZYMV. Strains from some regions, for example, Réunion Island and Singapore, are highly divergent molecularly, whereas isolates from other regions are more closely related, and fall into three main clusters without geographic structure; more recent sequence data available for Asian strains (China, Korea, Taiwan) indicate that some of those strains tend to locate between previously described groups or clusters. Molecular analyses have allowed, in a few cases, tracking the putative origins of ZYMV strains emerging in a new area.

Genetics and Evolution

WMV and ZYMV genome organization is very similar to that of other potyviruses sequenced so far. The single-stranded, positive-sense RNA genome (9.6 kbp for ZYMV, 10 kbp for WMV excluding the polyadenylated 3′ extremity) is translated as a single polyprotein that is self-cleaved in 10 functional proteins.

The full-length sequence revealed that WMV is related to the legume-infecting soybean mosaic virus (SMV); however, the P1 protein is 135 amino acids longer than that of SMV, and the N-terminal half of P1 shows no relation to SMV but is 85% identical to another legume-infecting potyvirus, bean common mosaic virus (BCMV). This suggests that WMV has emerged through an ancestral recombination event between an SMV-like and a BCMV-like potyvirus. SMV and BCMV have narrow host ranges, mostly restricted to legumes, while WMV is one of the potyviruses with the broadest host range, including monocots and dicots. The impact of its recombinant nature on WMV biological properties remains unknown and speculative.

Partial sequence data also indicate the presence of intraspecific recombinants in WMV: the molecular variability was structured in three groups based on the CP sequence; when other parts of the genome were considered, several isolates switched groups, indicating that recombination may have taken place. By sequencing the full-length genome of putative recombinants, recombination points were characterized in different parts of the genome.

In the case of ZYMV, sequence analysis softwares suggest that recombination might have taken place, but the situation is not as clear-cut as for WMV.

Besides recombination, viral genomes also evolve by mutations that take place during replication. In the case of ZYMV, the biological consequences of several point mutations that emerged in natural conditions were assessed by molecular studies: sequences of closely related strains with different biological properties were compared, and the mutations observed – particularly nonsilent mutations located in conserved domains – were introduced by site-directed mutagenesis in an infectious cDNA of ZYMV in order to check if the mutation alone is sufficient to induce the difference in biological properties.

Molecular studies have thus shown that single point mutations in HC-Pro or P3 of ZYMV were important for symptoms. Similarly, single mutations in HC-Pro or CP greatly affect aphid transmissibility. For unknown reasons, some ZYMV isolates such as ZYMV-E15 seem to be particularly prone to mutations and many variants (mild, aphid nontransmissible, aggressive, or virulent) have been derived from this isolate.

Biotechnological Application

The development of infectious cDNA clones of ZYMV brought the possibility of using ZYMV as a biotechnological tool to produce proteins of pharmaceutical or crop protection interest. The gene coding for the protein of interest can be inserted in the ZYMV genome either between the P1 and HC-Pro domains or between the NIb and CP domains. For optimal protein production, there is the possibility to use mild clones that have the mutation in the FRNK motif of HC-Pro so that they produce mild symptoms and do not affect plant growth. The protein may be produced in the edible cucurbit fruits, and therefore used directly for oral administration. Several molecules of pharmaceutical interest have been efficiently produced through this technology: the human interferon-alpha 2, antiviral and antitumor proteins MAP30 and GAP31, and a mite allergen. This technology also provided a way to produce large amounts of nucleocapsid proteins of five tospoviruses that could be used for immunization and production of specific antibodies. Finally, the *bar* gene coding for a phosphinothricin acetyltransferase that confers resistance to herbicides based on glufosinate ammonium has been inserted into the mild ZYMV expression vector. In weed-infested plots, this construct efficiently protected inoculated zucchini squash plants from damage caused by a herbicide treatment that completely destroyed the weeds. In addition, these plants were protected against severe ZYMV isolates. It is interesting to see that ZYMV, which emerged as the most damaging cucurbit virus in the last decades, can be manipulated and used for the benefit of human health and agriculture.

See also: Bean Common Mosaic Virus and Bean Common Mosaic Necrosis Virus; Diagnostic Techniques: Serological and Molecular approaches; Papaya Ringspot Virus; Plant Antiviral Defense: Gene Silencing Pathway; Plant Resistance to Viruses: Engineered Resistance; Plant Resistance to Viruses: Natural Resistance Associated with Dominant Genes; Plant Virus Diseases: Economic Aspects; Plant Virus Vectors (Gene Expression Systems); Polyomaviruses of Mice; Potyviruses; Recombination; Ribozymes; Vector Transmission of Plant Viruses; Vegetable Viruses.

Further Reading

Arazi T, Slutsky SG, Shiboleth YM, *et al.* (2001) Engineering zucchini yellow mosaic potyvirus as a non-pathogenic vector for expression of heterologous proteins in cucurbits. *Journal of Biotechnology* 87: 67–82.

Astier S, Albouy J, Maury Y, and Lecoq H (2001) *Principes de virologie végétale: Génome, pouvoir pathogène, écologie.* Paris: INRA Editions.

Desbiez C and Lecoq H (1997) Zucchini yellow mosaic virus. *Plant Pathology* 46: 809–829.

Desbiez C and Lecoq H (2004) The nucleotide sequence of watermelon mosaic virus (WMV, *Potyvirus*) reveals interspecific recombination between two related potyviruses in the 5′ part of the genome. *Archives of Virology* 149: 1619–1632.

Fuchs M, Ferreira S, and Gonsalves D (1997) Management of virus diseases by classical and engineered protection. *Molecular Plant Pathology On-Line.* http://www.bspp.org.uk/mppol/1997/0116fuchs (accessed October 2007).

Lecoq H (1998) Control of plant virus diseases by cross protection. In: Hadidi A, Kheterpal RK, and Koganezawa H (eds.) *Plant Virus Disease Control*, pp. 33–40. St. Paul, MN: APS Press.

Lecoq H (2003) Cucurbits. In: Loebenstein G and Thottapilly G (eds.) *Viruses and Virus-Like Diseases of Major Crops in Developing Countries*, pp. 665–687. Dordrecht: Kluwer Academic Publishers.

Lisa V and Lecoq H (1984) *Zucchini Yellow Mosaic Virus. CMI/AAB Descriptions of Plant Viruses No. 282.* Kew, UK: Commonwealth Mycological Institute.

Purcifull DE and Hiebert E (1979) Serological distinction of watermelon mosaic virus isolates. *Phytopathology* 69: 112–116.

Purcifull D, Hiebert E, and Edwardson J (1984) *Watermelon Mosaic Virus 2. CMI/AAB Descriptions of Plant Viruses No. 293.* Kew, UK: Commonwealth Mycological Institute.

West Nile Virus

L D Kramer, Wadsworth Center, New York State Department of Health, Albany, NY, USA

© 2008 Elsevier Ltd. All rights reserved.

Glossary

Bridge vector Competent mosquito vector that becomes infected following feeding on vertebrate hosts within the enzootic transmission cycle, and subsequently infects humans and other incidental hosts.

Enzootic A pathogen that is constantly present in an animal population, but usually only affects a small proportion of animals at any one time.

Epidemic Severe outbreak within a region or a group of humans.

Epizootic An outbreak of disease affecting many animals of one kind at the same time.

Extrinsic incubation period Period of time required for mosquito to transmit virus after ingesting infectious blood meal.

Neuroinvasive disease Disease caused by pathogen that infects nerve cells.

Nonviremic transmission Viral transmission to uninfected arthropod in the absence of detectable replication of virus in the vertebrate host.

Pathogenesis Origin and development of disease; more specifically the cellular events and reactions and other pathologic mechanisms occurring in the development of disease.

Phylogenetic analysis A method developed by biological systematists to reconstruct evolutionary genealogies of species based on nucleotide sequence relatedness.

Vertical transmission Viral transmission from adult female arthropod to her immediate progeny.

Viremia The presence of virus in the blood.

Zoonosis Also called zoonotic disease; refers to pathogens that can be transmitted from animals, whether wild or domesticated, to humans.

Classification

West Nile virus (WNV) is a member of the family *Flaviviridae*, genus *Flavivirus*, which is composed of approximately 70 members classified into 12 antigenic subgroups. The family *Flaviviridae* includes two additional genera,

Pestivirus (including veterinary pathogens such as bovine viral diarrhea viruses) and *Hepacivirus* (including the human pathogen hepatitis C virus). WNV belongs to the Japanese encephalitis (JE) antigenic complex, which includes the human pathogens St. Louis encephalitis, Rocio, and Ilheus viruses in the Americas, Japanese encephalitis virus in Asia, Murray Valley encephalitis virus in Australia, and other viruses, most of them not associated with encephalitis.

History and Geographic Distribution

WNV was first isolated in 1937 from the blood of a febrile woman in the West Nile district of Uganda. The next evidence of activity occurred when WNV was isolated from mosquitoes and birds and identified as the etiologic agent in sick children in North Africa and the Middle East in the 1950s. West Nile neuroinvasive disease (WNND) was first recognized during an outbreak in elderly patients in Israel in 1957. In the 1960s, WN encephalitis in equines was recognized in Egypt and France. The largest outbreak of WN fever occurred in 1974 in Cape Province, South Africa, leading to approximately 10 000 cases. Increasing frequency of severe outbreaks began occurring in 1996 in humans and horses, largely in the Mediterranean Basin. The range of WNV expanded suddenly in 1999 with the introduction of the virus into the New York City area, although the mode of introduction is unknown. The virus became successfully established as it expanded its range westward to encompass the contiguous United States, northward into Canada in 2001, and southward beginning in 2002 into Mexico, Central America, the Caribbean and parts of South America. From 1999 to 2006, approximately 22 790 cases of WN disease were reported in the US. The largest epidemics of neuroinvasive disease caused by WNV in North America occurred in 2002 and 2003 in the US, when 2946 and 2866 cases, respectively, of WNND were identified. WNV disease also has been noted in Cuba and Argentina in addition to the US and Canada in the Western Hemisphere, but scant evidence of morbidity and mortality has been observed in tropical America. Possible reasons for the lack of overt disease in tropical America are cross-protection from other flaviviruses circulating in tropical regions, less virulent strains of WNV, less-competent arthropod and avian hosts than in temperate regions, and greater diversity of host species in the tropics.

Molecular Epidemiology

WNV is the most widely distributed flavivirus in the world (**Figure 1**). Isolates comprise two main lineages based on nucleotide sequence homology (**Figure 2**). Lineage 1 includes three clades: (1a) isolates from Africa, Asia, Europe, Middle East, North America, and Russia; (1b) isolates from Australia (Kunjin); (1c) most Indian isolates. The strain of WNV introduced into the US in 1999 has 99.7% homology in both nucleic acid and amino acid sequence with a 1998 Israeli goose isolate. Lineage 2 contains strains isolated in sub-Saharan Africa and Madagascar, and includes the prototype 1937 Ugandan isolate. Severe WNND in humans has been associated only with lineage 1 strains. The strain responsible for the South African WN fever outbreak in 1974 is in lineage 2. Comparison of isolates from lineage 1 indicates differences in virulence between North American and Afro-European strains, but the degree of difference varies with the specific avian host. Enhanced virulence of the North American strains appears to be partially correlated with envelope (E) protein glycosylation. Strains within lineage 2, while not associated with overt human outbreaks, also vary phenotypically with respect to virulence in animal models. Two potential new lineages have been identified recently; a lineage 3 virus was isolated from *Culex pipiens* in the Czech Republic (Rabensburg virus) in 1997 and 1999, and a lineage 4 virus isolated from *Dermacentor marginatus* ticks in Russia in 1998 (LEIV-Krnd88–190). Rabensburg virus showed 77–78% identity to lineage 1 and 2 WNV strains, 77% identity to strain LEIV-Krnd88–190, and 71–76% identity to other representatives of the JE antigenic complex.

Phylogenetic analyses of WNV isolates from the US indicate that the virus remains highly conserved genetically. A single conserved amino acid change in the E gene (V159A), first detected in 2001, occurred with increasing frequency from 2002 through 2004 throughout the US (**Figure 3**), and became the dominant genetic variant circulating throughout North America. This genotype, 'North American dominant', probably resulted from a random mutation that became fixed in the viral RNA of the originally introduced genotype, 'Eastern US' strain. The dominant genotype disseminates 2–4 days earlier following feeding, in *Culex* spp. mosquito vectors, effectively reducing the extrinsic incubation period and thereby increasing the reproductive ratio or R_0 of WNV. This reproductive advantage, and a higher viremia observed in some avian species, may have contributed to displacement of the Eastern US genotype. Temporal and spatial clustering of isolates suggest the occurrence in some cases of localized virus spread by mosquitoes or resident birds, and in others long-distance dispersal by migratory birds or possibly by human transport. Nonetheless, it is clear from the phylogenetic evidence based on US isolates that there was a single introduction of WNV into North America from the Old World. This is in contrast to WNV in Israel, where both lineage 1 and 2 strains are frequently introduced, most likely during bird migration.

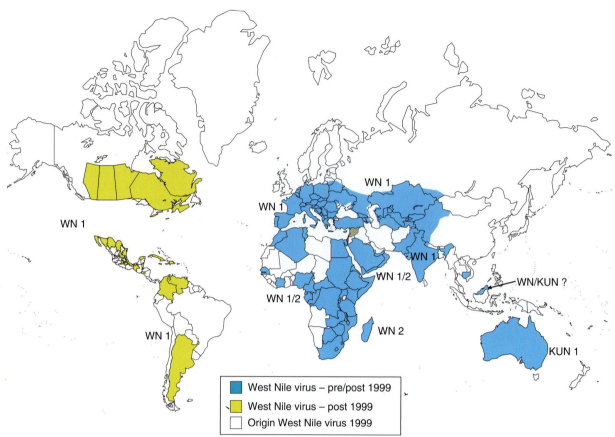

Figure 1 Global distribution of West Nile virus lineages 1 and 2. Areas in yellow have evidence of West Nile virus in the years (to 2006) following introduction of a lineage I (WN 1) virus strain in New York in 1999. This viral strain most likely originated in Israel, colored blue with yellow stripes. WN 1 viruses have been isolated from Africa, the Middle East, Europe, Russia, India, and North America. Isolations of lineage 2 WN viruses (WN 2) are found in Africa and the Middle East. Kunjin virus (KUN 1) is a subtype of lineage 1 isolated in Australia. A WN/KUN-like virus with a unique genotype (WN/KUN?) has been isolated from Sarawak, Malaysia. Modified from Scherret JH, Mackenzie JS, Hall RA, Duebel V, and Gould EA (2002) Phylogeny and molecular epidemiology of West Nile and Kunjin viruses. In: Mackenzie JS, Barrett ADT, and Duebel V (eds.) *Japanese Encephalitis and West Nile Viruses*, 379pp. Berlin, Heidelberg, and New York: Springer, with kind permission of Springer Science and Business Media.

The Virion and the Viral Genome

The West Nile virion is spherical, enveloped, approximately 50 nm in diameter, with relatively smooth surfaces and icosahedral symmetry (**Figure 4**). It contains a host-derived lipid bilayer surrounding a nucleocapsid that consists of the viral RNA complexed with multiple copies of the capsid protein. The viral genome is linear, positive-sense, single-stranded RNA, 11 029 bases in length. The 5′ noncoding region of WNV has a methylated type 1 cap (m^7GpppAmp); the 3′ noncoding region lacks a polyadenylated tail and in its place has a 5′-CU_{OH}-3′. These two regions contain conserved secondary structures critical to viral replication. The viral genome encodes three structural and seven nonstructural proteins in a single long open reading frame of 10 299 nt (nt 97–10 395), which is cleaved co- and post-translationally. The structural proteins (capsid (C), membrane (prM/M), and envelope (E)) are encoded at the 5′ end of the genome and the nonstructural proteins (NS1, NS2a, NS2b, NS3, NS4a, NS4b, and NS5) at the 3′ end. The E protein, the predominant protein of the virion, plays a major role in viral assembly, receptor binding, and membrane fusion, and is the principal target for neutralizing antibody. The NS proteins are responsible for viral RNA replication but also may function in viral assembly and in evasion of the host immune response. The putative function and nucleotide positions of the viral proteins are listed in **Table 1**.

Replication

WNV is able to replicate in a great variety of vertebrate cells. Replication is a complex process mediated in a controlled fashion through sequential interactions among viral RNA, proteins and host factors (**Figure 5**). The virion initially binds to the cell receptor and then enters via receptor-mediated endocytosis. Glycosaminoglycans have cleaved into the ten viral proteins. Genomic plus-sense RNA is transcribed into complementary minus-sense RNA at the endoplasmic reticulum membranes, forming replication complexes, which in turn serve as templates for the synthesis of new plus-sense RNA. The process is asymmetric, yielding 10- to 100-fold excess plus strands. The plus-sense RNA is packaged by viral C protein to form the nucleocapsid that is enclosed in an envelope consisting of a host-derived lipid bilayer and viral prM/M and E proteins. prM is cleaved in the *trans*-Golgi network, while M remains inserted in the envelope of the virion. Mature virions are released from the infected cell by exocytosis beginning 10–12 h after infection.

Ecology

Vectors

WNV is a zoonosis maintained in an enzootic cycle, transmitted primarily between avian hosts and ornithophilic mosquito vectors (**Figure 6**). Mosquitoes of the genus *Culex* are the predominant vectors in the enzootic cycle throughout the range of the virus' distribution, although the particular species of *Culex* varies according to geographic location. In the northeast, *Culex pipiens* and *C. restuans* may be important enzootic catalysts in the spring, while the former, an urban species that reaches high densities in midsummer, also may be the critical vector to humans in the northeastern and north central US as well as in recent outbreaks in Europe and Israel. *C. quinquefasciatus*, a member of the *C. pipiens* complex, is the primary vector in the southern US and, presumably, Latin America. Another important vector in these locations is *C. nigripalpus*, while *C. tarsalis* is the predominant vector in rural areas of western states in the US. Other *Culex* spp., such as *C. univitattus* and *C. antennatus* (Europe and Africa), *C. vishnui* complex (India), and *C. annulirostris* (Australia) have been implicated in the transmission cycle.

Vector competence varies between species and within populations of individual species. The *C. pipiens* complex contains two genetically distinct forms, that is, *C. pipiens* form 'pipiens' and form 'molestus'. The two forms differ in physiology and behavior with obvious implications to their epidemiological importance. Form 'pipiens' is thought to be exclusively ornithophilic while the urban form 'molestus' will feed on mammals. The two forms have been shown to not interbreed in northern Europe, in contrast to US populations, which contain individuals with hybrid genetic signatures (*pipiens* × *molestus*). Such hybridization, which also has been noted in southern Europe, may generate bridge vectors, disposed to feed on both birds and mammals. Indeed, US populations of *C. pipiens*, as well as *C. nigripalpus* and *C. tarsalis*, have been demonstrated to shift their feeding from birds to mammals in the late summer and early fall, and therefore may act as bridge vectors to infect equid and human hosts. It is

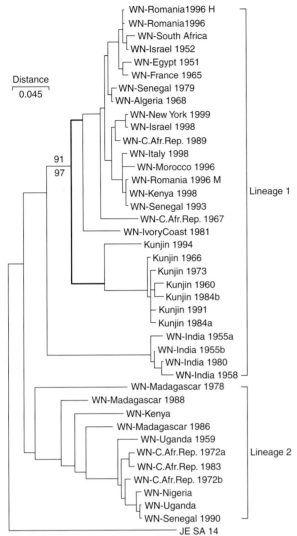

Figure 2 Phylogenetic tree based on E-glycoprotein nucleic acid sequence data (255 base pairs). The tree was constructed with the program MEGA by neighbor-joining with Kimura two-parameter distance (scale bar). Bootstrap confidence level (500 replicates) and a confidence probability value based on the standard error test were calculated using MEGA and are included on the tree (top and bottom values, respectively), illustrating support for the division between the lineage 1 WN virus group (not including the India isolates) and the KUN virus group. The best estimated length of the segment (bold line) separating these groups, in units of expected nucleotide substitutions per site, is 0.06928 and is statistically significantly positive ($P < 0.01$) by the likelihood ratio test (fastDNAml maximum likelihood program). An approximate 95% confidence interval for the true length of this segment is 0.03347, 0.10737. GenBank accession numbers can be found in Lanciotti RS, Roehrig JT, Deubel V, *et al.* (1999) Origin of the West Nile virus responsible for an outbreak of encephalitis in the northeastern United States. *Science* 286: 2333–2337, reproduced with permission.

been proposed as the putative receptors, although definitive proof is lacking. Following release of the nucleocapsid into the cytoplasm, the viral RNA is uncoated and translation proceeds, forming a single polyprotein which is

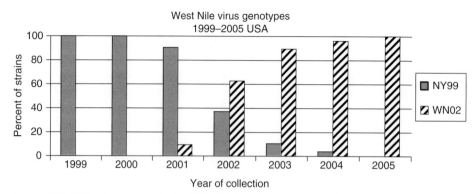

Figure 3 Prevalence of West Nile virus isolates in the United States characterized as Eastern US genotype (shaded box) and North American dominant genotype (striped box) from 2000 to 2005. Reproduced from Snapinn KW, Holmes EC, Young DS, Bernard KA, Kramer LD, and Ebel GD (2007) Declining growth rate of West Nile virus in North America. *Journal of Virology* 81(5): 2531–2534, with permission from American Society for Microbiology.

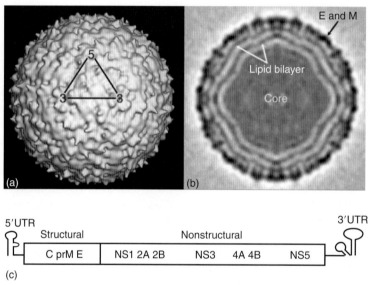

Figure 4 West Nile virion (a, b) and genome (c). WNV structure as reconstructed by cryo-EM. (a) A surface-shaded view with one asymmetric unit of the icosahedron indicated by the triangle. (b) Central section of the reconstruction, showing the concentric layers of mass density. (c) WNV genome, single-stranded positive-sense RNA, approximately 11 kb in length, consisting of a 5′ untranslated region (UTR), a single long open reading frame (ORF), and a 3′ UTR. The ORF encodes three structural and seven nonstructural proteins. (a, b) Modified from Mukhopadhyay S, Kim BS, Chipman PR, Rossmann MG, and Kuhn RJ (2003) Structure of West Nile virus. *Science* 302(5643): 248, as published in Kramer LD, Li J, and Shi PY (2007) West Nile virus: Highlights of recent advances. *Lancet Neurology* 6: 171–181, with permission from Science and Elsevier.

likely that other species of bridge vectors also become involved since virus has been isolated from mosquitoes of approximately 75 species of 10 genera worldwide; but the relative importance of each species in transmission to birds or humans must take into account factors of vectorial capacity, including field infection rates and population density. Many mosquito species actually may be incidental vectors of little epidemiologic significance, but further research is required to determine their importance in the ecology and epidemiology of WNV in North America.

The mechanism(s) of WNV perpetuation over adverse seasons and years may vary by region and country, but possible mechanisms include continuous low-level virus transmission, reinitiation after reintroduction of virus by migratory birds from locations where virus is active year-round, vertical transmission to females about to enter reproductive diapause in winter, and recrudescence of low levels of virus in chronically infected birds when mosquitoes are active (**Figure 6**). Viral RNA and infectious virus have been detected in experimentally infected wild birds more than 6 weeks after inoculation. Virus

Table 1 Functions and nucleotide positions of West Nile virus proteins

Viral protein	Function	Position in genome
Structural proteins		
C	Regulation of viral replication	97–465
(Pr)M	E folding and function	(466–7421) 742–966
E	Receptor binding	967–2469
	Membrane fusion	
	Viral assembly	
Nonstructural proteins (all NS proteins are required for viral replication)		
NS1	Early replication events (interaction with NS4A essential for replication)	2470–3525
NS2A[a]	Replication complex Interferon antagonist	3526–4218
NS2B[a]	Cofactor for viral protease Interferon antagonist assembly and release of virions	4219–4611
NS3	Serine protease	4612–6468
	NTPase and RTPase activity	
	RNA helicase Interferon antagonist	
NS4A[a]	Replication complex Interferon antagonist	6469–6915
NS4B[a]	Interferon antagonist Virion assembly	6916–7680
NS5	Methyl transferase	7681–10 395
	RNA dependent RNA polymerase Interferon antagonist	

[a]Functions not well characterized.

has been isolated from diapausing adult *C. pipiens* in the eastern US in the winter, from vertically infected male and female *C. univitattus* in Kenya, and from *C. quinquefasciatus* larvae in California during the summer transmission season. Vertical transmission has been demonstrated in the laboratory at rates varying with the population of mosquitoes tested.

Alternate modes of vector transmission have been observed in the laboratory and/or field. Nonviremic transmission has been demonstrated in the laboratory, with low rates of infection in co-feeding mosquitoes. This mode of transmission may expand the enzootic transmission cycle to involve mammals and birds that generally mount viremias too low to infect feeding mosquitoes. WNV has been isolated repeatedly in Russia from soft ticks (*Argasidae*), but their role in the transmission cycle is not clear. In addition, soft ticks have been demonstrated to transmit virus in the laboratory, and nonviremic transmission has been demonstrated. *Ixodidae* (hard ticks) allow the virus to pass transstadially, but are incompetent vectors. Other arthropods have been suggested as alternative vectors, including dermanyssoid mites, swallow bugs, and hippoboscid flies.

Vertebrates: Birds

Birds of more than 300 species in North America have been reported infected with WNV, as determined by virus isolation or antibody, confirming their role as the primary vertebrate in the enzootic cycle. Feeding by infected mosquitoes is the most common route of infection, but transmission to birds also has been demonstrated by direct contact, presumably via the fecal–oral route since there is significant shedding of virus, and by ingestion of infected mosquitoes or of carrion by omnivorous birds such as corvids and raptors. Experimental studies indicate that birds vary significantly in susceptibility and response to infection. The hallmark of WNV activity in North America is the significant morbidity and mortality in a wide range of species, although bird deaths also have been noted in the Middle East and recently in eastern Europe. An Egyptian strain isolated in the 1950s from a sick pigeon (*Columba livia*) is virulent in experimentally infected hooded crows (*Corvus corone*) and house sparrows (*Passer domesticus*). Corvids, especially crows, magpies, and jays, appear to be the most highly susceptible to disease, with 100% of American crows (*Corvus brachyrhynchos*) becoming sick and dying after infection with low doses of WNV. Nonetheless, corvids do not appear to be the primary amplifying hosts in the enzootic transmission cycle. In the northeastern and mid-Atlantic US, the American robin (*Turdus migratorius*) appears to be highly preferred by the predominant vector *C. pipiens*. Disproportionate numbers of mosquito blood meals are taken from robins and they have high levels of IgG antibody to WNV, indicating their importance in the enzootic cycle. In the laboratory, robins have average host competence determined by susceptibility, number of days infectious, and level of viremia or mean infectiousness. High levels of WNV IgG have been detected in other birds as well, including cardinals and wrens, and in some locations, house sparrows.

Vertebrates: Other

Thirty species of mammals and occasionally other vertebrates including reptiles and amphibians have been found

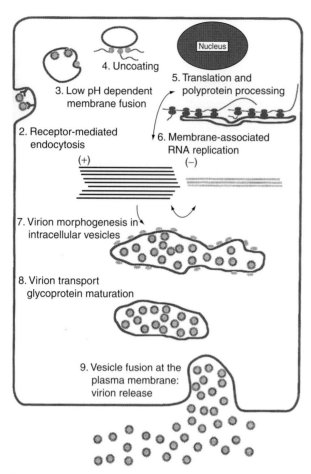

Figure 5 Flavivirus replication cycle. Highlights of the flaviviral replication cycle are diagrammed. Reproduced from Rice CM (1996) *Flaviviridae*: The viruses and their replication. In: Fields BN, Knipe DM, Howley PM, *et al.* (eds.) *Fields Virology*, 3rd edn., pp. 931–959. Philadelphia: Lippincott-Raven Publishers, with permission from Lippincott Williams and Wilkins.

infected with WNV. Their role in the transmission cycle is less significant than that of birds because of their generally low levels of viremia, but some small mammals, such as rabbits and chipmunks, have been demonstrated in the laboratory to mount sufficiently high levels of virus in the blood to infect a small proportion of feeding *Culex* spp. mosquitoes. In the US and Mexico, farmed alligators raised at high temperatures in crowded conditions demonstrate significant mortality and mount high viremias. Transmission appears to occur directly between alligators, as well as through ingestion of uncooked infected horse meat. Enzootics in equines have occurred in the US, France, Italy, Morocco, and Israel. Unvaccinated equines develop infections ranging from asymptomatic to encephalitic disease, and demonstrate a case–fatality rate of approximately 25%. Because of their low viremias, they likely are incidental hosts in the transmission cycle.

Low levels of WNV antibody in domestic pets have been detected in serosurveys in the US and in the Highveld region of South Africa. In the laboratory, dogs and cats are readily infected with WNV, but dogs do not develop detectable viremia and peak titers in cats are too low to infect most mosquitoes. Cats also may become infected by the oral route following feeding on dead infected mice. Sera from wild-caught vertebrates after intensive activity in 2003 in the US also demonstrated high levels of infection in raccoons, Virginia opossums, fox squirrels, and eastern gray squirrels. It is unclear how much morbidity and mortality is associated with WNV in wild mammals, but occasional cases of overt disease have been reported. Some small mammals, such as chipmunks and rabbits, have been shown experimentally to mount viremias sufficient to infect mosquitoes.

Vertebrate Pathogenesis

Infected mosquitoes expectorate virus with their saliva mostly intradermally but also intravascularly while probing and feeding, and their salivary secretions contain potent pharmacologic compounds that affect viral pathogenesis. The initial site of viral replication in the vertebrate is thought to be subdural Langerhans' dendritic cells at the site of inoculation. Activated dendritic cells migrate to draining lymph nodes where the virus replicates further, antigen processing begins, and an early immune response may become evident. Virus enters the blood by way of the efferent lymphatics and thoracic duct, resulting in a viremia that carries the virus to the visceral organs of the body, and possibly facilitates virus crossing the blood–brain barrier. However, the mechanism for the latter is unknown and may occur via infected inflammatory cells, retrograde transport along peripheral nerve axons, replication in epithelial cells, or by another route. Neurons, the main target cell in the central nervous system, suffer pathology, as do bystander nerve cells. In addition, there is immune-mediated tissue damage.

Factors such as age, immune status, and genetic susceptibility of the host, strain of virus, dose of inoculum, and route of infection, affect pathogenesis of WNV. In the mouse, variations in a single gene, Flv, on chromosome 5, determine the rodent's susceptibility to WN disease, but not infection. Susceptibility has been mapped to the gene encoding the L1 isoform of the interferon-inducible, antiviral effector enzyme $2'-5'$-oligoadenylate synthetase. An intact immune system is critical to prevention of disease, and both humoral and cellular factors play a role in protection against disease. Neutralizing antibody is the principal means of viral clearance. Interferon-dependent innate immune responses are essential in limiting infection of WNV, as is the adaptive immune response. Complement, dendritic cells and chemokines probably also play roles in viral clearance. The mechanism by which infection is cleared from neurons appears to involve CD8+ T cells in

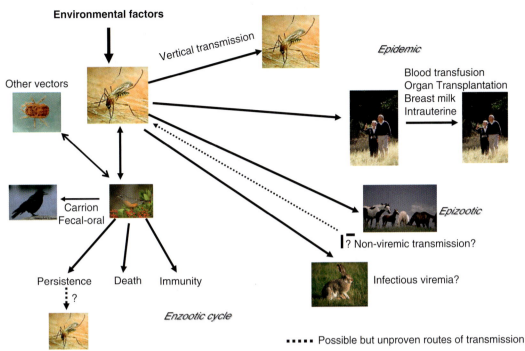

Figure 6 West Nile virus transmission cycle. The enzootic cycle is illustrated, as well as epidemic and epizootic hosts. Solid arrows indicate confirmed transmission pathways; dotted lines proposed pathways that have not been confirmed in nature.

the perforin-dependent class I major histocompatibility complex (MHC)-restricted system. Persistence of virus has been noted to occur in some vertebrates following natural and experimental infection, even in the presence of neutralizing antibody. Hamsters shed WNV in their urine for ≥ 8 months and virus can be isolated from their brains as long as 53 days after infection.

Clinical Disease

The range of symptoms associated with WNV infection extends from uncomplicated febrile illness to meningitis, neuropathies, muscle weakness, paralysis, and encephalitis. Symptoms generally become apparent 2–14 days after infection by mosquito bite, but 80% of infected individuals are asymptomatic. The majority of symptomatic patients present with flu-like symptoms, termed WN fever, which in the US has been associated with substantial morbidity. The duration of symptoms may be from days to months, and the patient may require hospitalization. WNND includes meningitis, encephalitis, and acute flaccid paralysis (AFP). From 1999 to 2007, 27 083 cases of WN disease were reported in the US, of which 57.3% were WN fever and 40.3% were WNND resulting in 1054 fatalities (3.9% of all, 9.7% of WNND) (**Table 2**). Given the proportion of symptomatic cases ($\sim 20\%$), approximately 135 415 individuals were infected with WNV over the 8-year period, and 8.1% of all infections were WNND. Males were more commonly infected than were females. Movement disorders, that is, dyskinesias, frequently characterized as parkinsonism are common. The constellation of fever, headache, and signs of meningeal irritation, cannot be easily discriminated from other flaviviral encephalitides. Patients with AFP experience loss of spinal anterior horn cells and have accompanying asymmetric weakness. Risk factors for WNND include age, diabetes, and possibly a history of hypertension or cardiovascular disease. The median age of WNND cases in the US since 1999 is 57 years (46 years for WNF), but there have been reports of WNND in children and AFP in young adults. Infection may persist in the central nervous system of immunologically compromised patients, leading to extended neurologic illness and sequelae. The mortality rate following WN encephalitis is greater than following WN meningitis. Immunosuppression, chronic renal disease, and hepatitis C infection have been recognized as risk factors for death after adjustment for age. Rare cases of hemorrhagic WNV disease have been noted.

The vast majority of human cases are transmitted by the bite of an infected mosquito. However, human-to-human transmission after blood transfusions, organ transplantation, and through mothers' milk have occurred in the US and there has been one instance of intrauterine transplacental transmission. High-throughput nucleic acid detection assays were developed and implemented by blood banks in response to transfusion transmission in order to protect the blood supply.

Table 2 Human West Nile virus disease cases, United States, 1999–2005

Year	Total cases	WNND cases (% of total)	WNF cases (% of total)	Other clinical/unspecified	Fatalities (% of WNND cases)
1999	62	59	3	0	7
2000	21	19	2	0	2
2001	66	64	2	0	9
2002	4 156	2 946	1 160	50	284
2003	9 856	2 860	6 830	166	264
2004	2 539	1 142	1 269	128	100
2005	3 000	1 294	1 607	99	119
2006	4 269	1 459	2 616	194	177
2007	3 114	1 059	2 026	29	92
Total	27 083	10 902 (40.3%)	15 515 (57.3%)	666	1054 (3.9%/9.7%)

WNND, West Nile neurologic disease; WNF, West Nile fever.
Reported to CDC, as of 06 November 2007.

Diagnostics

Human

Studies of infected blood donors indicated the presence of WNV IgM and IgA at the earliest on day 3 and in all cases by day 8–9 after RNA was detected; IgG antibody appeared about 1 day later (**Figure 7**). The detection of IgM antibody in the cerebrospinal fluid (CSF) by a monoclonal antibody (MAb) capture enzyme-linked immunosorbent assay (MAC ELISA) in conjunction with evidence of neurologic symptoms has been accepted as diagnostic of WNV disease. Presence of IgM antibody in the serum alone is strongly suggestive of recent infection but not definitive due to persistence for at least 16 months (199 days in the CSF) in patients with WNND, and to some cross-reactivity with antibody to other flaviviruses. The plaque reduction neutralization test (PRNT) in Vero cell culture remains the gold standard in diagnosis of flavivirus infections because of the extensive cross-reactions complicating diagnosis. A fourfold rise in neutralizing antibody titer between paired acute-phase and convalescent-phase sera confirms WNV infection, as does a fourfold or greater titer to WNV compared to related flaviviruses. However, interpretation is complicated in individuals who have had prior infection with another flavivirus. Microsphere immuno assay (MIA) using recombinant E protein has equivalent sensitivity to ELISA. In primary flavivirus infection, the specificity of the MIA procedure is high (>90%) compared to PRNT. However, as with PRNTs, the interpretation of WNV-E MIA is confounded when WNV is a secondary flavivirus infection. Sera from dengue (flavivirus) patients are highly likely to be positive in MIA procedures using recombinant WN E protein. Addition of WNV NS5 to a multiplex test panel increases specificity, since sera from patients with other past flavivirus infections are likely to be negative to WNV NS5 but reactive to WNV E. Serologic assays are more sensitive than detection of viral nucleic acid by reverse transcription-polymerase chain reaction (RT-PCR) once symptoms are evident because of the rapid clearance of virus by most individuals; however, detection of viral RNA is confirmatory.

Clinical findings also aid diagnosis. Neurologic symptoms indicate neuroinvasive disease. Electrophysiologic tests and magnetic resonance imaging (MRI) may be useful in cases of acute flaccid paralysis. Elevated lymphocyte counts and protein levels may be present in CSF during WNV infection. Epidemiological data also may be critical for diagnostic consideration because other flavivirus infections cause similar symptoms.

Surveillance Specimens

Submission of dead birds, particularly American crows, has provided an excellent sentinel system for WNV activity in the US, but depends upon participation by the public. In addition, over time this may not remain a reliable indicator of virus activity if selection for resistance to disease occurs. Tissues or swabs from dead birds and pooled mosquitoes are tested by real-time RT-PCR for viral nucleic acid, by cell culture procedures that detect live virus followed by confirmation using specific assays, and/or by procedures that detect viral antigen. High-throughput virus-specific molecular assays, such as real-time RT-PCR, have been developed to accommodate testing of large numbers of specimens in a timely manner. Rapid assays utilizing dip sticks to detect viral antigen also have been adopted for surveillance, but have low sensitivity. Detection of antibody in sera of live sentinel captive birds or wild-caught birds using an indirect or a competitive blocking ELISA, with confirmation by PRNT, generally has proven less useful as an early warning indicator of viral activity.

Prevention and Control

Vaccines and Therapeutics

Currently, the only available treatment for disease caused by WNV is supportive, given that no therapeutic

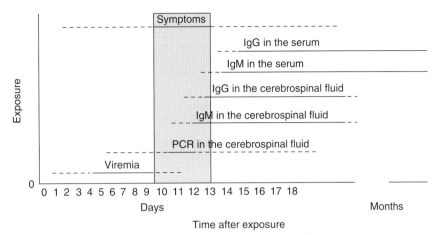

Figure 7 Schematic of virologic and serologic tests in West Nile virus encephalitis. Solid lines represent the more common results; broken lines represent reported ranges. The shaded box is an example of a typical patient. Reproduced from Gea-Banacloche J, Johnson RT, Bagic A, Butman JA, Murray PR, and Agrawal AG (2004) West Nile virus: Pathogenesis and therapeutic options. *Annals of Internal Medicine* 140: 545–553, with permission from American College of Physicians.

treatment has shown consistent clinical efficacy. Therapeutic interventions include passive administration of high-titer immune immunoglobulin. MAb, in particular humanized MAb against WNV E protein, also has shown some benefit. Interferon offers promise based on animal models but may not be helpful in treating the late-stage central nervous system disease. Finally, antiviral therapy using small sequence-specific nucleic-acid-based molecules to suppress WNV replication in tissue culture is being evaluated but the timing and high dose required remain issues of concern. Targets of intervention include viral RNA polymerase, RNA helicase, or other viral replication enzymes. WNV replicon-based high-throughput screening assays have been developed to aid in discovery of compounds with potential antiviral activity.

Advances also have been made in development and implementation of effective vaccines. Protection has been demonstrated in equids following immunization with inactivated cell culture-derived virus (Ft. Dodge). Recombinant canarypox virus containing the prM and E genes of WNV also has been licensed for use in equids. Two recombinant live virus vaccines are in clinical trial in humans. One uses a yellow fever 17D strain infectious clone, with the original yellow fever virus prM and E genes replaced by attenuated WNV genes (Acambis' Chimerivax-WN), and another uses attenuated dengue virus 4 with deletions in the $3'$ end as the backbone (NIH/Macrogenics). Both live, attenuated and chimeric vaccines have the advantage of eliciting humoral and cellular responses in the host after a single dose. Other approaches include DNA vaccines that encode prM, E, or C; subunit vaccines of purified recombinant prM and E; and cross-reactive flaviviruses or WNV variants.

Public Health Measures

Reduction of WNV transmission currently relies on reduction of mosquito populations or minimizing contact between vectors and humans. Mosquito control agencies in the US have undertaken extensive larviciding to kill immature stages of mosquito vectors in the aquatic environment. Source reduction, that is, removal of standing water, also has proven helpful. Personal protection measures, for example, wearing loose clothing, use of repellents, particularly 'DEET' (N,N-diethyl-m-toluamide or N,N-diethly-3-methylbenamide), or use of window screens, reduce direct contact with mosquitoes. Community action groups have worked to increase awareness and educate the public.

Summary

The introduction of WNV to the US and its subsequent spread and successful establishment in the Western Hemisphere alerted the world to the impact of globalization on pathogen dissemination. The number of WNND cases in 2002 and 2003 represent the largest outbreak of neurologic diseases in North America. Future research undoubtedly will address development of efficacious therapeutics, effective vaccines, novel methods to control mosquitoes, and accurate risk prediction. Each of these is dependent upon solid basic research on WNV ecology, pathogenesis, and immunology.

See also: Classical Swine Fever Virus; Dengue Viruses; Flaviviruses: General Features; Flaviviruses of Veterinary Importance; Japanese Encephalitis Virus; Yellow Fever Virus.

Further Reading

Beasley DW (2005) Recent advances in the molecular biology of West Nile virus. *Current Molecular Medicine* 5: 835–850.

Brinton MA (2002) The molecular biology of West Nile Virus: A new invader of the Western Hemisphere. *Annual Review of Microbiology* 56: 371–402.

Davis CT, Ebel GD, Lanciotti RS, et al. (2005) Phylogenetic analysis of North American West Nile virus isolates, 2001–2004: Evidence for the emergence of a dominant genotype. *Virology* 25: 252–265.

Diamond MS (2005) Development of effective therapies against West Nile virus infection. *Expert Review of Anti-Infective Therapy* 3: 931–944.

Gea-Banacloche J, Johnson RT, Bagic A, Butman JA, Murray PR, and Agrawal AG (2004) West Nile virus: Pathogenesis and therapeutic options. *Annals of Internal Medicine* 140: 545–553.

Hayes CG (1989) West Nile fever. In: Monath TP (ed.) *The Arboviruses: Epidemiology and Ecology*, vol. V, ch. 49, pp. 59–88. Boca Raton, FL: CRC Press.

Hayes EB and Gubler DJ (2006) West Nile virus: Epidemiology and clinical features of an emerging epidemic in the United States. *Annual Review of Medicine* 57: 181–194.

Kramer LD, Li J, and Shi PY (2007) West Nile virus: Highlights of recent advances. *Lancet Neurology* 6(2): 171–181.

Lanciotti RS, Roehrig JT, Deubel V, et al. (1999) Origin of the West Nile virus responsible for an outbreak of encephalitis in the northeastern United States. *Science* 286: 2333–2337.

Monath TP and Heinz FX (1996) Flavivirus. In: Fields BN, Knipe DM, Howley PM, et al. (eds.) *Fields Virology,* 3rd edn., pp. 961–1034. Philadelphia: Lippincott-Raven Publishers.

Mukhopadhyay S, Kim BS, Chipman PR, Rossmann MG, and Kuhn RJ (2003) Structure of West Nile virus. *Science* 302(5643): 248.

Rice CM (1996) *Flaviviridae*: The viruses and their replication. In: Fields BN, Knipe DM, Howley PM, et al. (eds.) *Fields Virology,* 3rd edn., pp. 931–959. Philadelphia: Lippincott-Raven Publishers.

Scherret JH, Mackenzie JS, Hall RA, Duebel V, and Gould EA (2002) Phylogeny and molecular epidemiology of West Nile and Kunjin viruses. In: Mackenzie JS, Barrett ADT, and Duebel V (eds.) *Japanese Encephalitis and West Nile Viruses*, 379pp. Berlin, Heidelberg, and New York: Springer.

Shi PY and Wong SJ (2003) Serologic diagnosis of West Nile virus infection. *Expert Review Molecular Diagnostics* 3: 733–741.

Smithburn KC, Hughes TP, Burke AW, et al. (1940) A neurotropic virus isolated from the blood of a native Ugandan. *American Journal of Tropical Medicine* 20: 471–492.

Snapinn KW, Holmes EC, Young DS, Bernard KA, Kramer LD, and Ebel GD (2007) Declining growth rate of West Nile virus in North America. *Journal of Virology* 81(5): 2531–2534.

White Spot Syndrome Virus

J-H Leu, J-M Tsai, and C-F Lo, National Taiwan University, Taipei, Republic of China

© 2008 Elsevier Ltd. All rights reserved.

Glossary

***In situ* hybridization (ISH)** A technique that uses a labeled complementary DNA or RNA probe to hybridize and detect a specific DNA or RNA sequence in a section of tissue (*in situ*).

Microarray A DNA microarray is a collection of microscopic dsDNA or oligonucleotide spots that are deposited on a solid glass slide at high density. Each spot commonly represents a single gene. DNA microarrays use DNA–DNA or DNA–RNA hybridization to perform simultaneous large-scale analyses of the expression levels of the corresponding genes.

RNAi (RNA interference) RNAi refers to the introduction of homologous double-stranded RNA to specifically interfere with the target gene's expression in the cells/organisms. It was originally discovered in *C. elegans*, and is now widely used in many organisms for null mutation. Recent studies suggest that RNAi can be applied in shrimp as well through intramuscular injection of dsRNA.

Introduction

White spot disease (WSD) is a highly contagious viral disease of penaeid shrimp. Its onset is rapid, and high levels of mortality can result within just a few days. Following the first outbreak of WSD in China and Taiwan in 1992–93, this viral disease spread quickly to other shrimp-farming areas, including Japan (1993), Thailand (1993), the United States (1995), Central and South America (1999), France (2002), and Iran (2002). Nowadays, WSD is considered endemic in almost all shrimp-producing countries in Asia and the Americas, causing serious economic damage to the shrimp culture industry. The causative agent of WSD is white spot syndrome virus (WSSV), which is a large (80–120 × 250–380 nm), non-occluded, rod-shaped to elliptical, double-stranded (ds) DNA virus. The virus has a remarkably broad host range among crustaceans, including many species of shrimp, crayfish, crab, and lobster. WSSV replication and virion assembly occur in the hypertrophied nuclei of infected cells without the production of occlusion bodies. The WSSV virion consists of a rod-shaped nucleocapsid surrounded by a trilaminar envelope and a unique, tail-like extension at one end. The virions contain a circular, supercoiled, dsDNA genome of about 300 kilobase pairs (kbp). Complete genome sequence analyses have shown that most of the WSSV open reading frames (ORFs) encode proteins bearing no homology to known proteins, while some identifiable genes are more homologous to eukaryotic genes than to viral genes. Based on these genetic analyses and on its unique morphological features, WSSV has been classified as the sole member of a new virus family.

Taxonomy

After its initial discovery, WSSV was classified as the genus *nonoccluded Baculovirus* (*NOB*) of the subfamily *Nudibaculovirinae* of family *Baculoviridae*. However, based on its unique morphological features, on its genomic structure and composition, and on phylogenetic analyses, *White spot syndrome virus 1* was reassigned by the International Committee on Taxonomy of Viruses in 2004 as the type species of the genus *Whispovirus* and the sole member within the new virus family *Nimaviridae*. The family name refers to the thread-like, polar extension on the virus particle (*nima* is Latin for 'thread'). Today, there is still only a single species within the genus *Whispovirus*. However, various geographical isolates with genotypic variability (variants) have been identified within this species. Other names that have been used in the literature for WSSV include: hypodermal and hematopoietic necrosis baculovirus (HHNBV), rod-shaped nuclear virus of *Penaeus japonicus* (RV-PJ), systemic ectodermal and mesodermal baculovirus (SEMBV), and white spot baculovirus (WSBV).

Transmission, Host Range, and Epidemiology Studies of WSSV

WSSV is contagious and highly virulent in penaeid shrimp. In cultured shrimp, WSSV infection can cause a cumulative mortality of up to 100% within 3–10 days of the first signs of disease. WSSV can be transmitted horizontally, either *per os* when the shrimp feed on diseased individuals, contaminated food or infected carcasses, or else through exposure to virus particles in the water, in which case the route of infection is primarily through the gills or other body surfaces. The virus is also transmitted vertically from brooder to offspring. However, transmission is not transovarial because the virus appears to attack only young developing oocytes, which die before reaching maturation. It is therefore more likely that vertical transmission is caused by contamination of the egg mass. Penaeid shrimp are highly susceptible to WSSV. Although there is evidence of resistance during the larval and early (younger than PL6) postlarval stages, WSSV can cause disease in shrimp at any growth stage.

WSSV has a remarkably broad host range. Almost every species of penaeid shrimp is susceptible to WSSV infection. Moreover, the virus can infect other marine, brackish, and freshwater crustaceans, including crayfish, crabs, and spiny lobsters. However, in contrast to penaeid shrimp, infection is often not lethal in these species, and consequently they may serve as reservoirs and carriers of the virus. Furthermore, at least one insect, the shore fly (a member of the family Ephyridae), as well as copepods collected from WSSV-affected farms, have been diagnosed as WSSV-positive by PCR, suggesting that they are also possible reservoir hosts.

In epidemiological studies of WSSV, variations in the number of the 54 bp tandem repeat located between the *rr1* and *rr2* genes (see the section titled 'Identification through homology comparison') have been used as a strain-specific, genetic marker. The number of repeats varies greatly when infected shrimps are collected from different ponds or from the same ponds at different times, but in almost every outbreak of WSD, shrimps from the same pond usually have the same number of repeats. This suggests that in each pond, a single WSSV isolate is the causative agent. High variations in the number of 54 bp repeats have been reported in strains of WSSV from Thailand, India, and Vietnam. In Vietnam, a particular repeat pattern was observed in shrimps from outbreak ponds, but no repeat pattern was found to be useful as a prediction of pathogenicity or geographical origin in any of these different strains. On the other hand, cultured shrimps and wild crustaceans from the same WSD outbreak pond had different 54 bp repeat patterns, suggesting that the wild crustaceans may not have been responsible for infecting the cultured shrimp. However, this interpretation has been questioned by a recent study showing that the number of the 54 bp repeats as well as the pathogenicity changed when a given WSSV isolate was passaged through different crustacean hosts. This study further implies that the variations in repeat number could result from host selection rather than geographical isolation.

Clinical Features and Pathology

The most commonly observed clinical sign of WSD in shrimp is white spots in the exoskeleton and epidermis. These may range in size from minute spots to disks several millimeters in diameter which may coalesce into larger plates. The spots may result from abnormal deposits of calcium salts by the cuticular epidermis or result from disruption to the transfer of exudates from the epithelial cells to the cuticle. Infected animals are lethargic, reduce their food consumption, and display a reddish to pink body discoloration due to expansion of cuticular chromatophores. Moribund shrimp exhibit systemic destruction of target tissues with many infected cells showing homogeneous hypertrophied nuclei. At advanced stages of infection, numerous virus particles are released into the hemolymph from the lesions, causing a general viremia.

It should be noted that although the white spots are a typical and characteristic clinical sign of WSD, white spots on the carapace of shrimp can also be caused by other environmental stress factors, such as high alkalinity or a bacterial shell disease. Conversely, moribund shrimp with WSD may have few, if any, white spots.

To date, no species of penaeid shrimp is known to show significant resistance to WSD.

Histopathology

The prime targets for WSSV replication are tissues of ectodermal (cuticular epidermis, fore- and hindgut, gills, and nervous tissue) and mesodermal (lymphoid organ, antennal gland, connective tissue, and hematopoietic tissue) origin. Tissues of endodermal origin (hepatopancreas and midgut) are not affected by the virus. Histopathological observation reveals similar cellular changes upon WSSV infection in all target tissues. In the early stage of infection, affected cells display nuclear hypertrophy (**Figure 1**), nucleoli dissolution, and chromatin margination, and the central area changes into a homogeneous eosinophilic region. The infected cells then proceed to develop an intranuclear eosinophilic Cowdry A-type inclusion, which subsequently becomes a light basophilic, denser inclusion separated by a transparent zone from the marginated chromatin. During this time, the cytoplasm becomes less dense and more lucent. In the late stage of infection, the nuclear membrane is disrupted, causing the intranuclear transparent zone to fuse with the lucent cytoplasm. At the end of cellular degeneration, the nucleus or whole cell disintegrates, leading to loss of cellular architecture. In moribund shrimp, most tissues and organs are heavily infected with the virus, and they exhibit severe multifocal necrosis.

Temporal studies coupled with *in situ* hybridization (**Figure 2**) have shown that the first WSSV-positive signals are detected in the stomach, gills, cuticular epidermis, and connective tissue of the hepatopancreas. At later stages of infection, the lymphoid organ, antennal gland, muscle tissue, hematopoietic tissue, heart, midgut, and hindgut also become positive. As infection proceeds, the stomach, gill, hematopoietic tissue, lymphoid organ, antennal gland, and cuticular epidermis become heavily infected with WSSV, which leads to serious damage and necrosis (and/or apoptotsis; see the section titled 'Viral–host relationships').

Viral Morphology and Morphogenesis

The WSSV virion is a nonoccluded particle with a trilaminar envelope. It is elliptical to bacilliform (olive-like) in shape and measures about 270 × 120 nm. A long, tail-like projection is often seen extending from the narrow end of the purified virion (**Figure 3(a)**). The cylindrical, rod-shaped nucleocapsid is about 300 × 65 nm, which is longer than the intact virion. The nucleocapsid displays a very distinctive pattern consisting of a stacked series of rings (about 16 in total) that run perpendicular to the longitudinal axis of the capsid. The thickness of these rings is quite constant, at about 20 nm. Each ring consists of two rows of 12–14 globular subunits, each approximately 10 nm in diameter (**Figure 3(b)**).

The entire WSSV replication cycle takes place in the nucleus and, in an acutely infected shrimp, can be completed within 24 h post infection. Initial signs of WSSV replication in the nucleus include nuclear hypertrophy and chromatin margination. Viral morphogenesis begins with *de novo* formation of the viral envelope, which can be seen at first as fibrillar fragments in the nucleoplasm. As infection progresses, these fragments form into the membranous and/or vesicular precursors of the viral envelope. The development of nucleocapsids begins with the formation of long empty tubules (**Figure 4(a)**) which have a diameter similar to that of empty nucleocapsids and a segmentation similar to that of nucleocapsids. They can exist individually or may be laterally aligned in groups of two or three to form a larger structure. The assembled tubules break into fragments of 12–14 rings to form empty naked capsids (**Figure 4(b)**) which are then surrounded by envelopes, leaving an opening at one end. Nucleoproteins, which have a filamentous appearance, enter the capsid through this open end, while

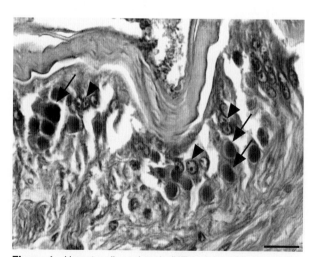

Figure 1 Hematoxylin and eosin (HE) staining of tissue section from *Penaeus monodon* cuticular epithelium of eyestalk infected with WSSV. Degenerated cells characterized by hypertrophied nuclei (black arrows) are readily seen. The cells with normal nuclei are also indicated with arrowheads. Scale = 20 μm.

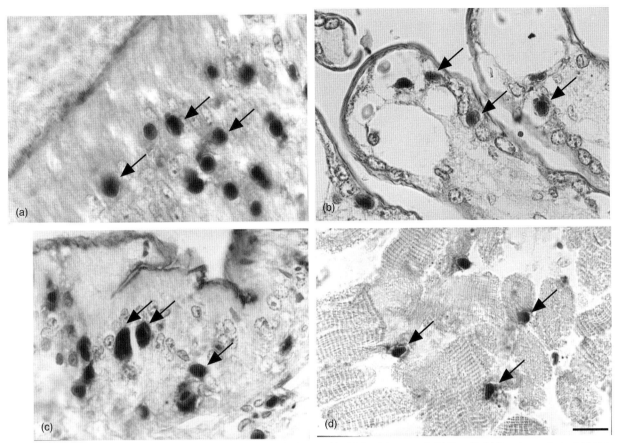

Figure 2 *In situ* hybridization analysis of WSSV positive cells (arrows) in (a) the integument, (b) the gill, (c) the stomach, and (d) the heart from experimentally infected *Penaeus monodon* at 60 hpi. The infected cells shown here in the heart are connective tissue cell; muscle cells are not usually targeted by WSSV. Scale = 20 μm.

simultaneously increasing the diameter of the virion. Finally, the open end of the nucleocapsid is closed, and the envelope narrows at the open end to form the apical tail of the mature virion, which has now obtained its characteristic olive-like shape.

Virion Structure and Composition

WSSV virions have a complex SDS-PAGE protein profile (**Figure 5(a)**). Early structural protein studies used SDS-PAGE coupled with Western blot analyses and/or N-terminal sequencing of proteins to identify at least six structural proteins, including VP35, VP28, VP26, VP24, VP19, and VP15. More recent studies have used mass spectrometry to analyze the protein profiles of purified WSSV virions separated by SDS-PAGE and 2-D electrophoresis, and more than 40 structural proteins have now been identified. The major structural proteins are VP664, VP28, VP26, VP24, VP19, and VP15 (**Figures 5** and **6**). A biochemical fractionation study using differential concentrations of salt and detergent has tentatively classified the WSSV structural proteins into envelope, tegument, and capsid proteins. VP15 is a basic capsid protein with *in vitro* DNA-binding activity, and it has high homology to the lysine- and arginine-rich DNA-binding proteins of the insect baculoviruses. This protein is therefore thought to be responsible for packing the WSSV DNA into the nucleocapsid. VP26, a tegument protein, can interact with actin, but the importance of this interaction remains unknown. VP19, VP24, and VP28 are envelope proteins, and an *in vitro* physical interaction between VP24 and VP28 has been reported. VP664, the major capsid protein, takes its name from its calculated molecular weight of 664 kDa, and it is not only the largest protein encoded by the WSSV genome, but also the largest viral structural protein ever found. This protein is encoded by the intron-less giant ORF (*wssv419*) of 18 234 nt. Immunoelectron microscopy of purified virions has shown that VP664 forms the globular subunits that are visible in the nucleocapsid ring structures (**Figure 6**). In addition to VP15 and VP664, the capsid contains at least four other minor proteins: VP160A (WSSV344), VP160B (WSSV94), VP60B (WSSV474), and VP51C (WSSV364) (**Figure 5**).

A number of viruses use the Arg-Gly-Asp (RGD) motif to bind to cellular integrins during infection and

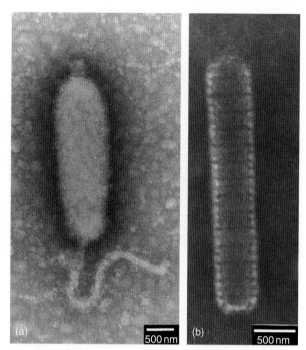

Figure 3 Morphology of the WSSV virions. (a) Intact WSSV virion showing the tail-like extension on one end. (b) A WSSV nucleocapsid showing the stacked ring structures, which are made up of two rows of globular subunits.

this cell attachment site signature can be identified in at least six WSSV structural proteins. It has been shown that one of these six structural proteins, VP110 (WSV035), can attach to shrimp host cells, and that this adhesion can be blocked by synthetic RGDT peptides, suggesting that the RGD motif in VP110 may play a role in WSSV infection. VP28 may also bind to host cells even though it lacks the RGD motif, and virus neutralization tests originally suggested that VP28 was critical for systemic WSSV infection. However, recent studies have shown that the observed phenomenon was, in fact, a nonspecific inhibitory effect of rabbit serum. Thus, to date, although SDS-PAGE indicates that VP28 is the most abundant virion protein, its role remains elusive. In addition to VP28, virus neutralization tests have been used to identify other structural proteins in the envelope and determine their role in WSSV infection. However, considering what is now known for VP28, the results of neutralization tests should be interpreted with care.

None of the five small major WSSV structural proteins (VP28, VP26, VP24, VP19, and VP15) appears to be glycosylated. However, increased migration in SDS-PAGE after N-linked glycosidase F treatment of structural protein VP180 (encoded by *wsv001*) suggests that this large collagen-like protein is N-glycosylated. This is the first reported evidence of an intact collagen gene in a virus genome.

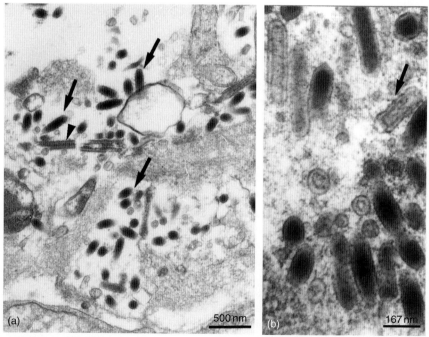

Figure 4 Transmission electron micrograph of WSSV-infected tissues from beneath the cephalothoracic exoskeletoal cuticle. (a) The viral particles spread among the necrotic area. The complete viral particles are indicated with arrows and the long empty tubules with an arrowhead. (b) High magnification of virus particles with rod-shaped morphology. A viral particle with an empty nucleocapsid is indicated with an arrow.

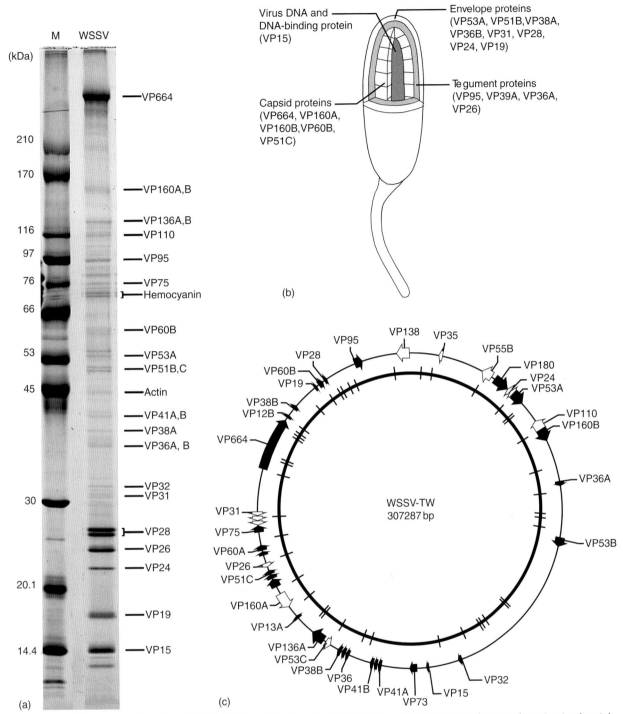

Figure 5 WSSV structural proteins. (a) SDS-PAGE profile of purified WSSV virions. The six major and some minor structural protein bands are indicated. The bands corresponding to two host proteins co-purified with the virion are also indicated. (b) A proposed schematic diagram to show the WSSV virion structure and the possible location of WSSV structural proteins and its DNA. (c) Distribution of 41 WSSV structural protein genes in the WSSV-TW genome. The inner circle shows predicted HindIII restriction enzyme cutting sites.

WSSV Genome Structure

The WSSV genome is a large ds circular DNA of about 300 kbp. The complete genome sequences of three isolates originating from China (WSSV-CN), Taiwan (WSSV-TW), and Thailand (WSSV-TH) indicated sizes of 305, 307, and 297 kbp, respectively. Genetic variations among these isolates include two major polymorphic loci

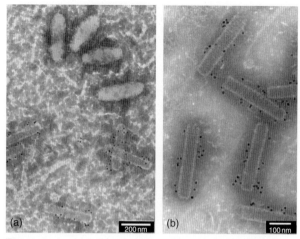

Figure 6 Immunoelectron microscopy analysis of purified WSSV virions detected with VP664 antibody followed by gold-labeled secondary antibody. (a) The VP664 antibody specifically binds to the nucleocapsid and not to the viral envelope. (b) Most of the gold particles are localized at the perimeter of the nucleocapsid.

(a 13 kbp deletion in WSSV-TH and a variable region prone to recombination), a transposase sequence that was only present in WSSV-TW, and variation in the number of repeats and single nucleotide mutations. Not including the two major polymorphic loci and the transposase sequence, these three isolates share an overall nucleotide sequence identity of 99.32%. It has been shown that differences in the major polymorphic 13 kbp deletion locus are related to differences in virulence of WSSV isolates.

The WSSV genome has a total G + C content of 41% uniformly distributed over the genome. Although most WSSV genome sequences are unique, 3% of the genome consists of highly repetitive sequences, which are organized into nine homologous regions (hrs). These are distributed throughout the genome, and are located largely in intergenic regions. The presence of hrs is a feature of many baculovirus genomes. Baculovirus hrs function as enhancers of early gene transcription and as initiation sites for DNA replication. They have also been implicated as sites of DNA recombination. It is likely that WSSV hrs will be shown to have similar functions.

A total of 532 putative ORFs consisting of a minimum of 60 codons have been identified by sequence analysis of the WSSV-TW isolate. Of these, 140 ORFs have a potential downstream polyadenylation site (AATAAA). The size of proteins encoded by these predicted ORFs ranges from 60 to 6077 amino acids. Only 30% of the predicted ORFs encode proteins homologous to any other known proteins or motifs. As about 40 structural proteins have been identified, approximately 8% of the ORFs encode structural proteins. Some of the predicted ORFs encode proteins that share significant similarities (40% or even higher) to each other and can therefore be classified into the same gene family. There are ten putative gene families in the WSSV genome, and it is possible that these families arose from gene duplication.

WSSV Genes

Identification through Homology Comparison

The WSSV genome contains several genes that can be identified unambiguously through homology searches. These genes are involved in nucleotide metabolism and DNA replication, and encode *thymidylate synthase*, *dUTPase*, *ribonucleotide reductases* (*rr1* and *rr2*; two separate ORFs encoding the two subunits), chimeric *thymidylate/thymidine kinase* (*TK-TMK*), and *DNA polymerase*. The chimeric *thymidylate/thymidine kinase* gene is a unique feature of WSSV, as these genes in other large DNA viruses are encoded in separate ORFs. The enzymatic activities of dUTPase, ribonucleotide reductase, and TK/TMK have been demonstrated by using purified recombinant proteins.

Some proteins from predicted ORFs show weak similarity to known proteins, including nonspecific nuclease (WSV191), TATA-box binding protein (TBP; WSV303), CREB-binding protein (CBP; WSV100), and helicase (WSV447). The activity of the nonspecific nuclease has been demonstrated using a purified recombinant protein produced in *Escherichia coli*.

Some proteins contain well-defined domains or motifs, and therefore their possible functions can be inferred. Two protein kinases, PK1 (WSSV482) and PK2 (ORF61), have been identified because they each contain the catalytic domain of serine/threonine protein kinase. Two proteins containing the EF-hand, calcium-binding motif have also been identified (WSV079 and WSV427; WSV427 is a latency-related gene), suggesting that WSSV might modulate calcium levels in infected cells. Four proteins with RING-H2 finger motifs have been annotated (WSV199, WSV222, WSV249, and WSV403). This motif has been implicated in ubiquitin-conjugating enzyme (E2)-dependent ubiquitination; many proteins containing a RING finger play a key role in the ubiquitination pathway. Ubiquitination activity has been demonstrated for WSV222 and WSV249, and their interacting host proteins have been identified using yeast-two hybrid screening. Some proteins contain conserved zinc finger or leucine zipper motifs. These motifs have been shown to be involved in DNA-protein interactions and in regulation of transcriptional activation. Of these, *wssv126* is one of three immediate early genes that have been identified to date.

Identification through Functional Studies

As most WSSV gene products show no significant homology to known proteins, *ab initio* research is required to identify their functions. For large dsDNA viruses, gene expression is usually categorized sequentially as immediate early, early, and late. Immediate early genes are expressed in the absence of *de novo* viral protein synthesis, early genes are involved in DNA synthesis, and late genes are expressed following viral DNA replication and encode the viral structural proteins and proteins required for assembly, maturation, and release of viral particles. Three immediate early genes, *ie1* (*wssv126*), *ie2* (*wssv418*), and *ie3* (*wssv242*), have been identified in WSSV-infected shrimp by showing that their transcription is not inhibited by cycloheximide. IE1 contains the Cys2/His2-type zinc finger motif, which is implicated in DNA-binding activity, suggesting that IE1 might function as a transcription factor. The gene products of *ie2* and *ie3* contain no recognizable functional motifs, and their functions remain unknown. Three latency–related genes have been identified: *wsv151*, *wsv427*, and *wsv 366*. These genes have been shown to be relatively highly expressed in asymptomatic carrier shrimp that tested negative for WSSV using a commercial PCR detection kit. When expressed as recombinant proteins in Sf9 insect cells, WSV151 is a nuclear protein of approximately 165 kDa, while WSV427 interacts *in vitro* with a novel shrimp serine/threonine protein phosphatase. A novel anti-apoptosis protein (ORF390) has also been identified. WSSV infection induces bystander, noninfected cells to undergo apoptosis, whereas WSSV-infected cells are not apoptotic. ORF390 has been shown to inhibit apoptosis induced either by infection with a mutant baculovirus or by treatment with actinomycin D. The protein has two putative caspase9 cleavage sites and a caspase3 cleavage site, and therefore might function like the AcMNPV P35 protein by directly binding to and inhibiting the activity of caspase9 and/or caspase3.

WSSV Gene Expression and Regulation

Since no suitable shrimp cell line is available, studies of WSSV gene expression have long been carried out *in vivo* in penaeid shrimp or crayfish by using RT-PCR. In the black tiger shrimp, *P. monodon*, WSSV immediate early and early genes (e.g., *rr1*, *rr2*, *pk1*, *tk-tmk*, *dnapol*, *ie1*, *ie2*, and *ie3*) are detected as early as 2–4 hpi (hours post infection) and their transcription levels either steadily and slowly increase till the end of the study (48 or 60 hpi) or else reach a plateau phase at 12–18 hpi. For late genes (the major viral structural protein genes), although very low levels could be detected as early as 2–4 hpi, transcription levels surged at 12–18 hpi and steadily increased thereafter. In crayfish, expression of the late genes was not observed until at least 18–24 hpi.

WSSV gene expression has been globally analyzed using microarrays. In one study, the results revealed that about 23.5% of WSSV genes began to express at 2 hpi, 4.2% at 6 hpi, 17.7% at 12 hpi, and 47.9% at 24 hpi. Most of the WSSV structural protein genes were expressed either at 12 hpi (9 WSSV structural protein genes) or at 24 hpi (14 WSSV structural protein genes). However, several structural protein genes were expressed at 2 hpi, suggesting that these proteins might also have important roles in the early infection stage. In another microarray study, clustering of the transcription profiles of the individual WSSV genes during infection showed two major classes of genes; the first class reached maximal expression at 20 hpi and the second class at 2 days post infection. Since most of the known WSSV early genes were found in the first class, the other genes in this class were also thought to be early genes. Conversely, the genes that clustered with the structural protein genes in the second class were likewise thought to be late genes. Therefore, RT-PCR and microarray studies suggest that, as for other large dsDNA viruses, the expression of WSSV genes is regulated in a coordinated series of cascades.

The promoter regions of WSSV early and late genes have been analyzed. The upstream region of WSSV early genes contains a TATA box and an initiator. This is similar to the *Drosophila* RNA polymerase II core promoter sequences, suggesting that the WSSV early genes are under the control of cellular transcription machinery. Alignment of the regions upstream of all the major structural protein genes has identified a degenerate motif (ATNAC) that could be involved in WSSV late gene transcription, while only one of these genes was found to contain a functional TATA box. These differences between the promoter regions of the early and late genes are sufficient to account for the differential transcription patterns between early and late genes, and they also suggest that WSSV uses at least two different classes of transcription machinery for gene expression. Since the WSSV genome contains no homologs of RNA polymerase subunits, at least one novel transcription factor (e.g., IE1, IE2, or IE3) would need to be induced at the early infection stage. These factors, which are presumably able to recognize a late gene-specific motif (such as the 'ATNAC' motif), could then recruit the host cellular transcription machinery to transcribe the late genes.

Viral–Host Relationships

Apoptosis

Apoptosis is a cell suicide program that enables multicellular organisms to direct and control cell numbers in tissues and to eliminate cells, including virus-infected cells that may be harmful to the survival of the organism. If apoptosis occurs during the early stage of infection, virus production is severely limited; because it inhibits the spread of progeny virus in the host, apoptosis is recognized as an antiviral defense. In consequence, most animal viruses have evolved strategies to evade or delay an early apoptosis response. The classical signs of apoptosis can be identified in WSSV-infected shrimp, including nuclear disassembly, fragmentation of DNA into a ladder, and increased caspase3 activity. Furthermore, there is a positive correlation between the severity of WSSV infection and the number of apoptotic cells. Among the WSSV target tissues, the subcuticlar and abdominal epithelia are the most seriously damaged, and these epithelial tissues exhibit the highest incidence of apoptotic cells. However, it is significant that cells displaying apoptotic characteristics do not contain virions, whereas those containing WSSV virions are not apoptotic. It is therefore reasonable to suppose that apoptosis is employed by shrimp as a protective response to prevent the spread of WSSV, while in the infected cells, the anti-apoptosis protein ORF390 blocks apoptosis and thus facilitates multiplication of the virus.

Hemocyte Responses to WSSV Infection

In crustaceans, the hemocytes (blood cells) mediate many defense-related activities that are essential for the nonspecific immunity, including phagocytosis, melanization, encapsulation, cytotoxicity, and clotting. They also produce a vast array of antimicrobial proteins, such as agglutinins, antimicrobial peptides, and lysozymes. Hemocyte responses to WSSV have been investigated thoroughly in penaeid shrimp and crayfish. WSSV infection decreases the total hemocyte count (THC) in shrimp but not in crayfish. The reduced THC in shrimp is probably due to lysis by the virus of infected hemocytes as well as due to virus-induced apoptosis in both circulating hemocytes and hematopoietic tissue. In addition, circulating hemocytes migrate to tissues that have a high number of virus-infected cells, although this is probably a general defensive response rather than a specific antiviral response. WSSV infection also induces apoptosis in crayfish hemocytes but the percentage of apoptotic hemocytes is much lower in crayfish (1.5%) than in shrimp (20%).

WSSV infection differentially affects the three morphologically and functionally distinct hemocyte types that have been identified in crustacean hemolymph, commonly referred to as semigranular cells (SGCs), granular cells (GCs), and hyaline cells (HCs). In shrimp and crayfish, both SGCs and GCs can be infected with WSSV but SGCs are the preferential target. This suggests that SGCs are more susceptible to WSSV and that the virus replicates more rapidly in SGCs than in GCs. However, the melanization activity of GCs in WSSV-infected crayfish is reduced. The third type of hemocyte (HC) is refractory to WSSV infection in shrimp but their role in antiviral defense is still not clear. On the other hand, GCs and SGCs each play a major defensive role and WSSV infection of these cell types is likely to weaken shrimp defenses and reduce shrimp health. Since hemocytes are also necessary for clotting, the low THC also accounts for the phenomenon that hemolymph withdrawn from WSSV-infected shrimp and crayfish always has a delayed (or sometimes completely absent) clotting reaction.

Protection of Shrimp Against WSSV Infection Using Vaccination and RNAi Strategies

Since the earliest outbreaks of WSD, several advances have been made in our understanding of how shrimp might be protected against WSSV infection. For example, although it is generally thought that shrimps lack the immunoglobulin-based adaptive immune system, recent studies have shown that when *Penaeus japonicus* shrimps survive either natural or experimental WSSV infections, they sometimes show resistance to subsequent challenge with WSSV. This 'quasi-immune response' suggests that shrimps may have an innate immune system that includes specific memory. Further research into this phenomenon has shown that intramuscular injection or oral administration of either inactivated WSSV virions or recombinant structural protein VP28 similarly provide shrimps and crayfish with some degree of protection against WSSV infection. Another potential means of limiting WSSV infection is to take advantage of the *in vivo* roles of RNA interference induced by dsRNA. In *Litopenaeus vannamei* shrimp, whereas long dsRNA was shown to induce both sequence-dependent and sequence–independent antiviral responses, WSSV gene-specific dsRNAs produce strong anti-WSSV activity. Similar results have also been reported for the gene-specific dsRNAs of another shrimp virus, yellow head virus. However, the anti-WSSV activity of WSSV gene-specific dsRNAs varies greatly from one WSSV gene to another, and the effectiveness of each candidate needs to be tested empirically. Further, the antiviral activity of WSSV gene-specific small interfering RNA (21–23 bp dsRNA) has not yet been conclusively demonstrated.

See also: Apoptosis and Virus Infection; Shrimp Viruses.

Further Reading

Leu JH, Tsai JM, Wang HC, *et al.* (2005) The unique stacked rings in the nucleocapsid of the white spot syndrome virus virion are formed by the major structural protein VP664, the largest viral structural protein ever found. *Journal of Virology* 79: 140–149.

Lo CF, Wu JL, Chang YS, *et al.* (2004) Molecular characterization and pathogenicity of white spot syndrome virus. In: Leung KY (ed.) *Molecular Aspects of Fish and Marine Biology, Vol. 3: Current Trends in the Study of Bacterial and Viral Fish and Shrimp Diseases*, pp. 155–188. Singapore: World Scientific.

Marks H, Ren XY, Sandbrink H, *et al.* (2006) In silico identification of putative promoter motifs of white spot syndrome virus. *BMC Bioinformatics* 7: 309–322.

Robalino J, Bartlett T, Shepard E, *et al.* (2005) Double-stranded RNA induces sequence-specific antiviral silencing in addition to nonspecific immunity in a marine shrimp: Convergence of RNA interference and innate immunity in the invertebrate antiviral response? *Journal of Virology* 79: 13561–13571.

Tsai JM, Wang HC, Leu JH, *et al.* (2004) Genomic and proteomic analysis of thirty-nine structural proteins of shrimp white spot syndrome virus. *Journal of Virology* 78: 11360–11370.

Tsai JM, Wang HC, Leu JH, *et al.* (2006) Identification of the nucleocapsid, tegument, and envelope proteins of the shrimp white spot syndrome virus virion. *Journal of Virology* 80: 3021–3029.

van Hulten MCW, Witteveldt J, Peters S, *et al.* (2001) The white spot syndrome virus DNA genome sequence. *Virology* 286: 7–22.

Witteveldt J, Cifuentes CC, Vlak JM, *et al.* (2004) Protection of *Penaeus monodon* against white spot syndrome virus by oral vaccination. *Journal of Virology* 78: 2057–2061.

Wongteerasupaya C, Pungchai P, Withyachumnarnkul B, *et al.* (2003) High variation in repetitive DNA fragment length for white spot syndrome virus (WSSV) isolates in Thailand. *Diseases of Aquatic Organisms* 54: 253–257.

Yang F, He J, Lin X, *et al.* (2001) Complete genome sequence of the shrimp white spot bacilliform virus. *Journal of Virology* 75: 11811–11820.

Zhang X, Huang C, and Hew CL (2004) Use of genomics and proteomics to study white spot syndrome virus. In: Leung KY (ed.) *Molecular Aspects of Fish and Marine Biology, Vol. 3: Current Trends in the Study of Bacterial and Viral Fish and Shrimp Diseases*, pp. 204–236. Singapore: World Scientific.

Zuidema D, van Hulte MCW, Marks H, *et al.* (2004) Virus–host interaction of white spot syndrome virus. In: Leung KY (ed.) *Molecular Aspects of Fish and Marine Biology, Vol. 3: Current Trends in the Study of Bacterial and Viral Fish and Shrimp Diseases*, pp. 237–255. Singapore: World Scientific.

Yatapoxviruses

J W Barrett and G McFadden, The University of Western Ontario, London, ON, Canada

© 2008 Elsevier Ltd. All rights reserved.

Introduction

The most recently accepted ICTV designation for the *Poxviridae* includes a new primate genus in the subfamily *Chordopoxvirinae*. The genus *Yatapoxvirus* contains two recognized members: the type species, *Yaba monkey tumor virus* (YMTV) and *Tanapox virus* (TPV).

History

Poxvirus members of the genus *Yatapoxvirus* have been identified relatively recently following infection of monkeys at primate centers in Nigeria (1957) and in several states of the USA (1967), and infection of native African populations of equatorial Africa (late 1950s and early 1960s). YMTV infection produces benign histiocytomas in man and monkeys that rapidly proliferate, leading to large subcutaneous masses that are considered benign tumors. TPV infection causes localized, dermal lesions which are located on the extremities and which resolve slowly. The yatapoxviruses have been shown to infect humans and other primates only (**Table 1**). Based on genomic sequence data and morphology studies, the yatapoxviruses are structurally similar to other members of the *Poxviridae*.

YMTV was first observed following a spontaneous outbreak of subcutaneous tumors in a colony of captive rhesus monkeys housed in open-air pens in June 1957 at Yaba, near Lagos, Nigeria. The superficial growths were first observed in a single rhesus monkey and subsequently spread to 20 of 35 rhesus monkeys from that colony over the following several weeks. Although this primate center housed several other species in close proximity, only a single other species of primate (dog-faced baboon) was affected during this outbreak. The lesions in all cases eventually regressed.

TPV was first identified (1957) following two outbreaks among local people living in villages along the Tana River of Kenya. The illness was characterized by acute febrile reaction and associated pox-like skin lesions. A more widespread illness of 50 cases occurred in 1962 that included both genders and a spread of ages. The observation that the infection occurred during years of dramatic floods and increased mosquito activity, and the fact that the lesions appeared only on areas of exposed skin suggested that a biting insect, most likely a mosquito, was the vector of transmission. Virus isolated from infected humans could only be grown *in vitro* on various human and monkey cell lines. In addition, further transmission of infection by the inoculation virus was successful only in man and monkey. Therefore, most likely, an undefined local monkey species acts as the reservoir species for TPV.

Supporting these conclusions was the occurrence of mild poxvirus infections in three primate centers in the USA in 1966. The affected monkeys, mainly rhesus macaques, exhibited lesions that were similar in appearance and histology to those of the native Kenyans infected with tanapox. The lesions were hypertropic, slow in developing, with little evidence of vesiculation, and these characteristics were quite different from the lesions produced by a similarly appearing nodule that arises from monkeypox infection of man. In addition, at least two handlers at these centers contracted an infection which produced symptoms and lesions that clinically and histologically resembled tanapox infection.

Taxonomy and Classification

The prefix 'yata' is a contraction from the names of the two recognized members of the yatapoxviruses: *ya*ba monkey tumor virus and *ta*napox virus. The members of the yatapoxviruses are closely related by sequence similarity, serology, and immunodiffusion tests. This genus is comprised of only two recognized members (YMTV and TPV) that so far have only been shown to infect primates, including humans. The members of this genus are immunologically distinct from members of the seven other *Chordopoxvirinae* genera. Vaccination by poxviruses from other genera, including vaccinia virus, will not protect from infection by

Table 1 Members of the *Yatapoxvirus* genus

Member	Abbreviation	Natural host	Major arthropod vector	Natural host disease	Length of infection
Yaba monkey tumor virus	YMTV	Monkeys of Africa and Malaysia	Probably mosquito	Large, multicellular masses (2–5 cm)	Spontaneously regresses 6–12 weeks
Tanapox	TPV	Humans	Probably mosquito	Individual, round, raised nodules	Resolution in 3–4 weeks
Yaba-like disease	YLDV	Captive primates/human handlers	Probably mosquito in monkeys, possibly scratches or bites to human handlers	Individual-few raised nodules	Resolution in 3–4 weeks

the yatapoxviruses and, conversely, a previous infection with a yatapoxvirus will not block other poxvirus infections.

A third member, yaba-like disease virus (YLDV), considered a strain of TPV, has been isolated from monkeys and their human handlers at several primate centers in the USA. Although ICTV does not recognize YLDV, the complete genomic sequence has been deposited in GenBank and provides scientists with genetic information about the closely related TPV.

Properties of the Virion

Morphologically, the yatapoxvirus viruses resemble typical oval-shaped poxvirus virions. YMTV and TPV particles contain a lipid outer membrane, inner membrane, and enclosed dumbbell-shaped core and lateral bodies. The virus measures between 250 and 300 nm on the long axis.

DNA Replication, Transcription, and Translation

Generally, viral DNA replication and protein synthesis of the yatapoxviruses is similar to, albeit slower than, the prototypical poxvirus, vaccinia virus. Yatapoxvirus DNA synthesis occurs in the cytoplasm of infected cells in 'virus factories'. Sequencing has confirmed that the genomes of the yatapoxviruses encode the standard suite of conserved poxvirus housekeeping genes, including transcription factors and enzymes required for viral replication, transcription, and translation.

One consistent difference between YMTV and other poxviruses has been the observation that YMTV replicates more efficiently *in vitro* at 35 °C rather than 37 °C. The reason for this temperature sensitivity is undefined; however, it may be the result of a required viral-encoded enzyme that is temperature sensitive or prolonged passage in a poikilothermic, nonmammalian host. Regardless, the cycle of replication follows standard poxvirus early gene expression, DNA synthesis, late gene expression, and morphogenesis. Viral DNA synthesis is detected as early as 3 h post infection (p.i.). Progeny virus first appears at 24 h p.i. *in vitro* and plateaus at 72 h p.i. This replication cycle is six to ten times longer than for vaccinia virus. YMTV has a narrow host range in tissue culture. In addition, *in vitro* infection involves a prolonged replication cycle and generally low virus yield. Continuous cells from cercopithecus monkey kidney cells (CV-1) are susceptible to YMTV infection; however, cell lines derived from other monkeys (BGMK, OMK) and other species (e.g., HeLa, RK13, and CEFs) do not support productive infection.

In contrast, TPV replication is most efficient at 37 °C. TPV replication follows the typical cascade of poxvirus DNA synthesis and production of infectious progeny but exhibits slower kinetics when compared to vaccinia virus, although it is much faster than YMTV. The eclipse period ranges from 24 to 48 h p.i.; however, this variation is dependent on the multiplicity of infection.

Transient expression of native yatapoxvirus genes from standard mammalian expression vectors results in low-to-no protein expression. However, if the viral coding sequence is optimized to favor the use of more commonly used human codons, the result is generally much higher protein expression levels from transfected human cells.

Properties of the Genome

The members of the yatapoxviruses encode the smallest characterized genomes within the *Poxviridae* (**Table 2**). Complete genomic sequencing information is now available for all yatapoxviruses. The genomes of the yatapoxvirus members are very A+T- rich, a characteristic shared by other members of the *Poxviridae* including members of the ortho-, sui-, capri-, and avipoxviruses. The YMTV genome is 70% A+T, 134.7 kbp long, and encodes for 140 genes. In contrast, TPV is 144.6 kbp, 73% A+T, and encodes for 155 genes (**Table 2**). Comparison between the complete YLDV sequence and the sequenced genomes of TPV indicates a high degree of sequence identity of approximately 98%, supporting the claim that TPV and YLDV are strains of the

Table 2 Features of the yatapoxvirus genomes

Member	Genome size (bp)	Single copy genes	Duplicated genes	Terminal inverted repeat (bp)	% A + T
TPV	144 565	155	1	1868	73
YMTV	134 721	139	1	1962	70.2
YLDV	144 575	150	1	1883	73

same virus. Comparison of the genomic sequences of YMTV and TPV indicates 78% identity. All of the genes identified in YMTV are also encoded by TPV; however, YMTV has lost 13 open reading frames (ORFs) found in TPV. The yatapoxviruses encode many of the same structural and housekeeping genes that are found in other poxviruses. The immunomodulatory genes of the yatapoxviruses also include a handful of novel genes predicted to be involved in regulation of immune response that are unique to the yatapoxvirus or are found in only a few other poxviruses. These include a new inhibitor of human tumor necrosis factor (TNF α (2L), virally encoded versions of chemokine receptors (7L and 145R), and TPV/YLDV (but not YMTV) encodes a viral IL-10 homolog (134R).

Immunity

YMTV and TPV show minimal to moderate cross-reactivity in collected sera. However, YMTV infection can protect nonhuman primates from TPV challenge. Monkeys infected with TPV and then challenged with YMTV exhibit reduced tumors and delayed symptoms. Comparison between YMTV and TPV infection identified both type-common and type-specific antigens. Circulating neutralizing antibodies, although present in the sera of several species of monkeys, are ineffective in preventing growth of YMTV tumors or reinfection. Immunity to superinfection is observed as long as tumors are present or regressing; however, following total tumor regression a new infection results in new tumor formation. Complement-fixing and complement-fixing-inhibiting antibodies are present in clinical and convalescent stages, respectively, of rhesus monkeys infected with either TPV or YMTV. The persistence of complement-fixing antibodies in monkeys infected with TPV is 10–12 weeks. Complement-fixing antibody was detected up to 35 weeks p.i. in monkeys infected with YMTV.

Tropism and Transmission – Humans

Naturally occurring YMTV infections have only been identified in nonhuman primates. However, accidental and volunteer infections have been established in humans. Infected humans develop lesions similar to those observed in monkeys although the proliferative responses are less pronounced and regression occurs earlier than in monkeys. Recovery of infectious virus from human lesions followed by serial passage and titration confirmed that YMTV was able to replicate productively in primary human tissue. Inoculation of rabbits, guinea pigs, hamsters, rats, mice, and dogs failed to produce proliferative lesions. In addition, no lesions were observed following inoculation of embryonated eggs. Testing of nonhuman primates suggests that rhesus and cynomolgus species are most susceptible as hosts.

TPV infections have only been identified in native populations of equatorial Africa or visitors to that region. The fact that infections appeared during years of extensive flooding, and that lesions are observed only on exposed areas of skin and occur on the extremities, suggests that transmission is via a biting arthropod vector (e.g., mosquitoes).

YLDV infection of monkey handlers at primate centers of the US exhibited a brief fever followed by production of a few raised, necrotic nodules that completely resolved.

Tropism – Nonhuman Primates

YMTV and YLDV have been identified from monkeys kept as research subjects at primate centers. Monkeys infected with YMTV were kept in open-air pens and infection spread from a single individual to other segregated monkeys, suggesting biting arthropods as the likely transmission vector. Research groups have screened sera collected from monkeys of various parts of the world for antibodies against either YMTV or TPV. Generally, antibodies against YMTV and TPV were identified from various monkey species from Africa, as well as cynomolgus monkeys from Malaysia, suggesting that these species suffer from infection, possibly subclinical, with TPV or YMTV. No antibodies to either virus were detected in Indian rhesus monkeys or from any of the monkey species tested from South America. Since rhesus monkeys are highly susceptible to clinical infection by both yatapoxviruses, it is surprising that none of the collected samples confirmed any previous infection. Although natural infection must be rare, the geographical location of the Indian primate species that are located between the African and Malaysian species (both of which were YMTV/TPV positive) suggests that negative serology results from primates in India may be misleading.

Pathogenesis

Infection of susceptible primates by YMTV targets histiocytes which rapidly divide and produce tumors. Histiocytes are cells, of either macrophage or Langerhans cell lineage, that migrate to areas of cellular disturbance. Following YMTV inoculation, histiocytes migrate to the site of virus inoculation during the first 48 h. The infiltrating histiocytes begin to exhibit cellular alterations, including enlargement of the nucleus and the prominence of nucleoli, by 72 h p.i. By day 5, the altered histiocytes become spindle shaped and exhibit mitotic activity. Also by day 5, inclusion bodies begin to become visible. Actively dividing cells begin to assume spiral forms between days 7 and 9 and acquire well-defined reticular patterns. By week 2, many degenerating cells along with other actively dividing cells become apparent. This scenario continues for the next 4–6 weeks. Eventually, the number of degenerating cells starts increasing and the number of dividing cells diminishes. Host immune response to infection is generally insignificant until regression of the lesion is nearly complete. At this point, there is evidence of lesion infiltration with host mononuclear cells. Necrosis of the center of the lesion is generally not part of the regression process; however, the tumor will exhibit surface ulceration. The beginning of degeneration is characterized by eosinophilic cytoplasmic mass, cytoplasmic lipid accumulation, vacuolization, and eventual cell dissolution. Tumor regression is not correlated to serum antibody titers. It is postulated that tumor regression is due to *in vivo* cytopathic effects rather than an innate immune response. However, the cytopathogenic effect of the virus eventually kills the cells and the tumor regresses. Natural regression can take up to several months, after which the individual is susceptible to reinfection.

TPV infection begins with a short febrile illness lasting several days that may include severe headaches, general fatigue, and body aches. This is followed by the appearance of a single, or a few, pock-like lesions that initially resemble those of variola. Tanapox lesions begin as a papule and develop into a hard, raised nodule and are normally located on the extremities. Nodules normally regress after 3–4 weeks.

Geographic and Seasonal Distribution

Naturally occurring yatapoxvirus infections have been extremely rare and geographically limited. TPV infections in human populations have only been observed in the equatorial belt of Africa between Zaire and Kenya. Recent TPV infections of humans have been limited to very few cases. Infection of human handlers with YLDV at primate centers in the US was effectively stopped by appropriate changes to biological safety conditions. YMTV infections have only been observed in nonhuman primates in Africa primate centers. Although a worldwide survey of primate sera has identified antibodies against YMTV in monkeys in both Africa and Malaysia, and TPV-neutralizing antibodies in native populations of equatorial Africa, the numbers of observed cases are rare.

The yatapoxviruses are probably transmitted by biting arthropod vectors (mosquitoes?) and therefore environmental conditions such as rainy seasons, excessive flooding, and natural occurrences that bring humans into closer contact with infected nonhuman primates will lead to increased observation of infections.

Immune Modulation

TPV is a self-limiting infection in humans and natural infection for YMTV has not been observed in humans. Only a small amount of research has been done on the molecular aspects of a TPV infection and none has been undertaken with YMTV. It is known that TPV-infected cells secrete a 38 kDa glycoprotein that has been shown to bind human interferon-γ, human IL-2, and human IL-5. In addition, the secreted, early protein 2L of TPV represents a new class of virally encoded tumor necrosis factor (TNF)-binding proteins found in the members of the yatapoxviruses, swinepox, and the unclassified deerpox virus. This 2L protein binds human TNFα with high affinity but is unable to bind other members of the human TNF superfamily. In addition, TPV 2L inhibits human TNFα from binding to TNF receptors I and II, as well as blocking TNF-induced cytolysis. However, 2L protein was shown to be unable to bind IFN-γ, IL-2, or IL-5 under experimental conditions.

TPV has been isolated from humans living in (or visiting) equatorial Africa and YLDV from workers at primate centers in the USA. The complete genomic sequence is presently available from TPV, YLDV and YMTV. Based on these sequences, it is clear that the TPV and YLDV genomes are 98% identical, which is clearly on the level of genetic similarity between strains of vaccinia virus. Based on the sequencing, a large number of host-modulating viral genes were predicted but only a handful have been studied. These include a new viral homolog of CCR8, a novel TNF inhibitor, and a viral-encoded member of the IL-10 family.

Epidemiology

YMTV is likely endemic in African monkeys. Neutralizing antibodies have been detected in monkeys from Africa and Malaysia, suggesting that YMTV possibly represents a latent infection in several monkey species. If this is true, it is surprising that monkeys from India have not yet been reported to contain the same neutralizing antibodies. Infection in man has only occurred through injection of isolated YMTV.

The reservoir host for TPV is thought to be wild monkeys in equatorial Africa, from which natives can be infected by mosquito transmission. TPV infection is more common in years of heavy flooding when high water forces monkeys and man into closer proximity, while at the same time mosquito populations are at their zenith.

Prevention and Control

Control of YMTV and YLDV infection in man requires normal biological safety protocols to be exercised by animal handlers, including the wearing of protective clothing. TPV infection in equatorial Africa will require mosquito control measures. Separation of Asian and African monkey species at primate centers and markets and protective clothing for merchants and animal handlers may also help.

Future Perspectives

The reservoir status of primates with the yatapoxviruses is not a major human health concern. However, the features of infection, including YMTV targeting of histiocytes and tumor formation, the lack of long-term protection following regression of the initial infection, the observation that early TPV infection is difficult to distinguish from monkeypox infection, and the observation that members of this genus of poxviruses do in fact infect humans without eliciting an effective immune response may offer some unique insights into the mechanism of human zoonotic infections. Given the current concerns for early diagnoses of any and all human poxvirus infections, research on this group of poxviruses is likely to continue.

See also: Poxviruses; Smallpox and Monkeypox Viruses; Vaccinia Virus; Zoonoses.

Further Reading

Brunetti CR, Amano H, Uedo Y, *et al.* (2003) Complete genomic sequence and comparative analysis of the tumorigenic poxvirus yaba monkey tumor virus. *Journal of Virology* 77: 13335–13347.
Brunetti CR, Paulose-Murphy M, Singh R, *et al.* (2003) A secreted high-affinity inhibitor of human TNF from tanapox virus. *Proceedings of the National Academy of Sciences, USA* 100: 4831–4836.
Downie AW and Espana C (1973) A comparative study of tanapox and yaba viruses. *Journal of General Virology* 19: 37–49.
Knight JC, Novembre FJ, Brown DR, Goldsmith CS, and Esposito JJ (1989) Studies on tanapox virus. *Virology* 172: 116–124.
Lee HJ, Essani K, and Smith GL (2001) The genome sequence of yaba-like disease virus. *Virology* 281: 170–192.
Nazarian SH, Barrett JW, Frace AM, *et al.* (2007) Comparative genetic analysis of genomic DNA sequences of two human isolates of *Tanapox virus*. *Virus Research* 129: 11–25.

Relevant Website

http://www.poxvirus.org – Poxvirus Bioinformatics Resource Center.

Yeast L-A Virus

R B Wickner, National Institutes of Health, Bethesda, MD, USA
T Fujimura and R Esteban, Instituto de Microbiología Bioquímica CSIC/University of Salamanca, Salamanca, Spain

© 2008 Elsevier Ltd. All rights reserved.

Glossary

Decapping Removal of the methylated GMP in 5′–5′ linkage at the 5′ end of most eukaryotic mRNAs.
Ribosomal frameshifting Ribosomes changing reading frame on an mRNA, in response to a special mRNA structure, to synthesize a protein from two overlapping reading frames.

Introduction

The L-A virus of bakers/brewers yeast *Saccharomyces cerevisiae* is one of several RNA viruses infecting this organism, each of which spreads via the cell–cell fusion of mating, rather than by the extracellular route. The totivirus L-A, like members of the very similar L-BC family of viruses, is a single-segment 4.6 kbp double-stranded RNA (dsRNA) virus encapsidated in icosahedral particles with a single major coat protein called Gag.

The 20S and 23S RNA replicons are naked cytoplasmic single-stranded RNA (ssRNA) replicons except for their bound RNA-dependent RNA polymerases.

The L-A virus serves as the helper virus for any of several smaller satellite dsRNAs, called M dsRNAs, each encoding a secreted protein toxin and immunity to that toxin, producing the 'killer' phenomenon. Killer strains can eliminate some of the competition by this means, although only about 10% of wild strains harbor a killer dsRNA, suggesting there are costs to carrying this replicon. The killer phenotype was used to study the genetics of M dsRNAs and the helper L-A. Several functional variants of L-A were defined based on their interactions with different M dsRNAs and with the host, and host mutants affected in virus expression or propagation were also examined.

History

In 1963, Makower and Bevan reported that some yeast strains secrete a toxin that kills other yeasts This led to the identification by Bevan and by Fink of cellular dsRNAs and later viral particles correlated with the killer phenomenon. Studies of the chromosomal genes affecting the killer system revealed that the Kex2 protease, identified by its requirement for toxin secretion, was the long-sought proinsulin-processing enzyme. The Mak3 N-acetyltransferase, whose acetylation of Gag is needed for viral assembly, established consensus sequences for such enzymes. The loss of the L-A virus in *mak3* mutants revealed a second dsRNA species of the same size, called L-BC, and unrelated to the killer system.

Virion Structure

L-A virions have icosahedral symmetry, but contain 120 Gag monomers per particle, in apparent violation of the rules of quasi-equivalence. In fact, each virion is composed of 60 asymmetric dimers of Gag (**Figure 1**), a feature that is common to the cores of all dsRNA viruses that have been characterized. One type of Gag molecule makes contact with the icosahedral fivefold and twofold axes, but not with the threefold axes. The second type of Gag finds itself in contact with the threefold axis, but not with either the five- or twofold axes (**Figure 1**). The two environments of Gag lead to two distinct conformations, suggesting that Gag may be more flexible than some other coat proteins. The L-A virion has more volume per nucleotide than do ssRNA or dsDNA viruses. These facts may reflect the requirement that the dsRNA moves inside the particle and is transcribed by the RNA-dependent RNA polymerase that is fixed to the inner virion wall (see below). Pores at the fivefold axes are assumed to allow entry of nucleotides and exit of (+) strand transcripts, but retention of the dsRNA genome. A trench on the outside contains His154, the central active site residue of the mRNA-decapping activity to which the 7-methyl-GMP structure becomes covalently attached (**Figure 2**). A

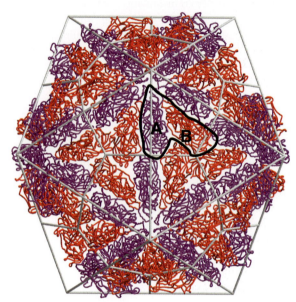

Figure 1 Wire diagram of the L-A virus capsid. The major coat protein, Gag, is found in two nonequivalent positions: 'A' molecules contact the fivefold and twofold axes, while 'B' molecules contact the threefold axes. Reproduced from Naitow I, Tang J, Canady M, Wickner RB, and Johnson JE (2002) L-A virus at 3.4 Å resolution reveals particle architecture and mRNA decapping mechanism. *Nature Structural Biology* 9: 725–728, with permission from Nature Publishing Group.

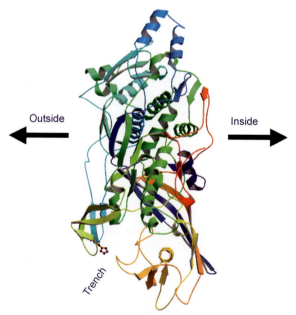

Figure 2 Ribbon diagram of a single Gag molecule. The trench on the outer surface includes His154, the Gag residue to which 7meGMP is covalently attached by the decapping activity.

layered density observed for the viral dsRNA may reflect the dsRNA's rigid structure and the fact that it is forced to press against the inner capsid wall.

Genome Organization

The single segment of the L-A genome is a 4.6 kbp dsRNA with two long open reading frames (ORFs) (**Figure 3**). The 5′ ORF encodes the major coat protein, called Gag in analogy with retroviruses, while the longer 3′ ORF, called Pol, encodes the RNA-dependent RNA polymerase, and has homology with similar enzymes of other ssRNA and dsRNA viruses. Pol is expressed only as a fusion protein whose amino end is a nearly complete Gag molecule. The fusion is carried out by a −1 ribosomal frameshift event in the region of overlap of the two ORFs. An RNA pseudoknot, in the region of overlap of the ORFs, slows ribosome progression at a point on the mRNA (of the form X XXY YYZ (0 frame indicated)) where bonding of the A-site and P-site tRNAs to the mRNA is nearly as good in the −1 frame as in the 0 frame. About 1% of ribosomes slip into the −1 frame, and, continuing translation, make the Gag–Pol fusion protein.

Replication Cycle

Transcription

Viral particles have an RNA-dependent RNA polymerase that is due to the Pol part of the Gag–Pol fusion protein. Transcription is a conservative reaction, meaning that the parental strands remain together after the synthesis of a new (+) strand. The L-A (+) strand transcripts are extruded from the particles into the cytoplasm where they serve as both mRNA and as the species that is packaged by coat proteins to make new viral particles. However, (+) strand transcripts of the smaller M dsRNA or of deletion mutants of L-A are often retained within the particle where they may be replicated. This is called 'headful replication' because the volume of the particle seems to be the determinant of how many dsRNA molecules may accumulate in each particle.

RNA Packaging

The RNA packaging site (**Figure 3**) is a stem–loop structure about 400 nucleotides from the 3′ end of the L-A (+) strand. This stem–loop has an essential A residue protruding on the 5′ side of the stem. The loop sequence is also important, but the stem sequence is not critical, as long as the stem structure can form. The packaging site on the RNA is recognized by the proximal part of the Pol domain of the Gag–Pol fusion protein. The Gag part of Gag–Pol is part of the capsid structure, so the Gag–Pol fusion protein structure assures packaging of viral (+) strands in new particles.

RNA Replication

The (−) strand synthesis reaction is called replication (**Figure 4**). This reaction involves recognition of the

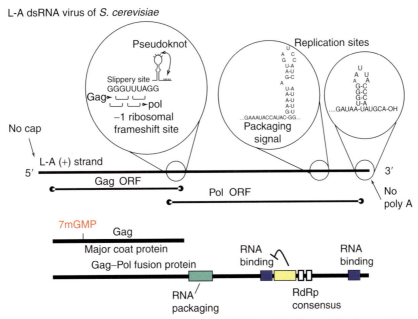

Figure 3 Genome organization. The L-A (+) strand encodes Gag and Pol in overlapping reading frames. The mRNA lacks 5′ cap and 3′ polyA structures. The location of the packaging site on the RNA and the region of the Pol protein segment (green box) that recognizes the packaging site are indicated. The RNA sites for replication, and the parts of Pol with homology to other RNA-dependent RNA polymerases (RdRp) are indicated. The cryptic *in vitro* RNA binding site (blue box) is inhibited by the area shown by the yellow box.

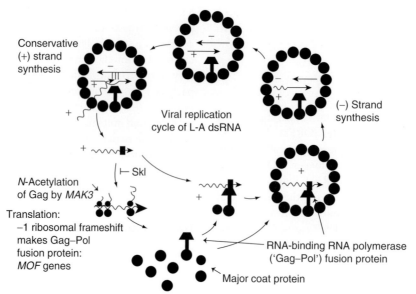

Figure 4 Replication cycle of the L-A virus. Both (+) and (−) RNA strand synthesis occur within the viral particles, but at different stages of the cycle. RNA (+) strands (transcripts) are extruded from the particles and serve as both mRNA and the species packaged to make new particles. Translation of viral mRNA is blocked by Ski proteins. N-terminal acetylation of Gag by Mak3p is necessary for viral assembly.

internal binding site on the L-A (+) strands by Pol, followed by interaction with the now nearby 3′ end and initiation of new RNA (−) chains. Multiple rounds of RNA synthesis can proceed within the viral particle in the 'headful replication' process discussed above.

Viral Translation

The translation apparatus is the battleground of RNA viruses with their hosts. Poliovirus protease cleaves eIF4G so that host-capped mRNAs cannot be translated but its own IRES-containing mRNAs are used. Influenza virus steals caps from host mRNAs.

L-A (+) strand transcripts lack both the 5′ cap and 3′ polyA structures typical of eukaryotic mRNAs, and so are at a distinct disadvantage. Mutations resulting in a relative deficiency of 60S ribosomal subunits selectively lose the ability to express cap-minus, polyA-minus mRNAs such as those of L-A. The *SKI* genes encode a series of proteins whose function is to specifically block the expression of non-polyA mRNAs. The Ski proteins have effects on both translatability of non-polyA mRNAs and on their stability. The *SKI1/XRN1* gene encodes a 5′→3′ exoribonuclease specific for uncapped mRNA. As a defense against its degradation of viral mRNA, L-A's Gag protein has a decapping activity that produces, from cellular mRNAs, 'decapitated decoys', uncapped RNAs that serve as alternative targets for the exonuclease action. In the absence of this decapping activity, viral mRNAs are not expressed, but deletion of the *SKI1/XRN1* gene restores viral mRNA expression.

L-A Genetics

Several natural variants of L-A have been described. The ability of an L-A variant to support M dsRNA replication was first called [HOK] (=helper of killer) and then H when it was found to be a property of L-As such as L-A-H, L-A-HN, or L-A-HNB. Making several chromosomal *MAK* genes dispensable for M propagation was named [B] (=bypass) and is found on L-A-HNB. Ability to exclude L-A-H was named [EXL] (=exclusion) and then just E on L-A-E. Insensitivity to the action of [EXL] was called [NEX] (=nonexcludable), and then shortened to N as in L-A-HN or L-A-HNB. The molecular basis of these interactions has resisted study because of the current inability to obtain L-A from a cDNA clone. M dsRNA can be supported from a cDNA clone of L-A, but the L-A virus has not been shown to be regenerated from these transcripts.

Other RNA Replicons in Yeast: L-BC, 20S RNA, 23S RNA

20S RNA was discovered as an RNA species which appeared when cells were placed under conditions that induce meiosis and spore formation, namely near-starvation for nitrogen and provision of acetate as a carbon source. It was shown that 20S RNA is an independent replicon, and can be made independent of the sporulation or meiosis processes. 20S RNA encodes its own RNA-dependent RNA polymerase which is bound to the otherwise naked cytoplasmic RNA. The mechanism of

20S RNA replication control by culture conditions remains to be elucidated.

23S RNA is a related, but independent yeast replicon which was found as a dsRNA form (called T). 23S RNA also encodes its RNA-dependent RNA polymerase. Both 20S RNA and 23S RNA can be launched from cDNA clones to form replicating virus.

L-BC is a totivirus, related to L-A but independent of it. Its dsRNA is essentially the same size as that of L-A although its copy number is usually about tenfold lower. Thus it was not detected until chromosomal mutants that lose L-A were examined. L-BC does not interact with the killer system, and confers no obvious phenotype, so its genetics has not been extensively explored.

See also: Narnaviruses; Totiviruses; Viral Killer Toxins.

Further Reading

Bevan EA, Herring AJ, and Mitchell DJ (1973) Preliminary characterization of two species of dsRNA in yeast and their relationship to the 'killer' character. *Nature* 245: 81–86.

Bostian KA, Elliott Q, Bussey H, Burn V, Smith A, and Tipper DJ (1984) Sequence of the preprotoxin dsRNA gene of type I killer yeast: Multiple processing events produce a two-component toxin. *Cell* 36: 741–751.

Dinman JD, Icho T, and Wickner RB (1991) A −1 ribosomal frameshift in a double-stranded RNA virus of yeast forms a gag-pol fusion protein. *Proceedings of the National Academy of Sciences, USA* 174–178.

Esteban R, Fujimura T, and Wickner RB (1989) Internal and terminal cis-acting sites are necessary for *in vitro* replication of the L-A double-stranded RNA virus of yeast. *EMBO Journal* 8: 947–954.

Fujimura T, Esteban R, Esteban LM, and Wickner RB (1990) Portable encapsidation signal of the L-A double-stranded RNA virus of *S. cerevisiae*. *Cell* 62: 819–828.

Fujimura T, Ribas JC, Makhov AM, and Wickner RB (1992) Pol of gag-pol fusion protein required for encapsidation of viral RNA of yeast L-A virus. *Nature* 359: 746–749.

Fuller RS, Brake A, and Thorner J (1989) Intracellular targeting and strructural conservation of a prohormone-processing endoprotease. *Science* 246: 482–486.

Icho T and Wickner RB (1989) The double-stranded RNA genome of yeast virus L-A encodes its own putative RNA polymerase by fusing two open reading frames. *Journal of Biological Chemistry* 264: 6716–6723.

Leibowitz MJ and Wickner RB (1976) A chromosomal gene required for killer plasmid expression, mating, and spore maturation in *Saccharomyces cerevisiae*. *Proceedings of the National Academy of Sciences, USA* 73: 2061–2065.

Masison DC, Blanc A, Ribas JC, Carroll K, Sonenberg N, and Wickner RB (1995) Decoying the cap-mRNA degradation system by a dsRNA virus and poly(A)-mRNA surveillance by a yeast antiviral system. *Molecular and Cellular Biology* 15: 2763–2771.

Naitow H, Canady MA, Wickner RB, and Johnson JE (2002) L-A dsRNA virus at 3.4 angstroms resolution reveals particle architecture and mRNA decapping mechanism. *Nature, Structural Biology* 9: 725–728.

Sommer SS and Wickner RB (1982) Yeast L dsRNA consists of at least three distinct RNAs; evidence that the non-Mendelian genes [HOK], [NEX] and [EXL] are on one of these dsRNAs. *Cell* 31: 429–441.

Tercero JC and Wickner RB (1992) *MAK3* encodes an N-acetyltransferase whose modification of the L-A *gag* N-terminus is necessary for virus particle assembly. *Journal of Biological Chemistry* 267: 20277–20281.

Vodkin MH, Katterman F, and Fink GR (1974) Yeast killer mutants with altered double-stranded ribonucleic acid. *Journal of Bacteriology* 117: 681–686.

Wickner RB (2006) Viruses and Prions of Yeast, Fungi and Parasitic Microorganisms. In: Knipe DM and Howley PM (eds.) *Fields Virology*, 5th edn. Philadelphia, PA: Lippincott Williams and Wilkins.

Yellow Fever Virus

A A Marfin, Centers for Disease Control and Prevention, Atlanta, GA, USA
T P Monath, Kleiner Perkins Caufield and Byers, Menlo Park, CA, USA

© 2008 Elsevier Ltd. All rights reserved.

Glossary

Aedes A genus of mosquitoes strongly involved in the transmission of yellow fever virus.
Aedes (Ae.) aegypti is a species belonging to the genus that breeds in close association with humans and is often responsible for epidemics of yellow fever.
Flavivirus A virus in the genus *Flavivirus*.
Genotype Genetic composition of an individual (virus), used to delineate relatedness of individuals.

Host A vertebrate species (including humans) infected by virus. An effective host is one that developes viremia sufficient to infect vectors and thus contributes to the virus transmission cycle.
Pathogenesis The process whereby virus infection proceeds in the host, leading to spread of the infection to vital organs and the development of damage to cells and tissues.
Transmission cycle The sequential infection of blood-feeding vectors and vertebrate hosts

responsible for maintaining and amplifying virus in nature.

Vector An arthropod (e.g., mosquito or tick) capable of becoming actively infected by blood feeding and of subsequently transmitting virus to a vertebrate host.

Introduction

Yellow fever virus (YFV) is the prototype species of the virus family *Flaviviridae* (and of the genus *Flavivirus*), which includes approximately 70 single-stranded RNA viruses, most of which are transmitted by mosquitoes or ticks. Other viruses in this family include the four dengue viruses, West Nile virus, Japanese encephalitis virus, and tick-borne encephalitis viruses.

History

Yellow fever (YF) is the original 'viral hemorrhagic fever', a systemic illness characterized by high viremia; hepatic, renal, and myocardial injury; hemorrhage; and high lethality. Genetic sequence analysis reveals that YFV diverged from the ancestral flaviviral lineage, roughly 3000 years ago, earlier than did other mosquito-borne flaviviruses. The first description of epidemic hematemesis occurred in Mexico in 1648, which suggests that the virus and its mosquito vector, *Aedes aegypti*, were introduced from Africa with the slave trade. Furthermore, monkeys of New World species are more susceptible to lethal infection than are African monkeys, also suggesting that YFV was recently introduced. The West African and South American genotypes of YFV are more closely related to each other than to other genotypes, indicating the probable source of introduction. During the eighteenth and nineteenth centuries, YF epidemics affected coastal cities in the Americas and Europe; the disease was likely introduced via ships infested with YFV-infected mosquitoes. In the mid- and late nineteenth century, physicians, most notably Carlos Findlay, suggested that YFV was transmitted by mosquitoes. In 1900, Walter Reed and colleagues demonstrated that the agent was a filterable virus transmitted by *Ae. aegypti* mosquitoes. In 1927, YFV was isolated from a Ghanaian man; efforts to establish continuous direct passage in animals resulted in establishment of the Asibi strain. Isolation of the Asibi strain and a contemporary ('French') strain isolated in Dakar, Senegal, enabled development of quantitative methods that allowed more precise scientific studies, as well as two live, attenuated vaccines.

Taxonomy, Classification, and Variation

YFV is antigenically distinguished from all other flaviviruses by neutralization test, and all strains of YFV belong to a single serotype. Among flaviviruses, YFV is more closely related to Wesselsbron, Sepik, Edge Hill, Bouboui, Uganda S, Banzi, Jugra, Saboya, and Potiskum viruses.

Wild-type YFV is genetically stable compared to other RNA viruses, likely due to restrictions imposed by replication in vertebrate and invertebrate hosts and maintained by a high-fidelity RNA-dependent RNA polymerase. Despite this stability, genomic sequence analyses of structural and nonstructural genes and of YFV-specific repeat sequences have identified at least seven distinct genotypes. Analyses indicate YFV arose in Africa and divided into West and East African genotypes prior to introduction into the Americas. Currently, five African genotypes are recognized – West Africa I, West Africa II, East/Central Africa, East Africa, and Angola. South American YFV can be divided into two genotypes – South America I (in Brazil) and South America II (in the western part of the continent). These genotypes demonstrate a high degree of homology but can frequently be clustered by time periods, geography, and by predominant transmission patterns.

Properties of the Virion

YFV virions are small (40–60 nm diameter), spherical, and have short surface projections. The icosohedral nucleocapsid contains the single-stranded RNA genome and a single core protein (C protein), and is surrounded by a lipid bilayer. Viral infectivity is rapidly inactivated by heat (56 °C for 30 min), ultraviolet radiation, and lipid detergents.

Properties of the Genome

The YFV genome is comprised of a single strand of plus-sense (i.e., infectious) RNA that contains 10 862 nt and is composed of a 118-nt 5′ noncoding region (NCR) preceded by a type 1 cap structure, a single open reading frame (ORF) of 10 233 nt that encodes 11 viral proteins, and a 511-nt 3′ NCR without a polyadenylated tail. The 5′ and 3′ NCRs have conformational structure and complementary sequences that are important in cyclization of the viral genome during encapsidation and replication and which function as promoters during replication. Mutations or deletions in these regions affect replication and virulence. The 3′ NCR contains conserved consensus sequences which pair with 5′ NCR sequences during cyclization and serve as the common recognition site for the viral polymerase. Directly upstream from these sequences, the 3′ NCR forms a 3′ terminal hairpin structure and pseudoknots to serve as promoters for genomic replication.

Viral Proteins

The ORF encodes three structural proteins at the 5′ end (capsid (C), pre-membrane (preM), and envelope (E) proteins), followed downstream by 7 nonstructural (NS) proteins (NS1–NS2A–NS2B–NS3–NS4A–NS4B–NS5). The mature virion includes these three structural proteins, while the NS proteins are responsible for replication and polyprotein processing including cleavage of the polyprotein into proteins.

The C protein (MW ∼ 11 kDa) interacts with genomic RNA to form the virion nucleocapsid, anchors the nucleocapsid to the endoplamic reticulum, and provides signal sequence to the preM protein. The preM glycoprotein (∼27 kDa) forms an intracellular heterodimer to stabilize the E protein during exocytosis. During exocytosis, the preM protein is processed by a furin-like protease leaving a small M structural protein (∼8 kDa) anchored in the virus envelope of the extracellular virion. The M protein spans the viral membrane and has exposed antigenic domains that may induce a minor immunologic response. Abnormal preM/M cleavage may result in incorporation of preM into mature virions that later affect E protein conformation, inhibit fusion, and reduce infectivity.

The E protein (∼50 kDa) is the major surface structure glycoprotein. The three-dimensional configuration of the E protein forms three domains, is determined by disulfide bonding, and is critical for its role in cell tropism, membrane fusion, virulence, and immunity. Strain-, type-, and flavivirus group-specific epitopes are present in this glycoprotein. The hydrophobic C-terminus of this protein is anchored to the lipid bilayer of the viral envelope by a 170-Å-long rod. The C-terminus is connected by a flexible region to domain I, the central part of the molecule with up-and-down topology having eight antiparallel β-strands and containing the N-terminus. Inhibition of this flexible region results in decreased infectivity. Two long loops in domain II extend laterally and are responsible for dimerization. A conserved stretch of 14 amino acids in these loops is the fusion domain that allows internalization of the nucleocapsid into the infected cell. Domain III contains sites involved in binding cell receptors. Conformational neutralization determinants are scattered on the outer surface of all three domains.

The NS1 glycoprotein is involved in RNA replication through its interaction with NS4A and is both expressed on the cell surface and released extracellularly. The extracellular form contains virus-specific and cross-reactive epitopes. Antibodies to NS1 do not neutralize virus infectivity but do provide protective immunity by complement-mediated lysis and rapid clearance of cells with NS1 on the surface. NS1 is highly conserved across all YF strains. NS2A is a small protein (∼22 kDa) that interacts with NS3, NS5, and 3′ NCR sequences and plays an important role in RNA replication and in the assembly or release of virions. NS3 (∼70 kDa) and NS2B (∼14 kDa) interact to form a complex with important enzymatic functions, including serine protease (responsible for post-translational cleavage of virus polyprotein), RNA helicase, and RNA triphosphatase activities. Because of its critical functions, the NS3 gene sequence is also highly conserved. NS3 is also present in cell membranes, contains virus-specific T-cell epitopes that are targets for cytotoxic T-cells. The NS4A (∼16 kDa) and NS4B (∼27 kDa) proteins are membrane associated and play a role in regulating RNA replication. The NS5 protein (∼103 kDa) is a highly conserved RNA-dependent RNA polymerase in virus replication and methyltransferase during 5′ cap methylation.

Assembly and Replication

Assembly of these proteins into viral particles and subsequent exocytosis of mature virions occur in close association with the endoplasmic reticulum of secretory cells (e.g., mosquito salivary gland and various mammal exocrine and endocrine glands). Virion assembly occurs in concert with host cell secretory components and virions are released in secretory granules without significantly disturbing host cellular function or macromolecular synthesis.

After release, virions enter uninfected host cells by attaching to still-undefined receptors and are taken up in clathrin-coated vesicles. Following an acid-mediated change in the E protein that allows fusion of the virus with the host's endosomal membrane, nucleocapsids are released from the vesicles into the cytoplasm. The full-genome-length, plus-sense RNA is translated to make complementary minus strands that serve as templates for progeny plus strands. These plus strands serve as mRNAs for translation of structural and NS proteins including enzymes required for continued virus production and post-translational cleavage of the polyprotein; as templates for the transcription of new minus-sense strands; and as genomes of newly produced mature virions.

Molecular Determinants of Virulence

The entire genomes of the 17D and Asibi strains and of the French neurotropic vaccine and French viscerotropic strains have been sequenced and compared. Comparing sequences and different biological properties of these pairs allows insight into molecular determinants of virulence. Although some potentially important specific sequences have been identified, virulence is clearly multigenic and determined by nonstructural and structural genes. Most studies on these molecular determinants have employed mouse models. Because mice manifest neurotropism and not viscerotropism, such studies are of limited value in dissecting pathogenesis. More recent studies in hamsters susceptible to lethal hepatic dysfunction and necrosis resembling wild-type YFV may allow identifying genetic sequences associated with viscerotropism, but YFV strains must be adapted by serial passage in hamsters before they elicit disease. Nonhuman primates are susceptible to wild-type YFV strains and develop a syndrome closely resembling that in humans.

The 17D and Asibi strains differ at four nucleotides in the 3′ NCR and 20 amino acids in the coding region – eight in the E gene; four in the NS2A gene; two in the NS5 gene; and one each in the M, NS1, NS2B, NS3, NS4A, and NS4B genes.

Given the functional importance of E protein in attachment of YFV to and entry into the cell, one or more of the amino acid differences in this protein likely plays a role in attenuation. Mutations in three areas of E protein alter virulence properties – the tip of the fusion domain in domain II, the molecular hinge between domains I and II, and the portion of domain III containing the putative receptor ligand. Of the eight amino acids that distinguish E proteins of Asibi and 17D viruses four are nonconservative, suggesting that one or more may be responsible for attenuation. Three of these nonconservative changes occur in the hinge region and likely alter the acid-dependent change in E protein conformation required for virus entry. Of nonconservative mutations outside of the hinge region, two (E305 and E380) are located in domain III, which contains determinants involved in tropism and cell attachment. The mutation at amino acid E380 of YF 17D occurs in the putative integrin cell receptor. Studies of this region in other flaviviruses also suggest that mutations in the putative cell receptor-binding region of domain III play a role in determining neurovirulence.

Other studies emphasize the multigenic nature of virulence and suggest that one or more of the 11 amino acid changes in NS proteins, or base changes in the 3′ NCR, may contribute to attenuation. Studies with other flaviviruses have shown that mutations in the NS proteins reduce neurovirulence, probably because NS proteins are critical to the virus replication mechanism. Eleven amino acid changes in the NS proteins of Asibi strain viruses occurred during derivation of the 17D vaccine strain. These changes in NS proteins may affect assembly or release of YFV particles, changes in the RNA helicase and triphosphatase enzymes for unwinding RNA during replication, and activity of the RNA-dependent RNA polymerase. Similarly, the 3′ NCR terminal region is involved in folding of the stem–loop structure which serves as a critical promoter during replication. Changes in this sequence or the number of repeat sequences may alter the stem–loop region, interfere with virus replication, and contribute to attenuation.

Little is known about the determinants of YFV viscerotropism (the ability of YFV to replicate and damage non-neural tissue, such as the liver) principally because assessing this property in nonhuman primates is difficult. Because neuro- and viscerotropic properties may reside in distinct regions, one cannot conclude that attenuation of one correlates with attenuation of the other. However, the hinge region of the E protein contains residues implicated in both neurotropism and viscerotropism. Studies in golden hamsters, which develop disease resembling human YFV, suggest that domain III sites implicated as cell receptor ligands are responsible for hepatotropism.

Host Range and Virus Propagation

A highly conserved region in the domain III of the E protein contains an arginine-glycine-aspartic acid (RGD) motif and is the likely site of virus-cell attachment. Because of the broad host range of flaviviruses, the receptors are highly conserved structures found in chordate and arthropod phyla. YFV replicates and produces cytopathic changes and plaque formation in a wide variety of cell types including primary chick and duck embryo cells; continuous porcine, hamster, rabbit, and monkey kidney cell lines; and cells of human origin such as HeLa, KB, Chang liver, and SW-13. The virus replicates in Fc receptor-bearing macrophages and macrophage cell lines; replication is enhanced by antibody. Mosquito cell lines, especially AP-16 cells, are highly susceptible and often used for primary isolation or efficient laboratory propagation.

Virus isolation is accomplished by intrathoracic inoculation of *Toxorhynchites* or *Ae. aegypti* mosquitoes, inoculation of mosquito cell line cultures (i.e., AP61 and C6/36 cells), intracerebral inoculation of suckling mice or inoculation of specific nonhuman primates. Intrathoracic inoculation of mosquitoes is the most sensitive method. Because infected mosquitoes show no signs of infection, virus is demonstrated by immunofluorescence, subsequent passage to a susceptible host, or amplification of YFV RNA sequences by reverse transcriptase-polymerase chain reaction (RT-PCR). AP61 cells are more sensitive than other *in vitro* methods for primary isolation and show cytopathic effects within 5–7 days of inoculation. Immunofluorescence assays to identify viral antigens and RT-PCR to identify RNA sequences are positive before development of cytopathic effects.

Susceptible vertebrate hosts that show signs of infection include infant mice, which develop fatal encephalitis when infected by peripheral or intracerebral routes; mice greater than 8 days old develop fatal encephalitis only following intracerebral challenge. Nonhuman primates of many species develop fatal hepatitis resembling human disease. The only nonprimate species that develop lethal hepatitis following infection are the European hedgehog and the golden hamster. Antibodies to YFV have been found in a wide variety of field-collected wild vertebrates and wild animals such as rodents and bats. With the exception of opossums in South America, there is no support for a role of nonprimate species in the transmission cycle.

Geographic and Seasonal Distribution

YF occurs in tropical South America and sub-Saharan Africa, where the enzootic transmission cycle involves tree-hole-breeding mosquitoes and nonhuman primates. *Aedes aegypti*-infested regions of Central and North America, the Caribbean, and Southern Europe had periodic outbreaks through the early 1900s and are considered at risk, should the virus be reintroduced. Despite the prevalence of *Ae. aegypti* in tropical Asia, YFV has not appeared. Cross-protection by immunity to dengue viruses may decrease the probability that YFV can be transported from inaccessible, endemic parts of Africa and South America. Low vector competence of Asian populations of *Ae. aegypti* may be responsible for diminishing the risk of virus transmission. Areas at highest risk for reintroduction of virus and secondary spread by re-emerging *Ae. aegypti* include coastal regions and interior towns throughout equatorial South America, Panama, Central America, the West Indies, Mexico, and the southern United States. All are areas affected by YF in the past.

Breeding of many mosquito species that transmit YFV occurs in tree holes and is dependent on rainfall; consequently, transmission increases during tropical rainy seasons and decreases or stops during dry seasons. The peridomestic vector *Ae. aegypti* breeds in receptacles used by humans for water storage, and is less dependent on rainfall. Where this mosquito is involved in virus transmission, YF may occur in the dry season in both rural and urban areas. Temperature also influences YFV transmission rates; a few degrees increase may shorten the extrinsic incubation period (a period of viral replication in a recently infected mosquito prior to the mosquito being capable of transmitting virus to a host) by days, resulting in a significantly increased rate of transmission. Warm temperature also increases biting and reproductive rates of *Ae. aegypti*.

Disease Incidence

Official notifications of disease likely underestimate disease incidence and case–fatality rate, although underreporting occurs to variable degree. During the 20-year period ending in 2004, 28 264 cases and 7880 deaths were reported to the World Health Organization (WHO). In Africa, which accounted for 24 684 cases (87%), the annual incidence varied between 1 and 5104 cases, suggesting highly variable reporting. Despite inconsistent reporting, enzootic and endemic transmission was known to occur continuously in affected regions. During this period, Africa reported 5815 or 74% of the worldwide deaths due to YF, with a case–fatality rate of 24%. The frequency and intensity of epidemics in Africa are due to interhuman transmission by mosquito vectors present in high-density, high-human populations, and low immunization coverage. West Africa appears to be at highest risk of YF emergences, probably reflecting the role of domestic *Ae. aegypti* mosquito vectors. In South America, YF occurs in the Amazon region and contiguous areas. Between 1985 and 2004, 3580 cases and 2074 deaths (case–fatality rate 58%) were reported to WHO by South American countries. The annual incidence varies by country due to fluctuating epizootic activity. The incidence of YF in South America is roughly 20% of that of Africa, due to transmission by enzootic as opposed to epidemic vectors, lower densities of vectors, monkeys and human hosts, and relatively high vaccination coverage. The higher case–fatality rate in South America probably reflects the reporting of more severe cases there, and the fact that surveillance there is based on death reports and postmortem liver examination, although it remains possible that the South American genotype(s) is more virulent than those in Africa.

Epidemiology and Transmission Cycles

YFV is present at high titers in the blood of infected, susceptible hosts (humans and nonhuman primates) for several days, during which mosquito vectors collecting a blood meal may become infected. Virus replicates sequentially in the midgut epithelium, body, and salivary glands of the mosquito. The period between the infecting blood meal and the point at which sufficient replication allows transmission to a vertebrate host is the 'extrinsic incubation period'. This is a temperature-dependent process and takes roughly a week to complete.

Transmission cycles differ according to the hosts and vectors involved. Although the resulting clinical disease does not differ, recognizing the type of cycle is important for disease control. The 'urban cycle' involves humans as the viremic host and *Ae. aegypti* breeding in peridomestic, man-made containers that hold unpolluted water. The 'sylvatic cycle', the predominant pattern in equatorial rain forests of Africa and South America, involves transmission between nonhuman primates and tree-hole-breeding mosquito vectors. In Africa, a third cycle, the 'savanna cycle', is recognized in moist grasslands bordering the rain forests where *Aedes* mosquitoes other than *Ae. aegypti* feed on both humans and nonhuman primates. In all of these settings, the rate of transmission varies widely depending on the density of vectors and the presence of susceptible primates. Primates of many nonhuman species are susceptible to YFV infection. Most African primates produce viremias sufficient to infect mosquitoes without developing illness, whereas many South American primates develop lethal infections. Depletion of vertebrate hosts through natural immunization and death during epizootic waves is a factor in the cyclic appearance of YF activity. In many areas, with environmental pressures markedly reducing monkey populations, human beings serve as hosts in the YFV transmission cycle. No nonprimate species are involved in enzootic transmission.

The risk of nonimmune persons becoming infected is determined by geographic location, season, activities that lead to exposure, duration of exposure to mosquito bites, and the intensity of YFV transmission. Although reported cases of human disease are an important guide to YFV activity, reported numbers may be low due to a high level of immunity in the population or to insensitive surveillance. In areas where vaccination is widely practiced, YFV may circulate between monkeys and mosquitoes, with few human cases occurring.

The means of survival of YFV across long dry seasons, when sylvatic mosquito vectors are absent, remains in question. Transovarial transmission to *Aedes* and *Haemagogus* mosquito eggs, which survive desiccation in tree holes and hatch when the rains return, likely play a major role. Low-level horizontal transmission by drought-resistant vectors, and alternative horizontal and vertical transmission cycles involving ticks, are theorized mechanisms. Although persistent infection of experimentally infected nonhuman primates has been documented, infections in these hosts generally are not accompanied by viremia sufficiently high to infect vectors.

The most alarming future prospect involves the reemergence of urban transmission in South America and the spread to heavily populated, *Ae. aegypti*-infested areas currently free from disease, such as the coastal regions of South America, the Caribbean, and North America, and regions where the virus has never become established but which have susceptible human populations and competent mosquito vectors, such as the Middle East, coastal East Africa, the Indian subcontinent, Asia, and Australia. Alterations in human demography and behavior, in virus activity, and in the distribution of *Ae. aegypti* underlie the potential for epidemiological change. Such alterations include: (1) increasing urban populations; (2) changes in commerce, transportation, and communication that allow human population expansion into previously remote endemic zones; (3) reinvasion of South America by the peridomestic *Ae. aegypti*; (4) presence of efficient jungle vectors in degraded periurban habitats; (5) introduction of new competent *Aedes* species (e.g., *Ae. albopictus*) that may bridge the jungle and urban transmission cycles; (6) relaxation of regulations and of enforcement of vaccination certification for travelers; and (7) possibly climate change.

Laboratory-acquired infections were common in the pre-vaccine era and remain of concern today, particularly where unvaccinated clinical laboratory personnel encounter blood from patients in early stages of illness. Some laboratory infections were probably acquired via bites of experimentally infected mosquitoes or of wild mosquitoes infected after feeding on experimental animals. Others resulted from direct contact with blood or aerosols of dried virus. The stability of YFV permits transmission within a short period after generation of an infectious aerosol.

Clinical Features

The broad clinical spectrum of YF includes abortive infection with nonspecific flu-like illness, and 'classic' YF – potentially lethal pansystemic disease with fever, jaundice, renal failure, and hemorrhage. Serosurveys to identify persons with asymptomatic infections estimate that only about 15–25% of infected persons develop classic YF. Diagnosis of sporadic cases without classic symptoms is difficult; as a result, often only cases of classic YF are notified, and deaths may be more frequently notified than cases. These features lead to an underestimate of morbidity and overestimate of case–fatality rate. The variability in clinical response to infection is multifactoral and includes intrinsic and acquired host resistance factors and differences in pathogenicity of virus strains.

Classic YF is a triphasic disease that begins with abrupt onset of fever, headache, myalgia, anorexia, and nausea 3–6 days after infection. This 'period of infection' is clinically nonspecific but corresponds to a viremic phase when a person is infectious to feeding mosquitoes. Fatal cases appear to have a longer duration of viremia than do survivors. After 3–4 days, a 'period of remission', lasting up to 48 h, begins with abatement of fever and symptoms. Cases of abortive infection simply recover at this stage. Because such cases remain anicteric with nonspecific symptoms, clinical diagnosis of YF is not possible. Roughly 15–25% of persons infected with YFV enter a 'period of intoxication', typically 3–5 days after onset, characterized by a moderate to severe disease characterized by jaundice, the return of fever, relative bradycardia, oliguria, hemetemesis, and other hemorrhagic diatheses. Virus disappears from blood and antibodies appear 7–10 days after infection. The subsequent course reflects dysfunction of the hepatic, renal, and cardiovascular systems. Central nervous system (CNS) signs include delirium, convulsions, and coma. In severe cases, the cerebrospinal fluid is under increased pressure and may contain elevated protein but no cells. No inflammatory changes have been found in brains of persons with severe illness; cerebral edema or metabolic factors apparently cause CNS signs. Between the fifth and tenth day of illness, the patient either dies or recovers rapidly. Terminal events are characterized by hypotension, shock, and respiratory distress.

Pathology and Pathogenesis

Two biological properties are inherent to all wild-type YFV strains: 'viscerotropism', the ability to cause viremia and subsequently to infect and damage liver, heart, and kidneys, and 'neurotropism', the ability to infect the brain parenchyma causing encephalitis. Wild-type YFV strains are predominantly viscerotropic in humans and in nonhuman primates. True YF viral encephalitis is rare; even after intracerebral inoculation of YFV, susceptible nonhuman primates die from hepatitis rather than from encephalitis.

The course of infection following inoculation of wild-type YFV has been partially investigated using nonhuman primates. Virus replicates at the site of skin inoculation and spreads to draining lymph nodes, then to central lymph nodes, and subsequently to visceral organs. Lymphoid cells are early viral replication sites, and in the liver Kupffer cells in the sinusoids are the gateway to subsequent hepatocyte infection. Results of prior research, primarily of dengue infections, makes clear that dendritic cells (DCs) play a primary role in the early stages of infection, and that these cells not only process viral antigens into immunogenic peptides and present them to T-cells, but also secrete a cascade of immunoregulatory cytokines that shape innate and adaptive immune responses.

In humans with viscerotropic disease, degeneration of hepatocytes occurs in the last phase of infection. In fatal cases, up to 100% of hepatocytes undergo coagulative necrosis characteristic of apoptosis. The mid-zone of the liver lobule is principally affected, with sparing of cells bordering the central vein and portal tracts. Viral antigen and viral RNA are demonstrable in cells undergoing pathological changes, suggesting that cytopathology is mediated by direct injury from virus replication and from accumulation of virions and viral protein in the endoplasmic reticulum. The reason for this peculiar distribution of hepatic injury is unknown; mid-zonal necrosis has been described in low-flow hypoxia, ATP depletion and oxidative stress of marginally oxygenated cells at the border between anoxic and normoxic cells. Eosinophilic degeneration with condensed nuclear chromatin (Councilman bodies), indicative of apoptosis, characterizes hepatocyte injury rather than the ballooning and rarefaction necrosis seen in hepatitis caused by other viruses. Inflammatory changes are minimal and persons who survive YF do not develop scarring or cirrhosis due to this infection. Eosinophilic degeneration without inflammation, along with fatty change of renal tubular epithelium, also characterizes renal pathology. Focal degeneration may be present in myocardial cells. Lymphoid follicles of lymph nodes and spleen show necrosis. The brain shows edema and petechial hemorrhage but viral invasion and encephalitis are rare. Decreased hepatic synthesis of clotting factors causes hemorrhagic manifestations.

The mediators of the profound shock of YF are unknown but cytokine dysregulation (systemic inflammatory response syndrome, SIRS) likely mediates this terminal stage as in other sepsis syndromes. The appearance of antibody and cellular responses coincides with onset of the 'period of intoxication', suggesting that immune clearance may be, in part, responsible for the cytokine storm events. Persons with fatal YF have elevated pro-inflammatory cytokines compared to persons with nonfatal cases of YF, resembling the profile seen in bacterial sepsis. Overall, these findings suggest that high levels of pro-inflammatory cytokines contribute to lethality.

Immune Responses

Prior to appearance of specific cytotoxic T-cells and immunoglobulins, innate immunity, comprised of natural killer (NK) cells, interferons (IFNs), and other pro-inflammatory cytokines, play a significant role in limiting viral replication during early infection. NK cells recognize and lyse cells displaying viral antigens; secrete IFN-α and pro-inflammatory cytokines that have direct antiviral activity; activate DCs and macrophages; and promote a Th1 adaptive immune response. YFV antigens directly activate DCs via type I IFN signal transduction and Toll-like receptors. Through Toll-like receptor activation, pro-inflammatory molecules (e.g., interleukins IL1-β, IL2, IL6, IL8, IL12) are expressed and further activate NK cells. This robust innate response to YFV likely underlies the rapid, strong, and durable adaptive responses to this virus. Flaviviruses have evolved weak mechanisms for avoiding these innate responses, including blocking the IFN signal transducer pathways by NS2A and NS4B. Those patients with a poor outcome may develop signs of a dysregulated pro-inflammatory cytokine response (SIRS).

Seven to ten days after infection, a specific immune response is detectable when immunoglobulin M (IgM; enzyme-linked immunosorbent assay, ELISA), hemagglutination-inhibiting, and neutralizing antibodies begin to appear. IgM antibodies peak during the second week and usually decline rapidly over the next 30–60 days while neutralizing antibody continues to increase. When YFV represents the patient's first flavivirus infection, the magnitude of the IgM response is significantly greater than in persons with prior flavivirus exposure. Although little is known about cytolytic antibodies against viral proteins on the surface of infected cells, antibody-dependent cell-mediated cytotoxicity (ADCC), and cytotoxic T-cells, these mechanisms, in addition to neutralizing antibody, presumably mediate clearance of primary infection. Based on immunologic studies of YFV vaccine recipients, neutralizing antibody generally peaks 4–6 weeks after infection but high titers of neutralizing antibodies persist for more than 10 years and provide complete protection against disease on reexposure to the virus. No documented case of a second clinical YFV infection has ever been reported. Although initial antibody response to infection is YFV antigen specific, with affinity maturation, specificity declines and cross-reactions with other flaviviruses develop during the subsequent several week of the immune response. Persons with prior heterologous flavivirus immunity develop broadly cross-reactive antibody responses during YFV infections. Previous infection with flaviviruses, such as Zika, dengue, or Wesselsbron viruses, provides partial cross-protection against YFV and may ameliorate the clinical severity of YF.

Prevention and Control

Domestic control of *Ae. aegypti* mosquitoes remains important but is difficult to sustain. Currently, the most effective approach to control of YF is by immunizing persons living in or traveling to endemic areas. 17D vaccine is the only strain currently used for human immunization against YFV. It is a live, attenuated vaccine produced in embryonated chicken eggs. 17D vaccines are not biologically cloned and are heterogeneous mixtures of multiple virion subpopulations ('genetic swarms'); differences in plaque size, oligonucleotide fingerprints, and nucleotide sequences have been found but do not appear to affect safety or efficacy. Currently, manufacturers in seven countries market YFV 17D vaccine. Three manufacturers in Brazil, France, and Senegal produce large amounts of vaccine for the Expanded Programme of Immunization and for mass vaccination campaigns. As of 2006, annual global vaccine production was approximately 60 million doses.

Monath *et al.* thoroughly review the development, immune response, and efficacy of the 17D vaccine strain.

As of 2006, over 400 million persons have been immunized with YFV vaccines. Over roughly 70 years of use, 17D vaccines have been acknowledged as one of the safest and most effective live vaccines in use. Recently, however, close scrutiny has been brought to bear as clinical and histopathological evidence has emerged linking 17D vaccines to severe and previously unrecognized adverse events, including viscerotropic disease closely resembling that caused by wild-type YFV. Although rare mutational events in 17D vaccine virus during replication in the host can alter pathogenicity, these recently reported serious adverse events were not associated with either mutations that change virulence or tropism of the virus or selection of virulent variants *in vivo*. Investigations suggest that host susceptibility, rather than a change in the virus, is responsible for these serious, adverse events. Advanced age and thymic disease appear to be risk factors for development of vaccine-associated viscerotropic disease. The incidence of this complication, which carries a case–fatality rate of approximately 50%, is believed to be 1:400 000.

See also: Yellow Head Virus.

Further Reading

Barrett ADT and Monath TP (2003) Epidemiology and ecology of yellow fever. *Advances in Virus Research* 61: 291–317.

Carter HR (1931) *Yellow Fever: An Epidemiological and Historical Study of Its Place of Origin.* Baltimore, MD: Williams and Wilkins.

Cordellier R (1991) The epidemiology of yellow fever in western Africa. *Bulletin of the World Health Organization* 69(1): 73–84.

Monath TP and Barrett ADT (2003) Pathogenesis and pathophysiology of yellow fever. *Advances in Virus Research* 60: 343–397.

Monath TP, Teuwen D, and Cetron M (in press) Yellow fever vaccine. In: Plotkin SA, Orenstein WA, and Offit PA (eds.) *Vaccines Expert Consult*, 5th edn. Philadelphia, PA: Saunders.

Strode GK (ed.) (1951) *Yellow Fever,* pp. 385–426. New York: McGraw-Hill.

Yellow Head Virus

P J Walker, CSIRO Australian Animal Health Laboratory, Geelong, VIC, Australia
N Sittidilokratna, Centex Shrimp and Center for Genetic Engineering and Biotechnology, Bangkok, Thailand

© 2008 Elsevier Ltd. All rights reserved.

Glossary

Antennal gland Complex excretory glands located behind the eyes on antenna on the head of decapods.

Hepatopancreas An organ of the digestive tract of arthropods and fish that provides the functions which are performed separately by the liver and pancreas in mammals.

Lymphoid organ Also known as the Oka organ, a component of the hematopoeitic system of penaeid shrimp consisting of lymphoid cells around the two subgastric arteries.
Pseudoknot A complex folded structure in an RNA molecule.
Slippery sequence A nucleotide sequence at which ribosomal slippage can occur during translation to cause a change in the reading frame.

Introduction

Yellow head virus (YHV) is a pathogen of the black tiger shrimp (prawn), *Penaeus monodon*, which is one of the world's major aquaculture species. Yellow head disease was first reported in central Thailand in 1990 from which it spread rapidly along the eastern and western coasts of the Gulf of Thailand to southern farming regions. Outbreaks of yellow head disease have since been reported from most of the major shrimp farming countries in Asia. It is suspected that the YHV (rather than monodon baculovirus, which is not usually pathogenic for juvenile shrimp) may have previously caused the crash of the shrimp farming industry in Taiwan during the late 1980s. Mortalities usually occur during the mid-late stages of grow-out in ponds with complete crop loss commonly occurring within 3 days of the first signs of disease. YHV is one genotype in a complex of closely related viruses infecting black tiger shrimp. Other genotypes include gill-associated virus (GAV) which has been associated with relatively less severe forms of disease in farmed shrimp in Australia, and at least four other genotypes for which no disease association has yet been established. YHV and the other genotypes are endemic throughout the Indo-Pacific region, occurring commonly as low-level chronic infections in healthy shrimp.

Taxonomy and Classification

YHV is a positive-sense single-stranded RNA (ssRNA) virus that shares aspects of genome organization, replication, and transcription with coronaviruses, toroviruses, and arteriviruses with which it is classified in the order *Nidovirales*. In 2002, the International Committee on Taxonomy of Viruses (ICTV) established the genus O*kavirus* in the new family *Roniviridae* to accommodate YHV and closely related GAV. *Okavirus* is derived from the Oka or lymphoid organ of penaeid shrimp in which the virus is commonly detected; *Roniviridae* is derived from the sigla rod-shaped nidovirus. *Gill-associated virus* was assigned as the type species of the genus because its biological and molecular characterization were more complete. YHV is currently classified as a member of the species *Gill-associated virus*. No virus other than those described in the yellow head complex is currently assigned to the *Roniviridae* but several viruses with similar morphology have been reported in crabs and fish. Roniviruses are the only members of the order *Nidovirales* that are currently known to infect invertebrates.

Virion Structure and Morphology

YHV virions are rod-shaped, enveloped particles ($\sim$50 nm $\times$ $\sim$175 nm) with prominent diffuse spikes ($\sim$8 nm $\times$ $\sim$11 nm) projecting from the surface (**Figure 1(a)**). Internal helical nucleocapsids are approximately 25 nm in diameter and have a periodicity of 5–7 nm. Filamentous nucleocapsid precursors, approximately 15 nm in diameter and of variable length ($\sim$80–450 nm), are observed in the cytoplasm, sometimes densely packed in paracrystalline arrays (**Figure 1(b)**). Nucleocapsids acquire trilamellar lipid envelopes by budding through membranes into intracytoplasmic vesicles or at the cell surface (**Figure 1(c)**). It has been reported that long nucleocapsid precursors generate elongated, enveloped structures that subsequently fragment into mature virions. The morphology of GAV virions is indistinguishable from that of YHV.

YHV virions contain a polyadenylated 26.6 kDa (+) ssRNA genome and three structural proteins. The nucleoprotein (p20) is a highly hydrophilic, basic protein that complexes with the genomic RNA in nucleocapsids. Transmembrane glycoproteins gp64 and gp116 are components of the envelope that form the visible projections on the virion surface. YHV infectivity can be at least partially neutralized by antibody to gp116 but not by antibody to gp64. It is reported that gp116 docks with a 65 kDa cell membrane protein (pmYRP65) that mediates YHV entry into susceptible shrimp cells. Knockdown of pmYRP65 expression has been reported to totally abrogate susceptibility of shrimp cells to YHV infection.

Genome Organization and Transcription Strategy

The 26 662 nt YHV genome comprises four long open reading frames (ORFs) designated ORF1a, ORF1b, ORF2, and ORF3 (**Figure 2**). ORF1a (12 216 nt) and ORF1b (7887 nt) encode all of the elements of a large replicase complex. ORF1a encodes a 4072 aa polyprotein (pp1a) that contains a 3C-like cysteine protease catalytic domain flanked by putative transmembrane domains. The pp1a protease has autolytic activity and appears to be involved in processing the replicase polyproteins. ORF1b overlaps ORF1a by 37 nt. Expression of ORF1b requires

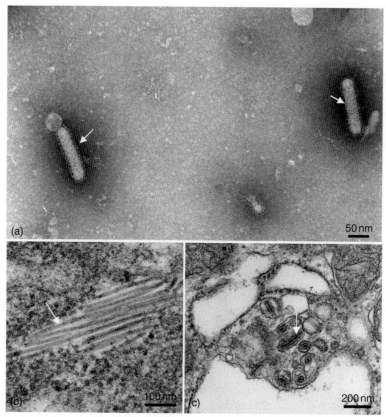

Figure 1 Transmission electron micrographs of YHV. (a) Image of virions stained with heavy metal salts showing the external appearance of the enveloped particles (arrows). (b) Image of an ultrathin section of helical nucleocapsids (arrow) within the gill of an infected shrimp. (c) Image of an ultrathin section of virions (arrow) within the gill of an infected shrimp. Kindly provided by Dr. Alex Hyatt, CSIRO, Australian Animal Health Laboratory, Geelong, Australia.

a −1 ribosomal frameshift at a slippery sequence (AAAUUUU) near a complex pseudoknot structure in the mRNA. The extended 6688 aa polyprotein (pp1ab) contains RNA-dependent RNA polymerase (RdRp), multinuclear zinc-binding (ZBD), helicase (HEL), 3′-5′ exoribonuclease (ExoN), uridylate-specific endoribonuclease (NendoU), ribose-2′-O-methyltransferase (O-MT) catalytic domains, and other cysteine/histidine-rich domains that are conserved in pp1ab of other nidoviruses. ORF2 encodes the 146 aa nucleocapsid protein (p20). Roniviruses are unique among known nidoviruses in that the nucleocapsid protein gene is located upstream rather than downstream of the glycoprotein genes. ORF3 encodes a 1666 aa polyglycoprotein that is processed to generate virion envelope glycoproteins gp64 and gp116. Proteolytic cleavage of ORF3 occurs at two [Ala-X-Ala] motifs immediately following predicted transmembrane domains that appear to function as signal peptides. The cleavage also generates a 228 aa (∼22 kDa) protein that contains triple membrane-spanning domains and resembles M-proteins in coronaviruses. The YHV M-like protein appears to be present in infected cells at relatively low levels. The YHV genome also features significant noncoding regions, including a 71 nt untranslated region (UTR) at the 5′-terminus, a 352 nt UTR between ORF1b and ORF2, and a 54 nt UTR between ORF2 and ORF3. The 677 nt region between ORF3 and the 3′-poly[A] tail contains no long ORFs (>65 nt) and so also appears to be a long UTR (**Figure 2**).

Much of our understanding of YHV molecular biology has been obtained by comparison with closely related GAV. The 26 235 nt GAV genome is similar in structural organization to YHV, varying principally in the size and structure of the UTRs. In GAV, the ORF1b–ORF2 UTR comprises only 93 nt. The 638 nt region downstream of GAV ORF3 encodes a 252 nt ORF (ORF4) that has potential to express an unidentified 83 aa polypeptide with a deduced molecular weight ∼9.2 kDa. A short ORF in the corresponding region of the YHV genome is truncated with a termination codon after only 20 aa and is unlikely to be expressed. Like other nidoviruses, the GAV genome is transcribed as a nested set of 3′-co-terminal mRNAs comprising the full-length genome and two subgenomic messenger RNAs (sg mRNAs) that initiate at conserved

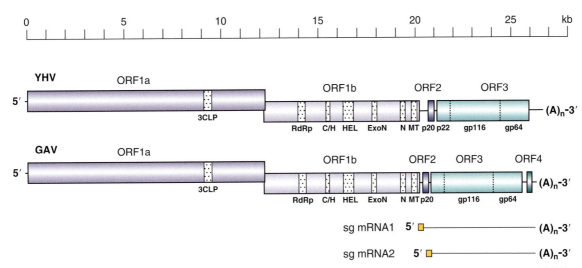

Figure 2 Genome organization of YHV and GAV indicating the locations of subgenomic mRNAs (sg mRNA1 and sg mRNA2) and functional domains in ORF1a (3CLP) and ORF1b (RdRp, C/H, HEL, ExoN, N, and MT). Also indicated are proteolytic cleavage sites in the ORF3 polyprotein that is processed post-translation to generate triple-membrane-spanning protein (p22) and transmembrane glycoproteins (gp116 and gp64).

transcription-regulating sequences (TRSs) in noncoding regions immediately upstream of ORF2 and ORF3. However, unlike coronaviruses and arteriviruses, conserved GAV (and YHV) TRSs are not present in the 5'-UTR of genomic RNA and so do not mediate splicing of common 5'-leader sequences on to the sg mRNAs. Sequences with partial identity to the conserved ORF2 and ORF3 TRSs occur upstream of GAV ORF4 (and the truncated YHV ORF4) but these do not appear to be functional.

Geographic Distribution and Host Range

Surveys for the presence of viral genomic RNA have indicated that YHV and other genotypes in the complex are endemic in black tiger shrimp populations across its natural geographic range throughout the Indo-Pacific. Yellow head disease has been reported in farmed tiger shrimp from Thailand, Taiwan, China, the Philippines, Vietnam, Malaysia, Indonesia, India, Sri Lanka, and Madagascar. Although natural infection and disease have been reported only in black tiger shrimp and kuruma shrimp (*Marsupenaeus japonicus*), YHV can cause high rates of mortality following experimental infection of most other farmed marine shrimp species, including Pacific white shrimp (*Litopenaeus vannamei*), Pacific blue shrimp (*Litopenaeus stylirostris*), brown tiger shrimp (*Penaeus esculentus*), white banana shrimp (*Fenneropenaeus merguiensis*), white shrimp (*Litopenaeus setiferus*), brown shrimp (*Farfantepenaeus aztecus*), hopper and brown-spotted shrimp (*Farfantepenaeus duorarum*), red endeavour prawn (*Metapenaeus ensis*), and Jungas shrimp (*Metapenaeus affinis*). Some species of palemonid shrimp and krill are also susceptible to experimental infection. Crabs appear to be refractory to YHV infection and disease.

GAV has been associated with a less-aggressive disease of juvenile black tiger shrimp in Australia called mid-crop mortality syndrome. However, several other viruses have also been detected in shrimp with this condition and the etiology remains uncertain. GAV does cause disease and mortalities following experimental infection of several farmed shrimp species, including black tiger, brown tiger, and kuruma shrimp. GAV and other genotypes in the YHV complex have also been detected in healthy black tiger shrimp from Taiwan, the Philippines, Malaysia, Brunei, Indonesia, Vietnam, Thailand, India, Mozambique, and Fiji. A very high prevalence of GAV infection has been reported in healthy black tiger shrimp from eastern Australia. Evidence of GAV infection has also been detected in mud crab (*Scylla serrata*) in an experimental aquaculture facility.

Pathology

Shrimp are susceptible to YHV infection from late post-larval stages but mass mortality in ponds usually occurs in early-to-late juvenile stages. Disease and mortalities usually occur within 2–4 days of a period of exceptionally high feeding activity followed by an abrupt cessation of feeding. Moribund shrimp congregate at pond edges near the surface and may exhibit a bleached overall appearance and discoloration of the cephalothorax caused by yellowing of the underlying hepatopancreas.

YHV infects tissues of ectodermal and mesodermal origin, including lymphoid organ, hemocytes, hematopoeitic tissue, gill lamellae, and spongy connective tissue of the subcutis, gut, antennal gland, gonads, nerve tracts, and ganglia. In severe infections, there is a generalized cell degeneration with prominent nuclear condensation, pyknosis and karyorrhexis, and basophilic, perinuclear cytoplasmic inclusions in affected tissues. There is evidence of apoptosis, including chromatin condensation and DNA fragmentation, in hemocytes, lymphoid organ, and gill tissues and it has been suggested that widespread apoptosis rather than necrosis is the cause of disease and mortalities.

YHV, GAV, and other viruses in the yellow head complex can also occur as low-level chronic infections in apparently healthy shrimp. Chronic infections have been observed in shrimp of all life stages collected from hatcheries and farms, and in the survivors of experimental infection. For GAV, the progression of infection following experimental challenge has been shown to be dose related. Shrimp infected with a high dose of GAV progress rapidly to disease with high viral loads and typical pathology leading to mortalities. Shrimp infected with a low dose do not develop disease and the virus persists as a low-level infection for at least 60 days. There is also evidence that stress can lead to rapid increases in viral load. For YHV, the onset of disease has been associated with the stress of molting. During chronic infections, there is little histopathology other than the accumulation of partitioned foci of cells with hypertrophic nuclei (spheroid bodies) in the lymphoid organ. Spheroid bodies appear to form in shrimp as part of a nonspecific defense mechanism for clearance of infectious agents and other foreign bodies.

Host Response to Infection

As invertebrates, shrimp lack antibodies, cytokines, T-lymphocytes, and other powerful components of the vertebrate immune system that allow a specific adaptive response to viral infection, clearance of virus and infected cells, and long-term immunological memory. There is also no evidence in shrimp of interferon, natural killer (NK) cells, or other key components of the vertebrate natural immune system that allow an immediate nonspecific defense against viruses. Nevertheless, shrimp do appear to have a capacity to respond to viral infection and highly pathogenic viruses are commonly present as low-level chronic infections in apparently healthy shrimp. For YHV, there is no evidence of an inflammatory response at the primary sites of infection. However, YHV accumulates in spheroid bodies in the lymphoid organ during chronic persistent infections, and it is thought that the lymphoid organ has an important role in filtering granulated hemocytes and the clearance of viruses from infected shrimp. It has been reported that cells within lymphoid organ spheroids become apoptotic during infection and may be cleared during molting. Apoptotic cells have been observed in lymphoid organs, hemocytes, and gills during acute YHV infections in what appears to be a fundamental host defensive reaction. It has also been reported that double-stranded RNA (dsRNA) corresponding to sequences in viral replicase and glycoprotein genes specifically inhibits YHV infection *in vitro* and *in vivo*, suggesting that RNA interference may play a role in the host response to infection.

Transmission

The natural transmission cycle of YHV has not been studied in detail. Experimentally, YHV infection and disease can be transmitted horizontally by injection, ingestion of infected tissue, immersion in membrane-filtered tissue extracts, or by cohabitation with infected shrimp. Transmission of disease by ingestion has been demonstrated from the late postlarval stages onward. Transmission has also been demonstrated by injection of black tiger shrimp with extracts of paste shrimp (*Acetes* sp.) and mysid shrimp (*Palaemon styliferus*) collected from infected ponds. For GAV, there is evidence that horizontal transmission can occur from chronically infected shrimp in the absence of disease.

There is no direct evidence of vertical transmission of YHV but it can be detected as a chronic infection in broodstock prior to spawning, and polymerase chain reaction (PCR) screening to eliminate infected broodstock and seed is increasingly being used to reduce risks of yellow head disease in ponds. GAV has been detected in spermatophores and mature ovarian tissue of broodstock, and in fertilized eggs and nauplii spawned from infected females. Examination by electron microscopy has revealed virions in seminal fluid but not in sperm cells. Artificial insemination of infected broodstock has shown that vertical transmission occurs efficiently from both male and female parents. Transmission is probably by surface contamination or infection of tissue surrounding the fertilized egg. The high prevalence of yellow head complex viruses in postlarvae collected from hatcheries in Australia and several Asian countries supports the view that vertical transmission has an important role in the infection cycle of all genotypes, particularly during propagation for aquaculture.

Genetic Diversity

YHV is one of several closely related genotypes that have been detected in black tiger shrimp in the Indo-Pacific region. Analysis of nucleotide and deduced amino

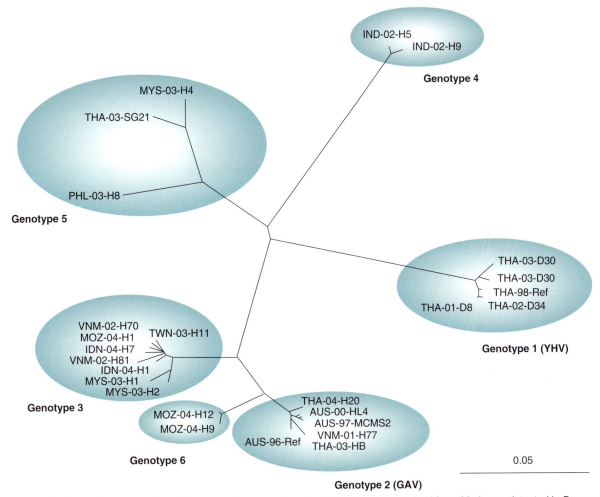

Figure 3 Phylogenetic tree constructed from an alignment of nucleotide sequences obtained from 26 viruses detected in *Penaeus monodon* shrimp from the Indo-Pacific region. The sequences correspond to a 671 nucleotide region of the ORF1b gene encompassing elements of the HEL domain. The alignment illustrates the six known genotypes in the yellow head complex including YHV (genotype 1), GAV (genotype 2), and four other genotypes (genotypes 3, 4, 5, and 6) that have been detected only in healthy shrimp. The viruses were obtained from shrimp collected in India (IND), Malaysia (MYS), Thailand (THA), the Philippines (PHL), Australia (AUS), Vietnam (VNM), Mozambique (MOZ), Indonesia (IDN), and Taiwan (TWN) between 1997 and 2004. Sequences kindly provided by Mrs Priyanjalie Wijegoonawardane, CSIRO Livestock Industries, St. Lucia, Australia.

acid sequences in a relatively conserved region of the ORF1b gene has identified at least six distinct genetic lineages in the complex (**Figure 3**). In pairwise alignments, nucleotide sequence identity between consensus sequences representing each genotype ranges from 80.3% to 96.5%. Variation within genotypes is generally low, with nucleotide sequence identities between isolates in the range 97.1–100%, except genotype 5 for which three available isolates have been reported to share 93.0–97.1% identity.

YHV (genotype 1) is the only genotype that has been detected in shrimp with typical signs of yellow head disease. Although the disease has been reported from many sites in Asia, isolates are currently available only from Thailand and Taiwan. Genotype 1 is the most distantly related to other lineages and appears to occur less commonly than other genotypes in healthy shrimp. GAV (genotype 2) is the only other lineage known to be associated with disease of any form. Analysis of complete genome sequences of prototype strains of YHV and GAV indicates similar nucleotide sequence identities for ORF1a (79.7%), ORF1b (82.3%), ORF2 (81.0%), and a slightly lower level of identity for ORF3 (73.2%). Amino-acid-sequence identities between YHV and GAV proteins are similar for the replicase pp1ab (84.9%), nucleoprotein p20 (84.4%), and glycoprotein gp64 (83.9%), and lower for the M-like protein p22 (74.8%) and glycoprotein gp116 (71.7%).

GAV has been detected in black tiger shrimp from Australia, Vietnam, and Thailand. Phylogenetic analysis of sequences analyzed to date suggests that all isolates may have originated from translocated Australian shrimp. Genotype 3 has been detected to date in Taiwan, Vietnam, Indonesia, Malaysia, Thailand, and Mozambique. It appears to be the most widely distributed and most frequently detected genotype. Genotype 4 has been detected only in India. Genotype 5 has been detected in the Philippines, Malaysia, and Thailand. As indicated above, genotype 5 is genetically the most diverse genotype and may be split into three distinct lineages as more isolates become available. Genotype 6 has been detected only in Mozambique.

Assignment of these genotypes has been based primarily on comparisons of sequences in a conserved region of the ORF1b gene. Analysis of nucleotide sequences in the 5'-terminal region of the ORF3 polyglycoprotein gene indicates more genetic variability and suggests that genetic recombination is contributing to diversity in the complex. Of 24 isolates examined recently, almost one-third were assigned to different genotypes in comparative phylogenetic analyses of nucleotide sequences in the ORF1b and ORF3 regions. Genetic recombination is a phenomenon known to occur commonly in other nidoviruses. It appears that the vast international trade in live shrimp broodstock and seed for aquaculture is providing adequate opportunities for recombination and diversification of the gene pool. This appears to confound the assignment of coherent genetic lineages and may have significant consequences for both the emergence and definitive diagnosis of disease.

Diagnosis and Disease Management

Gross clinical signs of YHV infection including yellowing of the carapace and erratic swimming behavior are not observed consistently and are not sufficiently pathognomonic to be useful for disease diagnosis. Histologically, moderate to large numbers of basophilic, spherical, cytoplasmic inclusions in tissues of ectodermal and mesodermal origin are indications of YHV infection and can be used for presumptive diagnosis. Confirmatory diagnosis of yellow head disease requires the use of electron microscopy or molecular methods such as the reverse transcriptase-polymerase chain reaction (RT-PCR) or *in situ* hybridization assays. Antibody-based tests such as western blotting and dot-blot nitrocellulose enzyme immunoassay (NC-EIA) are also available. Low-level chronic infections with YHV and other genotypes can be detected by nested RT-PCR or other highly sensitive molecular genetic tests such as real-time PCR or loop-mediated isothermal amplification (LAMP). Accurate genotype assignment can only be achieved by PCR, sequence analysis, and comparison with sequences of other known genotypes.

No effective vaccines or therapeutics are currently available for the control of YHV and no genetically resistant shrimp stocks have been reported yet. Disease management is primarily through pathogen exclusion by PCR screening of broodstock and/or seed, the application of on-farm biosecurity and sanitary measures, and stress reduction by careful management of water quality during grow-out.

Current Status

Key aspects of the biology of YHV infection are yet to be resolved and yellow head continues to be a disease of concern to aquaculture farmers. No direct link has been demonstrated between the presence of virus in infected broodstock and the appearance of disease on farms. Assumptions about vertical transmission come by analogy with GAV and may well be accurate. However, the prevalence of YHV in healthy shrimp appears to be far lower than for GAV and other genotypes, and it is unclear how it maintains a cycle of natural infection. It is possible that YHV is commonly introduced to ponds in healthy wild shrimp or other carrier crustaceans but surveys to date have not revealed a likely source. The host–viral interaction during the chronic phase of infection, the transition from chronic to acute phases, and the role of stress in disease emergence are also poorly understood, and there is little understanding of the molecular basis of virulence variations between YHV and other genotypes. A more comprehensive study of the sources of YHV infection and host and/or environmental factors leading to emergence of yellow head disease should be conducted. Emerging capabilities in shrimp genomics and proteomics will greatly facilitate this work.

There is an emerging understanding of RNA interference (RNAi) as a potentially powerful mechanism for the control of viral diseases. Inhibition of YHV infection in primary lymphoid organ cell culture has been demonstrated by treatment with dsRNA corresponding to YHV protease, polymerase, and helicase domains. Injection of shrimp with protease domain dsRNA has also been shown to inhibit YHV replication and mortalities. Knockdown of the shrimp dicer-1 endoribonuclease gene expression has demonstrated that the antiviral effects of dsRNA are caused by RNAi. RNAi technology has useful applications in studies of the molecular biology of YHV infection and, if delivered cost-effectively, could potentially find commercial application in the management of yellow head disease.

Roniviruses are also seen as important links in understanding the evolutionary biology of (+) ssRNA viruses. Considerations of virion structure and the size, complexity and structural organization of the genome suggest that roniviruses form a genetic lineage ancestral to coronaviruses and toroviruses. Studies of the ronivirus 3C-like cysteine protease encoded in ORF1a have also revealed structural similarities to coronaviruses in the catalytic site but substrate specificity and binding sites are more similar to those of potyviruses, suggesting that they bridge the gap between these distantly related proteases. A pseudoknot structure and slippery sequence at the ribosomal frameshift site is also distinct from the H-type structures characteristic of many vertebrate nidoviruses. Further molecular studies of ronivirus structure and function should provide insights into the evolution of these unusual viruses.

See also: Barley Yellow Dwarf Viruses; Tomato Yellow Leaf Curl Virus.

Further Reading

Assavalapsakul W, Smith DR, and Panyim S (2006) Identification and characterization of a *Penaeus monodon* lymphoid cell-expressed receptor for yellow head virus. *Journal of Virology* 80: 262–269.

Chantanachookin C, Boonyaratpalin S, Kasornchandra J, *et al.* (1993) Histology and ultrastructure reveal a new granulosis-like virus in *Penaeus monodon* affected by 'yellow-head' disease. *Diseases of Aquatic Organisms* 17: 145–157.

Cowley JA, Cadogan LC, Spann KM, Sittidilokratna N, and Walker PJ (2004) The gene encoding the nucleocapsid protein of gill-associated nidovirus of *Penaeus monodon* prawns is located upstream of the glycoprotein gene. *Journal of Virology* 78: 8935–8941.

Cowley JA, Dimmock CM, Spann KM, and Walker PJ (2000) Gill-associated virus of *Penaeus monodon* shrimp: An invertebrate virus with ORF1a and ORF1b genes related to arteri- and coronaviruses. *Journal of General Virology* 81: 1473–1484.

Cowley JA, Dimmock CM, and Walker PJ (2001) Gill-associated nidovirus of *Penaeus monodon* prawns transcribes 3′-coterminal subgenomic RNAs that do not possess 5′-leader sequences. *Journal of General Virology* 83: 927–935.

Cowley JA, Hall MR, Cadogan LC, Spann KM, and Walker PJ (2002) Vertical transmission of covert gill-associated virus (GAV) infections in *Penaeus monodon*. *Diseases of Aquatic Organisms* 50: 95–104.

Cowley JA and Walker PJ (2002) The complete genome sequence of gill-associated virus of *Penaeus monodon* prawns indicates a gene organisation unique among nidoviruses. *Archives of Virology* 147: 1977–1987.

Dhar AK, Cowley JA, Hasson KW, and Walker PJ (2004) Genomic organization, biology and diagnosis of Taura syndrome virus (TSV) and yellowhead virus (YHV) of penaeid shrimp. *Advances in Virus Research* 63: 353–421.

Gorbalenya A, Enjuanes L, Ziebuhr J, and Snijder EJ (2006) *Nidovirales*: Evolving the largest RNA virus genome. *Virus Research* 117: 17–37.

Jitrapakdee S, Unajak S, Sittidilokratna N, *et al.* (2003) Identification and analysis of gp116 and gp64 structural glycoproteins of yellow head nidovirus of *Penaeus monodon* shrimp. *Journal of General Virology* 84: 863–873.

Sittidilokratna N, Hodgson RAJ, Cowley JA, *et al.* (2002) Complete ORF1b-gene sequence indicates yellow head virus is an invertebrate nidovirus. *Diseases of Aquatic Organisms* 50: 87–93.

Sittidilokratna N, Phetchampai N, Boonsaeng V, and Walker PJ (2006) Structural and antigenic analysis of the yellow head virus nucleocapsid protein p20. *Virus Research* 116: 21–29.

Walker PJ, Bonami JR, Boonsaeng V, *et al.* (2005) Roniviridae. In: Fauquet CM, Mayo MA, Maniloff J, Desselberger U, and Ball LA (eds.) *Virus Taxonomy: Eighth Report of the International Committee on Taxonomy of Viruses*, pp. 973–977. San Diego, CA: Elsevier Academic Press.

Walker PJ, Cowley JA, Spann KM, *et al.* (2001) Yellow head complex viruses: Transmission cycles and topographical distribution in the Asia-Pacific region. In: Browdy CL and Jory DE (eds.) *The New Wave: Proceedings of the Special Session on Sustainable Shrimp Culture, Aquaculture*, pp. 292–302. Baton Rouge: World Aquaculture Society.

Ziebuhr J, Bayer S, Cowley JA, and Gorbalenya AE (2003) The 3C-like proteinase of an invertebrate nidovirus links coronavirus and potyvirus homologs. *Journal of Virology* 77: 1415–1426.

Zoonoses

J E Osorio and T M Yuill, University of Wisconsin, Madison, WI, USA

© 2008 Elsevier Ltd. All rights reserved.

Introduction

Zoonoses are diseases transmissible from vertebrate animals, other than humans, to people. Mammals, birds, reptiles, and probably amphibians are reservoirs or amplifier hosts for these viral zoonoses. Frequently, these viruses cause little or no overt disease in their nonhuman vertebrate hosts. Some zoonotic viruses have very limited host ranges; others may infect a wide range of vertebrates. Human infection may vary from unapparent to fatal disease. Both new and old viral zoonoses are especially important in emerging and re-emerging virus diseases. Transmission of zoonotic viruses may occur by a variety of routes. They include: 'direct' (rabies) or 'indirect' (hantavirus) contact; 'nosocomial' (arenavirus and filovirus); 'aerosol transmission' (SARS coronavirus); 'vertical' (*in utero*) (arenaviruses); and 'vector- or arthropod-borne' (yellow fever, YF). Viral zoonotic diseases occur on every continent except, perhaps Antarctica. Some are found around the world, in a variety of ecological settings. Others are found only in very limited ecologic and geographic foci. Although hundreds of viruses are zoonotic, the importance of many of these viruses has not yet been established. Some of the more important viral zoonoses will be discussed briefly.

Rabies Virus

Rabies is one of the oldest reported zoonoses. Rabies virus infection causes nervous system disease that ends in death. Animals can become infected without nervous system disease, develop antibodies, and survive, but play no role in transmission. Classical rabies is found all around the world except in Antarctica, Britain, the Hawaiian Islands, Australia, and New Zealand. Transmission occurs by the bite of an infected animal. Aerosol (droplet) transmission is rare. Dogs and cats are the main reservoirs in tropical developing countries where more than 99% of all human cases occur. In industrialized countries, wild mammals are the main reservoirs and the species involved vary from region to region. The principal species are as follows: in North America, skunks, raccoons, and foxes; in Europe, foxes; and in the Caribbean, mongooses. Bats in all enzootic regions harbor rabies with vampire bats especially important in the Neotropics, where they transmit rabies to cattle, horses, and other domestic animals, and, occasionally, to humans. Rabies virus is classified in the genus *Lyssavirus* of the family *Rhabdoviridae*. Genetic relationships between rabies isolates from different species and geographic areas have been established by genomic sequence analysis (**Table 1**).

Diagnosis is based on characteristic altered behavior of infected mammals, confirmed by either isolation of virus; demonstration of intracellular antigen by immunofluorescence; or of virus genomic sequences. Postexposure treatment is accomplished by thorough washing of the bite wound, administration of hyperimmune serum or globulin, and administration of antirabies vaccine. Dogs and cats in enzootic areas should be vaccinated. Other domestic animals and humans at high risk should also be vaccinated. Vaccination campaigns of free-ranging red fox populations in Europe and raccoons and coyotes in the USA have been carried out by oral administration of recombinant vaccinia-vectored vaccines in bait.

Hantavirus Hemorrhagic Fevers and Pulmonary Syndrome Viruses

Hantaviruses belong to the genus *Hantavirus* of the family *Bunyaviridae*. In the Americas, hantavirus can cause hantavirus pulmonary syndrome (HPS), an infectious disease typically characterized by fever, myalgia, and headache and followed by dyspnea, noncardiogenic pulmonary edema, hypotension, and shock. HPS has also been reported and confirmed in seven countries in South America: Argentina, Bolivia, Brazil, Chile, Paraguay, Uruguay, and Panama (**Table 2**). Hantaviruses are harbored by wild rodents which often live in close association with humans. Virus

Table 1 Rabies and related lyssaviruses

Virus name	Lyssavirus genotype	Location	Host
Rabies	1	Worldwide	Many wild and domestic mammals
Lagos bat	2	Africa	Bats, water mongoose (but no human disease)
Mokola	3	Africa	Several terrestrial mammals
Duvenhage	4	Africa	Bats
European bat-1	5	Europe	Bats
European bat-2	6	Europe	Bats
Australian bat	7	Australia	Bats
Aravan	New, proposed	Kyrgystan	Bats
Khujand	New, proposed	Tajikistan	Bats
Irkut	New, proposed	Russia	Bats
West Caucasian bat	New, proposed	Russia	Bats

Table 2 Hantaviruses that cause human disease

Virus name	Distribution	Rodent host	Human disease
Hantaan	Asia, Europe	*Apodemus agrarius*	Hemorrhagic fever with renal syndrome
Seoul	Worldwide	*Rattus* spp.	Hemorrhagic fever with renal syndrome
Dobrava-Belgrade	Europe, Middle East	*Apodemus flavicollis*	Hemorrhagic fever with renal syndrome
Puumala	Europe, Asia	*Clethrionomys glareolus*	Hemorrhagic fever with renal syndrome
Sin Nombre	North America	*Peromyscus maniculatus*	Pulmonary syndrome
New York	North America	*Peromyscus leucopus*	Pulmonary syndrome
Bayou	North America	*Oryzomys palustris*	Pulmonary syndrome
Black Creek Canal	North America	*Sigmodon hispidus*	Pulmonary syndrome
Andes	South America	*Oligoryzomys longicaudatus*	Pulmonary syndrome
Hu39694	South America	Unknown	Pulmonary syndrome
Juquitiba	South America	Unknown	Pulmonary syndrome
Laguna Negra	South America	*Calomys laucha*	Pulmonary syndrome
Lechiguanas	South America	*Calomys laucha*	Pulmonary sybdrome
Oran	South America	*Oligoryzomys longicaudatus*	Pulmonary syndrome
Choclo	Panama	*Zygodontomys brevicauda*	Pulmonary syndrome
Monogahela	North America	*Peromyscus maniculatus nubiterrae*	Pulmonary syndrome
Bermejo	South America	*Oligoryzomys chacoensis*	Pulmonary syndrome
Central Plata	South America	*Oligoryzomys flavescens*	Pulmonary syndrome
Araraquara	South America	*Bolomys lasiurus*	Pulmonary syndrome

is shed in urine and other excreta. Outbreaks of HPS have been associated with ecological changes and invasion of human habitations by expanding rodent populations. Diagnosis has been complicated by the lack of efficient and sensitive isolation and serological methods. Rodent control and avoidance of exposure to rodent excreta, especially in dust, are the only methods available currently for prevention of transmission to humans.

Arenavirus Hemorrhagic Fever Viruses

Arenaviruses are transmitted by the same kind of rodents that carry hantaviruses. They can also cause hemorrhagic fevers and, even though the prototype of this family has been long known (lymphocytic choriomeningitis virus, LCMV), viruses within this family are still being discovered. They produce human diseases in the Old World (Lassa fever in Africa) and New World (Junin, Machupo, and, later on, Guanarito and Sabia in South America). There are about 22 different arenaviruses in the Americas, but only four are associated with significant human disease. These pathogenic arenaviruses establish persistent infection in their rodent hosts, and the virus is shed in urine, infecting humans who live in close contact with these contaminated environments. Lassa fever is also transmitted nosocomially in rural hospitals to other people in contact with blood from viremic patients. Control of these diseases is attempted mainly by reduction of rodent populations. A live, attenuated vaccine has been developed for Argentine hemorrhagic fever, and a vaccinia-vectored vaccine has been developed for Lassa fever. Ribavirin is effective for treating arenavirus infection if administered early in the course of infection.

YF Virus

YF is a flavivirus that causes hemorrhagic disease with severe liver damage and death in up to half of the most acute cases. Humans or primates transport the virus from its sylvan cycle in forested areas to rural or urban areas, where other vector mosquitoes transmit it. YF remains a disease of significant public health importance, with an estimated 200 000 cases and 30 000 deaths annually. The disease is endemic in tropical regions of Africa and South America; nearly 90% of YF cases and deaths occur in Africa. It is a significant hazard to unvaccinated travelers to these endemic areas. Re-establishment of the major urban vector, *Aedes aegypti*, the recent spread of the Asian tiger mosquito (*Ae. albopictus*) as well as the rise in air travel has increased the risk of introduction and spread of the disease. YF is an acute infectious disease characterized by sudden onset with a two-phase development, separated by a short period of remission. The clinical spectrum of YF varies from very mild, nonspecific, febrile illness to a fulminating, sometimes fatal disease with pathognomic features. In severe cases, jaundice, bleeding diathesis, with hepatorenal involvement is common. There is no specific treatment for YF, making the management of YF patients extremely problematic. YF can be diagnosed by virus isolation, detection of circulating antigens, demonstration of a significant rise in specific YF virus antibodies, and microscopic detection of viral inclusion bodies or antigen in tissues taken at postmortem examination. Insecticide spraying and elimination of breeding sites in homes can be used for vector control in epidemic situations. Disease can be prevented in humans by vaccination.

West Nile Virus

West Nile virus (WNV) is an arthropod-borne RNA flavivirus that causes a mild infection to acute febrile disease with rash, and occasional encephalitis (mainly in the elderly) is produced in humans. The virus is widely spread, occurring from India and Pakistan westward through the Middle East and into Africa, and northward into Europe, the republics of the former USSR, and in the Western Hemisphere. WNV first appeared in North America in 1999 with an outbreak in New York City producing high mortality in crows and other birds. Subsequently, the virus has spread rapidly throughout North America, the Caribbean, Mexico, and into South America. WNV is now present in every state except Hawaii, and Alaska. It is believed that the NY99 virus has remained stable and unchanged. The ecology of WNV involves maintenance in a bird–mosquito–bird cycle and occasional infection of humans and horses. *Culex* spp. are the main vectors, and birds are the vertebrate hosts. There is no vaccine currently available for humans, although there is one for equine animals.

Chikungunya Virus

Chikungunya (CHIK) is an alphavirus of the family *Togaviridae* that has been responsible for acute febrile disease with rash and severe arthralgia in people in Africa and Asia. An outbreak of CHIK was reported on Reunion Island in March 2005 that resulted in >3500 confirmed cases and an estimated 250 000 suspected cases, affecting >25% of the island's inhabitants. CHIK virus is maintained in sylvan or savanna cycles involving wild primates and arboreal *Aedes* mosquitoes. In both Africa and Asia, the virus also has an urban cycle involving humans and *Ae. aegypti* mosquitoes that is more important from a public health standpoint.

Sindbis Virus

Sindbis (SIN) virus is one of the most widely distributed arthropod-borne viruses in the world, being found in Africa, Europe, Asia, and Australia. Disease in humans is usually mild, and is characterized by acute fever, with arthralgia, myalgia, and rash. There are periodic epidemics in Finland, where it is termed Podosta disease. SIN virus is maintained in wild bird populations, with transmission by *Culex* spp. mosquitoes. In Africa and the Middle East, SIN is often found in the same ecosystems where WN virus is being transmitted. The virus is an alphavirus of the *Togaviridae*. Phylogenetic analysis indicates that there is one major genetic cluster of western SIN virus strains in Africa and another in Australia and Asia. There is evidence of some geographic mixing of western strains of SIN virus that suggest long-distance transport via migrating birds. There is no vaccine available. Since many of the mosquito vectors breed in extensive rice fields, large-scale control would be expensive.

Crimean-Congo Hemorrhagic Fever Virus

Crimean-Congo hemorrhagic fever (CCHF) virus is very widely distributed, and is found from eastern Europe and the Crimean, eastward through the Middle East to western China, and southward through Africa to South Africa. CCHF is characterized by severe hemorrhagic fever with hepatitis, with case mortality of 10–50%. Maintenance of CCHF virus involves horizontal transmission from *Hyalomma* ticks to mammals, and vertical transmission in ticks through the eggs. *Hyalomma* ticks have also been found on birds migrating between Europe and Africa – a mechanism for long-distance dispersal of the virus. Human CCHF cases have occurred in workers handling livestock and their products in Saudi Arabia and the United Arab Emirates which have been attributed to importation of infected cattle and their ticks from Somalia and the Sudan. There are no vaccine or tick control measures available. CCHF virus belongs to the genus *Nairovirus* of the *Bunyaviridae*. Genetic analysis indicates that reassortment and recombination occur in nature.

Sandfly Fever Viruses

Sandfly fever (Sicilian, Naples, and Toscana) viruses are endemic in the Mediterranean area. They cause acute febrile disease in humans, with occasional aseptic meningitis. In central Italy, Toscana virus (TOSV) caused one-third of previously undiagnosed cases of aseptic meningitis. There are at least two genetic lineages of TOSV – Spanish and Italian. The viruses are members of the genus *Phlebovirus* of the *Bunyaviridae*. They are transovarially and horizontally transmitted by phlebotomine sandflies. Wild mammals are presumed reservoirs.

Viruses Occurring in the Americas

Encephalitis Viruses

Venezuelan equine encephalitis (VEE) viruses are made up of a closely related complex of subtypes with several varieties, which have differing epidemiology, geographic distributions, and disease importance. The epizootic/epidemic (VEE, IAB, and IC) virus variants are of greatest concern. In equine animals, the virus causes acute encephalitis, and case fatality may approach 80%. Survivors may have serious neurological deficits. Although the case–fatality rate in humans is low (less than 1%), the large numbers of acutely infected people that occur during an epidemic may completely overwhelm the local healthcare system. VEE, IAB, and IC viruses are maintained in northern South America, where they have periodically swept through Venezuela and Colombia in epidemic waves, with occasional extensions into Ecuador and massively through Central America into Mexico and South Texas. Epidemic spread depends on the availability of susceptible equine populations (the amplifying host) and abundant mosquito vectors of several species. The interepidemic maintenance systems remain undefined. There is evidence that the epizootic strains may arise by mutation of subtype ID enzootic virus. The enzootic strains are maintained in limited foci involving rodents and *Culex* (*Melanconion*) spp. mosquitoes from Florida to Argentina. With the exception of subtype IE, which has caused epizootics in horses in Mexico, these enzootic virus strains do not cause disease in equine animals, but can cause acute febrile illness in humans. The VEE complex viruses are in the genus *Alphavirus* of the *Togaviridae*. There is an effective live, attenuated vaccine for both human and equine use. Because the maintenance of equine herd immunity is costly, most animal health agencies do not carry out ongoing, intensive vaccination campaigns. Thus, the risk of reoccurrence of explosive outbreaks remains.

Eastern (EEE) and western (WEE) equine encephalitis viruses occur in epidemic form in North America, but have also been found in Central and South America. Generally, EEE is maintained in eastern North America but has caused scattered epizootics and cases in the Caribbean, and in Central and South America. EEE virus can be divided into a North American-Caribbean clade, an Amazon Basin clade, and a Trinidad, Venezuela, Guyana, Ecuador, and Argentina clade. During the past several years, there have been modest increases in several US states in the number of reported human cases of EEE. In North America, WEE occurs in western and prairie states and provinces and along the west coast. WEE has caused sporadic cases of encephalitis in equine animals, but not humans, in Argentina and Uruguay. Both involve wild birds and mosquito vectors, with spillover into equine population and humans, causing clinical encephalitis and death. Central nervous sequelae may occur among survivors. Effective vaccines are available commercially for equine animals, and experimental vaccines are used for laboratory personnel. Effective mosquito abatement to control vector populations has been carried out in the West for many years. Insecticide application is used for vector control in epidemic situations.

St. Louis encephalitis (SLE) virus occurs from Canada to Argentina and causes sporadic but extensive epidemics in the USA, with most epidemics occurring in the West, down the Ohio and Mississippi valleys into Texas, and in Florida. Wild passerine birds are amplifying hosts in North America, but in the Southeastern USA and the Neotropics mammals may play an epidemiologic role in virus maintenance and transmission. SLE virus is transmitted by *Culex* spp. mosquitoes in the USA. SLE virus is a flavivirus of the *Togaviridae*, and is closely related to Japanese encephalitis virus. In humans, SLE is characterized by febrile disease, with subsequent encephalitis or aseptic meningitis, and strikes older people more often than the young. Since no vaccine is available, SLE prevention and control relies on surveillance, vector control, and screening of dwelling windows and doors.

Powassan (POW) virus is a North American member of the flavivirus tick-borne encephalitis (TBE) complex. Although POW virus is widely distributed across the USA and Canada, and westward into far eastern Russia, disease (febrile, with encephalitis) has only been detected in the eastern states and provinces of North America. The transmission cycle involves small mammals and *Ixodes* ticks.

La Crosse (LAC) and other California serogroup encephalides are human pathogens in North America. Prior to the arrival of WNV, LACV was the most important cause of endemic arboviral encephalitis in the US, causing an estimated average of 80 cases per year, affecting mainly preschool-aged children. It is endemic in the Upper Midwest, but occasional cases occur elsewhere. Although fatality is uncommon, the disease is severe enough to cause prolonged hospitalization. LAC virus is maintained transovarially in treehole breeding by *Ochlerotatus* (formerly *Aedes*) *triseriatus* mosquitoes with horizontal transmission to small forest mammal reservoirs and to humans. The other California group viruses affecting people have similar epidemiologies, but do not cause disease as commonly. California encephalitis virus was isolated in California, and has occasionally caused human disease there. Snowshoe hare (SSH) virus occurs in the Northern USA and across Canada, and has caused human encephalitis in the eastern provinces. Jamestown Canyon (JC) virus is widely distributed across the USA, and has been shown to cause human disease, mainly in adults in the Midwest, and to infect deer. Like LAC virus, these other viruses have the same close epidemiological relationship with their *Ochlerotatus* vectors. These viruses are members of the California serogroup of the genus *Bunyavirus* of the family *Bunyaviridae*. SSH virus is an antigenic variant of LAC virus.

Colorado Tick Fever Virus

Colorado tick fever (CTF) is endemic in sagebrush–pine–juniper habitats of the higher elevations (over 1200 m) in the mountains of the western states and provinces of North America. Although seldom fatal, CTF can cause serious disease in humans (fever, chills, headache, retro-orbital pain, photophobia, myalgia, abdominal pain, and generalized malaise) with prolonged convalescence. CTF may present as hemorrhagic or central nervous system disease, and is most severe in preadolescent children. Males are infected over twice as frequently as are females. The virus is transmitted by and overwinters in *Dermacentor andersoni*. Wild rodents are the vertebrate hosts, and develop a prolonged viremia. CTF virus is classified in the genus *Coltivirus* of the family *Reoviridae* and is serologically related to Eyach virus from Germany. Avoidance of tick bites is the main preventive measure available, but control of rodents and the ticks that inhabit their burrows can be applied in foci of virus maintenance in the field.

Vesicular Stomatitis Virus

Vesicular stomatitis (VS) virus is endemic in Central and northern South America and in the southeastern USA, causing an acute, febrile vesicular disease in cattle, horses, and pigs. Sporadic VS epidemics occur in the southwestern states of the USA. Both of the major serotypes, VS-Indiana and VS-New Jersey, cause influenza-like illness in humans and are an occupational hazard to people handling cattle. The VS viruses comprise a complex of related serotypes and subtypes in the Americas, with related vesiculoviruses (family *Rhabdoviridae*) in Africa and Asia. Many of these viruses are transmitted horizontally and transovarially by phlebotomine sandflies, with evidence for infection of wild rodents and other small mammals. However, the role of these mammals in the epidemiology of VS viruses is unclear because they do not develop viremia. Grasshoppers have been shown to be susceptible experimentally, but their role as reservoirs or amplifiers in nature remains to be established.

Other Neotropical Viruses

Oropouche virus, a Simbu serogroup bunyavirus, causes epidemics, occasionally severe, or acute febrile disease with arthralgia and occasional aseptic meningitis in humans in the Brazilian and Peruvian Amazon as well as Surinam and Panama. During rainy season epidemics, the virus is transmitted by *Culicoides paraensis* biting midges. Enzootic maintenance cycles are believed to involve forest mammals and arboreal mosquitoes.

Mayaro (MAY) virus occurs epidemically in the Brazilian and Bolivian Amazon Basin, and has also been associated with human febrile disease in Surinam and Trinidad. In humans, the acute, nonfatal, febrile disease with rash is clinically similar to CHIK, an alphavirus to which it is antigenically and taxonomically related. MAY virus appears to be maintained in nature in a cycle similar to that of YF, with arboreal mosquito vectors and primate hosts, but also involving other mammals and birds.

Una virus is a close relative of MAY virus and causes human febrile disease also, but its natural history is not known. Una virus has been isolated from several mosquito species, and has been found at scattered sites from northern South America to Argentina. Antibodies have been found in humans, horses, and birds. Genetic analysis suggests that Una virus is maintained in discrete foci.

Rocio virus was first isolated from fatal human encephalitis cases during an explosive outbreak of acute febrile disease in coastal Sao Paulo State, Brazil in 1975, after which sporadic outbreaks have continued. This virus is an ungrouped flavivirus in the *Togaviridae* and is serologically related to Murray Valley encephalitis virus from Australia. The epidemiology is unclear but probably involves wild birds, and several mosquito species are suspected vectors.

Cowpox-Like Viruses

Cantagalo and related viruses are orthopoxviruses newly reported in Brazil. They can cause vesiculopustular lesions on the hands, arms, forearms, and face of dairy milkers.

Virus particles can be detected by either direct electron microscopy (DEM) in vesicular fluids and scab specimens or isolated in cell culture and embryonated chicken eggs. The epidemiological significance of these new vaccinia viral strains and their origins remains unknown.

Viruses Occurring in Europe

Tahyna (TAH) virus is widely distributed in Europe and has been reported in Africa. TAH virus produces an influenza-like febrile disease, with occasional central nervous system involvement. The virus is a bunyavirus of the California serogroup, in the *Bunyaviridae*. Like LAC virus, small forest mammals are TAH virus reservoirs, and the virus is horizontally and transovarially transmitted by *Ochlerotatus* mosquitoes. There are no effective control measures.

Omsk hemorrhagic fever occurs in a localized area of western Siberia. Disease can be severe, with up to 3% case fatality, and sequelae are common. This virus is a member of the TBE complex of the flaviviruses. The virus is epizootic in wild muskrats, which had been introduced into the area, and is associated with ixodid ticks. Muskrat handlers are at highest risk of infection. Water voles and other rodents are also vertebrate hosts of the virus. TBE virus vaccine is used in high-risk individuals to provide protection.

TBE virus is also a member of the TBE complex of flaviviruses. TBE virus has been classified into three subtypes: European, far eastern, and Siberian. Because recreation in wooded areas has increased in recent years, TBE has become the most frequent arthropod-borne disease in Europe. The virus occurs in deciduous forests in Western Europe from the Mediterranean countries, westward to France, northward to the Scandinavian countries, and eastward to Siberia. It is maintained in a transmission cycle involving small mammals and *Ixodes* spp. ticks. Human infection also occurs through the consumption of unpasteurized milk from infected cows and goats. Infection can be prevented by an inactivated vaccine and avoidance of tick bites.

Cowpox virus is an orthopoxvirus in the *Poxviridae*. It has a wide host range. Domestic cats are the most important source of human infection, transmitting the virus from wild rodent reservoirs to people. In addition to cattle, this virus has produced severe, generalized infections in a variety of incidental animal hosts in zoos and circuses, including elephants and large cats, which may die. Humans develop typical poxvirus lesions (vesicle and pustule formation), usually on the hands. Laboratory diagnosis (characterization of isolated virus) is required to differentiate cowpox from other nodule-forming zoonotic poxviruses such as orf virus, bovine papular stomatitis virus, and pseudocowpox virus, which are worldwide in distribution.

Viruses Occurring in Africa and the Middle East

Rift Valley Fever Virus

Rift Valley fever virus (RVF) is among the most serious arbovirus infections in Africa today. Repeated RVF epidemics in sub-Saharan Africa cause serious disease in small ruminant animals and humans. RVF disease has expanded its historical geographic range in the livestock-raising areas of eastern and southern Africa and into the Middle East (Saudi Arabia and Yemen) over the past 25 years, causing massive epidemics in Egypt, along the Mauritania–Senegal border and in Madagascar. A major outbreak in East Africa began in 2006 in northeastern Kenya, and spread into southern Somalia and Tanzania. Cattle, sheep, and humans are affected. Abortion storms with febrile disease and bloody diarrhea occur in ruminant animals, and mortality may be heavy in young stock. Most infected humans develop febrile disease, with prolonged convalescence. A few individuals develop more severe disease, with liver necrosis, hemorrhagic pneumonia, meningoencephalitis, and retinitis with vision loss. The human case–fatality rate is less than 1%. RVF virus is in the genus *Phlebovirus* of the *Bunyaviridae*. In sub-Saharan Africa, RVF virus is closely tied to its *Aedes* mosquito vectors. RVF vectors transmit the virus transovarially and horizontally. The virus persists in mosquito eggs laid around seasonally flooded pools and depressions. When these pools flood, the eggs hatch and infected mosquitoes emerge and begin transmission. The vertebrate reservoir hosts of RVF virus are unknown. Field and laboratory workers need to exercise caution to avoid becoming infected by exposure to the virus during postmortem examination of animals or processing materials in the laboratory. Both live, attenuated and inactivated vaccines are available for animals, but the unpredictability of scattered, sporadic RVF outbreaks across sub-Saharan Africa is a major obstacle for implementation of extensive, cost-effective vaccination programs.

Marburg and Ebola Viruses

The reappearance of epidemic Ebola disease in Kikwit, Democratic Republic of the Congo (formerly Zaire) in 1995 and Makokou, Gabon in 1996 again focused international attention on this hemorrhagic disease. Marburg and Ebola viruses have sporadically caused severe hemorrhagic fever in humans. Marburg virus, although of African origin, first appeared in laboratory workers in Germany who had handled cell cultures originating from African primates. Later, epidemics of severe hemorrhagic fever occurred in the Sudan and in Zaire, and Ebola virus was isolated. The first nonlaboratory epidemic of Marburg virus occurred in the Democratic Republic of the Congo from 1998 to 2000. A second, and more severe, outbreak occurred in Uige Province of Angola from 2004 to 2005, when 329 of 374 infected people died (case–fatality rate of 88%). These viruses produce hemorrhagic shock syndrome and visceral organ necrosis, and have the highest case–fatality rate (30–90%) of the hemorrhagic fevers. These viruses, with their bizarre filamentous, pleomorphic morphology, belong to the family *Filoviridae*. They are presumed to be zoonotic, but their hosts in nature and mechanisms of transmission in the field have not been determined. Most of the Makokou, Gabon patients had very recently butchered chimpanzees. A variant of Ebola virus has been isolated from chimpanzees from Côte D'Ivoire, but since wild primates suffer severe disease, they are unlikely to be maintenance reservoirs. Nosocomial transmission of Marburg and Ebola viruses has occurred frequently; a high level of patient isolation and biosafety containment are essential to avoid hospital- and laboratory-acquired infection. Serologic diagnosis is accomplished by means of indirect immunofluorescence or enzyme-linked immunosorbent assay (ELISA) test, with antigen specificity confirmed by western blot. No vaccines or control measures are available.

Monkeypox Virus

Human monkeypox is a severe, smallpox-like illness. Monkeypox belongs to the genus *Orthopoxvirus* of the *Poxviridae*. Monkeypox virus (MPXV) is endemic in rodents in West and Central Africa, with the occurrence of sporadic human cases. The case–fatality rate in humans appears to be higher in Central Africa than in West Africa, raising questions about possible difference in virulence in these two large geographic areas. The largest epidemic of human monkeypox ever documented occurred in the Katako-Kombe area of the Democratic Republic of the Congo (formerly Zaire) in 1996–97, with over 500 people becoming ill and five deaths. Rodent-to-human transmission occurred, as did subsequent secondary human-to-human spread. Vaccinia is protective against infection, but its use has been discontinued with the eradication of smallpox. In 2003, monkeypox emerged for the first time in the Western Hemisphere and caused an outbreak in the United States (Midwestern states) affecting 37 people exposed to ill prairie dogs purchased from pet stores. The virus entered the US upon the importation of exotic rodents from Ghana (West Africa). Recent nucleotide sequence analysis demonstrated the existence of two genetically distinct variants of the virus, called the West African and Congo Basin clades. The strain that caused the US outbreak belonged to the West African clade.

Semliki Forest Virus

Semliki Forest (SF) virus caused an extensive epidemic of human disease in Bangui, Central African Republic, in 1987. SF virus is an alphavirus in the *Togaviridae*. It occurs across East, Central, and West Africa, and has been isolated from various mosquitoes and from wild birds. Antibodies have also been found in wild mammals. The SF virus maintenance cycle probably involves *Ae. africanus* mosquitoes and vervet monkeys.

Orungo Virus

Orungo (ORU) virus caused mild epidemic disease (fever, nausea, headache, and rash) in Nigeria. The virus occurs in a band across Africa from Uganda to Sierra Leone. It is probably mosquito transmitted, but the species that transmit it in nature are not known. Although the vertebrate reservoir hosts are unknown, wild primates have antibody and are suspected to be involved in virus maintenance.

Alkhurma Virus

Alkhurma virus (a variant of Kysanur Forest disease virus, family *Flaviviridae*, genus *Flavivirus*) is an emerging pathogen responsible for hemorrhagic fever in the Middle East. This virus was isolated from hemorrhagic fever patients in Saudi Arabia in 1995. Transmission can occur from tick bites, handling carcasses of infected animals, or drinking unpasteurized milk. The case–fatality rate is 25%.

Viruses Occurring in Asia

Influenza Viruses

Influenza viruses belong to the family *Orthomyxoviridae*, which consists of five genera: influenza A, influenza B, influenza C, Isavirus, and Thogoto viruses. Influenza A viruses are widely distributed in nature and can infect a wide variety of birds and mammals. Influenza A virus subtypes are classified on the basis of the antigenicity of their surface glycoproteins hemagglutinin (HA) and neuraminidase (NA). To date, 16 HA and 9 NA genes are known to exist. Of these genes, only three HA (H1, H2, and H3) and two NA (N1 and N2) subtypes have circulated in the human population in the twentieth century. During the last 100 years, the most catastrophic impact of influenza was the pandemic of 1918, also known as the Spanish flu (H1N1), which resulted in the loss of more than 500 000 lives in the United States and caused about 40 million deaths worldwide.

The ability of influenza viruses to undergo antigenic changes is the cause of ongoing significant public health concern. New subtypes emerge when human virus captures genes from animal influenza viruses via reassortment; an event that can occur when both virus types simultaneously infect a host (antigenic shift). The threat imposed by influenza virus has been further elevated with the recent introductions of avian influenza viruses into the human population. Avian influenza viruses were initially considered nonpathogenic for humans. However, this perception has changed since 1997, when 18 Hong Kong residents were infected by an avian influenza virus of the H5N1 subtype that resulted in six deaths. Over the next few years, several other cases of direct avian-to-human transmission were reported, including the ongoing outbreak of highly pathogenic H5N1 influenza viruses in several Asian, African, and European countries. Migratory waterfowl – most notably wild ducks – are the natural reservoir of avian influenza viruses, and these birds are also the most resistant to infection. Domestic poultry, including chickens and turkeys, are particularly susceptible to epidemics of rapidly fatal influenza. Direct or indirect contact of domestic flocks with wild migratory waterfowl has been implicated as a frequent cause of epidemics. Live bird markets have also played an important role in the spread of epidemics. Viruses of low pathogenicity can, after circulation for sometimes short periods in a poultry population, mutate into highly pathogenic viruses. Quarantine of infected farms and destruction of infected or potentially exposed flocks are standard control measures aimed at preventing spread to other farms and eventual establishment of the virus in a country's poultry population.

Severe Acute Respiratory Syndrome

In February 2003, a new and previously unknown disease, severe acute respiratory syndrome (SARS), was reported to the World Health Organization (WHO). SARS originated in the province of Guangdong in southern China in November 2002 where it initially was thought to cause atypical pneumonia. However, within a short time the virus spread to Hong Kong, Singapore, Vietnam, Canada, the United States, Taiwan, and several European countries. A novel coronavirus (CoV) was identified as the etiological agent. The SARS-CoV affected more than 8000 individuals worldwide and was responsible for over 700 deaths during the first outbreak in 2002–03. For reasons unknown the SARS virus is less severe and the clinical progression a great deal milder in children younger than 12 years of age. In contrast, the mortality rate was highest among patients >65 years and can exceed 50% for persons at or above the age of 60 years. Farmed masked palm civets (*Paguma larvata*) and two other mammals in live animal markets in China were sources of SARS-CoV human infection. Three species of horseshoe bats (*Rhinolophus* spp.) are probable wildlife reservoirs in China.

Kyasanur Forest Disease

Kyasanur Forest disease (KFD) was first recognized in India in 1957, when an acute hemorrhagic disease appeared in wild monkeys and people frequenting forested areas. KFD has been slowly spreading in India. Human cases have increased from 1999 to 2005, with peak incidence in January and February. The cause of this increase is unknown. KFD virus is a member of the TBE complex of flaviviruses. The basic virus maintenance cycle involves forest mammals (primates, rodents, bats, and insectivores) and ixodid ticks, mainly *Haemaphysalis spinigera*. The virus can be isolated in mice and cell cultures, including tick cells. An inactivated vaccine provides some protection to people at risk of infection.

Japanese Encephalitis Virus

Japanese encephalitis (JE) virus is found in a broad area from far eastern Russia, northeastern Asia through China and Southeast Asia to Papua New Guinea and the Torres Strait Islands of Australia and westward into India. JE virus causes the greatest number of clinical human cases, thousands annually, predominantly in children. It produces encephalitis in humans and horses, and acute febrile disease with abortion in swine, an amplifying host. Herons and egrets are wildlife amplifying hosts. The virus is transmitted by *Culex* spp. mosquitoes. The over-wintering mechanism in temperate Asia is unknown. JE virus is a member of a complex of four related flaviviruses in the family *Flaviviridae*. Prevention of disease is mainly through vaccination of humans, horses, and swine. Insecticides and integrated pest control measures that include natural compounds (*Bacillus thurengiensis* toxins), larvicidal fish, and larval habitat modification have been successfully used in China. Use of pyrethroid-impregnated bed netting can also prevent transmission.

Chandipura

Chandipura virus is ubiquitous across the Indian subcontinent. It is a *Vesiculovirus* in the *Rhabdoviridae*. Chandipura has caused epidemics of febrile diseases, sometimes with encephalopathy. An outbreak occurred in 2004 with a case–fatality rate of 78.3% in children in India (Gujarat State). The virus is transmitted by *Phlebotomus* spp. Sergentomine sandflies and infects a variety of mammals. The virus has also been isolated in West Africa.

Nipah

Nipah virus (NiV) was first recognized in peninsular Malaysia in 1998, where it caused encephalitis and respiratory disease in commercially raised pigs, with transmission to humans in contact with them, with 40–76% case fatality. The virus was found in five species of giant fruit bats (*Pteropus* spp.) there and NiV was isolated from partially eaten fruit. Subsequently, there were five NiV outbreaks recognized in Bangladesh, also associated with *Pteropus* bats. Transmission in Bangladesh was directly from bats, via contaminated fruit and date palm sap.

Viruses Occurring in Australia

Murray Valley Encephalitis Virus and Kunjin Virus

Murray Valley encephalitis (MVE) virus and the closely related Kunjin virus are flaviviruses that cause encephalitis, although Kunjin virus more commonly produces a nonencephalitic illness with polyarthralgia. MVE virus is endemic in avian species and is found in humans in northern Western Australia, the Northern Territory, and Queensland. MVE virus is endemic in northern areas of Western and Northern Australia, and in New Guinea. Kunjin virus occurs over a much wider area, including most of tropical Australia, Sarawak, Borneo, Papua New Guinea, and Saibai Island in the Torres Strait. There is some evidence that infection with these viruses is increasing in incidence. In northern Australia, MVE cases occur predominantly between February and July, corresponding to the end of the monsoon season, when the mosquito vector (*Culex annulirostris*) proliferates in flooded environments. MVE and Kunjin viruses are flaviviruses, family *Flaviviridae*, of the Japanese encephalitis complex. Kunjin is a subtype of WNV. RNA sequencing indicates that the Australian strains of MVE virus are similar to, but different from Papua New Guinea isolates. No vaccine is available. Control is achieved through application of larvicides.

Ross River Virus

Ross River (RR) virus has caused annual epidemics of febrile disease with polyarthritis and rash with most cases occurring in November through April. It is the most commonly reported arthropod-borne virus disease in Australia. RR virus occurs in all Australian states and territories, but is most commonly found in the northern states and coastal areas. Within the past two decades, RR virus has spread through several Pacific Islands in epidemic form and appears to have become endemic in New Caledonia. Convalescence can be long. RR virus is an alphavirus of the family *Togaviridae*. The enzootic maintenance cycles of RR virus in Australia are not well defined, but wild and domestic mammals appear to be the reservoir hosts, and the principal mosquito vectors are salt marsh *Aedes* spp. and freshwater *Culex* spp. In the Pacific Islands outbreaks, the virus was probably transmitted from person to person by *Aedes* mosquitoes.

Barmah Forest Virus

Barmah Forest virus is the second most common mosquito-borne disease in Australia. It causes subclinical and clinical infections in humans, including fever, myalgia, polyarthralgia, and rash. It is an alphavirus in the *Togaviridae*. The virus appears to be endemic in eastern Australia. It has been isolated from 25 different mosquito species in five genera. *Ochlerotatus vigilax* (previously known as *Aedes vigilax*), is considered a major vector. Its vertebrate hosts have not been established, although marsupials are suspected.

Control

On-going disease surveillance supported by rapid, reliable diagnosis is critical for recognition and control of zoonotic diseases. Serological diagnosis by means of ELISA tests, virus neutralization, and other tests, coupled with detection of the virus itself by virus isolation, immunohistochemistry, immunofluoresence, or PCR techniques are standard laboratory approaches, but are not available in all countries. Timely laboratory results must reach clinicians treating infected individuals and become incorporated into epidemiological databases and early warning systems to assure rapid response by public health authorities to control outbreaks appropriately.

Control of zoonotic virus diseases is accomplished by breaking the cycle of transmission. This is usually achieved by eliminating or immunizing vertebrate hosts, and reducing vector populations. Reduction of reservoir host populations is usually not accomplished because it is too expensive, not environmentally safe, and not technically or logistically feasible. However, there have been some notable exceptions. Bolivian hemorrhagic fever, caused by Machupo virus, was controlled by reduction of its rodent hosts through intensive rodenticiding. The principal vampire bat reservoir of rabies, *Desmodus rotundus*, is being controlled by the application of warfarin-type anticoagulants. Control programs like these have to be continuous to be effective. Their reduction or discontinuation results in host population recovery through reproduction and immigration, which may result in re-emergence of disease sweeping through the increasing, susceptible cohort.

Immunization of hosts is another control approach. Safe and effective rabies vaccines are being used for immunization of humans, domesticated animals, and some wildlife species. The human diploid cell vaccine is extremely effective, free of adverse effects, and widely available but at a cost too high for use in many developing countries. Safe, effective animal vaccines of cell culture origin are on the market. After some initial public resistance, raccoon populations in the eastern USA and wild foxes in Europe are being successfully immunized by means of an oral, vaccinia-vectored recombinant vaccine.

This experience illustrates the need for public understanding, in order to counteract fear of dispersal of a genetically engineered virus. However, vaccines will not be developed for many zoonotic viral diseases that affect relatively few people and are of very limited concern geographically.

Vector control is another promising but difficult area of zoonoses reduction or elimination. Insecticide application has become more problematic because both vectors themselves, as well as public opinion, have become more resistant to their use. Integrated pest management techniques, well developed for the control of many crop insects, along with the use of natural pesticides such as *Bacillus thurengiensis* toxin, offers promise for the effective, environmentally safe control of dipterous vectors. Control of tick vectors is likely to remain a problem for some time to come.

Emerging and Re-emerging Zoonoses

Ecological Change

Human disturbance has become a feature of nearly every part of the planet. All too often these disturbances create habitats that favor increases in populations of key hosts and vectors, with subsequent increased transmission of viral zoonoses. Nowhere are ecological changes happening more rapidly and profoundly than in the world's tropics. Conversion of tropical forests to agricultural ecosystems simplifies diverse ecosystems and provides either native or introduced host or vector species the conditions necessary to become more abundant, and sustain intensified virus transmission in areas where people live and work. In Africa, recent YF epidemics have been increasing dramatically in agricultural areas. Some agricultural irrigation development projects have created extensive vector breeding habitats, with an increase in mosquito-transmitted disease. The extensive dams constructed in Senegal were followed by epidemics of RVF, with numerous cases of disease in humans and small ruminant animals. The public health consequences of development projects must never be overlooked.

Global climate change will also bring ecological changes and shifts of human populations that will affect the occurrence of viral zoonoses. There is general consensus that changes in the global climate will happen with unprecedented speed. With those changes will come alterations in the geography of natural and agricultural ecosystems, with corresponding changes in the distribution of zoonotic diseases and the intensity of their transmission. It is clear that El Niño southern oscillation phenomena have increased rainfall, with resulting increases in rodent populations and occurrence of HPS in the Southwestern USA, and increased breeding sites for mosquito vectors of RVF virus in Africa. While it is not possible to predict accurately what the world will be like in 100 years or what zoonotic diseases are likely to be most troublesome, it is certain that things will be different, and constant surveillance will be essential to avoid serious problems or deal promptly and effectively with the ones that arise.

Movement of zoonotic viruses can result from the displacement of infected animals, contaminated animal products, and virus-carrying arthropod vectors. Pets, sport, laboratory, and agricultural animals are moving around the world as never before. Although international and national regulations have been established to prevent movement of infected individuals, it is not possible to test for all possible zoonotic viruses, and prevent them from crossing international boundaries. Moreover, significant numbers of animals of high commercial value move illegally. The importation of highly virulent Newcastle disease (ND) virus has been occasionally linked to smuggled birds.

Zoonotic viruses may be transported by movement of arthropod vectors, too. Just as the YF mosquito, *Ae. aegypti*, moved around the world in water casks aboard sailing vessels, mosquitoes are transported around the world in international commerce. Ships still transport mosquito vectors. The Asian tiger mosquito, *Ae. albopictus*, has become established in the Western Hemisphere after multiple introductions in eggs deposited in used tires. This mosquito is capable of transmitting YF, VEE, JC, and LAC encephalitis viruses. Perhaps of greater concern, modern transport aircraft have been shown to move vector mosquitoes internationally.

Human activity alters animal populations, contact between wild and domestic animals, and human–animal interactions, changing the occurrence of zoonotic diseases and the risk of infection to humans. For example, emergence of new influenza strains is related to the interaction of populations of people, pigs, and aquatic birds.

Social Change

Increasing human populations place great demands on the public health and other government services, especially in developing countries where needs for zoonoses diagnosis, control, and prevention are greatest and resources are most limited. Some preventive measures could be implemented by the people who live in the affected areas themselves, and at minimal costs, if they knew why and how they needed to do it. Public education and information is essential for control and prevention of zoonotic diseases; however, it takes more than civic action to deal with them. Delivery of public education, disease surveillance and diagnosis, and the technical materials and logistical support for control or preventive programs depend on national or international scientific and financial support. International technical cooperation and financial support are imperative.

Summary

Zoonoses are diseases transmissible from animals, other than humans, to people. Both new and old viral zoonoses are important in emerging and re-emerging virus diseases. Some zoonotic viruses occur worldwide, in a variety of ecological settings. Others are found only in limited ecologic and geographic foci. Important worldwide zoonotic viruses include rabies, hantaviruses, arenaviruses, yellow fever virus, chikungunya virus, Sindbis virus, Crimean-Congo hemorrhagic fever virus, and the sandfly fever viruses. In the Americas, common zoonotic viruses include the encephalitis viruses, Colorado tick fever virus, vesicular stomatitis, and others. Zoonotic viruses in Europe include Tahyna virus, the tick-borne encephalitis viruses, and cowpox virus. There are several zoonotic viruses in Africa and the Middle East, including Rift Valley fever virus, Marburg and Ebola filoviruses, monkeypox virus, Semliki Forest virus, Orungo virus, and Alkhurma virus. Zoonotic viruses occurring in Asia include the influenza viruses, SARS coronavirus, Kyasanur Forest virus, Japanese encephalitis virus, Chandipura virus, and Nipah virus. Zoonotic viruses occurring in Australia include Murray Valley encephalitis virus, Kunjin virus, Ross River virus, and Barmah virus. The human and animal health importance of these viruses and their control are discussed. Because rapid ecological change is occurring worldwide, additional zoonotic viruses will emerge.

See also: Bunyaviruses: General Features; Emerging and Reemerging Virus Diseases of Vertebrates; Fish Rhabdoviruses; Hantaviruses; Lassa, Junin, Machupo and Guanarito Viruses; Marburg Virus.

Further Reading

Daszek P, Cunningham AA, and Hyatt AD (2000) Emerging infectious diseases of wildlife – Threats to biodiversity and human health. *Science* 287: 443–449.

Gritsun TT, Nuttall PA, and Gould EA (2003) Tick-borne flaviviruses. *Advances in Virus Research* 61: 317–371.

Karesh WB, Cook RA, and Newcomb J (2005) Wildlife trade and global disease emergence. *Emerging Infectious Diseases* 11: 1000–1002.

World Health Organization. *Avian Influenza: Assessing the Pandemic Threat.* Geneva: WHO/CDS/2005.29.

TAXONOMIC INDEX

A

Adenoviridae **1:1–9, 1:9–16, 1:17–23**, 2:120
Alfamovirus **3:46–56**
Allexivirus **1:96–98**
Alloherpesviridae 2:431
Alphacryptovirus **1:98–104**
Alphaherpesvirinae 1:363, 2:405–406, 2:431, **4:581–585**
Alphavirus **5:96–101**, 5:107
Amdovirus 4:95
Anellovirus **1:104–111**
Aphthovirus **2:265–274**
Aquabirnaviridae 3:83–84
Aquareovirus **1:163–169**
Arteriviridae **1:176–186**
Ascoviridae **1:186–193**
Ascovirus 1:188*t*
Asfarviridae 1:50
Astroviridae **1:204–210**, 2:121
Astrovirus 1:205
Atadenovirus 2:120
Aviadenovirus 1:27
Avsunviroidae 5:335
Avulavirus 4:41*t*

B

Baculoviridae **1:211–219, 1:219–225, 1:247–254, 1:254–265**, 1:265, 3:497–498
Barnaviridae **1:286–288**
Benyvirus **1:308–314**
Betacryptovirus **1:98–104**
Betaherpesvirinae 1:635–636, 2:431
Birnaviridae 1:321
Bocavirus 4:95
Bornaviridae **1:341–347**, 3:324
Bornavirus 1:341
Brevidensovirus 4:83, 4:96
Bromoviridae **1:386–390**, 2:255
Bunyaviridae 1:391, **1:399–401**, 3:479

C

Caliciviridae **1:410–419**, 2:117
Capillovirus 1:420
Capripoxvirus **1:427–432**
Cardiovirus **1:440–448**
Carlavirus **1:448–453**
Carmovirus **1:453–457**
Caudovirales 5:11*t*
Caulimoviridae **1:457–464, 1:464–469**
Caulimovirus 1:464
Chloriridovirus 3:155
Chordopoxvirinae 4:323
Chrysoviridae **1:503–513**
Chrysovirus 1:503–504
Circoviridae **1:513–520**
Closteroviridae 2:255
Coltivirus **1:533–541**, 4:536
Comoviridae 1:569
Coronaviridae **1:549–554, 1:554–563**, 2:122, 3:419–420, 5:151–152
Corticoviridae 3:1
Coxsackievirus **1:580–587**
Cucumovirus **1:614–620**
Curtovirus **1:301–307**
Cystoviridae 3:6
Cytomegalovirus **1:624–634, 1:634–642, 2:474–485**

D

Deltavirus 2:375
Densovirinae 4:76, 4:96
Densovirus 4:96

Dependovirus 4:93
Dicistroviridae **2:37–44**
Dinovernavirus 3:134

E

Ebolavirus 2:57
Endornavirus **2:109–116**
Enterovirus 2:65–66
Entomopoxvirinae **2:136–140**
Erythrovirus 4:95

F

Fijivirus 4:150
Filoviridae **2:198–205**, 3:324
Flaviviridae 2:6, **2:234–241, 2:241–253**
Flavivirus 2:6, 2:242
Flexiviridae **2:253–259**
Foveavirus 1:422
Furovirus **2:291–296**
Fuselloviridae **2:296–300**

G

Gammaherpesvirinae 1:363, 2:431, **4:585–594**
Geminiviridae **2:97–105, 4:164–170**
Giardiavirus **2:312–316**
Granulovirus **1:211–219**, 1:247, 1:265

H

Hantavirus **2:317–321**
Henipavirus **2:321–327**
Hepacivirus 2:249, 2:367
Hepadnaviridae **2:327–335, 2:335–343**
Hepatovirus 2:343
Herpesviridae **1:362–368, 2:205–212, 2:383–397, 2:405–411**, 2:431, **2:498–505**
Hordeivirus **2:459–467**
Hypoviridae 2:288, **2:580–585**

I

Ictalurivirus 2:431
Idaeovirus **3:37–42**
Iflavirus 3:42
Ilarvirus **3:46–56**
Influenzavirus A 3:96
Influenzavirus B 3:96
Influenzavirus C 3:96
Inoviridae 2:191
Inovirus 2:191, 3:117–118
Iridoviridae 1:50, **3:155–161, 3:161–167**
Iridovirus 3:155, 3:167
Isavirus 3:96
Iteravirus 4:83, 4:96

L

Lentivirus 4:594, 4:603
Leporipoxvirus **3:225–231**
Levivirus **1:381–386**
Luteoviridae **3:231–238**
Lymphocystivirus 3:155

M

Machlomovirus **3:259–263**
Maculavirus 5:206
Marafivirus 5:205
Marnaviridae **3:280–285**

Mastadenovirus 1:26–27
Megalocystivirus 3:155
Metapneumovirus 4:41*t*
Metaviridae **3:301–311**
Mimiviridae **3:311–319**
Mimivirus 3:312
Mononegavirales **3:324–334**, 5:11*t*
Morbillivirus **3:285–292**, 4:41*t*
Mycoreovirus **3:378–383**
Myoviridae **3:30–37**

N

Nanoviridae **3:385–391**
Narnaviridae **3:392–398**
Narnavirus 3:392
Necrovirus **3:403–405**
Nepovirus **3:405–413**
Nidovirales **3:419–430**, 5:11*t*
Nodaviridae **3:430–438**
Norovirus 2:117, 3:439
Novirhabdovirus 2:221–222
Nucleopolyhedrovirus 1:247, 1:254, 1:265

O

Ophiovirus 3:447
Orbivirus **3:454–466**
Orthobunyavirus **3:479–483**
Orthomyxoviridae 3:96, **3:483–489**
Orthoreovirus 4:383
Oryzavirus 4:150
Ourmiavirus **3:500–501**

P

Papillomaviridae **4:8–18, 4:26–34, 4:34–40**
Paramyxoviridae 2:321, 3:285, 3:324, **4:40–47, 4:52–57**
Paramyxovirinae 2:321
Parapoxvirus **4:57–63**
Parechovirus 2:122
Partitiviridae **4:63–68, 4:68–76**
Partitivirus 4:63, 4:68–69
Parvoviridae **4:76–85, 4:85–90, 4:90–97**
Parvovirinae **4:90–97**
Parvovirus 4:92
Pecluvirus **4:97–103**
Pefudensovirus 4:96
Pestivirus 2:246
Phlebovirus 4:490–491
Phycodnaviridae **4:116–125**
Phytoreovirus 4:150
Picobirnaviridae 1:327
Picornaviridae 1:542, 2:122, **4:129–140**, 4:243
Pneumovirus **1:614–620**, 4:41*t*
Podoviridae 3:31
Polerovirus **3:231–238**
Poliovirus 4:243
Polydnaviridae **4:256–261**
Polyomaviridae 4:34–35, 4:271, 4:277
Pomovirus **4:282–287**
Pospiviroidae 5:335
Potexvirus **4:310–313**
Potyviridae **4:313–322**
Poxviridae 4:323
Pseudoviridae **4:352–357**

R

Ranavirus 3:155
Reoviridae 3:133–134, **4:149–156, 4:382–390**
Respirovirus 4:41*t*, 4:47
Retroviridae **2:105–109**, 4:446, **4:459–467**

Rhabdoviridae **1**:111–121, **2**:221–227, 2:228, 3:324, **4**:187–197, **4**:576–581
Rhadinovirus 3:190
Rhinovirus 1:542, **4**:467–475
Roniviridae 3:419–420
Rotavirus 2:118
Rubivirus 5:107
Rubulavirus 4:19, 4:41*t*

S

Sadwavirus **4**:523–526
Sapovirus 3:439
Seadornavirus **4**:535–546
Sequiviridae **4**:546–551
Sequivirus **4**:546–551
Sobemovirus **4**:644–652
Suipoxvirus **3**:225–231

T

Tectiviridae 3:1
Tenuivirus **5**:24–27
Teschovirus 2:126–127
Tetraviridae **5**:27–37
Thogotovirus 3:96
Tobamovirus **5**:68–72
Tobravirus **5**:72–76
Togaviridae 1:174, 2:255, **5**:76–83, **5**:91–96, **5**:107–116
Tombusviridae 1:453, **5**:145–151
Torovirus **5**:151–157
Tospovirus **5**:157–163
Totiviridae **5**:163–174
Totivirus **5**:163–174
Trichovirus 1:424
Tymoviridae 2:255, **5**:199–207

U

Umbravirus **5**:209–214

V

Varicosavirus **5**:263–268
Vesiculovirus 1:497
Vitivirus 1:425

W

Waikavirus 4:547

Y

Yatapoxvirus **5**:461–465

SUBJECT INDEX

Notes

Cross-reference terms in italics are general cross-references, or refer to subentry terms within the main entry (the main entry is not repeated to save space). Readers are also advised to refer to the end of each article for additional cross-references - not all of these cross-references have been included in the index cross-references.

The index is arranged in set-out style with a maximum of three levels of heading. Major discussion of a subject is indicated by bold page numbers. Page numbers suffixed by T and F refer to Tables and Figures respectively. *vs.* indicates a comparison.

This index is in letter-by -letter order, whereby hyphens and spaces within index headings are ignored in the alphabetization. Prefixes and terms in parentheses are excluded from the initial alphabetization.

To save space in the index the following abbreviations have been used

CJD - Creutzfeldt–Jakob disease

CMV - cytomegalovirus

EBV - Epstein–Barr virus

HCMV - human cytomegalovirus

HHV - human herpesvirus

HIV - human immunodeficiency virus

HPV - human papillomaviruses

HSV - herpes simplex virus

HTLV - human T-cell leukemia viruses

KSHV - Kaposi's sarcoma-associated herpesvirus

RdRp - RNA-dependent RNA polymerase

RNP - ribonucleoprotein

SARS - severe acute respiratory syndrome

TMEV - Theiler's murine encephalomyelitis virus

For consistency within the index, the term "bacteriophage" has been used rather than the term "bacterial virus."

A

A_2 gene (bacteriophage Qβ), single gene-mediated host cell lysis 3:257–258
A151R polyprotein, African swine fever virus replication 1:46–47
A238L protein, African swine fever virus infection 1:47
A/Aichi/68 HA1 (influenza virus), antigenic drift 1:131–132
AalDNV *see* Aedes albopictus densovirus type 2 (AalDNV)
Abaca bunchy top virus (ABTV) 3:389t
Abacavir 2:510
Abiotic transmission, rice yellow mottle virus 4:488
Abomasal, definition 1:427
Abortion, equid herpesvirus 1 infection *see* Equid herpesvirus 1 infection
Abortive infection 4:110–111
Abortive transformation, murine polyomavirus infection 4:273

ABPV *see* Acute bee paralysis virus (ABPV)
A/B toxin(s), K28 virus toxin *vs.* 5:303–304
Abutilon mosaic virus (AbMV), DNA β satellite origin 1:320–321
AbV-1 *see* Agaricus bisporus virus 1 (AbV-1)
Abystoma tigrinum virus (ATV), genome sequencing 2:233
Ac32 gene, granuloviruses 1:217
Acanthamoeba polyphaga mimvirus 3:311–312
Acantholysis, definition 5:250
Acanthoma
  definition 3:319
  molluscum contagiosum virus infection 3:322
Acaricides, Crimean-Congo hemorrhagic fever prevention 1:602–603
Accessory protease(s) (PL^{pro}), nidoviruses, proteolytic processing 3:424
Accidental hosts, cowpox virus 1:577
AC genes, bean golden mosaic virus 1:298
Acherontia atropas virus (AaV) 5:28t
Acheta domesticus densovirus (AdDNV), infection 4:79

Acibenzolar-S-methyl, tomato spotted wilt virus management 5:137, 5:137f
Acidianus bottle-shaped virus (ABV) 1:590, 1:591f
Acidianus filamentous virus(es) (AFV), virion structure 1:589–590
  core 1:590
Acidianus filamentous virus 1 (AFV1) 1:593
Acidianus filamentous virus 2 (AFV2) 1:593
Acidianus two-tailed virus (ATVc)
  genome 1:593
  genome integration 1:591–592
  integrase 1:592–593
  host cell lysis 1:591
  virion morphology 1:587, 1:589f
  tail structure 1:587–589
Acid sphingomyelinase (ASM)
  definition 1:541
  rhinovirus infection prevention 1:548
ACLAME (A Classification of Genetic Mobile Elements) 5:357t
ACLSV *see* Apple chlorotic leaf spot virus (ACLSV)

499

Subject Index

AcMNPV *see* Autographa californica multiple nucleopolyhedrovirus (AcMNPV)
ACMV *see* African cassava mosaic virus (ACMV)
ACMVUG *see* African cassava mosaic virus (ACMV)
Acquired immunodeficiency syndrome (AIDS) *see* AIDS
Acquisition access period (AAP), mungbean yellow mosaic virus transmission 3:368
Acute bee paralysis virus (ABPV)
    host range 2:40
    IRES structure 2:41f
Acute coryza *see* Common cold
Acute flaccid paralysis (AFP) 2:130
Acute hemorrhagic conjunctivitis *see* Conjunctivitis
Acute infections
    cannulation model, Visna-Maedi virus infection 5:426
    persistent *vs.* 4:109
    see also specific viral infections
Acute respiratory disease (ARD) 2:554–555
Acute-T-cell lymphomas, Herpesvirus Ateles 4:586–587
Acyclovir 2:436–437, 2:557
    channel catfish virus treatment 2:211
    chronic suppressive therapy 1:145–146
    combination therapy 1:152
    EBV infection 2:165–166
    history 1:145–146
    HSV infection 2:394–395
        mode of action 2:394–395
        resistance 2:394–395
    Pacheco's disease 2:410
    selectivity 1:146
    varicella-zoster virus infection 5:255
        resistance 5:255
Ad5 E1A proteins 1:10f
Ad9, adenovirus malignant transformation 1:16
Ad12 E1A proteins 1:12–13
AD169 strain, HCMV 1:636, 2:486
Adansonian system, classification 5:19
Adaptation, bacteriophage population ecology 2:75–76
Adaptive immune system 3:60
    definition 1:121
    henipavirus infections 2:326
    hepatitis C 2:373
    persistent infection 4:112
    principal effector molecules 3:60–61
    T lymphocytes 3:78–79
    see also Cell-mediated immune response; *specific components*
Adaptive landscape, definition 4:359
Adaptor proteins, apoptosis 1:232
ADAR1 *see* Adenosine deaminase 1 (ADAR 1)
ADE *see* Antibody-dependent enhancement (ADE)
Adefovir 1:143t
    HAART-associated renal disease 1:56–57
    hepatitis B treatment 2:359, 2:365–366
    resistance development 2:359
Adelaide river virus (ARV) 4:184t
    genome 1:358f, 1:360
        α1 ORF 1:360
        structural protein genes 1:360
    transcription 1:360
        β gene 1:360
        β-L gene junction 1:360
Adeno-associated virus vectors 2:303
    dependovirus replication 4:93
    genome 4:93
        episomes 4:94–95
        recombinant genomes 4:94–95
    replication 2:303, 4:93, 4:94f
    structure 2:303
    see also Recombinant adeno-associated virus (rAAV)
Adenocarcinoma
    Lucke' tumor herpesvirus 2:206
Adenosine deaminase 1 (ADAR 1)
    interferon-induced antiviral response 3:115–116
    murine polyomavirus 4:273
Adenosine kinase, beet curly top virus pathogenesis 1:306

Adenoviridae 1:1, 1:2t, 1:10t, 1:17, 1:24, 2:120, 5:11t
Adenovirus(es) **1:1–9, 1:17–23**
    antigenic properties 1:5
        genus-specific antigens 1:5
        neutralization assays 1:5
    apoptosis inhibition 1:157–160, 1:158t
    biological properties 1:5
    bovine 1:26
    characteristics 2:117t
    co-evolution 1:1
    DNA replication 1:19, 1:21f
        origin of replication 1:20
        proteins associated 1:19–20
    early studies 2:457
    equine 1:26
    fiber genes 1:3, 1:6
        genome organization 1:3
    gene expression 1:19
    gene therapy vectors 2:303
        helper dependent vectors 2:303
        recombinant 2:303
        serotypes 2:303
    genome 1:3, 1:18
        fiber genes *see above*
        nucleotide composition 1:3, 1:4f
        splicing 1:3
    history 1:1, 1:17
    human *see* Human adenovirus(es)
    infection *see* Adenovirus infections
    morphology 2:554–555
    phylogeny 1:7f, 1:8
    proteins 1:4
    see also specific proteins
    replication 1:3, 4:409–410
        assembly 1:21
        DNA replication *see above*
        host cell attachment 1:18, 1:18f
        host cell entry 1:3–4, 1:18
        rate 1:22
        release 1:21
        transcription 1:3–4
        VAI RNA 1:3–4, 1:161
    serotypes 1:9
    simian 1:26
    stability 1:3, 1:25
    structure 5:391
    transmission 1:5
    vaccines/vaccination 1:29
    vectors 1:23, 2:303
        helper dependent vectors 2:303
        recombinant 2:303
        serotypes 2:303
    virions 1:3, 1:17, 2:78f
        capsid 1:17–18
        capsomers 1:3, 2:78
        electron microscopy 2:78, 2:85–86
        fiber unit 1:17–18
        penton base 1:17–18
        viral core 1:18
    see also specific viruses
Adenovirus(es), enteric (EAd) 2:120
    classification 2:120
    evolution 2:120
    genetics 2:120
    geographical distribution 2:120
    history 2:120
    infection
        clinical features 2:121
        diagnosis 2:120
        nonseasonal epidemics 2:121
        pathogenesis 2:121
    morphology 2:120f
    seasonal distribution 2:120
    see also specific viruses
Adenovirus death protein (ADP) 1:22
Adenovirus infections **1:24–30**
    acute hemorrhagic conjunctivitis 2:495
    cytopathic effects 1:22
    detection 1:28
    epidemic keratoconjunctivitis 2:495, 2:495f
    epidemiology 1:25

    HAdV-A 1:25
    HAdV-B1 1:25
    HAdV-B2 1:25
    HAdV-C 1:25
    HAdV-D 1:25
    HAdV-E 1:26
    HAdV-F 1:26
    tissue tropism 1:25
    epizootiology 1:26
        bovine 1:26
        canine 1:26
        equine 1:26
        frog 1:28
        ovine 1:26
        porcine 1:26
        reptilian 1:27
        simian 1:26
    history 1:24
    identification 1:28
    immune system evasion 1:22–23, 4:114t
    pathology 1:5, 1:22
    persistence 4:109t
    prevalence 1:25
    prevention 1:29
    simple follicular conjunctivitis 2:495
    therapy 1:29
    tissue tropism 1:22, 1:22f
    transmission 1:25
Adjunctive therapy, organ transplantation 3:471
Adjuvant(s) 3:68
    definition 1:69, 5:226
    HIV vaccine development *see* HIV vaccines/vaccination
Admin table(s), virus databases 5:352
Adoxophyes orana granulovirus (AdorGV) 1:212
ADP-ribose-1″-phosphatase (ADRP), nidoviruses 3:425
Adsorption, definition 5:365
Adsorption electron microscopy 2:19, 2:21f
Adult T-cell leukemia (ATLL) 2:561, 2:565, 5:198
    classification 2:565
        subtypes 2:565, 2:566t, 2:567t
    clinical features 2:561, 2:566, 2:567
        acute 2:561, 2:566
        chronic 2:561, 2:566
        hematopoietic precursor cell differentiation 2:567–568
        hypercalcemia 2:567–568
        onset 2:561
        skin lesions 2:568, 2:568f
        smoldering 2:561
    complications 2:567
        opportunistic infection 2:568, 2:569–570
        pulmonary 2:568
    cytological features 2:566, 2:567f
        cytogenetic abnormalities 2:566
        lymph node histology 2:566
    diagnosis 2:565, 2:566t
    epidemiology 2:564, 2:565
        incidence 2:565
    etiology 2:561
        carriers 2:561
    immuno-virological features 2:566
    pathogenesis 2:568, 2:569f
        leukemogenesis 2:561, 2:563
        tax protein 2:562, 2:568–569
    prognosis 2:569
    therapy 2:569
Adult T-cell lymphoma 2:565
    EBV associated 2:155
    HTLV-1 associated *see* Human T-cell leukemia virus 1 (HTLV-1)
Adventitious agents, vaccine safety 5:226, 5:228
Adverse event following immunization (AEFI) 5:226, 5:227–228, 5:230t, 5:231t, 5:232
    benefits 5:232, 5:232t
    classification 5:232
    definition 5:232
    evaluation 5:231
    public perceptions 5:230

vaccine safety assessment 5:232
  benefits 5:232, 5:232t
  public perceptions 5:230
Aedes
  breeding cycle 5:473
  definition 5:469
  see also Mosquito(es)
Aedes aegypti
  Chikungunya virus transmission 5:100
  Dengue virus transmission 2:5, 2:7, 2:9–10
    defective viruses 1:173
  infection control, Aedes aegypti densovirus 3:128, 4:79
  larval habitats 2:9–10
  RNAi in disease control 3:153
    O'nyong-nyong virus 3:154
  Venezuelan equine encephalitic virus transmission 5:99–100, 5:104
  yellow fever virus transmission 2:455
Aedes aegypti densovirus (AeDNV) 3:128, 4:79
Aedes albopictus
  Dengue virus transmission 2:5, 2:7
  eastern equine encephalitic virus transmission 5:103–104
Aedes albopictus densovirus type 2 (AalDNV) 4:83
  virion structure 4:78
Aedes polynesiensis 2:5
  Dengue virus transmission 2:7
Aedes pseudoscutellaris dinovernavirus (ApDNV-1) 3:135, 3:136, 3:136f
  genome 3:139t, 3:140–141
  infection 3:140
  infectivity 3:140
  protein functions 3:141, 3:141t
Aedes sollicitans, eastern equine encephalitic virus transmission 5:103–104
Aedes taeniorhynchus, Venezuelan equine encephalitic virus transmission 5:104
Aedes vexans arabiensis, Rift Valley fever virus transmission 4:491
AeMV see Arabis mosaic virus (ArMV)
Aerosol transmission
  common cold 4:472
  nepovirus transmission 3:411
  SARS CoV 4:557
  tobacco ringspot virus transmission 3:411
Aerosol vaccines, safety 5:229
Affinity maturation (immunoglobulins) 3:61
AFMoV see Angelonia flower mottle virus (AFMoV)
Africa
  banana bunchy top virus 1:277
  cotton leaf curl disease 1:563
African cassava mosaic disease **1:30–37**
  control 1:34, 2:101
    biological control 1:36
    crop isolation 1:35
    cultural methods 1:35
    East African cassava mosaic virus infection 1:36
    infected plant removal 1:35
    whitefly resistance 1:36
  economic importance 1:30, 1:34
    yield losses 1:34
  field-level epidemiology 1:33
  future work 1:36
  history 1:30
  host responses 1:33
    post-transcriptional gene silencing 1:33
  infection mechanisms 1:33
  pandemics 1:32, 2:100
    crop yield losses 2:101
    emergence mechanisms 2:104
    epidemiology 2:101
    geographical distribution 2:100, 2:100f
    symptoms 2:101
    transmission mechanisms 2:101
  plant resistance 1:34
    plant breeding 1:34–35
  regional epidemiology 1:33
  transgenic resistance 1:35
    DNA interference 1:35
    nonviral protein expression 1:35

RNA interference 1:35
viral protein expression 1:35
African cassava mosaic virus (ACMV) 1:31–32
  classification 2:101
  distribution 1:32f
  epidemiology 2:101
  infection see African cassava mosaic disease
  regional epidemiology 1:33–34
  transmission 2:104
  Ugandan strain 2:100–101
    epidemiology 2:101
    geographical distribution 2:100–101
    infection symptoms 2:101, 2:101f
    transmission 2:101
  virus-induced gene silencing 5:326
African green monkeys
  simian immunodeficiency virus infection see Simian immunodeficiency virus infection
African green monkey cytomegaloviruses (AgmCMV)
  genome 1:636–637
  infectivity 1:639f
  isolates 1:636
African green monkey kidney cells, SV40 infection 4:630
African horse sickness (AHS)
  cardiac (subacute) form 1:40
    macrolesions 1:41
  clinical signs 1:40
    cardiac (subacute) form 1:40
    horse sickness fever 1:40
    mixed form 1:40
    pulmonary form 1:40
  control 1:42
    movement restriction 1:42
  diagnosis 1:41
    clinical signs 1:41–42
    conformation 1:42
  differential diagnosis 1:42
    equine viral arteritis vs. 1:42
    purpura hemorrhagica vs. 1:42
  geographical distribution 1:37
  historical aspects 1:37
  macrolesions 1:41
    cardiac (subacute) form 1:41
    mixed form 1:41
    pulmonary interlobular edema 1:41
  microlesions 1:41
  mixed form 1:40
  mortality rate 3:458
  as notifiable disease 1:37
  pathogenesis 1:40
    cardio-vascular system 1:40–41
    lymphatic system 1:40–41
    regional lymph nodes 1:40
    viral organ accumulation 1:40
    viremia 1:40
  pathology 1:41
  pulmonary form 1:40
  treatment 1:42
African horse sickness virus (AHSV) **1:37–43**, 2:96
  capsid 3:462–463
  epidemiology 1:41
  genome 1:37, 1:39f
  heat resistance 1:38–40
  history 3:457
  host range 1:40, 3:457
  infection see African horse sickness (AHS)
  replication 1:39f
  serotypes 1:37, 1:38
  taxonomy 1:37
  transmission 1:41, 3:457
    Culicoides 1:41
  virion 1:37, 1:38f
    nonstructural protein 1 1:37, 1:39f
    nonstructural protein 2 1:37
    nonstructural protein 3 1:38, 1:39f
    nonstructural protein 3a 1:38, 1:39f
    VP1 1:37
    VP3 1:37
    VP4 1:37
    VP6 1:37
    VP7 trimers 1:37, 1:38f

African peanut clump virus see Peanut clump virus (PCV)
African swine fever virus (ASFV) **1:43–51**, 2:96
  assembly 1:46
  characteristics 1:171t
  classification 1:50
  distribution 1:43
  genome
    gene functions 1:49t
    isolate diversity 1:43
    structure 1:45, 1:48f, 1:170–172
  history 1:43
  infection see below
  multigene families 1:48
  replication 1:45, 1:46–47, 1:46f
    A151R polyprotein 1:46–47
    aggresomes 1:46
    DNA replication 1:46
    mRNA transcription 1:46
    virus entry into macrophages 1:45
  transmission 1:43, 1:44f
  virus assembly 1:46
  virus-encoded proteins 1:47, 1:49t
    A238L protein 1:47
    AP endonuclease 1:47
    DNA replication and repair 1:47
    immune system evasion 1:47
    mRNA transcription 1:47
  virus structure 1:45, 1:45f
    capsid 1:45
    stability 1:170–172
African swine fever virus infection
  apoptosis 1:44–45, 1:48
  characteristics 1:44
  host range 1:43
  immune evasion 1:47
    A238L protein 1:47
    CD2v protein 1:47–48
    macrophages 1:47
    NL-S protein 1:47–48
  macrophages 1:44
    immune system evasion 1:47
    replication 1:44
    virus entry 1:44, 1:45
  pathogenesis 1:44
  pigs 1:44
AFV see Acidianus filamentous virus(es) (AFV)
AG7088 1:548
Agalactia, definition 1:427
Agar double diffusion assays, plum pox virus strains/groups 4:239
Agaricus bisporus virus 1 (AbV-1) 1:510, 2:286–288
  genome 1:510
  virion structure 1:510
Agenerase 2:513
Ageratum yellow vein virus (AYVV)
  βC1 1:318
  begomovirus satellite identification 1:315
Age-relations
  Japanese encephalitis 3:184
  neural cell infection 1:473
  St Louis encephalitis virus transmission 5:273
  vector transmission, animal viruses 5:273
  western equine encephalomyelitis virus transmission 5:273
  zoster (shingles) 5:252, 5:252f
Aggresome(s), African swine fever virus replication 1:46–47
AgmCMV see African green monkey cytomegaloviruses (AgmCMV)
AgMNPV see Anticarsia gemmatalis multiple nucleopolyhedrovirus (AgMNPV)
Agnathans, immunoglobulin phylogeny 3:58, 3:60f
Agnoprotein (LP1), SV40 infection 4:634
AGO1, Argonaute proteins 3:151
AGO2, Argonaute proteins 3:151
Agonadal (AG) tissue, Hz-2V infection 3:144–145
Agraulis vanillae virus (AvV) 5:28t
Agriculture/farming
  bovine enteroviruses 2:123–124
  capripoxviruses 1:429

Agriculture/farming (continued)
    classical swine fever 1:529–530
    fish 2:207
    hepatitis E virus infection 2:382
    yellow fever virus epidemics 5:494
Agrobacterium tumefaciens-based expression system
    (T-DNA) 4:232f, 4:233
    definition 5:375
    Helicoverpa armigera stunt virus 5:36
    heterooligomeric protein expression 4:233
    plant-produced vaccines 5:224
    platforms 4:237f
    recombination 4:232f, 4:233
    virus-induced gene silencing 5:377
        limitations 5:379
Agroinfection 4:233
    definition 3:263
Agroinfiltration 1:81
Agroinoculation
    definition 1:457, 5:138
    viroid transmission 5:335
Agropyron mosaic virus (AgMV) 1:490
AHF *see* Argentine hemorrhagic fever (AHF)
AHS *see* African horse sickness (AHS)
AHSV *see* African horse sickness virus (AHSV)
Aichi virus, gastroenteritis 2:122
AIDS
    clinical features 4:599–600
    definition 1:69
        CD4+ T lymphocyte count 1:52
    etiology 2:143–144
    life expectancy 1:51–52
    medical achievements 2:535t
    ocular complications 2:496
        classification of 2:496
    pandemics 2:534
    pathogenesis 2:534–535, 2:535f, 2:538f
    related infections
        HCMV infection 2:439
        JCV reactivation 4:265–266
        KSHV infection *see* Kaposi's sarcoma-associated herpesvirus (KSHV)
        prevalence 2:505
    vaccine development *see* HIV vaccines/vaccination
AIDSinfo database 5:357t
AIDSVAX 1:70
AILV *see* Artichoke Italian latent virus (AILV)
Aino virus, Akabane virus *vs.* 1:80
Airborne transmission 2:140, 2:145–146
AK602 2:507t
Akabane virus 1:76–81, 1:391t, 3:480
    Aino virus *vs.* 1:80
    classification 1:76
    epidemiology 1:78
    genome 1:76
    geographical distribution 1:76, 3:480–481
    historical aspects 1:76, 3:480
    taxonomy 1:76
    tissue tropism 1:78
    transmission 1:78
    vaccines/vaccination 1:80
    vectors 3:480
        control 1:80
    virion properties 1:76
        stability 1:77
Akabane virus infection
    arthrogryposis 1:77
        epidemics 1:77–78
    calves
        clinical signs 1:77
        pathogenesis 1:78
        pathology 1:78
    clinical signs 1:77
        congenital defects 1:76, 1:77, 3:480–481
    control 1:80
    diagnosis 1:79
    differential diagnosis 1:80
        associated viruses 1:80
        noninfectious causes 1:80
    hydraencephaly 1:77
        epidemics 1:77–78

fetal infection stages 1:78
    pathology 1:78
immunocompetence 1:78
isolation 1:79
management strategies 1:80
pathogenesis 1:78
    fetal infection stages 1:78
    incubation period 1:78
pathology 1:78
prophylaxis 1:80
seasonal distribution 1:79
vertebrate hosts 1:77
Alcelaphine herpesvirus 1 (AlHV-1) 1:364t
    species barriers 2:428
Alcephaline herpesvirus 1 (AIHV1) 1:364t
Alcephaline herpesvirus 2 (AIHV2) 1:364t
Aleutian mink disease virus (AMDV) 4:95
    genome 4:94f, 4:95
    infection 4:90, 4:95
    virion structure 4:92f
Alfalfa cryptic virus (ACV), genome 1:99–100
Alfalfa mosaic virus (AMV) 1:81–87, 5:63
    Bacopa virus infections 4:215
    characteristics 5:63t
    classification 1:81
    coat protein 1:81, 1:85
        crystallization 1:83
        dependent infectivity 3:47
        eIF4G binding 1:84f, 1:85
        monomer numbers 1:81–82
        movement 1:85
        RNA interaction 1:82–83, 1:85–86
        RNA replication 1:84–85
        RNA translation 1:83–84, 1:85
        structure 1:81–82
    control 1:86
        coat protein transformation 1:86
        reservoir hosts 1:86
        virus-free seeds 1:86
    encapsidation 1:85
    epidemiology 1:86
    genome 1:81, 1:82, 1:82f, 5:63–64
        CP binding site 1:82–83
        movement protein 1:82–83
        replicase protein P1 1:82–83
        replicase protein P2 1:82–83
    historical aspects 1:81
    infection *see below*
    movement 1:85
        coat protein 1:85
        movement protein 1:85
    promoter 3:54
    replication 3:49
        coat protein–RNA complex dissociation 3:54
    reservoir hosts 1:86
    RNA replication 1:84
        coat protein 1:84–85
        minus-strand synthesis 1:84
        minus to positive strand switch 1:86
        P1 protein 1:85
        P2 protein 1:85
        plus-strand synthesis 1:84–85
        in tobacco 1:84
    RNA translation 1:83, 1:84f
        coat protein role 1:83–84
        poly(A) binding protein (PABP) 1:83–84
    structure 1:81, 1:82f, 5:391
        β-barrel fold 5:381, 5:382f
        bottom component 1:81–82
        coat protein *see above*
        middle component 1:81–82
        top component *a* 1:81–82
        top component *b* 1:81–82
    taxonomy 1:81
    transmission 1:86
        aphid vector 1:86
        seeds 1:86
Alfalfa mosaic virus (AMV) infection
    economic significance 1:81
    host range 1:81
    Impatiens 4:215

petunia 4:213, 4:214f
potato 4:300t
tobacco *see below*
tobacco 5:63
    symptoms 5:63
virus–host relationships 1:86
    symptoms 1:86
Alfamovirus 1:81, 1:387, 3:46, 3:47
Algal blooms, termination 4:124
Algal virus(es) 1:87–95, 1:88t
    associated families 1:87
    definition 1:87
    double-stranded RNA viruses 1:90
    ecological implications 1:94
    phylogeny 1:93f
        maximum likelihood trees 1:90f
    ssDNA viruses 1:92
    ssRNA viruses 1:87
    see also specific viruses
Alignment algorithms (classification) 5:342
    BLAST 5:343
    global 5:343
    local 5:343
    multiple sequence 5:343
    pairwise sequence comparison *see* Pairwise sequence comparison (PASC)
    phylogenetic analysis 5:343
    scoring schemes 5:342–343
Alkaliphile 5:411
Alkhurma virus 5:53
    zoonoses 5:491
Allele(s) 4:177
    dominant 4:177
    recessive 4:178
Alleloleviridae 3:21
Allexivirus 1:96–98
    genome 1:97, 1:97f
        coat protein 1:97–98
        open reading frames 1:97–98
        subgenomic RNA 1:98
        TGB protein 1:97–98
    geographical distribution 1:98
    history 1:96
    replication 1:97
        RNA polymerase 1:97–98
    transmission 1:97
    viral composition 1:96
        nucleic acid 1:97
        physical properties 1:97
        physicochemical properties 1:97
        proteins 1:97
    viral structure 1:96, 1:97f
        morphological types 1:96
Allexivirus 1:96, 1:96t, 1:98, 1:426f, 2:255
Alligator(s), West Nile virus infection 5:445–446
Allium virus X (AVX) 4:311t
Allograft transplantation, HCMV infection 2:483–484
Alloherpesviridae 2:431
Almpiwar Group, animal Rhabdoviruses 1:116, 1:119t
Almpiwar virus 4:184t
Alovudine 2:507t
Alphacryptovirus 1:98, 1:100t, 2:109–110
Alphacryptovirus(es) 1:98–104, 1:100t
    classification 1:99
    evolutionary relationships 1:103
    genome 1:100
    history 1:98
    molecular biology 1:100
    replication 1:100
    taxonomy 1:99
    transmission 1:99, 1:102
    virion properties 1:99
    virus–host relationships 1:102
    see also specific viruses
Alpha-dystroglycan (aDG), arenavirus cell
    attachment/entry 3:245
Alphaentomopoxvirus 4:323, 4:323t
Alphaentomopoxvirus 2:136
    characteristics 2:137t
    host range 4:323
    structure 4:325

Alphaentomopoxvirus 4:323
Alphaherpesvirinae 1:364t, 2:383, 2:405–406, 2:431, 2:432t, 4:341, 4:583, 5:11t, 5:251
Alphaherpesvirus 2:432, 5:251
Alphaherpesvirus(es)
 avian viruses *see* Avian alphaherpesvirus(es)
 discovery history 2:420
 equine viruses 2:412
 genital disease 1:366f
 host range 1:365
 latent infections 2:392f
 replication 1:363
 simian *see* Simian alphaherpesvirus(es)
 transmission 1:365
 see also Bovine herpesvirus(es); Herpes simplex virus(es) (HSV); Varicella-zoster virus (VZV); *specific viruses*
Alphanodavirus 3:430
 capsids 3:432, 3:432f
  alpha protein structure 3:432
  promoters 3:432
  quaternary structure 3:432f, 3:433
 geographical distribution 3:431–432
 infection 3:437
  insects 3:126t
 RNA packaging 3:432, 3:433
  molecular determinants 3:433
  random cellular RNA 3:434
 RNA replication 3:435–436
 virion assembly 3:433
 virion maturation 3:433
Alphanodavirus 3:431
Alphapapillomavirus
 carcinoma associated 4:15
 evolution 4:12–13
Alphapapillomavirus 4:9, 4:35
Alpha protein structure, alphanodavirus capsids 3:432
Alpharetrovirus 4:460
Alpharetrovirus(es)
 assembly 2:518
 cell-derived sequences 4:461
 classification 4:460
 morphology 4:461–462
 taxonomy 4:446
 see also specific viruses
Alphavirus 2:230, 4:514, 5:83, 5:91–92, 5:96, 5:101, 5:107, 5:108–110, 5:109t, 5:117
Alphavirus(es) 5:77, 5:83, **5:96–101**, 5:97t, 5:108–110, 5:109t
 antigenic groups 5:83
 assembly 1:195, 1:197f
  capsid dimerization 1:196
  genome encapsidation 1:196
  glycoproteins 1:197
  nucleocapsid 1:196–197
  in vitro assembly system 1:196
 bioterrorism 5:408
 budding 1:200
 definition 5:107
 epidemiology 5:100
 evolution
  arbovirus relationships 1:174
  relationships 1:174, 5:96, 5:98f
  RNA sequences 5:110–111
 genetics 5:103
 genome 5:98, 5:119f
 geographical distribution 5:77, 5:84f
 history 5:96
 infection *see below*
 life cycle 1:195, 5:102–103
  E1 glycoprotein function 1:195
  E2 glycoprotein function 1:195
  E3 glycoprotein function 1:195
  nucleocapsid release 1:195
  replication *see below*
 propagation 5:87
 replication 5:98, 5:120f
  assembly 1:196
  budding 1:196, 5:121
  host cell attachment 5:98, 5:118
  host cell entry 5:118–119

 nonstructural proteins 5:98, 5:119, 5:120
 protein processing 5:99f
 replicase complex 5:119
 rubella virus *vs.* 5:121
 sites 5:112–113
 structural protein maturation 5:121
 replicon systems 5:121–122
 RNA recombination 4:377
 species 5:93t
 transmission 5:92, 5:99, 5:111
  mosquito-borne 5:99
  non-mosquito-borne 5:99
  transovarial 5:111
  zoonotic 5:111–112
 type species 5:91–92
  Sindbis virus 5:91–92
 virion structure 1:195, 1:196f, 5:98
  E1 glycoprotein 1:195–196
  E2 glycoprotein 1:195–196
  nucleocapsid 5:98
 see also specific viruses
Alphavirus infections
 animal models 5:85
 arthritis 5:85, 5:87t
  treatment 5:88
 associated diseases 5:117
 clinical manifestations 5:86t
 diagnosis 5:87, 5:115
 disease control 5:116
 hosts 5:111
 pathogenesis 5:115
 pathology
  cell morphological changes 5:103–104
  cytopathic effects 5:103–104
  disease patterns 5:96
  molecular changes 5:103–104
 treatment 5:88
 zoonoses 5:111–112
 see also specific viruses
Alstroemeria virus X (AlsVX) 4:311t
Alternanthera mosaic virus (AltMV) 4:311t
Alzheimer's disease
 HSV associated 2:391
 viral etiology 1:475
AMA infectious diseases database 5:357t
α-Amanitin, viroid replication 5:339
Amantadine
 history 1:145
 influenza treatment 2:556, 3:102
 resistant variants 1:148
Amasya cherry disease (ACD) 1:509–510
Amasya cherry disease associated chrysovirus (ACDACV) 1:504t, 1:505–506
Ambisense genome(s)
 definition 1:390, 1:482, 5:157
 tenuivirus 5:25–26, 5:25f
Amdovirus 4:95
Amdovirus(es) 4:95
 classification 4:92t
 genome 4:94f, 4:95
 see also specific viruses
AMDV *see* Aleutian mink disease virus (AMDV)
American oyster reovirus (13p2RV) 1:164t
AMEV *see* Amsacta moorei entomopoxvirus (AMEV)
Amidase(s), endolysin function 3:251
Amino acid sequences/sequencing
 aquareovirus genetic diversity 1:167
 barley yellow dwarf viruses 1:279–280
 EBV 2:158–159
 seadornavirus 4:541–542, 4:541f
Amino acid substitutions, eukaryotic initiation factor-4E- mediated resistance 4:183–184
Amino acid transport protein
 Mason–Pfizer monkey virus cell receptor 4:627
 Simian retrovirus D cell receptor 4:627
Aminoacyl-tRNA synthetase(s)
 definition 3:311
 mimivirus gene content 3:316
Aminothiazole derivatives 1:149
Ammonium molybdate 2:81

Amphibian herpesvirus(es) **2:205–212**
 classification 2:206
 distribution 2:207
 evolution 2:209
 genetics 2:209
  genome characteristics 2:210t
 growth properties 2:209
 history 2:206
 infection
  clinical features 2:207
  control 2:211
  immune response 2:211
  pathogenesis 2:207
  pathology 2:207
  prevention 2:211
 see also specific viruses
Amplicons
 definition 2:93
 gene therapy 2:304, 2:396
Amprenavir 2:513, 2:514t
AMV *see* Alfalfa mosaic virus (AMV)
Amyloid
 prion formation
  self-propagating 4:337
  transfection with 4:339
 structure 4:338f, 4:339
Amyotrophic lateral sclerosis
 echoviruses 2:69
 viral etiology 1:475
Analgesics 5:83
Analytical tools, virus databases 5:357t, 5:364
Anamnesis 3:319
Anaphase-promoting complex (APC) 1:11
Anaphylaxis 5:226
Anastomosis
 definition 2:580
 hypoviruses 2:584
Ancestral retrotransposons 4:436
Ancillary viral protein(s) 3:348, 3:349
Andean potato latent virus (APLV) 4:298t
Andean potato mottle virus (APMV) 4:298t
Anellovirus 1:105, 1:106
Anellovirus(es) **1:104–111**
 distribution 1:107
 epidemiology 1:107
  global distribution 1:107
  mixed infections 1:109
  non-human primate infection 1:109
  prevalence rates 1:107
  risk exposure 1:107
 genome 1:105, 1:106f
 history 1:105
 infection *see below*
 phylogeny 1:106, 1:108f
  open reading frame 1 1:107
  sequence homology 1:107, 1:109f
 replication 1:105
 small
  history 1:105
  phylogeny 1:106, 1:107, 1:109f
 transmission 1:107
 virion properties 1:105
 see also specific viruses
Anellovirus infections
 clinical significance 1:110
  diseases associated 1:110
  hepatitis 1:110
  viral load *vs.* host immune status 1:110
 detection approaches 1:109
 isolation from biological samples 1:109
 mixed 1:109
 viral load 1:109
Anemia
 definition 3:155
 equine infectious anemia virus infection 2:173

Angel fish aquareovirus (AFRV) 1:164t
Angelonia flower break virus (AnFBV) 4:217, 4:220f
Angelonia flower mottle virus (AFMoV)
  host ranges 4:208
  verbenaceae infection 4:218, 4:220f, 4:221f
Angelonia virus infections 4:215
  see also specific viruses
Angiotensin converting enzyme 2 (ACE-2), SARS CoV host cell binding 4:555–557
Angola, Marburg virus outbreak 2:200, 3:272, 3:275, 3:276f
Anguilla herpesvirus (AngHV-1) 2:233
Animal contact, human monkeypox virus infections 4:642
Animal domestication, arbovirus transmission 1:176
Animal importation, infection control 5:431
Animal model(s)
  defective interfering viruses 2:3
  Eastern equine encephalitic virus 5:79
  Ebola viruses infection 2:202–203
  equine encephalitis 5:103
  filovirus infection 2:202–203
  hepatitis C virus 2:370
  herpes simplex viruses latency 2:437
  Kaposi's sarcoma 3:193
  Marburg virus infection 2:202–203
  multiple sclerosis 5:43–44
  seadornavirus infections 4:545
  severe acute respiratory syndrome 4:558
  simian immunodeficiency virus see Simian immunodeficiency virus (SIV)
  togovirus infection 5:103
  transgenic plant vaccine production 5:221
  varicella-zoster virus infection 5:251
Animal reservoir(s)
  definition 2:377
  orthopoxviruses 1:579–580
Annihilator mutant, definition 1:231
Annotation see Database(s)
Anogenital warts see Human papillomavirus infections
ANS (1-anilo-8-naphthalene sulfonate) 5:35
Antagonistic epistasis 4:359
Antagonistic pleiotropy 4:359
Antennal gland 5:476
Anterograde transport, CNS invasion pathways 1:470
Antheraea eucalypti virus (AeV) 5:28t
Anthoxanthum latent blanching virus (ALBV) 2:459
Anthroponoses 5:270, 632:270
  Dengue viruses 5:270–271
Anthroponosis 5:268
Anti-apoptotic proteins
  adenovirus E1B-19K gene/protein 1:13f
  HCMV immune system evasion 1:637
Antibodies 3:56–70
  antigen binding see Antibody–antigen binding
  antigenicity 1:138
  biological functions 3:62
  definition 3:78
  history 3:56, 3:57t
  immunoglobulin phylogeny 3:58
  inactivation, common cold 4:472–473
  infections
    henipavirus 2:326
    HSV 2:393
    mammalian reovirus 4:389
    murine CMV 1:630t, 1:631
    nervous system viruses 1:470
    persistent infection 4:112
    pseudorabies virus 4:349
    rotavirus 4:512
    togavirus 5:87–88
  mutational resistance 5:323–324
  neonatal immunization 2:53
  neutralizing see Neutralizing antibodies (NAbs)
  phage display function 2:198
  plant virus vectors, expression 3:64
  prevalence 2:343
  protective 3:64
  structure, complementarity-determining regions 1:138
  T-cells vs. 3:71
  therapeutic, Marburg virus infection 3:279

vaccine development strategies 5:236–237
  AIDS see HIV vaccines/vaccination
  variation, HIV 1:132
  see also Immunoglobulin(s)
Antibody–antigen binding 1:138, 1:139
  antibody concentration 3:414
  antibody-dependent disease enhancement 3:66
  conserved epitope binding 3:417–418
  dissociation constant 3:414–415
  envelope glycoproteins 3:414–415
  infection neutralization see Infection neutralization
  occupancy 3:414
  see also Immune complexes
Antibody-dependent cellular cytotoxicity (ADCC), yellow fever 5:475–476
Antibody-dependent enhancement (ADE) 3:66
  disease association 3:66
Antibody phage display 2:448–449
  antibody selection 2:198
  bacteriophage M13 2:197–198
  biomedical applications 2:198
  potato virus Y 4:290
Anticarsia gemmatalis multiple nucleopolyhedrovirus (AgMNPV) 3:130–131
  limitations 3:131
Antigen(s) 1:137–142
  antibody binding see Antibody–antigen binding; Immune complexes
  antigenicity, viral proteins 1:137
  associated antibodies 1:138
  epitopes 1:137
    continuous 1:138, 1:139f
    definition 3:413, 4:594
    detection, T-cells 3:73
    discontinuous 1:138, 1:139f, 1:141
    immune system evasion 3:77
    reactivity level 1:139, 3:73
    size 1:139
    structure 1:139
    see also B-cell epitope(s)
  exogenous 3:72–73
  histo-blood groups 3:444
  immune response 1:137, 3:62–64
  infection neutralization 3:414
  molecular parameters 1:140
  nonenveloped viruses, assembly 1:204
  processing 3:72
    interference 3:77
  recognition 3:65f
  variation see Antigenic variation;
  see also specific viruses
Antigenemia, definition 4:507
Antigenic drift
  definition 3:95, 3:483
  vaccine development 5:240
    drift rate 5:240
  see also specific viruses
Antigenic shift
  definition 3:95, 3:483
  vaccine development 5:240
  see also specific viruses
Antigenic variation 1:127–137, 2:146–147
  definition 1:127
  manifestations 1:127
  mutational mechanisms 1:127
  vaccines 1:127–128
    development strategies 5:240
    human papilloma viruses 1:127–128
  viral serotypes 1:128
  see also specific viruses
Antigenome
  definition 2:375, 3:324, 4:52
  hepatitis delta virus replication 2:376f
  mononegavirus replication 3:330–331
Antigen presentation 1:121–126, 3:72
  definition 3:70
  pathways 1:123
    MHC-I 1:123, 3:72
    MHC-II 1:124, 3:72
  viral subversion 1:125
  see also Antigen-presenting cells (APCs)

Antigen-presenting cells (APCs) 1:122, 1:122f, 1:620, 3:72
  definition 1:69
  maturation 1:123, 1:126
  types 1:122–123
  see also specific types
Anti-G protein human monoclonal antibodies 2:326
Anti-HBsAg antibodies 2:357
Antiholin(s)
  definition 3:248, 3:250
  S107 3:253
    holin S105 vs. 3:253–254
  lysis timing 3:253
  strategy diversity 3:255
Anti-poxvirus chemical agents 3:230
Antiretroviral therapy 1:143t, 1:146, 1:153, 1:353, 2:505–517, 2:518
  combination therapy 1:151, 2:516
    acyclovir 1:152
    regimens 2:516
    see also Highly active antiretroviral therapies (HAART)
  in development 2:507t
  drug resistance 1:148–149
  entry inhibitors 2:506
  HAART see Highly active antiretroviral therapies (HAART)
  integrase inhibitors 2:507t, 2:510
  maturation inhibitors 2:516
  mode of action 2:516
  nucleoside analogs see Nucleoside analogs
  nucleotide analogs see Nucleotide analogs
  protease inhibitors see Protease inhibitors
  reverse transcriptase inhibitors 2:509
    non-nucleoside reverse transcriptase inhibitors see Non-nucleoside reverse transcriptase inhibitors (NNRTIs)
    nucleoside reverse transcriptase inhibitors see Nucleoside reverse transcriptase inhibitors (NRTIs)
  see also specific drugs
Antisense binding, mRNA 1:148
Antiviral drug(s) 1:142–154, 1:143t
  action spectrum 1:147
    broad spectrum 1:147, 1:154
  aminothiazole derivatives 1:149
  combination therapy 1:151
  compound classes 1:148
    antisense binding, mRNA 1:148
    helicase inhibitors 1:149
    neuraminidase inhibitors 1:149
    polymerase inhibitors 1:149
    protease inhibitors 1:148
  compound toxicity 1:147
    long-term therapy 1:147
    nephrotoxicity 1:147
    tests 1:147
  development 1:146
    coronaviruses 1:561
  drug interactions 1:147
  drug resistance 1:150, 1:151
    genetic barrier 1:150
  history 1:145
    acyclovir see Acyclovir
    drug development 1:146
  mechanism of action 1:148f, 1:151
  mutational potential 1:150
  nonenveloped viruses, assembly 1:204
  organ transplantation 3:469–470
    adjunctive therapy 3:471
    drug toxicity 3:471
    duration 3:471
    post-transplant prophylaxis 3:471
    prophylaxis 3:471
  prodrugs 1:150
  selective activity index 1:146
    assays 1:146
    CC_{50} 1:146–147
    definition 1:146
    EC_{50} 1:146
    EC_{99} 1:146

IC$_{50}$ 1:146
  therapeutic ratio 1:147
target classes 1:148
  niche targets 1:153
  virus assembly 1:149
  virus attachment 1:148
toxicity 1:147
  tests 1:147
virus fitness 1:151
virus latency 1:152
see also specific drug classes
see also specific drugs
ANV see Avian nephritis virus (ANV)
APCs see Antigen-presenting cells (APCs)
ApDNV-1 see Aedes pseudoscutellaris dinovernavirus (ApDNV-1)
AP endonuclease, African swine fever virus 1:47
Aphid(s)
  alfalfa mosaic virus transmission 1:86
  banana bunchy top virus transmission 1:277
  barley yellow dwarf virus transmission 1:279, 1:282–283
  bean common mosaic virus transmission 1:291
  bean common necrosis virus transmission 1:291
  caulimovirus transmission 1:463
  citrus tristeza virus transmission 1:521, 1:524
  cucumber mosaic virus transmission 1:618, 1:619, 5:287–288
  luteovirus transmission 3:235–236, 3:237
  nanovirus transmission 3:386
  papaya ringspot virus transmission 4:1, 4:4
  plant rhabdovirus transmission 4:195
  see also specific viruses
  plum pox virus transmission 4:241
  potato virus A transmission 4:302
  potato virus Y transmission 4:289, 4:293, 4:303
  potyvirus transmission, proteins associated 4:320
  sowthistle yellow vein virus transmission 4:195–196
  umbravirus transmission 5:213
  watermelon mosaic virus transmission 5:436
  zucchini yellow mosaic virus transmission 5:436
Aphid transmission factor (ATF), caulimovirus transmission 1:465
Aphis gossypii, zucchini yellow mosaic virus transmission 5:436
Aphthovirus 2:266
Apipoxvirus 4:324t
Aplastic anemia 2:343
APOBEC3 (cytidine deaminases) 3:109
APOBEC3G gene/protein
  evolution 2:471–472
  HIV replication 2:521
  hypermutation 2:471–472, 3:109
  nonsynonymous polymorphisms 2:471–472
  retrovirus infection resistance 2:471
APOBECs, definition 2:467
Apodemus (woodmice), cowpox virus hosts 1:577, 1:578
Apoptosis 1:154–162, 1:156f, 1:232, 1:252
  adaptor proteins 1:232
  as antiviral response 1:236
    cell culture 1:237
  biochemical changes associated 1:155
  biochemical pathways 1:232
  caspase-activated DNase (CAD) 1:252
  caspases 1:232f
  cellular death pathways 1:155
  cross-talk 1:155–157
  DARK 1:232
  death receptor signaling 1:157
    FLIP proteins 1:160
    viral caspase regulators 1:160
    viral regulators 1:157
  definition 1:104, 1:186, 1:231–232, 1:247, 2:534, 3:111, 5:116
  DIAP1 1:232
  DRONC 1:232
  infections
    adenoviruses 1:157–160
    African swine fever virus 1:44–45, 1:48
    ascovirus 1:191, 1:191f
    HCMV 1:637

HSV 2:391–392
iridovirus 3:173
KSHV 3:200
mammalian reoviruses 4:387
nidovirus 3:429
pseudorabies virus 4:349, 4:350
white spot syndrome virus 5:457
yeast viruses 1:162
  inhibition
    adenoviruses 1:157–160
    baculovirus see Baculovirus apoptosis inhibition
    KSHV 3:200
  IPNV induced 3:86
  mitochondrial pathway, viral regulators 1:161
    Bcl-2 family proteins 1:155–157, 1:156f, 1:161–162
  morphological changes associated 1:155
  persistent infection 4:113
  regulation 1:232
  regulatory viral RNAs 1:161
    virus-encoded miRNA 1:161
  RNA-binding proteins 1:161
    pathogen recognition receptors 1:161
  SINV glycoproteins 5:122
Apoptosome 1:231
Apoptotic bodies, definition 1:186, 3:111
Apple chlorotic leaf spot virus (ACLSV) 4:204
  associated diseases 1:424, 4:205
  associated species 1:424
  genome 1:424
    movement protein 1:424
    open reading frames 1:424
  hosts 1:424
  particle structure 1:421f
Apple mosaic virus (ApMV) 4:205
Apple stem grooving virus (ASGV) 1:420t, 4:202f, 4:206
  affected hosts 1:420
  genome 1:420–421
    coat protein 1:421
    open reading frames 1:420–421
  occurrence 1:420
  particle structure 1:420, 1:421f
  transmission 1:420
Apple stem pitting virus (ASPV) 4:205
  genome 1:423
    analysis 1:423f, 1:424
    open reading frames 1:423
  particle structure 1:421f
Apple topworking disease 1:422
Apricitabine 2:507t, 2:510
Apricot latent ringspot virus (ALRSV) 3:406t
Apricots, Sharka disease 4:239
Aptivus® 2:516
Apurinic/apyrimidinic
  definition 4:436
  distribution 4:442t
Aquabirnavirus 1:321–322, 1:322t, 2:232, 3:83, 3:84
Aquaculture, definition 4:560
Aquareovirus 1:163, 2:231, 4:536t
Aquareovirus(es) 1:163–169
  genetic diversity 1:164t, 1:167
    amino acid sequences 1:167
    genogroups 1:167
    RNA electrophoresis 1:167
    RNA–RNA hybridization 1:167
  genome 1:165
    open reading frame 1:165
    sequencing 1:165, 2:232
  geographical distribution 1:166
    aquareovirus A 1:166
    aquareovirus B 1:166
    isolates 1:166
  host range 1:166
  morphogenesis 1:163
  proteins 1:165
    nonstructural proteins 1:165
    NS1 1:165–166
    NS2 1:165–166
    NS4 1:165–166
    structural proteins 1:165

VP1 1:165–166
VP2 1:165–166
VP3 1:165–166
VP4 1:165–166
VP5 1:165–166
VP6 1:165–166
VP7 1:165–166
replication 1:163
species 2:232t
  isolation 1:169
  type species 4:536t
taxonomy/classification 1:163, 1:164t, 1:169
  antibody-based cross neutralization 1:163
  RNA cross-hybridization 1:163
  species demarcation criteria 1:163
transmission 1:167
virion 1:163, 1:165f
  capsid 1:163
see also specific viruses
Aquareovirus A 1:164t, 1:166, 4:536t
Aquareovirus B 1:164t, 1:166
Aquareovirus C 1:164t
Aquareovirus D 1:164t
Aquareovirus E 1:164t
Arabidopsis thaliana
  beet curly top virus 1:307
  eIF4E-mediated resistance 4:183, 4:184–185
  gene silencing 4:141, 4:142
    Dicer-like proteins 4:145
    natural *cis*-antisense transcript-derived siRNA pathway 4:142
    RDR genes 5:326–327
    RdRp 6 4:146–147
    siRNA accumulation 4:145–146
  infection resistance 1:456
    HRT gene 4:174–175
    RCY1 gene 4:175
  transposons 3:309
Arabis mosaic virus (ArMV) 1:490, 3:406t
  diseases 3:411
  infectious DNA clones 3:408–409
  satellites 3:411, 4:531
Aracacha virus B-oca strain 4:298t
Arbovirus(es) 1:170–176
  classification 1:170
  definition 1:170, 1:399, 2:5, 3:454, 3:479, 4:490, 5:83, 5:101, 5:107, 5:268, 5:269
  evolutionary relationships 1:173
    alphaviruses vs. 1:174
    flaviviruses vs. 1:173
  genome 1:170
  geographical distribution 1:170
  infection 1:170, 1:175
    in animals 1:175
    CNS associated 1:175
    encephalitis 1:474
    febrile illness 1:175
    hemorrhagic fever 1:175
    neural cell infection 1:473
    persistence mechanisms 1:397–398, 2:235
  taxonomy 1:170, 1:171t
  transmission cycles 1:170, 1:176
    animal domestication 1:176
    host feeding range 1:176
    human reservoirs 1:176
  virion composition 1:170
  virion properties 1:170
see also specific viruses
Archaea
  definition 3:472
  lipid structure, related to evolution 1:400
  origin of 3:473
Arenaviridae 3:203, 3:239, 3:247–248
Arenavirus(es) 3:203–212, 3:204t, 3:243–248
  assembly/budding 3:247
    matrix protein 3:247
    Z protein 3:247
  bioterrorism 5:408–409
  cell attachment/entry 3:245
    alpha-dystroglycan 3:245
    GP-2 glycoprotein 3:245

Arenavirus(es) (continued)
  classification 3:203, 3:204t
  epidemiology 3:205
  evolution 3:205, 3:240
  gene products 3:244
    see also specific proteins
  genetics 3:205
    B-cell epitope variation 3:205
  genome 3:243
    intergenic regions 3:243, 3:244, 3:244f
    terminal nucleotide sequences 3:243
  geographical distribution 3:204
  glycoproteins 3:244
    GP-1 3:244
    GP-2 3:244
    posttranslational processing 3:244
    precursor 3:243
    signal peptide 3:244
  history 3:203
  host range 3:204
  infections see below
  life cycle 3:245
  L polymerase 3:244
    RNA replication/transcription 3:245
    sequence analysis 3:245
    structure 3:244–245
  molecular phylogeny 3:247
  nucleoprotein 3:243, 3:244
    RNA replication/transcription 3:245, 3:246
  propagation 3:204, 3:205
  RdRp 3:243
  reverse genetics 3:247
  RNA replication/transcription 3:245
    genomic promoters 3:246–247
    homotypic viral interference 3:246
    L polymerase 3:245
    nontemplated nucleotides 3:247
    nucleoprotein 3:245, 3:246
    regulation 3:246, 3:246f
    Z protein 3:245–246
    Z RING domain 3:245–246
  seasonal distribution 3:204
  tissue tropism 3:207
  transmission 3:207
  vaccines/vaccination 3:210
  virion properties 3:203
    genome 3:203–204
    glycoprotein precursor gene 3:203–204
    morphology 3:203–204
    nuclear ribosomal protein 3:203–204
    nucleoprotein gene 3:203–204
  virion structure 3:244
  Z protein 3:245
    RNA replication/transcription 3:245–246
  see also specific viruses
Arenavirus infections
  clinical features 3:208
  control 3:210
    drug prophylaxis 3:211
    rodents 3:210
    therapy 3:211
  immune response 3:210
  Lassa fever see Lassa fever
  pathogenesis 3:207
  prevention 3:210
    vaccines 3:210
  rodents 3:204
    coevolution 3:205
    control 3:210
    transmission 3:204, 3:206
  zoonoses 2:95
    hemorrhagic fever 5:486
    hosts 5:486
    transmission 5:486
  see also specific diseases/disorders
Argasids, definition 2:234
Argas pusillus, Keterah virus 1:400
Argentine hemorrhagic fever (AHF)
  clinical features 3:209
  epidemiology 3:206
  histopathology 3:209

pathogenesis 3:208
  capillary leakage 3:208
  hypovolemic shock 3:208
  proteinuria 3:208
pathology 3:209
therapy 3:211
Arginine-glycine-aspartic acid (RGD) motif 2:67
Argonaute (AGO) proteins 3:151, 4:141, 4:143t
  AGO1 3:151, 4:142
  AGO2 3:151
  PAZ domain 3:151
Arracacha virus A (AVA) 3:406t
  diseases 3:411
Arteriviridae 1:177, 3:419–420, 5:11t, 5:152
Arterivirus 1:177
Arterivirus(es) 1:176–186
  classification 1:177
  envelope protein 3:423
    M protein 3:423
    S protein 3:423
  evolution 1:182
    RNA recombination 1:182
    serologic relationships 1:182
    serologic variability 1:182
  genetics 1:181
  genome 1:177, 1:179, 1:179f, 3:424f
    frameshifting 1:179–180
    open reading frames 1:179–180
    proteolytic processing 3:424
    replicase 1:179–180, 3:423–424, 3:425f
    transcription regulation sequences 1:179–180, 1:180f, 3:423–424, 3:428
  geographical distribution 1:177
  history 1:176
  host range 1:177
  infection 3:421
    clinical features 1:183
    control 1:185
    histopathology 1:183
    immune response 1:184
    immune system evasion 1:185
    pathogenicity 1:183
    pathology 1:181, 1:183
    prevention 1:185
    tropism 1:181, 1:182
  propagation 1:177
  protein properties 1:180
    nonstructural proteins 1:179–180, 1:179f, 1:181
    nucleocapsid 1:180–181
    pp1a 1:178f, 1:180
    pp1ab 1:180
  replication 1:180f, 1:181
    cytoplasmic double-membrane vesicle formation 1:181
    genome encapsidation 1:181
    sites 1:181
    subgenomic RNA production 1:180f, 1:181
  taxonomy 1:177
  transmission 1:182
  virion properties 1:177
    morphology 1:177–179
    stability 1:177–179
  virion structure 1:178f
    capsid 1:179
    envelope proteins 1:179, 3:423
    M protein 1:180–181, 3:423
    S protein 3:423
  see also specific viruses
Arthralgia 1:399
  Ross River virus infection 5:114
Arthritis
  alphavirus infection see Alphavirus infections
  caprine arthritis encephalitis virus infection 5:430
  definition 5:83
  rubella virus infection 4:518–519
Arthrogryposis
  Akabane virus infections see Akabane virus
  definition 1:76, 1:328
Arthropod parvovirus(es) see Parvovirus(es), arthropod
Arthropods, definition 2:234

Arthroponoses 5:270, 5:270f
Artichoke Aegean ringspot virus (AARSV) 3:406t
Artichoke curly dwarf virus (ACDV) 4:311t
Artichoke Italian latent virus (AILV) 3:406t
  diseases 3:411
Artichoke yellow ringspot virus (AYRSV) 3:406t
  tobacco disease 5:66
Artiodactyl(s) 1:440
ARV see Adelaide river virus (ARV)
ARV1 (Rudivirus), morphology 1:589, 1:589f
ASBVd see Avocado sunblotch viroid (ASBVd)
Ascites 1:37
Ascomycete(s) 2:574
Ascomycota 4:419
  retrotransposons 4:423, 4:424t
Ascoviridae 1:186, 1:187, 1:188t, 5:11t
Ascovirus 1:188t
Ascovirus(es) 1:186–193
  ecology 1:189
  evolution 1:192
  genome 1:189, 1:190f
  geographical distribution 1:187
  history 1:187
  host range 1:189
  infection
    cell biology 1:191
    cytopathology 1:187f, 1:188f, 1:191
    disease symptoms 1:191
    insects 1:126t
    pathogenesis 1:191
    pathology 1:191
    rate 1:189
    tissue tropism 1:192
    virion-containing vesicles 1:191–192
  morphology 1:186
  origin 1:192
  replication 1:192
    virion assembly 1:188f, 1:192
  taxonomy 1:187, 1:188t
  variants 1:187
  transmission 1:186, 1:189
  virion composition 1:189
    lipid membranes 1:189
  virion structure 1:188f, 1:189
  see also specific viruses
Aseptic meningitis 4:389, 4:490
  coxsackieviruses 1:585, 1:585t
  disease surveillance 2:46
    notifications 2:45
  HIV infection 1:57
  sandfly fever virus infection 5:488
Asexual organisms, species classification 5:403
Asexual sporulation 2:576–577
Asfarviridae 1:50, 1:170, 5:11t
Asfarvirus(es)
  apoptosis inhibition 1:158t
  see also specific viruses
Asfivirus, virion structure 1:170–172
ASFV see African swine fever virus (ASFV)
ASGV see Apple stem grooving virus (ASGV)
AsiA, bacteriophage transcription 5:177
Asialoglycoprotein receptor (ASGPR), hepatitis A virus infection 2:12
Asian macaque, SIV disease model 2:539–540, 4:599
Asian tiger mosquito 5:494
Asparagus virus 3 (AV-3) 4:311t
Aspergillus ochraceous virus (AoV), antigenic properties 4:65
ASPV see Apple stem pitting virus (ASPV)
Assemblin, HCMV capsid formation 2:487
Assembly
  definition 4:406
  origin of see Origin of assembly
Assembly protein, HCMV capsid formation 2:487
Asthma, rhinovirus infection 1:546
Astrocyte(s), JCV infection 4:281–282
Astroviridae 1:205, 2:121, 5:11t
Astrovirus 1:205

Astrovirus(es) **1:204–210**
　characteristics 2:117t
　classification 1:205, 2:121
　　nomenclature 1:205
　detection 1:204–205
　epidemiology 1:205
　　animals 1:206
　　co-pathogens 1:206
　　distribution 1:206
　　infection rates 1:205
　　serotypes 1:206
　evolution 1:208, 2:121
　　interspecies transmission 1:208
　　mutation rates 1:208
　genetics 2:121
　history 1:205, 2:121
　host range 2:121
　infections *see below*
　molecular biology 1:207
　　cell entry 1:207
　　genome organization 1:207f
　　open reading frames 1:207
　　RdRp, expression 1:207–208
　morphology 1:204–205, 1:205f, 2:121f
　propagation 1:206
　protein expression 1:207
　stability 1:205, 1:210
　transmission 1:205
　see also specific viruses
Astrovirus infections 1:204–205
　clinical features
　　avian 1:209
　　mammalian 1:208
　control 1:210
　immune response 1:209
　　cellular 1:209
　　humoral 1:209
　　turkey model 1:209–210
　pathogenesis 2:121
　　avian 1:209
　　experimental models 2:121
　　mammalian 1:208
　pathology
　　avian 1:209
　　mammalian 1:208
　prevalence 1:204–205, 2:121
　prevention 1:210
　treatment 1:210
ASV database 5:357t
Aswan Dam project, Rift Valley fever virus epidemic 4:494
Atadenovirus 1:1, 1:2t, 1:17, 1:26–27, 1:27–28, 2:120
Atadenovirus(es) 1:6
　bovine 1:8
　genome
　　LH region 1:8
　　nucleotide composition 1:6–8
　　organization 1:18
　　RH region 1:8
　host range 1:6–8
　host switches 1:5, 1:8–9
　nomenclature 1:1
　phylogeny 1:7f, 1:8
　see also specific viruses
Atazanavir 2:513, 2:514t
Ateline herpesvirus 2 *see* Herpesvirus Ateles (HVA)
Ateline herpesvirus 3 *see* Herpesvirus Ateles (HVA)
Atherosclerosis, rat cytomegalovirus 1:633
Atkinsonella hypxylon virus (AhV-1) 4:69, 4:71f
Atlantic salmon aquareovirus (ASRV) 1:164t
Atlantic salmon paramyxovirus (ASPV) 2:229
ATLL *see* Adult T-cell leukemia (ATLL)
ATPases
　dsDNA bacteriophage genome packaging 2:310–311
　P-loop, Crenarchaeal viruses 1:594
ATP-dependent transporters-associated antigen - processing (TAP) 1:124
ATP-DNA complex, dsDNA bacteriophage genome packaging 2:309–310

Attachment
　baculoviruses 1:250
　definition 4:407
　see also specific viruses
Attachment glycoproteins, henipavirus 2:322–323
Attenuated virus cross-protection, Sharka disease control 4:242
Attenuation (vaccines) *see* Live-attenuated vaccines
A-type inclusions (ATIs)
　avipoxvirus 2:275
　cowpox virus *see* Cowpox virus (CPXV)
Aubian wheat mosaic virus (AWMV) 1:490
Aujeszky's disease (AD) 4:341, 4:349
　immunology 4:349
　see also Pseudorabies virus (PrV)
Aum Shinrikyo, bioterrorism 5:407
Aura virus (AURAV) 5:109t
Aureusvirus 5:145–146
Australia
　banana bunchy top virus epidemics 1:277
　Hendra virus 2:324
　rabbit hemorrhagic fever virus 1:411
Australian antigen *see* Hepatitis B surface antigen (HBsAg)
Australian bat lyssavirus 4:184t
Autism, vaccines safety 5:233
Autoantibodies
　EBV infection, response to 2:153
　see also specific antibodies
Auto-disable syringe
　definition 5:226
　vaccine safety 5:227, 5:229
Autographa californica multiple nucleopolyhedrovirus (AcMNPV) 1:238, 1:247
　apoptosis inhibition 1:233f
　BV envelope composition 1:263
　DNA replication 1:250
　evolution 2:182
　expression vector function 3:131, 3:132
　genome 1:256, 1:259f
　　homologous repeat regions 1:218, 1:256–257
　host range 3:129
　host range determinants 1:252–253
　ODV envelope composition 1:263
　replication cycle 1:239f
　　budded virus 1:238
　　occlusion-derived virus 1:238
　virion structure 1:223
Autoimmunity, Visna-Maedi virus infection 5:430
Autolysis 3:249
Autonomous retrotransposons 4:436
Autophagy 5:378
　MHC-II antigen presentation 1:125
AVA *see* Arracacha virus A (AVA)
Avenavirus 5:145–146
Aviadenovirus 1:1, 1:2t, 1:17, 1:27, 2:120
Aviadenovirus(es) 1:6
　disease patterns 1:6
　fiber genes 1:6
　genome 1:6, 1:18
　　dUTP pyrophosphatase 1:6
　host range 1:6
　see also specific viruses
Avian adenovirus(es) 1:27
　infections 1:27
　　fiber genes 1:6
　　GAM-1 protein 1:6
　see also specific viruses
Avian alphaherpesvirus(es) 2:405–406
　genetics 2:408
　see also specific viruses
Avian hepadnavirus(es) **2:327–335**
　classification 2:327
　genome 2:329
　history 2:334
　host range 2:329
　infection 2:341
　　disease control 2:332
　　hepatocytes 2:327, 2:329, 2:330f, 2:332, 2:334f

　　pathogenesis 2:332
　　sites 2:327
　　tissue tropism 2:329, 2:334f
　phylogeny 2:327, 2:332, 2:333f
　viruses associated 2:332
　replication 2:327, 2:329
　　budding 2:330f, 2:331
　　cccDNA 2:327, 2:329–330, 2:331
　　DNA synthesis 2:330
　　DR1 2:330–331
　　DR1* 2:330–331
　　DR2 2:330–331
　　genome integration 2:331–332
　　nuclear transport 2:331
　　pregenome packaging 2:330, 2:331f
　　receptor binding 2:329
　　sites 2:327, 2:329
　　transcription 2:328f, 2:330, 2:330f
　structure 2:328
　transmission 2:327
　see also specific viruses
Avian hepatitis E virus 2:377–378, 2:378f, 2:380
　hepatitis–splenomegaly syndrome 2:380
Avian herpesvirus(es) **2:405–411**
　genetics 2:408
　host range 2:406, 2:407
　infections
　　control 2:410
　　pathobiology 2:409
　　prevention 2:410
　morphology 2:408
　phylogeny 2:407f
　virus propagation 2:407
　see also specific viruses
Avian influenza virus(es)
　classification 3:101
　evolution 3:98–99
　H5N1 pandemic 3:100, 3:493, 5:491
　transmission 3:96
　see also specific viruses
Avian leukemia virus 4:466–467
Avian leukosis virus(es) (ALV)
　classification 4:455
　infection
　　host range 4:455
　　oncogenes/oncogenicity 4:457
　　replication 4:456
　see also specific viruses
Avian metapneumovirus 4:45
　genome 3:327f
Avian nephritis virus (ANV) 1:209
　pathogenesis 1:209
Avian poxvirus(es) 4:326–327
　see also specific viruses
Avian reovirus(es) 4:383
　see also specific viruses
Avian retrovirus(es) **4:455–459**
　cellular transformation 4:457
　classification 4:455
　distribution 4:455
　endogenous viruses 4:457
　evolution 4:458
　genetics 4:457
　genome 4:455
　　modifications 4:455–456
　　noncoding regions 4:456
　　oncogenes 4:455–456
　glycoproteins 4:456
　history 4:455
　host range 4:455
　infections
　　control 4:458
　　immune response 4:458
　　oncogenes/oncogenicity 4:448
　　pathogenicity 4:458
　　prevention 4:458
　physical properties 4:456
　propagation 4:455
　protein properties 4:456
　　Env precursor 4:456
　　Gag-Pro precursor 4:456

Avian retrovirus(es) (continued)
glycoproteins 4:456
Pol reading frame 4:456
replication 4:456
particle release 4:457
provirus transcription 4:457
receptors 4:456
reverse transcription 4:456–457
translation 4:457
viral DNA integration 4:457
taxonomy 4:455
tissue tropism 4:458
transmission 4:458
virion properties 4:455
morphology 4:455
see also specific viruses
Avibirnavirus 1:322t, 2:232
Avihepadnavirus 2:327, 2:351–352
Avipoxvirus 2:274
Avipoxvirus(es) **2:274–284**
antigens 2:275
classification 4:324t
epizootiology 2:278
extracellular enveloped virus 2:276–277, 2:277f
transmission 2:279
gene expression 2:275
protein homologs 2:275
genome 2:275
infection see below
intracellular mature virus 2:276–277, 2:277f
transmission 2:279
isolation 2:279
molecular phylogenetics 2:277, 2:278f
molluscum contagiosum virus vs. 2:276
replication 2:276
genetic reactivation 2:276
host range restriction 2:276
lipid metabolism 2:277
virion morphogenesis 2:276
survival factors 2:275
A-type inclusion protein 2:275
glutathione peroxidase 2:275
immunomodulators 2:276
photolyases 2:275
taxonomy 2:274
transmission 2:279
vaccines/vaccination
infection prevention 2:281
recombinant vaccines 2:283
virion structure 2:274
virus propagation 2:277
see also specific viruses
Avipoxvirus infections
clinical manifestations 2:279
control 2:281
detection 2:279
inter-strain differentiation 2:281
intra-species differentiation 2:281
monoclonal antibodies 2:280
PCR-based analysis 2:280
restriction enzyme digestion 2:280
serology 2:279
Southern blotting 2:280
diagnosis 2:279
histology 2:279
host range 2:278
prevention 2:281
vaccination 2:281, 2:283
reintroduction programs, birds 2:279
treatment 2:283
variants 2:283
antigenic 2:283
integrated reticuloendotheliosis sequences 2:283
tumorigenic 2:283
Avirulence genes (Avr)
definition 4:170
plant natural resistance 4:171, 4:172t, 4:173
mutation 4:173
recessive resistance 4:178
recognition evasion 4:176

Avirulence proteins
definition 4:173
recognition 4:175
Avirulent strains, cacao swollen shoot virus infection 1:403
Avocado sunblotch viroid (ASBVd)
host range 5:335
replication 5:339, 5:340f
Avsunviroidae 5:11t, 5:333t, 5:335
Avsunviroids
monomer circularization mechanism 4:479–480
pathogenicity determinants 5:339–341
replication 4:476, 4:478f
see also specific viruses
Avulavirus 4:41t
Avulavirus(es)
classification 4:41t
genome 3:327f
see also specific viruses
Axoplasmic flow, neural cell infection 1:473
AYRSV see Artichoke yellow ringspot virus (AYRSV)
AYVV see Ageratum yellow vein virus (AYVV)
5′Azacytidine, EBV infection control 2:155–156
AZT see Zidovudine (AZT)

B

B2 protein, RNAi suppression 3:152
B5 antigen, VACV 5:249–250
B10 virus 5:413t
B119L polyprotein 1:46–47
B602L protein 1:46–47
Baboon cytomegalovirus (BaCMV) 1:635t, 1:639f
Baboon reovirus (BRV) 4:383
Babuvirus
characteristics 3:389t
genome 3:387f
Babuvirus 1:274, 3:385
BAC see Bacterial artificial chromosome (BAC)
Bacilliform particles 5:391
Bacillus anthracis, bioterrorism 5:406, 5:408t
Bacillus subtilis, associated phages 4:401, 4:402–403
Back mutations
definition 4:359
quasispecies 4:361–362
BacMam see Baculovirus expression vector(s)
Bacmid shuttle baculovirus expression vectors, Bac-to-Bac 1:241
Bacopa virus infections 4:215, 4:218f
see also specific viruses
BacPAK6 DNA 1:240, 1:240f
Bacteria
detection, bacteriophage use 2:448–449
filamentous phage pathogens 3:123
superinfection, molluscum contagiosum virus 3:322
Bacterial artificial chromosome (BAC)
definition 1:237
flashBAC expression vector system 1:244
varicella-zoster virus 5:256
Bacterial cell wall
carbohydrates 5:366
definition 5:365
penetration see Bacteriophage(s), bacterial cell entry
Bacterial viruses see Bacteriophage(s)
Bacteriophage(s) **2:442–450**
capsid assembly see Bacteriophage(s), capsid assembly
characteristics 2:444–445
classification 5:342, 5:347
'bacteriophage proteomic tree' 5:343
coat protein
folding 1:436
procapsid maturation 1:436, 1:437f
conversion 2:445–446
definition 2:442, 5:365
double stranded DNA see Double stranded DNA (dsDNA) bacteriophages
double stranded RNA see Double stranded RNA (dsRNA) bacteriophages
dsDNA see Double stranded DNA (dsDNA) bacteriophages

dsRNA see Double stranded RNA (dsRNA) bacteriophages
ecology see Bacteriophage(s), ecology
evolution **5:370–375**
capsid proteins 5:374
co-infections 5:371
evolutionary connections 5:374
genome comparisons 5:372
genome organization 5:371
host cell co-evolution 5:374–375
mechanisms 5:371, 5:372
filamentous ssDNA see Filamentous ssDNA bacteriophages
genome packaging **2:306–312**, 2:307t
capsid assembly 2:306
genome replication 2:306
genome vs. chromosome 2:307
nucleotide selection 2:307
packaging motor complex 2:307
precursor capsid assembly 2:307
historical aspects 3:21
host cell infection see Bacteriophage(s), bacterial cell entry
host cell lysis see Bacteriophage(s), host cell lysis
host cell receptors 4:407–408
icosahedral dsDNA see Icosahedral dsDNA bacteriophages
icosahedral ssDNA see Icosahedral ssDNA bacteriophages
library 2:190
lysogeny 5:181
Lwoff, André experiments 2:445
research history 2:445
in molecular biology 3:21–22
nomenclature 2:442–443
nonhomologous recombination 5:371–372, 5:372–374, 5:373f
genome comparisons 5:372
horizontal transfer 5:372
novel gene acquisition 5:374
novel sequence joints 5:372
populations 5:371
prevalence 2:444
productive infections 5:371
recombination 4:374
replication **4:399–406**
research history 2:443, 2:447t
applications 2:448, 2:449t
bacteriophage therapy 2:444, 2:449t
bacteriophage typing 2:444
classification 2:446
discovery 2:443, 2:444t, 2:445t
diversification 2:446
ecology 2:444, 2:449
genome properties 2:446
genome sequencing 2:446, 2:448t
lysogeny 2:445
molecular biology tools 2:444
viral nature 2:444
'single-gene lysis' 3:256
gene A₂ 3:257
gene E 3:256–257
gene L 3:258
single-stranded RNA see Single-stranded RNA bacteriophages
ssRNA see Single-stranded RNA bacteriophages
structure 2:448f
T4-even phages 4:401
T4-like phages see T4-like bacteriophages
T7-like phages see T7-like bacteriophages
tailed see Caudovirus(es); Tailed bacteriophages
therapy 2:444, 2:448–449, 2:449t
transcription regulation **5:174–186**
AsiA 5:177
elongation 5:175–176
host RNAP-mediated elongation 5:175
initiation 5:175–176
isomerization 5:175–176
promoter binding 5:175
RNAP structure 5:175
see also specific bacteriophages

typing 2:444, 2:446
  bacteriophage research history 2:444
  definition 2:442
  see also specific types
Bacteriophage(s), bacterial cell entry **5:365–370**
  cell wall penetration 5:367
    lysozymes 5:367
    lytic transglycosylases 5:367
    peptidoglycan-hydrolyzing enzymes 5:366f, 5:367
    pre-existing channels 5:366f, 5:367
  energetics 5:370
    host cell activity 5:370
    sources 5:370
  genome delivery 5:367
    icosahedral vertex 5:367
    virion dissociation 5:367
    virion penetration 5:367
    see also specific viruses
  host cell barriers 5:365
    eubacteria 5:365
    Gram-negative bacteria 5:365
    Gram-positive bacteria 5:365
  host recognition 5:365, 5:366f
    bacterial pilus 5:366
    cell wall carbohydrates 5:366
    flagellar proteins 5:366
    lipopolysaccharides 5:366
    receptor-binding proteins 5:365–366
  virion structure 5:365
  viral genome effects 5:365
Bacteriophage(s), capsid assembly **1:432–440**, 1:434f, 2:306
  assembly pathways 1:434
    protein conformational changes 1:434–435
  catalysis 1:438
    chaperonin assembly 1:438
    enzyme assembly 1:438
    jigs 1:438
    scaffolds 1:438
    templates 1:438
  covalent protein modifications 1:439
    cross-linking 1:439
    lipid membrane acquisition 1:439
    proteolytic cleavage 1:439
  history 1:433
  pathways 1:434
  strategies/mechanisms 1:435
    branched pathways 1:435
    genome types 1:434
    nucleic acid encapsidation 1:439
    quality control 1:435
    subassemblies 1:435
  symmetrical protein arrays 1:436
    helical arrays 1:437
    icosahedral array maturation 1:436, 1:437f
    symmetry disruption 1:437
    symmetry mismatches 1:437
  types 1:435t
Bacteriophage(s), ecology **2:71–78**
  bacteria-like vs. virion like existence 2:73
    favorable growth conditions 2:73–74
    infecting states 2:73–74
  community ecology 2:76
    bacterial physiological burdens 2:76
    bacterial versatility 2:76
    bacteriophage-mediated genetic transduction 2:76–77
    bacteriophage-mediated population crashes 2:76
    bacteriophage resistance 2:76
    principles 2:76
  ecosystem ecology 2:77
    bacteriophage-induced bacterial lysis 2:77
    nutritional movement impacts 2:77
  existence within environments 2:72
    bacteriophage environmental diversity 2:72
    bacteriophage number determination 2:72, 2:72f
  landscape ecology 2:73
  organismal ecology 2:73
    growth parameters 2:73
    life cycle variations 2:73, 2:73f
    spatially structured environments 2:73

population ecology 2:74
  adaptation 2:75–76
  bacterial density 2:75
  bacteriophage decay rate 2:74–75
  broth culture 2:74
  burst size 2:75
  competition 2:75
  decay rate 2:74–75
  environmental conditions 2:74
  extracellular search 2:74–75
  growth inhibition 2:75
  selective pressures 2:75
Bacteriophage(s), host cell lysis **3:248–258**
  antiholins 3:255
  endolysins 3:251
  extrusion 3:249
  holin–endolysin lysis see Holin–endolysin lysis
  lysis modes 3:250
  progeny virion release 3:249
  rotting vs. 3:249
  'single-gene lysis' 3:256
    gene A_2 3:257
    gene E 3:257
    gene L 3:258
Bacteriophage 21 3:254–255
Bacteriophage Bam35
  genome 3:5
  host range 3:3
Bacteriophage Bxbl 5:185
Bacteriophage Cp-1 4:402–403
Bacteriophage CTXΦ
  genome 3:121
  virulence factors 3:123
Bacteriophage φ29
  evolution 5:374
  genome size 3:36
  replication 4:402
    DNA polymerase 4:403, 4:404f
    membrane-localized proteins 4:404
    replication forks 4:403–404, 4:404f
    sliding back mechanism 4:403
Bacteriophage fd 2:194
  composition 2:193–194
  genome packaging 2:193–194, 2:193f
  pIII 2:194
  pVIII
    central region 2:194
    composition 2:193–194, 2:193f
    N-terminal region 2:194
    pentapeptide motifs 2:194, 2:194t
    structure 2:192t, 2:193–194
Bacteriophage Ff 3:117, 3:118f
  assembly pathway 2:191
  capsid helical symmetry 3:121
  DNA conformation 3:121
  DNA synthesis 2:196
    host-encoded polymerases 2:196
    pII 2:196
  genome 2:192f, 3:120
  genome packaging 2:191
  infection 2:194
    F pilus 2:195–196
    Tol–Pal system 2:195–196, 2:195f
  plaque formation 3:118
  post-transcriptional modification 2:196
  protein synthesis 2:196
    host-encoded polymerases 2:196
    pIII maturation 2:196
    pIV 2:196–197
    pVIII 2:196
  structure 3:119–120, 3:120f
    gp7/9 3:119–120
  transcription 2:196
  translation 2:196
  virus assembly 2:197
    pIV channel conformations 2:197
    progeny elongation 2:197
    pV-coated genome 2:197
    sites 2:197
Bacteriophage fOg44 3:254–255
Bacteriophage HK022 5:185

Bacteriophage HK97
  coat protein folding 1:436
    cross-linking 1:439
    evolution 5:374
    procapsid maturation 1:436, 1:437f
  evolution 5:372, 5:373f
  phylogeny 3:36
  structure 5:388, 5:389f
Bacteriophage φ6 3:6
  structure 3:7f
Bacteriophage L2 1:439
Bacteriophage L5 5:185
Bacteriophage λ
  BaculoDirect expression vector system see Baculovirus expression vector(s)
  CII accumulation 5:183
  genome 4:400–401
  genome packaging 2:309, 2:309f
    integrase host factor 2:309
  holin–endolysin lysis 3:252
    antiholin S107 3:253
    cell energy metabolism 3:252
    dual start motif 3:252f
    holin S105 3:253
    lambda holin S105 3:253
    lysis cassette 3:252–253, 3:252f
    promoter activation 3:252–253
    Rz–Rz1 genes 3:254
  life cycle 5:182f
  lysis 5:181
  lysogeny 5:181, 5:182, 5:182f
  lytic growth 5:184
    persistence 5:184
    suppression mechanisms 5:185
    terminator role 5:184
    transcription regulation 5:184
  promoters 5:182–183, 5:182f
    P_{aQ} 5:183
    P_{Int} 5:183
    P_L 5:182–183, 5:184
    P_R 4:400–401, 5:182–183
    P_{RE} 5:182–183
  prophage excision 5:182f, 5:184
    Fis protein 5:184
    Int protein 5:184
    Xis protein 5:184
  RecF recombination 4:374
  regulatory strategies 5:181
  replication 4:400–401
    host proteins 4:400–401
    protein O 4:400–401
  site-specific prophage insertion 5:181–182
  structure 3:30–31, 3:31f, 3:32
  tail shaft assembly 1:438
  tape measure protein 1:438
Bacteriophage M13
  cloning vector function 2:197
  genome 2:192, 2:192f
  gene expression 2:192
  intergenic region 2:192
  phage display 2:197–198
Bacteriophage 'morons' 2:76
Bacteriophage Mu
  genome 4:404–405
  replication 4:404
    DnaB 4:404–405
    protein PriA 4:404–405
    transposome formation 4:404–405
Bacteriophage N4 5:181
  regulatory strategies 5:181
  transcription
    early genes 5:181
    late genes 5:181
    middle genes 5:181
    promoters 5:181
    RNAPII 5:181
    vRNAP 5:181
Bacteriophage P2 5:185
  antiholin gene 3:255
  associated proteins 3:250
  host cell lysis 3:250, 3:255

Bacteriophage P22
　evolution 5:374
　procapsid structure 1:611–612
　structure 1:434f
　　scaffold proteins 1:438
Bacteriophage Pf1
　capsid helical symmetry 3:121
　DNA conformation 3:121
　plaque formation 3:118
Bacteriophage Pf2 1:201–202
Bacteriophage Pf3
　capsid helical symmetry 3:121
　DNA conformation 3:121
Bacteriophage Pf4 3:123–124
Bacteriophage phi6 4:379–380
Bacteriophage φX174
　DNA packaging 3:19
　DNA replication 3:16, 3:17f
　evolution 3:20
　gene E 3:256–257
　　MraY resistance 3:257
　　Rz–Rz1 vs 3:257
　　single gene lysis 3:256–257
　genetic map 3:15f
　genome packaging 2:311
　history 3:13
　host cell attachment 3:15–16
　host cell lysis 3:256–257
　host cell penetration 3:16
　morphogenesis 3:19
　nomenclature 3:13
　replication 4:405
　　GpC protein 4:405
　　preprimosome assembly 4:405
　　protein A 4:405
　　SSB protein 4:405
　　stage I 4:405
　　stage II 4:405
　　stage III 4:405
　structure 1:434f, 3:14f, 5:384–385, 5:384f
　　procapsid 3:18f
Bacteriophage PM2, genome
　delivery 5:368
Bacteriophage PRD1
　genome 3:5
　　delivery 5:368
　　encoded proteins 3:2t
　　organization 3:5, 3:5f
　host range 3:3
　inverted terminal repeats 4:402–403
　life cycle 3:3
　　DNA entry 3:4
　　host cell lysis 3:4–5
　　receptor recognition 3:3–4
　　replication initiation 3:4, 4:403
　structural motif conservation 1:610
　structure 1:434f, 1:438, 3:1, 3:3f
　　membrane 3:2–3, 3:2t
　　vertices 3:4f
　　X-ray crystallography 3:3f
'Bacteriophage proteomic tree' 5:343
Bacteriophage Qbeta 4:380
　A$_2$ gene 3:257–258
Bacteriophage SPP1
　genome 4:401
　replication 4:401
　　associated host proteins 4:401
　　helicase 4:401
　　primase 4:401
Bacteriophage T2
　history 1:433
　structure 2:448f
Bacteriophage T3 5:178
　coupled transcription 5:179
　DNA entry 5:179
　early transcription 5:179
　　promoters 5:179
　　termination 5:179
　late transcription 5:178f, 5:179
　　class III genes 5:179
　　termination 5:179

middle transcription 5:179
　class II genes 5:179
　promoters 5:178f, 5:179
regulatory strategies 5:178
　coupled transcription 5:179
　DNA entry 5:179
　early transcription 5:179
　late transcription 5:179
　middle transcription 5:179
　temporal gene expression 5:179
　transcription regulation 5:180
temporal gene expression 5:179
　early genes 5:179
　early gene transcription 5:179
transcription regulation 5:180
Bacteriophage T4 3:30–31, 3:31f, 5:176
　assembly
　　gene products 1:438–439
　　jig fiber 1:438
　DNA injection 3:33–34
　early transcription
　　Alt protein 5:176–177
　　early promoters 5:176
　　σ^{70} 5:176
　evolution 2:181
　genome map 5:176f
　genome size 3:36
　host cell lysis
　　lysozyme E 3:251
　　RI protein 3:255
　late transcription 5:177
　　gp33 5:178
　　gp45 5:178
　　gp55 5:178
　middle transcription 5:177
　　ADP-ribosylation 5:177
　　promoters 5:177
　　RNA synthesis 5:176f, 5:177
　protein injection 3:34
　regulatory strategies 5:176
　　early transcription 5:176
　　late transcription 5:177
　　middle transcription 5:177
　replication 4:402f
　replication forks 4:401
　structure 1:433, 1:434f, 3:32, 3:33f
　　baseplate 1:608, 1:609f
　　tails/tail fibers 3:32
Bacteriophage T5 3:33–34
Bacteriophage T7 5:178
　coupled transcription 5:179
　DNA entry 5:179
　early transcription 5:179
　　promoters 5:179
　　termination 5:179
　genetic map 5:178f
　late transcription 5:178f, 5:179
　　class III genes 5:179
　　termination 5:179
　middle transcription 5:179
　　class II genes 5:179
　　promoters 5:178f, 5:179
　mutants 1:612
　open reading frames 5:178
　regulatory strategies 5:178
　　coupled transcription 5:179
　　DNA entry 5:179
　　early transcription 5:179
　　late transcription 5:179
　　middle transcription 5:179
　　RNA polymerase 5:180
　　RNA processing 5:180
　　temporal gene expression 5:179
　　transcription regulation 5:180
　RNA polymerase 5:180
　RNA processing 5:180
　temporal gene expression 5:179
　　early gene transcription 5:179
　transcription regulation 5:180
　　class II genes 5:180
　　late genes 5:180

RNAP phosphorylation 5:180
T7 DNA host cell entry 5:180
Bacteriophage Xf1 3:121
Bac-to-Bac *see* Baculovirus expression vector(s)
BaculoDirect *see* Baculovirus expression vector(s)
Baculoviridae 1:211, 1:220, 1:247, 1:254, 1:265, 3:129, 3:497–498, 4:567, 5:11t
Baculovirus(es) **1:247–254**
　apoptosis inhibition *see below*
　budded viruses 1:247
　classification 1:247, 3:129
　　genome sequencing 1:247–248
　envelope fusion protein 4:451, 4:454f
　errantiviruses 4:453
　evolution 1:249
　expression vectors *see* Baculovirus expression vector(s)
　F protein 4:454
　　cleavage 4:453, 4:454f
　　definition 4:451
　genomes 1:249
　　core genes 1:249
　　multiple homologous repeat regions 1:249
　　transcription 1:249
　historical perspective 1:247
　host specificity 1:252, 1:265, 3:129–130
　infection *see* Baculovirus infection
　insect pest control 1:253, 3:129, 3:130t
　　infectivity 3:131
　　limitations 3:129–130
　morphology 1:247
　mosquito *see* Mosquito baculovirus(es)
　nomenclature 1:247
　　host species 1:248
　nucleocapsids 1:248–249
　occlusion bodies 2:137, 4:570
　　formation 1:265–266
　　function 2:137
　phenotypes 1:265
　phylogeny 1:220, 3:497–498
　replication cycle 1:250
　　attachment 1:250
　　budding nucleocapsids 1:251
　　cathepsin 1:251–252
　　cell culture 1:250
　　cell entry 1:250
　　chitinase 1:251–252
　　DNA replication 1:250
　　early gene expression 1:250
　　late gene expression 1:251
　　morphogenesis 1:251
　　p10 gene/protein 1:251
　　polyhedrin gene/protein 1:251
　　promoters 1:251
　　release 1:251
　　structural protein expression 1:251
　　transcription regulation 1:250
　　uncoating 1:250
　　virus dissemination 1:251
　　virus-enhancing factor (VEF) 1:250
　replication levels 3:129
　spindle production 2:138
　structure 1:265
　taxonomy 1:247
　transmission 3:310
　virion structure 2:137
　　budded 3:129
　　occlusion-derived 3:129
　see also specific viruses
Baculovirus apoptosis inhibition 1:158t
　host specificity 1:253
　inhibitor of apoptosis proteins 1:252
　　BIR domain 1:236f
　　distribution 1:235
　　gene homology 1:235–236
　　mechanism 1:235–236
　nonpermissive/semipermissive infections 1:234
　P35 1:252
　　caspase inhibition 1:160–161
　　general apoptosis inhibition 1:234
　　gene transcription effects 1:235
　　structure 1:234f

timing 1:234
viral cell entry 1:234
Baculovirus expression vector(s) 1:253
　BacMam 1:245
　　advantages for use 1:245–246
　　cell lines 1:245
　　complement-mediated inactivation 1:246
　　entry mechanism 1:245–246
　　gene therapy use 1:246
　　hybrid promoters 1:245
　　transduction efficiency 1:245
　BacPAK6 DNA 1:240, 1:240f
　Bac-to-Bac 1:241
　　advantages 1:241
　　bacmid shuttle 1:241
　　E. coli transformation 1:241
　　insect cell transfection 1:241, 1:242f
　　site-specific transposition 1:241
　　Tn7 transposon 1:241
　　viral DNA extraction 1:241, 1:242f
　BaculoDirect 1:241
　　bacteriophage λ 1:241
　　entry clone generation 1:243–244
　　Gateway cloning vectors 1:241
　　recombinant virus purification 1:244
　　recombination reaction 1:243–244
　definition 1:238
　disadvantages 1:246
　flashBAC 1:244
　　advantages for use 1:244–245
　　automated high-throughput
　　　screening 1:245
　　bacterial artificial chromosome
　　　utilization 1:244
　　compatibility 1:244
　　homologous recombination 1:244
　gene delivery, mammalian cells 1:245
　insect cell culture 1:238
　　cell lines 1:238
　　culture media 1:238
　plaque purification 1:240–241
　polyhedrin promoter 1:253
　recombinant viruses 3:131
　　efficacy improvements 1:240
　　isolation 1:240
　systems 1:238
　see also specific systems
　transfer vectors 1:239–240
Baculovirus infection
　env-mediated 3:310
　host defence 1:267
　　developmental residence 1:267
　　F proteins 1:267
　　GP64 1:267
　　infected midgut cell shedding 1:267
　　M trait 1:267
　host defence modulation 1:252
　　apoptosis inhibition see Baculovirus apoptosis
　　　inhibition
　　ecdysteroid glucosyltransferase 1:252
　　molting inhibition 1:252
　manifestations 3:129, 3:129f
　pathogenesis 1:265–272
　　in vitro see below
　secondary 1:265–266, 1:266f
　systemic infection 1:267
　terminal disease 1:269, 1:269f
　　pupation 1:269
　　V-CATH 1:269
　virulence 3:129
　in vitro pathogenesis 1:269
　　BRO proteins 1:271
　　F-actin cables 1:270
　　GV-specific cell lines 1:270
　　host nuclear structure 1:270, 1:271f
　　host RNA polymerase II 1:270, 1:270f
　　late-stage infection 1:270
　　microtubules 1:270, 1:271f
　　NPV-specific cell lines 1:270
　　nuclear F-actin 1:270–271
　　nuclear structural change 1:271

　　nucleus infection 1:270
　　P6.9–DNA interactions 1:271
　　P78/83 1:271
　　very late factor-1 1:270–271
　　very late infection phase 1:272
　　wandering behavior 1:269
Badnavirus 1:403, 1:459t
Badnavirus(es) 1:459t
　host chromosome integration 1:462
　open reading frames 1:466f
　ornamental plant species infection 1:403
　species 1:403
　transmission 1:463
　virion structure 1:458–460
　see also specific viruses
Bafinivirus(es)
　genome 3:424f
　replicase genes 3:425f
　see also specific viruses
BAG-1 protein, JCV 4:264
Bahia Grand Group, animal rhabdoviruses 1:116
Bahia Grand virus (BGV) 1:116
Bak protein(s), apoptosis 1:155–157
BALB/c mice, MHV-68 pathogenesis 3:377
Balsaminaceae virus infections 4:214
　see also individual species
Bam35 bacteriophage see Bacteriophage Bam35
Bamboo mosaic virus (BaMV) 4:311t
　RNA satellites 4:310, 4:531
　　expression 4:535
　　variation/evolution 4:534
BamHI-A rightward transcripts (BARTs)
　see Epstein–Barr virus (EBV)
BaMMV see Barley mild mosaic virus (BaMMV)
BaMV see Bamboo mosaic virus (BaMV)
Banana bunchy top virus (BBTV) 1:272–279
　characteristics 3:389t
　coat protein 1:274
　detection assays 1:278
　　ELISA 1:278
　　PCR 1:273, 1:278
　genome 1:274, 1:275f, 3:387–388, 3:387f
　　coat protein 1:274
　　intergenic regions 1:274
　　movement protein 1:274
　　nuclear shuttle protein 1:274
　　Rep 1:274
　geographical distribution 1:274–275, 1:276
　　Africa 1:277
　　current 1:277
　　historical aspects 1:276
　historical aspects 1:274
　　geographical distribution 1:276
　host range 1:273, 3:386
　infection see below
　phylogenetics 1:274, 1:276f
　resistance 1:278
　　transgenic plants 1:278
　taxonomy 1:274
　transmission 1:277, 3:386
　　aphid vectors 1:277
　　circulative transmission 1:274
　　vegetative propagation 1:277
　viral particles 1:274, 1:274f
Banana bunchy top virus infection
　control 1:277
　　genetically-modified resistant plants 1:278
　　infection eradication 1:277–278
　　isolation 1:277–278
　economic importance 3:386
　epidemiology 1:277
　　Australian epidemics 1:277
　　Fijian epidemics 1:277
　　Pakistan epidemic 1:277
　eradication 1:277
　symptoms 1:273, 1:273f, 3:386
　　histology 1:273
　　phloem 1:273
Banana streak virus (BSV) 1:403
　host genome integration 1:462
Bangladesh, Nipah virus 2:324

Banna virus (BAV) 4:536t
　antigenic relationships 4:541
　capsid structure 4:538, 4:539f
　characteristics 1:171t
　diagnostic assays
　　molecular assays 4:544–545
　　serology 4:544–545
　epidemiology 4:537
　functional studies 4:542
　　vaccine development 4:542
　　VP9 expression 4:542
　genome 4:538, 4:540f
　　conserved sequences 4:541–542
　　sequencing 4:539
　host range 4:537
　infections
　　clinical features 4:544
　　as public health problem 4:537–538
　phylogenetics 4:540–541
　proteins, electrophoretic profiles 4:538f, 4:539
　rotaviruses vs.
　　evolutionary relationships 4:542
　　structural features 4:542, 4:545f
　transmission 4:537
　as type species 4:536
BARE1 element 4:435
Barfin flounder nervous necrosis virus (BFNNV)
　2:229–230
Barley (Hordeum vulgare)
　dsRNAs 2:112
　protoplasts, Brome mosaic virus
　　replication 1:382
　resistance genes, loci 4:182
　viruses 1:490–497, 1:491t
　see also specific viruses
Barley mild mosaic virus (BaMMV) 1:492
　infection 1:493f
　control 1:493
Barley stripe mosaic virus (BSMV) 1:490, 2:459
　coat proteins 2:459, 2:461–462
　genome 2:459–460, 2:460f
　infection
　　chlorophyll affects 2:463–464
　　chloroplast morphological
　　　change 2:463–464
　　virion accumulation 2:463–464
　pathogenesis 2:464
　　ND18 strain 2:464, 2:466
　　proteins 2:463, 2:464–466
　replicase proteins 2:461
　replication, gRNAs 2:461
　TGB1/RNP complex 2:462
　TGB3 movement protein 2:464
Barley virus B1 (BarV-B1) 4:311t
Barley yellow dwarf virus (BYDV) 1:279, 1:496
　disease 1:476
　　resistance 1:496
　　symptoms 1:476, 1:496
　genetically-modified resistant plants 1:285–286
　genome 1:280
　　coat protein 1:282
　　noncoding regions 1:281
　　open reading frames 1:280
　　ORF4 mutations 1:282
　MAV strain 1:279
　　epidemiology 1:285
　　genome 1:280
　　replication 1:283–284
　　species-specificity 1:283
　PAS strain 1:280
　PAV strain
　　epidemiology 1:285
　　genome 1:280, 1:281–282
　　host range 1:282–283
　　replication 1:283–284
　　species-specificity 1:283
　RMV strain 1:279
　　replication 1:283–284
　SGV strain 1:279
　　replication 1:283–284
　transmission 1:496

Barley yellow dwarf virus(es) 1:279–286
　classification 1:279
　　strains 1:279, 1:280f
　　see also specific strains
　control 1:285
　　insecticide applications 1:285–286
　　insecticide-treated seeds 1:285–286
　　planting times 1:286
　　resistant strains 1:285–286
　　transgenic plants 1:285–286
　diagnosis 1:285
　　control strategies 1:285
　effects of 1:279
　epidemiology 1:285
　genome 1:280
　　coat protein 1:280, 1:282
　　'leaky scanning' 1:281
　　long-distance frame shift element 1:281–282
　　noncoding regions 1:281
　　open reading fame 1:280
　　ORF4 mutations 1:282
　　protein functions 1:282
　　VPgs 1:282
　historical aspects 1:279
　host range 1:282
　RdRp 1:279–280, 1:280f
　replication 1:283
　　phloem 1:283–284, 1:284f
　taxonomy 1:279
　　amino acid sequences 1:279–280
　　RdRps 1:279–280, 1:280f
　transmission 1:282
　　aphids 1:279, 1:282–283
　　circulative strain-specific manner 1:283
　　species-specificity 1:283
　virion particle 1:280, 1:280f
　　coat proteins 1:280
　　VPg 1:280
　virus–host relations 1:284
　　chlorosis 1:284–285
　　stunting 1:284–285
　　yield losses 1:284–285
　see also specific viruses
Barley yellow mosaic virus (BaYMV) 1:492
　infection
　　control 1:493
　　symptoms 1:492, 1:492f
　　yield losses 1:492
Barley yellow streak mosaic virus (BaYSMV) 1:490
　structure 1:490
　transmission 1:492
Barley yellow striate mosaic virus (BYSMV) 1:492
　infection 1:492
　replication 4:193
　transmission 1:492
Barmah Forest virus (BFV) 5:88
　characteristics 5:109t
　geographical distribution 5:84f, 5:88
　host range 5:88
　infection
　　clinical features 5:83–84, 5:86t, 5:87t, 5:88
　　zoonosis 5:493
Barnard, J E, Ectromelia virus 3:342
Barnaviridae 1:287, 2:288, 5:11t
Barnavirus 1:287, 2:288
Barnavirus(es) 1:286–288
　evolutionary relationships 1:287, 1:288f
　expression 1:287
　genome 1:287
　　open reading frames 1:287, 1:287f
　historical aspects 1:286
　host range 1:288
　infections, fungi 2:287t
　structure 1:287
　transmission 1:288
　type virus 1:287
　virion properties 1:287
　virion structure 1:287f
　see also specific viruses
Barrier control methods, potato virus Y infections 4:295

BART gene, micro RNAs 5:327
Barur virus (BARV) 1:116
Basal cell(s)
　division 4:20, 4:21
　HPV see Human papillomavirus(es) (HPVs)
　papillomavirus life cycle 4:28–29
Basella alba (Malabar spinach), dsRNAs 2:112
Basidiomycota 4:419
　retrotransposons 4:423, 4:424t
Basidiospore(s), fungal virus transmission 2:285–286
Bat(s)
　filovirus host 2:200
　Hendra virus host 2:95
　paramyxovirus host 4:40–42, 4:45–46, 4:56
　rabies virus host 4:370–371, 5:485
　SARS CoV host 1:555, 4:555
BAV see Banna virus (BAV)
Bawden, Frederick, tobacco mosaic virus 3:399
Bax protein(s), apoptotic role 1:155–157
BaYMN see Barley yellow mosaic virus (BaYMV)
BaYSMV see Barley yellow streak mosaic virus (BaYSMV)
BBMV see Broad bean mottle virus (BBMV)
BBNV see Broad bean necrosis virus (BBNV)
BBSV see Beet black scorch virus (BBSV)
BBTV see Banana bunchy top virus (BBTV)
BBWW
　Bacopa virus infections 4:215
　nonspecific infections 4:219, 4:221f
βC1, Ageratum yellow vein virus 1:318
BC1 protein, mungbean yellow mosaic viruses 3:367
bc-3 resistance gene see Bean common mosaic virus (BCMV); Bean common necrosis virus (BCMNV)
B cell(s)
　activation 3:65f
　attributes 3:61
　clonal selection 3:61
　definition 3:78
　development 1:621
　　Burkitt's lymphoma 5:195
　differentiation 1:121, 3:61
　function 1:122–123
　Lassa fever infection 3:210
　localization 1:122–123
　persistent infection 4:112
　transformation, lymphocryptovirus 4:592–593
　virus latency 2:440–441
　　antigen receptor 2:441
　see also Antibodies; Immunoglobulin(s); specific viral infections
B-cell antigen receptor (BCR), EBV reactivation 2:441
B-cell epitope(s)
　definition 1:137
　variation 3:205
B-cell malignancies 5:196
　lymphomas, reticuloendotheliosis virus infection 4:416
　multicentric Castleman's disease 5:196
Bcl-2 family proteins
　adenovirus E1B-19K gene/protein vs. 1:13–14
　apoptosis 1:155–157, 1:156f, 1:161–162
　definition 1:154
BCMNV see Bean common necrosis virus (BCMNV)
BCMV see Bean common mosaic virus (BCMV)
BCTV see Beet curly top virus (BCTV)
BCV see Beet cryptic virus (BCV)
BdMV see Burdock mottle virus (BdMV)
BDV see Border disease virus (BDV)
Bean (Phaseolus vulgaris)
　dsRNAs 2:111–112
　　molecular evolution 2:115
　as food crop 1:296
　resistance genes, loci 4:182
Bean calico mosaic virus (BaCMV)
　identification/naming 1:298
　infection 4:167f
　　ELISA diagnosis 1:299
Bean common mosaic virus (BCMV) 1:288–295
　classification 1:289
　epidemiology 3:217

　genome 1:292
　　composition 1:291
　　encoded proteins 1:292, 1:292f
　　post-translational proteolytic processing 1:292
　　recombination 1:293, 1:294
　　variability 1:293
　geographical distribution 1:289
　　origin 1:289–291
　　peanut stripe isolates 1:289–291
　history 1:289, 1:290f, 1:292f, 3:215
　host range 1:291, 1:294
　　experimental hosts 1:291
　　peanut stripe isolates 1:291
　　soybean isolates 1:291
　replication 1:292
　serological relationships 1:291
　　broad-spectrum monoclonal antibodies 1:292
　serotypes 1:289
　taxonomy 1:289
　　phylogenetic analysis 1:290f
　　subgroups 1:289, 1:289t
　transmission 1:291
　　aphids 1:291, 3:214
　　seeds 1:291, 3:213f, 3:214, 3:215–216
　virion properties 1:291
Bean common mosaic virus infection
　control 1:294
　diagnosis 1:294
　　RT-PCR 1:294
　　serological techniques 1:294
　host resistance genes 1:292
　　bc-3 resistance gene 1:292–293
　　I gene 1:292, 1:293
　　peanut stripe isolates 1:293
　　transgenic resistance 1:293
　pathogenicity 1:292
　　determinants 1:292, 3:215
　　seed color-related susceptibility 1:293
　pathology 1:292
　prevention 1:294
　symptomatology 1:294
　　cytoplasmic inclusion morphology 1:294
Bean common necrosis virus (BCMNV) 1:288–295
　classification 1:289
　genome 1:292
　　composition 1:291
　　encoded proteins 1:292, 1:292f
　　P1 gene variability 1:294
　　post-translational proteolytic processing 1:292
　　recombination 1:293, 1:294
　geographical distribution 1:289
　　origin 1:291
　history 1:289
　host range 1:291, 1:294
　　NL-3 K strain 1:293
　recombination 1:293
　　NL-3 K strain 1:293
　　occurrence 1:294
　replication 1:292
　serological relationships 1:291
　　broad-spectrum monoclonal antibodies 1:292
　taxonomy 1:289
　transmission 1:291
　　aphids 1:291
　　Dendrobium mosaic virus 1:291
　　seeds 1:291
　virion properties 1:291
Bean common necrosis virus infection
　control 1:294
　diagnosis 1:294
　　RT-PCR 1:294
　　serological techniques 1:294
　host resistance genes 1:292
　　bc-3 resistance gene 1:292–293
　　I gene 1:292, 1:293
　　introgression 1:293
　pathogenicity 1:292
　　determinants 1:292
　　hypersensitivity 1:292

pathology 1:292
prevention 1:294
symptomatology 1:294
  cytoplasmic inclusion morphology 1:294
Bean dwarf mosaic begomovirus (BDMV) 3:218
Bean golden mosaic virus (BGMV) **1:295–301**
genome
  AC genes 1:298
  bean golden yellow mosaic virus vs. 1:298–299
  common region 1:298
  nuclear shuttle protein 1:298
  open reading frames 1:298
identification/naming 1:298
infection see Bean golden/yellow mosaic disease
molecular characteristics 1:298
vectors 1:296
Bean golden/yellow mosaic disease
diagnosis 1:299
  ELISA 1:299
  PCR 1:299
disease resistance 1:299
economic importance 1:296
etiology 1:297
  DNA probes 1:297–298
  electron microscopy 1:297
  first isolation 1:297
  PCR 1:297–298
history 1:296
host species 1:299–300
management 1:300
  resistant cultivars 1:300–301
  vector control 1:300–301
pathogenicity 1:299
  nucleolus hypertrophy 1:299
resistance
  dwf gene 1:300
  plant breeding 1:299–300
soybean infections 3:217, 3:218
symptoms 1:296
vector control 1:300–301
Bean golden yellow mosaic virus (BGYMV) 4:165
genome 1:298–299
identification/naming 1:298
infection see Bean golden/yellow
  mosaic disease
molecular characteristics 1:297
Bean-pod mottle virus (BPMV) 5:392
Bean yellow dwarf virus 4:236
Bean yellow mosaic virus (BYMV)
infectivity 3:214
  CMV interactions 3:214–215
transmission 3:216f
  seeds 3:215–216
Bebaru virus (BEBV) 5:109t
Bechhold, Heinrich 2:455
Bee(s)
iflavirus infection 3:42
  transmission 3:45
ilarvirus transmission 3:55
  control 3:55
Beet black scorch virus (BBSV) 3:403
satellite virus 3:404
Beet cryptic virus (BCV) 1:98
BCV1 vs. BCV2 1:102
mixed infections 1:102
phylogeny 4:74–75
transmission 1:102
  cross-pollination 1:102
Beet curly top virus (BCTV) **1:301–307**, 4:165
C1 protein 1:303
C2 protein 1:303
C3 protein 1:303
C4 protein 1:303
control 1:306
  defective interfering DNAs 1:307
  plant breeding 1:307
  vector physical barriers 1:307
genome 1:303, 1:303f
  open reading frames 1:303
geographical distribution 1:302
history 1:301

host range 1:306
  Arabidopsis thaliana 1:307
infection 4:167f
  potatoes 4:300t
pathogenesis 1:306
  adenosine kinase 1:306
  C2 proteins 1:306
  C4 proteins 1:306
  leaf curl 1:305f, 1:306
  transcriptional gene silencing 1:306
phylogeny 1:302
Rep 1:303–304
  retinoblastoma-related tumor-suppressor protein
    binding 1:307
replication 1:303
  deletions 1:304
  recombination-dependent replication 1:303–304
  rolling circle replication 1:303–304
  subgenomic viral DNAs 1:304
seasonal distribution 1:302
  Circulifer tenellus 1:302–303
transmission 1:304
  Circulifer tenellus 1:301–302, 1:302–303, 1:305,
    1:305f
  phloem limitation 1:306
as type species 1:302
V1 protein 1:303
V2 protein 1:303
V3 protein 1:303
virus movement 1:304
Beetle(s)
legume virus transmission 3:216–217
maize chlorotic mottle virus transmission 3:262
plant virus transmission 5:280
population management, oryctes virus 3:146
Rice yellow mottle virus transmission 1:488
Beet leaf curl virus (BLCV), vector
  relationships 4:195
Beet mild curly top virus (BMCTV) 1:302
geographical distribution 1:302–303
host range 1:306
replication 1:304
Beet necrotic yellow vein virus (BNYVV) 1:308
antigenic properties 1:310
control
  genetic modification 1:310
  resistant varieties 1:309–310
cytoplasmic localization 1:310
genome 1:311f, 1:312
  RdRp 1:312
  subgenomic RNAs 1:311f, 1:312
geographical distribution 1:309
infection (rhizomania) 1:308
  protein dependence 1:312–313
  yield losses 1:308
particle properties 1:310
RNA satellites
  expression 4:535
  large single-stranded RNA satellites 4:531
  satellite-like single-stranded RNAs 4:532
as type species 1:308
Beet ringspot virus (BRSV) 3:406t
diseases 3:411
genome 3:407–408
  sequence identity 3:408
  infectious DNA clones 3:408–409
Beet severe curly top virus (BSCTV) 1:302
geographical distribution 1:302–303
host range 1:306
replication 1:304
Beet soil-borne mosaic virus (BSBMV)
antigenic properties 1:310
genome 1:312
geographical distribution 1:309
infection 1:308
transmission 1:308
Beet soil-borne virus (BSBV) 4:283
CP proteins 4:284
genome, RNA2 4:284
geographical distribution 4:285–286
RT proteins 4:284

serological relationship 4:285f, 4:286
TGB protein 4:283–284
virus–host interactions 4:284
Beet virus Q (BVQ) 4:283
CP proteins 4:284
genome, RNA2 4:284
geographical distribution 4:285–286
RT proteins 4:284
serological relationship 4:285f, 4:286
TGB protein 4:283–284
virus–host interactions 4:284
Beet western yellows virus, associated RNAs 4:533
BEFV see Bovine ephemeral fever virus (BEFV)
Begomovirus 1:30, 1:298, 1:302, 2:98, 4:165
Begomovirus(es) 4:164, 4:198
  β ssDNA satellites 1:315, 1:316f
    cotton leaf curl disease 2:102
    function 1:318
    functions 1:318
    identification 1:315
    interaction complexity 1:320
  diversification, rapid evolution 4:166
  DNA satellites 1:315, 4:165, 4:528t, 4:530
    β ssDNA satellites see above
  'helper' virus infection 4:166
  evolution 4:166
  genome 4:165
  infection see below
  resistance to 4:178–182
  transmission 2:98, 4:165
    whitefly 2:98
  type species 4:165
  virus-induced gene silencing vectors 5:378
  see also specific viruses
Begomovirus infections 4:166, 5:65, 5:289
  βC1 pathogenicity 1:318
    gene silencing suppression 2:102
    helper begomoviruses 1:318
    nuclear localization 1:318
  sweetpotato 4:661t, 4:666
    geographical distribution 4:666
    host range 4:666
    infection symptoms 4:666
    phylogeny 4:666
  symptoms 4:166
  tobacco 5:65
    geographical distribution 5:65
Beijerinck, Martinus W 2:450–451, 5:54
  virus discovery 3:399
BEL, LTR retrotransposons 4:438
Bell's palsy 5:234
Bemisia tabaci see Whitefly
Benyvirus 1:308, 4:283
Benyvirus(es) **1:308–314**
  antigenic properties 1:310
  control 1:309
    chemical control 1:309–310
    genetic modification 1:310
    resistant varieties 1:309–310
  differentiation of 1:312
  epidemiology 1:309
  genome 1:311f, 1:312, 1:313
  geographical distribution 1:309
  host range 1:308
  infections 1:308
    diagnosis 1:313
  morphology 1:313
  nucleic acid properties 1:312
  other taxa vs. 1:313
  particle properties 1:310, 1:310f
  transmission 1:309
    Polymyxa zoospores 1:309
  type species 1:308
  see also specific viruses
Berne virus see Equine torovirus (EToV)
Berrimah virus 4:184t
β-amyloid plaque visualization, bacteriophage display
  2:198
β-barrel fold
  alfalfa mosaic virus structure 5:381, 5:382f
  nonenveloped virus structure 5:381, 5:382f

Subject Index

β-barrel fold (*continued*)
   nudaurelia capensis omega virus structure 5:381
   spherical viruses *see* Spherical virus(es)
βC1
β capsid protein, Providence virus 5:33
β-sheet(s)
   prion propagation 4:340
   prion structure 4:339
β ssDNA satellite(s) **1:314–321**
   comparisons, Indian tomato leaf curl viruses *vs.* 5:130
   cotton leaf curl disease *see* Cotton leaf curl disease (CLCuD)
   diversity 1:318
   geographical distribution 1:318
   host adaptation 1:319–320
   origin 1:317f, 1:319f, 1:320
   replication 2:102
   structure 1:316, 1:317–318, 1:317f
      satellite conserved region 1:317–318
   trans-replication 1:320
Betacryptovirus 1:99, 1:100t, 2:109–110
Betacryptovirus(es) **1:98–104**, 1:100t
   evolutionary relationships 1:103
   genome 1:100
   history 1:98
   molecular biology 1:100
   replication 1:100
   transmission 1:99, 1:102
   virion properties 1:99
   virus–host relationships 1:102
   see also specific viruses
Betaentomopoxvirus 4:323, 4:323t
Betaentomopoxvirus 2:136
   characteristics 2:137t
   structure 4:325
Betaherpesvirinae 1:625, 1:635–636, 1:635t, 2:431, 2:432t, 2:474, 5:11t
Betaherpesvirus(es)
   associated genera 1:635–636, 1:635t
   bigenic amplification 2:426f
   classification criteria 2:432
   evolution 2:478–482
   genome 2:475, 2:476f
      alphaherpesviruses *vs.* 1:636
      gammaherpesviruses *vs.* 1:636
      sequences 1:636
   latency 2:439
   pathogenicity 2:427–428
   see also specific viruses
Betanodavirus 3:431
Betanodavirus(es) 3:430
   capsids 3:433
   genotypes 3:431
   host range 3:432
   infection
      fish 2:229
      symptoms 3:437–438
      transmission 3:437–438
   type species 2:229
   see also specific viruses
Betapapillomavirus 4:9, 4:10f
Betapapillomavirus, carcinoma associated-infection 4:15
Betaretrovirus 3:175, 3:335, 4:460, 4:624
Betaretrovirus(es)
   assembly 2:518, 4:462
   genome 4:624, 4:624f
   morphology 4:461–462
   ruminants **3:175–182**
   see also specific viruses
Betatetravirus(es)
   insect infections 3:126t
   see also specific viruses
Bet gene/protein *see* Foamy viruses
Bevirimat 2:507t, 2:516
BFV *see* Barmah Forest virus (BFV)
BGMV *see* Bean golden mosaic virus (BGMV)
BGYMV *see* Bean golden yellow mosaic virus (BGYMV)

Bhanja virus 1:399
   distribution 1:399
   Haemaphysalis intermedia 1:399
BHF *see* Bolivian hemorrhagic fever (BHF)
BHV1 *see* Bovine herpesvirus 1 (BHV1)
BHV2 *see* Bovine herpesvirus 2 (BHV2)
Bicistronic, definition 3:263
Big liver and spleen disease virus (BLSV) 2:380
The Big Picture Book of Viruses 5:357t
Big-vein disease of lettuce *see* Lettuce big-vein disease
BILR 355 BS 2:507t
Binomial names 5:405
Bioassay(s)
   luteoviruses 3:237
   Taura syndrome virus 5:7–8
Bioavailability 4:552
Biocides, pecluvirus infection control 4:102
Biocrime
   definition 5:407
   see also Bioterrorism
BioCyc database 5:357t
Biodissemination, oryctes rhinoceros virus transmission 3:499
Biodistribution, definition 2:301
BioHealth Database 5:357t
Bioinformatics, microarrays and 2:16
Bioinformatics Resource Centers (BRCs) 5:363
Biological control 3:125
   African cassava mosaic disease 1:36
   agents 3:125
   baculoviruses *see* Baculovirus(es)
   densoviruses *see* Densovirus(es)
   strategies 3:125
   see also Insect pest control; *specific viruses*
Biological fitness 4:359
Biological warfare
   definition 5:407
   see also Bioterrorism
Biopesticides, tetraviruses 5:36
Biosystem 3:400
Biotechnology
   bacteriophages, filamentous ssDNA 2:197
   bacteriophage use 2:448–449
   watermelon mosaic virus 5:439–440
   zucchini yellow mosaic virus 5:439–440
Bioterrorism **5:406–411**
   alphaviruses 5:408
   Aum Shinrikyo 5:407
   Bacillus anthracis 5:406, 5:408t
   biotechnology 5:409
      mousepox virus 5:409
      pandemic influenza 5:410
      synthetic genomes 5:410
   definitions 5:407
   deterrence 5:410
   history 5:407
   recognition 5:410
   response 5:410
   virus classification 5:409t
Biotic transmission, rice yellow mottle virus epidemiology 4:488
Biotype(s)
   bovine viral diarrhea viruses 1:376
   definition 1:374
Bioweapon(s) 5:407
   associated factors 5:408
   hypothetical attack 5:407–408, 5:408t
   see also Bioterrorism
Bipartite genomes
   definition 1:295
   potyviruses 4:316–317
Bipartite nuclear targeting sequence, definition 3:447
Bipartite virus, definition 4:164
Biphasic milk fever, tick-borne encephalitis virus infection 5:53
Bird(s)
   Crimean-Congo hemorrhagic fever virus 1:597
   see also individual species
   see also individual viruses
'Bird flu' pandemic *see* Influenza
BIR domain, inhibitor of apoptosis proteins 1:236f

Birnaviridae 1:321, 1:322t, 2:232, 5:11t
Birnavirus(es) **1:321–328**
   characteristics 5:62t
   evolutionary relationships 1:326
      VPg-linked genome replication strategy 1:326
   fish *see* Fish birnavirus(es)
   gene expression 1:322
   genera 2:232
   genome 1:322, 1:323f
      open reading frames 1:322
   history 1:321
   nomenclature 1:321
   RdRp 1:322
      catalytic motifs 1:322–323
      palm domain 1:322–323
   replication 1:325
      morphogenesis 1:326
      transcription 1:325–326
      VLP formation 1:326
   shellfish 4:566
   virion composition 1:322
   virion proteins 1:322
      capsid protein VP2 1:322, 1:324
      internal protein VP3 1:325
      nonstructural protein VP5 1:325
      protease VP4 1:325, 1:325f
      pVP2 peptides 1:324
      RdRp VP1 1:322
   virion structure 1:322, 1:323f
   virus assembly 1:325
   see also specific viruses
Birth defects, definition 4:514
Bison, bovine herpesvirus 2 1:367
Bivalves, definition 4:560
BIV_{R29} infections (cattle)
   experimental infection 1:352
   serology 1:350
BK virus(es) (BKV) **4:261–271**
   associated disorders 4:269
   cytoplasmic entry
      erythrocyte membrane binding 4:281
      sialic acid receptors 4:280–281
   genome oncogenicity
      experimental animals 4:269
      human tumors 4:270
      mechanisms of 4:269
      strain potency 4:269
   host cell attachment
      phospholipase A2 activity 4:281
      receptors 4:280
   infection 4:269
      organ transplantation syndromes 3:470
      phospholipid bilayer role 4:281
      transplant-associated 3:468
   reactivation 4:269
   replication 4:279
   structural organization 4:269
   transmission 4:269
   see also specific viruses
Blackcurrant reversion virus (BRV) 3:406t
   diseases 3:411
   transmission 3:411
Black queen-cell virus (BQCV)
   host range 2:40
   infection 2:42
   transmission 2:42
Black raspberry necrosis virus (BRNV) 3:37, 4:523, 4:524t
   genome sequencing 4:524–525
   host range 4:525
   transmission 4:525
   vector control 4:525–526
Black-seeded bean cultivars, infection resistance 1:293
'Black spot' 1:515
BLAST 5:352
   output 5:354
   pairwise sequence comparison *vs.* 5:345
   sequence comparison methods 5:343
   viral evolution study 2:179–180
Blebbing, definition 1:231
Blepharoconjunctivitis 2:491

Blindness
　HCMV 2:496
　measles 2:494–495
BLMoV *see* Blueberry leaf mottle virus (BLMoV)
Blood, virus shedding 5:316–317
Blood-borne virus(es)
　CNS invasion pathways 1:470, 1:472*f*
　see also specific viruses
Blood–brain barrier
　CNS invasion pathways 1:471
　nervous system viruses 1:469–470
　SIV infection 4:600
　TMEV infection 1:447
　West Nile virus infection 5:446
Blood transfusions
　Crimean-Congo hemorrhagic fever treatment 1:603
　HCV infection 2:336
　hepadnavirus transmission 2:336
　hepatitis E virus transmission 2:336
　HTLV-1 transmission 2:559
　KSHV transmission 3:191
　vCJD risk 1:373
Blueberry leaf mottle virus (BLMoV) 3:406*t*
　diseases 3:411
　genome 3:408
Blueberry shoestring virus (BSSV) 4:646*t*
Bluetongue virus(es) (BTV) **1:328–335**
　capsid assembly 1:333
　　VP3 1:333
　　VP7 1:333
　capsid structure 3:459, 3:459*f*, 3:460*f*
　　modifications 3:459–460
　　outer capsid proteins 1:329
　　VP2 3:459, 3:461, 3:462–463, 3:463–464
　　VP5 3:459, 3:462–463, 3:463–464
　　VP7 3:459–460, 3:461
　cell egress 1:333
　　calpactin light chain 1:333
　　NS3/NS3A proteins 1:333
　　Tsg101 1:333
　core particle 1:329, 1:330*f*, 5:382*f*, 5:388–391, 5:392
　　trimeric proteins 1:329–330
　　X-ray crystallography 5:388–391
　epidemiology 1:334
　gene expression control 4:153–154
　genome 1:331, 1:331*t*, 3:460, 3:461*f*, 3:462*f*
　　noncoding regions 1:331
　　sequence conservation 1:331
　　terminal sequences 1:331
　geographical distribution 3:454–457, 3:458
　history 1:328, 3:454
　host cell attachment 1:331
　　VP2 1:331–332
　　VP5 1:332
　host cell entry 1:331
　infection
　　clinical features 1:333–334, 3:458
　　control 1:334, 3:458–459
　　economic losses 3:458–459
　　host range 3:457
　　immunity 1:334
　　outbreaks, Europe 3:458
　　pathogenesis 1:333
　　prevention 1:334
　　sheep 1:333–334
　　tissue tropism 3:458
　morphology 1:329, 1:329*f*
　mRNA production 4:152–153
　protein synthesis 1:332
　　NS1 1:332
　　NS2 1:332, 1:333
　　phosphorylation 1:332–333
　replicase/transcription complex 4:153–154
　replication 1:331, 4:152–153
　seasonal distribution 3:458
　serotypes 3:454–457, 3:455*t*, 3:459, 3:464*f*
　　BTV-8 3:458
　stability 3:459
　structure 1:329, 1:329*f*, 3:459, 3:459*f*, 5:382*f*
　　capsid *see above*

　　core particle *see above*
　　NS3 3:464
　transcription 1:332
　　VP4 1:332
　　VP6 1:332
　transmission 1:333, 1:334
　　temperature effects 3:457
　　vector competence 3:457
　as type species 4:536*t*
　vaccines/vaccination 1:334
　　inactivated 1:334–335
　　live attenuated 1:334
　viral core channels 4:152–153
　virion properties 1:329
　see also specific viruses
BMCTV *see* Beet mild curly top virus (BMCTV)
BmDNV-1 *see* Bombyx mori densovirus type 1 (BmDNV-1)
BmNPV infection, wandering behavior 1:269
BMS-378806 2:507*t*
BMV *see* Brome mosaic virus (BMV)
BNYVV *see* Beet necrotic yellow vein virus (BNYVV)
Bocavirus 4:95
Bocavirus(es) 4:95
　classification 4:92*t*
　definition 2:93
　genome 4:94*f*, 4:95–96
　see also specific viruses
Bohle iridovirus (BIV), host range 3:156
Boletus virus X (BolVX) 4:311*t*
Bolivian hemorrhagic fever (BHF)
　clinical features 3:209
　epidemiology 3:206
　histopathology 3:209
　pathogenesis 3:208
　pathology 3:209
　therapy 3:211
Bombyx mori cypovirus 1 (BmCPV-1) 3:136
Bombyx mori densovirus type 1 (BmDNV-1) 4:79
　genome 4:83
　infection 4:79
Bombyx mori densovirus type 2 (BmDNV-2) 4:76
Bone marrow
　transplantation, human herpesvirus infection 2:498–499, 2:502
　Visna-Maedi virus infection 5:428
Booster vaccination 2:54
Bootstrap analysis 4:127
Border disease 2:247
　control 1:340
　　vaccines/vaccination 1:340
　detection 1:340
　　at birth 1:340
　　enzyme-linked immunosorbent assay 1:340
　　reverse transcriptase-polymerase chain reaction 1:340
　fetal infection 1:337
　　immune-mediated damage 1:337
　historical aspects 1:336
　pathogenesis 1:336, 1:337*f*, 2:248
　　cell culture 1:337–338
　　fleece coarseness 1:336
　　immune-mediated damage 1:337
　　neurological signs 1:336
　persistent viremia 1:336
　　late-onset disease 1:337
　　myelin deficiency 1:336–337
　symptoms 1:337
Border disease virus (BDV) **1:335–341**, 1:526–527, 2:246
　genogroups 1:338
　genome 1:338
　　internal ribosomal entry site 1:339
　　open reading frame 1:338
　　sequence identity 1:339
　　sequencing 1:338
　heat inactivation 1:338
　historical aspects 1:336
　isolation 1:336
　infection *see* Border disease
　molecular biology 1:338

　nonstructural proteins 1:339
　　N^{pro} 1:339–340
　　NS2-3 1:339–340
　　NS3 protease 1:340
　　NS5A 1:340
　particle size/morphology 1:338
　physico-chemical properties 1:338
　proteins 1:339, 1:339*f*
　replication 1:340
　　attachment 1:340
　　clathrin-dependent endocytosis 1:340
　structural proteins 1:339, 1:339*f*
　transmission 2:247
Borna disease virus (BDV) **1:341–347**
　detection 1:346–347
　epidemiology 1:344
　　virus reservoirs 1:344
　genome 1:341, 1:342*f*, 3:326*f*
　　open reading frames 1:341
　geographical distribution 1:342
　history 1:341
　infection *see below*
　life cycle 1:342
　propagation 1:342
　protein properties 1:341
　　open reading frames 1:341–342
　replication 1:342
　　efficiency 1:342, 3:331
　　mechanisms 1:342
　　splicing 3:330
　synonyms 1:341
　variability 1:344
　virion properties 1:341
　　nucleocapsid protein structure 3:331
　　phosphoprotein 3:331
　　stability 1:341
Borna disease virus infection
　clinical features 1:345
　　disease characterization 1:345
　　incubation time 1:345
　control 1:346
　histopathology 1:345–346, 1:346*f*
　host range 1:342
　　experimental hosts 1:344
　immune response 1:345
　pathogenicity 1:345
　pathology 1:345
　prevention 1:346
　rats 1:344
　　clinical features 1:345
　　immune response 1:345
　　pathology 1:345–346
　serology 1:344
　tissue tropism 1:342, 1:344
　　neurotropism 1:344–345
　transmission 1:344
Bornaviridae 3:324, 5:11*t*
Bornavirus 1:321–322
Bornavirus(es)
　characteristics 5:62*t*
　genome 3:326*f*
　P mRNA transcripts 3:331
　transcription 3:330
　see also specific viruses
Bos, Lute 5:55
Botrytis virus F (BVF), classification 2:288
Botrytis virus X 2:255
Bottle-shaped virus(es) 1:590
　see also specific viruses
Bottom component (B), alfalfa mosaic virus 1:81–82
BoTV *see* Bovine torovirus (BoTV)
Bovine adenovirus(es) 1:26
　see also specific viruses
Bovine atadenovirus(es) 1:8
　see also specific viruses
Bovine enteric calicivirus(es) 1:411
　see also specific viruses
Bovine enterovirus(es) (BEV) 2:123
　classification 2:123
　genome 2:126

Bovine enterovirus(es) (BEV) (continued)
  virion structure 2:128, 2:128f
  see also specific viruses
Bovine enzootic fever see Bovine ephemeral fever virus (BEFV)
Bovine ephemeral fever virus (BEFV) 4:184t
  antigenic variation 1:360
    site G1 1:360–361
    site G2 1:360–361
    site G3 1:360–361
    site G4 1:360–361
  epizootiology 1:355
    vectors 1:355
  genome 1:357, 1:358f
  geographical distribution 1:355
  history 1:354–355
  infection
    immune response 1:361
    pathogenesis 1:356
    pathology 1:356
  taxonomy 1:355, 1:356f
  transcription 1:359
    α coding region 1:359–360
    β coding region 1:359–360
    γ coding region 1:359–360
    polyadenylation signal 1:359–360
  transmission 1:115
  vaccines/vaccination 1:361, 1:362
  virion morphogenesis 1:356
  virion structure 1:356, 1:357f
    G protein 1:356–357
    nucleocapsid 1:356–357
  virus encoded proteins 1:357
    G_{NS} protein 1:359
    G protein 1:358, 1:361f
    L protein 1:359
    M protein 1:358
    N protein 1:357
    P protein 1:358
    small nonstructural proteins 1:359
Bovine ephemerovirus, genome 3:326f
Bovine gammaherpesvirus(es)
  distribution 1:365
  history 1:363
  transmission 1:365
Bovine herpesvirus(es) 1:362–368, 1:364t
  antigenic relationships 1:365
  classification 1:363, 1:364t
  epidemiology 1:365
  geographical distribution 1:365
  history 1:362
  infection
    clinical disease 1:366
    control 1:367
    immune response 1:367
    latency 1:365
    pathogenesis 1:365
    prevention 1:367
  replication 1:363
  structure 1:363
  see also specific viruses
Bovine herpesvirus 1 (BHV1) 1:364t
  genital disease control 1:367
  history 1:362–363
  latency 1:365
  vaccines/vaccination 1:367
Bovine herpesvirus 2 (BHV2) 1:364t
  history 1:363
  mammilitis 1:366
Bovine herpesvirus 3 (BHV3) 1:363
Bovine herpesvirus 4 (BHV4) 1:363, 1:364t
Bovine herpesvirus 5 (BHV5) 1:364t
Bovine herpesvirus 6 (BHV6) 1:364t
Bovine immunodeficiency virus(es) (BIV) 1:347–354
  classification 1:347
  clinical features 1:351
  economic impact 1:353
  epidemiology 1:351
  evolution 1:350
  genetics 1:350
    variability 1:350

genome 1:349
  open reading frames 1:349–350
geographical distribution 1:348, 1:348f
  prevalence 1:348
history 1:347
host range 1:350
infection
  clinical features 1:351
  control 1:353
  histopathology 1:352
  immune response 1:352
  pathology 1:352
  prevention 1:352
  secondary diseases 1:352
propagation 1:350
  associated cell types 1:350
serology 1:350
  BIV_{R29}-infected cattle 1:350
taxonomy 1:347
tissue tropism 1:351
transmission 1:351
  iatrogenic 1:351
  PCR detection 1:351
variability 1:350
virion structure 1:349, 1:349f
  viral core 1:349
  viral enzymes 1:349
see also specific viruses
Bovine influenza see Bovine ephemeral fever virus (BEFV)
Bovine malignant catarrhal fever (MCF)
  causative agents 1:363, 1:365, 1:366–367
  clinical features 1:366–367
  control 1:367
  history 1:363
  hosts 1:363, 1:365
  pathogenesis 1:365
  prevention 1:367
Bovine papillomavirus 1 (BPV-1) 4:32
  genome 4:31f
  pathology 4:29–30
  research history 4:26
  structure 4:32f
  viral protein functions 4:31–32
    E1 4:22
    E2 4:22, 4:36
    E4 4:32
    E5 4:23
    E6 4:31–32
    E7 4:31–32
    L1 4:32
    L2 4:32
Bovine papillomavirus 4 (BPV-4) 4:32–33
Bovine papular stomatitis virus (BPSV) 4:57–58
  host range 4:58
Bovine pestivirus(es) 2:4
  see also specific viruses
Bovine respiratory syncytial virus
  genome 4:42, 4:43f
  reverse genetics 4:44–45
Bovine rhinovirus(es) (BRVs)
  classification 4:468
  see also specific viruses
Bovine spongiform encephalopathy (BSE) 1:368–374, 5:189
  associated cases 1:368, 1:369–370, 5:189, 5:189f
  disease characteristics 5:189
  epidemiology 1:369, 1:370f
    mathematical models 1:369–370
  geographical distribution 1:370, 1:371f, 5:190
  history 1:368, 5:189
    detection 5:189
  incubation period 1:368–369
  novel forms 1:370–371
  origin 1:368, 5:189
  prevention 5:190
    active testing 1:369, 1:370
    legislation 1:368
    passive surveillance 1:369, 1:370
    specified bovine offal ban 1:369
  species barrier/prion variant 4:332, 4:339

tissue distribution 1:368–369
transmission 1:370
vaccine safety concerns 5:228
see also Creutzfeldt–Jakob disease (CJD)
Bovine torovirus (BoTV) 5:151–152
  antigenic properties 5:155
  attachment/cell entry 5:153
  buoyant densities 5:153
  epidemiology 5:155
  genotypes 5:155
  historical aspects 5:151
  host range 5:154
  infection
    diarrhea 5:155–156
    pathogenesis 5:155–156
  tissue tropism 5:155
Bovine viral diarrhea virus(es) 1:374–381, 1:526–527
  characteristics 1:378
    virion inactivation 1:378
    virion structure 1:378
  genome recombination 2:248
  history 1:375, 2:246
    biotypes 1:376
    early reports 1:375
    genotypes 1:377
    isolation 1:376
    modified live virus release 1:376
    postvaccinal mucosal disease 1:376
    reduction and eradication programs 1:378
  infection see below
  molecular biology 1:379
    genome 1:377f, 1:378, 1:379
    viral proteins 1:379
    viral replication 1:380
  vaccines/vaccination 2:248–249
  see also specific viruses
Bovine viral diarrhea virus 1 (BVDV-1)
  antigenic/genetic relationships 1:338
  classification 1:336
  type 2 vs. 1:378, 1:379, 2:246
Bovine viral diarrhea virus 2 (BVDV-2)
  antigenic/genetic relationships 1:338
  classification 1:336
  prevalence 2:246
  type 1 vs. 1:378, 1:379, 2:246
Bovine viral diarrhea virus infection
  clinical features 1:375, 1:375f, 2:248
    cytopathic type 1:375f, 1:376
    noncytopathic type 1:375f, 1:378
  control 1:380
  detection 1:380, 2:248
  hemorrhagic syndrome 1:377–378
    BVDV1 vs. BVDV2 1:379
    clinical symptoms 1:377–378
  mucosal disease
    disease characteristics 1:375
    etiology 1:376
  pathogenesis 2:248
  persistent 1:376
  pregnancy 2:248
BPSV see Bovine papular stomatitis virus (BPSV)
BPV-1 see Bovine papillomavirus 1 (BPV-1)
BQVC see Black queen-cell virus (BQCV)
Bracovirus(es) (BV)
  definition 4:256
  gene families 4:250–251
  immunoevasive activity 4:255–256
  immunosuppressive effects 4:254
  infections 4:253
    insects 3:126t
  phylogeny 4:252f
  replication 4:260–261
  virion structure 4:257, 4:257f
    nucleocapsid 4:257
  see also specific viruses
Bradford coccus 3:312
Brain
  human herpesvirus infection 2:503
  JCV infection 4:262
  pseudorabies virus infection 4:349
  see also Central nervous system (CNS)

Branch length, phylogenetic trees 4:126
Brassica napus, cauliflower mosaic virus silencing suppression 1:468–469
Brazilian wheat spike virus (BWSpV) 1:492, 5:24
BRCs database 5:357t
Breakthrough infection 1:69
Breast cancer
　mouse mammary tumor virus model 4:274
　Simian retrovirus D 4:625, 4:629
Breastfeeding, HIV infection prevention 1:66
Breda virus 2:122
　diarrhea 2:122
Brevidensovirus 4:76, 4:83, 4:94t, 4:96
Brevidensovirus(es) 4:76, 4:96
　classification 4:94t
　genome 4:76–77, 4:83, 4:95f, 4:96–97
　　monosense genomes 4:83
　virion structure 4:78
　see also specific viruses
Bridge vector(s) 4:652
　definition 5:440
　vector transmission, animal viruses 5:270
　West Nile virus transmission 5:443–444
BRNV *see* Black raspberry necrosis virus (BRNV)
Broad bean mottle virus (BBMV)
　defective-interfering RNAs 1:389, 4:380
　recombination 4:380
Broad bean necrosis virus (BBNV) 4:283
　CP proteins 4:284
　genome
　　RNA2 4:284
　　untranslated regions 4:283
　geographical distribution 4:285
　particles 4:283
　RT proteins 4:284
　small open reading frame 4:284
Broad spectrum antiviral therapy 1:147, 1:154
Brome mosaic virus (BMV) 1:381–386, 1:492
　alphavirus-like superfamily 1:381
　coat protein translation 1:381
　　RNA3 1:384
　foreign gene expression 1:381
　　reverse genetics 1:382
　gene expression 1:381
　genome 1:381, 1:383f
　　cis-signals 1:383
　　noncoding regions 1:383
　　promoter 3:54
　　RdRp enzyme 1:389
　　RNA1 1:381, 1:388f
　　RNA2 1:381, 1:388f
　　RNA3 1:381, 1:383–384, 1:388f, 1:389
　　RNA4 1:381, 1:384, 1:388f
　host factors 1:384
　infectious *in vitro* transcripts 1:381
　as a model 1:381
　protein characterization 1:381
　RdRp 1:382, 1:389
　　promoter interactions 1:384
　recombination, template switching 1:385
　replication 1:382, 1:383f
　　associated proteins 1:382
　　conserved proteins 1:382
　　host factors 1:384
　　protein localization 1:382
　　recombination 1:389
　RNA encapsidation 1:385, 1:389
　RNA-protein interactions 1:383
　RNA recombination 1:385, 4:377–378
　　homologous recombination 4:378, 4:378f
　　host factors 4:379
　　replicases 4:378–379
　structure 1:387, 1:388f
　subgenomic mRNA synthesis 1:384, 1:389
　　core promoter 1:384
　　pathway elucidation 1:384
　transmission 1:390, 5:280–281
　virion
　　assembly polymorphism 1:385
　　classes 1:381

　size 1:385
　structure 1:385
Brome mosaic virus infection
　host range 1:385
　spread of 1:385
　　essential proteins 1:385
　　systemic spread alterations 1:385–386
5′-Bromo-4-chloro-3-indolylphosphate (BCIP) 2:24
Bromoviridae 1:81, 1:381, 1:387, 2:255, 3:46, 5:11t
Bromovirus 1:81, 1:381, 1:387, 3:46
Bromovirus(es) **1:386–390**
　cucumovirus *vs.* 1:387
　gene expression 1:388
　genome 1:387t, 1:388
　homologous recombination 1:389
　infections *see below*
　nonhomologous recombination 1:389
　phylogenetics 1:387
　RNA replication 1:389
　taxonomy 1:387, 1:387t
　transmission 1:390
　tRNA-like structure 1:81
　virion structure 1:387
　virus–host relationship 1:390
　see also specific viruses
Bromovirus infections
　associated host factors 1:389
　resistance genes 4:179t
　tobacco disease, economic importance 5:63
　　alfalfa mosaic virus 5:63
　　cucumber mosaic virus 5:63
　　disease outbreaks 5:63
　　tobacco streak virus 5:64
Bronchiolitis
　definition 2:551
　human respiratory syncytial virus infection 2:548
Bronchopneumonia, morbillivirus infection 4:504
BRO proteins, baculovirus 1:271
Broth culture, bacteriophages 2:74
Brown citrus aphid *(Toxoptera citricida),* citrus tristeza virus transmission 1:521, 1:524
Brown trout reovirus 1:164t
BRSV *see* Beet ringspot virus (BRSV)
BSBMV *see* Beet soil-borne mosaic virus (BSBMV)
BSBV *see* Beet soil-borne virus (BSBV)
BSCTV *see* Beet severe curly top virus (BSCTV)
BSE *see* Bovine spongiform encephalopathy (BSE)
BSMV *see* Barley stripe mosaic virus (BSMV)
*Bsu*361 restriction enzyme 1:240
BSV *see* Banana streak virus (BSV)
Bubaline herpesvirus 1 (BuHV1) 1:364t
Budded virion(s) (BV)
　baculoviruses 3:129
　definition 1:254
　nucleopolyhedrovirus structure 1:255–256, 1:263
　　AcMNPV 1:263
　　GP64 protein 1:263
Budding
　alphaviruses 1:196, 1:200, 5:121
　avian hepadnavirus 2:330f, 2:331
　Chandipura virus 1:503
　duck hepatitis B virus 2:330f, 2:331
　Ebola viruses 2:59–60
　equine infectious anemia virus 2:169–170
　flaviviruses 1:199
　foamy virus 2:264
　HTLV-1 2:561
　maize mosaic virus 4:193, 4:196
　mononegaviruses 3:333
　mumps virus replication 3:359–360
　nucleorhabdoviruses 3:333, 3:334
　paramyxoviruses 3:333–334
　plant rhabdoviruses 4:194–195
　Rift Valley fever virus 1:397
　rubella virus 5:121
　vesicular stomatitis virus 2:295
　virus membranes 5:313, 5:313f
　　M proteins 5:314
Buffalo, capripoxvirus infection 1:428
Bu gene 3:41
Buggy Creek virus 5:93t

Bunyamweravirus 3:480
Bunyamwera virus (BUNV)
　genome 1:394f
　replication 1:397
Bunyaviridae 1:170, 1:391, 1:399, 2:317–318, 3:479, 4:490–491, 5:11t, 5:27, 5:133, 5:157–158
Bunyavirus(es) **1:390–399**
　attenuation 5:317–318
　definition 1:390
　genomes 1:172, 3:482
　　coding strategies 1:393, 1:394f
　　reassortment 1:395, 1:396f
　　RNA segments 1:392, 2:317–318
　　S segment 1:395
　　terminal complementarity 1:392–393
　infections 1:391t
　　apoptosis inhibition 1:158t, 1:161
　　human respiratory infection 2:555
　　persistent 1:397
　　vectors 1:392
　replication cycle 1:395
　　host cell attachment 1:395
　　host cell entry 1:395
　　RNA replication 1:396f, 1:397, 1:398f
　　transcription 1:395, 1:396f
　　translation 1:396–397
　　uncoating 1:395
　　virion assembly 1:397
　　virion release 1:397
　reverse genetics 1:398
　transmission 1:391, 1:392f
　　vector competence 1:392
　'unassigned' **1:399–401**, 1:400t
　　definition 1:399
　virion proteins 1:392, 1:393f
　　glycoproteins 1:393
　　L protein 1:393
　　nucleocapsid 1:392
　virion structure 1:172, 1:393f
　　morphology 1:393f
　see also specific viruses
Burdock mottle virus (BdMV) 1:308–309
　genome 1:312
　transmission 1:309
Burkitt's lymphoma
　chromosome translocations 2:154
　EBV infection 2:154, 2:154f, 2:158–159, 4:115, 5:194
　　B-cell development 5:195
　　endemic 2:154, 5:194
　　epidemiology 5:195
　　seroepidemiological studies 2:154
　　sporadic 5:194
　history 2:149
　incidence 2:154
　Simian retrovirus D 4:629
Burley tobacco 5:61
　virus resistance 5:64
Burnett, F M, Ectromelia virus 3:342
Burst size
　bacteriophage population ecology 2:75
　definition 2:71, 5:411
　haloviruses 5:412
　methanogenic Archaea-infecting viruses 5:412
Bushmeat
　HIV origin 2:532, 4:604
　SIV cross-species transmission 2:532–533
Bushpig *(Potamochoerus porcus),* African swine fever virus 1:43
BVDV-1 *see* Bovine viral diarrhea virus 1 (BVDV-1)
BVDV-2 *see* Bovine viral diarrhea virus 2 (BVDV-2)
BVQ *see* Beet virus Q (BVQ)
Bxb1 bacteriophage 5:185
BYDV *see* Barley yellow dwarf virus (BYDV)
Bymovirus 4:315t
Bymovirus(es) 1:492, 1:493
　genome 1:492
　infection
　　control 1:493
　　plant resistance genes 4:179t
　　symptoms 1:492, 2:292
　　P2-2 protein 2:293–294, 4:317

Subject Index

Bymovirus(es) (continued)
 transmission 5:281
 variants 1:492–493
 see also specific viruses
BYMV see Bean yellow mosaic virus (BYMV)
BYSMV see Barley yellow striate mosaic virus (BYSMV)

C

C1 proteins, beet curly top virus 1:303
C2 proteins, beet curly top virus 1:303, 1:306
C3 proteins, beet curly top virus 1:303
C4 proteins, beet curly top virus 1:303, 1:306
C-6 virus, sweetpotato infections 4:661t
Cabassou virus (CABV) 5:92, 5:93t
 characteristics 5:109t
Cabbage, turnip mosaic virus infection 5:289
Cabbage leaf curl virus (CabLCV) 5:326
CABV see Cabassou virus (CABV)
Cacao swollen shoot virus (CSSV) 1:403–410, 4:200
 classification 1:403
 genome 1:405, 1:406f
  dsDNA 1:404–405
  N-terminal coat protein 1:405
  open reading frames 1:405
  ORF1 1:405
  ORF2 1:405
  ORF3 1:405–406
  ORF alignments 1:405, 1:407t
  variation 1:405
  viral homology 1:406, 1:408f
 geographical distribution 1:404
 host range 1:403
 infection see below
 molecular characterization 1:404
 molecular epidemiology 1:408
 phylogenetic tree 1:405–406, 1:408f
 transmission 1:403, 1:404, 4:200
  Pseudococcidae mealybugs 1:404, 1:404f
  reservoir hosts 1:404
 variability 1:405, 1:407t
  PCR diagnostic tests 1:405
 virion 1:404
Cacao swollen shoot virus (CSSV) infection
 control 1:404, 1:409, 4:200
  'cutting-out' campaigns 1:404, 1:409
  eradication policies 4:200
  isolation 1:409
  PCR diagnosis 1:409
  resistance breeding 1:409
  strain protection 1:409
 diagnosis 1:408
 see also specific methods
 economic importance 1:403, 1:404
  geographical distribution 1:404
  yield losses 4:200
 PCR diagnosis 1:409
  ORF3 1:409
  primers 1:409
  sequence alignment 1:409
 reservoir hosts 1:404
 serology 1:408
  ELISA 1:408–409
  immunocapture polymerase chain reaction 1:408
  strain variation 1:408–409
  variation 1:405
  virobacterial agglutination test 1:408
 symptoms 1:403, 1:404f
  avirulent strains 1:403
  defoliation 1:403
  red veins 1:403
  stem swellings 1:403
Cache valley virus 1:391t
Cactus virus X (CVX) 4:311t
Cadmium-induced glycine-rich protein (cdiGRP) 3:355
Caecum, definition 1:37
Caenorhabditis elegans, RNA interference assays 3:149
 FHV infection 3:153
CAEV see Caprine arthritis encephalitis virus (CAEV)
CaGMV see Cassava green mottle virus (CaGMV)
Caladium virus X (CalX) 4:311t

Calcineurin-dependent pathways, African swine fever virus immune system evasion 1:47
Calcium channels, KP4 toxin action 5:216
Calf serum, vaccine safety concerns 5:228
Calibrachoa mottle virus (CbMV)
 Calibrachoa infection 4:214, 4:214f
 petunia infection 4:213
Calibrachoa virus infections 4:214
 see also specific viruses
Caliciviridae 1:90f, 1:410, 1:411, 1:411t, 2:117, 2:230, 2:378, 3:439, 3:439t, 3:440f, 5:11t
Calicivirus(es) (CaCV) 1:410–419, 2:116
 characteristics 2:117t
 epidemiology 1:415
 fish see Fish calicivirus(es)
 genetics 2:118
 genome 1:412–413, 1:413f, 3:440–441
  NS proteins 3:440–441, 3:442t
 geographical distribution 1:413, 2:117
 history 1:410, 2:116
 infections 3:439t
  clinical signs 1:417
  control 1:418, 2:118
  host range 1:414, 2:116–117
  immune response 2:118
  immunity 1:417
  pathogenesis 1:416, 2:118
  prevalence 2:118
  prevention 1:418, 2:118
  symptoms 1:515
  veterinary pathogens 3:439
 morphology 1:411–412, 2:116–117, 2:117f
 phylogeny 1:413f
 propagation 1:414
 properties 1:411
  nonstructural proteins 1:413
 seasonal distribution 2:117
 serologic relationships 1:415
 structure 1:411–412, 1:412f, 1:414f, 2:117f
  capsid 2:118
 transmission 2:117–118
 see also specific viruses
Calimorpha quadripunctata virus (CqV) 5:28t
Callitrichine herpesvirus 3 (CalHV-3) 4:587t
Calomys
 arenavirus transmission 3:204
 Junin virus infection 3:204–205
Calves
 astrovirus pathogenesis model 1:208
 see also specific virus infections
Calyx cell, definition 4:256
Camelpox virus (CMPV), infection 4:329
Campoletis sonorensis ichnovirus (CsIV)
 genome 4:258, 4:258f
  coding density 4:258–259
  cys-motif 4:255, 4:258–259
  open reading frames 4:258–259
  rep genes 4:258–259
  segment nesting 4:258–259
  transposable elements 4:257–258
  vankyrin gene family 4:258–259
  vinnexin gene family 4:258–259
 host immune response abrogation 4:253–254
  ankyrin genes 4:255
  encapsulation 4:255
  hemocyte adhesion 4:255
  innexin expression 4:255
  melanization 4:255
Campyloganthia, definition 1:328
CaMV see Cauliflower mosaic virus (CaMV)
Canada, HIV epidemiology 1:65
Canarypox-like virus(es) 2:278
 see also specific viruses
Canarypox virus (CNPV) 2:275
Cancer
 Mason–Pfizer monkey virus 4:629
 see also specific cancers
Cancer therapy, echoviruses 2:70
Candlestick syndrome, African cassava mosaic virus 2:101
Canine adenovirus infection 1:5
Canine distemper virus (CDV) 4:502

cell entry 3:285
evolution 4:502
geographical distribution 4:498
host cell binding 4:500
host range 4:499–500
infection 4:498
 clinical features 4:504
vaccines/vaccination 4:506
Canine oral papillomavirus (COPV) 4:33
Canine parvovirus infection
 control 4:90
 pathogenesis 2:122, 4:89
Canker
 chestnut blight 2:575
 definition 2:574
Cannibalism, Taura syndrome virus transmission 5:7
Cantalago virus 5:489–490
Canyon hypothesis, picornavirus gene expression 4:131
Cap genes, porcine circoviruses 1:516
Capillary fragility, Crimean-Congo hemorrhagic fever 1:601–602
Capillary endothelia, CNS invasion pathways 1:471–472
Capillary leakage, Argentine hemorrhagic fever 3:208
Capillovirus 1:419–427
 associated species 1:420, 1:420t
 biological properties 1:420
 classification 1:420t
 genome 1:420
  heterogeneity 1:422
 particle structure 1:420, 1:421f
 phylogeny 1:426f
 replication 1:420
 serology 1:421
 strains 1:422
Capillovirus 1:420, 1:420t, 1:426f
Cap-independent RNA translation 4:133, 4:133f
 rhinoviruses 4:470
Cap protein(s), porcine circoviruses 1:517, 1:519
Caprine arthritis encephalitis virus (CAEV) 5:424
 infection 5:430
  macrophage infection 5:428–430
 vaccines/vaccination 5:431
Caprine herpesvirus 1 1:364t, 1:366
Caprine herpesvirus 2 1:364t, 1:367
Capripoxvirus 1:427
Capripoxvirus(es) 1:427–432
 classification 1:427, 4:324t
 distribution 1:428
 evolution 1:429
 genetics 1:428
  recombination 1:428
  sequence similarity 1:428, 1:432
  stability 1:428
  strain identification 1:428
 history 1:427
 infections see below
 isolation 1:428
 propagation 1:428
 serologic relationships 1:429
 taxonomy 1:427
 viability 1:430
 see also specific viruses
Capripoxvirus infections
 clinical features 1:430
  papules 1:430
 epizootiology 1:429
  morbidity rates 1:429
  mortality rates 1:429
 histopathology 1:431
 host range 1:428
 immune response 1:431
 incubation period 1:430
 pathogenicity 1:430
 pathology 1:431
 prevention and control 1:431
  climate 1:431–432
  vaccination 1:431–432
 tissue tropism 1:429
 transmission 1:429–430
  papules 1:429–430
Capsicum see Pepper (Capsicum)

Capsid(s)
    architecture 5:365
    β-barrel folding motif 1:202f
    definition 4:407
    evolutionary connections 5:374
    helical 1:200–201
    icosahedral 1:200–201
        definition 4:514
    picornaviruses 1:201, 1:201f
    structural polymorphism 1:611–612
    symmetry 1:433
        helical 1:200–201
        icosahedral 1:200–201, 4:68
        picornaviruses 1:201, 1:201f
        T=2, definition 4:68
    see also specific viruses
Capsid-associated transcriptase, Ustilago maydis viruses 5:217
Capsid protein (CP)
    see specific viruses
Cap-snatching
    bunyavirus transcription 1:395–396
    definition 1:390, 3:483
    infectious salmon anemia virus 3:93
    influenza virus transcription 3:487
    tenuivirus replication 5:24, 5:25f, 5:26
Capsomers
    adenovirus virion morphology 1:3, 2:78
    definition 3:264
Carboxypeptidase D, avian hepadnavirus replication 2:329
Carcinogenesis
    HPV infection see Human papillomavirus(es) (HPVs)
    SV40 infection 4:631, 4:636, 4:637–638
    see also Oncogene(s)
Carcinoma
    alphapapillomavirus associated 4:15
    HPV associated 4:14–15
Cardamine chlorotic fleck virus (CCFV) 1:454
Cardamom bushy dwarf virus (CBDV), characteristics 3:389t
Cardiac disease
    coxsackieviruses 1:586
    HIV infection see HIV infection
Cardio-vascular system, African horse sickness 1:40–41
Cardiovirus 1:441, 1:442, 5:38
Cardiovirus(es) **1:440–448**
    classification 1:442
    encephalomyocarditis virus
        see Encephalomyocarditis virus (EMCV)
    epidemiology 1:444
    evolution 1:443
    genome 1:442
    geographical distribution 1:444
    history 1:441
    host range 1:444
    infection
        clinical features 1:445
        pathogenicity 1:445
        pathology 1:445
        prevention 1:447
    propagation 1:442
    proteins 1:442, 1:442f
        leader 2:126
    replication 1:443
    taxonomy 1:441t, 1:442, 1:443f, 1:444f
    transmission 1:444
    virion properties 1:442
    see also specific viruses
Caribbean, HIV epidemiology 1:64
Carlavirus 1:426f, 1:442, 1:448, 1:449t, 1:451f, 4:306
Carlavirus(es) **1:448–453**
    classification 1:448, 1:451f
    gene expression 1:450
    genome 1:450, 1:451f
        open reading frames 1:451
    geographical distribution 1:452
    host range 1:452
    infections
        control 1:452
        cytopathology 1:452

epidemiology 1:452
sweetpotato 4:661t
symptomatology 1:452
physicochemical properties 1:450
serology 1:452
structure 1:450
taxonomy 1:448, 1:451f
transmission 1:449t, 1:452
see also specific viruses
Carmovirus 1:453, 1:454t, 5:145–146
Carmovirus(es) **1:453–457**, 1:454t
    defective-interfering RNAs 1:456, 5:150
    distribution 1:453
    evolutionary relationships 1:453
    gene expression 1:455
    genome 1:454f, 1:455
        open reading frames 1:455
    host range 1:453
    infections
        economic significance 1:453
        resistance genes 4:179t
    replication 1:455
        3' coterminal subgenomic RNAs 1:455
        associated structural elements 1:456
    satellite viruses 1:456
    transmission 1:453
    virion assembly 1:454
    virion structure 1:453, 1:454
    virus–host interactions 1:456
    see also specific viruses
Carnation cryptic virus (CarCV) 1:99
Carnation latent virus 1:448–450
Carnation mottle virus (CarMV) 5:145–146
Carnation mottle virus (CarMV), genome 5:147f
Carnation ringspot virus (CRSV) 5:145–146
    genome 5:147f
Carp pox herpesvirus 2:233
Carp pox herpesvirus infection 2:208
Carriers
    definition 3:6, 5:411
    icosahedral enveloped dsDNA bacteriophages 3:12
    infectious pancreatic necrosis virus 3:87
    Simian retrovirus D 4:627–628
Carrot red leaf virus (CRLV), associated RNAs 4:533
Case-control studies 2:143
    definition 2:140
    retrospective 2:143
Case-fatality rate, definition 2:44, 2:142
Caspase(s) 1:155, 1:156f
    apoptosis 1:232f
    definition 1:154, 1:186, 1:231, 1:247
    inhibition, P35 1:235
    regulators, death receptor signaling 1:160
Caspase-3 1:155
Caspase-8 1:155
    activation 1:156f, 1:160
Caspase-9 1:155–157
Casphalia extranea, biological control 3:128
Casphalia extranea densovirus (CeDNV) 4:79
Cassava (M*anihot esculenta*) 1:30
Cassava American latent virus (CsALV) 3:406t
Cassava common mosaic virus (CsCMV) 4:311t
    genome 4:312
Cassava green mottle virus (CaGMV) 3:406t
Cassava mosaic disease (CMD) 4:198
    control 4:198
    detection 4:198
    symptoms 4:198
    transmission 4:198
    yield losses 4:197, 4:198
Cassava mosaic geminivirus(es) (CMGs) **1:30–37**
    distribution 1:31, 1:32f
    diagnostic tests 1:31–32
    genome 1:30
        open reading frames 1:30–31
    recombination 1:32
    replication 1:31
    structure 1:30

transmission 1:32
    vegetative cuttings 1:32–33
    whitefly vector 1:32–33
see also specific viruses
Cassava virus C (CsVC) 3:500
    taxonomy 3:500
Cassava virus X (CsVX) 4:311t
Cat(s)
    West Nile virus infection 5:446
    see also under Feline
    see also specific viruses
Catalytic RNA 5:332
Catarrhines 4:611
Caterpillars, ascovirus infection 1:189
Cathepsin 1:215t, 1:216
    baculovirus expression vectors 1:251–252
Cattle
    Akabane virus infection 1:78
    bluetongue virus infection 1:334
    capripoxviruses 2:124
    import, virus panzootics 4:497
    parapoxviruses see Parapoxvirus(es) (PPVs)
    rinderpest viruses see Rinderpest virus(es) (RPVs);
    see also under Bovine
Cattle feed, BSE 5:189, 5:190
Cattle plague 4:497
Caudovirales 3:30–31, 3:32t, 5:11t, 5:23, 5:413t
Caudovirus(es) 3:30–31
    assembly
        procapsid assembly 1:435–436
        protein cleavage 1:439
        protein conformational changes 1:434–435
        protein synthesis 1:434–435
        symmetry disruptions 1:437
        tail shaft assembly 1:438
    classification 5:11t, 5:23
    DNA injection 3:33–34
    evolution 3:474–475
        cystoviruses vs. 5:374
        genome comparisons 5:372
        genome organization 5:371
        inoviruses vs. 5:374
        mechanisms 5:371, 5:372
        microviruses vs. 5:374
        populations 5:371
        types 5:371
    genome packaging 2:306
    protein injection 3:34
    structure 3:31
    see also Bacteriophage(s); Icosahedral tailed dsDNA bacteriophages; specific viruses
Cauliflower mosaic virus (CaMV)
    DNA recombination mechanisms 4:376
    gene map 1:461f, 1:467f
    genome 1:467f
        discontinuity 1:460
        ORF I 1:460, 1:468
        ORF II 1:468
        ORF III 1:468
        ORF IV 1:466
        ORF V 1:467
        ORF VI 1:468
        size 1:460
    history 1:457
    inclusion bodies 1:458f, 1:464
    movement within plants, virion-associated protein 3:353
    pathogenicity determinants 1:463–464
    precapsid protein 1:466
    replication 1:457
        terminal redundancy 1:461
        ribosomal shunt mechanism 1:460–461
    RNase H 1:467
    replication 1:462
    35S promoter 1:457–458
    vector recognition 5:279
    virus stability 1:467
    as a virus vector 1:457–458
Caulimoviridae 1:403, 1:457, 1:458, 1:459t, 1:464–465, 4:430, 4:451, 4:482, 5:11t
Caulimovirus 1:459t, 1:464–465, 1:465t

Subject Index

Caulimovirus(es) **1:457–464, 1:464–469**, 1:465t
  bacilliform virions 1:458, 1:464
  capsid properties 1:465, 1:468
  DNA recombination mechanisms 4:376
  genome 1:460, 1:465
    discontinuity 1:460
    DNA structure 1:457
    replication 1:467
    reverse transcriptase 1:460
    RNA processing 1:468
    transcription 1:467
    translation mechanism 1:468
  host range 1:463, 1:464–465
  icosahedral virions 1:458, 1:464
  inclusion bodies 1:458f, 1:464, 1:465, 1:466f
  open reading frames 1:465, 1:466f
  protein properties 1:465
    aphid transmission factor 1:465
    capsid protein 1:466
    movement protein 1:465
    POL polyprotein and cleavage products 1:467
    processing 1:468
    transactivator of translation/viroplasmin 1:467
    virion-associated protein 1:466
  replication 1:461, 1:462f
  silencing and silencing suppression 1:468
  sweetpotato infections 4:661t
  transmission 1:463
    aphids 1:463
    aphid transmission factor 1:465
  virion properties 1:465
  virion structure 1:458
  virus stability 1:467
  see also specific viruses
Causality assessment, definition 5:226
CAV *see* Chicken anemia virus (CAV)
Cavalon IR, EHV-4 treatment 2:419
Cavemovirus 1:459t
Cavemovirus(es) 1:459t
  host chromosome integration 1:462
  open reading frames 1:466f
  transmission 1:463
  virion structure 1:458–460
  see also specific viruses
CbMV *see* Calibrachoa mottle virus (CbMV)
CBP/p300 1:11
CcBV *see* Cotesia congregata bracovirus (CcBV)
CC chemokine(s) 1:623
CCHF *see* Crimean-Congo hemorrhagic fever (CCHF)
CCHFV *see* Crimean-Congo hemorrhagic fever virus (CCHFV)
CCMV *see* Cowpea chlorotic mottle virus (CCMV)
CC-NB-LRR protein, natural resistance 4:171
CCR2b receptor, SIV cell attachment 4:606–607, 4:618
CCR5 receptor
  definition 2:534
  inhibitors 2:507t
  see also HIV infection; Simian immunodeficiency virus (SIV)
CCR7 receptor 1:624
CCRS *see* Cherry chlorotic rusty spot disease (CCRS)
CCRV *see* Channel catfish reovirus (CCRV)
CCSV *see* Cynodon chlorotic streak virus (CCSV)
CCV *see* Channel catfish virus (CCV)
CD2v protein, African swine fever virus 1:47–48
CD4 receptor, HIV infection 2:506, 4:408
CD4+ T-cells 3:75
  antiviral memory 3:80
  CD8+ T cells *vs.* 3:72–73, 3:72f
  circulation 3:75
  cytokine secretion 3:80
  definition 1:69, 2:534
  Dengue fever 2:13
  HIV infection *see* HIV infection
  immunopathology 3:77
  murine CMV infection 1:630t
  persistent infection 4:112
  regulatory function 3:75
  SIV infection *see* Simian immunodeficiency virus (SIV)
  subtypes 3:80
  TMEV infection, response 5:43
  Visna-Maedi virus infection 5:430–431
  see also T helper cell(s)
CD8+ T-cells 3:74, 3:74f
  antigen processing 3:72
  antiviral memory 3:80, 3:116
  CD4 T cells *vs.* 3:72–73, 3:72f
  clonal expansion 3:73f, 3:74, 3:75f, 3:116
  cytokine/chemokine production 3:74–75
  Dengue fever 2:13
  HIV infection *see* HIV infection
  mechanism of action 3:80
  murine CMV infection 1:630t
  persistent infection 4:112
  SIV infection 4:600–601, 4:609, 4:621
  Visna-Maedi virus infection 5:427
  see also Cytotoxic T-cells (CTLs)
CD46, virus-host cell binding 5:324
CDC NCID (National Center for Infectious Diseases) 5:357t
CDD (Conserved Domain Database) 5:357t
CDV *see* Canine distemper virus (CDV)
Cebus cytomegalovirus (CeCMV) 1:635t, 1:639f
CeHV-17 *see* Ceropithecine herpesvirus 17 (CeHV-17)
Cell-associated virus (CEV), vaccinia virus assembly 5:248
Cell attachment/entry
  endocytosis 4:409
    process 4:409
  fusion 4:409
  replication 4:407
  translocation 4:409
  virus membranes 5:310
    binding specificity 5:310
  see also Receptor(s); specific viruses
Cell barriers, bacterial cell viral entry
  see Bacteriophage(s), bacterial cell entry
Cell–cell fusion mating, yeast L-A virus transmission 5:465–466
Cell culture
  aquareovirus infection diagnosis 1:167–168
  baculoviruses 1:250
  Border disease 1:337–338
  Herpesvirus Saimiri 4:586
  varicella-zoster virus propagation 5:251
Cell-cycle link protein, definition 1:272
'Cell-derived' genes, virulence 5:318
Cell envelope, definition 5:365
Cell fusion, Visna-Maedi virus infection 5:430–431
Cell line immortalization, EBV-encoded nuclear antigen 3 family 2:162
Cell lysis
  acidianus two-tailed virus 1:591
  icosahedral enveloped dsDNA bacteriophages 3:11
Cell-mediated immune response **3:70–77**
  antigen presentation 3:72
  antigen processing 3:72
  common cold 4:473
  Ectromelia virus infection 3:345
  human papillomavirus infection 4:17
  immune evasion 3:77
  immunodominance 3:73, 3:73f
    pattern changes 3:73
  immunopathology 3:76
  induction 3:73, 3:74f
  persistent infections 3:76
  T-cell memory 3:75
  varicella-zoster virus infection 5:254–255
  see also specific components
Cell membranes
  BEFV proliferation 1:357
  icosahedral enveloped dsDNA bacteriophages 3:11
Cell(s), origin
  parasitic RNA, evolution of 3:476
  viral roles 3:478
  virus-first hypothesis 3:475
Cell receptors, measles virus 3:287
Cell signaling, murine polyomavirus middle T 4:274, 4:274f, 4:275
Cell surface markers, rhadinovirus human T-cell transformation 4:591
Cell-to-cell transmission
  cowpea mosaic virus proteins 1:573
  grapevine fanleaf virus 3:410
  maize streak virus 3:268
  nepovirus 3:407f, 3:410
  neural cell infection 1:473
Cell tropism
  definition 1:554
  host cell attachment 4:408
Cellular cytidine deaminases *see* Retrovirus infections, resistance to
Cellular DNA synthesis, murine polyomavirus large T 4:276
CeMV *see* Cetacean morbillivirus (CeMV)
Centers for Disease Control and prevention (CDC) 2:94
Central Africa, HIV infection 1:63
Central America, HIV infection 1:65
Central nervous system (CNS)
  arbovirus infection 1:175
  coxsackievirus infection 1:585
  equid herpesvirus 1 infection 2:417–418
  equine encephalitic togavirus infection 5:106
  flavivirus infection 2:245
  HSV infection 2:304
  JC virus infection 4:262
  lactate dehydrogenase-elevating virus infection 1:184
  MHV-68 infection 3:376
  rabies virus infection 4:370
  yellow fever 5:474–475
  see also Nervous system virus(es)
Centrosema mosaic virus (CenMV) 4:311t
Cephalothorax, definition 4:567
Ceratitis capitata idnoreovirus 5 (CcIRV-5) 3:140
Cereal virus(es) 1:475–482, 1:490–497
  see specific viruses
Cereal yellow dwarf disease 1:476
Cereal yellow dwarf virus (CYDV) 1:279, 1:496
  genome
    coat protein 1:282
    noncoding regions 1:281
    open reading frames 1:280
    ORF4 mutations 1:282
    VPg 1:282
  RNA satellites
    replication 4:533, 4:534
    structure 4:534
  RPS strain 1:279
  RPV strain 1:279
    epidemiology 1:285
    genome 1:279
    host range 1:282–283
    replication 1:283–284
    structure 3:233
  transmission 1:496
Cerebrospinal fluid (CSF)
  alphavirus encephalitis 5:106
  echovirus detection 2:69
Ceropithecine herpesvirus 12 (CeHV-12) 4:587t
Ceropithecine herpesvirus 14 (CeHV-14) 4:587t
Ceropithecine herpesvirus 15 (CeHV-15) 4:587t
Ceropithecine herpesvirus 17 (CeHV-17) 4:587t
  discovery 4:586
Certification schemes, infection control measures 1:524
Cervical cancer
  Gardasil vaccine 5:224
  HSV-2 associated 2:394
  papillomavirus infection *see* Human papillomavirus infections
Cervical intraepithelial neoplasia (CIN)
  detection 4:19
  HPV types associated 4:15
Cervid herpesvirus 1 (CerHV1) 1:364t
Cervid herpesvirus 2 (CerHV2) 1:364t
Cesium chloride gradient purification
  cowpea mosaic virus purification 1:569–570
  rhinoviruses 4:469

Cetacean morbillivirus (CeMV) 4:498
  epizootiology 4:504
  evolution 4:502
CfMV *see* Cocksfoot mottle virus (CfMV)
cg4715 protein 4:452f, 4:454
Chaetoceros salsugineum, viral infection of 1:87
Chaetoceros salsugineum nuclear inclusion virus (CsNIV) 1:88t, 1:92, 4:116
  algal infections 1:92
  burst size 1:92–94
  genome 1:92–94, 1:94f
  host features 1:92–94
  phylogeny 1:93f
  virion structure 1:94f
Chalcone synthase gene, virus-induced gene silencing 5:326
Challenge models, infectious pancreatic necrosis virus infection 3:88
Chandipura virus (CPHV) 1:497–503
  budding 1:503
  classification 1:497
  composition 1:499
  gene expression 1:500
    mRNA modifications 1:501
    viral RdRp 1:500, 1:502
  gene structure 1:499
  genome 1:498f, 1:499
  genome encapsidation 1:502
    model 1:502–503
    N protein 1:502–503
  history 1:497
  infection *see below*
  life cycle 1:499, 1:500f
    glycoprotein G role 1:500
    matrix M protein role 1:500
    viral invasion 1:499
  phosphoprotein P 1:501
    dimerization 1:501
    phosphorylation 1:501
    unphosphorylated 1:501, 1:502
  replication 1:502
    model 1:502
    transcription–replication switch 1:500f, 1:502
  structure 1:498f, 1:499
  taxonomy 1:497
  transcription 1:500
    stop–start model 1:500, 1:501f
  transmission 1:498–499
  viral assembly 1:503
    matrix protein 1:503
Chandipura virus infection
  acute encephalitis 1:497–498
    fatality rate 1:497–498
    symptoms 1:498
  epidemiology 1:497
  zoonoses 5:492
Changuinola virus 3:464–465
Channel catfish reovirus (CCRV) 1:164t
  genome 1:165
Channel catfish virus (CCV)
  capsid structure 2:207f
  genetics
    gene families 2:209–211
    genome characteristics 2:210t
    genome structure 2:209
    koi herpesvirus *vs.* 2:211
    open reading frames 2:209–211, 2:210f
  growth properties 2:209
    protein expression 2:209
  history 2:206–207
  infection
    acyclovir treatment 2:211
    clinical features 2:208
    detection 2:208
    histopathology 2:208
    immune response 2:211
    occurrence 2:207
    prevention and control 2:211
  transmission 2:208
  virulence 2:208
Chapare virus, geographical distribution 3:204

Chaperone(s)
  plant virus movement 3:353
  prions and 4:337, 4:338f
  specificity 4:337–338
Chaperonin(s), bacteriophage assembly 1:438
Charge-coupled device (CCD) camera 1:605
Charleville virus (CHVV) 4:184t
Chelonus, BV-mediated immunosuppression 4:254–255
Chemical control, benyvirus 1:309–310
Chemical platforms, tetraviruses 5:36
Chemical tags, Nudaurelia ω virus 5:36–37
Chemokine(s) 1:620–624
  α family 1:623
  β family 1:623–624
    receptors 4:408
  definition 1:620, 3:70
  Dengue fever 2:13
  families 1:623
  function 1:623
  homologs, KSHV immune evasion 3:200
  ligands, HIV-1 cell entry inhibitors 2:468–469
  receptors 1:624
    definition 4:611
    HHV immune evasion 2:501
    HIV-host cell interactions 5:323
    see also specific receptors
  regulation 1:620
  rhinovirus infection 1:546
Chemosis, definition 1:580
Chenopodium mosaic virus X (ChMVX) 4:311t
Chenopodium necrosis virus (ChNV) 3:403
Cheravirus 3:405–407
Cherries
  Sharka disease 4:239
  see also specific viruses
Cherry chlorotic rusty spot disease (CCRS) 1:509–510, 1:511
  etiology 1:511
Cherry chlorotic rusty spot disease associated chrysovirus (CCRSDACV), genome 1:505–506
Cherry leafroll virus (CLRV) 3:406t
  diseases 3:411
  population structures 3:412
  transmission 4:205
  walnut tree infection 4:206
Cherry mottle leaf virus (CMLV) 4:204
  genome 1:425
  hosts 1:424
  replication 1:425
Cherry virus A 1:420–421
Chestnut blight
  biocontrol 2:584
  Cryphonectria parasitica 2:575, 2:580
  epidemic 2:580–581
  history 2:575
Chick embryo, animal virus cultivation 2:456
Chicken anemia virus (CAV)
  associated disease symptoms 1:514
  genome 1:516–517
    anellovirus *vs.* 1:107
    rolling circle replication 1:517
  taxonomy 1:513–514, 1:514f
  tissue tropism 1:514
  virion structure 1:514
Chicken influenza virus 2:4
Chicken inhibitor-of-apoptosis protein (ch-IAP1) 4:418
Chickenpox
  clinical features 5:253, 5:254f
    lesions 5:253
    complications 5:253
      Reye syndrome 5:253
    historical aspects 5:250–251
    incubation period 5:253
Chickpea chlorotic stunt virus 5:344
Chicory yellow mottle virus (ChYMV) 3:406t
Chikungunya virus (CHIKV) 5:88, 5:113
  characteristics 1:171t
  evolutionary relationships 1:175
  genetic lineages 5:88
  geographical distribution 5:84f, 5:88, 5:113

  infection *see below*
  transmission 5:88, 5:100, 5:111–112, 5:113
    Aedes aegypti 5:100
    mosquitoes 5:111–112
    sylvan 5:113
Chikungunya virus infection
  acute illness 5:89
  clinical features 5:86t, 5:87t, 5:88
    hemorrhagic manifestations 5:89
  Dengue virus infection *vs.* 2:5
  disease outbreaks 5:89, 5:113, 5:487
  disease surveillance 5:89, 5:487
    global programs 2:50
  morbidity rates 5:100
  mortality rates 5:89
  pregnancy 5:87
  zoonoses 5:487
CHIKV0 *see* Chikungunya virus (CHIKV)
Chilo iridescent virus (CIV) *see* Invertebrate iridescent virus-6 (IIV-6)
Chimera(s), definition 1:513
Chimeric transcript(s)
  definition 4:445
  retrovirus recombination 4:446–447
Chimeric virus(es)
  definition 1:410
  see also specific viruses
Chimerivax Dengue virus, vaccine development 5:238–239, 5:240f
Chimpanzee cytomegalovirus (CzCMV) 1:635t, 1:639f
Chimpanzee rhadinovirus 4:587t
Chinese hamster ovary (CHO) cells, cowpox virus growth 1:576
Chinese mitten crab reovirus 1:164t
Chinese wheat mosaic virus (CWMV) 1:493
  host range 2:292
  transmission 1:493
Chinook salmon aquareovirus (DRCRV) 1:164t
Chinook salmon reovirus (SCRV) 1:164t
Chironomus luridus entomopoxivirus 2:136
Chitinase, baculoviruses 1:251–252
Chlamydia
  icosahedral ssDNA bacteriophages 5:368
  viruses *vs.* 3:476
Chlorella virus(es) 4:123
  history 4:116
  see also specific viruses
Chloridovirus(es)
  insect infections 4:126
  see also specific viruses
Chloris striate mosaic virus (CSMV) 1:476
Chlorophyll defects, plant diseases, economic aspects 3:218
Chloroplast(s)
  rice yellow mottle virus 4:487–488
  tymovirus replication site 5:203
  vaccine production in plants 5:223
Chlorosis
  barley yellow dwarf viruses 1:284–285
  cucumber mosaic virus infection 1:617–618
  definition 1:614, 4:207
Chlorovirus 4:117t, 4:118
Chlorovirus(es)
  insect infections 3:126t
  see also specific viruses
Chlorpromazine, JCV 4:280
Chordopoxvirinae 1:427, 2:136, 3:226, 4:58, 4:323–325, 4:324t, 5:11t, 5:461–462
Chordopoxvirus(es) (CV) 2:136
  bacopa virus infections 4:215
  entomopoxviruses *vs.* 2:137
  genome 4:325–326
  phylogeny 4:326–327
  see also Entomopoxvirus(es); *specific viruses*
Chorioretinitis 2:388–390
Choristoneura fumiferana cypovirus 16 (CfCPV-16) 3:138, 3:139t
Choristoneura fumiferana multiple embedded nucleopolyhedrovirus (CfMNPV) 1:248
  host range determinants 1:252–253
Choroid plexus

Subject Index

Choroid plexus (continued)
  CNS invasion pathways 1:471–472
  definition 1:37
Chromosomes
  definition 2:306
  fungal retrotransposons translocation breakpoints 4:426
  hepatitis B virus integration 2:366
  icosahedral tailed dsDNA bacteriophages 3:35
Chronic bee-paralysis virus (CBPV) 4:527t, 4:530
Chronic bee-paralysis virus-associated satellite virus (CBVA) 4:527t
Chronic infections 4:114
  immune control 4:112
  initiation 4:109–110
  pathogenesis 4:111
    reactivation 4:111
  virus–host co-evolution 4:111
  yellow head virus infection 5:480
  see also Persistent infections; specific infections
Chronic suppressive therapy, acyclovir 1:145–146
Chronic wasting disease (CWD) 5:188
  clinical characteristics 5:188
  geographical distribution 5:188
  host range 5:188
  prevention 5:189
  transmission 5:188
    experimental 5:188
Chrysanthemum stunt viroid (CSVd) 4:223, 4:223f
Chrysanthemum virus B 4:223
Chrysanthemum virus infections 4:221
  see also specific viruses
Chrysoviridae 2:111, 2:286, 5:11t, 5:214
Chrysovirus 1:503–504
Chrysovirus(es) 1:503–513
  biological properties 1:509
    chryso-P3 1:509t
      functions 1:511
      'phytoreo S7 domain' 1:507
    chryso-P4 1:508
      functions 1:511
  classification 1:503–504, 1:504t
  evolution 1:508f, 1:510, 1:512f
  gene expression 1:507
    capsid proteins 1:504t, 1:507
    RdRp 1:504t, 1:507, 1:508f
  genome 1:504, 1:505f, 1:507t
    5′ untranslated regions 1:506–507, 1:506f
      CAA repeats 1:506–507
    dsRNA segments 1:504–505, 1:506f
  infections
    fungi 2:287t
    mixed infection 1:509
  occurrence 1:511
  replication 1:508
  transmission 1:509, 1:510–511
  virion properties 1:504
    RdRp activity 1:504
  virion structure 1:508, 1:509f
    capsid 1:508–509, 1:510f
  see also specific viruses
Chub reovirus (CHRV) 1:164t
Chum salmon aquareovirus (CSRV) 1:164t
CHV1 see Cryphonectria parasitica virus (CHV1)
CHV-EP713 virus 2:581
  field trials 2:584
  genetic organization 2:581, 2:583f
  p29 expression 2:583–584
  phenotypic change determinants 2:583
CHV-Euro7 virus, phenotypic change determinants 2:583
CHV-NB58 virus, genetic organization 2:582
Chytrid fungi
  legume virus transmission 3:217
  tombusvirus transmission 5:149–150
Chytridiomycota 4:419
  retrotransposons 4:423, 4:424t
Cicadellid leafhoppers, phytoreovirus transmission 4:154–155
CID see Cytomegalic inclusion disease (CID)
Cidofovir 2:557

adenovirus infection treatment 1:29
HSV treatment 2:395
mode of action 2:395
CiMV see Citrus mosaic virus (CiMV)
CIN see Cervical intraepithelial neoplasia (CIN)
CIPHI Fact Sheets 5:357t
CI protein, potyvirus protein properties 4:317
Circadulina, definition 3:264
Circoviridae 1:107, 1:513, 5:11t
Circovirus(es) 1:513–520
  genome 1:516
    associated proteins 1:517
    open reading frames 1:517
  infection
    associated diseases 1:514
    diagnosis 1:516
    immune system interactions 1:516
    pathogenesis 1:514
  nanoviruses vs. 3:391
  PCV transcription 1:517
  replication 1:518
  taxonomy 1:513, 1:514f
  transmission 1:514
  virion structure 1:514
  VP1 gene/protein
    antibody induction 1:517
    replication 1:517
  see also specific viruses
Circular dichroism (CD), inovirus structure 3:120–121
Circular genome replication, dsDNA bacteriophage genome packaging 2:308–309
Circular permutation
  definition 3:161, 5:411
  invertebrate iridoviruses 3:163–164
Circulating recombinant forms, definition 2:525
Circulating vaccine-derived poliovirus 4:246–248
Circulative strain-specific manner, barley yellow dwarf virus transmission 1:283
Circulative transmission
  banana bunchy top virus transmission 1:274
  definition 1:272
Circulative virus(es)
  definition 1:482
  see also specific viruses
Circulifer tenellus, beet curly top virus 1:301–302, 1:302–303, 1:305, 1:305f
Circumcision, HIV infection 1:60, 1:66
Cirrhosis
  definition 2:241
  HCV infection 2:249
cis-acting replicating elements (cre)
  definition 4:129
  narnavirus see Narnavirus
  picornavirus see Picornavirus(es)
  RNP complexes see Ribonucleoprotein (RNP)
cis-acting siRNA pathway, plant gene silencing 4:142
Citrus mosaic bacilliform virus (CMBV) 1:403
Citrus mosaic virus (CiMV) 4:523, 4:524t
  genome sequencing 4:524–525
  host range 4:525
  transmission 4:525
Citrus production 1:520
Citrus psorosis virus (CPsV) 3:447, 3:448t
  coat protein 3:450
  serology 3:450
  control 3:453
  genome 3:449, 3:450t
    RNA2 3:449
    sequence 3:448–449
  pathogenicity 3:450–451
  phylogenetic analysis 3:447
  transmission 3:453
Citrus root disease 1:520
Citrus sudden death (CSD) 2:91, 5:206
  causal agents 2:92
    classification 2:92
  disease symptoms 2:92
  epidemiology 2:92
  geographical distribution 2:91

infection control 2:92
transmission 2:92
yield losses 2:92
Citrus sudden-death-associated virus (CSDaV) 2:92
  classification 2:92
  genome 2:92, 5:205
Citrus tristeza virus (CTV) 1:520–525, 2:92, 4:197
  classification 1:521
  genome 1:522
    defective RNAs 1:523, 1:523f
    functions 1:522
    nontranslated regions 1:522, 1:522f
    open reading frames 1:522–523, 1:522f
    subgenomic RNAs 1:523
  geographical distribution 1:520, 1:521, 4:197–198
  history 1:520
  infection see below
  replication 1:523
  strains 1:521, 4:198
  taxonomy 1:521
  transmission 1:521, 1:524, 4:197–198
    aphids 1:521, 1:524
    experimental 1:520
  virion structure 1:521
    thread-like particles 1:520
Citrus tristeza virus infection 1:521
  control measures 1:524, 4:198
    certification schemes 1:524
    cross-protection 1:524, 4:198
    eradication 1:524, 4:198
    genetically-modified resistant plants 1:525
    grafting 1:524
    resistance 1:524
  cytopathology 1:521
  diagnosis 1:524
  economic impact 1:524
    crop yield losses 4:197–198
  host range 1:521
CIYMV see Clover yellow mosaic virus (CIYMV)
CJD see Creutzfeldt–Jakob disease (CJD)
C-Jun protein, JCV gene expression 4:264
Clades
  definition 4:436
  varicella-zoster virus 5:252
Classical swine fever virus (CSFV) 1:525–533
  antigenic relationships 1:338, 1:527
  classification 1:336, 1:526
  cytopathic isolates 1:526
  epidemiology 2:247
  genome 1:526
  history 2:246
  phylogeny 1:526–527, 1:527f
  properties 1:526
  protein expression 1:526
  replication 1:526
    tissue location 1:529
    virion assembly 1:526
  taxonomy 1:526
  transmission 1:528
    experimental 1:528
  vaccines/vaccination 1:531, 2:248–249
    booster vaccines 1:531–532
    marker vaccines 1:531–532
Classical swine fever virus infection 1:528, 2:247
  acute 1:528
  antibodies, neutralization 1:531
  chronic 1:528
  clinical manifestations 2:248
  diagnosis 1:530, 2:248
    antibody detection 1:531
    differential diagnosis 1:530
    direct antigen detection 1:530
    isolation in cell culture 1:530
    nucleic acid detection 1:530
  economic impact 1:526, 1:532
  epizootiology 1:529
  histopathology 1:529
  history 1:526
  host range 1:528
  immune response 1:531
  incubation period 1:528

late onset 1:528
pathology 1:529
pigs 1:528
prenatal 1:528
prevention and control 1:532
secondary infections 1:528
severity 1:528
virulence 1:528
Classification **5:9–24**
   Adansonian system 5:19
   databases 5:362
   definition 5:9
   demarcation criteria 5:19, 5:20*t*
   hierarchical 5:10, 5:22
   Linnaean system 5:19
   order of presentation 5:11*t*, 5:21
      criteria 5:21
   phylogeny 4:128
   polythetic 5:19
   sequence comparisons 5:19, 5:20, 5:342
      alignment algorithms 5:342
      BLAST 5:343
      global 5:343
      local 5:343
      multiple sequence 5:343
      pairwise sequence comparison *see* Pairwise sequence comparison (PASC)
      phylogenetic analysis 5:343
      scoring schemes 5:342–343
   species **5:401–406**
      abstract classes *vs.* real objects 5:402
      Aristotelian class 5:402
      binomial names 5:405
      classical 5:403
      criteria associated 5:403
      definitions 5:9, 5:21, 5:402, 5:403
      demarcation 5:404
      diagnostic properties 5:405
      evolutionary concept 5:403
      nomenclature 5:405
      phenotypic concept 5:404
      phylogenetic analysis 4:126–127
      polythetic classes 5:401
      qualitative properties 5:402
      taxonomic 5:404
      taxonomic classes *vs.* 5:404
      typography 5:405
      universal system 5:22
      virus identification 5:405
   universal system 5:11*t*, 5:22
      families 5:23
      genera 5:22
      orders 5:23
      species *see above*
   virus properties 5:342, 5:347
   see also Nomenclature; Taxonomy
Class-switching *see* Immunoglobulin(s)
Clathrin-dependent endocytosis
   Border disease virus 1:340
   yellow fever virus 5:471
CLCuD *see* Cotton leaf curl disease (CLCuD)
CLCuMV *see* Cotton leaf curl Multan virus (CLCuMV)
Cleave site mutations
   Nudaurelia ω virus 5:34*f*, 5:35–36
   tetraviruses 5:35
Clethrionomys, cowpox virus hosts 1:577, 1:578
Clevudine, hepatitis B treatment 1:152–153
Climate, vector transmission 5:273
Climate change, zoonotic effects 5:494
Clinical lung disease, Visna-Maedi viruses 5:424
Clinical trials, gene therapy *see* Gene therapy
Clonal selection, B-cells 3:61
Cloning vector(s)
   bacteriophage M13 2:197
   definition 2:190
Closed circular DNA, avian hepadnavirus 2:327
Closteroviridae 1:521, 2:255, 5:11*t*
Closterovirus 1:521
Clouds, quasispecies 2:178–179
Clover yellow mosaic virus (ClYMV) 3:447, 4:311*t*

verbenaceae infection 4:218
CLRV *see* Cherry leafroll virus (CLRV)
'Clump' disease, pecluvirus infections 4:97–98
CMD *see* Cassava mosaic disease (CMD)
CMLV *see* Cherry mottle leaf virus (CMLV)
CMV *see* Cucumber mosaic virus (CMV)
c-myc, reticuloendotheliosis virus infection 4:416
CO_2 (carbon dioxide) symptom *see* Sigma virus (SIGV)
Coagulopathy
   Ebola virus infection 2:64, 2:203
   Marburg virus infection 2:203, 3:277
Coastal Plains virus 4:184*t*
Coat protein (CP) 3:354
   definition 1:432
   plant engineered resistance 4:156–157, 4:157–160, 4:162, 5:439
   RNA complexes 3:24
   see also specific viruses
Cocksfoot mild mosaic virus (CMMV) 4:646*t*
Cocksfoot mottle virus (CfMV) 4:646*t*
   infection resistance 4:651–652
   virion structure 4:650*f*
Cocoa necrosis virus (CNV) 3:406*t*
Coding capacity, replication 4:407
Coding sequences, pecluvirus genome 4:99
Codon 129 polymorphism 1:371, 1:372, 4:330
   CJD 4:331–332
   prion strain propagation 4:335
Codons, definition 4:623
Co-evolution
   adenoviruses 1:1
   chronic infection 4:111
   definition 2:317
Cohabitation experiments
   definition 5:91
   salmon pancreas disease virus 5:94
Cohort studies 2:143
   definition 2:140
Coho salmon aquareovirus (CSRV) 1:164*t*
Coho salmon reovirus (CSRV) 1:164*t*
   protective effects 1:167
Coiled-coil motif, definition 4:97
Coital exanthema 1:362–363, 1:366
Coleus blumei viroid 1 (CbVd1), transmission 5:335
Coliphage(s)
   bacteriophage conversion 2:445–446
   propagation 2:444–445
Colorado tick fever (CTF) 1:533–534
   clinical features 1:538–539
   cytopathic effects 1:536
   diagnosis 1:538–539
   distribution 1:534, 1:534*f*
   incidence 1:533–534
   male *vs.* females 5:489
   neurological disorders 1:538–539
   symptoms 5:489
Colorado tick fever virus (CTFV) 1:533, 4:536, 5:489
   characteristics 1:171*t*
   cross-reactivity 1:536–537
   distribution 1:534, 1:534*f*
   genome 1:536, 1:537*f*
   geographical distribution 5:489
   history 1:533–534
   host range 1:534–535, 5:489
   infection *see above*
   serotypes 1:533
   transmission 1:535
      blood transfusion 1:535
   as type species 4:536*t*
   vectors 1:534
   zoonoses 5:489
Color-coded table(s), virus databases 5:352
Colostrum-deprived (CD) calves, definition 5:151
Coltivirus 1:533, 1:539*f*, 4:536, 4:536*t*
Coltivirus(es) **1:533–541**, 1:539*f*
   antigenic relationships 1:536
   classification 1:533, 1:539*f*
   distribution 1:534
   epidemiology 1:534

genetic relationships 1:536
genome 1:535, 1:535*f*
   dsRNA segments 1:536
history 1:533
host range 1:535
infection
   clinical features 1:538
   diagnostic assays 1:538
   immunity 1:540
   prevention 1:540
   treatment 1:540
prototype species 4:536
replication 1:535
transmission 1:535
   vectors 1:535
type species 4:536*t*
virion properties 1:535
   stability 1:535–536
virion structure 1:172, 1:535*f*
see also specific viruses
Comet assay, vaccinia virus spread 5:248
Commelina virus X (ComVX) 4:311*t*
Commelina yellow mottle virus (CoMYV), genome 1:458
   discontinuity 1:458, 1:461
   gene map 1:461*f*
   open reading frames 1:461
   size 1:461
Commercialization, transgenic plants 4:5
Common ancestry, icosahedral tailed dsDNA bacteriophages 3:36
Common cold 4:468
   clinical features 4:472
      lower respiratory tract 4:472, 4:474*f*
      sleep pattern disruption 4:472
   coronavirus infection 1:554–555
   definition 4:552
   epidemiology 4:471
   immune response 4:472
      antibody-mediated virus inactivation 4:472–473
      cellular immunity 4:473
      immunoglobulin A 4:472–473
      immunoglobulin G 4:472–473
   incubation period 4:472
   National Center for Health Statistics 1:541
   pathogenesis 4:472
      proinflammatory mediators 4:472, 4:473*f*
   prevention/control 4:473
      antiviral therapy 4:473–475
      vaccine development 4:473–475
   rhinovirus infection *see* Rhinovirus infections
   symptoms 1:541
   transmission 1:541, 4:472
      aerosol inhalation 4:472
      hand-to-hand contact 4:472
Common region (CR)
   bean golden mosaic virus 1:298
   mungbean yellow mosaic viruses 3:366
Common vehicle transmission, definition 2:140, 2:145
Common warts, papillomavirus associated 4:16
Community ecology
   bacteriophages *see* Bacteriophage(s), ecology
Comoviridae 1:90*f*, 1:569, 1:573, 3:405–407, 5:11*t*
Comovirus 1:569
Comovirus(es)
   plant resistance genes 4:179*t*
   see also specific viruses
   structure 5:391–392
   see also specific viruses
Compacted genomes, RNA virus experimental evolution 4:363
Comparative genome analysis, Infectious salmon anemia virus 2:229
Competition, bacteriophage population ecology 2:75
Complement 3:110, 3:112
   activation 3:79–80, 3:79*f*, 3:112
   antiviral properties 3:110
   control proteins
      HIV-1 infection 3:110
      receptor role 5:322
   definition 3:78, 3:104, 3:112

Complement (continued)
  infection neutralization role 3:414
  KSHV immune evasion 3:199–200
  opsonization 3:110
  pathways 3:110, 3:112
  regulation 3:110
Complementarity-determining region(s) (CDRs)
  antibody footprint 1:138
  IgG paratopes 1:138
Complementary sense, definition 1:295
Complementary sense genes, maize streak virus 3:267
Complement fixation assay(s)
  coltivirus infection diagnosis 1:540
  Japanese encephalitis diagnosis 3:186
  orthobunyavirus 3:482
  principles 2:31
Compositae virus infections 4:221
  see also individual species
  see also specific viruses
CoMYV see Commelina yellow mottle virus (CoMYV)
Concatamer(s)
  definition 2:306, 4:475, 5:411
  replication 2:308–309
Confocal laser scan microscopy (CLSM) 2:26–27
Conformational selection model 4:330
Congenital infections/disease
  Akabane virus 1:76
  definition 4:514
  HCMV infection 2:483
  herpesviruses 2:146
  rubella virus see Congenital rubella syndrome (CRS)
Congenital malformations, Akabane virus 3:480–481
Congenital rubella syndrome (CRS) 4:514, 4:518
  clinical manifestations 4:519
  diagnosis 4:520
  histopathology 4:519
  neuropathology 4:519
  transmission 5:111
  virus persistence 4:519
Conjunctiva 2:491
  infection 2:491
Conjunctival hyperemia, definition 1:580
Conjunctivitis
  acute hemorrhagic 2:495
    coxsackievirus infections see Coxsackievirus(es)
    definition 2:130
    enterovirus 70 2:132
    human adenoviruses 2:495
    human enterovirus D 2:124
  adenoviruses see Human adenovirus(es)
  associated RNA viruses 2:492
  EBV 2:496
Connecticut virus (CNTV) 1:116
Connochaetes gnu (wildebeest), bovine malignant catarrhal fever 1:363
Consensus sequence, definition 4:164
Conservation (insect pest control) 3:125
Conserved consensus sequences, geminivirus genome 4:168
Conserved epitope binding, antigen–antibody complexes 3:417–418
Conserved functional domains, three-hybrid assay 3:28–29
Conserved genes
  Herpesvirus Ateles 4:588
  Herpesvirus Saimiri 4:588
  tailed bacteriophage evolution 2:182f
cis-acting Conserved sequence elements (CSEs) 5:120–121
Constitutive transport element (CTE), Simian retrovirus D 4:625
Continuous cropping, pecluvirus infection control 4:102
Continuous epitopes 1:138, 1:139f
Contractile tails, definition 3:30
Contrast transfer function (CTF), cryo-electron microscopy 1:605–606

Convalescence stage, Crimean-Congo hemorrhagic fever 1:601
Conventional dendritic cell(s) (cDC)
  activation 3:106
  definition 3:104
Convergence model, epidemiology 2:141, 2:142f
Convergent evolution, definition 4:256
Convolvulaceae 4:659–660
Coprophagy, definition 3:89
Coquillittidia perturbans, eastern equine encephalitic virus transmission 5:103–104
Co-receptor(s)
  definition 5:319
  infection role 5:323
    Ab-mediated neutralization 3:415
  receptor vs. 5:323
  specificity 5:323
  see also specific viruses
Core genes, baculoviruses 1:249
C-ori, definition 3:264
Corky ringspot (potato disease) 4:307, 4:307f
Cornea
  immunology 2:492
  infection 2:491
Corn leaf necrosis disease complex 1:482
Corn lethal necrosis (CLN) 3:259
Corn stunt disease complex 1:482
  symptoms 1:482
  transmission 1:482
Coronaviridae 1:177, 1:549, 1:554–555, 2:122, 3:419–420, 4:553–554, 5:11t, 5:151–152
Coronavirus 1:549, 2:230
Coronavirus(es) 1:549–554, 1:554–563, 2:122
  classification 1:549
  evolution 1:550t, 1:552, 1:553f
  gene expression 1:550
  genera 1:549, 2:230
  genome 1:550, 1:555–556, 1:559–560, 1:559t, 3:424f
    accessory genes 3:426–427
    nonstructural proteins 3:423–424
    nucleotide cyclic phosphodiesterase 3:425–426
    open reading frames 1:550–551, 1:555–556
    proteolytic processing 3:424
    replicase genes 3:425f
    replication–transcription complex 1:551, 3:423–424, 3:428
    size 1:550–551
  human 1:553
  infection see below
  nomenclature 1:554–555
  recombination 1:552, 1:558–559, 1:560
  replication 1:551, 1:552, 1:557, 1:558f
    glycoprotein synthesis 1:551–552
    host cell attachment 1:551–552, 1:556–557
    host cell fusion 1:557
    ORF1b translation 1:557
    replicase function 1:557
    template switching 1:557–558
    transcription 1:557–558, 1:559f
    virus particle formation 1:552
    virus particle release 1:552
  reverse genetics 1:560
  RNA recombination 4:377, 4:379, 4:379f
  SARS CoV see Severe acute respiratory syndrome coronavirus (SARS CoV)
  taxonomy 1:549, 3:420–421
  virion properties 1:549
  virion structure 1:549, 1:555–556, 1:555f, 2:122f
    accessory proteins 1:559
    envelope protein 3:423
    membrane protein 1:557
    nucleocapsid protein 1:557, 3:421–423
    spike glycoprotein 1:556–557, 1:556f
    S protein 3:423
  see also specific viruses
Coronavirus infections
  antiviral drug development 1:561
  associated diseases 1:553
    co-infection 1:553
    persistent infection 1:553
  associated species 1:550t

  cellular effects 3:429
  common cold 1:554–555
  host range 1:553
    pigs 1:553
  respiratory infection 2:555
  target tissues 1:549
  transmission 2:122
  vaccines/vaccination 1:553
    development 1:561
  virus–host associations 4:128
Corticoviridae 3:1, 5:11t
Corticovirus(es)
  genome delivery 5:368
  see also Icosahedral dsDNA bacteriophages; specific viruses
Corticovirus PM2 3:1, 3:2t
  genome 3:5
  encoded proteins 3:2t
  structure 3:2
  membrane 3:3
Corvids, West Nile virus vectors 5:445
Coryza see Common cold
Cosmid transfection system, varicella-zoster virus 5:256
Cote d'Ivoire, HIV infection 1:63
Cote d'Ivoire Ebola virus (CIEBOV) 2:57
Cotesia congregata bracovirus (CcBV)
  BV-mediated immunosuppression 4:254–255
  C-type lectins 4:255
  gene expression 4:250–251
  genome 4:260
Cotesia kariyai, polydnavirus immunoevasive activity 4:255–256
Co-transfection, definition 1:237
Cotton, begomovirus resistance 2:104
Cotton leaf curl Alabad virus (CLCuAV) 1:566
Cotton leaf curl Burewala virus (CLCuBV) 1:566
Cotton leaf curl disease (CLCuD) 2:102
  begomovirus satellite identification 1:315–316
  causal agents 2:102
  see also β ssDNA satellite(s)
  control 1:568, 2:102
  crop yield losses 2:102
  DNA-1 component 1:565
    definition 1:565
    nanovirus Rep-1 satellites vs. 1:565
    Rep protein 1:565
    transmission 2:102
  DNA β components 1:564, 1:565f, 1:566
    begomovirus association 1:566
    identification 1:564–565
    replication 1:566
  engineered resistance 1:568
    post-transcriptional gene silencing 1:568
    transcriptional gene silencing 1:568
  epidemiology 2:102
  etiology 1:564
    elucidation 1:564
  geographical distribution 1:563, 2:102, 2:103f
  historical aspects 1:563, 2:102
    in Africa 1:563
    etiology elucidation 1:564
    in Southern Asia 1:563
  monopartite begomoviruses 1:566, 1:567f
    replication 1:566
  natural resistance 1:568
    breakdown of 1:568
  species 1:568
  symptoms 1:563, 1:564f, 2:102, 2:104f
  transmission 2:102–104
  trans-replication 1:320
Cotton leaf curl Gezira virus 1:566–568, 1:567f
Cotton leaf curl Kokhran virus (CLCuKV) 1:566
Cotton leaf curl Multan virus (CLCuMV) 1:564
  begomovirus satellite identification 1:315
  β ssDNA functions 1:318
Cotton leaf curl Rajasthan virus (CLCuRV) 1:566
Cotton leaf curl virus (CLCuV) 1:563–569
  classification 2:102
  DNA satellites 4:530

Subject Index

Cottontail rabbit *(Sylvilagus floridanus)*, Shope fibroma virus infection 3:225
Cottontail rabbit papillomavirus (CRPV) 4:33
  pathology 4:29–30
  vaccination 4:33
Councilman bodies, yellow fever 5:475
Counterterrorism, mathematic modeling in epidemiology 2:147–148
Courtois, Ghislaine, Crimean-Congo hemorrhagic fever 1:597
Covalently closed circular DNA (cccDNA)
  avian hepadnavirus replication 2:327, 2:329–330, 2:331
  hepatitis B virus genome 2:360–361
Covert infection, definition 3:161
Cowpea chlorotic mottle virus (CCMV)
  Brome mosaic virion *vs.* 1:385
  structure 1:387, 1:388*f*, 5:382*f*, 5:385, 5:386*f*, 5:391
    capsid 1:388
    hexamer units 5:385
    icosahedral asymmetric unit 5:385
    pentamer units 5:385
    pH-dependent structural transitions 1:388
  transmission 1:390
Cowpea mosaic virus (CPMV) **1:569–574**
  in biotechnology 1:573
    plant expression vectors 1:573–574, 1:573*f*
  genome 1:570
    in vitro translation 1:571
    open reading frame 1:571
    RNA-1 1:571, 1:572*f*
    RNA-2 1:571, 1:572*f*
    sequencing 1:570–571
  historical aspects 1:569
  movement in plants 3:352
  particle properties 1:569
    electrophoresis 1:570
    nonenveloped particles 1:569
    RNA sedimentation behavior 1:569, 1:570*f*
  phylogeny 1:573
  proteins 1:571
    cell-to-cell transmission 1:573
    electrophoresis 1:570
    RdRp 1:572–573, 1:572*f*
    RNA-1 encoded 1:572, 1:572*f*
    RNA-2 encoded 1:573
    small basic protein 1:570
  purification 1:569
    cesium chloride gradients 1:569–570
    sucrose density gradients 1:569
  replication 1:573
    membrane association 1:573
  structure 1:570
    X-ray crystallography 1:570
  transmission 1:569
  as type member 1:569
  vectors, transmission 1:569
Cowpea stunt 3:218
Cowpox
  Apodemus 1:577, 1:578
  cats 1:577
    clinical features 1:578–579
    tissue tropism 1:578, 1:579
    virus transmission 1:578
  clinical features 1:578
  histopathology 1:579
  human infections 1:578, 1:579
  immune response 1:579
  pathology 1:579
  prevention and control 1:579
  rodents
    clinical features 1:579
    epidemics 1:576
    as a host 1:577
    tissue tropism 1:577
  smallpox vaccination 4:640
Cowpox-like virus(es)
  zoonoses 5:489
  see also specific viruses
Cowpox virus (CPXV) **1:574–580**
  accidental hosts 1:577

A-type inclusions 1:574–575, 1:575*f*, 1:576
  associated proteins 1:576
Brighton Red strain 1:575
  evolution 1:577
  genome size 1:575
  protein coding 1:576
classification 1:574
distribution 1:576
epidemiology 1:577
evolution 1:577, 2:182
genetics 1:577
genome 1:575
  stability 1:575
history 1:574
host range 1:577
infection *see* Cowpox
physical properties 1:576
  inactivation 1:576
  rodent epidemics 1:576
propagation 1:577
protein properties 1:576
  immune modulation 1:576
replication 1:576
serology 1:577
smallpox vaccination 5:243
taxonomy 1:574
tissue tropism 1:578
transmission 1:578
virion properties 1:575, 1:575*f*
virulence 1:578
zoonoses 5:490
Coxsackie-adenovirus receptor (CAR) 1:18–19, 1:22, 2:129, 5:323
Coxsackie A virus infection, clinical manifestations 1:582
Coxsackie B virus
  capsid gene sequences 1:583
  host cell binding 5:322–323
  infection
    clinical manifestations 1:582
    myocarditis 2:68–69
    neonates 1:585
  RNA replication 4:136
Coxsackievirus 1:581*t*
Coxsackievirus(es) **1:580–587**
  classification 1:581
    typing systems 1:581
  distribution 1:582
  epidemiology 1:582
    common serotypes 1:582
  evolution 1:583
    nucleotide substitutions 1:583
  genetics 1:583
    diversity 1:583
    genome organization 1:584*f*
  history 1:580
  host cell attachment 1:543–544, 1:582
  host range 1:581
  infections *see below*
  propagation 1:581
    cell systems 1:582
    clinical specimens 1:582
  replication 2:67
  taxonomy 1:581, 1:581*t*
  transmission 1:582
  virion structure 1:582, 1:583*f*
  see also specific viruses
Coxsackievirus infections
  acute hemorrhagic conjunctivitis 1:585, 1:585*t*
    disease characterization 1:585
    incubation period 1:585
  clinical manifestations 1:584, 1:585*t*
    acute hemorrhagic conjunctivitis *see above*
    cardiac disease 1:586
    central nervous system disease 1:585
    diabetes 1:586
    febrile rash 1:585
    neonatal disease 1:585
    perinatal disease 1:585
    pleurodynia 1:585

    prenatal disease 1:585
    respiratory disease 2:552
  immunity 1:583
    heterotypic response 1:583–584
  pathogenesis 1:582, 1:583
    disseminated infection 1:583
    primary infection site 1:583
  prevention 1:586
  treatment 1:586
Cp-1 bacteriophage 4:402–403
CP binding site (CPB), alfalfa mosaic virus 1:82–83
CP gene/protein
  transgenic plants 4:5
CpGV *see* Cydia pomonella granulovirus (CpGV)
CPHV *see* Chandipura virus (CPHV)
CPMV *see* Cowpea mosaic virus (CPMV)
C (capsid) protein
  measles virus 3:287
  yellow fever virus 5:471
CPsV *see* Citrus psorosis virus (CPsV)
CPV-1 *see* Cypovirus 1 (CPV-1)
CPXV *see* Cowpox virus (CPXV)
CR2 (cell surface receptor)
  EBV binding 5:322
  structure 5:322
Cranefly *(Tipula paludosa)*, larval infection 3:161
cre see cis-acting replicating elements *(cre)*
Crenarchaeal virus(es) **1:587–596**
  diversity 1:588*t*
    classification 1:587
    evolutionary relationships 1:594–595
  gene functions 1:594
    archaeal homologs 1:594
    dUTPase 1:594
    glycotransferases 1:594
    P-loop ATPases 1:594
    transcription regulators 1:594
  genomes
    gene transcription 1:593
    inverted terminal repeats 1:593
    linear genomes 1:593
    orthologous genes 1:594
    replication 1:593
    structure 1:593
  genomics 1:592*f*, 1:594
  history 1:587
  morphotypes 1:587
    bottle-shaped viruses 1:590
    droplet-shaped viruses 1:590
    filamentous viruses 1:589
    rod-shaped viruses 1:589
    spherical viruses 1:590
    spindle-shaped viruses 1:587
  virion composition 1:590
    proteins 1:590–591
  virus–host interactions 1:591
    optimum growth temperatures 1:591
  see also specific viruses
Crenarchaeota 5:412
Creutzfeldt–Jakob disease (CJD) 1:474, 4:340, 5:190
  characterization 5:190–191
  diagnosis 4:334, 5:190–191
  familial 5:191, 5:192
  iatrogenic 4:332, 5:191
  mortality 5:190–191
  progression 5:190–191
  PrP^{Sc} tissue distribution 4:334
  sporadic 4:331–332, 5:191
    subtypes 5:191
  variant *see* Variant Creutzfeldt–Jakob disease (vCJD)
Cricket paralysis virus (CrPV)
  gene expression 5:5
  genetic diversity 5:6
  geographical distribution 2:43
  host range 2:40–41
  IRES structure 2:41*f*
  structure 2:39*f*
Crimean-Congo hemorrhagic fever (CCHF)
  at-risk groups 1:602
  clinical features 1:601

Crimean-Congo hemorrhagic fever (CCHF) (continued)
    convalescence stage 1:601
    hemorrhagic manifestations 1:601, 1:601f
    hemorrhagic stage 1:601
    incubation stage 1:601
    neuropsychiatric changes 1:601
    prehemorrhagic stage 1:601
  control 1:602
  diagnosis 1:602
    ELISA 1:602
    immunofluorescent assays 1:602
    RT-PCR 1:602
    virus isolation 1:602
  differential diagnosis 1:602
  historical aspects 1:596–597
    Courtois, Ghislaine 1:597
  mortality 1:601
  pathogenesis 1:601
    capillary fragility 1:601–602
    host-induced mechanisms 1:602
    thrombocytopenia 1:601–602
    virus localization 1:601–602
  prevention 1:602
    acaricides 1:602–603
    exposure minimization 1:602–603
    insect repellents 1:602–603
    tick control 1:602–603
  symptoms 1:596
  treatment 1:603
    blood transfusions 1:603
    fluid replacement 1:603
    ribavirin 1:603
    supportive 1:603
  zoonoses 5:487
Crimean-Congo hemorrhagic fever virus (CCHFV) 1:391t, 1:596–603
  characteristics 1:171t
  classification 1:597, 1:598t
  ecology 1:597
    birds 1:597
    reservoir species 1:597
  epidemiology 1:597
    seasonal aspects 1:597–598
  genome 1:394f, 1:598
    L segment 1:599
    M segment 1:599
    S segment 1:599
  geographical distribution 1:597, 5:487–488
  history 1:596
  infection see Crimean-Congo hemorrhagic fever (CCHF)
  L RNA segment 1:599
    phylogenetics 1:600–601
  phylogenetics 1:599, 1:600f
    L RNA segment sequence 1:600–601
    M RNA segment sequence 1:600–601
    S RNA segment sequence 1:600–601
  replication 1:599
  reservoir hosts 1:597
  taxonomy 1:597
  transmission 5:487–488
    Hyalomma ticks 1:596, 1:597–598, 1:599f
  virion structure 1:598, 1:599f
    envelope glycoproteins 1:598–599
    large polypeptide 1:598–599
    nucleocapsid protein 1:598–599
Crimean hemorrhagic fever (CHF) see Crimean-Congo hemorrhagic fever (CCHF)
Crimson clover latent virus (CCLV) 3:406t
Crinivirus 2:87
Crinivirus(es) 2:88
  infection symptoms 2:88
  sweetpotato infections 4:661t
  transmission 2:88
  see also specific viruses
Cripavirus 2:37, 2:38t
Cripavirus, insect infections 3:126t
Crixivan see Indinavir
crmA gene 1:576
Crocodile poxvirus(es) 4:326–327

Crop(s)
  disease resistance
    barley yellow dwarf virus control 1:285–286
    bean golden/yellow mosaic disease management 1:300–301
    beet necrotic yellow vein virus control 1:309–310
    benyvirus control 1:309–310
    maize chlorotic mottle virus 3:263
    maize streak virus control 3:270
    pomovirus control 4:287
    see also Plant(s), natural disease resistance
  engineered resistance 2:454, 4:157, 4:159t
  pest-induced losses
    studies 4:197
    vegetatively propagation 4:197
  viroid infections 5:337–338
    transfer from wild species 5:338
  see also specific crops
  see also specific viruses
Cro protein 5:184
Cross-breeding, ornamental plant species virus infections 4:208, 4:226–229
Cross-linking, bacteriophage assembly 1:439
Cross-protection 2:454, 5:438f
  citrus tristeza virus infection control 1:524
  cucumber mosaic virus 1:615
  definition 1:520, 1:614, 3:46, 5:433
  papaya ringspot virus 4:5
  plant gene silencing 4:145
  vegetable viruses 5:286–287
  watermelon mosaic virus 5:437
  zucchini yellow mosaic virus 5:437
Cross-reactivity
  antibodies 5:87–88
  antigenic variation 1:127
  psittacinepox virus 2:278–279
Cross-species transmission
  hepatitis E virus 2:381
  HIV see HIV
  paramyxoviruses 4:40–42
  SIV see Simian immunodeficiency virus (SIV)
Cross-talk, apoptosis 1:155–157
Croup, definition 2:551
CRPV see Cottontail rabbit papillomavirus (CRPV)
CRS see Congenital rubella syndrome (CRS)
CRSV see Carnation ringspot virus (CRSV)
'Crumbly fruit' Raspberry bushy dwarf virus infection 3:38, 3:38f
Crustaceans, aquareovirus infections 1:167
Cryo-electron microscopy (cryoEM) 1:603–614
  advantages 2:81f, 2:82
  biological insights 1:610
  definition 1:193, 4:68
  history 1:603–604
  icosahedral reconstruction 1:605, 1:605f, 1:608, 1:608f
    algorithms 1:606
    contrast transfer function 1:605–606
    deconvolution 1:606
    power spectrum 1:605–606, 1:606f
    procedure steps 1:605–606, 1:605f
    signal-to-noise ratio 1:605–606
  image analysis 2:84–85
    resolution 2:84–85
  limitations 2:82
    radiation damage 2:82
  low-dose cryo-tomographic imaging 1:608, 1:609f
    resolution 1:609–610
  low-dose single particle imaging 1:604
    defocus values 1:605
    electron dose 1:605
    recording 1:605
    sample storage 1:604–605
  nonenveloped virus structure 5:380, 5:381
  sample preparation 1:604, 2:81, 2:81f
    concentration 2:81–82
    temperature 2:81–82
    vitrification 1:604, 2:84
  single-particle density maps 1:606, 1:607f
  structural motif conservation 1:610, 1:611f
  structural polymorphism in maturation 1:611, 1:612f

technological advances 2:82
  viral genome 1:612, 1:613f
  virus–antibody complex 1:612
  virus–cell receptor complex 1:612
  virus particle modeling, density map constraints 1:608
    docking 1:608
  see also specific viruses
Cryo-sectioning 2:83
  vitrification 2:83–84
  methods 2:83–84
Cryo-tomographic imaging (cryo-ET)
  analysis 2:84–85
    resolution 2:84–85
  CEMOVIS 2:84
  cryo-electron microscopy 1:604, 1:608
  sample preparation 2:82
Cryparin 2:578, 2:578f
Cryphonectria parasitica virus (CHV1)
  infection
    chestnut blight, 2:580
  transcriptional profiling studies, 2:583
Cryphonectria hypovirus 1 - EP713 (CHV-1/EP713)
  host range, 2:285
  infection, 3:379
  MyRV-1/Cp9B21 double infection, 3:381
Cryphonectria parasitica, 3:378
  CHV-1, biological control, 2:285
  mycoreovirus infection, 3:379f
Cryphonectria parasitica virus (CHV1) 2:289, 2:576
  classification 2:576
  gene expression 2:577–578
  genome 2:576, 2:576f
  history 2:576
  hypovirulence see Hypovirulence
  infection
    chestnut blight 2:575, 2:575f, 2:580
    fungal mating, effects 2:578
    signal transduction 2:577
    slow growth mutations 2:577
    symptoms 2:575f, 2:577
    vesicle accumulation 2:577, 2:577f
  protein expression studies 2:578, 2:579f
    processing signals 2:578
    protease p29 2:579
    vesicle accumulation 2:579
  structure 2:576
  transcriptional profiling studies 2:582–583
  transmission 2:576
  viruses associated 2:575–576
Cryphonectria hypovirus 1 - EP713 (CHV-1/EP713)
  host range 2:285
  infection 3:383
  MyRV-1/Cp9B21 double infection 3:381
Cryphonectria parasitica 3:378
  CHV-1, biological control 2:291
  mycoreovirus infection 3:379f
'Cryptic epitopes' HIV vaccine development 5:242
Cryptotope(s)
  definition 1:137–138
  location 1:137–138
  TMV antigenicity 5:58
Cryptovirus(es) 4:68–69
  associated plant hosts 1:99, 1:100t
  infection, viral concentrations 4:69
  mixed infections 1:102
  RdRp 1:99
    molecular mass 1:101
    phylogeny 1:103f
    replication 1:101–102
    sequences 1:101
    translation 1:101
  serology 1:102
  symptoms of infection 1:102
  transmission 4:69
  see also Alphacryptovirus(es); Betacryptovirus(es); specific viruses
Crystallization, alfalfa mosaic virus coat protein 1:83
CsCMV see Cassava common mosaic virus (CsCMV)
CSDaV see Citrus sudden-death-associated virus (CSDaV)

CSEI (Center for the Study of Emerging Infections) 5:357t
CSFV *see* Classical swine fever virus (CSFV)
CSL binding
　EBV-encoded nuclear antigen 2 2:161
　EBV-encoded nuclear antigen 3 family 2:162
CsIV *see* Campoletis sonorensis ichnovirus (CsIV)
CSSV *see* Cacao swollen shoot virus (CSSV)
CsVC *see* Cassava virus C (CsVC)
C-terminal domain, *TRIM5* gene/protein 2:469–470
CTF *see* Colorado tick fever (CTF)
CTFV *see* Colorado tick fever virus (CTFV)
'CTL exhaustion' viral titer effects 3:82
CTLs *see* Cytotoxic T-cells (CTLs)
CTV *see* Citrus tristeza virus (CTV)
CTXΦ bacteriophage *see* Bacteriophage CTXΦ
Cucumber mosaic virus (CMV) 1:614–620, 5:287
　bacopa virus infections 4:215
　characteristics 5:63t
　classification 1:615
　control, vector control 1:619
　ecology 1:618
　epidemiology 1:618
　evolution 1:619
　genetics 1:617
　　2b protein 1:617
　　3a protein 1:617
　　3b protein 1:617
　　hypersensitivity response 1:617
　　RNA functions 1:617
　　variation 1:619
　genome 1:616, 5:63–64
　　nontranslated regions 1:616
　　RNA species 1:616
　geographical distribution 1:615
　　subgroup IB 1:615
　　subgroup II 1:615
　history 1:615
　infection
　　chlorosis 1:617–618
　　control 1:619
　　cucurbits 5:287–288, 5:288f
　　cytopathology 1:618
　　dahlia 4:223, 4:223f
　　economic impact 1:618–619
　　genetically-modified resistant plants 1:619
　　host range 1:615, 4:208
　　Impatiens 4:215, 4:217f
　　mixed 5:287–288
　　necrosis 5:287–288
　　Osteospermum 4:221, 4:222f
　　pathology 1:617
　　pelargonium 4:211, 4:211f
　　pepper plants 5:287–288
　　petunia 4:213, 4:213f
　　potato 4:300t
　　sweetpotato 4:661t, 4:668
　　tobacco 5:63
　movement 1:617
　　2b protein 1:617
　　3a protein 1:617
　propagation 1:615
　replication 1:617
　　1a protein 1:617
　　2a protein 1:617
　　initiation 1:617
　　virion assembly 1:617
　RNA satellites 1:616
　　control 1:619
　small, linear single stranded RNA satellites 4:531–532
　　origin 4:534
　　replication 4:533
　　structure 4:534
　　variation/evolution 4:534
　synergism 5:435
　taxonomy 1:615
　transgenic plants 4:157, 4:158f
　　gene silencing suppression 5:331
　transmission 1:617, 1:618
　　aphids 1:618, 5:287–288

　　efficiency 1:618
　　horizontal 1:618
　　seed 1:618
　　vector recognition 5:279
　virus properties 1:616
　　stability 1:616
　virus structure 1:616
Cucumber soil-borne virus (CSBV) 1:453
Cucumber vein yellowing virus (CVYV) 2:90
　classification 2:90
　epidemiology 2:90
　genome 2:90
　geographical distribution 2:90
　host range 2:90–91
　infection
　　control 2:91
　　disease symptoms 2:90
　　yield losses 2:90
　P1 protein 4:317
　transmission 2:90
Cucumis melo (melon), dsRNAs 2:112
Cucumovirus 1:81, 1:387, 1:615, 3:46
Cucumovirus(es)
　disease resistance genes 4:179t
　petunia infection 4:213, 4:213f
　sweetpotato infections 4:661t
　tRNA-like structure 1:81
　see also specific viruses
Cucurbit aphid-borne yellows virus
　classification 5:344
　synergism 5:435
Cucurbit yellow stunting disorder virus (CYSDV) 2:88
　classification 2:89
　divergence 2:89
　epidemiology 2:89
　genome 2:89
　　encoded proteins 2:89
　　open reading frames 2:89
　geographical distribution 2:88
　infection
　　control 2:89
　　disease symptoms 2:88
　　host range 2:89
　　yield losses 2:88
　transmission 2:89
　virion structure 2:89
Culex
　EEEV transmission 5:99–100
　Saint Louis encephalitis virus transmission 4:655, 4:657–658
　West Nile virus transmission 5:443–444
Culex molestus, West Nile virus transmission 5:443–444
Culex nigripalpus, West Nile virus transmission 5:443
Culex nigripalpus nucleopolyhedrovirus (CuniNPV) 1:220
　auxiliary functions, associated genes 1:224
　DNA replication 1:223
　　bro proteins 1:223
　　essential genes 1:223
　genome 1:222
　　homology 1:222
　infection 1:222
　　histopathology 1:221f
　　pathology 1:220–221
　　rate 1:222
　life cycle 1:221
　phylogeny 1:224–225, 1:224f
　RNA transcription 1:222–223
　　species-specific differences 1:222–223
　　transcription factors 1:222–223
　transmission 1:221–222
　　salt effects 1:222
　virion functions 1:223–224
　　phenotypes 1:223
　virion occlusion 1:223–224
　　associated genes 1:223–224
　　occlusion body morphology 1:223–224
Culex pipiens, West Nile virus transmission 5:443
Culex pipiens densovirus (CpDNV) 4:82–83
Culex taeniopus, eastern equine encephalitis virus transmission 5:103–104

Culex tarsalis
　Saint Louis encephalitis virus transmission 4:655
　western equine encephalitic virus transmission 5:104
　West Nile virus transmission 5:443
Culicoides
　African horse sickness virus transmission 1:41
　definition 1:37, 3:454
　orbivirus transmission 3:457
Culicoides brevitarsis, Akabane virus transmission 1:78–79
Culicoides paraensis, Oropouche virus vector 3:482
Culicoid midges, definition 2:234
Culiseta melanura
　eastern equine encephalitic virus transmission 5:103–104
　EEEV transmission 5:99–100
　highlands J virus transmission 5:103–104
Cullin 5-based E3 ubiquitin ligase, in malignant transformation 1:14f, 1:15
Cultivar decline, definition 4:659
Cultural practices
　maize streak virus 3:270
　mungbean yellow mosaic virus management 3:369–370
　ornamental plant species virus infections 4:208
　pecluvirus infection control 4:102
CuniNPV *see* Culex nigripalpus nucleopolyhedrovirus (CuniNPV)
Curation, definition 5:348
Curtovirus 1:302, 2:98, 4:165
Curtovirus infections
　resistance genes 4:179t
　symptomatology 4:166
Curvularia protuberata, tropical panic grass association 2:290
Curvularia thermal tolerance virus (CThTV) 2:290
CUS-1 prophage 3:123
CUS-2 prophage 3:123
Cutaneous adult T-cell lymphoma (CTCL) 2:568, 2:568f
Cutaneous human papilloma virus(es)
　detection 4:14
　distribution 4:14
　transmission 4:10
Cutaneous lesions, varicella-zoster virus infection 5:254
Cut-like protein1/CCAT displacement protein 3:340
'Cutting-out' campaigns, cacao swollen shoot virus infection 1:404, 1:409
CVYV *see* Cucumber vein yellowing virus (CVYV)
CWD *see* Chronic wasting disease (CWD)
CWMV *see* Chinese wheat mosaic virus (CWMV)
CWP virus 5:413t
CXC chemokine(s) 1:623
CXCR4 chemokine receptor
　HIV infection 2:506, 2:518–519, 2:535–536, 5:323
　SIV-host cell interactions 4:618
CXCR5 chemokine receptor 1:624
Cyanosis, definition 1:37
Cycas necrotic shunt virus (CNSV) 3:406t
Cyclins, retroviral 2:214–215
Cyclosporine A (CsA)
　HCV replication 2:373
　HIV infection resistance 2:470
Cydia pomonella granulovirus (CpGV) 1:247
　gene content 1:214, 1:215f, 1:215t
　imperfect palindromes 1:218
　inhibitors of apoptosis 1:216–217
　insect pest control 3:130–131
CYDV *see* Cereal yellow dwarf virus (CYDV)
Cylindrical inclusions, definition 2:18
Cymbidium mosaic virus (CYMMV) 4:225–226, 4:225f, 4:226f, 4:227f, 4:311t
Cynodon chlorotic streak virus (CCSV) 1:477
　infection, symptoms 1:477, 1:478f
　transmission 1:477
Cynomolgus EBV 4:587t
Cynosurus mottle virus (CnMoV) 4:646t
Cypovirus 3:134, 3:134t, 4:536t
Cypovirus(es)
　antigenic relationships 3:141

Cypovirus(es) (continued)
see also specific viruses
genetic relationships 3:141
genome 3:136
segments 3:138, 3:139t
size distribution 3:138
termini 3:138, 3:139t
history 3:135
host range 3:135
infections see below
infectivity 3:137
insect pest control 3:126t
nomenclature 3:141
polyhedra 3:136, 3:137–138
replication 3:136
assembly 3:139
uptake 3:138–139
transmission 3:135
virion properties 3:134, 3:136
capsid 3:136
proteins 3:136
stability 3:137–138
see also specific viruses
Cypovirus 1 (CPV-1) 3:141
as type species 4:536t
Cypovirus 12 (CPV-12) 3:141
Cypovirus 14 (CPV-14) 3:141
Cypovirus infections 3:135
host range 3:135
insect pest control 3:127
polyhedra 3:135
composition 3:138
infectivity 3:137–138
pupae 3:135
symptoms 3:127, 3:135
tissues associated 3:135
Cyprinid herpesvirus 1 (CyHV-1) see Carp pox herpesvirus
Cyprinid herpesvirus 3 see Koi herpesvirus (KHV)
Cyprus, sandfly fever viruses 4:495
CYSDV see Cucurbit yellow stunting disorder virus (CYSDV)
Cys-motif, Campoletis sonorensis ichnovirus 4:258–259
Cysteine protease(s) 1:155
Cystic fibrosis transmembrane conductance regulator (CFTR), definition 4:507
Cystoviridae 3:7, 5:11t
Cystovirus(es)
evolution 5:374
see also specific viruses
genome delivery 5:369
genome packaging 2:306, 2:307t, 2:311
nucleotide selection 2:307
historical aspects 3:6
replication 4:406
L segment mRNA 4:406
pac sequence 4:406
see also Icosahedral enveloped dsDNA bacteriophages; specific viruses
Cytidine deaminases (APOBEC)
definition 4:611
innate immunity 3:109
SIV infection 4:622
Cytochrome c, apoptosis 1:155–157
Cytogenetic abnormalities, adult T-cell leukemia 2:566
Cytokine(s) 1:620–624
definition 1:620, 3:70, 3:78, 3:104, 3:112
Ectromelia virus infection 3:345
effects 1:620
families 1:621
function 1:620
interferon type I 5:136
murine CMV infection 1:630t, 1:632
networks 1:620
parapoxvirus infection 4:61
persistent infections 4:112
pro-inflammatory
common cold 4:473f
yellow fever 5:475
receptors 1:620
regulation 1:620
rhinovirus infection 1:546

yellow fever 5:475
see also specific cytokines
'Cytokine storm' 3:99
Cytomegalia 1:624, 2:490
Cytomegalic inclusion disease (CID) 2:474
cytopathology 2:474
diagnosis 2:474
Cytomegalovirus 1:625, 1:625t, 1:635–636, 1:635t, 2:432t, 2:474, 2:485–486
Cytomegalovirus(es) (CMV)
genomic comparisons 1:636–637
infections
chronic disease 4:114–115
measles vs. 3:289–290
replication
egress 1:628f, 1:629
genome packaging 1:628–629
virion morphology 1:625
virion structure 1:625
see also specific viruses
Cytopathic effect(s)
Colorado tick fever virus infection 1:536
definition 1:335, 1:374, 1:580, 2:212, 2:221, 3:161, 4:497, 5:116
togaviruses 5:122
Cytopathic vacuoles (CPVs), togavirus replication sites 5:121
Cytopathology
potato virus Y see Potato virus Y (PVY)
potexvirus 4:312
sadwavirus infection 4:525
tenuivirus infection 5:26
varicella-zoster virus infection 5:261
Cytoplasmic exchange, Cryphonectria parasitica virus transmission 2:576
Cytoplasmic life cycle, RNA viruses 4:377
Cytoplasmic male sterility (CMS), definition 2:109
Cytoplasmic polyhedrosis, cypovirus infection 3:135
Cytoplasmic spherical bodies, edornaviruses 2:110
Cytoplasmic targeting/retention signal (CTRS), retroviral replication 4:466
Cytorhabdovirus 1:119t, 3:160f, 4:187, 4:188t
Cytorhabdovirus(es)
classification 1:119t
see also specific viruses
genome 3:326f
host cell entry 3:333
replication 4:193, 4:194f
virus-encoded proteins 3:332–333
see also specific viruses
Cytoskeleton, intracellular movement, role 3:352
50% Cytotoxic concentration (CC_{50}), antiviral therapy 1:146–147
Cytotoxic T-cells (CTLs)
death receptor activation 3:345
definition 4:382
Ectromelia virus infection 3:345
granule exocytosis pathway 3:345
hepatitis A virus infection 2:12
HIV infection see HIV infection
mechanism of action 3:80
murine CMV infection 1:630t, 1:631, 1:632f
varicella-zoster virus infection 5:254–255
VSV infection 5:298
yellow fever 5:475
Cytotoxic test(s), antiviral therapy 1:147

D

D13 protein, vaccinia virus 5:248
Dacryoadenitis 2:491, 2:495
DAdV-1 see Duck adenovirus 1 (DAdV-1)
DAF see Decay-accelerating factor (DAF)
Dahlia, virus infections 4:221
see also specific viruses
Dahlia mosaic virus (DMV) 4:221, 4:222f
Daily survival (P), vectorial capacity 5:271
Dairy industry, economic losses due to phages 2:448–449
Dali 5:357t
Dama trima virus (DtV) 5:28t
Damom bushy dwarf virus (CBDV), characteristics 3:389t

Dane particles 2:335–336
hepatitis B virus 2:352
Daphne virus X (DVX) 4:311t
DARK, apoptosis 1:232
Darunavir 1:143t
Dasychira pudibunda virus (DpV) 5:28t
Database(s) 5:348–365, 5:357t
access 5:353
analytical tools 5:357t, 5:364
annotation 5:353
definition 5:348
Inferred from Electronic Annotation 5:353
Standard Operating Procedures 5:353
data associated 5:349
classification 5:362
functional motifs 5:363
major repositories 5:356
nucleotide sequence data 5:362
orthologous clusters 5:363
protein sequence data 5:363
structural information 5:363
taxonomy 5:362
virus-specific sequence data 5:363
definition 5:348
errors 5:355
formats 5:349, 5:350f
output 5:354, 5:354f
BLAST 5:354
web-based 5:354–355
search tools, web-based 5:356
tables
admin 5:352
BLAST 5:352
color-coding 5:352
gene properties 5:352
gene segments 5:352
genome_molecule 5:352–353
taxonomy 5:352
Database management system(s) (DBMS) 5:349
definition 5:348
Database schema, definition 5:348
Data protection act(s), HIV/AIDS surveillance 2:46
DB52 protein, HCMV 2:487
DCL1 protein, short interfering RNAs 5:326
DCL2 protein, short interfering RNAs 5:326
DCL3-dependent RNA silencing 4:145–146
DCL3 protein, short interfering RNAs 5:326
DCL4 protein 4:147–148, 5:326
DCV see Drosophila C virus (DCV)
DDBJ Sequence Database 5:357t
DDE transposases 4:436
Death certification, disease surveillance 2:45
Death-inducing signaling complex (DISC) 1:157
'Death rafts' lambda holin S105 3:253
Death receptor signaling 1:157
FLIP proteins 1:154, 1:160
apoptosis 1:160
viral caspase regulators 1:160
viral regulators 1:157
Decay-accelerating factor (DAF)
echovirus replication 2:67
enterovirus 70 2:132
enterovirus binding 2:129
virus binding 5:322–323
coxsackie-adenovirus receptor 5:323
coxsackievirus 1:543–544
enteroviruses 5:322–323
DED1 protein 1:384–385
Deer tick virus 2:239
DE expression, nucleopolyhedrovirus 1:262
Defective interfering DNAs (DI DNA) 2:1
beet curly top virus control 1:307
recombination see Recombination
Defective interfering particles
definition 1:211
equine herpesviruses 2:416
granuloviruses 1:218
Defective interfering RNAs (DI RNA) 2:1
definition 1:520
geminiviruses 1:315–316
hepatitis A virus 2:4

hepatitis B virus 2:4
  production 2:2f
 tombusviruses *see* Tombusvirus(es)
 tospoviruses 5:159
Defective interfering virus(es) **2:1–4**
 assays 2:3
 biological effects 2:3
 cyclic variations 2:2, 2:3f
 defectiveness 2:1
 defective viruses *vs.* 2:2
 experimental animal models 2:3
 genomes 2:1
 history 2:1
 interference 2:2
 matrix protein role 2:3
 natural infections 2:3
   detection 2:3–4
 structure 2:1
 see also specific viruses
Defective viruses, defective interfering
   viruses *vs.* 2:2
Defoliation, cacao swollen shoot virus
   infection 1:403
Deformations, definition 4:207
Deformed wing virus (DWV) 3:42
 genome 3:45
 tissue tropism 3:45
Degenerate primers, PCR *see* Polymerase chain
   reaction (PCR)
Degenerative quasispecies 4:359
Delavirdine 2:510, 2:511t
Delayed type hypersensitivity (DTH), TMEV
   infection 5:43
Delbrück, Max, bacteriophage infection kinetics
   2:444–445
Delphacid planthoppers *see* Planthoppers
Delta antigen 2:375
Deltabaculovirus 1:220
Delta protein(s) 2:376
Deltaretroviridae 2:558
Deltaretrovirus 4:460
Deltavirus *see* Hepatitis delta virus (HDV)
Deltavirus 2:375
Dementia
 HIV-associated *see* HIV infection, dementia
 nervous system viruses 1:474
Democratic Republic of Congo
 Ebola virus outbreaks 2:199
 Marburg virus outbreaks 2:200, 3:272
 monkeypox virus outbreak 5:491
Demyelination
 definition 5:37
 see also specific infections
Dendritic cells (DC)
 conventional *see* Conventional dendritic cell(s) (cDC)
 Dengue fever 2:13
 differentiation 3:112
 Ebola virus infection 2:202–203
 efficiency 1:123
 function 1:122–123
 HIV pathogenesis 2:538–539, 2:539f
 induction 3:73, 3:74f
 innate immunity 3:112
 localization 1:122–123
 maturation 3:109
 migration 1:125
 murine CMV infection 1:630t
 parapoxvirus infections 4:61
 persistent infections 4:112
 subsets 1:123
 types 3:112
 Visna-Maedi virus infection 5:425
 West Nile virus infection 5:446
Dendrobium mosaic virus, transmission 1:291
 bean common necrosis virus 1:291
Dendrograms, definition 2:174
Dendrolimus punctatus tetravirus (DpTV) 5:28t
 capsid protein precursors 5:31–32, 5:32t
 discovery 5:28
 genomes 5:31
   open reading frames 5:31
   RNA2 length 5:31

replicases 5:31
Dendrolimus spectabilis, biological control 3:127
Dengue fever 2:5
 bioterrorism 5:407
 clinical diagnosis 2:12
 clinical features 2:11
   adults 2:11
   children 2:11
   Dengue hemorrhagic fever/Dengue shock
     syndrome *see below*
   hemorrhagic fever 2:245
 control/prevention 2:13
   vaccines/vaccination 2:13
 diagnosis
 see specific methods
 disease severity 2:9
 epidemics 2:5, 2:6
   severity 2:9
 histopathology 2:12
 immune response 2:12
 immunoglobulin G diagnosis 2:12
   cross reactivity 2:12
   persistence 2:12
   secondary infection 2:12
 incidence 2:5–6, 2:6f
 incubation period 2:11
 pathology 2:12
   immunopathogenic mechanisms 2:13
 prevalence 2:243
 prognosis 2:12
 rate 2:243
 serology
   serologic relationships 2:8
   serotypes 2:5
   subcomplexes 2:6
   variability 2:8
 treatment 2:12
Dengue hemorrhagic fever/Dengue shock syndrome
   (DHF/DSS) 2:5, 2:11, 2:245
 characterization 2:10, 2:11
 control 2:13
 critical stage of disease 2:11
 epidemiology 2:9, 2:9f
 history 2:5
 infection prevention 2:13
 pathogenetic mechanism 2:9f, 2:10
Dengue virus(es) (DENVs) **2:5–14**, 2:96
 anthroponoses 5:270–271
 antigenic variation, yellow fever virus *vs.* 1:127
 associated proteins 2:6
 characteristics 1:171t
 classification 2:6
 cross reactivity 2:12
 DENV-1 serotype 2:5
   associated epidemics 2:10–11
 DENV-2 serotype 2:5
   American 2:7–8
   antigenic variation 2:9
   Cuba 2:10–11
   Southeast Asia 2:7–8
   vector modifications 3:153
   Vietnam 2:7–8
 DENV-3 serotype 2:5
   antigenic variation 2:9
   isolation 2:7
 DENV-4 serotype 2:5
   antigenic variation 2:9
 DNA vaccination 2:53
 epidemiology 2:9
   maintenance cycles 2:8f, 2:9
 etiology 2:5
 evolution 2:8, 2:8f
   theories 2:8
 genetics 2:7
   diversity 2:7
   genetic drift 2:8
   intraserotypic recombination 2:8
   partial nucleotide sequencing 2:7–8
 genetic variation studies 2:7
   partial nucleotide sequencing 2:7–8
 genome 2:244f
 geographical distribution 2:6

history 2:5
host range 2:7
life cycle 2:9–10
pathogenicity 2:10, 2:11f
replication sites 2:10
seasonal distribution 2:6
 influencing factors 2:7
serotypes 2:6
see also specific serotype
size 2:6
structure 2:243, 2:244f
taxonomy 2:6
tissue tropism 2:10
transmission 2:5, 2:6, 2:10, 2:243
 Aedes aegypti 1:173, 2:5, 2:7, 2:10
 Aedes albopictus 2:5, 2:7
 Aedes polynesiensis 2:5, 2:7
 defective virus 1:173
 peak transmission 2:7
 vectorial capacity 5:271
virus propagation 2:7
in vitro studies 2:10–11
see also specific viruses
Density gradient centrifugation 2:452
Densovirinae 4:76, 4:77t, 4:86, 4:92t, 4:94t, 4:96, 5:11t
Densovirus 4:94t, 4:96
Densovirus(es)
 ambisense genomes 4:77, 4:77t, 4:81
   subgroup A 4:81, 4:81f
   subgroup B 4:81
   subgroup C 4:81f, 4:82
   see also specific viruses
 biological control 4:79
 biological properties 4:78
 capsid proteins 1:228
 classification 4:76, 4:92t, 4:94t
 gene cassettes 4:95f, 4:96
 gene expression 4:79
   nonstructural proteins 4:81
   structural proteins 4:81
 genome 4:76–77, 4:79, 4:81f, 4:95f, 4:96
   ambisense *see above*
   leaky scanning 4:79–81
   monosense genomes 4:83
   phospholipase A2 motif 4:81
   sequence motif conservation 4:82f
 host range 4:79
 infection
   insects 3:126t
   pathology 4:79
   symptoms 4:79
   tissue tropism 4:79
 insect pest control 3:126t, 3:127, 4:79
 phylogenetic comparisons 4:82f, 4:84
 nonstructural motifs 4:84, 4:84f
 unclassified 4:83
 virion structure 4:78
   capsid 4:78
 see also specific viruses
Deoxythymidine monophosphate (dTMP) 3:477–478
Deoxyuridine triphosphatase 1:47
Dependovirus 4:93
Dependovirus(es)
 classification 4:92t
 see also specific viruses
 genome 4:94f
 infection 4:94–95
 replication 4:93
 see also specific viruses
Dera Ghazi Khan virus 1:598t
Dermacentor andersoni (tick), Colorado tick fever virus
   transmission 1:534, 1:535
Dermacentor variabilis (dog tick), Colorado tick fever
   virus transmission 1:534
Dermatitis, contagious pustular 4:57–58
Desomotubule(s), plant virus movement 3:352
Desquamation 4:552
Deubiquitinating enzyme(s) (DUB), SARS CoV 1:562
Developed countries, KSHV transmission 3:190
Developmental resistance, baculovirus 1:267
Development perturbation, hypovirulence 2:575, 2:577
Dexmethasone, BHV1 genital disease 1:367

DHBcAg (duck hepatitis B virus c antigen) 2:328–329
DHBeAg (duck hepatitis B virus e antigen) 2:328–329
DHBsAg (duck hepatitis B virus surface antigens) 2:328
DHBV *see* Duck hepatitis B virus (DHBV)
D'Herelle, Fêlix Hubert 2:439
   bacteriophage therapy 2:443
Diabetes mellitus, coxsackieviruses 1:586
Diabetes mellitus, type I, human enterovirus B 2:69
Diadromus pulchellus ascovirus (DpAV) 1:189
Diadromus pulchellus idnoreovirus
   (DpIRV-1) 3:135
   genome segments 3:140
   proteins 3:140, 3:140t
Diadromus pulchellus reovirus 4:536t
Diagnostic techniques
   equine infectious anemia virus 2:173
   molecular approaches 2:29–37
   multianalyte methods 2:35–36
   neglected methods 2:28
   new methods 2:14, 2:28
   organ transplantation, infections 3:470
   viroid infection 5:338
   see also specific methods
   see also specific viruses
Dianthovirus 5:145–146
Dianthovirus(es)
   genome 5:148
   see also specific viruses
   geographical distribution 5:150
   taxonomy 5:145–146
   tombusvirus taxonomy 5:145–146
   see also specific viruses
DIAP1, apoptosis 1:232
Diapause 4:652, 5:268
Diaporthe RNA virus (DRV) 2:288–289
Diarrhea
   adenovirus-induced 2:121
   astrovirus-induced 2:121
   bovine torovirus 5:155–156
   Breda virus 2:122
   coronavirus-induced 2:122
   human torovirus infection 5:156
   torovirus-induced 2:122
   see also specific virtal infections
Diascia, virus infections 4:215
   see also specific viruses
Diatom(s), viral infection 1:95
Dicer(s) 3:149
   definition 5:325
   domains 3:149–150
   gene evolution 3:154–155
   miRNA biogenesis 3:150, 5:326
   nodaviruses 3:436–437
   organismal distribution 3:149–150
Dicer-2 3:149–150
   R2D2 protein interactions 3:150–151
Dicer-like proteins
   engineered resistance in plants 4:161
   plant gene silencing 4:143t, 4:145
      antiviral defense 4:145
      Arabidopsis thaliana 4:145
      DCL3-dependent RNA silencing 4:145–146
Diceromyia, Dengue virus transmission 2:7
Dichlorodiphenyltrichloroethane (DDT), Dengue
   virus control 2:97
Dicistronic mRNA, definition 3:46
Dicistroviridae 1:90f, 2:37, 4:567, 5:11t
Dicistrovirus(es) 2:37–44
   biophysical properties 2:38, 2:39f
      morphology 2:38
      structural proteins 2:38, 2:40f
   classification 2:37
   gene expression 2:39
   genome 2:38, 2:39t
      intergenic regions 2:41f
      open reading frames 2:38–39
      untranslated regions 2:38–39
   geographical distribution 2:43
   history 2:37
   host range 2:40
   infection
      insects 3:126t

      pathology 2:42
   marnaviruses *vs.* 3:281, 3:281f
   phylogeny 2:43, 2:43f
   replication 2:39, 2:42t
      IRES elements 2:39, 2:41f
      translation initiation 2:39
   strain variation 2:43
   transmission 2:42
   see also specific viruses
Dicotyledon(s)
   definition 1:533
   recessive resistance genes 4:178
Didanosine 2:510
Diet, prion diseases 1:372, 4:332
Differential hosts, definition 4:659
Differentiating infected from vaccinated animals
   (DIVA), pseudorabies virus infection 4:350–351
Diffusion, lipid rafts 5:309
Dihydrofolate reductase, murine polyomavirus large
   T 4:276
Dimorphotheca virus infections 4:221
   see also specific viruses
Dinovernavirus 3:134, 3:134t
Dinovernavirus(es)
   antigenic relationships 3:142
   see also specific viruses
   distribution 3:136
   genetic relationships 3:142, 3:142f
   genome 3:140
   host range 3:136
   infection 3:136
   replication 3:140
   transmission 3:136
   virion properties 3:134, 3:140
   see also specific viruses
Dioscorea alata bacilliform virus (DaBV) 1:403
Diptera 1:111, 2:37, 3:133
Direct chemical control, ornamental plant species
   virus infections 4:226–229
Direct contact transmission 2:141, 2:145
Directed evolution, first-generation plant virus
   vectors 4:230
DIRS1 *see* LTR-retrotransposons
DIRS retrotransposons 4:433
Discistroviridae 4:567
Discorea latent virus (DLV) 4:311t
Disease(s)
   antibody-dependent enhancement 3:66
   causation criteria 2:144t
   definition 2:44
   prevention
      eradication by vaccination 2:51
      implications 2:148
   see also specific diseases
Diseases Conditions database 5:357t
Disease surveillance 2:44–51
   data analysis 2:48
      age distribution 2:48
      occupational groups 2:48
      person 2:48
      place 2:49
      seasonal trends 2:48
      secular trends 2:48
      time 2:48
      travel history 2:48
   data collection 2:45
      accuracy 2:47
      completeness 2:47
      consistency 2:48
      death certification 2:45
      environmental sources 2:46
      general practitioner systems 2:45
      hospital administrations 2:46
      laboratory data 2:45, 2:46
      mass control programs 2:47f
      nonhuman sources 2:46
      notifications 2:45
      outbreaks 2:46
      purpose 2:44–46
      representativeness 2:47
      serological surveillance 2:45
      sickness absence 2:46

      survey *vs.* 2:45
      system attributes 2:47
      system evaluation 2:50
      timeliness 2:47
   data interpretation 2:49
      data origin 2:49
      infection stage 2:49
   feedback 2:49
   food poisoning 2:46
   global programs 2:50
   international programs 2:50
   see also specific diseases
Disseminated intravascular coagulation (DIC) 2:198
   filovirus infection 2:203
Dissociation constant 3:413
   antibody–antigen binding 3:414–415
Distal symmetric sensory polyneuropathy *see* HIV
   infection
Distemper
   clinical features 4:504
   control 4:505
   diagnosis 4:506
   histopathology 4:504
   immune response 4:505
   pathology 4:504
   prevention 4:505
   serology 4:502
   vaccination 4:506
Distemper virus(es) **4:497–507**
   classification 4:498
   epizootiology 4:502
   evolution 4:501
   genome 4:500
   geographical distribution 4:498
   history 4:497
   host range 4:499
   infection *see* Distemper
   propagation 4:500
   taxonomy 4:498
   transmission 4:504
   see also specific viruses
Distinct genes, phylogenetic analysis 4:127
DmGypV *see* Drosophila melanogaster Gypsy virus
   (DmGypV)
DNA
   binding proteins
      nucleopolyhedroviruses 1:257–260
      SV40 T-ag 4:633
   concatamers 3:35–36
   injection
      bacteriophage T4 3:33–34
      icosahedral tailed dsDNA bacteriophages 3:33
   origin of 3:477
   packaging 1:432
   reassortment 4:166
   repeat regions 1:229
   replication 3:478
      African swine fever virus 1:47
      baculoviruses 1:250
   restriction digestion 2:456
   shuffling
      first-generation plant virus vectors 4:230
      recombinant adeno-associated virus 2:304
   substrate recruitment 2:309
   translocation 2:311
DNA-1 component *see* Cotton leaf curl disease
   (CLCuD)
DNA-A satellite(s), cotton leaf curl disease
   transmission 2:102
DNA bacteriophages 3:117
DNA β-components *see* β ssDNA satellite(s)
DNA β satellite, abutilon mosaic virus 1:320–321
DNA-B satellite(s), Indian tomato leaf curl viruses 5:130
DNA chromosome assembly, dsDNA bacteriophages
   2:308–309
DNA-dependent DNA polymerase, DNA virus
   evolution 2:182
DNA encapsidation 1:432
DNA helicase
   African swine fever virus 1:47
   origin of 3:478
DNA interference, African cassava mosaic disease 1:35

DNA ligase
    Amsacta moorei entomopoxivirus 2:139
    Melanoplus sanguinipes entomopoxivirus 2:139
DNA microarrays see Microarrays
DNA polymerase(s)
    African swine fever virus 1:47
    bacteriophage φX174 4:405
    DNA satellite replication 4:533
    HSV replication 2:401
    origin 3:478
DNA primase, African swine fever virus 1:47
DNA probes 1:295
    bean golden/yellow mosaic disease 1:297–298
DNA translocase(s), bacteriophage assembly 1:439
DNA transposon(s) 4:428
DNA vaccine(s) 2:51–55
    advantages 2:52
    antigen expression 2:52
    clinical applications 2:54
    composition 2:52
        EMCV IRES promoter insertion 2:53
        two gene expression cassettes 2:53
    concept 2:52
    delivery 2:52–53
    development strategies 5:236
        HIV 1:75
    improvements 2:54–55
        DNA uptake 2:54–55
    limitations 2:54
        transfection efficiency 2:54
    malaria 2:53
    multi-component 2:53
    neonatal immunization 2:53
        effectiveness 2:53
        memory induction 2:53
    plasmids 2:52
        co-administration of immune stimulators 2:54
        integration into cells 2:52–53
        interference 2:53
        production 2:52
        purification 2:52
    production 2:52
        timescale 2:52
    safety issues 2:52
DNA virus(es)
    antigenic variation 1:128
    evolution of 2:181
        DNA-dependent DNA polymerase comparisons 2:182
    fish viruses 2:227
    genome 1:612
    insect pest control 3:127
    nucleotide sequence analysis 2:21
    ocular disease 2:495
    recombination 4:374
        cellular machinery 4:374
        eukaryotic viruses 4:375
        functional marker mutations 4:374–375
        host genome acquisition 4:374
        host methods vs. 4:375f
        plant viruses 4:376
        prokaryotic viruses 4:374
        recA 4:374
        three-factor cross method 4:374–375
        viral DNA-repair systems 4:374
        virally-encoded systems 4:374
        see also specific viruses
    replication 4:407
        double-stranded DNA viruses 4:410
        RNA intermediates 4:411
        single-stranded DNA 4:410
    RNA viruses vs. 3:477
    see also specific viruses
dNTP-binding site, Maize streak virus rep gene 3:268
Doctor fish virus 16 (DFV-16) 2:233
Dogs
    rabies virus 4:372
    see also under Canine
Dolichos yellow mosaic virus 3:370–371
Domestic animals
    rabies virus vaccination 4:373
    see also specific animals

Dopamine receptor(s), JCV host cell attachment 4:280
'Dose sparing' vaccine development strategies 5:241–242
Double membrane vesicles (DMV)
    definition 1:554
    nidovirus replication 3:426
Double stranded DNA (dsDNA) 3:148
Double stranded DNA (dsDNA) bacteriophages
    genome packaging 2:308f
        ATPase complex 2:310–311
        ATP-DNA complex 2:309–310
        capsid architecture 2:307–308
        circular genome replication 2:308–309
        concatamer replication 2:308–309
        DNA chromosome assembly 2:308–309
        DNA substrate recruitment 2:309
        DNA translocation 2:311
        energy dependency 2:310f
        integrated segment replication 2:308–309
        linear DNA replication 2:308–309
        nucleotide selection 2:307
        precapsid assembly 2:307
        'riddle-of-the-ends' 2:308
        termination 2:310
    replication 4:400
        5' DNA ended terminal protein phages 4:402
        bacteriophage Mu 4:404
        bacteriophage SPP1 4:401
        φX174 phages 4:405
        lambdoid phages 4:400
        T4-like phages 4:401
        T7-like phages 4:401
    see also specific viruses
Double-stranded DNA (dsDNA) virus(es)
    classification 5:11t
    replication 4:410
        cytoplasmic 4:410
        nuclear 4:410
    RNA intermediates, replication 4:411
    transcription 4:410
        icosahedral enveloped bacteriophages 3:11
        RNA intermediate 4:411
    see also specific viruses
Double stranded RNA (dsRNA) bacteriophages
    genome packaging 2:306
    replication 4:405
    see also specific viruses
Double-stranded RNA (dsRNA) virus(es)
    classification 5:11t
    replication 4:410
    see also specific viruses
Double-subgenomic promoter virus(es) (DP-virus)
    infectious clone technology 5:121–122
    see also specific viruses
DpIRV-1 see Diadromus pulchellus idnoreovirus (DpIRV-1)
DpTV see Dendrolimus punctatus tetravirus (DpTV)
DPVweb (Descriptors of Plant Viruses) 5:357t
DR1, avian hepadnavirus 2:330–331
DR2, avian hepadnavirus 2:330–331
Dragon grouper nervous necrosis virus (DGNNV) 3:434
DRONC, apoptosis 1:232
Droplet-shaped virus(es) 1:590
    see also specific viruses
Drosophila C virus (DCV)
    gene expression 5:5
    geographical distribution 2:43
    host range 2:40–41
    transmission 2:42
Drosophila melanogaster
    gypsy retrotransposon 3:306
        env-mediated infection 3:310
        follicle cell, expression 3:302–304
        open reading frames 1:466f
        reverse transcription priming 3:306
        target site specificity 3:309
        transposition rates 3:302–304
    immune response
        Dicer gene evolution 3:154–155
        parasitoid attack 4:252–253
        protein tyrosine phosphatase function 4:252–253
        siRNA 3:149

retroelements associated 3:302
    flamenco 3:302–304
    gypsy see above
    transposition levels 3:302–304
Drosophila melanogaster Gypsy virus (DmGypV) 4:451
    cg4715 protein 4:454
    env protein cleavage 4:453
    monoclonal antibodies, infectivity 4:452
Drosophila melanogaster idnoreovirus-5 (DmIRV-5) 3:140
Drosophila melanogaster zam virus (DmZamV) 4:452–453
Drosophila S virus 3:135
Drug(s)
    antiretroviral see Antiretroviral therapy
    prophylaxis, arenavirus infection control 3:211
    resistance 1:151
        genetic barrier 1:150
        hepatitis B virus see Hepatitis B virus (HBV)
        replication 1:151
Dry eye, HIV infection 2:144
Dual-start motif 3:248
    antiholin diversity 3:252f
Duck adenovirus 1 (DAdV-1)
    classification 1:27
    egg drop syndrome 1:27
Duck hepatitis B virus (DHBV) 2:336
    classification 2:327
    genome 2:328f
        open reading frames 2:329
    infection 2:330f
        congenital 2:332
        histology 2:334f
        persistent 2:334f
        tissue tropism 2:332
    replication 2:327
        budding 2:330f
        cccDNA 2:327
        DNA synthesis 2:330
        DR1 2:330–331
        DR1* 2:330–331
        DR2 2:330–331
        nuclear transport 2:331
        pregenome packaging 2:331f
        receptors 2:329
        reverse transcription 2:328f
        sites 2:327
        transcription 2:328f
    structure 2:328
        antigens 2:328
        envelope 2:328
        nucleocapsid 2:328f
        polymerase 2:328
Ducklings, astrovirus infection 1:209
Duck virus enteritis (DVE) 2:406
Dugbe virus 1:597
dUTPase 1:594
dUTP pyrophosphatase 1:6
Duvenhage virus 4:184t
dwf gene 1:300
DWV see Deformed wing virus (DWV)
Dysesthesia, definition 5:250
Dysplasia 2:212
Dyspnea 5:423

E

E1A gene/protein see Adenovirus infections
E1B-19K gene/protein see Adenovirus infections
E1B-55K, adenovirus E4orf6 gene/protein interactions 1:14–15
E1B-AP5 gene/protein, adenovirus malignant transformation 1:15
E1B gene/protein, adenovirus malignant transformation 1:13
E1E4 see Human papillomavirus(es) (HPVs)
E1 glycoprotein function, alphavirus life cycle 1:195
E1 proteins
    alphavirus replication 5:99
    HPV see Human papillomavirus(es) (HPVs)
E2F/DP complex, HPV replication 4:25

E2 glycoprotein function, alphavirus life cycle 1:195
E2 protein
  alphavirus replication 5:99
  bovine viral diarrhea viruses 1:379
  HPV see Human papillomavirus(es) (HPVs)
E3 gene, siadenoviruses 1:8
E3 glycoprotein function, alphavirus life cycle 1:195
E4 gene/protein
  adenovirus malignant transformation 1:13
  adenovirus MRE11-RAD50-NSB1 complex 1:15
  orf6 see Adenovirus infections
  papillomaviruses 4:36
E5 gene/protein, HPV 4:23
E6 gene/protein see Human papillomavirus(es) (HPVs)
E6 promoter, papillomavirus gene expression 4:37–38
E7 gene/protein see Human papillomavirus(es) (HPVs)
E120R polyprotein, African swine fever virus replication 1:47
EACMCV see East African cassava mosaic Cameroon virus (EACMCV)
EACMKV see East African cassava mosaic Kenya virus (EACMKV)
EACMV see East African cassava mosaic virus (EACMV)
EACMZV see East African cassava mosaic Zanzibar virus (EACMZV)
Early genes
  baculoviruses 1:250
  Herpesvirus Saimiri 4:588–589
  icosahedral tailed dsDNA bacteriophages 3:34
Early proteins, murine polyomavirus 4:273
East Africa, HIV infection 1:62f
East African cassava mosaic Cameroon virus (EACMCV) 1:31–32
  distribution 1:32f
East African cassava mosaic Kenya virus (EACMKV) 1:31–32
  distribution 1:32f
East African cassava mosaic Malawi virus (EACMMV) 1:31–32
  distribution 1:32f
East African cassava mosaic virus (EACMV) 1:31–32
  regional epidemiology 1:33–34
East African cassava mosaic Zanzibar virus (EACMZV) 1:31–32
  distribution 1:32f
Eastern equine encephalitic virus (EEEV) 5:113
  animal models 5:79
  characteristics 5:109t
  epidemiology 5:104f
  evolutionary relationships 1:174
  geographical distribution 5:102
  host range 5:102–103
    North America 5:102–103
  infection
    case–fatality rates 5:105
    clinical features 5:105
    diagnosis 5:105
    encephalitis 5:78
    epidemic 5:101
    epizootics 5:103–104
    incidence 5:103–104
    incubation period 5:78
    pathogenesis 5:115
    pathology 5:106
    recovery 5:105
    treatment 5:82
    vaccines 5:488
  pathogenicity 5:105
  replication 5:105
  subtypes 5:101
  tissue tropism 5:104
  transmission 5:104
    vector control 5:111–112
    vectors 5:103–104
  virulence 5:105
Eastern Europe, HIV infection 1:64
EAV see Equine arteritis virus (EAV)
EBNA see Epstein–Barr virus (EBV)
EBNA-1 see Epstein–Barr virus (EBV)

EBNA-2 see Epstein–Barr virus (EBV)
EBNA-3 see Epstein–Barr virus (EBV)
EBNA-LP see Epstein–Barr virus (EBV)
Ebola hemorrhagic fever (EHF) 2:57
  diagnosis 2:63
  history 2:57
Ebolavirus 2:57, 2:199
Ebola virus(es) (EBOV) **2:57–65**
  bioterrorism 5:407
    fatality rates 5:408–409
    transmission 5:408–409
  cell tropism 2:61
  classification 2:57
  disease surveillance 2:199
  genetic stability 2:179
  genome 3:327f
    Marburg virus vs. 2:201–202
  history 5:490–491
  immune evasion 3:108
  inactivation 2:57–58
  life cycle 2:61f
    mRNA transcripts 2:61–62
    replication 2:61–62
    transcription 2:61–62
    virus assembly 2:62
  matrix proteins 2:59f
    L protein 2:59–60
    NP 2:59–60
    VP24 2:59–60
    VP30 2:59–60
    VP35 2:59–60
    VP40 2:59–60
  proteins 2:60t
    budding 2:59–60
    fusion protein biosynthesis 3:293
    glycoproteins 3:332
    matrix proteins see above
    RNP complex 2:59
  structure 3:329f
  taxonomy 2:57
  transmission 5:490–491
  viral receptors 2:61
  virion properties
    morphology 2:59f
    nucleocapsid 2:57–58
  see also specific viruses
Ebola virus infection
  animal models 2:202–203
  clinical features 2:62
  coagulopathy 2:203
  Democratic Republic of Congo 2:199
  dendritic cells 2:202–203
  diagnosis 5:490–491
  endothelial cell activation 2:203
  fatality rates 2:57
  immune response 2:203
    adaptive 3:108–109
    evasion 3:108
  incubation period 2:202
  mortality rate 2:199
  nonhuman primates 2:203
  pathogenesis 2:63
    Marburg virus vs. 3:278
  pathogenicity 2:63f
  symptoms 2:62
  treatment 2:204
    antiviral therapy 2:64
    experimental strategies 2:64
  vaccines 2:65
  viremia 2:62
  zoonoses 5:490
  see also Ebola hemorrhagic fever (EHF)
EBV see Epstein–Barr virus (EBV)
Ecdysteroid, definition 4:256
Ecdysteroid glucosyltransferase, baculovirus infection 1:252
Echinochloa hoja blanca virus (EHBV) 1:487, 5:24
  phylogenetics 5:26f
  species discrimination 5:26
Echinochloa ragged stunt virus (ERSV), characteristics 4:150t

Echovirus(es) **2:65–71**
  cancer therapy 2:70
  classification 2:65
  epidemiology 2:69
  genome 2:66–67
    5′ noncoding region 2:67
  history 2:65
  infections see below
  laboratory diagnosis 2:69
  molecular characteristics 2:66
    capsid 2:66–67
    structural studies 2:67
  mutation rate 2:68
    genetic lineages 2:70
    recombination 2:70
  replication 2:66
    associated proteins 2:67–68
    cellular effects 2:68
    cycle 2:67
    translation 2:67
  serotypes 2:65–66
    3D structures 2:66–67
  transmission 2:69
    children 2:69–70
  see also specific viruses
Echovirus 4, disease surveillance 2:45
Echovirus infections
  amyotrophic lateral sclerosis 2:69
  clinical manifestations 2:68
    respiratory disease 2:552
  human respiratory infection 2:552
  vaccines/vaccination 2:70
Ecological niche(s), definition 5:401
Ecology
  bacteriophages see Bacteriophage(s), ecology
  definition 3:212
  HIV cross-species transmission 2:532–533
  zoonoses 5:494
Economic importance
  alfalfa mosaic virus 1:81
  fish viruses 2:227
  infectious hematopoietic necrosis virus 2:228
  ornamental plant species virus infections 4:208
  potato mop-top virus infections 4:286–287
  viral hemorrhagic septicemia virus 2:227
Ecosystem, definition 2:71
Ectocarpus virus(es) (EsV) 4:116
  genome 4:119
    open reading frames 4:122f
  replication 4:122
  see also specific viruses
Ectothermic, definition 3:155
Ectromelia virus 3:342
  biotechnology vs. bioterrorism 5:409
  enzootic 3:343
  epidemiology 3:342
    inoculation routes 3:342–343
  evolution 2:182
  future work 3:346
  genetics 3:342
  geographical distribution 3:342
  historical aspects 3:342
    Barnard, J E 3:342
    Burnett, F M 3:342
    Marchal, J 3:342
  host range 3:342
  infection see below
  mouse strain susceptibility 3:343
  propagation 3:342
  protein p28 3:346
  seasonal distribution 3:342
Ectromelia virus infection
  clinical features 3:344f
    age effects 3:343
  future work 3:346
  histopathology 3:343
  immune response 3:345
    cell-mediated 3:345
    host-response modulation 3:346
    humoral immunity 3:345
  intracytoplasmic inclusion bodies 3:344f

lesions
  intranasal inoculation 3:345
  intraperitoneal inoculation 3:345
  kidneys 3:345
  liver lesions 3:344
  lymph nodes 3:345
  skin lesions 3:344
  spleen lesions 3:345
 mousepox 4:329
 pathogenesis 3:343
  incubation period 3:343
  as model system 3:343
  pathology 3:343
Ectropis obliqua picorna-like virus (EoPV) 3:45
Edible vaccine(s) 5:221
 polio vaccine *see* Polio vaccines/vaccination
 proof of concept 5:222–223
 rabies virus vaccination 4:373
Edornavirus(es) 2:109–116
 cytoplasmic spherical bodies 2:110
 genome 2:110f
  open reading frames 2:110–111
  RdRp 2:112
 historical aspects 2:110
 large dsRNAs 2:110
 members 2:113t
 molecular evolution 2:115
  host species dependence 2:115
  vertical transmission 2:115
 nonplant 2:111
  genomes 2:110f
  members 2:111
  Phytophthora 2:111
 phylogenetic trees 2:112f
 proteins 2:112
  uridine diphosphate glycosyltransferase 2:112–113
 RdRp 2:110–111
 replication 2:115
 rice 2:110
 RNA helicase 2:110–111
 site-specific coding strand nicks 2:110f
 tentative members 2:111
  phylogenetic trees 2:112f
 vertical transmission 2:113
  molecular evolution 2:115
 see also specific viruses
EDS1 protein, plants natural resistance 4:176
EEEV *see* Eastern equine encephalitic virus (EEEV)
Eel herpesvirus 2:233
Eel virus American (EVA) 2:228
Eel virus Europe X (EVEX) 2:228
EEV *see* Extracellular enveloped virus (EEV)
Efavirenz 2:510
50% Effective concentration (EC$_{50}$), antiviral therapy 1:146
Effective population size 4:359
Effector memory T-cell(s) 3:70
Effector T-cell(s) 1:620–621
 definition 3:70
 population expansion 3:74
Efuviritide 1:143t
Egf gene family 4:254
EGFR *see* Epidermal growth factor receptor (EGFR)
Egg drop syndrome (EDS) 1:27
Eggplant mosaic virus-tobacco strain (EMV-T) 5:67
 characteristics 5:63t
Eggplant mottled dwarf virus (EMDV) 5:67
 associated disease symptoms 5:67
 characteristics 5:63t
 incidence 5:67
 potato infections 4:300t
EHDV *see* Epizootic hemorrhagic disease virus (EHDV)
EHF *see* Ebola hemorrhagic fever (EHF)
EHNV *see* Epizootic hematopoietic necrosis virus (EHNV)
EHV-1 *see* Equid herpesvirus 1 (EHV-1)
EHV-2 *see* Equid herpesvirus 2 (EHV-2)
EHV-3 *see* Equid herpesvirus 3 (EHV-3)
EHV-4 *see* Equid herpesvirus 4 (EHV-4)
EHV-6 *see* Equid herpesvirus 5 (EHV-5)

EHV-9 *see* Equid herpesvirus 9 (EHV-9)
EIA *see* Equine infectious anemia (EIA)
EIAV *see* Equine infectious anemia virus (EIAV)
EICP0 protein, equid herpesvirus 1 replication 2:415
eIF4G, alfalfa mosaic virus coat protein binding 1:84f
Eigen, Manfred, quasispecies 2:177–178, 4:359–360
Eighth ICTV Report on Virus Taxonomy 5:10
 species demarcation criteria 5:21
Eimeria brunetti RNA virus 1 (EbRV-1) 5:173–174
ELDKWAS peptide *see* HIV vaccines/vaccination
Electron(s), penetrating power 2:78
Electron density
 single-particle density maps 1:607f
  resolution 1:607f
 virus particle modeling 1:607f
Electron microscopy (EM) 2:78–86
 adenovirus virion morphology 2:78
 adsorption EM 2:21f
 cell preparation 2:82
  antibody labelling 2:83
  classical thin sectioning 2:84f
  cryo-EM, vitreous sections 2:84
  cryo-sectioning *see* Cryo-sectioning
 cryo-transmission 4:68
 hepatitis B virus 2:351
 icosahedral tailed dsDNA bacteriophages 3:31
 image analysis 2:84
 immunoelectron microscopical decoration 2:21f
 immunosorbent EM 2:21f
 large particle preparation 2:82
 plant virus detection 2:27
  bean golden/yellow mosaic disease 1:297
  cytological alterations 2:27f
  tobacco mosaic virus 2:452
  virus particle aggregates 2:27f
 single particle imaging 2:78
  cryo-electron microscopy *see* Cryo-electron microscopy (cryoEM)
  cryo-electron tomography *see* Cryo-tomographic imaging (cryo-ET)
  metal shadowing *see* Metal shadowing
  negative staining *see* Negative staining
  sample contamination 2:78–79
 taxonomy 5:9–10
 ultrastructural studies 2:27
 varicella-zoster virus assembly 5:261
Electroporation
 definition 2:51
 DNA vaccine delivery 2:55
Elford, William, virus discovery 3:399
ELISA *see* Enzyme-linked immunosorbent assay (ELISA)
Elk herpesvirus 1 (EIHV1) 1:364t
Elongation terminator(s) 5:176
Elvucitabine 2:507t
EMBL Sequence Database 5:357t
EMCV *see* Encephalomyocarditis virus (EMCV)
EMDV *see* Eggplant mottled dwarf virus (EMDV)
Emerging viral diseases 2:93–97
 contributing factors 2:93
 definition 4:40
 ecological factors 2:96
 history 2:93
 human behavior 2:94
 human demographics 2:94
 recognition of 2:93
 vector control 2:97
 zoonoses 2:95
 see also specific viruses
Emiliania huxleyi virus (EhV) 4:116
 genome 4:119
  open reading frames 4:122f
  RNA polymerase 4:122
Emtricitabine 2:510
EMV-T *see* Eggplant mosaic virus-tobacco strain (EMV-T)
Enamovirus 3:231, 3:232f, 3:232t
Enanthem, definition 4:639
Enation, definition 1:301
Encapsidated RNA2, pomovirus genome 4:284

Encapsidated RNA3, pomovirus genome 4:284
Encapsidation
 alfalfa mosaic virus 1:85
 bacteriophages 1:439
 Brome mosaic virus assembly 1:385
 DNA 1:432
 foamy viruses life cycle 2:264
 genome 1:196
 heterologous *see* Hetero-encapsidation
Encephalitis 5:101
 Akabane virus infection 1:77
 alphavirus infection 5:115
 arbovirus infection 1:474
 bovine herpes virus 1 infection 1:366
 bovine herpes virus 5 infection 1:365
 bovine immunodeficiency viruses infection 1:352
 coxsackievirus infection 1:585t
 definition 1:474
 equid herpesvirus-9 infection 2:418
 HIV infection 1:57
 HSV infection 1:474
 human herpesvirus-6 infection 2:503
 Japanese encephalitis virus infection 2:236
 Langat virus infection 5:53
 mumps 3:361
 Powassan virus infection 2:239
 rabies 3:466
 St. Louis encephalitis virus infection 2:236
 TMEV infection 5:41–42
 zoonoses 5:488
 see also specific viruses
Encephalomyelitis
 definition 1:474
 TMEV infection 5:41–42
Encephalomyocarditis virus (EMCV)
 history 1:441
 infection 1:446f
  clinical features 1:445
  clinical outbreaks 1:441–442
  diagnosis 1:447
  histopathology 1:446f
  human 1:445
  pathogenicity 1:445
  pathology 1:445
  persistent 1:445–446
  pigs 1:445–446
  prevention 1:447–448
  route 1:445
  tissue tropism 1:445
  vaccines/vaccination 1:447–448
 internal ribosome entry sites 4:134–135
Encephalomyocarditis virus (EMCV) 1:443f
Encephalopathy
 betanodavirus fish infections 2:229–230
 definition 5:186
 Dengue virus infection 2:10
Endemic(s) 2:5
 definition 2:142
Enders, John Franklin, animal virus cultivation 2:456
Endive, tomato spotted wilt virus infection 5:288–289
Endocytosis
 definition 5:423
 host cell entry 4:409
 process 4:409
 virus membranes 5:310
 virus uncoating 4:409
Endogenous pararetroviral sequences (EPRVs) 4:430
Endogenous retrovirus(es) (ERV) 2:105–109
 classification 2:107
 definition 2:105
 discovery 2:106
 distribution 2:106
 evolution 2:106–107
 host benefits 2:107
 nomenclature 2:108
 origins 2:107
  mechanisms 2:107
 recombination 2:106
 structure 2:106f
 taxonomy 2:108f
 see also Provirus(es)

Endolysins 3:251
    amidases 3:251
    definition 3:248
    icosahedral tailed dsDNA bacteriophages 3:35
    see also Holin–endolysin lysis
Endoparasitic wasp(s)
    ascovirus infection 1:189
    definition 1:186
Endopeptidase(s), endolysin function 3:251
Endoplasmic reticulum (ER)
    Brome mosaic virus replication in yeast 1:383
    rotavirus replication 4:510
    yellow fever virus assembly 5:471
Endoplasmic reticulum-associated degradation (ERAD)
    definition 5:299
    K28 virus toxin translocation 5:304–305
Endoreduplication, definition 1:301
Endornavirus 2:288
Endosomes
    MHC-II antigen presentation 1:124–125
    proteases 1:124–125
Endothelial cell(s)
    Ebola virus infection 2:203
    Lassa fever pathogenesis 3:207
    Marburg virus infection 2:203
Endothelial syncytial cells, measles 3:289
Endotheliotropic elephant herpesvirus 2:428
Energetics, bacterial cell viral entry see Bacteriophage(s), bacterial cell entry
Energy dependency, dsDNA bacteriophage genome packaging 2:310f
Enfuvirtide 2:506
Engineered resistance, crops 2:454
Enhancin 1:256
Entecavir 1:143t
Enteric cytopathogenic human orphan viruses see Echovirus(es)
Enteric infection(s)
    astroviruses 1:204–205
    barriers against 5:315
Enteric virus(es) 2:116–123
    acute gastroenteritis 2:116
    see also specific viruses
Enterobacteria, icosahedral ssDNA bacteriophages 5:368
Enterobacteria phage M13 2:192t
    pVIII 2:193
Enterovirus 1:581, 1:581t, 2:65–66, 2:130, 4:243
Enterovirus(es) 2:130
    bovine see Bovine enterovirus(es) (BEV)
    classification 2:123
    genome 2:125f
        2A protein 2:126
        5′ nontranslated region 2:126
        leader protein 2:126
    host cell entry 5:315
    human see Human enterovirus(es)
    occurrence 2:70
    porcine 2:124
    replication 2:127
    simian see Simian enteroviruses (SEV)
    stability 2:124
    swine vesicular disease virus 2:124
    vaccines/vaccination 2:70
    virion properties 2:128
        stability 2:128
    virion structure 2:128f
        capsid 2:128
    see also Human enterovirus(es)
Enterovirus 68 2:130
    cellular interactions 2:131
    history 2:130
    infection
        clinical presentation 2:131
        epidemiology 2:131
        infectivity, loss 2:131
        pathogenesis 2:131
        transmission 2:131
Enterovirus 69 2:131
Enterovirus 70 2:132
    cellular interactions 2:132
    history 2:132
    infection
        acute hemorrhagic conjunctivitis 2:132
        clinical presentation 2:132
        control 2:133
        diagnosis 2:12
        epidemiology 2:132
        eye infection 2:132
        infectivity 2:133
        neurological symptoms 2:133
        pandemics 2:132
        pathogenesis 2:132
        pathology 2:132
        prevention 2:133
        treatment 2:133
    origin 2:132
    transmission 2:132
Enterovirus 71 2:133
    cellular interactions 2:133
    genetic characteristics 2:133
        genogroups 2:133
        subgroups 2:133
    history 2:133
    infection
        Asian epidemics 2:133
        clinical presentation 2:134
        control 2:134
        diagnosis 2:134
        epidemiology 2:133
        immune response 2:133
        pathogenesis 2:133
        pathology 2:134
        prevention 2:134
        treatment 2:134
    novel types 2:134
Enterovirus 94 2:135
    genome sequence comparisons 2:135
Entomobirnavirus 1:321–322, 1:322t, 2:232
Entomopathogenic nematodes, iridovirus infection 3:166
Entomopoxvirinae 2:136, 4:323, 5:11t
Entomopoxvirus(es) 2:136–140
    fusolin/spindles 2:138
    genome 2:138
        gene content 4:326
        genome size 2:138
    infections 2:137
        insects 3:126t
        insect pest control 3:128
        occlusion bodies 3:128
    molecular biology 2:139
        inhibitor of apoptosis genes 2:139–140
        proteolytic processing 2:139
        RNA polymerase complex 2:139
        transcription factor homologs 2:139
    phylogeny 2:136
        unclassified viruses 2:136
    replication 2:138
        life cycle 2:138–139
        RNA production 2:139
    virion structure 2:137f
        chordopoxviruses vs. 2:137
        poxviruses vs. 3:128–129
    see also specific viruses
ENTV see Enzootic nasal tumor virus (ENTV)
Envelope, definition 4:407
Enveloped virus(es)
    antiviral therapy 1:147
    see also specific viruses
    assembly see below
    families associated 5:308
    host cell attachment 5:308
    lipid composition 5:308
    membrane fusion 3:292
    membrane proteins 5:309
    replication
        assembly 4:411
        release 4:412
    see also Virus membrane(s)
Enveloped virus(es), assembly 1:193–200
    envelope 1:194
        acquisition 1:194
        glycoproteins 1:194
        origin 1:194
    host factor utilization 1:200
    icosahedral viruses 1:195
        alphaviruses 1:195
        flaviviruses 1:195
    see also specific viruses
Envelope fusion protein, insect retroviruses 4:454f
Envelope glycoprotein(s)
    antigen–antibody complexes 3:414–415
    Crimean-Congo hemorrhagic fever virus 1:598–599
    definition 1:127
    flaviviruses 1:198
    yellow head virus 5:477–478
    see also specific viruses
Envelope (env) protein
    definition 5:157
    see also specific viruses
Envelope (E) proteins
    Japanese encephalitis virus 3:183
    see also specific proteins
env gene/protein
    see specific viruses
ENVID (European Network for Diagnostics of "Imported" Viral Diseases) 5:357t
Environment, maize streak virus 3:270
Env-like domains (retrotransposons) 4:451
    definition 4:428
    LTR-retrotransposons 4:435
env-like gene
    definition 4:352
    functions 4:434
    metaviruses 3:309–310
    pseudoviruses 4:355
    SIRE1 virus 4:355
Env precursor protein, avian retroviruses 4:456
Enzootic disease
    definition 2:377
    fowlpox 2:274
Enzootic nasal tumor virus (ENTV) 3:181
    cell transformation 3:181–182
    classification 3:175
    infection 3:181
    Jaagsiekte sheep retrovirus vs. 3:181–182
Enzootic viruses
    see specific viruses
Enzyme(s), prions as 4:340
Enzyme immunoassay(s) (EIA) 2:32f
    definition 2:29
Enzyme-linked immunosorbent assay (ELISA)
    banana bunchy top virus 1:278
    bean calico mosaic virus infection 1:299
    bean golden mosaic virus infection 1:299
    bean golden yellow mosaic virus infection 1:299
    benyvirus infection diagnosis 1:313
    Border disease detection 1:340
    cacao swollen shoot virus infection 1:408
    citrus tristeza virus detection 4:198
    classical swine fever virus detection 1:530
    Crimean-Congo hemorrhagic fever 1:602
    definition 2:29
    HCV detection 2:374
    henipavirus infections 2:326
    Japanese encephalitis diagnosis 3:186
    Marburg virus detection 3:278–279
    measles diagnosis 3:289–290
    mungbean yellow mosaic viruses 3:364–365
    orthobunyavirus infection diagnosis 3:482–483
    papillomavirus infection 4:17
    plant reovirus detection 4:155
    plant viruses 2:24f, 2:453
    see also specific viruses
    potato virus Y 4:290
    pseudorabies virus detection 4:350f
    rhinovirus detection 4:471
    rice yellow mottle virus diagnosis 4:488
    squash leaf curl virus infection 1:299
    Taura syndrome virus, infection diagnosis 5:8
    varicella-zoster virus 5:252

virus isolation 4:494
VLP-based 4:17
VSV detection 5:298
Eorf6, adenovirus E1B-55K gene/protein interaction 1:14–15
Eosinophilic degeneration, yellow fever 5:475
Eosinophilic inclusions, varicella-zoster virus infection 5:254f
EpCV *see* Epirus cherry virus (EpCV)
Ependymal cells, nervous system viruses 1:470
Ephemerovirus 1:355
Ephemerovirus(es)
  classification 1:119t
  see also specific viruses
  genome 3:326f
  see also specific viruses
Ephrin B2, henipavirus tissue tropism 2:322–323
Epicrisis, definition 3:319
Epidemic(s) 2:525
  Chikungunya virus 5:89
  concentrated 1:58
  definition 2:141
  generalized 1:58
  groundnut rosette, Nigeria 3:217
  keratoconjunctivitis, adenovirus infection 2:495f
  SARS 4:552
  viral adaptation 2:533
  Visna-Maedi virus infection 5:424
  see also specific infections
'Epidemic curves' 2:147
Epidemic-level transmission, subsidence 5:272–273
Epidemiology **2:140–148**
  convergence model 2:142f
  definition 2:141
  disease occurrence assessment 2:141
    associated attributes 2:142
  disease prevention implications 2:148
  Henle-Koch postulates 2:143
  history 2:141
  mathematical modeling 2:147
    definition 2:141
    uses 2:148
  studies 2:142
    case-control studies 2:143
    cohort studies 2:143
    etiologic studies 2:143
    molecular epidemiologic studies 2:143
    parameters 2:142–143
    sentinel studies 2:144
    seroepidemiologic studies 2:144
    vaccine trials 2:144
  virus transmission 2:146
    among individuals 2:144
    arthropod transmission cycles 2:147
    common patterns 2:146
    host clinical status 2:146
    host population immunity 2:146
    modes 2:145
    population size 2:147
    virus virulence 2:146
    zoonotic transmission cycles 2:147
  see also specific diseases/disorders
Epidermal growth factor receptor (EGFR)
  activation by HPV E5 4:23
  HCMV-host cell binding 1:626
  molluscum contagiosum virus infection 3:320
Epidermal hyperplasia, fish adenoviruses 2:234
Epidermodysplasia verruciformis (EV) 4:16
  epithelial restriction barriers 4:35–36
  HPV infection 4:20
Epifluorescent microscopy
  bacteriophage counts 2:72
  definition 2:71
Epirus cherry virus (EpCV) 3:500
  taxonomy 3:500
Episomes
  definition 2:301
  EBV infection 2:158f
Epistasis, definition 4:359
Epitheliotropic 3:155

Epithelium
  morbillivirus infection 4:504–505
  papillomavirus infection 4:32
Epitope mapping, plant virus infection diagnosis 2:26
Epitopes *see* Antigen(s)
Epivir® *see* Lamivudine
Epizootic hematopoietic necrosis, fish iridoviruses 2:233
Epizootic hematopoietic necrosis virus (EHNV) 2:233, 3:155–156
  infection 3:158–159
Epizootic hemorrhagic disease virus (EHDV)
  capsid 3:462–463
  history 3:457
Epizootic infections 4:490
  avian adenoviruses 1:27
  avipoxviruses 2:278
  bluetongue virus 3:457
  definition 4:497, 4:567
  lumpy skin disease virus 1:427
  shellfish herpesviruses 4:563
Epizootic transmission, subsidence 5:272–273
EPR (Epidemic and Pandemic Alert and Response) 5:357t
E (envelope) proteins, yellow fever virus 5:471
Epsilonretrovirus 2:231
Epsilonretrovirus(es) 2:220f
  see also specific viruses
Epsilonretroviruses 2:219, 4:460
Epstein-Barr virus (EBV) **2:148–157**
  *Bam*HI-A rightward transcripts 2:157–158
    expression patterns 2:165
    Notch signaling pathway 2:163–164
    open reading frames 2:159f
  defective interfering viruses 2:4
  distribution 2:149
  EBV-encoded nuclear antigens 2:157–158
    latency 2:440–441
    see also specific types
  EBV-encoded nuclear antigen 1 2:161
    latency 2:160
  EBV-encoded nuclear antigen 2 2:161
    CSL targeting 2:161
    EBV-encoded nuclear antigen 3 , regulation by 2:162
    EBV-encoded nuclear antigen-LP, regulation by 2:162
    gene expression effects 2:161
    latency 2:160–161
    PU-1 targeting 2:162
    transcription transactivator 2:161
  EBV-encoded nuclear antigen 3 2:162
    cell line immortalization 2:162
    CSL binding 2:162
    EBV-encoded nuclear antigen 2 regulation 2:162
    gene duplication 2:162
    transactivation domains 2:162
  EBV-encoded nuclear antigen-LP 2:161
    EBV-encoded nuclear antigen 2 regulation 2:162
    latency 2:160–161
  EBV-encoded small RNA 2:164
    expression patterns 2:165
    RNA-protein complexes 2:164
    structure 2:164
    tumorigenic potential 2:164
  evolution 2:151
  genetic diversity 2:158
    amino acid sequence identity 2:158–159
    sequence polymorphisms 2:159
    strains 2:158–159
  genetics 2:156f
    diversity *see above*
    intertypic recombination 2:150
  genome 2:148
    coding potential 2:157
    linearity 2:157
    open reading frames 2:157
    organization 2:157
    terminal repeats 2:157
    see also specific proteins
  historical aspects 2:149
  infection *see below*

latent membrane proteins 2:157–158
  oncogenesis 5:194
  see also specific types
latent membrane protein 1 2:162
  latency 2:160–161
  phenotypic changes 2:162
  signaling 2:162–163
  structure 2:162–163
  TRAF signaling 2:163
  transformation effector sites 2:162–163
latent membrane protein 2A 2:163
  Hodgkin lymphoma 2:163
  latency 2:160–161
  signaling 2:163
  structure 2:163
latent membrane protein 2B 2:163
  signaling 2:163
  structure 2:163
latent membrane protein 3 2:160–161
miRNAs 2:164
molecular biology **2:157–167**
  DNA recombination mechanisms 4:375
  persistence 2:152f
  propagation 2:150
  proteins 2:157
  see also specific proteins
replication 2:160
  cell entry 2:160
  induction efficiency 2:160
strain variation 2:150
  LMP1 variant 2:150
  P3HR-1 virus 2:150
  prototype strain 2:150
  strain 1 2:158–159
  strain 2 2:158–159
taxonomy 2:148
vaccines/vaccination 2:155–156
virion properties 2:157
  morphology 2:157
virion structure 2:149f
Epstein-Barr virus-encoded nuclear antigen 1 (EBNA-1) *see* Epstein-Barr virus (EBV)
Epstein-Barr virus-encoded nuclear antigen 2 (EBNA-2) *see* Epstein-Barr virus (EBV)
Epstein-Barr virus-encoded nuclear antigen 3 (EBNA-3) *see* Epstein-Barr virus (EBV)
Epstein-Barr virus-encoded nuclear antigen-LP (EBNA-LP) *see* Epstein-Barr virus (EBV)
Epstein-Barr virus-encoded small RNA (EBERs) *see* Epstein-Barr virus (EBV)
Epstein-Barr virus infection 2:164
  apoptosis inhibition 1:157–160
  associated carcinomas 5:195
    nasopharyngeal carcinoma 5:195
  B cells 2:151
    persistence 2:151–152
    propagation 2:150
  Burkitt's lymphoma *see* Burkitt's lymphoma
  carcinomas 5:195
    gastric 2:155
    nasopharyngeal *see below*
  conjunctivitis 2:496
  control 2:155
    5' azacytidine 2:155–156
  gene expression patterns 2:165
  Hodgkin's lymphoma 2:154f
    disease characterization 5:194
    incidence 2:154–155
    viral gene expression 5:194
  host range 2:150
  immune response 2:153
    infectious mononucleosis 2:153
  latency 2:157–158
    associated genes/proteins 5:194
    associated tumors 2:441
    EBV-encoded nuclear antigen 1 2:160
    EBV-encoded nuclear antigen 2 2:160–161
    EBV-encoded nuclear antigen-LP 2:160–161
    EBV nuclear antigens 2:440–441
    episome generation 2:158f
    gene expression 2:160

Epstein–Barr virus infection (*continued*)
    latent membrane protein 1 2:160–161
    latent membrane protein 2A 2:160–161
    latent membrane protein 3 2:160–161
    latent membrane proteins 2:151–152
    reactivation 2:165–166
    site 2:151
  lymphoproliferative malignancies 5:194
    adult T-cell lymphoma 2:155
    Burkitt's lymphoma *see* Burkitt's lymphoma
    Hodgkin's lymphoma *see above*
    in immunosuppressed 2:153
    lymphocyte transformation *in vitro* 5:194
    natural killer-cell lymphoma 2:155
    natural killer cells 5:195
    post-transplantation lymphoma 2:152*f*
    T cell 2:155
  molecular biology 2:159
  nasopharyngeal carcinoma 2:155
    viral gene expression 2:155
  oncogenesis 5:316
  persistent 2:165
    resting memory B-cells 2:166*f*
  prevention 2:155
  primary infection 2:152*f*
  reactivation 2:441*f*
  replication role 2:165
    acyclovir treatment experiments 2:165–166
    host, effects on 2:166–167
    latent infection reactivation 2:165–166
    Rta protein 2:166
    Zta protein 2:166
Epstein–Barr virus (EBV) mononucleosis syndrome 2:555
Equid herpesvirus 1 (EHV-1)
  defective interfering particles 2:416
    DNA structure 2:416
    open reading frame 2:416
    sequence conservation 2:416
  distribution 2:412
  DNA homology 2:413
  epidemiology 2:417
  experimental hosts 2:413
  gene cloning 2:413
  genome 2:413–414
    functional analyses 2:414
    gp2 2:414
    IR genes 2:414
    regulatory genes 2:414*t*
  histopathology 2:418
  history 2:411–412
  infection *see below*
  replication
    defective mutants 2:416
    early proteins 2:415
    entry receptor 2:414
    IE gene 2:415
    initiation 2:415
    IR3 gene 2:415
    late proteins 2:415
    transcription regulation 2:415
  tissue tropism 2:413
  virion structure 2:416
    capsid proteins 2:416
Equid herpesvirus 1 infection
  clinical features 2:411–412
    incubation period 2:417
    pregnancy 2:417
    recovery 2:417
  pathology 2:418
    experimental infections 2:418
  prevention and control 2:419
    bacterial complications 2:419
    quarantine 2:419
    vaccination 2:419
  spontaneous abortion 2:411–412
Equid herpesvirus 2 (EHV-2)
  distribution 2:412
  epidemiology 2:417
  history 2:412
  infection 2:418

Equid herpesvirus 3 (EHV-3)
  distribution 2:412
  DNA homology 2:413
  epidemiology 2:417
  history 2:412
  infection 2:418
    histopathology 2:418
Equid herpesvirus 4 (EHV-4)
  DNA homology 2:413
  epidemiology 2:417
  history 2:412
  infection 2:417
  pathology 2:418
Equid herpesvirus 5 (EHV-5)
  epidemiology 2:417
  infection 2:418
Equid herpesvirus 9 (EHV-9)
  DNA homology 2:413
  history 2:412
  infection 2:418
Equine adenovirus infection 1:26
Equine arteritis virus (EAV)
  genome 1:179–180
  geographical distribution 1:177
  history 1:176
  host range 1:177
  infection
    clinical features 1:183
    focal myometritis 1:183
    neutralizing antibodies 1:184
    pathogenicity 1:183
    pathology 1:183
    pregnancy 1:183
    tissue tropism 1:182–183
    vaccination 1:185
  structure 1:178*f*
Equine encephalitic togavirus(es) **5:101–107**
  classification 5:101
  conserved proteins 5:103
  distribution 5:102
  epidemiology 5:103
  evolution 5:103
  genetics 5:103
  genome 5:101
  history 5:101
  infection
    brain examination 5:106
    cellular immunity 5:106
    clinical features 5:105
    control 5:106
    horses 5:106
    host range 5:102
    immune response 5:106
    pathogenicity 5:105
    pathology 5:106
    prevention 5:106
  laboratory animals, infection 5:103
  life cycle 5:102–103
  nomenclature 5:101
  propagation 5:102
  protein properties 5:102
  replication 5:102
    cell attachment 5:102
    protein transport 5:102
  taxonomy 5:101
  tissue tropism 5:104
  transmission 5:104
  vaccines/vaccination 5:106
  virion properties 5:101
    sensitivity 5:101
    size 5:101
  virion structure 5:101
  in vitro propagation 5:103
  see also specific viruses
Equine herpesvirus(es) **2:411–420**
  classification 2:412
  defective interfering particles 2:416
  distribution 2:412
  epidemiology 2:417
  evolution 2:417
  genetics 2:413

  genome 2:413
    functional analyses 2:414
    gp2 2:414
    IR genes 2:414
  history 2:411
  host range 2:413
  infection
    clinical features 2:417
    histopathology 2:418
    immune response 2:418
    pathogenicity 2:417
    pathology 2:418
    persistent infection 2:416
    prevention and control 2:419
  morphology 2:412
  propagation 2:413
  replication 2:414
  serology 2:417
  taxonomy 2:412
  tissue tropism 2:417
  transmission 2:417
  vaccines/vaccination 2:419
    pregnancy 2:419
    side effects 2:419
  see also specific viruses
Equine infectious anemia (EIA) 2:167
  acute 2:172
  chronic 2:172
  clinical features 2:172*f*
  control 2:173
  histopathology 2:172
    hepatic necrosis 2:172–173
    lymphocytic infiltrations 2:172–173
  immune response 2:173
  inapparent carriers 2:172
  pathogenicity 2:171
  pathology 2:172
  prevention 2:173
  rate 2:170
  tissue tropism 2:171
Equine infectious anemia virus (EIAV) **2:167–174**
  classification 2:167
  diagnostic assays 2:173
  epidemiology 2:170
    incidence 2:169–170
  evolution 2:171*f*
    envelope variants 2:171
  genetics 2:170
    reverse transcription 2:170
  genome 2:169*f*
    env gene 2:168
    gag gene 2:168
    pol gene 2:168
    rev gene 2:168
    tat gene 2:168
  geographical distribution 2:170
  history 2:167
  host range 2:170
  infection *see* Equine infectious anemia (EIA)
  infectivity 2:171
  propagation 2:170
  replication 2:169*f*
    assembly 2:169–170
    budding 2:169–170
    genome integration 2:169
    host cell attachment 2:169
    polyprotein processing 2:169
    transcription 2:169
  seasonal distribution 2:170
  serological relationships 2:171
  transmission 2:170
    control 2:173
  vaccines/vaccination 2:173–174
  viral protein properties 2:168*f*
    capsid 2:168
    envelope glycoproteins 2:168
    RNP complex 2:168
  virion properties 2:167
    morphology 2:168*f*
Equine influenza virus(es) 3:101
  see also specific viruses

Equine torovirus (EToV) 5:151–152
  antigenic properties 5:155
  buoyant densities 5:153
  genotypes 5:155
  geographical distribution 5:154
  historical aspects 5:151
  host range 5:154
  infection, seroconversion 5:155–156
  replication
    attachment/cell entry 5:153
    transcription 5:153–154
Equine viral arteritis, African horse sickness vs. 1:42
Equine virus(es)
  see specific viruses
ER see Endoplasmic reticulum (ER)
ERAD see Endoplasmic reticulum-associated degradation (ERAD)
Eradication programs, rabies virus infection 4:372
ERCC4-like nuclease 1:47
E^{rns} protein
  bovine viral diarrhea viruses 1:378
  classical swine fever virus 1:526
Errantivirus 3:301–302, 4:451
Errantivirus(es) 4:453
  baculovirus vs. 4:453
  distribution 4:452
  envelope fusion protein 4:453
    cleavage 4:453
  infectivity 4:451
    env role 4:452
  retrotransposons 4:433
  see also specific viruses
Error(s), virus databases 5:355
Error catastrophe see Evolution (of viruses)
Error-prone replication 4:360
  tailed bacteriophage evolution 2:181
  viral evolution 2:177
Error threshold 4:359
Erythema infections 4:95
Erythrocyte(s)
  BK virus membrane binding 4:281
  human B19 parvovirus 4:89
Erythrocyte necrosis virus 3:155
Erythrovirus 4:95
Erythrovirus(es)
  classification 4:92t
  see also specific viruses
  genome 4:94f
  see also specific viruses
Escape hypothesis, viral origins 3:476f
  bacteriophages 3:473
  virus-encoded proteins 3:473–474
Escape mutants
  definition 1:127
  HIV 1:74–75
  passive immunization 3:67
  rate 3:66–67
  Visna-Maedi virus infection 5:430–431
Escherichia coli
  associated phages 4:400–401
    bacteriophage T4 5:175f
    ssRNA bacteriophage 4:405
  DNA vaccine production 2:52
  transcription, bacteriophage N4
    gyrase role 5:181
    SSB role 5:181
  transcription elongation terminators 5:176
  transformation, baculovirus expression vectors 1:241
Escherichia coli 018:K1:H7, virulence factors 3:123
Ethiopia, HIV infection 1:63
Ethylene, plant resistance to viruses 4:176–177
Etiologic studies, epidemiology 2:143
Etiology, definition 1:301
EToV see Equine torovirus (EToV)
Etravirine 2:510
Eubacteria, bacterial cell viral entry 5:365
Eucocytis meeki virus (EmV) 5:28t
euHCVdb 5:357t
Eukaryotic initiation factor-2α (eIF2α) 3:173
Eukaryotic initiation factor-3 (eIF-3) 1:384

Eukaryotic initiation factor-4E (eIF4E) 4:183f
  amino acid substitutions 4:183–184
  natural resistance vs. 4:184
  role 4:183f
  subfamily 4:183
  VPg interactions 4:185
Eukaryotic initiation factor-4F (eIF-4F)
  picornavirus replication 4:133
  plant resistance genes 4:185
  role 4:183f
Eukaryotic initiation factor-4G (eIF-4G)
  enterovirus translation 2:126
  foot-and-mouth disease virus 2:269
  picornavirus 2:269
  plant resistance genes 4:185
Eukaryotic virus(es)
  see specific viruses
Euploea corea virus (EcV) 5:28t
European bat lyssavirus 4:184t
European catfish virus (ECV) 2:233
European Center for Disease Prevention and Control 5:357t
European Molecular Biology Laboratory (EMBL) 5:356
European rabbit (Oryctolagus cuniculus), calicivirus infection 1:410
European sheatfish virus (ESV) 2:233
European spruce sawfly (Gilpinia hercyniae)
  biological control 3:125
  history 1:225–226
European wheat striate mosaic virus (EWSMV) 5:24
Euryarchaea-infecting virus(es) 5:411–423
  extreme halophilic Archaea-infecting viruses 5:415
    haloalkaliphilic virus 5:418
    head-and-tail viruses 5:415
    salterprovirus 5:418
    spherical halovirus SH1 5:420
  methanogenic Archaea-infecting viruses 5:412
    burst size 5:412
    latent period 5:412
    research 5:412
    history 5:412
  thermococcales-infecting virus (PAV1) 5:421
    genome 5:422
    virion composition 5:422
  see also specific viruses
EV see Epidermodysplasia verruciformis (EV)
Everglades virus (EVEV) 5:109t
Everted DNA
  definition 3:117
  inovirus DNA conformation 3:121
Evolution (of viruses) 2:174–184
  as basic science 2:175
  cellular vs. acellular 3:475
    metabolism 3:475
  continuous 2:180f
    genetic stability vs. 2:179
  error catastrophe 2:178
    definition 2:178–179, 4:359
    genome size limitation 2:178–179
    quasispecies see Quasispecies
  error prone replication 2:177
  experiments in see below
  historical aspects 2:177
    fitness definition 2:177
    mutation rates 2:177
    non-species definitions 2:177
  host evolution vs. 2:176
    non-extinction 2:176
    polyphyletic origins 2:176
    species jumping 2:176
  modular 5:145
  mutations 4:125
  quasispecies 2:177
    clouds 2:178–179
    definition 2:177–178
    Eigen, Manfred 2:177–178
    Shuster, Peter 2:177–178
    swarms 2:178–179
  in real time 2:175
  RNA stability 2:179

RNA virus–host interactions 3:154
sequence space 2:178
  fitness landscape 2:178f
  genetic bottlenecks 2:179
  similarity tools 2:179
    BLAST 2:180–181
    Markov chain Monte Carlo method 2:180–181
species classification 5:403
species jumping 2:175
specific features 2:175
virulence variation 2:175
virus–host congruence 2:179
see also Origin(s) of viruses; specific viruses
Evolution (of viruses), RNA virus experiments 4:362
  evolutionary properties 4:363
    evolutionary convergencies 4:364
    evolutionary parallelisms 4:364
    fitness tradeoffs 4:364
    transmission bottlenecks 4:364
  parameters 4:363
    compacted genomes 4:363
    high per-base mutation rate 4:363
    rapid growth 4:363
  quasispecies theory relevance 4:364
    error catastrophe vs. lethal mutagenesis 4:365
  robustness evolution 4:364
Evolutionary convergencies, RNA virus experimental evolution 4:364
Evolutionary parallelisms, RNA virus experimental evolution 4:364
Exanthem, definition 1:580
Exanthema subitum 2:498
Exaptation, definition 4:436
Exogenous antigen(s) 3:72–73
3'-5' Exonuclease (ExoN) 5:477–478
Exophthalmia, definition 3:89
3'-5' Exoribonuclease 3:424–425
Expanded Programme Immunization (EPI), hepatitis B virus vaccination 2:364
ExPASy 5:357t
Exposed-uninfected (EU) individual(s), definition 1:69
Extensible markup language (XML), virus database formats 5:350f
Extracellular antibody neutralization, neural cell infection 1:473
Extracellular enveloped virions, variola virus replication 4:639
Extracellular enveloped virus (EEV)
  African horse sickness vs. 1:42
  cowpox virion types 1:575
  see also specific viruses
Extracellular search
  bacteriophage population ecology 2:74–75
  definition 2:71
Extrahepatic biliary atresia (EHBA) 4:382
Extra small virus (XSV) 4:527t
  genome 4:574
  taxonomy 4:575
Extreme halophilic Archaea-infecting virus(es)
  see Euryarchaea-infecting virus(es);
  see specific viruses
Extreme resistance (ER) (plants) 4:174
Extrinsic incubation period see Vector(s)
Extrinsic pathway (apoptosis) 1:155
Eyach virus (EYAV) 1:533
  cross-reactivity 1:536–537
  distribution 1:534f
  evolution 1:537
  genome 1:540f
Eye
  anatomy 2:492f
  infections see Eye infection(s)
  lens 2:491
  physiology 2:491
  see also entries beginning ocular
Eye infection(s) 2:491–498
  adenoviruses 2:495
  associated viruses 2:68
  see also specific viruses
  hepatitis C virus 2:491–492
  molluscum contagiosum virus 2:496

Eye infection(s) (continued)
　necrotizing inflammation 2:492
　rubella virus 2:492–494
　transmission 2:491
　varicella zoster virus 2:496
　see also entries beginning ocular

F

Faba bean necrotic stunt virus (FBNSV) 3:386
　characteristics 3:389t
Faba bean necrotic yellows virus (FBNYV) 3:218f
　characteristics 3:389t
　epidemic, Egypt 3:218
　genome 3:387f
　　integral genome segments 3:388
　　satellite-like rep DNA 3:388–390
　host range 3:386
　infection
　　agroinoculation experiments 3:386
　　economic importance 3:386
　transmission 3:386
　　aphids 3:386
Fabavirus 1:573
F-actin cables, baculovirus in vitro
　pathogenesis 1:270
Famciclovir 1:150
　HSV infection 2:394–395
Families 5:9
Famvir® 1:143t
Fancy carp reovirus 1:164t
Farming see Agriculture/farming
Farr, William 2:141
Fascia 5:83
　togavirus infection 5:84
Fascial 1:37
Fascitis 5:83
　togavirus infection 5:84
Fas ligand(s)
　apoptotic role 1:157
　Fas interactions 3:80
Fatal familial insomnia (FFI) 5:192
　associated genetic mutation 5:192–193
　clinical characteristics 5:192–193
Fatal hemorrhagic disease 1:636
FBL-2 protein 2:373
FBNSV see Faba bean necrotic stunt virus (FBNSV)
FBNYV see Faba bean necrotic yellows virus (FBNYV)
fd bacteriophage see Bacteriophage fd
FDV see Fiji disease virus (FDV)
Febrile rash, coxsackieviruses 1:585
Fecal–oral cycle
　hepatitis E virus 2:380
　virus transmission patterns 2:146
Feldmannia virus(es) (FsV) 4:116
　see also specific viruses
Feline calicivirus(es) 1:410
　antigenic variation 1:415
　geographical distribution 1:413–414
　host range 1:414–415
　infection
　　clinical features 1:417
　　immunity 1:417
　propagation 1:415
　transmission 1:416
　vaccines/vaccination 1:418
　see also specific viruses
Feline immunodeficiency virus (FIV) 1:347–354
　classification 1:347
　diagnostics 1:353–354
　epidemiology 1:351
　　dual infections 2:187
　evolution 1:350
　　pathogenic 1:353–354
　genetics 1:350
　　subtypes 1:350
　　variability 1:350
　genome 1:349
　　open reading frames 1:349–350
　geographical distribution 1:348f
　　prevalence 1:348–349

history 1:347
host range 1:350
infection see below
propagation 1:350
serology 1:350
　cross-reactivity 1:350
taxonomy 1:347
　subtypes 1:348
tissue tropism 1:351
transmission 1:351
　transplacental 1:351
variability 1:350
virion structure 1:349f
　viral core 1:349
　viral enzymes 1:349
Feline immunodeficiency virus infection
　clinical features 1:351
　　CNS disease 1:352
　　stages 1:351–352
　　viral load 1:351–352
　histopathology 1:352
　immune response 1:352
　pathology 1:352
　　neurological abnormalities 1:352
　prevention and control 1:353
　　antiretroviral drugs 1:353
　　vaccine trials 1:353
Feline leukemia virus (FeLV) 2:107, 2:185–190
　classification 2:185
　control 2:189
　defective interfering viruses 2:4
　envelope gene variation 2:188
　　FeLV-A 2:188
　　FeLV-B 2:188
　　FeLV-C 2:188
　epidemiology 2:186
　　viral reactivation 2:186
　evolution 2:188
　genome 2:185
　geographical distribution 2:186
　history 2:185
　host range 2:187
　infection see below
　pathogenicity 2:188
　replication 2:185
　　life cycle 2:186f
　　virion assembly 2:186
　subgroups 2:187
　taxonomy 2:185
　transmission 2:189
　　age-related resistance 2:189
　vaccines/vaccination 2:189
　virion structure 2:185
Feline leukemia virus infection
　clinical features 2:187
　　mortality 2:187
　　neoplasms 2:187–188
　dual infections 2:187
　immune response 2:189
　immunosuppression 2:188
　oncogenesis 2:188
　pathology 2:187
　prevention 2:189
　transient infection 2:186
Feline panleukopenia virus (FPV) 4:85
　infection pathogenesis 4:89
Feline parvovirus(es) (CPV) 4:90
　see also specific viruses
Feline sarcoma virus(es) (FeSV) 2:185–190
　classification 2:185
　envelope gene variation 2:188
　epidemiology 2:186
　genome 2:185
　geographical distribution 2:186
　history 2:185
　infection
　　clinical features 2:187
　　control 2:189
　　immune response 2:189
　　pathology 2:187
　　prevention 2:189
　　vaccines/vaccination 2:189

pathogenicity 2:188
recombination 2:185
replication 2:185
taxonomy 2:185
transmission 2:189
virion structure 2:185
see also specific viruses
Ferret(s), influenza virus infection 3:101
FeSV see Feline sarcoma virus(es) (FeSV)
Fetal infection(s)
　see specific infections
Ff bacteriophage see Bacteriophage Ff
FFI see Fatal familial insomnia (FFI)
FG loops, translational repression/assembly initiation
　complex 3:28
F glycoprotein, measles virus 3:287–288
FHV see Flock house virus (FHV)
Fiber genes
　adenoviruses 1:3
　aviadenoviruses 1:6
　fowl adenoviruses 1:6
Fibrillarin 5:209
　umbravirus infection 5:212
Fibrin degradation products (FDP) 2:199
Fibroblast(s), CMV infection 1:639f, 3:321–322
Fibroblast growth factors (FGF), granuloviruses 1:217
Fibropapilloma-associated turtle herpesvirus
　(FPTHV) 2:428
Fibropapillomatosis 2:428
Fibropapillomavirus(es)
　life cycle 4:20
　see also specific viruses
　tissue tropism 4:30f
　see also specific viruses
Field-level epidemiology, African cassava mosaic
　disease 1:33
Field trials, papaya ringspot virus resistance 4:5
Fiji, banana bunchy top virus epidemics 1:277
Fiji disease virus (FDV) 4:149–150
　characteristics 4:150t
　genome 4:151f
　as type species 4:536t
Fiji leaf gall disease 4:149–150
Fijivirus 1:484, 4:150, 4:536t
Fijivirus(es)
　characteristics 4:150t
　classification 4:150
　　characteristics 4:150
　genome 4:150
　　dsRNA segments 4:151f
　　terminal sequences 4:151t
　host range 4:154
　transmission 4:154
　virion structure 4:151
　　capsid 4:151
　see also specific viruses
Filamentous brown algae, associated viruses 4:123
　ecological impact 4:124
Filamentous ssDNA bacteriophages 2:190–198,
　3:117–118, 3:119t
　bacterial hosts 3:123
　biotechnology 2:197
　classification 2:191
　composition 2:193
　genome 2:192f
　genome delivery 5:369
　　pili 5:369
　　receptor-binding proteins 5:369
　Gram-negative bacteria infection 3:117–118, 5:369
　life cycle 2:195
　　DNA synthesis 2:196
　　infection 2:195
　　protein synthesis 2:196
　　virus assembly 2:191–192
　phage display 3:124
　physical properties 2:191
　receptor complexes 3:118
　structure 2:191
　　pVIII 2:193f
　taxonomy 2:191
Filiformism 1:614
Filoviridae 2:57, 2:199, 3:273, 3:276f, 3:324

Filovirus(es) **2:198–205**
  bioterrorism 5:408–409
  characteristics 5:62t
  gene expression 2:201
  genome 2:202f
  history 2:199
  hosts 2:200
  transmission 2:200
    amplification 2:200
  vaccines/vaccination 2:204
  virion composition 2:200
  virion structure 2:201f
    glycoproteins 2:200–201
    receptors, putative 2:200–201
    RNP complex 2:201
    VP24 2:201
    VP30 2:201
    VP35 2:201
    VP40 2:201
  see also specific viruses
Filovirus infections
  animal models 2:202
  apoptosis inhibition 1:158t
  clinical features/presentation 2:202
  disseminated intravascular coagulation 2:203
  host responses 2:203
  incubation period 2:202
  prevention 2:204
    antisense technology 2:204–205
    candidate vaccines 2:204
    RNA interference 2:204–205
  treatment 2:204
    interferon type I 2:205
    recombinant nematode anticoagulant protein c2 2:204–205
Filtration experiments, virus discovery 3:398–399
Fingerlings 3:83
Finlaya, Dengue virus transmission 2:7
First-generation vectors *see* Plant virus(es)
Fish adenovirus(es) 1:28
  infections 1:28, 2:234
    epidermal hyperplasia 2:234
  see also specific viruses
Fish birnavirus(es) 2:232
  transmission 2:232
  vaccines 2:232
  vaccines/vaccination 2:232
  see also specific viruses
Fish calicivirus(es) 2:230
  phylogenetic analysis 2:230–231
  see also specific viruses
Fish herpesvirus(es) **2:205–212**
  classification 2:206
  distribution 2:207
  evolution 2:209
  genetics 2:209
  genomes 2:210t
  growth properties 2:209
  history 2:206
  infection 2:233–234
    clinical features 2:207
    host species 2:206
    immune response 2:211
    pathogenesis 2:207
    pathology 2:207
    prevention and control 2:211
  see also specific viruses
Fish iridovirus(es) 2:233
  epizootic hematopoietic necrosis 2:233
  genome sequencing 2:233
  virion morphology 2:233
  see also specific viruses
Fish nodavirus(es) 2:229
  affected species 2:229–230
Fish paramyxovirus(es) 2:229
  historical aspects 2:229
  see also specific viruses
Fish reovirus(es) 2:231
Fish retrovirus(es) **2:212–221**, 2:231
  historical aspects 2:231
  nucleotide sequences 2:231

phylogeny 2:220f
  see also specific viruses
Fish rhabdovirus(es) **2:221–227**, 2:228
  affected species 2:228
  classification 2:223f
  phylogenetic analysis 2:228
  reverse genetic analysis 2:228
  transmission 2:221
  see also specific viruses
Fish rhabdovirus infection
  challenge models 2:221
  control 2:225
    management practices 2:226
    transmission 2:226
    vaccination 2:226
  cytopathic effects 2:221
  detection 2:225
    serological methods 2:225
  mortality 2:221
  symptoms 2:221
  vaccines/vaccination 2:226
Fish virus(es) **2:227–234**
  DNA viruses 2:227
  economic impacts 2:227
  as reportable diseases 2:227
  RNA viruses 2:227
  see also specific viruses
Fitness 5:274
  viral evolution 2:177
Fitness landscape
  definition 2:175
  sequence space 2:178f
Fitness tradeoffs, RNA viruses 4:364
5′-nontranslated region, Pepino mosaic virus 4:105
Flagellar proteins, bacterial cell viral entry 5:366
flamenco (transcription factor) 3:302–304
Flanders virus (FLANV) 4:184t
*flash*BAC *see* Baculovirus expression vector(s)
Flat-type smallpox 4:640
Flaviviridae 1:170, 1:526–527, 2:6, 2:235, 2:242, 2:367, 3:183, 5:11t, 5:45, 5:440–441
Flavivirus 2:6, 2:242, 5:45, 5:108, 5:440–441
Flavivirus(es) **2:241–253**
  antigenic determinants 2:8–9
  apoptosis inhibition 1:158t
  assembly 1:198
    capsid 1:199
    glycoproteins 1:198
    pre-membrane protein cleavage 1:199
    vesicle packets 1:199
  budding 1:199
  classification 2:242
    re-classification 5:20
  definition 5:469
  epidemiology 2:243
    human activity 2:243
  evolution 1:174
    arboviruses *vs.* 1:173
    mutation frequency 1:173–174
  fusion proteins 3:297
  genome 1:172
    encoded proteins 2:235
  geographical distribution 1:174
  global distribution 2:243
  history 2:242
  host cell exit strategies 1:200
  human pathogens 1:198
  infections *see below*
  life cycle 1:198
  maturation 2:243–245
  physical properties 2:243
    stability 1:172
  polyprotein processing 1:198
  replication 2:243
  structure 2:242
  taxonomy 2:242t
  transmission 1:170
    mosquito-borne *vs.* tick-borne 1:173–174
    resistance 2:245
    tick-borne 2:6

  veterinary important **2:234–241**
    mosquito-borne 2:236
    tick-borne 2:239
    virion morphology 2:235
  virion structure 1:172
    capsid 1:198
    envelope 1:198
    immature virion 1:198
    mature virion 1:198
    nucleocapsid 1:199
    pre-membrane protein 1:198
    proteins 2:243–245
  see also specific viruses
Flavivirus infections
  animal disease 1:175
  cell tropism 2:245
  central nervous system 2:245
  clinical disease 2:243
  diagnosis 2:245
    immunoglobulin G 2:245–246
  host cell entry 2:242
    candidate receptors 2:243
  immunomodulation 4:114t
  pathogenesis 2:245
  persistence 4:109t
  prevention 2:245
  secondary infection 2:245
Fleck disease complex 4:203
  tymoviruses 4:203
Fleece coarseness, Border disease 1:336
Flexiviridae 1:448–450, 2:254t, 2:255, 2:288
Flexivirus(es)
  biological properties 2:254
  classification 2:253t
  genome 2:255f
  phylogenetic relationships 1:426f
    analysis limitations 2:257–258
    coat protein 2:258f
    movement protein 2:256
    replication protein 2:256f
    RNA-binding protein 2:258f
    TGB protein 2:257f
  replication 2:253
    translation 2:253
  transmission 2:254
  virion morphology 2:253
  see also specific viruses
FLIP proteins *see* Death receptor signalling
Flock house virus (FHV)
  host range 3:431–432
  RNAi suppression 3:152
  RNA replication
    cis-acting signals 3:436
    long-distance base paring 3:436
  structure 5:382f
Floral homeotic gene(s), virus-induced gene silencing 5:378–379
Flow cytometry 2:71
Flower breaking, definition 4:207
Fluid replacement, Crimean-Congo hemorrhagic fever treatment 1:603
Fluorescence antibody to membrane antigen (FAMA), varicella-zoster virus 5:252
6-Fluoro-3-hydroxy-3-pyrazine carboxamide (T-705) 1:153
Flv gene, West Nile virus 5:446–447
FMV *see* Fort Morgan virus (FMV)
Foamy viruses (FV) **2:259–265**
  associated hosts 2:259
  Bet gene/protein 2:471
    overexpression 2:263
    synthesis 2:263
  classification 2:260
  DNA integration
    life cycle 2:264
    vectors for gene therapy 2:264–265
  env gene/protein 2:262
  epidemiology 2:259
    at risk populations 2:259–260
  evolution 2:264

Foamy viruses (FV) (continued)
  gag gene/protein 2:262
    synthesis 2:262
  gene therapy vectors 2:264
    DNA integration 2:264–265
  genetics 2:264
    gene expression 2:262
    see also specific genes
  genome 2:262f
  infection 2:260
    tissue cultures 2:261f
  life cycle 2:263f
    assembly 2:264
    budding 2:264
    DNA integration 2:264
    encapsidation 2:264
    host cell entry 2:263–264
    packaging 2:264
    reverse transcription 2:264
    transcription 2:264
    translation 2:264
    uncoating 2:264
  morphology 2:261f
  nomenclature 2:259
  pol proteins 2:262
    enzymatic activity 2:262–263
    synthesis 2:262
  protein organization 2:262
    Bet 2:263
    env 2:262
    gag 2:262
    pol 2:262
    Tas 2:263
  taxonomy 2:260
  transmission 2:259–260
  virion properties 2:261
  see also specific viruses
Focal adhesion kinase (FAK), KSHV host cell entry 3:197
Focal myometritis 1:183
Foetal bovine serum, insect cell culture 1:238
fOg44 bacteriophage 3:254–255
Foliage, Sharka disease 4:238
Follicle cell(s), gypsy RNA transcripts, expression 3:302–304
Follicular dendritic cell(s) (FDC), prion replication 4:334
Fomite(s) 2:141
Fomite transmission, respiratory viruses 2:556
Food hygiene, norovirus infection 1:418
Food poisoning, disease surveillance systems 2:46
Foorkey disease 1:275–276
Foot-and-mouth disease
  clinical features 2:273f
  control 2:273
  immune response 2:273
  pathogenicity 2:272
  pigs 2:270–271
    vaccination 2:273
  prevention 2:273
  tissue tropism 2:272
Foot-and-mouth disease virus(es) (FMDV) 2:265–274
  animal enterovirus associated 2:124
  classification 2:266
  epidemiology 2:272
  evolution 2:272f
  genetics 2:271
  genome 2:267f
    internal ribosome entry sites 2:268f
    sequencing 2:143
    untranslated region 2:267
  geographical distribution 2:270f
    nonendemic areas 2:272
  history 2:266
  host range 2:270
  neutralization escape mutants 3:418
  propagation 2:270
  proteins 2:267
    antigenic sites 2:271
    3C^{pro} 2:268
    3D^{pol} 2:268

  leader 2:268
  P1 region 2:268
  P2 region 2:268
  P3 region 2:268
  structural 2:266
  truncations 2:266–267
  VP1 2:266–267
  VPg 2:267
  receptor binding 5:321
  replication
    host cell attachment 2:270
    host cell entry 2:270
    post-translational processing 2:267f
    RNA replication 2:268
    translation 2:269
    virus assembly 2:269
    virus release 2:269
  serological relationships 2:272f
  taxonomy 2:266
  transmission 2:272
  vaccines/vaccination 2:273
    formalin inactivation 2:273
  virion properties 2:266
    structure 2:266f
  see also specific viruses
Foreign gene delivery, giardiavirus vectors 2:314
Fort Morgan virus (FMV) 5:92
  characteristics 5:109t
Fosamprenavir 1:143t
Foscarnet 1:143t
Foscavir® 1:143t
Foveavirus 1:420t, 1:422, 1:426f
Foveavirus(es) 1:419–427
  associated species 1:422
  biological properties 1:422
  classification 1:420t
  genome 1:423
    heterogeneity 1:424
    open reading frames 1:423
  particle structure 1:421f
  phylogeny 1:426f
  replication 1:423
  strains 1:424
  tissue tropism 1:423–424
  see also specific viruses
Fowlpox
  cutaneous form 2:280f
  detection
    inter-strain differentiation 2:281
    intra-species differentiation 2:281
    monoclonal antibodies 2:280
    PCR 2:282f
  diphtheritic form 2:281f
  histology 2:279
  variants 2:283
Fowlpox-like virus(es) 2:278
  see also specific viruses
Fowlpox virus (FWPV) 2:274–284
  antigens 2:275
  history 2:274
  infection see Fowlpox
  lipid metabolism, host effects 2:277
  survival factors 2:275–276
  taxonomy 2:274
  transmission 2:279
  vaccines/vaccination 2:282f
    contamination 2:283
    virus propagation 2:277
  variants
    antigenic 2:283
    tumorigenic 2:283
Foxtail mosaic virus (FoMV) 4:311t
F-pilus, filamentous ssDNA bacteriophage infection 2:191
F proteins
  baculovirus infection resistance 1:267
  henipavirus 2:322–323
  measles virus 3:286
  nucleopolyhedrovirus 1:263
  paramyxovirus classification 3:285
FPV see Feline panleukopenia virus (FPV)

FpV-1 see Fusarium poae virus (FpV-1)
Fractalkine 2:542
FrAdV-1 see Frog adenovirus 1 (FrAdV-1)
Fragaria chiloensis latent virus (FCiLV) 3:47
Fragaria vesca, Raspberry bushy dwarf virus infection 3:39
Frameshifts, mouse mammary tumor virus replication 3:337f
Francisella tularensis, bioterrorism evaluation 5:408t
Frankliniella occidentalis, tomato spotted wilt virus association 5:136
Fraser's hypotheses, plant recessive resistance genes, mechanisms 4:182
Freesia sneak virus (FSV) 3:448
  coat protein 3:450
  geographical distribution 3:452–453
  transmission 3:453
Freeze-dried vaccine(s), safety 5:229
Freund's incomplete adjuvant, Visna-Maedi virus vaccination 5:431
Frog adenovirus(es)
  infection pathogenesis 1:28
  see also specific viruses
Frog adenovirus 1 (FrAdV-1) 1:28
Frog herpesvirus(es)
  genetics 2:211
  see also specific viruses
Frog virus 3 (FV3) 2:233
  replication cycle 3:171f
    cellular receptor, host cell entry 3:170
    DNA methylation 3:170
    DNA synthesis 3:170
    RNA synthesis, early 3:170
    RNA synthesis, late 3:170
    virion assembly 3:171
    sites of 3:172f
Frog virus 4 (FV4), genome 2:209
Frosch, Paul 5:55
  virus discovery 3:399
Fruit drop, Sharka disease 4:239
Fruit trees, diseases 4:201–207
  pome fruits 4:205
    causal agents 4:206
    control 4:206
    geographical distribution 4:206
    symptoms 4:205
    transmission 4:206
    yield losses 4:205
  stone fruits
    causal agents 4:205
    control 4:205
    geographical distribution 4:205
    nematode-borne 4:198
    Sharka 4:201
    symptoms 4:204
    transmission 4:205
    yield losses 4:204
Fry 3:83
FSV see Freesia sneak virus (FSV)
Fukuoka virus (FUKAV) 4:184t
Full length genome sequencing, plum pox virus 4:240
'Full virus vector' strategy see Plant virus(es)
Fulminant hepatitis 2:357
Functional genomics 5:375
Functional marker mutations, DNA virus recombination 4:374–375
Functional motifs, virus databases 5:363
Fungal virus(es) 2:284–291
  classification 2:287t
  double-stranded DNA viruses 2:286
    genomes 2:286
    morphology 2:286
    replication 2:289
  gene expression 2:289
  genomes 2:574
    ssRNA viruses 2:288
    unassigned ssRNA viruses 2:288
  historical aspects 2:284
  host range 2:285
    experimental 2:285
  hypovirulence see Hypovirulence

infections
  host defenses 2:289
  latency 2:285
  mixed 2:286
  plant–fungal mutualistic associations 2:291
  symptom expression 2:285
replication 2:289
  yeast as a model host 2:290
taxonomy 2:286
transmission 2:285
virological technical advances 2:289
see also Mycovirus(es)
Fungi
  classification 4:419
  definition 4:419
  filamentous, killer phenomenon 5:164
  infections
    biocontrol of 2:580
    symptom classification 2:574
    see also specific infections
  plant virus transmission 5:281
  retrotransposons *see* Retrotransposons, fungi
Furin
  definition 2:542
  henipavirus 2:322–323
Furin-like cleavage, baculovirus F protein 4:454f
Furovirus 1:308, 2:291–292, 2:295, 4:283
Furovirus(es) **2:291–296**
  classification 4:283
  definition 2:291
  genome 2:294f
  geographical distribution 2:292
  history 2:291
  infections
    control 2:292
    diagnosis 2:295
    diseases associated 2:292
    host range 2:292
    symptoms 2:295
  movement, long distance 2:292
  phylogeny 2:294f
  proteins 2:293
    coat protein 2:293–294
    movement 2:292
  transmission 2:292
  virus differentiation 2:293
  virus properties 2:293
    antigenic 2:293
    nucleic acids 2:293
    structure 2:293f
  see also specific viruses
Furovus 1:308
Fusarium poae virus (FpV-1) 4:73
  morphology 4:73
Fuselloviridae 1:587, 2:296–297, 5:11t
Fusellovirus(es) **2:296–300**
  biochemical characterization 2:297
  Crenarchaea infections 1:588t
  culture-independent studies 2:299
  definition 2:296–297
  distribution 2:297
  genomes 1:593
    conservation 1:593
  history 2:297
  integrases 2:299
    genome integration 2:299
    tRNA genes 2:299
    in vitro studies 2:299
  isolates 2:298
  plasmid virus hybrids 2:299
  structural genomics 2:300
  vectors 2:299
  virion morphology 1:587
  virion production 2:296–297
  see also specific viruses
'Fusiform' virus(es) 2:300
  see also specific viruses
Fusion, host cell entry 4:409
Fusion-associated small transmembrane (FAST) proteins 4:383–384
Fusion inhibitors, HIV antiviral therapy 1:148
Fusion loop(s), membrane fusion 3:300

Fusion peptide(s), membrane fusion 3:296
Fusion pore(s) 3:292
  formation 3:294f
Fusion protein(s)
  class I 3:293
    biosynthesis 3:293
    cooperativity 3:296
    general fusion model 3:297f
    hemagglutinin 3:293
    see also Hemagglutinin (HA)
    membrane anchors 3:295–296
    refolding 3:297f
  class II 3:297
    activated structure 3:297
    biosynthesis 3:297
    cooperativity 3:299
    native structure 3:297
    trimerization 3:299f
  class III 3:300f
    biosynthesis 3:299
    fusion loops 3:300
  definition 4:407
Fusogenic orthoreovirus(es) 4:383
  see also specific viruses
Fusolin/spindles, entomopoxviruses 2:138
FV *see* Foamy viruses (FV)
Fv1 (Friend virus susceptibility factor 1) 2:469
  saturation of 2:469
  strain specificity 2:469
FV3 *see* Frog virus 3 (FV3)
Fv4, receptor-envelope interaction 2:468–469
FWPV *see* Fowlpox virus (FWPV)

G

Gag (group-specific antigen)
  LTR retrotransposons 4:438–439
  Visna-Maedi virus infection, immune response 5:427
  yeast L–A virus virion structure 5:466f
gag gene/protein
  definition 4:428
  LTR retrotransposons 4:420
  pseudovirus organization 4:353
  structural similarities 2:520
  see also specific viruses
Gag-Pro precursor protein 4:456
Gait spasticity, TMEV infection 5:42
Galleria mellonella (wax moth), parvovirus infection 4:76
Galleria mellonella densovirus (GmDNV) 4:78
Gallid herpesvirus 1 (GaHV-1) 2:406
Gallid herpesvirus 2 (GaHV-2) 2:406
GALT *see* Gut-associated lymphoid tissue (GALT)
GAM-1 protein, fowl adenoviruses 1:6
Gammaentomopoxvirinae 4:323
Gammaentomopoxvirus
  characteristics 2:137t
  structure 4:325
Gammaentomopoxvirus 2:136, 4:323
Gammaherpesvirinae 1:364t, 2:431, 5:11t
Gammaherpesvirus(es)
  bigenic amplification 2:426f
  bovine *see* Bovine gammaherpesvirus(es)
  classification 2:432
  equine viruses 2:412
  latency 2:440
  lymphoproliferative disease 2:428
  pathogenicity 2:427–428
  simian **4:585–594**
    future work 4:593
    members 4:587t
  virus-encoded Bcl-2 proteins 1:162
  see also specific viruses
Gammapapillomavirus 4:9
  see also specific viruses
Gammaretrovirus 4:460
Gammaretrovirus(es)
  classification 4:460
  detection 2:17
  morphology 4:461–462
  taxonomy 4:446
  see also specific viruses

Ganciclovir (GCV) 2:557
  BaculoDirect expression vector system 1:244
  history 1:146
  prodrug 1:146
Gan Gan virus 1:400
Ganglia 2:436
Ganglioside(s), Gt1b 4:280
Ganjam virus 1:597f
Gardasil anti-HPV L1 major capsid protein VLP vaccine 5:224
  yield 5:224
Garlic common latent virus (GarCLV) 5:288
Garlic dwarf virus (GDV) 4:150t
Garlic virus(es) 5:288
  infection 5:288
Gastric acid, host defence 4:512
Gastric carcinoma(s)
  distribution 2:155
  EBV associated 2:155
Gastroenteritis
  associated viruses 2:116
  see also specific viruses
  deaths 2:116
  definition 3:438
  symptoms 2:116
Gastrointestinal tract
  adult T-cell leukemia 2:568
  enteric viruses 2:116
  MMR vaccine safety 5:233
  mouse mammary tumor virus transmission 3:337
  SIV replication 4:608
  togavirus infection 5:84
  villi 3:443
  virus entry 5:315
Gastrointestinal virus(es)
  organ transplantation 3:470
  see also specific viruses
Gateway cloning vectors, BaculoDirect expression system 1:241
GAV *see* Gill-associated virus (GAV)
gB gene, herpesvirus 2:426–427
GBLV *see* Grapevine Bulgarian latent virus (GBLV)
GBP-i protein, JCV 4:264
GB virus(es) 2:252
  structure 2:252
  see also specific viruses
GCMV *see* Grapevine chrome mosaic virus (GCMV)
GCRV *see* Grass carp aquareovirus (GCRV)
Geminiviridae 1:298, 1:302, 2:98, 4:165, 4:169f, 5:11t
Geminivirus(es) **4:164–170**
  defective interfering RNAs 1:315–316
  definition 2:97
  diversification 4:166
    DNA reassortment 4:166
    rapid evolution 4:166
  DNA recombination mechanisms 4:376
  DNA satellites 4:530
  economic importance 4:166
    distribution effects 4:167
    yield losses 4:166–167
  emerging **2:97–105**
  evolution 4:166
  genera 4:165
    discrimination 4:168
  genome 4:165
    conserved consensus sequences 4:168
    monopartite 4:165
  host shifts 4:166
  infections *see below*
  molecular diagnosis 4:167
    PCR 4:167
    phylogenetic analysis 4:169f
    replicase-associated protein 4:168
    serology 4:167
    viral coat protein DNA sequence 4:168
  movement 3:350
  phylogenetics 4:169f
    nanoviruses *vs.* 3:391
  proteins 4:165
  replication 1:317–318
  satellite viruses 4:534

Geminivirus(es) (continued)
    structure 4:165f
    transmission 4:165
    virus-induced gene silencing vectors 5:378
    see also specific viruses
Geminivirus infections 4:167f
    management 4:168
        variant detection 4:169
        vector population reduction 4:168–169
    sweetpotatoes 4:661t
    symptomatology 4:167f
    tobacco diseases 5:65
GenBank 5:350f
    sequence data 5:19f
        accuracy 5:362–363
    taxonomic data 5:19f
Gene(s)
    delivery, baculovirus expression vectors 1:245
    diversity, retrovirus evolution 2:183–184
    duplication, EBV-encoded nuclear antigen 3 2:162
    nonchromosomal, definition 4:336
Gene A$_2$, bacteriophage-mediated host cell lysis 3:257
Gene E, bacteriophage-mediated host cell lysis 3:257
Gene expression
    definition 4:407
    varicella-zoster virus latency 5:261
Gene expression profiling 2:14–15
    probes 2:15
        sequence specificity 2:15
Gene expression systems
    limitations 4:229
    plant virus vectors see Plant virus(es)
Gene-for-gene resistance 4:170
    history 4:171
Gene guns, DNA vaccine delivery 2:54–55
Gene L, bacteriophage-mediated host cell lysis 3:258
Gene Ontology (GO) Consortium 5:353
Gene parity plots 1:225
Gene property tables 5:352
Genera 5:9
General fusion model, class I fusion proteins 3:295–296
General practitioners (GPs), disease surveillance systems 2:45
Generative propagation, definition 4:207
Gene regulation
    varicella-zoster virus latency 5:262
Gene segment table(s) 5:352
Gene silencing 4:141–149
    cis-acting siRNA pathway 4:142
    core reactions 4:141
    definition 1:296
    environmental influences 4:148
    heterochromatin siRNA pathway 4:142
    natural cis-antisense transcript-derived siRNA pathway 4:142
    plant disease resistance see Genetically-modified plants, disease resistance
    plants see Plant(s)
    RdRp 6 4:146
    trans-acting siRNA pathway 4:142
    virus-induced see Virus-induced gene silencing (VIGS)
    see also MicroRNAs (miRNAs)
Gene tagging, fungal retrotransposons 4:427
Gene therapy 2:301–306
    biodistribution
        adeno-associated virus vectors 2:304
        clinical trials 2:305
    clinical trials 2:302
        administration route 2:304–305
        HSV 2:304
        lentivirus 2:304–305
        onco-retrovirus 2:304–305
        phases 2:305
        preclinical safety 2:305
        recombinant adeno-associated virus 2:303–304
        recombinant adenovirus 2:304–305
        special considerations 2:305
        tumors 2:304–305
        vector genome biodistribution 2:305
    considerations 2:301
    definition 2:301
    requirements 2:301
    vectors
        design 2:302
        effectivity 2:301
        host range 2:302
        toxicity 2:301
        types 2:302
        see also specific viruses
Genetically modified organism (GMO), definition 1:296
Genetically-modified plants
    antibody expression 3:64
    disease resistance see below
    exogenous protein expression 4:235
        bean yellow dwarf virus 4:236
        chemical induction 4:235
        estrogen-inducible promoters 4:236
        ethanol induction 4:236
        hybridization 4:235
        inducing treatments 4:235
        magnifection 4:234
        strategies 4:235
        transgenic systems 4:235
    gene silencing see below
    heterooligomeric protein expression 4:233
        'noncompeting viral vectors' 4:233
        transient systems 4:233
        yields 4:233
    industrial-scale protein expression 4:234
        Agrobacterium-mediated delivery 4:232f, 4:234
        genome modifications 4:234
        recombinant protein yield 4:234–235, 4:235f, 4:236f
    vaccine production see below
    virus vectors see below
Genetically-modified plants, disease resistance 4:156–164, 4:159t
    associated species 4:157, 4:159t
    benefits of 4:160
    broad spectrum resistance 4:163
    commercial release 4:160–161, 4:163
    disease control 4:157, 4:162
    economic impact 4:160–161, 4:163
    environmental safety 4:161
    evaluation 4:157
        field evaluation 4:163
    gene flow 4:162
    gene silencing 4:145, 4:147f, 5:331
        cross-protection 4:145
        HEN1 4:146
        plant recovery 4:145
        RISC 4:146
        SDE3 4:147
    global availability 4:157–160
    heterologous encapsidation 4:162
    mechanisms 4:161
        antiviral factors 4:157
        gene silencing see above
        post-transcriptional silencing 4:161
        protein-mediated 4:161
        RNA silencing 4:161, 4:162, 5:64
    plant development 4:156
        associated antiviral factors 4:157
        benefits of 4:160
        history 4:156
        transgenic-wild type hybrids 4:162
    vectors see Genetically-modified plants, virus vectors
        watermelon mosaic virus 4:162
        zucchini yellow mosaic virus 4:162
    pollen 4:162
    recombination 4:162
    risks, real vs. perceived 4:163
    RNA interference 3:153
    strategies 4:156
    target traits 4:157
    see also Virus-induced gene silencing (VIGS)
Genetically-modified plants, vaccine production 5:221–225
    animal models 5:221
    chloroplast expression 5:223
    contamination, vaccine safety 5:221
    cost 5:221
    'guinea pig' argument 5:224
    gut-associated lymphoid tissue 5:222
    history 5:222
    HPV 5:223
    limitations 5:223–224
    maize 5:223–224
    mucosa-associated lymphoid tissue 5:222
    Newcastle disease virus 5:224
    Norwalk virus 5:223
    obstacles 5:224
    plant tissues used 5:223
    ProdiGene 5:223–224
    public perceptions 5:221
    soybean plants 5:223–224
    technological advances 5:222
Genetically-modified plants, virus vectors 4:229–238, 5:274–282, 5:275t
    applications 4:233
        nanoscale material manufacture 4:236
        replication cycle manipulation 4:229–230
        research tools 4:233
    classification 5:276f, 5:277t
        'non-persistent' viruses 5:276–277
        persistent viruses 5:277
        semipersistent viruses 5:276–277
    cowpea mosaic virus 1:573–574, 1:573f
    first-generation vectors 4:230
        CP fusions 4:230, 4:232f
        directed evolution 4:230
        DNA shuffling 4:230
        'full virus' strategy 4:230
        host improvement 4:230
        large scale production 4:230
        limitations 4:230
        N-glycosylation 4:231–232
        post-transcriptional silencing see Post-transcriptional gene silencing (PTGS)
        recombinant protein yields 4:230
        replicon delivery 4:230
        TMV-based vectors 4:230
        viral promoters 4:230
    second-generation vectors 4:232
        viral constituents 4:232
Genetic bottlenecks, viral evolution 2:179
Genetic diversity
    gill-associated virus 5:481f
    potato virus Y tomato infections 4:295
Genetic drift
    definition 4:359
    Dengue viruses 2:8
    hantaviruses 2:318
Genetic reactivation, avipoxvirus replication 2:276
Genetic shift, hantaviruses 2:318
Gene transfer
    Herpesvirus Saimiri 4:591
    rhadinovirus human T-cell transformation 4:591
Genital infection(s)
    alphaherpesviruses 1:366f
    alphaherpesviruses, bovine 1:366f
    herpesviruses 2:436–437
    human papillomaviruses see Human papillomavirus(es) (HPVs)
Genitourinary medicine clinic(s) (GUM) 2:44
Genogroups
    aquareovirus 1:167
    Border disease virus 1:338
    definition 2:221
Genome(s)
    bacterial cell viral entry see Bacteriophage(s), bacterial cell entry
    co-packaging 1:133
    definition 1:520
    integration 2:331–332
    retrotransposons, effects of see Retrotransposons
    size
        error catastrophe limitation 2:178–179
        retrotransposon content 4:443
    see also specific viruses
Genome molecule tables, virus databases 5:352–353

Genome replication, definition 4:407
Genomes of the T4-like Phages database 5:357t
Genomic precursor 3:6
Genomics, phylogenetic analysis 4:127
Genomic segment 3:6
Genotypes/genotyping 5:469
　　definition 4:507
　　see also specific viruses
GEO (Gene Expression Omnibus) 5:357t
Geoduck clam aquareovirus (CLV) 1:164t
Geraniums
　　see under Pelargonium
German measles see Rubella
Germline(s), endogenous retroviral origins 2:107
Germplasm, infection control 3:219
Gerstmann–Sträussler–Scheinker syndrome (GSS) 5:192
　　clinical characteristics 5:192
　　genetic mutations 5:192
　　geographical distribution 5:192
　　illness duration 5:192
Getah virus (GETV) 5:92
　　characteristics 5:109t
GETV see Getah virus (GETV)
GFkV see Grapevine fleck virus (GFkV)
GFLV see Grapevine fanleaf virus (GFLV)
GFP see Green fluorescent protein (GFP)
gH/gL complex(es), human herpes virus replication 2:501
Giant cell disease, definition 4:611
Giant cell pneumonia, measles 3:289
Giant mimivirus, evolution 3:477
Giardia lamblia 2:312
Giardiasis 2:315
Giardiavirus 2:286, 2:313, 5:164, 5:165t
Giardiavirus(es) (GLV) **2:312–316**
　　biochemical characteristics 2:313
　　classification 2:313
　　evolution 2:315
　　as gene therapy vector 2:314
　　genome 2:314f, 5:167f
　　　　mRNA structure 2:314
　　　　mRNA translation 5:167
　　　　open reading frames 2:313
　　　　untranslated regions 2:314
　　geographical distribution 2:313
　　history 2:312
　　host range 2:313
　　infection 2:315f, 5:169
　　physical characteristics 2:313f
　　　　RdRp activity 2:313
　　replication 2:314
　　taxonomy 2:313
　　ScV vs. 2:315
　　see also specific viruses
Giardivirus 5:164
Gill-associated virus (GAV) 5:477
　　classification 4:567
　　genetic diversity 5:481f
　　genome 5:478–479
　　　　open reading frame 5:478–479
　　　　transcription-regulating sequences 5:478–479
　　infection
　　　　chronic infections 5:480
　　　　mortality 5:479
　　　　transmission 5:480
　　see also Yellow head virus (YHV)
Gill disease (fish), KHV infection 2:208
Gill necrosis disease 4:561
　　histopathology 4:562
　　manifestations 4:561–562
Gills, definition 4:560
Gilpinia hercyniae (European spruce sawfly)
　　biological control 3:125
　　history 1:225–226
Gilthead seabream reovirus 1:164t
Ginger chlorotic flack virus (GCFV) 4:646t
GINV see Grapevine berry inner necrosis virus (GINV)
Gladiavirus 5:165t
Glc1.8 (mucin), BV-mediated immunosuppression 4:254

Global Advisory Committee on Vaccine Safety (GACVS) 5:232
Global Rinderpest Eradication Programme (GREP) 4:505
Global vaccine coverage, measles 3:290f
Glomeromycota 4:419
　　retrotransposons 4:423
Glomerulonephritis
　　definition 5:423
　　immune complex deposition 3:79f
GLRaV-2 see Grapevine leafroll associated virus-2 (GLRaV-2)
Glucocorticoids, virus transcriptional regulation 3:340
Glut-1 receptor, HTLV-1 infection 2:559
Glutathione peroxidase, avipoxvirus survival factors 2:275
GLV see Giardiavirus(es) (GLV)
Glycan shields, infection neutralization 3:66–67
Glycine decarboxylase, avian hepadnavirus replication 2:329
Glycine max see Soybean (*Glycine max*)
Glycoprotein(s)
　　enveloped viruses 3:292
　　membrane fusion
　　　　biosynthesis 3:293
　　　　class I 3:293
　　　　class II 3:297
　　　　class III 3:299
　　see also specific glycoproteins
Glycoprotein G role, Chandipura virus life cycle 1:500
Glycosaminoglycan(s) 2:542
Glycosaminoglycan(s) (GAGs)
　　associated viruses 5:320–321
　　binding domains 5:321
　　receptor role 5:320
　　see also Heparan sulfate (HS)
　　VACV-host cell entry 5:245
Glycosidase(s), endolysin function 3:251
Glycosylation
　　HCMC modifications 2:489–490
　　plant virus vectors 4:231–232
　　varicella-zoster virus proteins 5:261
Glycotransferases 1:594
Gnasthostome immunoglobulins
　　diversity 3:58
　　phylogeny 3:60f
Gnotobiotic (Gn) calves 5:151
GO (Gene Ontology Database) 5:357t
Goatpox 1:427
　　epidemiology 1:427
Goatpox virus (SPPV), geographical distribution 4:329
Goatpox virus (SPPV), transmission 4:329
Gokushovirus(es)
　　genetic map 3:15f
　　genome content 3:14–15
　　microviruses vs. 3:20
　　morphogenesis 3:19
　　virion morphology 3:14–15
　　　　procapsid assembly 3:15
　　see also specific viruses
Gokushovirus φMH2K 3:20
Golden ide reovirus (GIRV) 1:164t
Golden shiner reovirus (GSRV) 2:231
　　genome 1:165
　　infection 1:166
Gold-handed tamarin lymphocryptovirus (SmiLHV1) 4:587t
Golgi apparatus
　　bunyavirus assembly 1:397
　　varicella-zoster virus assembly 5:261
Gonads, definition 3:495
Gonad-specific virus see *Helicoverpa zea*-2 virus (Hz-2V)
Gonotrophic cycles 4:652
　　definition 5:268
　　horizontal transmission 5:269
Gorilla(s) 2:529, 2:530f, 2:533
Gorilla rhadinovirus 4:587t
GP-1 glycoprotein, arenavirus 3:244
GP-2 glycoprotein, arenavirus 3:244, 3:245

gp41 gene/protein
　　definition 1:127
　　nucleopolyhedrovirus 1:264
GP64 protein
　　baculovirus infection resistance 1:267
　　nucleopolyhedroviruses 1:263, 1:264
　　BV envelope structure 1:263
gp120 see HIV
GPCMV see Guinea pig cytomegalovirus (GPCMV)
GpC protein, bacteriophage φX174 4:405
G protein
　　Hendra virus 2:323
　　henipavirus 2:323
G protein-coupled receptor(s) (GPCR), HCMV immune system evasion 1:637, 1:638f
Grafting 4:201–202
　　grapevine diseases see Grapevine(s), diseases
　　infection control measures 1:524
Gramineae
　　fijivirus hosts 4:154
　　oryzavirus hosts 4:154
　　phytoreovirus hosts 4:154–155
Gram-negative bacteria
　　bacterial cell viral entry 5:365
　　filamentous ssDNA bacteriophages 5:369
Gram-positive bacteria, bacterial cell viral entry 5:365
Gram staining, mimivirus structure 3:312f, 3:316–317
Granulin, definition 1:247
granulin gene, granulovirus 1:265
Granulocyte-macrophage colony-stimulating factor inhibitory factor (GM-CSF)
　　parapoxviruses 4:62
　　Visna-Maedi virus infection 5:428–430
Granulocytes, innate immunity 3:112
Granulovirus 1:211, 1:220, 1:247, 1:254, 1:265
Granulovirus(es) **1:211–219**
　　A-T content 1:216
　　classification 1:211, 1:220
　　fast-killing 1:212
　　gene expression 1:216
　　　　early genes 1:216
　　　　fibroblast growth factors 1:217
　　　　inhibitors of apoptosis 1:214, 1:216
　　　　late genes 1:216
　　　　metalloproteinase 1:216
　　　　polyhedron envelope/calyx protein 1:217
　　　　very late genes 1:216
　　genome 1:213, 1:213f, 1:213t, 1:214f
　　host range 1:265
　　infections
　　　　cycle 1:211
　　　　insects 3:126t
　　morphology 1:212f
　　nucleocapsids 1:249
　　occlusion bodies 1:211
　　packaging 1:265
　　　　granulin gene 1:265
　　pathogenesis 1:268
　　replication 1:212f, 1:214
　　　　origins of 1:218
　　slow-killing 1:212
　　taxonomy 1:211
　　virion structure 1:212f
　　　　budded virions 1:212f
　　　　occlusion-derived virions 1:211, 1:212f
　　see also specific viruses
Granzymes 3:74–75
Grapevine(s), diseases **4:201–207**
　　causal agents 4:203
　　classification 4:203
　　see also specific viruses
　　control 4:204
　　geographical distribution 4:203
　　graft incompatibility 2:91
　　　　causal agents 2:91
　　　　see also specific viruses
　　　　disease symptoms 2:91
　　　　geographical distribution 2:91
　　　　infection control 2:91
　　　　syrah decline 2:91

Grapevine(s), diseases (continued)
    transmission 2:91
    yield losses 2:91
  infectious decline 4:203
    transmission 4:203
  infectious degradation 4:203
  symptoms 4:202, 4:202f
  transmission 4:203
  yield losses 4:202
  see also specific viruses
Grapevine berry inner necrosis virus (GINV) 1:424
  genome 1:424
Grapevine Bulgarian latent virus (GBLV) 3:406t
  diseases 3:411
Grapevine chrome mosaic virus (GCMV) 3:406t
  diseases 3:411
  genome 3:408
Grapevine fanleaf virus (GFLV) 3:406t
  cell-to-cell movement 3:410
  control, transgenic plants 3:412
  diseases 3:411
  genome 3:407–408
  infectious DNA clones 3:408–409
  large single-stranded RNA satellites 4:531
  movement 3:352
  population structures 3:412
  regulated polyprotein processing 3:409
  replication 3:409
  satellites 3:410–411
  transmission, soil nematodes 3:411
Grapevine fleck virus (GFkV)
  characteristics 5:206
  genome 5:202f, 5:206
    coat protein 5:206
    replication polyprotein 5:206
Grapevine leafroll associated virus-2 (GLRaV-2) 2:91
  molecular variants 2:91
  transmission 2:91
Grapevine rootstock stem lesion-associated virus (GRSLaV) 2:91
Grapevine rupestris stem pitting-associated virus (GRSPaV) 2:91
Grapevine Tunisian ringspot virus (GTRSV) 3:406t
  diseases 3:411
Grapevine virus A (GVA)
  genome 1:425–426
  particle structure 1:421f
  replication 1:426
Grapevine virus B (GVB)
  genome 1:425–426
  replication 1:426
Grass carp aquareovirus (GCRV) 1:164t
  genome 1:165
  infection 1:166
  vaccine 1:167, 1:168–169
Grass carp virus (GCV) 2:231
Grasses, maize streak virus 3:265, 3:271
Great island virus 3:465
Green fluorescent protein (GFP)
  definition 5:325
  plant virus distribution 2:26–27
  plant virus movement 3:349, 3:354
Green fluorescent protein-transgene silencing, virus-induced gene silencing 5:326
Groundnut rosette 5:209–210
  causative agents 3:214–215
  Nigerian epidemic 3:217, 3:218
Groundnut rosette assistor virus (GRAV) 5:211
Groundnut rosette virus (GRV)
  cell-to-cell movement 5:212, 5:212–213, 5:212f
  infection see Groundnut rosette
  ORF3 5:212f
  ORF4 5:210f
  phloem-dependent long-distance movement 5:210f
  resistance 5:213
  satellite-like single-stranded RNAs 4:532
  transmission, RNA satellites 5:211
Group C virus(es), orthobunyavirus reassortment 3:482
  see also specific viruses
GRV see Groundnut rosette virus (GRV)

GS-9137 2:510–513
GS9148 (antiviral prodrug) 1:150
GSK-364735 2:507t
GSRV see Golden shiner reovirus (GSRV)
GSS see Gerstmann–Sträussler–Scheinker syndrome (GSS)
Gt1b ganglioside(s), JCV infection 4:280
GTRSV see Grapevine Tunisian ringspot virus (GTRSV)
Guanarito virus 2:95
  geographical distribution 3:204
  history 3:203
Guanidine, prion inhibition 4:337–338
Guanidine hydrochloride (Gu-HCl), picornavirus replication 4:136–137
Guanyltransferase, mycoreovirus structure–function relationships 3:378–379
Guard hypothesis, plant natural resistance 4:175
Guenon, definition 4:611
Guillain–Barré syndrome
  definition 5:226
  HIV infection 1:474–475
  swine flu vaccine safety 5:233
Guinea pig(s)
  foot-and-mouth disease virus infection 2:271
  Marburg virus infection 3:277–278
  varicella-zoster virus infection 5:251
  Zaire Ebola virus infection 2:62
Guinea pig cytomegalovirus (GPCMV) 1:633
  genome 1:633
  infection
    associated tissues 1:634
    transplacental infection 1:634
  vaccines 1:634
Gumboro disease see Infectious bursal disease virus (IBDV)
Guppy aquareovirus (GRV) 1:164t
Guppy virus 6 (GV-6) 2:233
Gut-associated lymphoid tissue (GALT)
  edible vaccine development 5:222
  HIV infection 1:71
  resident memory CD4 T cells 1:71
Guttaviridae 1:587, 5:11t
GVA see Grapevine virus A (GVA)
GVB see Grapevine virus B (GVB)
GV-specific cell lines, baculovirus in vitro pathogenesis 1:270
Gypcheck, insect pest control 3:130
gypsy retrotranspon see *Drosophila melanogaster*
Gyrovirus(es)
  associated diseases 1:514
  see also specific viruses

H

HA see Hemagglutinin (HA)
HAART see Highly active antiretroviral therapies (HAART)
Haddock reovirus 1:164t
Haemaphysalis intermedia, Bhanja virus 1:399
Hairpin ribozyme(s) 4:476
  ligation mechanism 4:479
    structural dynamics 4:479
  rate of folding 4:476
  schematic 4:478f
  virus-induced gene silencing 5:379
Haldane–Muller principle, quasispecies 4:361
Halibut reovirus 1:164t
Haloalkaliphilic virus(es) 5:418
  see also specific viruses
Haloarcula hispanica, SH1 virus infection 5:420
Halobacterium salinarum, associated viruses 5:412
Halophile, definition 5:411
Halovirus(es) 5:415
  burst size 5:412
  haloalkaliphilic viruses 5:418
  head-and-tail viruses 5:415
  latent period 5:412
  salterprovirus 5:418
  spherical halovirus SH1 5:420
  see also specific viruses

HAM see Tropical spastic paraparesis/HTLV-1 associated myelopathy (TSP/HAM)
Hammerhead ribozyme(s) 4:476
  classification 4:476
  definition 5:332
  ligation mechanism 4:479
    structural dynamics 4:479
  structure and folding 4:476
    catalytic core 4:476–478
    magnesium ions 4:478
    secondary structure 4:480f
Hand, foot and mouth disease (HFMD)
  coxsackieviruses 1:585
  enterovirus 71 infection 2:134
  neurological complications 2:134
  symptoms 2:134
Hand-to-hand contact, common cold 4:472
Hanseniaspora uvarum, killer phenomenon 5:300
Hantaan virus 1:391t
  genomes 1:394f
Hantavirus 1:391, 1:391t, 2:317–318, 3:480, 4:490–491
Hantavirus(es) 1:399, **2:317–321**
  diagnostics 2:320
  ecology 2:318
    European pathogens 2:318
  epidemiology 2:318
  genome 2:317
  history 2:317
  incidence 2:317
  infections see below
  phylogeny 2:318–319
  replication, host cell entry 1:395
  seroprevalence 2:320
  transmission 1:391, 2:318
  types 2:319t
  vaccines/vaccination 2:320
  virus structure 2:317
  see also specific viruses
Hantavirus infections
  clinical features 2:319
  emerging diseases 2:95
  hemorrhagic fever with renal syndrome 2:317, 5:485
    associated species 2:317, 2:318
    incidence 2:317, 2:318
    pathogenesis 2:320
    risk factors 2:318
  humans 2:317
    respiratory infection 2:555
  pathogenesis 2:319
  prevention 2:320
  vaccines 2:320
  zoonoses 5:486t
    geographical distribution 5:485–486
    hemorrhagic fever 5:485
    rodents 2:318
Hantavirus pulmonary syndrome (HPS) 2:317, 2:555
  pathogenesis 2:320
  zoonoses 5:485
Hare fibroma virus (HFV) 3:225
  characteristics 3:225t
  infection 3:229–230
HaRNAV see *Heterosigma akashiwo* RNA virus (HaRNAV)
HaRNAV virus 4:116
Harrisinia brillians granulovirus (HabrGV) 1:269
Hart Park Group, animal rhabdoviruses 1:115, 1:119t
Hart Park virus 4:184t
HASTY, microRNA pathway in gene silencing 4:141
Hatchery, definition 4:560
HAV see Hepatitis A virus (HAV)
Hazelnut trees(s), diseases 4:206
  causal agents 4:207
    classification 4:207
  geographical distribution 4:206
  symptoms 4:206
  transmission 4:207
  yield losses 4:206
  see also specific viruses
HBeAg see Hepatitis B e antigen (HBeAg)
HBoV see Human bocavirus (HBoV)
HBsAg see Hepatitis B surface antigen (HBsAg)

HBV *see* Hepatitis B virus (HBV)
HCC *see* Hepatocellular carcinoma (HCC)
HCMV *see* Human cytomegalovirus (HCMV)
HcRNAV *see* Heterocapsa circularisquama
 RNA virus (HcRNAV)
HCV *see* Hepatitis C virus (HCV)
HCV Database Project 5:357t
HCV pseudoparticles (HCVpp) 2:372
H DNA 4:588
H (heavy) dsRNAs, Ustilago maydis
 viruses 5:215
HDV *see* Hepatitis delta virus (HDV)
Head, definition 3:30
Head-and-tail virus(es) 5:415
 morphology 5:412
Headful packaging 5:411
Heat inactivation, Border disease virus 1:338
Heat resistance, African horse sickness
 virus 1:38–40
Heat-shock experiments, Sigma virus
 replication 4:579
Heat shock protein 70 homolog (Hsp70h) 3:353
HEF *see* Hemagglutinin-esterase-fusion (HEF)
HeLa cells, JC virus infection 4:280
Helical arrays
 bacteriophage assembly 1:437
 bacteriophages 1:437
 inovirus protein coat assembly 1:438
 myovirus 1:437–438
Helical particles 5:383f, 5:385
Helical symmetry 5:393
Helical virus(es) 1:201
 see also specific viruses
Helicase(s)
 RNA virus evolution 2:180–181
 yellow head virus 5:477–478
 see also DNA helicase
Helicase inhibitors 1:149
Helicoverpa amigera stunt virus (HaSV) 5:28t
 Agrobacterium-mediated transgenesis 5:36
 capsid 5:33
  assembly 5:33
  maturation 5:34–35
  morphology 5:31f
  protein precursors 5:31–32
 discovery 5:28
 genomes 5:29, 5:31
  RNA2 length 5:31
 infection 5:36
  histopathology 5:36
 insect pest control 3:127
 replicases 5:31
Helicoverpa zea virus(es), oryctes rhinoceros virus *vs.*
 3:496
 see also specific viruses
Heliothis zea-1 virus (Hz-1V)
 genome 3:145f
 history 3:144
 morphology 3:145
 replication 3:146
  stages 3:146
  transcription 3:146
Heliothis zea-2 virus (Hz-2V)
 genome 3:145f
 history 3:145f
 infection
  agonadal tissue 3:144–145
  pathology 3:146–147
  symptoms 3:147
 morphology 3:145
  envelope 3:148f
 replication 3:147
Helminthosporium victoriae 145S virus (Hv145SV)
 1:503–504
 classification 1:503–504
 genome 1:505–506
 infection
  disease characteristics 5:169–170, 5:169f
  transformation vector construction 5:169–170
 structure, coat protein 1:509
 transmission 5:169–170

Helminthosporium victoriae 190S virus (Hv190SV)
 2:285, 5:164, 5:165t, 5:166
 capsids 5:166
 genome 5:165t
  open reading frames, overlap 5:166–167, 5:167f
  RdRp 5:167
  structural features 5:167
 Hv-p68 gene/protein 5:171f
 infection
  disease characteristics 5:169–170, 5:169f, 5:171
  symptom severity 5:169–170, 5:169f, 5:170f
  transformation vector construction 5:169–170
 replication 5:168f
  phosphorylation 5:168–169, 5:168f, 5:171
  proteolytic processing 5:168–169, 5:168f, 5:171
  RNA polymerase activity 5:167–168
 taxonomy 5:164, 5:165t
 transmission 5:169–170
 yeast L–A virus *vs.* 5:165
Helper adenovirus(es) *see* Adeno-associated
 virus vectors
Helper component protein (HC-Pro) 2:579
 definition 4:546
 potato virus Y 4:288
 potyviruses *see* Potyvirus(es)
 RNAi suppression 3:152
 virus transmission 5:280f
  watermelon mosaic virus 5:436
  zucchini yellow mosaic virus 5:436
Helper component (HC)-transcomplementation
 definition 5:274
 plant virus transmission 5:281–282
Helper-dependent mini-replicons, human respiratory
 syncytial virus 2:547
Helper T-cells *see* T helper cell(s)
Helper virus(es)
 begomovirus infection 4:166
 see also specific viruses
 β ssDNA satellites 1:314–315
 definition 4:526
 yeast L–A virus 5:465–466
 see also specific viruses
Hemadsorption 1:43
Hemagglutination-inhibition (HAI) test(s)
 cowpox virus infection 1:579
 influenza virus 1:129–131, 2:31
 Japanese encephalitis diagnosis 3:186
 orthobunyavirus reassortment 3:482
 principles 2:31
 togavirus detection 5:107–108
Hemagglutination-neuraminidase protein (HN)
 see Paramyxovirus(es)
Hemagglutinin (HA)
 definition 1:127, 4:497
 HA2
  chain reversal 3:294, 3:295f
  loop region 3:294
  thermostability 3:294–295
 henipavirus 2:322–323
 human parainfluenza viruses 4:47–48
 influenza viruses *see* Influenza virus(es)
 Japanese encephalitis virus 3:183
 mumps virus 3:356
Hemagglutinin-esterase-fusion (HEF) 1:559–560,
 3:492
 definition 3:89
 function 3:485
 infectious salmon anemia virus 3:93
 structure 3:492
Hemagglutinin-neuraminidase (HN), henipavirus
 2:322–323
Hematophagous arthropod vectors
 definition 1:111, 1:170
 orbivirus transmission 3:457
Hematopoietic necrosis herpesvirus 2:208, 2:233
Hematopoietic precursor cells, adult T-cell leukemia
 2:567–568
Hemifusion 3:292
Hemifusion diaphragm 5:311f
Hemiptera 2:37, 4:164
Hemivirus 4:353t

Hemivirus(es) 4:352
 see also specific viruses
Hemocoel 4:250
 definition 3:231
Hemoconcentration 1:328
Hemocyte(s)
 definition 4:250, 4:560
 insect immune responses
  adhesion factors 4:252–253
  differentiation 4:251–252, 4:253f
  proliferation 4:251–252, 4:253f
 types 4:252
 white spot syndrome virus infection *see* White spot
  disease (WSD)
Hemocytic infection virus disease 4:561
Hemolymph 1:301
 definition 3:231
 plant virus transmission 5:278
Hemolysis
 equine infectious anemia virus infectious 2:173
 mumps virus infection 3:356
Hemorrhage
 acute CSFV infection 1:528
 Chikungunya virus infection 5:89
 Marburg virus infection 3:278
Hemorrhagic fever
 African swine fever virus infection 1:44
 arbovirus infection 1:175
 arenavirus infection 5:486
 bioweapons 5:408–409
 see also specific hemorrhagic fevers
Hemorrhagic fever with renal syndrome (HFRS)
 see Hantavirus infections
Hemorrhagic kidney syndrome (HKS) 3:94
Hemorrhagic pock(s), cowpox 1:577, 1:579
Hemorrhagic smallpox 4:640
Hemorrhagic syndrome
 bovine viral diarrhea viruses *see* Bovine viral
  diarrhea virus infection
 definition 1:374
HEN1, plant gene silencing 4:146
Henderson–Patterson bodies *see* Molluscum lesion
Hendra virus (HeV) 4:40, 4:52
 Australia 2:324
 fusion proteins, proteolytic processing 4:54
 genetic stability 2:179
 genome 3:327f
 global distribution 4:45
 historical aspects 2:94, 2:321
 infection 2:554
  clinical features 2:325
  direct transmission 2:324
  G protein 2:323
  host cell attachment 4:44, 4:53–54
  host range 2:554
  pathology 2:325
  zoonoses 2:95
Henipavirus 2:321, 4:41t
Henipavirus(es) 2:321–327, 2:322f
 cell attachment
  attachment glycoproteins 2:322–323
  ephrin B2 2:322–323
  glycoprotein attachment proteins 2:322–323
  hemagglutinin 2:322–323
  hemagglutinin-neuraminidase 2:322–323
  sialic acid-independent mechanisms 2:322–323
 classification 4:41t
 epidemiology 2:324
 evolution 2:323
 future work 2:326
 genetics 2:323
 genome 3:327f
 geographical distribution 2:323
 history 2:321
 infections *see below*
 molecular biology 2:322
  F protein 2:322–323, 2:326
  furin 2:322–323
  G protein 2:326
  interferon inhibition 2:323, 2:326
  P gene 2:323, 2:326

Subject Index

Henipavirus(es) (continued)
    V protein 2:323, 2:326
    W protein 2:323
  molecular features 2:321–322
  propagation 2:323
  transmission 2:324
    direct vs. indirect transmission 2:324
  see also specific viruses
Henipavirus infections
  clinical features 2:325
    neurological 2:325
    prolonged infection 2:325
    respiratory problems 2:325
  control 2:326
    ant-G protein human monoclonal antibodies 2:326
    ribavirin 2:326
  ELISA 2:326
  histopathology 2:325
  host range 2:323
  immune responses 2:326
    adaptive 2:326
    antibodies to 2:326
    innate 2:326
  morbidity 2:324–325
  mortality 2:324–325
  pathogenicity 2:324
  pathology 2:325
    natural hosts 2:326
  prevention 2:326
  serology 2:323
  tissue tropism 2:324
    ephrin B2 expression 2:324
  vaccines/vaccination 2:326
  see also specific viruses
Henivirus 4:352
Henle-Koch postulates, epidemiology 2:143, 2:144t
Hepacivirus 2:242, 2:249, 2:367, 5:440–441
Hepacivirus(es) 2:249
  history 2:249
  infection
    clinical disease 2:249, 2:251
    diagnosis 2:252
    epidemiology 2:249
    pathogenesis 2:251
    treatment 2:252
  replication 2:249
  structure 2:249
  see also specific viruses
Hepadnaviridae 2:327, 2:351–352, 2:360, 4:460, 5:11t
Hepadnavirus(es) 2:335–343
  avian see Avian hepadnavirus(es)
  epidemiology 2:337
  evolution 3:477
  genetic variation 2:337, 2:339f, 2:340f
    common determinants 2:337–338
    open reading frames 2:338
    subtypes 2:337–338
  genome 2:336, 2:338f
    open reading frames 2:336
  geographical distribution 2:338
  infection
    apoptosis inhibition 1:158t
    avian see Avian hepadnavirus(es)
    clinical features 2:340
    immunomodulation 4:114t
    pathology 2:340
    persistent 4:109t
    prevention 2:341
    rodents 2:341
    treatment 1:143t, 2:341
  LTR retrotransposons 4:438
  phylogeny 2:337f
  structure 2:336
    surface proteins 2:336, 2:338f
  transmission 2:335
    parenteral 2:335
    post-transfusion 2:336
  see also specific viruses
Heparan sulfate (HS), receptor role 5:320
  alphavirus 5:118
  foot-and-mouth disease virus 5:324

  porcine respiratory and reproductive syndrome virus 1:181
  pseudorabies virus 4:346
  tick-borne encephalitis viruses 5:46
Hepatic necrosis, equine infectious anemia virus infection 2:172–173
Hepatitis 1:104
Hepatitis A
  asialoglycoprotein receptor 2:12
  clinical features 2:347
    biochemical parameters 2:347
    children 2:347
    extrahepatic manifestations 2:348
  control 2:350
    disease surveillance 2:50
  diagnosis 2:349
    immunoglobulin G 2:12
  epidemics 2:343
  epidemiology 2:349
    at risk individuals 2:350
    seasonal distribution 2:350
    seroprevalence 2:349–350
  fulminant 2:348
    fatality rate 2:348
  host range 2:346
  immune response 2:349
    adaptive 2:349
    innate 2:349
  infection course 2:348f
    incubation period 2:347
    reconvalescence period 2:347
  pandemics 2:343
  pathology 2:347
  prevention 2:350
  prolonged 2:348
  relapsing 2:348
    frequency 2:348
    pathogenesis hypotheses 2:348
  tissue tropism 2:346
Hepatitis A virus (HAV) 2:343–350
  cell culture 2:347
    cytopathic variants 2:347
  defective interfering RNAs 2:4
  gene expression 2:344
    proteolytic processing 2:344, 2:346f
    RNA transcription 2:344
    translation 2:344
  genome 2:344, 2:345t
    internal ribosome entry site 4:130, 4:134–135
  growth characteristics 2:347
  history 2:343
  infection see Hepatitis A
  morphogenesis 2:344, 2:346f
  morphology 2:344
  physicochemical properties 2:344
  replication cycle 2:347
  stability 2:344, 2:350
  taxonomy 2:343
    genotypes 2:343
  transmission 2:343, 2:346
  vaccines/vaccination 2:350
    development 5:236–237
Hepatitis B 2:333–334
  acute infection 2:357
  antiviral therapy 1:143t, 1:152
    clevudine 1:152–153
    drug resistance 1:151, 2:359, 2:365
    genotype effects 2:355–357
    lamivudine 1:152, 2:359, 2:365
    nucleoside analogs 2:359
    nucleotide analogs 2:359
    telbivudine 1:152
  children 2:340
  chronic infection 2:357
    course of 2:357, 2:358f
    HBeAg negative group 2:358
    immune clearance phase 2:357–358
    mutant viruses 2:358
    nonreplicative phase 2:357–358
    reactivation phase 2:357–358
    treatment 2:364

    drug resistance 2:342
    epidemiology 2:352
      prevalence 2:352
  hepatitis delta virus co-infection 2:342, 2:375, 2:376
  hepatocarcinogenesis 5:197
  hepatocellular carcinoma see Hepatocellular carcinoma (HCC)
  infection markers 2:357
    anti-HBsAg antibodies 2:357
  natural history 2:357
    fulminant hepatitis 2:357
  organ transplantation 3:469
  persistent 4:109t
  treatment 2:359
    antiviral therapy see above
    HBeAg-negative patients 2:359
    HBeAg-positive patients 2:359
    immunomodulators 2:359
    protracted 2:359
Hepatitis B core antigen (HBcAg) 2:360
Hepatitis B e antigen (HBeAg) 2:352–353, 2:364
  mutation 2:364
  negative patients, treatment 2:359
  positive patients, treatment 2:359
  seroconversion 2:364
  tolerogenic effects 2:352–353
  transmission, placental 2:474
Hepatitis B immune globulin (HBIG) 2:364
Hepatitis B surface antigen (HBsAg) 2:360, 2:362
  antibody response 2:362
    infection marker 2:357
  discovery 2:350–351
  hepatitis delta virus co-infection 2:342
  phylogenetic tree 2:351f
  research history 2:335–336
  structure 2:362, 2:363f
  transgenic plant vaccine production 5:223
    limitations 5:223–224
    virus yield 5:223
  vaccination programs 2:363, 5:197
  virion structure 2:352
Hepatitis B virus (HBV) 2:350–360, 2:360–367
  antiviral therapy 1:152
  chromosomal integration 2:366
  defective interfering RNAs 2:4
  distribution 2:352
  drug resistance 2:342, 2:359
    molecular biology 2:359
  duck hepatitis B virus vs. 2:334
  epidemiology 2:352
    prevalence 2:352
  genome 2:329, 2:336, 2:338f, 2:352, 2:353f, 2:360–361, 2:361f
    cccDNA 2:360–361
    nucleocapsid protein 2:352–353
    open reading frames 2:336–337, 2:361
    pre-core/core open reading frames 2:352–353
    pre-core mRNA 2:352–353
    pregenomic RNA 2:352–353, 2:353f
  genotypes 2:355, 2:356t
    antiviral response effects 2:355–357
    geographical distribution 2:355
    pathogenesis effects 2:355–357
  geographical distribution 2:340, 2:341f, 2:355
  hepatitis B e antigen see Hepatitis B e antigen (HBeAg)
  hepatitis B surface antigen see Hepatitis B surface antigen (HBsAg)
  hepatitis delta virus association 4:532–533
  historical aspects 2:350
    discovery 2:350–351
    electron microscopy 2:351
    hepatocellular carcinoma connection 2:351
    PCR 2:351
    plasma-derived vaccine development 2:351
    recombinant vaccine development 2:351
  history 2:336
  infection see Hepatitis B
  phylogenetic trees 2:351f

proteins
    L (large) protein 2:352
    M (middle) protein 2:352
    S (small) protein 2:352
replication 2:337, 2:353, 2:354f, 2:360, 2:362f
    direct repeats 2:361
    hepatocyte receptors 2:353
    host cell entry 2:360
    polymerase 2:364–365, 2:365f
    pregenomic RNA 2:353–354, 2:355f, 2:361, 2:362f, 2:365f
    RNase H 2:353
    templates 2:353
    transcription 2:361
subtypes 2:354, 2:356t
    S protein sequence 2:354–355
taxonomy/classification 2:351
    phylogenetic tree 2:351f
transmission 2:334, 2:335–336
vaccines/vaccination 2:51–52, 2:341–342, 2:358, 2:363, 4:108–109
    beneficial aspects 2:358–359
    composition 5:237
    efficacy 2:358–359
    'plasma-derived' vaccine 2:363–364
    reactions to 5:230t, 5:231t
    recombinant vaccines 2:358
    safety 5:233
    vaccine escape mutants 2:359
variants 2:357
    evolution 2:355–357
virion structure 2:352, 2:352f, 2:360, 2:362f, 2:375
    Dane particles 2:352
    lipid bilayer 2:352
virus entry 5:315
Hepatitis C
    acute 2:251
    chronic 2:251
    clinical manifestation 2:251, 2:368, 4:115
        disease progression 2:368, 2:368f
        extrahepatic symptoms 2:368
        incubation period 2:368
        viremia 2:368
    diagnosis 2:93, 2:251, 2:374
    eye 2:491–492
    histopathology 2:368
    host range 2:370
    immune response
        adaptive 2:373
        innate 2:373
        T-cell responses 3:76
    immune system evasion 3:108
    liver 2:341
        biopsy 2:252
        cirrhosis 2:249
        disease mechanism 2:251–252
        hepatocarcinogenesis 5:197
    organ transplantation 3:469
    pathogenesis 2:251
    pathology 2:368
    persistent 2:251
    risk factors 2:368
    screening 2:368, 2:374
    transmission 2:368
    treatment 2:252, 2:374
        antiviral therapy 1:153
Hepatitis C virus (HCV) 2:367–375
    evolution 2:177f
    experimental systems 2:370
        animal models 2:370
        cell culture 2:372, 2:372f
    genome 2:249–250, 2:250f, 2:369, 2:369f
        core+1 proteins 2:370
        IRES structure 2:249–250
        nonstructural proteins 2:369–370
    genotypes 2:249, 2:367–368
    geographical distribution 2:367
    history 2:249, 2:367
    host cell entry 2:251
    infection see Hepatitis C
    prevalence 2:249, 5:198

protein functions 5:198
    cellular proteins 2:250–251
    E1 2:250
    E2 2:250
    NS5A 2:251
    p7 2:250
    proteases 2:250–251
    RNA-binding core protein 2:250
replication cycle 2:370, 2:371f
    host cell binding 2:370
    rate 1:150
    replicon system 1:153
    virus release 2:370
taxonomy 2:367
tissue tropism 2:370
virion properties 2:249–250, 2:370
    polyprotein processing 2:250
virion structure 2:371f
    pseudoparticles 2:372
virus-host interactions 2:373
    replication factors 2:373
Hepatitis delta virus (HDV) 2:342, 2:375–377, 4:532
    classification 2:375
    epidemiology 2:376
    genome 2:342, 2:375–376
    infection
        clinical features 2:377
        control 2:377
        hepatitis B co-infection 2:342, 2:375, 2:376
    properties 2:375
    replication 2:375, 2:376f, 4:530
        antigenome 2:376f
        mechanism 2:376
        site 2:376
    ribozyme 4:480
        replication 4:480
        structure 4:481f
    satellite viruses 2:375
    structure 2:375
    transmission 2:336, 2:376
Hepatitis E virus (HEV) 2:377–383
    avian see Avian hepatitis E virus
    classification 2:378
        phylogeny applications 4:128
    gene expression 2:378
    genome 2:378, 2:379f
        ORF3 2:380
        transcription 2:380
        translation 2:380
    history 2:377–378
    infection see Hepatitis E virus infection
    natural history 2:382
    taxonomy 2:378, 2:379f
    transmission 2:377–378, 2:380, 2:382
        intravenous infection 2:382
    vaccines/vaccination 2:382–383
    virion properties 2:378
        sedimentation coefficient 2:378
        stability 2:378
    virion structure 2:378
        capsid 2:378–380
Hepatitis E virus infection
    cross-species 2:381
    diagnosis, immunoglobulin G 2:382
    epidemiology 2:380
        mortality rate 2:380, 2:381
        seroprevalence 2:380
    host range 2:381
    immunity 2:382
    pathogenesis 2:381
    prevention and control 2:382
    rodents 2:381
    tissue tropism 2:381
    zoonoses 2:381
    swine 2:380
Hepatitis–splenomegaly (HS) syndrome, avian hepatitis E virus 2:380
Hepatitis Virus Database 5:357t
Hepatocellular carcinoma (HCC) 2:341, 5:197
    hepatitis B 2:351, 2:366, 5:197
        immunization 2:341–342

hepatitis C 5:197
    transcription transactivation 5:197
Hepatocellular necrosis, Marburg virus infection 3:278
Hepatocyte(s)
    avian hepadnavirus infection see Avian hepadnavirus(es)
    degeneration, yellow fever 5:475
    hepatitis B 2:341, 2:353
    hepatitis C 2:368
Hepatopancreas 4:567, 5:476
Hepatopancreatic parvovirus (HPV)
    genome 4:572, 4:572f
    geographical distribution 4:572
    host range 4:572
    infection 4:571
    taxonomy 4:572
Hepatovirus 2:343
Herd immunity 2:141, 2:146–147
L'He´ritie´r, P, Sigma virus discovery 4:576
Heron hepatitis B virus 2:327
Herpangina, coxsackieviruses 1:585, 1:585t
Herpes folliculitis 2:388–390
Herpes simplex virus(es) (HSV) 2:383–397, 2:397–405, 4:582, 4:582f
    associated proteins 2:399
        cytopathological involvement 2:403
        DNA replication 2:401
        nomenclature 2:399, 2:402t
        pathogenesis involvement 2:403
        regulatory 2:399
        structural 2:401
        see also specific proteins
    binding, glycoproteins 5:321
    classification 2:383
    DNA recombination mechanisms 4:375
    envelope glycoproteins
        function 2:397, 2:403
        structure 2:397
    epidemiology 2:387
    gene expression 2:399
        enhancers 2:399
        phases 2:398
        promoters 2:399, 2:401f
        transcript properties 2:399
    gene therapy vectors 2:396
        brain tumors 2:304
        clinical trials 2:304
        genome 2:304
        infected cell types 2:304
        therapeutic genes 2:396
    genome 2:383, 2:398, 2:400f
        inversions 2:398
        map 2:400f
        open reading frames 2:399
        regions 2:398
    history 2:383
    host cell entry 2:398
    molecular epidemiological studies 2:143
    oncolytic viruses 2:396
    replication 2:384, 2:385f, 2:398
        assembly 2:386, 2:399
        DNA 2:385–386, 2:398
        egress 2:386, 2:399
        glycoproteins 2:398
        host cell effects 2:386
        host cell entry 2:384
        post-translational processing 2:398
        release 2:399
        RNA export 2:398
        transcription 2:385, 2:398
        translation 2:398
    research 2:404
    transmission 2:387
    vaccines/vaccination 2:395–396
        clinical trials 2:396
    virion structure 2:383, 2:384f, 2:397
        capsid 2:207f, 2:383–384, 2:397, 2:399
        envelope 2:397
        tegument 2:397
    see also specific viruses

Herpes simplex virus 1 (HSV-1)
  apoptosis activation 2:388
  evolution 2:181
  genome, varicella-zoster virus vs. 5:252
  host cell entry 2:388
  infection
    children 2:388–390
    encephalitis 2:387
    manifestations 2:390f
    pathogenicity 2:389f
    persistence 4:109t
    rhinitis 2:388–390
    seroprevalence 2:387
  reactivation 2:388–390, 2:405f
Herpes simplex virus 2 (HSV-2)
  acquisition, influencing factors 2:387
  animal studies 2:388
  genome, varicella-zoster virus vs. 5:252
  HIV infection association 1:66
    clinical trials 1:66
    pre-exposure prophylaxis 1:66
  infection 2:390–391, 2:390f
    clinical features 2:390–391
    females 2:390–391
    fetal risks 2:387
    males 2:390–391
    pathogenicity 2:389f
    persistent 4:109t
    pregnancy 2:387
    proliferative lesions 2:395f
    seroprevalence 2:387
    squamous cervical cancer 2:394
    treatment 2:395t
  replication 2:386f
  transmission 2:388
    children 2:388
    neonatal 2:388
Herpes simplex virus-associated erythema multiforme (hem) 2:391
Herpes simplex virus infections
  antibodies 2:393
  apoptosis inhibition 1:157–160
  associated Alzheimer's disease 2:391
  associated risk 2:387
  clinical features 2:388, 4:115
    associated disease 2:388
    children 2:388–390, 2:555
    immunocompromised patients 2:391
    incubation period 2:388
  diagnosis 2:393
  encephalitis 1:474, 2:387
    neonatal mortality 2:388–390
    transmission 2:388
  eye 2:384, 2:388–390, 2:491–492, 2:495–496
    neonates 2:495–496
  host range 2:384
  immune evasion 1:125, 2:144–145, 2:393–394, 2:394f, 3:108–109
  immune response 2:393
    children 2:393
    naturally acquired immunity 2:393–394
  latency 2:391, 2:392f, 2:404, 2:436–442, 2:439f
    animal models 2:437
    establishment of 2:437
    gene expression 2:438
    immunological studies 2:438
    latent phase transcripts 2:404
    maintenance of 2:438
    mechanism 2:391–392
  neonatal/fetal disease 2:393
    mortality rates 2:393
  neoplasia 2:394
    Ras/MEK/ERK pathway 2:394
  nervous system infection 1:469
  organ transplantation 3:469
  pathogenicity 2:388, 2:389f
  pathology 2:393
  pharyngitis 2:555
  pregnancy 2:393
  prevention 2:394

reactivation 2:386, 2:405f, 2:438, 2:439f
  associated stimuli 2:391
  transplant-associated 3:467
serology 2:387
therapy 2:394
  acyclovir see Acyclovir
  foscarnet 2:395
  helicase inhibitors 1:149
  long-term daily suppressive therapy 2:394–395
tissue tropism 2:387
Herpestidae, definition 4:367
Herpesviridae 1:363, 1:635–636, 2:148–149, 2:206, 2:233, 2:405–406, 2:412, 2:423, 2:424t, 2:431, 2:432t, 2:474, 2:498, 4:341, 5:11t, 5:251, 5:347
Herpesvirus(es) 2:383, 2:430–436, 2:474
  amphibian see Amphibian herpesvirus(es)
  avian see Avian herpesvirus(es)
  bovine see Bovine herpesvirus(es)
  caudoviruses vs. 1:610–611
  classification 2:423, 2:431, 2:432t
    biological 2:432
    criteria 2:423
    genera 2:233, 5:251
    genomic 2:432
    host specificity 2:423–424
    morphological 2:431
    phylogenetic trees 2:425f, 2:426–427
    sequence 2:432, 2:433f
    serological 2:432
    subfamilies 1:363, 1:635–636, 2:423, 2:424t, 2:431, 4:581–582
  definition 2:431
  discovery 2:420–430
    degenerate primer PCR 2:420
    detection methods 2:420
    history 2:420
    numbers discovered 2:421, 2:424t
    pathogenic impact 2:427
  DNA polymerase gene 2:423
  equine see Equine herpesvirus(es)
  evolution 2:435
    co-adaptation 2:435
    timescale 2:435
  fish see Fish herpesvirus(es)
  gB structure 3:300
  gene content 2:434
    arrangement 2:434–435
    complements 2:434
    core genes 2:435
    sequence comparisons 2:435
    transcript functions 2:434–435
  gene therapy vectors 2:304, 2:396
    structure 2:304
  genome 2:433, 2:434f
    5′ CG deficiency 2:433
    comparisons 1:636
    structure types 2:433–434
  genotypes 2:378
  percentage identity 5:347
  replication 2:430, 4:409–410
    DNA replication 2:431
    expression 2:431
    host cell attachment 3:310
    host cell entry 2:430
    maturation 2:431
    release 2:431
  virion structure 2:430, 2:431f
    capsid 2:430
    core 2:430
    envelope 2:430
    tegument 2:430
  see also specific viruses
Herpesvirus 8 (HHV-8) see Kaposi's sarcoma-associated herpesvirus (KSHV)
Herpesvirus Ateles (HVA) 4:586, 4:587t
  classification 4:585–586
  gene content 4:588
    conserved genes 4:588
  genome 4:588
    H DNA 4:588
    L DNA 4:588

historical aspects 4:586
infection
  acute-T-cell lymphomas 4:586–587
  affected species 4:586
  host range 4:586
  pathology 4:586
  transfected T-cell lines 4:587–588
isolation 4:586
replication 4:588
strain classification 4:588
transmission 4:586
Herpes virus entry mediator (HVEM), viral replication 2:398
Herpesvirus infections
  antiviral therapy 1:143t, 1:146, 1:152
    acyclovir see Acyclovir
    drug resistance 1:151
    HSV see Herpes simplex virus(es) (HSV)
  apoptosis inhibition 1:158t
  congenital disease 2:146
  HCMV coinfection 2:485–486
  hosts 2:421–422, 2:424t
  human respiratory infection 2:555
  immune evasion 1:125
  immunomodulation 4:114t
  latency 2:436–442
  organ transplantation 3:467, 3:469
  pathogenic impact, species barriers 2:428
  persistence 4:109t
  RNAi therapeutics 5:327
  shellfish see Shellfish virus(es)
Herpesvirus papio 2 (HVP-2), detection 4:582–583
Herpesvirus Saimiri (HVS) 2:428, 4:586, 4:587t
  affected species 4:586
  cell culture 4:586
  classification 4:585–586
  gene content 4:588
    conserved genes 4:588
    early infection genes 4:588–589
    host nucleus localization 4:589
  genome 4:588
    H DNA 4:588
    L DNA 4:588
  historical aspects 4:586
    discovery 4:586
  host range 4:586
  human T-cell transformation 4:593
    gene transfer 4:591
  oncogenic signaling 4:589
  pathology 4:586
  replication 4:588
  sequestered cellular genes 4:589
  Stp oncoproteins 4:589
    human T-cell transformation 4:591
    sequence homology 4:589
    signaling 4:589
    T-cell leukemia 4:589
    transcription 4:589
  strain classification 4:586, 4:588, 4:588f
  Tio oncoprotein 4:590
    Lck phosphorylation 4:590
    Src phosphorylation 4:590
  Tip oncoproteins 4:589
    human T-cell transformation 4:591
    Lck binding 4:590
    signaling 4:590
    Src kinase homology 4:589–590
    T-cell proliferation 4:589–590
  transformation 4:589
  transmission 4:586
Herpes zoster ophthalmicus, HIV infection 2:497
Herpetic whitlow 2:388–390
Herring aquareovirus (HRV) 1:164t
HERVd (Human Endogenous Retrovirus Database) 5:357t
Heterobasidion annosum, partitivirus transmission 4:67
Heterocapsa circularisquama, viral infection of 1:87

Subject Index

Heterocapsa circularisquama RNA virus (HcRNAV) 1:88t, 1:89
  ecotypes 1:89, 1:90
  genome 1:89, 1:91f
    open reading frames 1:89, 1:89f
  host range 1:89–90
  host-virus dynamics 1:95
  phylogeny 1:89, 1:93f
  virion structure 1:89, 1:91f
Heterochromatin protein 1 (HP1) 2:439
Heterochromatin siRNA pathway 4:142
Hetero-encapsidation 4:97
  zucchini yellow mosaic virus transmission 5:436
Heterogeneous nuclear ribonucleoprotein (hnRNP), nidovirus replication 3:427
Heterokaryon 5:299
Heterooligomeric protein expression, transgenic plants 4:233
Heterosigma akashiwo 3:280
  infection
    cytopathology 3:280
    susceptibility 3:280
Heterosigma akashiwo RNA virus (HaRNAV) 3:280, 3:281
  genome 3:280–281, 3:281f
  phylogenetic analyses
    polymorphisms 3:283, 3:284t
    RdRp domains 3:281, 3:283
  virion structure 3:280, 3:280f
  proteins associated 3:281, 3:281f
Heterotelomeric parvovirus(es), replication 4:90–91
  see also specific viruses
Heterotypic immune response, coxsackieviruses 1:583–584
Het-S prion 4:337, 4:340
HeV *see* Hendra virus (HeV)
HEV-A *see* Human enterovirus(es) (HEV)
HEV-B *see* Human enterovirus B (HEV-B)
HEV replicon system 2:380
Hexamer units, cowpea chlorotic mottle virus structure 5:385
Hexons
  adenovirus antigens 1:5
  HSV capsid structure 2:397
HF1 virus 5:413t, 5:415
  genome 5:417, 5:417f
  host range 5:415
HF2 virus 5:413t, 5:415
  genome 5:417, 5:417f
  structure 5:417f
HFMD *see* Hand, foot and mouth disease (HFMD)
H glycoprotein, measles virus 3:287–288
HGS004 2:507t
Hh-1 virus 5:413t
Hh-3 virus 5:413t
HHV-4 *see* Epstein–Barr virus (EBV)
HHV-6 *see* Human herpesvirus-6 (HHV-6)
HHV-7 *see* Human herpesvirus-7 (HHV-7)
HHV-8 *see* Kaposi's sarcoma-associated herpesvirus (KSHV)
Hibiscus chlorotic ringspot virus (HCRSV) 1:453
Hibiscus-infecting tobamovirus(es) 5:69
  see also specific viruses
Hibiscus latent ringspot virus (HLRSV) 3:406t
High-affinity laminin receptor (HALR), Sindbis virus 5:118
Highlands J virus 5:81
  characteristics 5:109t
  evolutionary relationships 1:174
  geographical distribution 5:102
  infection
    encephalitis 5:81
    epidemiology 5:81–82
    host range 5:102–103
  transmission, *Culiseta melanura* 5:103–104
Highly active antiretroviral therapies (HAART) 1:66–68, 1:353, 2:506
  associated renal disease 1:56–57
  drug resistance 2:516, 2:518
  HCMV infection 2:484
  HIV infection *see* HIV infection

JCV reactivation 4:265–266
KSHV infection 3:192, 3:194
Highly exposed seronegative individual(s) (ES)
  see HIV infection
Highly polymorphic region, infectious salmon anemia virus 3:93
High pathogenicity avian influenza (HPAI) virus(es) 3:101
  HA cleavage sites 3:101–102
  see also specific viruses
High per-base mutation rate, RNA virus 4:363
High Plains virus (HPV) 1:477
  infection 1:477
High-pressure freezing, cryo-electron microscopy 2:84
High-throughput screening 5:375
Hind III enzyme, cowpox virus genome digestion 1:574, 1:575
Hippocampus sclerosis, Borna disease virus associated 1:342–344, 1:343f
Hirame rhabdovirus (HIRRV) 2:221–222, 2:228, 4:184t
Hirst, George, influenza virus 2:456
His1 virus 5:413t, 5:418
  cell exit 5:420
  DNA polymerase 5:420
  genome 5:420, 5:421f
  structure 5:418, 5:420f
His2 virus 5:413t, 5:418
  cell exit 5:420
  DNA polymerase 5:420
  genome 5:420, 5:421f
    open reading frames 5:418–420
  structure 5:418–420, 5:420f
Histiocytes, Yaba monkey tumor virus infection 5:464
Histo-blood groups 1:410, 3:444
Histo-group antigen(s) (HBGAs), receptor role 5:321
Histone(s)
  acetylation 1:11
  EBV latency, role 2:442
Histopathology
  Helicoverpa armigera stunt virus infection 5:36
  henipavirus infections 2:325
  Providence virus 5:36
  rabbitpox 3:347
  tetravirus infections 5:36
'Hit and run' transformation, adenovirus malignant transformation 1:16
HIV 2:517–525
  antibody variation 1:132
  antigenic variation 1:128, 1:132
    circulating recombinant forms 1:59, 1:132–133
    envelope glycoprotein *see below*
    genome co-packaging 1:133
    global variation 1:132
    intra-patient variation 1:132
    population genetics 1:133
    replicase 1:133, 1:133f
    unique recombinant forms 1:59
  biocrime 5:407
  classification 1:132–133, 2:526
    serogroups 1:59, 2:526
    subtypes 1:59, 1:60f, 2:524, 2:526
  cross-species transmission 2:525–526, 2:529, 2:536–537, 4:604–606
    ecological change 2:532–533
    evidence 4:612
    geographic location 2:529
    HIV-1 2:529
    HIV-2 2:531
    method 2:532
    origins *see below*
    pathogenic potential 2:533
    risks 2:532
    timeline 2:532
  defective interfering viruses 2:4
  envelope glycoproteins 1:132, 1:134
    antibody recognition 1:136
    binding 5:322
    CD4 binding site 1:136
    glycosylation 1:136
    structure 1:134
    see also specific types

env gene/protein
  CD4 T-cell response 1:71
  consensus sequence 1:73–74
  immunogenicity 1:75
  infection pathogenesis 2:538f, 2:540
  membrane fusion 3:296–297
  neutralizing antibodies 1:72–73
  retrovirus organization 2:519–520
  sequence evolution 1:73
escape mutants 2:541
etiology 2:143–144
evolution 2:175, 2:176f, 2:527
  HIV-1/SIV$_{cpz}$ 2:528, 2:530f
  HIV-2/SIV$_{gor}$ 2:529, 2:530f
  HIV-2/SIV$_{smm}$ 2:527
  ZR59 2:532
expression regulation 3:304–306
gag gene/protein
  expression 1:73–74
  retrovirus organization 2:519–520
  vaccine developments 1:73–74
  genome 2:518, 2:519f, 2:526, 2:528f
gp41
  antigenic variation 1:136
  host cell entry 2:506
gp120 2:519–520, 2:538, 5:310, 5:323
  antigenic variation 1:134, 1:135–136
  definition 1:127
  glycosylation 1:136
  host cell attachment/entry 2:506, 2:538, 4:408, 5:323
  hypervariable loops 1:135–136
gp160
  neotopes 1:137–138
  neutralizing antibodies 1:137–138
historical aspects 2:517–518, 2:525, 2:534, 4:460
  discovery 4:625
host cell attachment, receptors 4:408
infection *see* HIV infection
life cycle 2:505–506
neutralizing antibodies 1:72–73, 1:137–138
nomenclature 4:594–595
origin 1:59, 2:525–534, 4:595f, 4:611
  bushmeat 2:532
  chimpanzees 2:528, 2:530f
  evolution *see above*
  gorillas 2:533
  polio vaccine, human-induced theory 2:530f
  subspecies identification 2:528, 2:529
  vertical transmission 2:528–529
regulatory proteins 2:521
  Nef 2:523
  Rev 2:522
  Tat 2:522
  Vif 2:523, 4:598–599
  Vpr 2:522
  Vpu 2:523
replication 2:518, 2:520f
  activation 2:539–540
  APOBEC3G 2:521, 3:109, 4:598–599
  cellular factors, role 2:521
  control 2:535
  co-receptors 1:134–135, 2:518–519
  early 2:537
  gp41 2:519–520
  host cell entry 1:134–135, 2:506, 2:519
  primary receptor 2:518–519
  rate 1:150
  transcription 2:519
  translation 2:519–520
  viral release 2:520–521
  virus entry 5:315
  virus shedding 5:316–317
Rev protein 2:522
  binding activity 2:522
  splicing 2:522
sensitivity, *TRIM5* gene/protein 2:470
taxonomy 2:526
transmission 1:59, 1:60, 1:60t, 1:73, 2:94, 4:601, 4:602
  cross-species *see above*
  determinants 2:537

HIV (continued)
    influencing factors 2:538
    risk factors 1:60
  vaccines/vaccination see HIV vaccines/vaccination
  see also AIDS
HIV-1
  cell entry
    CCR5 co-receptor 2:468
    chemokine ligands as inhibitors 2:468–469
  origin 2:528, 2:530f
    subspecies identification 2:528, 2:529
    vertical transmission 2:528–529
  vaccines/vaccination
    AIDSVAX trials 1:70
    ELDKWAS peptide 1:141
  virus protein U 2:523
    definition 2:467
    functions 2:523
    lentiviruses 2:473
    structure 2:523
  ZR59 2:532
HIV-1, Human Protein Interaction Database 5:357t
HIV-2
  associated serogroups 2:533
  cross-species transmission 2:531, 2:531f
  evolution 2:527
  geographical distribution 1:59
  origin 2:531, 2:531f
    geographic location 2:529
    timeline 2:532
  prevalence 1:59, 2:524
HIV Drug Resistance Database 5:357t
Hivid® 2:510
HIV infection 1:51–58, 3:195
  acute renal failure 1:55
    causes 1:55–56
    incidence 1:55–56
  antiretroviral therapy 1:146, 1:153
    fusion inhibitors 1:148
    see also below highly-active antiretroviral therapy
  associated infections 2:94
    HPV 4:14
    human respiratory infections 2:556
    Japanese encephalitis concurrent infection 3:184
    JCV reactivation 4:262, 4:265–266
    Kaposi's sarcoma see Kaposi's sarcoma (KS)
    KSHV see Kaposi's sarcoma-associated herpesvirus (KSHV)
    measles 3:289
  associated malignancies 1:52
  cases 1:51–52
  CCR5 receptor 2:468
    Δ32 genetic mutation 2:506, 2:537, 5:318–319
    HHV co-infection 2:503
    host-cell interactions 2:506, 2:518–519, 5:318–319, 5:323
    pathogenesis 2:535–536
  CD4+ T cells 1:52, 2:534–535
    attachment 1:136, 2:518–519, 5:322
    bystander activation 2:540–541
    cell count 1:52
    co-receptors 2:535–536
    depletion 2:535, 2:535f
    downregulation 1:72, 2:524
    gp120 conformation 2:540, 5:323
    MALT-associated 2:537t, 2:539–540
    proliferation 2:535
  CD8+ (cytotoxic) T-cells 1:71
    acute-to-chronic phase transition 2:539f, 2:540
    bystander activation 2:540–541
    highly exposed seronegative individuals 1:70–71
    vaccination strategies 1:69–70
  dementia 1:57, 1:474
    disease progression 1:57
    manifestations 1:57
  disease surveillance
    data collection sources 2:45, 2:46
    data interpretation 2:49
    data protection acts 2:46
    death certification 2:46
    risk factors 2:46
    screening systems 2:47
  distal symmetric sensory polyneuropathy 1:58
    disease characterization 1:58
    opportunistic infection treatment effects 1:58
    prevalence rate 1:58
  eye 2:144, 2:146
    herpes zoster ophthalmicus 2:497
    retinopathy 2:146
  gp160, neutralizing antibodies 1:137–138
  Guillain–Barré syndrome 1:474–475
  highly active antiretroviral therapies
    associated renal disease 1:56, 1:57
    immune reconstitution inflammatory syndrome
        see Immune reconstitution inflammatory syndrome (IRIS)
    mortality rate 1:51–52
    neurological manifestations 1:57
  highly exposed seronegative individuals 1:70–71
    CTL responses 1:70–71
    seroconversion 1:71
  history 1:59, 2:505
  HLA effects
    progression indicator 1:70–71, 2:538
    vaccine development 1:74–75
  host cell interactions 4:598, 5:321
  immune complex disease 1:56
    categories 1:56
    clinical presentation 1:56
  immune evasion 1:72t, 1:74t
    Ab-mediated neutralization 2:540, 3:418, 3:418f
    Ab-mediated neutralization mutants 3:418, 3:418f
    APOBEC3G 3:109
    virus infectivity factor 3:109
  immune response
    Ab-dependent disease enhancement 3:66
    CD4+ T cells see above
    CD8+ (cytotoxic) T-cells see above
    immune evasion see above
    monoclonal antibodies 3:64–65
    neutralizing antibodies see below
    T-cell responses 3:76
  intravenous drug abuse 1:60, 1:64
    treatment 1:66
  kidney disease 1:54
    acute renal failure 1:55
    chronic renal failure 1:56
    mortality rates 1:54–55
    nephropathy see below
    thrombotic microangiopathy 1:56
  lipodystrophy 1:53
    fat accumulation sites 1:53, 1:54f, 1:55f
    management 1:53–54
    thymidine nucleoside analog treatment 1:53
  metabolic syndrome 1:53, 1:53t
    cardiovascular morbidity 1:53
    cardiovascular mortality 1:53
    exacerbation 1:53
    management 1:53
    manifestations 1:53
    prevalence 1:53
  mortality rates, influenza vs. 3:96
  nephropathy 1:56
    ethnicity 1:56
    microalbuminuria 1:56
    prevalence 1:56
    treatment 1:56
  neurological complications 1:57, 1:472f, 1:473–474, 1:474–475
    encephalitis 1:57
  neutralizing antibodies 1:132, 3:64–65
    acute-to-chronic phase transition 2:540
    env protein 1:72–73
    gp160 1:137–138
  neutralizing monoclonal antibodies 1:69–70, 1:71
    properties 1:73t
  opportunistic infections 1:52–53, 1:52t, 1:68
    human herpesviruses 2:503
    Pneumocystis jirovecii 1:52
  organ transplantation, risks 3:470
  pathogenesis 2:534–542, 2:538f, 4:115
    activation-induced apoptosis 2:536
    acute phase 2:537, 2:539
    acute-to-chronic transition 2:539f, 2:540
    bystander activation 2:540–541
    bystander cell death 2:534, 2:536
    cell turnover rate 2:536
    chronic phase 2:537, 2:539f, 2:540
    CTL-mediated response 2:540
    dendritic cells 2:538–539, 2:539f
    disease progression, markers 2:535
    env protein 2:538f, 2:540
    immunopathogenesis 2:536, 4:115
    mucosa-associated lymphoid tissue, role 2:538f, 2:539–540
    progression duration 2:536
    replication control 2:535
    therapy implications 2:541
  patient care 1:66
  persistence 4:109t
  prevention 1:59, 1:65
    see also Antiviral drug(s)
  primary isolates 1:69
  protease inhibitors 1:148
    drug resistance 1:148–149
    resistance to, cyclosporine A 2:470
  as sexually transmitted disease 1:60, 4:601
    risk factors 1:60
    SIV vs. 4:601
  susceptibility, Δ32 mutation 2:506, 2:537, 5:318–319
  treatment 1:66, 1:67t
    highly active antiretroviral therapies see above
    protease inhibitors see above
  zoonoses 2:95
HIV infection epidemiology 1:58–68, 1:60f, 1:61f
  Asia 1:63, 1:64
    prevalence 1:63
    Southeast 1:63–64
  Brazil
    prevalence 1:64–65
    risk factors 1:64–65
    treatment 1:66–68
  Caribbean 1:64
  China
    IV drug abuse 1:64
    plasma donation 1:64
    prevalence 1:64
    transmission 1:64
  developing countries 2:534
  Eastern Europe 1:64
    IV drug abuse 1:64
  global prevalence 1:61
  global response 1:65
  HIV-2 see HIV-2
  Latin America 1:64
  Middle East 1:65
  molecular 1:59
  new infections 1:61
  North Africa 1:65
  North America 1:65
    Canada 1:65
    new infections 1:65
    risk factors 1:65
  Oceania 1:65
  sub-Saharan Africa 1:61, 1:62, 1:62f
    new infections 1:62–63
    prevalence 1:61–62, 1:62f
    risk factors 1:62
    treatment 1:66–68
  transmission 1:60, 1:60t
    risk factors 1:60
  Western Europe 1:65
  see also specific countries
HIV Sequence Database 5:357t
HIV vaccines/vaccination 1:66, 1:69–76, 5:242
  adjuvants 1:75
    co-administration 1:75
    delivery systems 1:75
    immune potentiators 1:75
  AIDSVAX 1:70

antibodies 1:72
    antigen optimization 1:73
    env evolution 1:72, 1:72t
    in vivo role 1:72
antigenic drift 5:240
    drift rate 5:240
antigenic shift 5:240
antigens 5:242
Banna virus 4:542
cell-mediated immune response 1:74
    antigen target selection 1:74
cellular response induction 5:242
clinical trials 1:70
    CTL responses 1:70
    tat protein neutralization 1:72
common cold 4:473–475
coronaviruses 1:561
delivery methods 1:75
ELDKWAS peptide 1:141
envelope glycoproteins 1:134, 1:134f, 1:135f, 1:136
env gene/protein 1:72–73
glycoproteins 5:242
history 1:69–70
immune response induction 1:75
    adjuvants 1:75
    HLA genotype 1:70–71
    immunogen class 1:75
    mutation rate 1:70
implications 2:541
    CTL-based vaccines 2:542
neutralizing antibodies 1:141
    HIV gp41 1:141
    induction 2:542, 5:242
protection correlates 1:70
    SIV studies 1:71
safety 5:234
    disease course alterations 5:234
    immune response alterations 5:234
    transmission 5:234
studies 1:70
    killed microorganisms 1:70
    live-attenuated microorganisms 1:70
    subunit variations 1:70
VaxGen 1:70
vectors 1:75, 5:242
HK022 bacteriophage 5:185
HK97 bacteriophage *see* Bacteriophage HK97
HMPV *see* Human metapneumovirus
Hodgkin's and Reed–Sternberg (HRS) cells 2:154–155
Hodgkin's lymphoma
    EBV associated *see* Epstein–Barr virus infection
    latent membrane protein 2A 2:163
Hog cholera virus *see* Classical swine fever virus (CSFV)
Holin–endolysin lysis 3:250
    bacteriophage fOg44 3:254–255
    bacteriophage λ *see* Bacteriophage λ
    definition 3:248
    pinholins 3:254
    Rz–Rz1 genes 3:252
        definition 3:249
        gene diversity 3:254
        gene product function 3:254
        reading frames 3:254
    SAR endolysins 3:254
    signal sequences 3:254
Holins 3:250
    definition 3:248, 3:250
    homology 3:250, 3:251f
    icosahedral tailed dsDNA bacteriophages 3:35
    lambda holin S105 3:253
    mutational effects 3:249–250
    S^{21} function 3:254–255
    topology 3:250
        transmembrane helical domains 3:250, 3:251f
    triggering 3:248
Homologous recombination
    bacteriophages 4:402f
        evolution 5:371–372
    baculovirus expression vectors 1:239–240, 1:244
        *flash*BAC 1:244

brome mosaic virus 4:378, 4:378f
bromoviruses 1:389
    definition 1:238
    hantaviruses 2:318
    phylogenetic analysis 4:127
    RNA viruses 4:377–378, 4:378f
Homologous repeat region(s)
    definition 1:254
    granuloviruses 1:218
    nucleopolyhedroviruses 1:256–257
Homoplasty 4:436
Homopterans, definition 1:111
Homotelomeric parvovirus(es)
    replication 4:90–91
    see also specific viruses
Homotypic viral interference, arenavirus replication 3:246
Hong Kong influenza (1968) pandemic 3:100
    genome reassortment 3:99
Hop stunt viroid (HSVd), host range 5:335
Hordeivirus(es) **2:459–467**
    cereal genomic analysis applications 2:466
    coat protein 2:461–462
    cytopathology 2:463
    gene expression 2:459
    genome 2:459
        gRNA-encoded proteins 2:460–461
        gRNAs 2:460
        sgRNAβ1 2:461
        sgRNAβ2 2:461
    host range 2:459
    infection
        chloroplast morphological changes 2:463–464
        TGB movement protein role 2:464
    pathogenesis 2:464
    proteins 2:461
        coat proteins 2:461
        pathogenesis proteins 2:463
        replicase 2:461
        TGB movement proteins *see below*
    RdRp 2:461
    replication 2:463, 2:464, 2:465f
    TGB movement proteins 2:462
        chloroplasts, effects on 2:464
        expression 2:461
        TGB1 2:462
        TGB2 2:462
        TGB3 2:463
    transmission 2:464
    see also specific viruses
Hordeum vulgare see Barley (*Hordeum vulgare*)
Horizontal gene transfer
    bacteriophage evolution 5:372
    definition 3:30
    icosahedral tailed dsDNA bacteriophages 3:36
    partitivirus evolution 4:74–75
Horizontal transmission 5:274
    definition 2:141
    mouse mammary tumor virus 3:337
Hormone-regulated gene expression, virus transcriptional regulation 3:338f, 3:340
Hormone response elements (HREs), definition 3:334
Horse(s)
    West Nile virus infection 2:237
    see also under equine
Horsegram yellow mosaic virus 3:370–371
Horsepox, smallpox vaccination 4:640
Horsepox virus (HPXV), vaccinia virus origin 5:243–244
Horseradish curly top virus (HrCTV) 1:302
    geographical distribution 1:302–303
    host range 1:306
Horse sickness fever 1:40
Hospital administrations, disease surveillance 2:46
Host(s)
    clinical status related to epidemiology 2:146
    definition 5:469
    genome
        DNA virus recombination 4:374
        mouse mammary tumor virus replication 3:336

resistance
    genes 4:156
    RNA silencing 4:161, 4:162
    see also specific viruses
Hosta virus X (HVX) 4:311t
Host cell(s) 2:505
Host cell factor-1 (HCF-1), varicella-zoster virus gene expression 5:259
Host specificity, baculoviruses *see* Baculovirus(es)
House sparrows, Saint Louis encephalitis virus transmission 4:655
HPIVs *see* Human parainfluenza virus(es) (hPIVs)
HPS *see* Hantavirus pulmonary syndrome (HPS)
HPV *see* Hepatopancreatic parvovirus (HPV); High Plains virus (HPV)
HPV16 *see* Human papilloma virus 16 (HPV16)
HPV31 *see* Human papilloma virus 31 (HPV31)
HrCTV *see* Horseradish curly top virus (HrCTV)
HRSV *see* Human respiratory syncytial virus (HRSV)
HRT gene, *Arabidopsis thaliana* 4:174–175
HRV2 *see* Human rhinovirus (HRV2)
HRV16 *see* Human rhinovirus 16 (HRV16)
Hs1 virus 5:413t
HSV-1 *see* Herpes simplex virus 1 (HSV-1)
HTLV-1 *see* Human T-cell leukemia virus 1 (HTLV-1)
HTLV-1 associated myelopathy (HAM) *see* Tropical spastic paraparesis/HTLV-1 associated myelopathy (TSP/HAM)
HTLV-1-associated myelopathy (HAM) *see* Tropical spastic paraparesis/HTLV-1 associated myelopathy (TSP/HAM)
HTLV-2 *see* Human T-cell leukemia virus 2 (HTLV-2)
HToV *see* Human torovirus (HToV)
Hughes virus 1:598t
Human adenovirus(es)
    classification, serotypes 1:17
    see also specific viruses
    epidemiology
        HAdV-A 1:25
        HAdV-B1 1:25
        HAdV-B2 1:25
        HAdV-C 1:25
        HAdV-D 1:25
        HAdV-E 1:26
        HAdV-F 1:26
    gene expression 1:20f
        coding potential 1:19
        MLTU coding potential 1:19
    genome 1:18, 1:18f
    infections *see below*
    taxonomy 1:1
Human adenovirus infections 1:26, 1:28
    acute hemorrhagic conjunctivitis 2:495
    classification 1:24, 1:24t
    epidemic keratoconjunctivitis 2:495, 2:495f
    experimental infection 1:28
    human respiratory infection 2:554
        vaccines 2:557
    immune response 1:28
    immunocompromised patients 1:28
    incubation period 1:28
    malignant transformation *see below*
    pathogenesis 1:28
        classification 1:24, 1:24t
        experimental infection 1:28
        immune response 1:28
        immunocompromised patients 1:28
        incubation period 1:28
        infection sites 1:28
        pathology 1:28
        serotypes 1:24, 1:24t
    pathology 1:28
    serotypes 1:24, 1:24t
    simple follicular conjunctivitis 2:495
Human adenovirus infections, malignant transformation **1:9–16**
    Ad9 1:16
    anaphase-promoting complex 1:11

Human adenovirus infections, malignant
  transformation (continued)
  clinical manifestations 1:9
  E1A gene/protein 1:10, 1:10f, 1:12, 1:19
    Ad5 E1A proteins 1:10f
    Ad12 E1A proteins 1:12–13
    CBP/p300 1:11
    exon 2 1:12
    expression 1:10–11
    MHC downregulation 1:12–13
    N-terminal region 1:11–12
    oncogenicity of 1:12
    p400 binding 1:11–12
    PML-containing nuclear oncogenic bodies
      interactions 1:12
    pRB interactions 1:12
    promyelocytic leukemia interactions 1:12
    protein expression 1:10f
    transforming properties of 1:10
    TRRAP binding 1:11–12
  E1B-19K gene/protein 1:13
    anti-apoptotic signaling role 1:13f
    BCL-2 vs. 1:13–14
    transforming properties of 1:13
  E1B-55K gene/protein 1:13, 1:14, 1:14f
    Eorf6 interactions 1:14–15
  E1B-AP5 gene/protein 1:15
  E1B gene/protein 1:13
  E4 gene/protein 1:13
  E4orf1 gene/protein 1:16
  E4orf3 gene/protein 1:14, 1:15
  E4orf6 gene/protein 1:14
    C-terminal region 1:14–15
    E1B-55K interactions 1:14–15
  histone acetylation 1:11
  history 1:9
  'hit and run' transformation 1:16
  major histocompatibility complex downregulation
    1:12–13
  MRE11-RAD50-NSB1 complex 1:14, 1:15
    E4orf3 1:15
  NFκB 1:12–13
  nuclear export signals 1:14
  properties 1:9
  transfection studies 1:9–10
  tumorigenic potential 1:9
Human B19 parvovirus(es) 4:85–86
  tissue tropism 4:89
  see also specific viruses
Human behavior, emerging viral diseases 2:94
Human bocavirus (HBoV) 2:94
  infection 2:555
Human coronavirus(es) 1:553
Human cytomegalovirus (HCMV) 2:474–485,
  2:485–491
  classification 2:474
  evolution 2:476f, 2:478
  genetics 2:475, 2:479t
    sequence identity 2:477
    variants 2:475
  genome 1:637, 2:476f, 2:486
    HSV-1 vs. 2:486
    origin of DNA replication 2:486
    variable genes 1:637
  geographical distribution 2:475
  glycoproteins 2:487–488
  history 2:474
  laboratory strains 1:639
  NS5A protein 2:369–370
  physical properties 2:489
    infectivity 2:489
  prevalence 2:475
  propagation 2:475
  proteins 2:486
    early 2:486
    glycosylation 2:489–490
    immediate early 2:486
    late 2:486
    nonvirion 2:486
    phosphorylation 2:489–490
    posttranslational modification 2:489
    proteolytic cleavage 2:489–490
    virion 2:487, 2:488t
  replication 1:626, 2:475, 2:478f, 2:489
    DE phase 2:477
    DNA replication 2:489
    egress 2:477–478, 2:478f
    encapsidation 2:477–478
    host cell entry 1:627–628, 2:477
    IE phase 1:627–628, 2:477
    L phase 1:628, 2:477
    metabolic changes in infected cells 2:489
    regulation 2:489
    tethering 1:626
    transcription characterization 2:489
    translation characterization 2:489
  seasonal distribution 2:475
  transmission 2:439, 2:482
    horizontal 2:482
    placental 2:474
    transplacental 2:474, 2:482–483
  virion properties 2:475–477, 2:485
    capsid 2:485–486, 2:487, 2:488t, 2:490
    capsomeres 2:485–486
    envelope 2:485–486, 2:487–488, 2:488t, 2:490
    nucleocapsid 2:475–477
    tegument 2:475–477, 2:485–486, 2:487, 2:488t
Human cytomegalovirus infection 2:474
  AIDS 2:439
  antiviral drugs 2:474, 2:484
    HAART 2:484
  blindness 2:496
  clinical features 2:483, 4:114–115
    chronic disease 4:114–115
    congenital disease 2:483
  control 2:484
  cytopathology 2:490
    HCMV release 2:490
  diagnostic tests 3:471
  host range 2:475
    cell tropism 2:475
  immune evasion 1:125, 1:637, 2:484, 2:487, 3:110
    anti-apoptotic proteins 1:637
    apoptosis 1:637
    G protein-coupled receptors 1:637, 1:638f
  immunocompromised hosts 2:474
  infectious retinitis 2:496, 2:496f, 2:497
  latency 2:439, 2:440f, 2:482
    chromatin structure 2:439
    genome 2:440
    major immediate early promoter 2:439
  newborns 2:483
  pathogenesis 2:482
  pathology 2:483
  persistence 4:109t
  pregnancy 2:482–483, 4:114–115
  prevention 2:484
  reactivation 2:439, 2:440, 2:440f, 2:483
    JCV/HIV infection 4:266
    transplant-associated 3:467–468, 3:468t
  transplantation 2:483–484, 3:469
    allograft 2:483–484
    solid organ 2:484
    universal prophylaxis 3:471
  virus shedding 2:482
Human demographics, emerging viral
  diseases 2:94
Human diploid cell vaccine 5:493–494
Human endogenous retrovirus(es) (HERV) 2:108
  see also specific viruses
Human enterovirus(es) (HEV) 2:130–136
  classification 1:581, 2:130
  features 2:130
    structure 2:130
  history 2:130
  infection
    disease surveillance 2:46
    respiratory disease 2:552
  novel types 2:131t, 2:134
    history 2:134
  transmission 2:130
  see also specific viruses
Human enterovirus B (HEV-B) 2:65–66
  enterovirus 69 2:131
  infection
    clinical manifestations 2:68
    diabetes, type I 2:69
    neonates 2:68
  novel types 2:135
Human enterovirus D (HEV-D) 2:124
Human extrahepatic biliary atresia (EHBA),
  mammalian reovirus infection 4:389
Human herpesvirus(es)
  see specific viruses
Human herpesvirus 4 (HHV-4) see Epstein–Barr
  virus (EBV)
Human herpes virus 5 see Human cytomegalovirus
  (HCMV)
Human herpesvirus-6 (HHV-6) 2:498–505
  classification 2:498
  genome 2:499, 2:500f
  geographical distribution 2:498
  HHV-6A 2:498, 2:502
  HHV-6B 2:498, 2:502
  history 2:498
  infection see below
  laboratory culture 2:499
  replication 2:501
  transmission 2:502
Human herpesvirus-6 infection
  antiviral therapy 2:504–505
  cell tropism 2:499
  congenital 2:502
  control 2:504
  diagnosis 2:504
  disease associations 2:502
  immune response 2:504
    protective immunity 2:504
  immunocompromised patients 2:502
    bone marrow transplantation 2:498–499, 2:502
    solid organ transplantation 2:498–499, 2:502,
      3:469
  immunomodulation 2:501
  infant fever 2:502
  latency 2:499
  neuroinflammatory disease 2:503
    multiple sclerosis 2:504
  pathogenesis 2:502
  persistence, brain 2:503
Human herpesvirus-7 (HHV-7) 2:498–505
  classification 2:498
  genome 2:499
  geographical distribution 2:498
  history 2:498
  infection see below
  laboratory culture 2:499
  replication 2:501
  transmission 2:502
Human herpesvirus-7 infection
  antiviral therapy 2:504–505
  cell tropism 2:499
  control 2:504
  cytopathic effects 2:499
  diagnosis 2:504
  disease associations 2:502
  immune response 2:504
    protective immunity 2:504
  immunocompromised patients 2:502
    bone marrow transplantation 2:502
    solid organ transplantation 2:502
  immunomodulation 2:501
  infant fever 2:502
  latency 2:499
  neuroinflammatory disease 2:503
    multiple sclerosis 2:504
  pathogenesis 2:502
  persistence, brain 2:503
Human herpesvirus-8 (HHV-8) see Kaposi's
  sarcoma-associated herpesvirus (KSHV)
Human herpesvirus infections
  immunocompromised patients
    bone marrow transplantation 2:502
    HIV/AIDS 2:503

solid organ transplantation 2:502
  persistence 4:109t
Human immunodeficiency virus see HIV
Humanized antibodies 2:543
Human lymphocyte antigen (HLA), HIV infection
  see HIV infection
Human metapneumovirus infection 2:553
  clinical features 2:553
  pediatric 2:553
Human monkeypox virus (MPXV) 4:329
Human papillomavirus(es) (HPVs) **4:8–18, 4:18–26**
  capsid structure 4:25–26
  classification 4:8
  E1^E4 gene/protein 4:22
    cornified cell envelope association 4:23
    expression 4:22–23
    motifs 4:23
    overexpression 4:23
    role 4:22
  E1 gene/protein 4:21
    associated cellular factors 4:21–22
    role 4:21, 4:36
  E2 gene/protein 4:22
    binding sites 4:21–22, 4:36
    Brd4 binding disruption 4:22
    role 4:22
    transcriptional regulation 4:22
  E5 gene/protein 4:23
  E6 gene/protein 4:24
    cellular targets 4:37
    immortalization 4:24
    motifs 4:24
    protein interactions 4:24
    role 4:24, 4:37
  E7 gene/protein 4:25
    associated cellular factors 4:25
    pocket protein-binding motif 4:25
    pRB interactions 4:25
    role 4:25, 4:37
    structure 4:25
  evolution 4:10f, 4:12
    mechanisms 4:13
  gene functions 4:21
  see also specific genes
  genome 4:8, 4:19
    early region 4:8
    late region 4:8
    oncogenes 4:19
    open reading frames 4:8
    post-transcriptional modifications 4:19
    regions 4:19, 4:19f
    transcription 4:19
    upstream regulatory region 4:8
  genotypes 4:8–9, 4:18–19
    genital high-risk types 4:17
    variants 4:9–10
  history 4:8
  L1 4:13, 4:25
  L2 4:25
  life cycle 4:20, 4:20f
    basal cell 4:20
    cellular location 4:20
    DNA integration 5:196
    host cell entry 4:20
    nonproductive infection establishment 4:20
    nonproductive infection maintenance 4:21
    productive infection 4:10–12, 4:21
    productive stage 4:21
    progeny production 4:21
    suprabasal cell 4:20
  replication 4:10
    gene expression 4:10–12
    genome integration 4:12
    host cell entry 4:10–12
    release 4:10–12, 4:12f
    vegetative 4:10–12
  taxonomy 4:8, 4:10f
  transmission 4:10, 4:38
    cutaneous types 4:10
    genital infections 4:38
    genital types 4:10

risk factors 4:38
vaccines/vaccination 2:51–52, 4:19, 4:39, 5:223, 5:225
  antigenic variation 1:127–128
  bivalent vaccine 4:18
  development 5:235
  strategies 4:17, 5:241
  tetravalent vaccine 4:18
  virion structure 4:8, 4:9f
    capsids 4:13
  see also specific viruses
Human papillomavirus infections
  anogenital warts 4:14–15, 4:16
    intraepithelial neoplasia 4:16
  carcinogenesis/neoplasia 4:8, 4:10–12, 4:38–39
    anogenital cancer 4:19
    anogenital warts see above
    cervical cancer see below
    conjunctival tumors 2:496
    diagnosis 4:39
    squamous cell carcinoma 2:492f, 2:496, 4:20, 5:197
  cervical cancer 4:8, 4:18–19, 4:38–39, 5:196
    causative mechanism 4:15
    Papanicolaou test 4:16, 4:17, 4:39
    risk factors 4:15–16
    transformation zone 4:38–39
  clinical features 4:16
  control 4:17
  diagnosis 4:39
  as emerging disease 2:93
  epidemiology 4:14
    Hybrid Capture 2 test 4:14
  epidermodysplasia verruciformis 4:20, 5:197
  eye infections 2:491–492
  genital infections
    anogenital warts see above
    cell immortalization 4:12
    clinical features 4:16
    detection 4:14
    distribution 4:14
    immune response 4:17
    incidence 4:14
    transformation zone 4:10
    transmission 4:10
  histopathology 4:14, 4:16
  host range 4:10
  immortalization potential 4:21
    E5 4:23
  immune response 4:16, 4:39
  immunodeficiency 4:17
  immunomodulation 4:114t
  infection sites 5:197
  pathogenesis 4:19, 5:196
  pathogenicity 4:11t, 4:14, 4:38
  pathology 4:16
  persistent 4:109t, 4:115
  prevention 4:17
  renal transplant recipients 3:470
  serological relationships 4:13
  tissue tropism 4:10
  treatment 4:17, 4:39
  vaccines/vaccination 4:19, 4:108–109
  warts 4:19
    anogenital see above
    common 4:16
    experimental transmission 4:8, 4:26
    genital 4:16
    incidence 4:16
    palmoplantar myrmecia 4:16
    pathogenesis 4:19
    pathogenicity 4:14–15
    treatment 4:39
Human papilloma virus 5 infection,
  epidermodysplasia verruciformis 4:8
Human papilloma virus 6 infection, anogenital warts
  4:14–15
Human papilloma virus 8 (HPV8), carcinogenic
  potential 5:197
Human papilloma virus 11 infection, anogenital warts
  4:14–15
Human papilloma virus 16 (HPV16) 4:14
  gene expression 5:196

E2 binding sites 4:36
E5 role 4:23–24
E6 role 4:24–25
E7 role 4:25f
genome 4:35f
infection see below
taxonomy 4:35
Human papilloma virus 16 infection
  cancer 4:16, 4:115
    epithelial restriction barriers 4:35–36
    related histological abnormalities 5:196
    tumor development 4:10
  immune system evasion 4:17
  nonproductive infection 4:20–21
Human papilloma virus 18 (HPV18), E7 role 4:25,
  4:115
Human papilloma virus 31 (HPV31)
  E5 role 4:23–24
  E6 role 4:24–25
  E7 role 4:25
  L2 role 4:26
Human parainfluenza virus(es) (hPIVs) **4:47–52**
  classification 4:47
  evolution 4:50
  genetics 4:50
  genome 4:47
  geographical distribution 4:49
  hemagglutinin–neuraminidase 4:47–48
    cleavage 4:48
    morphology 4:48–49
  historical aspects 4:47
  infection 2:553
    clinical features 4:50t, 4:51
    control 4:51
    diagnosis 4:51
    epidemiology 4:50, 4:50t
    host range 4:50
    immune response 4:51
    pathogenicity 4:50
    pathology 4:51
    prevention 4:51
    transmission 4:50
  propagation 4:50
  proteins 4:47
    attachment protein 4:48
    fusion protein 4:49
    hemagglutinin–neuraminidase see above
    matrix protein 4:47–48, 4:49
    nucleocapsid 4:47–48
    P protein 4:47–48
    structural proteins 4:47–48
  replication 4:49
  serology 4:50
  taxonomy 4:47
  transcription 4:49
  translation 4:49
  types 4:47
  virion structure 4:47, 4:48f
  see also specific viruses
Human polyomavirus see BK virus; JC virus; KI virus;
  WU virus
Human parainfluenza virus-4 2:17
Human reservoirs
  see specific viruses
Human respiratory syncytial virus (HRSV)
  2:542–551
  antigens 2:547
    dimorphism 2:553
    monoclonal antibodies against 2:547
    subgroups 2:547
  classification 2:543
  genetics 2:547
    recombinant virus 2:547
  genome 2:545f, 2:546
  history 2:543
  infection see below
  replication 2:546
    antigenome 2:546
    host cell attachment 2:547
    progeny release 2:546
    transcription 2:546

Human respiratory syncytial virus (HRSV) (continued)
  stability 2:547
  transmission 2:548
  vaccines/vaccination 2:550
    disease enhancement 2:550, 2:557
  virion proteins 2:543
    F glycoprotein 2:543, 2:544
    G glycoprotein 2:543
    L protein 2:543
    M2-2 protein 2:545
    M protein 2:545
    NS1 protein 2:545–546
    NS2 protein 2:545–546
    nucleocapsid 2:543
    nucleocapsid phosphoprotein 2:543
    small hydrophobic protein 2:543, 2:544–545
    transcriptional processivity factor 2:543
  virion structure 2:543, 2:545*f*
Human respiratory syncytial virus infection 2:553
  cell culture 2:546
  children 2:553
  clinical features 2:548
  complications 2:548
  diagnosis 2:548
  elderly patients 2:548
  epidemiology 2:548, 2:549*f*, 2:553
    incidence 2:548
  experimental animals 2:546
  immunity 2:549
  immunoprophylaxis 2:549
  incubation period 2:548
  pathogenesis 2:548*f*, 2:549
  re-infection 2:548
  risk factors 2:548
  treatment 2:549
Human respiratory tract 2:551
Human respiratory virus(es) **2:551–558**
  antiviral drugs 2:556
  co-infections 2:556
    transmission 2:556
  disease syndromes 2:551
  immunocompetent hosts 2:551
  immunocompromised hosts 2:552
  vaccines 2:557
  see also specific viruses
Human rhinovirus(es) 5:391–392, 5:392*f*
  see also specific viruses
Human rhinovirus 2 (HRV2)
  antigenicity 1:545
  host cell attachment 1:543–544
Human rhinovirus 14 (HRV14) 1:545
Human rhinovirus 16 (HRV16)
  infection 1:546
  macrophage activation 1:547
Human rhinovirus 87 (HRV87) 1:545–546
Human rhinovirus A (HRV-A) 4:468
Human rhinovirus B (HRV-B) 4:468
Human subacute sclerosing panencephalitis (SSPE) 2:4
Human T-cell leukemia virus 1 (HTLV-1) **2:558–564**, 5:198
  basic leucine zipper protein 2:563
  classification 2:558
  evolution 2:561
  genetics 2:560
    clonal integration 2:566–567, 2:567*f*
    stability 2:565
    transcriptional regulation 2:213–214
  genome 2:559, 2:560*f*
    pX region 2:558, 2:560
    regulatory factors 2:560
    variability 2:561, 2:565
  history 2:558, 2:564
  receptor 2:559
  replication 2:559, 2:560*f*
    budding 2:561
    integration 2:560
    maturation 2:561
    reverse transcription 2:560
    splicing 2:560
    transcription 2:560

Tax protein 2:562
  cell cycle checkpoint attenuation 2:562
  transcriptional activation 2:562
  transcriptional repression 2:562
  tumor suppressor protein inactivation 2:562
transmission 2:559
  across species 2:561
  blood transfusion 2:559, 2:564–565
  breastfeeding 2:559, 2:563, 2:564–565
  clustering 2:559
  sexual transmission 2:559, 2:564–565
  zoonosis 2:559
Human T-cell leukemia virus 1 infection 2:558, 2:559, **2:564–574**, 2:565*t*
  adult T-cell leukemia 5:198
  adult T-cell lymphoma 2:565, 5:198
    cell morphology 5:198–199
    characterization 5:198–199
    classification 2:565
    clinical features 2:566, 2:567
    complications 2:567
    cytological features 2:566
    diagnosis criteria 2:565
    epidemiology 2:565
    immuno-virological features 2:566
    pathogenesis 2:568
    symptoms 5:198–199
    therapy 2:569
  associated cell types 2:559
  associated myelopathy *see* Tropical spastic paraparesis/HTLV-1 associated myelopathy (TSP/HAM)
  cell transformation 2:559
    T-cell transformation 5:198
  control 2:563
  epidemiology 2:564
    incidence 2:564–565
  geographical distribution 2:559
    age-dependent infection 2:559
    clustering 2:559
    endemic areas 2:564
    familial aggregation 2:559
  host range 2:559
  infective dermatitis 2:571
    children 2:572
    differential diagnosis 2:572
    disease characterization 2:572, 2:572*f*
    geographical distribution 2:571–572
    history 2:571–572
    occurrence 2:572
    pathology 2:572
    treatment 2:572
  kerato-conjunctivitis 2:573
  myositis 2:572
    disease characterization 2:572
    histology 2:572–573
    musculo-skeletal symptoms 2:572–573
    sporadic inclusion-myositis 2:573
    Tax protein 2:572–573
  pathogenicity 2:561
  persistence 4:109*t*
  prevention 2:563
  tropical spastic paraparesis *see* Tropical spastic paraparesis/HTLV-1 associated myelopathy (TSP/HAM)
  uveitis 2:573
    prognosis 2:573
    symptoms 2:573
    therapy 2:573
  zoonoses 2:559
Human T-cell leukemia virus 2 (HTLV-2) 2:559
  genetic stability 2:179
  infection, persistence 4:109*t*
Human torovirus (HToV) 5:151–152
  antigenic properties 5:155
  buoyant densities 5:153
  genotypes 5:155
  geographical distribution 5:154
  historical aspects 5:151
  infection, diarrhea 5:156

Humoral immune system
  Ectromelia virus infection 3:345
  varicella-zoster virus infection 5:254
  see also Antibodies; B cell(s); Cytokine(s); Immunoglobulin(s)
Humpty Doo virus 4:184*t*
Humulus japonicus latent virus (HJLV) 3:47
Hv145SV *see* Helminthosporium victoriae 145S virus (Hv145SV)
Hv190SV *see* Helminthosporium victoriae 190S virus (Hv190SV)
HVA *see* Herpesvirus Ateles (HVA)
Hv-p68 gene/protein
  activity 5:171–172
  overexpression 5:171, 5:171*f*
  upregulation 5:171, 5:171*f*
HVS *see* Herpesvirus Saimiri (HVS)
Hyalomma marginatum, Wanowrie virus 1:400
Hyalomma ticks, Crimean-Congo hemorrhagic fever virus transmission 1:596, 1:597–598, 1:599*f*
Hybrid Capture 2 test, human papillomavirus screening 4:14
Hybridization
  industrial-scale protein production 4:235
  microarray design 2:16
Hybrid promoters, BacMam baculovirus expression vectors 1:245
Hydraencephaly (HE), definition 1:76, 1:328
Hydrangea ringspot virus (HdRSV) 4:311*t*
Hydropericardium 1:37, 1:328
Hydrophobin(s) 2:578
Hydropic, definition 1:427
Hydrothorax 1:37, 1:328
Hydroxyurea, EBV infection control 2:155–156
Hymenoptera 1:225–226
  definition 2:37, 3:133
  iflavirus infection 3:42
Hymenopteran parasitoids, iridovirus infection 3:166
Hypercalcemia, adult T-cell leukemia 2:567–568
Hyperendemicity 2:5
  Dengue virus transmission 2:5, 2:10–11, 2:11*f*
Hypergammaglobulinemia
  definition 4:594
  SIV infection 4:601
Hypermutation, APOBEC3G 3:109
Hyperplasia
  definition 1:301, 2:212, 3:155, 4:594, 5:423
  papillomavirus-induced 4:31–32
Hypersensitivity reactions 4:173–174
  Arabidopsis 4:174–175
  bean common necrosis virus infection 1:292
  cucumber mosaic virus 1:617
  definition 4:170, 4:173, 5:375
  potato virus Y potato infections 4:291, 4:292*f*
  Rx genes 4:174
  Sharka disease control 4:242
  tobacco mosaic virus 4:173, 4:174*f*
Hyperthermophiles
  definition 1:587, 5:411
  optimum growth temperatures 1:591
Hypertrophy 1:301
Hyphae 2:574
Hyphal anastomosis
  Cryphonectria parasitica virus transmission 2:576
  definition 2:284, 4:68
  Hv145SV transmission 5:169–170
  Hv190SV transmission 5:169–170
  mycoreovirus movement 3:380
Hypocritae jacobeae virus (HjV) 5:28*t*
Hypogammaglobulinemia, post-transplant 3:471
Hyposoter exiguae ichnovirus 4:257*f*
Hyposoter exiguae idnoreovirus 2 (HeIRV-2) 3:140
Hypotension 1:328
Hypothetical taxonomy units (HTUs) 4:126
Hypotonic hyporesponsive episode 5:226
Hypoviridae 2:111, 2:288, 2:581, 5:11*t*
Hypovirulence **2:574–580**
  Cryphonectria parasitica virus 2:576
  definition 2:284, 2:574–575, 2:580

development perturbation 2:575, 2:577
fungal viruses 2:285, 2:574
virulence perturbation 2:575
Hypovirus(es) **2:580–585**
　anastomosis 2:584
　biological control 2:584
　　transgenic hypovirulent strains 2:584
　double stranded RNA satellites 4:531
　fungal dsRNA viruses 2:111
　gene expression 2:581
　genetic organization 2:581, 2:582f
　host range 2:585
　infections, fungi 2:287t
　replication 2:289
　taxonomy 2:581
　virulence attenuation 2:583
　virus–host interactions 2:582
　　phenotypic changes 2:582
　　replication element mapping 2:583
　　symptom determinants 2:583
　see also specific viruses
Hypovirus p29, RNA silencing 2:290
Hypovolemic shock, Argentine hemorrhagic fever 3:208
Hz-1V see Helicoverpa zea-1 virus (Hz-1V)
HZ-2v see Helicoverpa zea-2 virus (Hz-2V)

I

I-309 1:623–624
Iatrogenic transmission, definition 2:141, 2:146
ICAM-1, rhinovirus cell entry 4:470, 4:471
Ichneumonoid wasp(s), polydnavirus phylogeny 4:261
Ichnovirus(es) (IV)
　definition 4:256
　see also specific viruses
　gene families 4:250–251
　infection
　　host tissue 4:253
　　immunoevasive activity 4:255–256
　　immunosuppressive effects 4:255
　　insects 3:126t
　phylogeny 1:192, 4:252f
　replication 4:260–261
　virion structure 4:257, 4:257f
　see also specific viruses
Icosahedral, definition 1:533
Icosahedral array maturation 1:436, 1:437f
Icosahedral asymmetric unit (IAU), cowpea chlorotic mottle virus 5:385
Icosahedral dsDNA bacteriophages **3:1–6**
　classification 3:1
　enveloped see Icosahedral enveloped dsDNA bacteriophages
　genomes 3:5
　genomics 3:5
　life cycle 3:3
　　DNA entry 3:4
　　genome replication 3:4
　　host cell lysis 3:4
　　particle assembly 3:4
　　receptor recognition 3:3
　virion properties 3:1
　virion structure 3:1
　　DNA 3:2
　　membrane 3:2
Icosahedral enveloped dsDNA bacteriophages **3:6–13**
　applications 3:12
　　polymerase 3:12
　carrier state 3:12
　classification 3:7
　genetic tools 3:12
　　reverse genetics 3:12
　genome 3:9, 3:10f
　　nonstructural proteins 3:10
　　sequence identity 3:10
　　structural proteins 3:9
　　structure 3:6
　genome delivery 5:366f, 5:369
　　lipopolysaccharide binding 5:369
　　pilus binding 5:369
　　spike proteins 5:369

historical aspects 3:6
host range 3:7
with internal membrane 5:368
nucleocapsid 3:7
　assembly 3:8
physico-chemical properties 3:7
polymerase complex 3:7–8, 3:8f
　conformational changes 3:8
RdRp 3:6, 3:7–8, 3:8f
replication 3:10, 3:10f
　host cell attachment 3:11
　host cell lysis 3:11
　membrane acquisition 3:11
　nucleocapsid maturation 3:11
　outer membrane fusion 3:11
　packaging 3:11
　plasma membrane penetration 3:11
　procapsid assembly 3:11
　recombination 3:11
　transcription 3:11
structural proteins 3:8, 3:9
　P1 protein 3:7–8, 3:8f, 3:9f
　P2 protein 3:8, 3:9f
　P4 protein 3:7–8, 3:9f
　P7 protein 3:7–8
　P8 protein 3:8, 3:9f
structure 3:7, 3:7f
　membrane envelope 3:8
in vitro assembly/packaging/replication 3:7, 3:12
see also specific viruses
Icosahedral particles
　cryo-electron microscopy 1:605, 1:605f
　P=3 5:391, 5:392f
　T=1 5:381, 5:384f
　T=3 5:385, 5:386f
　T=4 5:385, 5:387f
　T=7 icosahedral particles 5:388, 5:389f
　T=13 5:382f, 5:388
　T=25 5:390f, 5:391
Icosahedral ssDNA bacteriophages **3:13–20**
　cell lysis 3:19
　　protein E 3:19
　DNA
　　binding protein 3:19
　　packaging 3:19
　　replication 3:16
　evolution 3:19
　　studies 3:19
　gene expression 3:14t, 3:16
　genome content 3:14
　genome delivery 5:368
　　lipopolysaccharide-binding 5:368–369
　　spike complexes 5:368–369
　history 3:13
　host cell
　　attachment 3:15
　　penetration 3:15
　　recognition 3:15
　hosts 5:368
　morphogenesis 3:17
　　termination of 3:19
　virion morphology 3:14, 3:14f
　see also specific bacteriophages
　see also specific viruses
Icosahedral ssRNA bacteriophages **3:21–30**
　applications 3:28
　capsid 3:22, 3:23f
　　conformations 3:22–24
　　maturation protein 3:24
　gene expression control 3:24, 3:25f
　　maturation process 3:26
　　Min Jou sequence 3:26
　　RdRp 3:26
　　replicase 3:25f, 3:26
　genome delivery 5:369
　　pilus binding 5:369–370
　life cycle 3:22
　　assembly initiation 3:24
　　cell entry 3:24
　　coat protein–RNA complexes 3:24
　　RdRp 3:24
　RdRp 3:24, 3:26

three-hybrid assay 3:28, 3:29f
　conserved functional domains 3:28–29
　nicotine vaccine 3:28–29
　non-bacteriophage protein encapsidation 3:29
　vaccine identification 3:28–29
translational repression/assembly initiation complex 3:23f, 3:26, 3:27f
　coinfection 3:28
　FG loops 3:28
　mechanism 3:28
　RNA-binding site 3:28
two-hybrid assay 3:29
Icosahedral symmetry 5:394
　cryo-electron microscopy 1:604, 1:606
Icosahedral tailed dsDNA bacteriophages **3:30–37**
　capsids 3:31, 3:31f, 3:33f, 3:35
　　stability 3:31–32
　DNA injection 3:33
　gene expression 3:34
　　early genes 3:34
　　gene classes 3:34
　　late genes 3:34
　　RNA polymerase 3:34
　genomes 3:35
　　chromosome delivery 3:35
　　common ancestry 3:36
　　common structural themes 3:36
　　delivery 5:367
　　diversity 3:36
　　DNA concatamers 3:35–36
　　horizontal exchange 3:36
　　replication 3:35
　historical aspects 3:30–31
　host interaction 3:34
　lysis 3:35
　　endolysin 3:35
　　Holin proteins 3:35
　protein injection 3:34
　replication 3:35
　structure 3:31, 3:31f, 3:32t
　　electron microscopy 3:31
　tails/tail fibers 3:32, 3:33f
　　binding to bacterium 3:32
　　long contractile tails 3:31, 3:31f, 5:366f, 5:367
　　long noncontractile tails 3:31, 3:31f, 5:367–368
　　short tails 3:31, 5:367, 5:368
　　temperate vs. virulent 3:33
　virion assembly 3:34
　　capsid 3:35
　　pathways 3:34
　　procapsids 3:35
　　tail 3:34
Icosahedral vertex, bacterial cell viral entry 5:367
ICP0, HSV 2:385, 2:401
ICP22, HSV 2:385, 2:403–404
ICP27, HSV 2:400–401
Ictalurid herpesvirus 1 see Channel catfish virus (CCV)
Ictalurivirus 2:233, 2:431, 2:432t
Icterus, definition 2:350
ICTVdb 5:357t
ICTVNet 5:357t
Idaeovirus 3:37
Idaeovirus(es) **3:37–42**
　geographical distribution 3:37
　　black raspberry isolates 3:37
　　resistance-breaking isolates 3:37
　　Scottish isolates 3:37
　infection see below
　nucleotide sequences 3:39
　protein sequences 3:40
　　39kDa protein 3:40
　　190kDa protein 3:40
　　coat protein 3:40
　transmission 3:38
　viral relationships 3:41
　virion composition 3:39
　　RNA 3:39
　virion properties 3:39, 3:39f
　see also specific viruses
Idaeovirus infections
　control 3:41
　detection 3:41

Idaeovirus infections (continued)
　disease symptoms 3:38
　　experimentally infected plants 3:39
　　naturally infected plants 3:38
　resistance genes 4:179t
　therapy 3:41
Idnoreovirus 3:134, 3:134t, 4:536t
Idnoreovirus(es)
　antigenic relationships 3:142
　see also specific viruses
　distribution 3:135
　genetic relationships 3:142
　genome 3:139, 3:139t
　　segments 3:139
　host range 3:135
　infection 3:135
　replication 3:139
　transmission 3:135
　virion properties 3:134, 3:139
　　stability 3:139
　　structure 3:139
　see also specific viruses
IE-1/IE-0 complex 1:261–262
IE1 protein, HCMV 1:640, 2:486
IE2 protein, HCMV 1:640, 2:486
IE62, varicella-zoster virus 5:259
IE63, varicella-zoster virus 5:261
IEDB (Immune Epitope Database and Analysis Resource) 5:357t
If1 bacteriophage 3:121
Iflavirus 1:90f, 3:42, 3:44, 3:44f
Iflavirus(es) **3:42–46**, 3:44f
　capsid proteins 3:42, 3:43f
　　dicistroviruses *vs.* 3:44
　　host cell entry, role 3:43
　　maturation 3:43
　　VP4 3:43
　characteristics 3:42, 3:42f
　detection methods 3:45
　genome 3:43, 3:43t
　　nonstructural proteins 3:43
　　polyprotein cleavage 3:43, 3:43f
　infection
　　pathology 3:44
　　tissue tropism 3:44
　replication 3:42f, 3:45
　　IRES function 3:45
　　translation 3:45
　　untranslated regions 3:44f, 3:45
　　VPg protein 3:45
　transmission 3:44
　see also specific viruses
Igbo-Ora virus 5:89
IKe bacteriophage, capsid helical symmetry 3:121
Ilarvirus 1:81, 1:387, 3:46, 4:205
Ilarvirus(es) **3:46–56**
　coat protein
　　functional domain 3:52
　　genome activation role 3:49–52
　　RNA binding 3:52–53, 3:53f
　　virus structure 3:47
　genome 3:47
　host range 3:47, 3:55
　infections *see below*
　replication 3:49
　　conserved motifs 3:49–52, 3:52f
　　genome activation 3:49
　　poly A tail 3:52
　　RNA–coat protein complex 3:52
　　transcription 3:53–54
　taxonomy 3:46
　　genus-distinguishing features 3:46
　　strains 3:48t
　　subgroups 3:47, 3:48t, 3:50f
　transmission 3:54, 4:205
　　pollen-borne 3:54–55
　virus structure 3:47, 3:50f
　see also specific viruses
Ilarvirus infections
　control 3:54

　epidemiology 3:54
　vegetative propagation 3:54
　shock phase 3:47
　sweetpotatoes 4:661t
ILTAB (International Laboratory for Tropical Agricultural Biotechnology) 5:357t
Iltovirus 2:406, 2:432t
Iltovirus infections 2:409
　prevention 2:410
Imaging techniques
　see specific techniques
Immediate early proteins (IE1), sawfly baculovirus 1:228
Immune adherence hemagglutination (IAHA), varicella-zoster virus 5:252
Immune clearance phase, chronic hepatitis B 2:357–358
Immune complexes 1:139
　antibody concentration 3:414
　conserved epitope binding 3:417–418
　definition 3:78
　dissociation constant 3:414–415
　envelope glycoproteins 3:414–415
　formation 3:79–80
　occupancy 3:414
Immune evasion 2:144–145, 3:60, 3:66, 3:77
　antigen processing interference 3:77
　insects 4:255–256
　interferons 3:107, 3:109
　　adaptive immune response consequences 3:109
　invertebrate **4:250–256**
　　parasitoid attack 4:251
　iridovirids 3:168–170
　pattern recognition receptors 3:108
　RIG-I pathway 3:108
　RNA dependent protein kinase 3:108–109
　T-cell epitope mutation 3:77
　virus-mediated, RNA dependent protein kinase 3:108–109
　see also specific viruses
Immune globulin, varicella-zoster virus 5:254
Immune homeostasis 3:71
Immune-mediated damage, Border disease 1:337
Immune potentiators, HIV vaccine development 1:75
Immune reconstitution inflammatory syndrome (IRIS) 1:52–53
　opportunistic infections 1:52–53
Immune response 2:457
　adaptive *see* Adaptive immune system
　antibody-mediated *see* Antibodies; Humoral immune system
　antigenicity, viral proteins 1:137, 3:62–64
　cell-mediated *see* Cell-mediated immune response
　definition 1:296
　DNA vaccines 2:54
　evasion *see* Immune evasion
　gene therapy 2:303
　heterotypic 1:583–584
　innate *see* Innate immune system
　insects 4:251, 4:255
　modulation
　　antigen presentation 1:125
　　cowpox virus 1:576
　　KSHV latency 3:199
　persistent infection 4:112
　　dysregulation 4:113
　　pathogen–host interactions 4:112
　regulation 4:113
　virus clearance *vs.* pathology 3:78, 3:79f
　see also specific diseases/viral infections
Immune stimulator molecules, co-administration in DNA vaccination rephrase 2:54
Immunization
　definition 5:226
　hepatitis A 2:350
　hepatitis B 2:341–342
　　hepatitis D virus coinfection 2:377
　neonates 2:53
　　effectiveness 2:53
　passive
　　definition 3:414
　　Ebola virus infection treatment 2:64

　　immunoglobulins 3:67
　　mucosal applications 3:67
　　post-exposure 3:67
　　pre-exposure 3:67
　SARS 4:559
　T-cell response 3:77
　see also Inoculation; Vaccines/vaccination
Immunoassay(s)
　principles 2:32
　　enzyme labels 2:32
　　solid-phase 2:32, 2:32f
　sensitivity 2:33
　specificity 2:33
　see also specific assays
Immunoblotting, principles 2:33
Immunocapture polymerase chain reaction (IC-PCR) 1:408
Immunodeficiency
　human papillomavirus infection 4:17
　rhadinovirus human T-cell transformation 4:590
　rotavirus infection 4:512
　see also Immunosuppression; *specific diseases/disorders*
Immunodeficient vaccine-derived poliovirus (IVDPV) 4:248–249
Immunodominance
　cell-mediated immunity 3:73, 3:73f
　definition 3:71
　pattern changes 3:73
Immunoelectron microscopy, benyvirus infection diagnosis 1:313
Immunoevasive ovarial protein(s) (IEP) 4:255–256
Immunofluorescent assays
　Crimean-Congo hemorrhagic fever 1:602
　Dengue virus 2:8–9
　Japanese encephalitis diagnosis 3:186
　principles 2:32f, 2:33
Immunogenicity **1:137–142**
　definition 1:140
　peptide conformation 1:140–141
Immunoglobulin(s)
　affinity maturation 3:61
　assembly 3:61
　　adaptive immunity 3:61
　class switching
　　phylogeny 3:58
　　role 3:66f
　complement activation 3:79–80, 3:79f
　expression 3:79–80
　hypermutation 3:61
　immunomodulatory role 3:67
　isotypes 3:63f
　　switching 3:66f
　passive immunization 3:67
　　in clinical use 3:67
　　isolation 3:67
　　purification 3:67
　phylogeny 3:58, 3:59f, 3:60f
　　agnathans 3:58, 3:60f
　　antibody-mediated immunity 3:58
　　somatic variation 3:58
　production 1:621
　　sites 3:61
　prophylaxis, varicella-zoster virus infection 5:255
　receptor role 5:321
　structure 3:62f
　virus neutralization 3:79–80
　　antigen recognition 3:79–80
　see also Antibodies; *specific types*
Immunoglobulin A (IgA)
　common cold 4:472–473
　distribution 3:64f
　functions 3:64f
　hepatitis A virus infection 2:12
　mumps virus infection 3:362–363
　rotavirus infection 4:512
　structural organization 3:63f, 3:64f
　subclasses 3:61
Immunoglobulin D (IgD)
　distribution 3:64f
　function 3:64f
　structural organization 3:63f

Immunoglobulin E (IgE)
    distribution 3:64f
    function 3:64f
    structural organization 3:63f
Immunoglobulin G (IgG)
    common cold 4:472–473
    measles 3:289
    serological assays 2:30
        distribution 3:64f
        function 3:64f
        paratopes 1:138
        structural organization 3:63f
        see also specific viruses/infections
    SIV infection 4:600–601
Immunoglobulin M (IgM)
    alphavirus infection 5:87
    Dengue fever 2:12
    distribution 3:64f
    equine encephalitic togaviruses 5:82, 5:106
    function 3:64f
    hepatitis A 2:12
    measles 3:289
    rubella 4:520
    serological assays 2:30, 2:144
        complement fixation 2:31
        see also specific viruses/infections
    structural organization 3:63f
    West Nile virus 5:448
    yellow fever 5:475–476
Immunoglobulin Y (IgY) 3:58
Immunogold labeling 2:20
Immunohistochemistry
    emerging viral disease detection 2:94
    human respiratory syncytial virus
        infection 2:549
Immunological memory
    B cells 3:61
    definition 3:70, 3:71
    DNA vaccines 2:53
    latency 4:109–110
    persistent infection 4:109–110, 4:111f
    regulation 4:113
Immunomodulation
    avipoxvirus survival factors 2:276
    hepatitis B treatment 2:359
    persistent infection 4:113, 4:114t
        apoptosis 4:113
        cell stress 4:113
        TLR signaling 4:113
Immunopathology 3:78–83
    associated host factors 3:81
        age 3:81
        genetics 3:82
        site of infection 3:82
    associated virus characteristics 3:81
        genetics 3:82
        viral titer 3:82
        virus strain 3:82
    cell-mediated immunity 3:76
    definition 3:71, 3:78
    infection 4:113
    Visna-Maedi virus infection 5:428
    see also specific viral infections
Immunoprecipitation
    papaya ringspot virus proteins 4:2
    RNP complexes 3:396
Immunosorbent electron microscopy (ISEM)
    definition 4:103
    plant virus infection diagnosis 2:19, 2:21f, 2:23f
Immunosorbent techniques, definition 2:18
Immunosuppression
    KSHV infection 3:190
    lymphoma 5:194
    measles 3:288
    organ transplantation, infections 3:466
    persistent infection 4:115
    reticuloendotheliosis virus infection 4:415–416
    TMEV infection 5:44
    see also Immunodeficiency; specific viral infections
IMNV see Infectious myonecrosis
    virus (IMNV)

Impatiens infections 4:214
    unknown 4:215, 4:218f
    see also specific viruses
Impatiens necrotic spot virus (INSV)
    Bacopa virus infections 4:215
    chrysanthemum infections 4:224
    dahlia infections 4:221–222
    geographical distribution 5:158
    host ranges 4:208
    Impatiens infections 4:214–215, 4:215f
    Osteospermum infection 4:221
Importins 5:299
INCB9471 2:507t
Incidence rate 1:59, 2:141, 2:142
Inclusion bodies 1:482
Inclusion body myositis, mumps 3:363
Inclusions
    potato virus Y 4:291
    tenuivirus infection 5:24, 5:25f
Incubation, Crimean-Congo hemorrhagic
    fever 1:601
Incubation, extrinsic 2:5
Incubation period
    chickenpox 5:253
    common cold 4:472
    definition 1:482
    Ectromelia virus 3:343
    hepatitis C 2:368
    Japanese encephalitis 3:185–186
Indexing, definition 4:659
India, HIV infection 1:64
Indian cassava mosaic virus (ICMV) 1:31–32
Indian peanut clump virus (IPCV) 1:494, 4:98
    genome 4:99
        open reading frames 4:101, 4:101t
        peanut clump virus vs. 4:101
        RNA2 4:101
        sequence comparisons 4:101
    geographical distribution 4:98
    infection
        diagnosis 4:98
        serological diagnosis 4:98–99
        reservoirs 4:98
        transmission 4:98–99
        virus particles 4:99
Indinavir 1:143t
    HAART-associated renal disease 1:56–57
Indirect contact transmission 2:145
Inducible NADPH-dependent nitric oxide synthase
    4:387–388
Infected midgut cell shedding, baculovirus infection
    resistance 1:267
Infection(s)
    abortive 4:110–111
    acute
        definition 2:146
        diagnosis 2:30, 2:30f
        T-cell memory 3:75f, 3:76
    control 5:493
    enteric 2:146
    immune response phases 3:61f
    immune system regulation 4:113
    immunomodulation 4:113
    immunopathology 4:113
    localization 5:316
        cellular susceptibility 5:316
        regulation 5:316
    organ transplantation see Organ transplantation,
        infection risks
    persistent
        adaptive immunity 4:112
        antibodies 4:112
        cell-mediated immunity 3:73, 3:73f,
            3:75f, 3:76
        definition 1:374, 4:514
        virus perpetuation 2:146
    risk 2:141
    spread 5:315
        blood-tissue barrier 5:316
        neural 5:316
    subclinical 2:146

systemic 4:514
see also specific infections
see also specific viruses
Infection neutralization 3:413–419
    antibody–antigen binding 3:414
        antibody concentration 3:414
        dissociation constant 3:414–415
        envelope glycoproteins 3:414–415
        occupancy 3:414
    antigens 3:414
    definition 3:413–414
    experimental measurement 3:414
    glycan shields 3:418–419
    kinetics 3:416
    mechanism 3:64–65, 3:415
    molecularity 3:416
    resistance 3:417
    tissue culture 3:418
    virion epitope occupancy 3:414
    virion surface area–Ab relationship 3:417
    virus escape 3:417
Infectious bronchitis virus (IBV)
    gene expression 1:551f
    infection
        associated disease 1:553
        host range 1:553
        persistent infection 1:553
        tissue targets 1:553
    structure 1:550f
    virion structure 1:549, 1:550f
Infectious bursal disease virus (IBDV)
    infection 1:326
        immunosuppression 1:326–327
        mortality rate 1:326–327
        very virulent strains 1:326–327
    structure 1:323f
        pVP2 1:324–325
        VP2 1:324
        VP3 1:325
        VP4 1:325
        VP5 1:325
    VP2 gene/protein
        base domain 1:324
        projection domain 1:324
        shell domain 1:324
Infectious clone(s)
    definition 1:549, 4:514
    genetically modified coronaviruses 1:553
Infectious degradation, grapevine disease,
    transmission 4:203
Infectious DNA clones
    Grapevine fanleaf virus 3:408–409
    nepovirus 3:408–409
Infectious flacherie virus (IFV)
    classification 5:2–3
    genome 3:45, 5:3
Infectious hematopoietic necrosis virus (IHNV) 2:228,
    4:184t
    economic effects 2:228
    genome 2:223–224, 2:224f, 3:326f
    reverse genetic analysis 2:228
    salmonid infection 2:221–222
Infectious hypodermal and hematopoietic necrosis
    virus (IHHNV)
    genome 4:571–572, 4:572f
    geographical distribution 4:572
    host range 4:572
    infection 4:571
    taxonomy 4:572
    virion structure 4:78–79, 4:571
Infectious laryngotracheitis (ILT)
    antibody detection 2:410
    clinical features 2:406, 2:409
        incubation period 2:409
        symptom severity 2:409
    pathogenesis 2:409
    pathology 2:409
Infectious laryngotracheitis (ILT) virus
    cytopathology 2:407
    genetics 2:408
    history 2:406–407

Infectious laryngotracheitis (ILT) virus (*continued*)
  host range 2:407
  latency 2:409
  morphology 2:408
  propagation 2:407
  reactivation 2:409
  vaccination 2:406
Infectious mononucleosis (IM) 2:151, 2:153
  immune response 2:153
    lymphocytosis 2:153
  incidence 2:153
  measles *vs.* 3:289–290
  see also Epstein–Barr virus infection
Infectious myonecrosis virus (IMNV) 4:572
  genome 4:573, 4:573*f*, 5:167
  geographical distribution 4:573
  host range 4:573
  infection 5:164, 5:169
  properties 4:572
  taxonomy 4:573, 5:164
Infectious pancreatic necrosis virus (IPNV) 2:232, **3:83–89**, 3:85*f*
  genome 3:84, 3:86*f*
    segment A 3:84
    segment B 3:84, 3:86*f*
  host cell attachment/entry 3:84–86
  infection *see below*
  phylogenetic analysis 3:84, 3:85*f*
  prevalence 3:88
  propagation 3:84
  structure 3:84, 3:86*f*
    antigens 3:84, 3:86*t*
    neutralization sites 3:84
    VP1 3:84–86
    VP2 1:324, 3:84
    VP3 3:84
    VP4 1:325, 3:84
    VP5 1:325
  transmission 2:232, 3:88
  vaccines/vaccination 2:233, 3:88
  virulence determinants 3:87, 3:88*t*
Infectious pancreatic necrosis virus infection 1:327
  acute 3:86
    characteristics 3:86
    histopathology 3:86
  challenge models 3:88
  clinical manifestations 1:327, 3:83
  control
    breeding 3:89
    management 3:88
    vaccination 3:88
  cytopathic effects 3:84
  disease patterns 3:83
  epidemiology 3:88
  mortality 1:327, 3:83, 3:87
  persistent 3:86
    carrier fish 3:87
    virus location 3:87
  serological 3:84
  serotype Sp 3:87
Infectious pseudo-recombinants 1:296
Infectious pustular balanoposthitis 1:362–363
Infectious pustular vulvovaginitis (IPV) 1:362–363
Infectious salmon anemia virus (ISAV) 2:229, **3:89–95**
  detection 3:91
  economic impacts 2:227
  genetics 3:91, 3:92*t*
    cap-stealing 3:93
    groups 3:93
    hemagglutinin-esterase gene 3:93
    highly polymorphic region 3:93
    reassortment 3:93
    transcription 3:93
    virulence 3:93
  genome 2:229, 3:90
    comparative analysis 2:229
    encoded proteins 3:91, 3:92*f*
    overlapping reading frames 3:91
  geographical distribution 3:90
  history 3:90
  host range 3:90
    reservoirs 3:91
  infection 2:229
    clinical features 3:90, 3:94
    control 3:95
    diagnosis 3:95
    disease course 3:93
    hemorrhagic kidney syndrome 3:94
    histopathology 3:94
    immune response 3:95
    incubation period 3:93
    mortality 3:90, 3:93
    pathogenesis 3:93
    pathology 3:94
    prevention 3:95
    shedding 3:91
    target cells 3:94
  infectivity 3:91
  replication 3:90
  structure 3:90, 3:90*f*
  surface glycoproteins 3:90
  target cells 3:94
  transmission 3:91
    vertical 3:91
  vaccines/vaccination 3:95
  vectors 3:91
    sea louse 3:91
Infectious spleen and kidney necrosis virus (ISKNV) 2:233
Infective dermatitis, HTLV-1 associated *see* Human T-cell leukemia virus 1 infection
Infectivity
  Aedes pseudoscutellaris dinovernavirus 3:140
  seadornaviruses 4:538–539
Infectivity factors, nucleopolyhedrovirus primary infection 1:266–267
Inferred from Electronic Annotation (IEA) 5:353
Inflammasome(s) 1:620, 1:621–622
Inflammation
  HSV infection 2:393
  measles 3:289
  MHV-68 infection 3:375
  necrotizing, eye infections 2:492
  yellow head virus infection 5:480
Influenza 2:554
  antiviral therapy 1:145, 1:153, 3:102
    amantadine 3:102
    channel-blocking compounds 1:148
    resistant variants 1:148
  Asian pandemic (1957) 3:100
    mortality 3:100
  control 5:491
  disease 3:96
    epidemics 3:96
    HIV *vs.* 3:96
    symptoms 3:96
  disease surveillance 2:50
  H5N1 pandemic 2:554, 3:100
    avian influenza viruses 3:100, 3:493, 5:491
    human infections 3:100
    mortality rate 3:100
    transmission 3:100
    vaccination development 5:241–242
  hemagglutination inhibition tests 2:31
  human 3:100
    replication 3:101
  immunity 3:101
  ocular diseases 2:492
  pandemics 2:97, 3:100
    associated pig infections 3:101
    avian 3:493
    bioterrorism 5:410
    H1N1 pandemic 3:100, 5:491
    H2N2 pandemic 3:100
    H3N2 pandemic 3:100
    H5N1 *see above*
  pigs 3:101
  symptoms 2:554
  zoonoses 5:491

Influenza A virus
  antigen structure variation 1:128
    antigenic drift 3:100
    sequence identity 3:96–97, 3:493
    subtypes 1:128, 3:96, 3:489, 3:493, 5:491
  birds 3:101
    replication 3:101
  evolution 2:180*f*
  genome 3:99*f*, 3:483–484, 3:485
    associated genes/proteins 3:485*t*
  hemagglutinin 3:485, 3:490
    function 3:485
    neuraminidase ratio 3:97
    post-translational modifications 3:485
  host range 3:96–97
    natural host 3:101
  immune evasion, NS1 protein 3:109
  infection 2:554
    pandemics 3:96, 3:100
  matrix proteins 3:485
  neuraminidase
    hemagglutinin ratio 3:97
    inhibitors of 3:493
    structure 3:493
  replication
    maturation 4:411–412
    uncoating 4:409
  sialic acid 5:320
  vaccines/vaccination 3:103
    development 5:241
    development strategies 5:241
    NS1 gene 5:238
  see also Influenza
Influenza B virus 2:554
  antigenic variation 3:96–97
  genome 3:483–484
  hemagglutinin 3:490
  NB protein 3:97–98
  neuraminidase 3:492
  vaccination 3:103
Influenza C virus
  antigenic variation 3:96–97
  genome 3:483–484
    noncoding sequences 3:484
    RNP complex 3:484
  hemagglutinin-esterase-fusion 3:484, 3:492
  matrix proteins 3:485
Influenza Genome Sequencing Project 5:357*t*
Influenza Sequence Database 5:357*t*
Influenza vaccine(s) 2:458, 2:557, 3:103
  antigenic variation, H2 displacement 5:240
  inactivated 3:103
    manufacture 3:103
  intranasal, Bell's palsy *vs.* 5:234
  live-attenuated virus 3:103
    cold-adapted 3:103, 5:241–242
    efficacy 3:103
  production timescale 5:241–242, 5:241*f*
  reactions to 5:230*t*
  safety 5:228
  strain selection 5:241–242
Influenza virus(es) **3:483–489**
  antibody binding 3:493
  antigenic drift 1:128, 1:131, 1:131*f*, 3:100, 3:493
    A/Aichi/68 HA1 1:131–132
    antigenic sites 1:131*f*
    mechanism 1:128, 1:131
    neuraminidase 1:128, 1:132, 1:132*f*
    Type A influenza viruses 1:128
  antigenic shift 1:128, 1:129*f*, 3:101
    Type A influenza viruses 1:128, 1:129*f*
  antigenic variation 1:128, 1:131*f*, 3:493
    antigenic drift *see above*
    antigenic shift *see above*
    epitopes 1:129, 1:131*f*
    H3 variants 1:131–132
    neuraminidase 1:128, 1:132*f*
    neutralization mechanism 1:129
  antigens **3:489–495**
    see also Hemagglutinin (HA); Neuraminidase (NA)
  apoptosis inhibition 1:158*t*

avian *see* Avian influenza virus(es)
classification 3:483, 3:489
evolution 3:98
    mutation rates 1:150, 3:98–99
genetic reassortment 3:99
genome 3:97–98, 3:484
    RNP 3:97
    H5N1 3:491
    H7N7 2:554
hemagglutinin *see below*
host range
    animal 3:101
    molecular determinants 3:101
isolation 3:96
matrix proteins 3:485
neuraminidase *see below*
neutralization 1:129
    epitopes 1:129, 1:131f
    mechanism 1:129
nomenclature 3:96, 3:483, 3:489–490
NP protein 3:485
NS1 protein 3:102
    function 3:102
    immune system evasion 3:109
    viral pathogenicity 3:102
NS proteins 3:485–486
PB2 protein 3:102
    mRNA cap binding 3:102
    virulence determinance 3:102
physical properties 3:484
propagation 3:96
protein properties 3:484
recombination 3:99
release 3:487
    NA function 3:488
replication 3:486
    assembly 3:487
    host cell attachment 3:486, 3:486f
    host cell entry 3:486
    RNA synthesis 3:487
    sites 3:96
    transcription 3:487
reverse genetics 3:99, 3:488, 3:488f
    vaccines 3:488
    virus reconstruction 3:488
structure 3:96, 3:97f, 3:98f
    matrix protein function 3:97–98
    nonstructural proteins 3:97–98
    receptor binding site 1:203
taxonomy 3:96
tissue tropism 3:96
transmission 3:96
virion composition 3:484
virion packaging 3:487
    associated proteins 3:487
    random incorporation model 3:487–488
    selective incorporation model 3:487–488
virion structure 3:484, 3:484f
see also individual types
Influenzavirus A 3:96
Influenzavirus B 3:96
Influenzavirus C 3:96
Influenza virus hemagglutinin
    Ab-mediated neutralization 3:415
    antigenic variation 1:139f, 3:418, 3:493
        amino acid substitutions 3:493
        glycosylation 3:494
        H3 HA subtype 1:128, 3:493
    cleavage 3:101
    definition 3:489
    evolution 3:98–99
    function 3:485, 3:490
        influenza A virus 3:490
            receptor binding activity 3:97–98
        influenza B virus 3:490
    mutation rates 3:98–99, 3:100
    receptor specificity 3:102
    structure 3:293, 3:490, 3:491f, 3:493
        antibody complexes 3:494, 3:494f
        HA0 3:294, 3:294f, 3:491
        HA1 3:101–102, 3:293

        HA2 3:101–102, 3:294, 3:295f, 3:491, 3:492f
        pH effects 3:491, 3:492f
        receptor binding domain 3:490
        sialic acid 3:486, 3:490–491
    synthesis 3:490
        post-translational modifications 3:485, 3:490
Influenza virus neuraminidase 3:102, 3:485, 4:412
    active site 3:492
    antigenic drift 1:128, 1:132, 1:132f
    definition 3:489
    enzymatic activity 3:488, 5:320
    influenza A virus 3:492
        receptor destroying activity 3:97–98
    influenza B virus 3:492
    inhibitors 1:149, 3:102, 3:488
        resistant variants 3:103
    mutation rates 3:100
    structure 3:492, 3:493f
        antibody complexes 3:494f
        sialic acid 3:492
    synthesis 3:492
    WSN strain 3:102
Influenza Virus Resources 5:357t
'In-group' phylogenetic analysis 4:127
inhibitor of apoptosis (iap) genes 2:139–140
Inhibitors of apoptosis (IAPs) 1:160–161
    baculovirus *see* Baculovirus apoptosis inhibition
    chicken 4:418
    definition 1:154
    granulovirus gene expression 1:214, 1:216–217
    sawfly baculovirus genome 1:226
50% Inhibitory concentration (IC_{50}), antiviral therapy 1:146
Initiator caspases, definition 1:231
Innate immune system 3:58, **3:104–111**, **3:111–117**
    antiviral effector protein targeting 3:108, 3:109
    cellular responses 3:112
        dendritic cells 3:112
        granulocytes 3:112
        macrophages 3:112
        monocytes 3:112
        natural killer cells 3:110, 3:112
    clearance 4:114
    complement *see* Complement
    control genes 3:286
    cytidine deaminases 3:109
    definition 1:121, 3:78, 3:111
    henipavirus infections 2:326
    hepatitis C 2:373
    interferons *see* Interferon(s)
    pathogen-associated molecular patterns 3:106
    pattern-recognition receptors 3:106, 3:107
    persistent infection 4:112
    PRR component targeting 3:108
    RNA interference *see* RNA interference (RNAi)
Innovator, EHV-4 treatment 2:419
Inoculation
    definition 1:296
    routes 3:342–343
    see also Immunization; Vaccines/vaccination
Inoculation access period (IAP) 3:368
Inoculative release (insect pest control) 3:125
Inoviridae 2:191, 3:117–118, 5:11t
Inovirus 2:191, 2:192t, 3:117–118
Inovirus(es) **3:117–125**
    capsid helical symmetry 3:121
    definition 3:117
    discovery 3:117
    DNA conformations 3:118f, 3:121
    ecology 3:117
    evolution 5:374
    genome 3:121
        delivery 5:369
        packaging 2:306
    host range 3:117
    life cycles 3:122
        assembly 3:123
        bacteriophage attachment 3:122
        DNA penetration 3:122
        DNA replication 3:122

        export 3:123
        gene expression 3:122
        prophage carrier states 3:123
        virion extrusion 3:122
    morphology 3:118, 3:118f
        flexibility 3:118–119
        length 3:118–119
    protein subunit conformation 3:120
    structure 3:117
        coat proteins 3:119
        DNA content 3:119
        helical arrays 1:438
    virion stoichiometry 3:121
    see also Filamentous ssDNA bacteriophages
Insect(s)
    associated viruses 3:126
    control of *see* Insect pest control
    crop losses 3:125
    infections, insects 3:126t
    innate immune response 3:149
    parasitoid attack
        immune response 4:251
        polydnavirus immune evasion 4:255
    pierce-sucking 5:274, 5:275–276
    polydnavirus infection 4:253
        virus-mediated immune abrogation 4:253
    RNA viruses 5:3
    vectors *see* Insect vectors
    viruses
    see specific viruses
Insect cell culture
    culture media 1:238
    growth temperature range 1:238
Insecticides 2:454
    barley yellow dwarf virus control 1:285–286
    biological control *vs.* 3:129–130
    cucumber mosaic virus control 1:619
    Indian tomato leaf curl viruses 5:133
    maize streak virus control 1:481, 3:270
    Rift Valley fever virus control 4:496
    watermelon mosaic virus infection 5:437
    zucchini yellow mosaic virus infection 5:437
Insect pest control **3:125–133**
    Anticarsia gemmatalis multiple nucleopolyhedrovirus 3:130–131
    chemical *vs.* biological control 3:129–130
    DNA viruses 3:127
        baculoviruses *see* Baculovirus(es)
        densoviruses 3:127, 4:79
        entomopoxviruses 3:128
        Oryctes rhinoceros virus 3:127
    host specificity 3:125–126, 3:129
    RNA viruses 3:127
        reoviruses 3:127
        tetraviruses 3:127
    strategies 3:125
        biological control 3:125
        conservation 3:125
        inoculative release 3:125
        inundation 3:125–126
Insect reovirus(es) **3:133–144**
    antigenic relationships 3:141
    distribution 3:135
    genetic relationships 3:141
    genome 3:136
    history 3:135
    host range 3:135
    infection 3:135
    replication 3:136
    taxonomy 3:134t
    transmission 3:135
    virion properties 3:136
    see also specific viruses
Insect repellents, Crimean-Congo hemorrhagic fever prevention 1:602–603
Insect retrovirus(es) **4:451–455**
    envelope fusion protein 4:451
    cellular homolog 4:454
    LTR retroelements, classification 4:451
    see also LTR-retroelements

Insect retrovirus(es) (continued)
 origin 4:451, 4:452f
 retroelements 4:451
 see also specific viruses
Insect vectors 2:458
 capripoxvirus 1:429–430, 1:432
 hematophagous
  definition 1:170
  orbivirus transmission 3:457
 luteovirus 3:237–238
 potato virus Y transmission 4:289, 4:289f
 role in viral zoonoses 2:147
 virus replication 3:147
 see also individual species
Insect virus(es)
 nonoccluded see Nonoccluded insect virus(es) (NOIV), unassigned;
 see specific viruses
Insertional mutagenesis
 feline leukemia viruses 2:188, 2:189t
 fungal retrotransposons 4:427
 gene therapy 2:302
 mouse mammary tumor virus infection 3:339
 retroviral integration 2:105
Insertion sequences, definition 5:411
Instar, definition 2:234
Institute of Medicine of the National Academy of Sciences, USA 2:93
INSV see Impatiens necrotic spot virus (INSV)
IntAct (Protein Interactions) 5:357t
Integrase(s) 5:411
 attachment site, fuselloviruses 2:299
 definition 4:428, 5:411
 genome integration, host cell 1:592–593, 2:510
 inhibitors 2:507t, 2:510
 LTR retrotransposons 4:438–439
 metaviruses 3:306, 3:307–309
 mouse mammary tumor virus 3:335
 pseudoviruses 4:354–355
 Simian retrovirus D 4:627
Integrase host factor, bacteriophage λ genome packaging 2:309
Integrated pest management (IPM) schemes 3:125
Integrated reticuloendotheliosis sequences, avipoxvirus variants 2:283
Integrated segment replication, dsDNA bacteriophage genome packaging 2:308–309
Integration 3:301
Integration sites, mouse mammary tumor virus infection 3:339–340
Integrative bacteriophage, definition 3:117
Integrin(s), viral cell entry
 hantaviruses 1:395
 HPV 4:20
 KSHV 3:197
 rotaviruses 4:510
 see also specific types
Integrin $\alpha_2\beta_1$, echovirus 1 replication 2:67
Integrin $\alpha_3\beta_1$, KSHV host cell entry 3:197
Integrin $\alpha_6\beta_4$, HPV replication 4:20
Integrin $\alpha_v\beta_3$, foot-and-mouth disease virus-host cell interactions 2:537, 5:324
Integrin β_1, mammalian reoviruses, host cell entry 4:387
Intelence® 2:510
Intentional infection
 HIV 5:407
 see also Bioterrorism
Intercellular adhesion molecule 1 (ICAM-1)
 coxsackievirus attachment 1:543–544
 rhinovirus infection 1:546–547
Intercellular movement (plant virus(es)) 3:348
Interference, antigen processing 3:77
Interfering RNAs see RNA interference (RNAi)
Interferon(s) 3:105, 3:105f
 activation 3:105–106
 adenosine deaminase 1 3:115–116
 antiviral protein induction 3:107
 definition 3:104, 3:111, 5:117
 interference/evasion 3:107, 3:109
  component targeting 3:108, 3:109
  henipavirus 2:323, 2:326

 influenza infection 3:102
 paramyxovirus 3:108
 porcine circovirus 1:516
 mammalian reovirus infection 4:390
 murine CMV infection 1:632
 persistent infection 4:112
 production signals 3:107
 signaling 3:107
 therapy
  filovirus infection 2:205
  hepatitis B 2:342, 2:364
  hepatitis delta virus treatment 2:377
  phlebovirus infection 4:495–496
  rhinovirus infection 2:556–557
  side effects 3:116
  treatable diseases 3:116
 togaviruses 5:123
 type I see below
 type II 3:429–430
 varicella-zoster virus infection 5:255
 VSV infection 5:298
 see also specific types
Interferon(s), type I 3:112
 arterivirus infection 3:429–430
 coronavirus infection 3:429–430
 definition 3:104
 direct antiviral effects 3:115
 filovirus infection treatment 2:205
 indirect antiviral effects 3:116
 induction 3:113, 3:113f
  IRF-7 3:114–115
  pattern recognition receptors 3:113
 rhinovirus infection, immune response 1:547–548
 signaling 3:115
  defects 3:112
  JAK–STAT pathway 3:115
 Visna-Maedi infection 5:430
 see also Interferon-α (IFN-α)
Interferon-α (IFN-α)
 lymphocytic choriomeningitis virus infection 3:242
 togavirus infection 5:80, 5:123
Interferon-β (IFN-β)
 gene expression 3:114–115, 3:114f
 hepatitis A 2:349
 hepatitis C 2:373
 togavirus infection 5:80, 5:123
Interferon-γ (IFN-γ) 1:621
 active form 1:621
 arterivirus infection 3:430
 Ectromelia virus infection 3:345–346
 function 1:621
 hepatitis A 2:349
 receptors 1:621
Interferon regulatory factor-3 (IRF-3) 3:107
Interferon regulatory factor-4 (IRF-4) 4:417–418
Interferon regulatory factor-7 (IRF-7) 3:107, 3:114–115
Interferon stimulated gene(s) (ISG)
 definition 5:117
 togavirus infection 5:123
Intergenic region(s) (IGR)
 arenavirus genome 3:243, 3:244, 3:244f
 banana bunchy top virus 1:274
 definition 2:37, 5:157
 dicistrovirus genome 2:41f
 Pepino mosaic virus 4:105, 4:106
Interleukin-1 (IL-1) 1:621–622
 receptors 1:621–622
Interleukin-2 (IL-2) 1:622
 receptors 1:622
 signal transduction 1:622
Interleukin-4 (IL-4) 1:622
 cellular effects 1:622
 Ectromelia virus infection 3:345–346
Interleukin-5 (IL-5) 1:622
Interleukin-6 (IL-6) 1:622
 KSHV homolog 3:193–194, 3:200
Interleukin-7 (IL-7) 1:622
 receptor 1:622
Interleukin-8 (IL-8), Dengue fever 2:13

Interleukin-10 (IL-10) 1:622–623
 parapoxviruses 4:62
 receptor 1:622–623
Interleukin-12 (IL-12) 1:623
 function 1:620
 receptor 1:623
Interleukin-13 (IL-13), cellular effects 1:622
Interleukin-15 (IL-15) 1:623
Interleukin-17 (IL-17) 1:623
Interleukin-18 (IL-18) 1:622
Interleukin-26 (IL-26) 4:591
Intermediate subviral particles (ISVPs) 4:386
Internal ribosome entry sites (IRES)
 Border disease virus 1:339
 classical swine fever virus 1:526
 definition 1:541, 3:42, 4:129, 5:1
 echoviruses 2:67
 enteroviruses 2:126
 picornaviruses see Picornavirus(es)
 rhinovirus genome 4:469
 swine vesicular disease virus 2:127
Internal sequence (INT), LTR retrotransposons 4:438
International Agricultural Research Institutes (IARCs) 3:219
International Catalog of Arboviruses Including Certain Viruses of Vertebrates 1:170
International Code for Virus Classification and Nomenclature 5:10–19
International Committee on Taxonomy of Viruses (ICTV) 5:9, 5:362
 database 5:10–19
 Eighth ICTV Report on Virus Taxonomy 5:10
 historical aspects 5:10
 organization 5:10
 species, definition 5:403–404
 structure 5:10
 subcommittees 5:10–19
 taxonomic proposals 5:10
International Working Group on Legume Viruses (IWGLV) 3:213
International Working Group on Vegetable Viruses, merger 3:213–214
InterPro 5:357t
Intracellular enveloped virus (IEV) 4:328
Intracellular mature virus (IMV)
 avipoxviruses see Avipoxvirus(es)
 cowpox virion types 1:575
 vaccinia virus see Vaccinia virus
 variola virus replication 4:639
Intracellular movement
 cytoskeleton role 3:352
 definition 3:348
Intracytoplasmic A particles, mouse mammary tumor virus replication 3:336–337
Intracytoplasmic inclusion bodies, Ectromelia virus infection 3:343, 3:344f
Intraepithelial neoplasia, papillomaviruses 4:16
Intranasal infection, Borna disease virus 1:344–345
Intranasal inoculation, Ectromelia virus infection lesions 3:345
Intraocular infection 2:491
Intraperitoneal inoculation, Ectromelia virus infection lesions 3:345
Intraserotypic recombination, Dengue viruses 2:8
Intraspecies barrier(s), fungal virus recognition 2:289
Intra-species differentiation, avipoxviruses 2:281
Introgression 5:138
Introns, group I
 catalytic core 4:478–479
 mimivirus gene content 3:317
 ribozyme function 4:475
 structure 4:475
Intussusception, definition 4:507
Inundation (insect pest control) 3:125–126
Invertebrate iridescent virus(es) (IIVs) 3:161
 taxonomy 3:168
Invertebrate iridescent virus-1 (IIV-1) 3:161

Invertebrate iridescent virus 3 (IIV-3)
　genome 3:164
　　unique genes 3:164
　infection 3:166f
　structure 3:161
Invertebrate iridescent virus-6 (IIV-6) 3:161
　genome 3:164, 3:165f
　　ribonucleotide reductase 3:164
　genome sequencing 2:233
　infection 3:165
　structure 3:162–163, 3:163f
Invertebrate iridovirus(es) **3:161–167**
　classification 3:161, 3:162t
　　tentative species 3:161
　ecology 3:166
　　genotypic variation 3:166
　　seasonal transmission 3:166
　evolution 3:164
　genome 3:163, 3:165f
　　circular permutation 3:163–164
　　core genes 3:164
　　methylation levels 3:164
　　organization 3:165f
　　putative genes 3:164
　　repetitive DNA 3:164
　geographical distribution 3:161
　history 3:161
　infections *see below*
　propagation 3:162
　replication 3:164
　structure 3:161, 3:162–163, 3:163f
　　capsid 3:162–163
　　capsomers 3:162–163
　　core 3:162–163
　transmission 3:166
　virion properties 3:162
　see also specific viruses
Invertebrate iridovirus infections 3:166f
　cell–cell fusion 3:166
　covert infection 3:161
　　detection 3:161
　　insects affected 3:165
　disease characteristics 3:164
　disease signs 3:164
　economic importance 3:167
　host range 3:162
　patent disease 3:161, 3:166
　pathology 3:165
　route 3:162
　tissue tropism 3:165–166
Inverted terminal repeats (ITRs), Crenarchaeal
　viruses 1:593
Invirase® 1:143t
in vitro assembly system, alphavirus 1:196
in vitro cell tropism, Simian retrovirus D 4:628
in vitro evolution, gene therapy
　viral vectors 2:304
In vitro polymerase assays, leishmaniaviruses 3:221
in vitro replication, Providence virus 5:32
in vitro translation, cowpea mosaic virus
　genome 1:571
in vivo cell tropism, Simian retrovirus D
　4:626f, 4:628
Iontophoresis, definition 2:436
IPCV *see* Indian peanut clump virus (IPCV)
IPNV induced apoptosis 3:86
Ipomoea batacus (sweetpotato(es)) *see* Sweetpotato
　(*Ipomoea*)
Ipomoea crinkle leaf curl virus (ICLCV) 4:666
Ipomoea vein mosaic virus 4:665
Ipomoea yellow vein virus (IYVV) 4:666
Ipomovirus 4:315t
IR2 protein, equid herpesvirus 1 replication 2:415
IR3 gene, equid herpesvirus 1 replication 2:415
IR4 protein, equid herpesvirus 1
　replication 2:415
Iranian wheat stripe virus (IWSV) 5:24
I resistance gene
　bean common mosaic virus 1:292
　bean common necrosis virus 1:293
IRF-3, innate immunity 3:107
IRF-7, innate immunity 3:107

IR gene(s), equid herpesvirus 1 2:414
Iridocyclitis 2:491
Irido-like virus(es)
　shellfish viruses *see* Shellfish virus(es);
　　see specific viruses
Iridovirid(s)
　associated proteins 3:168–170
　　immune evasion 3:168–170
　genome 3:168
　　coding potential 3:169t
　　DNA repeats 3:170
　　inversions 3:168–170
　　open reading frames 3:168–170
　　size 3:168–170
　host range 3:170
　replication **3:167–174**
　taxonomy 3:167
　　MCP analyses 3:170
　　phylogenetic trees 3:169f
　virion structure 3:167
Iridoviridae 1:50, 2:233, 3:158t, 3:167, 3:168t
Iridovirus 2:233, 3:155, 3:161, 3:168t
Iridovirus(es) **3:155–161**
　apoptosis, effects on 3:173
　apoptosis inhibition 1:158t
　control 3:159
　emerging pathogens 3:159
　fish *see* Fish iridovirus(es)
　genome 3:168
　geographical distribution 3:156
　historical aspects 3:155–156
　host range 3:156
　infections
　　clinical features 3:156
　　disease characteristics 3:164–165
　　effects on host cell 3:173
　　insects 3:126t
　　pathology 3:156
　invertebrates *see* Invertebrate iridovirus(es)
　phylogeny 1:192
　replication **3:167–174**
　　DNA methylation 3:170
　　DNA synthesis 3:170
　　RNA synthesis, early 3:170
　　RNA synthesis, late 3:171
　　virion assembly 3:172
　　virus entry 3:170
　structure 3:157f
　transmission 3:159
　virion structure 3:167
　　capsid 3:167
　see also specific viruses
Iris diaphragm 2:491
IR-PTGS 4:142–145
　post-transcriptional gene silencing 4:142–145
Isavirus 3:96
ISEM *see* Immunosorbent electron microscopy
　(ISEM)
Isogenic line, definition 2:109
Isolates, potato virus Y potato infections 4:291
Isolation
　banana bunchy top virus 1:277
　cacao swollen shoot virus infection 1:409
　varicella-zoster virus infection 5:255
Isometric, definition 1:296
Israel turkey meningoencephalomyelitis
　virus 2:237
　mosquito-borne 2:237
Iteravirus 4:76, 4:83, 4:96
Iteravirus(es)
　classification 4:94t
　see also specific viruses
　genome 4:76–77, 4:95f
　monosense genomes 4:83
　virion structure 4:78
　see also specific viruses
IUMS-Virology 5:357t
Ivanovsky, Dmitri 5:54
　virus discovery 3:398–399
IVDB (Influenza Virus Database) 5:357t
Ivory Coast Ebola virus (ICEBOV) 2:199
Ixodid, definition 2:234

J

Ja.1 virus 5:413t
Jaagsiekte sheep retrovirus (JSRV) **3:175–182**
　cell transformation 3:180
　　Env-induced transformation 3:181
　　oncogene activation 3:180–181
　　phosphatidylinositol 3-kinase
　　　docking 3:181
　　proto-oncogene transduction 3:180–181
　classification 3:175
　endogenous 3:178
　　Env protein 3:179
　　expression regulation 3:178–179
　　JSRV homology 3:178
　　long terminal repeats 3:179
　　replication defects 3:179
　　tissue tropism 3:178
　genome 3:177f
　　env gene 3:176
　　gag gene 3:175–176
　　noncoding regions 3:176
　　orf-x 3:176
　　pol gene 3:175–176
　　pro gene 3:175–176
　　sequence variability 3:176
　history 3:175
　oncogenesis 2:219
　ovine pulmonary adenocarcinoma *see* Ovine
　　pulmonary adenocarcinoma (OPA)
　particle assembly interference 2:473
　replication cycle 3:176
　　assembly 3:178
　　expression 3:177
　　hyaluronidase 2 cellular receptor 3:176
　　long terminal repeats 3:176
　virion proteins 3:175
　　capsid protein 3:175–176
　　matrix protein 3:175–176
　　reverse transcriptase 3:175–176
JAK–STAT pathway, interferon type I
　signaling 3:115
Jamestown Canyon (JC) virus 5:489
Japan, adult T-cell leukemia incidence 2:565
Japanese eel reovirus 1:164t
Japanese encephalitis
　affected species 3:185
　case fatality rates 2:236
　clinical symptoms 3:185
　　incubation period 3:185–186
　control 3:187
　　vector control 3:187
　definition 3:182–183
　diagnosis 3:186
　　complement fixation 3:186
　　enzyme-linked immunosorbent assay 3:186
　　hemagglutination inhibition 3:186
　　immunofluorescence 3:186
　epidemiology 3:184
　　age-related 3:184
　　HIV concurrent infection 3:184
　　mosquito vectors 3:184
　　seasonality 3:184
　　zoonoses 3:184–185
　livestock 3:182–183
　pathogenesis 2:245
　prevention 5:492
　treatment 3:186
　vaccines 3:187
　　adverse events 3:187
　　live vaccine 3:187
　　protective immunity mechanisms 3:187
　　recommendations 3:187
　　success of 3:187
　zoonoses 3:184–185
Japanese encephalitis (JE) antigenic complex, West
　Nile virus classification 5:440–441
Japanese encephalitis virus (JEV) **3:182–188**
　characteristics 1:171t
　epidemiology 2:243
　genotypes/genotyping 3:185
　geographical distribution 3:184

Japanese encephalitis virus (JEV) (continued)
   historical aspects 3:182–183
   host range 2:236
   infection see Japanese encephalitis
   molecular epidemiology 3:185
     antigenic groups 3:185
     genotypes 3:185
     strain variation 3:185
     studies 3:185
   pathogenicity 3:185
     host determinants 3:185
     neuroinvasiveness variation 3:185
     viral determinants 3:185
   physical properties 3:183
     hemagglutinin 3:183
     soluble complement-fixing antigen 3:183
     virion morphology 3:183
   RdRp 3:183–184
   replication 3:183
     nonstructural proteins 3:183–184
     RdRp 3:183–184
     receptor-mediated endocytosis 3:183–184
     RNA synthesis 3:183–184
     virion assembly 3:184
   structural proteins 3:183
     capsid protein 3:183
     envelope protein 3:183
     membrane protein 3:183
   transmission 3:186
     mosquito vectors 3:186
   vaccine, reactions to 5:231t
   virulence 3:185
Jaundice
   hepadnaviruses 2:340
   hepatitis A infection 2:343
JC virus(es) (JCV) **4:261–271**
   cellular transcription factors 4:263
   classification 4:261–262
   genome 4:263f
     coding regions 4:262
     regulatory regions 4:264f
   history 4:262
   infection see below
   latency 4:261–262
   life cycle 4:265f
     capsid proteins 4:265
     LT-Ag role 4:265
   LT-Ag protein
     gene expression 4:263–264
     life cycle 4:265
     oncogenic role 4:268
   Mad-1 strain
     genome organization 4:262
     oncogenic potential 4:267–268
     regulatory region 4:264f
     sequence repeats 4:263
   oligodendrocyte infection 1:473
   phylogenetics 4:128
   reactivation 4:265
     HIV infection 2:94
   receptors 4:280
   replication
     agnoprotein 4:278
     chlorpromazine 4:280
     cytoplasmic entry 4:280
     nuclear entry 4:280
     serotonergic receptors 4:280
     sialic acid receptor function 4:280
   tissue tropism 4:261–262
   viral interactions 4:266
     BK viruses 4:267
     HHV-6 4:266
     human CMV 4:266
   virion structure 4:261–262
JC virus infections
   associated cells 4:262
   astrocytes, host cell susceptibility 4:281–282
   demyelination 4:267
   glial cells 4:280
   Gt1b gangliosides 4:280
   HeLa cells 4:280
   host cell attachment 4:280
   host cell susceptibility
     astrocytes 4:281–282
     NF1X expression 4:281
   lytic infection vs. tumor induction 4:267
     oncogenic potential 4:262
     PML 4:267
   persistence 4:109t
   prevalence 4:261–262
   renal allograft recipients 3:470
   at risk patients 4:262
JDV see Jembrana disease virus (JDV)
Jembrana disease virus (JDV)
   genome 1:349
     open reading frames 1:349–350
   geographical distribution 1:348f
   infection 1:351
   taxonomy 1:348
   virion structure 1:349
Jenner, Edward 1:574
   smallpox 4:639
Jeryl Lynn strain (mumps virus), vaccination 3:363
Jet injection, DNA vaccine delivery 2:55
JEV see Japanese encephalitis virus (JEV)
JFH-1 HCV isolate 2:372–373
JGMV see Johnson grass mosaic virus (JGMV)
Jigs, bacteriophage assembly 1:438
Johnson grass mosaic virus (JGMV) 1:477
   host range 1:477
   infection
     control 1:477
     resistance 1:477
     symptoms 1:477
'Join-cut-copy recombination' 4:401
Joinjakaka virus 1:118
J protein(s), microvirus DNA packaging 3:19
JSRV see Jaagsiekte sheep retrovirus (JSRV)
Junction adhesion molecule-A (JAM-A), mammalian reoviruses 4:387
Junin virus
   geographical distribution 3:204
   history 3:203
   host range 3:206
   infection see Argentine hemorrhagic fever (AHF)
   transmission 3:204–205
   vaccine against 3:210–211

K

K1 virus toxin 5:306–307
K28 virus toxin
   A/B toxins vs. 5:303–304
   α-component 5:305–306
   β-subunit 5:303–304
   cell binding 5:306–307
   cellular effects
     apoptosis 5:305
     cell cycle progression 5:305
     DNA synthesis 5:305
     immunity 5:307
     lethality 5:306f
     resistance mechanisms 5:305–306
   concentration effects 5:306f
   endocytosis 5:303
   endoplasmic reticulum exit 5:305f
     Sec61 complex 5:304–305
   intracellular transport 5:303
   nuclear entry 5:304
   preprotoxin processing 5:301–303
     Kex1 cleavage 5:303–304
Kadam virus group 2:239
Kadipiro virus (KAD) 4:536
   diagnostic assays 4:544–545
   epidemiology 4:537
   genome
     conserved sequences 4:541–542
     sequencing 4:539
   host range 4:537
   phylogenetics 4:540–541
   proteins 4:538f
   transmission 4:537
Kakugo virus (KV) 3:45
Kamese virus (KAMV) 4:184t
Kaplan criteria, norovirus infection diagnosis 3:444
Kaposin A, KSHV latency 3:199
Kaposi's sarcoma 5:195
Kaposi's sarcoma (KS) 5:195
   animal model 3:193
   antiherpesvirus drugs 3:193
   antiviral drugs 3:193
   cellular immune status 3:193
   clinical subtypes 5:195–196
   definition 2:93
   history 3:195
   KSHV detection 3:193
   latent genes 3:200–201
   see also specific genes
   ocular associated 2:497
   pathology 3:193
   terminal repeat lengths 3:200–201
   tumor growth 3:193
   tumor morphology 5:195–196
   viral gene expression 5:195–196
Kaposi's sarcoma-associated herpesvirus (KSHV) 4:587t
   apoptosis inhibition 3:200
   characterization 3:189–190
   classification 4:585–586
   evolution 3:190
     lineages 3:190f
     recombination 3:197
     subtypes 3:197
   genetics 3:191
     LUR-encoded proteins 3:191
     Rhesus rhadinovirus vs. 4:592
     terminal repeats 3:191
   genome 3:196f
     gene functions 3:196
     miRNAs 3:195–196
     open reading frames 3:195–196
     origins of replication 3:196
   geographical distribution 3:189
   history 3:189
   host cell entry 3:197
     associated cellular signal activation 3:197
     gB glycoprotein 3:197
     receptors 3:191
   infection see Kaposi's sarcoma;
   see below
   inhibitory proteins 3:189–190
   kaposin B
     immune evasion role 3:200
     latency 3:199
   molecular piracy 3:192
     viral cyclins 3:192
   replication 3:197
     lytic 3:191
   sequestered cellular genes 4:589
   taxonomy 3:190
   tissue tropism 3:197
     associated cell lines 3:197
   transmission 3:190
     blood transfusion 3:191
     casual transmission 3:190–191
     developed countries 3:190
     immunosuppression 3:190
     maternal-child 3:190–191
     transplantation 3:191
   vIRF proteins
     apoptosis inhibition 3:200
     latency 3:199
     primary effusion lymphoma 3:201
   virion structure 3:195
Kaposi's sarcoma-associated herpesvirus (KSHV)
   complement-control protein (KCP) 3:199–200
Kaposi's sarcoma-associated herpesvirus infection 3:200
   associated autoimmune disease 3:194
   associated malignant disease see Kaposi's sarcoma (KS)
   B-lymphoid malignancies 5:196
   diagnosis 3:194
   epidemiology 3:190
   host range 3:191
   immune evasion 3:199
     chemokine homologs 3:200
     complement control protein 3:199–200

interleukin-6 homolog 3:200
kaposin B 3:200
latency 3:199
modulator of immune recognition 3:199
immune response 3:192
incidence 3:190
incubation period 3:193
latent persistence 3:197, 3:198f
associated genes 3:198f
associated promoters 3:198f
organ transplantation 3:469
prevention 3:194
primary effusion lymphoma 3:193
apoptosis inhibition 3:200
latency 3:197–198
reactivation 3:469
treatment 3:194
Karolysis, definition 3:155
Karyorrhexis, definition 3:155
Kashmir bee virus 2:40
Kasokero virus 1:400
infection 1:400
transmission 1:400
Kawasaki disease 2:93
KEGG (Kyoto Encyclopedia of Genes and Genomes) 5:357t
Kelp fly virus (KFV) 3:44f
Kemerovo virus
geographical distribution 3:465
infection 3:465
Kenya, HIV infection 1:63
Keratinocyte(s)
human papillomavirus replication 4:10–12
immortalization 4:23
MCV infection 3:321
parapoxvirus infection 4:61
Kerato-conjunctivitis, HTLV-1 associated 2:573
Kern Canyon Group, animal rhabdoviruses 1:116
Kern Canyon virus 1:116
Keterah virus 1:400
Argas pusillus 1:400
global distribution 1:400
histological studies 1:400
Scotophilus temmencki 1:400
Keuraliba virus 4:184t
Kex2 processing, Cryphonectria parasitica virus 2:579f
KFDV *see* Kyasanur Forest disease virus (KFDV)
KHV *see* Koi herpesvirus (KHV)
Kidney(s)
Ectromelia virus infection lesions 3:345
HIV infection *see* HIV infection
infectious salmon anemia virus infection 3:94
see also under Renal
Killed vaccines
development 5:236
neonates 2:53
safety issues 5:228
Simian retrovirus D 4:629
Killer toxin(s) **5:299–308**
adsorption 5:307
dsRNA viruses 5:300
immunity 5:307
lethality 5:306
Killer virus system 5:299
Kimberley virus 4:184t
Kinetic neutralization curves 3:416
KI polyomavirus (KIPyV) 4:262
Koch's postulates 1:314
Koi herpesvirus (KHV)
detection 2:208
distribution 2:207
genome 2:209, 2:210t
growth properties 2:209
history 2:207
infection 2:208
mortality 2:207
Kolongo Group, animal rhabdoviruses 1:118
Kolongo virus (KOLV) 1:118
Kontonkan virus 1:112–114
Koolpinyah virus (KOOLV) 1:118
Koplik's spots, measles 3:289–290
Korean hemorrhagic fever 2:317

KP1 toxins, *Ustilago maydis* viruses 5:215–216
KP4 toxin, *Ustilago maydis* viruses 5:216f
KP6 toxin, *Ustilago maydis* viruses 5:217f
KP1461 2:507t
Kre1p toxin 5:306–307
Kakugo virus (KV) 3:45
KS *see* Kaposi's sarcoma (KS)
KSHV *see* Kaposi's sarcoma-associated herpesvirus (KSHV)
Kunjin virus, zoonoses 5:493
geographical distribution 5:493
Kununurra virus 1:118
Kuru 1:474
codon polymorphism 5:190
disease transmission 4:332
experimental 5:190
history 5:190
incubation period 4:332
Kuruma shrimp *(Penaeus japonicus)*, infection 4:568
Kyasanur Forest disease 5:53
human cases 2:239
zoonoses 5:492
Kyasanur Forest disease virus (KFDV) 2:239
characteristics 1:171t
tick transmission 2:239
vectors 2:239
Kyphosis 1:76

L

L1 endonuclease, LINE-like retrotransposons 4:439–440
L1 proteins
HPV 4:25
HPV, vaccination 4:39
LINE-like retrotransposons 4:439–440
L2 bacteriophage 1:439
L2 proteins, HPV 4:25
L5 bacteriophage 5:185
Laboratory-acquired infections, yellow fever 5:474
Laboratory Response Network (LRN) 5:410
Laccase(s) 2:578
La Crosse (LAC) virus 5:489
characteristics 1:171t
classification 3:480
infections 5:489
mortality 3:480
morphology 1:393f
prevalence 3:480
reassortment 3:482
transmission 5:489
venereal transmission 5:269
vertical transmission 5:269
Lactase deficiency, rotavirus-induced diarrhea 4:511
Lactate dehydrogenase-elevating virus (LDV)
genome 1:179–180
geographical distribution 1:177
history 1:176
host range 1:177
infection *see below*
LDV-C 1:181
replication 1:181
host cell entry 1:181
transmission 1:182
Lactate dehydrogenase-elevating virus infection
cell tropism 1:177
clinical features 1:183
histopathology 1:184
immune response 1:184–185
cytotoxic T cells 1:185
neutralizing antibodies 1:184
immune system evasion 1:185
pathogenicity 1:183
pathology 1:184
lacZ gene/protein, baculovirus expression vector systems 1:240f
La France disease 1:510
historical aspects 1:286
La France infectious virus (LFIV) 1:286
La France isometric virus (LFIV) *see* Agaricus bisporus virus 1 (AbV-1)
Lagging strand 4:399

Lagomorph 1:533
Lagos bat virus 4:184t
Lagovirus 1:411t, 3:439
Lagovirus(es)
history 1:410
infections 3:439t
nonstructural proteins 1:412–413
see also specific viruses
Lake Victoria marburgvirus 3:273
Lambda holin S105 3:253
antiholin S107 vs. 3:253–254
membrane potential collapse 3:253
missense mutations 3:253
topology 3:253
λ1 protein, reovirus structure 4:394f
λ2 protein, reovirus structure 4:393
λ3 protein, reovirus structure 4:394f
λ bacteriophage *see* Bacteriophage λ
Lambda R, bacteriophage-mediated cell lysis 3:251
Lambdoid phages *see* Bacteriophage λ
Lamivudine 1:143t
hepatitis B therapy 1:152
LANA-1 (latency-associated nuclear antigen) 3:196
binding sites 3:197–198
KSHV infection, immune response 3:192
RTA inhibition 3:198
Lancisi, Giovanni, rinderpest epizootics 4:505
Landlocked salmon virus 1:164t
Langat virus
encephalitis 5:53
vaccines/vaccination 5:53
Laodelphax striatellus, rice black streaked dwarf virus transmission 1:484
Large dsRNAs, edornaviruses 2:110
Largemouth bass iridovirus (LMBV) 2:233
Large polypeptide (L), Crimean-Congo hemorrhagic fever virus 1:598–599
Large retrotransposon derivatives (LARDs), LTR-retrotransposons, unclassified 4:434
Large T *see* Polyomavirus(es)
Larvae
Aedes aegypti 2:9–10
definition 4:560
virus control
nucleopolyhedrovirus 1:266
Saint Louis encephalitis virus 4:659
Lassa fever
clinical features 3:208
children 3:208
complications 3:209
proteinuria 3:208–209
white blood cell count 3:208–209
control 5:486
epidemiology 3:206
histopathology 3:209
pathogenesis 3:207
endothelial cell dysfunction 3:207
incubation period 3:207
organ damage 3:207
pathology 3:209
risk factors 3:206
therapy 3:211
transmission 3:206
vaccination 3:210–211
Lassa virus
genetics 3:205
geographical distribution 3:204
history 3:203
host range 3:205
infection *see* Lassa fever
lymphocytic choriomeningitis virus cross-reactivity 3:240
tissue tropism 3:207
Last universal common ancestor (LUCA)
definition 3:473
viral origins 3:474
Late gene expression, baculoviruses 1:251
Latency **2:436–442**
antiviral therapy 1:152
characteristics 2:436
common themes 2:442
dynamic model 2:436

Latency (continued)
  establishment 2:404
  gene expression 2:438
    detection 2:438
  immunological memory 4:109–110
  immunological studies 2:438
  maintenance 2:404
  persistent vs. 4:110f
  post-transplantation reactivation 3:467
    antirejection therapy 3:467
    associated viruses 3:468
    cellular pathways 3:467
    herpesviruses 3:467
    risk factors 3:467
  reactivation 2:405f
  respiratory tract infection 2:552
  static model 2:436
  see also specific viral infections
Latency-associated nuclear antigen see LANA-1 (latency-associated nuclear antigen)
Latency-associated transcript (LAT) 1:152
  function 2:391–392
  HSV detection 2:437
  HSV infection 2:391–392
  pseudorabies virus infection 4:343f
  structure 2:404f
Latent membrane protein 1 (LMP-1) see Epstein–Barr virus (EBV)
Latent membrane protein 2A (LMP-2A) see Epstein–Barr virus (EBV)
Latent membrane protein 2B (LMP-2B) see Epstein–Barr virus (EBV)
Latent membrane proteins (LMPs) see Epstein–Barr virus (EBV)
Lateral flow tests, principles 2:33
Latex tests, principles 2:33
Latin America
  HIV epidemiology 1:64
  potato virus infection 4:297
  see also individual countries
Latinized binomials, nomenclature 5:22
Lck
  phosphorylation, Tio oncoprotein 4:590
  Tip oncoprotein binding 4:590
L (light) dsRNAs, *Ustilago maydis* viruses 5:215
Le+, VSV replication 5:294
Leader proteins
  apthoviruses 2:126
  enteroviruses 2:126
Leading strand, definition 4:399
Leaf curl, beet curly top virus pathogenesis 1:305f
Leafhopper(s)
  Maize streak virus transmission 3:264f
  plant rhabdovirus transmission 4:195
  rice tungro disease transmission 4:485
Leafroll disease
  grapevines 4:202f
    causal agents 4:203
    transmission 4:203–204
  potato virus infection 4:297
Leaky scanning 4:644
  barley yellow dwarf viruses 1:281–282
  definition 4:97
  densovirus genome 4:79–81
Leaves
  Maize streak virus 3:270
  potato virus Y 4:291
Lectins, C-type
  alphavirus host cell attachment 5:118
  BV-mediated immunosuppression 4:255
Le Dantec virus (LDV) 4:184t
Leek white stripe virus (LWSV) 3:403
  geographical distribution 3:404
Leek yellow stripe virus 5:288
LEF-3 protein 1:257–260
LEF11 protein 1:257–260
Legumes 3:212
  associated viruses see Legume virus(es)
  definition 3:212–213
  seeds 3:212–213
  see also Crop(s)

Legume virus(es) 3:212–220
  ecology 3:215
    infection sources 3:215
  economic importance 3:218
  epidemiology 3:218f
    disease development 3:218f
  genetically-modified resistant plants 3:217, 3:219
  history 3:213f
  host range 3:213
  resistant cultivars 3:217
  seed transmission
    infection control 3:219
    transmission 3:215–216
  transmission 3:215
    direct contact 3:216
    fungi 3:217
    human 3:217
    living organisms 3:216
    nematodes 3:217
    noncirculative 3:216–217
    nonpersistent 3:216–217
    persistent 3:214f
    seed-borne see above
    semipersistent 3:216–217
  variation 3:215
  vectors 3:216
    infection control 3:219
  see also specific viruses
Legume virus infections
  control 3:219
    germplasm 3:219
    resistance 3:219
    vectors 3:219
  economic aspects
    early infections 3:218
    late infections 3:214f, 3:218
  symptoms 3:214f
Leiomyosarcoma, definition 2:212
Leishmania RNA virus (LRV)
  evolution 5:173–174
  LRV1-4, endoribonuclease activity 3:222
  taxonomy 3:220
  viral persistence
    5' UTR function 3:223
    transcript cleavage 3:223
Leishmaniavirus 2:286, 3:220, 5:164, 5:165t
Leishmaniavirus(es) 3:220–225
  endoribonuclease activity 3:222
  5' UTR role 3:222
  genome 3:222f
    open reading frames 3:222f
  history 3:220
    RNA protection assays 3:221
    In vitro polymerase assays 3:221
  polyprotein processing 3:222
    infection maintenance, role 3:222
  properties 3:220
  RdRp 3:221
  replication 3:221
    model 3:224f
    RdRp activity 3:221
  structure 3:221f
  viral persistence mechanisms 3:223
    5' UTR functions 3:223
    LRV1-4 transcript cleavage 3:223
    ribosomal RNA 3:223
    translation 3:223
  see also specific viruses
Lembombo virus 3:465
Lens, eye see Eye
Lentiferon (lentivirus-induced interferon) 5:428–430
Lentivirus 1:347–348, 2:167, 2:526, 4:460, 4:594, 4:603, 5:424
Lentivirus(es)
  classification 5:424
    sequence homologies 5:21
  evolution 4:595f
    error-prone replication 4:596–597
    recombination 4:596–597
  gene therapy
    psychology 2:302–303
    vectors 2:302

  genome 4:605f
    group A 4:595–596
    group B 4:595–596
    group C 4:595–596
  morphology 4:603
  nomenclature 4:597f
  phylogeny 4:605f
  post-cell entry resistance 2:469
  taxonomy 4:595
  transmission 4:594
  vif gene/protein 2:471
  Vpu protein 2:473
  see also specific viruses
Lentivirus protein R (Vpr) 2:522
  G2 cell cycle arrest 2:522–523
  PIC translocation function 2:522–523
Lentivirus-induced interferon (lentiferon) 5:428–430
Lepeophtheirus salmonis (sea louse), infectious salmon anemia virus vector 3:91
Lepidoptera
  ascovirus infection see Ascovirus(es)
  definition 2:37
  granulovirus infection 1:211–212
  iflavirus infection 3:42
  nucleopolyhedrovirus infection 1:258f
  polydnavirus life cycle 4:260–261
Lepidopteran nucleopolyhedrovirus(es) (NPV) 1:219–220
  classification 1:220
  occlusion bodies 1:220
  phylogeny 1:260f
Leporipoxvirus 3:225t, 3:226
Leporipoxvirus(es) 3:225–231
  classification 3:226
  DNA properties 3:226
  evolution 3:228
  genetic variability 3:228
  geographical distribution 3:227
  history 3:225
  host range 3:227
  infection
    clinical features 3:228
    control 3:230
    cutaneous tumors 3:228–229
    histopathology 3:229
    immune response 3:229
    pathogenesis 3:227
    pathology 3:229
    prevention 3:230
    subcutaneous tumors 3:229
  propagation 3:227
  protein properties 3:226
  replication
    DNA replication 3:226
    sites 3:227–228
    transcription 3:226
    translation 3:226
  seasonal distribution 3:227
  taxonomy 3:226
  tissue tropism 3:228
  transmission 3:228
    efficiency 3:228
    vectors associated 3:228
  virion properties 3:226
    morphology 3:226
  see also Myxoma viruses (MYX)
Lepus europaenus, lagovirus infection 1:410
Lesions
  bovine herpes viruses 1:365
  chickenpox 5:253
  cowpox infections 1:579
  parapoxvirus infections 4:57
Lettuce, tomato spotted wilt virus infection 5:288–289
  see also specific viruses
Lettuce big-vein disease
  definition 5:263
  historical aspects 5:263
Lettuce big-vein virus (LBVaV)
  coat protein 5:266
    phylogenetics 5:266
  evolution 5:265f

genome 5:264f
geographical distribution 3:452f
historical aspects 5:263
host range 5:267
L protein 5:265
  evolution 5:265f
  phylogenetics 5:265–266
phylogenetics 5:265–266
protein 5 5:266–267
RdRp 5:265
transcription 5:267
transmission 5:264
  Olpidium brassicae vector 5:264
virus particles 5:264
Lettuce necrotic yellows virus (LYNV) 4:193
  genome 3:326f
Lettuce ring necrosis virus (LRNV) 3:447
  coat protein 3:450
  control, soil sterilization 3:453
  genome 3:449
    RNA4 3:449
  geographical distribution 3:452
  transmission, *Olpidium brassicae* 3:453
Leucania separata nucleopolyhedrovirus (LeseNPV)
  1:256
Leukemic cells, HTLV-1 infection 2:561
Leukemogenesis *see* T-cell leukemia
Leukoencephalitis 5:424
Leukosis virus(es) (ALV), avian *see* Avian leukosis
  virus(es) (ALV)
Leviviridae 2:288, 3:21, 5:11t
Levivirus(es)
  gene L 3:257f
  see also specific viruses
  see also specific viruses
Lexiva® 1:143t
L'Héritier, P, Sigma virus discovery 4:576
Liao ning virus (LNV) 4:536
  epidemiology 4:537
  genome
    conserved sequences 4:541–542
    sequencing 4:539
  host range 4:537
  phylogenetics 4:540–541
  proteins 4:538f
  replication
    mammalian cell lines 4:539
    mosquito cell lines 4:539
  transmission 4:537
Lice, definition 5:91
Life, definition 3:399
Life expectancy, AIDS 1:51–52
Life-long immunity, varicella-zoster virus infection
  5:254–255
Ligases *see* DNA ligase
Liliaceae infections 4:224
  see also individual species
Lilium infections 4:224
  see also specific viruses
Lily mottle virus (LMoV) 4:225f
Lily symptomless virus (LSV) 4:225f
Lily virus X (LVX) 4:225
LINE-1 clade 4:439
Lineage-specific retrotransposon 4:436
Linear DNA replication 2:308–309
LINE-like retrotransposons
  mechanism of action 4:440f
    L1 proteins 4:439–440
  open reading frames 4:439
  origins 4:441
  structure 4:439
  untranslated region 4:439
Linnaean system, classification 5:19
Lipid bilayers/membranes 5:308
  acquisition, bacteriophage assembly 1:439
  bacteriophages, assembly 1:439
  composition 5:308
  enveloped viruses
    assembly 1:194
    membrane fusion 3:292
    SFV E1 insertion 3:299

function 5:308
hepatitis B virus virion 2:352
lipid arrangements 5:308–309
lipid composition 5:309
measles virus 3:286
Lipid metabolism, avipoxviruses 2:277
Lipid rafts
  definition 3:292
  Ebola virus assembly 2:62
  measles virus assembly 3:333
  membrane fusion 3:296–297
  replication 4:411
  virus membranes
    composition 5:309
    diffusion 5:309
Lipodystrophy
  antiviral therapy 1:147
  HIV infection *see* HIV infection
Lipopolysaccharide (LPS)
  bacterial cell viral entry 5:366
  bacteriophage ϕX174 cell attachment 3:15–16
  definition 5:365
  icosahedral enveloped dsDNA bacteriophages 5:369
  icosahedral ssDNA bacteriophages 5:368–369
Liposomes
  cypovirus uptake 3:138–139
  DNA vaccine delivery 2:54
Lipothrixviridae 1:587, 5:11t
Lipothrixvirus(es)
  Crenarchaea infections 1:588t
  see also specific viruses
  envelope composition 1:591
  genomes 1:593
  virion structure 1:589
  see also specific viruses
Lipovnik virus
  geographical distribution 3:465
  infection 3:465
Liquid arrays, plant virus detection 2:28
Live-attenuated vaccines 5:236
  definition 4:514
  development 5:236
    history 5:238
  Japanese encephalitis 3:187
  neonates 2:53
  safety issues 2:52–53
Liver
  anellovirus infection 1:110
  avian hepadnavirus infection 2:327
  cancer 5:197
  hepatitis A virus infection 2:346
  hepatitis B virus infection 5:197
  hepatitis C virus infection 2:368
  infectious salmon anemia virus infection 3:94
  Marburg virus infection 3:278
  transplantation 3:469
  see also under Hepatic
Liver damage, Lassa fever 3:209
Liver lesions, ectromelia virus infection lesions 3:344
Loeffler, Friedrich 5:55
  virus discovery 2:455
Long contractile tails, icosahedral tailed dsDNA
  bacteriophages 5:366f
Long-distance frame shift element (LDFE)
  1:281–282
Long-distance movement
  maize streak virus 3:268
  nepovirus 3:410
Long distance RNA/RNA interactions, RNP
  complexes 3:397f
Long intergenic region (LIR), definition 3:264
Long interspersed nuclear elements (LINEs) 4:439
  definition 4:437
  LINE-1 clade 4:439
  retrotransposons 4:429–430
  see also LINE-like retrotransposons
Long noncontractile tails, icosahedral tailed dsDNA
  bacteriophages 5:367
Long terminal repeats (LTRs)
  definition 4:428
  mouse mammary tumor virus 3:335–336

pseudovirus genome 4:355
retrovirus evolution 2:183–184
Simian retrovirus D 4:625
Long-term nonprogressors (LTNP)
  definition 1:69
  HIV infection 1:71
Long-term therapy 1:147
Loop-mediated isothermal amplification (RT-LAMP)
  2:25
Lopinavir 2:513
Louping ill virus 2:240
  infection 2:240–241
    clinical manifestations 5:52–53
  transmission 2:240
    ticks 2:240
Low density lipoprotein (LDL) receptors,
  coxsackievirus attachment 1:543–544
Lower respiratory tract
  common cold 4:474f
  definition 2:551
  infection causes 2:551–552
  structures associated 2:551
Low pathogenicity avian influenza (LPAI) virus(es)
  3:101
  HA cleavage sites 3:101–102
L polymerase
  arenaviruses *see* Arenavirus(es)
  rhabdovirus phylogeny 1:112–114
L protein
  definition 5:263
  Lettuce big-vein virus *see* Lettuce big-vein virus
    (LBVaV)
  L RNA segment *see* Crimean-Congo hemorrhagic
    fever virus (CCHFV)
LRV1-4 *see* Leishmania RNA virus
LSM1-7/PAT/DHH1 protein complex 1:384–385
LT-Ag protein, BKV 4:269
LTR-retroelements
  classification 4:451
  definition 3:301
  reverse transcriptase 4:452f
LTR-retrotransposons 4:421f
  BEL 4:438
  classification 4:429–430
  definition 4:420
  DIRS1 4:438
    genome distribution 4:443
  discovery 3:302
  distribution 4:442t
  evolution 4:441
    quasispecies-like populations 4:432
  genera 4:433
    metaviruses 4:433
    pseudoviruses 4:433
  genes, effects on 4:444
  hepadnaviruses 4:438
  history 4:437
  life cycle 4:429f
    amplification 4:432f
    LC8 proteins 4:430–431
  lineage 3:305t
  mechanism of action 4:439
  members 4:437
  open reading frames 4:420
  origins, acquisition 4:441
  phylogenetic studies 4:441
  replication 4:424
    gag gene 4:420
    pol gene 4:420
    primer-binding site 4:425
    reverse transcription 4:425
    variations 4:425
  retroelements vs. 4:451
  retroviruses vs. 4:431–432
  reverse transcription 3:307f
  structural diversity 4:429f
  structure 4:438f
    Gag 4:438–439
    integrase 4:438–439
    internal sequence 4:438
    mutations 4:439

LTR-retrotransposons (continued)
    open reading frames 4:438–439
    Pol 4:438–439
        primer binding site 4:439
        reverse transcriptase 4:438–439
        ribonuclease H 4:438–439
    Tnt1 elements 4:432
    transcription 4:430
    Ty3/gypsy see Ty3/gypsy retrotransposon(s)
    unclassified 4:434
        large retrotransposon derivatives 4:434
        terminal-repeat retrotransposons in miniature 4:434
Lucerne Australian latent virus (LALV) 3:406t
Lucerne Australian symptomless virus (LASV) 4:523
    geographical distribution 4:523
    host range 4:525
Lucerne transient streak virus (LTSV) 4:647
Lucke' tumor herpesvirus (LTHV)
    adenocarcinoma 2:206
    cellular morphology 2:206f
    distribution 2:207
    genome 2:209
    history 2:206
    tumor induction 2:209
Lumpy skin disease (LSD) virus 1:427
    epidemiology 1:429
Lungs
    MHV-68 infection 3:375
    SARS infection 4:557–558
    see also under Pulmonary
Luteoviridae 1:279, 3:231, 3:232f, 3:232t, 5:209
    see also specific viruses
Luteovirus 3:231
Luteovirus(es) **3:231–238**
    biological properties 3:231
    epidemiology 3:237
    evolutionary relationships 3:235
    gene expression 3:233
        post-transcriptional gene silencing 3:234–235
        strategies 3:233
        VPg 3:235
    genome 3:234f
        open reading frames 3:231
        subgenomic RNAs 3:233–234
    geographical distribution 3:237
    history 3:231
    host range 3:235
    infection
        co-infection 3:237
        control 3:237
        crop yield losses 3:231
        diagnosis 3:237
        host–virus relationships 3:236
        plant resistance genes 4:179t
        symptoms 3:236–237
    phylogeny 3:232f
    replication 3:236
    taxonomy 3:231
    tobacco disease, economic importance 5:66
    transmission 3:235
        aphids 3:235–236
        capsid RTD function 3:236
        circulative 3:235
        initiation 3:236f
        nonpropagative 5:278f
    virion composition 3:233
    virion properties 3:231
        stability 3:231–233
    virion structure 3:233f
        capsids 3:233
    see also specific viruses
Luteovirus-associated RNAs 4:533
Lwoff, André, bacteriophage lysogeny experiments 2:445
Lychnis ringspot virus (LRSV) 2:459
    coat proteins 2:459
    genome 2:459–460
Lychnis symptomless virus (LycSLV) 4:311t
LydA holin, inhibition 3:255–256
Lymantria dispar multiple nucleopolyhedrovirus (LdMNPV) 1:217
    virion structure 1:223

Lymantria ninayi virus (LnV) 5:28t
Lymphatic system, African horse sickness 1:40–41
Lymph nodes
    adult T-cell leukemia histology 2:566
    bovine immunodeficiency virus infection 1:352
    dendritic cell activation 3:73–74
    Ectromelia virus infection lesions 3:345
Lymphadenopathy
    MHV-68 infection 3:375–376
    SIV infection 4:601
Lymphoblastoid cell lines (LCLs)
    EBV evolution 2:151
    EBV propagation 2:150
Lymphocryptovirus 2:148–149, 2:432t
Lymphocryptovirus (LCV) 4:592
    B-cell transformation 4:592–593
    classification 4:585
    distribution 4:593
    genome sequence 4:592
    phylogenetic trees 2:427f
    taxonomy 2:426–427
    EBV 2:148–149
Lymphocystis disease virus 1 (LCDV-1) 2:233
Lymphocystivirus 2:233, 3:155, 3:168t
Lymphocystivirus(es)
    as emerging pathogen 3:159
    see also specific viruses
    geographical distribution 3:156
    host range 3:156
    infections, fish 3:156–158
    phylogeny 3:160f
    transmission 3:159–160
    see also specific viruses
Lymphocyte(s) 3:78–79
    apoptosis in African swine fever infection 1:44–45
    apoptosis in filovirus infection 2:203
    infiltrations, equine infectious anemia virus infection 2:172–173
    in vitro transformation 5:194
    see also B cell(s); CD4+ T-cells; CD8+ T-cells
Lymphocytic choriomeningitis virus (LCMV) **3:238–243**
    classification 3:239
    distribution 3:239
    epidemiology 3:205–206
    evolution 3:240
    genetics 3:240
        polymerase-binding sites 3:240
        recombination 3:240
        variants 3:240
    genome, intergenic regions 3:244
    history 3:238
    host range 3:204–205
    infection see below
    phylogenetics 3:247–248
    propagation 3:239
    replication, respiratory syncytial virus vs. 3:246
    reverse genetics 3:247
    RNA replication/transcription 3:245–246
    serology 3:240
    strains 3:238
        pathogenicity 3:241
    structure 3:81
    taxonomy 3:239
    tissue tropism 3:239–240
    transmission 3:239–240
    transplant-related transmission 3:466–467
    Z protein 3:245–246
        promyelocytic leukemia protein interaction 3:245
        RNA replication/transcription 3:245–246
Lymphocytic choriomeningitis virus infection
    animal models 3:239f
        adult infection 3:81
        CNS infection 3:82
        host genetics 3:82
        infection site 3:82
        neonatal infection 3:81
        survival-determining factors 3:81
    clinical features 3:241
        grippe-like form 3:241
        incubation period 3:241
        meningeal form 3:241

meningoencephalomyelitic form 3:241
        mortality 3:241
    cytotoxic T-cells 3:238–239
        initial site of infection 3:82
    histopathology 3:241
    immune evasion 1:125
    immune response 3:242
        early stage 3:242
        late stage 3:242
        survival-determining factors 3:81
        T cells see above
    pathogenesis 3:239f
    pathogenicity 3:241
        persistent infection 3:241
        strain variant differences 3:241
    pathology 3:241
    persistent, terminal genomic deletions 3:247
    prevention and control 3:242
Lymphocytosis, SIV infection 4:601
Lymphoid follicles
    definition 5:424
    yellow fever 5:475
Lymphoid organs, definition 5:477
Lymphoid tissue, SIV infection 4:601
Lymphomas
    feline immunodeficiency viruses 1:351–352
    fish retroviruses 2:213
    immunosuppressed patients, EBV infections 5:194
    Pacheco's disease 2:410
    SIV infection 4:601
Lymphopenia 4:552
Lymphotoxins 1:620
Lyon-4 virus, antigenic properties 5:155
Lysis
    definition 3:248
    icosahedral tailed dsDNA bacteriophages
        see Icosahedral tailed dsDNA bacteriophages
    rotting vs. 3:249
    triggers 3:249–250
Lysogen 5:411
Lysogenic coliphage λ 2:445–446
Lysogeny 2:445
    definition 2:442
Lysosomal protease(s) 1:157
Lysozymes, bacterial cell viral entry 5:367
Lyssavirus(es) 4:367–368
    genome 3:326f
    zoonoses 5:486t
    see also specific viruses
Lytic transglycosylases, bacterial cell viral entry 5:367
Lytic virus(es) 5:411
    replication 4:412
    see also specific viruses

M

M1 virus, apoptosis 1:162
M2 gene
    B cell specificity 3:375
    latency 3:376–377
M2 ion channel blocker(s), influenza treatment 3:102
M2 virus, apoptosis 1:162
M13 bacteriophage see Bacteriophage M13
MAb 2F5, HIV-1 vaccine development 1:141
Macaca fasicularis gamma virus 4:587t
Macaca mulatta rhadinovirus see Ceropithecine herpesvirus 17 (CeHV-17)
Macaca nemestrina rhadinovirus 2 (MnRRV) 4:587t
Macaques see Rhesus macaques
Machlomovirus 5:145–146
    evolution 3:263
    genetics 3:261
    genome 3:259
    geographical distribution 3:261
    historical aspects 3:259
    infections
        control 3:263
        epidemiology 3:262
        histopathology 3:262
        host range 3:261

pathogenicity 3:262
pathology 3:262
prevention 3:263
propagation 3:261
replication 3:260
seasonal distribution 3:261
serotypes 3:261
transmission 3:262
variation 3:261
virion structure 3:259
see also specific viruses
Machlomovirus 3:259
Machupo virus
geographical distribution 3:204
history 3:203
host range 3:204–205
infection 3:210
Macluravirus 4:315t
Macrobrachium rosenbergii nodavirus (MrNV)
extra small virus complex 4:573
genome organization 4:573
geographical distribution 4:575
host range 4:575
infection *see* White tail disease
properties 4:574f
taxonomy 4:575
virus relationships 4:575
genome 4:574f
taxonomy 4:575
Macrobranchium rosenbergii nodavirus (MrNV) 4:527t
Macrolesions, African horse sickness *see* African horse sickness (AHS)
Macromolecular Structure Database 5:357t
Macrophages
African swine fever virus *see* African swine fever virus
astrovirus infection, role 1:209–210
Ebola virus infection 2:63–64
filovirus infection 2:203
function 1:122–123
immune control 4:112
innate immunity 3:112
JCV infection 4:267
localization 1:122–123
murine CMV infection 1:630t
rhinoviral infections 1:547
Maculavirus 5:200, 5:200t, 5:206
Maculavirus(es) 5:206
characteristics 5:206
properties 5:206
see also specific viruses
Maculopapular rash
definition 5:83
togavirus infection 5:84
Mad-1 strain *see* JC virus
Mad-4 strain (JC virus), genome 4:262
Mad cow disease *see* Bovine spongiform encephalopathy (BSE)
Maedi–Visna virus (MV) 1:126
Magnesium ion(s)
CuniNPV transmission 1:221–222
ribozyme folding 4:475–476
Main protease (M^pro), nidoviruses 3:424
Maintenance of killer genes (MAK), *Saccharomyces cerevisiae* 5:170–171
Maintenance transmission 5:272
seasonal changes 5:272
West Nile virus 5:272
Maize *(Zea mays)*
barley yellow dwarf disease 1:476
cereal yellow dwarf disease 1:476
corn stunt complex disease 1:482
diseases associated 1:476
vaccine production 5:223–224
virus infections **1:475–482**
control 1:475–476
crop yields 1:475–476
symptoms 1:476
transmission 1:475–476
see also specific viruses

Maize chlorotic dwarf virus (MCDV) 4:546–547
genome 4:548
geographical distribution 1:477
isolate variation 4:550–551
protein properties
leader protein 4:549
proteases 4:548–549
RdRp 4:549
Maize chlorotic mottle virus (MCMV) 3:259
discovery 3:259
epidemiology 3:262
genetics 3:261
RdRp 3:261
genome 5:147f
geographical distribution 3:261
infection *see* Maize mottle chlorotic stunt disease
pathogenicity 3:262
synergistic relationships 3:262
yield reduction 3:262
RdRp 3:261
replication 3:260–261
RNA strand templates 3:260–261
resistant plant lines 3:263
seasonal distribution 3:261
serotypes 3:261–262
structure 3:260f
transmission 3:262
beetles 3:262
soil 3:262
water 3:262
Maize dwarf mosaic virus A (MDMV-A) 1:479
infection
resistance 1:479
symptoms 1:479
yield losses 1:479
synergistic relationships, maize chlorotic mottle virus 3:262
transmission 1:479
Maize fine streak virus (MFSV) 1:479
infection, symptoms 1:478f
transmission 1:479
Maize Indian Fiji-like virus 1:479
maize infections 1:479
Maize Iranian mosaic virus (MIMV) 1:479
host range 1:479
Maize mosaic virus (MMV) 1:479
geographical distribution 1:479
infection
resistance 1:479
symptoms 1:479
replication, budding 4:193
transmission 4:195f
insect tissues associated 4:195–196
Maize mottle chlorotic stunt disease 1:479
geographical distribution 1:479–480
symptoms 1:479–480, 3:262–263
transmission 1:479–480
yield losses 1:478
Maize necrotic streak virus (MNeSV) 1:480
Maize rayado fino virus (MRFV) 1:480
geographical distribution 1:480
infection
control 1:480
cytological symptoms 5:205–206
incidence 1:480
replication, sites 5:205–206
transmission 5:205–206
Maize rough dwarf virus (MRDV) 1:480
characteristics 4:150t
geographical distribution 1:480
infection
control 1:480
symptoms 1:478f
transmission 1:480
Maize sterile stunt virus (MSSV) 1:480
Maize streak virus (MSV) 4:165
B strain 1:494
coat protein 3:266
diversity 3:265
genome sequence 3:265–266
relation to other mastreviruses 3:265f

evolution 3:265
future work 3:271
genome 3:266
coat protein 3:266
complementary sense genes 3:267
historical aspects 3:264
infection *see below*
long intergenic region 3:266
origin of replication 3:266
molecular biology 3:268
movement 3:270
cell-to-cell 3:268
leaves 3:270
long-distance movement 3:268
shoot apex 3:270
movement proteins 3:266
posttranslational modifications 3:268
particle assembly 3:268
particle structure 3:264f
plant retinoblastoma-related protein 3:267
repA (regulatory protein) 3:266
transactivation domain 3:268
translation 3:267
rep gene 3:266
dNTP-binding site 3:268
replication 3:268
rolling-circle replication 3:269f
short intergenic region 3:265f
complementary strand synthesis 3:266–267
strains 3:265f
transmission 1:481
leafhopper vectors 3:264f
virion sense genes 3:268
Maize streak virus infection 4:167f
control 1:481
insecticides 3:270
resistant crops 3:270
epidemiology 3:270
cultural practices 3:270
environment 3:270
host range 3:265
grasses 3:265
symptoms 1:481
yield losses 3:265
Maize stripe virus (MSpV) 5:24
host range 1:481
infection, symptoms 1:481
species discrimination, rice stripe virus 5:26
transmission 1:481
Maize white line mosaic virus (MWLMV) 4:527t
Major histocompatibility complex (MHC)
Ab-mediated immunity 3:56
allelic variation 1:123
assembly 1:123
co-stimulatory molecules 1:122
definition 3:78
downregulation
adenovirus infection 1:12–13
KSHV infection 3:193
expression 1:122
HCV infection response 2:373–374
host susceptibility to infection 5:318
restriction 3:72f
T-cell receptors 3:71–72
SIV infection 4:600–601
translocation 1:123
Major histocompatibility complex (MHC) class I,
antigen presentation 1:620–621
cross-presentation 1:124
inflammatory form 1:123–124
loading complex 1:124
molecular mechanism 1:124
peptide binding 1:123–124
peptide degradation 1:123–124
structure 1:123
Major histocompatibility complex (MHC) class II
antigen presentation 1:620–621
autophagy 1:125
endosome proteases 1:124–125
peptide editing 1:124–125

Major histocompatibility complex (MHC) class II (*continued*)
  invariant chain association 1:124
    pseudo-peptide 1:124
  mouse mammary tumor virus transmission 3:337
  structure 1:124
    peptide binding cleft 1:124
Malabaricus grouper nervous necrosis virus (MGNNV) 3:433f
  capsid 3:433
Malarial catarrhal fever *see* Bluetongue virus(es) (BTV)
Malaysia, Nipah virus 2:323–324
Mal de Rio Cuarto virus (MRCV) 4:150t
  host range 1:481
  infection
    control 1:481
    symptoms 1:478f
  transmission 1:481
Malignant rabbit fibroma virus (MRV) 3:225–226
Malignant transformation
  adenoviruses *see* Adenovirus(es)
  avian leukosis virus 4:458
  HIV associated 1:52
  reticuloendotheliosis viruses
    see Reticuloendotheliosis viruses (REV)
Malpighian cell(s) 2:212
Malva veinal necrosis virus (MVNV) 4:311t
Mamestra brassicae multiple nucleopolyhedrovirus (MbMNPV) 3:129
Mammalian cells
  B-tropic mouse leukemia virus infection 2:469
  Liao ning virus replication 4:539
  N-tropic mouse leukemia virus infection 2:469
Mammalian orthoreovirus 4:392
Mammalian orthovirus 4:536t
Mammalian reovirus(es) 4:384
  apoptosis 4:387
  distribution 4:385
  epidemiology 4:385
    age-related susceptibility 4:385
  genome 4:384t
    reassortants 4:385
    S1 segment 4:385f
  host cell attachment 4:387
    junction adhesion molecule-A 4:387
    σ1 protein function 4:387
  infection *see below*
  intermediate subviral particles 4:386
  phylogeny 4:385f
  serotypes 4:384
  strains 4:384
  transmission 4:386f
    L2 gene function 4:385–386
    S1 gene function 4:385–386
  virus–cell interactions 4:387
    apoptosis 4:388f
    cellular signaling pathways associated 4:387
    inducible NADPH-dependent nitric oxide synthase 4:387–388
    M2 gene function 4:387
    S1 gene function 4:387
  see also specific viruses
Mammalian reovirus infections
  clinical features 4:388
    aseptic meningitis 4:389
    children 4:388–389
    human extrahepatic biliary atresia 4:389
  host range 4:385
  immune response 4:389
    antibodies 4:389
    interferon 4:390
    passive protection 4:389–390
    T cells 4:389
  pathogenesis 4:386
  spread 4:386
  tissue tropism 4:386f

Mammary adenocarcinomas, mouse mammary tumor virus infection 3:338–339
Mammary gland enhancer (MGE), mouse mammary tumor virus 3:341
Mandarivirus 1:426f, 2:255
Mansonia dubitans, Venezuelan equine encephalitic virus transmission 5:104
Mantle, definition 4:561
Marafivirus 5:200, 5:200t, 5:205
Marafivirus(es) 5:205
  characteristics 5:205
  genome 5:205
    marafibox 5:205
  infection 5:201
  phylogenetic relationships 5:204f
  properties 5:205
    polyprotein processing 5:205
  replication cycle 5:205
  transmission 5:205
  virion structure 5:205
    coat proteins 5:205
  see also specific viruses
Maraviroc 2:506
Marburg hemorrhagic fever (MHF) 3:272, 3:277
  genome 3:327f
Marburg virus **3:272–280**
  bioweapons 5:408
  classification 3:273
  composition 3:273
  evolutionary relationships 3:275
    Ebola *vs.* 3:275
    phylogenetic analysis 3:276f
  gene expression 3:273
  genetic stability 2:179
  genome 3:274f
    Ebola virus *vs.* 2:201–202
    polymerase complex 3:274–275
  history 5:490–491
  inactivation 3:273
  infection *see below*
  infectivity 3:273
  occurrence 3:273f
  replication 3:275
  structure 3:274f
    glycoprotein spikes 3:275
    L protein 3:274–275
    membrane-anchored spikes 3:273
    RNP complex 3:273
    VP24 3:275
    VP30 3:274
    VP40 3:275
  taxonomy 3:273
  transmission 3:276
Marburgvirus 2:199, 3:273
Marburg virus infection
  animal models 2:202–203
  clinical features 3:277f
  coagulopathy 2:203
  Democratic Republic of Congo 2:200
  diagnosis 5:490–491
    enzyme-linked immunosorbent assay 3:278–279
    RT-PCR 3:278–279
  disease outbreaks 3:272
  disease surveillance 2:199
    Democratic Republic of Congo 3:272
    Uíge Province, Angola 3:276f
  endothelial cell activation 2:203
  fatality rate 3:272
  hepatocellular necrosis 3:278
  host range 3:276
    animal reservoir 3:276–277
  immune response 2:203
  incubation period 2:202
  liver function impairment 3:278
  neutralization antibodies 3:279
  nonfatal cases 3:277f
  pathogenesis 3:277
    Ebola virus *vs.* 3:278
    guinea pigs 3:277–278

  pathology 3:277
    hepatocellular necrosis 3:278
    parenchymal cells 3:278
  prevention 3:279
  transmission 5:490–491
  treatment 3:279
    supportive therapy 3:279
    therapeutic antibodies 3:279
  Uíge Province, Angola 2:200
  vaccines/vaccination 3:279
  zoonoses 5:490
Marchal, J, Ectromelia virus 3:342
Mardivirus
  genetics 2:409
  nomenclature 2:406
Mardivirus 2:432t
Marek's disease
  pathobiology 2:409
  prevention 2:411
Marek's disease virus (MDV)
  genetics 2:408–409
    genome integration 2:409–410
  history 2:407
  infection
    associated T-cell lymphoproliferative syndromes 2:406
    clinical features 2:409
    control 2:411
    differential diagnosis 2:410
    economic impact 2:411
    host range 2:407–408
    latency 2:409–410
    pathobiology 2:411
    pathogenesis 2:409
    pathology 2:409
    vaccination 2:406
  morphology 2:408
  propagation 2:407–408
  replication 2:409–410
  transmission 2:409–410
Marine bacteriophage(s), ecology 4:124
Marine birnavirus (MABV) 4:566
Markov chain Monte Carlo method, viral evolution 2:180–181
Marmoset animal model, varicella-zoster virus infection 5:251
Marnaviridae 1:87, 1:90f, 3:280, 3:281
Marnavirus(es) **3:280–285**
  phylogenetic analyses
    capsid proteins 3:284f
    RdRp domains 3:282f
  see also specific viruses
Maruca vitrata nucleopolyhedrovirus (MaviNPV) 1:256
MAS, definition 1:296
Mason–Pfizer monkey virus (MPMV)
  cell receptor, amino acid transport protein 4:627
  classification 4:624
  historical aspects 4:624–625
  human infections 4:629
    cancer 4:629
    schizophrenia 4:629
    serological surveys 4:629
  natural hosts 4:627
  pathogenesis 4:628
  transmission 4:624
  viral assembly 2:264
Mass vaccination programs, disease surveillance 2:47f
Mastadenovirus 1:1, 1:17, 1:26–27, 2:120
Mastadenovirus(es) 1:17
  biological properties 1:5
    genome 1:6
    infectivity 1:5–6
    proteins 1:6
  bovine 1:27
  genome 1:18
  host cell entry 1:6
  replication 1:4–5
  virion morphology 1:3
  see also specific viruses

Master sequence 4:359
Mastitis, MuV infection 3:361
Mastomys, Lassa fever epidemiology 3:206
Mastrevirus 1:296, 2:98, 4:165
Mastrevirus(es)
    infections 5:65
        monocots 4:166
    Rep proteins 1:518
    tobacco diseases, economic importance 5:65
    type species 4:165
    see also specific viruses
Maternal immunity, vertebrate parvoviruses 4:89
Maternally derived antibodies, neonatal immunization 2:53
Maternal milk, mouse mammary tumor virus transmission 3:337
Maternal transmission 1:376
Matrix metalloproteinase(s) (MMPs) 1:216
Matrix protein, definition 4:407
Matrix (M) proteins
    arenavirus assembly/budding 3:247
    Chandipura virus 1:503
    Ebola virus *see* Ebola virus(es) (EBOV)
    influenza A virus 3:485
    influenza C virus 3:485
    influenza viruses 3:485
    influenza viruses structure 3:97
    mononegaviruses 3:331
    orthoretroviruses 4:463–464
    paramyxoviruses 3:333
    vesicular stomatitis viruses 5:295
Matsukemin, *Dendrolimus spectabilis* control 3:127
Maturation
    definition 4:407
    icosahedral ssRNA bacteriophages 3:26
    replication 4:411
        proteases 4:411–412
        protein cleavage 4:411
Maturation inhibitors, antiretroviral therapy 2:516
Maturation (A) protein 3:24
Maus Elberfeld (ME) virus 1:441–442
MAV strain *see* Barley yellow dwarf virus (BYDV)
Maximum likelihood (ML), phylogenetic analysis 4:126–127
Maximum parsimony (MP), phylogenetic analysis 4:126–127
Mayaro virus 5:489
    characteristics 1:171*t*
    geographical distribution 5:113
    infection 5:113–114
        clinical features 5:87*t*
        transmission 5:89
Mayer, Adolf 5:54
MCF *see* Bovine malignant catarrhal fever (MCF)
MDA-5 helicases 3:59–60
    structural rearrangement 3:107
M-dsRNA, *Saccharomyces cerevisiae* L-A virus 5:300
M (medium) dsRNAs, Ustilago maydis viruses 5:215
M (middle) dsRNAs, Ustilago maydis viruses 5:217–218
Mealie blight *see* Maize streak virus (MSV)
Mealie yellow *see* Maize streak virus (MSV)
Mealybug(s), badnavirus transmission 1:463
Measles
    blindness 2:494–495
    clinical features 3:288
        Koplik's spots 3:289–290
    complications 3:288
        giant cell pneumonia 3:289
        HIV infection 3:289
        immunosuppression 3:288
        latent tuberculosis reactivation 3:288
        measles inclusion body encephalitis 3:289
        subacute sclerosing panencephalitis 3:289
    control 3:289
    diagnosis 3:289
        enzyme-linked immunosorbent assay 3:289–290
        reverse transcription PCR 3:289–290
    differential diagnosis 3:289–290
        cytomegalovirus infections *vs.* 3:289–290
        infectious mononucleosis *vs.* 3:289–290
        rubella *vs.* 3:289–290
        scarlet fever *vs.* 3:289–290
        toxoplasmosis *vs.* 3:289–290
        typhus *vs.* 3:289–290
    disease surveillance 2:47*f*
        age distribution 2:48
    elimination 5:232
    fatality 3:288
    histopathology 3:289
        endothelial syncytial cells 3:289
        Warthin–Finkeldy giant cells 3:289
    historical aspects 3:288
    immune response 3:289
        immunoglobulin G 3:289
        immunoglobulin M 3:289
        T-cell responses 3:289
    keratopathy 2:494–495
    mortality rate 4:52
    ocular complications 2:494–495
    pathology 3:289
        inflammation 3:289
        necrosis 3:289
        secondary infections 3:289
    persistent infection 2:147
    prevention 3:289
        global vaccine coverage 3:290*f*
        vaccination 3:290
    protective immunity 3:288
    secondary infections 3:289
    transmission 3:288
    vaccines/vaccination 3:290
        atypical measles 3:289
Measles inclusion body encephalitis 3:289
Measles/mumps/rubella (MMR) vaccine 4:57
    reactions to 5:230*t*
    safety 5:228
        autism *vs.* 5:233
        bowel disease *vs.* 5:233
        public perceptions 5:231
Measles virus 3:285–292
    assembly 3:333
    cellular receptors 3:287
        membrane cofactor protein 3:288
        signaling lymphocyte activation molecule 3:285
    classification 3:285
    C protein 3:287
    envelope proteins 3:287
    F glycoprotein 3:287–288
    F protein spike 3:286
        particle structure 3:286
    genome 3:287*f*
        innate immunity control genes 3:286
        nucleocapsid protein 3:286
    host cell binding 5:322
    H protein 3:287–288
        host cell binding 4:53–54
    human respiratory infection 2:553
    human subacute sclerosing panencephalitis association 2:4
    infection *see* Measles
    L protein 3:286
        particle structure 3:286
    M protein 3:287
    multivalent vaccines 3:291
        hepatitis B surface antigen 3:291
    nucleocapsid protein 3:286
    oncolysis 3:291
    particle structure 3:286*f*
        lipid bilayer 3:286
        nucleocapsid core 3:286
        nucleocapsid protein 3:286
        polymerase cofactor 3:286
    P protein 3:286
        function 3:287
    RdRp 3:287
        stop–start mechanism 3:287
    replication 3:286
    RNP complex 3:287
    transmission 2:553–554
    vaccines/vaccination
        Ab-dependent disease enhancement 5:231
        definition 5:226
        efficacy *vs.* potency 5:228
        multivalent *see above*
        see also Measles/mumps/rubella (MMR) vaccine
    virus perpetuation 2:147
Mechanical transmission
    Pepino mosaic virus 4:108
    Taura syndrome virus 5:7
    vegetable viruses 5:284–285
Mediastinal lymph node (MLN), MHV-68 infection 3:375
Mediterranean crab *(Portunus depurator)*, shrimp virus infection 4:568
Mediterranean shorecrab reovirus (RC84) 1:164*t*
Mediterranean swimming crab reovirus (P) 1:164*t*
MedScape 5:357*t*
Megalocystivirus 2:233, 3:155, 3:167–168, 3:168*t*
Megalocystivirus(es)
    geographical distribution 3:156
    see also specific viruses
    host range 3:156
    infections, fish 3:156–158
    phylogeny 3:160*f*
    transmission 3:159–160
    vaccines/vaccination 3:160
Meibomian gland 2:491
Melaka virus 2:95
Melanization
    IV-mediated immunosuppression 4:255
    parasitoids 4:253*f*
Melanoplus sanguinipes entomopoxvirus (MSEV) 2:136
    DNA ligase 2:139
    genome 2:138
    insect pest control 3:128*f*
Melon, resistance genes 4:182
    quantitative resistance 4:182
Melon necrotic spot virus (MNSV)
    eIF-4E-mediated resistance to 4:186
    transmission 1:453
Membrane cofactor protein (MCP; CD46), measles virus cell entry 3:288
Membrane fusion 3:292–301
    fusion loops 3:300
    fusion peptide 3:296
    fusion protein
        activated structure 3:297
        cooperativity 3:296
        native structure 3:297
    glycoproteins
        biosynthesis 3:293
        class I 3:293
        class II 3:297
        class III 3:299
        HA0 3:294
        HA2, low pH-activated conformation 3:294
        hemagglutinin structure 3:293
        herpesvirus gB structure, low pH-activated conformation 3:300
        VSV G structure, low pH-activated conformation 3:300
    lipid role 3:296
    protein cooperativity 3:296
    sites 3:292
    stalk hypothesis 3:292
    see also PH-mediated fusion
Membrane fusion protein, sawfly baculovirus genome 1:228
Membrane proliferation, nepovirus 3:409
Membrane proteins 5:310*f*
    aggregation 5:309
    enveloped viruses 5:309
    functions 5:308
    integral 5:309
        composition 5:309
        functions 5:309
    Japanese encephalitis virus 3:183
    peripheral 5:309
    viroporins 5:309
    see also specific viruses

570 Subject Index

Memory B-cell(s), EBV persistence 2:166f
Memory T-cell(s) 3:80
  acute infection, response 3:76
  classification 3:75–76
Menangle virus 2:95
Mengo virus 1:443–444
  virion structure 5:38–39
Meningitis
  aseptic *see* Aseptic meningitis
  echovirus 30 2:70
  human enterovirus B associated 2:68
  lymphocytic choriomeningitis virus associated 3:238
  Toscana virus associated 4:495
Meningococcal disease-associated prophage (MDA phage) 3:123
Meningoencephalitis, lymphocytic choriomeningitis virus associated 3:238–239
Merck Manual of Diagnosis and Therapy, Infectious Diseases 5:357t
Meristem
  definition 5:282
  virus-induced gene silencing 5:378–379
Meristem tip culture 2:454
Messenger RNA (mRNA)
  definition 4:407
  giardiaviruses 2:314
  monocistronic 1:513
  narnavirus host persistence 3:397f
  polycistronic 1:513
Metabolic syndrome, HIV infection *see* HIV infection
Metabolism
  cellular *vs.* acellular evolution 3:475
  thiomersal safety in vaccines 5:233
Metalloproteinase, granulovirus gene expression 1:216
Metal shadowing
  helix handedness determination 2:82
  sample preparation 2:83f
Metapneumovirus
  avian 4:45
  classification 4:41t
  genome 3:327f
Metapneumovirus 4:41t
Metaviridae 3:301, 4:430, 4:451, 5:11t
Metavirus(es) 3:301–311
  env gene/protein 3:309
    homology 3:301–302
    infection 3:309
    expression 3:302
    transposition levels 3:302–304
  genome 4:429f
  host range 3:302
  infections, fungi 2:287t
  integration 3:301
    target site specificity 3:309
  LTR-retrotransposons 4:433
  reverse transcriptase homology 3:301
  reverse transcription 3:307f
    genome circularization 3:306
    priming 3:306
    RNase H activity 3:306
  transposition
    flamenco, transcription factor 3:302–304
    levels 3:302–304
  see also specific viruses
Metazoan, definition 2:105
Methanococcus voltae, 'fusiform' viruses 2:300
Methanogenic Archaea-infecting virus(es)
  see Euryarchaea-infecting virus(es)
Methanogens, definition 5:411
Methanothermobacter marburgensis virus (ψM1) 5:412
  genome 5:415
    open reading frames 5:412–415
    pseudomurein endoisopeptidase 5:415
  related viruses 5:412
  structure 5:415f
Methionine homozygotes (PRNP) *see* Codon 129 polymorphism
Methyl-β-cyclodextrin
  definition 1:541
  HRV2 uptake 1:543–544

Methyl-transferase(s), gene silencing 4:142
mf-1 gene, *Cryphonectria parasitica* mating 2:578
mf-2 gene, *Cryphonectria parasitica* mating 2:578
M glycoprotein, toroviruses 5:153
MHC *see* Major histocompatibility complex (MHC)
MHV-68 *see* Murine gammaherpesvirus 68 (MHV-68)
MIANS (maleimide-1-anilo-8-naphthalene sulfonate) 5:35
Mice *see* Mouse/mice
Microangiopathy-associated renal disease 1:57
Microarrays 2:35–36
  bioinformatics 2:16
  clinical applications 2:17
    DFA testing *vs.* 2:17
    panviral arrays 2:17
  definition 5:450
  design, viral detection 2:15
    parameters 2:15
  diagnostic techniques 2:35–36
  DNA 2:14
  implications
    public health 2:18
    scientific 2:18
  limitations 2:17
  plant virus detection 2:26
  research applications 2:17
  sample amplification 2:15
    panviral 2:16
  sample complexity 2:16
  sample processing 2:15
  sample origin 2:15–16
Microfilaments
  definition 3:348
  plant virus movement 3:352
Microhabitat, definition 2:234
Microlesions, African horse sickness 1:41
Micromonas pusilla, viral infection of 1:87
Micromonas pusilla reovirus (MpRV) 1:90
  biological properties 1:90–91
    virion structure 1:92f
  classification 1:90f
  genome 1:92f
  host features 1:95
  phylogeny 1:93f
  viral co-existence 1:95
Microplitis demolitor bracovirus (MdBV)
  genome 4:259f
    coding density 4:259–260
    egf-motif genes 4:259–260
    glc gene family 4:259–260
    ptp family 4:259–260
    segment O gene family 4:259–260
    transposable elements 4:257–258
    vankyrin gene family 4:259–260
  host immune abrogation 4:254
    gene families 4:254
Micropropagation, definition 4:207
MicroRNAs (miRNAs) 4:141
  apoptosis 1:161
  BART gene 5:327
  biogenesis, Dicers 3:150
  definition 1:154
  EBV 2:164
  plant gene silencing 4:141
    methylation 4:141
  viral suppression 5:328f
  virus interactions 3:152
    SV40 3:152–153
Microsphere immuno assay (MIA), West Nile virus detection 5:448
Microsporidia 4:419
  retrotransposons 4:423
Microtubules
  baculovirus in vitro pathogenesis 1:271f
  definition 3:348
  plant virus movement 3:350–352
Microtus, cowpox virus hosts 1:577
Microviridae 5:11t
Microvirus(es) 3:13
  cell lysis 3:19
    protein E 3:19

  DNA packaging 3:19
    J proteins 3:19
  evolution 5:374
    caudoviruses *vs.* 5:374
    Gokushoviruses *vs.* 3:20
  gene expression 3:16
    mRNA stability 3:16–17
  genome content 3:14
  genome packaging 2:306
    nucleotide selection 2:307
  morphogenesis 3:17
    termination of 3:19
  structure 1:434f
    scaffold proteins 1:438
  virion morphology 3:14f
  see also Bacteriophage φX174
Middelburg virus (MIDV) 5:92
  characteristics 5:109t
  evolutionary relationships 5:97
Middle component (M), alfalfa mosaic virus 1:81–82
Middle East, HIV epidemiology 1:65
Midgut accessibility, nucleopolyhedrovirus primary infection 1:266
MIE transcription units, cytomegaloviruses 1:640
  promoter/enhancer regions 1:641f
Migration
  Dengue virus transmission 2:8
  rabies virus host range 4:372–373
Mild virus strains, papaya ringspot virus 4:4
Milk vetch dwarf virus (MDV)
  characteristics 3:389t
  genome 3:387f
  host range 3:386
  transmission 3:386
Milot-strain cross-pollination, definition 5:433
Milt, definition 3:83
Mimiviridae 3:312
Mimivirus 3:311–319
Mimivirus(es) 3:311–319
  amino acid composition 3:313–314
  classification 3:312
  evolution 3:318f
    gene homologs 3:318
  gene content 3:313–314
    duplications 3:314–315
    NCLDV-associated core genes 3:315
  genome 3:314f
    DNA repair-related genes 3:316
    DNA synthesis genes 3:316
    host signaling interfering pathway genes 3:316
    intein domains 3:317
    introns 3:317
    metabolic pathway-associated genes 3:316
    promoter structure 3:315
    translation-related genes 3:316
  history 3:312
  host range 3:312
  pathogenicity 3:312
  replication cycle 3:313
    infection scenario 3:313f
    transcription 3:313–314
  virion structure 3:317
    Gram staining 3:312f
    morphology 3:312f
    particle-associated mRNAs 3:318
    proteomics 3:317
  see also specific viruses
Mimotopes 1:140
  cross-reactivity 1:140
  definition 1:137
  screening 1:140
Min Jou sequence, icosahedral ssRNA bacteriophages 3:26
Mint virus X (MVX) 4:311t
Minus-strand synthesis, alfalfa mosaic virus RNA replication 1:84
Minus to positive strand switch, alfalfa mosaic virus RNA replication 1:86
Mirafiori lettuce virus (MiLV) 3:447
  coat protein 3:450
  phylogenetics 3:450

control, soil sterilization 3:453
genome 3:449
RNA2 3:449
RNA4 3:449
sequence 3:448–449
transmission, *Olpidium brassicae* 3:453
miRNAs *see* MicroRNAs (miRNAs)
Mites
mycoreovirus transmission 3:382–383
plant virus transmission 5:280
Mitochondria, nodavirus infection 3:436f
Mitochondrial apoptosis pathway
caspase activation 1:155–157
viral regulators 1:161
Bcl-2 family proteins 1:156f
Mitovirus 2:288
Mixed infections
anellovirus 1:109
aquareovirus 1:166
Pepino mosaic virus 4:107
see also specific infections
MK-0518 2:510
M-like protein, yellow head virus 5:477–478
MM virus 1:441–442
MMWR (Morbidity and Mortality Weekly report) 5:357t
ModA/ModB, bacteriophage T4 transcription 5:177
Modified live virus (MLV) vaccine
bovine viral diarrhea viruses 1:376
classical swine fever virus 1:531
infectious laryngotracheitis virus 2:406
Modular evolution model, bacteriophage evolution 5:372
Modulator of immune recognition (MIR), KSHV immune evasion 3:200
Mokola virus 4:184t
Molecular assays
Kadipiro virus 4:544–545
pecluvirus infections 4:99
seadornavirus infections 4:544–545
Molecular biology, techniques 2:456
see also specific methods
Molecular chaperones 3:348
Molecular clock models
HIV origin 2:532
phylogenetic analysis 4:127
phylogeny 4:125
Molecular epidemiologic studies 2:143
definition 2:143
Molecularity
definition 3:413
infection neutralization 3:416
stoichiometry *vs.* 3:416
multihit 3:417
Poisson distribution 3:417f
Molecular piracy, KSHV 3:192
Molecular strain typing, definition 4:330
Molluscipoxvirus 2:136, 4:326–327
Molluscs, definition 4:561
Molluscum contagiosum virus (MOCV) 3:319–324
avipoxviruses *vs.* 2:276
classification 3:319
epidermal growth factor receptor expression 3:320
evolution 3:320
genetic subtypes 3:320
genome 3:321f
poxvirus homology 3:320
history 3:319
infection
acanthoma 3:322
bacterial superinfection 3:322
clinical features 3:322
diagnosis 3:323
differential diagnosis 3:323
eye 2:496
host range 3:321
immune response 3:323
incidence 3:322
outbreaks 3:322
pathology 3:322
prevention 3:323

transmission 3:322
treatment 3:323
nonstructural proteins 3:320
propagation 3:321
survival factors, glutathione peroxidase 2:275–276
variola virus *vs.* 2:276
virion morphology 3:320f
Molluscum lesion 3:319
appearance 3:322
composition 3:322
persistence 3:322
Molting inhibition, baculovirus infection 1:252
Mongooses, rabies virus host range 4:373
Monitoring, phylogeny 4:127–128
Monkey(s)
SV40 hosts 4:636
see also specific types
Monkeypox virus (MPXV) 4:641
classification 4:641
ecology 4:643
evolution 4:643
future work 4:644
genetics 4:643
geographic expansion 4:643
Sudan 4:643
USA 4:643
historical aspects 4:641
infection *see below*
transmission 5:491
virion structure 4:642f
Monkeypox virus infection
disease surveillance 5:491
epidemics 5:491
fatality rate 5:491
host range 4:643
new hosts 4:643
humans 4:641
animal contact 4:642
clinical features 4:641
diagnostics 4:644
epidemiology 4:641
misdiagnosis 4:641–642
person-to-person transmission 4:642f
smallpox *vs.* 4:641
treatment 4:643
zoonoses 5:491
Monocistronic mRNA, definition 4:623
Monoclonal antibodies (MAbs)
avipoxvirus detection 2:280
Monoclonal antibodies, bean common mosaic virus 1:292
definition 4:623
Ebola virus infection treatment 2:64
HIV-1 vaccine development 1:141
passive immunization 3:67
potato virus Y 4:290
TMV structure elucidation 5:58
transgenic plant expression 4:232f
TMV utilization 4:236–237
use in epidemiology 2:143
watermelon mosaic virus 5:436
zucchini yellow mosaic virus 5:436
Monocotyledons
definition 1:533
mastrevirus infections 4:166
resistance genes, recessive 4:178
Monocyte chemotactic protein(s) (MCP) 1:623–624
Monocytes
Ebola virus infection 2:202–203
innate immunity 3:112
murine CMV infection 1:630t
rhinoviral infections 1:547
Monomer circularization mechanism, avsunviroids 4:479–480
Mononegavirales 3:325, 4:40, 5:291
Mononegavirus(es) 4:47
assembly 3:333
budding 3:333
definition 3:447
genome 3:329
gene end signals 3:329
gene start signals 3:329

intergenic regions 3:330
promoter sequences 3:329
glycoproteins 3:332
host cell entry 3:333
mRNA splicing 3:330
phylogenetic tree 3:328f
replication 3:330–331
RNA replication 3:330
transcription 3:330
initiation 3:330
polymerase 3:330
termination 3:330
viral proteins 3:331
glycoproteins 3:332
L protein 3:332
matrix proteins 3:331
nonstructural proteins 3:332
nucleocapsid 3:331
phosphoprotein 3:331
P mRNA encoded proteins 3:331
virus structure 3:329f
nucleocapsid complex 3:325–329
see also specific viruses
Mononuclear phagocytes, Dengue virus infection 2:10
Monopartite genomes
geminiviruses 4:165
potyviruses 4:314, 4:316–317
Monopartite virus(es) 4:164
see also specific viruses
Monophyletic group, definition 2:105
Mopeia virus 3:244
Morbillivirus 3:285, 4:41t, 4:498
Morbillivirus(es) 3:285, 4:47, 4:498
classification 4:41t
evolution 4:501–502, 4:501f
genome 3:327f, 4:500–501, 4:500f
encoded proteins 4:42–43, 4:500–501
genome length requirement 4:501
mRNA synthesis 4:500–501
nonstructural proteins 4:501
host cell receptors 4:500
infection
cytopathic effects 4:500
incubation period 4:504
route 4:45–46
sites 4:45–46
phylogeny 4:45
propagation 4:40–42, 4:500
replication, host cell attachment 4:44
serology, H protein 4:502
signaling lymphocyte activation molecule 3:285
transmission 4:45–46
see also specific viruses
Morphogenesis, baculoviruses 1:251
Mosquito(es) 2:235–236
cell lines 4:539
cg4715 protein, lineage 4:454
disease control methods 5:116
larval control 2:13
vertical transmission 2:8
Mosquito(es), as vectors
Akabane virus 1:77, 1:78–79
alphavirus 5:99, 5:111–112
bovine ephemeral fever virus 1:355
Chikungunya virus 5:111–112
control 2:13, 5:449
Dengue virus 2:5, 2:7
eastern equine encephalitic virus 5:78, 5:101
Eastern equine encephalitis virus 5:111–112
equine encephalitic togaviruses 5:102, 5:105, 5:106
flaviviruses 2:236
Israel turkey meningoencephalomyelitis virus 2:237
Japanese encephalitis virus 2:236, 3:184, 3:186
Saint Louis encephalitis virus 4:656
seadornavirus 4:537
Sitiawan virus 2:237
St. Louis encephalitis virus 2:236
Tataguine virus 1:400
Venezuelan equine encephalitic virus 5:111–112, 5:112f

Mosquito(es), as vectors (*continued*)
    Wesselsbron virus 2:237
    western equine encephalitic virus 5:77–78, 5:101
    West Nile virus 2:236, 5:443–444, 5:445
    yellow fever virus 2:238
    zoonoses 5:494
    see also individual species
Mosquito baculovirus(es)
    auxiliary functions 1:224
    see also specific viruses
    classification 1:220, 1:220t
    DNA replication 1:223
    epizootiology 1:222
    genome 1:222
    histopathology 1:220
    history 1:219–220
    host range 1:221
    life cycle 1:221
    occlusion bodies 1:220
        properties 1:220
        structure 1:221f
    pathology 1:220
    phylogeny 1:224, 1:224f
    RNA transcription 1:222
    structural functions 1:223
    transmission 1:221
    virion functions 1:223
    virion occlusion 1:223
    virion properties 1:220
Mosso das Pedras virus (MDPV) 5:109t
Mossuril virus (MOSV) 4:184t
Most recent common ancestor (MRCA), phylogenetic trees 4:126
MotA, bacteriophage transcription role 5:177
Moth(s), Hz-2V infection 3:147
Mother-to-child transmission, HIV infection 1:60–61
Mount Elgon Bat Virus (MEBV) 1:118
Mouse/mice
    lactate dehydrogenase-elevating virus infection 1:183
    models
        coxsackievirus isolation 1:582
        Ebola virus infection 2:62
        rhabdovirus isolation 1:111
        SARS infection 4:558
        VEEV infection 5:115
    strains, Ectromelia virus susceptibility 3:343
    transgenic *see* Transgenic mice;
    see also under Murine
Mouse encephalomyelitis virus(es) *see* Theiler's murine encephalomyelitis virus (TMEV)
Mouse-encephalomyelitis virus (ME virus) 1:441–442
Mouse hepatitis virus (MHV)
    replication 3:427
    RNA recombination 1:560
Mouse leukemia virus (MLV)
    mammalian cell infection 2:469
    mammalian cell resistance 2:469
    resistance 2:469
Mouse mammary tumor virus (MMTV) 3:334–342
    capsid protein 3:335
    cell entry, transferrin receptor 1 3:336
    classification 3:335
    endoribonuclease activity 3:222
    future work 3:341
    genetic experimentation 3:339
    genetics 3:338
    genome 3:335, 3:335f
        long terminal repeats 3:335–336
        precursor polypeptides 3:335–336
        primer-binding site 3:335–336
        repeat regions 3:335–336
        splice acceptor 3:335–336
        splice donor 3:335–336
    history 3:335
    infection *see below*
    integrase 3:335
    middle T, breast cancer model 4:274
    nomenclature 3:335
    nucleocapsid protein 3:335
    replication 3:336
        frameshifts 3:336–337, 3:337f
        host genome integration 3:336
        intracytoplasmic A particles 3:336–337
        long terminal repeats 3:336
        precursor protein translation 3:336–337
        preintegration complex 3:336
        Rem protein 3:336–337
        reverse transcriptase 3:336
        RNA polymerase II 3:335f, 3:336
        sag gene 3:335f, 3:336
        termination 3:336
    reverse transcriptase 3:335
    RNP 3:335
    susceptibility 3:338
    taxonomy 3:335
    tissue tropism 3:337
        Cut-like protein1/CCAT displacement protein 3:340
        mammary gland enhancer 3:341
        negative regulatory elements 3:340
        special AT-rich binding protein 1 3:340
    transcriptional regulation 3:340
        glucocorticoids 3:340
        hormone-regulated gene expression 3:338f, 3:340
        transcription factor IID 3:340
    transmission 3:337, 3:338f
        horizontal 3:337
        intestinal tract 3:337
        maternal milk 3:337
        MHC class II 3:337
        T-cell receptor 3:337
        vertical 3:337
    virion 3:335
        capsid protein 3:335
Mouse mammary tumor virus infection
    immune response 3:341, 3:341f
        Sag proteins 3:341, 3:341f
        T cells 3:341
    mammary adenocarcinomas 3:338–339
    pathogenicity 3:339
        insertional mutagenesis 3:339
        integration sites 3:339–340
        T-cell lymphomas 3:339–340
    resistance to 3:339
        neutralizing antibodies 3:339
        Toll-like receptor 4 signaling 3:339
    susceptibility 3:338
        cellular factors 3:339
        host factors 3:339
        MHC class II subtypes 3:339
        mouse strains 3:338–339
Mouse minute virus (MMV), evolution 2:182–183
Mousepox virus *see* Ectromelia virus
Movement disorders, West Nile virus infection 5:447
Movement proteins (MPs) 3:349, 3:350t, 5:145
    alfalfa mosaic virus 1:82–83, 1:85
    banana bunchy top virus 1:274
    caulimovirus 1:465
    definition 3:264, 3:348
    flexivirus phylogenetics 2:256
    functional characteristics 3:349
    furovirus 2:292
    Maize streak virus *see* Maize streak virus (MSV)
    potexvirus 4:312
    tombusvirus replication 5:148–149
    tospoviruses 5:134–135
Movement restriction, African horse sickness control 1:42
MP-12 vaccine, Rift Valley fever virus 4:496
M protein
    measles virus 3:287
    Sigma virus structure 4:576–577
MPyV, receptors 4:279
MRC-5 vaccine, Rift Valley fever virus 4:496
MRE11-RAD50-NSB1 (MRN) complex *see* Adenovirus infections
M RNA segment sequence, Crimean-Congo hemorrhagic fever virus 1:599, 1:600–601
M trait, baculovirus infection resistance 1:267
Mu bacteriophage *see* Bacteriophage Mu
Mucambo virus (MUCV), characteristics 5:109t
Mucins
    definition 2:543
    Glc1.8 4:254
Mucosa-associated lymphoid tissue (MALT)
    definition 2:534
    edible vaccine development 5:222
    HIV infection, role 2:538f, 2:539–540, 2:541
Mucosal disease (MD) 1:374
Mucosal edema 1:328
Mucosal immunity, vaccination 5:223
Mulberry ringspot virus (MRSV) 3:406t
Muller's ratchet 2:179, 4:359
    definition 2:175
    VSV evolution 5:297
Multianalyte diagnostic techniques 2:35–36
Multicentric Castleman's disease (MCD) 3:194, 3:195, 3:201, 5:196
Multicentric fibrosarcoma(s), feline leukemia viruses 2:188
Multi-dose vials, vaccine safety 5:229
Multigene families 1:43
    African swine fever virus 1:48
Multinuclear zinc-binding protein (ZBD), yellow head virus 5:477–478
Multiorgan failure, yellow fever vaccine 5:234
Multiple DNA-B components, mungbean yellow mosaic viruses 3:366
Multiple embedded nucleopolyhedrovirus (MNPV) 1:248
Multiple homologous repeat regions, baculoviruses 1:249
Multiple nucleocapsid per envelope (MNPVs), nucleopolyhedrovirus packaging 1:265, 1:266f
Multiple sclerosis (MS)
    animal model 5:43–44
    hepatitis B vs. 5:233
    human herpesvirus infection 2:504
    viral etiology 1:475
Multiple serotype vaccines 5:240
Multiple transcription initiation sites, mungbean yellow mosaic viruses 3:367
Multiplicative landscape
    definition 4:359
    error catastrophe 4:362
    quasispecies 4:362
Multiplying (M-type) infection, tobraviruses 5:73
Multivalent vaccines, measles virus *see* Measles virus
μ1 protein, reovirus structure, outer capsid 4:394f, 4:396
μ2 protein, reovirus structure, inner capsid 4:393–394
Mumps
    clinical features 3:356, 3:361
        associated organs 3:362t
        incubation period 3:361
        prognosis 3:362t
    disease surveillance 2:45
        data collection sources 2:45
    epidemiology 3:361
        acquired immunity 3:361
        genotypes 3:361
        history 3:361
        incidence 3:362t
    histopathology 3:357f, 3:361
    host range 3:360
        experimental 3:360
    immune response 3:362
        cell-mediated 3:362–363
        humoral 3:362–363
    nervous system infection 1:469
    ocular complications 2:495
    pathogenicity 3:362
        experimental infection 3:362
        neurovirulence 3:362
    pathology 3:361, 3:362t
        isolation 3:361
    prevention and control 3:363
    serology 3:360
Mumps virus (MuV) 3:356–363
    chemical sensitivity 3:356
    classification 3:356
    distribution 3:360
    evolution 3:360

genetics 3:360
  reverse 3:363
genome 3:327f, 3:357, 3:358f
history 3:356
  epidemiology 3:361
inclusion body myositis and 3:363
infection see Mumps
propagation 3:360
  cytopathic effect 3:360
protein properties 3:357, 3:358t
  small hydrophobic protein 3:357, 3:360
RdRp 3:358–359, 3:359–360
replication 3:357
  budding 3:359–360
  co-transcriptional editing 3:359
  host cell attachment 3:357–358
  Ig sequences 3:359, 3:359f
  site of 3:361
  transcription 3:358–359
  translation 3:359
SH protein 4:55
taxonomy 3:356
tissue tropism 3:361
transmission 3:361
vaccines/vaccination 3:361, 3:363
  Jeryl Lynn strain 3:363
  see also Measles/mumps/rubella (MMR) vaccine
virion properties 3:356
  physical 3:357
  structure 3:356, 3:357f
Mungbean yellow mosaic India virus (MYMIV) 3:364
  DNA β component associated 1:320
  genome 3:365–366, 3:365f
  phylogenetic relationships 3:370–371
  replication, Rep protein 3:366
  transmission 3:368
  as type species 3:364
Mungbean yellow mosaic virus(es) 3:364–372
  classification 3:364
  coat protein 3:364–365
  ecology 3:368
  ELISA 3:364–365
  epidemiology 3:368
  gene expression regulation 3:366
    BC1 protein 3:367
    multiple transcription initiation sites 3:367
    Rep protein 3:367
  genome 3:365–366, 3:365f
    common region 3:366
    multiple DNA-B components 3:366
    open reading frames 3:365–366
  geographical distribution 3:364
  infection see below
  infectivity 3:367
    Agrobacterium tumifaciens experiments 3:367
  phylogenetic relationships 3:368t, 3:370–371
    nucleotide sequence 3:369t, 3:371
    phylogenetic tree 3:371, 3:371f
  phylogenetic trees 3:371f
  replication 3:366
    Rep protein 3:366
    rolling circle replication 3:366
  taxonomy 3:364
  transcript analysis 3:366–367
    RT-PCR 3:366–367
  transmission 3:368
    acquisition access period 3:368
    inoculation access period 3:368
    vector lifecycle 3:368
    whitefly 3:366, 3:368
  as type species 3:364
  virion structure 3:364
    coat protein 3:364–365
  see also specific viruses
Mungbean yellow mosaic virus infections
  host range 3:367
  management 3:369
    cultural practices 3:369–370
    disease resistance 3:369
    RNAi 3:370
  symptoms 3:364, 3:365f

Mupapillomavirus 4:9, 4:10f, 4:11t
MurA
  inhibition 3:256–257, 3:256f, 3:257–258
  missense mutation 3:257–258
Murein, synthesis inhibition 3:256–257, 3:256f
Murine cytomegalovirus(es) (MCMV) **1:624–634**
  classification 1:635–636, 1:635t
  genome 1:626, 1:627f, 1:636–637
    Smith strain 1:626
  laboratory strains 1:639
  latency 1:629, 2:440
    transcription regulation 1:629
  reactivation 1:629
  replication 1:627–628, 1:628f
    capsid formation 1:628–629
    origin of replication 1:627f, 1:628
    transcription regulation 1:629
  virion morphology 1:625
  virion structure 1:625
    capsid 1:625
    envelope 1:626
    glycoproteins 1:626t
    tegument 1:625–626
Murine cytomegalovirus infections
  antibodies 1:631
  associated diseases 1:629
    affecting factors 1:629
  fibroblasts 3:321–322
  host immune response 1:629, 1:630t
    associated genes 1:630
    immune system evasion 1:629, 1:630t
  pathogenesis 1:629
  seroprevalence 1:629
Murine gammaherpesvirus 68 (MHV-68) **3:372–378**
  ecology 3:372
  egress 3:374
  genome 3:373, 3:373f
    open reading frames 3:373
  host cell entry 3:375
  infection see below
  origins 3:372
  replication 3:374
    essential genes 3:374
    nonessential gene expression 3:374
    transcription 3:374
  spread 3:375, 3:376
  strains 3:373
  structure 3:373
  unique viral genes, expression 3:373, 3:374
    clusters 3:373–374
    M1 3:374–375
    M2 3:374–375
    M3 3:374–375
    M4 3:374–375
    vtRNA 3:374–375
Murine gammaherpesvirus 68 infection 3:374
  associated cell types 3:375
  B cells 3:374
    lymphomas 3:377
    M2 gene 3:375
    mediastinal lymph node 3:375
    S11 cell line 3:377
  inflammatory response 3:375
  influencing host factors 3:376
  latency 3:376
    associated cell types 3:376
    M2 gene 3:376–377
    NS0 cell line 3:374
    ORF73 3:376–377
  pathogenesis 3:376f, 3:377
    lymphoma 3:377
    splenomegaly 3:377
Murine hepatitis virus (MHV) 1:553
Murine leukemia virus(es) (MLVs)
  defective interfering viruses 2:4
  evolution 2:183–184
  xenotropic behavior 2:468
Murine norovirus(es) 1:418
  see also specific viruses
Murine polyomaviruses see Polyomavirus(es), murine
Muromegalovirus 1:635–636, 2:432t

Murray Valley encephalitis (MVE) virus
  characteristics 1:171t
  geographical distribution 5:493
  seasonal distribution 5:493
  zoonoses 5:493
Muscle
  Akabane virus infection 1:78
  togavirus infection 5:83–84
Mushroom bacilliform virus (MBV) 1:287, 1:510, 2:288
  evolutionary relationships 1:287, 1:288f
  host range 1:288
  phylogeny 4:652
  transmission 1:288
  virion properties 1:287
  virion structure 1:287, 1:287f
Muskrats, Omsk hemorrhagic fever virus 2:239–240
Mutagenesis, insertional see Insertional mutagenesis
Mutagens, fungal retrotransposons 4:426
Mutation(s)
  antiviral therapy 1:150
  chronic hepatitis B 2:358
  evolution 4:125
  LTR retrotransposons 4:439
Mutational resistance, antibodies 5:323–324
Mutational robustness 4:359
  quasispecies see Quasispecies
Mutation rates
  rhinoviruses 4:470
  RNA viruses 4:377
  viral evolution 2:177
Mutation–selection balance 4:359
Mx proteins, interferon-induced antiviral response 3:107, 3:115–116
Myalgia 1:399
  definition 1:580
Mycoreovirus 2:286, 3:378, 3:379f, 3:380, 4:536t
Mycoreovirus(es) **3:378–383**
  gene expression, fungal 3:383
  genome 3:380
    encoded proteins, functions 3:380
    segments 3:380, 3:381f
  genomic relationships 3:380
  infection 3:379–380
    asexual spores 3:380
    egress 3:379–380
    μ1 protein homologs 3:379–380
    mixed 3:381
    movement, hyphal anastomosis 3:380
  nomenclature 3:380
  structure–function relationships 3:378
    guanyltransferase 3:378–379
  taxonomy 3:378, 3:379f, 3:380
  transmission 3:380, 3:382–383
  see also specific viruses
Mycoreovirus-1/Cp9B21 2:285, 3:381
  double infection 3:381
  gene product expression 3:381
  host, effects 3:383
  transmission 3:381
  as type species 4:536t
  virulence 3:381
Mycoreovirus-2/CpC18 3:382
  host, effects 3:383
  infected culture phenotype 3:379f, 3:382
  MyRV-1/Cp9B21 vs. 3:382
Mycoreovirus-3/RnW370 3:382
  composition 3:382
  transmission 3:382–383
  white rot control 3:382
Mycovirus(es) 4:63, 4:68–69
  classification 4:63
  definition 2:109, 2:284, 4:68
  dsRNA genomes 2:286
  infectivity 2:290
  ssRNA genomes 2:288
  transfection 4:66
  unassigned ssRNA viruses, genomes 2:288
  see also Chrysovirus(es); specific viruses
Myelin deficiency, Border disease 1:336–337
Myeloid dendritic cells (mDC) 3:112
  type I interferon induction 3:115

Myiasis, definition 1:427
Myocardiotropism 1:440
Myocarditis
  coxsackie B virus 2:68–69
  definition 1:440, 1:533
Myositis *see* Human T-cell leukemia virus 1 infection
Myoviridae 3:30–31, 5:11*t*, 5:413*t*
Myovirus(es) 3:32*t*
  Euryarchaeota infections 5:413*t*
  genome packaging 2:306, 2:307*t*
  long contractile tails 5:367
  structure 1:434*f*, 3:31
    baseplates 1:437–438
    tails 1:437–438
  tails/tail fibers 3:32
  see also specific viruses
Myristic acid, capsid stabilization 2:128
Myrobalan latent ringspot virus (MLRSV) 3:406*t*
Myxomatosis 4:329
  clinical features 3:228–229
    primary tumors 3:228–229
    secondary tumors 3:229
  mortality rates 3:225, 3:228
  pathogenicity 3:228
  rabbit eradication programs 3:228
Myxoma viruses (MYX) 4:329
  characteristics 3:225*t*
  history 3:225
  infection
    apoptosis inhibition 1:161
    immune modulation 3:227
    immune response 3:230
    viral migration 3:229
    see also Myxomatosis
  virulence 2:146, 3:228

N

N1a gene/protein 4:3
N1b gene/protein 4:3
N4 bacteriophage *see* Bacteriophage N4
NADL strain, bovine viral diarrhea viruses 1:376–377
Nairobi sheep disease virus (NSDV) 1:391*t*, 1:596
  historical aspects 1:596, 1:597
  Rhipicephalus appendiculatus tick vector 1:598
Nairovirus 1:391, 1:391*t*, **1:596–603**, 3:480, 4:490–491
Nairovirus(es) 1:399
  history 1:596
  M segment-encoded polyprotein 1:395
  species 1:596, 1:598*t*
Nanoscale material manufacture, plant virus vector applications 4:236
Nanoviridae 1:274, 3:385, 5:11*t*
Nanovirus 3:385
Nanovirus(es) **3:385–391**
  aphid transmission 3:386
  associated DNAs 4:533
  biological properties 3:386
    host range 3:386
    tissue tropism 3:386
  characteristics 3:389*t*
  economic importance 3:386
  genome 1:315–316, 3:387, 3:387*f*
    DNA structure 3:387
    DNA types 3:388, 3:389*t*
    integral genome segments 3:388
    open reading frames 3:390
    satellite-like rep proteins 3:388, 3:390*f*
  geographical distribution 3:385
  history 3:385
  phylogenetic relationships 3:391
    circoviruses *vs.* 3:391
    geminiviruses *vs.* 3:391
  proteins 3:387*f*, 3:389*t*, 3:390
    DNA-C 3:390
    DNA-M 3:390
    DNA-N 3:390
    functions 3:387
    nonstructural 3:390
    nuclear shuttle protein 3:390
    rep 3:390

Rep-1 satellites, DNA-1 component *vs.* 1:565
replication 3:390–391
  DNA replication 3:390–391
  dsDNA synthesis 3:390–391
  mRNA transcription 3:390–391
transmission 3:386
  aphids 3:386
trans-replication 1:315–316
virion properties 3:385
  serology 3:386
  structure 3:386*f*
see also specific viruses
Narcissus mosaic virus (NMV) 4:311*t*
  genome 4:312
Narnaviridae 2:288, 3:392, 5:11*t*
Narnavirus 2:288, 3:392
Narnavirus(es) **3:392–398**
  cis-acting replication signals 3:394
    negative strands 3:395–396
    positive strands 3:394–395, 3:395*f*
  double stranded RNA satellites 4:531
  gene expression 2:289
  genome 2:288, 3:392, 3:393*f*
  historical background 3:392
  infections *see below*
  in vivo generation 3:394
  proteins 3:393, 3:393*f*
  replication intermediates 3:394
    RdRp 3:394
  virion lack 3:392
  see also Ribonucleoprotein (RNP)
Narnavirus infections
  fungi 2:287*t*
  host persistence 3:392, 3:396
    mRNA degradation 3:396, 3:397*f*
    'top-half' domain 3:395*f*, 3:398
Nasal infection route 5:155
Nasal T-cell lymphoma 5:195
Nasopharyngeal carcinoma 5:195
  anaplastic 5:195
  incidence 2:155, 5:195
  tumor characterization 2:155
  see also specific virus infections
National Bioforensics Analysis Center 5:410
National Center for Biomedical Ontology 5:357*t*
National Center for Biotechnology Information (NCBI) 5:356
  Assembly Archive 5:357*t*
  database 5:357*t*
  Entrez search engine 5:356–362
  pairwise sequence comparison, implementation 5:344
    PASC interface 5:344
  PubMed search engine 5:356–362
  Reference Sequence database 5:362–363
  Trace Archive 5:357*t*
  Viral Genome Resource 5:357*t*
  Viral Genomes Project 5:362
  Viral Genotyping Tool 5:357*t*
National Center for Health Statistics, common cold 1:541
National Immunization Days (NIDs), polio vaccination 4:245–246
Natrialba magadii, φCh1 infection 5:418
Natsudaidai dwarf virus (NDV) 4:523, 4:524*t*
  genome sequencing 4:524–525
  host range 4:525
  transmission 4:525
Natural *cis*-antisense transcript-derived siRNA pathway 4:142
Natural killer (NK) cells 3:110, 3:112
  definition 3:78
  EBV-associated lymphoma 2:155, 5:195
  function 3:110
  lymphocytic choriomeningitis virus infection 3:242
  murine CMV infection 1:630–631, 1:630*t*
  regulation 3:110
  yellow fever 5:475
Natural resistance, vegetable viruses 5:286–287, 5:287*t*
Natural selection 3:402

Navel orange infectious mottling virus (NIMV) 4:523, 4:524*t*
  genome sequencing 4:524–525
  host range 4:525
  transmission 4:525
NB-LRR protein, definition 4:170
NCBI *see* National Center for Biotechnology Information (NCBI)
ND10/POD domains, KSHV replication 3:199
Ndumu virus (NDUV) 5:109*t*
Necrosis
  apoptosis *vs.* 1:157
  definition 1:154–155, 4:207
  measles 3:289
Necrovirus 3:403, 5:145–146
Necrovirus(es) **3:403–405**
  genome 3:403, 3:403*f*
    open reading frames 3:403
    positive-sense RNA strand 3:403
  geographical distribution 3:404
  infection, host range 3:404
  interspecific relationships 3:404
    serology 3:404
  necrovirus, tobacco disease 5:66
  satellite viruses 3:404, 4:527*t*
  structure 3:403
  transmission 3:404
    seeds 3:404
    soil 3:404
    vectors 3:404
  type species 3:403
  virion particles 3:403
  virus–host relationships 3:404
  see also specific viruses
Needleman–Wunsch algorithms (classification) 5:343
nef gene, simian immunodeficiency virus 4:622
Nef protein (HIV) 2:523
  expression 2:524
  functions 2:523–524
Negative regulatory elements, mouse mammary tumor virus 3:340
Negative sense genome, definition 5:157
Negative staining 2:79, 2:80*f*
  advantages 2:81
  heavy metal salts 2:79
  image analysis 2:85–86
  limitations 2:81
  positive staining *vs.* 2:79
  sample preparation 2:79, 2:80*f*
    concentration 2:79
  stains 2:79
    ammonium molybdate 2:81
    phosphotungstic acid 2:81
    sodium silicotungstate 2:79, 2:80*f*
    uranyl acetate 2:79, 2:80*f*
  support films 2:79
Negative-strand RNA virus(es)
  recombination 4:380
  see also specific viruses
Negishi virus 2:241
Negro coffee mosaic virus (NeCMV) 4:311*t*
Neighbor-joining methods
  definition 2:525
  phylogenetic analysis 4:126–127
Neisseria meningitidis, virulence factors 3:123
Nelfinavir 2:514*t*
Nelson Bay virus (NBV) 4:383
Nematicides, nepovirus control 3:412
Nematodes, plant virus transmission 5:280*f*, 5:281
  grapevine disease 4:203
  legume virus 3:217
  pea-early browning virus 3:217, 3:218
  tobacco rattle virus 4:308
Nemesia ring necrosis virus (NeRNV) 4:215, 4:219*f*, 5:200*t*, 5:204
  nemesia infections 4:215–217, 4:220*f*
  verbenaceae infection 4:218
Nemesia virus infections 4:215
  see also specific viruses
NendoU, nidovirus genome 3:425
Neocheck-S, insect pest control 3:130

Neodiprion abietis nucleopolyhedrovirus
  (NeabNPV) 1:225–226
Neodiprion lecontei nucleopolyhedrovirus
  (NeleNPV) 1:226
  gene parity plots 1:228, 1:229f
  genome 1:226, 1:227f
    auxiliary genes 1:228
    DNA repeat regions 1:229
    inhibitors of apoptosis 1:226
    membrane fusion proteins 1:228
    nonsyntenic region 1:226, 1:227f
    regulator of chromosome condensation 1:228
    trypsin-like serine protease 1:226–227
    zinc finger protein 1:228
  phylogeny 1:229
Neodiprion sertifer nucleopolyhedrovirus
  (NeseNPV) 1:226
  gene parity plots 1:228, 1:229f
  genome 1:226
    auxiliary genes 1:228
    DNA repeat regions 1:229
    inhibitors of apoptosis 1:226
    membrane fusion proteins 1:228
    nonsyntenic region 1:226, 1:227f
    regulator of chromosome condensation 1:228
    trypsin-like serine protease 1:226–227
    zinc finger protein 1:228
  phylogeny 1:229
Neonatal disease(s)
  coxsackieviruses 1:585
  HCMV 2:483
  see also specific infections
Neonatal immunization, maternally derived
  antibodies 2:53
Neoplasia
  definition 2:212
  reticuloendotheliosis virus infection 4:416
  simian immunodeficiency virus infection 4:608–609
  see also Cancer; Carcinogenesis
Neotopes
  definition 1:137–138
  TMV antigenicity 5:58
Neotropical virus(es) 5:489
Nephotettix cincticeps, rice dwarf virus transmission 1:483
Nephotettix impicticeps, rice tungro virus transmission 1:485
Nephropathia epidemica (NE) 2:317, 2:318
  clinical features 2:319–320
Nephropathy, polyomavirus-associated 4:269
Nephrotoxicity
  antiretroviral treatment 1:54–55
  antiviral therapy 1:147
Nepovirus 1:573, 3:405, 4:523
Nepovirus(es) 3:405–413
  cell-to-cell movement 3:407f, 3:410
  future research 3:412
  genera 3:406t
  genome 3:407, 3:408f
    sequence identity 3:408, 3:408f
    untranslated regions 3:408
    viral protein linkage 3:407–408
  historical perspective 3:405
  infections see below
  long distance movement 3:410
  particle structure 3:407, 3:407f
  population structures 3:412
    sequence diversity 3:412
  regulated polyprotein processing 3:409
    NTB domain 3:409
    proteinases 3:409
  replication 3:409
    membrane proliferation 3:409
    NTB protein 3:409–410
    VPg domain 3:409–410
  satellites 3:410
    type D 3:411
    VPg links 3:410–411
  transmission 3:411
    aerial vectors 3:411
    seed transmission 3:411
    soil nematodes 3:411
    vector species specificity 3:411
  see also specific viruses

Nepovirus infections 3:411
  control 3:412
    infected plant removal 3:412
    nematicides 3:412
    transgenic plants 3:412
  economic impact 3:411
  host range 3:410
  infectious DNA clones 3:408–409
  petunia infection 4:213
  plant post-transcriptional gene silencing 3:410
  symptomatology 3:410
  tobacco disease 5:65
Nerine virus X (NVX) 4:311t
Nerve terminals, CNS invasion pathways 1:470
Nervous system virus(es) 1:469–475
  acute infections 1:474
  see also Encephalitis
  anatomic considerations 1:469
    blood–brain barrier 1:469–470
    ependymal cells 1:470
    extracellular environment 1:469–470
    neuron membranes 1:470
  cell damage mechanisms 1:473
    immune responses 1:473
    lysis 1:473
  clinical features 1:474
  CNS invasion pathways 1:470, 1:471f
    along nerves 1:470
    anterograde transport 1:470
    blood-bourne 1:470, 1:472f
    blood–brain barrier 1:471
    capillary endothelia 1:471–472
    choroid plexus 1:471–472
    experimental systems 1:470
    nerve terminals 1:470
    olfactory spread 1:470–471
    particle size vs. 1:471
    reticuloendothelial cells 1:471
  neural cell infection 1:473
    age-relation 1:473
    axoplasmic flow 1:473
    cell-to-cell spread 1:473
    extracellular antibody neutralization 1:473
    oligodendrocytes 1:473
    vascular endothelial cells 1:473
  physiologic considerations 1:469
    antibodies 1:470
    T cells 1:470
  slow infections 1:474
    dementia 1:474
  see also Neuroinvasive disease; specific viruses
Neuraminidase (NA)
  definition 1:127
  function 1:149
  human parainfluenza viruses 4:47–48
  infection neutralization 3:413–414
  influenza viruses see Influenza virus
    neuraminidase
  mumps virus 3:356
Neuraminidase inhibitors 1:149
Neurodegenerative disease, JCV
  infection 4:262
Neuroinvasive disease 4:652
  definition 1:469, 5:440
  Japanese encephalitis virus 3:185
  West Nile virus 3:466
Neurological disease
  Border disease 1:336
  equid herpesvirus 1 infection 2:417–418
  henipavirus infections 2:325
  Nipah virus infection 2:325
Neuronal cell death, prions 4:336
Neuronal inclusions, rabies virus 4:371
Neuropsychiatric changes, Crimean-Congo
  hemorrhagic fever 1:601
Neurospora crassa VS ribozyme 4:480
Neurotropic virus(es)
  definition 1:441, 1:469
  see also specific viruses
Neurovirulence test(s)
  definition 5:226
  mumps virus pathogenicity 3:362

Neurovirulent virus(es)
  definition 1:469
  see also specific viruses
Neutralization see Infection Neutralization
Neutralization assays
  adenovirus 1:5
  antigen production measurements 3:414
  defective viruses 3:414
  echovirus 2:69
  human papillomavirus 4:17
  orthobunyavirus 3:482
  quantal assays 3:414
  steps 3:414
  West Nile virus 5:448
  yellow fever virus 5:470
Neutralization escape 3:417
  foot-and-mouth disease virus 3:418
  HIV-1 3:418
Neutralization resistance see Resistance
Neutralizing antibodies (NAbs) 3:64
  avian retroviruses 4:458
  classical swine fever virus 1:531
  definition 1:127, 2:534
  effectiveness 3:64
  experimental measurement 3:414
  flavivirus infection 3:66
  HIV infection see HIV infection
  human herpesvirus infection 2:504
  human papillomavirus infection 4:13–14
  immunoglobulin isotypes 3:65–66
  influenza viruses 1:129–131
  Marburg virus infection 3:279
  mechanism, replication blocking 3:415
  mouse mammary tumor virus infection 3:339
  picornaviruses 1:138
  residual infectivity 3:415
    'persistent fraction' 3:415
  Tat protein 1:72
  tests for
    principles 2:31
    residual infectivity 2:31
  vaccine development 1:141
    HIV gp41 1:141
  virus binding 3:64–65
Neutralizing monoclonal antibodies (NmAbs),
  definition 1:69
Nevirapine 1:143t
  limitations 2:510
Newcastle disease virus 5:316
  genome 3:327f
  infection 4:45–46
  tissue tropism 4:46
  vaccine development
    glycoproteins 5:224
    plant-produced vaccine 5:224
New England Primate Research Center (NEPRC) 4:603
New Minto virus (NMV) 1:116
NF-1 protein 4:264, 4:281
NF-1X protein 4:281
NFκB
  adenovirus malignant transformation 1:12–13
  apoptotic role 1:156f, 1:157–160
  BV-mediated immunosuppression 4:254, 4:255
  definition 1:154
  JCV gene expression 4:264
  KSHV latency 3:198–199
  oncogenicity role 4:417
Ngaingan virus 4:184t
Ngari virus
  genome segment reassortment 1:395
  strain reassortment 3:482
N gene(s)
  plant resistance 4:173
  tobacco resistance 4:176
NIa protease 4:321
NIa protein 4:317–319
NIb protein 4:319
Nicotiana benthamiana
  satellite virus pathogenicity 1:318
  virus-induced gene silencing 5:376, 5:378
    gene silencing 4:146–147, 4:148
  virus vector research 4:231

Nicotiana glutinosa 5:68
Nicotiana rustica 5:60
Nicotiana tabacum see Tobacco *(Nicotiana tabacum)*
Nicotiania glutinosa, viral assays 5:68
Nicotine vaccine, three-hybrid assay 3:28–29
Nidovirales 1:177, 1:549, 2:230, 3:419, 3:420f, 5:11t, 5:23, 5:152
Nidovirus(es) 1:177, **3:419–430**, 3:420f, 5:11t, 5:23
  definition 1:549
  evolution 1:182
  families 1:549, 4:553–554
  fish 2:230
  gene expression 3:423
  gene transcription 3:428
    sgRNAs 3:428
    transcription–regulation sequences 3:428
  genome 3:419–420, 3:423, 3:424f
    accessory protein genes 3:426
    ADP-ribose-1″-phosphatase 3:425
  infection
    apoptosis 3:429
    disease associated 3:421
    host cell affects 3:429
  nucleocapsid protein 3:421–423
  origin 3:428
  phylogeny 3:419, 3:421f
  proteolytic processing 3:423
    3C-like protease 3:424
    accessory proteases 3:424
  replicase 3:423, 3:425f
    3′–5′ exoribonuclease 3:424–425
    ADP-ribose-1″-phosphatase 3:425
    NendoU 3:425
    nucleotide cyclic phosphodiesterase 3:425
    ribose-2′-O-methyltransferase 3:424–425
  replication 3:427
    cis-acting sequences 3:427
    double membrane vesicles 3:426
    host factors associated 3:427
  structural protein genes 3:426
  virus structure 3:421, 3:422f
    structural proteins 3:423
  see also specific viruses
Niemann–Pick disease, definition 1:541
NIH Vaccine Research Center 5:357t
Nilaparvarta lugens reovirus (NLRV)
  genome 4:151, 4:152t
  pathogenicity 4:155
Nilaparvata bakeri, rice ragged stunt virus transmission 1:485
Nilaparvata lugens, rice ragged stunt virus transmission 1:485
Nilaparvata lugens reovirus (NLRV) 4:150t
Nimaviridae 1:247, 4:567, 5:11t, 5:451
Nipah virus (NiV) 4:40, 4:52
  epidemiology 2:324
  evolution
    Bangladesh 2:324
    Malaysia 2:323–324
  fusion proteins 4:54
  genetic stability 2:179
  global distribution 4:45
  historical aspects 2:321
    classification 4:128
  host cell attachment 4:44, 4:53–54
  host range 2:554
  infection *see below*
  transmission 2:324
Nipah virus infection
  clinical features 2:325, 4:56–57
    neurological 2:325
    prolonged infection 2:325
  human respiratory infection 2:554
  immune system evasion, IFN targeting 3:108
  mortality rate 4:56–57
  pathology 2:325
  risk factors 2:324
  serology 2:323
    antibodies to 2:326
  zoonoses 2:95, 5:492
Nitazoxanide 4:512

Nitrocellulose enzyme immunoassay (NC-EIA) 5:482
4-Nitrotetrazoliumchloride blue (NTB) 2:24
Nkolbisson virus (NKOV) 1:116
NKV virus(es) 2:235
NL-S protein 1:47–48
Nodamura virus (NoV) 3:431
  host range 3:431–432
  pathogenesis 3:437
Nodaviridae 2:229, 3:430, 3:431, 3:431t, 4:567, 5:11t
Nodavirus(es) **3:430–438**
  apoptosis inhibition 1:158t
  B2 protein 3:152
  capsids 3:432
  gene expression regulation 3:435
    protein synthesis 3:435
    RNA3 3:435
    RNA synthesis 3:435
  genetics 3:436
    recombination 3:436
  genome 3:430, 3:434, 3:434f
    packaging 3:433
    RNA1 3:430, 3:434
    RNA2 3:430, 3:434
    RNA3 3:434–435
  geographical distribution 3:431, 3:431t
  infections
    epidemiology 3:437
    host range 3:431, 3:431t
    insects 3:126t
    mitochondrial changes 3:436f, 3:437
    pathogenesis 3:437
    virus–host interactions 3:437
  protein B2 3:152, 3:436, 3:437f
  RdRp 3:436
  RNA recombination 4:379
  RNA replication 3:435
    cis-acting sequences 3:436
    protein A 3:435
    replication complexes 3:435, 3:436f
    RNA1 3:435
  RNA silencing suppression 3:152, 3:436, 3:437f
  satellite viruses 4:527t
  taxonomy 3:431, 3:431t
  virion assembly 3:433
  virion properties 3:432
  see also specific viruses
NOD-like receptor(s) (NLRs) 1:155
Nomenclature **5:9–24**
  definition 5:9
  latinized binomials 5:22
  nonlatinized binomials 5:22
  orthography 5:9, 5:21–22
  retrotransposons 4:437
  uniform system 5:21
  see also Classification
Non-β-barrel fold 5:381, 5:383f
Nonchromosomal gene(s), definition 4:336
Noncoding regions (NCRs)
  see specific viruses
Noncoding RNA, RNA satellites replication 4:533–534
'Noncompeting viral vectors' 4:233
Nondefective 'parent' virus(es) (St particles) 2:1, 2:3
Nonenveloped virus(es) 2:321
Nonenveloped virus assembly **1:200–204**
  antigenic sites 1:204
  antiviral agents 1:204
  architecture 1:200
    capsid 1:200–201, 1:201f
    genome size 1:200–201
  atomic structure 1:201
  evolution 1:203
  host receptor recognition site 1:203
  nucleic acid–protein interaction 1:202
  spherical viruses, atomic structure 1:202
  structure determination methods 1:201
Nonenveloped virus structure **5:380–393**
  β-barrel fold 5:381, 5:382f
  cryo-electron microscopy 5:380, 5:381
  historical aspects 5:380–381
  non-β-barrel fold 5:381, 5:383f
  nuclear magnetic resonance 5:381

  pseudo-symmetric particles 5:391
    P=3 icosahedral particles 5:391, 5:392f
  quasi-symmetric particles 5:385
    bacilliform particles 5:391
    $T=3$ icosahedral particles 5:385, 5:386f
    $T=4$ icosahedral particles 5:385, 5:387f
    $T=7$ icosahedral particles 5:388, 5:389f
    $T=13$ icosahedral particles 5:382f, 5:388
    $T=25$ icosahedral particles 5:390f, 5:391
  symmetric capsules 5:381
    helical particles 5:383f, 5:385
    $T=1$ icosahedral particles 5:381, 5:384f
  X-ray crystallography 5:380, 5:381
  X-ray fiber diffraction 5:380
Non-extinction, viral evolution 2:176
Non-Hodgkin's lymphoma
  incidence in AIDS patients 2:153–154
  ocular associated 2:497
Nonhomologous recombination
  bacteriophages *see* Bacteriophage(s)
  bromoviruses 1:389
  definition 2:301
  RNA viruses 4:377–378
Nonhuman primates *see* Primates, nonhuman
Nonlatinized binomials, nomenclature 5:22
Non-LTR retrotransposons 4:428–429, 4:438f, 4:439
  distribution 4:442t
  members 4:439
  phylogenetic studies 4:441
  see also Long interspersed nuclear elements (LINEs); Penelope-like elements (PLEs); Short interspersed repetitive elements (SINEs); SVA retrotransposons; Target-primed retrotransposons
Nonmultiplying (NM-type) infection, tobraviruses 5:73
Non-nucleoside reverse transcriptase inhibitors (NNRTIs) 1:149, 2:507t, 2:509
  see also specific drugs
  drug structure 2:510
  genetic barriers 1:150–151
  limitations 2:510
Nonoccluded insect virus(es) (NOIV), unassigned **3:144–148**
  genome 3:145
  history 3:144
  infection
    pathology 3:146
    persistence 3:147
  physical properties 3:145
  replication 3:145
  transmission 3:146
Nonpermissive cells, rhinoviruses 4:471
Non-phage protein encapsidation, three-hybrid assay 3:29
Nonplant edornavirus(es) *see* Edornavirus(es)
Nonprimate cytomegalovirus(es) **1:624–634**
Nonpropagative transmission, definition 1:301
Nonreplicative phase, chronic hepatitis B 2:357–358
Nonsegmented genomes, single-stranded negative sense RNA virus replication 4:410
Non-species definitions, viral evolution 2:177
Nonspecific flu-like illness, yellow fever 5:474
Nonsteroidal anti-inflammatory agents (NSAIDs), definition 5:83
Nonstructural (NS) protein(s)
  see specific viruses
Nonsynonymous polymorphisms, *APOBEC3G* 2:471–472
Nontemplated nucleotides, arenavirus replication 3:247
Nontranslated regions, satellite viruses 4:530
Nonviral protein expression, African cassava mosaic disease 1:35
Nonviremic transmission 5:270
  definition 5:440
Norovirus 1:411–412, 1:411t, 2:117, 3:439, 3:439t
Norovirus(es) **3:438–445**
  antigenic variation 1:415
  characteristics 2:117t
  classification 3:438–439, 3:440f, 3:441t

composition 3:440
culture 1:411
evolution 3:442
  genetic recombination 3:442–443
gene expression 3:440
genetics 2:118
genogroups 2:117
genome 2:117, 3:440, 3:442f
  open reading frames 3:440
geographical distribution 1:414
history 1:411
infections *see below*
prevalence 2:118
recombination 1:415
replication 3:441, 3:443f
  receptors associated 1:416–417
  site 3:443
structure 1:411–412, 2:117f, 3:440, 3:442f
  VP1 3:440
  VP2 3:440
  VPg 3:440
taxonomy 3:439
transmission 1:416, 2:117–118, 3:443
vaccines/vaccination 1:418, 3:444
see also specific viruses
Norovirus infection 3:439t
  clinical features 1:417, 3:443
    gastroenteritis 1:411, 2:117–118, 3:438–439, 3:443
  complications 1:418
  control 1:418, 3:444
  diagnosis 3:444
  disease surveillance 2:46
    data collection sources 2:46
    gastroenteritis 2:46
  food hygiene 1:418
  host range 1:414–415, 3:443
  immune response 1:417–418, 3:444
    histo-blood groups 3:444
  Norwalk virus 1:411
  pathogenesis 1:416
    incubation 1:416
  pathogenicity 3:443
    murine virus 3:443
  prevention 3:444
  treatment 3:443–444
North Africa, HIV epidemiology 1:65
North America, ophiovirus 3:450–451
Northern cereal mosaic virus (NCMV) 1:494, 4:184t
  host range 1:494
  infection, yield losses 1:494
Norvir® 1:151
Norvirhabdovirus, genome 1:112
'Norwalk agent' 2:117
Norwalk virus
  cDNA clones 1:415
  detection 1:411
  gastroenteritis 1:411
  history 3:438
  plant-produced vaccine 5:223
  structure 1:412f
Norwegian salmonid alphavirus (NSAV/SAV-3) 2:230, 5:93t, 5:94, 5:95
Nosocomial transmission, definition 1:59, 2:141, 2:146
Nosology, definition 1:533
Notch signaling pathway, *Bam*HI-A rightward transcripts 2:163–164
Notifications, disease surveillance
  data collection, sources 2:45
  meningitis, aseptic 2:45
Novirhabdovirus 1:119t, 2:221–222, 2:222t, 2:228, 3:160f
Novirhabdovirus(es) 2:221
  classification 1:119t, 2:221–222, 2:222t
  definition 2:221
  genetic diversity 2:224
  genome 2:223–224, 2:224f, 3:326f
  transcription 2:223–224
  virion structure 2:222–223, 2:224f
    glycoprotein 2:222–223
    matrix protein 2:222–223
    nonvirion protein 2:222–223
    nucleocapsid 2:222–223

phosphoprotein 2:222–223
polymerase 2:222–223
see also specific viruses
N-pilus, filamentous ssDNA bacteriophage infection 2:191
N^{pro} protein
  Border disease virus 1:339–340
  bovine viral diarrhea viruses 1:378, 1:379
  classical swine fever virus 1:526
N protein
  bacteriophage lambda 5:184
  Chandipura virus genome encapsidation 1:502–503
NPV-specific cell lines, baculovirus in vitro pathogenesis 1:270
NS1 protein
  aquareovirus 1:165–166
  flaviviruses 3:66
  influenza viruses *see* Influenza virus(es)
  parvoviruses 4:88
NS2-3 protein
  Border disease virus 1:339–340
  bovine viral diarrhea virus biotypes 1:376–377, 1:377f, 1:379–380
NS2 protein
  aquareovirus 1:165–166
  parvoviruses 4:88
NS3/4A protease, apoptosis inhibition 1:161
NS3 protease, Border disease virus 1:340
NS3 protein, bovine viral diarrhea virus biotypes 1:376–377, 1:377f
NS4A protein, bovine viral diarrhea virus 1:379–380
NS4B protein, bovine viral diarrhea virus 1:379–380
NS4 protein, aquareovirus 1:165–166
NS5A protein, Border disease virus 1:340
NS5B protein, HCMV 2:369–370
NSDV *see* Nairobi sheep disease virus (NSDV)
NSm protein 5:134–135
nsP1 5:98
nsP2 5:98
nsP3 5:98
nsP4
  alphavirus replication 5:98
  viral enterotoxic role 4:511
NTB domain, nepovirus 3:409
NTB protein, nepovirus 3:409–410
N-terminal coat protein, cacao swollen shoot virus 1:405
Nuclear export signals (NES)
  adenovirus malignant transformation 1:14
  Borna disease virus replication 1:342
  umbravirus infection 5:212
Nuclear F-actin, baculovirus in vitro pathogenesis 1:270–271
Nuclear hypertrophy
  definition 3:495
  Penaeus duorarum single nuclear polyhedrosis virus 4:571
Nuclear localization signals (NLS)
  Borna disease virus replication 1:342
  ophiovirus 3:450
  porcine circoviruses 1:519
  umbravirus infection 5:212
Nuclear magnetic resonance (NMR)
  inovirus structure 3:120–121
  nonenveloped virus structure 5:381
Nuclear shuttle protein (NSP)
  banana bunchy top virus 1:274
  bean golden mosaic virus 1:298
  definition 1:272
Nuclear transport, avian hepadnavirus replication 2:331
Nuclease(s), polymerase chain reaction 2:35
Nucleic acid amplification techniques *see* Polymerase chain reaction (PCR)
Nucleic acid detection 2:33
Nucleic acid probes, pecluvirus 4:99
Nucleic Acid Research (NAR) 5:356–362, 5:357t
Nucleocapsid(s)
  alphavirus assembly 1:196–197
  baculoviruses 1:248–249
  definition 4:407
  maturation, icosahedral enveloped dsDNA bacteriophages 3:11

nucleopolyhedrovirus primary infection 1:267
potexvirus 4:310
varicella-zoster virus 5:256
yellow head virus 5:477
Nucleocapsid core
  alphavirus assembly 1:195
  definition 1:194
  measles virus 3:286
  orthoretrovirus structure 4:463–464
Nucleocapsid (N) protein
  Crimean-Congo hemorrhagic fever virus 1:598–599
  hepatitis B virus 2:352–353
  measles virus 3:286, 3:287
  mouse mammary tumor virus 3:335
  tenuivirus 5:25
  toroviruses 5:153
Nucleocytoplasmic large DNA virus (NCLDV) superfamily 1:50
  associated core genes 3:315
  classification 3:315, 3:315f
  mimiviruses 3:312, 3:315
  phycodnavirus 4:118
  definition 3:311
  evolution 3:474–475, 3:478
Nucleolus
  definition 5:209
  hypertrophy, bean golden/yellow mosaic disease 1:299
  umbravirus infection 5:212
Nucleolytic ribozyme(s)
  chemical reaction 4:475–476, 4:477f
  definition 4:475
Nucleopolyhedrovirus 1:220, 1:247, 1:254, 1:260f, 1:265
Nucleopolyhedrovirus(es) (NPVs) 1:225–226, **1:254–265**
  classification 1:254
  DNA replication 1:250–251, 1:257
  genes 1:257–260, 1:259f, 1:261t, 1:264
    DE expression 1:262
    IE-1/IE-0 complex 1:261–262
    late expression factors 1:262
    oral infectivity 1:264
  genome 1:255, 1:256
  group I 1:263–264
  group II 1:263
  history 1:225–226
  infection *see below*
  life cycle 1:255
    host cell entry 1:256
    morphogenesis 1:262
    nucleocapsids 1:249
    occlusion bodies 1:264
  occlusion-derived virions 1:255, 1:255f, 1:263
    *Ac*MNPV replication 1:238
    function 1:223
    life cycle 1:221
  packaging 1:265
    multiple nucleocapsid per envelope 1:265
    polyhedrin gene 1:265
    single nucleocapsid per envelope 1:265, 1:266f
  proteins 1:257–260
    F proteins 1:263
    GP41 1:264
    GP64 1:263, 1:264
    P6.9 1:262–263
    VP39 1:262–263
  transcription 1:256
    early gene expression 1:261
    late gene expression 1:262
    temporal regulation 1:261
  virus structure
    BV envelopes 1:255–256, 1:257f, 1:263
    nucleocapsids 1:255, 1:255f, 1:257f, 1:262
    ODV envelopes 1:255–256, 1:255f, 1:263
    polyhedra 1:264
    tegument proteins 1:264
  see also Lepidopteran nucleopolyhedrovirus(es) (NPV); *specific viruses*
Nucleopolyhedrovirus infections
  DNA infectivity 1:261
  host range 1:265

Nucleopolyhedrovirus infections (continued)
  infection cycle 1:256, 1:258f
  insects 3:126t
  pathogenesis 1:266
  primary infection 1:266
    infectivity factors 1:266–267
    larval protection strategies 1:266
    midgut accessibility 1:266
    nucleocapsids 1:267
    occlusion derived virus 1:266–267
    receptor binding 1:266–267
  secondary infection 1:267
  tracheal cells 1:267
Nucleoprotein (NP)
  arenaviruses 3:243, 3:244
  yellow head virus 5:477
Nucleorhabdovirus 1:119t, 1:488, 3:160f, 4:187, 4:188t
Nucleorhabdovirus(es)
  budding 3:333, 3:334
  classification 1:119t
  genome 3:326f
  host cell entry 3:333
  infection model 4:193–194, 4:194f
  tobacco diseases, economic importance 5:67
  virus-encoded proteins 3:332–333
    nucleocapsid 3:331
    phosphoprotein 3:331
  see also specific viruses
Nucleoside analogs 1:149
  channel catfish virus treatment 2:211
  hepatitis B treatment 2:342, 2:359, 2:365
  history 1:145
  see also specific drugs
Nucleoside diphosphate kinase (NDK) 3:316
Nucleoside reverse transcriptase inhibitors (NRTIs) 2:507t, 2:509, 2:511t
  drug resistance 1:150–151
  side effects 2:509
Nucleotide analogs 1:149
  hepatitis B treatment 2:359
  see also specific drugs
Nucleotide cyclic phosphodiesterase (CPD), nidoviruses 3:425
Nucleotide selection
  bacteriophage genome packaging 2:307
  cystovirus genome packaging 2:307
  dsDNA bacteriophage genome packaging 2:307
Nucleotide sequences
  fish retroviruses 2:231
  mungbean yellow mosaic viruses 3:369t, 3:371
  virus databases 5:362
Nucleotide substitution rate, definition 4:436
Nuclear pyknosis, definition 1:580
Nucleus infection, baculovirus in vitro pathogenesis 1:270
Nudaurelia capensis β virus (NβV) 5:28t
  capsid morphology 5:31f
  capsid protein precursors 5:29, 5:32t
  discovery 5:27–28
  genomes 5:29
  infection 5:36
Nudaurelia capensis omega virus (NωV) 5:28t
  capsid assembly 5:33
  capsid maturation 5:34–35
  capsid morphology 5:31f
  capsid protein precursors 5:31–32, 5:32t
  capsid structure 5:33, 5:34f
  chemical tags 5:36–37
  cleave site mutations 5:34f, 5:35–36
  discovery 5:28
  genomes 5:31
    RNA2 length 5:31
  structure 5:382f, 5:385–388, 5:387f
    β-barrel fold 5:381
Nudaurelia ε virus (NεV) 5:28t
Nupapillomavirus 4:9, 4:10f, 4:11t
Nupapillomavirus 4:9
Nursery, definition 4:561
Nus factors, bacteriophage lambda 5:184
Nymph, definition 1:533
Nystagmus, definition 1:76

O

Oak Vale virus 4:184t
Oat blue dwarf virus (OBDV)
  infection 5:205–206
  replication 5:205–206
Oat chlorotic stunt virus (OCSV) 5:145–146
  genome 5:147f
Oat golden stripe virus (OGSV) 2:292
Oat sterile dwarf virus (OSDV) 4:150t
OBDV *see* Oat blue dwarf virus (OBDV)
Obligate intracellular parasites, icosahedral ssDNA bacteriophages 5:368, 5:369
Obodhang virus 4:184t
Obuda pepper virus (ObPV) 4:173
Occlusion bodies (OB)
  baculovirus *see* Baculovirus(es)
  definition 1:211, 1:219, 1:254
  insect pest control 3:128
  nucleopolyhedrovirus 1:264
Occlusion-derived virion(s) (ODV)
  baculoviruses 3:129
  definition 1:254
  nucleopolyhedroviruses *see* Nucleopolyhedrovirus(es) (NPVs)
Occupancy, antigen–antibody complexes 3:414
Occupations
  St Louis encephalitis virus 5:273
  vector transmission 5:273
Oceania, HIV epidemiology 1:65
Ockelbo disease 5:90
  seasonal distribution 5:90
Oct-1 recognition sequence, varicella-zoster virus 5:259
Ocular adnexae 2:491
Ocular diseases
  AIDS *see* AIDS
  DNA viruses 2:494t, 2:495
  influenza virus 2:492
  mumps virus 2:495
  RNA viruses 2:492, 2:493t
  uveitis 2:573
Ocular immunology 2:492
  tears 2:492
Odonstyle, definition 3:405
Odontoglossum ringspot virus (ORSV) 4:225–226, 4:228f
Odontophore, definition 3:405
Office Internationale des Epizooties (OIE) 5:1
Ogyia pseudosugata multiple nucleopolyhedrovirus (MNPV) 3:125
OHMV *see* Omsk hemorrhagic fever virus (OHFV)
Oil palm(s), infection control 3:128, 4:79
Oka vaccine strain, varicella-zoster virus 5:253, 5:255
Okavirus 4:567, 5:477
Okavirus(es)
  classification 4:567
  genome 3:424f
  replicase genes 3:425f
Okazaki fragment 4:400
OLE1 protein 1:384–385
Oleavirus 1:81, 1:387, 3:46
2′–5′ Oligoadenylate (2–5 OAS), interferon-induced antiviral response 3:107, 3:115–116
Oligodendrocytes
  demyelination in JCV infection 4:267
  neural cell infection 1:473
  TMEV infection 5:42–43
Oligonucleotides, use in microarrays 2:15
Olpidium brassicae
  Lettuce big-vein virus 5:264, 5:268
  Lettuce ring necrosis virus transmission 3:453
  Mirafiori lettuce virus transmission 3:453
  Ophiovirus transmission 3:453
  Tulip mild mottle mosaic virus transmission 3:453
Olive latent ringspot virus (OLRSV) 3:406t
Olive latent virus 1 (OLV-1) 3:403
  geographical distribution 3:404
  virus–host relationship 3:404–405

Olive mild mosaic virus (OMMV) 3:403
  geographical distribution 3:404
OLV-1 *see* Olive latent virus 1 (OLV-1)
Omegatetravirus(es)
  capsid protein precursors 5:31–32, 5:31f, 5:32t
  insect infections 3:126t
Omsk hemorrhagic fever 5:53, 5:490
Omsk hemorrhagic fever virus (OHFV) 2:239, 5:53
  muskrats 2:239–240
  vector 2:239–240, 5:53
Oncogene(s)
  adenoviral 1:10–11, 1:13
  see also Adenovirus infections
  definition 2:213
  HPV 4:19
  retroviral *see* Retrovirus(es)
Oncogenesis 2:457, 5:316
  hepatitis B 2:366
  Herpesvirus Saimiri 4:589
  HPV E6 4:24
  Jaagsiekte sheep retrovirus 2:219
Oncolysis, measles virus 3:291
Oncolytic virotherapy 3:285, 3:291
One-step growth 2:442
Onion yellow dwarf virus (OYDV) 5:288
ONNV *see* O'nyong nyong virus (ONNV)
Ontology, definition 5:348
O'nyong nyong virus (ONNV) 5:89, 5:114
  characteristics 1:171t
  evolutionary relationships 1:175
  geographical distribution 5:84f, 5:89, 5:114
  infection
    clinical features 5:86t, 5:87t, 5:89
    epidemic 5:89
    RNAi role 3:154
  transmission 5:89, 5:100, 5:114
Oomycetes 2:288
Open reading frames (ORFs)
  aquareovirus 1:165
  *Bam*HI-A rightward transcripts 2:159f, 2:163–164
  barley yellow dwarf viruses 1:280
  barnaviruses 1:287, 1:287f
  bean golden mosaic virus 1:298
  beet curly top virus 1:303
  Border disease virus 1:338
  cacao swollen shoot virus 1:405, 1:407t
  cassava mosaic geminiviruses 1:30–31
  cereal yellow dwarf virus 1:280
  cowpea mosaic virus genome 1:571
  definition 1:296, 1:482, 1:533, 5:199
  Dendrolimus punctatus tetravirus 5:31
  EBV 2:157
  edornaviruses 2:110–111, 2:112, 2:113
  giardiavirus 2:313
  gill-associated virus 5:478–479
  hepatitis E virus 2:378
  Indian peanut clump virus 4:101
  LINE-like retrotransposons 4:439
  LTR retrotransposons 4:420, 4:438–439
  mungbean yellow mosaic viruses 3:365–366
  necrovirus 3:403
  omega tetraviruses 5:31
  ophiovirus 3:449
  Peanut clump virus 4:101t
  pecluvirus 4:99–100, 4:101t
  Pepino mosaic virus 4:105
  pomovirus 4:283
  retrotransposons 4:437
  rice yellow mottle virus 4:486
  sadwaviruses 4:524, 4:525f
  target-primed retrotransposons 4:420–421, 4:425
  tetraviruses 5:29
  toroviruses 5:152–153
  yeast L–A virus 5:467
  yellow fever virus genome 5:470–471
  yellow head virus 5:477–478, 5:482
  see also under ORF
Operational taxonomy units (OTUs) 4:126
Ophidium brassicae 5:263
Ophioviridae 3:447
Ophiovirus 3:447, 5:263

Ophiovirus(es) **3:447–454**
  Asia 3:450–451
  classification 3:447, 3:449f
    coat protein sequences 3:447
    RT-PCR 3:448
    serology 3:448
    species demarcation 3:447
  genome 3:449, 3:449f, 3:450t
    nuclear localization sequence 3:450
    open reading frames 3:449
    RNA2 3:449
    RNA4 3:449
    subgenomic RNAs 3:449–450
  geographical distribution 3:450
    Asia 3:450–451
    Mediterranean 3:450–451
    North America 3:450–451
    South America 3:450–451
  host ranges 3:453
  infection
    control 3:453
    cytopathology 3:453
    pathogenicity 3:450, 3:451f
    prevention 3:453
    symptoms 3:453
  members 3:448t
  proteins 3:450
    coat protein 3:450
  RdRp 3:447
  taxonomy 3:447
    RdRp 3:447
  transmission 3:453
    Olpidium brassicae 3:453
  virion particles 3:448
  see also specific viruses
Opportunistic infections
  adult T-cell leukemia complications 2:568, 2:569–570
  HIV infection *see* HIV infection
Opsonization 3:110
  definition 3:104
Opt-out HIV testing 1:66
Opuntia virus X (OPX) 4:311t
Or-1V *see* Oryctes virus (Or-1V)
Oral bioavailability, antiviral prodrugs 1:150
Oral focal epithelial hyperplasia 4:16
Oral infection route 5:155
Oral infectivity, nucleopolyhedrovirus genes 1:264
Oral vaccination *see* Edible vaccine(s)
Orbivirus 3:454, 3:455t, 4:536, 4:536t
Orbivirus(es)
  animal pathogens associated 3:454
  classification 3:454, 3:455t
  evolutionary relationships 3:461, 3:463f
  genome 1:172, 3:460, 3:462t
    open reading frames 3:460
    untranslated regions 3:460
  history 3:454
  host range 3:457
  infection 3:458
    epidemiology 3:458
    human 3:464
  replication 3:460
    cell tropism 3:461
  transmission 3:454, 3:457
  type species 4:536t
  viral RNAs 3:459
  virion properties 3:459
  virion proteins 3:459
  virion structure 1:172, 3:459
  see also specific viruses
Orchidaceae virus infections 4:225
  see also specific viruses
Orchitis 1:533
Order(s) 5:10
  definition 5:9
  universal system 5:23
ORF1, cacao swollen shoot virus 1:405
ORF2, cacao swollen shoot virus 1:405
ORF3, cacao swollen shoot virus 1:405–406
ORF4 mutations
  barley yellow dwarf virus 1:282
  cereal yellow dwarf virus 1:282

ORF29, varicella-zoster virus latency 5:262
ORF1629, baculovirus expression vectors 1:244
ORF d63, fusellovirus protein structure 2:300
ORF f93, fusellovirus protein structure 2:300
ORFV *see* orf virus (ORFV)
orf virus (ORFV) 4:57–58
  genome 4:59, 4:60f
    pathogenesis-related genes 4:59–60
  host range 4:58
  inactivation 4:58
  infection
    clinical features 4:58
    immune response 4:61
    lesion progression 4:58
    reinfection 4:58, 4:61
    teat lesions 4:59
  morphology 4:59f
  synonyms 4:58
  vaccines/vaccination 4:62–63
  virus shedding 4:58
Organism(s), definition 3:400
  cluster concept 3:400
  properties 3:400
  reproductive lineage 3:400
Organismal ecology
  bacteriophages *see* Bacteriophage(s), ecology
Organotypic raft(s) 4:32
Organ transplantation, infection risks **3:466–472**
  antiviral therapy 3:469–470
    adjunctive therapy 3:471
    drug toxicity 3:471
    duration 3:471
    post-transplant prophylaxis 3:471
    prophylaxis 3:471
  associated viruses 3:466, 3:466t
    HCMV infection 2:483–484
    KSHV transmission 3:191
    rabies virus 4:370
  clinical consequences 3:466
  diagnostic assays 3:470
  donor-derived 3:466
  effects 3:468, 3:468t
  immunocompromised patients 3:466
  latent infection reactivation 3:467
    antirejection therapy 3:467
    associated viruses 3:467
    cellular pathways 3:467
    herpesviruses 3:467
    Kaposi's sarcoma 3:469
    risk factors 3:467
  novel infections 3:468
  screening 3:466, 3:467
  therapy 3:471
  vaccination 3:471, 3:472t
  virus-specific syndromes 3:469
  see also specific transplants
Organ transplantation rejection 3:468–469
Oriental tobacco 5:61
Origin of assembly sequence (OAS)
  definition 5:68, 5:145
  pecluvirus 4:101–102
Origin of replication (ori)
  definition 1:211
  HSV virus vectors 2:396
  maize streak virus 3:266
Origin(s) of viruses 3:402, **3:472–479**, 3:474f
  ancient origin 3:474
  cellular *vs.* acellular evolution 3:475
  classical view 3:473
    consequences of 3:473
  DNA, origin of 3:477
  escape hypothesis 3:473, 3:476, 3:476f
    bacteriophages 3:473
    virus-encoded proteins 3:473–474
  genome plasticity 3:477–478
  modern cells, relationship with 3:473
  phylogenetic analysis 3:402
  reduction hypothesis 3:473, 3:476, 3:476f
  RNA *vs.* DNA viruses 3:477
  trinity concept 3:473–474
  viral eukaryogenesis hypothesis 3:478
  virus-first hypothesis 3:473, 3:475, 3:476f

  acellular machinery, coevolution of 3:475
  see also Evolution (of viruses)
ori-Lyt, HCMV 1:640, 2:486
Ornamental plant species virus infections **4:207–229**, 4:216t
  control 4:226
    cross-breeding 4:226–229
    direct chemical control 4:226–229
    guidelines 4:226–229
  cross-breeding 4:208
  cultural practice 4:208
  economic importance 4:208
  host ranges 4:208
  risk of 4:208
  species introduction 4:208
  vector introduction 4:208
  worldwide trade 4:208
  see also specific species
Ornithodoros moubata (soft tick vector) 1:43, 1:44f
Oropouche virus 1:391t, 3:481, 5:489
  vector 3:482
Orthobunyavirus 1:76, 1:391, 1:391t, 3:480, 4:490–491
Orthobunyavirus(es) 1:399, **3:479–483**
  definition 3:479
  genome segment reassortment 1:393f, 1:395
  historical aspects 3:479
  host species 3:480
  infection 3:480
    diagnosis 3:482–483
    treatment 3:482
  morphology 1:393f
  reassortment 3:482
    complement fixation tests 3:482
    group C viruses 3:482
    hemagglutination inhibition tests 3:482
    neutralization tests 3:482
  replication, host cell entry 1:395
  RNA structure–function relationships 3:482
  species 3:480, 3:481t
  transmission 1:392, 1:392f, 3:480
  vectors 3:480
  virion proteins
    NSm protein 1:393
    NSs protein 1:395
  see also specific viruses
Orthography, nomenclature 5:9, 5:21–22
Orthohepadnavirus 2:351–352
Orthohepadnavirus, species infected 2:351–352
Orthologous clusters, virus databases 5:363
Orthomyxoviridae 1:170, 3:96, 3:483–484, 3:489, 5:11t
Orthomyxovirus(es)
  fish 2:229
  genera 3:96, 3:483–484, 3:489, 3:490
  human respiratory infection 2:554
  virion structure 1:172
  see also specific viruses
Orthopoxvirus 3:342, 4:324t, 4:639, 4:641, 5:244–245
Orthopoxvirus(es) 5:489–490
  animal reservoirs 1:579–580
  classification 4:324t
  conserved genes 2:138
  molluscipoxvirus 2:136
  phylogeny 4:326–327
  see also specific viruses
Orthoptera, definition 1:111, 2:37
Orthoreovirus 4:383, 4:391, 4:392, 4:536t
Orthoreovirus(es) 3:378–379
  fusogenic 4:383
    FAST proteins 4:383–384
  genome 4:383, 4:384t
  infection
    clinical features 4:388
    immune response 4:389
  type species 4:536t
  see also specific viruses
Orthoretrovirinae 2:231, 2:260t, 4:412–413, 4:430, 4:460, 5:11t
Orthoretrovirus(es)
  assembly 2:518
  classification 4:460

Orthoretrovirus(es) (*continued*)
    life cycle 2:263
    viral proteins 4:463–464
    see also specific viruses
Orungo virus 3:465
    zoonoses 5:491
OrV *see* Oryctes rhinoceros virus (OrV)
Oryctalagus, myxoma infection *see* Myxoma viruses (MYX)
Oryctes rhinoceros virus (OrV) **3:495–500**
    assembly 3:495
    biocontrol 3:499
    genome 3:497
        genetic variation 3:497
        restriction map 3:497, 3:497*f*
        size 3:497
    history 3:495
    infection *see below*
    insect pest control 3:127, 3:132
    nomenclature 3:495
    phylogeny 3:497, 3:498*f*
    replication 3:495
        cell entry 3:495
    taxonomy 3:497
    transmission 3:132, 3:499
        adult beetles 3:499
        larvae 3:499
    virion properties 3:496, 3:496*f*
        associated proteins 3:496
        size 3:496
        tail-like appendage 3:496
Oryctes rhinoceros virus infection 3:495, 3:496*f*
    insect pest control 3:132
    midgut epithelium 3:497*f*
    pathology 3:495
        adult beetles 3:495
        larval morphology 3:495
Oryctes virus (Or-1V) 3:144
    beetle population management 3:146
    infection 3:144
        beetle population management 3:146
        pathology 3:146
        persistence 3:147
        replication 3:146
        symptoms 3:146
        transmission 3:146
    morphology 3:145, 3:147*f*
    replication 3:145–146
        structural proteins 3:146
        virion assembly 3:145–146
Orygia psuedotsugata multiple neucleopolyhedrovirus (MNPV) 3:125
Oryza rufipogon endornavirus (OrEV) 2:110–111
    genome 2:110*f*
Oryza sativa see Rice *(Oryza sativa)*
Oryza sativa endornavirus (OsEV) 2:110–111
    copy number regulation 2:114
        in vitro *vs.* in vivo 2:114
        pollen grains 2:114
    genome 2:110*f*
    vertical transmission 2:113–114
    host species 2:114
Oryzavirus 4:150–151, 4:536*t*
Oryzavirus(es)
    characteristics 4:150*t*
    classification 4:150
    genome 4:151
        dsRNA segments 4:151–152, 4:152*f*, 4:153*t*
        size 4:150–151
    host range 4:154
    transmission 4:154
    type species 4:536*t*
    virion structure 4:151
        capsid 4:151–152
    see also specific viruses
Oseltamivir 3:102–103, 3:492, 3:493*f*
    influenza virus infection 2:556
OsEV *see* Oryza sativa endornavirus (OsEV)
Osloss strain, bovine viral diarrhea viruses 1:376–377
O-some 4:400
Ostelmavir 1:143*t*

Osteospermum virus infections 4:221
    see also specific viruses
Ostreid herpesvirus 1 (OsHV-1) 2:429, 4:563, 4:564*f*
    genome 4:564*f*
    genes 4:564–565
    ORFs 4:564–565
    taxonomy 4:565
Ourmia melon virus 3:500
    pathogenicity 3:501
    taxonomy 3:500
Ourmiavirus 3:500
Ourmiavirus(es) **3:500–501**
    control 3:501
    cytopathology 3:501
    genome 3:500
        RNA-1 3:500
        RNA-2 3:500
        RNA-3 3:500
    taxonomy 3:500
    geographical distribution 3:501
    morphology 3:500
    other viruses *vs.* 3:500
    particles 3:500
    pathogenicity 3:501
    prevention 3:501
    proteins 3:501
    transmission 3:501
    see also specific viruses
Outbreak control, phylogeny 4:127
Outer membrane fusion, icosahedral enveloped dsDNA bacteriophages 3:11
'Outgroup' phylogenetic analysis 4:127
Output (information utilization), virus databases 5:354, 5:354*f*
    BLAST 5:354
OvHV-2 *see* Ovine herpesvirus 2 (OvHV-2)
Ovine adenovirus(es) 1:26
Ovine herpesvirus 1 (OvHV-1) 1:364*t*
Ovine herpesvirus 2 (OvHV-2) 1:364*t*
Ovine herpesvirus 2 (OvHV-2) infection
    clinical features 1:367
    species barriers 2:428
Ovine pulmonary adenocarcinoma (OPA) 3:179
    disease characterization 3:179–180, 3:180*f*
    experimental model 3:180
Oviposition, definition 3:264
Oyster(s), norovirus infection 1:418
Oyster velar virus (OVV) 4:561
Oyster velar virus disease (OVVD)
    histopathology 4:562
    manifestations 4:561–562

P

P1 gene/protein
    alfalfa mosaic virus 1:85
    bean common necrosis virus 1:294
    icosahedral enveloped dsDNA bacteriophages 3:7–8, 3:8*f*, 3:9*f*
    potyvirus 4:317
    rhinovirus 4:469
    rice yellow mottle virus 4:486–487
    tobacco etch virus 4:317
P2 bacteriophage *see* Bacteriophage P2
P2 gene/protein
    alfalfa mosaic virus 1:85
    bacteriophage PRD1 3:1–2
    caulimovirus 5:279
    icosahedral enveloped dsDNA bacteriophages 3:8, 3:9*f*
    rhinovirus 4:469
P3-6K1 protein, tobacco etch virus 4:317
P3 gene, zucchini yellow mosaic virus 5:437–438
P3 gene/protein
    rhinovirus 4:469
    watermelon mosaic virus 5:437–438
P4 gene/protein
    icosahedral enveloped dsDNA bacteriophages 3:7–8, 3:9*f*
    ssDNA bacteriophages 2:311–312
P5CDH gene, gene silencing 4:142

P6.9–DNA interactions, baculoviruses 1:271
P7 gene/protein, icosahedral enveloped dsDNA bacteriophages 3:7–8
P8 gene/protein, icosahedral enveloped dsDNA bacteriophages 3:8, 3:9*f*
p10 gene/protein, baculoviruses 1:251
p19 gene/protein, RNAi suppression 3:152
P22 bacteriophage *see* Bacteriophage P22
P23 protein, alphavirus 5:98
p29 protease, CHV-EP713 virus 2:583–584
P35 *see* Baculovirus apoptosis inhibition
p37 protein, cauliflower mosaic virus 1:466
p44 protein, cauliflower mosaic virus 1:466
p53 gene/protein
    apoptosis, small T protein murine polyomavirus 4:275
    HPV E6 4:24
    HTLV-1 2:562
        tax protein function *see* Tax protein
    mutation rates 5:198
    papillomavirus 4:37
p54 gene/protein, African swine fever virus 1:46–47
p72 gene/protein, African swine fever virus 1:46–47
P78/83, baculoviruses 1:271
P90 gene/protein, togavirus 5:119
P123 gene/protein, alphavirus 5:98
P150 gene/protein, togavirus gene/protein 5:119
p220 polyprotein, African swine fever virus 1:46–47
p400 gene/protein, adenovirus E1A 1:11–12
P1234 gene/protein
    alphavirus 5:98
    togavirus 5:120
Pacheco's disease 2:406
    acyclovir 2:410
    clinical signs 2:409
    history 2:406–407
    immune prophylaxis 2:410
    incubation period 2:409
    psittacid herpesvirus 1 2:406
Packaging
    definition 3:6
    icosahedral enveloped dsDNA bacteriophages 3:11
Packaging ATPase, definition 2:306
Packaging motor complex, bacteriophages 2:307
Pac site 3:6
PAD4 protein, plant natural resistance 4:176
Padlock probes, plant virus detection 2:28, 2:28*f*
Pairwise sequence comparison (PASC) 5:19, **5:342–348**, 5:357*t*
    applications 5:344
        boundaries 5:344
        taxonomy demarcations 5:344, 5:345*f*
    BLAST *vs.* 5:345
    considerations 5:344
    identity distribution 5:343–344
    limitations 5:347
        parameter variability 5:347
        programs 5:347
    NCBI implementation 5:344
        PASC interface 5:344
    principles 5:343
Pakistan, banana bunchy top virus epidemic 1:277
Palindromes 1:211
Palivizumab 2:550, 2:556
Palm civets, SARS CoV transmission 1:555, 4:555, 4:559–560
Palmoplantar myrmecia, papillomavirus associated warts 4:16
PAMPs *see* Pathogen-specific molecular patterns (PAMPs)
Pandemics 2:525
    definition 1:30, 1:59, 2:44, 2:97, 2:142, 2:517, 2:525, 3:95
    see also specific infections
    see also specific viruses
Panencephalitis, subacute sclerosing (SSPE) 2:494–495
Pangola stunt virus (PaSV) 4:150*t*
Panicavirus 5:145–146, 5:147*f*
Panicovirus 5:145–146
Panicum mosaic satellite virus (PMSV) 4:527*t*
    host virus accumulation interference 4:530
    variation/evolution 4:534
    virion structure 4:530

Panicum mosaic virus (PMV) 5:145–146, 5:147f
  genome 5:147f
Panleukopenia 4:85
Panviral microarrays
  bioinformatics 2:16
  clinical applications 2:17
  sample amplification 2:16
Papanicolaou test ('pap test'), cervical papillomavirus infection 4:16, 4:17, 4:39
Papaya, engineered resistance 4:160–161
Papaya leaf curl virus (PaLCuV) 1:566
Papaya leaf distortion mosaic virus (PLDMV), papaya ringspot virus vs. 4:1
Papaya lethal yellowing virus (PLYV) 4:645
  infection 4:651
Papaya mosaic virus (PapMV) 4:311t
Papaya ringspot virus (PRSV) **4:1–8**
  classification 4:1
  CP gene/protein 4:2–3, 4:4
  evolution 4:2
  genome 4:1, 4:2f, 4:3t
  geographical distribution 4:3
  historical aspects 4:1
  host range specificity determinants 4:3
  infection *see below*
  papaya leaf distortion mosaic virus vs. 4:1
  particle size 4:1
  phylogenetics 4:3
  proteins 4:3t
    immunoprecipitation 4:2
    proteinases 4:2
    see also specific proteins
  sequence diversity 4:2
    CP gene/protein 4:2–3
    isolates 4:3
    management proteins 4:2–3
    phylogenetics 4:3
  suppression of gene silencing 4:4
  transmission 4:4
    aphid vectors 4:1
Papaya ringspot virus infection
  host range 4:1
    P-type 4:1, 4:3
    W-type 4:1, 4:3
  local lesion formation determinants 4:4
  pathogen-derived resistance 4:4, 4:6f
    cross-protection 4:5
    mild virus strains 4:4
    tolerant plant breeding 4:5
  symptoms 4:1, 4:2f
  symptom severity determinants 4:4
    CP gene/protein 4:4
    strain-specific crossprotection 4:4
  transgenic resistance 4:5, 4:157–160, 4:321
    commercialization 4:5
    CP gene 4:5
    field trials 4:5
    'parasite-derived protection' 4:5
    resistance to 4:6f, 4:7
Papillomatosis 4:16
Papillomaviridae 4:8–9, 4:27, 4:34
Papillomavirus(es) **4:26–34**
  apoptosis inhibition 1:158t
  classification 4:27
  evolution 4:28f
    L1 nucleotide sequences 4:35
  gene expression 4:30
    chromatin structure 4:38
    E6 promoter 4:37–38
    early region 4:30
    late region 4:30
  genome 4:32f
    open reading frames 4:34
  history 4:26
  infections *see below*
  propagation 4:32
  proteins 4:31
    E1 4:31
    E2 4:31
    E2TA 4:31
    E4 4:32

E4 ORF 4:36
E5 4:31
E6 4:31–32
E7 4:31–32
E8 4:30–31
L1 4:32
L2 4:32
  replication 4:30
    cell surface receptors 4:30
    mutation rates 4:27
    virion assembly 4:30
  taxonomy 4:27
  tissue tropism 4:30f
  transmission 4:27
  types 4:34
  virion properties 4:32
  virion structure 4:32f
  see also specific viruses
Papillomavirus infections
  clinical features 4:27
  control 4:33
  detection 4:33
  diagnosis 4:33
  epidemiology 4:38
  host range 4:27
    species specificity 4:27
  immune response 4:33
    humoral 4:33
    regression 4:33
  pathogenicity 4:27
  pathology 4:30f
  prevention 4:33
Papova-like virus(es), shellfish viruses 4:566
Papovaviridae 4:277
Papovavirus(es)
  human respiratory infection 2:556
  infections, persistence 4:109t
  see also specific viruses
Papua New Guinea, HIV epidemiology 1:65
Papules
  definition 4:639
  smallpox 4:639–640
Parainfluenza virus(es) **4:40–47**
  classification 4:40
  epidemiology 4:45
  genetic relationships 4:45
  genome 4:43f
    leader region 4:42
    transcriptional control signals 4:42
  host range 4:40
  infection 2:553
    control 4:45
    diseases 4:45
    immune response 4:45
    pathogenesis 4:45
    prevention 4:45
  propagation 4:40
  proteins 4:42
  replication 4:44
    host cell attachment 4:44
    M protein 4:44
    RNA synthesis 4:44
    transcription 4:44
    virus assembly 4:44
  reverse genetics 4:44
  serologic relationships 4:45
  taxonomy 4:40
  transmission 4:45
  virion properties 4:42f
Paralogs
  cytomegaloviruses 1:638f
  definition 1:634
  mimivirus genome 3:314–315
Paralysis
  coxsackieviruses 1:580–581
  poliovirus 4:242–243
Paralytic rabies *see* Rabies virus (RABV)
Paramecium bursaria, chlorella virus infection 4:116
Paramecium bursaria chlorella 1 virus (PBCV-1), structure 5:391

Paramyxoviridae 2:321, 3:285, 3:324, 3:356, 4:40, 4:41t, 4:47, 4:52
Paramyxovirinae 2:321, 2:543, 2:544f, 4:40, 4:41t, 4:52, 5:11t
Paramyxovirus(es) 3:285
  animal pathogens 4:41t
  apoptosis inhibition 1:158t
  assembly 3:333
  attachment 4:42–43
    sialic acid 4:53–54
  budding 3:333–334
  characteristics 5:62t
  classification 4:52
    F protein 3:285
    subfamilies 4:40
  fish *see* Fish paramyxovirus(es)
  F protein 3:285
  fusion 4:42–43
    conformational changes 4:54
    proteolytic processing 4:54
  genome 4:42
    overlapping reading frames 4:52–53
    P gene 4:44
  glycoprotein 4:42–43
  glycoproteins 3:332
    hemagglutination activity 3:332
  hemagglutination-neuraminidase protein
    function 3:358t
    proteolytic cleavage 3:359
    structure 3:356
  infection *see below*
  nucleocapsid 4:54
    structure 3:356
  P mRNA transcripts 3:331
  polymerase 4:54
    abundance 4:54–55
    P–L protein complex 4:54–55
  proteins 4:42
    matrix 4:54
    phosphoprotein 4:54
    small hydrophobic proteins 4:42–43
    viral degradation complex 4:55
    V protein 4:44
    see also specific types
  replication 4:55f
    assembly 4:55–56
    glycoprotein modifications 4:55–56
    transcription 4:55–56
    translation 4:55–56
  reverse genetics 4:56
  ribonucleoprotein 3:356
  rinderpest virus relationship 4:497
  taxonomy 2:321
  transmission 4:56
  as vaccine vectors 4:44–45
  viral degradation complex 4:55
  virion structure 4:498
    nucleocapsid 3:356
    ribonucleoprotein structure 3:356
  virus structure 3:329f
  see also specific viruses
Paramyxovirus infections 4:56
  host range 4:56
  human respiratory infection 2:553
  immune response
    evasion 3:108
    MDA-5 3:106
  persistence 4:109t
  prevention 4:46
    vaccination 4:46
  treatment 4:56
  vaccines/vaccination 4:46–47
    live attenuated vs. inactivated 4:46
Parapoxvirus 4:58, 4:324t
Parapoxvirus(es) (PPVs) **4:57–63**
  classification 4:58
  genomes 4:59
    recombination 4:60
    transcription 4:60
  history 4:57–58
  host range 4:58

Parapoxvirus(es) (PPVs) (continued)
   inections *see below*
   physical properties 4:59
   propagation 4:58
   protein properties 4:60
      structural proteins 4:60
   replication 4:60
   taxonomy 4:58
   transmission 4:59
      vectors 4:62
   virion properties 4:59f
   see also specific viruses
Parapoxvirus infections
   clinical features 4:58
   epidemiology 4:58
   immune response 4:61
   immunomodulation 4:61
      chemokine-binding protein 4:62
      granulocyte-macrophage colony-stimulating factor inhibitory factor 4:62
      interferon resistance protein 4:62
      viral interleukin-10 4:62
      viral vascular endothelial growth factor 4:61
   pathogenesis 4:61
   serologic relationships 4:58
Pararetroviridae 4:430
Pararetroviruses, plants *see* Plant pararetrovirus(es)
'Parasite-derived protection' transgenic plants 4:5
Parasitoids
   definition 4:250
   insect immune responses 4:251
      encapsulation 4:253f
      hemocyte differentiation 4:251–252
      hemocyte proliferation 4:251–252
      polydnavirus immune evasion 4:255
      signaling pathways 4:252–253
   life cycle 4:251f
   wasps 4:250
Paratopes
   definition 1:137
   IgG 1:138
   infection neutralization 3:416
Parechovirus 2:122
Parechovirus, respiratory disease 2:553
Parenchyma, rice yellow mottle virus 4:487–488
Parenchymal cells, Marburg virus infection 3:278
Pariacoto virus (PaV)
   host range 3:431–432
   nucleic acid–protein interaction 1:203f
   RNA packaging 3:432f
   structure 3:432f
Paris polyphylla virus X (PPVX) 4:311t
Parkinson's disease, viral etiology 1:475
Parotitis, MuV infection 3:356
Parr 3:83
Parrots, Pacheco's disease 2:406
Parry's Creek virus (PCV)
   morphology 1:112f
Parsley virus 5 (PaV-5) 4:311t
Parsnip virus 3 (ParV-3) 4:311t
Parsnip virus 5 (ParV-5) 4:311t
Parsnip yellow fleck virus (PYFV)
   genome 4:548f
   isolate variation 4:550
   transmission 4:551
Partial genome sequencing, epidemiology 2:143
Particle assembly
   maize streak virus 3:268
   retrovirus infection resistance 2:473
Particle maturation, virus structure 5:399
Particle purification, seadornavirus 4:538
Particles **5:393–401**
   maize streak virus 3:264f
   Simian retrovirus D 4:626f
Partitiviridae 1:99, 1:103f, 1:104f, 1:503–504, 2:109–110, 2:286, 4:68–69, 5:11t, 5:214
Partitivirus 4:63, 4:64t, 4:68–69
Partitivirus(es) 1:100t
   biological properties 4:69
   classification 2:109–110
      fungal dsRNA viruses 2:111

double stranded RNA satellites 4:531
   genera 1:99
   genome 1:100
   infections
      fungi 2:287t
      host range 4:75
      viral concentrations 4:69
   phylogeny 1:103f
      CP cluster 4:74–75
      horizontal transfer 4:74–75
      RdRp cluster 4:74–75
   RdRp 4:63, 4:66–67, 4:69–71
      replicase activity 4:69
      transcriptase activity 4:69–71
   replication 4:71f
   taxonomy 4:74
   transmission 4:69
   virion properties 4:69
   virion structure 4:72
      capsid 4:72
   virus characterization 1:99
   see also specific viruses
Partitivirus(es), fungal **4:63–68**
   antigenic properties 4:65
   classification 4:63
   dsRNA viruses 2:111
   evolution 4:65
   gene expression 4:65
   genome 4:65
      dsRNA1 4:65
      dsRNA2 4:65
   infections
      mixed 4:69
      pathogenesis 4:67
   infectivity assays 4:75
   multiplication 4:66f
   phylogenetic relationships 4:67f
   taxonomy 4:63
   tentative members 1:100t
   transfection 4:66
   transmission 4:67
      asexual spores 4:69
   virion properties 4:64
   virion structure 4:64
Parvoviridae 2:555, 4:76, 4:86, 4:92, 4:92t, 4:567, 4:572, 5:11t
Parvovirinae 4:86, 4:92
Parvovirus 4:92
Parvovirus(es) 2:122
   arthropod **4:76–85**
      capsid structure 4:76
      classification 4:76
      history 4:76
   capsid 4:86
      assembly 4:86
      cell receptor binding 4:87
      genome packaging 4:86
      role 4:86
      structure 4:87f
   classification 4:92t
      subfamilies 4:92
   control 4:89
   evolution 4:88
      rate 4:88–89
   genome 4:94f
      DNA replication initiator proteins 4:90
      gene cassettes 4:90
      hairpins 4:90
      inverted terminal repeats 4:88f
      NS1 protein 4:88
      NS2 protein 4:88
      open reading frames 4:87–88
      packaging 4:86
      promoters 4:87–88
      sequence variation 4:88
   geographical distribution 4:86
   history 4:85
   infections *see below*
   propagation 4:86
   proteins
      antigenic sites 4:91–92

      polypeptides 4:91
      SF3 helicases 4:91–92
      VP1 4:91
   replication 4:90
      DNA amplification 4:90–91
      heterotelomeric parvoviruses 4:90–91
      homotelomeric parvoviruses 4:90–91
      phases 4:90
      site-specific single-strand nicks 4:91
   seasonal distribution 4:86
   sequence motif conservation 4:82f
   shrimp 4:571
      genome organization 4:571
      geographical distribution 4:572
      host range 4:572
      taxonomy 4:572
   taxonomy 4:86
   tissue tropism 4:86
      B19 virus 4:86
   transmission 4:86
   vaccines/vaccination 4:89
   virion structure 4:92f
   see also specific viruses
Parvovirus B19 4:95
   aplastic anemia 4:115
   transplant-associated transmission 3:468
      diagnostic assays 3:470–471
      syndromes associated 3:470
Parvovirus infections 2:555
   age-related disease 4:86
   diagnosis 4:89
   endosome role 4:87
   host range 4:86
   immune response 4:89
      maternal immunity 4:89
   insects 3:126t
   nuclear entry 4:87
   pathogenesis 4:89
   persistence 4:109t
   prevention 4:89
PASC *see* Pairwise sequence comparison (PASC)
Pasteur, Louis, animal virus cultivation 2:456
Patas monkeys, simian hemorrhagic fever virus infection 1:183
Patchouli virus X (PatVX) 4:311t
Patent infection 3:161
   NPV pathology 1:220
Pathogen-associated molecular patterns (PAMPs) 1:123
   definition 3:104
Pathogen-derived resistance (PDR) 4:156
   definition 1:563
   history 4:156
   papaya ringspot virus *see* Papaya ringspot virus (PRSV)
   RNAi role 3:152
Pathogenesis (of viruses) **5:314–319**
   definition 5:314
   entry 5:315
   infection 5:314
   localization 5:316
   resistance 5:318
   shedding 5:316
   spread 5:315
   susceptibility 5:318
      genetic loci 5:318
      mapping 5:318
   tissue tropism 5:316
   transmission 5:315, 5:316
   virulence 5:317
      genetic determinants 5:317
      measurement of 5:317
      pathogenic mechanisms 5:318
   see also specific infections
   see also specific viruses
Pathogenesis proteins (γb), hordeiviruses 2:463
Pathogen exclusion
   aquareovirus infection management 1:168–169
   yellow head virus infection management 5:482

Pathogenicity
    see also specific infections
Pathogenicity determinants, avsunviroids 5:339–341
Pathogen-specific molecular patterns
        (PAMPs) 5:119f
    dsDNA 3:114
    dsRNA 3:113
    envelope proteins 3:114
    ssRNA 3:113–114
Patois group viruses, reassortment 3:482
PATRIC 5:357t
Pattern-recognition receptors (PRRs) 3:106
    apoptosis 1:161
    innate immunity 3:107
    interferon type I 5:119f
    virus-mediated immune evasion 3:108
PAV1 virus 5:421
    genome 5:422
    virion composition 5:422
PAV strain *see* Barley yellow dwarf virus (BYDV)
PAZ domain, Argonaute proteins 5:151
PB2 protein *see* Influenza virus(es)
PBCV-1 virion
    genome 4:120f
        methylation 4:119
        open reading frames 4:122f
        polyamine biosynthesis 4:124
    replication 4:119
        assembly 4:119–122
        DNA synthesis 4:119–122
        host cell attachment 4:119
        host cell entry 4:119
        release 4:119–122
        transcription 4:119
    structure 4:118f
        capsid 4:118
        glycoproteins 4:118–119
        lipid membrane 4:118
PBFDV *see* Psittacine beak and feather disease virus
        (PBFDV)
PCPV *see* Pseudocowpox virus (PCPV)
PDB (Protein DataBank) 5:363
PDV *see* Prune dwarf virus (PDV)
PDZ domain(s), HPV E6 interactions 4:24
PE2 polyprotein, alphavirus assembly 1:197
Pea *(Pisum)*, resistance genes, loci 4:182
Peaches *(Prunus persica)*, Sharka disease 4:239
Peach mosaic virus (PcMV) 4:204
Peach rosette mosaic virus (PRMV) 3:406t
Pea early-browning virus (PEBV)
    disease recurrence 3:217
    movement studies 5:73
    transmission 5:75
        2b protein, role 5:75
        2c protein role 5:75
        nematodes 3:217
        seeds 5:74–75
Pea enation mosaic disease, prevention 5:213
Pea enation mosaic virus-1 (PEMV-1), transmission
        3:236
Pea enation mosaic virus-2 (PEMV-2)
    movement 5:212–213
        ORF3 5:212
    RNA satellites function 5:211
Peanut clump virus (PCV) 4:98
    genome 4:100f
        Indian peanut clump virus *vs.* 4:101
        intergenic regions 3:244
        open reading frames 4:101t
        RNA2 4:101
    geographical distribution 4:98
    infection
        diagnosis 4:98
        economic impact 4:98
        symptoms 4:98
    serological diagnosis 4:98–99
    transmission 4:98–99
    virus particles 4:99
Pea seed-borne mosaic potyvirus (PSbMV)
    3:215–216
    transmission 3:216

Pecluvirus 1:308, 4:98, 4:283
Pecluvirus(es) **4:97–103**
    biological properties 4:98
    furovirus *vs.* 2:295–296
    future work 4:103
    genome 4:99
        coding sequences 4:99
        noncoding sequences 4:100f
        open reading frames 4:99–100
        post-transcriptional gene silencing 4:99–100
        RNA1 4:99–100
        RNA2 4:99–100
        sequence comparisons 4:100
        t-RNA-like structure 4:100
    geographical distribution 4:98
    historical aspects 1:308
    infections *see below*
    molecular diagnosis 4:99
        nucleic acid probes 4:99
        reverse transcription PCR 4:99
    molecular properties 4:99
    particle assembly 4:101
    reservoirs 4:98
    seasonal distribution 4:98
    transmission 4:98
        Polymyxa vectors 4:98–99
        sap 4:98–99
        seeds 4:98–99
    virus-like particles 4:102
    virus particles 4:99
        origin of assembly sequence 4:101–102
    see also specific viruses
Pecluvirus infections
    'clump' disease 4:97–98
    control 4:102
        biocides 4:102
        continuous cropping 4:102
        cultural practices 4:102
        host plant resistance 4:102
    epidemiology 4:101
        host life cycle 4:102
    host range 4:98
    molecular diagnosis 4:99
    serological diagnosis 4:99
    symptoms 4:98
Pefudensovirus 4:76, 4:96
Pefudensovirus(es)
    classification 4:94t
    genome 4:95f
Pegylated interferon α-2, hepatitis B treatment 2:359
Pelargonium chlorotic ring pattern virus (PCRPV)
        4:210f
Pelargonium flower break virus (PFBV) 4:209f
    host ranges 4:208
Pelargonium leaf curl virus 4:210f
Pelargonium line pattern virus (PLPV) 4:209f
Pelargonium ring spot virus (PelRSV) 4:209
Pelargonium vein clearing virus (PelVCV) 4:210
Pelargonium virus infections 4:208
    nonspecific 4:210
    specific 4:208
    yellow net vein disease 4:210f
    see also specific viruses
Pelargonium zonate spot virus (PZSV) 4:210
Penaeid shrimps 2:37
Penaeus duorarum single nuclear polyhedrosis virus
        (PdSNPV)
    genome 4:570
    infection cycle 4:571
    nuclear hypertrophy 4:571
Penaeus japonicus (kuruma shrimp), infection 4:568
Penaeus monodon densovirus (PmDNV) 4:83–84
Penaeus monodon-type baculovirus (MBV) 4:570
    geographical distribution 4:571
    host range 4:571
    transmission 4:571
Penaeus vannamei singular nucleopolyhedrosis virus
        (PvSNPV) 4:570
Penelope-like elements (PLEs) 4:421–423
    distribution 4:442t
    history 4:437

    origins 4:441
    replication 4:426
    structure 4:440
    transposition 4:441
Penetration, definition 4:407
Penicillium chrysogenum virus (PcV) 1:503–504
    chryso-P3 protein 1:507
    evolutionary relationships 1:511
    genome 1:505f
        5' untranslated regions 1:506f
    virion structure 1:509f
        capsid 1:510f
        coat proteins 1:509
Penicillium stoloniferum virus F (PsV-F)
    genome 4:73
    virion morphology 4:73
Penicillium stoloniferum virus S (PsV-S)
    genome 4:72–73
    replication 4:66f
    virion morphology 4:72f
        capsid 4:73f
        dimeric protrusions 4:72–73
Pennisetum mosaic virus (PenMV) 1:481
Pennisetum mosaic virus, maize infections 1:481
Pentamer units, cowpea chlorotic mottle virus 5:385
Penton bases
    adenoviruses 1:17–18
    HSV capsid structure 2:397
Pepino mosaic virus (PepMV) 2:87
    antiserum 4:104–105
    capsid protein 4:104
    classification 2:87
    coat protein 4:106
    genome 2:87
        coat protein 4:106
        5'-nontranslated region 4:105
        intergenic region 4:105
        open reading frames 2:87
        TGB protein 1 4:105
        TGB protein 2 4:106
        TGB protein 3 4:106
        3'-nontranslated region 4:106
        triple gene block 4:105
    geographical distribution 2:87
    historical aspects 4:104
    infection *see below*
    isolates/strains 4:106
        full-length sequences 4:106
        relationships between 4:107
        restriction fragment length polymorphism
                4:106–107
        RT-PCR 4:106
        sequence data 4:106
        single-stranded conformation polymorphism
                pattern 4:106
    phylogeny 2:88f
    structure 2:87
    transmission 2:87
        mechanical transmission 4:108
        seed transmission 4:108
    viability 2:87
    virions 4:104
        capsid protein 4:104
        stability 4:104–105
Pepino mosaic virus infection
    control 2:88
    damage 4:107
    environmental conditions 4:108
    epidemiology 2:87
        geographical distribution 4:107
        mixed infections 4:107
        population studies 4:107
        strain dependency 4:107
    host range 2:87
        alternative hosts 2:87
        experimental 2:87
        natural hosts 4:104
        weed hosts 4:104
    mixed infections 4:108
    symptoms 2:87
        inoculation studies 4:104

Pepino mosaic virus infection (*continued*)
  tomato 2:87
    symptoms 2:87
    yield losses 2:87
Pepper *(Capsicum)*
  cucumber mosaic virus infection 5:287–288
  potato virus Y infections *see* Potato virus Y (PVY)
  resistance genes, loci 4:182
Pepper mild mottle virus (PMMoV) 5:288
Pepper ringspot virus (PeRSV)
  geographical distribution 5:76
  host range 5:76
Pepper veinal mottle virus (PVMV), resistance to 4:184
Pepper vein banding virus (PVBV), nuclear inclusion a protein 4:317–319
Pepsin, host defence against rotaviruses 4:512
Peptide editing, MHC-II antigen presentation 1:124–125
Peptidoglycan A 5:365
Peptidoglycan-hydrolyzing enzymes, bacterial cell viral entry 5:366*f*
Peptidyl-prolyl cis transferase-isomerase (PPIase), role in cell lysis 3:19
Perforin(s) 3:80
Pericarditis 1:533
Perina nuda virus (PnV) 3:45
Perinatal disease, coxsackieviruses 1:585
Perinet virus 4:184*t*
Periplaneta fuliginosa densovirus (PfDNV)
  classification 4:77
  infection 4:79
Persistent infections **4:108–116**
  acute *vs.* 4:110*f*
  apoptosis 4:113
  chronic disease manifestations 4:114
    pregnancy 4:114–115
  concepts 4:108
  definition 1:335
  immune control 4:112
    adaptive response 4:112
    cytokines 4:112
    dendritic cells 4:112
    dysregulation 4:113
    immunological memory 4:111*f*
    interferons 4:112
    macrophages 4:112
    pathogen–host interactions 4:112
  immunocompromised patients 4:115
  immunomodulation 4:113
    apoptosis 4:113
    cell stress 4:113
    TLR signaling
  immunopathology 4:113
  latent *vs.* 4:109
  pathogenesis 4:111*f*
    reactivation 4:111
  vaccines/vaccination 4:108–109
    booster 4:109–110
  viruses associated 4:112
Persistently associated transcript (PAT1), Hz-1V replication 3:146
Person-to-person transmission, monkeypox virus 4:642*f*
Pertussis vaccine, safety 5:228
Peste des petits ruminants virus (PPRV)
  epidemiology 4:45
  epizootiology 4:503*f*
  geographical distribution 4:499*f*
  host range 4:499
  propagation 4:500
Pestivirus 1:336, 2:242, 2:246, 5:108, 5:440–441
Pestivirus(es) 2:246
  antigenic relationships 1:338
  bovine viral diarrhea viruses 1:378
  classification 1:336
  E1 protein 2:247–248
  E2 protein 2:247–248
  genetic relationships 1:338
  genome 2:247*f*
    ORF processing 2:247–248

history 2:246
infection 2:246
  control 2:248
  diagnosis 2:248
  disease manifestations 2:246–247
  epidemiology 2:246
  host range 2:246
  pathogenesis 2:248
  persistence 2:246
  prevention 2:248
physical properties 2:247
replication 2:247
structure 2:247
transmission 2:246
see also specific viruses
Petechiae, definition 3:454
Petunia vein banding virus (PetVBV) 4:211
Petunia vein clearing virus (PVCV) 4:212*f*
  transmission 1:462
  host genome integration 1:462
Petunia virus infections 4:211
  nonspecific 4:211
  specific 4:211
  see also specific viruses
Petuvirus
  host chromosome integration 1:462
  open reading frames 1:466*f*
  virion structure 1:458–460
Petuvirus 1:459*t*
PEV *see Porcine enterovirus(es)* (PEV)
Peyer's patches, poliovirus infection 4:243
Pf1 bacteriophage *see* Bacteriophage Pf1
Pf2 bacteriophage 1:201–202
Pf3 bacteriophage *see* Bacteriophage Pf3
Pf4 bacteriophage 3:123–124
PFAM database 5:357*t*
P gene, henipavirus 2:323
PG virus 5:413*t*
PHABULOSA (PHB) 4:142
PHABVOLUTA (PHV) 4:142
Phacochoerus aethiopicus (warthog), African swine fever virus 1:44*f*
Phaeovirus 4:117*t*
Phage *see* Bacteriophage(s)
Phage display *see* Antibody phage display
Pharyngoconjunctival fever 2:554–555
Phaseolus vulgaris see Bean *(Phaseolus vulgaris)*
Phase transitions 4:359
Phenotypes 4:40
  species classification 5:404
Phi6 bacteriophage 4:379–380
Φ29 bacteriophage *see* Bacteriophage ϕ29
ϕCh1 virus 5:418
  genome 5:419*f*
    methylation 5:418
    open reading frames 5:418
  structure 5:418*f*
ϕF1 virus 5:413*t*
ϕF3 virus 5:413*t*
ϕH virus 5:413*t*
  genome 5:412
  morphology 5:412
  transfection 5:412
ϕN virus 5:412
ϕX174 bacteriophage *see* Bacteriophage ϕX174
Philosamia cynthia X ricini virus (PxV) 5:28*t*
Φ_L promoter 5:179
Φ_M promoter 5:179
Phlebotomine flies, definition 2:234
Phlebotomus fever virus(es) 4:490
  see also specific viruses
Phlebotomus perniciosus, Toscana virus transmission 4:491
Phlebovirus 1:391, 3:480, 4:490–491, 5:27
Phlebovirus(es) **4:490–497**
  control 4:496
  definition 4:490
  G2 proteins 3:310
  genome 4:492*f*
  history 4:491
  infections *see below*

NSs protein 1:395
species 4:493*t*
structure 4:490–491
transmission prevention 4:496
vaccines/vaccination 4:496
see also specific viruses
Phlebovirus infections
  diagnostic procedures 4:494
    isolate identification 4:494
    serologic tests 4:494
    virus detection 4:494
    virus recovery 4:494
  epidemiology 4:494
  prevention 4:496
  therapy 4:495
Phloem
  banana bunchy top virus 1:273
  barley yellow dwarf virus replication 1:284*f*
  definition 5:209
pH-mediated fusion
  herpesvirus gB structure 3:300
  influenza-host cell entry 3:486*f*
  VSV G structure 3:300
Phocine distemper virus (PDV)
  epizootiology 4:502–504
  geographical distribution 4:498–499
  history 4:498
  propagation 4:500
Phosphatidylserine, VSV-host cell attachment 5:292–293
Phosphoenolpyruvate carboxykinase, reticuloendotheliosis virus 4:416
Phospholipase A2
  BK virus-host cell attachment 4:281
  densovirus 4:81
  parvovirus 4:78
Phospholipid bilayer, BK virus infection, role 4:281
Phosphoprotein P *see* Chandipura virus
Phosphoproteins
  HCMC 2:489–490
  murine polyomavirus middle T 4:274–275
  VSV 5:295
Phosphorylation, varicella-zoster virus proteins 5:261
Phosphotransferase, sawfly baculovirus genome 1:227
Phosphotungstic acid, negative staining 2:81
Photolyases, avipoxvirus survival factors 2:275
Phthorinaea operculella granulovirus (PhopGV) 1:214*f*
Phycodnaviridae 1:87, 4:118
Phycodnavirus(es) **4:116–125**
  composition 4:118
  ecology 4:124
  evolution 2:181–182
  genomes 4:119
  history 4:116
  replication 4:119
    assembly 4:119–122
    DNA synthesis 4:119–122
    host cell attachment 4:119
    host cell entry 4:119
    release 4:119–122
    transcription 4:119
  structure 4:118
  viral encoded enzymes 4:123
    sugar metabolism 4:123
  see also specific viruses
Phylogenetics 4:126
  alignment algorithms 5:343
  bootstrap analysis 4:127
  definition 2:130
  distinct genes 4:127
  fungal retrotransposons 4:422*f*
  genomics 4:127
  homologous recombination 4:127
  'in-group' 4:127
  LTR retrotransposons 4:441
  maximum likelihood 4:126–127
  maximum parsimony 4:126–127
  molecular clock models 4:127
  neighbor-joining methods 4:126–127
  non-LTR retrotransposons 4:441

'outgroup' 4:127
satellite viruses 4:534
sequence alignment 4:126, 5:343
species differences 4:126–127
unweighted pair group method with arithmetic means 4:126–127
see also specific viruses
Phylogenetic trees 4:126
branch length 4:126
definition 2:105
hypothetical taxonomy units 4:126
most recent common ancestor 4:126
operational taxonomy units 4:126
topology 4:126
see also specific viruses
Phylogeny 4:125–129
antibody-mediated immunity 3:58
applications 4:127
classification 4:128
definition 1:441
diversity 4:125–126
geographical distribution 4:128
goals of 4:125
identification 4:127
molecular clock model 4:125
monitoring 4:127–128
new virus identification 4:128
outbreak control 4:127
virus–host associations 4:128
zoonotic reservoir identification 4:128
see also specific viruses
Physical contact, Simian retrovirus D transmission 4:627–628
Phytoene desaturase (PDS), virus-induced gene silencing role 2:466
Phytophthora 2:288
edornaviruses 2:111
Phytophthora endornavirus 1 (PEV1) 2:111
Phytoplankton
ecological dynamics 1:94–95
viral infection of 1:87
Phytoreovirus 1:483, 1:484, 4:150–151, 4:536t
Phytoreovirus(es) 4:536t
characteristics 4:150t
classification 4:150
genome 4:152
dsRNA segments 4:153f
size 4:150–151
host range 4:154
transmission 4:154
type species 4:536t
virion structure 4:152
capsid 4:152
see also specific viruses
Phytoreovirus P7 proteins 1:507
Phytosanitation, cassava mosaic disease control 4:198
PI-3 kinase, Jaagsiekte sheep retrovirus infection 3:181
Picobirnavirus(es) 1:321–328
definition 2:93
detection 1:327
genome 1:327
HIV co-infection 2:94
mutation 1:327
RdRp 1:327
RdRp structure 1:327
virion structure 1:327
see also specific viruses
Picorna-like capsids 5:398
Picorna-like virus(es) 2:37
shellfish viruses 4:565
see also specific viruses
Picornaviridae 1:90f, 1:442, 1:542, 1:573, 1:580–581, 2:65–66, 2:266, 2:343, **4:129–140**, 4:130t, 4:243–244, 4:244t, 4:468, 5:2–3, 5:11t, 5:20, 5:38
Picornavirus(es) 2:65–66
apoptosis inhibition 1:158t
capsid 4:132f
canyon region 4:131
VP1 4:131

VP2 4:131
VP3 4:131
VP4 4:131
capsid structure 1:201f
classification 4:130t
3D^{pol} 4:137
fish 2:229
gene expression 4:131
genomes 4:131f
internal ribosome entry sites 4:134f
noncoding regions 4:130
infections
antibody-mediated neutralization 3:416
insects 3:126t
respiratory infection 2:552
internal ribosome entry sites 4:130, 4:133, 4:133f, 4:138–139
hepatitis A virus replication 2:344
hepatitis C virus replication 2:370
PABP interactions 4:138–139
PCBP interactions 4:135
PTB interactions 4:134f
replication 4:133
type I 4:133
type II 4:133
type vs. type II 4:134–135
neutralizing antibodies 1:138
polyprotein processing 4:130
process 5:20
RdRp 4:137
replication 2:127
cis-acting replicating elements 4:139
host cell entry 4:131
internal ribosome entry sites *see above*
polyprotein processing 4:136f
RNA replication 4:138f
RNA synthesis 4:139
translation 4:132
uncoating 4:409
VPg role 4:132–133
RNA recombination 4:377
structure 5:391–392
viral proteinase substrates 4:135
capsid proteins 4:135
protein cleavage shutoff 4:137
RNA replication proteins 4:136
translation inhibition 4:137
see also specific viruses
Picornavirus-like superfamily 3:44
Pig(s)
lymphoproliferative disease 2:428
post-weaning multisystemic wasting syndrome 1:515–516
vesicular exanthema 1:410
see also under Porcine
Pigeon pea sterility mosaic virus (PPSMV) 3:216–217
Pigeonpox virus(es) 2:278
pIII, bacteriophage fd 2:194
Pike fry rhabdoviruses (PFRV) 2:228
Pili 3:21
filamentous ssDNA bacteriophage infection 5:369
F-type 2:191
N-type 2:191
type IV 2:191
icosahedral enveloped dsDNA bacteriophages 5:369
icosahedral ssRNA bacteriophages 5:369–370
viral entry 5:366
Pilin, filamentous ssDNA bacteriophage infection 2:191
Pineapple bacilliform virus (PBV) 1:403
Pinholins
definition 3:249
holin–endolysin lysis 3:254
Pinniped calicivirus infection, pigs 1:414–415
Pinnipeds
calicivirus infection 1:414–415
definition 1:410
Pinwheel, definition 1:483
Pirie, Norman, tobacco mosaic virus 3:399
Pixuna virus (PIXV), characteristics 5:109t

Plant(s)
development 5:330
post-transcriptional gene silencing 3:410
removal, disease control 1:35, 3:412
Plant(s), natural disease resistance (dominant genes) **4:170–177**
avirulence genes 4:173
NB-LRR protein 4:171, 4:172t, 4:175
coiled-coil structure 4:171
divergence 4:171
TIR-NB-LRR 4:171
R gene-*Avr* recognition mechanisms 4:175
guard hypothesis 4:175
products 4:171
responses to viruses 4:173
R genes
durability 4:175–176
polymorphism 4:171
products 4:171
responses to viruses 4:173
signaling mechanisms 4:176
Sw-5 gene 4:174
Plant(s), natural disease resistance (recessive genes) **4:177–187**
eukaryotic initiation factor-4E *see* Eukaryotic initiation factor-4E (eIF4E)
gene types 4:178
mechanisms 4:182
Fraser's hypotheses 4:182
translation initiation factors, roles 4:183
molecular structure 4:182
pathotypes 4:182
phenotypes 4:182
prevalence 4:178, 4:179t
VPg role 4:178–182
Plantago asiatica mosaic virus (PlAMV) 4:311t
Plantago severe mottle virus (PlSMoV) 4:311t
Plantain virus X (PlVX) 4:311t
Plant breeding
bean golden/yellow mosaic disease resistance 1:299–300
beet curly top virus resistance 1:307
papaya ringspot virus resistance 4:5
potato virus Y resistance 4:293, 4:294f
Sharka disease resistance 4:242
Plant chrysovirus(es)
occurrence 1:511
see also specific viruses
Plant–fungal mutualistic associations 2:291
Planthoppers
fijivirus transmission 4:150t, 4:154
oryzavirus transmission 4:150t, 4:154
rice ragged stunt virus transmission 1:485
tenuivirus transmission 5:26
Planting times, barley yellow dwarf virus control 1:286
Plant pararetrovirus(es)
endogenous 1:464–465
genera 1:464–465
see also specific viruses
Plant-produced vaccines *see* Vaccines/vaccination
Plant reovirus(es) **4:149–156**, 4:150t
classification 4:150
distribution 4:154
gene expression 4:152
genome 4:151
host range 4:150, 4:154
infection 4:149–150
control 4:155
diagnosis 4:155
symptoms 4:155
pathogenicity 4:155
replication 4:152
taxonomy 4:150
transmission 4:150t, 4:154
virion structure 4:151
see also specific viruses
Plant retinoblastoma-related protein (pRBR), maize streak virus 3:267
Plant rhabdovirus, transmission 4:195
Plant viroids/virusoids, ribozymes of 4:476

Subject Index

Plant virus(es)
  history 2:450–455
    biochemical age 2:451
    biological age 2:451
    biophysical age 2:451
    crop resistance 2:454
    enzyme-linked immunosorbent assays 2:453
    heat treatments 2:454
    meristem tip culture 2:454
    molecular biology age 2:453
    nucleotide sequence information 2:453
    prehistory 2:450
    purification 2:453
    replication 2:453
    vectors 2:453–454
    viral entity recognition 2:450
    yeast systems 2:453
  host cell receptors 4:408
  infections *see below*
  morphologies 2:22f, 2:23f
  recombination 4:376
  replication 2:453
  vectors *see* Plant virus vectors;
  see also specific viruses
Plant virus infections
  diagnosis 2:18–29
    applications 2:19
    functional studies 2:26
    host plant reactions 2:21, 2:24f
    immunoelectron microscopy *see below*
    see also specific techniques
  economic aspects 4:156, **4:197–201**
    chlorophyll defects 3:218
    climatic regions 4:197
    faba bean necrotic yellows nanovirus infection 3:218
    transmission 4:197
    yield losses 4:197
  emerging/re-emerging viruses **2:86–93**
  immunoelectron microscopy 2:20, 2:21f, 2:23f
    antibody reactivities 2:20–21
    mixed infections 2:20–21
  movement **3:348–356**
    host factors 3:349
    intercellular 3:348, 3:352
    intracellular 3:348, 3:349, 3:351f
    RNA viruses 3:350
    systemic 3:348, 3:353
    transport/accumulation regulation 3:354
    tubule forming viruses 3:353
    see also Movement proteins (MPs)
  nucleotide sequence analysis 2:21
    DNA viruses 2:21
    RNA viruses 2:21
  physical studies 2:26
  primary diagnosis 2:19
    symptom inspection 2:19, 2:20f
  routine detection 2:23
  structural property studies 2:26
  vectors *see below*
  virus-encoded proteins 3:332
  see also specific viruses
Plant virus vectors
  associated viruses 4:229
  deconstruction 4:229, 4:230
  definition 2:18, 5:275–276
  evolution 5:281
    viral variant distribution 5:281
    virus acquisition patterns 5:281
  history 4:230
  insects *see below*
  non-insect vectors associated 5:277t
    fungi 5:281
    mites 5:280
    nematodes 5:281
  vegetable virus transmission 5:284–285
    control 5:286–287
  see also specific vectors
Plant virus vectors, insects 5:276
  beetle-mediated transmission 5:280
  circulative transmission 5:276f, 5:277, 5:277t

    nonpropagative 5:278f
    propagative 5:277
  classification 5:276
  noncirculative transmission 5:276f
    capsid strategy 5:280f
    helper strategy 5:279
    see also Helper component
    host cell attachment 5:279
    retention sites 5:279
  see also specific insects
Plaque assays 2:72f
  fusellovirus detection 2:297–298
  history 2:456
Plaque-forming unit (Pfu) 3:21
Plaque reduction (PRNT) 4:494
Plasmacytoid dendritic cells (pDC) 3:112
Plasmalemma, ascovirus infection 1:191
Plasmaviridae 5:11t
Plasmavirus(es), genome delivery 5:368
Plasmids, DNA vaccines *see* DNA vaccine(s)
Plasmodesmata
  caulimovirus infection 1:464
  definition 1:457
  plant virus movement 3:348
  TMV-associated changes 5:71
Plasmodiophorid fungi *see Polymyxa* (plasmodiophorid fungi)
Plasmodiophorid fungi (*Polymyxa graminis*)
  furovirus transmission 2:292
  legume virus transmission 3:217
  pomovirus transmission 4:285f
  rice necrosis mosaic virus transmission 1:489
  rice stripe necrosis virus transmission 1:489
Plastic mulches, infection control 5:437
Plastochron 3:264
Platelet-derived growth factor receptor β (PDGFβ), HPV E5 4:23
Platforms, *Agrobacterium tumefaciens*-based expression system 4:237f
Plautia stali intestine virus (PSIV), gene expression 5:5
Pleconaril 2:70
  rhinovirus infection 2:556–557
Pleiotropic mutations 4:177
Pleomorphic dsDNA bacteriophages, genome delivery 5:368
Pleurodynia
  coxsackieviruses 1:585
  definition 1:580
'Plowright vaccine' rinderpest viruses 4:506
Plum (*Prunus*)
  raspberry bushy dwarf virus infection 3:39
  Sharka disease 4:239
  see also entries under Prunus
  see also specific viruses
Plum pox virus (PPV) 4:201
  genome 4:239
    noncoding region 4:317
  geographical distribution 4:238
  historical aspects 4:201
  infections *see* Sharka disease
  strains/groups 4:240f
    agar double diffusion assays 4:239
    evolution 4:240–241
    full length genome sequencing 4:240
    PAGE 4:239
    PPV-C (cherry) 4:240
    PPV-D (*Dideron*) 4:240
    PPV-EA (El Amar) 4:240
    PPV-M (*Marcus*) 4:239–240
    PPV-Rec (*recombinase*) 4:240
    PPV-W (Winona) 4:240
    RFLP 4:241
    serology 4:241
  symptoms 4:201
  transmission 4:238
    aphid vectors 4:241
    efficiency 4:241
    nonpersistent infections 4:241
    seed transmission 4:241
  virion 4:239
Plus-strand synthesis, alfalfa mosaic virus 1:84–85

PM2 (corticovirus) *see* Corticovirus PM2
PM2 bacteriophage, genome delivery 5:368
PML-containing nuclear oncogenic bodies, adenovirus E1A gene/protein interactions 1:12
PMS1 virus 5:413t
Pneumabort K 2:419
Pneumocystis jirovecii
  definition 4:623
  HIV-associated opportunistic infection 1:52
Pneumonia
  cats, cowpox virus infection 1:579
  definition 2:551
  HSV associated 2:391
  human respiratory syncytial virus infection 2:548
  mimivirus infection 3:312
  SARS 4:552
Pneumonitis 5:424
Pneumovirinae 4:40, 4:41t, 4:47, 4:52
Pneumovirus 4:41t
Pneumovirus(es)
  classification 4:41t
  genome 3:327f
    leader region 4:42
  M2 gene 3:332
  nonstructural proteins 3:332
  nucleocapsid 3:325–329
  propagation 4:40–42
  see also specific viruses
Poa semilatent virus (PSLV) 2:459
  coat proteins 2:459
  genome 2:459–460
  pathogenesis proteins 2:463
  TGB3 movement protein 2:464
Pock 4:639
Pocket factor, enterovirus capsid stabilization 2:128
Pocket protein-binding motif, HPV E7 4:25
Podospora anserina, prion 4:337
Podosta disease *see* Sindbis virus (SINV)
Podoviridae 3:31, 3:32t, 5:11t
Podovirus(es)
  DNA injection 3:33–34
  genome packaging 2:306
  short tails 5:368
  structure 1:434f
Pogosta disease, seasonal distribution 5:90
Poinsettia latent mottle virus (PnLMoV) 4:651f
Poinsettia mosaic virus (PnMV) 5:204f
  characteristics 5:206
  genome 5:202f
  host range 5:206
  properties 5:206
  transmission 5:206
Point mutations, bacteriophage evolution 5:371
Point-of-care tests (POC), principles 2:33
Poisson distribution, infection neutralization 3:416–417
Pol (polymerase), LTR retrotransposons 4:438–439
Polerovirus 3:231, 3:232f, 3:232t
pol gene/protein
  LTR retrotransposons 4:420
  Simian retrovirus D 4:627
  SIV serological relationships 4:607–608
polh promoter, baculovirus 1:238
  *flash*BAC 1:244
Poliomyelitis 2:65–66
  acute 1:474
  disease surveillance 2:48f
    data collection sources 2:48f
  immune system evasion 3:108
  pathogenesis 4:242
  Peyer's patches 4:243
  systemic infection 4:243
  TMEV infection 5:42
  transmission 4:243
  vaccine-derived 4:246
Polio vaccines/vaccination 2:458
  disease eradication 5:232
    developing countries 4:245–246
    Northern Europe 4:245–246
    tropical climates 4:245–246
  HIV-1 origin, human-induced theory 2:530f

inactivated 4:244
monovalent 4:245–246
National Immunization Days 4:245–246
Nigeria
   circulating vaccine-derived poliovirus 4:246–248
   disease eradication 4:246
   National Immunization Days 4:246
oral 4:244
post-vaccination virus excretion 4:245
Sabin strains 4:244
   mutation 5:317
   type 1 4:244–245
   type 2 4:244–245
   type 3 4:245
safety
   adverse reactions 5:230t
   SV40 contamination 4:630
   vaccine-associated paralytic poliomyelitis 4:246
vaccination program cessation 4:249
Poliovirus 4:129
  Ab-mediated neutralization 3:416
  cell surface binding 2:129
  classification 4:129
  genome 4:243
    noncoding regions 4:139
    sequencing 2:143
  geographical clustering 4:244
  host cell receptor 4:131–132
  infection see Poliomyelitis
  neural cell infection 1:473
  protein 3A 2:579–580
  proteinase substrates
    protein 2C 4:136
      structure 4:136–137
    protein 3AB 4:137
  proteolytic processing 2:127
  RNA translation 2:457
  serotypes 4:243
  vaccines/vaccination see Polio vaccines/vaccination
  viremia 4:243
  virulence 5:318
    vaccine reversion 4:246
  virus structure 4:132f
Poliovirus 4:243
Poliovirus type 3, structure 5:382f
Pollen
  gene flow in transgenic plants 4:162
  Oryza sativa endornavirus 2:114
POL polyprotein, caulimoviruses 1:467
pol proteins
  see specific viruses
Pol reading frame, avian retroviruses 4:456
Polyacrylamide gel electrophoresis (PAGE), plum pox virus strains/groups 4:239
Polyactide-L-glycolide (PLG), DNA vaccine delivery 2:54
Polyadenylation
  cauliflower mosaic virus transcription 1:467–468
  definition 1:296
Poly-A retrotransposons *see* Target-primed retrotransposons
Polyarthritis 1:399
Poly(A) binding protein (PABP)
  alfalfa mosaic virus coat protein 1:83–84
  picornaviruses 4:138–139
Polycistronic mRNA, single-stranded positive sense RNA viruses 4:410
Polycistronic RNA 5:145
Polyclonal antiserum 1:296
Polydnaviridae 1:247, 4:256–257, 5:11t
Polydnavirus(es) (PDV) **4:256–261**
  apoptosis inhibition 1:158t
  classification 4:256
  definition 4:256
  genome 4:257
    gene expression 4:258
    sequence analysis 4:257–258
    transposable elements 4:257–258
  infection *see below*
  life cycle 4:260f
    assembly 4:260–261
    virion accumulation 4:260–261

phylogeny 4:261
replication 4:260
  gene expression 4:260–261
taxonomy 4:256
virion structure 4:257
  brachovirus 4:257
  bracovirus 4:257
  ichnovirus 4:257
virulence genes 4:250
see also specific viruses
Polydnavirus infections 4:253
  immune evasion 4:253
    virulence genes 4:250
  insects 3:126t
  virus-mediated immune abrogation 4:253
  wasp–virus mutualism 4:250
Polyhedra *see* Occlusion bodies
Polyhedrin 1:238
  baculoviruses 1:251
  definition 1:254
  gene 1:265
  molecular arrangement 3:136
  structure 3:136
Polyhedron envelope/calyx protein (PEP/P10) 1:217
Polylactosaminoglycan 2:543
Polymerase chain reaction (PCR) 2:34f
  advantages 2:35
  definition 2:29
  degenerate primers 2:420, 2:421f
    detection spectrum 2:423
    method variations 2:421
    primer binding sites 2:422f
  detection/diagnosis
    adenovirus 1:29
    anellovirus 1:109
    avipoxviruses 2:280
    banana bunchy top virus 1:273
    bean golden mosaic virus infection 1:299
    bean golden/yellow mosaic disease 1:297–298
    bean golden yellow mosaic virus 1:299
    benyvirus 1:313
    cacao swollen shoot virus 1:405
    Dengue virus 2:8–9
    emerging viral diseases 2:93
    fusellovirus 2:299
    geminiviruses 4:167
    hepatitis B virus 2:351
    herpesviruses 2:421f
    orthobunyavirus 3:482–483
    plant viruses 2:25f
    potyvirus 4:316
    pseudorabies virus 4:350
    rabies virus 4:372
    rhinovirus 4:471
    Taura syndrome virus 5:8
    VSV 5:298
  internal controls 2:35
  limitations 2:33–34
  multiplexing 2:25
  principles 2:33–34
    amplification 2:34
    product visualization 2:34f
    sequencing 2:35
  reverse transcription *see* Reverse transcription polymerase chain reaction (RT-PCR)
  vaccine safety 5:228
Polymerase complex (PC)
  definition 3:6
  icosahedral enveloped dsDNA bacteriophages 3:8f
Polymerase inhibitors 1:149
Polymerases
  discovery 2:456–457
  icosahedral enveloped dsDNA bacteriophages 3:12
  large 5:296
    conserved domains 5:296
  see also specific types
Polymerase stuttering, mumps virus 3:358–359
Polymorphisms
  definition 5:186
  nonsynonymous 2:471–472

Polymyxa (plasmodiophorid fungi)
  benyvirus transmission 1:309
  furovirus transmission 2:292
  legume virus transmission 3:217
  pecluvirus transmission 4:98–99
  rice necrosis mosaic virus transmission 1:489
  rice stripe necrosis virus transmission 1:489
Polymyxa graminis see Plasmodiophorid fungi *(Polymyxa graminis)*
Polyoma enhancer, polyomavirus 4:272
Polyomaviridae 4:34–35, 4:277, 4:631
Polyomavirus 4:631
Polyomavirus(es) **4:277–282**
  apoptosis inhibition 1:158t
  classification 4:277
    papillomaviruses *vs.* 4:277
  evolution 2:182–183
  genome 4:278
    functional regions 4:278
  host cell susceptibility 4:281
    DNA-binding proteins 4:281
    non-coding region hypervariability 4:281
  infections
    associated nephropathy 4:269
    respiratory 2:556
  murine *see below*
  physical properties 4:277
    stability 4:278
  physicochemical properties 4:277
    virion assembly 4:278
  proteins 4:278
    agnoprotein 4:278
    capsid 4:278
    T proteins 4:278
  replication 4:409–410
    control 4:281
    cytoplasmic entry 4:279
    nuclear entry 4:279
    receptors 4:279
  structure 5:388
  taxonomy 4:34–35
  virion morphology 4:277
    capsomers 4:277
    structural proteins 4:277
  see also specific viruses
Polyomavirus(es), murine **4:271–277**
  cytoplasmic entry 4:279
    endocytosis 4:279
    host cell receptor binding 4:279
    large plaque isolates 4:279
    small plaque isolates 4:279
  genome 4:272f
    genetic analysis 4:272
    polyoma enhancer 4:272
    T antigens 4:271–272
    see also specific antigens
  historical aspects 4:271
  infection 4:273
    abortive transformation 4:273
    entry processes 4:273
    kinetics of 4:273
    stable transformation 4:273
  large T 4:275
    cellular DNA synthesis 4:276
    dihydrofolate reductase activation 4:276
    infection kinetics 4:273
    replication 4:273
    retinoblastoma gene product binding 4:276
    structure 4:276f
    thymidine kinase activation 4:276
    transcription 4:273
    transformation 4:276
  middle T 4:274
    cell signaling 4:274f
    membrane association 4:274
    phosphorylation 4:274–275
    PTK complex 4:274
    small T complementation 4:275
    small T *vs.* 4:273–274
    src family tyrosine kinases 4:274
    transcription 4:273

588 Subject Index

Polyomavirus(es), murine (continued)
    transformation 4:274
        transgenic models 4:274
    as models 4:271
    particle structure 4:271
        VP1 structural protein 4:271
        VP2 structural protein 4:271
        VP3 structural protein 4:271
    replication 4:272
    transcription 4:272
    RNA transcription 4:273
        ADAR enzyme 4:273
        early proteins 4:273
        T proteins 4:273
    small T 4:275
        cellular distribution 4:275
        cellular functions 4:275
        middle T complementation 4:275
        middle T vs. 4:273–274
        p53-induced apoptosis 4:275
        transcription 4:273
        transformation role 4:275
    transformation 4:273
        host effects 4:277
    tumorigenesis 4:271
Polyphyletic origins, viral evolution 2:176
Poly(A) polymerase, Amsacta moorei entomopoxivirus 2:139
6K Polyprotein, alphavirus 1:197
Polyprotein(s) 4:129
    alphavirus assembly 1:197
    definition 2:343, 4:407
    leishmaniaviruses 3:222
    rice yellow mottle virus 4:487
Polypurine tract (PPT) 4:428
Polypyrimidine tract binding protein (PTB), IRES interactions 4:134f
Polythetic class(es) (classification)
    definition 5:401
Pome fruits, viral infections see Fruit trees, diseases; see specific viruses
Pomovirus 1:308, 4:283
Pomovirus(es) **4:282–287**
    classification 4:283
    control 4:286
        genetic modification 4:287
        resistant cultivars 4:287
    CP gene/protein 4:284
    genome 4:283
        encapsidated RNA2 4:284
        encapsidated RNA3 4:284
        open reading frames 4:283
        RT proteins 4:284
        TGB proteins 4:283–284
        untranslated regions 4:283
        virus movement proteins 4:283–284
    geographical distribution 4:284
    historical aspects 1:308
    infection
        control 4:286
        diagnosis 4:286
        host range 4:284
        virus–host interactions 4:284
    movement 4:285f
    particles 4:283
    reverse-phase PCR 4:286
    serological relationship 4:285f
    taxonomy 4:283
    transmission 4:284
        plasmodiophorid vectors 4:285f
    see also specific viruses
Pongine herpesvirus 1 (PoHV-1) 4:587t
Pongine herpesvirus 2 (PoHV-2) 4:587t
Pongine herpesvirus 3 (PoHV-3) 4:587t
Population ecology, bacteriophages see Bacteriophage(s), ecology
Population genetics
    fungal retrotransposons 4:427
    HIV antigenic variation 1:133
Population studies, Pepino mosaic virus 4:107
Porcine adenovirus(es) 1:26

Porcine circovirus (PCV)
    chimeras 1:516
    genome 1:515f
        conserved regions 1:516
        open reading frames 1:517
    proteins, functional analyses 1:518
        DNA binding 1:518
        localization 1:519
        P_{rep} repression 1:518–519
        protein–protein interactions 1:519
        replication 1:518
    replication 1:518
        conserved elements 1:518
        origin of replication 1:518
    structure 1:514f
    transcription 1:517
        P_{cap} 1:517–518
        P_{rep} 1:517–518
    type 1 1:516
    type 2 1:516
        host cell attachment 1:517–518
        immune system interactions 1:516
        post-weaning multisystemic wasting syndrome 1:515–516
        tissue tropism 1:518
Porcine enterovirus(es) (PEV) 2:124
    genome 2:126
Porcine lymphotropic gammaherpesvirus 1 (PLHV-1) 2:428
Porcine lymphotropic herpesvirus 3 (PLHV-3) 2:421
Porcine parvovirus(es) (PPV) 4:89
Porcine respiratory and reproductive syndrome virus (PRRSV) 1:176–177
    geographical distribution 1:177
    infection see below
    structure 1:178f
Porcine respiratory and reproductive syndrome virus infection
    cell tropism 1:182–183
    clinical features 1:183
    host range 1:177
    immune response 1:184
        neutralizing antibodies 1:184
    pathogenicity 1:183
    vaccination 1:185
Porcine torovirus (PToV) 5:151–152
    antigenic properties 5:155
    classification 5:151–152
    genotypes 5:155
Porencephaly, definition 1:76
por mutation (MurA) 3:257–258
Portal 3:30
Portal protein 1:433
    bacteriophage structure 1:437
Porthos latent virus (PLV), genome 5:147f
Portunus depurator (Mediterranean crab), shrimp virus infection 4:568
Positive-sense RNA strand, necrovirus 3:403
Positive staining, negative staining vs. 2:79
Positive strand RNA 4:129
Pospiviroidae 5:11t, 5:333t, 5:335
Pospiviroids
    classification 5:335
    monomer circularization mechanism 4:479–480
    replication 4:478f
    see also specific viruses
Post-attachment neutralization (PAN) 3:415
    kinetics 3:416
Postexposure treatment, rabies virus 4:373
Postherpetic neuralgia (PHN) 5:253, 5:262
Postlarvae, definition 4:567
Postmarketing surveillance (vaccine safety) 5:231
Post-transcriptional gene silencing (PTGS) 5:375–376
    African cassava mosaic disease 1:33
    βC1 role 1:318
    cis-acting siRNA pathway 4:142–145
        IR-PTGS 4:142–145
        S-PTGS 4:142–145
    cotton leaf curl disease 1:568

    definition 4:97
    exploitation of 5:376
    luteoviruses 3:234–235
    pecluvirus 4:99–100
    potato virus Y helper component protein 4:288
    resistance, engineered 4:161
    tombusviruses 5:148–149
    vector competence 5:271
    VIGS system development 5:375–376
Post-transcriptional modifications
    arenavirus glycoproteins 3:244
    toroviruses 5:154
Post-transplantation lymphomas (PTLs), EBV associated 2:152f
Post-weaning multisystemic wasting syndrome (PMWS) 1:515–516
    diagnostic criteria 1:516
    experimental disease reproduction 1:515–516
    mortality rate 1:515–516
Potamochoerus porcus (bushpig), as an African swine fever virus host 1:43
Potato (Solanum), infection resistance
    breeding 4:297–302
    extreme resistance 4:303f
    gene duplication 4:172–173
    N genes 4:303
    R genes 4:303
    types 4:297–302
Potato 14R virus (P14V) 4:298t
Potato aucuba mosaic virus (PAMV) 4:300t, 4:311t
    transmission 4:310
Potato black ringspot virus (PBRSC) 3:406t
    diseases 3:411
Potato deforming mosaic virus (PDMV) 4:298t
Potato latent virus (PotLV) 4:300t
Potato leafroll virus (PLRV) 4:297t
    economic impact 4:199
    engineered resistance 4:157–160
    transmission 4:302
Potato mop-top virus (PMTV) 4:283
    CP gene/protein 4:284
    geographical distribution 4:286
    infection
        control 4:308
        detection 4:308
        economic importance 4:286–287
        symptoms 4:286f
    movement in plants 4:285f
    particles 4:283
    RNA3 genome 4:284
    RT proteins 4:284
    serological relationship 4:285f
    small open reading frame 4:284
    TGB protein 4:283–284
    tissue distribution 4:286
    transmission 4:287
    virus–host interactions 4:284
Potato rough dwarf virus (PRDV) 4:298t
Potato spindle tuber disease (PSTVd) 5:332
    epidemiology 5:338
    gene silencing suppression 5:330
    host range 5:335
    propagation 5:338
    replication 5:339
        RNA cleavage 5:339
    symptoms 5:337f
    tobacco 5:341
    viral movement 5:339
        phloem 5:339
Potato virus(es) 4:199
    control 4:199
        genetically-modified resistant plants 4:303
        measures used 4:302
        quarantine 4:303
        resistance 4:297–302
        seed potatoes 4:302
    disease traits 4:302
        damage associated 4:302
        symptoms 4:302
        yield losses 4:302
    history 4:297

pathogens 4:303
  transmission 4:302
    aphid 4:302
  see also specific viruses
Potato virus A (PVA) 4:306
  geographical distribution 4:306
  host range 4:306
  infection
    diagnosis 4:306
    symptoms 4:306
    yield losses 4:306
  propagation 4:306
  serogroups 4:306
Potato virus M (PVM) 4:307f
  genome 1:451f
Potato virus P (PVP) 4:298t
Potato virus S (PVS) 4:297t
  genome variability 4:306
  host range 4:306–307
  infection 4:306f
  transmission 4:302
Potato virus T (PVT) 4:298t
Potato virus U (PVU) 3:406t
Potato virus V (PVV) 4:300t
Potato virus X (PVX) 4:297t
  Avr gene expression 4:172–173
  genome 4:312f
  history 2:451
  infection
    control 4:306
    diagnosis 4:305–306
    Rx gene resistance 4:174
    transgenic resistance 5:325–326
    yield losses 4:199
  strains 4:305–306
  transmission 4:302
  as type species 4:310
  virulence 4:305–306
  virus-induced gene silencing 5:327–328
Potato virus Y (PVY) 4:297t
  characteristics 5:63t
  coat protein 4:288
  control 4:287
  evolution 4:296
  genome 4:288f
    coat protein 4:288
  historical aspects 4:287
    Smith, Kenneth Manley 2:451
  infection see below
  phylogenetics, potyvirus relationships 4:291
  RNA synthesis 4:288
  serology 4:290
    ELISA 4:290
      monoclonal antibodies 4:290
      phage display antibody technology 4:290
      strain discrimination 4:290–291
  significance 4:287
  strains 4:303
  taxonomy 4:296
  transmission 4:199
    aphid 4:303
    aphid vectors 4:289
    determination 4:289
    insect vectors 4:289f
    vegetative propagation 4:288–289
  viral particle 4:288
    helper component protein 4:288
    viral genome linked protein 4:288
  virus-induced gene silencing 5:327–328
Potato virus Y infection 4:215
  cytopathy 4:291
    inclusions 4:291
    leaves 4:291
  disease symptoms 5:64
  economic impact, yield losses 4:199
  genetically-modified resistant
    plants 4:157–160
  host plants 4:289
    specificity 4:289
    weeds 4:289
  see also individual species

peppers 4:289
  barrier control methods 4:295
  control 4:293–294
  isolate effects 4:293
  pvr^2 gene 4:294
  pvr^4 gene 4:294–295
  resistance genes 4:293–294
  see also specific genes
  symptoms 4:294f
  yield effects 4:293
petunia 4:213f
potato 4:289, 4:291
  aphid transmission 4:293
  breeding 4:294f
  control methods 4:293
  hypersensitivity reactions 4:292f
  isolates 4:291
  quarantine 4:293
  strains 4:291
  symptoms 4:291f
  unclassified isolates 4:291–292
  yield reduction 4:291
tobacco 4:289, 4:295
  as diagnostic species 4:290f
  economic importance 5:64
  resistance breeding 4:295
  strains 4:295
  symptoms 4:295f
  yield losses 4:295
tomato 4:289, 4:295
  economic impact 5:64
  genetic diversity 4:295
  resistance 4:295
  strains 4:295
Potato yellow dwarf virus (PYDV) 4:196
Potato yellowing virus (PYV) 4:298t
Potato yellow mosaic virus (PYMV) 4:298t
Potato yellow vein virus (PYVV) 4:298t
Potexvirus 1:426f, 2:255, 4:310–313, 4:311t
Potexvirus(es) 4:310–313
  CP gene/protein 4:312
    degradation 4:312
  cytopathology 4:312
  genome 4:312
    CP 4:312
    movement proteins 4:312
    replicase 4:312
    sequences 4:311t
    untranslated region 4:312f
  geographical distribution 4:313
  infections see below
  phylogenetics 4:310
  physicochemistry 4:310
    CP degradation 4:312
    sedimentation coefficients 4:310–312
    ultraviolet absorbance spectra 4:310–312
  serology 4:313
  structure 4:310
    nucleocapsids 4:310
    virions 4:310
  symptomatology 4:313
  taxonomy/classification 4:310
    phylogenetic analysis 4:310
    species demarcation 4:310
  transmission 4:310
  type species 4:310
  see also specific viruses
Potexvirus infections 4:310
  control 4:313
    transgenic resistant plants 4:313
  epidemiology 4:313
  host range 4:310
Potyviridae 1:289, 1:573, 4:314, 4:315f, 4:315t, 5:11t
Potyvirus 1:289, 4:313, 4:314, 4:314f, 4:315f, 4:315t,
  5:433–434, 5:434f
Potyvirus(es) 4:313–322
  biotechnological applications 4:321
    vectors 4:321
  classification 4:313
  evolution 4:320
    speciation events 4:321

genomes 4:316
  5' noncoding region 4:317
  bipartite structure 4:316–317
  monopartite structure 4:314
helper component protein
  functions 4:317
  pathogenicity 4:320
  protein properties 4:317
  RNAi suppression 3:152
  transmission 5:279
infections see below
legume viruses 3:215
propagation 4:319
proteins 4:317
  CI protein 4:317
  coat proteins 4:316
  CP gene/protein 4:319
  helper component 4:317
  NIa protein 4:317
  NIb protein 4:319
  P1 protein 4:317
  pathogenicity 4:320
  proteolytic processing 4:318f
  VPg protein 4:317–319
replication 4:319f
  host-virus interactions 4:320
  RdRp 4:319
  replication complex 4:319–320
  RNA replication 4:320
  siRNA accumulation 4:319–320
  virus transport 4:320
RNA recombination 4:377
serology 4:316
  CP core 4:316
taxonomy 4:314
  criteria 4:313
  phylogenetic analysis 4:315f
  standards 4:314f
transmission 4:320
  aphids 4:320
  vectors 4:321
virion properties 4:316
  structure 4:316
virus-induced gene silencing, viral synergism
  5:327–328
see also specific viruses
Potyvirus infections
  control 4:321
  diagnostics 4:316
    RT-PCR 4:316
  epidemiology 4:321
  host range 4:313–314
  see also individual species
  outbreak severity 4:321
  pathogenicity 4:320
  plants
    orchids 4:225–226
    petunia 4:213
    sweetpotato 4:663
    tobacco 5:64
  resistance to 4:178–182
    genes 4:179t
  symptoms 4:313
Powassan virus (POWV) 5:488
  encephalitis 2:239
  host 2:239
  tick-borne 2:239
  vector 2:239
Poxviridae 1:427, 1:574–575, 3:226, 3:342, 4:323, 5:11t,
  5:244–245, 5:347
Poxvirus(es) 4:322–330
  apoptosis inhibition 1:158t
  avian 4:326–327
  classification 4:323
  genome size 5:347
  DNA recombination mechanisms 4:375
  entomopoxvirus vs. 3:128–129
  gene content 4:326
    DNA replication associated genes 4:326
    nucleotide composition 4:326
    transcription associated genes 4:326

Poxvirus(es) (continued)
    genome 4:325
        central region 4:325–326
        inverted terminal repeats 4:325–326
        molluscum contagiosum virus vs. 3:320
        size 5:347
    immunomodulation 4:114t
    infections see below
    membrane development 5:308
    phylogeny 4:326f
    replication 4:407
        DNA replication 4:327
        gene expression 4:327
        host cell entry 4:327
        morphogenesis 4:322–323
        transcription 4:326
        uncoating 4:409
    virion structure 4:325
    see also specific viruses
Poxvirus Bioinformatics Resource Center 5:357t
Poxvirus infections
    disease associated 4:322–323
    immune evasion 4:329
        cellular protein mimics 4:329–330
    insects 3:126t
    pathogenesis 4:328
    persistence 4:109t
    prevention 3:230
Pecluvirus, genome 4:100f
PPL-100 2:507t
P protein, Sigma virus structure 4:576–577
PPV-C (cherry), plum pox virus strains/groups 4:240
PPV-D (Dideron), plum pox virus strains/groups 4:240
PPV-EA (El Amar), plum pox virus strains/groups 4:240
PPV-M (Marcus), plum pox virus strains/groups 4:239–240
PPV-Rec (recombinase), plum pox virus strains/groups 4:240
PPV-W (Winona), plum pox virus strains/groups 4:240
Pradimicin A (PRM-A) 1:153
Prasinovirus 4:117t
PRD1 bacteriophage see Bacteriophage PRD1
Precapsid assembly, dsDNA bacteriophage genome packaging 2:307
Precursor capsid assembly, bacteriophages 2:307
Precursor polypeptides, mouse mammary tumor virus 3:335–336
Precursor protein translation, mouse mammary tumor virus replication 3:336–337
Pre-existing channels, bacterial cell viral entry 5:366f
Pregenome packaging, avian hepadnavirus 2:331f
Pregenomic RNA (pgRNA)
    hepatitis B virus 2:353f
    hepatitis B virus replication 2:355f
Pregnancy
    Akabane virus infection 1:77
    bovine viral diarrhea virus infection 2:248
    chronic infections 4:114–115
    enterovirus 71-induced hand, foot and mouth disease 2:134
    HCMV infection 2:482–483
    rubella virus infection 4:518
Prehemorrhagic stage, Crimean-Congo hemorrhagic fever 1:601
Preintegration complex (PIC)
    HIV 2:519
    retroviral replication 4:465
PRE-M gene(s), vaccine vectors 5:238–239
PreM (pre-membrane) protein, yellow fever virus 5:471
Prenatal disease, coxsackieviruses 1:585
Preprimosome
    bacteriophage φX174 4:405
    definition 4:400
Preprotoxin
    apoptotic role 1:162
    definition 5:299
    processing 5:303f
Prestige 2:419

Prezista 1:143t
Prevaccinol 2:419
Prevalence rate, definition 2:141
Prevention of mother-to-child transmission (PMTCT), HIV 1:66
Primary effusion lymphoma (PEL) 5:196
    KSHV see Kaposi's sarcoma-associated herpesvirus (KSHV)
Primary host defence, baculovirus infection see Baculovirus infection
Primates, nonhuman
    anellovirus infection 1:109
    Dengue virus transmission 2:7
    Ebola virus infection 2:203
    filoviruses 2:200
    hepatitis B virus 2:338
    HIV cross-species transmission 4:612
    influenza virus infection 3:101
    KSHV-like viruses 3:190
    SIV infection 4:612–617, 4:620
    SV40 hosts 4:636
    yellow fever 5:475
    see also individual species
    see also specific viruses
Primer-binding site (PBS)
    definition 4:428
    LTR retrotransposons 4:425, 4:430, 4:439
    mouse mammary tumor virus 3:335–336
Primitive forest cycle, Dengue virus 2:8f, 2:9
Primosome 4:400
Principle effector molecules, adaptive immunity 3:60–61
PRINTS database 5:357t
Prion(s) 2:457, **4:330–336**
    classification 5:11t
    composition 5:187
    conformational selection model 4:330
    definition 4:330, 4:337, 5:186
    diseases see Prion diseases
    domains 4:338f
        shuffling 4:339
    formation, amyloid 4:337, 4:339
    function 5:187
    history 4:331
    infectivity 1:369
    molecular strain typing 4:330
    nomenclature 4:331, 4:335
    prion protein (PrP^C)
        conformational states 4:331
        definition 4:330, 5:186
        depletion 4:336
        expression 4:332–333
        knock out mice 4:333
        polymorphism 4:340
        structure and function 4:331, 4:332
    PRNP gene see PRNP gene
    protein-only hypothesis 4:332, 4:334
        C-terminal domain 4:333
        N-terminal domain 4:333
        PrP conformational states 4:332
    PrP^L, neurodegeneration 4:336
    replication 4:334
    scrapie isoform (PrP^{SC})
        detection 1:369
        protein conformation 4:333–334
        proteolysis 4:333, 4:333f
        structure and function 4:333
    strains 4:334
        biological properties 4:334
        detection 4:335
        structure and function 4:332
    transmission 4:332
        barrier 4:331
Prion(s), yeast and fungal **4:336–341**
    amyloid structure 4:339
    biological role 4:340
    chaperones and 4:338f
    domain shuffling 4:339
    enzymes as 4:340
    formation 4:339
    generation 4:339

genetic signature 4:337, 4:337f
history 4:337
self-propagating amyloid 4:337
species barriers, variants and 4:338
transfection with amyloid 4:339
Prion diseases 4:337
    chronic wasting disease 5:188
    diagnosis 4:334
    etiological agent 5:186
    genetic mutation associated 5:192
        familial CJD 5:192
        fatal familial insomnia 5:192
        Gerstmann–Sträussler–Scheinker syndrome 5:192
    genetic susceptibility 4:331–332
    host range 5:188t
    incubation period 4:330, 4:332, 4:334, 5:186
    neuropathology 5:186
    pathogenesis 4:334
        histopathological changes 4:334, 4:335f
        neuronal cell death 4:336
        tissue distribution 4:334
    PrP^C conversion 5:187
    subclinical infection 4:330
    see also Bovine spongiform encephalopathy (BSE); Creutzfeldt–Jakob disease (CJD); Kuru; Scrapie; Variant Creutzfeldt–Jakob disease (vCJD)
PRNP gene 4:330
    codon 129 polymorphism 1:371, 1:372, 4:330
    prion strain propagation 4:335
    sporadic CJD 4:331–332
    strain propagation 4:335
    variant CJD 4:331–332
    knock-out mice 4:331, 4:333
    pathogenic mutations 4:332
PRO140 2:507t
Proboscivirus 1:636
    classification 1:635t
Procapsids
    definition 1:433, 2:306, 3:6, 3:13, 3:30
    icosahedral enveloped dsDNA bacteriophages 3:11
    icosahedral tailed dsDNA bacteriophages 3:35
    maturation 1:436, 1:437f
Processive polymerization 4:400
Processivity factor 4:400
ProdiGene 5:223–224
Prodrugs, antiviral therapy 1:150
pro gene/protein, Simian retrovirus D 4:624f, 4:625
Prognathia 1:328
Programmed cell death see Apoptosis
Programmed cell death-1 (PD-1) 1:126
Progressive multifocal leukoencephalopathy (PML)
    associated immunosuppressive conditions 4:262
    clinical signs 4:267
    histopathology 4:267f
    JCV infection vs. tumor induction 4:267
Prohead 3:30
Pro-inflammatory cytokines see Cytokine(s)
Pro-integration complex (PIC), mouse mammary tumor virus 3:336
Projection mapping, icosahedral particles 1:608
Promoters, baculoviruses 1:251
Promyelocytic leukemia (PML)-containing nuclear oncogenic bodies (PODs) 1:12
Promyelocytic leukemia (PML) protein 3:245
Propagative infection 1:483
Prophages
    bacteriophages 2:445
    definition 2:442, 5:411
Pro-phenoloxidases (pro-POs) 4:251–252, 4:252–253, 4:254
Prophylaxis, organ transplantation 3:471
Pro proteins, retroviruses 4:446, 4:461
PROSITE 5:357t
Protapanteles paleacrita brachovirus 4:257f
Prot, definition 4:428
Protease(s)
    2A^{pro} 1:542
    arteriviruses 1:180
    mode of action 2:513

nidoviruses 3:424
pseudovirus gene expression 4:354–355
SARS CoV therapeutic targets 1:561–562, 1:561f
virus maturation 4:411–412
see also specific types
Protease inhibitors 1:148, 2:507t, 2:513, 2:514t
HCV therapy 1:153
Protease p29, Cryphonectria parasitica virus 2:579
Protective antibodies 3:64
Protective immunity
definition 5:226
Japanese encephalitis vaccines 3:187
measles 3:288
Protein(s)
cross-linking, bacteriophage assembly 1:439
evolution of 3:478
sequence data, virus databases 5:363
see also specific proteins
Protein 5, lettuce big-vein virus 5:266–267
Protein A, bacteriophage φX174 4:405
Protein antibiotics 3:249, 3:256f
Protein arrays, symmetrical 1:436
helical arrays 1:437
icosahedral array maturation 1:436, 1:437f
symmetry disruption 1:437
symmetry mismatches 1:437
Proteinases
nepovirus 3:409
papaya ringspot virus 4:2
see also specific types
Protein DataBank (PDB) 5:363
Protein kinases, plant resistance 4:176–177
Protein methylesterase (PME) 3:355
Protein priming replication 5:411
Protein tyrosine phosphatase(s) (PTPs)
immunosuppressive effects 4:254, 4:255
Microplitis demolitor bracovirus 4:259–260
Proteinuria, Argentine hemorrhagic fever 3:208
Proteolytic cleavage
arterivirus genome 3:424
bacteriophage assembly 1:439
fusion protein biosynthesis 3:293
Hendra virus fusion proteins 4:54
Nipah virus fusion proteins 4:54
posttranslational modifications in HCMC 2:489–490
PrP^{SC} 4:333, 4:333f
Sendai virus fusion proteins 4:54
Proteomics 3:311
Protomer, definition 1:580
Proto-oncogenes 2:213
definition 2:105
Protoplasts 5:157
Brome mosaic virus replication 1:382
fungal virus transfer 2:290
Providence virus (PrV) 5:28t
assembly 5:33
β capsid protein 5:33
capsid structure 5:31f, 5:33
genome 5:29
infection, histopathology 5:36
in vitro replication 5:32
morphology 5:31f
protein precursors 5:29–31, 5:30f
transmission electromicroscopy 5:33
Provirus(es)
definition 2:505
feline leukemia viruses 2:185–186
primers 2:105–106
retroviral integration 2:105, 4:446–447
selective forces 2:107
structure 2:105–106
PrP^C *see* Prion(s)
PrP^{SC} *see* Prion(s), scrapie isoform; Scrapie (PrP^{SC})
PRRs *see* Pattern-recognition receptors (PRRs)
PRRSV *see* Porcine respiratory and reproductive syndrome virus (PRRSV)
Prune dwarf virus (PDV) 4:204
mixed infections 4:239

Prunus necrotic ringspot virus (PNRSV) 4:204
coat protein 3:53, 3:53f
mixed infections 4:239
transmission 3:55
Pruritus, pseudorabies virus infection 4:348
PrV *see* Providence virus (PrV)
Prymnesiovirus 4:117t
Pseudococcidae (mealybugs), cacao swollen shoot virus transmission 1:404, 1:404f
Pseudocowpox virus (PCPV) 4:57–58
host range 4:58
spread 4:58
synonyms 4:58
Pseudogenes 2:360
Pseudoknots
definition 5:68
hepatitis delta virus 4:480
VS RNA 4:480
yellow head virus 5:477–478
Pseudolysogeny 2:71
Pseudomonas aeruginosa, virulence factors 3:123–124
Pseudomurein endoisopeptidase, Methanothermobacter marburgensis virus 5:415
Pseudoparticles, hepatitis C virus 2:372
Pseudoplusia includens virus (PiV) 5:28t
Pseudorabies virus (PrV) 4:341–352
Bartha strain 4:351
capsid 4:342
assembly 4:346
replicative role 4:346
UL17 4:342
UL25 4:342
envelope 4:346
glycoproteins 4:344t, 4:346
nonglycosylated membrane proteins 4:346
genome 4:342, 4:343f
gene functions 4:344t
genes 4:344t
inversions 4:342
history 4:341
infection *see below*
infectivity 4:348
morphology 4:341, 4:342f
neurobiological tool function 4:351
replication cycle 4:346, 4:347f
capsid assembly 4:346
DNA replication 4:346
egress 4:346–347
host cell attachment 4:346
host cell entry 4:346
IE180 function 4:346
nucleotide metabolism 4:346
transcription 4:346
UL3 protein kinase 4:346–347
UL31/UL34 complex formation 4:346–347
virion morphogenesis 4:347–348
taxonomy 4:341
tegument 4:342
capsid-proximal tegument 4:342–346
envelope-proximal tegument 4:342–346
formation 4:347–348
UL41 4:342–346
UL48 4:342–346, 4:347–348
tissue tropism 4:349
transmission 4:348–349
Pseudorabies virus infection
antibodies 4:349
apoptosis 4:349, 4:350
clinical features 4:341, 4:348
control 4:350
ELISA 4:350, 4:350f
screening 4:350
vaccination 4:349, 4:350
epidemiology 4:350
histology 4:349
host range 4:341, 4:348
immune system evasion 4:349
immunology 4:349
antibodies 4:349
infected vs. vaccinated animals 4:350–351, 4:350f

latency 4:343f, 4:349
neural 4:348–349
pathology 4:348
pigs 4:348f
mortality rate 4:348
neonates 4:348
pregnancy 4:348
vaccines/vaccination 4:349, 4:350
Pseudorecombination
definition 1:30, 1:615, 5:124
Indian tomato leaf curl viruses 5:128–129, 5:131–132
Pseudotype virions, virus membranes 5:313
Pseudoviridae 4:352, 4:430, 4:451, 5:11t
Pseudovirus 4:352, 4:353t
Pseudovirus(es) 4:12, **4:352–357**
classification 4:353t
env-like gene 4:355
evolutionary relationships 4:355
phylogenetic tree 4:356f
reverse transcriptase 4:355–356
gene expression 4:353
integrase 4:354–355
protease 4:354–355
genome 4:353, 4:354f
gag/pol ratio 4:353–354
long terminal repeats 4:355
pol coding region 4:354–355
primer binding sites 4:355
hemiviruses *vs.* 4:433
infections, fungi 2:287t
LTR-retrotransposons 4:433, 4:451
organization 4:433
phylogeny, genome sequencing 4:357
replication cycle 4:355
gag translation 4:355
LTR regions 4:355
pol translation 4:355
viral cDNA integration 4:355
viral cDNA synthesis 4:355
virion properties 4:352
virus-like intermediate 4:352–353
virus-like particles 4:352–353
see also specific viruses
PSI-inducibility (PIN) prion 4:337, 4:339
φ6 bacteriophage *see* Bacteriophage φ6
PSI prion 4:337
isogenic strains 4:340
species barrier 4:338
Psittacid herpesvirus 1 (PsHV-1)
genetics 2:408
host range 2:407
infection *see* Pacheco's disease
propagation 2:407
replication sites 2:409
transmission 2:409
virus shedding 2:409
Psittacine beak and feather disease virus (PBFDV)
disease symptoms 1:514–515
virion structure 1:514
Psittacinepox virus(es) 2:278, 2:278f
antigenic cross-reactivity 2:278–279
ψM1 virus 5:412
genome 5:415
open reading frames 5:412–415
pseudomurein endoisopeptidase 5:415
related viruses 5:412
structure 5:412–415, 5:415f
ψM2 virus 5:412
genome 5:415, 5:416f
ψM100 virus 5:412
genome 5:416f
homology 5:415
pseudomurein endoisopeptidase 5:415
Psorophora confinnis, Venezuelan equine encephalitic virus transmission 5:104
PSV *see* Pyrobaculum spherical virus (PSV)
PTK complex, murine polyomavirus middle T 4:274
PToV *see* Porcine torovirus (PToV)
PU-1 targeting, EBV-encoded nuclear antigen 2 2:162
PubMed 5:356–362, 5:357t

Pulau virus 2:95
Pulmonary complications, adult T-cell leukemia 2:568
Pulmonary edema
  definition 1:328
  enterovirus 71 infection 2:134
Pulmonary interlobular edema, African horse sickness 1:41
Pulsed-field electrophoresis 2:71
Pupation, baculovirus terminal disease 1:269
Purpura hemorrhagica 1:37
  African horse sickness vs. 1:42
Pustules 4:639
Puumala virus 1:391t
pVII see Bacteriophage fd
pVIII gene/protein
  bacteriophages, filamentous ssDNA structure 2:193, 2:193f
  structure 2:193, 2:193f
pvr^2 gene/protein
  eIF4E-mediated resistance, plants 4:184
  potato virus Y infections 4:294
pvr^4 gene/protein, potato virus Y infections 4:294–295
pvr6 gene/protein, eIF4E-mediated resistance, plants 4:184
Pyknosis 3:342
Pyrobaculum spherical virus (PSV)
  envelope composition 1:591
  morphology 1:590, 1:590f
  protein composition 1:591
Pyrococcus abyssi
  'fusiform' viruses 2:300
  PAV1 infection 5:421–422

Q

Qalyub virus 1:598t
Qβ RNA polymerase, quasispecies 2:177–178
Q protein, bacteriophage lambda 5:184, 5:185
Quantitative real-time PCR, rhinovirus detection 4:471
Quantitative taxonomy see Taxonomy
Quantitative trait loci (QTL)
  definition 4:178
  polygenic resistance in plants 4:178
    gene mapping 4:182
Quarantine
  equid herpesvirus 1 infection 2:419
  potato virus infection control 4:303
  potato virus Y potato infections 4:293
  Sharka disease control 4:241
Quasi-equivalence
  definition 1:433, 5:145
  icosahedral array assembly 1:433
  tombusviruses 5:146
  virus structure 5:395
Quasi-equivalent conformers, definition 3:21
Quasispecies 4:359–365
  common assumptions 4:361
    relaxation of 4:361
  definition 5:274
  error catastrophe 4:362
    multiplicative landscape 4:362
  fundamental model 4:360
    equation solutions 4:360
    error-prone replication 4:360
  historical context 4:359
    Eigen, Manfred 4:359–360
  HIV 1:128, 1:133
  mutation robustness, development of 4:361
    back mutations 4:361–362
    Haldane–Muller principle 4:361
    secondary mutations 4:361–362
    'survival of the flattest' 4:362, 4:361–362
  relevant predictions 4:361
  scientific context 4:360
    as population genetics 4:360
    variability 4:360, 4:360t
Taura syndrome virus 5:6
viral evolution see Evolution (of viruses)

R

R2D2 protein (RNA interference) 3:150
  Dicer-2 interactions 3:150–151
Rabbit(s)
  CSFV vaccine production 1:528
  economic impact 1:411
  HSV reactivation 2:437
  see also specific infections
Rabbit fibroma virus (SFV) 3:225t
Rabbit hemorrhagic disease 1:416
  clinical features 1:417
  control 1:418
  vaccines/vaccination 1:417, 1:418
Rabbit hemorrhagic fever virus 1:410
  antigenic variation 1:415
  Australia 1:411
  geographical distribution 1:413–414
  host range 1:414–415
  infection see above
  replication, receptors associated 1:417
  stability 1:412
  structure 1:412f
  transmission 1:416
Rabbitpox
  clinical features 3:347, 3:347f
  future work 3:347
  histopathology 3:347
  immune response 3:347
  pathogenesis 3:346
  pathology 3:347
Rabbitpox virus 3:346
  classification 3:346
  epidemiology 3:346
  genetics 3:346
  history 3:346
Rabies virus (RABV) 2:96, 4:184t, 4:367–373
  cell tropism 4:370
  classification 4:367–368, 5:485
  epidemiology 4:372
  evolution 4:368
  genome 3:326f, 4:367, 4:368f
    G gene 4:367–368
    M gene 4:367–368
    P gene 4:367–368
  geographical distribution 2:96
  infection see below
  infectivity 4:367–368
  morphology 4:367, 4:368f
  phosphoprotein 3:331
  phylogeny 4:368, 4:369f
    Arctic lineage 4:368, 4:372
    'cosmopolitan' lineage 4:368, 4:372
    mutations 4:368
  proteins 4:367–368
  replication 4:370
  transmission 4:370
  vaccines/vaccination 2:96, 4:373, 5:237, 5:485, 5:493–494
    domestic animals 4:373
    humans 4:373
    potency 4:373, 5:237
    wild animals 4:373
Rabies virus infection
  acute 1:474
  clinical spectrum 4:371
    encephalitic form 4:371
    incubation period 4:371–372
    neural cell infection 1:473
    paralytic form 4:371
    symptoms 4:371
  control 4:373
  diagnosis 4:371, 5:485
    cell culture 4:371–372
    neuronal inclusions 4:371
    PCR 4:372
    virus isolation 4:371–372
  eradication 4:372
  host range 2:96, 4:370–371, 4:372
    bats 4:370–371, 5:485
    dogs 4:372
    migratory species 4:372–373
    mongooses 4:373
    see also Zoonoses, below
  mortality 2:96
  organ transplantation risks 4:370
    encephalitis 3:466
  pathogenesis 4:370
    neural involvement 4:370
  postexposure treatment 4:373, 5:485
  prevention 4:373
  survival 4:370–371
  zoonoses 5:485, 5:486t
    animal reservoirs 5:485
    geographical distribution 5:485
    transmission 5:485
Racecadotril 4:511, 4:512
Racivir 2:507t
Radiological abnormalities 4:552
Ralstonia solanacearum, resistance in rice 4:175
Raltegravir 2:510
Raman spectroscopy 2:190
Ramsay–Hunt syndrome, zoster (shingles) 5:253
Rana catesbeiana virus Z (RCV-Z) 3:172f
Rana pipiens, Lucke' tumor herpesvirus infection 2:209
Ranavirus 2:233, 3:155, 3:168t
Ranavirus(es)
  geographical distribution 3:156
  host range 3:156
  infection 3:158–159
  phylogeny 3:160f
  transmission 3:159–160
Ranid herpesvirus 1 see Lucke' tumor herpesvirus (LTHV)
RANTES 1:623–624
Ranunculus white mottle virus (RWMV) 3:447, 3:448t
  coat protein 3:450
  genome 3:449
  geographical distribution 3:451
  phylogenetic analysis 3:447
  transmission 3:453
Raphidovirus 4:117t
Rapid growth, RNA virus experimental evolution 4:363
Rashes, smallpox 4:639–640
Raspberry bushy dwarf virus (RBDV) 3:37, 3:41
  geographical distribution 3:37
    black raspberry isolates 3:37
    resistance-breaking isolates 3:37
    Scottish isolates 3:37
  ilarvirus vs. 3:41
  infection
    control 3:41
    detection 3:41
    disease symptoms 3:38, 3:38f
    naturally infected plants 3:38
  nucleotide sequences 3:39, 3:39f
    cucumber mosaic virus 2b gene vs. 3:40
    non-coding sequences 3:40
    RNA-1 3:40
    RNA-2 3:39–40
  resistant cultivars 3:41
  virion composition 3:39
  virion properties 3:39, 3:39f
Raspberry ringspot virus (RpRSV) 3:406t, 4:204
  diseases 3:411
  infectious DNA clones 3:408–409
  transmission 3:411
Rat1 mutation (MurA) 3:257–258
Rat cytomegalovirus (RCMV) 1:632
  atherosclerosis 1:633
  infection
    atherosclerosis model 1:633
    immune system evasion 1:633
    immunocompromised animals 1:633
    pathogenesis 1:633
    reactivation 1:633
  Maastricht strain 1:632–633
    genome 1:633
Rategravir 2:510
RDR genes, *Arabidopsis thaliana* 5:326–327
Reactivation 2:405f
  antigen detection 2:33
  chronic hepatitis B 2:357–358
  chronic infection pathogenesis 4:111

HCMV infection 2:483
HHV-6 2:502–503
HHV-7 2:502
latency 4:110–111
persistent infection pathogenesis 4:112
Reactome database 5:357t
Readthrough domain(s) (RTD), luteovirus transmission 3:236
Reassortment
  cucumber mosaic variation 1:619
  definition 1:390, 3:479, 4:164, 4:391
  Ngari virus strains 3:482
  orthobunyavirus *see* Orthobunyavirus(es)
  Patois group viruses 3:482
  reovirus genome 4:392
recA, DNA virus recombination 4:374
Receptor(s) 5:310, **5:319–324**
  clinical isolates, laboratory adaptation 5:323
    antibody resistance 5:323–324
  competition, retrovirus infection resistance 2:468–469
  complement control proteins 5:322
  definition 4:407
  glycosaminoglycan 5:320
  heparan sulfate 5:320
  histo-group antigens 5:321
  host cell attachment 4:407–408, 5:319
    binding specificity 4:408–409
    cell tropism 4:408
    plant viruses 4:408
  identification 5:319
    direct 5:320
    indirect 5:320
  immunoglobulin superfamily proteins 5:321
  inactivation 2:468
  infection neutralization role 3:416
  sialic acid 5:320
  specificity 5:320
  see also specific receptors
  see also specific viruses
Receptor activated pathway (apoptosis) 1:155
Receptor-binding proteins
  bacterial cell viral entry 5:365–366
  definition 5:365
  filamentous ssDNA bacteriophages 5:369
Recombinant virus(es)
  definition 3:95, 3:483, 5:124
  HIV vaccine development 1:75
  see also specific viruses
Recombinant adeno-associated virus (rAAV), gene therapy vector 2:303
  DNA shuffling 2:304
  peptide insert size 2:304
  serotypes 2:304
  in vitro evolution 2:304
Recombinant adenovirus (rAd), gene therapy vector 2:303
  first generation 2:303
  second generation 2:303
Recombinant herpes simplex virus, gene therapy vector 2:303
Recombinant nematode anticoagulant protein c2
  Ebola virus infection treatment 2:64, 2:204–205
  Marburg virus infection treatment 2:204–205
Recombinant vaccines
  avipoxviruses 2:283
  development 5:237
    immune response 5:237–238
    virus structure 5:237–238
  hepatitis B virus 2:351, 2:358
  TMV 5:59–60
  vaccinia virus vaccines 4:629
Recombinase *see* Integrase(s)
Recombination **4:374–382**
  Agrobacterium tumefaciens-based expression system 4:232f, 4:233
  bacteriophage evolution 5:371–372
  baculovirus expression vectors 1:243–244
  Brome mosaic virus 1:389
  Campoletis sonorensis ichnovirus 4:258–259
  capripoxviruses 1:428
  coronaviruses 1:552, 1:558–559, 1:560

cucumber mosaic virus 1:619
defective-interfering RNAs 4:380
  historical aspects 4:380
  negative-strand RNA viruses 4:380
  single internal deletions 4:380
definition 1:410, 4:164
echoviruses 2:70
endogenous retroviruses 2:106
enteroviruses 2:127–128
feline sarcoma viruses 2:185
heterologous encapsidation 4:162
homologous *see* Homologous recombination
icosahedral enveloped dsDNA bacteriophages 3:11
influenza viruses 3:99
intertypic 2:150
intraserotypic 2:8
KSHV 3:197
nodaviruses 3:436
nonhomologous *see* Nonhomologous recombination
noroviruses 1:415, 3:442–443
retroviruses 4:446–447
RNA viruses 4:377
sapoviruses 3:442–443
T-cell receptors 3:71–72
tomato leaf curl viruses, Indian 5:131
watermelon mosaic virus 5:439
zucchini yellow mosaic virus 5:439
Recombination-dependent replication, beet curly top virus 1:303–304
Red clover mottle comovirus (RCMV) 3:216
Red clover necrotic mosaic dianthovirus (RCNMV) 3:216
Red clover necrotic mottle virus 3:354–355
Red grouper reovirus 1:164t
Red grouse, louping ill virus infection 2:240–241
Red Queen hypothesis 2:175, 2:179, 5:297
Red raspberry (*Rubus idaeus*), idaeovirus infection 3:37
Red sea bream iridovirus (RSIV), infected cell morphology 3:173
Red-spotted grouper nervous necrosis virus (RGNNV) 2:229–230
Reduction hypothesis, viral origins 3:473, 3:476, 3:476f
Red veins, cacao swollen shoot virus infection 1:403
Reed, Walter, Yellow Fever Virus 2:455
Reemerging diseases
  ecological factors 2:96
  history 2:93
  recognition of 2:93
  vertebrate **2:93–97**
    contributing factors 2:93
    human behavior 2:94
    human demographics 2:94
    zoonoses 2:95, 5:494
  see also specific infections
ref genes/proteins 4:579–580, 4:580f
RegB endoribonuclease, bacteriophage T4 5:177
Regional lymph nodes, African horse sickness 1:40
Regulated polyprotein processing
  Grapevine fanleaf virus 3:409
  nepovirus *see* Nepovirus
Regulator of chromosome condensation (RCC1) protein 1:228
Regulator of complement activity (RCA) 5:322
  γHV68 3:110
Regulator of expression of virion proteins (Rev) *see* HIV
Regulator of transcriptional activation (RTA), KSHV 3:195, 3:199
Regulatory viral RNAs, apoptosis 1:161
Reintroduction program(s) (birds), poxvirus infection 2:279
'Relational database management systems' (RDBMSs) 5:351–352
Release
  baculoviruses 1:251
  definition 4:407
Relenza *see* Zanamivir
Relenza® 1:143t
Rem A 3:334
Rem protein, mouse mammary tumor virus replication 3:336–337

Renal failure, chronic, HIV infection 1:56
Renal transplantation
  immunosuppression 3:471
  JC polyomavirus infection 3:470
  papillomavirus infection 3:470
  therapy 3:471
Reniform 1:186
Reo-like virus(es), shellfish viruses 4:566
Reoviridae 1:93f, 1:163, 1:170, 2:118–119, 2:231, 2:286, 3:133–134, 3:134t, 3:454, 4:149–150, 4:383, 4:384t, 4:391, 4:392t, 4:509, 4:536, 5:11t, 5:344–345, 5:346f, 5:346t
Reovirus(es) 3:454, 4:149–150, **4:382–390, 4:390–399**, 4:392t
  characteristics 2:117t
  classification 4:384t
  fungal dsRNA viruses 2:111
  fusogenic orthoreoviruses 4:383
  genera 2:231, 4:536, 4:536t
  genome 1:163, 4:391, 4:391t, 4:394f
    large segments 4:391–392
    medium segments 4:391–392
    nontranslated region 4:391–392, 4:394f
    RdRp conservation 1:537
    reassortants 3:454, 4:392
    small segments 4:391–392
  history 4:383
  infections
    apoptosis inhibition 1:158t
    clinical features 4:388
    fish 2:231
    fungi 2:287t
    immune response 4:389
    insects 3:126t
    respiratory 2:556
  inner capsid 4:392, 4:393f
    λ1 protein 4:393, 4:394f, 4:395t
    λ2 protein 4:393, 4:394f
    λ3 protein 4:393, 4:394f
    μ2 protein 4:393–394
    σ2 protein 4:393f, 4:394
  insect pest control 3:127
  mammalian *see* Mammalian reovirus(es)
  outer capsid 4:394
    μ1 protein 4:394f, 4:396
    σ1 protein 4:396
    σ3 protein 4:396
    σ3 protein cleavage 4:396–398
  plant *see* Plant reovirus(es)
  RdRp
    classification 5:344–345, 5:346f, 5:346t
    gene conservation 1:537, 1:538t, 1:539f
    sequence comparison 5:344–345, 5:346f, 5:346t
  replication 4:396, 4:397f
    assembly 4:398
    assortment process 4:398
    cap formation 4:398
    host cell binding 4:396–398, 4:397f
    host cell entry 4:393f, 4:396–398
    λ2 proteins 4:398–399, 4:399t
    σ3 protein 4:398
    σNS protein 4:398
    tissues associated 4:396
    transcription 4:397f, 4:398
  strain nomenclature 4:383
  taxonomy/classification 1:163, 3:454, 4:383
  virulence determinants 5:318
  virus structure 4:383, 4:391
  see also specific viruses
Reovirus type 1 Lang, virulence determinants 5:318
REP 9, antiviral therapy, enveloped viruses 1:147
RepA, definition 3:264
repA (regulatory protein), maize streak virus *see* Maize streak virus (MSV)
Repeat (R) regions, mouse mammary tumor virus 3:335–336
Repetitive DNA, iridoviruses 3:164
rep gene/protein
  adeno-associated virus vectors 2:303
  bacteriophage φX174 3:16
  banana bunchy top virus 1:274
  beet curly top virus *see* Beet curly top virus (BCTV)

rep gene/protein (*continued*)
  Campoletis sonorensis ichnovirus 4:258–259
  definition 3:264, 3:385
  Maize streak virus *see* Maize streak virus (MSV)
  mungbean yellow mosaic India virus replication 3:366
  mungbean yellow mosaic viruses 3:366, 3:367
  nanoviruses 3:388, 3:390, 3:391
  porcine circoviruses 1:516, 1:517–518, 1:518–519
  tomato leaf curl viruses Indian replication 5:132
  tomato yellow leaf curl virus 5:144
Replicase(s)
  alfalfa mosaic virus 1:82–83
  arteriviruses 3:423–424, 3:425*f*
  brome mosaic virus 4:378–379
  coronaviruses 1:552, 1:552*f*
  Dendrolimus punctatus tetravirus 5:31
  engineered resistance in plants 4:156–157
  Helicoverpa armigera stunt virus 5:31
  hordeiviruses 2:461
  icosahedral ssRNA bacteriophages 3:25*f*, 3:26
  omega tetraviruses 5:31
  potexvirus 4:312
  RNA virus evolution 2:180–181
  Thosea asigra virus 5:29
  tobravirus replication 5:74–75
Replicase-associated protein (REP), geminiviruses 4:168
Replication 1:147, **4:406–412**, 4:408*f*
  antiviral therapy *see* Antiviral drug(s)
  assembly 4:411
    sites 4:411
  blocking, Ab-mediated neutralization 3:415
  coding capacity 4:407
  definition 1:296
  double-stranded DNA viruses *see* Double stranded DNA (dsDNA) bacteriophages
  host cell attachment 4:407
    receptors *see* Receptor(s)
  maturation *see* Maturation
  release 4:412
  stages 4:407, 4:408*f*
  temperature effect 5:315
  transcription *see* Transcription
  uncoating 4:409
    endocytic pathway 4:409
  see also specific viruses
Replication competent, definition 2:301
Replication competent retrovirus (RCR) 2:302
'Replication complex' ribonucleoprotein 3:393–394
Replication initiator proteins (rep), nanoviruses 3:388, 3:390, 3:391
Replication protein, flexiviruses 2:255, 2:256*f*
Replication–transcription complex, arterivirus genome 3:423–424
Replicon systems
  alphaviruses 5:121–122
  definition 5:117
  HCV 1:153
  hepatitis E virus 2:380
  industrial-scale protein production 4:235
  togaviruses 5:121–122
Replisome 4:400
Reportable diseases, fish viruses 2:227, 2:228*t*
Reptilian adenovirus(es) 1:27
Rescriptor® 2:510
Resequin, EHV-4 treatment 2:419
Reservoir hosts
  see specific viruses
Residence, vector transmission 5:273
Residual infectivity, neutralizing antibody tests 2:31
Resins
  antibody labelling 2:83
  classical thin sectioning 2:83
Resistance
  antibody-mediated infection neutralization 3:417
    conserved epitope binding 3:417–418
    mutations 3:417
  breeding
    cacao swollen shoot virus infection 1:409
    potato virus Y tobacco infections 4:295
    rice yellow mottle virus control 4:488–489
  definition 1:296

gene-for-gene
  definition 4:170, 4:171
  history 4:171
  partial 4:178
  potato virus Y tomato infections 4:295
  virus-induced gene silencing 5:378
  Visna-Maedi virus infection 5:430
  watermelon mosaic virus 5:437–438
  zucchini yellow mosaic virus 5:437–438
  see also Plant(s), natural disease resistance (dominant genes); Plant(s), natural disease resistance (recessive genes)
Resistance genes, definition 4:170
Respiratory distress syndrome, hantavirus infection 2:317
Respiratory infections
  adenoviruses 1:25
  anelloviruses 1:110
  barriers against 5:315
  disease surveillance 2:46
  echoviruses 2:68
  EHV-4 2:418
  equid herpesvirus 1 2:418
  henipavirus infections 2:325
  human respiratory syncytial virus *see* Human respiratory syncytial virus (HRSV)
  lymphocytic choriomeningitis virus *see* Lymphocytic choriomeningitis virus infection
  organ transplantation 3:469–470
  virus entry 5:315
  see also specific infections
Respiratory syncytial virus 3:246
  genome 3:327*f*
Respiratory virus(es)
  see specific viruses
Respirovirus 4:41*t*, 4:47
Respirovirus(es)
  classification 4:41*t*
  genome 3:327*f*
'Resting complex' ribonucleoproteins 3:393–394
Resting memory B-cells, EBV infection 2:165, 2:166*f*
Reston Ebola virus (REBOV)
  genome 2:58–59
  geographical distribution 2:62
  history 2:57
Restriction enzyme digests 2:456
  avipoxvirus detection 2:280
  varicella-zoster virus 5:252
Restriction factor 2:467
Restriction fragment length polymorphisms (RFLPs)
  apple stem pitting virus strains 1:424
  definition 1:301, 4:103
  Pepino mosaic virus 4:106–107
  plum pox virus strains/groups 4:241
Restrictive infection, Visna-Maedi virus infection 5:425–426
Reticulated evolution, definition 2:175
Reticuloendothelial cells, CNS invasion pathways 1:471
Reticuloendotheliosis viruses (REV) **4:412–419**
  classification 4:412
  genome 4:413, 4:414*f*
  geographical distribution 4:415
  history 4:412
  infection *see below*
  protein properties 4:413, 4:414*f*
    envelope protein 4:413
    v-Rel 4:413
  replication 4:413
    genome integration 4:415
    nucleocapsid assembly 4:415
    RNA transcription 4:414
    subgenomic RNAs 4:414–415
    viral maturation 4:415
  taxonomy 4:412
  transmission 4:415
  vaccine contamination 2:283
  virion properties 4:413
    morphology 4:413

v-Rel complexes
  mutations 4:417, 4:417*f*
  target genes 4:417
  transformation 4:417, 4:418*t*
  see also specific viruses
Reticuloendotheliosis virus infections
  host range 4:415
  immune response 4:415
  pathogenesis 4:415
    Gag proteins 4:416
    neoplastic diseases 4:416
    non-neoplastic diseases 4:415
    transformation by REV-T 4:416
    v-Rel complexes *see* Reticuloendotheliosis viruses
  runting syndrome 4:416
Retina 2:491
  acute necrosis 2:497
  betanodavirus fish infections 2:229–230
  infection 2:491
Retinitis, HCMV infection 2:496, 2:496*f*, 2:497
Retinoblastoma (RB) cell cycle regulator
  function 4:37
  murine polyomavirus large T 4:276
Retinoblastoma-related tumor-suppressor protein (pRBR) 1:307
Retinochoroiditis 2:491
  HCMV 2:496
Retinoic acid inducible protein (RIG-I) pathway 3:59–60, 3:106
  viral evasion 3:108
Retrales 4:430
Retriposons *see* Target-primed retrotransposons
Retroelements 4:451
  definition 4:430
  LTR retrotransposons *vs.* 4:451
Retroperitoneal fibromatosis-associated herpesvirus 4:587*t*
Retroperitoneal fibromatosis herpesvirus (RFHV) 4:628–629
Retroperitoneal fibrosis, rhadinovirus infection 4:592
Retrotransposineae 4:430
Retrotransposition, definition 4:419
Retrotransposons 4:352, **4:436–445**
  definition 4:352, 4:419, 4:437
  errantiviruses 4:433
  evolutionary features 4:441
  functions 4:437
  fungi *see below*
  genome effects 4:443
    content 4:443
    genes, effects on 4:444
    genomic distribution 4:443
    mammalian 4:443
  history 4:437
    whole genome sequencing 4:437
  host responses 4:444
  life cycle 3:304*f*
  nomenclature 4:437
  origins 4:441
  as phylogenetic markers 4:442
  plants *see below*
  retroviruses *vs.* 4:352
  structure 4:437, 4:438*f*
    open reading frames 4:437
  target-primed *see* Target-primed (TP) retrotransposons
  see also LINE-like retrotransposons*see also* LTR-retroelements*see also* Metavirus(es)*see also* Non-LTR retrotransposons*see also* Penelope-like elements (PLEs)
  see also SINE retrotransposons
Retrotransposons, fungi **4:419–427**
  distribution 4:423
  diversity 4:423–424
  genome effects 4:426
    chromosomal translocation breakpoints 4:426
    mutagens 4:426
    proportion from retrotransposons 4:426
  host effects 4:426–427
    RIP 4:426–427

phylogenetic analysis 4:420, 4:422f
replication 4:424
types 4:420, 4:421f, 4:422f
uses of 4:427
gene tagging 4:427
insertional mutagens 4:427
population genetics 4:427
variation 4:427
see also specific types
Retrotransposons, plants **4:428–436**
classification 4:428, 4:431t
env-like functions 4:434
LTR-retrotransposons see LTR-retroelements
nomenclature 4:435
non-LTR-retrotransposons 4:428–429
classification 4:429–430
structural variants 4:428
transmission 4:428
Retrovir® see Zidovudine (AZT)
Retrovirales 4:430
see also specific viruses
Retroviridae 1:347–348, 2:107, 2:108f, 2:185, 2:213, 2:231,
2:260, 2:260t, 2:518, 2:526, 3:175, 3:335, 4:412–413,
4:430, 4:446, 4:451, 4:459, 4:460, 4:461f, 4:603,
4:604t, 4:624, 5:11t
Retrovirus(es) **4:459–467**
acute transforming 4:466
apoptosis inhibition 1:158t
avian see Avian retrovirus(es)
classification 2:213, 2:260t, 4:459, 4:460
endogenous see Endogenous retrovirus(es) (ERV)
env gene/protein 4:446
deletions 4:447
surface protein 4:463
transmembrane protein 4:463
evolution 2:183
gene diversity 2:183–184
long terminal repeats 2:183–184
fish see Fish retrovirus(es)
gag gene/protein 4:446, 4:461
deletions 4:447
fusions 4:447–448
virion assembly 4:466
gene therapy vectors 2:302
life cycle 2:302
replication competent 2:302
types 2:302
genome 2:213, 4:446, 4:459, 4:460, 4:462f, 4:594
history 4:459
infection see below
life cycle 2:213, 2:467–468, 2:468f
infection initiation 2:213
LTR-retrotransposons vs. 4:431–432
morphology 4:461, 4:463f, 4:594
pol proteins 4:446
deletions 4:447
proteins 4:463
receptors 4:464
replication 4:464
assembly 4:466
attachment 4:464
genomic mRNA transcription 4:465
integration 4:465
maturation 4:411–412
penetration 4:464
polyprotein processing 2:213, 4:466
reverse transcription 4:464
translation 4:465–466
virion release 4:466
retrotransposons vs. 4:352
RNA recombination 4:380
structure 2:213
taxonomy 4:446, 4:460, 4:461f
transporters 4:464
see also specific viruses
Retrovirus infections
antiviral therapy 1:143t
immunomodulation 4:114t
integration 3:307–309
proviruses 2:105, 4:446–447
selective forces 2:107

oncogenes/oncogenicity 2:213, 2:457, **4:445–450**
cellular sequence acquisition 4:446, 4:447f
classes of 4:449t
co-infection 4:447
cooperation of 4:448
defective replication 4:447, 4:466–467
functional classes of 4:448, 4:449t
fusion proteins 4:447
gain of function 4:448
historical perspective 4:446
overexpression 4:448
taxonomy 4:446
pathogenesis 4:466
acute transforming viruses 4:466
persistence 4:109t
resistance to see below
Retrovirus infections, resistance to **2:467–474**
acquisition 2:467–468
cellular cytidine deaminases 2:471, 2:472f
APOBEC3G 2:471
specificity 2:471
future work 2:473
post-entry 2:469
see also Fv1 (Friend virus susceptibility factor 1);
TRIM5 gene/protein
specificity 2:469
virus entry 2:468
receptor competition 2:468–469
receptor-envelope interaction 2:468–469
receptor inactivation 2:468
route importance 2:471
xenotropism 2:468
virus generation 2:471
particle assembly 2:473
zinc finger antiviral protein 2:473
Reverse genetics
arenaviruses 3:247
Brome mosaic virus BMV 1:382
bunyaviruses 1:398
coronaviruses 1:560
definition 1:390, 3:6, 3:96, 3:483, 4:40, 5:375
fish rhabdoviruses 2:228
icosahedral enveloped dsDNA bacteriophages 3:12
Infectious hematopoietic necrosis virus 2:228
influenza viruses 3:99, 3:488
lymphocytic choriomeningitis virus 3:247
parainfluenza viruses 4:44–45
paramyxovirus characterization 4:40
RNAi 3:151–152
Viral hemorrhagic septicemia virus 2:228
Reverse-phase polymerase chain reaction, pomovirus
4:286
Reverse transcriptase(s)
active site 4:464–465
antiretroviral drug mechanism of action 2:509
definition 1:457, 4:419, 4:428
HIV antigenic variation 1:133
inhibitors 2:509
LTR-retrotransposons 3:301, 4:438–439
mouse mammary tumor virus 3:335, 3:336
pseudovirus evolution 4:355–356
retrotransposons 4:428–429
Reverse transcriptase-endonuclease, Simian
retrovirus D 4:627
Reverse transcription 3:307f, 4:409–410, 4:464, 4:465f
double-stranded DNA virus replication 4:411
equine infectious anemia virus genetics 2:170
LTR retrotransposons 4:425
priming
gypsy retrotransposon 3:306
Tf1 retrotransposon 3:306–307
Ty3 retrotransposon 3:306
single-stranded RNA virus replication 4:410–411
virus classification 5:11t
Reverse transcription polymerase chain reaction (RT-PCR)
aquareovirus infection diagnosis 1:167–168
Border disease diagnosis 1:340
classical swine fever virus diagnosis 1:530–531
Crimean-Congo hemorrhagic fever diagnosis 1:602
definition 2:29, 4:103

echovirus infection diagnosis 2:69
furovirus infection diagnosis 2:295
iflavirus infection diagnosis 3:45
Marburg virus infection diagnosis 3:278–279
measles diagnosis 3:289–290
monkeypox virus infection diagnosis 4:644
mungbean yellow mosaic virus infection diagnosis
3:366–367
ophiovirus classification 3:448
pecluvirus infection diagnosis 4:99
Pepino mosaic virus infection diagnosis 4:106
phlebovirus infection diagnosis 4:494
plant virus infection diagnosis 2:21–23, 2:26
principles 2:35f
rice yellow mottle virus diagnosis 4:488
tospovirus infection diagnosis 5:162
yellow fever diagnosis 5:472
yellow head virus infection diagnosis 5:482
Rev protein see HIV
REV-T-mediated transformation,
reticuloendotheliosis virus infection 4:416
Rex protein, HTLV-1 2:560–561
Reyataz® 2:513
Reye syndrome, chickenpox 5:253
Rhabdoviridae 1:111, 1:170, 1:488, 1:497, 2:228, 3:324,
4:187, 4:367–368, 4:576, 5:11t, 5:291
Rhabdovirus(es) 1:111, 3:324
animal see below
apoptosis inhibition 1:158t
associated genera 1:497
characteristics 5:62t
genome 1:173, 3:326f
glycoproteins 3:332
fusion protein biosynthesis 3:300
host cell entry 3:333
morphology 1:497
plant see below
virion structure 1:173
virus structure 3:325, 3:329f
see also specific viruses
Rhabdovirus(es), animal **1:111–121**
Almpiwar Group 1:116, 1:119t
classification 1:111, 1:119t
immune reagents 1:111
evolutionary relationships 1:112, 1:114f
L polymerase, block III 1:112–114
fish see Fish rhabdovirus(es)
genetics 1:112
genome 1:112, 1:113f
open reading frames 1:112
history 1:111
host range 1:115
infection
clinical features 1:115
insects 3:126t
pathology 1:115
isolate CS1056 1:118
isolate CSIRO75 1:118
isolate DP1163 1:118
Kolongo Group 1:118
morphology 1:112, 1:112f
Mount Elgon Bat Virus Group 1:118
nonassigned serogroups 1:113t, 1:115
Bahia Grand Group 1:116
Hart Park Group 1:115, 1:119t
Kern Canyon Group 1:116, 1:119t
Le Dantec Group 1:116, 1:119t
Sawgrass Group 1:116
Timbo Group 1:116
propagation 1:111, 1:115
Sandjimba Group 1:118
Tibrogargan Group 1:117, 1:119t
transmission 1:115
virion properties 1:112
Rhabdovirus(es), plant **4:187–197**
classification 4:187
distribution 4:195
evolution 4:195, 4:196
genome 4:189, 4:190f
gene-junction sequences 4:189
noncoding sequences 4:189

Subject Index

Rhabdovirus(es), plant (continued)
    open reading frames 4:189
    structure 4:189
host range 4:196
    specificity determinants 4:196
infections
    barriers 4:196
    cytopathology 4:193
    orchid infections 4:225–226
matrix protein 4:188–189
nucleocapsid core 4:188–189
nucleocapsid protein 4:189f, 4:190
    nuclear localization signals 4:190
pathogens associated 4:187
polymerase 4:193
    activity 4:193
proteins 4:189
    glycoprotein 4:189f, 4:192
    matrix 4:192
    phosphoprotein 4:190
    position X proteins 4:192
    position Y orfs 4:189f, 4:192
    replication 4:193
    see also specific proteins
replication 4:193
    nuclear entry 4:194–195
    protein interactions 4:194–195
    virion-associated polymerase 4:193–194
    viroplasm formation 4:194–195
taxonomy 4:187, 4:188t
transmission 4:187
vector relationships 4:195
virion composition 4:188, 4:189
    lipids 4:189
    matrix protein 4:188–189
    nucleocapsid core 4:188–189
virion morphology 4:188, 4:189f
Rhadinovirus 2:432t, 3:190, 3:195, 4:585–586
Rhadinovirus(es) 3:190, 3:195, 4:585–586, 4:587t
    distribution 4:593
    evolution 3:197
    gene vectors 4:593
    human T-cell transformation 4:585–586, 4:590, 4:593
        cell surface markers 4:591
        generation 4:590
        gene transfer 4:591
        immunodeficiency research 4:590
        interleukin-26 discovery 4:591
        phenotypes 4:590
    molecular piracy 3:192
    murine gammaherpesvirus 68 3:372
    retroperitoneal fibrosis 4:592
    sequestered cellular genes 4:589
    subgroups 2:424–426
    see also specific viruses
Rhesus macaque cytomegalovirus (RhCMV) 1:635t, 1:639f
Rhesus macaque rhadinovirus see Ceropithecine herpesvirus 17 (CeHV-17)
Rhesus macaques
    simian hemorrhagic fever virus infection 1:183
    SIV infection see Simian immunodeficiency virus infection
    yatapoxvirus infection 5:461
Rhesus papillomavirus 1 (RhPV) 4:33
Rhesus rhadinovirus (RRV) 2:428, 4:591
    discovery 4:586
    genome 4:592
        KSHV vs. 4:592
    historical aspects 4:591–592
    natural occurrence 4:591
    pathology 4:591
    replication 4:592
Rhinitis 1:104
    HSV-1 associated 2:388–390
Rhinomune 2:419
Rhinopneumonitis, EHV infections 2:417
    clinical management 2:419
Rhinorrhea 2:543
Rhinovirus 1:542

Rhinovirus(es) **4:467–475**
    antigenicity 1:204, 1:545
    antigenic variability 4:471
        cross-reactivity 4:471
    assembly 4:470
    capsid composition 1:542
    capsid proteins 4:468–469
    cell entry 4:470
        ICAM-1 4:470, 4:471
    cesium chloride gradient purification 4:469
    classification 4:468
    detection 4:471
        enzyme-linked immunosorbent assays 4:471
        PCR 4:471
        quantitative real-time PCR 4:471
        serology 4:471
    genome 1:542, 1:542f, 2:125, 4:469
        5' nontranslated region 2:126, 4:139
        internal ribosome entry sites 4:469
        IRES sequence 1:542
        P1 4:469
        P2 4:469
        P3 4:469
    history 4:468
    host cell attachment 1:543, 1:543f, 1:544f, 5:321–322
        antirhinoviral strategies 1:547–548, 1:547t
        canyon hypothesis 1:544, 1:545f, 1:545t, 1:612–613, 5:321–322
        fatty acid function 1:544–545
        HRV2 1:543–544
        ICAM-1 1:546
        LDL receptors 1:546
        methyl-β-cyclodextrin effects 1:543–544
        VCAM-1 1:546
    host cell entry 1:543
    infection see Common cold;
        see below
    life cycle 4:470
    phylogeny 1:545
    propagation 4:471
    proteins 4:469
        VPg protein 4:469–470
    RdRp 4:470
    replication 1:543, 4:470
        cap-independent translation 4:470
        capsid maturation 1:542
        mutation rate 4:470
        polyprotein processing 1:542
        polyprotein translation 1:542
        RdRp 4:470
        VPg 3AB priming 1:543
    serotypes 4:468, 4:468t, 4:471
    survival features 4:469
    taxonomy 1:542, 4:468
    transmission 1:541
    virion 4:468, 4:469f
        structure 1:544
    see also specific viruses
Rhinovirus infections 2:552, 4:129
    antirhinoviral strategies 1:547
        acid sphingomyelinase 1:548
        cross-reacting antibodies 1:548
        vaccine development 5:240–241
    asthma 1:546
    COPD 1:546
    cycle 1:543–544
    detection 1:546
    host cell reactions 1:546
        cytokine production 1:546
    host range 4:471
        nonpermissive cells 4:471
        in tissue culture 4:471
    lower respiratory tract illness 2:552
    models 1:546
    PI-3 kinase
        host cell reactions 1:546–547
        inhibition 1:547
    prevention, acid sphingomyelinase 1:548
    in serious illness 4:468
    serology 4:471

symptoms, molecular origin 1:546
treatment 2:556–557
see also Common cold
Rhizidiovirus, fungal infections 2:287t
Rhizoctonia solani virus 717 (RhsV 717)
    antigenic properties 4:65
    genome 4:65t
        homology 4:65f
    structure 4:64f
Rhizomania see Beet necrotic yellow vein virus (BNYVV)
Rhizosolenia setigera, viral infections 1:87
Rhizosolenia setigera RNA virus (RsRNAV) 1:87, 1:88t
    biological properties 1:87
    genome 1:87–88
    phylogenetic analysis 3:281–282, 3:283t
        RdRp domains 3:283
    phylogeny 1:93f
    propagation 1:87–88
    virion structure 1:89f
Rhododendron necrotic ringspot virus (RoNRSV) 4:311t
Rhubarb virus 1 (RV-1) 4:311t
Ribavirin 1:143t
    arenaviruses 3:211
    Crimean-Congo hemorrhagic fever treatment 1:603
    definition 2:543
    henipavirus infection treatment 2:326
    human respiratory syncytial virus 2:549–550
    Lassa fever treatment 3:211
    phleboviruses 4:495
    respiratory virus infection treatment 2:556
Ribonuclease H, LTR retrotransposons 4:438–439
Ribonuclease P (RNaseP), structure 4:475
Ribonucleoprotein (RNP) 3:393
    cis-acting signals 3:396
        immunoprecipitation 3:396
        long distance RNA/RNA interactions 3:396, 3:397f
    definition 4:129, 5:157
    mouse mammary tumor virus 3:335
    RdRps 3:393–394
    'replication complex' 3:393
    'resting complex' 3:393
    see also specific viruses
Ribonucleotide reductase
    African swine fever virus 1:47
    Invertebrate iridescent virus-6 3:164
Ribose-2'-O-methyltransferase (O-MT)
    nidoviruses genome 3:424–425
    yellow head virus 5:477–478
Ribosomes
    frameshifting 1:549, 3:220, 4:623, 4:644
        Simian retrovirus D 4:627
        Ustilago maydis viruses 5:214–215
    RNA, evolution of 3:478
    RNA 23S, ribozyme function 4:475
    shunt
        cauliflower mosaic virus 1:460–461
        definition 1:457
Ribozyme(s) **4:475–481**
    chemical reactions 4:475
        ligation mechanism 4:479
        monomer circularization mechanism 4:479
    definition 2:375, 4:475, 5:332
    group I introns 4:475
    hammerhead see Hammerhead ribozyme(s)
    Neurospora crassa 4:480
    nucleolytic 4:475
        chemical reaction 4:475–476, 4:477f
    plant viroids/virusoids 4:476
        classification 4:477f
        hairpin see Hairpin ribozyme(s)
        replication 4:476, 4:478f
    satellite viruses 4:480
        classification 4:477f
    Varkud satellite 4:475
Rice *(Oryza sativa)* **1:482–489**
    edornaviruses 2:110
    infection 1:483
    production 1:483

viruses
see under Oryza
see specific viruses
Rice black streaked dwarf virus (RBSDV) 1:481, 1:484
  characteristics 4:150*t*
  classification 1:484
  genome 1:484
  history 1:484
  host range 1:484
  infection 1:484
    maize 1:481
    resistance 1:484
  proteins 1:484
  transmission 1:484
Rice dwarf virus (RDV) 1:483, 4:149–150
  characteristics 4:150*t*
  disease 1:483, 4:149–150
    symptoms 1:483–484
  genome 4:152, 4:153*f*, 4:154*t*
  history 1:483
  host range 1:483–484
  transmission 1:483
  vectors 1:483
Rice gall dwarf virus (RGDV) 1:484
  characteristics 4:150*t*
  classification 1:484
  disease 1:484
    symptoms 1:484
  host range 1:484
  tissue tropism 1:484
  transmission 1:484
  vectors 1:484
Rice grassy stunt virus (RGSV) 5:24
  disease 1:486
    resistance 1:487
    symptoms 1:487
  genome 1:486–487
  history 1:486–487
  transmission 1:487
Rice hoja blanca virus (RHBV) 1:487, 5:24, 5:26, 5:26*f*
  classification 1:487
  genome 1:487
  geographical distribution 1:487
  history 1:487
  infection 1:487, 5:26
    epidemics 5:24
    host range 1:487–488
    incubation within vector 1:487
    resistance 1:487–488
    symptoms 1:487–488
    transmission 1:487
Rice necrosis mosaic virus (RNMV)
  classification 1:489
  disease 1:489
  transmission 1:489
Rice ragged stunt virus (RRSV) 1:484–485, 4:149–150
  characteristics 4:150*t*
  classification 1:484–485
  genome 4:152*f*, 4:153*t*
  history 1:484–485
  infection 1:484
    host range 1:485
    pathogenicity 4:155
    resistance 1:485
  structure 1:484–485
  transmission 1:485
  as type species 4:536*t*
Rice stripe necrosis virus (RSNV) 1:308, 1:488–489
  antigenic properties 1:310
  genome 1:312, 1:489
  geographical distribution 1:309, 1:489
  history 1:488–489
  infection 1:308, 1:488
    resistance 1:489
    symptoms 1:488–489
  structure 1:488–489
  transmission 1:488–489
  soil 1:309
Rice stripe virus (RSV) 1:486
  classification 1:486
  genome 1:486

geographical distribution 1:486
history 1:486
infection 1:486
  epidemics 5:24
  host range 1:486
  resistance 1:486
  symptoms 1:486
species discrimination, maize stripe virus 5:26
transmission 1:486
as type species 5:24
vectors 1:486
Rice tungro bacilliform virus (RTBV) 4:482
  classification 4:482
  genome 1:485, 4:482, 4:484*f*
    transcription 4:482
  host range 1:485–486
  infection *see below*
  rice tungro spherical virus coupling 1:458, 4:551
    transmission 1:463
  strains 4:482–484
  structure 4:484*f*
  transmission 4:484*f*
  vectors 1:485
Rice tungro disease 1:458, 1:485, **4:481–485**
  control 4:485
  geographical distribution 4:482, 4:483*f*
  history 1:485
  plant resistance 1:486, 4:485
  severity 4:482
  subgroups 1:485
  symptoms 1:485–486, 4:482, 4:483*f*
  transmission 4:482, 4:483*f*, 4:551
  vectors associated 4:482
Rice tungro spherical virus (RTSV) 4:484, 4:546–547
  classification 4:484
  genome 1:485, 4:485, 4:548
    polyprotein processing 4:485
    structure 4:548*f*
  host range 1:485–486, 4:482
  isolate variations 4:550
  plant resistance 1:486
  proteins
    coat proteins 4:549–550
    leader protein 4:549
    proteases 4:548–549
    RdRp 4:549
  rice tungro bacilliform virus coupling 1:458, 4:551
    transmission 1:463, 1:485
  structure 4:484, 4:484*f*, 4:547*f*
    capsid 4:484–485
  transmission 4:484*f*
  vectors 1:485
Rice wilted stunt virus 5:24
Rice yellow mottle virus (RYMV) 4:646*t*
  biological properties 4:487
    chloroplasts 4:487–488
    cytology 4:487–488
    host range 4:487
    parenchyma 4:487–488
    plant, distribution in 4:487
    vector transmission 4:488
    xylem 4:487–488
  coat protein 4:486, 4:487
  diversity 4:489
  evolution 4:489
  genome 1:488, 4:486, 4:487*f*
    coat protein 4:487
    open reading frames 4:486
    P1 protein 4:486–487
    polyproteins 4:487
    RNA satellites 4:486
    VPg 4:487
  historical aspects 1:488, 4:485–486
  identification 4:488
  infection *see below*
  phylogenetic relationships 4:487
  RNA satellites 4:486
  transmission 1:488
    abiotic transmission 4:488
    biotic transmission 4:488

vectors 4:488
  control 4:488
virion particle 4:486, 4:486*f*
  coat protein 4:486
  X-ray crystallography 4:486
virion structure 4:649, 4:650*f*
  P1 protein 4:647
Rice yellow mottle virus infection 1:488
  control
    genetically-modified resistant plants 1:488
    resistance breeding 4:488–489
    transgenic plants 4:488–489
    vector control 4:488
  crop yield losses 4:651
  diagnosis 4:488
    ELISA 4:488
    RT-PCR 4:488
    symptoms 4:488
  epidemiology 4:488, 4:489*f*
    abiotic transmission 4:488
    biotic transmission 4:488
    as emergent disease 4:488
    symptoms 4:488
  host range 4:488
  pathology 4:650–651
  resistance 1:488, 4:651–652
  symptoms 1:488
Rice yellow stunt virus (RYSV) 1:488, 4:184*t*
  disease 1:488
    resistance 1:488
  genome 1:488
  history 1:488
  movement hypothesis, P3 protein 4:192
  P6 protein 4:192
  transmission 1:488
Rickettsia, neural cell infection 1:473
'Riddle-of-the-ends' dsDNA bacteriophage genome packaging 2:308–309
Rift Valley fever **4:490–497**
  affected species 4:492
  bioterrorism evaluation 5:407–408, 5:408*t*
  clinical features 4:491, 4:492–493
  disease manifestations 5:490
  epidemics
    Aswan Dam project 4:494
    Egypt 4:492–493
    insecticides 4:496
  epidemiology 4:494
    epizootics 4:494, 5:490
  incidence 4:492
  mortality 4:492–493, 4:494
  zoonoses 5:490
Rift Valley fever virus (RVFV) 1:391*t*
  characteristics 1:171*t*
  genome 4:491
  genomes 1:394*f*
  infection *see above*
  isolation 4:491
  maturation 4:492
  morphology 4:491
  protein translation 4:491
  replication strategy 4:491
    ambisense replication 4:492
    budding 1:397
  transmission 4:491, 4:492, 5:490
    Aedes vexans arabiensis 4:491
    prevention 4:496
  vaccines/vaccination 4:496
    formalin-inactivated 4:496
    inactivated 4:496
    live vaccine candidates 4:496
  virion structure 4:491
RIG-like helicase(s) 1:155
Rilpivirine 2:507*t*
Rimantadine
  history 1:145
  influenza treatment 3:102
  resistance 1:148, 3:102
RIN4 protein, plant viral resistance 4:175

Rinderpest virus(es) (RPVs) **4:497–507**
  associated animal pathogens 4:497
  classification 4:498
  epizootiology 4:502
  evolution 4:501
  genome 4:500
  geographical distribution 4:498
  history 4:497
    cattle imports 4:497
    panzootics 4:497
  host cell binding 4:500
  host range 4:499
    susceptibility 4:499*t*
  infection
    clinical features 4:504
    control 4:505
    diagnosis 4:506
    histopathology 4:504
    immune response 4:505
    pathology 4:504
    prevention 4:505
    serology 4:502
    vaccination 4:506
  propagation 4:500
  signaling lymphocyte activation molecule 3:285
  taxonomy 4:497, 4:498
  transmission 4:504
  vaccines/vaccination 4:506
    egg-adapted vaccines 4:506
    'Plowright vaccine' 4:506
  see also specific viruses
RING domain, definition 1:231
Rio Negro virus (RNV), characteristics 5:109*t*
RIP (repeat-induced point mutation), fungal retrotransposons 4:426–427
RI protein, T4 bacteriophage host cell lysis 3:255
RIRIS *see* Immune reconstitution inflammatory syndrome (IRIS)
Ritonavir 1:151, 2:513, 2:514*t*
Rmcf, receptor-envelope interaction 2:468–469
RMV strain, barley yellow dwarf virus *see* Barley yellow dwarf virus (BYDV)
RNA
  chemical modification in evolution 3:477–478
  folding 4:475–476
  hypoviruses 2:581
  influenza virus replication 3:487
  microarray use 2:15–16
  packaging, yeast L–A virus 5:467, 5:467*f*
  poliovirus translation 2:457
  synthesis
    Japanese encephalitis virus replication 3:183–184
    potato virus Y 4:288
RNA-1
  cowpea mosaic virus genome 1:571, 1:572*f*
  ourmiavirus 3:500
  pecluvirus genome 4:99–100
RNA-2
  beet soil-borne virus genome 4:284
  beet virus Q genome 4:284
  broad bean necrosis virus genome 4:284
  citrus psorosis virus 3:449
  cowpea mosaic virus 1:571, 1:572*f*
  Dendrolimus punctatus tetravirus genomes 5:31
  Helicoverpa armigera stunt virus genomes 5:31
  Indian peanut clump virus 4:101
  mirafiori lettuce virus 3:449
  Nudaurelia ω virus 5:31
  omega tetraviruses 5:31
  ophiovirus 3:449
  ourmiavirus 3:500
  peanut clump virus 4:101
  pecluvirus genome 4:99–100
RNA-3
  ourmiavirus 3:500
  potato mop-top virus genome 4:284
RNA-4
  lettuce ring necrosis virus 3:449
  mirafiori lettuce virus 3:449
  ophiovirus 3:449

RNA-binding proteins
  apoptosis 1:161
  flexiviruses phylogenetics 2:257, 2:258*f*
  pathogen recognition receptors 1:161
RNA-binding site, translational repression/assembly initiation complex 3:28
RNA cross-hybridization, aquareovirus taxonomy/classification 1:163
dsRNA dependent protein kinase (PKR)
  HSV infection 2:403
  interferon-induced antiviral effects 3:107, 3:115–116
  iridovirus infection 3:173
  togavirus infection 5:123
  virus-mediated immune system evasion 3:108–109
RNA-dependent RNA polymerase (RdRp) 1:99
  cellular 5:326–327
  definition 4:52, 5:117
  gene silencing 4:141, 4:142
  see also specific viruses
RNA-dependent RNA polymerase 6 4:146
RNA-directed DNA methylation (RdDM) 4:141
RNA editing
  definition 2:375, 3:285
  hepatitis delta virus replication 2:376
RNA electrophoresis, aquareovirus genetic diversity 1:167
RNA genome(s)
  positive sense 5:117
  see also specific viruses
RNA helicase 2:110–111
RNA-induced silencing complex (RISC) 5:376
  associated proteins 3:149, 5:326–327
  definition 3:148
  nodavirus RNA silencing 3:436–437
  plant gene silencing 4:146
  resistance engineering in plants 4:161
  tombusvirus replication 5:148–149
RNA interference (RNAi) **3:148–155**, 3:150*f*, 5:327
  Argonaute protein interactions 3:151
  components 3:149
  definition 1:154, 2:199, 3:149, 4:310, 5:325, 5:450
  Dicers *see* Dicer(s)
  disease control 3:153
  function 3:149
  mechanism of action 3:149
    effector phase 3:149
    initiator phase 3:149
  micro RNAs *see* Micro RNAs (miRNAs)
  properties 3:151
    animals 3:151
    plants 3:151
  R2D2 protein interactions 3:150
  small interfering RNA *see* Small-interfering RNAs (siRNAs)
  transitive RNAi 3:151
  vector competence 5:271
  viral encoded suppressors 3:152
  virus interactions 3:152
    African cassava mosaic disease 1:35
    B2 protein, Flock house virus 3:152
    filovirus infection prevention 2:204–205
    model systems 3:153
    mosquitoes–arbovirus interactions 3:154
    mungbean yellow mosaic virus management 3:370
    RNA virus–host evolution 3:154
    white spot disease 5:458
    yellow head virus infection management 5:482
  see also specific viruses
  see also Post-transcriptional gene silencing (PTGS)
RNA polymerase
  antigenic variation, role 1:127
  elongation 5:175–176
  elongation termination 5:176
  entomopoxviruses 2:139
  icosahedral tailed dsDNA bacteriophages 3:34
  isomerization 5:175–176
  modifications in T4 infection 5:176–177
  origin of 3:478
  promoter binding 5:175
  structure 5:175
  transcription initiation 5:175–176

RNA polymerase II
  coxsackievirus 1:583
  HSV transcription 2:398, 2:399
  influenza virus replication 3:487, 3:493
  iridovirus 3:170, 3:171–172
  mouse mammary tumor virus 3:335*f*, 3:336
RNA protection assays, leishmaniaviruses 3:221
RNA-protein complexes, EBV-encoded small RNA 2:164
RNA pseudoknot, yeast L–A virus 5:467
RNA recombination
  arterivirus evolution 1:182
  mouse hepatitis virus 1:560
RNA replication
  alfalfa mosaic virus *see* Alfalfa mosaic virus (AMV)
  yeast L–A virus 5:467, 5:468*f*
RNA–RNA crossover events, RNA virus recombination 4:377
RNA–RNA hybridization
  aquareovirus genetic diversity 1:167
  aquareovirus infection diagnosis 1:167–168
RNAs and Proteins of dsDNA Viruses database 5:357*t*
RNA sedimentation behavior, cowpea mosaic virus 1:569, 1:570*f*
RNase H
  hepatitis B virus replication 2:353
  metavirus 3:306
  see also specific viruses
RNase III
  RNA interference 3:149–150
  T7 RNA processing 5:180
RNase protection assays, plant virus detection 2:29
RNA silencing 5:376
  DNA viruses 4:146
  engineered plant resistance 4:161, 4:162
  fungal virus infection defence 2:290
  hordeiviruses 2:463
  nodaviruses 3:436
  see also RNA interference (RNAi); Virus-induced gene silencing (VIGS)
RNA-specific adenosine deaminase 1 (ADAR 1) *see* Adenosine deaminase 1 (ADAR 1)
RNA stability, viral evolution 2:179
RNA strand templates, maize chlorotic mottle virus 3:260–261
RNA structure–function relationships, orthobunyavirus 3:482
RNA translation, alfalfa mosaic virus *see* Alfalfa mosaic virus (AMV)
RNA virus(es)
  antiviral therapy 1:143*t*
  DNA viruses *vs.* 3:477
  evolution of 2:180
    helicases 2:180–181
    reduction hypothesis 3:476
    replicases 2:180–181
  fish viruses 2:227, 2:228
  fusion proteins 3:293*t*, 3:297
  genome 1:612
  insect pest control 3:127
  nucleotide sequence analysis 2:21
  ocular disease 2:492, 2:493*t*
  positive strand 2:378
    replication 1:382
  recombination 4:377
    cytoplasmic life cycle 4:377
    homologous recombination 4:377–378, 4:378*f*
    with host-derived sequence 4:377
    molecular mechanisms 4:377
    mutation rates 4:377
    nonhomologous recombination 4:377–378
    recombination by reassortment 4:377
    RNA–RNA crossover events 4:377
  replication 4:407
    double-stranded RNA 4:410
    single-stranded negative sense RNA 4:410
    single-stranded positive sense RNA 4:410
    single-stranded RNA viruses, DNA intermediate 4:410
  research history 2:452
  see also specific viruses

'RNA world' 3:473
  virus evolution 3:475
Rnq1 amyloid, prion generation 4:339
Robovirus(es) 2:317
Rochambeau virus (RBUV) 1:118
Rocio virus 5:489
  characteristics 1:171t
Rodents
  cardiovirus infection 1:444–445
  Saint Louis encephalitis virus transmission 4:656
  see also specific viruses
Roguing 5:282
  vegetable viruses 5:286–287
Rolling-circle replication (RCR)
  beet curly top virus 1:303–304
  circoviruses 1:517
  definition 1:314, 1:513, 2:375, 5:138
  equine herpesviruses 2:415
  hepatitis delta virus 2:376
  KSHV 3:199
  maize streak virus 3:268, 3:269f
  mungbean yellow mosaic virus replication 3:366
  ribozymes 4:476, 4:478f
  RNA satellites replication 4:534
  varicella-zoster virus 5:258
Rolling-hairpin replication (RHR) 4:90
Roniviridae 3:419–420, 4:567, 5:11t, 5:152, 5:477
Ronivirus(es)
  genome 3:424f
    accessory proteins 3:426
    structural proteins 3:426
  infection 3:421
  structure 3:421–423
  see also specific viruses
Root beardness see Beet necrotic yellow vein virus
  (BNYVV)
Root madness see Beet necrotic yellow vein virus
  (BNYVV)
Rosellinia necatrix, Mycoreovirus 3 infection 3:380
Rosellinia necatrix virus (RnV-1)
  infectivity assays 4:75
  latency 4:67–68
Roseola infatum 2:502
Roseolovirus 1:635–636, 2:432t
Roseolovirus(es)
  classification 1:635–636, 1:635t
  genome
    direct repeats 2:499
    origin of lytic replication 2:499–500
  infection 2:498
  see also specific viruses
Ross River virus (RRV) 5:89, 5:114
  characteristics 1:171t
  genotypes 5:89
  geographical distribution 5:84f, 5:89
  host range 5:90
  infection see below
  seasonal distribution 5:90
  transmission 5:90, 5:100, 5:114
Ross River virus infection
  clinical features 5:87t
  clinical manifestations 5:83–84, 5:86t, 5:114
    arthralgia 5:114
    arthritis 5:85, 5:114
    gastrointestinal 5:84
    pregnancy 5:87
    rash 5:84
  host factors 5:87
  incidence 5:90
  zoonoses 5:493
    geographical distribution 5:493
Rotarix 4:513
RotaShield 4:513
RotaTeq 4:513
Rotavirus 2:118–119, 4:509, 4:536t
Rotavirus(es) 2:118, **4:507–513**
  characteristics 2:117t
  classification 2:118, 4:509
    genotypes 2:118–119, 4:509
    serotypes 2:118–119, 4:509
    strain variability 4:509

genetics 2:119
genome 4:509
  functional analysis 4:510
  genetic diversity mechanisms 4:510
  geographical distribution 2:119
  global distribution 4:509
  group A, genome reassortment 4:509
history 2:118, 4:507
infections see below
morphology 4:508, 4:508f
  associated proteins 4:508
mutation rate 4:510
replication 4:509
  host cell entry 4:510
  nonstructural proteins 4:510
  sites of 4:510–511
transmission 4:511
vaccines/vaccination 2:120, 4:512–513
  development 5:235, 5:238
  dose regimen 5:238
  efficacy vs. potency 5:228
  reassortant strains 5:238
Rotavirus A 4:536t
Rotavirus infections
  acute gastroenteritis 2:118, 2:119–120, 4:507, 4:512
  antibodies 4:512
  children 2:118, 2:119–120, 4:507, 4:512
  clinical characteristics 2:119, 4:511
    diarrhea see below
    immunodeficiency 4:512
    virus excretion 4:511
  control 2:120
  diagnosis 4:511
    ELISA 4:511
    RNA analysis 4:511
  diarrhea 2:118, 4:507
    children 2:119–120, 4:511
    mechanisms 2:119, 4:511, 4:511t
    pigs 4:511
  epidemiology 4:509
    temperate zones 4:509
  host range 2:119, 4:509
  human pathogens 4:509
  immune response 2:120, 4:512
    antibodies 4:512
    children 4:512
    neonates 4:512
  mortality 2:118, 4:507, 4:511–512
    children 4:507
  pathogenesis 2:119, 4:510, 4:511t
  prevalence 4:507, 4:508t, 4:509, 4:511–512
  prevention 2:120
  re-infection 4:508t, 4:512–513
  respiratory 2:555
  seasonal distribution 2:119
  treatment 2:119–120, 4:511
  virus-host interactions 4:507
Rottboelia yellow mottle virus (RoMoV) 4:646t
Rous sarcoma virus (RSV)
  expression regulation, frameshifts 3:304–306
  oncogenesis studies 4:446, 4:459–460
  propagation 4:446
  reverse transcription priming 3:308f
  transformation-defective 4:446
Route importance, retrovirus infection resistance
  2:471
Row covers, infection control 5:437
RPV strain, cereal yellow dwarf virus see Cereal yellow
  dwarf virus (CYDV)
RSP2 protein 4:175
Rta protein, EBV infection 2:166
RT proteins 4:284
Rubella 5:117
  affected organs 4:519
  clinical features 4:518
    arthritis 4:518–519
    eyes 2:492–494
    rash 4:518–519
    rashes 4:518–519
    vascular lesions 4:519
  complications 5:111

congenital see Congenital rubella syndrome (CRS)
control 4:520
cytopathology 4:517, 5:103–104
diagnosis 4:520
  immunoglobulin M 4:520
  serodiagnosis 4:519
epidemic 4:514
fetal 4:518, 4:519, 4:520
fetal infections 4:518, 4:519, 4:520
histopathology 4:519
host range 4:514
immune response 4:519
incubation period 4:518
measles vs. 3:289–290
pathogenesis 4:519
pathology 4:519
persistence 4:519
pregnancy 4:518, 4:520
prevention 4:520
re-infection 4:518
Rubella virus (RUBV) **4:514–522**, 5:83
  classification 4:514, 5:83
  epidemiology 4:518
    vaccination programs 4:518
  evolution 4:517, 5:111
  genetics 4:517
  genome 4:515, 4:516f, 5:117, 5:119f
    infectious clone development 4:515
    open reading frames 4:515
    P150 protein 4:515
    proteolytic cleavage 4:515
  infection see Rubella
  propagation 4:514
  protein processing, alphavirus vs. 5:121
  replication 4:516
    budding 5:121
    capsid morphogenesis 4:517
    cellular site 4:517, 4:518
    glycan modifications 4:517
    host cell attachment 5:118
    host cell entry 4:516, 5:118–119
    regulation 4:517
    replication complex 5:119
    RNA synthesis 4:516–517
    structural protein maturation 5:121
    translation 4:516–517
  replicon systems 5:121–122
  serological relationships 4:518
  taxonomy 5:117
  tissue tropism 4:518
  transmission 4:518, 5:113
  vaccines/vaccination 4:520
    arthritis risk 4:520–521
    developing countries 4:521
    history 4:521
    modifications 4:521
    pregnancy 4:520–521
    programs 4:518
  virion properties 4:515, 4:515f
    C protein 4:515
    glycoprotein spikes 4:515
    stability 4:515, 4:517–518
  virion structure 4:514, 5:108
  virus perpetuation 2:146
Rubivirus 2:230, 4:514, 5:91–92, 5:107, 5:108, 5:117
Rubulavirus 4:41t, 4:47
  associated species 3:356
  classification 4:41t
  genome 3:327f
    P gene 4:44
  glycoproteins 3:332
  immune evasion, STAT protein targets 3:108
Rubulavirus 4:19, 4:41t
Rubus Chinese seed-borne virus (RCSV)
  4:523, 4:524t
  geographical distribution 4:523
  host range 4:525
Rubus idaeus (red raspberry)
  idaeovirus infection 3:37
  Raspberry bushy dwarf virus infection 3:38
Rudiviridae 1:587, 5:11t

Subject Index

Rudivirus(es) 1:587
  Crenarchaea infections 1:588t
  DNA-binding protein 1:591
  gene functions 1:594
  genome 1:593
  morphology 1:589
Rugose mosaic disease 3:55
Rugose wood disease complex 4:202–203, 4:202f
  causal agents 4:203
  transmission 4:203–204
Runting syndrome, reticuloendotheliosis virus infection 4:416
Rupestris stem pitting-associated virus (RSPaV) 1:423
Ruprintivir 1:548
Rural cycle, Dengue virus 2:8f, 2:9
Ruska, Helmut, physical studies of viruses 2:455–456
Russian influenza (1977) pandemic 3:100
Rv-cyclin(s) 2:214–215
  localization 2:216
  transcription activation 2:216
  transcription inhibition 2:216
Rx2 gene, plant resistance 4:174
Rx genes
  hypersensitive response 4:174
  plant resistance 4:174
    signaling mechanisms 4:176
Ryegrass mottle virus (RGMoV) 4:646t, 4:650–651
Rymovirus 4:315t
Rz-Rz1 genes *see* Holin–endolysin lysis

S

S41 virus 5:413t
S45 virus 5:413t
S50.2 virus 5:413t
S107 antiholin 3:253
  holin S105 *vs.* 3:253–254
  lysis timing 3:253
S4100 virus 5:413t
S5100 virus 5:413t
Sabia virus 2:95
  history 3:203
Sabin, Albert Bruce 4:383
Sacbrood virus (SBV) 3:42
  classification 5:2–3
  genome 3:45, 5:3
Saccharomyces cerevisiae
  Brome mosaic virus replication studies 1:382
  killer phenomenon 5:164, 5:170–171, 5:300
    maintenance of killer genes 5:170–171
    phenotype inheritance 5:300
    preprotoxin processing 5:301–303
    superkiller genes 5:170–171
  PIN prion 4:337
  retrotransposons 4:420
  Ty elements 4:420
  vacuolar protease B 4:340
  viral replication model 2:290
  virus transmission 5:300
Saccharomyces cerevisiae L-A virus (ScV-L-A) *see* Yeast L–A virus (ScV-L-A)
Saccharomyces cerevisiae M virus(es) (ScV-M)
  killer phenomenon 5:300–301
  M28 virus 5:301
    preprotoxin processing 5:304f
  replication 5:301
Sadwavirus 3:405–407, 4:523, 4:524t
Sadwavirus(es) **4:523–526**
  epidemiology 4:525
  genome 4:524
    open reading frame 4:524, 4:525f
  geographical distribution 4:523
  host range 4:525
  infections *see below*
  phylogenetics 4:524
  satellites 4:525
  sequence comparisons 4:524
  strains 4:523
  taxonomy/classification 4:523, 4:524t
  type species 4:523
  virion 4:523, 4:524f
    sedimentation rates 4:523–524
    size 4:523–524
  virus propagation 4:525
  see also specific viruses
Sadwavirus infections
  cytopathology 4:525
  prevention/control 4:525
    detection 4:525–526
    vector control 4:525–526
'Safety box(es)', vaccine safety 5:229–230
sag gene/protein, mouse mammary tumor virus 3:335f, 3:336, 3:341, 3:341f
Saimirine herpesvirus 2 *see* Herpesvirus Saimiri (HVS)
Saint Louis encephalitis virus (SLEV) 2:236, **4:652–659**
  classification 4:653
  distribution 2:236, 4:653, 4:654f
  epidemiology 4:653, 4:654f, 4:656
    overwintering 4:656
    virus reintroduction 4:657
  geographical distribution 5:488
  geographic variation 4:653–655
    clusters 4:653–655
  history 4:653
  infection *see below*
  lineages 4:653–655
  multiple strain circulation 4:653–655
  pathogenicity 4:656
  replication 4:656
    conditions 4:657–658
  structure 4:653–655
  subsidence 4:658
  transmission 2:236, 4:657–658, 5:488
    cycles 4:658
  virulence changes 4:653–655
  virus amplification 4:657
Saint Louis encephalitis virus infection
  asymptomatic 2:236
  case fatality rates 2:236
  clinical features 4:656, 5:488
  control 4:659
  epidemics 4:653
    seroprevalence rates 4:658
  host range 4:655
    arthropods 4:655
    birds 4:655
    domestic animals 4:655
    humans 4:655
    wild mammals 4:656
  persistence 4:657
  prevention 4:659
  reintroduction via migration 4:657
  risk factors
    age 4:658
    occupation 4:658
    residence 4:657f, 4:658
    socioeconomic status 4:658
    weather 4:659
Sakhalin virus 1:598t
Salicylic acid (SA)
  definition 4:170
  plant signaling mechanisms 4:176–177
Saliva, West Nile virus transmission 5:446
Salivary gland virus(es) *see* Human cytomegalovirus (HCMV)
Salmonid alphavirus(es) (SAVs) 2:230, 5:92
  see also specific viruses
Salmonid herpesvirus(es) 2:208
  genetics 2:211
  growth properties 2:209
Salmonid herpesvirus 1 (SalHV-1) 2:233
Salmonid herpesvirus 2 (SalHV-2) 2:233
Salmon pancreas disease virus (SPDV) 2:230, 5:93t, 5:94
  characteristics 5:109t
  cohabitation experiments 5:94
  disease 5:94
  genome 5:94
  identification 5:94
  infection 2:230
Salmon swimbladder sarcoma-associated retrovirus (SSSV) 2:217
  genome 2:217–218, 2:218f
    central DNA flap 2:217–218
  infection
    mortality 2:219
    symptoms 2:217, 2:218f
  isolation 2:217
  pathogenesis 2:219
    insertional activation 2:219
  phylogeny 2:219–220, 2:220f
  prevalence 2:218
  seasonality 2:218
  sequence comparisons 2:218
  transmission 2:218
Salterprovirus 5:418
  Euryarchaeota infections 5:413t
  'fusiform' viruses 2:300
Salterprovirus 5:413t
Sandfly 4:490
Sandfly fever virus(es) 4:490, 4:493, 1:391
  epidemiology 4:495
  infection
    aseptic meningitis 5:488
    clinical features 4:493
  isolation 4:491
  transmission 5:488
  zoonoses 5:488
  see also specific viruses
Sandjimba Group, animal rhabdoviruses 1:118
Sandjimba virus (SJAV) 1:118
San Miguel sea lion virus (SMSV) 1:410, 2:230
  geographical distribution 1:413–414
  structure 1:412f
S-antigen, Borna disease virus 1:341–342
Sap, pecluvirus transmission 4:98–99
Sapelovirus 2:126–127
Sapovirus 1:411, 1:411t, 2:117, 3:438–439, 3:439t, 3:440f, 3:441t
Sapovirus(es) **3:438–445**
  antigenic variation 1:415
  characteristics 2:117t
  classification 3:438–439, 3:440f, 3:441t
  composition 3:440
  evolution 3:442
    genetic recombination 3:442–443
  gene expression 3:440
  genome 3:440, 3:442f
    open reading frames 3:440
  history 1:411
  infection *see below*
  nonstructural proteins 1:412–413
  replication 3:441, 3:443f
  structure 3:440, 3:442f
    VP1 3:440
    VP2 3:440
    VPg 3:440
  taxonomy 3:439
  transmission 1:416, 3:443
  vaccines/vaccination 3:444
  see also specific viruses
Sapovirus infections 3:439t
  clinical features 3:443
    pediatric gastroenteritis 3:439
  complications 1:418
  control 1:418, 3:444
  diagnosis 3:444
  gastroenteritis, clinical signs 1:417
  host range 1:414–415, 3:443
  immune response 1:417–418, 3:444
    histo-blood groups, antigens 3:444
  pathogenicity 3:443
  prevention 3:444
Saquinavir 1:143t
Sarcolemma 1:533
SARS *see* Severe acute respiratory syndrome (SARS)
SARS-CoV RNA SSS Database 5:357t
Satellite(s)
  Arabis mosaic virus 3:411
  classification 5:11t
  definition 1:314, 2:97, 4:164, 4:526

Satellite DNAs **4:**526–535
  definition 4:526
  replication 4:533
    host DNA polymerase 4:533
    single-stranded DNA satellites 4:530
    see also β ssDNA satellite(s)
  structure 4:533
    sequence comparisons 4:533
  see also specific types
Satellite dsRNA, definition 5:299
Satellite-like RNA
  classification 4:529*t*
  definition 4:526
Satellite maize white line mosaic virus (SMWLMV) 4:527*t*
Satellite nucleic acids
  classification 4:527, 4:528*t*
  definition 4:526
  evolution 4:534
  geographical distribution 4:527
  historical aspects 4:527
  sequence expression 4:535
  sequence variation 4:534
  viruses, control of 4:535
  see also Satellite DNAs; Satellite RNAs
Satellite Panicum mosaic virus (SPMV) 4:527*t*
Satellite-related nucleic acids 4:529*t*, 4:532
  satellite-like ssRNAs 4:532
  virus-associated DNAs 4:529*t*, 4:533
  virus-associated nucleic acids 4:532
Satellite RNAs **4:**526–535
  definition 1:615, 4:526, 4:644
  double stranded 4:528*t*, 4:531
  replication 4:526, 4:533
    helper virus *see* Helper virus(es)
    host virus replication machinery 4:533
    noncoding RNA 4:533–534
    rolling-circle mechanism 4:534
  single stranded 4:531
    large 4:528*t*, 4:531
    small circular 4:528*t*, 4:532
    small linear 4:528*t*, 4:531
  structure 4:533
    in vitro models 4:534
  see also specific viruses
Satellite tobacco mosaic virus (STMV) 4:527*t*
  structure 5:381–384, 5:384*f*
Satellite viruses 4:480, **4:**526–535
  β ssDNA satellites *see* β ssDNA satellite(s)
  classification 4:527, 4:527*t*
  definition 3:385, 4:526
  evolution 4:534
  geographical distribution 4:527
  historical aspects 4:527
  host virus accumulation interference 4:530
  nontranslated regions 4:530
  particle structure 4:530
  sequence variation 4:534
    phylogenetic analysis 4:534
  structure 1:314–315
  viruses, control of 4:535
  see also specific viruses
Satellite virus-induced gene silencing (VIGS)
  see Virus-induced gene silencing (VIGS)
Satsuma dwarf virus (SDV) 4:524*t*
  genome sequencing 4:524–525
  geographical distribution 4:523
  host range 4:525
  transmission 4:525
  as type species 4:523
  vector control 4:525–526
'Savanna type' yellow fever transmission 5:473–474
Sawflies 1:225–226, 1:229–230, 1:230*f*
Sawfly baculovirus(es) **1:**225–231
  gene content 1:226
  gene parities 1:228
  genome 1:226
    auxiliary genes 1:228
    densovirus capsid protein 1:228
    DNA repeat regions 1:229
    immediate early proteins 1:228
    inhibitors of apoptosis 1:226
    membrane fusion protein 1:228
    open reading frames 1:226
    phosphotransferase 1:227
    regulator of chromosome condensation protein 1:228
    trypsin-like serine protease 1:226
    zinc finger protein 1:228
  infection 1:229–230
    phylogeny 1:228
    tissue tropism 1:229–230
Sawgrass Group, Rhabdoviruses, animal 1:116
SBV *see* Sacbrood virus (SBV)
Scab material, orf virus shedding 4:58, 4:63
Scaffold proteins
  bacteriophage assembly 1:438
  capsid formation in HCMV 2:487
  definition 1:433, 1:438, 3:13
  microviruses
    functions 3:15
    morphogenesis role 3:17–18
Scallion virus X (ScaVX) 4:311*t*
Scarlet fever, measles *vs.* 3:289–290
Schistosoma mansoni, ribozyme structure 4:478–479
Schizophrenia
  Borna disease virus associated 1:342–344
  Mason–Pfizer monkey virus 4:629
  viral etiology 1:475
Schlumbergera virus X (SCX) 4:311*t*
SCID mice, definition 4:382
Sclera 2:491
Sclerophthora macrospora virus A (SmV A) 2:289
Sclerophthora macrospora virus B (SmV B) 2:289
Sclerotinia sclerotiorum debilitation-associated, replicase 2:288
Scoliosis 1:76
Scoring schemes, alignment algorithms 5:342–343
Scotophilus temmencki, Keterah virus 1:400
Scrapie (PrP^{SC}) 1:474, 4:338, 5:187
  history 5:186
  occurrence 5:187–188
  prion function 4:333
  prion properties 5:186–187
  prion structure 4:333–334
  susceptibility 5:188
  transmission 5:187
  see also Bovine spongiform encephalopathy (BSE)
Scrophulariaceae virus infections 4:215
  see also specific viruses
ScV, giardiavirus *vs.* 2:315
ScV-L-A *see* Yeast L–A virus (ScV-L-A)
SDE3, antiviral defense 4:147, 4:147*f*
SDE5, virus-induced gene silencing function 5:326–327
SDI RNA *see* Defective interfering RNAs (DI RNA)
SDV *see* Satsuma dwarf virus (SDV)
Seadornavirus 1:533, 4:536, 4:536*t*, 4:540, 4:540*t*
Seadornavirus(es) **4:**535–546, 4:536*t*
  antigenic relationships 4:540
    amino acid-sequencing 4:541–542, 4:541*f*
    conserved sequences 4:542
    RdRp 4:542, 4:543*t*, 4:544*f*
    VP7 4:541–542
  capsid structure 4:538, 4:538*f*
    electron microscopy 4:539*f*
  definition 4:536
  distribution 4:537, 4:537*f*
  epidemiology 4:537
  genome 1:172, 4:538
    sequencing 4:539
  historical overview 4:536
  infections *see below*
  infectivity 4:538–539
  nomenclature 4:536
  particle purification 4:538
  phylogenetic relationships 4:540, 4:540*t*
  proteins 4:538
    electrophoresis 4:538*f*, 4:539
    electrophoretic profiles 4:538*f*, 4:539
    VP1 4:538
    VP2 4:538
    VP3 4:538
    VP4 4:538
    VP8 4:538
    VP9 4:538
    RdRp 4:542, 4:543*t*, 4:544*f*
  replication 4:538, 4:539
    mosquito cell lines 4:539
  transmission 4:537
    mosquito vector 4:537
  type species 4:536, 4:536*t*
  vectors 4:537
  virion properties 4:538
  see also specific viruses
Seadornavirus infections
  animal models 4:545
  clinical features 4:544
  diagnostic assays 4:544
    molecular assays 4:544–545
    serology 4:544–545
  host range 4:537
  immunity 4:546
  treatment 4:546
    vaccines 4:546
  vaccines/vaccination 4:546
Seagulls, Taura syndrome virus transmission 5:7
Sea louse (*Lepeophtheirus salmonis*), infectious salmon anemia virus vector 3:91
Seals
  canine distemper virus 4:498
  phocine distemper virus 4:498
Seasonal trends
  disease surveillance 2:48
  maintenance transmission 5:272
  subsidence 5:273
  see also specific viral infections
Sec61 complex, K28 virus toxin 5:304–305
Secondary infections
  measles 3:289
  nucleopolyhedrovirus 1:267
Secondary mutations, quasispecies 4:361–362
Second-generation vectors, plant viruses 4:232
Secretory cells, yellow fever virus assembly 5:471
Secretory component (SC), adaptive immunity role 3:61
Secular trends, disease surveillance 2:48
Seed(s), viral transmission 5:284–285
  alfalfa mosaic virus 1:86
  bean common mosaic virus 1:291, 1:293
  bean common necrosis virus 1:291
  control 5:285*t*, 5:286–287
  cucumber mosaic virus 1:618
  necrovirus 3:404
  nepovirus 3:411
  pecluvirus 4:98–99
  Pepino mosaic virus 4:108
  plum pox virus 4:241
  zucchini yellow mosaic virus 5:436
Seed coats, infection susceptibility 1:293
Seedling-yellow reaction, citrus tristeza virus infection 1:521
Seed potato production 4:297–302
  certification schemes 4:199
Seed programs (plants), infection control 1:295, 3:219
Segmented genomes
  double-stranded RNA virus replication 4:410
  single-stranded negative sense RNA virus replication 4:410
Segment nesting, Campoletis sonorensis ichnovirus 4:258–259
Selection coefficient 4:359
Selective activity index *see* Antiviral drug(s)
Self antigens 1:124
Selonopsis invicta virus-1 (SINV-1) 2:40
Semipersistent transmission 1:457, 1:483, 4:546
Semliki Forest virus (SFV) 5:90
  antigenic group 5:83
  characteristics 5:109*t*
  geographical distribution 5:84*f*, 5:90
  host protein synthesis shut-off 5:123
  replication 5:118
  zoonoses 5:491
Semotivirus 4:451

Semotivirus(es)
    classification 4:451
    distribution 4:452f, 4:453
    see also specific viruses
Sendai virus 4:54
    genome 3:327f
Sentinel studies, epidemiology 2:144
Seoul virus 1:391t
    associated disease 2:318, 2:319–320
Sequence(s) 1:296
    alignment, phylogenetic analysis 4:126
    alignment algorithms see Alignment algorithms (classification)
    analysis, arenavirus L polymerase 3:245
    Blueberry leaf mottle virus 3:408
    comparisons, pecluvirus genome 4:100
    Grapevine chrome mosaic virus 3:408
    icosahedral enveloped dsDNA bacteriophages 3:10
    nepovirus 3:408, 3:408f, 3:412
Sequence homology, Stp oncoproteins 4:589
Sequence polymorphisms, EBV 2:159
Sequence space 4:359
    definition 2:175
    viral evolution see Evolution (of viruses)
Sequestered cellular genes
    Herpesvirus Saimiri 4:589
    Human herpesvirus 8 4:589
    rhadinovirus 4:589
Sequiviridae 1:90f, 4:546–547, 4:547t, 5:11t
Sequivirus 4:547, 4:550
Sequivirus(es) 4:546–551
    genome 4:548, 4:548f
    infection, tissue tropism 4:547
    isolate/strain variation 4:550
    phylogenetic relationships 4:550
        coat proteins 4:550
        NTP-binding domains 4:550
        RdRp 4:550
    protein properties 4:548
        proteases 4:548
        proteolytic processing 4:548
    taxonomy 4:547
    viral interactions 4:551
    virion properties 4:547
    see also specific viruses
Serine-like protease 2:343
Seroconversion, equine torovirus 5:155–156
Seroepidemiologic studies 2:144
Serological assay(s)
    applications 2:31
    clinical specimens used 2:31
    coltivirus infection 1:540
    definition 4:207, 4:514
    limitations 2:31
    principles 2:30, 2:31
        complement fixation tests 2:31
        hemagglutination inhibition tests 2:31
        immunoassays 2:32
        immunoblotting 2:33
        isotype-specific markers 2:30
        lateral flow tests 2:33
        latex tests 2:33
        neutralizing antibody tests 2:31
        point-of-care tests 2:33
    rubella virus infection 4:519
Serological differentiation index (SDI) 5:59
Serology
    avipoxviruses 2:279
    citrus psorosis virus 3:450
    disease surveillance 2:45, 2:46
    geminivirus molecular diagnosis 4:167
    Indian peanut clump virus 4:98–99
    Mason–Pfizer monkey virus 4:629
    ophiovirus classification 3:448
    orthobunyavirus infection diagnosis 3:482–483
    peanut clump virus 4:98–99
    pecluvirus infections 4:99
    plum pox virus strains/groups 4:241
    potato virus Y see Potato virus Y (PVY)
    rhinovirus detection 4:471
    rhinoviruses 4:468, 4:468t, 4:471

seadornavirus infections 4:544–545
Simian retrovirus D 4:629–630, 4:629f
Seroprevalence
    epidemiology 2:142
    simian alphaherpesviruses infection 4:582, 4:584
Serosal hemorrhage, definition 1:328
Serotonin receptor 2A (5HT$_{2A}$), JCV cell attachment 4:265, 4:280
Serotypes, definition 4:507
Serp1 protein 3:227
Serpentine chemokine receptor(s) 2:276
Seryl–threonyl protein kinase 5:180
Sesbania mosaic virus (SeMV) 4:646t, 4:649, 4:650f
Severe acute respiratory syndrome (SARS) 1:549, 1:553, 2:94, 4:552–560
    animal models 4:558
    cellular effects 4:558
    classification 4:553–554
    clinical features 2:555, 4:557
        children 4:557
        extrapulmonary manifestations 4:557
    diagnosis 2:17, 4:558
        sero-conversion 4:558
        viral load 4:558
    disease surveillance 2:50, 4:552
        data collection sources 2:50
    epidemic 1:555, 2:96, 4:552, 4:553t
    geographical distribution 5:492
    history 4:552
    host range 1:553, 2:94
        zoonoses 1:555, 4:555, 5:492
    immune response 4:559
    incubation period 4:557
    mortality rate 4:552, 4:557, 5:492
    organ transplantation 3:470
        diagnosis 3:470
    pathogenesis 4:558
    pathology 4:557–558
    re-emergence 4:559
    treatment 4:558
        antiviral drugs 4:558
        antiviral therapy 4:558
        neutralizing monoclonal antibodies 1:561
        virus targets 1:561–562, 1:561f
    virus–host associations 4:128
Severe acute respiratory syndrome coronavirus (SARS CoV) 2:555, 4:553
    ecology 4:555
    endolysins, definition 3:249
    genome 3:426f, 4:554, 4:554f
        accessory protein production 4:555
        open reading frames 4:554
        polyprotein production 4:554–555
    historical aspects, classification 4:128
    holin–endolysin lysis 3:254
        membrane depolarization 3:255
        RI protein 3:255
    host cell binding, angiotensin converting enzyme 2 4:555–557
    infection see Severe acute respiratory syndrome (SARS)
    origin 5:492
    phylogeny 4:555, 4:556f
    receptors 4:555
    reverse genetics 1:560–561
    structure 4:553–554
    transmission 4:557
        aerosol transmission 4:557
        host factors 4:557
        interspecies 4:560
        zoonosis 5:492
    Urbani strain 4:559
    vaccines/vaccination 1:561, 4:559, 5:237
        passive immunization 4:559
        timescale 5:237
    virulence, protein 6 1:559–560
Sexually transmitted disease(s)
    HCMV infection 2:482
    HPV infection 4:19, 4:38
    HTLV-1 2:559
    see also Venereal transmission; specific infections

SFV E1, membrane fusion 3:297, 3:298f, 3:299
SGV strain see Barley yellow dwarf virus (BYDV)
SH1 virus 5:413t, 5:420
    genome 5:421, 5:422f
    stability 5:420–421
    structure 5:420, 5:422f
SH3BGRL protein 4:418
Shallot latent virus (SLV) 5:288
Shallot virus X (ShVX) 1:96
Sharka disease 4:200, 4:201, 4:202f, 4:204, 4:321
    control 4:241
        attenuated virus cross-protection 4:242
        hypersensitive response development 4:242
        plant breeding programs 4:242
        quarantine measures 4:241
        tolerant varieties 4:241
        transgenic plants 4:242
    distribution 4:239
    ecology 4:241
    economic importance 4:238
        crop yield losses 4:201
        stone fruit crops 4:238
    host range 4:238
        Prunus 4:238
    mixed infections 4:239
        prune dwarf virus 4:239
        prunus necrotic ringspot virus 4:239
    symptomatology 4:238
        affecting factors 4:238–239
        apricots 4:239
        cherries 4:239
        foliage 4:238
        fruit drop 4:239
        ornamental trees 4:239
        peach 4:239
        plum 4:239
Shedding 2:144
    animal viruses 5:316
    persistent viruses 5:316–317
Sheep
    astrovirus pathogenesis model 1:208
    capripoxvirus transmission 1:429
    Jaagsiekte sheep retrovirus infection see Jaagsiekte sheep retrovirus (JSRV)
    louping ill virus infection 2:240–241
    ovine pulmonary adencarcinoma see Ovine pulmonary adenocarcinoma (OPA)
    parapoxvirus infection see Parapoxvirus(es) (PPVs)
    prion transmission 4:338
    see also under Ovine
Sheeppox 1:427
    clinical features 1:430f, 1:431f
    epidemiology 1:427, 1:429
    epizootiology 1:429
    eradication of 1:428
Sheeppox virus (SPPV)
    geographical distribution 4:329
    infection see above
    transmission 4:329
Sheep red cells, complement fixation tests 2:31
Shellfish
    cultivation 4:561
    definition 4:561
Shellfish virus(es) 4:560–567
    birnaviruses 4:566
    herpesviruses 4:563
        diagnosis 4:565
        epidemiological surveys 4:565
        epizootiology 4:563
        evolution 4:565
        genome 4:563
        hosts 4:563
        infections 4:563
        locations 4:563
        taxonomy 4:565
        vertebrate viruses vs. 4:563–564
        virus ultrastructure 4:563
    historical aspects 4:561
    information lack 4:561
    irido-like viruses 4:561
        epizootiology 4:561

histopathology 4:562
hosts 4:561
infections 4:561
locations 4:561
morphology 4:562f
see also Gill necrosis disease
nonspecific viral infections 4:561
papova-like viruses 4:566
picorna-like viruses 4:565
reo-like viruses 4:566
Shine–Dalgarno sequence 3:220
Shingles see Varicella-zoster virus (VZV)
Shoot apex, maize streak virus 3:270
Shope fibroma virus (SFV) infection
cottontail rabbit 3:225
immune response 3:229–230
pathogenicity 3:228
tumor development 3:228
Short intergenic region (SIR)
definition 3:264
maize streak virus see Maize streak virus (MSV)
Short interspersed repetitive elements (SINEs) 4:439
definition 4:437
distribution 4:442t
genes, effects on 4:444
origins 4:441
retrotransposons 4:429–430
Short inverted terminal repeats, bacteriophages 4:402–403
Short tails, icosahedral tailed dsDNA bacteriophages 3:31, 5:367, 5:368
Shrimp baculovirus(es) 4:569
characteristics 4:570
morphology 4:570
occlusion bodies 4:570
geographical distribution 4:571
host range 4:571
infection cycle 4:571
polyhedra 4:570, 4:570f
Shrimps, penaeid
definition 2:37
infection see Shrimp virus(es)
Shrimp virus(es) 4:567–576, 4:568t
baculoviruses see Shrimp baculovirus(es)
disease control 4:575
disease emergence, factors associated 4:569
farmed shrimp production, impacts 4:569
annual production losses 4:569
history 4:567
infectious myonecrosis virus infection see Infectious myonecrosis virus (IMNV)
Taura syndrome disease see Taura syndrome disease (TSD)
white spot syndrome virus see White spot syndrome virus;
see also specific viruses
Shuster, Peter, quasispecies 2:177–178
Siadenovirus 1:1, 1:2t, 1:17
Siadenovirus(es) 1:1, 1:8
evolutionary history 1:8
genome 1:8, 1:18
E3 gene 1:8
host switches 1:5, 1:8–9
see also specific viruses
Sialic acid
BK viruses 4:280–281
influenza viruses 1:149
paramyxovirus–host cell binding 4:53–54
porcine respiratory and reproductive syndrome 1:181
α-2,3-Sialic acid
BK virus-host cell binding 4:280–281
influenza virus-host cell binding 3:486
α-2,6-Sialic acid
associated viruses 5:320
influenza virus-host cell binding 3:102
JCV cell attachment 4:265, 4:280
receptor role 5:320
structure 5:320
Sialic acid-independent mechanisms (SLAM), henipavirus cell attachment 2:322–323

Sialoadhesin 1:181
Sibine fusca, biological control 3:128
Sibine fusca densovirus. (SfDNV) 4:79
Sickness absence, disease surveillance 2:46
Sida yellow mosaic virus 1:318–319, 1:319f
Sieve-tube cells 1:296
σ1 protein
function 4:387
outer capsid 4:396
σ2 protein, inner capsid 4:393f, 4:394
σ3 protein, outer capsid 4:396–398
σ^{70} 5:175
bacteriophage λ 5:185
bacteriophage T4 5:177
AsiA binding 5:177
σ-factor(s) 5:175
bacteriophage λ 4:400–401
definition 4:400
see also specific factors
Sigma virus (SIGV) 1:118, **4:576–581**
classification 4:576
phylogenetic tree 4:576, 4:577f
definition 4:576
discovery 4:576
L'He´ritie'r, P 4:576
Tessier, G 4:576
genome 4:576, 4:577f
infection see below
phylogenetic trees 4:577f
replication 4:579
early mutants 4:579
heat-shock experiments 4:579
late phase mutants 4:579
middle-stage mutants 4:579
ts mutants 4:579
transmission 4:576, 4:578
efficiency 4:579
females 4:579
males 4:579
vertical 4:578
virion structure 4:576, 4:577f
M protein 4:576–577
P protein 4:576–577
Sigma virus infection
CO_2 sensitivity 4:576, 4:578
specificity 4:578
thoracic ganglia invasion 4:578
detection 4:578
host immunity 4:579
ref genes/proteins 4:579–580, 4:580f
host range 4:578
Sigmodontinae 2:93
Signaling lymphocyte activation molecule (SLAM), cell entry
canine distemper virus 3:285
measles virus 3:285, 3:288
morbillivirus 3:285
rinderpest virus 3:285
Signal peptide (SP), arenavirus glycoproteins 3:244
Signal transducers and activators of transcription (STAT), henipavirus 2:326
Silencing-suppressor proteins
definition 5:145
resistance engineering in plants 4:161
Simian adenovirus(es)
pathogenesis 1:26
see also specific viruses
Simian agent 8 (SA8) 4:582–583
Simian alphaherpesvirus(es) **4:581–585**
classification 4:583, 4:583t
epidemiology 4:584
evolution 4:583
history 4:581
infection 4:584
cross-reactivity 4:582–583
cytopathic effects 4:582f, 4:584
immune response 4:584
reactivation 4:584
seroprevalence 4:582, 4:584
physical properties 4:584
replication 4:584

transcription 4:584
translation 4:584
transmission 4:584
see also specific viruses
Simian cytomegalovirus(es) **1:634–642**
classification 1:635
epidemiology 1:638
evolution 1:635
gene expression control 1:640
genome 1:636
infection, persistent 1:638
physical properties 1:638
replication strategies 1:638
virion structure 1:636
see also specific viruses
Simian enteroviruses (SEV) 2:124
A-2 plaque virus 2:124
genome 2:127
see also specific viruses
Simian foamy virus (SFV)
genetic stability 2:179
human infection 2:532
Simian hemorrhagic fever virus (SHFV) 1:177
genome 1:179–180
geographical distribution 1:177
host range 1:177
infection
clinical features 1:183
immune response 1:185
pathogenicity 1:183
Simian immunodeficiency virus (SIV) **4:603–611**
agm 4:621
infectivity 4:621
classification 2:526, 2:527t, 4:603, 4:604t, 4:612, 4:613t
host dependent 2:526
serology 2:526
coreceptor usage 4:618
cpz (chimpanzee) 2:528, 2:530f, 2:533
CCR5 receptor expression 4:620
cross-species transmission 2:533, 4:611, 4:612
evolutionary history 2:528
geographical foci 4:618
hunting 2:533
origin 2:530f, 2:533
subspecies identification 2:528
vertical transmission 2:528–529
cross-species transmission 4:618
bushmeat 2:532–533, 4:604
risks 2:532
SIVagm 4:618
SIVcpz origin 2:533
SIVsmm 4:611
definition 2:534
discovery 4:594
epidemiology 4:608
evolution 4:596, 4:605f, 4:607
primate hosts 2:533
genetics 4:607
diversity 4:607
DNA synthesis 4:607
genome composition 4:617
genome 4:595, 4:605f, 4:617
accessory genes 4:607, 4:617
auxiliary genes 4:606t, 4:607
open reading frames 4:607
patterns 4:617, 4:617f
recombination 4:596, 4:617
structural genes 4:607, 4:617
geographical distribution 4:604
gor (gorilla) 2:530f
history 4:603, 4:604t
infection see below
mac (macaque)
evolution 2:527–528
history 4:594, 4:612
nef gene 4:622
origin 4:596
molecular biology 4:597
host cell receptors 4:598, 4:618
life cycle 4:597–598
propagation 4:598

Subject Index

Simian immunodeficiency virus (SIV) (continued)
  nomenclature 4:594, 4:597f
  origin 4:595f
  phylogenetic relationships 4:612, 4:615t, 4:616f
  propagation 4:598, 4:606, 4:618, 4:619t
  receptors associated 4:605f, 4:606, 4:606t
    co-receptors 4:606–607
  replication
    natural host 4:620, 4:621f
    replicative capacity 4:599
    sites 4:608
  serologic relationships 4:607
  Simian retrovirus D vs. 4:628
  spike protein structure 1:609–610, 1:610f
  taxonomy 4:594, 4:603, 4:612
  transmission 4:599, 4:608, 4:620
    blood 4:601
    cross-species see above
    experimental 4:601–602
    interspecies 4:595f, 4:596–597, 4:597f
    mother–infant 4:602
    sexual 4:601
  variants 4:600
  virulence 4:622
Simian immunodeficiency virus infection **4:611–623**
  African green monkeys
    immune response 4:622
    infectivity 4:621
  animal models **4:594–603**
    clinical features 4:601
    dissemination 4:599
    histopathology 4:601
    immune response 4:600
    pathogenesis 4:599
    pathology 4:601
    vaccine testing 4:602
  blood–brain barrier 4:600
  CCR5 receptor
    host-cell interactions 4:606–607, 4:618
    receptor expression 4:620
    natural hosts 4:620
  CD4+ T cells 4:599–600, 4:603–604
    host cell attachment 4:598, 4:603–604, 4:618
    receptor expression, natural hosts 4:620
  CD8+ T cells 4:600–601
  cell tropism 4:608–609, 4:620
  clinical features 4:601, 4:608, 4:622
  control 4:610
  diagnosis 4:618
    antibody detection 4:618
    cell culture, assay 4:618
    propagation 4:618
  dissemination 4:599, 4:599f
  dual/multiple infections 4:597
  epidemiology 4:618
  histopathology 4:601
  history 4:612
  HIV infection animal model 4:601
  HIV vs. 4:601
  host range 2:525–526, 2:536–537, 4:600, 4:604, 4:618
    AIDS resistance 2:537–538
    genetic resistance 4:622
  immune response 4:600, 4:609, 4:610f, 4:620
    antibodies 4:620–621
    CD4+ T-cell depletion 4:620
    CD8+ lymphocytes 4:609
    hypergammaglobulinemia 4:601
    IL-7 role 4:620
    immunoglobulin G 4:600–601
    natural host 4:620
    T-cell regenerative capacity 4:622
  lymphadenopathy 4:601
  lymphocytosis 4:601
  lymphoid tissues 4:601
  lymphomas 4:601
  major histocompatibility complex 4:600–601
  opportunistic infections 4:608–609
  pathogenesis 2:536–537, 2:537t, 4:599, 4:620, 4:621f
  pathology 4:601, 4:620, 4:622

  persistence 4:609, 4:620
    natural host 4:620
  prevalence 4:618
  prevention 4:610
    vaccines/vaccination 4:610–611
  rhesus macaques 4:612
    clinical presentation 4:608, 4:622
    history 4:603
    virulence 4:622
  terminal stages 4:608–609
  tissue tropism 4:620
  transmission 4:618
  viral load 4:608–609
Simian retrovirus D **4:623–630**
  capsid structure 4:626f, 4:627
  cell receptor 4:627
  classification 4:624
  env gene/protein 4:627
  gag gene/protein 4:625
  genome 4:624f
    long terminal repeats 4:625
  historical aspects 4:624
    discovery 4:625
  human infections 4:629
    breast cancer association 4:625, 4:629
    Burkitt's lymphoma 4:629
    serological surveys 4:629–630, 4:629f
    tissue culture 4:629
  in vitro cell tropism 4:628
  in vivo cell tropism 4:626f, 4:628
  natural hosts 4:627
  pathogenesis 4:628
  replication 4:625
    constitutive transport element 4:625
    env gene/protein 4:627
    gag gene/protein 4:625
    integrase 4:627
    particles 4:626f
    pol gene/protein 4:627
    pro gene/protein 4:624f, 4:625
    reverse transcriptase-endonuclease 4:627
    ribosomal frameshifts 4:627
    spike protein 4:627
    surface membrane receptor 4:627
    transmembrane protein 4:627
  serotypes 4:624
  simian immunodeficiency virus vs. 4:628
  transmission 4:624, 4:627
    carriers 4:627–628
    physical contact 4:627–628
  vaccines 4:629
    killed-virus 4:629
    recombinant vaccinia virus 4:629
Simian T-cell leukemia virus(es) (STLV), isolation 2:558
Simian T-cell leukemia virus III (STLV-III) 4:603
Simian varicella virus (SVV) 4:584, 5:251
Simian virus 5 (SV5)
  isolation 3:356
  SH protein 4:55
Simian virus 40 (SV40) **4:630–638**
  classification 4:631
  distribution 4:636
  DNA recombination mechanisms 4:375–376
  epidemiology 4:636
  genetics 4:636
    restriction map 4:632, 4:632f
    strain origins 4:637t
  genome 4:631, 4:632f
  history 4:630
  host range 4:636
  infection
    agnoprotein LP1 4:634
    carcinogenesis 4:631, 4:636, 4:637–638
    immune response 4:638
    pathogenicity 4:637
    pathology 4:637
    prevention and control 4:638
  microRNA 3:152–153, 5:327
  oligodendrocyte infection 1:473
  poliovirus vaccine 4:630, 4:637
    safety events 5:232

  propagation 4:636
  proteins 4:633
    capsid 4:634
    T-antigens 4:633, 4:634–635
  receptors 4:279
  replication 4:634
    cytoplasmic entry 4:635
    MHC class I receptor 4:633
    microRNA 4:635
    nucleic acid replication 4:635
    post-translational processing 4:635
    time course 4:634–635
    transcription 4:635
    virion uptake and release 4:635
  serology 4:636, 4:637, 4:638
  structure 5:382f, 5:388, 5:391
  T-antigens
    cellular protein targets 4:633–634
    functional domains 4:634f
    properties/functions 4:633, 4:633t
    replication role 4:634–635
    transcription role 4:635
    variable domain 4:634
  taxonomy 4:631
  tissue tropism 4:637
  transmission 4:637
  virion properties 4:631, 4:631t
    DNA packaging 4:631, 4:632
    maturation 4:636
    stability 4:631
Simple follicular conjunctivitis, adenovirus infection 2:495
Simplexvirus 2:432t
Sindbis virus (SINV) 5:90
  antigenic group 5:83
  characteristics 1:171t
  genetic lineages 5:90
  geographical distribution 5:90
  glycoproteins, apoptosis 5:122
  host range 5:90
  infection see below
  replication 5:118
    high-affinity laminin receptor 5:118
  structure 5:118f
  transcriptional downregulation 5:122
  transmission 5:90
  as type species 5:91–92
Sindbis virus infection
  apoptosis 5:122
  cellular responses 5:122
    IFN-mediated antiviral response 5:123–124
  clinical features 5:86t, 5:87t
  disease characterization 5:487
  epidemics 5:90
  zoonoses 5:487
    strain mixing 5:487
SINE retrotransposons
  mechanism of action 4:440
  structure 4:440
Singapore grouper iridovirus (SGIV) 2:233
Single-dose vials, vaccine safety 5:229
'Single-gene lysis'
  definition 3:249
  see also Bacteriophage(s)
Single internal deletions, defective-interfering RNAs 4:380
Single nucleocapsid per envelope (SNPVs), nucleopolyhedrovirus 1:265, 1:266f
Single-step growth
  bacteriophage organismal ecology 2:73
  definition 2:71
Single-stranded conformation polymorphism (SSCP) pattern 4:106
Single-stranded DNA satellites see β ssDNA satellite(s)
Single-stranded DNA virus(es)
  algal viruses 1:92
  classification 5:11t
  filamentous see Filamentous ssDNA bacteriophages

genome packaging 2:306, 2:307t, 2:312f
P4 enzyme 2:311–312
translocating NTPase 2:311–312
replication 4:410
Single-stranded genome 1:483
Single-stranded RNA bacteriophages
genome 1:439, 2:306, 2:311, 4:405
external protein components 2:306
replication 4:405
replicase composition 4:405–406
Single-stranded RNA virus(es)
algal viruses 1:87
fungal viruses 2:288
negative sense
classification 5:11t
replication 4:410
positive sense
classification 5:11t
replication 4:410
replication 4:410
see also specific viruses
Sin Nombre virus 1:391t
detection 2:93–94
transmission 2:95
SINV *see* Sindbis virus (SINV)
Siphoviridae 3:30–31, 5:11t, 5:181, 5:413t
Siphovirus(es)
DNA injection 3:33–34
Euryarchaeota infections 5:413t
genome packaging 2:306, 2:307t
protein cross-linking 1:439
structure 3:31
tail shaft assembly 1:438
see also specific viruses
SIRE1 virus, *env*-like gene 4:355
Sirevirus 4:352, 4:353t
Sirevirus(es) 4:431t, 4:433
genome 4:354
see also specific viruses
siRNA *see* Small-interfering RNAs (siRNAs)
Site-specific coding strand nicks, edornaviruses 2:110f, 2:113
Sitiawan virus 2:237
SIV *see* Simian immunodeficiency virus (SIV)
Sixth disease 2:498
Skin
tumors, fish 2:213
viral infection 5:315
barrier against 5:315
papillomaviruses 4:27–28, 4:30f
see also Epithelium
Skin lesions
adult T-cell leukemia 2:568
clinical features 2:568, 2:568f
cowpox infections 1:579
Ectromelia virus infection lesions 3:344
Ski proteins, yeast L–A virus 5:468
SLAM protein, virus-host cell binding 5:324
morbillivirus replication 4:44, 4:500
Sleeping disease virus (SDV) 5:93t, 5:94
characteristics 5:109t
classification 5:95
genotypes/genotyping 5:95
genotyping 5:95
glycoproteins 5:95
historical aspects 5:94
isolation 5:95
Sleep pattern disruption, common cold 4:472
Sliding back mechanism, definition 4:399
Slippery sequence, definition 5:477
slyD gene 3:257
Small-angle X-ray scattering (SAXS) 5:34–35
Small basic protein (VPg), cowpea mosaic virus 1:570
Small-interfering RNAs (siRNAs) 4:141, 5:326, 5:376
Arabidopsis thaliana gene silencing 4:145–146
definition 3:149, 5:325
engineered resistance in plants 4:161
mammalian cells 3:153

production 3:149
RNA silencing role 5:376
signal amplification 3:151
viroid pathogenicity 5:341
Smallpox
Ab-mediated immunity 3:56
antiviral drugs 5:249
bioterrorism 5:407, 5:408
clinical features 4:639
eye infections 2:496
flat-type 4:640
hemorrhagic 4:640
papules 4:639–640
rash 4:639–640
vesicles 4:639–640
deaths 1:574
eradication 2:458, 4:640, 5:244
World Health Organization 4:640
historical aspects 4:639, 5:244
Jenner, Edward 4:639, 4:640
variola minor 4:639
human monkeypox virus infections *vs.* 4:641
mortality rate 4:329
pathogenesis 4:329
transmission 4:640
vaccines/vaccination 1:574, 1:579, 2:458, 4:640, 5:244
B5 antigen 5:249–250
complications 4:640, 5:244
cowpox 4:640
horsepox 4:640
immune response 5:249
variolation 4:640
disease impact 4:640
Small ruminant lentiviruses (SRLV) 5:424
Small T, murine polyomavirus *see* Polyomavirus(es)
Smear test(s) 4:16
Smelt aquareovirus (SRV) 1:164t
Smith, Kenneth Manley, potato viruses 2:451
Smithiantha latent virus (SmiLV) 4:311t
Smith–Waterman algorithms (classification) 5:343
SMK toxin, *Ustilago maydis* viruses 5:215–216, 5:216f
Smoldering adult T-cell leukemia 2:561
Smolts, definition 3:83, 3:90, 5:91
SMSV *see* San Miguel sea lion virus (SMSV)
Smut killer toxins 2:285
Snakehead reovirus (SKRV) 1:164t
Snakehead retrovirus (SnRV) 2:216, 2:231
genome 2:217, 2:217f
open reading frames 2:217
nucleotide sequence 2:231
phylogeny 2:219, 2:220f
Snakehead rhabdovirus (SHRV) 2:221–222, 2:228, 4:184t
reverse genetic analysis 2:228
Snow, John, epidemiology 2:141
Snowshoe hare (SSH) virus 5:489
classification 3:480
Sobemovirus 1:488, 4:645
Sobemovirus(es) **4:644–652**
classification 4:645, 4:646t
genome 4:645
open reading frames 4:645, 4:648f
satRNA 4:645
sgRNA 4:645
geographical distribution 4:646t
host range 4:645
infection
economic importance 4:650
pathology 4:650
resistance 4:179t, 4:650
symptoms 4:650–651
movement within host 4:649
tissues associated 4:649
phylogenetic relationships 4:651f, 4:652
protein functions 4:647
P1 4:647
polyprotein 4:648
serine protease 4:648
VPg 4:648

replication 4:645
mutation rates 4:647
RdRp expression 4:645, 4:648
satRNA 4:647
translation 4:645
satellite viruses 4:527t
stability 4:649
subcellular localization 4:649
transmission 4:645, 4:646t
virion structure 4:649, 4:650f
capsid 4:649
coat protein 4:649
see also specific viruses
Social change, zoonoses 5:494
Socioeconomic status, vector transmission, animal viruses 5:273
Sodium silicotungstate, negative staining 2:79, 2:80f
Soft tick vector *(Ornithodoros moubata)*, African swine fever virus host 1:43, 1:44f
Soil
maize chlorotic mottle virus transmission 3:262
necrovirus transmission 3:404
rice stripe necrosis virus 1:309
sterilization
infection control measures 5:286–287
lettuce ring necrosis virus control 3:453
mirafiori lettuce virus control 3:453
tombusvirus transmission 5:149–150
Soil-borne cereal mosaic virus (SBCMV) 1:493, 1:494f
geographical distribution 1:493
infection, control 1:493
Soil-borne wheat mosaic virus (SBWMV) 1:308, 1:493, 2:291–292
genome 2:293
geographical distribution 1:493
infection 2:292
resistance 1:493
movement, via xylem 2:292
properties
antigenic 2:293
structure 2:293f
strains 2:292
transmission 1:309
zoospores 2:292
Soil nematodes
Grapevine fanleaf virus transmission 3:411
nepovirus transmission 3:411
tobacco ringspot virus transmission 3:411
tomato ringspot virus transmission 3:411
Solanaceae
Tnt1 elements 4:432
virus infections 4:211
see also Potato *(Solanum)*; Tomato *(Solanum esculentum)*
Solanum apical leaf curl virus (SALCV) 4:298t
Solanum nodiflorum mottle virus (SNMoV) 4:646t
Solenopsis invicta virus-1 (SINV-1) 2:40
Solid organ transplantation, human herpesvirus infection 2:498–499, 2:502
Solid-phase immunoassay(s) 2:32, 2:32f
Soluble complement-fixing antigen, Japanese encephalitis virus 3:183
Sonchus yellow net virus (SYNV) 4:184t
genome 3:326f
glycoprotein function 4:192–193
nucleocapsid protein expression 4:190, 4:191f
phosphoprotein function 4:190
complex formation 4:190–191
importin binding 4:191
nuclear export signals 4:191
phosphorylation sites 4:190
polymerase protein 4:193
Sooty mangabeys, HIV-2/SIVsmm evolution 2:527, 4:604–606
Sorghum chlorotic spot virus (SrCSV) 2:292
Sour orange rootstock, citrus tristeza virus pandemic 1:520
South African cassava mosaic virus (SACMV) 1:31–32
distribution 1:32f
South America, ophiovirus 3:450–451
Southern Asia, cotton leaf curl disease 1:563

Southern bean mosaic virus (SBMV) 4:646t
Southern blotting, avipoxviruses 2:280
Southern cowpea mosaic virus (SCPMV) 4:646t, 4:649, 4:650f
　　P1 protein 4:647
Southern elephant seal virus (SESV) 5:92, 5:93t
　　characteristics 5:109t
　　genetics 5:92–94
　　isolation 5:92–94
　　tick transmission 5:92–94
Soviet Union, bioweapons 5:408
Sowbane mosaic virus (SoMV) 4:646t
Sowthistle yellow vein virus 4:195–196
Soybean *(Glycine max)*
　　bean golden mosaic virus infection 3:217
　　infection resistance 4:175
　　　　I gene 4:175
　　vaccine production 5:223–224
　　see also specific viruses
Soybean mosaic virus (SMV)
　　epidemiology 3:217
　　pathogenicity 4:320
　　resistance evasion 4:175
　　transmission 3:214
　　　　seeds 3:215–216
Soymovirus
　　open reading frames 1:466f
　　transmission 1:463
　　virion structure 1:458–460
Soymovirus 1:459t
Spanin, definition 3:249, 3:250
Spanish influenza pandemic (1918-1919) 3:100, 5:491
　　bioterrorism 3:99
　　mortality rate 3:100
　　virus pathogenicity 3:99
Spatial structure, definition 2:71
Spawner mortality virus (SMV) 4:571
SPCFV *see* Sweetpotato chlorotic fleck virus (SPCFV)
SPCSV *see* Sweetpotato chlorotic stunt virus (SPCSV)
SPDV *see* Salmon pancreas disease virus (SPDV)
Special AT-rich binding protein 1 (SATB1) 3:340
Species barriers
　　alcelaphine herpesvirus 1 2:428
　　prion variants and 4:338
Species definition *see* Classification
Species jumping
　　definition 2:141
　　viral evolution 2:175
Specific pathogen free (SPF) lines
　　definition 5:1
　　Taura syndrome virus 5:3
Spectroscopy, inovirus structure 3:120–121
SPFMV *see* Sweetpotato feathery mottle virus (SPFMV)
Spherical halovirus SH1 5:420
　　genome 5:421, 5:422f
　　stability 5:420–421
　　structure 5:420, 5:422f
Spherical virus(es) 1:202
　　β-barrel fold 1:202, 1:202f
　　　　drug-induced modifications 1:204
　　　　evolution 1:203
　　capsid protein structure 1:202
　　　　β-barrel folding motif *see above*
　　Crenarchaeal virus morphotypes 1:590
　　see also specific viruses
Spheroidin 2:137
Spheroplast, definition 5:299
Spherules, Brome mosaic virus replication in yeast 1:383
Spike complexes, icosahedral ssDNA bacteriophages 5:368–369
Spike (S) proteins
　　coronavirus structure 1:556–557
　　　　conformational change 1:557
　　　　therapeutic developments 1:557
　　icosahedral enveloped dsDNA bacteriophages 5:369
　　Simian retrovirus D 4:627
　　toroviruses 5:153, 5:154
Spill-back events, definition 2:221
Spill-over events, definition 2:221

Spinach curly top virus (SCTV) 1:302
　　geographical distribution 1:302–303
　　host range 1:306
　　replication 1:304
Spinal leptomeninges, inflammation 5:42
Spindle-shaped sulfolobus virus(es) (SSVs), morphology 1:587, 1:589f
Spindle-shaped sulfolobus virus 1 (SSV1)
　　replication induction 1:592–593
　　transcription 1:593
　　　　open reading frames 1:593–594
　　　　promoters 1:594
Spindle-shaped virus(es)
　　Crenarchaeal viruses 1:587
　　see also specific viruses
SpLCV *see* Sweetpotato leaf curl virus (SpLCV)
Spleen lesions, Ectromelia virus infection lesions 3:345
Splenomegaly
　　definition 1:410
　　lactate dehydrogenase-elevating virus infection 1:184
　　MHV-68 infection 3:375, 3:377
　　SIV infection 4:601
Splice acceptor (SA) 3:335–336
Splice donor (SD) 3:335–336
Splicing
　　adenoviruses 1:3
　　influenza virus 3:487
Spodoptera frugiperda, insect cell culture 1:238
Sporadic Creutzfeldt–Jakob disease (sCJD) 1:371
Sporadic inclusion-myositis (sIBM) 2:571f, 2:573
Spore balls 4:282
Spores, asexual, partitivirus transmission 4:69
SPP1 bacteriophage *see* Bacteriophage SPP1
SPPV *see* Goatpox virus (SPPV)
Spraing (potato disease) 4:282, 4:307, 4:307f
Spring viremia of carp virus (SVCV) 2:224–225, 2:228, 4:184t
　　phylogeny 2:223f, 2:224–225
Sprot, definition 4:428
S protein sequence, hepatitis B virus subtypes 2:354–355
S-PTGS, post-transcriptional gene silencing 4:142–145
Spumaretrovirinae 2:231, 2:260t, 4:460, 5:11t
Spumavirus 4:460
Spumavirus(es) 4:460
　　classification 4:460
　　gag protein 4:464
　　gene therapy vectors 2:302–303
　　replication characteristics 4:460
　　see also specific viruses
Squamous cell carcinoma (SCC)
　　HSV-2 associated 2:394
　　human papilloma virus infection 2:492f, 2:497
Squamous intraepithelial lesions (SIL), human papillomavirus infection 4:16
Squash leaf curl virus (SLCV)
　　ELISA diagnosis 1:299
　　identification/naming 1:298
Squash plants
　　genetically-modified resistant plants 4:158f
　　　　CP genes 4:157–160
　　　　gene flow 4:162
　　viral disease control 4:157, 4:158f
　　see also specific viruses
Squirrel fibroma virus (SqFV) 3:225–226
　　characteristics 3:225t
　　infection 3:229–230
Squirrel monkey lymphocryptovirus (SscLHV1) 4:587t
src family tyrosine kinases
　　murine polyomavirus 4:274
　　phosphorylation, Tio oncoprotein 4:590
　　Tip oncoprotein homology 4:589–590
Sri Lankan cassava mosaic virus (SLCMV) 1:31–32
　　DNA satellites 4:530–531
S RNA segment sequence, Crimean-Congo hemorrhagic fever virus 1:599, 1:600–601
SRO5 gene silencing 4:142

SSB protein, bacteriophage φX174 replication 4:405
SSSV *see* Salmon swimbladder sarcoma-associated retrovirus (SSSV)
SSV-1
　　detection 2:297, 2:297f
　　fuselloviruses 2:297
　　genome sequence 2:297, 2:298f
SSV-2, fuselloviruses 2:298
SSV-3, fuselloviruses 2:298
SSV-K1, fuselloviruses 2:298
SSV-RH, fuselloviruses 2:298
ST-246, antiviral function 1:149
Stable transformation, murine polyomavirus infection 4:273
Stalk hypothesis, membrane fusion 3:292
Stalk-pore pathway, virus fusion 5:310–311, 5:311f, 5:313f
Standard Operating Procedures (SOPs), database annotation 5:353
Stanley, Wendel 5:55
　　tobacco mosaic virus 2:452, 3:399
Stavudine 2:510
Stealth virus 1:636
　　definition 1:634
Stegomya, Dengue virus transmission 2:7
Stem cell(s), HSV infection 2:388
Stem pitting 4:204
　　citrus tristeza virus infection 1:521
Stem swellings, cacao swollen shoot virus infection 1:403
Stiffsiekte *see* Bovine ephemeral fever virus (BEFV)
St Louis encephalitis virus
　　transmission
　　　　age 5:273
　　　　mosquito-borne 2:236
　　　　occupation 5:273
　　as zoonosis 5:270, 5:270f
STMV *see* Satellite tobacco mosaic virus (STMV)
Stocrin® 2:510
Stoichiometry, definition 3:413
Stone fruits, viral infections *see* Fruit trees, diseases; *see specific viruses*
Stop-and-start model 5:263
St particles (nondefective 'parent' virus(es)) 2:1, 2:3
Stp oncoproteins *see* Herpesvirus Saimiri (HVS)
Strains
　　cacao swollen shoot virus infection 1:408–409
　　Japanese encephalitis virus 3:185
　　potato virus Y 4:290–291, 4:292–293
　　tobacco infections 4:295
　　varicella-zoster virus 5:251–252
Stratified squamous epithelia, HPV life cycle 4:20
Strawberry crinkle virus 4:184t
Strawberry latent ringspot virus (SLRSV) 4:523, 4:524t
　　genome sequencing 4:524–525
　　geographical distribution 4:523
　　host range 4:525
　　RNA satellites 4:525
　　transmission 4:525
　　vector control 4:525–526
Strawberry mild yellow edge virus (SMYEV) 4:311t
　　genome 4:312
Strawberry mottle virus (SMoV) 4:523, 4:524t
　　genome sequencing 4:524–525
　　geographical distribution 4:523
　　host range 4:525
　　transmission 4:525
　　vector control 4:525–526
Stress, viral reactivation 2:438
Stress kinases, togavirus infection response 5:123
Striped bass aquareovirus (SBRV) 1:164t
Striped jack nervous necrosis virus (SJNNV) 2:229–230
　　genome 2:229
Striped leaf disease *see* Maize streak virus (MSV)
Stromal keratitis 2:388–390
Strongyloides stercoralis, HTLV-1 associated infection 2:568
Structural information, virus databases 5:363
Structural proteins
　　see specific viruses

Structure determination methods, nonenveloped viruses 1:201
Structured query language (SQL) 5:353–354
Stunt disease of tobacco 5:263–264
  see also Lettuce big-vein virus (LBVaV)
Stunting, barley yellow dwarf viruses 1:284–285
Subacute sclerosing panencephalitis (SSPE) 1:474, 2:494–495, 3:289
  definition 3:285
Subclinical infection, definition 2:146
Subcutaneous fibromatosis 3:225–226
Subgenomic DNAs, beet curly top virus 1:304
Subgenomic RNA (sgRNA)
  beet necrotic yellow vein virus 1:311f, 1:312
  carmovirus 1:455
  definition 1:520, 3:46, 4:207, 5:145, 5:200
  nidoviruses 3:428
  ophiovirus 3:449–450
  tombusvirus 5:149
Sub-Saharan Africa, HIV epidemiology see HIV infection
Subsidence
  seasonal changes 5:273
  vector transmission 5:272
Substilisin-like protease(s), fusion protein synthesis 3:293
Subterranean clover mottle virus (SCMoV) 4:646t, 4:651
Subterranean clover stunt virus (SCSV)
  characteristics 3:389t
  economic importance 3:386
  genome 3:387f, 3:388
  host range 3:386
  transmission 3:386
    aphids 3:386
Subtractive hybridization
  definition 4:445
  Rous sarcoma virus, transformation-defective 4:446
Subunit vaccine(s) 2:52, 5:236
Subviral agents, classification 5:11t
Subviral RNA Database 5:357t
Suc-HSA 2:507t
Sucrose density gradients, cowpea mosaic virus purification 1:569
Sudan
  HIV epidemiology 1:65
  monkeypox virus 4:643
Sudan Ebola virus (SEBOV)
  fatality rates 2:57, 2:199
  genome 2:58–59
  host range 2:62
Sugarcane bacilliform virus (SCBV) 1:403
  host range 1:463
Sugarcane mosaic virus (SCMV) 1:481
  maize chlorotic mottle virus synergy 3:262
Suid herpesvirus 1 see Pseudorabies virus (PrV)
Suipoxvirus 3:225t, 3:226, 4:324t
Suipoxvirus(es) 3:225–231
  classification 3:225t, 3:226, 4:324t
  DNA properties 3:226
  evolution 3:228
  genetic variability 3:228
  geographical distribution 3:227
  history 3:225
  infection
    clinical features 3:228
    control 3:230
    histopathology 3:229
    host range 3:227
    immune response 3:229
    pathogenesis 3:227
    pathogenicity 3:228
    pathology 3:229
    prevention 3:230
    tissue tropism 3:228
  propagation 3:227
  protein properties 3:226
  replication
    DNA replication 3:226
    sites 3:227–228
    transcription 3:226
    translation 3:226
  seasonal distribution 3:227
  taxonomy 3:226
  transmission 3:228
  virion properties 3:226
    morphology 3:226
  see also specific viruses
Sulfolobus, fuselloviruses 2:297
Sulfolobus islandicus filamentous virus (SIFV) 1:589–590
Sulfolobus islandicus rod-shaped virus(es) (SIRV)
  genome 1:593
  morphology 1:589, 1:589f
  replication 1:593
Sulfolobus neozealandicus droplet-shaped virus (SNDV)
  genome modifications 1:593
  morphology 1:590, 1:591f
Sulfolobus solfataricus, fuselloviruses 2:297–298
  vectors 2:299
Sulfolobus tengchongensis spindle-shaped virus 1 (STSV1)
  gene functions 1:594
  genome 1:593
  genome modifications 1:593
  morphology 1:589
Sulfolobus turreted icosahedral virus (STIV)
  morphology 1:590, 1:590f
  structure 5:382f, 5:390f, 5:391
Sup35p
  domain shuffling 4:339
  formation 4:337, 4:339
Superantigens, definition 3:334
Superkiller genes (SKI), Saccharomyces cerevisiae 5:170–171
Supportive therapy
  Crimean-Congo hemorrhagic fever 1:603
  Marburg virus infection 3:279
Suppression of gene silencing, papaya ringspot virus 4:4
Suppressor of virus-induced gene silencing 3:46
Suprabasal cell(s) see Human papillomavirus(es) (HPVs)
Supramolecular adhesion complex (SMAC) 1:122f
Supraorbital fossae 1:37
Surveillance measures, transmission risk forecasting 5:274
'Survival of the flatest' quasispecies 4:361–362, 4:362f
Sustiva® 2:510
SV40 see Simian virus 40 (SV40)
SVA retrotransposons 4:439
  genes, effects on 4:444
  mechanism of action 4:440
  structure 4:440
    variable number tandem repeat 4:440
Sw-5 gene, plant natural resistance 4:174
Swarms, quasispecies 2:178–179
Sweetpotato (Ipomoea) 4:659
  production 4:659–660, 4:660f
  vegetative propagation 4:660
Sweetpotato virus(es) 4:659–669, 4:661t
  average yields 4:199
  begomoviruses 4:666
  control 4:200
    breeding programs 4:200
  diseases 4:199
    yield losses 4:200
  grafting 4:660, 4:664f
  infections 4:667, 4:668f, 4:669f
    multiple infections 4:667–668
  potyviruses 4:663, 4:665
    geographical distribution 4:665
    phylogeny 4:665, 4:665f
  propagation 4:200
  transmission 4:200
  see also specific viruses
Sweetpotato chlorotic fleck virus (SPCFV) 4:661t, 4:667
  genome 4:667
  host range 4:667
Sweetpotato chlorotic stunt virus (SPCSV) 4:200, 4:661t, 4:667
  genome 4:667
  geographical distribution 4:667
  host range 4:667
  infections
    chlorotic dwarf 4:668
    multiple infections 4:667–668
  serotypes 4:667
  structure 4:667
  transmission 4:667
Sweetpotato feathery mottle virus (SPFMV) 4:661t, 4:663
  host range 4:663–664
  infection rate 4:200
  multiple infections 4:667–668
  phylogeny 4:663–664, 4:665f
  structure 4:663–664
  transmission 4:663–664
Sweetpotato latent virus (SwPLV) 4:661t, 4:665
  infection rate 4:200
Sweetpotato leaf curl Georgia virus (SPLCGV) 4:661t, 4:666
Sweetpotato leaf curl virus (SpLCV) 1:302, 4:661t, 4:666
  phylogeny 4:666
Sweetpotato leaf speckling virus (SPLSV) 4:661t, 4:667
Sweetpotato mild mottle virus (SPMMV) 4:661t, 4:665–666
Sweetpotato mild speckling virus (SPMSV) 4:661t, 4:665
Sweetpotato severe mosaic disease (SPSMD) 4:668–669
Sweetpotato vein mosaic virus (SPVMV) 4:661t, 4:665
Sweetpotato virus 2 4:661t
Sweetpotato virus G (SPVG) 4:661t, 4:665
Sweetpotato virus Y (SPVY) 4:665
Sweetpotato yellow dwarf virus (SPYDV) 4:661t
Swimbladder sarcoma virus (SSSV) 2:231
Swine cytomegalovirus 1:634
Swine fever
  bioterrorism 5:407
  see also African swine fever virus (ASFV); Classical swine fever virus (CSFV)
Swine flu (vaccination) 5:233
Swine hepatitis E virus 2:377–378
Swine pestivirus(es) 2:4
Swinepox virus (SPV) 3:226
  characteristics 3:225t
Swine vesicular disease virus (SVDV) 2:124
  genome 2:127
  history 2:124
  structure 2:128–129
  virion structure 2:128–129
  virulence 2:127
SwPLV see Sweetpotato latent virus (SwPLV)
syk tyrosine kinase 1:546–547
Sylvatic circulation
  definition 5:83
  yellow fever virus 2:238–239, 5:473–474
Sylvilagus floridanus (cottontail rabbit)
  leporipoxvirus infection 3:229
  Shope fibroma virus infection 3:225
Symbionin, luteovirus transmission 5:278
Sympatric 4:611
Sympatrically 3:479
Syncitin protein, retroviral origin 2:107
Syncytia
  channel catfish virus infection 2:209
  definition 3:285
  foamy virus infection 2:260
Synergism
  definition 5:124
  Indian tomato leaf curl viruses 5:128–129
Synovial fluid 5:83
Synovial membranes 5:83
Synteny 1:225
Synthetic genome(s), bioterrorism 5:410
Syphovirus(es) 5:367–368
Systemic immune response syndrome (SIRS), yellow fever 5:475
Systemic movement (plant viruses), definition 3:348

T

T2 bacteriophage *see* Bacteriophage T2
T3 bacteriophage *see* Bacteriophage T3
T4 bacteriophage *see* Bacteriophage T4
T4-even phages 4:401
T4-like bacteriophages
  genome 4:401
  replication 4:401, 4:402f
    leading strand synthesis 4:401
    origin of replication 4:401
T5 bacteriophage 3:33–34
T7 bacteriophage *see* Bacteriophage T7
T7 DNA polymerase 4:401–402
T7 helicase/primase 4:401–402
T7-like bacteriophages
  genome 4:401–402
  replication 4:401
    bacteriophage-encoded proteins 4:401–402
    concatamer formation 4:402
    origin of replication 4:401–402
    promoters 4:401–402
    terminal repeat sequence duplication 4:402, 4:403f
T7 lysozyme 3:251
T7 RNA polymerase 4:401–402, 5:180
  consensus sequence 5:180
  efficiency 5:180
  primase initiated synthesis 4:402
  promoters 5:180
  reverse genetics 4:44–45
  termination efficiency 5:180
T7 ssDNA binding protein 4:401–402
T-705 (6-fluoro-3-hydroxy-3-pyrazine carboxamide) 1:153
Taastrup virus 4:184t
Tacaribe complex virus(es), classification 3:203
Tacaribe virus (TACV) 3:247–248
Tachykinin 2:543
TAdv-3 *see* Turkey adenovirus 3 (TAdV-3)
Tahyna virus 1:391t, 5:490
Tail, definition 3:30
Tailed bacteriophages 2:448t
  assembly 1:439
  evolution 2:181
    conserved genes 2:182f
    error prone replication 2:181
    genome comparisons 5:372
    genome organization 5:371
    types 5:371
  populations 5:371
    co-infections 5:371
    productive infections 5:371
  structure 1:433
    symmetry disruptions 1:437
  see also specific types
Tail fiber, definition 3:30
TAK-220 2:507t
TAK-652 2:507t
Tamarin, definition 1:104
Tamdy virus 1:400
Tamiflu *see* Oseltamivir; Oseltamivir
Tamiflu® 1:143t
Tamus red mosaic virus (TRMV) 4:311t
Tanapox virus (TPV) 5:462, 5:462t
  distribution 5:464
  genome 5:462–463
  history 5:461
  host range 5:465
  infection
    control 5:465
    immune modulation 5:464
    immunity 5:463
    pathogenesis 5:464
    prevention 5:465
  replication 5:462
  tropism
    humans 5:463
    nonhuman primates 5:463
T-antigens
  murine polyomavirus 4:271–272
  SV40 proteins *see* Simian virus 40 (SV40)

Tape measure protein, bacteriophage λ 1:438
TAP/NXF1 mRNA export pathway 2:398
TaqMan PCR 2:26
Targeting induced local lesions in genomes (TILLING) 4:186
Target-primed (TP) retrotransposons 4:421f
  distribution 4:423
  open reading frames 4:420–421
  protein coding regions 4:420–421
  replication 4:425
    open reading frame 4:425
Taro bacilliform virus (TaBV) 1:403
TAS genes, plant gene silencing 4:142
Tasmanian Atlantic salmon reovirus (TSRV) infection 1:168f
Tas protein, foamy virus 2:263
TATA-binding protein (TBP), varicella-zoster virus 5:259
TATA box
  HSV replication 2:401f
  JCV genome 4:264f
Tataguine virus 1:400
  global distribution 1:400
  infection 1:400
  transmission 1:400
Tat protein
  antibody neutralization 1:72
  expression 2:522
  function 2:522
  JCV reactivation 4:266
Taura syndrome disease (TSD) 5:1
  clinical signs 5:3f
  control 5:8
  diagnosis 5:7
    bioassay 5:7–8
    ELISA 5:8
    histology 5:8
    RT-PCR 5:8
  economic impact 5:1
  farming practices 4:569
  geographical distribution 5:7
  host range 2:40–41
  mortality rate 5:3–4
  pathogenesis 5:4
    acute phase 5:4
    chronic phase 5:4
    histopathology 5:3f
    target tissues 5:4
    transition phase 5:4
  prevention 5:8
  resistance 5:3
Taura syndrome virus (TSV) 5:1–9
  classification 4:567
  epizootics 5:1
  evolution 5:6
    quasispecies 5:6
  as fungicides 5:1–2
  gene expression 5:5
    encoded proteins 5:5
    IGR region 5:5
    inhibition of apoptosis 5:5
    IRES 5:5
    protease motif 5:5
    transcription 5:5
  genetics 5:5
    Belize isolates 5:6
    diversity 5:7f
    Mexican isolates 5:5–6
    Texas isolates 5:6
  genetic variation 2:43
  genome 5:1
    open reading frames 5:5
  history 5:1
  infection *see* Taura syndrome disease (TSD); *see above*
  transmission 5:6
    cannibalism 5:7
    mechanical 5:7
    seagulls 5:7
  virion structure 2:38
TAV *see* Tomato aspermy virus (TAV)

TaV *see* Thosea asigra virus (TaV)
Taxa 5:19
  definition 1:399
Taxonomy 5:9–24
  definition 5:9
  historical aspects 5:9
    electron microscopy 5:9–10
    pathogenic properties 5:9–10
  nomenclature 5:21
  PASC applications 5:344
  potato virus Y 4:296
  quantitative 5:20
  species classification vs.demarcation 5:404
  taxa descriptions 5:19
  viral species 5:21
  virus databases 5:352
  see also Classification; *specific viruses*
Tax protein 5:198
  adult T-cell leukemia pathogenesis 2:559–560
  HTLV-1 infection *see* Human T-cell leukemia virus 1 (HTLV-1)
  myositis 2:572–573
TBE E, membrane fusion role 3:298f
TBRV *see* Tomato black ring virus (TBRV)
TBSV *see* Tomato bushy stunt tombusvirus (TBSV)
TC-10 GTPase 4:418
TC-83 live attenuated vaccine, Venezuelan equine encephalitic virus 5:106
T cell(s) 3:61–62
  activation 1:122
    HIV infection 2:524
  antibodies *vs.* 3:71
  Borna disease virus infection 1:345
  classes 1:121
  cytotoxic *see* CD8+ T-cells
  definition 3:78
  Dengue fever 2:13
  downregulation 1:126
  exhaustion 3:76
    definition 3:71
  HCV infection 2:373–374
  HIV infection 3:76
  HSV infection 2:393
  HTLV-1 infection *see* Human T-cell leukemia virus 1 (HTLV-1)
  immunopathology role 3:80f
  mammalian reovirus infection 4:389
  measles 3:289
  memory *see* Memory T-cell(s)
  mouse mammary tumor virus infection 3:341
  nervous system viruses 1:470
  persistent infection 3:75f
  proliferation, Tip oncoproteins 4:589–590
  recognition 3:72f
  subpopulations 3:80
  tolerance 1:124
  transformation, HTLV-1-associated malignancies 5:198
  viral load, response 3:76
  Visna-Maedi virus infection 5:427
  see also CD4+ T-cells
T-cell immunoglobulin mucin 1 (TIM1), hepatitis a virus detection 2:346
T-cell leukemia 2:561
  adult *see* Adult T-cell leukemia (ATLL)
  HTLV basic leucine zipper protein 2:563
  Stp oncoproteins 4:589
  tax protein 2:562
T-cell lymphomas
  adult *see* Adult T-cell lymphoma
  EBV-associated 5:195
  mouse mammary tumor virus infection 3:339–340
  reticuloendotheliosis virus infection 4:416
T-cell lymphosarcoma, feline leukemia viruses 2:187–188
T-cell receptor (TCR) 3:71
  definition 3:71
  engagement 3:73f
  MHC-restriction 3:71–72
  mouse mammary tumor virus transmission 3:337
  recombination 3:71–72

TCID$_{50}$, definition 1:37
TCV see Turnip crinkle virus (TCV)
T-DNA see *Agrobacterium tumefaciens*-based expression system (T-DNA)
Tear(s), ocular immunology 2:492
Tectiviridae 1:1, 3:1, 5:11t
Tectivirus(es)
   genome delivery 5:368
   genome packaging 2:306
   structure 1:434f
      symmetry disruptions 1:437
   see also Icosahedral dsDNA bacteriophages; *specific viruses*
Tegument
   HCMV structure 2:485–486
   HSV structure 2:397
Telbivudine 1:152
   resistance development 2:359
Telomerase 4:418
Telomeric repeats, mardivirus 2:409
Temperate virus, definition 5:411
Temperature limits, extrinsic incubation period 5:271–272
Templates
   bacteriophage assembly 1:438
   Brome mosaic virus, RNA recombination 1:385
Tench reovirus (TNRV) 1:164t
Tenofovir 2:510
   HAART-associated renal disease 1:56–57
Tenosynovitis, definition 5:83
'Tentative Species Names' 5:21
Tenuivirus 5:24
Tenuivirus(es) **5:24–27**
   antigenic properties 5:26
   biological properties 5:26
   genome 5:25f
      ambisense arrangement 5:25f
      nucleocapsid protein 5:25
      RNPs 5:25
   historical aspects 5:24
   infection
      cytopathology 5:26
      host range 5:26
      inclusions 5:25f
      symptoms 5:26
   phylogenetics 5:26
   relation to other taxa 5:27
   replication 5:25
      cap-snatching 5:25f
      transcription 5:25f
   transmission 5:26
      plant hopper vector 5:26
   type species 5:24
   virion 5:24
      components 5:24
      see also *specific components*
      morphology 5:25f
      physical properties 5:24
   see also *specific viruses*
Teratogens, definition 3:479
Terminal nucleotide sequences, arenavirus genome 3:243
Terminal protein(s) (TP) 4:402
   classification 4:402–403
   TP–dAMP complex 4:403
Terminal redundancy
   cauliflower mosaic virus transcription 1:467–468
   definition 5:411
Terminal-repeat retrotransposons in miniature (TRIMs), LTR-retrotransposons, unclassified 4:434
Terminal repeats, EBV 2:157
Termination, dsDNA bacteriophage genome packaging 2:310
Teschovirus 2:126–127
Tessier, G, Sigma virus discovery 4:576
Tetraviridae 5:11t, 5:27, 5:28t
Tetravirus(es) **5:27–37**
   β-like genome 5:30f
      capsid protein precursor 5:31f
      omegatetraviruses vs. 5:29
      open reading frames 5:29
      RdRp 5:29

biotechnology applications 5:36
   biopesticides 5:36
   chemical platforms 5:36
capsid assembly 5:33
capsid maturation 5:34
   labeling experiments 5:35
   small-angle X-ray scattering 5:34–35
capsid morphology 5:31f
capsid proteins 5:29
   precursors 5:31f
capsid structure 5:34f
   cryo-electron microscopy 5:33
   high-resolution 5:33
   X-ray crystallography 5:33
cleave site mutations 5:35
definition 5:27
genomes 5:29
   delivery 5:368
   two RNA species 5:29
   see also *specific genome types*
historical aspects 5:27–28
infections 5:36
   histopathology 5:36
   host range 5:36
   insects 3:126t
   pathogenicity 3:127
   sites 3:127
   symptoms 3:127
   transmission 5:36
   tropism 5:36
insect pest control 3:127
ω-like genome 5:31
   capsid protein precursors 5:31f
   open-reading frames 5:31
   replicases 5:31
   RNA2 5:31
RdRp 5:29
structure 5:387f
see also *specific viruses*
TEV movement (RTM) proteins 3:355
Tf1 transposon (yeast)
   expression 3:306
   integrase 3:309
      target site specificity 3:309
   reverse transcription priming 3:307f
Tf2 transposon (yeast)
   frameshifts 3:304–306
   induction, stress 3:304
   translation 3:304–306
TFIIIB (transcription factor) 3:309
TFIIIC (transcription factor) 3:309
TGB movement protein(s) 3:352
   allexivirus genome 1:97–98
   beet soil-borne virus 4:283–284
   beet virus Q 4:283–284
   flexiviruses 2:257f
   hordeiviruses 2:462
      TGB1 see TGB1 movement protein
      TGB2 see TGB2 movement protein
      TGB3 see TGB3 movement protein
   pomovirus genome 4:283–284
   potato mop-top virus 4:283–284
   see also *specific proteins*
TGB1 movement protein 2:462
   expression 2:462
   Pepino mosaic virus 4:105
   RNA binding activity 2:462
   RNA helicase activity 2:462
   TGBp1 helicase 3:353
TGB2 movement protein 2:462
   Pepino mosaic virus 4:106
   TGBp2 3:352
TGB3 movement protein 2:463
   domains 2:463
   Pepino mosaic virus 4:106
   translation 2:463
Thailand, HIV infection 1:63–64
Theiler's murine encephalomyelitis virus (TMEV) **5:37–45**
   classification 5:38f
   final gene products 5:39

genome 5:40f
   coding limits 5:39
geographical distribution 5:37
history 1:441
infection *see below*
physical properties 5:38
polyprotein processing 5:40f
   capsid proteins 5:40
   cleavage sites 5:39
   L* protein 5:40
RNA replication 5:40
   internal ribosome entry site 4:134–135
serologic relationships 5:38
taxonomy 1:443f
tissue tropism 5:41
transmission 5:41
virion structure 5:39f
   capsid proteins 5:38–39
   Mengo virus vs. 5:38–39
Theiler's murine encephalomyelitis virus infection
   blood–brain barrier 1:447
   clinical features 1:446
      gait spasticity 5:42
      late disease 5:42
      poliomyelitis 5:42
      spinal leptomeninges, inflammation 5:43f
   demyelination 5:43f
      incidence 5:44
   diagnosis 1:447
   genomic determinants 5:41
   histopathology 1:447
   host range 5:37
   immune response 5:43
      demyelination 5:43f
      IgG response 5:43
      survival-determining factors 3:82
   macrophages
      pathogenesis 5:43f
      persistence 5:42–43
   neurovirulence 1:446–447
      capsid proteins 5:41
      L* protein 5:41
   pathogenesis 1:446
   pathology 1:446
   persistence sites 5:42
Theiler's virus see Theiler's murine encephalomyelitis virus (TMEV)
T helper cell(s)
   cytokine/chemokine production 1:621
   differentiation 1:621
   type 1
      definition 2:51
      immune response functions 3:80
   type 2
      cytokine production 1:621
      definition 2:51
   see also CD4+ T-cells
Therapeutic ratio, antiviral therapy 1:147
Thermococcales-infecting virus (PAV1)
   see Euryarchaea-infecting virus(es)
Thermoproteus tenax spherical virus 1 (TTSV1) 1:590
Thermoproteus tenax virus 1 (TTV1)
   host cell lysis 1:591
   virion core structure 1:590
θ replication mode
   bacteriophage λ 4:400–401
   definition 4:400
Thiafora virus 1:598t
Thiomersal 5:228
Thogoto virus
   characteristics 1:171t
   virion structure 1:172
Thogotovirus 3:96
Thoracic ganglia invasion, Sigma virus 4:578
Thosea asigra virus (TaV) 5:28t
   capsid assembly 5:33
   capsid protein precursor 5:29–31
   capsid protein precursors 5:29–31
   genome 5:29
   RdRp 5:29
   replicase 5:29

Thread-like particles, citrus tristeza virus 1:520
3′-nontranslated region, Pepino mosaic virus 4:106
Three-day sickness *see* Bovine ephemeral fever virus (BEFV)
Three-factor cross method, DNA virus recombination 4:374–375
Three-hybrid assay *see* Icosahedral ssRNA bacteriophages
Thrip vectors
  tobacco streak virus transmission 3:55
    Tomato spotted wilt virus *see* Tomato spotted wilt virus
  tospovius *see* Tospovirus(es)
Thrombocytopenia
  classical swine fever 1:529
  Crimean-Congo hemorrhagic fever 1:601–602
  definition 2:241
  Dengue fever symptoms 2:11–12
Thrombosis, definition 1:328
Thrombotic microangiopathy, HIV infection 1:56
Thymic lymphosarcoma(s), feline leukemia viruses 2:189
Thymidine kinase activation, murine polyomavirus large T 4:276
Thymidine nucleoside analogs 1:53
Thymidylate kinases, virus-encoded proteins in African swine fever virus 1:47
Tibrogargan virus (TIBV) 4:184t
Tick(s) 2:235
  as an African swine fever virus host 1:44f
  coltivirus transmission *see* Coltivirus(es), transmission
  control, Crimean-Congo hemorrhagic fever prevention 1:602–603
  flavivirus transmission 2:239
  life cycle 1:173–174
  vertical transmission 5:269
  West Nile virus transmission 5:445
Tick-borne encephalitis virus (TBEV) 2:240
  bioterrorism evaluation 5:407–408
  European subtype 2:240
  Far Eastern subtype 2:240
  geographical distribution 5:490
  host cell attachment 5:48f
  host cell entry 5:46
    cellular receptors 5:46
    E protein 5:46
  infectivity 5:45
  life cycle 5:46
  nomenclature 5:45
  nonviremic transmission 5:270
  phylogeny 5:46f
  protein processing 5:48
  RNA replication 5:49
    associated proteins 5:49–50
    replication complex formation 5:51f
    replicative intermediates 5:49
  RNA translation 5:50f
  Siberian subtype 2:240
  taxonomy 5:45
  vaccines/vaccination 5:53
  virion assembly 5:48f
  virion maturation 5:50
  virion release 5:50
  virion structure 5:45
    capsid 5:48f
    envelope glycoprotein 5:47f
      conformational changes 5:48f
    M protein 5:46
Tick-borne encephalitis virus infection
  biphasic milk fever 5:53
  clinical manifestations 5:52
  control 5:53
  encephalitis 2:240
    incidence 5:51
  incidence 5:51
  pathogenesis 5:51
  prevention 5:53
  transmission 5:490
    vertical transmission 5:269
  zoonoses 5:490

Tick-borne flavivirus(es) 2:239
  see also specific viruses
TICV *see* Tomato infectious chlorosis virus (TICV)
Tiger frog virus (TFV) 2:233
Tiger prawn reovirus 1:164t
Tiger puffer nervous necrosis virus (TPNNV) 2:229–230
Tiger shrimp reovirus 1:164t
Timbo Group, rhabdoviruses 1:116
Tind, fusellovirus promoter 2:299
Tioman virus 2:95
Tip oncoproteins *see* Herpesvirus Saimiri (HVS)
Tipranavir 2:516
Tipula paludosa (cranefly), larval infection 3:161
TIR-NB-LRR protein, plant natural resistance 4:171
Tissue culture
  neutralization sensitivity 3:418
  Simian retrovirus D 4:629
Tissue factor (TF) expression
  definition 2:199
  Ebola virus infection 2:64
Tissue print immunoblotting, plant virus infection diagnosis 2:25f
Tissue tropism
  definition 4:407
  see also specific viruses
TLC1 element 4:435
TLCNDV *see* Tomato leaf curl New Dehli virus (TLCNDV)
TLCV *see* Tobacco leaf curl virus (TLCV)
TLRs *see* Toll-like receptor(s) (TLRs)
T lymphocytes *see* T cell(s)
Tm-2 genes 4:176
TM Biocontrol-1 3:125
TME *see* Transmissible mink encephalopathy (TME)
TMEV *see* Theiler's murine encephalomyelitis virus (TMEV)
TMGMV *see* Tobacco mild green mosaic virus (TMGMV)
TMMMV *see* Tulip mild mottle mosaic virus (TMMMV)
TMV *see* Tobacco mosaic virus (TMV)
Tn7 transposon, Bac-to-Bac expression vectors 1:241
TNSV *see* Tobacco necrosis satellite virus (TNSV)
Tnt1 elements 4:432
  U3 regions 4:432
T-number, definition 3:30
TNV *see* Tobacco necrosis virus (TNV)
TNV-A *see* Tobacco necrosis virus A (TNV-A)
TNV-D *see* Tobacco necrosis virus D (TNV-D)
TNX-355 2:507t
  HIV entry inhibitors 2:506
Tobacco (*Nicotiana tabacum*) 5:60
  history 5:60
  infections 5:61
    alfalfa mosaic virus 5:63
    artichoke yellow ringspot virus 5:66
    economic importance 5:61
    potato spindle tuber disease 5:341
    see also specific diseases/viruses
  plant virus movement 3:350–352, 3:351f, 3:353–354
  satellite virus-induced silencing system 5:378
  types 5:61
  virus propagation, associated species 5:61
  virus resistance 5:61
    TBRV-acquired immunity 5:325–326
    TRSV-acquired immunity 5:65
Tobacco apical stunt virus (TASV) 5:62t
Tobacco bushy top disease 5:67
Tobacco bushy top virus (TBTV) 5:62t
Tobacco curly shoot virus (TbCSV) 1:320
  characteristics 5:62t
Tobacco etch virus (TEV)
  characteristics 5:62t
  genome 5:64
  geographical distribution 5:64
  infection
    economic importance 5:64
    symptoms 5:64
  NIa protease 4:321
  P1 protein 4:317

P3-6K1 protein 4:317
  transgenic resistance 5:64
  transmission 5:64
Tobacco leaf curl disease 5:65
Tobacco leaf curl-like virus(es) 5:65
  genome 5:65
  geographical distribution 5:65
Tobacco leaf curl virus (TLCV) 5:65
  characteristics 5:62t
  geographical distribution 5:65
Tobacco mild green mosaic virus (TMGMV) 5:61–63
  characteristics 5:62t
Tobacco mosaic satellite virus (TMSV) 4:535
  characteristics 5:63t
Tobacco mosaic virus (TMV) 5:62t
  antigenicity 5:58
  assembly 5:57f
    disk aggregate 5:56
    elongation 5:68–69
    initiation 5:56–57
  biotechnology applications 5:59
    engineered resistance 4:156–157
    experimental vaccines 5:223
    magnifection 4:234
    nanoscale material production 4:236
    promoter utilization 4:230
  capsid protein 4:236
    monoclonal antibody production 4:236–237
    protein fusion 4:236
  cell-cell movement 5:59
  characteristics 5:62t
  chemical properties 5:55
  coat proteins 5:68–69
    aggregation forms 1:201–202
    disks 1:201–202
    epitopes 5:58
    symmetry 1:201–202
  crystallization 3:399
  disassembly 5:57
  gene silencing 5:376
  genome 5:385
  history 5:61
  hypersensitive response 4:174f
  infection
    Impatiens infections 4:215
    nonspecific infections 4:222f
    petunia infection 4:212f
    potato 4:300t
    resistance to 4:173, 5:68
    sweetpotatoes 4:661t
  movement within plants 3:350–352
    movement protein 3:351f
  mutants 5:55–56
  physical properties 5:55
  replication 5:59
    translation 5:59
  satellite viruses 4:527t
  structure 5:68–69
    capsid protein-nucleic acid interactions 5:385
    historical aspects 5:380–381
    origin of assembly 5:69
    X-ray crystallography 5:380–381
  transient vector construction 5:61
  transmission 5:285
    capsid strategy 5:280f
Tobacco mottle virus (TMoV) 5:66–67
  characteristics 5:62t
Tobacco necrosis satellite virus (TNSV) 4:527
  characteristics 5:63t
  noncoding regions 4:530
  properties 4:530
  virion structure 4:530
Tobacco necrosis virus (TNV) 5:66
  characteristics 5:62t
  genome 5:147f
  legume infection 3:214f
  petunia infection 4:213
  satellite viruses 4:527
  transmission via chytrid fungi 3:214f

Subject Index 611

Tobacco necrosis virus A (TNV-A) 5:145–146
   geographical distribution 3:404
   satellite virus 3:404
   virus–host relationship 3:404–405
Tobacco necrosis virus D (TNV-D) 3:403
   geographical distribution 3:404
   satellite virus 3:404
   tulip infections 4:224
   virus–host relationship 3:404–405
Tobacco necrotic dwarf virus (TNDV) 5:66
   characteristics 5:62t
Tobacco rattle virus (TRV) 5:72
   associated vectors 5:66
   characteristics 5:62t
   genome 5:376–377
   geographical distribution 5:76
   host range 5:76
   infection
     detection 4:308
     petunia infection 4:213
     symptoms 5:76
     tulips 4:224
   morphology 4:307–308
   movement studies 5:73
     RNA1 5:73
     RNA2 5:73
   recombination 5:74
   transmission 4:307–308
     2b protein, role 5:75
   virus-induced gene silencing role 5:376
Tobacco ringspot virus (TRSV) 3:406t
   associated disease symptoms 5:65
   cell-to-cell movement 3:410
   characteristics 5:62t
   diseases 3:411
   infections
     pelargonium infections 4:210
     petunia infection 4:213
   infectious DNA clones 3:408–409
   particle structure 3:407
   ribozyme 4:476
     structural dynamics 4:479
   satellite viruses 3:411
     RNA satellites 4:534
   transmission 5:65
     aerial vectors 3:411
     soil nematodes 3:411
Tobacco rosette disease 5:66
Tobacco streak virus
   characteristics 5:62t
   economic importance 5:63
   genome 3:50f
     open reading frames 3:47
   infections
     dahlia 4:223
     host range 3:55
     Impatiens 4:215
     petunia 4:213
     potato 4:300t
     sweetpotato 4:661t
   transmission 3:55
Tobacco stunt virus (TSV) 5:66
   characteristics 5:62t
Tobacco vein clearing virus (TVCV), transmission 1:462
   host genome integration 1:462
Tobacco vein distorting virus (TVDV) 5:66–67
   characteristics 5:62t
Tobacco vein mottling virus (TVMV)
   characteristics 5:62t
   geographical distribution 5:64–65
   transmission 5:64
   virion structure 4:316
Tobacco virus(es) **5:60–68**
   associated satellite viruses 5:63t
   naturally occurring 5:63t
   satellite viruses 5:61
   see also specific viruses
Tobacco wilt virus (TWV) 5:62t
Tobacco yellow dwarf virus (TYDV) 5:65
Tobacco yellow vein assistor virus 5:62t

Tobacco yellow vein virus (TYVV) 5:67
Tobamovirus 5:59, 5:68
Tobamovirus(es) **5:68–72**
   antigenic relationships 5:59
   cis-acting sequences 5:69
   classification 5:68
   composition 5:68
   genome 5:70f
     hibiscus-infecting viruses 5:69
     open reading frames 5:69
   history 5:68
   infection
     Calibrachoa 4:214
     Impatiens 4:214–215
     petunia 4:213
     resistance 4:173
     symptoms 5:288
     tobacco 5:61
     virus-host interactions 5:70
   replication 5:71
     rate 5:71
     translational disassembly 5:71
   satellite TMV 5:71
   structure 5:68
   taxonomy 5:68
   transmission 5:288
   viral proteins 5:70
     coat protein 5:70
     movement protein 5:70
   see also specific viruses
Tobravirus 5:72
Tobravirus(es) **5:72–76**
   characteristics 5:72
   coat protein 5:72–73
     antigenic regions 5:72–73
   furovirus *vs.* 2:295–296
   gene expression 5:73
     RNA1 5:74
     RNA2 5:74
     subgenomic RNA 5:74
   gene silencing 5:75
   genome 5:73
   infection
     diseases associated 5:76
     M-type 5:73
     NM-type 5:73
     recovery 5:76
     resistance 5:76
     tobacco 5:66
   movement studies 5:73
   proteins encoded 5:74
     cell-to-cell movement 5:74–75
     coat proteins 5:75
     replicase 5:74–75
     RNA1 5:73
     RNA2 5:73
     subgenomic RNA 5:74
   RNA sequences 5:73f
     recombination 5:74
     RNA1 5:72–73
     RNA2 5:72–73
     subgenomic RNA 5:72–73
   satellite viruses 4:527t
   taxonomy 5:72
   transmission 5:72
   virion production 5:72
   virion structure 5:72f
   see also specific viruses
ToCGV *see* Tomato leaf curl Gujarat virus (ToCGV)
ToCV *see* Tomato chlorosis virus (ToCV)
Togaviridae 1:174, 1:195, 2:230, 2:255, 4:514, 5:11t, 5:77, 5:83, 5:91–92, 5:96, 5:101, 5:108, 5:117
Togavirus(es) **5:107–116**
   cell viability 5:122
   evolution 5:108
   fish 2:230
   genera 2:230
   genome 1:173
   history 5:107
   infections *see below*
   infectious clone technology 5:121

   pathogens associated 5:113
   see also specific viruses
   phylogenetic relationships 1:174
   RdRp 5:119
   replication 5:120f
     genome packaging 5:121
     host cell entry 5:118
     sites 5:121
     specificity 5:120–121
     structural protein maturation 5:121
     transcription 5:119
     translation 5:119
     uncoating 5:118
     virion egress 5:121
     virus attachment 5:118
   systematics 5:108
   transmission cycles 5:111
   vectors 5:121–122
   virion morphology 2:230
   virion structure 1:173
     E1 envelope glycoprotein 5:108
     envelope spikes 5:108
     glycoproteins 5:117
     nucleocapsid 5:108
     stability 1:173
   see also specific viruses
Togavirus infections 5:121
   2–5 A synthase/RNaseL pathway 5:123–124
   clinical manifestations 5:83
     acute 5:83
     children 5:83
     chronic illness 5:85
     encephalitis *see below*
     joint involvement *see below*
     pregnancy 5:85
     rashes 5:84
   cytopathic effects 5:122
   diagnosis 5:87
     antibodies 5:87–88
     hemagglutination inhibition tests 5:107–108
   diseases associated 5:111
     New World viruses 5:111
     Old World viruses 5:111
   double-subgenomic promoter viruses 5:121–122
   encephalitis **5:76–83**
     diagnosis 5:82
     epidemiology 5:80f
     geographical distribution 5:79f
     treatment 5:82
   host cell interactions 5:122
     protein shut-off 5:122
   hosts 5:111
   interferon-mediated antiviral response 5:123
   joints 5:83–84
     pattern of involvement 5:84
     swelling 5:84
   replicon systems 5:121–122
   treatment 5:88
ToLCD *see* Tomato leaf curl disease (ToLCD)
Toll-like receptor(s) (TLRs)
   apoptosis inhibition 1:161
   cellular distribution 3:106
   definition 3:104
   function 1:123
   innate immune response 3:149
     expression 3:58–59
     nucleic acid recognition 3:58–59
     signal cascades 3:58–59
     viral protein recognition 3:106
   research history 3:58–59
   signaling
     parasitoid attack 4:252–253
     persistent infection 4:113
Toll-like receptor 4 (TLR4)
   adapter proteins 3:107
   mouse mammary tumor virus infection 3:339
Tol QRA proteins, bacteriophages 3:122
Tomato (*Solanum esculentum*)
   Sw-5 gene 4:174
   see also specific viruses
Tomato apical stunt viroid (TASVd) 5:339–341

Subject Index

Tomato aspermy virus (TAV)
  chrysanthemum infections 4:224
  petunia infection 4:213
  small, linear single stranded RNA satellites 4:531–532
Tomato black ring virus (TBRV) 3:406t
  infections 3:411
    peach tree 4:198
    pelargonium 4:211f
    petunia 4:213
    potato 4:300t
  large single-stranded RNA satellites 4:531
  transgenic resistance 5:325–326
Tomato bushy stunt tombusvirus (TBSV) 3:216
  genome 5:147f
  RNA recombination 4:379
  structure 1:454–455
Tomato chlorosis virus (ToCV) 2:90
  epidemiology 2:90
  genome 2:90
  geographical distribution 2:90
  infection
    classification 2:90
    control 2:90
    disease symptoms 2:90
    yield losses 2:90
  virion morphology 2:90
Tomato infectious chlorosis virus (TICV) 2:89
  classification 2:89
  epidemiology 2:89
  genome 2:89
  geographical distribution 2:89
  host range 2:89
  infection
    control 2:90
    disease symptoms 2:89
    yield losses 2:89
  transmission 2:89–90
  virion morphology 2:89
Tomato leaf curl Bangalore virus (ToLCBV) 1:567f
Tomato leaf curl disease (ToLCD)
  epidemiology 5:125
    relative humidity 5:126
    vector movement 5:125–126
    virus persistence 5:126
    weather, effects 5:125–126
  history 5:125
  yield losses 5:125
Tomato leaf curl Gujarat virus (ToCGV) 5:127
  DNA satellites 5:127
Tomato leaf curl New Delhi virus (TLCNDV) 5:127
  DNA β component association 1:320
  DNA satellites 4:530–531
  infection 5:132
    movement protein 5:132
    nuclear shuttle protein 5:132
  replication 1:566
Tomato leaf curl virus (TLCV)
  DNA satellites 4:530
  satellite viruses 4:527
Tomato leaf curl viruses (Indian) 5:124–133
  distribution 5:128f
  diversity 5:126
  DNA-A satellites
    comparisons 5:129
    pseudo-recombination 5:131–132
    recombination 5:131
  epidemiology 5:125
  genome 5:126
  infections
    host range 5:126
    management 5:133
    pathogenesis 5:132
    resistance 5:133
    symptoms 5:125f
  molecular relationships 5:129f
    common region sequence comparisons 5:130
    DNA-A sequence comparisons 5:129
    DNA-B sequence comparisons 5:130
    DNA-β sequence comparisons 5:130
  phylogeny 5:129f
  pseudo-recombination 5:128–129
    DNA-A 5:131–132
  recombination 5:131f
    DNA-A 5:131
  replication 5:132
    Rep protein 5:132
  synergism 5:128–129
  tomato leaf curl Karnataka virus 5:128
  tomato leaf curl Pune virus 5:128
  transmission 5:125
  see also specific viruses
Tomato leaf curl virus-satellite (ToLCV-sat) 1:315
Tomato mild green mosaic virus (TMGMV)
  Impatiens infections 4:217f
  petunia infection 4:213f
Tomato mosaic virus (ToMV) 5:61–63
  bacopa virus infections 4:215
  Calibrachoa infection 4:214f
  characteristics 5:63t
  petunia infection 4:211–212
  phloem-dependent long-distance movement 5:212
  potato infection 4:300t
Tomato pseudo-curly top virus (TPCTV) 4:165
  infections 4:166
Tomato ringspot virus (ToRSV) 3:406t
  diseases 3:411
  genome 3:407–408
    sequence identity 3:408
  infectious DNA clones 3:408–409
  pelargonium infections 4:210
  petunia infection 4:213
  regulated polyprotein processing 3:409
  replication 3:409
    NTB protein 3:409–410
  transmission, soil nematodes 3:411
Tomato spotted wilt virus (TSWV) 5:133–138
  biology 5:133
  epidemiology 5:136
  genome 5:134f
    L RNA 5:134
    M RNA 5:134
    N gene variability 5:133–134
    RdRp 5:134
    ribonucleoproteins 5:134
    S RNA 5:134
  history 5:157
  infection 5:133–134
    acibenzolar-S-methyl 5:137f
    bacopa virus infections 4:215
    chrysanthemum infections 4:224f
    dahlia infections 4:223f
    diagnosis 5:136
    economic impact 5:136
    host ranges 5:133–134
    Impatiens infections 4:214
    management 5:137
    nemesia infections 4:220f
    NSm protein role 5:134–135
    Osteospermum infection 4:222f
    pelargonium infections 4:211f
    potato 4:308–309
    symptoms 5:134f
    tobacco 5:65
  molecular biology 5:134
  morphology 5:134
  RdRp 5:134
  resistance
    Tm-2 genes 4:176
    transgenic 5:325–326
  structure 5:288–289
  taxonomy 5:134f
    classification descriptors 5:133
  transmission see below
  virus structure 5:65
Tomato spotted wilt virus transmission 5:133–134
  thrips 5:157
    acquisition 5:159–160
    control 5:162–163
    infection effects 5:161
    management 5:137
Tomato yellow leaf curl China virus (TYLCCNV)
  βC1 role 1:318
  DNA satellites 4:535
Tomato yellow leaf curl disease 2:98
  crop yield losses 2:98
  geographical distribution 2:99f
  symptoms 2:98
Tomato yellow leaf curl Sardinia virus (TYLCSV)
  begomovirus satellite identification 1:315
  geographical distribution 5:140–141
Tomato yellow leaf curl virus (TYLCV) 5:289
  begomovirus satellite identification 1:315
  DNA β component associated 1:320
  classification 2:98
  epidemiology 2:98
  gene expression 2:100
  genome 5:289
    open reading frames 2:101
  geographical distribution 5:289
  history 5:138
  host range 5:139
  infection see Tomato yellow leaf curl disease; see below
  structure 5:289
  transmission 5:289
    disease incidence 5:139
    sexual 5:143
    vector relationships 5:141
    host fertility 5:143
    host life expectancy 5:143
    host tissues associated 5:141
    virus retention 5:141
    whitefly see below
  virus reservoir 2:98
  whitefly 5:139
    biotypes 5:139
    disease incidence 5:139
    fertility, host effects 5:143
    life expectancy, host effects 5:143
    sexual transmission 5:143
    tissues associated 5:141
    virus retention 5:141
Tomato yellow leaf curl virus infection 4:167f
  control 2:100
    breeding programs 4:200–201
    chemical 4:200–201
    genetic engineering 5:144
    physical barriers 4:200–201
  diagnosis 5:140–141
  genetically-modified resistant plants 5:144
    hybrid plants 5:143
    marker-assisted breeding 5:144
    Rep gene 5:144
  symptoms 5:290f
  tomato 5:139
    breeding programs, viral resistance 5:144f
    yield losses 4:200
  see also Tomato yellow leaf curl disease
Tombusviridae 1:453, 3:403, 5:145
Tombusvirus 5:145–146
Tombusvirus(es) 1:453
  defective interfering RNAs
    pathogenesis 5:150
    recombination 4:380
    replication 5:149
  epidemiology 5:149
  evolution 5:145
  genome 5:146
  geographical distribution 5:150
  host range 5:149
  infection see below
  phylogeny 5:145
  RdRp 1:453, 5:146, 5:148–149
  replication 5:148
    defective-interfering RNA 5:149
    movement protein expression 5:148–149
    post-transcriptional gene silencing 5:148–149
    termination 5:149
    translation 5:147f

RNA satellites
  genome organization 5:148
  pathogenesis 5:150
satellite viruses 5:148
stability 5:148
transmission 5:145
virion structure 5:148f
  CP subunits 5:148f
  P domain 5:146
  R domain 5:146
  S domains 5:146
see also specific viruses
Tombusvirus 19 5:328
Tombusvirus infections
  cytopathology 5:150
  pathogenesis 5:150
    defective interfering RNAs 5:150
    recombination 5:150
    RNA satellites 5:150
  resistance genes 4:179t
ToMV see Tomato mosaic virus (ToMV)
Tonate virus (TONV) 5:109t
Tonoplast, definition 1:615
Tonsil(s)
  EBV replication 2:441
  vCJD diagnosis 4:334
Top component *a* (Ta) 1:81–82
Top component *b* (Tb) 1:81–82
'Top-half' domain, narnavirus host persistence 3:395f
Topocuvirus 1:296, 2:98, 4:165
Topoisomerases, mimivirus gene content 3:316
Topology, phylogenetic trees 4:126
Torovirus 1:549, 2:230, 5:151
Torovirus(es) 2:122
  antigenic properties 5:155
  classification 5:151
  envelope protein
    M protein 3:423
    S protein 3:423
  epidemiology 5:155
  gene expression, transcription 3:428
  genetic relationships 5:155
  genome 5:152f
    nucleotide cyclic phosphodiesterase 3:425f
    open reading frames 5:152–153
    proteolytic processing 3:424
    RdRp 5:152–153
    replicase genes 3:425f
  geographical distribution 5:154
  history 5:151
  infection
    clinical features 5:155
    gastroenteritis 3:421
    histopathology 5:156
    host range 5:154
    immune response 5:156
    pathogenesis 5:155
    pathology 5:156
    prevention/control 5:156
    tissue tropism 5:155
    transmission 2:122
  physical properties 5:153
    buoyant densities 5:153
  proteins 5:153
    envelope associations 5:153
    M glycoprotein 5:153
    nucleocapsid protein 5:153
    spike protein 5:153
  RdRp 5:152–153
  replication 5:153
    assembly 5:154
    attachment 5:153
    budding 5:154
    egress 5:154
    entry 5:153
    maturation 5:154
    post-translational processing 5:154
    transcription 5:153
    translation 5:154
    uncoating 5:153

  taxonomy 5:151
  transmission 5:155
    oral/nasal route 5:155
  virions 5:152f
  see also specific viruses
Torque teno mini virus (TTMV) 1:105, 1:108f
  discovery 1:105
  genome 1:105
    open reading frame 2 1:106
  infection
    diseases associated 1:110
    mixed 1:109
    viral load *vs.* host immune status 1:110
  morphology 1:105
Torque teno virus (TTV) 1:105, 1:108f
  epidemiology 1:107
    global distribution 1:107
    prevalence rates 1:107
    risk exposure 1:107
  genome 1:106f
  history 1:105
  infection
    hepatitis 1:110
    mixed 1:109
    viral load *vs.* host immune status 1:110
  morphology 1:105
  replication 1:105–106
    open reading frame 1 1:106
    untranslated region 1:105–106
  sequence homology 1:109f
ToRSV see Tomato ringspot virus (ToRSV)
Torticollis, definition 1:76
Toscana virus (TOSV) 4:493
  epidemiology 4:495
  infection
    clinical features 4:493
    disease duration 4:494
    seroprevalence studies 4:494
    transmission 4:493
  Phlebotomus perniciosus 4:491
Tospovirus 1:391, 4:490–491, 5:133, 5:157–158
Tospovirus(es) **5:157–163**
  classification 5:157–158
  definition 1:399
  epidemiology 5:161
  evolution 5:162
  genetics 5:162
    genetic drift 5:162
    genome reassortment 5:162
    haplotypes 5:162
  genome 5:157–158
    defective interfering RNA 5:159
    IGRs 5:159
    L RNA 5:160f
    M RNA 5:160f
    S RNA 5:160f
    transcription termination 5:159
  geographical distribution 5:158
  history 5:157
  infection
    chrysanthemum infections 4:224
    control 5:162
    cytopathology 5:161
    detection 5:162
    diagnosis 5:162
    economic aspects 5:158
    host range 5:158
    nemesia infections 4:220f
    NSm protein role 1:393–395
    Osteospermum infection 4:221
    pathogenicity 5:162
    petunia infection 4:213
    prevention 5:162
    resistance genes 4:179t
    symptoms 5:162
  propagation 5:158
  protein properties 5:159
    glycoprotein function 5:159–160
    N protein 5:159–160
    NSm protein 5:159–160
    NSs protein 5:159–160

  polyprotein cleavage 5:159–160
  tubule induction 5:159–160
  replication 5:160f
  resistance 5:162
  taxonomy 5:157
  tobacco disease, economic importance 5:65
  transmission 5:158
    mechanical 5:162
    seed/pollen 5:162
    thrips 5:158
    weeds 5:162
  virion properties 5:159f
    nucleocapsids 5:158
    ribonucleoprotein 5:158
  see also specific viruses
TOSV see Toscana virus (TOSV)
Totiviridae 2:286, 2:313, 3:220, 4:567, 5:164, 5:214
Totivirus 2:286, 5:164, 5:165t, 5:214
Totivirus(es) 2:111, 4:567
  biological properties 5:169
  classification 5:164
  double stranded RNA satellites 4:531
  evolution 4:65
  evolutionary relationships 5:172
    capsid sequences 5:173f
    RdRp sequences 5:172f
  gene expression 5:166
  genera 5:214
  genome 5:164
    open reading frames 5:167f
    RdRp expression 5:164
  infection
    fungi 2:287t
    mixed 5:169
  killer phenomenon 5:164
    host genes associated 5:170–171
    preprotoxin processing 5:301
    toxin maturation 5:301
  phylogenetic analysis 2:286
  RdRp expression 2:286
  replication cycle 5:167
    RNA polymerase activity 5:167–168
  taxonomy 5:164
  transmission 5:169
  virion composition 5:165
  virion properties 5:164
  virion structure 5:166f
    capsids 5:165
  virus–host relationships 5:170
  see also specific viruses
Toxic shock syndrome, vaccine contamination 5:229
Toxin(s), bacterial, bacteriophage-mediated cell lysis 3:249–250
Toxoplasmosis, measles *vs.* 3:289–290
Toxoptera citricida (brown citrus aphid), citrus tristeza virus transmission 1:521
T phage(s), morphotypes 2:448f
TΦ promoter, T7 bacteriophage transcription 5:179
T proteins, murine polyomavirus 4:273
TPV see Tanapox virus (TPV)
Tracheal cells, nucleopolyhedrovirus 1:267
TRAF signaling, latent membrane protein 1 2:163
Trager duck spleen necrosis virus (TDSNV) 4:413–414
Tranosema rostrale ichnovirus (TrIV) 4:255
Trans-acting siRNA pathway, plant gene silencing 4:142
  TAS genes 4:142
Transactivation domains
  EBV-encoded nuclear antigen 3 family 2:162
  maize streak virus 3:268
Transactivation response (TAR) element
  JCV reactivation in HIV infection 4:266
  RNA interference in HIV infection 1:161
Transactivator of translation/viroplasmin (TAV), cauliflower mosaic virus 1:467
Transcriptase, bunyavirus replication 1:396f
Transcription 4:410
  discontinuous 1:554
  double-stranded RNA viruses 4:410
  immediate early 2:437

Transcription (continued)
  single-stranded DNA viruses 4:410
  single-stranded negative sense RNA 4:410
  single-stranded positive sense RNA 4:410
  single-stranded RNA viruses, DNA intermediate 4:410
  Stp oncoproteins 4:589
  see also specific viruses
Transcriptional gene silencing (TGS) 5:326
  beet curly top virus pathogenesis 1:306
  cotton leaf curl disease 1:568
  definition 5:325
Transcriptional regulatory sequence(s) (TRSs) 1:554
Transcription factor IID (TFIID) 3:340
Transcription factors, plant resistance to viruses 4:176–177
Transcription-regulating sequences (TRSs)
  arterivirus genome 1:180f
  coronaviruses 1:551
  crenarchaeal viruses 1:594
  gill-associated virus 5:478–479
  nidoviruses 3:428
Transcription transactivator, EBV-encoded nuclear antigen 2 2:161
Transcytosis, definition 4:507
Transduction
  definition 5:411
  feline leukemia viruses 2:188
Trans-envelope network, definition 2:191
Transesterification, definition 4:475
Transfected T-cell lines, Herpesvirus Ateles 4:587–588
Transfection, definition 5:411
Transfection studies, adenovirus malignant transformation 1:9–10
Transferrin receptor 1 (Tfr1), mouse mammary tumor virus cell entry 3:336
Transfer RNA (tRNA)
  nonsense suppressor 4:337
  pseudovirus taxonomy 4:352
  reverse transcription priming 3:306
Transfer RNA-like structure
  bromovirus 1:81
  cucumovirus 1:81
  definition 4:97
  pecluvirus genome 4:100
Transfer RNA (Uracil-5)-methyltransferase 3:316
Transformation, malignant see Malignant transformation
Transformation effector sites (TES), latent membrane protein 1 2:162–163
Transgene(s)
  definition 1:457
  gene therapy 2:301
Transgenic mice
  HCV infection 2:370–372
  influenza virus infection 3:101
  JCV oncogenic potential 4:268
  SV40 pathogenicity 4:637–638
Transgenic pathogen-derived resistance, definition 1:520
Trans-Golgi vesicles, Cryphonectria parasitica virus infection 2:579
Translation 4:183f
  maize streak virus 3:267
  simian alphaherpesviruses 4:584
  togavirus infection 5:122
Translational codons, definition 1:296
Translational repression
  definition 3:21
  icosahedral ssRNA bacteriophages see Icosahedral ssRNA bacteriophages
Translation enhancer elements, SINV infection 5:122
Translation initiation factors, mimivirus gene content 3:316
Translocating NTPase, ssDNA bacteriophage genome packaging 2:311–312
Translocation, host cell entry 4:409
Transmembrane glycoproteins, yellow head virus 5:477
Transmembrane (TM) protein, Simian retrovirus D 4:627

Transmissible gastroenteritis virus (TGEV) 1:550t
  M glycoprotein 1:549–550
  porcine 1:553
Transmissible mink encephalopathy (TME)
  occurrence 5:187
  prion strains 4:334–335
Transmissible spongiform encephalopathies (TSEs)
  see Prion diseases
Transmission 5:315
  airborne 2:140
  barriers 4:331
  bottlenecks 4:364
  definition 2:141
  see also specific transmission modes
  epidemiology 2:144
    common patterns 2:146
    modes 2:145
    virus entry 2:144
    virus shedding 2:144
  horizontal see Horizontal transmission
  maternal 1:376
  nonviremic 5:440
  routes 5:315
  transovarial 1:170, 1:533, 5:111
    definition 1:186
  transplacental 1:351
    definition 2:141
  transstadial 1:533
  vertical see Vertical transmission;
  see also specific viruses
Transmission cycle, definition 5:469
Transmission electron microscopy, Providence virus 5:33
Transmission propensity, definition 5:433
Transovarial transmission see Transmission
Transplacental transmission see Transmissible gastroenteritis virus (TGEV)
Transposable elements, retrotransposons
  see Retrotransposons
Transposition, Bac-to-Bac expression vector 1:241
Transposons
  Arabidopsis thaliana 3:309
  definition 4:437
Transpososome, definition 4:400
Transpososomes, bacteriophage Mu 4:404–405
Transtadial infection, definition 1:483
Tree shrew herpesvirus 1:634
T regulatory cells (Tregs), HSV infection 2:393–394
Triangulation number
  definition 1:433
  icosahedral phage assembly 1:436
Triatoma virus (TrV) 2:42
Tribec virus
  geographical distribution 3:465
  infection 3:465
Trichomonas vaginalis virus(es) 5:172f
Trichoplusia ni, insect cell culture 1:238
Trichoplusia ni granulovirus (TnGV) 1:268–269
Trichoplusia ni TED virus (TniTedV) 4:453
  definition 4:451
Trichoplusia ri virus (TrV) 5:28t
  infection 5:36
Trichovirus 1:420t, 1:424, 1:426f
Trichovirus(es) 1:419–427
  associated species 1:424
  biological properties 1:424
  classification 1:420t
  genome 1:424
    open reading frames 1:424
  particle structure 1:421f
  phylogeny 1:426f
  replication 1:424
  serology 1:425
  strains 1:425
  transmission 1:424
  see also specific viruses
Trigeminal ganglia, HSV latency 2:391
TRIM5α, HIV replication, role 2:521
TRIM5 gene/protein 2:469–470
  evolution 2:470
  HIV sensitivity 2:470

  sequences 2:470
  structure 2:470f
    C-terminal domain 2:469–470
TRIM-Cyp, HIV inhibition 2:470
Triple gene block (TGB)
  definition 4:310
  Pepino mosaic virus 4:105
Triradiate lumen, definition 3:405
Triskelions, definition 1:328
Triticum mosaic virus (TriMV) 1:494
Tritimovirus 4:315t
tRNA see Transfer RNA (tRNA)
Trocara virus (TROV) 5:109t
Tropical spastic paraparesis/HTLV-1 associated myelopathy (TSP/HAM) 1:475
  clinical features 2:563
  diagnosis 2:570
  disease characterization 2:571f
  etiology 2:563
  geographical distribution 2:570
  incidence 2:563
  incubation period 2:570
  neurological features 2:570
  pathogenesis 2:571
  prognosis 2:571
  therapy 2:571
TRRAP, adenovirus E1A gene/protein binding 1:11–12
TRSV see Tobacco ringspot virus (TRSV)
Trubanaman virus 1:400
TRV see Tobacco rattle virus (TRV)
Trypsin-like serine protease 1:226
TSD see Taura syndrome disease (TSD)
TSeq_orgname, virus database formats 5:349–351
ts mutants, Sigma virus replication 4:579
Tst-1 protein, JCV 4:264
TSV see Taura syndrome virus (TSV)
TTV see Torque teno virus (TTV)
Tuberculosis (TB), HIV related infection 1:68
Tubules, bluetongue virus replication 1:332
Tulip(s), infection history 2:450
Tulipa virus infections 4:224
  see also specific viruses
Tulip breaking virus (TBV) 4:224f
Tulip mild mottle mosaic virus (TMMMV) 3:447
  coat protein 3:450
  disease-resistant cultivars 3:453
  genome 3:449
  geographical distribution 3:452f
  Olpidium brassicae transmission 3:453
'Tulipomania' 2:450
Tulip virus X (TVX) 4:225
Tumor(s) 2:457
  avipoxviruses infection 2:283
  brain
    BKV associated 4:270
    JCV associated 4:268
  EBV-encoded small RNA 2:164
  FeLV associated 2:187–188
  JCV infection 4:267
  MHV-68 associated 3:377
  murine polyomavirus 4:271
  see also Cancer; Oncogene(s)
Tumor necrosis factor-α (TNF-α) 1:621
  apoptotic role 1:157
  function 1:621
  Marburg virus infection 3:278
  receptor binding 1:621
  synthesis 1:621
Tumor necrosis factor binding protein (vTNF-BP) 1:157
Tumor necrosis factor receptor(s) (TNFR) 1:621
  apoptotic role 1:157
Tumor suppressor(s)
  adenoviral targets 1:9–10
  E1A/pRB interactions 1:12
  E7/pRB interactions 4:25
Tumor virus(es), human 5:193–199
  oncogenic virus features 5:193
  see also specific viruses
Tungrovirus
  open reading frames 1:466f
  resistance genes 4:179t

virion structure 1:458–460
see also specific viruses
Tungrovirus 1:459t, 4:482
Tupaia, definition 1:104
Turbot aquareovirus (TRV) 1:164t
Turkey adenovirus 3 (TAdV-3)
　classification 1:27
　genome 1:4f
　pathogenicity 1:8
Turkeypox virus 2:277
Turkeys, astrovirus infection 1:206
Turkish tobacco 5:61
Turnip crinkle virus (TCV) 1:454f
　assembly 1:455
　　origin of 1:455
　CP, role 1:456
　defective-interfering RNAs 1:456
　gene expression 1:455
　gene silencing suppression 5:330
　genome, intergenic regions 3:244
　host range 1:454
　resistance to 1:456
　　Arabidopsis thaliana 1:456
　RNA recombination 4:379
　RNA satellites 1:456
　transmission 1:454
　virion structure 1:454–455
　　CP subunit 1:454–455
Turnip mosaic virus (TuMV) 5:289
　host range 5:289
　Impatiens infections 4:218f
　infection symptoms 5:289
　petunia infection 4:213
　structure 5:289
Turnip rosette virus (TRoV) 4:646t
Turnip vein clearing virus (TVCV)
　Impatiens infections 4:214–215
　petunia infection 4:211–212
Turnip yellow mosaic virus (TYMV) 5:200
TVCV *see* Tobacco vein clearing virus (TVCV)
TVL, host range 5:154
TVMV *see* Tobacco vein mottling virus (TVMV)
20S RNA 3:392
22S RNA 3:392
2A protein, enteroviruses 2:126
2B protein, picornavirus 4:136
2–5 A synthase/RNaseL pathway, togavirus infection 5:123–124
Twort, Frederick William, bacteriophage research 2:443
Ty1-Copia retrotransposon(s) 4:433
　history 4:433
Ty1 gene/protein
　replication cycle 4:355
　virion structure 4:354f
Ty3/gypsy retrotransposon(s) 4:433
　distribution 4:443
　expression regulation 3:304
　　frameshifts 3:304–306
　host genome integration
　　target site specificity 3:309
　　TFIIIB, transcription factor 3:309
　　TFIIIC, transcription factor 3:309
　induction 3:304
　reverse transcription priming 3:306
Ty elements, *Saccharomyces cerevisiae* 4:420
TYLCCNV *see* Tomato yellow leaf curl China virus (TYLCCNV)
Tymoviridae 2:255, 5:11t, 5:200
Tymovirus 5:200, 5:200t
Tymovirus(es) **5:199–207**
　capsid structure 5:201f
　　empty capsids 5:201f
　characteristics 5:201
　classification 5:200t
　fleck disease complex 4:203
　genome 5:202f
　　coat protein 5:203
　　open reading frames 5:201
　　overlapping protein 5:203
　　replication polyprotein 5:203
　　tRNA-like structure 5:202–203
　　tymobox 5:203
　infection 5:203
　phylogenetic relationships 5:204f
　properties 5:201
　　stability 5:201–202
　replication 5:203
　　encapsidation 5:202
　　RNA replication 5:203
　　sites 5:201
　　vesicle formation 5:203
　species demarcation 5:204
　tobacco diseases 5:67
　transmission 5:203
Typhus, measles *vs.* 3:289–290
Typography, species nomenclature 5:405
Tyrosine recombinase (YR) retrotransposons 4:421f
　distribution 4:423
　replication 4:426

U

U3 regions, Tnt1 elements 4:432
Ubiquitination, killer toxin immunity 5:307
Uganda, HIV infection 1:63
UHBV *see* Urochloa hoja blanca virus (UHBV)
Ukraine, HIV infection 1:64
UL5 protein 2:415
UL112 phosphoproteins 2:486–487
Ulcerative disease rhabdovirus (UDRV) 2:228
Ultraviolet absorbance spectra, potexvirus 4:310–312
Ultraviolet resonance Raman spectroscopy, definition 2:191
Umbravirus 5:209
Umbravirus(es) **5:209–214**
　assistor virus interactions 5:212
　　groundnut rosette assistor virus 5:211
　classification 5:209
　gene expression 5:210
　genome 5:211f
　　frameshifting 5:210–211
　　open reading frames 5:210–211
　host range 5:213
　infection 5:209–210
　　control 5:213
　　disease associations 5:209
　　fibrillarin 5:212
　　nucleolus involvement 5:212
　　prevention 5:213
　　resistance genes 4:179t
　movement
　　cell-to-cell 5:211
　　phloem-dependent long-distance 5:212
　nomenclature 5:209
　replication 5:211
　RNA satellites 5:211
　tobacco disease, economic importance 5:66
　transmission 5:209
　　aphids 5:213
　virus properties 5:210f
Unassigned virus(es) 5:21–22
Una virus (UNAV) 5:489
Uncoating 4:409
　baculoviruses 1:250
　definition 4:407
　endocytic pathway 4:409
　products 4:409
　replication 4:409
UniProt 5:363
United States of America (USA)
　bioweapons 5:408
　monkeypox virus 4:643
Universal tree of life 3:473
Untranslated regions (UTRs)
　broad bean necrosis virus genome 4:283
　giardiavirus genomes 2:314
　LINE-like retrotransposons 4:439
　nepovirus 3:408
　pomovirus genome 4:283
　potexvirus 4:312f
　yellow head virus 5:479f
Unweighted pair group method with arithmetic means (UPGMA) 4:126–127

Upper respiratory tract, structure 2:551
Uranotaenia sapphirina nucleopolyhedrovirus (UrsaNPV) 1:220
　occlusion bodies 1:220
　transmission 1:221–222
Uranyl acetate stains 2:80f
　limitations 2:79
Urbani, Carlo 4:552
Urbanization, Dengue virus 2:8f
'Urban type' yellow fever transmission 5:473–474
Ure2p prions
　domain shuffling 4:339
　essential sequences 4:339
　formation 4:337
URE3 prion 4:337
Urethritis, HSV-2 infection 2:390–391
Uridine diphosphate glycosyltransferase (UGT) 2:112–113
Uridylate-specific endoribonuclease (NendoU) 5:477–478
Uridylylation
　definition 4:129
　picornaviruses 4:139–140
Urochloa hoja blanca virus (UHBV) 5:24
　phylogenetics 5:26f
　species discrimination 5:26
Urodeles, definition 3:155
UrsaNPV *see* Uranotaenia sapphirina nucleopolyhedrovirus (UrsaNPV)
Urticarial 5:83
Ustilago maydis virus(es) **5:214–219**
　definition 5:214
　genetic engineering 5:218
　genome 5:214
　　H (heavy) dsRNAs 5:215
　　isolates 5:215
　　L (light) dsRNAs 5:215
　　M (medium) dsRNAs 5:215
　　ribosomal frameshifting 5:214–215
　RdRp 5:214, 5:218
　replication 5:217
　　capsid-associated transcriptase 5:217
　　L (light) dsRNAs 5:217
　　M (middle) dsRNAs 5:217–218
　structure 5:218
　taxonomy 5:214
　toxins 5:215
　　KP1 5:215–216
　　KP4 5:216f
　　KP6 5:217f
　　SMK 5:216f
　　structure 5:216f
　transcription 5:217
Ustilago maydis virus H1 (UmV-H1), genome 5:166
Uukuniemi virus, structure 4:490–491
Uveitis 2:573
　HTLV-1 associated 2:573
　　prognosis 2:573
　　symptoms 2:573
　　therapy 2:573

V

V1 protein, beet curly top virus 1:303
V2 protein, beet curly top virus 1:303
V3 protein, beet curly top virus 1:303
Vaccine-associated paralytic poliomyelitis (VAPP) 4:246
Vaccines/vaccination 2:458
　administration methods 3:68
　　aerosol 5:229
　　injection *see below*
　　intradermal *vs.* intramuscular 2:54
　animal viruses 5:225
　　development strategies 5:235
　antigenic variation 1:127–128
　autogenous 1:440
　components, reactions to 5:230
　development **5:235–243**
　　animal models 5:236
　　antigenic variation 5:240
　　candidates 5:235–236

Vaccines/vaccination (continued)
　　DNA 5:239
　　　multiple serotypes 5:240
　　　principles 5:236
　　　recommended cell cultures 5:237
　　　technological advances 5:235–236
　　　vectors 5:238
　　　see also Vector(s)
　　　viral epitopes 1:141
　　　virus properties 5:235–236
　　diluent 5:226
　　history 2:51
　　human viral diseases 3:68
　　　categories 3:68
　　　effectiveness 3:68
　　injection 5:229
　　　auto-disable syringe 5:229–230
　　　unsafe practices 5:229
　　niche targets 1:153–154
　　organ transplantation, infections 3:471
　　persistent viral infections 4:108–109
　　plasma-derived 2:351
　　production in plants see Genetically-modified
　　　plants
　　safety see below
　　T-cell responses 3:76
　　technological advances 2:51–52
　　trials, epidemiology 2:144
　　types 5:236
　　see also specific types
　　vectors 4:44–45
　　veterinary see Veterinary vaccine(s)
　　Zaire Ebola virus infection 2:65
　　see also Inoculation
Vaccines/vaccination, safety 5:226–235
　　administration 5:229
　　　aerosol 5:229
　　　auto-disable syringe 5:227
　　　injectable 5:229
　　　multidose vials 5:229
　　　oral 5:229
　　　single-dose vials 5:229
　　adventitious agent 5:226
　　adverse event following immunization see Adverse
　　　event following immunization (AEFI)
　　AEFI surveillance 5:232
　　aerosol 5:229
　　causality assessment 5:227–228
　　classical events 5:232
　　　SV40 5:232
　　　swine flu 5:233
　　contamination 5:229
　　　plant produced vaccines 5:221
　　　public perceptions 5:231
　　　vial type 5:229
　　design 5:228
　　　considerations 5:227f
　　functions 5:227
　　manufacture 5:228
　　　killed 5:228
　　　live 5:228
　　　potency 5:228
　　monitoring and assessment 5:231
　　presentation 5:229
　　programmatic factors 5:227
　　　freeze-dried 5:229
　　public perceptions 5:230
　　recipient reactions 5:230
　　　common/mild 5:230
　　　rare/severe 5:230
　　SV40 contamination 4:630, 4:637, 5:228
　　thermostability 5:230
　　thiomersal 5:233
　　vaccine-associated paralytic poliomyelitis 4:246
Vaccinia-like virus(es) 5:244
Vaccinia virus (VV) 4:325, 5:243–250
　　antiviral therapy 5:249
　　apoptosis inhibition 1:161
　　attenuated strains 5:244
　　　extracellular enveloped virus 5:244
　　　intracellular mature virus 5:244

　　cell culture 5:248
　　DNA recombination mechanisms 4:375
　　expression vectors 5:249
　　genome 2:275, 5:245
　　host cell entry 5:245, 5:247f
　　host range 5:249
　　immune evasion 1:125
　　intracellular mature virus 4:325, 5:245
　　　assembly 4:327, 5:248
　　　attenuation 5:244
　　　export pathways 4:328
　　　proteinaceous bodies 4:327
　　　structure 4:325
　　morphogenesis 4:328
　　origin 5:243
　　rabbits see Rabbitpox
　　replication 1:576, 5:245
　　　assembly 4:327, 5:248, 5:248f
　　　associated proteins 5:246–247
　　　DNA replication 5:246–247
　　　export pathways 4:328
　　　host cell entry 4:327, 5:245
　　　morphogenesis 4:327, 4:328f
　　　transcription 5:245
　　taxonomy 5:244
　　　strains 5:244–245
　　vectors 5:249
　　virion structure 4:325, 5:245
　　　core 4:325, 5:245
　　　extracellular enveloped virus 4:325
　　　intracellular mature virus 4:325
　　　morphogenesis 5:248f
　　virulence 5:249
　　　animal models 5:249
　　　associated proteins 5:249
　　　attenuation 5:249
　　　immunomodulators 5:249
　　　intradermal infection 5:249
　　　intranasal infection 5:249
Vacuolar protease B (Prb1p), prion function 4:340
Vacuoles, sobemovirus localization 4:649–650
Valaciclovir 1:143t
　　antiviral prodrugs 1:150
　　HSV infection 2:394–395
Valcyte® 1:143t
Valency, definition 3:413
Valganciclovir (VGCV) 1:143t, 1:146
Valtrex® 1:143t
Vankyrin gene family
　　Campoletis sonorensis ichnovirus 4:258–259
　　Microplitis demolitor bracovirus 4:259–260
　　phylogeny 4:261
Variable number tandem repeat (VNTR), SVA
　　retrotransposons 4:440
Variant Creutzfeldt–Jakob disease (vCJD) 1:371,
　　4:331–332, 4:333f, 4:335f, 5:191
　　associated cases 1:372, 5:191
　　classical Creutzfeldt–Jakob disease vs. 5:191t
　　clinical and subclinical infection 1:371
　　　characteristics 5:191t
　　　diagnosis 5:191–192
　　　mortality rate 1:372
　　　prevalence 1:371
　　epidemiology 1:372
　　　associated deaths 1:372, 1:372f
　　　valine homozygotes 1:372
　　geographical distribution 1:372–373, 1:373t
　　history 1:368
　　infectivity 1:371
　　mortality rate 1:368
　　origin 1:371
　　pathogenesis 1:371
　　prevention and control 1:372
　　public health implications 1:373
　　　blood transfusions 1:373
　　sporadic vs. 1:371
　　transfusion associated 4:334, 5:192
Varicella-zoster virus (VZV) 5:250–256, 5:254f
　　assembly 5:261
　　　electron microscopy 5:261
　　　Golgi apparatus 5:261

　　bacterial artificial chromosome 5:256
　　classification 5:251
　　cosmid transfection system 5:256
　　cytopathology 5:261
　　epidemiology 5:252
　　　geographical distribution 5:251
　　　seasonal distribution 5:251
　　evolution 5:252
　　　clades 5:252
　　future work 5:255
　　　wide-spread vaccination 5:255
　　gene expression 5:259
　　　host cell transcription apparatus 5:259
　　　see also specific genes
　　genetics 5:251
　　　strains 5:251–252
　　　see also specific genes
　　genome 5:252, 5:256
　　　HSV vs. 5:252
　　　open reading frames 5:256
　　genome replication 5:258
　　　bidirectional 5:258
　　　rolling-circle mechanism 5:258
　　historical aspects 5:250–251
　　　first cultivation 5:251
　　host range 5:251
　　immediate early (IE) gene expression 5:259
　　　host cell factor-1 (HCF-1) 5:259
　　　IE62 5:259
　　　Oct-1 recognition sequence 5:259
　　infection see Chickenpox; Zoster (shingles);
　　see below
　　molecular biology 5:256–263
　　nucleocapsid 5:256
　　post-translational processing 5:261
　　　glycosylation 5:261
　　　phosphorylation 5:261
　　productive infections 5:256
　　　cell attachment 5:256–258
　　promoters 5:259
　　　classification 5:259
　　　TATA-binding protein 5:259
　　propagation 5:251
　　　cell culture 5:251
　　protein translation 5:260
　　release 5:261
　　　tissue culture studies 5:261
　　reservoir 5:251
　　serology 5:252
　　　enzyme-linked immunosorbent assay 5:252
　　　fluorescence antibody to membrane antigen 5:252
　　　immune adherence hemagglutination 5:252
　　taxonomy 5:251
　　transmission 5:253
　　vaccines/vaccination 4:108–109, 5:255
　　　attenuation mechanisms 5:262
　　　effects of 5:255
　　　Oka vaccine strain 5:253, 5:255
　　variability 5:252
　　　restriction enzyme digests 5:252
　　virion structure 5:256
　　　morphology 5:256
Varicella-zoster virus (VZV), infection
　　animal models 5:251
　　eye 2:496
　　histology 5:253
　　　cutaneous lesions 5:254
　　　eosinophilic inclusions 5:254, 5:254f
　　immune response 5:254
　　　cellular immune responses 5:254–255
　　　cytotoxic T-cells 5:254–255
　　　humoral immune responses 5:254
　　　life-long immunity 5:254–255
　　latency 2:438, 5:253
　　　gene expression 5:261
　　　gene regulation 5:262
　　　host cell factor-1 5:262
　　　IE63 5:261
　　　ORF29 5:262
　　management 5:255
　　　immune globulin 5:254

organ transplantation 2:496
pathogenicity 5:253
pathology 5:253
prevention 5:255
 acyclovir 5:255
 exposure 5:255
 immunoglobulin prophylaxis 5:255
 isolation 5:255
 leukocyte interferon 5:255
primary 5:253
tissue tropism 5:253
see also Chickenpox; Zoster (shingles)
Varicellovirus 5:251, 5:347
Varicosavirus 5:263, 5:264
Varicosavirus(es) **5:263–268**
 evolution 5:264
 genetics 5:264, 5:264f
 geographical distribution 5:267
 historical aspects 5:263
 host range 5:267
 tobacco disease, economic importance 5:66
 transcription 5:267
 transmission 5:268
 vectors 5:268
 see also Lettuce big-vein virus (LBVaV); *specific viruses*
Variolae vaccinae, history 1:574
Variola major, bioterrorism 5:407
 biotechnological impacts 5:409–410
 as a bioweapon 5:408
Variola minor 4:639
Variolation *see* Smallpox
Variola virus (VARV) 4:322–323, 4:639
 classification 4:639
 evolution 4:643
 genetics 4:643
 historical aspects 4:639
 infection *see* Smallpox
 molluscum contagiosum virus *vs.* 2:276
 person-to-person transmission 4:643
 replication
  extracellular enveloped virions 4:639
  intracellular mature virions 4:639
Varkud satellite (VS) ribozyme(s) 4:475
 secondary structure 4:480f
VA RNA genes, adenovirus replication 1:3–4
 immune system evasion 1:22–23
Varroa destructor virus 1 (VDV-1) 3:42
 genome 3:45
 tissue tropism 3:45
Varroa mite(s), iflavirus transmission 3:42, 3:45
Vascular choroid 2:491
Vascular dysfunction
 Ebola virus infection 2:62, 2:63, 2:203
 HCMV infection 2:474
 Marburg virus infection 2:203, 3:278
Vascular endothelial cells, neural cell infection 1:473
Vascular endothelial growth factor (vEGF) 4:61–62
Vascular lesions, rubella virus infection 4:519
Vascular permeability, Dengue fever 2:12
VaxGen, HIV-1 vaccine trials 1:70
VBRC database 5:357t
V-CATH, baculovirus terminal disease 1:269
vcyv protein, KSHV latency 3:198
Vector(s)
 animal viruses *see below*
 bridge vectors *see* Bridge vector(s)
 competence 4:653, 5:271
  definition 1:390, 5:269
  post-transcriptional gene silencing 5:271
  RNA interference 5:271
 definition 2:141, 2:146, 2:234, 2:301, 3:212, 4:164, 4:208, 5:60, 5:107, 5:269, 5:274, 5:282, 5:470
 extrinsic incubation period 5:271–272
 feeding 5:269
  blood-meal host-selection patterns 5:271
 gene silencing
  DNA viruses 5:378
  RNA viruses 5:376
 infection control 1:295, 5:494
 introduction 4:208

maintenance 4:652
plant viruses *see* Plant virus(es)
transient 5:60
vaccine development strategies 5:236, 5:238
see also specific vectors
Vectorial capacity 5:271
 daily survival 5:271
 definition 5:269
 Dengue viruses 5:271
 vector blood-meal host-selection patterns 5:271
 zoonoses 5:271
Vectors, animal virus(es) **5:268–274**
 bridge vectors 5:270
 cycle types 5:270, 5:270f
  arthroponoses 5:270, 5:270f
  zoonoses 5:270
 definition 5:269, 5:269f
 enabling factors 5:271
  vector competence 5:271
  vectorial capacity 5:271
 extrinsic incubation period 5:271
  definition 5:271–272
  temperature limits 5:271–272
  vector behavior 5:271–272
 forecasting risk 5:274, 5:274f
  surveillance measures 5:274
 horizontal 5:269
  gonotrophic cycle 5:269
  vector feeding 5:269
 nonviremic transmission 5:270
 phases of transmission 5:272
  amplification 5:272
  maintenance 5:272
  subsidence 5:272
 risk factors 5:273
  age 5:273
  climate 5:273
  occupation 5:273
  residence 5:273
  socioeconomic status 5:273
 venereal transmission 5:270
 vertical 5:269
 ticks 5:269
 see also specific viruses
Vegetable(s) 5:282
 losses 5:283
 see also Crop(s); *individual species*
Vegetable virus(es) 2:87, **5:282–291**
 ecology 5:284
 epidemiology 5:284
 garlic viruses 5:288
 hosts 5:284–285
 infection control measures 5:285t, 5:286
  cross-protection 5:286–287
  natural resistance 5:286–287, 5:287t
  roguing 5:286–287
  soil contamination 5:286–287
 pathogenicity 5:284
  genetic variability 5:284–285
 taxonomy 5:283, 5:283t, 5:284t
 transmission
  control 5:286
  environmental factors 5:285
  mechanical 5:284–285
  seeds *see* Seed(s), viral transmission
  vectors 5:284–285
  vegetative propagation 5:284–285, 5:286t
 see also specific viruses
Vegetative incompatibility *(vic)* genes 2:576
 definition 2:284, 2:580
Vegetative propagation
 banana bunchy top virus transmission 1:277
 cassava mosaic geminiviruses transmission 1:32–33
 definition 4:208
 Ilarviruses 3:54
 potato virus Y transmission 4:288–289
 vegetable virus transmission 5:286t
Vein-clearing, definition 5:200
Veins, plant virus movement 3:353–354, 3:354f
Velum, definition 4:561

Velvet tobacco mosaic virus (VTMoV) 4:646t
Venereal transmission
 LaCrosse virus 5:269
 vector transmission, animal viruses 5:270
 see also Sexually transmitted disease(s)
Venezuelan equine encephalitic virus (VEEV) 5:79, 5:111–112, 5:112f, 5:114
 bioterrorism evaluation 5:407–408, 5:408t
 characteristics 1:171t
 geographical distribution 5:77, 5:102, 5:488
 host cell attachment 5:118
 host range
  enzootic 5:79–80, 5:103, 5:104
  epizootic 5:79–80, 5:103, 5:104
 structure 5:118f
 subtypes 5:80
 tissue tropism 5:104
 transmission 5:79–80, 5:104, 5:111–112, 5:112f
  Aedes 5:99–100, 5:104, 5:112
  enzootic 5:80, 5:114
  epizootic 5:114
  minimum infectious dose 5:112
 vaccines/vaccination 5:488
Venezuelan equine encephalitic virus (VEEV) infection 5:488
 clinical features 5:105
  adults 5:80
  children 5:105
 diagnosis 5:82
 epidemic 5:80, 5:101, 5:488
 epidemiology 5:104
 experimental 5:105
 fatality rates
  horses 5:80, 5:488
  humans 5:488
 immune response 5:80–81
  blood-brain barrier 5:81
  cell-mediated cytotoxicity 5:81
  γβ T-cell deficiency 5:81
  vaccinated animals 5:81
 murine model 5:80–81, 5:115
 pathogenesis 5:115
 pathogenicity 5:105
 pathology 5:80
 symptoms 5:114
 treatment 5:82
Venezuelan hemorrhagic fever 3:207
 clinical features 3:209
 epidemiology 3:207
 histopathology 3:210
 pathology 3:210
Venturia canescens, polydnavirus immunosuppressive activity 4:255–256
Verbenaceae virus infections 4:218
 nonspecific infections 4:219, 4:221f
 see also specific viruses
Vero cells
 definition 2:93
 phlebovirus isolation 4:494
Vertebrates
 Akabane virus infections 1:77
 as viral reservoirs in zoonoses 2:147
Vertebrate virus(es) **2:455–459**
 see also specific viruses
Vertex, definition 5:365
Vertical transmission 2:5
 definition 2:37, 2:141, 2:146, 5:274, 5:440
 edornaviruses 2:113, 2:115
 LaCrosse virus 5:269
 mouse mammary tumor virus 3:337
 Oryza sativa endornavirus *see Oryza sativa* endornavirus (OsEV)
 Sigma virus 4:578
 tick-borne encephalitis virus 5:269
 yellow head virus 5:480
Very late factor-1 (VLF-1) 1:270–271
Very late infection phase, baculovirus in vitro pathogenesis 1:272
Vesicle-associated membrane protein-associated protein A (VAP-A) 2:373
Vesicle packet(s), flaviviruses 1:199

Vesicles
　　definition 4:639
　　smallpox 4:639–640
Vesicular disease, definition 1:410
Vesicular exanthema 1:410
　　clinical features 1:417
　　control 1:418
　　epidemiology 1:415–416
　　incubation period 1:417
　　pathogenesis 1:416
Vesicular exanthema virus infection *see* Vesicular exanthema
Vesicular lesions, foot-and-mouth disease virus 2:272
Vesicular stomatitis, disease history 1:497
Vesicular stomatitis virus(es) (VSV) 5:291–299
　　assembly 3:333
　　classification 5:291
　　diagnosis 5:298
　　distribution 5:298
　　evolution 5:297
　　　　ecological factors 5:297
　　　　mutations 5:297
　　　　origins 5:297
　　fusion loops 3:300
　　genome 3:326f, 5:292, 5:292f
　　glycoproteins
　　　　function 5:292–293, 5:296
　　　　glycoprotein G 3:299–300, 5:312–313
　　　　structure 5:296
　　history 5:291
　　host range 5:296
　　immune response 5:298
　　infection
　　　　incubation period 5:298
　　　　pathogenesis 5:298
　　　　pathology 5:298
　　　　prevention and control 5:298
　　　　zoonoses 5:489
　　matrix (M) proteins 5:295
　　　　domains 5:295
　　　　function 5:295–296
　　　　kinetics 5:295
　　　　structure 5:295–296
　　nucleocapsid core
　　　　immune response involvement 5:298
　　　　structure 5:295
　　particle structure 5:291
　　　　RNP core 5:291–292
　　protein functions 5:295
　　　　glycoprotein 5:292–293, 5:296
　　　　large polymerase 5:296
　　　　matrix protein 5:295
　　　　nucleocapsid 5:295
　　　　phosphoprotein 5:295
　　replication 5:292, 5:293f
　　　　assembly 5:295
　　　　attachment 5:292
　　　　budding 5:295
　　　　cell culture 5:296–297
　　　　gene expression 5:293
　　　　host cell entry 5:292
　　　　uncoating 5:292
　　RNA polymerase II 5:293–294
　　　　error rate 5:297
　　　　functionally distinct pools of 5:294
　　　　Le+ role 5:294
　　　　mRNA cap formation 5:294
　　　　mRNA synthesis 5:292f
　　RNP core
　　　　function 5:295
　　　　structure 5:291–292
　　structure 5:291
　　taxonomy 5:291
　　transmission 1:115, 5:291, 5:297
　　　　experimental 5:296
　　see also specific strains
Vesicular stomatitis Algoas virus (VSAV) 5:297–298
Vesicular stomatitis Indiana virus (VSIV) 4:184t
　　characteristics 1:171t
　　history 5:291
　　structure 5:292f

Vesicular stomatitis New Jersey virus (VSNJV) 4:184t
　　distribution 5:297–298
　　history 5:291
　　transmission 5:296
Vesiculovirus 1:497, 1:498t, 5:291, 5:291t
Vesiculovirus(es) 1:119t
　　definition 2:221
　　genome 3:326f
　　geographical distribution 5:291t
Vesiculovirus-like virus(es) 2:222t, 2:224
　　genome 2:224f, 2:225
　　　　intergenic regions 2:225
　　phylogeny 2:225
　　virion structure 2:224f, 2:225
Vesivirus 1:411t, 3:439
Vesivirus(es)
　　genome 1:413
　　history 1:410
　　infections 3:439t
　　stability 1:412
　　see also specific viruses
Veterinary vaccine(s) 3:68
　　aims 3:68
　　research and development 3:68
　　development strategies 3:235
VFLIP
　　apoptosis 1:160
　　KSHV 3:198–199, 3:200
Vibrio parahaemolyticus 3:123–124
Vicia cryptic virus (VCV) 1:99
　　cultivar analysis 1:103
　　genome 1:100, 1:101f
　　　　5' nontranslated region 1:101, 1:101f
　　　　open reading frames 1:100
Vicia faba endornavirus (VfEV) 2:110–111
　　genome 2:110f
　　molecular evolution 2:115
Vicriviroc 2:507t
　　HIV entry inhibitors 2:506
VIDA database 5:357t
VIDE (Virus Identification Data Exchange) 5:357t
Videx® 2:510
Vietnam, DENV-2 2:7–8
vif gene/protein
　　definition 2:467
　　lentiviruses 2:471
Vinnexin gene family 4:258–259
Viola mottle virus (VMoV) 4:311t
VIPERdb database 5:357t, 5:363
VIPER-like elements 4:421–423
　　replication 4:426
Viracept® 2:514t
Viral binding sites (VBS), *Saccharomyces cerevisiae* L-A virus replication 5:301, 5:302f
Viral Bioinformatics - Canada 5:357t
Viral chemokine-binding protein (vCBP), parapoxviruses 4:62
Viral chemokine receptors (vCCRs)
　　HHV infection 2:501
　　see also specific receptors
Viral 'cross-talk' transplant rejection 3:468–469
Viral degradation complex (VDC) 4:55
"Viral eukaryogenesis," 3:478
Viral Genome Database (VGD) 5:351–352, 5:351f, 5:355f
Viral genome linked protein (VPg) 4:288
Viral Genomes Project 5:362
Viral hemorrhagic septicemia virus (VHSV) 2:228, 4:184t
　　economic effects 2:227, 2:228
　　host range 2:228
　　reverse genetic analysis 2:228
　　salmonid infection 2:221–222
Viral inclusion bodies (VIBs) 1:332
Viral infectivity factor (Vif) 2:523, 4:598–599
　　mechanism of action hypotheses 2:523
　　SIV infection 4:622
Viral inhibitor of apoptosis (vIAP) 3:200
Viral interferon resistance protein (OVIFNR) 4:62

Viral load
　　definition 2:505
　　T cells, response effects 3:76
Viral mitochondria-localized inhibitor of apoptosis (vMIA) 1:158t, 1:162
Viral nervous necrosis 2:229–230
'Viral ocean' 3:474
Viral protein genome-linked (VPg) protein 2:125
　　definition 2:37, 4:129, 4:644
　　nepovirus 3:407–408
　　picornavirus replication 4:132–133
　　potyvirus protein properties 4:317–319
　　recessive plant resistance genes 4:178–182
　　replication strategy, birnavirus evolutionary relationships 1:326
　　viral resistance 4:184
Viral vascular endothelial growth factor (vEGF) 4:61
Viramune® *see* Nevirapine
Viread® *see* Tenofovir
Viremia 1:399, 4:653
　　active 5:315–316
　　African horse sickness 1:40
　　African swine fever virus 1:44
　　cell-associated 5:316
　　definition 4:40, 5:269, 5:440
　　Dengue virus 2:9
　　Ebola virus infection 2:62
　　hepatitis C 2:368
　　hepatitis C virus 2:368
　　immunoglobulin role 3:79–80
　　phleboviruses 4:494
　　plasma 5:316
　　　　transit time 5:316
　　poliovirus 4:243
　　termination 5:315–316
Viremogenicity 4:653
vIRF protein(s) *see* Kaposi's sarcoma-associated herpesvirus (KSHV)
Virginia tobacco 5:61
Virion(s)
　　adsorption, bacteriophages 2:73
　　cacao swollen shoot virus 1:404
　　definition 1:433, 1:483, 2:234, 2:306, 2:505, 3:30, 4:407, 4:623, 5:157, 5:365
　　dissociation, bacterial cell viral entry 5:367
　　enveloped 4:514
　　epitope occupancy
　　　　definition 3:413
　　　　infection neutralization 3:414
　　immature 1:193
　　iridoviruses 2:233
　　Japanese encephalitis virus 3:183, 3:184
　　lysis-induced release 3:249
　　Maize streak virus 3:268
　　necrovirus 3:403
　　ophiovirus 3:448
　　penetration 5:367
　　potexvirus 4:310
　　production 3:401–402
　　receptor-binding site 3:401
　　structure 2:457
　　　　hepatitis B surface antigen 2:352
　　togaviruses 2:230
　　toroviruses 5:152, 5:152f
　　virus *vs.* 3:400
Virion-associated transcriptional transactivator (VATT) 3:170
Virion-containing vesicle(s) 1:186
　　see also specific viruses
Virion-induced transcriptional transactivator (VITT) 3:170
Virobacterial agglutination test (VBA) 1:408
Viroceptors 5:318
　　leporipoxvirus infection 3:227
Viroid(s) 5:332–342
　　classification 5:11t, 5:333t, 5:335, 5:336f
　　definition 2:375, 3:348, 5:332
　　evolution 3:475, 5:336f, 5:341
　　genome 5:332, 5:333t, 5:334f
　　　　domains 5:333
　　　　loop E motif 5:333–335

properties 5:332
ribozyme-mediated cleavage 5:333–335, 5:334f
geographical distribution 5:337
history 5:332
host range 5:335, 5:336f, 5:341
infections *see below*
movement 5:339
   cell-cell 5:339
origin 5:341
replication 5:339
   α-amanitin 5:339
   error rate, evolution related 5:341
   host-encoded enzymes associated 5:339
   RNA cleavage 5:340f
structure
   stability 5:341
   tertiary 5:333–335
transmission 5:335
   agroinoculation 5:335
   experimental 5:335
   wild species *vs.* crops 5:338
virus-induced gene silencing 5:330
Viroid infections
control 5:338
cytopathic effects 5:335
diagnostic tests 5:338
epidemiology 5:338
pathogenicity 5:339
   small interfering RNA 5:341
symptomatology 5:335, 5:337f
Virokines 5:318
leporipoxvirus infection 3:227
see also Cytokine(s)
VirOligo database 5:357t
Viroplasm(s) 4:194–195, 4:194f
definition 2:18
Viroporin(s) 5:309
membrane proteins 5:309
picornavirus replication 2:127
Virosomes 5:310
VirP1 protein 5:339
Virucidal agent(s) *see* Antiviral drug(s)
Virulence 5:317
attenuating mutations, genetic determinants 5:317, 5:318
definition 4:178
genetic determinants 5:317
   attenuating mutations 5:317, 5:318
   'cell-derived' genes 5:318
   leporipoxviruses 3:227
   phenotype 5:317
   principles 5:317
   suipoxviruses 3:227
measurement of 5:317
pathogenic mechanisms 5:318
vaccine safety 5:228–229
vaccinia virus 5:249
variation, viral evolution 2:175
see also specific viruses
Virulence factors 3:123
Avr proteins 4:173
transmission 3:123
yellow fever virus *see* Yellow fever (YF) virus
Virulence gene(s) 4:170
Virulence perturbation, hypovirulence 2:575
Virulent-systemic (VS) feline calicivirus disease 1:416
Virus(es) 3:398–403
antigen detection 2:33
assembly
   antiviral therapy 1:149
   virus membranes 5:312
associated nucleic acids 4:526
attachment 1:148
bacteria *vs.* 2:456
binding neutralization 3:64–65
biochemistry 2:456
caspase regulators 1:160
chemical composition 2:456
Chlamydia vs. 3:476
coat protein DNA sequence 4:168
core 3:6

'cross-talk' apoptosis 1:155–157
cultivation 2:456
definition 2:451, 2:455, 3:400
discovery 3:398
   history 3:398
diseases 3:66
see also specific diseases
DNA-repair systems 4:374
DNA virus recombination 4:374
envelope 5:365
evolution *see* Evolution (of viruses)
fitness 3:402
   antiviral therapy 1:151
   definition 3:402
fusion 5:310
   hemifusion diaphragm 5:310–311, 5:311f
   stalk-pore pathway 5:310–311, 5:311f, 5:313f
   virosomes 5:310
fusion proteins 5:310
   activation 5:312, 5:313f
   properties 5:311, 5:313f
   structural features 5:312
genomes 2:457
   evolution of 3:477–478
   integration 3:401
historical aspects **2:455–459**
immune system evasion 1:637, 2:144–145, 2:487
induced cell transformation 3:180
infection neutralization 3:417
isolates 3:402
isolation 1:602
latency *vs.* active replication 3:401–402
life cycle 3:400
membranes *see* Virus membrane(s)
mitochondrial apoptosis 1:161
movement proteins 4:283–284
origins *see* Origin(s) of viruses
perpetuation in nature 2:146
   arthropod transmission cycles 2:147
   host clinical status 2:146
   host population immunity 2:146
   population size 2:147
   virus virulence 2:146
   zoonotic transmission cycles 2:147
progeny 3:402
proteins 3:401, 3:401f
   escape hypothesis 3:473–474
   expression 1:35
   immunogenicity *see* Immunogenicity
   purification
      BaculoDirect expression vector system 1:241
      X-gal 1:240–241
release 4:412
replication *see* Replication
serotypes 3:402
   antigenic variation 1:128
species classification 5:405
specific sequence data 5:357t, 5:363
strains
   characteristics 3:402
   definition 3:402
structure **5:393–401**
   helical symmetry 5:393
   icosahedral symmetry 5:394
   particle maturation 5:399
   picorna-like capsids 5:398
   quasi-equivalent capsids 5:395
structures **5:393–401**
subversion 1:125
transfer RNA 3:375
transmission *see* Transmission;
see specific transmission modes
tropism 2:505
vaccines *see* Vaccines/vaccination
virion *vs.* 3:400
virus particle *vs.* 3:400
see also specific topics
see also specific viruses
Virus-associated interfering RNA (VAI RNA) 1:3–4, 1:161
Virus attachment protein, definition 4:407

Virus-enhancing factor (VEF), baculoviruses 1:250
Viruses: From Structure to Biology 5:357t
Virus-first hypothesis 3:473, 3:475, 3:476f
Virus-free seeds, alfalfa mosaic virus 1:86
Virus Identification Data Exchange project (VIDE) 3:213
Virus-indexed propagation program(s) 4:200
Virus-induced gene silencing (VIGS) 4:148, **5:325–332, 5:375–380**
Agrobacterium tumefaciens-based transformation vectors 5:377, 5:379
Arabidopsis thaliana 5:326–327
barley stripe mosaic virus 2:466
definition 5:325, 5:375–376
food crops, vectors 5:378
functional genomics tool 5:376
   advantages 5:379
   DNA virus-based vectors 5:326, 5:377t, 5:378
   limitations 5:379
   RNA virus-based vectors 5:376, 5:377t
   VIGS-related phenotype 5:378
gene characterization applications 5:378
   resistance gene pathways 5:378
GFP-transgene silencing 5:326
history 5:325
induction 5:325
overexpression 4:148
phytoene desaturase gene 5:326
principles 5:376
solanaceous hosts 5:378
suppression **5:325–332**, 5:329t
   cross-kingdom activity 5:328–330
   HC-Pro 5:327–328, 5:328f, 5:330
   immune evasion 5:327
   miRNA pathway 5:328f, 5:331
   plant development, affects 5:330
   reversal of silencing assays 5:328
   transgenic plant grafting 5:328
   viral synergism 5:327–328
tobacco mosaic virus 5:376
transgenic resistance, plants
   cross-protection 5:326
   tobacco 5:325–326
   tomato 5:325–326
viral-derived siRNAs 5:326
   antiviral therapies 5:327
Virus infectivity factor (Vif)
HIV-1 infection 4:598–599
immune system evasion 3:109
SIV infection 4:622
Virus-like particles (VLPs) 1:98
algal infections 1:87
birnavirus replication 1:326
definition 4:97, 4:352, 4:428
LTR-retrotransposon life cycle 4:430
pecluvirus 4:102
pseudoviruses 4:352–353
salterprovirus 5:418
Virus membrane(s) **5:308–314**
acquisition 5:308
budding 5:313, 5:313f
   M proteins 5:314
fusion *see below*
ligand role 5:310
   endocytosis 5:310
   host cell attachment 5:310
lipid bilayers 5:308
   lipid arrangements 5:308–309
   lipid composition 5:309
   lipid rafts 5:309
proteins 5:309, 5:310f
   aggregation 5:309
   functions 5:308
   integral 5:309
   peripheral 5:309
   viroporins 5:309
pseudotype virions 5:313
synthesis 5:310f, 5:312
   glycosylation 5:312
viral assembly 5:312

Virus membrane fusion 5:310
   hemifusion diaphragm 5:310–311, 5:311f
   proteins 5:310
      activation 5:312, 5:313f
      properties 5:311, 5:313f
      structural features 5:312
   stalk-pore pathway 5:310–311, 5:311f, 5:313f
   virosomes 5:310
Virus nervous necrosis disease 3:432
Virus neutralization test (VNT) 1:531
Virus World 5:357t
Viscerotropism see Yellow fever
Visna 4:460
Visna-Maedi virus(es) (VMV) **5:423–432**
   animal importation 5:431
   cell tropism 5:424, 5:425
      cellular receptors 5:426
   env gene/protein 5:431
   genome 5:424, 5:425f, 5:425t
   geographical distribution 5:430
   history 5:424
   infections see below
   replication cycle 5:424
      macrophage infection 5:425–426
      provirus integration 5:424
   synonyms 5:424
   transmission 5:426
      oral 5:426
      respiratory 5:426
   vaccines/vaccination 5:431
      Freund's incomplete adjuvant 5:431
   virion structure 5:424
Visna-Maedi virus infection
   acute infection cannulation model 5:426
   clinical lung disease 5:424
   control 5:431
      animal importation 5:431
      eradication 5:431
      serological tests 5:431
   dendritic cells 5:425
   escape mutants 5:430–431
   granulocyte monocyte colony stimulating factor 5:428–430
   Icelandic epidemic 5:424
   immune response 5:426
      abnormalities 5:428
      acute infection cannulation model 5:426, 5:427f
      antibody response 5:427
      CD8+ T cells 5:427
      ENV antigens 5:427
      GAG antigens 5:427
      genetics, role 5:430
      macrophages see below
      neutralizing antibodies see below
      protective immune response 5:427
      seroconversion 5:426–427
      T-lymphocyte response 5:427
   immune system evasion 5:430
      CD4+ T cells 5:430–431
      cell fusion 5:430–431
      genome integration 5:430–431
   interferons, type I 5:430
   latency 5:425–426
   macrophages 5:425
      immune system evasion 5:430–431
      pathogenesis 5:428–430
      phenotypic changes 5:428–430
   neutralizing antibodies 5:430–431
      vaccination 5:431
   pathogenesis 5:428–430, 5:429t
      associated tissues 5:428
      autoimmunity 5:430
      clinical signs 5:428
      cytopathic effects 5:428
      host cell entry 5:428
      immunopathology 5:428
      lesion composition 5:428
      lymphocytes, role 5:430
   persistent 5:428
   resistance 5:430
   treatment 5:431

Visnia 1:474
Vitivirus 1:420t, 1:425, 1:426f
Vitivirus(es) **1:419–427**
   associated species 1:425
   biological properties 1:425
   classification 1:420t
   disease characterization 1:425
   genome 1:425
   hosts 1:425
   particle structure 1:421f, 1:425
   phylogeny 1:426, 1:426f
   replication 1:425
   serology 1:426
   transmission 1:425
   see also specific viruses
Vitrification, cryo-electron microscopy 1:604, 2:83–84
   CEMOVIS 2:84
   methods 2:83–84
   sample concentration 1:604
VLF-1 protein 1:257–260
VLPA3 virus 5:413t
VLP-based, papillomavirus infection 4:17
VOCs database 5:357t
VOGs database 5:357t
Voles
   cowpox virus hosts 1:577, 1:578
   murine gammaherpesvirus 68 3:372
Vomiting, rotaviruses 4:512
V-ori 3:264
VP1 gene/protein
   African horse sickness virus 1:37
   aquareovirus 1:165–166
   circovirus see Circovirus
   echovirus detection 2:69
   foot-and-mouth disease virus 2:266–267
   micromonas pusilla reovirus 1:91–92
   murine polyomavirus 4:271
   seadornavirus 4:538
   swine vesicular disease virus structure 2:128–129
VP2 gene/protein
   aquareovirus 1:165–166
   circoviruses 1:517
   infectious bursal disease virus see Infectious bursal disease virus
   murine polyomavirus 4:271
   parvovirus capsid structure 4:86
   poliovirus 2:129
   seadornavirus 4:538
   swine vesicular disease virus structure 2:129
VP3 gene/protein
   African horse sickness virus 1:37
   aquareovirus 1:165–166
   birnaviruses 1:325
   murine polyomavirus 4:271
   poliovirus 2:129
   seadornavirus proteins 4:538
   swine vesicular disease virus structure 2:129
VP4 gene/protein
   African horse sickness virus 1:37
   aquareovirus 1:165–166
   birnaviruses 1:325, 1:325f
   poliovirus 2:129
   seadornavirus proteins 4:538
   swine vesicular disease virus structure 2:129
VP5 gene/protein
   aquareovirus 1:165–166
   birnaviruses 1:325
VP6 gene/protein
   African horse sickness virus 1:37
   aquareovirus 1:165–166
VP7 gene/protein
   African horse sickness virus 1:37, 1:38f
   aquareovirus 1:165–166
   seadornavirus 4:541–542
VP8 gene/protein, seadornavirus proteins 4:538
VP9 gene/protein
   Banna virus 4:542
   seadornavirus proteins 4:538
VP16 gene/protein, HSV latency 2:437

VPg protein
   barley yellow dwarf viruses 1:280, 1:282
   cereal yellow dwarf virus 1:282
   nepovirus 3:409–410
   rhinoviruses 4:469–470
   rice yellow mottle virus 4:487
Vpr see Lentivirus protein R (Vpr)
V protein, henipavirus 2:323, 2:326
Vpu gene/protein see HIV-1
v-Rel complexes see Reticuloendotheliosis viruses
VTA virus 5:413t

W

Waikavirus
   genome 4:548
   tissue tropism 4:547
Waikavirus 4:547, 4:547t
Walleye dermal sarcoma (WDS) 2:214f, 2:215
   appearance 2:214
   incidence 2:214
   viral accessory protein roles 2:215–216
Walleye dermal sarcoma virus (WDSV) 2:231
   accessory proteins, pathogenic role 2:215
      OrfB protein 2:216
      OrfC protein 2:216
      origins 2:216
   disease cycle control 2:216
   genome
      alternative splicing 2:215f, 2:216
      open reading frames 2:214–215
   infection see above
   isolation 2:214, 2:215f
   molecular biology 2:231
   transmission 2:214
Walleye discrete epidermal hyperplasia (WEH) 2:214, 2:214f
   tumor incidence 2:214
Walleye epidermal hyperplasia virus(es) (WEHV)
   genome, open reading frames 2:214–215
   infection see above
   isolation 2:214, 2:215f
   transmission 2:214
Walleye epidermal hyperplasia virus type 1 (WEHV-1) 2:231
Walleye epidermal hyperplasia virus type 2 (WEHV-2) 2:231
Walleye fish, skin tumors 2:214
Walleye herpesvirus 2:208–209
Walnut blackline 4:206
Walnut trees(s), diseases 4:206
   causal agents 4:206
   control 4:206
   geographical distribution 4:206
   symptoms 4:206
   transmission 4:206
   yield losses 4:206
Wandering behavior
   baculovirus infection 1:269
   BmNPV infection 1:269
Wanowrie virus 1:400
   Hyalomma marginatum 1:400
Warthin–Finkeldy giant cells, measles 3:289
Warthog (*Phacochoerus aethiopicus*), as an African swine fever virus host 1:43, 1:44f
Warts see Papillomavirus infections
Water
   hepatitis E virus transmission 2:380, 2:382
   Maize chlorotic mottle virus transmission 3:262
Watermelon mosaic virus (WMV) **5:433–440**
   biotechnological applications 5:439
   classification 5:433
   efficient vectors 5:437
   epidemiology 5:436
      virus sources 5:436
   evolution 5:439
   genetics 5:439
   geographical distribution 5:435
   history 5:433
   host range 5:435

infection *see below*
  spread 5:437
  strains 5:433
  taxonomy 5:433, 5:434f
  use in transgenic plants 4:157, 4:158f
    heterologous encapsidation 4:162
  variability 5:438
  vector relationships 5:436
Watermelon mosaic virus infection
  control methods 5:437
    cross-protection 5:437, 5:438f
    genetically-modified resistant plants 5:438
    prophylaxis 5:437
    resistant cultivars 5:437
  diagnostic methods 5:435
  symptoms 5:434, 5:435f
  synergism 5:435
Watson, James
  bacteriophage structure, research history 1:433
  virus protein coats 2:452
Wax moth *(Galleria mellonella)*, parvovirus infection 3:127–128, 4:76
  symptoms 3:127–128
Weed(s)
  cucumber mosaic virus transmission 1:618–619
    control 1:619
  potato virus Y 4:289
  tobravirus transmission 5:76
  zucchini yellow mosaic virus transmission 5:436–437
WEEV *see* Western equine encephalitis virus (WEEV)
Well boats 3:90
Wesselsbron virus 2:237
  geographical distribution 2:238
  infection 2:238
  Rift Valley fever virus vaccine 2:237–238
  transmission 2:238
    mosquitoes 2:237
  vaccine 2:237–238
West Africa, HIV infection 1:63
Western blotting
  definition 4:623
  plant virus detection 2:26
  yellow head virus infection 5:482
Western equine encephalitis virus (WEEV) 5:77, 5:114
  Ab-mediated neutralization 3:416
  characteristics 1:171t
  evolutionary relationships 1:174, 5:96–97
  geographical distribution 5:77, 5:102, 5:114–115
  infection *see below*
  isolation 5:82
  phylogeny 5:77–78
  subtypes 5:101
  tissue tropism 5:105
  transmission 5:77–78, 5:99–100, 5:101, 5:115
    age 5:273
    forecasting risk 5:274
Western equine encephalitis virus infection 5:77, 5:82
  case–fatality rates 5:77, 5:105
  clinical features 5:77, 5:105, 5:115
    encephalitis characterization 5:77
  diagnosis 5:82
  epidemiology 5:104, 5:104f
  host range 5:102–103
  incubation period 5:77
  pathogenicity 5:105
  pathology, brain examination 5:106
  treatment 5:82
Western Europe, HIV infection 1:65
  transmission 1:65
West Nile (WN) encephalitis 5:447
West Nile (WN) fever 5:441
  outbreaks 5:441
  symptoms 5:447
West Nile neuroinvasive disease (WNND) 5:441
  epidemics 5:441
  fatalities 5:447
  risk factors 5:447
  symptoms 5:447
  virus lineage associated 5:441

West Nile virus (WNV) 2:96, 2:236, **5:440–450**
  bioterrorism 5:407
  bird vectors 5:445
    American robin 5:445
    corvids 5:445
    morbidity/mortality rates 5:445
    surveillance specimens 5:448
  characteristics 1:171t
  classification 5:440
  ecology 5:443
  epidemiology 2:243, 2:247f, 5:441
    'North American dominant' genotype 5:441
  genome 5:442
    noncoding regions 5:442
    nonstructural proteins 5:442
    proteins encoded 5:445t
    structural proteins 5:442
  geographical distribution 4:128, 5:441, 5:443f, 5:487
    lineages 2:237, 5:441
  history 5:441
  phylogenetic analysis 5:441, 5:443f
  replication 5:442, 5:446f
    host cell attachment 5:442–443
    RNA synthesis 5:442–443
    transcription 5:442–443
  transmission 2:236, 5:447
    nonviremic 5:445
    perpetuation mechanisms 5:444–445
    vectors *see below*
    vertical 5:444–445, 5:447f
  vectors 2:236, 5:443, 5:445
    American robin 5:445
    birds *see above*
    competence 5:443–444
    mosquitoes 2:236
    vertebrates 5:445
  virion structure 5:442, 5:444f
West Nile virus infection
  alligators 5:445–446
  clinical features 2:236, 5:447
    movement disorders 5:447
  control 5:448
    public health measures 5:449
    therapeutics 5:448
    vaccination 5:448
    vaccines/vaccination 5:448
  diagnostics 5:448
    human 5:448
    surveillance specimens 5:448
  disease surveillance 5:487
  epidemics 2:236
  horses 2:237
  host range 2:237
  immune response 5:446–447
  immune system evasion 3:110
    complement 3:110
  incidence 5:448t
  maintenance transmission 5:272
  neurological 2:237, 5:82
  organ transplantation, risks 3:466, 3:470
  pathogenesis 5:446
    host status 5:446–447
  zoonosis 5:443, 5:447f, 5:487
  see also West Nile neuroinvasive disease (WNND)
Whatroa virus (WHAV) 5:92, 5:93t
  characteristics 5:109t
Wheat American striate mosaic virus (WASMV) 1:495
Wheat dwarf virus (WDV) 1:495
  infection 1:495
  strains 1:495
  transmission 1:495
Wheat eqlid mosaic virus (EqMV) 1:495
Wheat spindle streak mosaic virus (WSSMV) 1:493, 1:493f
Wheat streak mosaic virus (WSMV) 1:495
  host range 1:495
  infection
    prevention 1:495
    resistance 1:495
    symptoms 1:495, 1:496f

  synergistic relationships, maize chlorotic mottle virus 3:262
  transmission 1:495, 5:280–281
Wheat virus(es) **1:490–497**, 1:491t
  see also specific viruses
Wheat yellow head virus (WYHV) 1:495
Wheat yellow leaf virus 1:495
Whispovirus 4:567, 5:451
White bream virus (WBV) 2:230
White clover cryptic virus(es) (WCCV) 1:99
  phylogeny 4:67f
White clover mosaic virus (WClMV) 4:311t
  genome 4:312
  transmission 4:310, 4:312
White-faced saki lymphocryptovirus (PpiLHV1) 4:587t
Whitefly *(Bemisia tabaci)*
  African cassava mosaic virus transmission 1:36, 2:101
  cassava mosaic geminiviruses transmission 1:32–33
  cotton leaf curl disease transmission 2:102–104
  cucumber vein yellowing virus transmission 2:90
  cucurbit yellow stunting disorder virus transmission 2:89
  definition 5:138
  mungbean yellow mosaic virus transmission 3:366, 3:368
  population control 2:104
  tomato infectious chlorosis virus transmission 2:89–90
  tomato yellow leaf curl virus *see* Tomato yellow leaf curl virus
  vilruliferous, definition 5:138
Whitefly-transmitted geminivirus(es) (tomato) (WTG) 4:200, 5:125
White rot control, mycoreovirus-3/RnW370 3:382
White spot disease (WSD) **5:450–451**
  clinical features 5:451
  farming practices 4:569
  geographical distribution 5:450–451
  hemocytes
    host responses 5:458
    semigranular cells 5:458
  histopathology 5:452
  host range 5:450–451
  mortality rate 5:451
  pathology 5:451
  RNAi strategies 5:458
  shrimp production losses 4:569
  vaccines/vaccination 5:458
White spot syndrome virus (WSSV) **5:450–459**
  apoptosis 5:457
  classification 4:567
  clinical features 5:451
  epidemiology 5:451
  evolution 2:182
  gene expression 5:457
    early genes 5:457
    immediate early genes 5:457
    promoter regions 5:457
    regulation 5:457
    structural protein genes 5:457
  genes 5:456
    encoded proteins 5:456
    expression *see above*
    identification, comparative homology 5:456
    identification, functional studies 5:456
    immediate early 5:456–457
    latency-related 5:456–457
  genome 5:455
    tandem repeats 5:451
  geographical distribution 5:450–451
  histopathology 5:452, 5:452f, 5:453f
  host cell attachment 5:453–454
  host range 5:451
  host relationships 5:457
    apoptosis 5:457
    hemocyte responses 5:458
  infection *see* White spot disease (WSD)
  life cycle 5:452–453
  morphogenesis 5:452, 5:454f

White spot syndrome virus (WSSV) (continued)
  morphology 5:452, 5:454f
  pathology 5:451
  structural proteins 5:453, 5:455f, 5:456f
    capsid 5:453
    envelope 5:453–454
  synonyms 5:451
  taxonomy 5:451
  transmission 5:451
  virion assembly 5:452–453
  virion composition 5:450–451, 5:453
  virion structure 5:453, 5:455f
White sturgeon iridovirus (WSIV) 3:155
  infection 3:156–158
White tail disease 4:573
  affected tissues 4:573
  geographical distribution 4:575
  host range 4:575
Wild animals
  rabies virus vaccination 4:373
  see also specific animals
Wild boar
  arterivirus evolution 1:182
  classical swine fever 1:530, 1:532
Wildebeest (Connochaetes gnu), bovine malignant catarrhal fever 1:363
Wild potato mosaic virus (WPMV) 4:298t
Wild type, definition 4:359
WIN compounds, antirhinoviral strategies 1:547–548
Winter wheat mosaic virus 5:24
Woese, Carl 3:474
  trinity concept of evolution 3:473–474
Wongabel virus (WONV) 4:184t
Woodchuck hepatitis virus (WHV) 2:329, 2:366–367
Woodmice (Apodemus), cowpox virus hosts 1:577, 1:578
World Health Organization (WHO)
  Dengue hemorrhagic fever/Dengue shock syndrome diagnostic criteria 2:11–12, 2:12t
  smallpox eradication 4:640
  yellow fever incidence 5:473
Worldwide trade, ornamental plant species virus infections 4:208
Wound tumor virus (WTV) 3:382–383
  characteristics 4:150t
  genome 4:152, 4:153f
  as type species 4:536t
  virion structure 4:152
W protein, henipavirus 2:323
WSN influenza strain, neuraminidase 3:102
WU virus (WUV) 4:262

X

Xenotransplantation
  papillomavirus propagation 4:32
  retroviral expression 2:108–109
Xenotropic behavior
  definition 2:467
  murine leukemia viruses 2:468
Xenotropism, retrovirus infection resistance 2:468
Xf1 bacteriophage 3:121
X-gal, baculovirus expression vectors, plaque purification 1:240–241
X-linked lymphoproliferative syndrome (XLP) 2:153
  mortality rate 2:153
XLP see X-linked lymphoproliferative syndrome (XLP)
X protein (Borna disease virus) 3:331
X-ray crystallography 1:603, 2:453
  bluetongue virus core particle structure 5:388–391
  cowpea mosaic virus 1:570
  nonenveloped virus structure 5:380, 5:381
  rice yellow mottle virus 4:486
  tetravirus capsid structure 5:33
  tobacco mosaic virus 2:452, 5:380–381
X-ray diffraction 1:201
  nonenveloped virus structure 5:380
XSV see Extra small virus (XSV)
Xylem
  definition 3:348
  rice yellow mottle virus 4:487–488

Y

Yaba-like disease virus (YLDV) 5:462, 5:462t
  distribution 5:464
Yaba monkey tumor virus (YMTV) 5:462t
  distribution 5:464
  genome 5:462–463
  history 5:461
  infection 5:461
    control 5:465
    immunity 5:463
    pathogenesis 5:464
    prevention 5:465
    tumor regression 5:464
  tropism
    humans 5:463
    nonhuman primates 5:463
  virion properties 5:462
Yatapoxvirus 4:324t, 5:461, 5:462t
Yatapoxvirus(es) 5:461–465, 5:462t
  classification 4:324t
  epidemiology 5:464
  genome 5:462, 5:463t
  geographical distribution 5:464
  history 5:461
  infection
    control 5:465
    immune modulation 5:464
    immunity 5:463
    pathogenesis 5:464
    prevention 5:465
  replication
    DNA replication 5:462
    transcription 5:462
    translation 5:462
  seasonal distribution 5:464
  transmission 5:463
  tropism
    humans 5:463
    non-human primates 5:463
  virion properties 5:462
  see also specific viruses
YDJ1 protein, Brome mosaic virus 1:384–385
Yeast(s)
  killer phenomenon 2:285, 5:164
    ion channel formation 5:307
    maintenance of killer genes 5:170–171
    superkiller genes 5:170–171
    toxin immunity 5:307
  prions see Prion(s), yeast and fungal
  Tf1 transposon see Tf1 transposon (yeast)
  Tf2 transposon see Tf2 transposon (yeast)
  Ty3 retrotransposon see Ty3/gypsy retrotransposon(s)
  viruses
    apoptosis 1:162
    replication models 2:290
    see also specific viruses
  see also under Saccharomyces
Yeast L-A virus (ScV-L-A) 5:465–469
  apoptosis 1:162
  capsid structure 5:165
  evolution 5:173–174
  genetics 5:468
  genome 5:467, 5:467f
    open reading frame 5:467
    open reading frames 5:166, 5:300–301
    RNA pseudoknot 5:467
  Helminthosporium victoriae 190S virus vs. 5:165
  as helper virus 5:465–466
  history 5:466
  killer phenomenon 5:170–171
    host genes associated 5:170–171
    satellite dsRNAs associated 5:300
  replication 5:301, 5:467
    RNA packaging 5:467, 5:467f
    RNA polymerase activity 5:167–168
    RNA replication 5:467, 5:468f
    transcription 5:467
    viral binding sites 5:301, 5:302f
  translation 5:468
    positive strand 5:468
    Ski proteins 5:468
  transmission, cell–cell fusion
    mating 5:465–466
  virion structure 5:466, 5:466f
    Gag 5:466–467, 5:466f
Yellow dwarf virus(es)
  infection
    risk assessment 1:496
    symptoms 1:496
    transmission 1:496
  see also specific viruses
Yellow fever
  cases associated 5:487
  clinical features 5:474
    classic disease 5:474–475
    CNS involvement 5:474–475
    fatal cases 5:474–475
    nonspecific flu-like illness 5:474
    spectrum 2:243
  diagnosis 2:243
  epidemics 5:494
    agriculture 5:494
  epidemiology 2:243, 5:473
    virus survival 5:474
  geographical distribution 5:473
  historical aspects 5:470
    vector identification 5:470
  immune response 5:475
    antibody-dependent cellular cytotoxicity 5:475–476
    cytotoxic T-cells 5:475
    immunoglobulin M 5:475–476
    natural killer cells 5:475
    pro-inflammatory cytokines 5:475
  incidence 5:473
    WHO figures 5:473
  mortality rate 5:487
  neurotropism 5:475
  pathogenesis 2:245, 5:475
    cytokine dysregulation 5:475
    primate models 5:475
    systemic immune response syndrome 5:475
  pathology 5:475
  prevention/control 5:476
    vaccines/vaccination see below
  seasonal distribution 5:473
    Aedes breeding cycle 5:473
  transmission cycles 5:473
    laboratory-acquired infections 5:474
    'savanna type' 5:473
    'sylvatic type' 5:473
    'urban type' 5:473, 5:474
  vaccines/vaccination 5:476
    17D vaccine 5:476
    safety 2:246, 5:234
  vaccine/vaccination 2:239
    reactions to 5:231t
    safety 2:246, 5:234
    strain 17D 5:238–239, 5:240f
  vectors, identification 5:470
  viscerotropism 5:475
    Councilman bodies 5:475
    eosinophilic degeneration 5:475
    hepatocyte degeneration 5:475
    lymphoid follicles 5:475
Yellow fever (YF) virus 2:238, 5:469–476
  assembly 5:471
    endoplasmic reticulum 5:471
    secretory cells 5:471
  cell entry, clathrin-coated vesicles 5:471
  characteristics 1:171t
  circulation, sylvatic 2:238–239
  classification 5:470
    neutralization tests 5:470
    RdRp 5:470
  genome 5:470
    noncoding region 5:470–471
    open reading frames 5:470–471
    reverse transcriptase-PCR 5:472

history 2:238, 2:242, 2:455
host range 2:238–239, 5:472
    vertebrates 5:472–473
infection *see* Yellow fever
isolation 5:472
propagation 5:472
proteins 5:471
    C (capsid) protein 5:471
    E (envelope) proteins 5:471, 5:472
    nonstructural proteins 5:471, 5:472
    preM (pre-membrane) protein 5:471
RdRp 5:470
replication 5:471
taxonomy 5:470
transmission 2:238, 5:487
    Aedes aegypti 2:455
    mosquitoes 2:238
variation 5:470
    Dengue virus *vs.* 1:127
    genotypes 5:470
virion 5:470
virulence determinants 5:471
    E protein 5:472
    nonstructural proteins 5:472
    sequences 5:471–472
    viscerotropism 5:471–472
zoonoses 5:487
Yellow head virus (YHV) **5:476–483**
    classification 4:567
    genetic diversity 5:480, 5:481f
        open reading frames 5:482
    genome 5:477
        open reading frames 5:477–478, 5:482
        pseudoknot structure 5:477–478
        untranslated region 5:477–478, 5:479f
    genotypes 5:479
    geographical distribution 5:479
    historical aspects 5:477
    host range 5:479
    infection *see below*
    proteins 5:477–478
    RdRp 5:477–478
    structural proteins 5:477
    taxonomy/classification 5:477
    transmission 5:480
        vertical transmission 5:480
    virion 5:477, 5:478f
        nucleocapsids 5:477
        nucleoprotein 5:477
        structural proteins 5:477
        transmembrane glycoproteins 5:477
    see also Gill-associated virus (GAV)
Yellow head virus infection
    cell entry 5:477
    chronic infections 5:480
    diagnosis 5:482
        nitrocellulose enzyme immunoassay 5:482
        RT-PCR 5:482
        Western blotting 5:482
    host response 5:480
        inflammation 5:480
    management 5:482
        pathogen exclusion 5:482
        RNA interference 5:482
    mortality 5:479
    pathology 5:479
    tissue types 5:480
Yellow mosaic-affected vines 4:202, 4:202f
Yellows disease, raspberry bushy dwarf virus 3:38, 3:38f
Yersinia pestis, virulence factors 3:123
YGEV *see* Transmissible gastroenteritis virus (TGEV)
YHV *see* Yellow head virus (YHV)

Yield losses
    African cassava mosaic disease 1:34
    barley yellow dwarf viruses 1:284–285
    beet necrotic yellow vein virus infection 1:308
    geminiviruses 4:166–167
    maize chlorotic mottle virus 3:262
    potato virus Y potato infections 4:291
YLDV *see* Yaba-like disease virus (YLDV)
ψM1 virus 5:413t
ψM2 virus 5:413t
ψM100 virus 5:413t
YMTV *see* Yaba monkey tumor virus (YMTV)
Young pigeon disease syndrome (YPDS) 1:515
Y-transposase 4:436

Z

Zaire Ebola virus (ZEBOV)
    classification 2:57
    endemic 2:57
    genome 2:58
    geographical distribution 2:62
    history 2:57
    host range 2:62
        experimental models 2:62
    infection
        animal models 2:202–203
        clinical features 2:62
        mortality rate 2:199
        pathogenicity 2:63
        treatment 2:64
    life cycle 2:61
    protein properties 2:59
        glycoproteins 2:60
        matrix proteins 2:59
        RNP complex 2:59
    taxonomy 2:57
    vaccines 2:65
        recombinant VSV 2:65
    virion properties 2:57
Zalcitabine 2:510
ZAM elements 4:433
Zanamivir 1:143t
    influenza virus infection treatment 1:153, 2:556, 3:102–103, 3:492
Zea mays see Maize *(Zea mays)*
ZEBOV *see* Zaire Ebola virus (ZEBOV)
Zebrafish endogenous retrovirus 2:219
    phylogeny 2:219–220, 2:220f
Zerit® 2:510
Ziagen® 2:510
Zidovudine (AZT) 2:510, 2:511t
    adult T-cell leukemia therapy 2:569–570
Zimbabwe, HIV epidemiology 1:65–66
Zinc finger antiviral protein (ZAP), retrovirus infection resistance 2:473
Zinc finger proteins
    Ilarvirus replication 3:52–53
    sawfly baculovirus genome 1:228
Zoo animals, cowpox infection 1:577
Zoonoses 5:270, **5:485–495**
    arthropod vectors 2:147
        Asian tiger mosquito 5:494
        sandflies 5:488
    control 5:493
    definition 1:170, 2:141, 2:147, 2:241, 2:377, 2:517, 4:40, 4:52, 4:490, 4:624, 5:107, 5:269, 5:440
    diagnosis 5:493
    ecological change 5:494
    as emerging diseases 2:95, 5:494
    organ transplant infections 5:490
    reservoir identification, phylogeny 4:128

social change 5:494
transmission 5:485
    control 5:493
vectorial capacity 5:271
vector transport 5:494
see also specific diseases/disorders
see also specific viruses
Zoospore(s)
    furovirus transmission 2:292
    plant virus transmission 5:281
Zoster (shingles)
    clinical features 5:253, 5:254f
    complications 5:253
        immunocompromised 5:253
        postherpetic neuralgia 5:253
        Ramsay–Hunt syndrome 5:253
    epidemiology 5:252
        age-relation 5:252, 5:252f
        immunocompromised 5:252
    etiology 5:252
    virus reactivation 5:253
    immunocompromised patients 5:252, 5:253
    pathology 5:253–254
Zostera marina (seagrass), dsRNAs 2:112
Z protein
    arenavirus assembly/budding 3:247
    lymphocytic choriomeningitis virus replication/transcription 3:245–246
Z RING domain, arenavirus replication 3:245–246
Zta protein, EBV infection 2:166
Zucchini yellow mosaic virus (ZYMV) 5:290, **5:433–440**
    biotechnological applications 5:439
    classification 5:433
    efficient vectors 5:437
    epidemiology 5:436
        spread 5:437
        virus sources 5:436
    evolution 5:439
    genetics 5:439
    geographical distribution 5:435
    history 5:433
    host range 5:435
    infection *see below*
    taxonomy 5:433, 5:434f
    transmission 5:290
        Aphis gossypii 5:436
        via seed 5:436
        weeds 5:436–437
        ZYMV-NAT 5:436
    use in transgenic plants 4:157, 4:158f
        heterologous encapsidation 4:162
    variability 5:438
    vector relationships 5:436
Zucchini yellow mosaic virus (ZYMV) infection
    control methods 5:437
        cross-protection 5:437, 5:438f
        genetically-modified resistant plants 5:438
        prophylaxis 5:437
        resistant cultivars 5:437
    diagnostic methods 5:435
    epidemics 5:290
    symptoms 5:290, 5:434, 5:435f
    synergism 5:435
Zygocactus symptomless virus (ZSLV) 4:311t
Zygocactus virus X (ZVX) 4:311t
Zygocin 5:307
    structure 5:307
Zygomycota 4:419
    retrotransposons 4:423, 4:424t
Zygosaccharomyces bailii
    killer spectrum 5:307
    phenotype inheritance 5:300
ZYMV *see* Zucchini yellow mosaic virus (ZYMV)

Set
978-0-12-373935-3

Volume 5 of 5
978-0-12-374116-5